Lecture Notes in Computer Science 4052

Commenced Publication in 1973
Founding and Former Series Editors:
Gerhard Goos, Juris Hartmanis, and Jan van Leeuwen

Michele Bugliesi   Bart Preneel
Vladimiro Sassone   Ingo Wegener (Eds.)

Automata, Languages and Programming

33rd International Colloquium, ICALP 2006
Venice, Italy, July 10-14, 2006
Proceedings, Part II

Volume Editors

Michele Bugliesi
Università Ca'Foscari
Dipartimento di Informatica
Via Torino 155, 30172 Venezia-Mestre, Italy
E-mail: bugliesi@dsi.unive.it

Bart Preneel
Katholieke Universiteit Leuven
Department of Electrical Engineering-ESAT/COSIC
Kasteelpark Arenberg 10, 3001 Leuven-Heverlee, Belgium
E-mail: Bart.Preneel@esat.kuleuven.be

Vladimiro Sassone
University of Southampton
School of Electronics and Computer Science
SO17 1BJ, UK
E-mail: vs@ecs.soton.ac.uk

Ingo Wegener
Universität Dortmund
FB Informatik, LS2
Otto-Hahn-Str. 14, 44221 Dortmund, Germany
E-mail: ingo.wegener@uni-dortmund.de

Library of Congress Control Number: 2006928089

CR Subject Classification (1998): F, D, C.2-3, G.1-2, I.3, E.1-2

LNCS Sublibrary: SL 1 – Theoretical Computer Science and General Issues

ISSN 0302-9743
ISBN-10 3-540-35907-9 Springer Berlin Heidelberg New York
ISBN-13 978-3-540-35907-4 Springer Berlin Heidelberg New York

Springer is a part of Springer Science+Business Media

springer.com

Printed in Germany

Typesetting: Camera-ready by author, data conversion by Scientific Publishing Services, Chennai, India
Printed on acid-free paper SPIN: 11787006 06/3142 5 4 3 2 1 0

Preface

ICALP 2006, the 33rd edition of the International Colloquium on Automata, Languages and Programming, was held in Venice, Italy, July 10–14, 2006. ICALP is a series of annual conferences of the European Association for Theoretical Computer Science (EATCS) which first took place in 1972. This year, the ICALP program consisted of the established track A (focusing on algorithms, automata, complexity and games) and track B (focusing on logic, semantics and theory of programming), and of the recently introduced track C (focusing on security and cryptography foundation).

In response to the call for papers, the Program Committee received 407 submissions, 230 for track A, 96 for track B and 81 for track C. Out of these, 109 papers were selected for inclusion in the scientific program: 61 papers for Track A, 24 for Track B and 24 for Track C. The selection was made by the Program Committee based on originality, quality, and relevance to theoretical computer science. The quality of the manuscripts was very high indeed, and several deserving papers had to be rejected.

ICALP 2006 consisted of four invited lectures and the contributed papers. This volume of the proceedings contains all contributed papers presented at the conference in Track A, together with the paper by the invited speaker Noga Alon (Tel Aviv University, Israel). A companion volume contains all contributed papers presented in Track B and Track C together with the papers by the invited speakers Cynthia Dwork (Microsoft Research, USA) and Prakash Panangaden (Mc Gill University, Canada). The program had an additional invited lecture by Simon Peyton Jones (Microsoft Research, UK), which does not appear in the proceedings.

ICALP 2006 was held in conjunction with the Annual ACM International Symposium on Principles and Practice of Declarative Programming (PPDP 2006) and with the Annual Symposium on Logic-Based Program Synthesis and Transformation (LOPSTR 2006). Additionally, the following workshops were held as satellite events of ICALP 2006: ALGOSENSORS 2006 - International Workshop on Algorithmic Aspects of Wireless Sensor Networks; CHR 2006 - Third Workshop on Constraint Handling Rules; CL&C 2006 - Classical Logic and Computation; DCM 2006 - 2nd International Workshop on Developments in Computational Models; FCC 2006 - Formal and Computational Cryptography; iETA 2006 - Improving Exponential-Time Algorithms: Strategies and Limitations; MeCBIC 2006 - Membrane Computing and Biologically Inspired Process Calculi; SecReT 2006 - 1st Int. Workshop on Security and Rewriting Techniques; WCAN 2006 - 2nd Workshop on Cryptography for Ad Hoc Networks.

We wish to thank all authors who submitted extended abstracts for consideration, the Program Committee for their scholarly effort, and all referees who assisted the Program Committees in the evaluation process.

Thanks to the sponsors for their support, to the Venice International University and to the Province of Venice for hosting ICALP 2006 in beautiful S. Servolo. We are also grateful to all members of the Organizing Committee in the Department of Computer Science and to the Center for Technical Support Services and Telecommunications (CSITA) of the University of Venice. Thanks to Andrei Voronkov for his support with the conference management software EasyChair. It was great in handling the submissions and the electronic PC meeting, as well as in assisting in the assembly of the proceedings.

April 2006

Michele Bugliesi
Bart Preneel
Vladimiro Sassone
Ingo Wegener

Organization

Program Committee

Track A

Harry Buhrman, University of Amsterdam, The Netherlands
Mark de Berg, TU Eindhoven, The Netherlands
Uriel Feige, Weizmann Institute, Isreal
Anna Gal, University of Texas at Austin, USA
Johan Hastad, KTH Stockholm, Sweden
Edith Hemaspaandra, Rochester Institute of Technology, USA
Kazuo Iwama, Kyoto University, Japan
Mark Jerrum, University of Edinburgh, UK
Stefano Leonardi, Università di Roma, Italy
Friedhelm Meyer auf der Heide, Universität Paderborn, Germany
Ian Munro, University of Waterloo, Canada
Sotiris Nikoletseas, Patras University, Greece
Rasmus Pagh, IT Univerisy of Copenhagen, Denmark
Tim Roughgarden, Stanford University, USA
Jacques Sakarovitch, CRNS Paris, France
Jiri Sgall, Academy of Sciences, Prague, Czech Republic
Hans Ulrich Simon, Ruhr-Universität Bochum, Germany
Alistair Sinclair, University of Berkeley, USA
Angelika Steger, ETH Zürich, Switzerland
Denis Thérien, McGill University, Canada
Ingo Wegener, Universität Dortmund, Germany (Chair)
Emo Welzl, ETH Zurich, Switzerland

Track B

Roberto Amadio, Université Paris 7, France
Lars Birkedal, IT University of Copenhagen, Denmark
Roberto Bruni, Università di Pisa, Italy
Mariangiola Dezani-Ciancaglini, Università di Torino, Italy
Volker Diekert, University of Stuttgart, Germany
Abbas Edalat, Imperial College, UK
Jan Friso Groote, Eindhoven University of Technology, The Nederlands
Tom Henzinger, EPFL, Switzerland
Madhavan Mukund, Chennai Mathematical Institute, India
Jean-Éric Pin, LIAFA, France
Julian Rathke, University of Sussex, UK
Jakob Rehof, Microsoft Research, Redmont, USA

Vladimiro Sassone, University of Southampton, UK (Chair)
Don Sannella, University of Edinburgh, UK
Nicole Schweikardt, Humboldt-Universität zu Berlin, Germany
Helmut Seidl, Technische Universität München, Germany
Peter Selinger, Dalhousie University, Canada
Jerzy Tiuryn, Warsaw University, Poland
Victor Vianu, U. C. San Diego, USA
David Walker, Princeton University, USA
Igor Walukiewicz, Labri, Université Bordeaux, France

Track C

Martín Abadi, University of California at Santa Cruz, USA
Christian Cachin, IBM Research, Switzerland
Ronald Cramer, CWI and Leiden University, The Netherlands
Ivan Damgrd, University of Aarhus, Denmark
Giovanni Di Crescenzo, Telcordia, USA
Marc Fischlin, ETH Zürich, Switzerland
Dieter Gollmann, University of Hamburg-Harburg, Germany
Andrew D. Gordon, Microsoft Research, UK
Aggelos Kiayias, University of Connecticut, USA
Joe Kilian, Rutgers University, USA
Cathy Meadows, Naval Research Laboratory, USA
John Mitchell, Stanford University, USA
Mats Näslund, Ericsson, Sweden
Tatsuaki Okamoto, Kyoto University, Japan
Rafael Ostrovksy, University of California at Los Angeles, USA
Pascal Paillier, Gemplus, France
Giuseppe Persiano, University of Salerno, Italy
Benny Pinkas, HP Labs, Israel
Bart Preneel, Katholieke Universiteit Leuven, Belgium (Chair)
Vitaly Shmatikov, University of Texas at Austin, USA
Victor Shoup, New York University, USA
Jessica Staddon, PARC, USA
Frederik Vercauteren, Katholieke Universiteit Leuven, Belgium

Organizing Committee

Michele Bugliesi, University of Venice (Conference Chair)
Andrea Pietracaprina, University of Padova (Workshop Co-chair)
Francesco Ranzato, University of Padova (Workshop Co-chair)
Sabina Rossi, University of Venice (Workshop Co-chair)
Annalisa Bossi, University of Venice
Damiano Macedonio, University of Venice

Referees

Martín Abadi
Masayuki Abe
Zoe Abrams
Gagan Aggarwal
Mustaq Ahmed
Cagri Aksay
Tatsuya Akutsu
Susanne Albers
Eric Allender
Jean-Paul Allouche
Jesus Almansa
Helmut Alt
Joel Alwen
Andris Ambainis
Elena Andreeva
Spyros Angelopoulos
Elliot Anshelevich
Lars Arge
Stefan Arnborg
Sanjeev Arora
Vincenzo Auletta
Per Austrin
David Avis
Moshe Babaioff
Michael Backes
Evripides Bampis
Nikhil Bansal
Jeremy Barbay
Paulo Baretto
David Barrington
Roman Bartak
Adam Barth
Paul Beame
Luca Becchetti
Amos Beimel
Elizabeth Berg
Robert Berke
Piotr Berman
Thorsten Bernholt
Ivona Bezakova
Laurent Bienvenu
Mathieu Blanchette
Yvonne Bleischwitz
Isabelle Bloch
Avrim Blum
Liad Blumrosen
Alexander Bockmayr
Alexandra Boldyreva
Beate Bollig
Giuseppe Battista
Nicolas Bonichon
Vincenzo Bonifaci
Joan Boyar
An Braeken
Justin Brickell
Patrick Briest
Gerth Brodal
Gerth Stlting Brodal
Gerth S. Brodal
Peter Buergisser
Jonathan Buss
Gruia Calinescu
Christophe De Cannire
Flavio d'Alessandro
Alberto Caprara
Iliano Cervesato
Timothy M. Chan
Pandu Rangan Chandrasekaran
Krishnendu Chatterjee
Arkadev Chattopadhyay
Avik Chaudhuri
Kamalika Chaudhuri
Chandra Chekuri
Zhi-Zhong Chen
Joseph Cheriyan
Benoit Chevallier-Mames
Markus Chimani
Christian Choffrut
Marek Chrobak
Fabian Chudak
David Clarke
Andrea Clementi
Bruno Codenotti
Richard Cole
Scott Contini
Ricardo Corin
Graham Cormode
Jose Correa
Veronique Cortier
Stefano Crespi
Maxime Crochemore
Mary Cryan
Felipe Cucker
Artur Czumaj
Sanjoy Dasgupta
Ccile Delerable
Xiaotie Deng
Jonathan Derryberry
Jean-Louis Dessalles
Nikhil Devanur
Luc Devroye
Florian Diedrich
Martin Dietzfelbinger
Jintai Ding
Irit Dinur
Benjamin Doerr
Eleni Drinea
Petros Drineas
Stefan Droste
Laszlo Egri
Friedrich Eisenbrand
Michael Elkin
Leah Epstein
Kimmo Eriksson
Thomas Erlebach
Peter Gacs
Rolf Fagerberg
Ulrich Faigle
Piotr Faliszewski
Pooya Farshim
Arash Farzan
Serge Fehr
Sandor Fekete
Rainer Feldmann
Stefan Felsner
Coby Fernadess
Paolo Ferragina
Jiri Fiala

Faith Fich
Matthias Fitzi
Abraham Flaxman
Lisa K. Fleischer
Rudolf Fleischer
Fedor Fomin
Lance Fortnow
Pierre Fraignaud
Paolo Franciosa
Matt Franklin
Eiichiro Fujisaki
Satoshi Fujita
Toshihiro Fujito
Stanley Fung
Martin Furer
Jun Furukawa
Martin Frer
Bernd Gaertner
Martin Gairing
Steven Galbraith
Clemente Galdi
G. Ganapathy
Juan Garay
Naveen Garg
Ricard Gavalda
Dmitry Gavinsky
Joachim Gehweiler
Stefanie Gerke
Abhrajit Ghosh
Oliver Giel
Reza Dorri Giv
Andreas Goerdt
Eu-Jin Goh
Leslie Goldberg
Mikael Goldmann
Aline Gouget
Navin Goyal
Gregor Gramlich
Robert Granger
Alexander Grigoriev
Martin Grohe
Andre Gronemeier
Jiong Guo
Ankur Gupta
Anupam Gupta
Venkatesan Guruswami
Inge Li Grtz
Robbert de Haan
Torben Hagerup
Mohammad Taghi Hajiaghayi
Michael Hallett
Dan Halperin
Christophe Hancart
Goichiro Hanaoka
Dan Witzner Hansen
Sariel Har-Peled
Thomas Hayes
Meng He
Lane Hemaspaandra
Javier Herranz
Jan van den Heuvel
Martin Hirt
Michael Hoffmann
Dennis Hofheinz
Thomas Hofmeister
Christopher M. Homan
Hendrik Jan Hoogeboom
Juraj Hromkovic
Shunsuke Inenaga
Piotr Indyk
Yuval Ishai
Toshimasa Ishii
Hiro Ito
Toshiya Itoh
Gabor Ivanyos
Riko Jacob
Jens Jaegerskuepper
Sanjay Jain
Kamal Jain
Klaus Jansen
Thomas Jansen
Jesper Jansson
Stanislaw Jarecki
Wojciech Jawor
David Johnson
Tibor Jordan
Philippe Jorrand
Jan Juerjens
Valentine Kabanets
Jesse Kamp
Sampath Kannan
Haim Kaplan
Sarah Kappes
Bruce Kapron
Juha Karkkainen
Julia Kempe
Johan Karlander
Howard Karloff
Jonahtan Katz
Akinori Kawachi
Claire Kenion
Krishnaram Kenthapadi
Rohit Khandekar
Subhash A. Khot
Samir Khuller
Eike Kiltz
Guy Kindler
Valerie King
Lefteris Kirousis
Daniel Kirsten
Bobby Kleinberg
Adam Klivans
Johannes Koebler
Jochen Koenemann
Petr Kolman
Guy Kortsarz
Michal Koucky
Elias Koutsoupias
Dexter Kozen
Darek Kowalski
Matthias Krause
Hugo Krawczyk
Klaus Kriegel
Alexander Kroeller
Piotr Krysta
Ludek Kucera
Noboru Kunihiro
Eyal Kushilevitz
Minseok Kwon
Shankar Ram Lakshminarayanan
Joseph Lano
Sophie Laplante
Christian Lavault

Ron Lavi
Thierry Lecroq
Troy Lee
Hanno Lefmann
Francois Lemieux
Asaf Levin
Benoit Libert
Christian Liebchen
Andrzej Lingas
Helger Lipmaa
Sylvain Lombardy
Alex Lopez-Ortiz
Zvi Lotker
Laci Lovasz
Chris Luhrs
Rune Bang Lyngs
Peter Mahlmann
John Malone-Lee
Heikki Mannila
Alberto Marchetti-Spaccamela
Martin Marciniszyn
Gitta Marchand
Stuart Margolis
Martin Mares
Russell Martin
Toshimitsu Masuzawa
Jiri Matousek
Giancarlo Mauri
Alexander May
Elvira Mayordomo
Pierre McKenzie
Kurt Mehlhorn
Aranyak Mehta
Nele Mentens
Mark Mercer
Ron van der Meyden
Ulrich Meyer
Peter Bro Miltersen
Dieter Mitsche
Shuichi Miyazaki
Burkhard Monien
Cris Moore
Thomas Moscibroda
Mitsuo Motoki
Ahuva Mu'alem Kamesh Munagala
Ian Munro
Kazuo Murota
Petra Mutzel
Hiroshi Nagamochi
Shin-ichi Nakano
Seffi Naor
Gonzalo Navarro
Frank Neumann
Antonio Nicolosi
Rolf Niedermeier
Jesper Buus Nielsen
Svetla Nikova
Karl Norrman
Dirk Nowotka
Robin Nunkesser
Regina O'Dell
Hirotaka ONO
Mitsunori Ogihara
Kazuo Ohta
Chihiro Ohyama
Yusuke Okada
Yoshio Okamoto
Christopher Okasaki
Ole Østerby
Janos Pach
Anna stlin Pagh
Anna Palbom
Konstantinos Panagiotou
Leon Peeters
Derek Phillips
Toniann Pitassi
Giovanni Pighizzini
Wolf Polak
Pawel Pralat
Pavel Pudlak
Prashant Puniya
Yuri Rabinovich
Jaikumar Radhakrishnan
Stanislaw Radziszowski
Harald Raecke
Prabhakar Ragde
Zia Rahman
S. Raj Rajagopalan
V. Ramachandran
Dana Randall
S. Srinivasa Rao
Ran Raz
Alexander Razborov
Andreas Razen
Ken Regan
Ben Reichardt
Christophe Reutenauer
Eleanor Rieffel
Romeo Rizzi
Martin Roetteler
Phillip Rogaway
Amir Ronen
Dominique Rossin
Peter Rossmanith
Joerg Rothe
Arnab Roy
Leo Ruest
Milan Ruzic
Kunihiko Sadakane
Cenk Sahinalp
Kai Salomaa
Louis Salvail
Peter Sanders
Mark Sandler
Rahul Santhanam
Palash Sarkar
Martin Sauerhoff
Daniel Sawitzki
Nicolas Schabanel
Christian Schaffner
Michael Schapira
Dominik Scheder
Christian Scheideler
Christian Schindelhauer
Katja Schmidt-Samoa
Georg Schnitger
Henning Schnoor
Uwe Schoening
Gunnar Schomaker
Eva Schuberth
Andreas Schulz

Nathan Segerlind
Meinolf Sellmann
Pranab Sen
Hadas Shachnai
Ronen Shaltiel
Abhi Shelat
Bruce Sheppard
Oleg M. Sheyner
David Shmoys
Detlef Sieling
Jiri Sima
Mohit Singh
Naveen Sivadasan
Matthew Skala
Steve Skiena
Michiel Smid
Adam Smith
Shakhar Smorodinsky
Christian Sohler
Alexander Souza
Paul Spirakis
Michael Spriggs
Reto Sphel
Rob van Stee
Stamatis Stefanakos
Daniel Stefankovic
Cliff Stein
Bernhard von Stengel
Tobias Storch
Madhu Sudan
Dirk Sudholt
Mukund Sundararajan
Koutarou Suzuki
Maxim Sviridenko
Tibor Szabo
Zoltan Szigeti
Troels Bjerre Sørensen
Asano Takao
Hisao Tamaki
Akihisa Tamura
Eva Tardos
Sebastiaan Terwijn
Pascal Tesson
Prasad Tetali
Ralf Thoele
Karsten Tiemann
Yuuki Tokunaga
Takeshi Tokuyama
Eric Torng
Patrick Traxler
Luca Trevisan
Tatsuie Tsukiji
Gyorgy Turan
Pim Tuyls
Ryuhei Uehara
Chris Umans
Falk Unger
Takeaki Uno
Pavel Valtr
Sergei Vassilvitskii
Ingrid Verbauwhede
Kolia Vereshchagin
Damien Vergnaud
Adrian Vetta
Ivan Visconti
Berthold Voecking
Heribert Vollmer
Sergey Vorobyov
Sven de Vries
Stephan Waack
Uli Wagner
Michael Waidner
Bogdan Warinschi
Osamu Watanabe
Brent Waters
Kevin Wayne
Stephanie Wehner
Ralf-Philipp Weinmann
Andreas Weissl
Tom Wexler
Erik Winfree
Peter Winkler
Kai Wirt
Carsten Witt
Ronald de Wolf
Stefan Wolf
David Woodruff
Mutsunori Yagiura
Hiroaki Yamamoto
Go Yamamoto
Shigeru Yamashita
Takenaga Yasuhiko
Yiqun Lisa Yin
Filip Zagorski
Guochuan Zhang
Yuliang Zheng
Ming Zhong
Hong-Sheng Zhou
Xiao Zhou
Wieslaw Zielonka
Eckart Zitzler
David Zuckerman
Philipp Zumstein
Uri Zwick

Sponsoring Institutions

IBM Italy
Venis S.P.A - Venezia Informatica e Sistemi
Dipartimento di Informatica, Università Ca' Foscari
CVR - Consorzio Venezia Ricerche

Table of Contents – Part II

Invited Papers

Zero-Knowledge and Signatures

Cryptographic Protocols

Secrecy and Protocol Analysis

Cryptographic Primitives

Bounded Storage and Quantum Models

Foundations

Multi-party Protocols

Games

Semantics

Automata I

Models

Equations

Logics

Automata II

Table of Contents – Part I

Formal Languages

Approximation Algorithms I

Approximation Algorithms II

Graph Algorithms I

Algorithms I

Complexity I

Data Structures and Linear Algebra

Graphs

Complexity II

Game Theory I

Algorithms II

Game Theory II

Networks, Circuits and Regular Expressions

Fixed Parameter Complexity and Approximation Algorithms

Graph Algorithms II

Differential Privacy

Cynthia Dwork

Microsoft Research
dwork@microsoft.com

Abstract. In 1977 Dalenius articulated a desideratum for statistical databases: nothing about an individual should be learnable from the database that cannot be learned without access to the database. We give a general impossibility result showing that a formalization of Dalenius' goal along the lines of semantic security cannot be achieved. Contrary to intuition, a variant of the result threatens the privacy even of someone not in the database. This state of affairs suggests a new measure, *differential privacy*, which, intuitively, captures the increased risk to one's privacy incurred by participating in a database. The techniques developed in a sequence of papers [8, 13, 3], culminating in those described in [12], can achieve any desired level of privacy under this measure. In many cases, extremely accurate information about the database can be provided while simultaneously ensuring very high levels of privacy.

1 Introduction

A statistic is a quantity computed from a sample. If a database is a representative sample of an underlying population, the goal of a privacy-preserving statistical database is to enable the user to learn properties of the population as a whole, while protecting the privacy of the individuals in the sample. The work discussed herein was originally motivated by exactly this problem: how to reveal useful information about the underlying population, as represented by the database, while preserving the privacy of individuals. Fortuitously, the techniques developed in [8, 13, 3] and particularly in [12] are so powerful as to broaden the scope of private data analysis beyond this orignal "representatitive" motivation, permitting privacy-preserving analysis of an object that is itself of intrinsic interest. For instance, the database may describe a concrete interconnection network – not a sample subnetwork – and we wish to reveal certain properties of the network without releasing information about individual edges or nodes. We therefore treat the more general problem of *privacy-preserving analysis of data*.

A rigorous treatment of privacy requires definitions: What constitutes a failure to preserve privacy? What is the power of the adversary whose goal it is to compromise privacy? What auxiliary information is available to the adversary (newspapers, medical studies, labor statistics) even without access to the database in question? Of course, utility also requires formal treatment, as releasing no information or only random noise clearly does not compromise privacy; we

M. Bugliesi et al. (Eds.): ICALP 2006, Part II, LNCS 4052, pp. 1–12, 2006.

will return to this point later. However, in this work privacy is paramount: we will first define our privacy goals and then explore what utility can be achieved given that the privacy goals will be satisified[1].

A 1977 paper of Dalenius [6] articulated a desideratum that foreshadows for databases the notion of semantic security defined five years later by Goldwasser and Micali for cryptosystems [15]: access to a statistical database should not enable one to learn anything about an individual that could not be learned without access[2]. We show this type of privacy cannot be achieved. The obstacle is in *auxiliary information*, that is, information available to the adversary other than from access to the statistical database, and the intuition behind the proof of impossibility is captured by the following example. Suppose one's exact height were considered a highly sensitive piece of information, and that revealing the exact height of an individual were a privacy breach. Assume that the database yields the average heights of women of different nationalities. An adversary who has access to the statistical database and the auxiliary information "Terry Gross is two inches shorter than the average Lithuanian woman" learns Terry Gross' height, while anyone learning only the auxiliary information, without access to the average heights, learns relatively little.

There are two remarkable aspects to the impossibility result: (1) it applies regardless of whether or not Terry Gross is in the database and (2) Dalenius' goal, formalized as a relaxed version of semantic security, cannot be achieved, while semantic security for cryptosystems can be achieved. The first of these leads naturally to a new approach to formulating privacy goals: the risk to one's privacy, or in general, any type of risk, such as the risk of being denied automobile insurance, should not substantially increase as a result of participating in a statistical database. This is captured by *differential privacy.*

The discrepancy between the possibility of achieving (something like) semantic security in our setting and in the cryptographic one arises from the utility requirement. Our adversary is analagous to the eavesdropper, while our user is analagous to the message recipient, and yet there is no decryption key to set them apart, they are one and the same. Very roughly, the database is designed to convey certain information. An auxiliary information generator knowing the data therefore knows much about what the user will learn from the database. This can be used to establish a shared secret with the adversary/user that is unavailable to anyone not having access to the database. In contrast, consider a cryptosystem and a pair of candidate messages, say, $\{0, 1\}$. Knowing which message is to be encrypted gives one no information about the ciphertext; intuitively, the auxiliary information generator has "no idea" what ciphertext the eavesdropper will see. This is because by definition the ciphertext must have no utility to the eavesdropper.

---

[1] In this respect the work on privacy diverges from the literature on secure function evaluation, where privacy is ensured only modulo the function to be computed: if the function is inherently disclosive then privacy is abandoned.

[2] Semantic security against an eavesdropper says that nothing can be learned about a plaintext from the ciphertext that could not be learned without seeing the ciphertext.

In this paper we prove the impossibility result, define differential privacy, and observe that the interactive techniques developed in a sequence of papers [8, 13, 3, 12] can achieve any desired level of privacy under this measure. In many cases very high levels of privacy can be ensured while simultaneously providing extremely accurate information about the database.

Related Work. There is an enormous literature on privacy in databases; we briefly mention a few fields in which the work has been carried out. See [1] for a survey of many techniques developed prior to 1989.

By far the most extensive treatment of disclosure limitation is in the statistics community; for example, in 1998 the *Journal of Official Statistics* devoted an entire issue to this question. This literature contains a wealth of privacy supportive techniques and investigations of their impact on the statistics of the data set. However, to our knowledge, rigorous definitions of privacy and modeling of the adversary are not features of this portion of the literature.

Research in the theoretical computer science community in the late 1970's had very specific definitions of privacy compromise, or what the adversary must achieve to be considered successful (see, eg, [9]). The consequent privacy guarantees would today be deemed insufficiently general, as modern cryptography has shaped our understanding of the dangers of the leakage of partial information. Privacy in databases was also studied in the security community. Although the effort seems to have been abandoned for over two decades, the work of Denning [7] is closest in spirit to the line of research recently pursued in [13, 3, 12].

The work of Agrawal and Srikant [2] and the spectacular privacy compromises achieved by Sweeney [18] rekindled interest in the problem among computer scientists, particularly within the database community. Our own interest in the subject arose from conversations with the philosopher Helen Nissenbaum.

2 Private Data Analysis: The Setting

There are two natural models for privacy mechanisms: interactive and non-interactive. In the non-interactive setting the data collector, a trusted entity, publishes a "sanitized" version of the collected data; the literature uses terms such as "anonymization" and "de-identification". Traditionally, sanitization employs techniques such as data perturbation and sub-sampling, as well as removing well-known identifiers such as names, birthdates, and social security numbers. It may also include releasing various types of synopses and statistics. In the interactive setting the data collector, again trusted, provides an interface through which users may pose queries about the data, and get (possibly noisy) answers.

Very powerful results for the interactive approach have been obtained ([13, 3, 12] and the present paper), while the non-interactive case has proven to be more difficult, (see [14, 4, 5]), possibly due to the difficulty of supplying utility that has not yet been specified at the time the sanitization is carried out. This intuition is given some teeth in [12], which shows concrete separation results.

3 Impossibility of Absolute Disclosure Prevention

The impossibility result requires some notion of utility – after all, a mechanism that always outputs the empty string, or a purely random string, clearly preserves privacy[3]. Thinking first about deterministic mechanisms, such as histograms or k-anonymizations [19], it is clear that for the mechanism to be *useful* its output should not be predictable by the user; in the case of randomized mechanisms the same is true, but the unpredictability must not stem only from random choices made by the mechanism. Intuitively, there should be a vector of questions (most of) whose answers *should* be learnable by a user, but whose answers are not in general known in advance. We will therefore posit a *utility vector*, denoted w. This is a binary vector of some fixed length κ (there is nothing special about the use of binary values). We can think of the utility vector as answers to questions about the data.

A *privacy breach* for a database is described by a Turing machine $\mathcal{C}$ that takes as input a description of a distribution $\mathcal{D}$ on databases, a database DB drawn according to this distribution, and a string – the purported privacy breach– and outputs a single bit[4]. We will require that $\mathcal{C}$ always halt. We say the adversary *wins*, with respect to $\mathcal{C}$ and for a given $(\mathcal{D}, DB)$ pair, if it produces a string s such that $\mathcal{C}(\mathcal{D}, DB, s)$ accepts. Henceforth "with respect to $\mathcal{C}$" will be implicit.

An auxiliary information generator is a Turing machine that takes as input a description of the distribution $\mathcal{D}$ from which the database is drawn as well as the database DB itself, and outputs a string, z, of auxiliary information. This string is given both to the adversary and to a *simulator*. The simulator has no access of any kind to the database; the adversary has access to the database via the privacy mechanism.

We model the adversary by a communicating Turing machine. The theorem below says that for any privacy mechanism San() and any distribution $\mathcal{D}$ satisfying certain technical conditions with respect to San(), there is always some particular piece of auxiliary information, z, so that z alone is useless to someone trying to win, while z in combination with access to the data through the privacy mechanism permits the adversary to win with probability arbitrarily close to 1. In addition to formalizing the entropy requirements on the utility vectors as discussed above, the technical conditions on the distribution say that learning the *length* of a privacy breach does not help one to guess a privacy breach.

Theorem 1. *Fix any privacy mechanism* San() *and privacy breach decider* $\mathcal{C}$. *There is an auxiliary information generator* $\mathcal{X}$ *and an adversary* $\mathcal{A}$ *such that for all distributions* $\mathcal{D}$ *satisfying Assumption 3 and for all adversary simulators* $\mathcal{A}^*$,

$$\Pr[\mathcal{A}(\mathcal{D}, \mathrm{San}(\mathcal{D}, DB), \mathcal{X}(\mathcal{D}, DB)) \text{ wins}] - \Pr[\mathcal{A}^*(\mathcal{D}, \mathcal{X}(\mathcal{D}, DB)) \text{ wins}] \geq \Delta$$

where Δ *is a suitably chosen (large) constant. The probability spaces are over choice of* $DB \in_R \mathcal{D}$ *and the coin flips of* San, $\mathcal{X}$, $\mathcal{A}$, *and* $\mathcal{A}^*$.

[3] Indeed the height example fails in these trivial cases, since it is only through the sanitization that the adversary learns the average height.

[4] We are agnostic as to *how* a distribution $\mathcal{D}$ is given as input to a machine.

The distribution $\mathcal{D}$ completely captures any information that the adversary (and the simulator) has about the database, prior to seeing the output of the auxiliary information generator. For example, it may capture the fact that the rows in the database correspond to people owning at least two pets. Note that in the statement of the theorem all parties have access to $\mathcal{D}$ and may have a description of $\mathcal{C}$ hard-wired in; however, the adversary's strategy does not use either of these.

Strategy for $\mathcal{X}$ and $\mathcal{A}$ when all of w is learned from San(DB)*:* To develop intuition we first describe, slightly informally, the strategy for the special case in which the adversary always learns all of the utility vector, w, from the privacy mechanism[5]. This is realistic, for example, when the sanitization produces a histogram, such as a table of the number of people in the database with given illnesses in each age decile, or a when the sanitizer chooses a random subsample of the rows in the database and reveals the average ages of patients in the subsample exhibiting various types of symptoms. This simpler case allows us to use a weaker version of Assumption 3:

Assumption 2. *1.* $\forall 0 < \gamma < 1 \; \exists n_\gamma \; \Pr_{DB \in_R \mathcal{D}}[|DB| > n_\gamma] < \gamma$*; moreover* n_γ *is computable by a machine given* $\mathcal{D}$ *as input.*
 2. There exists an ℓ *such that* both *the following conditions hold:*
 (a) Conditioned on any privacy breach of length ℓ*, the min-entropy of the utility vector is at least* ℓ*.*
 (b) Every $DB \in \mathcal{D}$ *has a privacy breach of length* ℓ*.*
 3. $\Pr[\mathcal{B}(\mathcal{D}, San(DB))$ wins$] \leq \mu$ *for all interactive Turing machines* $\mathcal{B}$*, where* μ *is a suitably small constant. The probability is taken over the coin flips of* $\mathcal{B}$ *and the privacy mechanism* San()*, as well as the choice of* $DB \in_R \mathcal{D}$*.*

Intuitively, Part (2a) implies that we can extract ℓ bits of randomness from the utility vector, which can be used as a one-time pad to hide any privacy breach of the same length. (For the full proof, ie, when not necessarily all of w is learned by the adversary/user, we will need to strengthen Part (2a).) Let ℓ_0 denote the least ℓ satisfying (both clauses of) Part 2. We cannot assume that ℓ_0 can be found in finite time; however, for any tolerance γ let n_γ be as in Part 1, so all but a γ fraction of the support of $\mathcal{D}$ is strings of length at most n_γ. For any fixed γ it is possible to find an $\ell_\gamma \leq \ell_0$ such that ℓ_γ satisfies both clauses of Assumption 2(2) on all databases of length at most n_γ. We can assume that γ is hard-wired into all our machines, and that they all follow the same procedure for computing n_γ and ℓ_γ. Thus, Part 1 allows the more powerful order of quantifiersd in the statement of the theorem; without it we would have to let $\mathcal{A}$ and $\mathcal{A}^*$ depend on $\mathcal{D}$ (by having ℓ hard-wired in). Finally, Part 3 is a nontriviality condition.

The strategy for $\mathcal{X}$ and $\mathcal{A}$ is as follows. On input $DB \in_R \mathcal{D}$, $\mathcal{X}$ randomly chooses a privacy breach y for DB of length $\ell = \ell_\gamma$, if one exists, which occurs with probability at least $1 - \gamma$. It also computes the utility vector, w. Finally, it chooses a seed s and uses a strong randomness extractor to obtain from w

[5] Although this case is covered by the more general case, in which not all of w need be learned, it permits a simpler proof that exactly captures the height example.

an ℓ-bit almost-uniformly distributed string r [16, 17]; that is, $r = \text{Ext}(s, w)$, and the distribution on r is within statistical distance ϵ from U_ℓ, the uniform distribution on strings of length ℓ, even given s and y. The auxiliary information will be $z = (s, y \oplus r)$.

Since the adversary learns all of w, from s it can obtain $r = \text{Ext}(s, w)$ and hence y. We next argue that $\mathcal{A}^*$ wins with probability (almost) bounded by μ, yielding a gap of at least $1 - (\gamma + \mu + \epsilon)$.

Assumption 2(3) implies that $\Pr[\mathcal{A}^*(\mathcal{D}) \text{ wins}] \leq \mu$. Let d_ℓ denote the maximum, over all $y \in \{0,1\}^\ell$, of the probability, over choice of $DB \in_R \mathcal{D}$, that y is a privacy breach for DB. Since $\ell = \ell_\gamma$ does not depend on DB, Assumption 2(3) also implies that $d_\ell \leq \mu$.

By Assumption 2(2a), even conditioned on y, the extracted r is (almost) uniformly chosen, independent of y, and hence so is $y \oplus r$. Consequently, the probability that $\mathcal{X}$ produces z is essentially independent of y. Thus, the simulator's probability of producing a privacy breach of length ℓ for the given database is bounded by $d_\ell + \epsilon \leq \mu + \epsilon$, as it can generate simulated "auxiliary information" with a distribution within distance ϵ of the correct one.

The more interesting case is when the sanitization does not necessarily reveal all of w; rather, the guarantee is only that it always reveal a vector w' within Hamming distance κ/c of w for constant c to be determined[6]. The difficulty with the previous approach is that if the privacy mechanism is randomized then the auxiliary information generator may not know which w' is seen by the adversary. Thus, even given the seed s, the adversary may not be able to extract the same random pad from w' that the auxiliary information generator extracted from w. This problem is solved using *fuzzy extractors* [10].

Definition 1. *An* $(\mathcal{M}, m, \ell, t, \epsilon)$ fuzzy extractor *is given by procedures* (Gen,Rec)*.*

1. Gen *is a randomized generation procedure. On input* $w \in \mathcal{M}$ *outputs an "extracted" string* $r \in \{0,1\}^\ell$ *and a public string* p*. For any distribution* W *on* $\mathcal{M}$ *of min-entropy* m*, if* $(R, P) \leftarrow \text{Gen}(W)$ *then the distributions* (R, P) *and* (U_ℓ, P) *are within statistical distance* ϵ*.*
2. Rec *is a deterministic reconstruction procedure allowing recovery of* $r = R(w)$ *from the corresponding public string* $p = P(w)$ *together with any vector* w' *of distance at most* t *from* w*. That is, if* $(r, p) \leftarrow \text{Gen}(w)$ *and* $\|w - w'\|_1 \leq t$ *then* $\text{Rec}(w', p) = r$*.*

In other words, $r = R(w)$ looks uniform, even given $p = P(w)$, and $r = R(w)$ can be reconstructed from $p = P(w)$ and any w' sufficiently close to w.

We now strenthen Assumption 2(2a) to say that the entropy of the source San(W) (vectors obtained by interacting with the sanitization mechanism, all of

[6] One could also consider privacy mechanisms that produce good approximations to the utility vector with a certain probability for the distribution $\mathcal{D}$, where the probability is taken over the choice of $DB \in_R \mathcal{D}$ and the coins of the privacy mechanism. The theorem and proof hold *mutatis mutandis.*

distance at most κ/c from the true utility vector) is high even conditioned on any privacy breach y of length ℓ **and** $P = \text{Gen}(W)$.

Assumption 3. *For some ℓ satisfying Assumption 2(2b), for any privacy breach $y \in \{0,1\}^\ell$, the min-entropy of* $(\text{San}(W)|y)$ *is at least $k+\ell$, where k is the length of the public strings p produced by the fuzzy extractor*[7].

Strategy when w need not be fully learned: For a given database DB, let w be the utility vector. This can be computed by $\mathcal{X}$, who has access to the database. $\mathcal{X}$ simulates interaction with the privacy mechanism to determine a "valid" w' close to w (within Hamming distance κ/c). The auxiliary information generator runs $\text{Gen}(w')$, obtaining $(r = R(w'), p = P(w'))$. It computes n_γ and $\ell = \ell_\gamma$ (as above, only now satisfying Assumptions 3 and 2(2b) for all $DB \in D$ of length at most n_γ), and uniformly chooses a privacy breach y of length ℓ_γ, assuming one exists. It then sets $z = (p, r \oplus y)$.

Let w'' be the version of w seen by the adversary. Clearly, assuming $2\kappa/c \leq t$ in Definition 1, the adversary can reconstruct r. This is because since w' and w'' are both within κ/c of w they are within distance $2\kappa/c$ of each other, and so w'' is within the "reconstruction radius" for any $r \leftarrow \text{Gen}(w')$. Once the adversary has reconstructed r, obtaining y is immediate. Thus the adversary is able to produce a privacy breach with probability at least $1-\gamma$. It remains to analyze the probability with which the simulator, having access only to z but not to the privacy mechanism (and hence, not to any w'' close to w), produces a privacy breach.

In the sequel, we let $\mathcal{B}$ denote the *best* machine, among all those with access to the given information, at producing producing a privacy breach ("winning").

By Assumption 2(3), $\Pr[\mathcal{B}(\mathcal{D}, \text{San}(DB)) \text{ wins}] \leq \mu$, where the probability is taken over the coin tosses of the privacy mechanism and the machine $\mathcal{B}$, and the choice of $DB \in_R \mathcal{D}$. Since $p = P(w')$ is computed from w', which in turn is computable from $\text{San}(DB)$, we have

$$p_1 = \Pr[\mathcal{B}(\mathcal{D}, p) \text{ wins}] \leq \mu$$

where the probability space is now also over the choices made by Gen(), that is, the choice of $p = P(w')$. Now, let U_ℓ denote the uniform distribution on ℓ-bit strings. Concatenating a random string $u \in_R U_\ell$ to p cannot help $\mathcal{B}$ to win, so

$$p_2 = \Pr[\mathcal{B}(\mathcal{D}, p, u) \text{ wins}] = p_1 \leq \mu$$

where the probability space is now also over choice of u. For any fixed string $y \in \{0,1\}^\ell$ we have $U_\ell = U_\ell \oplus y$, so for all $y \in \{0,1\}^\ell$, and in particular, for all privacy breaches y of DB,

$$p_3 = \Pr[\mathcal{B}(\mathcal{D}, p, u \oplus y) \text{ wins}] = p_2 \leq \mu.$$

[7] A good fuzzy extractor "wastes" little of the entropy on the public string. Better fuzzy extractors are better for the adversary, since the attack requires ℓ bits of residual min-entropy after the public string has been generated.

Let W denote the distribution on utility vectors and let $\mathrm{San}(W)$ denote the distribution on the versions of the utility vectors learned by accessing the database through the privacy mechanism. Since the distributions $(P, R) = \mathrm{Gen}(W')$, and (P, U_ℓ) have distance at most ϵ, it follows that for any $y \in \{0,1\}^\ell$

$$p_4 = \Pr[\mathcal{B}(\mathcal{D}, p, r \oplus y) \text{ wins}] \leq p_3 + \epsilon \leq \mu + \epsilon.$$

Now, p_4 is an upper bound on the probability that the *simulator* wins, given $\mathcal{D}$ and the auxiliary information $z = (p, r \oplus y)$, so

$$\Pr[\mathcal{A}^*(\mathcal{D}, z) \text{ wins}] \leq p_4 \leq \mu + \epsilon\,.$$

An $(\mathcal{M}, m, \ell, t, \epsilon)$ fuzzy extractor, where $\mathcal{M}$ is the distribution $\mathrm{San}(W)$ on utility vectors obtained from the privacy mechanism, m satisfies: for all ℓ-bit strings y which are privacy breaches for some database $D \in DB$, $\mathrm{H}_\infty(W'|y) \geq m$; and $t < \kappa/3$, yields a gap of at least

$$(1 - \gamma) - (\mu + \epsilon) = 1 - (\gamma + \mu + \epsilon)$$

between the winning probabilities of the adversary and the simulator. Setting $\Delta = 1 - (\gamma + \mu + \epsilon)$ proves Theorem 1.

We remark that, unlike in the case of most applications of fuzzy extractors (see, in particular, [10, 11]), in this proof we are not interested in hiding partial information about the source, in our case the approximate utility vectors W', so we don't care how much min-entropy is used up in generating p. We only require sufficient residual min-entropy for the generation of the random pad r. This is because an approximation to the utility vector revealed by the privacy mechanism is not itself disclosive; indeed it is by definition safe to release. Similarly, we don't necessarily need to maximize the tolerance t, although if we have a richer class of fuzzy extractors the impossibility result applies to more relaxed privacy mechanisms (those that reveal worse approximations to the true utility vector).

4 Differential Privacy

As noted in the example of Terry Gross' height, an auxiliary information generator with information about someone not even in the database can cause a privacy breach to this person. In order to sidestep this issue we change from absolute guarantees about disclosures to relative ones: any given disclosure will be, within a small multiplicative factor, just as likely whether or not the individual participates in the database. As a consequence, there is a nominally increased risk to the individual in participating, and only nominal gain to be had by concealing or misrepresenting one's data. Note that a bad disclosure can still occur, but our guarantee assures the individual that it will not be the presence of her data that causes it, nor could the disclosure be avoided through any action or inaction on the part of the user.

Definition 2. *A randomized function $\mathcal{K}$ gives* ϵ-differential privacy *if for all data sets D_1 and D_2 differing on at most one element, and all $S \subseteq Range(\mathcal{K})$,*

$$\Pr[\mathcal{K}(D_1) \in S] \leq \exp(\epsilon) \times \Pr[\mathcal{K}(D_2) \in S] \tag{1}$$

A mechanism $\mathcal{K}$ satisfying this definition addresses concerns that any participant might have about the leakage of her personal information x: even if the participant removed her data from the data set, no outputs (and thus consequences of outputs) would become significantly more or less likely. For example, if the database were to be consulted by an insurance provider before deciding whether or not to insure Terry Gross, then the presence or absence of Terry Gross in the database will not significantly affect her chance of receiving coverage.

This definition extends to group privacy as well. A collection of c participants might be concerned that their collective data might leak information, even when a single participant's does not. Using this definition, we can bound the dilation of any probability by at most $\exp(\epsilon c)$, which may be tolerable for small c. Note that we specifically aim to disclose aggregate information about large groups, so we should expect privacy bounds to disintegrate with increasing group size.

5 Achieving Differential Privacy

We now describe a concrete interactive privacy mechanism achieving ϵ-differential privacy[8]. The mechanism works by adding appropriately chosen random noise to the answer $a = f(X)$, where f is the *query function* and X is the database; thus the query functions may operate on the entire database at once. It can be simple – eg, "Count the number of rows in the database satisfying a given predicate" – or complex – eg, "Compute the median value for each column; if the Column 1 median exceeds the Column 2 median, then output a histogram of the numbers of points in the set S of orthants, else provide a histogram of the numbers of points in a different set T of orthants."

Note that the complex query above (1) outputs a vector of values and (2) is an adaptively chosen sequence of two vector-valued queries, where the choice of second query depends on the *true answer* to the first query. Although complex, it is soley a function of the database. We handle such queries in Theorem 4. The case of an adaptively chosen series of questions, in which subsequent queries depend on the *reported* answers to previous queries, is handled in Theorem 5. For example, suppose the adversary first poses the query "Compute the median of each column," and receives in response noisy versions of the medians. Let M be the reported median for Column 1 (so M is the true median plus noise). The adversary may then pose the query: "If M exceeds the true median for Column 1 (ie, if the added noise was positive), then ... else ..." This second query is a function not only of the database but also of the *noise* added by the privacy mechanism in responding to the first query; hence, it is adaptive to the behavior of the mechanism.

[8] This mechanism was introduced in [12], where analagous results were obtained for the related notion of ϵ-indistinguishability. The proofs are essentially the same.

5.1 Exponential Noise and the $L1$-Sensitivity

We will achieve ϵ-differential privacy by the addition of random noise whose magnitude is chosen as a function of the largest change a single participant could have on the output to the query function; we refer to this quantity as the *sensitivity* of the function[9].

Definition 3. *For $f : \mathcal{D} \rightarrow R^d$, the* $L1$-sensitivity *of f is*

$$\Delta f = \max_{D_1, D_2} \|f(D_1) - f(D_2)\|_1 \tag{2}$$

for all D_1, D_2 differing in at most one element.

For many types of queries Δf will be quite small. In particular, the simple counting queries ("How many rows have property P?") have $\Delta f \leq 1$. Our techniques work best – ie, introduce the least noise – when Δf is small. Note that sensitivity is a property of the function alone, and is independent of the database.

The privacy mechanism, denoted $\mathcal{K}_f$ for a query function f, computes $f(X)$ and adds noise with a scaled symmetric exponential distribution with variance σ^2 (to be determined in Theorem 4) in each component, described by the density function

$$\Pr[\mathcal{K}_f(X) = a] \propto \exp(-\|f(X) - a\|_1/\sigma) \tag{3}$$

This distribution has independent coordinates, each of which is an exponentially distributed random variable. The implementation of this mechanism thus simply adds symmetric exponential noise to each coordinate of $f(X)$.

Theorem 4. *For $f : \mathcal{D} \rightarrow R^d$, the mechanism $\mathcal{K}_f$ gives $(\Delta f/\sigma)$-differential privacy.*

Proof. Starting from (3), we apply the triangle inequality within the exponent, yielding for all possible responses r

$$\Pr[\mathcal{K}_f(D_1) = r] \leq \Pr[\mathcal{K}_f(D_2) = r] \times \exp(\|f(D_1) - f(D_2)\|_1/\sigma)\,. \tag{4}$$

The second term in this product is bounded by $\exp(\Delta f/\sigma)$, by the definition of Δf. Thus (1) holds for singleton sets $S = \{a\}$, and the theorem follows by a union bound.

Theorem 4 describes a relationship between Δf, σ, and the privacy differential. To achieve ϵ-differential privacy, one must choose $\sigma \geq \epsilon/\Delta f$.

The importance of choosing the noise as a function of the sensitivity of the entire complex query is made clear by the important case of *histogram queries*, in which the domain of data elements is partitioned into some number k of classes, such as the cells of a contingency table of gross shoe sales versus geographic

[9] It is unfortunate that the term *sensitivity* is overloaded in the context of privacy. We chose it in concurrence with *sensitivity analysis*.

regions, and the true answer to the query is the k-tuple of the exact number of database points in each class. Viewed naïvely, this is a set of k queries, each of sensitivity 1, so to ensure ϵ-differential privacy it follows from k applications of Theorem 4 (each with $d = 1$) that it suffices to use noise distributed according to a symmetric exponential with variance k/ϵ in each component. However, for any two databases D_1 and D_2 differing in only one element, $||f(D_1) - f(D_2)||_1 = 1$, since only one cell of the histogram changes, and that cell only by 1. Thus, we may apply the theorem once, with $d = k$ and $\Delta f = 1$, and find that it suffices to add noise with variance $1/\epsilon$ rather than d/ϵ.

Adaptive Adversaries. We begin with deterministic query strategies F specified by a set of query functions f_ρ, where $f_\rho(X)_i$ is the function describing the ith query given that the first $i-1$ (possibly vector-valued) responses have been $\rho_1, \rho_2, \ldots, \rho_{i-1}$. We require that $f_\rho(X)_i = f_{\rho'}(X)_i$ if the first $i-1$ responses in ρ and ρ' are equal. We define the sensitivity of a query strategy $F = \{f_\rho : \mathcal{D} \to (R^+)^d\}$ to be the largest sensitivity of any of its possible functions, ie: $\Delta F = \sup_\rho \Delta f_\rho$.

Theorem 5. *For query strategy $F = \{f_\rho : \mathcal{D} \to R^d\}$, the mechanism $\mathcal{K}_F$ gives $(\Delta F/\sigma)$-differential privacy.*

Proof. For each $\rho \in (R^+)^d$, the law of conditional probability says

$$\Pr[\mathcal{K}_F(X) = \rho] = \prod_{i \leq d} \Pr[\mathcal{K}_F(X)_i = \rho_i | \rho_1, \rho_2, \ldots \rho_{i-1}] \tag{5}$$

With $\rho_1, \rho_2, \ldots, \rho_{i-1}$ fixed, $f_\rho(X)_i$ is fixed, and the distribution of $\mathcal{K}_F(X)_i$ is simply the random variable with mean $f_\rho(X)_i$ and exponential noise with variance σ^2 in each component. Consequently,

$$\Pr[\mathcal{K}_F(X) = \rho] \propto \prod_{i \leq d} \exp(-\|f_\rho(X)_i - \rho_i\|_1/\sigma) \tag{6}$$

$$= \exp(-\|f_\rho(X) - \rho\|_1/\sigma) \tag{7}$$

As in Theorem 4, the triangle inequality yields $(\Delta F/\sigma)$-differential privacy.

The case of randomized adversaries is handled as usual, by fixing a "successful" coin sequence of a winning randomized strategy.

Acknowledgements. Kobbi Nissim introduced me to the topic of interactive privacy mechanisms. The impossibility result is joint work with Moni Naor, and differential privacy was motivated by this result. The definition, the differential privacy mechanism, and Theorems 4 and 5 are joint work with Frank McSherry. The related notion of ϵ-indistinguishable privacy mechanisms was investigated by Kobbi Nissim and Adam Smith, who were the first to note that histograms of arbitrary complexity have low sensitivity. This example was pivotal in convincing me of the viability of our shared approach.

References

[1] N. R. Adam and J. C. Wortmann, Security-Control Methods for Statistical Databases: A Comparative Study, *ACM Computing Surveys* 21(4): 515-556 (1989).
[2] R. Agrawal and R. Srikant. Privacy-preserving data mining. In *Proc. ACM SIGMOD International Conference on Management of Data*, pp. 439–450, 2000.
[3] A. Blum, C. Dwork, F. McSherry, and K. Nissim. Practical privacy: The SuLQ framework. In *Proceedings of the 24th ACM SIGMOD-SIGACT-SIGART Symposium on Principles of Database Systems*, pages 128–138, June 2005.
[4] S. Chawla, C. Dwork, F. McSherry, A. Smith, and H. Wee. Toward privacy in public databases. In *Proceedings of the 2nd Theory of Cryptography Conference*, pages 363–385, 2005.
[5] S. Chawla, C. Dwork, F. McSherry, and K. Talwar. On the utility of privacy-preserving histograms. In *Proceedings of the 21st Conference on Uncertainty in Artificial Intelligence*, 2005.
[6] T. Dalenius, Towards a methodology for statistical disclosure control. *Statistik Tidskrift 15*, pp. 429–222, 1977.
[7] D. E. Denning, *Secure statistical databases with random sample queries*, ACM Transactions on Database Systems, 5(3):291–315, September 1980.
[8] I. Dinur and K. Nissim. Revealing information while preserving privacy. In *Proceedings of the 22nd ACM SIGMOD-SIGACT-SIGART Symposium on Principles of Database Systems*, pages 202–210, 2003.
[9] D. Dobkin, A.K. Jones, and R.J. Lipton, Secure databases: Protection against user influence. *ACM Trans. Database Syst. 4*(1), pp. 97–106, 1979.
[10] Y. Dodis, L. Reyzin and A. Smith, Fuzzy extractors: How to generate strong keys from biometrics and other noisy data. In *Proceedings of EUROCRYPT 2004*, pp. 523–540, 2004.
[11] Y. Dodis and A. Smith, Correcting Errors Without Leaking Partial Information, In *Proceedings of the 37th ACM Symposium on Theory of Computing*, pp. 654–663, 2005.
[12] C. Dwork, F. McSherry, K. Nissim, and A. Smith. Calibrating noise to sensitivity in private data analysis. In *Proceedings of the 3rd Theory of Cryptography Conference*, pages 265–284, 2006.
[13] C. Dwork and K. Nissim. Privacy-preserving datamining on vertically partitioned databases. In *Advances in Cryptology: Proceedings of Crypto*, pages 528–544, 2004.
[14] A. Evfimievski, J. Gehrke, and R. Srikant. Limiting privacy breaches in privacy preserving data mining. In *Proceedings of the 22nd ACM SIGMOD-SIGACT-SIGART Symposium on Principles of Database Systems*, pages 211–222, June 2003.
[15] S. Goldwasser and S. Micali, Probabilistic encryption. *Journal of Computer and System Sciences 28*, pp. 270–299, 1984; prelminary version appeared in *Proceedings 14th Annual ACM Symposium on Theory of Computing*, 1982.
[16] N. Nisan and D. Zuckerman. Randomness is linear in space. *J. Comput. Syst. Sci.*, 52(1):43–52, 1996.
[17] Ronen Shaltiel. Recent developments in explicit constructions of extractors. *Bulletin of the EATCS*, 77:67–95, 2002.
[18] Sweeney, L., Weaving technology and policy together to maintain confidentiality. J Law Med Ethics, 1997. 25(2-3): p. 98-110.
[19] L. Sweeney, Achieving k-anonymity privacy protection using generalization and suppression. *International Journal on Uncertainty, Fuzziness and Knowledge-based Systems, 10* (5), 2002; 571-588.

The One Way to Quantum Computation

Vincent Danos[1], Elham Kashefi[2], and Prakash Panangaden[3]

[1] Université Paris 7 & CNRS
vincent.danos@pps.jussieu.fr
[2] IQC - University of Waterloo & Christ Church - Oxford
ekashefi@iqc.ca
[3] School of Computer Science, McGill University, Montréal, Québec, Canada
prakash@cs.mcgill.ca

Abstract. Measurement-based quantum computation has emerged from the physics community as a new approach to quantum computation where measurements rather than unitary transformations are the main driving force of computation. Among measurement-based quantum computation methods the recently introduced one-way quantum computer [RB01] stands out as basic and fundamental.

In this work we a concrete syntax and an algebra of these patterns derived from a formal semantics. We developed a rewrite theory and proved a general standardization theorem which allows all patterns to be put in a semantically equivalent standard form.

1 Introduction

The emergence of quantum computation has changed our perspective on many fundamental aspects of computing: the nature of information and how it flows, new algorithmic design strategies and complexity classes and the very structure of computational models [NC00]. New challenges have been raised in the physical implementation of quantum computers. This paper is a contribution to a nascent discipline: quantum programming languages.

This is more than a search for convenient notation, it is an investigation into the structure, scope and limits of quantum computation. The main issues are questions about how quantum processes are defined, how quantum algorithms compose, how quantum resources are used and how classical and quantum information interact.

In the mid 1980s Deutsch [Deu87] showed how to use superposition – the ability to produce linear combinations of quantum states – to obtain computational speedup. The most dramatic results were Shor's celebrated polytime factorization algorithm [Sho94] and Grover's sublinear search algorithm [Gro98]. Remarkably one of the problematic aspects of quantum theory, the presence of non-local correlation – an example of which is called "entanglement" – turned out to be crucial for these algorithmic developments.

Only recently has there been significant interest in quantum programming languages; i.e. the development of formal syntax and semantics and the use of

M. Bugliesi et al. (Eds.): ICALP 2006, Part II, LNCS 4052, pp. 13–21, 2006.

standard machinery for reasoning about quantum information processing. The first definitive treatment of a quantum programming language was the flowchart language of Selinger [Sel04]. It was based on combining classical control, as traditionally seen in flowcharts, with quantum data.

So far the main framework to explore quantum computation has been the circuit model [NC00], based on unitary evolution. This is very useful for algorithmic development and complexity analysis. Recently physicists have introduced novel ideas based on the use of measurement and entanglement to perform computation [GC99, RB01, RBB03]. This is very different from the circuit model where measurement is done only at the end to extract classical output. In measurement-based computation the main operation to manipulate information and control computation is measurement. This is surprising because measurement creates indeterminacy yet it is used to express deterministic computation defined by a unitary evolution.

A computation consists of a phase in which a collection of qubits are set up in a standard entangled state. Then measurements are applied to individual qubits and the outcomes of the measurements may be used to determine further measurements. Finally – again depending on measurement outcomes – local unitary operators, called corrections, are applied to some qubits; this allows the elimination of the indeterminacy introduced by measurements. The phrase "one-way" is used to emphasize that the computation is driven by irreversible measurements.

Our approach to understanding the structural features of measurement-based computation is to develop a formal calculus. One can think of this as an "assembly language" for measurement-based computation. Ours is the first programming framework specifically based on the one-way model. We first develop a notation for such classically correlated sequences of entanglements, measurements, and local corrections. Computations are organized in patterns[1], and we give a careful treatment of the composition and tensor product (parallel composition) of patterns. We show next that such pattern combinations reflect the corresponding combinations of unitary operators. An easy proof of universality follows.

The idea of computing based on measurements emerged from the teleportation protocol [BBC+93]. The goal of this protocol is for an agent to transmit an unknown qubit to a remote agent without actually sending the qubit. This protocol works by having the two parties share a maximally entangled state called a Bell pair. The parties perform *local* operations – measurements and unitaries – and communicate only classical bits. Remarkably, from this classical information the second party can reconstruct the unknown quantum state. In fact one can actually use this to compute via teleportation by choosing an appropriate measurement [GC99]. This is the key idea of measurement-based computation.

It turns out that the above method of computing is actually universal. This was first shown by Gottesman and Chuang [GC99] who used two-qubit

[1] We use the word "pattern" rather than "program", which is what they are, because this corresponds to the commonly used terminology in the physics literature.

measurements and given Bell pairs. In the one-way computer, invented by Raussendorf and Briegel [RB01], one uses only single-qubit measurements with a particular multi-party entangled state, the cluster state.

There are at least two reasons to take measurement-based models seriously: one conceptual and one pragmatic. The main pragmatic reason is that the *one-way* model is believed by physicists to lend itself to easier implementations, see, for example [Nie04]. Conceptually the measurement-based model highlights the role of entanglement and separates the quantum and classical aspects of computation; thus it clarifies in particular the interplay between classical control and the quantum evolution process.

The main point of this paper is to introduce alongside our notation, a calculus of local equations over patterns that exploits some special algebraic properties of the entanglement, measurement and correction operators. More precisely, we use the fact that that 1-qubit XY measurements are closed under conjugation by Pauli operators and the entanglement command belongs to the normalizer of the Pauli group. We show that this calculus is sound in that it preserves the interpretation of patterns. Most importantly, we derive from it a simple algorithm by which any general pattern can be put into a standard form where entanglement is done first, then measurements, then corrections. We call this *standardization.*

The consequences of the existence of such a procedure are far-reaching. Since entangling comes first, one can prepare the entire entangled state needed during the computation right at the start: one never has to do "on the fly" entanglements. Furthermore, the rewriting of a pattern to standard form reveals parallelism in the pattern computation. In a general pattern, one is forced to compute sequentially and obey strictly the command sequence, whereas after standardization, the dependency structure is relaxed, resulting in lower depth complexity.

The full paper develops the one-way model *ab initio* but there may be certain concepts with which the reader might not be familiar: qubits, unitaries, measurements, Pauli operators and the Clifford group; these are all readily accessible through the excellent book of Nielsen and Chuang [NC00].

2 Measurement Patterns

We first develop a notation for 1-qubit measurement based computations. The basic commands one can use in a pattern are:

- 1-qubit auxiliary preparation N_i
- 2-qubit entanglement operators E_{ij}
- 1-qubit measurements M_i^α
- and 1-qubit Pauli operators corrections X_i and Z_i

The indices i, j represent the qubits on which each of these operations apply, and α is a parameter in $[0, 2\pi]$. Expressions involving angles are always evaluated modulo 2π. These types of command will be referred to as N, E, M

and C. Sequences of such commands, together with two distinguished – possibly overlapping – sets of qubits corresponding to inputs and outputs, will be called *measurement patterns*, or simply patterns. These patterns can be combined by composition and tensor product.

Importantly, corrections and measurements are allowed to depend on previous measurement outcomes. We prove in the full paper that patterns without those classical dependencies can only realize unitaries that are in the Clifford group. Thus dependencies are crucial if one wants to define a universal computing model and it is also crucial to develop a notation that will handle these dependencies.

Commands. Preparation N_i prepares qubit i in state $|+\rangle_i$. The entanglement commands are defined as $E_{ij} := \wedge Z_{ij}$ (controlled-Z), while the correction commands are the Pauli operators X_i and Z_i.

Measurement M_i^α is defined by orthogonal projections on

$$|+_\alpha\rangle := \frac{1}{\sqrt{2}}(|0\rangle + e^{i\alpha}|1\rangle)$$
$$|-_\alpha\rangle := \frac{1}{\sqrt{2}}(|0\rangle - e^{i\alpha}|1\rangle)$$

followed by a trace out operator. The parameter $\alpha \in [0, 2\pi]$ is called the *angle* of the measurement. For $\alpha = 0$, $\alpha = \frac{\pi}{2}$, one obtains the X and Y Pauli measurements. Operationally, measurements will be understood as destructive measurements, consuming their qubit. The *outcome* of a measurement done at qubit i will be denoted by $s_i \in \mathbb{Z}_2$. Since one only deals here with patterns where qubits are measured at most once (see condition (D1) below), this is unambiguous. We take the specific convention that $s_i = 0$ if under the corresponding measurement the state collapses to $|+_\alpha\rangle$, and $s_i = 1$ if to $|-_\alpha\rangle$.

Outcomes can be summed together resulting in expressions of the form $s = \sum_{i \in I} s_i$ which we call *signals*, and where the summation is understood as being done in $\mathbb{Z}_2$. We define the *domain* of a signal as the set of qubits on which it depends.

Dependent corrections will be written X_i^s, Z_i^s and $Z_i^{\alpha,s}$ and dependent measurements will be written ${}^t[M_i^\alpha]^s$, where $s, t \in \mathbb{Z}_2$ and $\alpha, \beta \in [0, 2\pi]$. The meaning of dependencies for corrections is straightforward: $X_i^0 = Z_i^0 = Z_i^{\alpha,0} = I$ (no correction is applied), while $X_i^1 = X_i$, $Z_i^1 = Z_i$ and $Z_i^{\alpha,1} = Z_i^\alpha$. In the case of dependent measurements, the measurement angle will depend on s, t and α as follows:

$${}^t[M_i^\alpha]^s := M_i^{(-1)^s\alpha + t\pi} \tag{1}$$

so that, depending on the parities of s and t, one may have to modify the α to one of $-\alpha$, $\alpha + \pi$ and $-\alpha + \pi$. These modifications correspond to conjugations of measurements under X and Z:

$$X_i M_i^\alpha X_i = M_i^{-\alpha} \tag{2}$$
$$Z_i M_i^\alpha Z_i = M_i^{\alpha+\pi} \tag{3}$$

and so we will refer to them as the X and Z-actions. Note that these two actions are commuting, since $-\alpha + \pi = -\alpha - \pi$ up to 2π, and hence the order in which one applies them doesn't matter.

As we will see later, relations (2) and (3) are key to the propagation of dependent corrections, and to obtaining patterns in the standard entanglement, measurement and correction form. Since the measurements considered here are destructive ones, the above equations actually simplify to

$$M_i^\alpha X_i = M_i^{-\alpha} \tag{4}$$
$$M_i^\alpha Z_i = M_i^{\alpha-\pi} \tag{5}$$

Patterns

Definition 1. *Patterns consists of three finite sets V, I, O, and a finite sequence of commands $A_n \dots A_1$, read from right to left, applying to qubits in V in that order, i.e. A_1 first and A_n last, such that:*

(D0) *no command depends on an outcome not yet measured;*
(D1) *no command acts on a qubit already measured;*
(D2) *no command acts on a qubit not yet prepared, unless it is an input qubit;*
(D3) *a qubit i is measured if and only if i is not an output.*

The set V is called the pattern *computation space*, and we write $\mathfrak{H}_V$ for the associated quantum state space $\otimes_{i\in V}\mathbb{C}^2$. The sets I, O will be called respectively the pattern *inputs* and *outputs*, and we will write $\mathfrak{H}_I$, and $\mathfrak{H}_O$ for the associated quantum state spaces. The sequence $A_n \dots A_1$ will be called the pattern *command sequence*, while the triple (V, I, O) will be called the pattern *type*.

To run a pattern, one prepares the input qubits in some input state $\psi \in \mathfrak{H}_I$, while the non-input qubits are all set in the $|+\rangle$ state, then the commands are executed in sequence, and finally the result of the pattern computation is read back from outputs as some $\phi \in \mathfrak{H}_O$. Clearly, for this procedure to succeed, we had to impose the (D0), (D1), (D2) and (D3) conditions. Indeed if (D0) fails, then at some point of the computation, one will want to execute a command which depends on outcomes that are not known yet. Likewise, if (D1) fails, one will try to apply a command on a qubit that has been consumed by a measurement (recall that we use destructive measurements). Similarly, if (D2) fails, one will try to apply a command on a non-existent qubit. Condition (D3) is there to make sure that the final state belongs to the output space $\mathfrak{H}_O$, *i.e.*, that all non-output qubits, and only them, will have been consumed by a measurement when the computation ends.

We will write (D) for the conjunction of our definiteness conditions (D0), (D1), (D2) and (D3). Whether a given pattern verifies (D) or not is statically verifiable on the pattern command sequence. Here is a concrete example:

$$\mathcal{H} := (\{1,2\}, \{1\}, \{2\}, X_2^{s_1} M_1^0 E_{12} N_2)$$

with computation space $\{1,2\}$, inputs $\{1\}$, and outputs $\{2\}$. To run $\mathcal{H}$, one first prepares the first qubit in some input state ψ, and the second qubit in state $|+\rangle$, then these are entangled to obtain $\wedge Z_{12}(\psi_1 \otimes |+\rangle_2)$. Once this is done, the first qubit is measured in the $|+\rangle$, $|-\rangle$ basis. Finally an X correction is applied on the output qubit, if the measurement outcome was $s_1 = 1$.

A last thing to note, is that one does not require inputs and outputs to be disjoint subsets of V. This seemingly innocuous additional flexibility is actually quite useful to give parsimonious implementations of unitaries [DKP05].

Pattern combination. We are interested now in how one can combine patterns into bigger ones.

The first way to combine patterns is by composing them. Two patterns $\mathcal{P}_1$ and $\mathcal{P}_2$ may be composed if $V_1 \cap V_2 = O_1 = I_2$. Provided that $\mathcal{P}_1$ has as many outputs as $\mathcal{P}_2$ has inputs, by renaming the pattern qubits, one can always make them composable.

Definition 2. *The composite pattern $\mathcal{P}_2\mathcal{P}_1$ is defined as:*
— $V := V_1 \cup V_2$, $I = I_1$, $O = O_2$,
— *commands are concatenated.*

The other way of combining patterns is to tensor them. Two patterns $\mathcal{P}_1$ and $\mathcal{P}_2$ may be tensored if $V_1 \cap V_2 = \varnothing$. Again one can always meet this condition by renaming qubits in a way that these sets are made disjoint.

Definition 3. *The tensor pattern $\mathcal{P}_1 \otimes \mathcal{P}_2$ is defined as:*
— $V = V_1 \cup V_2$, $I = I_1 \cup I_2$, *and* $O = O_1 \cup O_2$,
— *commands are concatenated.*

In contrast to the composition case, all unions involved here are disjoint. Therefore commands from distinct patterns freely commute, since they apply to disjoint qubits, and when we say that commands have to be concatenated, this is only for definiteness.

It is routine to verify that the definiteness conditions (D) are preserved under composition and tensor product. These details as well as the operational semantics and denotational semantics and the proof of universality are described in the full paper [DKP]

3 The Measurement Calculus

We turn to the next important matter of the paper, namely standardization. The idea is quite simple. It is enough to provide local pattern rewrite rules pushing Es to the beginning of the pattern, and Cs to the end.

The equations. A first set of equations give means to propagate local Pauli corrections through the entangling operator E_{ij}. Because $E_{ij} = E_{ji}$, there are only two cases to consider:

$$E_{ij}X_i^s = X_i^s Z_j^s E_{ij} \tag{6}$$
$$E_{ij}Z_i^s = Z_i^s E_{ij} \tag{7}$$

These equations are easy to verify and are natural since E_{ij} belongs to the Clifford group, and therefore maps under conjugation the Pauli group to itself.

A second set of equations give means to push corrections through measurements acting on the same qubit. Again there are two cases:

$$ {}_t[M_i^\alpha]^s X_i^r = {}_t[M_i^\alpha]^{s+r} \tag{8} $$

$$ {}_t[M_i^\alpha]^s Z_i^r = {}_{t+r}[M_i^\alpha]^s \tag{9} $$

These equations follow easily from equations (4) and (5). They express the fact that the measurements M_i^α are closed under conjugation by the Pauli group, very much like equations (6) and (7) express the fact that the Pauli group is closed under conjugation by the entanglements E_{ij}.

Define the following convenient abbreviations:

$$ [M_i^\alpha]^s := {}_0[M_i^\alpha]^s, \; {}_t[M_i^\alpha] := {}_t[M_i^\alpha]^0, \; M_i^\alpha := {}_0[M_i^\alpha]^0, $$
$$ M_i^x := M_i^0, \; M_i^y := M_i^{\frac{\pi}{2}} $$

The rewrite rules. We now define a set of rewrite rules, obtained by orienting the equations above:

$$
\begin{array}{lll}
E_{ij}X_i^s & \Rightarrow X_i^s Z_j^s E_{ij} & EX \\
E_{ij}Z_i^s & \Rightarrow Z_i^s E_{ij} & EZ \\
{}_t[M_i^\alpha]^s X_i^r & \Rightarrow {}_t[M_i^\alpha]^{s+r} & MX \\
{}_t[M_i^\alpha]^s Z_i^r & \Rightarrow {}_{r+t}[M_i^\alpha]^s & MZ
\end{array}
$$

to which we need to add the *free commutation rules*, obtained when commands operate on disjoint sets of qubits:

$$
\begin{array}{ll}
E_{ij}A_{\boldsymbol{k}} \Rightarrow A_{\boldsymbol{k}}E_{ij} & \text{where } A \text{ is not an entanglement} \\
A_{\boldsymbol{k}}X_i^s \Rightarrow X_i^s A_{\boldsymbol{k}} & \text{where } A \text{ is not a correction} \\
A_{\boldsymbol{k}}Z_i^s \Rightarrow Z_i^s A_{\boldsymbol{k}} & \text{where } A \text{ is not a correction}
\end{array}
$$

where $\boldsymbol{k}$ represent the qubits acted upon by command A, and are supposed to be distinct from i and j. Clearly these rules could be reversed since they hold as equations but we are orienting them this way in order to obtain termination.

Condition (D) is easily seen to be preserved under rewriting.

Under rewriting, the computation space, inputs and outputs remain the same, and so do the entanglement commands. Measurements might be modified, but there is still the same number of them, and they still act on the same qubits. The only induced modifications concern local corrections and dependencies. If there was no dependency at the start, none will be created in the rewriting process.

In this conference version of the paper we omit all proofs.

Standardization. Write $\mathcal{P} \Rightarrow \mathcal{P}'$, respectively $\mathcal{P} \Rightarrow^\star \mathcal{P}'$, if both patterns have the same type, and one obtains the command sequence of $\mathcal{P}'$ from the command sequence of $\mathcal{P}$ by applying one, respectively any number, of the rewrite rules of the previous section. We say that $\mathcal{P}$ is *standard* if for no $\mathcal{P}'$, $\mathcal{P} \Rightarrow \mathcal{P}'$ and the procedure of writing a pattern to standard form is called standardization[2].

[2] We use the word "standardization" instead of the more usual "normalization" in order not to cause terminological confusion with the physicists' notion of normalization.

One of the most important results about the rewrite system is that it has the desirable properties of determinacy (confluence) and termination (standardization). In other words, we will show that for all $\mathcal{P}$, there exists a unique standard $\mathcal{P}'$, such that $\mathcal{P} \Rightarrow^\star \mathcal{P}'$. It is, of course, crucial that the standardization process leaves the semantics of patterns invariant. This is the subject of the next simple, but important, proposition,

Proposition 4. *Whenever* $\mathcal{P} \Rightarrow^\star \mathcal{P}'$, $[\![\mathcal{P}]\!] = [\![\mathcal{P}']\!]$.

We now state the main results. First, we state termination.

Proposition 5 (Termination). *For all* $\mathcal{P}$, *there exists finitely many* $\mathcal{P}'$ *such that* $\mathcal{P} \Rightarrow^\star \mathcal{P}'$.

The next theorem establishes the important determinacy property and furthermore shows that the standard patterns have a certain canonical form which we call the NEMC form. The precise definition is:

Definition 6. *A pattern has a* NEMC *form if its commands occur in the order of* N*s first,* E*s ,* M*s, and then* C*s.*

We will usually just say "EMC" form since we can assume that all the auxiliary qubits are prepared in the $|+\rangle$ state we usually just elide these N commands.

Theorem 1 (Confluence). *For all* $\mathcal{P}$, *there exists a unique standard* $\mathcal{P}'$, *such that* $\mathcal{P} \Rightarrow^\star \mathcal{P}'$, *and* $\mathcal{P}'$ *is in EMC form.*

We conclude this subsection by emphasizing the importance of the EMC form. Since the entanglement can always be done first we can always derive the entanglement resource needed for the whole computation right at the beginning. After that only local operations will be performed. This will separate the analysis of entanglement resource requirements from the classical control.

4 Conclusion

We have presented a calculus for the one-way quantum computer. We have developed a syntax of patterns and, much more important, an *algebra* of pattern composition. We have seen that pattern composition allows for a structured proof of universality, which also results in parsimonious implementations. We develop an operational and denotational semantics for this model; in this simple first-order setting their equivalence is clear.

We have developed a rewrite system for patterns which preserves the semantics. We have shown further that our calculus defines a quadratic-time standardization algorithm transforming any pattern to a standard form where entanglement is done first, then measurements, then local corrections.

We feel that our measurement calculus has shown the power of the formalisms developed by the programming languages and logics community to analyze quantum computations. The ideas that we use: rewriting theory, (primitive) type theory and above all, the importance of reasoning compositionally, locally and modularly, are standard for the analysis of traditional programming languages. However, for quantum computation these ideas are in their infancy. It is not merely a question of adapting syntax to the quantum setting; there are fundamental new ideas that need to be confronted. What we have done here is to develop such a theory in a new, physically-motivated setting.

Acknowledgments

We have benefited from discussions with Hans Briegel, Anne Broadbent, Dan Browne, Ellie D'Hondt, Philippe Jorrand, Harold Olivier, Simon Perdrix and Marcus Silva. Prakash Panangaden thanks EPSRC and NSERC for support and Samson Abramsky and the Oxford University Computing Laboratory for inspiration and hospitality at Oxford.

References

[BBC+93] C Bennett, G Brassard, C Crepeau, R Jozsa, A Peres, and W Wootters. Teleporting an unknown quantum state via dual classical and EPR channels. *Phys Rev Lett*, pages 1895–1899, 1993.

[Deu87] D. Deutsch. Quantum computers. *Computer Bulletin*, 3(2):24, June 1987.

[DKP] Vincent Danos, Elham Kashefi, and Prakash Panangaden. The measurement calculus. Available from `www.cs.mcgill.ca/~prakash/pubs.html`.

[DKP05] V. Danos, E. Kashefi, and P. Panangaden. Robust and parsimonious realisations of unitaries in the one-way model. *Phys. Rev. A*, 72, 2005.

[GC99] D. Gottesman and I. L. Chuang. Quantum teleportation is a universal computational primitive. *Nature*, 402:390, 1999.

[Gro98] Lov K. Grover. A framework for fast quantum mechanical algorithms. In *Proceedings of STOC'98 – Symposium on Theory of Computing*, pages 53–62, 1998.

[NC00] M.A. Nielsen and I.L. Chuang. *Quantum Computation and Quantum Information*. Cambridge University Press, Cambridge, 2000.

[Nie04] M. A. Nielsen. Optical quantum computation using cluster states. *Phys. Rev. Lett.*, 93(4):040503, 2004.

[RB01] R. Raussendorf and H.-J. Briegel. A one-way quantum computer. *Physical Review Letters*, 86(5188), 2001.

[RBB03] R. Raussendorf, D. E. Browne, and H.-J. Briegel. Measurement-based quantum computation on cluster states. *Phys. Rev. A*, 68(022312), 2003.

[Sel04] P. Selinger. Towards a quantum programming language. *Mathematical Structures in Computer Science*, 14(4):527, 2004.

[Sho94] P.W. Shor. Algorithms for quantum computation: Discrete logarithms and factoring. In *Proceedings of FOCS'94 – Symposium on Foundations of Computer Science*, page 124, 1994.

Efficient Zero Knowledge on the Internet

Ivan Visconti

Dip. di Informatica ed Appl., Università di Salerno
Via S. Allende, 84081 Baronissi (SA) - Italy
visconti@dia.unisa.it

Abstract. The notion of concurrent zero knowledge has been introduced by Dwork et al. [STOC 1998] motivated by the growing use of asynchronous networks as the Internet.

In this paper we show a transformation that, for any language L admitting a Σ-protocol, produces a 4-round concurrent zero-knowledge argument system with concurrent soundness in the bare public-key (BPK, for short) model. The transformation only adds $O(1)$ modular exponentiations, and uses standard number-theoretic assumptions and polynomial-time simulation.

A tool that we construct and use for our main result is that of efficient concurrent equivocal commitments. We give an efficient construction of this gadget in the BPK model that can be of independent interest.

1 Introduction

In several settings the original notion of zero knowledge [1] (which only considers one prover and one verifier that run the proof in isolation) was insufficient. The notion of *concurrent zero knowledge* [2] formalizes security in a scenario in which several verifiers interact concurrently with the same prover and maliciously coordinate their actions so to extract information from the proofs. This notion is being studied in the plain model where there is no additional set-up infrastructure or network assumption. In [3] it has been showed that in the plain model constant-round black-box concurrent zero knowledge is impossible for non-trivial languages. In the plain model, the most efficient concurrent zero-knowledge proof systems has been presented in [4] on top of a more general result [5]. In [4], any language L that admits an efficient Σ-protocol[1] is transformed in a concurrent zero knowledge proof system. Unfortunately both the round complexity and the number of modular exponentiations required by the resulting protocol are $\omega(\log n)$. Other models are being studied to achieve efficient, and, in particular, constant-round concurrent zero-knowledge protocols. Specifically, the timing model [2] makes other assumptions on the network asynchronousity; the preprocessing model [6] requires an interactive preprocessing

[1] The transformation can be applied to a more general class of protocols but in this paper we focus on Σ-protocols.

M. Bugliesi et al. (Eds.): ICALP 2006, Part II, LNCS 4052, pp. 22–33, 2006.

stage involving all parties; the common random/reference string models [7, 8], require a trusted third party or a physical assumption; the single-prover model [9] assumes the existence of only one stateful prover.

The model that seems to have the minimal set-up or network assumptions is the bare public-key (BPK) model [10], where verifiers register their public keys in a public file during a set-up stage. There is no interactive preprocessing stage, no trusted third party, no physical assumption, no assumption on the asynchronousity of the network. In this model concurrent soundness is harder to achieve than sequential soundness, as noted in [11], who discussed four distinct and increasingly stronger soundness notions. Indeed, the constant-round concurrent zero-knowledge (in fact, resettable zero-knowledge, a stronger notion from [10]) protocols in the BPK model presented in [10, 11] only enjoy sequential soundness. In [12] a constant-round concurrently sound concurrent zero-knowledge argument system in the BPK model is presented under non-standard assumptions on the hardness of computational problems against sub-exponential-time adversaries. The use of such non-standard assumptions is referred to as "complexity leveraging" and is very related to the notion of super-polynomial-time simulation used in [13] and both correspond to relaxed notions of security.

Equivocal commitment schemes. A commitment scheme is a two-phase protocol between two polynomial-time Turing machines `sen` and `rec`. The security of this primitive is based on the following properties: 1) hiding, i.e., a cheating `rec` can not guess with probability significantly better than 1/2 which message has been committed over any possible pair of different messages; 2) binding, i.e., a cheating `sen` should be able to open a commitment (i.e., to decommit) with both m and $m' \neq m$ only with very small (i.e., negligible) probability.

An equivocal commitment scheme is a special commitment scheme. It allows an efficient algorithm, referred to as equivocator, to violate at its wish the binding property and at the same time, no efficient malicious receiver $\texttt{rec}^\star$ detects this cheating behavior with respect to commitments and decommitments of honest senders. Obviously any equivocator needs a special feature that is not available to any malicious sender $\texttt{sen}^\star$, otherwise the existence of the equivocator contradicts the binding property. Several special features for the equivocator have been proposed in the past as knowledge of an auxiliary information [14] (i.e., so called "trapdoor commitments"), knowledge of the description of the adversarial receiver [15], rewinding capabilities [6]. A constant-round equivocal commitment scheme in the BPK model was presented in [16]. As for the case of the notion of soundness, the notion of binding of an equivocal commitment scheme in the BPK is subtle. The authors of [16] showed that a concurrent malicious sender could succeed in a protocol that instead is secure with respect to sequential adversarial senders. The construction given in [16] is a *concurrent* equivocal commitment scheme in the BPK model and thus is secure with respect to concurrent malicious senders. In the proposed scheme, the commitment phase needs 3-round while the decommitment phase is non-interactive. Unfortunately the construction is not practical since the number of modular exponentiations it needs is linear in the length of the security parameter.

Concurrent zero knowledge from concurrent equivocal commitments. In [8] efficient concurrent equivocal commitments are used to achieve efficient concurrent zero-knowledge argument systems in both the common reference string and the shared random string models. More concretely, given a language L with an efficient Σ-protocol and an efficient concurrent equivocal commitment scheme, the results of [8] produce an efficient 3-round concurrent zero-knowledge argument system. The transformation only adds a few modular exponentiations to the computations required by the Σ-protocol. Proof (in contrast to argument) systems with similar properties have been recently showed in [17]. However the common reference string and the shared random string models need the existence of a trusted third party or a physical assumption.

Another efficient transformation was presented in [6]. It adds only $O(1)$ rounds and $O(1)$ modular exponentiations to the computations of the Σ-protocol but unfortunately they require a strong set-up assumption. [6] needs an interactive preprocessing for each proof that has to be run later (this seems to be very problematic in practice).

A very challenging open question is therefore the possibility of constructing a transformation as efficient as the ones of [6, 8] but that works with a seemingly better set-up assumption.

In [16], the constructions given in [8, 6] are implemented in the BPK model under standard number-theoretic assumptions and polynomial-time simulation using concurrent equivocal commitment schemes in the BPK model. Unfortunately, in contrast with the efficient transformations of [8, 6], the transformation of [16] adds a number of modular exponentiations that is linear in the size of the challenge of the Σ-protocol. This overhead is added by their implementation of the concurrent equivocal commitment scheme in the BPK model.

1.1 Our Results

In this paper we show a more efficient transformation that only adds $O(1)$ rounds and $O(1)$ modular exponentiations to the ones required by the Σ-protocol and that works in the BPK model.

More precisely, we show a transformation that, for any language L admitting a Σ-protocol, produces a 4-round (i.e., the round complexity is optimal [11]) concurrent zero-knowledge argument of knowledge with concurrent soundness in the BPK model that only adds $O(1)$ modular exponentiations and uses standard number-theoretic assumptions and polynomial-time simulation. This improves all previous results since they either were in models with stronger set-up or network assumptions, or they were not efficient, or they were not fully concurrently secure, or they were using non-standard intractability assumptions.

Following the previous approaches, a tool that we construct and use is that of efficient concurrent equivocal commitments in the BPK model. We give an efficient construction of this gadget in the BPK model (it needs only $O(1)$ modular exponentiations) and that can be of independent interest.

2 The BPK Model and Its Players

In the BPK model verifiers have to announce their public keys and provers have to download the file with all public keys before any protocol starts. No public key is certified and the directory containing the registered users can even be completely controlled by the adversary. Therefore the BPK model is a relaxed version of known set-up assumptions in cryptography as the public-key infrastructure, and the interactive preprocessing. The BPK model does not assume the existence of any trusted third party neither of any physical assumption. It is therefore more practical than the common reference and shared random string models (see [8] for efficient concurrent zero knowledge in these models). Since the BPK model does not need interaction during the preprocessing, it is more practical then the interactive preprocessing used in [6] (see Section 5.2 of [6]). Moreover, when protocols start there is no assumption on the asynchronousity of the network in contrast to the timing model [2]. When the first stage is completed, only a bounded number of verifiers can play in the second stage, this is more attractive than the single-prover requirement of [9].

The BPK model for commitment schemes. The definitions for argument systems in the BPK model can be found in [11, 16]. Here we give the definitions of commitment schemes in the BPK model. In particular we consider the notion of a concurrent equivocal commitment scheme. This definition was implicit in [16]. In our notation we use "for all x" meaning for all values x depending on the security parameter n and of length polynomial in n. Formally, there exists a public file F that is a collection of records, each containing a public key.

An (honest) sender $\mathtt{sen}$ is an interactive deterministic polynomial-time Turing machine that takes as input a security parameter 1^n, a file F, a string m (i.e., the message to be committed), a reference $\mathtt{pk}$ to an entry of F and a random tape. $\mathtt{sen}$ after running an interactive protocol with a receiver $\mathtt{rec}$ outputs $\mathtt{aux}$ or the special symbol $\perp$; later $\mathtt{sen}$ uses $\mathtt{aux}$ to compute $\mathtt{dec}$ and sends the pair $(\mathtt{dec}, m)$ to $\mathtt{rec}$ in order to open the committed message m.

An (honest) receiver $\mathtt{rec}$ is an interactive deterministic polynomial-time Turing machine that works along with $\mathtt{sen}$ in the following two stages: 1) on input a security parameter 1^n and a random tape, $\mathtt{rec}$ generates a key pair $(\mathtt{pk}, \mathtt{sk})$ and stores the public key $\mathtt{pk}$ in one entry of the file F; this stage is executed only once by each receiver; 2) $\mathtt{rec}$ takes as input the secret key $\mathtt{sk}$, and a random string, and outputs a message $\mathtt{com}$ or the special symbol $\perp$ after performing an interactive protocol with a sender $\mathtt{sen}$; later $\mathtt{rec}$ receives the pair $(\mathtt{dec}, m)$ from $\mathtt{sen}$ and verifies that the pair $(\mathtt{com}, \mathtt{dec})$ corresponds to a message m. The interaction between senders and receivers start after all receivers have completed their first stage.

Malicious senders in the BPK model. We say that $\mathtt{sen}^\star$ is an *s-concurrent malicious* sender if it is a probabilistic polynomial-time Turing Machine that, on input 1^n and PK, can perform the $s(n)$ interactive protocols with a receiver $\mathtt{rec}$ as follows: 1) if $\mathtt{sen}^\star$ is already running i protocols $0 \leq i < s(n)$ he can start a new protocol with rec; 2) he can send a message for any running protocol, receive immediately the response from rec and continue.

Given an s-concurrent malicious sender sen* and a honest receiver rec, an *s-concurrent attack* of sen* is performed as follows: 1) the first stage of rec is run on input 1^n and a random string to obtain a pair (pk, sk); 2) sen* is run on input 1^n and pk to start a new protocol; 3) whenever sen* starts a new protocol, rec uses a new random string r and sk, and interacts with sen*.

Malicious receivers in the BPK model. We say that rec* is an *s-concurrent malicious* receiver if it is a probabilistic polynomial-time Turing Machine that, on input 1^n and pk, can perform the following $s(n)$ interactive protocols with a sender sen: 1) if rec* is already running i protocols $0 \leq i < s(n)$ he can decide the i-th protocol to be started with sen; 2) he can output a message for any running protocol, receive immediately the next message from sen and continue[2].

Given an s-concurrent malicious receiver rec* and a honest sender sen, an *s-concurrent attack* of rec* is performed as follows: 1) in its first stage, rec*, on input 1^n and a random string, generates a public file F; 2) rec* is run on input 1^n and F so to start the first protocol with sen; 3) whenever rec* starts a new protocol, sen uses a new random string, and interacts with rec*.

We now define concurrent equivocal commitments in the BPK model. We stress that we give a definition that works with an interactive commitment phase and a non-interactive decommitment phase since these are the properties of the commitment scheme that we will construct. We assume that parties use n as security parameter.

Definition 1. (sen, rec, Ver) *is a concurrent equivocal* commitment scheme *(CS, for short) in the BPK model if:*

- **correctness:** *for all m, let* sen *be a honest sender that receives m as input in the game described above, then: 1)* rec *outputs* com *2)* sen *outputs* dec *and 3)* Ver *is an efficient algorithm such that* $\mathtt{Ver}(\mathtt{com}, \mathtt{dec}, m) = 1$.
- **binding:** *for all sufficiently large n, for any s-concurrent malicious sender* sen* *that runs the game described above with a honest receiver* rec, *there is a negligible function ν such that for all* com *given in output by* rec *the probability that* sen* *outputs a pair* $(\mathtt{dec}_0, \mathtt{dec}_1)$ *such that* $\mathtt{Ver}(\mathtt{com}, \mathtt{dec}_0, m_0) = 1 \wedge \mathtt{Ver}(\mathtt{com}, \mathtt{dec}_1, m_1) = 1 \wedge m_0 \neq m_1 \wedge |m_0| = |m_1|$ *is less than $\nu(k)$.*
- **hiding:** *for all sufficiently large n, for any pair of same-length vectors $\bar{m}_0$, $\bar{m}_1$ of* POLY(n)*-bit messages, and for any s-concurrent malicious receiver* rec* *that runs the game described above with a honest sender* sen, *the view of* rec* *when interacting with* sen *on input $\bar{m}_0$ is computationally indistinguishable from the one when interacting with* sen *on input $\bar{m}_1$, where* sen *in the i-th session commits to the i-th element of the vector received as input.*
- **equivocality:** *there exists an efficient equivocator M such that for any s-concurrent malicious receiver* rec* *it holds that:*

[2] The message that follows the last message of the commitment phase is the decommitment dec sent by sen.

- *M and* $\mathtt{rec}^\star$ *output respectively* $\mathtt{aux}_i$ *and* $\mathtt{com}_i$ *after the commitment phase of the i-th session; then M on input* $\mathtt{aux}_i$ *and any message m, outputs* $\mathtt{dec}_i$ *such that* $\mathtt{Ver}(\mathtt{com}, \mathtt{dec}_i, m) = 1$.
- *the distribution of the view of* $\mathtt{rec}^\star$ *when interacting with M is computationally indistinguishable from the one when interacting with a honest sender* $\mathtt{sen}$.

3 Equivocal Commitments in the BPK Model

In this section we show an efficient construction for concurrent equivocal commitments in the BPK model that only needs a few modular exponentiations. For the sake of simplifying the notation, sometimes we omit the modulus from modular operations.

SimC, simulatable commitment of a message. Consider (p, q, h) such that p, q are primes, $p = 2q + 1$, h is a generator of the only subgroup G_q of $Z_p^\star$ that has order q. In [4] Micciancio and Petrank presented the following perfectly binding commitment scheme. In order to commit to $z \in Z_q$, the sender chooses a random generator g of G_q and computes $\hat{g} = g^r \bmod p$, $\hat{h} = h^{r+z} \bmod p$ where $r \in_R Z_q$. $\mathtt{sen}$ sends g and $\mathtt{com} = (\hat{g}, \hat{h})$. The corresponding decommitment is the pair (r, z) and it can be verified that $\mathtt{com} = (g^r, h^{r+z})$. As discussed in [4], this commitment scheme that we refer to as SimC, is perfectly binding since $\hat{g}$ and $\hat{h}$ uniquely determine r and $r + z$. Computational hiding follows from the DDH assumption. Moreover, in [4] the authors show that such a computed commitment $\mathtt{com}$ is *simulatable*, in the sense that it admits an efficient 3-round public-coin honest-verifier zero-knowledge proof system for proving that $\mathtt{com}$ is a commitment of z. Their proof system enjoys optimal soundness as there exists at most one challenge that allows an adversarial prover to succeed in proving a false statement. Moreover the simulator perfectly simulates true statements.

SimDlogC, simulatable commitment of a discrete logarithm. A variation of the proof system for SimC can be used to prove that $\mathtt{com}$ is the commitment of the discrete logarithm in base h of an element h' of G_q (actually, $\mathtt{com}$ is an El Gamal encryption of h' and thus it uniquely determines its discrete logarithm in base h). We refer to this scheme as SimDlogC. The proof system works as follows. First the prover computes the pair $(\bar{g}, \bar{h}) = (g^s, h^s)$ for a randomly chosen s in Z_q and sends this pair to the verifier. The verifier answers sending a random challenge c in Z_q. The prover computes $a = cr + s$ and sends it to the verifier. Finally the verifier accepts the proof if and only if $\hat{g}^c\bar{g} = g^a$ and $(\hat{h}/h')^c\bar{h} = h^a$. The only variation with respect to the proof system given in [4] is that here the verifier checks that $(\hat{h}/h')^c\bar{h} = h^a$ while in [4], the verifier has to check that $(\hat{h}/h^z)^c\bar{h} = h^a$. The simulator on input $(\hat{g}, \hat{h})$ randomly chooses a and c and sets $\bar{g} = g^a/\hat{g}^c, \bar{h} = h^a/(\hat{h}/h')^c$. Again, the only update with the simulator of [4] is that h' is used instead of h^z. All properties enjoyed by SimC are obviously enjoyed by SimDlogC.

The proof system of SimDlogC is not a proof of knowledge. Note that in order to compute a commitment com of the discrete logarithm of h' in base h, knowledge of this discrete logarithm is not necessary since it is possible to compute $\mathtt{com} = (\hat{g}, \hat{h}) = (g^r, h^r h')$ with $r \in_R Z_q$. It is possible to honestly run the public-coin honest-verifier zero-knowledge proof system discussed above on such a computed commitment. Indeed, notice that the discrete logarithm z of h' in base h is never used in the proof. Since we have that $\hat{h} = h^r h' = h^{r+z}$, the prover only needs to know r.

OR-composition of SimDlogC. Given $\mathtt{pk}_b = (p, q, h, h_b)$ for $b = 0, 1$, the public-coin honest-verifier zero-knowledge proof system of SimDlogC can be used to prove that a commitment $\mathtt{com}_b$ is the commitment of the discrete logarithm of h_b in base h, where h_b is an element of G_q.

Let com be a commitment of a message z computed using SimDlogC. If z is either the discrete logarithm of h_0 in base h or of h_1 in base h, it is possible to OR-compose [18, 19] the public-coin honest-verifier zero-knowledge proof system of SimDlogC thus proving that com is either a commitment of the discrete logarithm of h_0 in base h or the commitment of the discrete logarithm of h_1 in base h.

Σ-protocols. A Σ-protocol is a 3-round interactive protocol between a PPT honest prover P and a PPT honest verifier V. P and V receive as common input a statement "$x \in L$". P has as auxiliary input a witness y for $x \in L$ (L is an $\mathcal{NP}$-language). At the end of the protocol V decides whether the transcript is accepting with respect to the statement or not. Σ-protocols have the following properties: 1) completeness, means that V always accepts when interacting with P; 2) public coin, means that V sends random bits only; 3) special soundness, means that given two accepting transcripts (a, c, z) and (a, c', z') for a statement "$x \in L$", if $c \neq c'$ then $x \in L$ (i.e., the statement is true) and there exists an efficient extractor, that on input (x, a, c, z, c', z') outputs a witness y for $x \in L$; 4) special honest-verifier zero knowledge, means that there is an efficient algorithm S, referred to as simulator, that on input a true statement "$x \in L$" outputs for any c a pair (a, z) such that the triple (a, c, z) is indistinguishable from the transcript of a conversation between P and V.

$\Sigma^{\mathtt{com}}$, AND-composition of the Σ-protocol on $(\hat{g}, \hat{h})$. When the discrete logarithm z of h' is known and the commitment $\mathtt{com} = (\hat{g}, \hat{h}) = (g^r, h^{r+z})$ of z is computed using the regular procedure, it is possible to use the Σ-protocol of Schnorr [20] composed by means of [18, 19] for proving knowledge of either the discrete logarithm of $\hat{g}$ in base g and the discrete logarithm of $\hat{h}$ in base h. In the example described above one can therefore prove knowledge of both r and $r + z$. We refer to $\Sigma^{\mathtt{com}}$ as this AND-composed Σ-protocol.

Efficient equivocal commitments from any efficient Σ-protocol. In [21, 22] transformations that output equivocal commitments from Σ-protocols were presented. The message space for the resulting equivocal commitment scheme is exactly the challenge space of the considered Σ-protocol. The resulting scheme is non-interactive and works in the common reference string model. We will use it as an ingredient in our construction in the BPK model. The commitment scheme is based on the fact that given a Σ-protocol for proving "$x \in L$", the simulator of the special honest-verifier zero-knowledge property, can output for any m a triple (a, m, z) that is indistinguishable from a real transcript. The sender can thus use this simulator to commit to m. More precisely, the sender sends a to the receiver and then opens the commitment sending the pair (m, z). The receiver outputs m if and only if (a, m, z) is an accepting transcript and $\perp$ otherwise. Knowledge of a witness y for "$x \in L$" allows an equivocator to first send a and then for any m, to compute z such that (a, m, z) is accepting. The binding property crucially needs that the prover does not know y. The hiding property is perfect if the output of the simulator is perfectly indistinguishable from a real conversation.

For the case of $\Sigma^{\mathtt{com}}$, using this transformation we obtain an efficient equivocal commitment scheme $\mathtt{Com}_{\Sigma^{\mathtt{com}}}$ since a few modular exponentiations achieve the commitment and decommitment of an element of Z_q.

3.1 Achieving Concurrent Equivocality

We now show our construction of efficient concurrent equivocal commitments in the BPK model. Our protocol uses SimDlogC and combines it with $\mathtt{Com}_{\Sigma^{\mathtt{com}}}$.

The basic idea is that the receiver $\mathtt{rec}$ during the preprocessing stage generates the public key $\mathtt{PK} = (\mathtt{pk}_0 = (p, q, h, h_0), \mathtt{pk}_1 = (p, q, h, h_1))$ and keeps secret one of the two discrete logarithms of h_0 and h_1 in base h. Then during the protocols he first uses the efficient and special 3-round witness-indistinguishable proof of knowledge (WI-PoK, for short) for proving knowledge of one of the two secret keys. This can be obtained by composing the protocol of Schnorr [20] with the techniques of [18, 19]). This is special since knowledge of the witness (i.e., the discrete logarithm) is only needed for computing the third message. Let $\mathtt{PoK}_1, \mathtt{PoK}_2$ and $\mathtt{PoK}_3$ be the messages played in these 3 rounds. $\mathtt{PoK}_1$ is the only message played in the first round of the resulting concurrent equivocal commitment scheme in the BPK model.

The sender $\mathtt{sen}$ on input m and $\mathtt{PK}$ uses SimDlogC for computing the commitment $\mathtt{com} = (\hat{g}, \hat{h})$ that is the commitment of the discrete logarithm of h_0 in base h (we stress that knowledge of this discrete logarithm is not required for computing this commitment and for running the corresponding proof). Then he uses the sender algorithm of $\mathtt{Com}_{\Sigma^{\mathtt{com}}}$ on input m and $\mathtt{com}$ obtaining a pair (a, z) such that (a, m, z) is an accepting transcript for this Σ-protocol. Then $\mathtt{sen}$ uses the OR-composition of SimDlogC on input $\mathtt{com}$ to compute the first message $\hat{a}$ for proving that $\mathtt{com}$ is either a commitment of the discrete logarithm of h_0 in base h or a commitment of the discrete logarithm of h_1 in base h (we stress that this can be done without knowing the discrete logarithm). The message sent by $\mathtt{sen}$ in the second round is thus $(g, \mathtt{com}, \hat{a}, a, \mathtt{PoK}_2)$ and its output includes the decommitment information $(m, z, \mathtt{aux}_z)$. In the third round the receiver sends $\mathtt{PoK}_3$ and a challenge $\hat{c}$.

$\mathtt{sen}$ uses $(\mathtt{aux}_z, \hat{c})$ to compute $\hat{z}$ and opens the commitment sending $(\hat{z}, m, z)$. The receiver verifies the correctness of the opening by checking that (a, m, z) and $(\hat{a}, \hat{c}, \hat{z})$ are accepting for their corresponding statements.

The key idea of this new commitment scheme is that once the sender has played the second round, in order to decommit, $\mathtt{com}$ must be a commitment of either the first secret key or the second secret key (this follows from the optimal soundness of the OR-composition of SimDlogC). Moreover since $\mathtt{com}$ is perfectly binding, the witness (that therefore corresponds to either $\mathtt{sk}_0$ or $\mathtt{sk}_1$) extracted from two different openings of $\mathtt{Com}_{\Sigma}\mathtt{com}$ is fixed and therefore can not be changed by exploiting concurrent man-in-the-middle attacks[3]. This is a crucial property for proving the binding property with respect to s-concurrent malicious senders. Finally we stress that an equivocator can extract the secret key of the receiver running the extractor of the witness-indistinguishable proof of knowledge. Then he can freely equivocate in a straight-line fashion since knowledge of this secret key allows it to run the equivocator of $\mathtt{Com}_{\Sigma}\mathtt{com}$ and the honest prover algorithm of the OR-composed honest-verifier zero-knowledge proof of SimDlogC. Notice that the number of extraction procedures that the equivocator has to run is bounded by the size of the public file.

Theorem 1. Assuming the intractability of the DDH assumption modulo integers of the form $p = 2q + 1$, for p, q primes, the previously described protocol is an efficient concurrent equivocal commitment scheme in the BPK model.

Proof. Completeness can be verified by inspection. The hiding property follows by the perfect honest-verifier zero-knowledge property of the Σ-protocols and the proof system of SimDlogC.

Assume by contradiction that an s-concurrent malicious sender $\mathtt{sen}^\star$ succeeds in computing two decommitments to different messages of the same commitment. This means that $\mathtt{sen}^\star$ outputs two different messages m, m' and two string z, z' such that for the same message a sent during the second round of the protocol, both transcripts (a, m, z) and (a, m', z') are accepting. By the special-soundness property of $\Sigma^{\mathtt{com}}$ and the optimal soundness of SimDlogC either $\mathtt{sk}_0$ or $\mathtt{sk}_1$ is extracted. This can be used by an algorithm $\mathcal{A}$ to break the discrete logarithm assumption as follows. $\mathcal{A}$ receives a discrete logarithm challenge $\mathtt{pk}$ and then generates the entry of the public file as $\mathtt{PK} = (\mathtt{pk}_0, \mathtt{pk}_1)$ where either $\mathtt{pk}_0$ or $\mathtt{pk}_1$ is equal to $\mathtt{pk}$ (the choice is random) while for the other entry $\mathcal{A}$ knows the secret key. $\mathcal{A}$ runs the protocol with $\mathtt{sen}^\star$ and as discussed above $\mathcal{A}$ extracts one of two secret keys. If the extracted secret key corresponds to $\mathtt{pk}$ then $\mathcal{A}$ breaks the challenge. Assume therefore that $\mathcal{A}$ always extracts the already known secret key. This means that in case $\mathcal{A}$ knows $\mathtt{sk}_0$, he uses it in the concurrent protocols and always extracts $\mathtt{sk}_0$ from the output of $\mathtt{sen}^\star$. The opposite case happens when

[3] The capability of an s-concurrent malicious sender $\mathtt{sen}^\star$ that runs many concurrent interactions with the same receiver allows him to mount a man-in-the-middle attack between the proof of knowledge he receives in a session j and a commitment that he computes and open in a different session j'. This attack could allow the man-in-the-middle to equivocate using the same secret key used in the WI-PoK.

$\mathcal{A}$ always uses $\mathtt{sk}_1$ and extracts $\mathtt{sk}_1$. We can apply the same hybrid arguments of [16] and thus an efficient algorithm $\mathcal{A}'$ obtains from the output of $\mathtt{sen}^\star$ two different openings of the commitment of the i-th session. Moreover, from these openings $\mathcal{A}'$ extracts the same secret key he uses in a given session j. We now distinguish two cases. 1) The WI-PoK given by $\mathcal{A}'$ is session j is completed before the second round of session i is played. This case can not happen otherwise A' by relaying the messages can break the witness indistinguishability of the WI-PoK. 2) The WI-PoK given by $\mathcal{A}'$ in session j is completed after the second round of session i is played. In this case, since $\mathcal{A}'$ has not decided yet which witness has to be used (note that the WI-PoK is special), the probability that the secret key extracted is the equal to the used one is only $1/2$. This contradicts the previous assumption that $\mathcal{A}$ always extract the same witness used in session j.

The equivocality property can be proved as follows. The equivocator runs the extractor of the WI-PoK given by the receiver and obtains a secret $\mathtt{sk}$. Then the equivocator uses SimDlogC for computing a commitment of $\mathtt{sk}$. Now the equivocator can run the honest prover algorithm of $\Sigma^{\mathtt{com}}$ obtaining a and computes $\hat{a}$ using the OR-composed proof system of SimDlogC. Later for any message m he can compute z such that (a, m, z) is accepting for $\Sigma^{\mathtt{com}}$. The same can be done for computing $\hat{a}$ and then $\hat{z}$ such that $(\hat{a}, \hat{c}, \hat{z})$ is accepting for proving that $\mathtt{com}$ is a commitment of one of two secret keys. Note that the equivocator crucially uses knowledge of the secret key and therefore knowledge of the discrete logarithms of the pair computed by means of SimDlogC to commit to the secret key. The running time of the equivocator is polynomial, this is the major benefit of the BPK model. Indeed the equivocator is required to run only one extraction procedure for each entry of the public file. The remaining part of the work is straight-line. Since the size of the public file is polynomial, so is the running time of the equivocator.

The indistinguishability of the equivocator with respect to a real sender can be proved using the following standard hybrid arguments. The game played by the prover is modified by letting it to commit to the same secret key used by the simulator. This change is not noticeable, otherwise the hiding property of SimDlogC is broken[4]. The game of this modified prover differs from the simulation only because the prover uses the simulator of $\Sigma^{\mathtt{com}}$ while the simulator uses the honest prover of $\Sigma^{\mathtt{com}}$. However, the two distributions are perfectly indistinguishable.

4 Efficient Concurrent Zero Knowledge

We now briefly show how to obtain a 4-round concurrently sound concurrent zero-knowledge argument of knowledge in the BPK model for any language L that admits a Σ-protocol. We show an efficient transformation that only adds one round and $O(1)$ modular exponentiations. For the definition of concurrent zero-knowledge with concurrent soundness in the BPK model, please see [16, 11].

[4] Note that this step actually requires additional hybrid arguments since the prover (potentially) commits to a different secret key in a polynomial number of sessions.

Let Σ_L be a Σ-protocol for the language L. The basic idea is that first the prover computes the first message a_L of Σ_L, then he uses the concurrent equivocal commitment scheme constructed in Section 3.1 to compute a commitment $\mathtt{com}_a$ of a_L that is sent to the verifier. Note this commitment is executed running the 3-round protocol discussed in Section 3.1. The verifier appends to the third round of this interactive commitment scheme a challenge c_L of Σ_L. In the fourth message the prover simply opens the commitment $\mathtt{com}_a$ of a_L and computes and sends the third message z_L of Σ_L. The verifier accepts if and only if both the commitment has been correctly opened and (a_L, c_L, z_L) is accepted by the verifier of Σ_L.

Completeness is straightforward. Concurrent zero knowledge is obtained as in the previous works of [6, 8]. The simulator uses the equivocator of the commitment scheme in order to compute commitments that can be equivocated. Consider session i, let $\mathtt{com}_{a_i}$ be the commitment computed by the simulator and let c_i be the challenge of Σ_L sent by $V^\star$. The simulator runs the simulator of the special honest-verifier zero-knowledge property of Σ_L for obtaining an accepting transcript (a_i, c_i, z_i). Then still working as the equivocator it opens $\mathtt{com}_{a_i}$ as a_i and completes the proof. Since the equivocator can compute any polynomial number of equivocal commitments in polynomial time, and since the additional work of S is straight-line, the resulting running time of S is still polynomial in the security parameter. Witness extraction and concurrent soundness follow from the special soundness of Σ_L and the (concurrent) binding of the commitment scheme.

Alternative construction. The previously discussed construction is modular on top of a concurrent equivocal commitment scheme. It is also possible to give a direct construction in which $\mathtt{Com}_\Sigma\mathtt{com}$ is not used. Indeed, after using SimDlogC for computing $\mathtt{com} = (\hat{g}, \hat{h})$, the standard technique proposed by [18, 19] can be used directly. Here the prover has to 1) prove that "com is the commitment of the discrete logarithm of either h_0 or h_1" using SimDlogC, 2) prove knowledge of a) a witness for $x \in L$ or b) the discrete logarithms of $\hat{g}$ in base g and either the one of $\hat{h}$ in bases h_0 or the one of $\hat{h}$ in bases h_1. The honest prover obviously can complete the protocol using the witness y for "$x \in L$". Instead, the simulator will use the secret key committed in com to indistinguishability complete the proof. Concurrent soundness still holds since the commitment com can correspond to at most one secret key.

Acknowledgments

I would like to thank Pino Persiano and Yunlei Zhao for many useful discussions on zero knowledge in the BPK model and the anonymous reviewers of ICALP '06 for their comments. This work has been supported in part by the European Commission through the IST program under Contract IST-2002-507932 ECRYPT and in part through the FP6 program under contract FP6-1596 AEOLUS.

References

1. Goldwasser, S., Micali, S., Rackoff, C.: The Knowledge Complexity of Interactive Proof-Systems. In proc. of (STOC '85). (1985) 291–304
2. Dwork, C., Naor, M., Sahai, A.: Concurrent Zero-Knowledge. In proc. of STOC '98, ACM (1998) 409–418
3. Canetti, R., Kilian, J., Petrank, E., Rosen, A.: Black-Box Concurrent Zero-Knowledge Requires $\omega(\log n)$ Rounds. In proc. of STOC '01, ACM (2001) 570–579
4. Micciancio, D., Petrank, E.: Simulatable Commitments and Efficient Concurrent Zero-Knowledge. In proc. of Eurocrypt '03. Vol. 2045 of LNCS, Springer-Verlag (2003) 140–159
5. Prabhakaran, M., Rosen, A., Sahai, A.: Concurrent Zero-Knowledge with Logarithmic Round Complexity. In proc. of FOCS '02. (2002) 366–375
6. Di Crescenzo, G., Ostrovsky, R.: On Concurrent Zero-Knowledge with Pre-processing. In proc. of Crypto '99. Vol. 1666 of LNCS, 485–502
7. Blum, M., De Santis, A., Micali, S., Persiano, G.: Non-Interactive Zero-Knowledge. SIAM J. on Computing **20** (1991) 1084–1118
8. Damgård, I.: Efficient Concurrent Zero-Knowledge in the Auxiliary String Model. In proc. of Eurocrypt '00. Vol. 1807 of LNCS, Springer-Verlag (2000) 418–430
9. Persiano, G., Visconti, I.: Single-Prover Concurrent Zero Knowledge in Almost Constant Rounds. In proc. of ICALP 05. Vol. 3580 of LNCS, Springer-Verlag (2005) 228–240
10. Canetti, R., Goldreich, O., Goldwasser, S., Micali, S.: Resettable Zero-Knowledge. In proc. of STOC '00, ACM (2000) 235–244
11. Micali, S., Reyzin, L.: Soundness in the Public-Key Model. In proc. of Crypto '01. Vol. 2139 of LNCS, Springer-Verlag (2001) 542–565
12. Di Crescenzo, G., Persiano, G., Visconti, I.: Constant-Round Resettable Zero Knowledge with Concurrent Soundness in the Bare Public-Key Model. In proc. of Crypto '04. Vol. 3152 of LNCS, Springer-Verlag (2004) 237–253
13. Pass, R.: Simulation in Quasi-Polynomial Time and Its Applications to Protocol Composition. In proc. of Eurocrypt '03. Vol. 2045 of LNCS, 160–176
14. Brassard, J., Chaum, D., Crepéau, C.: Minimum Disclosure Proofs of Knowledge. Journal of Computer and System Science **37** (1988) 156–189
15. Pass, R., Rosen, A.: New and Improved Constructions of Non-Malleable Cryptographic Protocols. In proc. of STOC '05, ACM (2005) 533–542
16. Di Crescenzo, G., Visconti, I.: Concurrent Zero Knowledge in the Public-Key Model. In proc. of ICALP 05. Vol. 3580 of LNCS, Springer-Verlag (2005) 816–827
17. Catalano, D., Visconti, I.: Hybrid Trapdoor Commitments and Their Applications. In proc. of ICALP 05. Vol. 3580 of LNCS., Springer-Verlag (2005) 298–310
18. Cramer, R., Damgård, I., Schoenmakers, B.: Proofs of Partial Knowledge and Simplified Design of Witness Hiding Protocols. In proc. of Crypto '94. Vol. 839 of LNCS., Springer-Verlag (1994) 174–187
19. De Santis, A., Di Crescenzo, G., Persiano, G., Yung, M.: On Monotone Formula Closure of SZK. In: Proc. of FOCS '94. (1994) 454–465
20. Schnorr, C.P.: Efficient Signature Generation for Smart Cards. Journal of Cryptology **4** (1991) 239–252
21. Feige, U., Shamir, A.: Zero-Knowledge Proofs of Knowledge in Two Rounds. In proc. of Crypto '89. Vol. 435 of LNCS, Springer-Verlag (1990) 526–544
22. Damgård, I.: On the Existence of Bit Commitment Schemes and Zero-Knowledge Proofs. In proc. of Crypto '89. Vol. 435 of LNCS, Springer-Verlag (1990) 17–27

Independent Zero-Knowledge Sets*

Rosario Gennaro[1] and Silvio Micali[2]

[1] IBM T.J.Watson Research Center, Yorktown Heights, NY, USA
rosario@watson.ibm.com
[2] MIT CSAIL, Cambridge, MA, USA

Abstract. We define and construct *Independent Zero-Knowledge Sets (ZKS) protocols.* In a ZKS protocols, a Prover commits to a set S, and for any x, proves non-interactively to a Verifier if $x \in S$ or $x \notin S$ without revealing any other information about S. In the *independent* ZKS protocols we introduce, the adversary is prevented from successfully correlate her set to the one of a honest prover. Our notion of independence in particular implies that the resulting ZKS protocol is non-malleable.

On the way to this result we define the notion of *independence* for commitment schemes. It is shown that this notion implies non-malleability, and we argue that this new notion has the potential to simplify the design and security proof of non-malleable commitment schemes.

Efficient implementations of ZKS protocols are based on the notion of *mercurial commitments.* Our efficient constructions of independent ZKS protocols requires the design of *new* commitment schemes that are simultaneously independent (and thus non-malleable) and mercurial.

1 Introduction

The notion of Zero Knowledge Sets (ZKS) was introduced by Micali, Rabin and Kilian in [17]. In these protocols, one party (Alice) holds a secret database Db which can be accessed by another party (Bob) via queries. When Bob queries the database on a key x, Alice wants to make sure that nothing apart from $Db(x)$ is revealed to Bob, who at the same time wants some guarantee that Alice is really revealing the correct value.

Micali *et al.* presented a very ingenious solution to this problem, based on a new form of commitment scheme (later termed *mercurial commitments* in [14]). In a nutshell, Alice first commits to the entire database in a very succinct way, and then when Bob queries a given key x, Alice answers with a "proof" π_x that $Db(x) = y$ according to the original commitment. Their solution is efficient and based on the discrete logarithm assumption.

A construction based on general assumptions, and allowing more general queries on the database, was presented in [19]. However their construction required generic ZK proofs, based on Cook-Levin reductions and thus was less efficient than [17]. The original construction in [17] has been generalized to hold under various assumptions in [14] and [5].

* Extended Abstract. An extended version, which contains all formal definitions and proofs, is available at the IACR Eprint Archive: http://eprint.iacr.org/2006/155

M. Bugliesi et al. (Eds.): ICALP 2006, Part II, LNCS 4052, pp. 34–45, 2006.

MALLEABILITY. ZKS protocols guarantee simply that when Bob queries x, only the value of $D(x)$ is disclosed. However, this is only one of possible attacks that can be carried on a cryptographic protocol. It is well known that proving confidentiality may not be sufficient, in an open network like the Internet, where an Adversary can play the role of "man-in-the-middle" between honest parties.

First formalized in [10], the notion of malleability for cryptographic protocols describes a class of attacks in which the adversary is able to correlate her values to secret values held by honest players. In a ZKS protocol, for example, this would take the form of the adversary committing to a set somewhat related to the one of a honest player and then using this to her advantage.

The confidentiality property of ZKS protocols does not prevent such an attack from potentially taking place. Indeed such an attack could be devised against the protocol from [17]. What we need is an enhanced definition of security, to make sure that databases committed by one party are independent from databases committed to by a different party.

NON-MALLEABLE COMMITMENTS. The first non-malleable commitment scheme was presented in [10], but it required several rounds of communication. A breaktrough result came with a paper by Di Crescenzo, Ishai and Ostrovsky (DIO) [8] which constructed a non-interactive and non-malleable commitment scheme. Following the DIO approach several other commitment schemes were presented with improved efficiency or security properties (e.g. [9, 7, 15, 12]).

The DIO approach has a very interesting feature: non-malleability is proven by showing that the commitment satisfies a basic "independence" property (though this property is not formally defined as such), and then it is shown that this property implies non-malleability. All the commitment schemes that followed the DIO approach have a proof of security structured in a similar way. However the only "original" part of the proof in each scheme is the proof that the commitment satisfies this "independence" property. The second part of the proof is basically identical in all the proofs.

OUR CONTRIBUTION

- We define the notion of *Independent Zero Knowledge Sets* which enforces the independence of databases committed by various parties. We also define the notion of independence for *commitment schemes*. This definition captures the crucial notion of security in a DIO-like commitment.
  Once this notion of independence is formalized we restate the second part of the DIO proof as a formal theorem that shows once and for all that independent commitments are non-malleable.
  We believe that isolating the notion of independence has the potential to simplify the design and security proof of non-malleable commitments in the future.
- We present efficient independent ZKS protocols. These protocols are enabled by the efficient constructions of new commitment schemes that are simultaneously independent (and thus non-malleable) and mercurial.
- Finally we define various notions of non-malleability for ZKS protocols. We then ask if the DIO theorem (that independence implies non-malleability for

commitments) holds for ZKS protocols as well. Surprisingly the answer is not that simple. We show under which conditions independent ZKS protocols are also non-malleable.

2 Zero-Knowledge Sets

ZKS DEFINITION. An *elementary database* Db is a subset of $\{0,1\}^* \times \{0,1\}^*$ such that if $(x, v) \in Db$ and $(x, v') \in Db$ then $v = v'$. With $[Db]$ we denote the *support* of Db, i.e. the set of $x \in \{0,1\}^*$ for which $\exists v$ such that $(x, v) \in Db$. We denote such unique v as $Db(x)$; if $x \notin [Db]$ then we also write $Db(x) = \perp$. Thus Db can be thought of as a partial function from $\{0,1\}^*$ into $\{0,1\}^*$.

In a ZKS protocol we have a Prover and Verifier: the Prover has as input a secret database Db. The Prover runs in time polynomial in $||[Db]||$ (the cardinality of the support) and the number of queries, while the Verifier runs in time polynomial in the maximal length of $x \in [Db]$, which we assume to be publicly known. They also have a common input string σ, which can be a random string (in which case we say that we are in the *common random string* model) or a string with some pre-specified structure (in which case we say we are in the *common parameters* model).

The Prover first commits in some way to the database Db. This commitment string is then given as input to the Verifier. Then the Verifier asks a query x and the Prover replies with a string π_x which is a proof about the status of x in Db. The Verifier after receiving π_x outputs a value y (which could be $\perp$) which represents his belief that $Db(x) = y$, or *bad* which represents his belief that the Prover is cheating.

A ZK Set protocol must satisfy completeness, soundness and zero-knowledge. Informally completeness means that if $Db(x) = y$ the Prover should always be able to convince the verifier of this fact. Soundness means that no efficient Prover should be able to produce a commitment to Db, a value x and two proofs π_x, π'_x that convince the Verifier of two distinct values for $Db(x)$. Finally zero-knowledge means that an efficient verifier learns only the values $Db(x)$ from his interaction with the Prover, and nothing else. In particular the Verifier does not learn the values $Db(x')$ for an x' not queried to the Prover (following [13] this is stated using a simulation condition).

2.1 Mercurial Commitments

A mercurial commitment scheme [14] is a commitment with two extra properties:

1. On input a message m, the sender can create two kinds of commitments: a *hard* and a *soft* commitment.
2. There are two kinds of openings: a regular opening and a partial opening or *teasing*.

The crucial properties of a mercurial commitment are: (i) both hard and soft commitments preserve the secrecy of the committed message (semantic security);

(ii) hard commitments are indistinguishable from soft ones; (iii) soft commitments cannot be opened, but can be teased to any value (even without knowing any trapdoor information); (iv) hard commitments can be opened or teased only to a single value (unless a trapdoor is known).

A construction of mercurial commitments was implicitly presented in [17] based on discrete log. More constructions were presented in [14, 5], including one based on general assumptions. Let us recall the discrete log construction and a new one based on RSA[1].

MERCURIAL COMMITMENTS BASED ON DISCRETE LOG. This commitment is based on [1, 18]; the mercurial property was introduced in [17]. The public information is a cyclic group G of prime order q, where multiplication is easy and the discrete log is assumed to be hard. Also two generators g, h for G.

To hard commit to M, choose $\rho, R \in_r Z_q$: let $h_\rho = g^\rho h$ and commit using Ped-Com with bases g, h_ρ i.e. compute $C = g^M h_\rho^R$. The hard commitment is h_ρ, C. The opening is M, R, ρ and the verification of a hard commitment is to check the above equations. To soft commit, choose $\rho, R \in_r Z_q$: let $h_\rho = g^\rho$ and commit to 0 using Ped-Com with bases g, h_ρ i.e. compute $C = h_\rho^R$. The soft commitment is h_ρ, C Notice that in a soft commitment, one actually knows the discrete log of h_ρ with respect to g, while in a hard-commitment computing such discrete log is equivalent to computing $\log_g h$. Thus to tease the above soft commitment to M', one produces R' with $R' = R - M'\rho^{-1} \bmod q$. The verification of such a teasing consists in checking that $g^{M'} h_\rho^{R'} = C$.

MERCURIAL COMMITMENTS BASED ON RSA. This commitment is based on [6]; the mercurial property is an original contribution of this paper. The public information is an RSA modulus N, a prime e, such that $GCD(e, \phi(N)) = 1$; and $s \in_R Z_N^*$. To hard commit to M, choose $\rho, R \in_R Z_N^*$: let $s_\rho = s\rho^e \bmod N$ and commit using RSA-Com with base s_ρ i.e. compute $C = s_\rho^M R^e$. The hard commitment is s_ρ, C. The opening is M, R, ρ and the verification of a hard commitment is to check the above equations. To soft commit, choose $\rho, R \in_r Z_N^*$: let $s_\rho = \rho^e$ and commit to 0 using RSA-Com with base s_ρ i.e. compute $C = R^e$. The soft commitment is s_ρ, C Notice that in a soft commitment, one actually knows e-root of s_ρ, while in a hard-commitment computing such root is equivalent to computing the e-root of s. Thus to tease the above soft commitment to M', one produces R' with $R' = R\rho^{-M'} \bmod N$. The verification of such a teasing consists in checking that $s_\rho^{M'}(R')^e = C \bmod N$.

2.2 Constructing ZK Sets

Using any mercurial commitment it is possible to construct a ZKS protocol as shown in [17].

Let l to denote the maximal length of an input $x \in [Db]$. As we said above we assume this to be a publicly known value. The Prover uses a variation of a

[1] This construction of mercurial commitments based on RSA was independently discovered in [14].

Merkle tree [16]. The Prover builds a tree of height l and stores a commitment to $Db(x)$ in the x-leaf (notice that if $x \notin [Db]$ then $Db(x) = \perp$). Then the Prover stores in each internal node a commitment to the contents of its two children: this is done by hashing the values of the two children using a collision-resistant hash function and then committing to the resulting value. The final commitment to Db is the value stored at the root. To prove the value of $Db(x)$ the Prover just decommits all the nodes in the path from the root to x (in particular this means that he reveals the values stored at their siblings, but without decommitting them), thus providing a Merkle-authentication path from the leaf to the root. The Verifier checks this path by checking the all the decommitments are correct.

Unfortunately the above algorithm runs in time 2^l, no matter what the size of the database is. In order to have the Prover run in time polynomial in $|[Db]|$, a pruning step is implemented as follows. First of all, we use mercurial commitments, to compute the commitments. In the above tree, we consider all the maximal subtrees whose leaves are *not* in $[Db]$. We store a soft-commitment in the roots of those trees. The rest of the tree is computed as above, using the hard commitments. Now the running time of the Prover is at most $2l|[Db]|$ since it is only computing full authentication paths for the leaves inside Db.

The question is now how do you prove the value of $Db(x)$. If $x \in [Db]$ then you just decommit (open) the whole authentication path from its leaf to the root, as before.

Let x be the a query such that $x \notin [Db]$, i.e. $Db(x) = \perp$. Let y be the last node on the path from the root to x that has a commitment stored in it. We associate soft-commitments to the nodes on the path from y to x and their siblings, including x. Then we compute an authentication path from the root to x, except that we *tease* (rather than open) each commitment to the hash of the commitment of the children. Notice that we can seamlessly do this from the root to x. Indeed from the root to y these are either hard or soft commitments, and we only tease the hard ones to their real opening. From y to the leaf those are soft commitments that can be teased to anything.

3 Independent Zero-Knowledge Sets

INDEPENDENT COMMITMENTS. As we said in the introduction, our starting point was the DIO approach [8] to build non-malleable commitments. In order to prove the non-malleability of their commitment scheme they first proved the following property.

Consider the following scenario: ℓ honest parties[2] commit to some messages and the adversary, after seeing their commitment strings, will also produce a commitment value. We require that this string must be different from the commitments of the honest parties (otherwise the adversary can always copy the

[2] To be precise in [8] only the case $\ell = 1$ is considered, and suggested how to easily extend it to constant ℓ. The case of arbitrary ℓ (polynomial in the security parameter) is presented by Damgård and Groth in [7] to construct *reusable* non-malleable commitments.

behavior of a honest party and output an identical committed value). At this point the value committed by the adversary is *fixed*, i.e. no matter how the honest parties open their commitments the adversary will always open in a unique way.

In [8] this property is not formally defined but it is used in a crucial way in the proof of non-malleability. We put forward a formal definition for it (see full version) and we say that such a commitment scheme is *ℓ-independent.* If it is ℓ-independent for any ℓ (polynomial in the security parameter) we say that is simply *independent.*

As mentioned in the Introduction, following the DIO approach, several other non-malleable commitments were presented (e.g. [9, 7, 15, 12]). *All* these schemes are independent according to our definition. Moreover their non-malleability proofs all share the same basic structure: an "original" part which proves that they are independent (once one formalizes the notion, as we did) – and a second part (common to all proofs and which basically goes back to DIO [8]) that the independence property implies non-malleability.

By formalizing the notion of independence we can then rephrase this second part of the DIO proof as a separate theorem:

Theorem 1 (DIO [8]). *If an equivocable commitment scheme is ℓ-independent then it is (ℓ, ϵ)-non-malleable with respect to opening, for any ϵ. As a consequence, if an equivocable commitment scheme is independent then it is ϵ-non-malleable with respect to opening, for any ϵ.*

3.1 Defining Independence for ZK Sets

Let us consider a man-in-the-middle attack for a ZK Sets protocol. In such an attack, the Adversary would interact with the Verifier but while on the background is acting as a verifier himself with a real Prover. Of course we can't prevent the Adversary from relaying messages unchanged from the Prover to the Verifier and vice versa. But we would like to prevent an adversary to commit to a related database to the one committed by the real Prover and then manage to convince the Verifier that $Db(x)$ is a value different than the real one. When we define *independence* for ZKS our goal will be to prevent this type of attacks.

A WEAK DEFINITION. A first approach is to treat ZKS protocols in a similar way as commitments. Then the definition of independence would go as follows. The adversary commits to a set after seeing the commitment of the honest prover, but *before* making any queries about the set committed by the honest prover. What we would like at this point is that the set committed by the adversary is *fixed*, i.e. it will not depend on the answers that the honest prover will provide on queries on his own committed set.

We call the above *weak independence.* This property is easily achieved by combining any ZKS protocol with any independent commitment.

A STRONGER DEFINITION. It may not be reasonable to assume that the adversary does not query the honest provers before committing. Thus a stronger definition of independence allows such queries. However once the adversary has seen the value $Db(x)$ of x in the database Db held by the honest prover, it can always commit to a set which is related to Db by the mere fact that the adversary *knows* something about Db (for example the adversary could committ to Db' where $Db'(x) = Db(x)$).

The idea is to make sure that the set committed by the adversary is *independent from the part of the honest prover's set that the adversary has not yet seen.* Here is how we are going to formalize this.

Consider an adversary $\mathcal{A} = (\mathcal{A}_1, \mathcal{A}_2)$ which tries to correlate its database to the one of a honest prover. $\mathcal{A}_1$ sees the commitments of the honest provers, queries them (concurrently) on some values, and then outputs a commitment of its own. $\mathcal{A}_2$ is given concurrent access to the provers to whom he can ask several database queries while answering queries related to $\mathcal{A}_1$'s commitment. We would like these answers to be independent from the answers of the honest provers *except the ones provided to* $\mathcal{A}_1$ *before committing.*

In other words $\mathcal{A}_1$, after seeing the commitments $Com_1, \ldots, Com_\ell$ of ℓ honest provers, does the following: (i) queries Com_i on some set Q_i of indices, which are answered according to some database Db_i, and then (ii) outputs a commitment Com of its own.

We now run two copies of $\mathcal{A}_2$, in the first we give him access to provers that "open" the Com_i according to the databases Db_i; in the second instead we use some different databases Db'_i. However the restriction is that Db'_i *must agree with* Db_i *on the set of indices* Q_i. At the end $\mathcal{A}_2$ outputs a value x and the corresponding proof π_x with respect to Com. We require that the database value associated to x in the two different copies of $\mathcal{A}_2$ must be the same, which implies that it is "fixed" at the end of the committing stage.

Of course we must rule out the adversary that simply copies the honest provers, as in that case achieving independence is not possible (or meaningful). Thus we require that $\mathcal{A}_1$ output a commitment Com different from the honest provers' Com_i. A formal definition follows.

Given two databases Db, Db' and a set of indices Q we define the operator $\dashv$ as follows: $Db' \dashv_Q Db$ is the database that agrees with Db' on all the indices except the ones in Q where it agrees with Db.
We say that a ZKS protocol is ℓ*-independent* if the following property holds (where Q_i is the list of queries that $\mathcal{A}_1$ makes to the oracle $\mathsf{Sim2}^{Db_i(\cdot)}(\omega_i, Com_i)$):

ZKS ℓ-independence. For any adversary $(\mathcal{A}_1, \mathcal{A}_2)$ and for any pair of ℓ-tuple of databases $Db_1, \ldots, Db_\ell$ and $Db'_1, \ldots, Db'_\ell$ the following probability

$$Pr\left[\begin{array}{c} (\sigma, \omega_0) \leftarrow \mathsf{Sim0}(1^k)\ ;\ (Com_i, \omega_i) \leftarrow Sim1(\omega_0)\ \forall\, i = 1, \ldots, \ell\ ; \\ (Com, \omega) \leftarrow \mathcal{A}_1^{\mathsf{Sim2}^{Db_i(\cdot)}(\omega_i, Com_i)}(\sigma, \omega)\ \text{ with }\ Com \neq Com_i\ \forall i\ ; \\ (x, \pi_x) \leftarrow \mathcal{A}_2^{\mathsf{Sim2}^{Db_i(\cdot)}(\omega_i, Com_i)}(\sigma, \omega)\ ; \\ (x, \pi'_x) \leftarrow \mathcal{A}_2^{\mathsf{Sim2}^{Db'_i \dashv_{Q_i} Db_i(\cdot)}(\omega_i, Com_i)}(\sigma, \omega)\ : \\ bad \neq \mathsf{V}(\sigma, Com, x, \pi_x) \neq \mathsf{V}(\sigma, Com, x, \pi'_x) \neq bad \end{array}\right]$$

is negligible in k.

The above notion guarantees independence only if the adversary interacts with a bounded (ℓ) number of honest provers. We say that a ZKS protocol is *independent* if it is ℓ-independent for any ℓ (polynomial in the security parameter). In this case independence is guaranteed in a fully concurrent scenario where the adversary can interact with as many honest parties as she wants.

THE STRONGEST POSSIBLE DEFINITION. A stronger definition allows $\mathcal{A}_1$ to copy one of the honest provers' commitments, but then restricts $\mathcal{A}_2$ somehow. Namely, we say that either $Com \neq Com_i$ for all i, or if $Com = Com_i$ for some i, the answer of $\mathcal{A}_2$ must be "fixed" on all the values x which she does not ask to the ith prover. We call this *strong independence.*

We say that a ZKS protocol is *strongly ℓ-independent* if the following property holds, where Q_i (resp. Q'_i) is the list of queries that $\mathcal{A}_1$ (resp. $\mathcal{A}_2$) makes to the oracle $\mathsf{Sim2}^{Db_i(\cdot)}(\omega_i, Com_i)$:

ZKS strong ℓ-independence. For any adversary $(\mathcal{A}_1, \mathcal{A}_2)$ and for any pair of ℓ-tuple of databases $Db_1, \ldots, Db_\ell$ and $Db'_1, \ldots, Db'_\ell$ the following probability

$$Pr\left[\begin{array}{c} (\sigma,\omega_0) \leftarrow \mathsf{Sim0}(1^k)\ ;\ (Com_i,\omega_i) \leftarrow Sim1(\omega_0)\ \forall\, i=1,\ldots,\ell\ ; \\ (Com,\omega) \leftarrow \mathcal{A}_1^{\mathsf{Sim2}^{Db_i(\cdot)}(\omega_i,Com_i)}(\sigma,\omega)\ ; \\ (x,\pi_x) \leftarrow \mathcal{A}_2^{\mathsf{Sim2}^{Db_i(\cdot)}(\omega_i,Com_i)}(\sigma,\omega)\ ; \\ (x,\pi'_x) \leftarrow \mathcal{A}_2^{\mathsf{Sim2}^{Db'_i \dashv_{Q_i} Db_i(\cdot)}(\omega_i,Com_i)}(\sigma,\omega)\ : \\ bad \neq \mathsf{V}(\sigma,Com,x,\pi_x) \neq \mathsf{V}(\sigma,Com,x,\pi'_x) \neq bad \text{ AND} \\ ((Com \neq Com_i\ \forall i) \text{ OR } (\exists i : Com = Com_i \text{ AND } x \notin Q'_i)) \end{array}\right]$$

is negligible in k.

Again we say that a ZKS protocol is (strongly) independent if it is (strongly) ℓ-independent for any ℓ (polynomial in the security parameter).[3]

3.2 Constructing Independent ZKS

In this section we show how to modify the original protocol presented in [17] (recalled in Section 2.2) using a different type of commitment which will yield

[3] Note that since we need to open the same database commitment according to two different databases, the definition is stated in terms of the simulated provers. But simulated executions are indistinguishable from real ones, so that independence property holds in real life too.

This is the reason why we restrict the database Db'_i to agree with Db_i on the queries that were asked by the adversary before committing. In our proofs of security this requirement does not matter (i.e. the adversary would not be able to output (x, π_x, π'_x) such that V outputs different values for $Db(x)$ depending on which proof, π_x or π'_x, is provided, *even if Db'* does not agree with Db_i on the set Q_i). But the simulated execution is indistinguishable from the real one only if the answers are consistent. Thus after $\mathcal{A}_1$ has seen a given value for $Db_i(x)$, we need to make sure that in both copies of $\mathcal{A}_2$ the same value appears for $Db_i(x)$, in order to make the simulated run indistinguishable from a real one.

strong independence. This new commitment schemes that we introduce are simultaneously independent and mercurial.

STRONG 1-INDEPENDENCE BASED ON DISCRETE LOG. The starting point of this protocol is Pedersen's commitment, modified as in [17] to make it mercurial. In order to achieve independence we modify the commitment further, using techniques inspired by the non-malleable scheme in [9]. We are going to describe the protocol that achieves strong independence and later show how to modify it if one is interested in just independence.

DLSI-ZKS

- *CRS Generation.* On input 1^k selects a cyclic group G of order q, a k-bit prime, where the discrete logarithm is assumed to be intractable and multiplication is easy. It also chooses three elements $g_1, g_2, h \in_R G$. Finally it selects a collision-resistant function H with output in Z_q. The CRS is $\sigma = (G, q, g_1, g_2, h, H)$
- *Prover's Committing Step.* On input *Db* and the CRS σ. Choose a key pair sk,vk for a signature scheme. Let $\alpha = H(\mathsf{vk})$ and $g_\alpha = g_1^\alpha g_2$. Run the prover's committing step from [17] on *Db* and the mercurial commitment defined by $\sigma_\alpha = (G, q, g_\alpha, h)$ to obtain *Com*, *Dec*. Output *Com*, vk.
- *Prover's Proving Step.* On input x compute π_x with respect to σ_α, Com, Dec using the prover's proving step from [17]. Then output Com, π_x and sig_x a signature on (Com, x) using sk.
- *Verifier.* Check that sig_x is a valid signature of (Com, x) under vk; if yes, compute $\alpha = H(\mathsf{vk})$ and $g_\alpha = g_1^\alpha g_2$ and run the [17] Verifier on $(\sigma_\alpha, Com, x, \pi_x)$, otherwise output *bad*.

Theorem 2. *Under the discrete logarithm assumption,* DLSI-ZKS *is a strong 1-independent zero-knowledge set protocol.*

STRONG INDEPENDENCE UNDER THE STRONG RSA ASSUMPTION. We are going to use the mercurial RSA commitment described in Section 2.1. In order to achieve independence we are going to modify it further using techniques inspired by [7, 12], which require the Strong RSA assumption [2, 20]. Here is a description of the protocol.

SRSA-ZKS

- *CRS Generation.* The key generation algorithm chooses a k-bit modulus, N as the product of two large primes p, q and a random element $s \in_R Z_N^*$. Also selects a collision-resistant hash function H which outputs prime numbers $> 2^{k/2}$. Notice that such primes are relatively prime to $\phi(N)$. The CRS is $\sigma = (N, s, H)$. [4]

[4] We can use the techniques from [12] to implement H efficiently. Also if we choose N as product of two *safe* primes, then it is sufficient for H to outputs primes smaller than $2^{k/2-1}$, speeding up all computations.

- *Prover's Committing Step.* On input *Db* and the CRS σ. Choose a key pair sk,vk for a signature scheme. Let $e = H(\mathsf{vk})$. Run the Prover's committing step from [17] on *Db* and the mercurial commitment defined by $\sigma_e = (N, s, e)$ to obtain Com, Dec. Output Com, vk.
- *Prover's Proving Step.* On input x compute π_x with respect to σ_e, Com, Dec using the Prover's proving step from [17]. Then output Com, π_x and sig_x a signature on (Com, x) using sk.
- *Verifier.* Check that sig_x is a valid signature of (Com, x) under vk; if yes, compute $e = H(\mathsf{vk})$ and run the [17] Verifier on $(\sigma_e, Com, x, \pi_x)$, otherwise output *bad.*

Theorem 3. *Under the Strong RSA Assumption,* SRSA-ZKS *is a strong independent zero-knowledge set protocol.*

If one is interested in simple independence (rather than strong independence) both of the above protocols can be simplified by using more efficient one-time signature schemes for vk and just sign *Com*. Even more efficiently, to obtain independence, one can use a message authentication code in place of a signature scheme (the original idea in [9]). Informally, the basic idea is to commit to a random MAC key a using a basic trapdoor commitment: call this commitment A. Set $\alpha = H(A)$ (resp. $e = H(A)$) and now use it the same way we used α (resp. e) in DLSI-ZKS (resp. SRSA-ZKS). To answer a query x, open A as a, produce π_x and a MAC of *Com* under a. However note that both these variations (one-time signatures or MAC) cannot be used for strong independence as we need to sign several messages (C, x_i) with the same key.

It is possible to obtain strong independence under the newly introduced Strong DDH assumption over Gap-DDH groups [3]. This approach uses the multi-trapdoor commitment from [12] based on this assumption, modified it to make it both mercurial and independent. Details appear in the final version.

4 Independence Versus Non-malleability for ZKS

In the previous section we showed that independence implies non-malleability for commitments. Does this implication extend to the case of ZKS protocols as well? The answer, surprisingly, is not that simple.

The first thing to clarify, of course, is a definition of non-malleability for ZKS protocols. Informally in the commitment case [10], a non-malleable commitment satisfies the following property. An adversary $\mathcal{A}$ is fed with a commitment to a message m, and she outputs another commitment to a message m'. If $\mathcal{A}$ manages to commit to a message m' related to m then there is another machine $\mathcal{A}'$ that outputs a commitment to m' without ever seeing a commitment to m. So in other words the commitment is not helping $\mathcal{A}$ in committing to related messages.

Our definition of non-malleability for ZKS follows the same paradigm. Except that, as in the case of independence, we have to deal with the fact that a ZKS commitment is a commitment to a large string and that the adversary may receive partial openings before creating her own commitment. For this reason we present three separate definitions, each stronger than the previous one and investigate their relationship with our notion of ZKS independence.

ZKS WEAK NON-MALLEABILITY. A first attempt would be to consider ZKS protocols simply as commitments to large strings. In other words, as in the case of weak ZKS independence, the adversary commits *before* querying the honest provers. In this case the definition of ZKS non-malleability would be basically identical to non-malleability for commitment schemes.

Corollary 1. *If a ZKS protocol is weakly ℓ-independent then it is weakly (ℓ, ϵ)-non-malleable with respect to opening.*

ZKS NON-MALLEABILITY. We can strengthen the above definition by allowing the adversary to query the committed databases before producing its own commitment, which must be different from the ones of the honest provers.

However now we are faced with a "selective decommitment problem" [11]. A ZKS commitment is a commitment to a large set of strings: by allowing the adversary to query some keys in the database we are basically allowing a selective decommitment of a subset of those strings (some points in the database).

Thus to obtain this form of ZKS non-malleability we need a commitment scheme which is secure against the selective decommitment problem. We do not know if independent or non-malleable commitments are secure in this sense. Universally composable (UC) commitments [4], on the other hand, are secure in the selective decommitment scenario.

However to obtain an efficient ZKS protocol, such UC commitments would have to be used inside the [17] construction, and thus would have to be mercurial as well. Unfortunately we do not know any commitment that is simultaneously mercurial and UC (not to mention also non-interactive).

Another approach is to restrict the distribution of the committed databases. Under this assumption we can prove that independence will suffice.

Let $\mathcal{IDB}$ be the family of distributions over databases where each distribution can be efficiently sampled conditioned on the value of some points in the database. In other words a distribution $\mathcal{DB} \in \mathcal{IDB}$ if after sampling $Db \in \mathcal{DB}$ and a set of points x_i it is possible to efficiently sample $Db' \in \mathcal{DB}$ such that Db, Db' agree on x_i. An example of such a class of distributions is the one in which the value of each element in the database is independent from the others.

Theorem 4. *If a ZKS protocol is ℓ-independent then it is (ℓ, ϵ)-non-malleable with respect to opening, with respect to the distribution class $\mathcal{IDB}$.*

ZKS STRONG NON-MALLEABILITY. In this definition we allow the adversary to copy one of the commitments, of the honest provers. Now recall that when she is queried on her committed database, she can query the honest provers in the background on their databases. Since she copied the (say) i^{th} committed database, a distinguisher can always detect a correlation between the adversary's and P_i's answer to the same query x. But we require that this must be *all* that the distinguisher can see. In other words, the distinguisher cannot see any correlation between the answers of $\mathcal{A}$ and the answers of all the other P_j's; and cannot see any correlation between the answers of $\mathcal{A}$ and the answers of P_i unless it queries them on the same value.

Theorem 5. *If a ZKS protocol is strongly ℓ-independent then it is strongly (ℓ, ϵ)-non-malleable with respect to opening, with respect to the distribution class $\mathcal{IDB}$.*

References

1. J. Boyar, S.A. Kurtz and M.W. Krentel. *A Discrete Logarithm Implementation of Perfect Zero-Knowledge Blobs.* J. Cryptology 2(2): 63-76 (1990).
2. N. Barić, and B. Pfitzmann. *Collision-free accumulators and Fail-stop signature schemes without trees.* EUROCRYPT '97, Springer LNCS 1233, pp.480-494.
3. D. Boneh and X. Boyen. *Short Signatures without Random Oracles.* EUROCRYPT'04, Springer LNCS 3027, pp.382–400.
4. R. Canetti and M. Fischlin. *Universally Composable Commitments.* CRYPTO'01, Springer LNCS 2139, pp.19-40.
5. D. Catalano, Y. Dodis and I. Visconti. *Mercurial Commitments: Minimal Assumptions and Efficient Constructions.* TCC'06, Springer LNCS 3876, pp.120–144.
6. R. Cramer and I. Damgård. *New Generation of Secure and Practical RSA-based signatures.* CRYPTO'96, Springer LNCS 1109, pp.173-185.
7. I. Damgård, J. Groth. *Non-interactive and reusable non-malleable commitment schemes.* STOC'03, pp.426-437.
8. G. Di Crescenzo, Y. Ishai, R. Ostrovsky. Non-Interactive and Non-Malleable Commitment. STOC'98, pp.141–150.
9. G.Di Crescenzo, J. Katz, R. Ostrovsky, A. Smith. *Efficient and Non-interactive Non-malleable Commitment.* EUROCRYPT 2001, Springer LNCS 2045, pp.40-59.
10. D. Dolev, C. Dwork and M. Naor. *Non-malleable Cryptography.* SIAM J. Comp. 30(2):391–437, 200.
11. C. Dwork, M. Naor, O. Reingold and L. Stockmeyer. *Magic Functions.* FOCS'99.
12. R. Gennaro. *Multi-trapdoor Commitments and Their Applications to Proofs of Knowledge Secure Under Concurrent Man-in-the-Middle Attacks.* CRYPTO'04, Springer LNCS 3152, pp.220–236.
13. S. Goldwasser, S. Micali and C. Rackoff. *The Knowledge Complexity of Interactive Proof Systems.* SIAM J. Comput. 18(1): 186-208 (1989)
14. M. Chase, A. Healy, A. Lysyanskaya, T. Malkin and L. Reyzin. *Mercurial Commitments and Zero-Knowledge Sets based on general assumptions.* EUROCRYPT'05, Springer LNCS 3494, pp.422-439.
15. P. MacKenzie and K. Yang. *On Simulation-Sound Commitments.* EUROCRYPT'04, Springer LNCS 3027, pp.382-400.
16. R.C. Merkle. *A Digital Signature Based on a Conventional Encryption Function.* CRYPTO'87, Springer LNCS 293, pp.369–378.
17. S. Micali, M.O. Rabin and J. Kilian. *Zero-Knowledge Sets.* FOCS'03, pp.80-91.
18. T. Pedersen. *Non-interactive and information-theoretic secure verifiable secret sharing.* CRYPTO'91, Springer LNCS 576, pp.129–140.
19. R. Ostrovsky, C. Rackoff and A. Smith. *Efficient Consistency Proofs for Generalized Queries on a Committed Database.* ICALP 2004.
20. R. Rivest, A. Shamir and L. Adelman. *A Method for Obtaining Digital Signature and Public Key Cryptosystems.* Comm. of ACM, 21 (1978), pp. 120–126

An Efficient Compiler from Σ-Protocol to 2-Move Deniable Zero-Knowledge

Jun Furukawa[1], Kaoru Kurosawa[2], and Hideki Imai[3]

[1] NEC Corporation, Japan
j-furukawa@ay.jp.nec.com
[2] Ibaraki University, Japan
kurosawa@mx.ibaraki.ac.jp
[3] Research Center for Information Security,
National Institute of Advanced Industrial Science and Technology, Japan
h-imai@aist.go.jp

Abstract. Pass showed a 2-move deniable zero-knowledge argument scheme for any $\mathcal{NP}$ language in the random oracle model at Crypto 2003. However, this scheme is very inefficient because it relies on the cut and choose paradigm (via straight-line witness extractable technique). In this paper, we propose a very efficient compiler that transforms any Σ-protocol to a 2-move deniable zero-knowledge argument scheme in the random oracle model, which is also a resettable zero-knowledge and resettably-sound argument of knowledge. Since there is no essential loss of efficiency in our transform, we can obtain a very efficient undeniable signature scheme and a very efficient deniable authentication scheme.

Keywords: deniable, efficient, constant-round, resettable zero-knowledge, the random oracle model, resettably-sound argument of knowledge, Σ-protocol.

1 Introduction

Zero-knowledge interactive proof systems, first proposed by Goldwasser, Micali and Rackoff [19], have the significant property that they leak no knowledge other than the validity of the proven assertion. It has been shown in [22] that every NP-statement can be proved in zero-knowledge if one-way functions exist. Because of these properties, these proof systems have been found to be very important tools in many cryptographic applications.

The original definition of zero-knowledge considered the setting in which a single prover and a verifier execute only one instance of a protocol. However, in more realistic settings, where many computers are connected through the Internet and protocols may be concurrently executed, many verifiers may interact with the same prover simultaneously. Proof systems that are zero-knowledge even in such a setting are called concurrent zero-knowledge (cZK).

The term "concurrent zero-knowledge" was coined by Dwork, Naor, and Sahai in [13], and they observed that the zero-knowledge property does not necessarily carry over to the concurrent setting. Indeed, Goldreich and Krawczyk showed

M. Bugliesi et al. (Eds.): ICALP 2006, Part II, LNCS 4052, pp. 46–57, 2006.

in [21] the existence of protocols that are ordinary zero-knowledge and yet fail dramatically to be zero-knowledge in the concurrent scenario. Moreover, Kilian, Petrank, and Rackoff showed a negative result in [26] such that any language that has a 4-move black-box cZK proof argument is in BPP. Canetti, Kilian, Petrank, and Rosen proved in [7] that black-box cZK proof systems for any non-trivial language require a non-constant number ($\tilde{\Omega}(\log k)$) of rounds.

Despite these negative results, many protocols that achieve round efficiency and adequate security in concurrent settings have been presented under some additional assumptions. These include cZK under the timing assumption [13], resettable zero-knowledges (rZK) in the public-key model [6] and the weak public-key model [29], cZK in the auxiliary string model [11], universally composable zero-knowledge in the common reference string (CRS) model [5], etc.

Now even though the notions of cZK in the CRS model and the auxiliary string model achieve a kind of zero-knowledgeness, they lose some of the spirit of the original definition. In particular, as is mentioned in [13], these models are not sufficient to yield the property of deniability. An interactive protocol is called deniable zero-knowledge if the transcript of its interaction does not leak any evidence of interaction. For example the simulators in the CRS model and the auxiliary string model are powerful enough to control their strings, while the verifiers in these models are never able to control them. As a result, a verifier interacting with a prover in these models are able to output a transcript that cannot be generated by the verifier alone. Hence, the verifier's possession of such a transcript is an evidence of its interaction with the prover.

The question of whether or not there exists a constant-round deniable cZK argument under additional assumptions was studied by Pass [34]. He showed that no black-box constant-round deniable cZK argument for non-trivial language exists in the CRS model. It is also shown there that a 2-move constant-round straight-line witness extractable deniable cZK argument exists for any $\mathcal{NP}$-language in the random oracle (RO) model. However, this argument system is inefficient since it relies on the cut and choose technique. Fischlin also proposed in [18] a communication efficient straight-line witness extractable zero-knowledge proof that can be applied to deniable cZK argument. However, this argument still requires rather large computational complexity.

Besides proving the existence of a certain kind of zero-knowledge protocol for every language in $\mathcal{NP}$, it is also important for practical applications to construct a compiler which transforms Σ-protocols to certain kind of zero-knowledge proof systems or arguments. Σ-protocols are 3-move special honest verifier zero-knowledge protocols with special soundness property. We call such a compiler, a *Σ-compiler.* Σ-compilers are useful in practical point of view since many efficient Σ-protocols for many relations are proposed until now. Σ-compilers for cZK argument and that for rZK and concurrently sound protocol with small overhead are proposed by [11] and [37], respectively. The results of Pass [34] and Fischlin [18] mentioned above are indeed proposal of Σ-compiler for 2-move straight-line witness extractable deniable zero-knowledge argument in the random oracle model. However, their Σ-compiler have large overheads.

Our Contribution
In this paper, we propose a more efficient Σ-compiler for deniable zero-knowledge argument *with very small overhead* in the random oracle model. We also prove that our Σ-compiler simultaneously provides the following properties:

- The resulting protocol is **deniable** resettable zero-knowledge (rZK).
- The resulting protocol is 2-move (constant-round).
- The computational and communication overhead of compilation is very small.
- The resulting protocol is resettably-sound argument of knowledge (RSAK).

We assume here the existence of an efficient invulnerable generator for some NP language.

We note that the result of Pass also provides resettable soundness and can, with slight modification, provide rZK property. (These facts have not been shown before.) In this sense, the essential improvement in our scheme is with respect to efficiency.

The overhead of our compiler is computation, verification, and transmission of Fiat-Shamir transformations of any Σ-protocol, i.e., NIZK-argument. This Σ-protocol is chosen independently to the proven statement and we choose the most efficient one within an allowed assumptions. On the other hand, the compiler in [34] requires the verifier to generate the corresponding NIZK argument by Cut & Choose method instead of Fiat-Shamir transformation. Thus, its overhead will be larger than that of ours in the proportion of the security parameter to one. Fischlin [18] improved efficiency of communication complexity but its computational overhead is till large.

Although our protocol itself is efficient, it does not provide a straight-line simulator. It only provides a rather complicated but still polynomially bounded simulator. This can be compared to cZK protocols under the timing assumption [13] and rZK protocols in the public-key models [6, 29]. All types of protocols achieve deniability with efficient non straight-line simulator in different models. Our protocol is the first that provides an efficient non straight-line simulator in the random oracle model for what ?

Here, rZK and RSAK [2] are, respectively, stronger notions of cZK and argument of knowledge. The requirement for rZK is more restricting than that of cZK in the sense that proof systems or arguments must be cZK even if verifiers in these protocols are able to reset provers. Meanwhile, protocols which are still argument of knowledge against provers who can reset verifiers are called RSAK.

The notion of RSAK was proposed by Barak et al. in [2]. As is pointed out in [6], rZK arguments of knowledge are impossible to achieve for non-trivial languages as long as the ability of knowledge extractor is limited to black-box oracle access to the prover. By exchanging the roles of provers and verifiers, and those of simulators and knowledge extractors, this impossibility holds for zero-knowledge RSAK. However, the negative result are only with respect to the standard definition and may not hold in the random oracle model where simulators and knowledge extractors are more powerful than, respectively, the provers and verifiers in the sense that they are able to control random oracles.

Applications

The security of Chaum's undeniable signature scheme was recently formally proved in the random oracle model [33], where the confirmation protocol and the disavowal protocol are both 4-move.

Now by directly applying our Σ-compiler, we can obtain a very efficient 2-move confirmation protocol and a very efficient 2-move disavowal protocol which are concurrent deniable zero-knowledge as well. It means that our variant of Chaum's undeniable signature scheme is not only more efficient but also secure even against concurrent attacks.

Further, our protocols are resettably-sound argument of knowledge. Hence, our protocols remain secure in a setting where parties in protocols are implemented by devices, which cannot reliably keep state (e.g., smart card), being maliciously reset to prior state. And, the resulting protocols are available in the setting when it is impossible or too costly to generate fresh randomness on the fly.

Another application is deniable identification. In Schnorr's identification scheme, a cheating verifier (Bob) will compute his challenge as a hashed value of the first message of the prover (Alice). Then the transcript of the protocol is an evidence of the fact that Alice executed the protocol with Bob. So the privacy of Alice is not protected. In this sense, Schnorr's identification scheme is not deniable. Now by applying our Σ-compiler, we can obtain a very efficient 2-move deniable identification scheme.

Organization

Our paper is organized as follows. Section 2 describes the basic concepts involved in constructing the proposed compiler in the random oracle model. Section 3 describes our approach and then proposes our Σ-compiler for 2-move deniable rZK that is also RSAK with no essential loss of efficiency in the random oracle model. Section 4 describes the main idea of the reason why the result of our proposed compilation is deniable. Section 5 discusses the efficiency of our compiler.

2 Preliminaries

2.1 Notation

For a random oracle RO, $RO(x)$ denotes its output on input x.

Definition 1. *A function $f(n)$ is* negligible *if $\forall c > 0 \; \exists N \; \forall n > N, \;\; f(n) < \frac{1}{n^c}$.*

Definition 2. *Let $R \subset \{0,1\}^* \times \{0,1\}^*$ be a relation. We say that (x, w)* satisfies *R if $(x, w) \in R$, where x is called an* common input *and w is called a* witness. *Define $L_R = \{x | \; \exists w \; s.t., (x, w) \in R\}$. Also let $R_n = R \cap (\{0,1\}^n \times \{0,1\}^n)$.*

Definition 3. *A* generator *for a relation R is a deterministic polynomial time Turing machine G_R which outputs $(x, w) \in R_n$ on input a random string $r_G \in \{0,1\}^{Q(n)}$, where $Q(\cdot)$ is some polynomial.*

G_R is called an invulnerable *generator for R if for any polynomial time nonuniform algorithm A, $\Pr[(x, A(x)) \in R]$ is negligible in n, where $(x, w) \leftarrow G_R$ and the probability is taken over r_G.*

For example, let $R = \{((p, g, y = g^w \bmod p), w)$, where p is a prime, $g \in Z_p^*$ has a prime order$q$ which is close to p and $w \in Z_q$. Then G_R is an invulnerable generator for R under the discrete log assumption if it outputs a random $((p, g, y = g^w \bmod p), w) \in R$.

2.2 Deniable ZK in the Random Oracle Model

We consider the random oracle model, where a prover P and a (malicious) verifier V^* have access to a random oracle O. In the definition of zero-knowledge, however, a distinguisher D does not have access to O. Hence S has only to generate a view of V^* by providing V^* with a fake random oracle O' which S can manipulate arbitrarily.

Therefore, V^* cannot necessarily generate his view by himself in the *real* world, where D has access to O. This means that V^* can use the view as an evidence of the fact that P executed the protocol, and P cannot deny it. Indeed, V^* can show the view as an evidence to the third party who has access to O.

On the other hand, in the definition of *deniable* zero-knowledge, D has access to the random oracle O. So S must be able to generate a view of V^* which cannot be distinguished from the real one by D who has access to O. Therefore, in a *deniable* zero-knowledge protocol, there is no evidence of the fact that P executed the protocol because V^* can generate his view. Hence P can deny that fact.

2.3 Concurrent ZK and Resettable ZK

A concurrent zero-knowledge (cZK) protocol is a zero-knowledge proof system that withstand malicious verifiers who can interact for polynomial times with the prover in an "interleaved way" about the same theorem. In a resettable zero-knowledge (rZK) protocol, a malicious verifier may not only interact for polynomial times with the prover in an "interleaved way", but also enforce that, in each such interaction, the prover has the same initial configuration (and thus use the same random tape) [6].

Here, we introduce a deniable variant of rZK in the random oracle model. Without loss of generality, we assume that each message of the verifier contains the entire communication history up to that point. Furthermore, we assume that the prover is memoryless: it responds to each message based solely on the input, the random input and the received message.

Definition 4. *An interactive protocol (P, V) for a relation R is said to be* (black-box) deniable rZK in the random oracle model *if, there exists a probabilistic polynomial time simulator S such that, for every probabilistic polynomial time adversary V^*, the following two distribution ensembles are computational indistinguishable by every probabilistic polynomial time distinguisher who can* *access a random oracle RO: Let each distribution be indexed by a sequence of common inputs $\bar{x} = (x_i)_{i=1,...,poly(n)}$ and the corresponding sequence of prover's auxiliary-inputs $\bar{w} = (w_i)_{i=1,...,poly(n)}$ such that $(x_i, w_i) \in R_n$ for all i.*

Distribution 1: *This is defined by the following random process which depends on P and V^*.*

1. *Randomly select and fix RO and $t = poly(n)$ random-tapes, $\{r_i\}_{i=1,...,t}$, for P, resulting in deterministic strategies $P^{(i,j)} = P_{x_i,w_i,r_j}$ defined by $P_{x_i,w_i,r_j}(\alpha) = P(x_i, w_i, r_j, \alpha)$, for $i, j \in \{1, \ldots, t\}$. Each $P^{(i,j)}$ is called an incarnation of P. P is allowed to access the random oracle RO.*
2. *Machine V^* is allowed to arbitrarily interact with all the incarnations of P (i.e., V^* sends arbitrary messages to each of the $P^{(i,j)}$ and obtains the responses of $P^{(i,j)}$ to such messages) and the random oracle RO. Once V^* decides it is done interacting with the $P^{(i,j)}$'s, it (i.e., V^*) produces an output based on its view of these interactions.*

Distribution 2: *The output of $S(\bar{x})$. RO is randomly selected and fixed at first. S has black-box access to the random oracle RO and V^{**} and is able to control a random oracle RO*. V^{**} is the same as V^* except that the random oracle that V^{**} accesses is RO* rather than RO.*

It is important to notice that the simulator is able to control the random oracle, i.e., choose the outputs of the random oracle, that the verifier accesses but is unable to control the one that the distinguisher accesses. The latter property is the key feature of deniability. The former property comes from the fact that the simulator can black-box access the verifier and is not essential for deniability. Our simulator leverages this property for simulation while the simulator of Pass only uses the property that it can catch random oracle queries of the verifier but does not fully leverage the former property. To leverage this property, our simulator rewinds the verifier for polynomial times.

Barak et al. defined the notion of RSAK in [2]. Since it is easy to know its random oracle variant by analogy, we omit to present it here.

2.4 Σ-Protocols and Σ-Compilers

Σ-protocols were introduced by Cramer, Damgård and Schoenmakers in [9]. Informally, a Σ-protocol is a 3-round public-coin special honest verifier zero-knowledge protocol which satisfies special soundness in the knowledge-extraction sense. They are widely used in numerous important cryptographic applications including digital signatures by using the famous Fiat-Shamir methodology [17].

For a relation R, let $L_R = \{x \mid (x, w) \in R\}$.

Definition 5. A *Σ-protocol for relation R*, denoted by (A_R, C_R, Z_R, V_R), is a 3-round protocol (P, V) as follows, where P is a prover and V is a verifier. Let x be a common input and w be the private input to P, where $(x, w) \in R$. Let $r_P \in \{0, 1\}^n$ denote the random input of P.

In the first round, P sends a to V, where a is generated by computing a function A_R on input x, w, r_P. In the second round, V sends e to P, where e is randomly chosen from a set C_R. (C_R is implicitly indexed by n.) In the third round, P sends z to V, where z is generated by computing a function Z_R on input x, w, e, r_P. Finally, V computes $V_R(a, e, z) = 1/0$, where 1 means accept and 0 means reject.

We require that the following three conditions are satisfied.

- Completeness. If P and V follow the protocol, V always accepts.
- Special soundness. From any common input x and any pair of accepting conversations (a, e, z) and (a, e', z') with $e \neq e'$, one can efficiently compute w such that $(x, w) \in R$.
- Special honest verifier zero-knowledge (SHVZK). There exists a probabilistic polynomial-time Turing machine S_R, called a simulator, as follows. For any $x \in L_R$, on input x and a random challenge string e, S_R outputs an accepting conversation (a, e, z) which follows the same probability distribution as the real conversation between the honest P and V.

A Σ*-compiler* is a transformation which transforms a Σ-protocol for a relation R to a zero-knowledge proof system or argument for the same R which satisfies a certain property.

3 Proposed Σ-Compiler in the Random Oracle Model

3.1 Our Approach

In the model of *deniable* zero-knowledge protocols, a prover P, a (malicious) verifier V^*, and a distinguisher D have access to the same random oracle O as shown below.

$$(O \leftrightarrow P) \leftrightarrow (V^* \leftrightarrow O) \text{ and } D \leftrightarrow O. \qquad (1)$$

In this model, P cannot see the queries of V^* to O nor control the answers of O.

In the simulation of S, however, the simulator S can provide V^* with a fake random oracle O' as follows.

$$O \leftrightarrow S \leftrightarrow (V^* \leftrightarrow O') \text{ and } D \leftrightarrow O. \qquad (2)$$

In the simulated world, S can see the queries of V^* to O' and control the answers of O'.

In the methods of Pass and Fischlin, S sees the queries of V^* to O', but does not control the answers of O'. On the other hand, we construct S which both sees the queries of V^* and controls the answers of O'. This is a critical part of our approach. We use a similar technique for knowledge extractor as well.

3.2 Proposed Compiler

Now we present our Σ-compilers which output 2-move deniable rZK and RSAK protocols in the random oracle model. respectively. Suppose that there exists a Σ-protocol for a relation R. Let P and V be a prover and a verifier, respectively, and let x be a common input and w be the private input to P, where $(x, w) \in R$. Then our 2-move protocol proceeds as follows.

1. V chooses a random $(\bar{x}, \bar{w}) \in R$ and sends $\bar{x}$ to P. V then proves that he knows $\bar{w}$ non-interactively.
2. P proves that she knows w or $\bar{w}$ non-interactively.

We use Fiat-Shamir transformation [17] to construct non-interactive arguments, and use the technique of [9] to construct the P's message (OR protocol).

Although our protocol is very simple, it has not been known that it is deniable zero-knowledge. This is probably because many people believed that the straight-line extractability of the witness of the verifier is necessary. Our main contribution is then to proof that the above simple construction is indeed deniable zero-knowledge.

We first present Construction 1 which is deniable zero-knowledge only, but not rZK. We next show Construction 2 which is rZK, where Construction 2 is obtained by applying a known technique to Construction 1.

Suppose that there exists a Σ-protocol (A_R, C_R, Z_R, S_R) for a relation R, and a Σ-protocol $(A_{\bar{R}}, C_{\bar{R}}, Z_{\bar{R}}, S_{\bar{R}})$ for a relation $\bar{R}$ which has an invulnerable generator $G_{\bar{R}}$, where $\bar{R}$ can be the same as R. We assume that $C_R = C_{\bar{R}}$ and $\frac{1}{|C_R|} = \frac{1}{|C_{\bar{R}}|}$ is negligible in the security parameter n.

Let x be a common input and w be the private input to P, where $(x, w) \in R$. P has random tapes $r_E, r_S, r_P \in \{0,1\}^n$, and V has random tapes $r_V, r_G \in \{0,1\}^n$. They are allowed to have access to a random oracle RO whose output is uniformly distributed over $C_R = C_{\bar{R}}$.

Construction 1. 1. V sends $(\bar{x}, \bar{a}, \bar{z})$ to P which are generated as follows.
   (a) V generates $(\bar{x}, \bar{w}) \in \bar{R}$ by running $G_{\bar{R}}$ on input r_G.
   (b) V computes $\bar{a} = A_R(\bar{x}, \bar{w}, r_V)$.
   (c) V queries $(\bar{x}, \bar{a})$ to RO, and RO returns $\bar{e}$ to V.
   (d) V computes $\bar{z} = Z_R(\bar{x}, \bar{w}, \bar{e}, r_V)$.
2. P sends $((a, e, z), (\bar{a}', \bar{e}', \bar{z}'))$ to V which are computed as follows.
   (a) P computes $\bar{e} = RO(\bar{x}, \bar{a})$, and verifies that $V_R(\bar{x}, \bar{a}, \bar{e}, \bar{z}) = 1$.
   (b) P generates a random $\bar{e}' \in C_{\bar{R}}$ by using r_E. P then generates a simulated view $(\bar{x}, \bar{a}', \bar{e}', \bar{z}')$ by running $S_{\bar{R}}$ on input $(\bar{x}, \bar{e}', r_S)$.
   (c) P computes $a = A_R(x, w, r_P)$.
   (d) P queries $(x, a, \bar{x}, \bar{a}')$ to RO, and RO returns d.
   (e) P computes $e = d \oplus \bar{e}'$
   (f) P computes $z = Z_R(x, w, e, r_P)$.
3. V accepts iff

$$RO(x, a, \bar{x}, \bar{a}') = e \oplus \bar{e}', \ V_{\bar{R}}(\bar{x}, \bar{a}', \bar{e}', \bar{z}') = 1 \text{ and } V_R(x, a, e, z) = 1.$$

Theorem 1. *The above protocol is RSAK for relation R in the random oracle model.*

The proof is given in the full paper [1].

Theorem 2. *The above protocol is deniable cZK in the random oracle model.*

The idea of simulation is given in Section 4 and the proof is given in the full paper [1].

Construction 2. The construction is the same as Construction 1 except the following changes.

1. P has a pseudorandom function $f_k : (\{0,1\}^n)^4 \rightarrow (\{0,1\}^n)^3$, where the index k is randomly chosen by P.
2. P generates its random tapes as $(r_E, r_S, r_P) = f(x, \bar{x}, \bar{a}, \bar{z})$.

Theorem 3. *Construction 2 is rZK as well as RSAK and cZK.*

Proof. Construction 1 is admissible and hybrid deniable zero-knowledge, which implies that Construction 2 is deniable rZK from [6]. See [6] or the full paper [1] for the definitions of admissible protocols and hybrid deniable zero-knowledge and the validity of the transformation from Construction 1 to Construction 2. □

4 Idea of Simulation

This section shows an idea of the proof of Theorem 2. To prove deniable concurrent zero-knowledgeness, we need to construct a simulator S for any adversary V^* who creates verifiers $V_1, V_2, \cdots$, where $V_1, V_2, \cdots$ run our protocol concurrently with a single prover P, and all of P, V^*, S and distinguishers D have access to the same random oracle O.

In our protocol, V^* sends α to P at step 1 and P sends β to V^* at step 2, where α and β are described in Construction 1. We say that $C = (\bar{a}, \bar{e}, \bar{z})$ and $C' = (\bar{a}, \bar{e}', \bar{z}')$ are a matching pair on $\bar{x} \in L_{\bar{R}}$ if they are accepting conversations of the Σ-protocol on input $\bar{x}$ and $\bar{e} \neq \bar{e}'$. Then the basic idea is that:

1. S can compute $\bar{w}$ such that $(\bar{x}, \bar{w}) \in \bar{R}$ if S somehow obtains a matching pair (C, C') on $\bar{x}$. This is due to the special soundness.
2. S can complete the simulation in polynomial time if S uses a fake random oracle O'. S uses this type of simulation for obtaining (C, C').
3. P proves that she knows w or $\bar{w}$. Hence S can simulate, with respect to the real random oracle O, the role of P if S knows $\bar{w}$.

For simplicity, suppose that if V_i queries $(\bar{x}_1, \bar{a}_1)$ to O, then V_i always sends some $\alpha_i = (\bar{x}_1, \bar{a}', \bar{z}')$ to P (and never aborts), where $\bar{a}'$ may not be the same as $\bar{a}_1$. Then S behaves as follows. Fix the random oracle O, and the random tapes of P and A. Suppose that V_i queries $(\bar{x}_i, \bar{a}_i)$ to O at time t_i, where $t_1 < t_2 < \cdots$.

1. S runs V^* by using O, and finds that V_1 queries $(\bar{x}_1, \bar{a}_1)$ to O at time t_1.
2. S repeats the following until S gets a matching pair on $\bar{x}_1$: By using a fake random oracle O', S runs V^* from t_1 until V_1 sends some α_1 to P.
3. S computes $\bar{w}_1$ from the matching pair. S can, from now on, (as P) computes β_1 for any α_1 (sent by V_1) which includes $\bar{x}_1$ because she has $\bar{w}_1$.

4. S runs V^* by using O from the beginning, and finds that V_2 queries $(\bar{x}_2, \bar{a}_2)$ to O at time t_2.
5. S repeats the following until S gets a matching pair on $\bar{x}_2$: By using a fake random oracle O', S runs V^* from t_2 until V_2 sends some α_2 to P.
6. S computes $\bar{w}_2$ from the matching pair. S can now (as P) computes β_2 for any α_2 (sent by V_1) which includes $\bar{x}_2$ because she has $\bar{w}_2$,
7. and so on.

If S succeeds in the above simulation, then S can output a transcript which is the same as the original one. In particular, all the participants use the same random oracle O. Hence our protocol is deniable cZK.

In the general case where some V_i may abort without outputting $\bar{z}$, the simulation gets to be much complicated. In this case, S repeats simulation with chosen output values of random oracles for many times to obtains two $\bar{z}$ for each $(\bar{x}, \bar{a})$ but gives up if its number of repetitions exceeds a certain number of times $(2n^2Q(n))$. Such a multiple trial is required since there may be (with some probability) a case when some V_i aborts without outputting $\bar{z}$ when S is simulating with chosen output values of random oracles but outputs $\bar{z}$ when S is simulating with the output values of real random oracle. The number of times the S tries to obtain $\bar{z}$ is $2n^2Q(n)$. Here, $Q(n)$ is the running time of P^*.

It turns out that such a simulation can be successful with the probability larger than $1/2$. Hence, repeating it for polynomial times enable the successful simulation with overwhelming probability. The number of time S repeats this simulation is n. The total running time of the simulator is $n^3Q(n)^2$.

The above simulator is not black-box simulator since it needs to know the running time of verifier. However, it is easy to construct a black-box simulator from the proposed simulator. The black-box simulator executes the proposed simulator repeatedly until it complete simulation, by, in each execution, it increase the order of time that it assumes as the running time of verifier.

5 Efficiency

Our protocol (which is illustrated at the beginning of Sec.3) is almost as efficient as the underlying Σ-protocol because Fiat-Shamir transformation and OR-protocol have very small overhead. Moreover, efficient Σ-protocols are known for many useful relations. Hence our construction will find a lot of applications.

On the other hand, the compiler of Pass [34] requires a Cut & Choose method which is very inefficient. Indeed, its overhead is proportion to the security parameter n while ours is only a small constant. Fischlin [18] proposed a straight-line witness extractable proof that has smaller communication complexity than the method of Pass. However, its overhead still depends on the security parameter n. Hence, its communication/computation complexity is still larger than that of ours.

As an example, let us consider the case when our compiler is applied to Schnorr's identification protocol illustrated in the following.

Let p, q be primes such that $q|p-1$ and g be a generator of the order q subgroup of $(\mathbb{Z}/p\mathbb{Z})^*$. The secret key for the prover P is $w \in_R \mathbb{Z}/q\mathbb{Z}$ and the public key is $x = g^w \bmod p$. Let V denote the verifier.

1. P chooses $r \in \mathbb{Z}/q\mathbb{Z}$ randomly and sends $a = g^r \bmod p$ to V.
2. V sends a random $c \in \mathbb{Z}/q\mathbb{Z}$ to P.
3. P sends $z = r + cw \bmod q$ to V.
4. V accepts iff $g^z = ax^c \bmod p$.

The communication cost and computational cost are roughly 1/50 and 1/32, respectively, of those of Pass' scheme and are roughly 1/3 and 1/36 [1], respectively, of those of Fischlin's scheme.

References

1. An Efficient Compiler from Σ-Protocol to 2-move Deniable Zero-Knowledge (full version with examples and proofs). Manuscript.
2. Boaz Barak, Oded Goldreich, Shafi Goldwasser, Yehuda Lindell: Resettably-Sound Zero-Knowledge and its Applications. FOCS 2001: 116-125
3. Mihir Bellare, Adriana Palacio: GQ and Schnorr Identification Schemes: Proofs of Security against Impersonation under Active and Concurrent Attacks. CRYPTO 2002: 162-177
4. Jan Camenisch, Victor Shoup: Practical Verifiable Encryption and Decryption of Discrete Logarithms. CRYPTO 2003: 126-144 2002
5. Ran Canetti: Universally Composable Security: A New Paradigm for Cryptographic Protocols. FOCS 2001, pp. 136-145
6. R. Canetti, O. Goldreich, S. Goldwasser, and S. Micali: Resettable Zero-Knowledge. Proc. of STOC 2000.
7. R. Canetti, J. Kilian, E. Petrank, and A. Rosen: Black-box concurrent zero-knowledge requires Omega (log n) rounds. ACM Symposium on Theory of Computing, pp. 570-579, 2001.
8. David Chaum, Hans Van Antwerpen: Undeniable Signatures. CRYPTO 1989: 212-216
9. Ronald Cramer, Ivan Damgård, Berry Schoenmakers: Proofs of Partial Knowledge and Simplified Design of Witness Hiding Protocols. CRYPTO 1994: 174-187
10. Giovanni Di Crescenzo, Giuseppe Persiano, Ivan Visconti: Constant-Round Resettable Zero Knowledge with Concurrent Soundness in the Bare Public-Key Model. CRYPTO 2004: 237-253
11. I. Damgård: Efficient Concurrent Zero-Knowledge in the Auxiliary String Model. LNCS 1807, Proc. Eurocrypt 2000, pp.419-430
12. Vergnaud Damien: Private communication.
13. C. Dwork, M. Naor, and A. Sahai: Concurrent zero-knowledge: Proc. of STOC 30, pp. 409–428, 1998
14. Cynthia Dwork, Moni Naor: Zaps and Their Applications. Electronic Colloquium on Computational Complexity (ECCC)(001): (2002)
15. C. Dwork, A. Sahai: Concurrent Zero-Knowledge: Reducing the Need for Timing Constraints. LNCS 1462, pp.442- 1998.

[1] Operations required in our scheme and those in Fischlin's are very different. Hence this comparison is really rough.

16. U. Feige and A. Shamir: Zero Knowledge Proofs of Knowledge in Two Rounds. Crypto 89, pp.526-544, 1990
17. Amos Fiat, Adi Shamir: How to Prove Yourself: Practical Solutions to Identification and Signature Problems. CRYPTO 1986: 186-194.
18. Marc Fischlin: Communication-Efficient Non-Interactive Proofs of Knowledge with Online Extractors. CRYPTO 2005:
19. S. Goldwasser, S. Micali, and C. Rackoff: The Knowledge Complexity of Interactive Proof Systems. SIAM J. Comput. Vol. 18, No. 1, pp. 186-208 1989.
20. Oded Goldreich, Ariel Kahan: How to Construct Constant-Round Zero-Knowledge Proof Systems for NP. J. Cryptology 9(3): 167-190 (1996)
21. O. Goldreich and H. Krawczyk: On the Composition of Zero Knowledge Proof Systems. SIAM J. on Computing, Vol.25, No.1, pp.169-192,1996
22. O. Goldreich, S. Micali, and A. Wigderson: A Proof that Yields Nothing but Their Validity or All Languages in NP Have Zero-Knowledge Proof System. ACM, 38, 1, pp. 691-729 1991.
23. O. Goldreich and Y. Oren: Definitions and properties of Zero-Knowledge proof systems. Journal of Cryptology, Vol. 7, No. 1 pp. 1-32 1994.
24. R. Impagliazzo, L.Levin and M.Luby: Pseudo-random Generation from one-way functions. STOC 1989, pp.12–24 (1989)
25. J. Kilian and E. Petrank: Concurrent zero-knowledge in poly-logarithmic rounds. STOC 2001.
26. J. Kilian, E. Petrank, and C. Rackoff: Lower Bounds for Zero Knowledge on the Internet. FOCS 1998: 484-492.
27. J. Kilian, E. Petrank, R. Richardson: On Concurrent and Resettable Zero-Knowledge Proofs for NP.
28. Kaoru Kurosawa, Swee-Huay Heng: 3-Move Undeniable Signature Scheme. EUROCRYPT 2005: 181-197
29. S. Micali1 and L. Reyzin: Min-Round Resettable Zero-Knowledge in the Public-Key Model. Eurocrypt 2001, LNCS 2045, pp.373-393, 2001
30. Silvio Micali, Leonid Reyzin: Soundness in the Public-Key Model. CRYPTO 2001: 542-565
31. M. Michels and M. Stadler: Efficient Convertible Undeniable Signature Schemes. Proc. SAC 1997 pp. 231-244 1997.
32. M. Naor: Bit Commitment Using Pseudo-Randomness. Journal of Cryptology, vol.4, 1991,pp.151-158
33. W. Ogata, K. Kurosawa and S.H. Heng, The Security of the FDH Variant of Chaum's Undeniable Signature Scheme, Accepted by IEEE Trans. on IT.
34. R. Pass: On Deniability in the Common Reference String and Random Oracle Model. CRYPTO 2003, pp. 316-337.
35. R. Richardson and J. Kilian: On the Concurrent Composition of Zero-Knowledge Proofs. EUROCRYPT 1999, pp. 415-431, 1999.
36. Amit Sahai: Non-Malleable Non-Interactive Zero Knowledge and Adaptive Chosen-Ciphertext Security. FOCS 1999: 543-553
37. Yunlei ZHAO: Concurrent/Resettable Zero-Knowledge With Concurrent Soundness in the Bare Public-Key Model and Its Applications. Cryptology ePrint Archive, Report 2003/265.

New Extensions of Pairing-Based Signatures into Universal Designated Verifier Signatures*

Damien Vergnaud

Laboratoire de Mathématiques Nicolas Oresme
Université de Caen, Campus II, B.P. 5186,
14032 Caen Cedex, France
Damien.Vergnaud@math.unicaen.fr

Abstract. The concept of universal designated verifier signatures was introduced by Steinfeld, Bull, Wang and Pieprzyk at Asiacrypt 2003. We propose two new efficient constructions for pairing-based short signatures. The first scheme is based on Boneh-Boyen signatures and, its security can be analyzed in the standard security model. We reduce its resistance to forgery to the hardness of the strong Diffie-Hellman problem, under the knowledge-of-exponent assumption. The second scheme is compatible with the Boneh-Lynn-Shacham signatures and is proven unforgeable, in the random oracle model, under the assumption that the computational bilinear Diffie-Hellman problem is untractable. Both schemes are designed for devices with constrained computation capabilities since the signing and the designation procedure are pairing-free.

1 Introduction

Recently many *universal designated verifier signature* protocols have been proposed (*e.g.* [13, 17, 18]). The present paper focuses on the proposal of two new efficient constructions for pairing-based short signatures [3, 5]. The resistance to forgery of the first scheme relies on the hardness of the strong Diffie-Hellman problem, under the knowledge-of-exponent assumption, in the standard security model, and the one of the second scheme relies, in the random oracle model, on the hardness of a new computational problem (not easier than the widely used computational bilinear Diffie-Hellman problem).

Related Work. The concept of *designated verifier signatures* (DVS, for short) was introduced by Jakobsson, Sako and Impagliazzo in 1996 [10]. These signatures are intended to a specific and unique designated verifier, who is the only one able to check their validity. Motivated by privacy issues associated with dissemination of signed digital certificates, Steinfeld, Bull, Wang and Pieprzyk [17] defined, in 2003, a new kind of signatures called *universal designated-verifier signatures* (UDVS, for short). This primitive can function as a standard publicly-verifiable digital signature scheme but has an additional functionality which allows any holder of a signature to designate the signature to any verifier. Again,

* Dedicated to my wife Juliette, on her 30th birthday.

M. Bugliesi et al. (Eds.): ICALP 2006, Part II, LNCS 4052, pp. 58–69, 2006.

the designated-verifier can check that the message was signed by the signer, but is unable to convince anyone else of this fact. Steinfeld *et al.* proposed an efficient UDVS scheme constructed using any bilinear group-pair and Laguillaumie and the author suggested in [13] a variant which significantly improves this protocol. Both schemes are compatible with the key-generation, signing and verifying algorithms of the Boneh-Lynn-Shacham [5] signature scheme (BLS). In [3], Boneh and Boyen proposed efficient pairing-based short signatures (BB) whose security can be analyzed in the standard security model. A UDVS scheme compatible with a variant of Boneh and Boyen's scheme has been proposed by Zhang, Furukawa and Imai [18].

Contributions of the Paper. The main contribution of the paper is to provide a new efficient UDVS protocol compatible with the *original* Boneh-Boyen scheme. The idea underlying our design relies on the homomorphic properties of BB signatures. The new scheme, called UDVS-BB, is unforgeable in the standard security model assuming the hardness of the strong Diffie-Hellman problem [3], under the *knowledge-of-exponent assumption (KEA)* [1,7]. The protocol proposed by Zhang *et al.* is proven unforgeable assuming the hardness of the same algorithmic problem, but under a stronger assumption. The computational workload of UDVS-BB amounts to three exponentiations over bilinear groups for designating a signature and four pairing evaluations to verify it, and moreover, the size of the signatures is much smaller than the one of Zhang *et al.*'s signatures.

Using the same design principle, we found a new UDVS protocol compatible with the BLS signatures which is well-suited for devices with constrained computation capabilities and low bandwidth. Indeed the designation procedure of the signatures is pairing-free and the resulting size is comparable to the length of DSA signatures. The security analysis for this scheme, called UDVS-BLS, takes place in the random oracle model [2]: we show that this scheme is unforgeable with respect to a new computational assumption weaker than the widely used computational bilinear Diffie-Hellman assumption.

2 Definitions

2.1 Notations

The set of n-bit strings is denoted by $\{0,1\}^n$ and the set of all finite binary strings is denoted by $\{0,1\}^*$. Let $\mathcal{A}$ be a probabilistic Turing machine running in polynomial time (a PPTM, for short), and let x be an input for $\mathcal{A}$. The probability space that assigns to a string σ the probability that $\mathcal{A}$, on input x, outputs σ is denoted by $\mathcal{A}(x)$. The support of $\mathcal{A}(x)$ is denoted by $\mathcal{A}[x]$.

Given a probability space S, a PPTM that samples a random element according to S is denoted by $x \xleftarrow{R} S$. For a finite set X, $x \xleftarrow{R} X$ denotes a PPTM that samples a random element uniformly at random from X.

2.2 Universal Designated Verifier Signatures

In this subsection, we recall the definition of UDVS schemes [12, 17].

Definition 1. *A* universal designated verifier signature scheme Σ *is an 8-tuple* $\Sigma = (\mathsf{Setup}, \mathsf{SKeyGen}, \mathsf{VKeyGen}, \mathsf{Sign}, \mathsf{Verify}, \mathsf{Designate}, \mathsf{Fake}, \mathsf{DVerify})$ *such that*

- $(\mathsf{Setup}, \mathsf{SKeyGen}, \mathsf{Sign}, \mathsf{Verify})$ *is a signature scheme:*
  - $\Sigma.\mathsf{Setup}$ *is a PPTM which takes an integer* 1^k *as input. The output are the* public parameters cp *which contain a description* $D_{\mathcal{M}}$ *of a set* $\mathcal{M} \subseteq \{0,1\}^*$ *called the* message space. k *is called the* security parameter *and an element of* $\mathcal{M}$ *is called a* message.
  - $\Sigma.\mathsf{SKeyGen}$ *is a PPTM which takes the public parameters as input. The output is a pair* $(\mathsf{sks}, \mathsf{pks})$ *where* sks *is called a* signing secret key *and* pks *a* signing public key.
  - $\Sigma.\mathsf{Sign}$ *is a PPTM which takes the public parameters, a message, and a signing secret key as inputs and outputs a bit string.*
  - $\Sigma.\mathsf{Verify}$ *is a PPTM which takes the public parameters, a message* m, *a bit string* σ *and a signing public key* pks. *It outputs a bit. If the bit output is 1 then the bit string* σ *is said to be a* signature *on* m *for* pks.
- $\Sigma.\mathsf{VKeyGen}$ *is a PPTM which takes the public parameters as input. The output is a pair* $(\mathsf{skv}, \mathsf{pkv})$ *where* skv *is called a* verifying secret key *and* pkv *a* verifying public key.
- $\Sigma.\mathsf{Designate}$ *is a PPTM which takes the public parameters, a message* m, *a signing public key* pks, *a signature* σ *on* m *for* pks *and a verifying public key as inputs and outputs a bit string.*
- $\Sigma.\mathsf{Fake}$ *is a PPTM which takes the public parameters, a message, a signing public key and a verifying secret key as inputs and outputs a bit string.*
- $\Sigma.\mathsf{DVerify}$ *is a deterministic PTM which takes the public parameters, a message* m, *a bit string* τ, *a signing public key* pks, *a verifying public key* pkv *the matching verifying secret key* skv *as inputs. It outputs a bit. If the bit output is 1 then the bit string* τ *is said to be a* designated verifier signature *on* m *from* pks *to* pkv.

Σ *must satisfies the following properties, for all* $k \in \mathbb{N}\backslash\{0\}$, *all* $cp \in \Sigma.\mathsf{Setup}[1^k]$, *all* $(\mathsf{pks}, \mathsf{sks}) \in \Sigma.\mathsf{SKeyGen}[cp]$, *all* $(\mathsf{pkv}, \mathsf{skv}) \in \Sigma.\mathsf{VKeyGen}[cp]$ *and all messages* m:

- CORRECTNESS OF SIGNATURE:

$$\forall \sigma \in \Sigma.\mathsf{Sign}[cp, m, \mathsf{sks}], \quad \Sigma.\mathsf{Verify}[cp, m, \sigma, \mathsf{pks}] = \{1\}.$$

- CORRECTNESS OF DESIGNATION:

$$\forall \sigma \in \Sigma.\mathsf{Sign}[cp, m, \mathsf{sks}], \quad \forall \tau \in \Sigma.\mathsf{Designate}[cp, m, \mathsf{pks}, \sigma, \mathsf{pkv}], \\ \Sigma.\mathsf{DVerify}[cp, m, \tau, \mathsf{pks}, \mathsf{pkv}, \mathsf{skv}] = \{1\}.$$

- SOURCE HIDING:

$$\Sigma.\mathsf{Designate}(cp, m, \mathsf{pks}, \Sigma.\mathsf{Sign}(cp, m, \mathsf{sks}), \mathsf{pkv}]) = \Sigma.\mathsf{Fake}(cp, m, \mathsf{pks}, \mathsf{skv}).$$

The correctness properties insure that a properly formed (designated verifier) signature is always accepted by the (designated) verifying algorithm. The source hiding property states that given a message m, a signing public key pks, a verifying public key pkv and a DVS τ on m from pks to pkv it is unconditionally infeasible to determine if τ was produced by $\Sigma.\mathsf{Designate}$ or $\Sigma.\mathsf{Fake}$.

For digital signatures, the *de facto* standard notion of security was defined in [9], as *unforgeability against chosen message attacks* (EF-CMA). UDVS scheme must satisfy a similar property which was formally defined in [12, 13, 17]:

Unforgeability (UDVS-EF-CMA): given a signing public key pks and a verifying public key pkv, it should be computationally infeasible for an adversary which engages in polynomially many runs of the protocol with the signer, interleaved at its own choosing, to produce a DVS from pks to pkv on a *new* message.

This definition does not capture that the adversary cannot generate a new signature on a previously signed message (the so-called *strong unforgeability*).

2.3 Bilinear Maps and Computational Assumptions

The security of asymmetric cryptographic tools relies on assumptions about the hardness of certain algorithmic problems. Bilinear maps such as Weil or Tate pairing on elliptic curves and hyperelliptic curves have found various applications in cryptography (*e.g.* [4, 3, 5]). In the following, we briefly review the basic definitions about bilinear maps and in order to highlight that our schemes apply to any secure instanciation of BLS and BB signatures, we do not pin down any particular generator, but instead parameterize definitions and security results by a choice of generator.

Definition 2. *A* prime-order-BDH-parameter-generator *is a PPTM that takes as input* $k \in \mathbb{N} \setminus \{0\}$ *and outputs a tuple* $(q, \mathbb{G}_1, \mathbb{G}_2, \mathbb{G}_3, \langle\cdot,\cdot\rangle, \psi)$ *satisfying the following conditions:*

1. q *is a prime with* $2^{k-1} < q < 2^k$*;*
2. $(\mathbb{G}_1, +)$, $(\mathbb{G}_2, +)$ *and* $(\mathbb{G}_3, \times)$ *are groups of order* q*;*
3. $\psi : \mathbb{G}_2 \longrightarrow \mathbb{G}_1$ *is an isomorphism s.t. there exists a PPTM to compute* ψ*;*
4. $\langle\cdot,\cdot\rangle : \mathbb{G}_1 \times \mathbb{G}_2 \longrightarrow \mathbb{G}_3$ *satisfies the following properties:*
   (a) $\langle [a]Q, [b]R\rangle = \langle Q, R\rangle^{ab}$ *for all* $(Q, R) \in \mathbb{G}_1 \times \mathbb{G}_2$ *and all* $(a, b) \in \mathbb{Z}^2$*;*
   (b) $\langle\cdot,\cdot\rangle$ *is non degenerate (*i.e. $\langle\psi(P), P\rangle \neq 1_{\mathbb{G}_3}$ *for some* $P \in \mathbb{G}_2$*);*
   (c) *there exists a PPTM to compute* $\langle\cdot,\cdot\rangle$.

Notations: In the following, we denote by $\mathsf{E}_{\mathbb{G}}$ (*resp.* T) the time complexity for evaluating exponentiation in a group $\mathbb{G}$ (*resp.* a pairing) and by $\ell_i(k)$ the bit-length of the representation of elements of a group $\mathbb{G}_i$ of k-bit order q.

Let $(q, \mathbb{G}_1, \mathbb{G}_2, \mathbb{G}_3, \langle\cdot,\cdot\rangle, \psi)$ be as above, $P_2 \in \mathbb{G}_2$ and let $P_1 = \psi(P_2)$. In margin to the classical Diffie-Hellman problems in the groups $\mathbb{G}_1$, $\mathbb{G}_2$ and $\mathbb{G}_3$, the introduction of bilinear maps in cryptography gives rise to new algorithmic problems. The unforgeability of UDVS-BLS is based on a new algorithmic problem, that we call the *strong computational bilinear Diffie-Hellman problem*:

Strong Computational Bilinear Diffie-Hellman (SCBDH): let $(x, y, z) \in \mathbb{N}^3$. Given $([x]P_2, [y]P_2, [z]P_1)$, compute $(Q, R) \in \mathbb{G}_1 \times \mathbb{G}_2$ such that $\langle Q, R\rangle = \langle P_1, P_2\rangle^{xyz}$.

This problem is not easier than the computational bilinear Diffie Hellman problem which has already been widely used (*e.g.* [4, 13, 17]). In particular, the unforgeability of UDVS-BLS reduces to a weaker assumption than all UDVS schemes compatible with BLS proposed up to now [13, 17].

To analyze the security of their signatures, Boneh and Boyen [3] introduced a new computational problem, on which relies also the unforgeability of our scheme UDVS-BB:

ℓ-Computational Strong Diffie-Hellman (ℓ-CSDH): let $x \in \mathbb{N}$. Given $\ell \in \mathbb{N}$ and $([x]P_2, \ldots, [x^\ell]P_2) \in \mathbb{G}_2^\ell$, compute a pair $\big([(x+h)^{-1}]P_1, h\big)$ in $\mathbb{G}_1 \times [\![1, q-1]\!]$.

3 Description of the New Schemes

In this section, we describe our new UDVS schemes. We give in details the ideas underlying their design, since we are convinced that they may be of independent interest. The general principle is based on an elegant technique proposed by Damgård [7] and aimed at making public-key encryption scheme secure against chosen ciphertext attacks.

3.1 Damgård's Encryption Scheme and KEA

Let $(\mathbb{G}, +)$ be a group of prime order q, let k be the bit size of the representation of elements of $\mathbb{G}$ and let P be a generator of G. In 1991, Damgård [7] presented a simple variant of the El Gamal encryption scheme in $\mathbb{G}$. In his proposal, Alice publishes two public keys $A_1 = [a_1]P$ and $A_2 = [a_2]P$ and keeps secret their discrete logarithms a_1 and a_2. When Bob wants to privately send a message $m \in \{0,1\}^k$ to Alice, he picks uniformly at random an integer $r \in [\![1, q-1]\!]$ and transmits the triple (Q_1, Q_2, C) where $Q_1 = [r]P$, $Q_2 = [r]A_1$ and $C = m \oplus ([r]A_2)$. When she receives the ciphertext (Q_1, Q_2, C), Alice checks whether the equality $Q_2 = [a_1]Q_1$ holds: if it is the case, she retrieves the message m, as $m = C \oplus ([a_2]Q_1)$, otherwise she rejects the ciphertext. Damgård proved that if the decisional Diffie-Hellman problem is hard in $\mathbb{G}$, then this scheme is semantically secure against (non-adaptive) chosen ciphertext attacks, if we assume the knowledge-of-exponent assumption [1].

Intuitively this assumption states that, without the knowledge of a_1, the only way to generate couples $(Q_1, Q_2) \in \mathbb{G}^2$, verifying $Q_2 = [a_1]Q_1$, is to choose an integer $r \in [\![1, q-1]\!]$ and to compute $Q_1 = [r]P$ and $Q_2 = [r]A_1$. There are many ways in which the formulation of KEA can be varied to capture this intuition that the only way to generate a Diffie-Hellman triple is to know the corresponding exponent. Usually, this is done by saying that for any PPTM outputting such a triple, there is an "extractor" than can return this exponent.

In the following definition, we propose a new variant of KEA in the bilinear setting (which reduces to the classical KEA, when Gen is a so-called *symmetric* prime-order-BDH-parameter-generator). For our purposes, it is necessary to allow the adversary to be randomized (in that case, it is important that the extractor gets the coins of the adversary as an additional input, since otherwise the assumption is clearly false).

Definition 3. *Let* Gen *be a prime-order-BDH-parameter-generator and let* $\mathbf{A}$ *and* $\overline{\mathbf{A}}$ *be two PPTM's. We consider the following random experiments, where* $k \in \mathbb{N} \setminus \{0\}$ *is a security parameter:*

Experiment $\mathbf{Exp}^{kea}_{\mathsf{Gen},\mathbf{A},\overline{\mathbf{A}}}(k)$

$(q, \mathbb{G}_1, \mathbb{G}_2, \mathbb{G}_3, \langle\cdot,\cdot\rangle, \psi) \xleftarrow{R} \mathsf{Gen}(k)$; $x \xleftarrow{R} [\![1, q-1]\!]$; $P_2 \xleftarrow{R} \mathbb{G}_2 \setminus \{\mathbb{O}_{\mathbb{G}_2}\}$
$(R, S) \leftarrow \mathbf{A}((q, \mathbb{G}_1, \mathbb{G}_2, \mathbb{G}_3, \langle\cdot,\cdot\rangle, \psi), P_2, [x]P_2; \varpi)$
$r \leftarrow \overline{\mathbf{A}}((q, \mathbb{G}_1, \mathbb{G}_2, \mathbb{G}_3, \langle\cdot,\cdot\rangle, \psi), P_2, [x]P_2, \varpi; \overline{\varpi})$
Return 1 *if* $(R, S) \in \mathbb{G}_1^2$, $S = [x]R$ *and* $R \neq [r]\psi(P_2)$, 0 *otherwise*

Let $\varepsilon \in [0,1]^{\mathbb{N}}$. *We define the* advantage *of* $\mathbf{A}$ *relative to* $\overline{\mathbf{A}}$ via

$$\mathbf{Adv}^{kea}_{\mathsf{Gen},\mathbf{A},\overline{\mathbf{A}}}(k) = \Pr\left[\mathbf{Exp}^{kea}_{\mathsf{Gen},\mathbf{A},\overline{\mathbf{A}}}(k) = 1\right].$$

1. $\overline{\mathbf{A}}$ *is a* ε-kea-extractor *for* $\mathbf{A}$ *if for all* $k \in \mathbb{N} \setminus \{0\}$, $\mathbf{Adv}^{kea}_{\mathsf{Gen},\mathbf{A},\overline{\mathbf{A}}}(k) \leq \varepsilon(k)$
2. *We say that the knowledge-of-exponent assumption holds for* Gen *if for every PPTM* $\mathbf{A}$, *there exists a PPTM* $\overline{\mathbf{A}}$ *and a negligible*[1] *function* ε *such that* $\overline{\mathbf{A}}$ *is a* ε-KEA-extractor *for* $\mathbf{A}$.

3.2 Description of the Protocol UDVS-BB

Boneh-Boyen's Signatures. In 2004, Boneh and Boyen [3] proposed a new application of bilinear structures to construct efficient short signatures. Their idea is to plug the message to be signed in the exponent and, in order to avoid trivial "homomorphic" forgeries, to do so in a non-linear way.

Let Gen be a prime-order-BDH-parameter-generator. Let $k \in \mathbb{N} \setminus \{0\}$, let $(q, \mathbb{G}_1, \mathbb{G}_2, \mathbb{G}_3, \langle\cdot,\cdot\rangle, \psi)$ be some output of $\mathsf{Gen}(1^k)$ and let $P_2 \in \mathbb{G}_2 \setminus \{\mathbb{O}_{\mathbb{G}_2}\}$ and $P_1 = \psi(P_2)$. Alice's signing secret/public keys are pairs $(u_a, v_a) \in [\![1, q-1]\!]^2$ and $(U_a, V_a) = ([u_a]P_2, [v_a]P_2) \in \mathbb{G}_2^2$ (respectively) and the signatures on a message $m \in [\![1, q-1]\!]$ for these keys are pairs $(r, [(u + m + rv)^{-1}]P_1)$ in $[\![1, q-1]\!] \times \mathbb{G}_1$.

The unforgeability of the scheme BB reduces to the ℓ-CSDH problem in the standard security model.

The Scheme UDVS-BB. The principle underlying the UDVS scheme UDVS-BB is based on Damgård's idea. Let us suppose that Bob has published a public key $U_b = [u_b]P_2$ and that the pair $\sigma = (r, S)$ in $[\![1, q-1]\!] \times \mathbb{G}_1$ is a BB signature produced by Alice, on a message m. If Cindy wants to designate σ to Bob, she picks uniformly at random an integer $t \in [\![1, q-1]\!]$ and sets $Q_1 = [t]S$, $Q_2 = [t]\psi(U_b)$ and $Q_3 = [t]P_1$. The quadruple $\tau = (r, Q_1, Q_2, Q_3)$ is the resulting DVS on m. The protocol UDVS-BB is described with all the details in figure 1.

The following simple observations are intuitive arguments in favor of the security of the protocol.

[1] *i.e.* $\forall c \geq 0, \exists K_c \in \mathbb{N}, \forall k \in [\![K_c, +\infty[\![, \varepsilon(k) \leq k^{-c}$.

1. Under KEA, the equality $\langle Q_3, U_b\rangle = \langle Q_2, P_2\rangle$ **[1]** insures Bob that Cindy knows the value t such that $Q_2 = [t]\psi(U_b)$ and $Q_3 = [t]P_1$.
2. If **[1]** is satisfied, Bob is convinced that Cindy knows the group element $S = [t^{-1}]Q_1$. The BB verification equality $\langle S, U_a + [m]P_2 + [r]V_a\rangle = \langle P_1, P_2\rangle$, holds if and only if the equality $\langle Q_1, U_a + [m]P_2 + [r]V_a\rangle = \langle Q_3, P_2\rangle$ **[2]** does. Therefore, if the equalities **[1]** and **[2]** are true, the quadruple τ proves to Bob that Alice has actually signed the message m.
3. However, this quadruple cannot convince anyone else, since it could have been produced by Bob himself. Indeed, if Bob samples uniformly at random $(r, \tilde{t})$ in $[\![1, q-1]\!]^2$ and computes the group elements: $Q_1 = [\tilde{t}]P_1$, $Q_3 = [\tilde{t}]\psi(U_a) + [\tilde{t}\cdot m]P_1 + [\tilde{t}\cdot r]\psi(V_a)$ and $Q_2 = [u_b]Q_3$, he produces quadruples which verify **[1]** and **[2]** and follow the same distribution as those produced by Cindy (namely with $t \equiv_q \tilde{t}(a_1 + m + a_2 r)$).

Algorithm UDVS-BB.Setup
Input: $k \in \mathbb{N}$
Output: cp

$(q, \mathbb{G}_1, \mathbb{G}_2, \mathbb{G}_3, \langle\cdot,\cdot\rangle, \psi) \xleftarrow{R} \mathsf{Gen}(k)$
$P_2 \xleftarrow{R} \mathbb{G}_2 \setminus \{\mathbb{O}_{\mathbb{G}_2}\}$
$P_1 \leftarrow \psi(P_2)$, $g = \langle P_1, P_2\rangle$
$D_{\mathcal{M}} \leftarrow$ " $[\![1, q-1]\!]$ "
$\mathsf{cp} = ((q, \mathbb{G}_1, \mathbb{G}_2, \mathbb{G}_3, \langle\cdot,\cdot\rangle, \psi), P_1, P_2, g, D_{\mathcal{M}})$

Algorithm UDVS-BB.Sign
Input: cp, m, $\mathsf{sk} = (u, v)$
Output: $\sigma = (r, S)$

$r \xleftarrow{R} [\![1, q-1]\!]$
$S \leftarrow [(u + m + vr)^{-1}]P_1$

Algorithm UDVS-BB.VKeyGen
Input: cp
Output: (sk, pk)

$u \xleftarrow{R} [\![1, q-1]\!]$
$\mathsf{sk} \leftarrow u$, $\mathsf{pk} \leftarrow [u]P_2$

Algorithm UDVS-BB.Fake
Input: cp, m, $\mathsf{skv} = u_b$, $\mathsf{pks} = (U_a, V_a)$
Output: $\tau = (r, Q_1, Q_2, Q_3)$

$(r, t) \xleftarrow{R} [\![1, q-1]\!]^2$
$R \leftarrow [t]\psi(U_a) + [t\cdot m]P_1 + [t\cdot r]\psi(V_a)$
$Q_1 \leftarrow [t]P_1$, $Q_2 \leftarrow [u_b]R$, $Q_3 \leftarrow R$

Algorithm UDVS-BB.SKeyGen
Input: cp
Output: (sk, pk)

$(u, v) \xleftarrow{R} [\![1, q-1]\!]^2$
$\mathsf{sk} \leftarrow (u, v)$; $\mathsf{pk} \leftarrow ([u]P_2, [v]P_2)$

Algorithm UDVS-BB.Verify
Input: cp, m, $\sigma = (r, S)$, $\mathsf{pk} = (U, V)$
Output: $b \in \{1, 0\}$

$s \leftarrow \langle S, U + [m]P_2 + [r]V\rangle$
If $s = g$ then $b \leftarrow 1$ else $b \leftarrow 0$

Algorithm UDVS-BB.Designate
Input: cp, m, $\mathsf{pks} = (U_a, V_a)$, $\mathsf{pkv} = U_b$, $\sigma = (r, S)$
Output: $\tau = (r, Q_1, Q_2, Q_3)$

$t \xleftarrow{R} [\![1, q-1]\!]$
$Q_1 \leftarrow [t]S$, $Q_2 \leftarrow [t]\psi(U_b)$, $Q_3 \leftarrow [t]P_1$

Algorithm UDVS-BB.DVerify
Input: cp, m, $\mathsf{skv} = u_b$, $\mathsf{pks} = (U_a, V_a)$, $\tau = (r, Q_1, Q_2, Q_3)$
Output: $b \in \{1, 0\}$

$\alpha_1 \leftarrow \langle Q_1, U_a + [m]P_2 + [r]V_a\rangle$
$\alpha_2 \leftarrow \langle Q_3, P_2\rangle$
$\beta_1 \leftarrow \langle Q_3, [u_b]P_2\rangle$, $\beta_2 \leftarrow \langle Q_2, P_2\rangle$
If $\alpha_1 = \alpha_2$ and $\beta_1 = \beta_2$ then $b \leftarrow 1$ else $b \leftarrow 0$

Fig. 1. Description of the protocol UDVS-BB(Gen)

Remark 1. Given a UDVS produced by UDVS-BB, it is easy, by random scalar multiplication, to produce a new signature on *the same* message for the *same* public keys. It is admitted that weak forgery is no real threat whatsoever.

Remark 2. The computational workload of UDVS-BB.DVerify for the designated verifier can be reduced to only two pairing evaluations and one exponentiation thanks to the knowledge of u_b by checking that $Q_2 = [u_b]Q_3$ instead of $\beta_1 = \beta_2$.

3.3 Description of the Protocol UDVS-BLS

Boneh-Lynn-Shacham's Signatures. In [5], Boneh *et al.* presented the signature scheme BLS that works in any bilinear cryptographic context. The scheme can be seen as a variant of the FDH signature scheme [2]. The protocol BLS is efficient, produces short signatures (for carefully chosen parameters), and is reducible in the random oracle model to the co-CDH problem [5].

The Scheme UDVS-BLS. Let Gen be a prime-order-BDH-parameter-generator. Let $k \in \mathbb{N} \setminus \{0\}$, let $(q, \mathbb{G}_1, \mathbb{G}_2, \mathbb{G}_3, \langle\cdot,\cdot\rangle, \psi)$ be some output of $\mathsf{Gen}(1^k)$ and let

Algorithm UDVS-BLS.Setup
Input: $k \in \mathbb{N}$
Output: cp
$(q, \mathbb{G}_1, \mathbb{G}_2, \mathbb{G}_3, \langle\cdot,\cdot\rangle, \psi) \xleftarrow{R} \mathsf{Gen}(k)$
$P_2 \xleftarrow{R} \mathbb{G}_2 \setminus \{\mathbb{O}_{\mathbb{G}_2}\}$
$D_{\mathcal{M}} \leftarrow$ " $\{0,1\}^*$ "
$\mathcal{H} \xleftarrow{R} (\mathbb{G}_1)^{\{0,1\}^*}$
$\mathsf{cp} \leftarrow ((q, \mathbb{G}_1, \mathbb{G}_2, \mathbb{G}_3, \langle\cdot,\cdot\rangle, \psi), P_2, D_{\mathcal{M}}, \mathcal{H})$

Algorithm UDVS-BLS.Verify
Input: cp, m, $\mathsf{pk} = U$, $\sigma = S$
Output: $b \in \{1, 0\}$
$H \leftarrow \mathcal{H}(m)$
$s \leftarrow \langle H, U\rangle$
If $s = \langle S, P_2\rangle$ then $b \leftarrow 1$ else $b \leftarrow 0$

Algorithm UDVS-BLS.Fake
Input: cp, m, $\mathsf{skv} = u_b$, $\mathsf{pks} = U_a$
Output: $\tau = (Q_1, Q_2)$
$t \xleftarrow{R} [\![1, q-1]\!]$
$Q_1 \leftarrow [t^{-1}]H(m)$
$Q_2 \leftarrow [t \cdot u_b]U_a$

Algorithms UDVS-BLS.SKeyGen UDVS-BLS.VKeyGen
Input: cp
Output: (sk, pk)
$\mathsf{sk} = u \xleftarrow{R} [\![1, q-1]\!]$
$\mathsf{pk} \leftarrow [u]P_2$

Algorithm UDVS-BLS.Sign
Input: cp, $m \in \{0,1\}^*$, $\mathsf{sk} = u$
Output: $\sigma = S$
$H \leftarrow \mathcal{H}(m)$, $S \leftarrow [u]H$

Algorithm UDVS-BLS.Designate
Input: cp, m, $\mathsf{pkv} = U_b$, $\sigma = S$
Output: $\tau = (Q_1, Q_2)$
$t \xleftarrow{R} [\![1, q-1]\!]$
$Q_1 \leftarrow [t]S \quad Q_2 \leftarrow [t^{-1}]U_b$

Algorithm UDVS-BLS.DVerify
Input: cp, m, $\mathsf{skv} = u_b$, $\mathsf{pks} = U_a$, $\tau = (Q_1, Q_2)$
Output: $b \in \{0, 1\}$
$H \leftarrow H(m)$
$s \leftarrow \langle [u_b]H, U_a\rangle$
If $s = \langle Q_1, Q_2\rangle$ then $b \leftarrow 1$ else $b \leftarrow 0$

Fig. 2. Description of the protocol UDVS-BLS(Gen)

$P_2 \in \mathbb{G}_2 \setminus \{\mathbb{O}_{\mathbb{G}_2}\}$. Let $U_a = [u_a]P_2$ (*resp.* $U_b = [u_b]P_2$) be Alice's (*resp.* Bob's) public key. Alice's signatures are elements $S = [u_a]H \in \mathbb{G}_1$, where the group element H is the hash value of the signed message m. The discrete logarithm of H is unknown to all users, therefore, whence the signature S is randomized as above: $Q_1 = [t]S$ for some $t \in [\![1, q-1]\!]$, it suffices to reveal the element $Q_2 = [t^{-1}]U_b$ to prove, in a non-transferable way, to Bob that Alice actually signed the message m. The quadruple $(U_a, U_b, H, \langle Q_1, Q_2\rangle)$ is indeed a bilinear Diffie-Hellman quadruple which could have been produced by using secret information from Alice or Bob, but not otherwise under the assumption that the SCBDH problem is intractable. The protocol UDVS-BLS is described with all the details in figure 2.

4 Security Results

In this section, we state that our schemes resist existential forgeries. The proofs are more or less routine and, due to lack of space, they are only sketched.

4.1 Unforgeability of the Scheme UDVS-BB

The following lemma state that, under KEA, the scheme UDVS-BB(Gen) is UDVS-EF-CMA-secure if and only if the scheme BB(Gen) is EF-CMA-secure.

Lemma 1. *Let* Gen *be a prime-order-BDH-parameter-generator and let $\mathcal{A}$ be a polynomial time UDVS-EF-CMA-adversary against* UDVS-BB(Gen). *Assuming KEA, there exist a polynomial time EF-CMA-adversary $\mathcal{B}$ against* BB(Gen) *such that the difference*

$$\left|\mathbf{Succ}^{efcma}_{\mathsf{BB(Gen)},\mathcal{B}} - \mathbf{Succ}^{\mathsf{UDVS\text{-}EF\text{-}CMA}}_{\mathsf{UDVS\text{-}BB(Gen)},\mathcal{A}}\right|$$

is a negligible function of the security parameter.

Proof. The algorithm $\mathcal{B}$ takes as inputs public parameters cp and a signing public key pks. It computes a verifying pair of keys (u_b, U_b) by running the algorithm UDVS-BB.VKeyGen(cp) and then executes $\mathcal{A}$ on inputs cp, pks and U_b. It simply forwards the $\mathcal{A}$'s signature queries to its own signing oracle and the simulation of the verifying oracle is trivial since the protocol UDVS-BB is publicly verifiable.

Let us denote $\mathcal{C}$ the algorithm whose execution is identical to the one of $\mathcal{A}$, except that it returns the couple (Q_3, Q_2), when $\mathcal{A}$ returns the quadruple $\tau^\star = (r, Q_1, Q_2, Q_3)$. If $\tau^\star$ is a valid forgery then the quadruple (P_2, U_b, Q_3, Q_2) is a valid Diffie-Hellman quadruple. Under KEA, there exists a PPTM $\overline{\mathcal{C}}$ which, given as inputs $\mathcal{C}$'s random tape and $\mathcal{C}$'s inputs (*i.e.* cp, pks and U_b), outputs $t \in [\![1, q-1]\!]$ such that $Q_3 = [t]P_1$ and $Q_2 = [t]\psi(U_b)$ with probability negligibly close to $\mathbf{Succ}^{\mathsf{UDVS\text{-}EF\text{-}CMA}}_{\mathsf{UDVS\text{-}BB(Gen)},\mathcal{A}}$.

$\mathcal{B}$ runs the algorithm $\overline{\mathcal{C}}$ to get this value $t \in [\![1, q-1]\!]$ and outputs the pair $\sigma^\star = (r, [t^{-1}]Q_1)$ which is valid forgery for the signature scheme BB if $\tau^\star$ is a valid forgery and $Q_3 = [t]P_2$. $\mathcal{B}$ is therefore an EF-CMA-polynomial time adversary whose success probability against BB(Gen) is negligible close to $\mathbf{Succ}^{\mathsf{UDVS\text{-}EF\text{-}CMA}}_{\mathsf{UDVS\text{-}BB(Gen)},\mathcal{A}}$. □

Combining this lemma with the unforgeability result of [3] we get:

Theorem 1. *Let* Gen *be a prime-order-BDH-parameter-generator and let $\mathcal{A}$ be a polynomial time UDVS-EF-CMA-adversary against* UDVS-BB(Gen). *Under KEA, there exist a polynomial time CSDH-adversary $\mathcal{B}$ against* Gen *such that*

$$\left|\mathbf{Succ}^{csdh}_{\mathsf{Gen},\mathcal{B}} - \mathbf{Succ}^{UDVS\text{-}EF\text{-}CMA}_{\mathsf{UDVS\text{-}BB(Gen)},\mathcal{A}}\right|$$

is a negligible function of the security parameter. □

Remark 3. Since KEA is a somewhat strange and impractical assumption, it would be better if we could do without it, as it has been recently done by Gjøsteen [8] for Damgård's encryption scheme. In [16], we reduce (without using any non-black-box assumption, such as KEA) the unforgeability of UDVS-BB to a well-defined (though *ad hoc*) computational problem:

Problem $\mathcal{P}(\ell)$: let $(x, y) \in \mathbb{N}^2$. Given $\ell \in \mathbb{N}$, $([x]P_2, [x^2]P_2, \dots, [x^\ell]P_2) \in \mathbb{G}_2^\ell$ and $([y]P_2, [(xy)]P_2, \dots, [(x^\ell y)]P_2) \in \mathbb{G}_2^{\ell+1}$, compute a quadruple (R_1, R_2, R_3, h) in $\mathbb{G}_1^3 \times [\![1, q-1]\!]$ such that $[(x+h)]R_2 = R_1$ and $R_3 = [y]R_1$lem.

4.2 Unforgeability of the Scheme UDVS-BLS

The theorem below states that UDVS-BLS(Gen) is UDVS-EF-CMA-secure in the random oracle model assuming the intractability of the SCBDH problem in Gen. It is worth noting that this security result *does not* depend on KEA.

Theorem 2. *Let* Gen *be a prime-order-BDH-parameter-generator and let $\mathcal{A}$ be a (τ, q_S, q_V)-UDVS-EF-CMA-adversary against* UDVS-BLS(Gen) *in the q_H-random oracle model. There exist a τ'-SCBDH-adversary $\mathcal{B}$ against* Gen *such that*

$$\begin{cases} \tau' = \tau + (q_H + 2q_S)(\mathsf{E}_{\mathbb{G}_1} + O(1)) + q_V(\mathsf{E}_{\mathbb{G}_3} + O(1)) \\ \mathbf{Succ}^{SCBDH}_{\mathsf{Gen},\mathcal{B}} \geq \mathbf{Succ}^{UDVS\text{-}EF\text{-}CMA}_{\mathsf{UDVS\text{-}BLS(Gen)},\mathcal{A}} / q_S(q_V + 1). \end{cases}$$

Proof (Sketch). Thanks to the random oracle model assumption, the proof is completely similar to the proof of security of the schemes proposed in [12, 17]. Our exact security reduction relies on two clever techniques from [6, 15]:

- Following a well-known technique due to Coron [6], a random coin decides whether $\mathcal{B}$ introduces the challenge in the answer to the random oracle or an element with a known preimage. This introduce the (small) loss factor q_S in the success probability.
- Using an approach due to Ogata, Kurosawa and Heng [15], introduced to analyze the security of Chaum's undeniable signatures, we do not need a decisional oracle to simulate the verification queries. The idea is that, unless UDVS-BLS is not unforgeable, all verification queries necessarily involve DVSs that were obtained from signing oracles (and can be readily checked) or that are invalid. $\mathcal{B}$'s strategy is to guess which verification query involves a forged signature and reject signatures involved in all other queries. This is done at the expense of losing the factor q_V in $\mathcal{B}$'s probability of success.

Due to space constraints, details are left to the reader. □

5 Efficiency Issues and Additional Properties

In the table 1, we compare the performance of all pairing-based UDVSs proposed up to now. For concreteness, we assume that all schemes are instantiated with the Tate pairing on a supersingular elliptic curve of MOV degree 6 on a ground base field of size 171 bits and that computing this bilinear map is 10 times more expensive than computing a scalar multiplication on the curve (whose computation time is arbitrarily set to 1). The new schemes compare very favorably in performance with respect to systems proposed so far and they can be used over a low bandwidth channel (UDVSs are longer than those produced by DVSBMH [13], but this scheme is not well-suited for devices with constrained computation capabilities since the designation procedure is much more costly).

Table 1. Efficiency comparison of pairing-based UDVSs

Protocol	DVSBM [17]	DVSBMH [13]	UDVS-BLS § 3.3	ZFI [18]	UDVS-BB § 3.2
Signatures	BLS			Variant of BB	BB
Model	Random Oracle Model			Standard Model	
Sign	$1\ \mathsf{E}_{\mathbb{G}_1}$			$1\ \mathsf{E}_{\mathbb{G}_1}$	$1\ \mathsf{E}_{\mathbb{G}_1}$
Verify	$2\ \mathsf{P}$			$1\ \mathsf{P} + 2\ \mathsf{E}_{\mathbb{G}_2}$	$1\ \mathsf{P} + 2\ \mathsf{E}_{\mathbb{G}_2}$
Designate	$1\ \mathsf{P}$	$1\ \mathsf{P}$	$1\ \mathsf{E}_{\mathbb{G}_1} + 1\ \mathsf{E}_{\mathbb{G}_2}$	$1\ \mathsf{P} + 2\ \mathsf{E}_{\mathbb{G}_2}$	$1\ \mathsf{E}_{\mathbb{G}_1} + 2\ \mathsf{E}_{\mathbb{G}_2}$
(in practice)	10	10	**2**	12	**3**
DVerify	$1\ \mathsf{P}$	$1\ \mathsf{P}$	$2\ \mathsf{P} + 1\ \mathsf{E}_{\mathbb{G}_1}$	$2\ \mathsf{P} + 2\ \mathsf{E}_{\mathbb{G}_2}$	$2\ \mathsf{P} + 3\ \mathsf{E}_{\mathbb{G}_2}$
(in practice)	10	10	31	22	23
Size	$\ell_3(k)$	k	$\ell_1(k) + \ell_2(k)$	$\ell_1(k) + \ell_2(k) + \ell_3(k)$	$2\ell_1(k) + \ell_2(k)$
(in bits)	1024	80	**342**	1366	**513**

Finally, it is worth mentioning that our schemes have additional properties, for instance:

- UDVS-BB can be extended to give the *first* efficient construction of universal multi-DVS [12, 14] in the standard security model. The multi-user scheme inherits the efficiency properties of UDVS-BB with the same DVS size (which, in particular, does not grow with the number of verifiers).
- In some cases [10, 13] it may be desirable that UDVSs provide a stronger notion of privacy: the *privacy of signer's identity* [13]. The scheme UDVS-BLS provides this security requirement assuming the hardness of the so-called xyz-decisional co-Diffie Hellman problem [11].

Details and additional extensions will be given in [16].

Acknowledgements. It is a pleasure to acknowledge Fabien Laguillaumie and Benoît Libert for their great comments and simplifying suggestions on a preliminary version of this paper. The author is grateful to Willy Susilo and Rui Zhang for providing a copy of their papers [14, 18].

References

1. M. Bellare & A. Palacio – "The Knowledge-of-Exponent Assumptions and 3-Round Zero-Knowledge Protocols." in *Advances in Cryptology* - Crypto *2004*, LNCS vol. 3152, Springer, 2004, p. 273–289.
2. M. Bellare & P. Rogaway – "Random Oracles are Practical: A Paradigm for Designing Efficient Protocols." in *Proceedings of the First ACM Conference on Computer and Communications Security*, ACM Press, 1993, p. 62–73.
3. D. Boneh & X. Boyen – "Short Signatures Without Random Oracles." in *Advances in Cryptology* - Eurocrypt *2004*, LNCS vol. 3027, Springer, 2004, p. 56–73.
4. D. Boneh & M. Franklin – "Identity-Based Encryption from the Weil Pairing. " *SIAM J. Comput.* **32** (2003), no. 3, 586–615.
5. D. Boneh, B. Lynn & H. Shacham – "Short Signatures from the Weil Pairing." *J. Cryptology* **17** (2004), no. 4, p. 297–319.
6. J.-S. Coron – "On the Exact Security of Full Domain Hash." in *Advances in Cryptology* - Crypto *2000*, LNCS vol. 1880, Springer, 2000, p. 229–235.
7. I. B. Damgård – "Towards Practical Public Key Systems Secure Against Chosen Ciphertext Attacks." in *Advances in Cryptology* - Crypto*'91*, LNCS vol. 576, Springer, 1992, p. 445–456.
8. K. Gjøsteen – "A New Security Proof for Damgård's ElGamal." in *Topics in Cryptology* - CT-RSA *2006*, LNCS vol. 3860, Springer, 2006, p. 150–158.
9. S. Goldwasser, S. Micali & R. L. Rivest – "A Digital Signature Scheme Secure Against Adaptive Chosen-Message Attacks." *SIAM J. Comput.* **17** (1988), no. 2, p. 281–308.
10. M. Jakobsson, K. Sako & R. Impagliazzo – "Designated Verifier Proofs and Their Applications." in *Advances in Cryptology* - Eurocrypt*'96*, LNCS vol. 1070, Springer, 1996, p. 143–154.
11. F. Laguillaumie, P. Paillier, & D. Vergnaud – "Universally Convertible Directed Signatures." in *Advances in Cryptology* - Asiacrypt *2005*, LNCS vol. 3788, Springer, 2005, p. 682–701.
12. F. Laguillaumie & D. Vergnaud – "Multi-designated Verifiers Signatures." in *Information and Communications Security, 6th International Conference, ICICS 2004*, LNCS vol. 3269, Springer, 2004, p. 495–507.
13. — "Designated Verifier Signatures: Anonymity and Efficient Construction from *any* Bilinear Map." in *4th Conference on Security in Communication Networks, SCN 2004*, LNCS vol. 3352, Springer, 2005, p. 107–121.
14. C. Y. Ng, W. Susilo & Y. Mu – "Universal Designated Multi Verifier Signature Schemes" in *International Workshop on Security in Networks and Distributed Systems, SNDS 2005*, IEEE Press, 2005, p. 305–309.
15. W. Ogata, K. Kurosawa & S.-H. Heng – "The Security of the FDH Variant of Chaum's Undeniable Signature Scheme." in *8th International Workshop on Practice and Theory in Public Key Cryptography - PKC 2005*, LNCS vol. 3386, Springer, 2005, p. 328–345.
16. D. Vergnaud – "New Extensions of Pairing-based Signatures into Universal (Multi) Designated Verifier Signatures. " *submitted,* 2006.
17. R. Steinfeld, L. Bull, H. Wang & J. Pieprzyk – "Universal Designated-Verifier Signatures." in *Advances in Cryptology* - Asiacrypt *2003*, LNCS vol. 2894, Springer, 2003, p. 523–542.
18. R. Zhang, J. Furukawa & H. Imai – "Short Signature and Universal Designated Verifier Signature without Random Oracles." in *Applied Cryptography and Network Security, ACNS 2005*, LNCS vol. 3531, Springer, 2005, p. 483–498.

Corrupting One vs. Corrupting Many: The Case of Broadcast and Multicast Encryption*

Daniele Micciancio and Saurabh Panjwani

University of California, San Diego
http://www-cse.ucsd.edu/users/{daniele, spanjwan}

Abstract. We analyze group key distribution protocols for broadcast and multicast scenarios that make blackbox use of symmetric encryption and a pseudorandom generator (PRG) in deriving the group center's messages. We first show that for a large class of such protocols, in which each transmitted ciphertext is of the form $E_{K_1}(K_2)$ (E being the encryption operation; K_1, K_2 being random or pseudorandom keys), security in the presence of a single malicious receiver is equivalent to that in the presence of collusions of corrupt receivers. On the flip side, we find that for protocols that *nest* the encrytion function (use ciphertexts created by enciphering ciphertexts themselves), such an equivalence fails to hold: there exist protocols that use nested encryption, are secure against single miscreants but are insecure against collusions.

Our equivalence and separation results are first proven in a symbolic, Dolev-Yao style adversarial model and subsequently translated into the computational model using a general theorem that establishes soundness of the symbolic security notions. Both equivalence and separation are shown to hold in the computational world under mild syntactic conditions (like the absence of encryption cycles).

We apply our results to the security analysis of 11 existing key distribution protocols. As part of our analysis, we uncover security weaknesses in 7 of these protocols, and provide simple fixes that result in provably secure protocols.

Keywords: Broadcast Encryption, Multicast Encryption, Group Key Distribution, Collusion-resistance.

1 Introduction

Private communication in dynamic groups is a cryptographic task of significant practical import. The problem, in a nutshell, is to enable an information provider to broadcast data to a large, dynamic set of "priveleged" receivers, while ensuring that at every instant, receivers outside this set are unable to procure the data. Two different models have been used in the literature to study this problem—one, known as *broadcast encryption* [8], assumes that all receivers are stateless

* This material is based upon work supported by the National Science Foundation under ITR Grant CCR-0313241 and Cyberturst Grant CCR-0430595. A full version of the paper can be downloaded from the second author's webpage.

M. Bugliesi et al. (Eds.): ICALP 2006, Part II, LNCS 4052, pp. 70–82, 2006.

and so, individual broadcasts are decipherable independently of other transmissions; and the other, usually referred to as *multicast encryption* or *multicast key distribution* [18], allows receivers to maintain state and, thus, decrypt the current data based on all past transmissions. Applications (of both models) have a wide spectrum, ranging from secure pay-per-view services over the Internet to protection mechanisms for digital media.

From a security perspective, privacy over broadcast channels raises new and challenging issues, not easily addressable using techniques for conventional point-to-point privacy. The multi-receiver setting opens up a new avenue for attacks—numerous malicious receivers can now potentially *collude* with each other and combine their secret information to decrypt the transmissions of the sender (even when they are not part of the priveleged set). Furthermore, miscreants can exploit past transmissions of the sender to recover current classified information (or, possibly, use future transmissions to do so later on). Proving security of protocols in the presence of such adversarial behavior is a difficult (and cumbersome) task, and so, protocol designers tend to rely more on intuition, rather than mathematical rigor, in making security arguments. Protocols are typically analyzed using a symbolic model of computation, one in which malicious behaviour is specified using fixed symbolic rules, often referred to as the Dolev-Yao rules. While the Dolev-Yao model enables simple and tractable security proofs, the question of whether such proofs imply security in the face of arbitrary computational attacks, is quite often left unresolved.

The general tendency to ignore and "shortcut" security analysis of protocols has the consequence that a bulk of multicast and broadcast encryption protocols exist in the literature without any meaningful proofs for (or against) their security claims—out of thirteen (symmetric-key) protocols that we surveyed from the literature, we found only three to have been *correctly* proven secure using strong computational definitions of security (one was claimed, though not proven, to be secure). For most of the remaining protocols, the security proofs provided, if any, involved only informal, Dolev-Yao style security arguments. Some protocols were not even accompanied with any security argument.

Our Contribution. In this paper, we concern ourselves with the provable security of broadcast and multicast encryption protocols that make use of symmetric-key cryptography. Instead of studying broadcast/multicast "encryption" directly, we focus on the related problem of *group key distribution (GKD)*, where the goal is to enable the sender to establish a shared secret key among a group of priveleged receivers on a broadcast channel (while keeping it secret from the rest of the receivers). A secure protocol for this task, coupled with a secure symmetric-key encryption scheme, naturally yields a solution to the group privacy problem[1].

We analyze GKD protocols that make blackbox use of a symmetric-key encryption scheme and a pseudorandom generator (PRG), in generating the center's messages, invoking both these primitives in an arbitrary, intermingled

[1] Indeed, all broadcast/multicast encryption protocols we know of involve group key distribution.

fashion. Our first finding is an equivalence relation between two security notions for these protocols—we show that for a large class of such protocols, in which encryption is not nested (that is, each ciphertext is of the form $E_{K_1}(K_2)$) security against multiple corrupt receivers is equivalent to security against a single corrupt receiver. This equivalence holds both in the symbolic (Dolev-Yao) model (that is, symbolic security against single corruptions implies symbolic security against multiple corruptions), and, subject to some mild syntactic conditions (e.g., absence of encryption cycles), also in the computational model. The equivalence in the computational setting is, in fact, of a very strong flavor: if one can prove a protocol (within the said class) computationally secure against single corruptions for *some* implementation of the cryptographic primitives, then it is collusion-resistant for *every* implementation of the primitives satisfying standard security properties (semantic security against chosen plaintext attacks for the former and computational indistinguishability for the latter).

We exemplify the significance of this equivalence result by applying it to the security analysis of various existing protocols. (See Table 1 in Sect. 4.) Most protocols (11 out of 13) surveyed by us don't make use of nested encryption and, as such, a proof of security against single corruptions for such protocols automatically implies collusion-resistance. As a part of our analysis, we uncover security weaknesses in 7 of the surveyed protocols, and provide simple fixes that result in protocols that are provably secure against arbitrary (polynomial-time) computational attacks.

Our techniques to prove this equivalence result don't generalize to capture protocols that use nested encryption (that is, transmit ciphertexts created by iterative encryption of a key using *multiple* other keys), and, in fact, they cannot do so. We demonstrate this by constructing a protocol that uses nesting (in fact, at most two iterations of E per ciphertext suffice), is secure against single corruptions but is totally broken by malicious coalitions (of size as small as two). As with the equivalence result, our separation holds both in the symbolic and computational models of security.

Protocols like the one used in our separation result have already been known to exist [5, 7]. (We remark that both these protocols require stateful receivers while ours does not.) Such protocols have a significant advantage over collusion-resistant protocols in terms of communication efficiency (constant versus logarithmic) and, in fact, they *beat* known lower bounds on the communication cost of GKD protocols [12]. Our results provide a precise explanation for this anomaly: although the bound of [12] applies to nested-encryption protocols, it holds only when collusion-resistance is satisfied. In fact, from our equivalence theorem (and the result of [12]), it follows that the efficiency of [5, 7] is unachievable using single encryption alone; it is precisely the use of nesting (and relaxation of the security requirements) that provide the efficiency gain.

OUR APPROACH. Our equivalence and separation results for GKD protocols are obtained using a modular two-stage approach. We first prove the results in the Dolev-Yao model (Sect. 2), treating encryption and PRGs as abstract

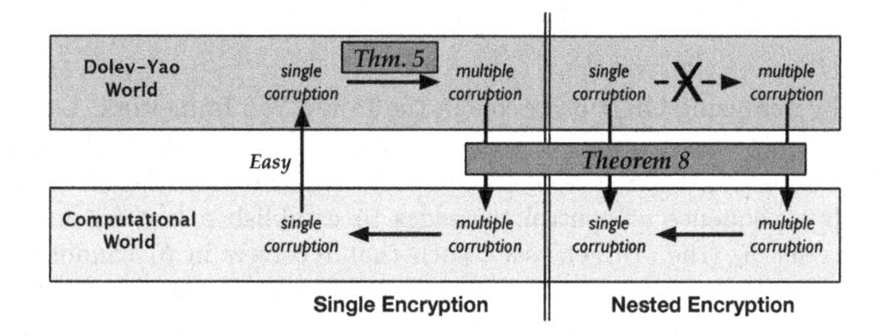

Fig. 1. Our results are proved first in the Dolev-Yao model (the solid arrow show implication, crossed-out one shows separation) and then interpreted in the computational model by proving soundness of the Dolev-Yao definitions (Thm. 8)

operators with perfect security properties. All proofs in this world are quite simple and intuitive, owing to the symbolic treatment of the primitives. As a second step (Sect. 3), we translate these results into the standard framework of computational cryptography by proving that our symbolic security notions are sound in a strong computational sense (provided some syntactic restrictions are obeyed by the protocol). This is achieved via an extension of a computational soundness theorem proven by us in [13]. Our extension incorporates the use of PRGs that can be applied in an arbitrary, nested manner (with polynomially-many nestings per seed) and greatly increases the applicability of the originally theorem. (9 out of 11 protocols we apply our results to make use of PRGs.)

This "two-step" approach not only makes the proof of our equivalence theorem simpler, but also alleviates much of the trouble in analyzing protocol security. A similar approach had already been taken in the seminal work of Abadi and Rogaway [1], and subsequent extentions of the same, with applications to multicast key distribution [13] and security of XML data [2]. In these papers, computational soundness theorems were used to translate security definitions from the symbolic to the computational setting. In this paper, we take the approach one step further, using it to translate (from the symbolic to the computational setting) not just security notions, but *relations* among these notions. The extension is not completely trivial, as it involves both soundness [1] and completeness [14] considerations, which are implicit in the protocol partitioning method underlying our computational equivalence proof. The soundness theorem itself (given in the full version) is of independent interest and could be applicable in other settings where encryption and PRGs are the only used primitives.

In our discussions on the computational security of GKD protocols, we focus on a scenario in which the dynamics of group membership are adversarially chosen in an adaptive way, but the decision of whether a receiver is malicious or not is made at the outset (non-adaptively). Dealing with adaptive corruptions is an important problem by itself, but is largely out of the scope of this paper. (See the full version for some partial results that address adaptive corruptions.)

2 The Result in the Dolev-Yao Model

We begin by analyzing GKD protocols in the Dolev-Yao framework. Let $[N] := \{1, \cdots, N\}$ denote a set of receivers having access to a broadcast channel. For any subset S of $[N]$, let $\overline{S}$ denote $[N] \setminus S$. At any instant t, a central authority $\mathcal{C}$ sends a sequence of control messages to establish a key K_t among receivers in a set S_t (the "target" set), such that receivers in $\overline{S_t}$ cannot recover K_t. We assume that all receivers and the center $\mathcal{C}$ have blackbox access to three cryptographic operations: a pair of functions, (E, D), modelling symmetric encryption, and a PRG G. The encryption pair satisfies the obvious correctness criterion: for any key K and message M, $D_K(E_K(M)) = M$. G models a length-doubling PRG[2]; it takes as input a key K and outputs two keys, $G_0(K)$ and $G_1(K)$. All information stored/exchanged during protocol execution is modeled using abstract expressions derived from the variable M in the following grammar:

$$\begin{aligned} \mathsf{M} &\rightarrow \mathsf{K} \mid E_{\mathsf{K}}(\mathsf{M}) \\ \mathsf{K} &\rightarrow \mathsf{Rand} \mid G_0(\mathsf{K}) \mid G_1(\mathsf{K}) \end{aligned} \tag{1}$$

Here K is a variable for keys, which can either be purely random (derived via the symbol $\mathsf{Rand} \rightarrow R_1|R_2|R_3|\cdots$) or pseudorandom (obtained by applying G_0 or G_1 on other keys). Some example expressions that can be obtained from this grammar are R_1, $G_0(G_1(R_2))$ (keys) and $E_{R_1}(R_2)$, $E_{G_0(R_2)}(E_{R_3}(R_4))$ (ciphertexts). We say that a key K_2 is *derived* from K_1, denoted $K_1 \Rightarrow_g K_2$, if $K_2 = G_{b_1}(\cdots G_{b_l}(K_1)\cdots)$ for some bits $b_1, \cdots, b_l$ and $l \geq 0$.

A GKD protocol has three components: Setup, Send and Decrypt. The first one, Setup, initializes the states of all receivers and of the center. The center's initial state, Δ_0, is an arbitrary set of keys, Keys, obtained from variable K above, and that of the ith receiver, (for any $i \in [N]$) is a set $\mathsf{Keys}[i]$, each such set being derivable from Keys. We use $\mathsf{Keys}[S]$ to denote $\bigcup_{i \in S} \mathsf{Keys}[i]$.

The other two algorithms Send and Decrypt are used for key updates. For any $t > 0$, Send takes a set $S_t \subseteq [N]$, and the current state of the center, Δ_{t-1} as input and outputs a set of messages $\mathsf{Msgs}(S_t)$ (to be sent to all receivers), while also updating $\mathcal{C}$'s state to Δ_t. Depending upon the manner in which messages are created, two protocol classes can be defined: Protocols in which every message in $\mathsf{Msgs}(S_t)$ is an arbitrary expression derived from variable M above are called *nested-encryption GKD (*N-GKD*) protocols.* A special case is one where protocols don't nest the encryption function for creating ciphertexts (this corresponds to replacing the rule $\mathsf{M} \rightarrow E_{\mathsf{K}}(\mathsf{M})$ with $\mathsf{M} \rightarrow E_{\mathsf{K}}(\mathsf{K})$); such protocols are called *single-encryption GKD (*S-GKD*) protocols.* Most protocols in the literature belong to this special case.

[2] Note that a PRG with an arbitrary expantion factor—the ratio between output length and input length—can be easily implemented using a length-doubling PRG. We use the latter for simplicity of analysis.

The target receivers decrypt messages using the following function:

Definition 1. (Key Recovery) For any set of keys, KSet, and any set of ciphertexts, CSet, the set of keys that can be recovered from CSet given KSet, denoted $\mathbf{Rec}(\mathsf{KSet}, \mathsf{CSet})$, is the smallest set R satisfying:

1. $\mathsf{KSet} \subseteq \mathsf{R}$.
2. If $K \in \mathsf{R}$, then $G_0(K), G_1(K) \in \mathsf{R}$.
3. If $K_1, \cdots, K_m \in \mathsf{R}$ and $E_{K_1}(E_{K_2} \cdots E_{K_m}(K) \cdots) \in \mathsf{CSet}$, then $K \in \mathsf{R}$

Roughly speaking, the DECRYPT algorithm applies the above function per target receiver, with KSet being equal to the keys available to that receiver and CSet the set of transmitted ciphertexts. For any sequence of target sets, $\widetilde{S}_t = (S_1, \cdots, S_t)$, we let $\mathsf{Msgs}(\widetilde{S}_t)$ be the set of all messages output by SEND when given this sequence as input, i.e., $\mathsf{Msgs}(\widetilde{S}_t) = \bigcup_{t'=1}^{t} \mathsf{Msgs}(S_{t'})$. Messages in $\mathsf{Msgs}(S_t)$ (resp. $\mathsf{Msgs}(\widetilde{S}_t)$) can be partitioned into keys $\mathsf{MKeys}(S_t)$ (resp. $\mathsf{MKeys}(\widetilde{S}_t)$) sent in clear, and ciphertexts $\mathsf{Ciph}(S_t)$ (resp. $\mathsf{Ciph}(\widetilde{S}_t)$).

Definition 2. (Correctness) A GKD protocol is called *stateless* if for all t, for all $S_t \subseteq [N]$, $\exists K$ s.t. $\forall i \in S_t$, $K \in \mathbf{Rec}(\mathsf{Keys}[i] \cup \mathsf{MKeys}(S_t), \mathsf{Ciph}(S_t))$. It is called *stateful* if for all sequences, $\widetilde{S}_t = (S_1, S_2, \cdots, S_t) \subseteq (2^{[N]})^*$, $\exists K$ s.t. $\forall i \in S_t$, $K \in \mathbf{Rec}(\mathsf{Keys}[i] \cup \mathsf{MKeys}(\widetilde{S}_t), \mathsf{Ciph}(\widetilde{S}_t))$.

Stateless GKD protocols (corresponding to the broadcast encryption model) are a special case of stateful ones (which correspond to multicast encryption). Any key satisfying the above criterion is called a *group key* at time t. We assume that for every t there is a distinguished group key that is used in applications like broadcast encryption at time t and denote it by K_t.

Security. Security of a GKD protocol $\Lambda = (\text{SETUP}, \text{SEND}, \text{DECRYPT})$ in the Dolev-Yao model refers to incapability of non-target receivers to recover group keys using our symbolic recovery rules. This can be formalized in two ways:

Definition 3. A GKD protocol Λ is secure against single corruptions (in the Dolev-Yao model) if for all t, for all sequences of target sets, $\widetilde{S}_t = (S_1, \cdots, S_t)$, for every $t' \leq t$ and every $i \notin S_{t'}$, $K_{t'} \notin \mathbf{Rec}(\mathsf{Keys}[i] \cup \mathsf{MKeys}(\widetilde{S}_t), \mathsf{Ciph}(\widetilde{S}_t))$.

Definition 4. A GKD protocol Λ is collusion-resistant(in the Dolev-Yao model) if for all t, for all sequences of target sets, $\widetilde{S}_t = (S_1, \cdots, S_t)$, for every $t' \leq t$, $K_{t'} \notin \mathbf{Rec}(\mathsf{Keys}[\overline{S_{t'}}] \cup \mathsf{MKeys}(\widetilde{S}_t), \mathsf{Ciph}(\widetilde{S}_t))$[3].

Note that we enforce that non-target receivers not be able to procure $K_{t'}$ even *after* viewing future transmissions of the center (a requirement often called *backward secrecy*). The definitions are common to both stateless and stateful protocols. Our first result is the equivalence between these definitions for the case of S-GKD protocols.

[3] It is not hard to verify that our definition of collusion-resistance is equivalent to one in which a Dolev-Yao adversary *adaptively* corrupts an arbitrary *subset* S of receivers, and then computes the group key K_t, by evaluating $\mathbf{Rec}(\mathsf{Keys}[S] \cup \mathsf{MKeys}(\widetilde{S}_t), \mathsf{Ciph}(\widetilde{S}_t))$, for some t such that $S_t \cap S = \emptyset$.

Theorem 5. An S-GKD protocol is secure against single corruptions (satisfies Defn. 3) if and only if it is collusion-resistant (satisfies Defn. 4).

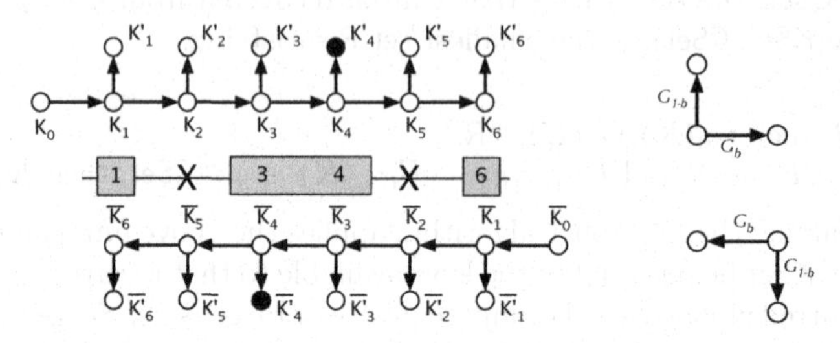

Fig. 2. An illustration of the separation protocol for $N = 6, S_t = \{1, 3, 4, 6\}$

On the flip side, we show that it is possible to design an N-GKD protocol, even for stateless receivers, that is secure against solitary malicious receivers but not collusion-resistant. Our protocol involves the use of, what we call here, *fully-pseudorandom chains (FPCs)* of keys, a notion similar to that of forward-secure PRGs [3]. Let G be a length-doubgling PRG. Let $b \in \{0, 1\}$. An FPC of length N, built from a random key K_0 (the *seed*), is a sequence of N key pairs $((K_i, K_i'))_{i \in [N]}$ such that $\forall i \in [N]$, $K_i = G_b(K_{i-1})$ and $K_i' = G_{1-b}(K_i)$. The K_i''s in this chain are "fully" pseudorandom in the sense that it is computationally infeasible to distinguish between them and a sequence of N independent random keys. In our protocol, SETUP creates two FPCs of length N using two different seeds K_0 and $\overline{K}_0$ (one called the *forward chain* and the other *backward chain*) and gives the keys $(K_i, \overline{K}_{N-i+1})$ to receiver i. Note that given this, receiver i can derive the key pairs $(K_i, K_i'), \cdots, (K_N, K_N')$ in the forward chain and $(\overline{K}_{N-i+1}, \overline{K}'_{N-i+1}), \cdots, (\overline{K}_N, \overline{K}'_N)$ in the backward chain.

To transmit a key K_t secretly to a set S_t, SEND divides the sequence $(1, \cdots, N)$ into the smallest possible set of intervals such that every $i \in S_t$ is contained in exactly one interval and no $i \in \overline{S_t}$ is contained in any interval. For e.g., if $N = 6$ and the target set is $\{1, 3, 4, 6\}$ (Fig. 2), these intervals would be $(1), (3, 4), (6)$. Let $I_1, \cdots, I_{r+1}$ denote these intervals with r being $|\overline{S_t}|$. For each interval $I_j = (j_1, \cdots, j_m)$, SEND outputs a ciphertext $E_{\overline{K}'_{N-j_1+1}}(E_{K'_{j_m}}(K_t))$. This ciphertext can be decrypted only by the receivers who know both K'_{j_m} and $\overline{K}'_{N-j_1+1}$, which is exactly the receivers in I_j. (In the figure, the black keys denote the keys used to encrypt K_t so as to transmit it to receivers $(3, 4)$.) Receiver $i \in S_t$ determines which interval I_j it belongs to and decrypts the corresponding ciphertext. It is not hard to verify that this protocol is secure against single corruptions (satisfies Defn. 3) but is not collusion-resistant (fails Defn. 4).

3 Interpretation in the Computational Setting

A natural question to ask at this point is whether our results in the Dolev-Yao model apply to practical implementations where the cryptographic operations are real programs satisfying computational security properties and the adversary is an arbitrary (polynomially-bounded) entity. In this section, we provide sufficient conditions under which this is true. We first define a class of GKD protocols, called *safe protocols*, which satisfy certain syntactic conditions on the way keys are used in the protocol. These safety conditions implement well-established and commonly-accepted (computational) cryptography practices. Then, via a generalization of the computational soundness theorems of [1,13], we show that any safe GKD protocol secure in the Dolev-Yao model is also secure (under the corresponding security notion) in the computational setting, provided the underlying primitives satisfy standard security properties. Combining this result with a case analysis, we show that for all safe GKD protocols, our equivalence and separation results of Sect. 2 are also true in the computational setting.

Definition 6. (Safe Protocols) A GKD protocol $\Lambda = (\textsc{Setup}, \textsc{Send}, \textsc{Decrypt})$ is called safe if for any input, $\widetilde{S_t} = (S_1, \cdots, S_t)$, given to it, the following conditions are satisfied:

1. *Proper key usage:* Any key K is used by at most one cryptographic primitive. This is a well established cryptography/security practice, which, in our case, means two things: (a) Encryption keys (i.e., keys K occurring in subexpressions of the form $E_K(M)$) are never used as input to the PRG (i.e., in subexpressions of the form $G_b(K)$); (b) a group key K_t (which can potentially be used for keying another primitive in an application, e.g. broadcast encryption), is used neither as an encryption key nor as an input to the PRG.
2. *No Encryption Cycles:* Define a relation $\rightarrow$ over keys such that $K_1 \rightarrow K_2$ if K_1 encrypts K_2 at any instant in the protocol. We require that the composition of the relations $\rightarrow$ and $\Rightarrow_g$, be acyclic. For e.g., messages like $E_{G_0(K_1)}(K_1)$ or message pairs like $(E_{K_1}(K_2), E_{G_1(K_2)}(K_1))$ are disallowed.
3. *Key Deployment after Key Distribution:* For any two (not necessarily distinct) keys, K_1, K_2, such that $K_1 \Rightarrow_g K_2$, if K_1 is used as a message (either in clear or encrypted under other keys) at time t_1, and K_2 is used as an encryption key at time t_2, then $t_2 \geq t_1$. In other words, once a key K_2 has been deployed for encrypting messages, the protocol can no longer distribute it (not even can it distribute a pseudorandom preimage K_1 of K_2, from which K_2 can, quite easily, be recovered).

These conditions are essential for the application of our computational soundness theorem to GKD protocols. (In fact, condition 1 is necessary to guarantee computational security of any GKD protocol in the sense we define below.) Notice that both our equivalence and separation results in the Dolev-Yao model also hold when restricted to safe protocols: The former follows from the fact that the definition of safety is independent of the Dolev-Yao adversarial model and for the latter, observe that our separation protocol satisfies all safety conditions.

Let $\Lambda = (\textsc{Setup}, \textsc{Send}, \textsc{Decrypt})$ be an N-receiver GKD protocol in the Dolev-Yao model. In the computational interpretation of Λ, $\Lambda^{\Pi,\mathcal{G}}$, all messages and keys are bitstrings corresponding to the "computational evaluation" of symbolic expressions used in Λ based on a (computational) encryption scheme $\Pi = (\mathcal{E}, \mathcal{D})$[4] and a (computational) length-doubling PRG $\mathcal{G}(\cdot) = \mathcal{G}_0(\cdot) \| \mathcal{G}_1(\cdot)$ ($\|$ denotes bitstring concatenation). The setup program, $\text{SETUP}^{\Pi,\mathcal{G}}$, takes a security parameter η as input and sets the initial states of all parties. The initial state of the center (resp. the ith receiver), denoted $\delta_0^{\mathcal{C}}$ (resp. δ_0^i), is the computational evaluation of the set Keys (resp. $\mathsf{Keys}[i]$). $\text{SEND}^{\Pi,\mathcal{G}}$, as before, receives a set S_t and the current center state $\delta_{t-1}^{\mathcal{C}}$ as input, and outputs a set of messages $\mathsf{msgs}(S_t)$ (the evaluation of all expressions in $\mathsf{Msgs}(S_t)$) and $\mathcal{C}$'s updated state $\delta_t^{\mathcal{C}}$. Finally, $\text{DECRYPT}^{\Pi,\mathcal{G}}$ takes a receiver index i, the corresponding current state δ_{t-1}^i, the current transmission $\mathsf{msgs}(S_t)$, and outputs the updated state δ_t^i and either a key k_t^i or `fail`. Correctness, now, means that for all sequences $\widetilde{S}_t = (S_1, \cdots, S_t)$, for every $i, j \in S_t$, the keys k_t^i and k_t^j output after running $\text{DECRYPT}^{\Pi,\mathcal{G}}$ with inputs $\mathsf{msgs}(S_1), \cdots, \mathsf{msgs}(S_t)$ are the same (equal to the bitstring group key k_t). In stateless protocols, δ_t^i equals δ_0^i for all i and t.

Security. Security of GKD protocols is defined using a game played between an adversary $\mathcal{A}$ and a challenger $\mathcal{B}$. Both are given a security parameter η as input (and $\mathcal{A}$ must run in time polynomial in η). First, the challenger invokes $\text{SETUP}(\eta)$ to generate the initial states, $\delta_0^{\mathcal{C}}, \delta_0^1, \cdots, \delta_0^N$. It also generates a uniformly random challenge bit b and initializes a protocol counter t to 1. The adversary first specifies a set of corrupt receivers $C(\mathcal{A})$ (in return for which it is given $\{\delta_0^i\}_{i \in C(\mathcal{A})}$), and then makes several queries, each query being of one of two types:

- `send`(S_t) (for some $S_t \subseteq [N]$): $\mathcal{B}$ runs SEND, returns $\mathsf{msgs}(S_t)$ to $\mathcal{A}$, updates states of all parties and sets $t \leftarrow t + 1$; or
- `challenge`(t') (for some $t' \leq t$ such that $S_{t'} \cap C(\mathcal{A}) = \emptyset$): If $b = 0$, $\mathcal{B}$ returns the group key at time t', $k_{t'}$; else, it generates a fresh random key $r_{t'}$ and returns it.

The advantage of $\mathcal{A}$ in the game, denoted $\mathbf{Adv}^{gkd}_{\Lambda^{\Pi,\mathcal{G}}}(\mathcal{A}, \eta)$, is the absolute difference between the probability that $\mathcal{A}$ outputs 1 when $b = 1$ and the probability of the same event when $b = 0$.

Definition 7. A GKD protocol Λ is *secure against single (resp. multiple) corruptions* (in the computational model) if for any adversary $\mathcal{A}$ satisfying $|C(\mathcal{A})| = 1$ (resp. $|C(\mathcal{A})| \geq 0$), $\mathbf{Adv}^{gkd}_{\Lambda^{\Pi,\mathcal{G}}}(\mathcal{A}, \eta)$ is a negligible function of η.

For the case of stateless GKD protocols, our definition parallels definitions of broadcast encryption already existing in the literature [15] with the major difference that we focus on defining group key distribution rather than the problem of group privacy. (More discussion on this issue appears in the full version.)

Theorem 8. (Security Theorem) Let Λ be any safe GKD protocol. Let Π be any `ind-cpa` secure encryption scheme (i.e., satisfying semantic security against chosen plaintext attacks) and $\mathcal{G}$ any secure pseudorandom generator. If

[4] We consider encryption schemes where key generation involves picking a uniformly random bitstring of length equal to the security parameter.

Λ is secure without collusions (resp. collusion-resistant) in the Dolev-Yao model, that is Λ satisfies Defn. 3 (resp. Defn. 4) then $\Lambda^{\Pi,\mathcal{G}}$ is secure against single (resp. multiple) corruptions in the computational model.

Equivalence and Separation. We now use Theorem 8 to translate of our equivalence result for S-GKD protocols from Sect. 2 into the computational setting. Consider the class of safe S-GKD protocols implemented with an encryption scheme Π and a PRG $\mathcal{G}$. We first partition this class into two sub-classes—protocols in the first catagory are secure without collusions in the Dolev-Yao model (satisfy Defn. 3) while those in the second class are not. Based on the equivalence in the Dolev-Yao model (Thm. 5) and the computational soundness of Dolev-Yao collusion-resistance (Thm. 8), we conclude that protocols in the first catagory are secure against multiple corruptions *for any secure instantiations of* Π *and* $\mathcal{G}$. Second, observe that if a protocol (whether safe or not) does not satisfy security without collusions in the Dolev-Yao setting (fails Defn. 3), then it is trivially insecure: it can be broken for *every implementation of* Π *and* $\mathcal{G}$, by executing the Dolev-Yao attack in the computational setting. Thus, we have:

Theorem 9. Let Λ be a safe S-GKD protocol. If $\Lambda^{\Pi,\mathcal{G}}$ is secure against single corruptions for some encryption scheme Π and some PRG $\mathcal{G}$, then $\Lambda^{\Pi,\mathcal{G}}$ is secure against multiple corruptions for any `ind-cpa` secure encryption scheme Π and any secure PRG $\mathcal{G}$.

For protocols that use nested encryption, we have the following theorem:

Theorem 10. There exists an N-GKD protocol that is secure against single corruptions for any (computationally secure) implementation of Π and $\mathcal{G}$, but is not collusion-resistant.

This separation is demonstrated by the protocol of Sect. 2. (The single-corruption security of the protocol follows from its Dolev-Yao security and Thm. 8.)

4 Analysis of Known Protocols

In this section, we summarize our analysis of various existing GKD protocols based on the results of Sect. 3. Out of 13 protocols for symmetric-key broadcast/multicast encryption that we surveyed from the literature, we found only 3 protocols to have been accompanied with proofs that establish computational security of the protocol. The security analysis of the remaining protocols, if undertaken at all, has hitherto been restricted to Dolev-Yao style arguments, and without any computational justification of such analysis. Indeed, we find that 7 of these protocols have weaknesses in their design which make them vulnerable to attacks by computational adversaries. For example, the broadcast encryption protocols of [9, 17, 6] make use of "key chains" generated by applying a cryptographic function f iteratively on a random key K to produce a sequence of values $f(K), f(f(K)), f(f(f(K))), \cdots$. The values in the chain are subsequently used as keys to encrypt other keys during key distribution and it is claimed that an adversary can decrypt a ciphertext $E_{K'}(M)$ for some K' in the key chain only if it knows K' or some other value *preceding* K' in the chain. [9,17] suggest implementing f with a one-way hash function and [6] proposes to do so with a one-way permutation or a PRG. We note that none of these implementations are sufficient to guarantee security of the respective protocols in a computational sense. Technically, the deployment of key chains in all these protocols conflicts with our first safety condition: keys are used for encryption, and as inputs to another cryprographic primitive (that is, f), which, as already discussed, is bad cryprographic practice. In principle, this could lead to complete recovery of all values in such key chains (and, consequently, of all group keys) even by a passive observer of the protocol.

In the same vein, various multicast encryption protocols [18, 4, 7, 16] are computationally insecure. For a receiver "join" event (at time t), they either recommend encrypting the new group key K_{t+1} with K_t [18, 4], or deriving K_{t+1} from K_t via a PRG [16]. ([7] encrypts K_{t+1} under K_t during both "leave" and "join" events.) Such usage of group keys can, in principle, compromise their pseudorandomness and thus render the resulting encryption protocol totally insecure.

Table 1. Analysis of 11 GKD protocols surveyed by us. *Comp-single* and *Comp-multiple* refer to security against single and multiple corruptions respectively in the computational model. The last column shows which protocols use nested encryption.

Protocol	*Safe?*	*Fixed*	*Security from our results*	*Nesting?*
LKH [18]	No	Yes	Comp-multiple	No
LKH+ [4]	No	Yes	Comp-multiple	No
Subset Diff. (SD) [15]	Yes	—	Comp-multiple	No
ELK [16]	No	Yes	Comp-multiple	No
LSD [10]	Yes	—	Comp-multiple	No
Stratified SD (SSD) [9]	No	Yes	Comp-multiple	No
DDKC [17]	No	Yes	Comp-multiple	No
Skip. Chains (SC) [6]	No	Yes	Comp-multiple	No
Improved SSD/SC [11]	Yes	—	Comp-multiple	No
Boolean Func. Min. [5]	Yes	—	Comp-single	Yes
LOR [7]	No	Yes	Comp-single	Yes

Fortunately, these weaknesses are quite straightforward to fix in most cases. For the protocols that make use of (insecure) key chains, the fix simply involves replacing the chain with an FPC (Sect. 2). The protocols that use group keys for keying primitives within the GKD protocol can be patched in the following manner: if K_t is the group key distributed in the original protocol for time t, then the patched protocol instead uses $G_0(K_t)$ as the group key and any encryption/pseudorandom generation that was previously done using K_t is now performed using $G_1(K_t)$. (For e.g., the control message $E_{K_t}(K_{t+1})$ gets substituted by $E_{G_1(K_t)}(K_{t+1})$.) These modifications make the protocols compatible with our safety requirements and enable application of our security theorem (Thm. 8). Table 1 presents the results from our analysis of 11 GKD protocols, including the protocols fixed in the above manner. 9 of the listed protocols don't make use of nested encryption and for these it suffices to verify Dolev-Yao security of the protocol against single corruptions (defn. 3), and subsequently to invoke our equivalence theorem (Thm. 9) in order to establish collusion-resistance. The 2 protocols that use nested encryption [5, 7] were already known to be collusion-insecure, and for these a Dolev-Yao proof of security against single corruptions implies the corresponding computational criterion.

References

1. M. Abadi and P. Rogaway. Reconciling two views of cryptography (the computational soundness of formal encryption). *J. of Crypt.*, 15(2):103–127, 2002.
2. M. Abadi and B. Warinschi. Security analysis of cryptographically controlled access to XML documents. In *Proc. of 24th ACM Symposium on Principles of Database Systems (PODS)*, pages 108–117, 2005.
3. M. Bellare and B. Yee. Forward security in private key cryptography. In *Topics in Cryptology – CT-RSA 2003*, pages 1–18, 2003.
4. R. Canetti, J. A. Garay, G. Itkis, D. Micciancio, M. Naor, and B. Pinkas. Multicast security: A taxonomy and some efficient constructions. In *IEEE INFOCOM'99*, pages 708–716, 1999.
5. I. Chang, R. Engel, D. Kandlur, D. Pendarakis, and D. Saha. Key management for secure internet multicast using boolean function minimization techniques. In *IEEE INFOCOMM '99*, pages 689–698, 1999.
6. J. H. Cheon, N. su Jho, M.-H. Kim, and E. S. Yoo. Skipping, cascade, and combined chain schemes for broadcast encryption. Cryptology ePrint Archive, Report 2005/136. Prelim. version in Eurocrypt 2005.
7. J. Fan, P. Judge, and M. H. Ammar. Hysor: Group key management with collusion-scalability tradeoffs using a hybrid structuring of receivers. In *Proc. of the IEEE International Conference on Computer Communications Networks*, 2002.
8. A. Fiat and M. Naor. Broadcast encryption. In *Advances in Cryptology - CRYPTO '93*, pages 480–491, 1993.
9. M. T. Goodrich, J. Z. Sun, and R. Tamassia. Efficient tree-based revocation in groups of low-state devices. In *Advances in Cryptology - CRYPTO '04*, pages 511–527, 2004.
10. D. Halevy and A. Shamir. The LSD broadcast encryption scheme. In *Advances in Cryptology: CRYPTO 2002*, pages 47–60, 2002.
11. J. Y. Hwang, D. H. Lee, and J. Lim. Generic transformation for scalable broadcast encryption schemes. In *Advances in Cryptology - CRYPTO '05*, pages 276–292, 2005.
12. D. Micciancio and S. Panjwani. Optimal communication complexity of generic multicast key distribution. In *Advances in Cryptology - Eurocrypt '04*, pages 153–170, 2004.
13. D. Micciancio and S. Panjwani. Adaptive security of symbolic encryption. In *Theory of Cryptography Conference, TCC 2005*, pages 169–187, 2005.
14. D. Micciancio and B. Warinschi. Completeness Theorems for the Abadi-Rogaway Logic of Encrypted Expressions. *Journal of Computer Security*, 12(1):99–129, 2004.
15. D. Naor, M. Naor, and J. Lotspiech. Revocation and tracing schemes for stateless receivers. In *Advances in Cryptology – CRYPTO 2001*, pages 41–62, 2001.

16. A. Perrig, D. Song, and D. Tygar. ELK, a new protocol for efficient large-group key distribution. In *IEEE Symposium on Security and Privacy*, 2001.
17. P. Wang, P. Ning, and D. Reeves. Storage-efficient stateless group key distribution. In *7th International Information Security Conference (ISC)*, pages 25–38, 2004.
18. C. K. Wong, M. Gouda, and S. S. Lam. Secure group communications using key graphs. *IEEE/ACM Transactions on Networking*, 8(1):16–30, Feb. 2000.

Cryptographically Sound Implementations for Communicating Processes
(Extended Abstract)

Pedro Adão[1,⋆] and Cédric Fournet[2]

[1] Center for Logic and Computation, IST, Lisboa, Portugal
[2] Microsoft Research

Abstract. We design a core language of principals running distributed programs over a public network. Our language is a variant of the pi calculus, with secure communications, mobile names, and high-level certificates, but without any explicit cryptography. Within this language, security properties can be conveniently studied using trace properties and observational equivalences, even in the presence of an arbitrary (abstract) adversary.

With some care, these security properties can be achieved in a concrete setting, relying on standard cryptographic primitives and computational assumptions, even in the presence of an adversary modeled as an arbitrary probabilistic polynomial-time algorithm. To this end, we develop a cryptographic implementation that preserves all properties for all safe programs. We give a series of soundness and completeness results that precisely relate the language to its implementation.

1 Secure Implementations of Communications Abstractions

When designing and verifying security protocols, some level of idealization is needed to provide manageable mathematical treatment. Accordingly, two views of cryptography have been developed over the years. In the first view, cryptographic protocols are expressed algebraically, within simple languages. This formal view is suitable for automated computer tools, but is also arguably too abstract. In the second view, cryptographic primitives are probabilistic algorithms that operate on bitstrings. This view involves probabilities and limits in computing power; it is harder to handle formally, especially when dealing with large protocols. Getting the best of both views is appealing, and is the subject of active research that aims at building security abstractions with formal semantics and sound computational implementations.

In this work, we develop a first sound and complete implementation of a distributed process calculus. Our calculus is a variant of the pi calculus; it provides name mobility, reliable messaging and authentication primitives, but neither explicit cryptography nor probabilistic behaviors. Taking advantage of concurrency theory, it supports simple reasoning, based on labeled transitions and observational equivalence. We precisely define its concrete implementation in a computational setting. We establish general soundness

⋆ Partially supported by FCT grant SFRH/BD/8148/2002, FEDER/FCT project Fiblog POCTI/2001/MAT/37239, and FEDER/FCT project QuantLog POCI/MAT/55796/2004.

M. Bugliesi et al. (Eds.): ICALP 2006, Part II, LNCS 4052, pp. 83–94, 2006.

and completeness results in the presence of active adversaries, for both trace properties and observational equivalences, essentially showing that high level reasoning accounts for all low-level adversaries. We illustrate our approach by coding security protocols and establishing their computational correctness by simple formal reasoning.

We implement high-level functionalities using cryptography, not high-level views of cryptographic primitives. Following recent related works, we could instead have proceeded in two steps, by first compiling high-level communications to an intermediate calculus with ideal, explicit cryptography (in the spirit of [3, 2]), then establishing the computational soundness of this calculus with regards to computational cryptography. However, this second step is considerably more delicate than our present goal, inasmuch as one must provide a sound implementation for an arbitrary usage of ideal cryptography. In contrast, for instance, our language keeps all keys implicit, so no high-level program may ever leak a key or create an encryption cycle. (We considered targeting existing idealized cryptographic frameworks with soundness theorems, but their reuse turned out to be more complex than a direct implementation.)

Our concrete implementation relies on standard cryptographic primitives, computational security definitions, and networking assumptions. It also combines typical distributed implementation mechanisms (abstract machines, marshaling and unmarshaling, multiplexing, and basic communications protocol.) This puts interesting design constraints on our high-level semantics, as we need to faithfully reflect their properties and, at the same time, be as abstract as possible. In particular, our high-level environments should be given precisely the same capabilities as low-level probabilistic polynomial-time (PPT) adversaries. For example, our language supports abstract reliable messaging: message senders and receivers are authenticated, message content is protected, and messages are delivered at most once. On the other hand, under the conservative assumption that the adversary controls the network, we cannot guarantee message delivery, nor implement private channels (such that some communications may be undetected). Hence, the simple rule $\overline{c}\langle M\rangle.P \mid c(x).Q \rightarrow P \mid Q\{M/x\}$, which models silent communication "in the ether" for the pi calculus, is too abstract for our purposes. (For instance, if P and Q are implemented on different machines connected by a public network, and even if c is a restricted channel, the adversary can simply block all communications.) Instead, we design high-level rules for communications between explicit principals, mediated by an adversary, with abstract labels that enable the environment to perform traffic analysis but not forge messages or observe their payload. Similarly, process calculi feature non-deterministic infinite computations, and we need to curb these features to meet our low-level complexity requirements.

Contents. This extended abstract is organized as follows. Section 2 defines our low-level target model. Section 3 presents our high-level language and semantics. Section 4 defines and illustrates high-level equivalences. Section 5 outlines our concrete implementation. Section 6 states our soundness and correctness theorems. Section 7 concludes.

A technical report [6] provides additional details and definitions, including the definition of our cryptographic implementation, examples and applications, and all proofs.

Related Work. Within formal cryptography, process calculi are widely used to model security protocols. For example, the spi calculus of Abadi and Gordon [4] neatly models secret keys and fresh nonces using names and their dynamic scopes. Representing active attackers as pi calculus contexts, one can state (and prove) trace properties and observational equivalences that precisely capture the security goals for these protocols. Automated provers (e.g. [10]) also help verify these goals.

Abadi, Fournet, and Gonthier develop distributed implementations for variants of the join calculus, with high-level security but no cryptography, roughly comparable to our high-level language. Their implementation is coded within a lower-level calculus with formal cryptography. They establish full abstraction for observational equivalence [3, 2]. Our approach is similar, but our implementation is considerably more concrete. Also, due to the larger distance between high-level processes and low-level machines, our soundness results are more demanding. Abadi and Fournet also propose a labeled semantics for traffic analysis, in the context of a pi calculus model of a fixed protocol for private authentication [1].

The computational soundness of formal cryptography is an active area of research, with many recent results for languages that include selected cryptographic primitives. Abadi and Rogaway initially consider formal encryption against passive attackers [5] and establish the soundness of indistinguishability. Backes, Pfitzmann and Waidner [8] achieve a first soundness result with active attackers, initially for public-key encryption and digital signatures. They extend their result to symmetric authentication [9] and encryption [7]. Micciancio and Warinschi [16] also establish soundness in the presence of active attacks, under different simpler assumptions.

Other works develop computationally sound implementations of more abstract security functions on top of cryptography. For example, Canetti and Krawczyk build computational abstractions of secure channels in the context of key exchange protocols, with modular implementations, and they establish sufficient conditions to realize these channels [11]. Targeting the idealized cryptographic model of Backes et al. [8], Laud [14] implements a deterministic process calculus and establishes the computational soundness of a type system for secrecy.

Another interesting approach is to supplement process calculi with concrete probabilistic or polynomial-time semantics. Unavoidably, reasoning on processes becomes more difficult. For example, Lincoln, Mitchell, Mitchell, and Scedrov [15] introduce a probabilistic process algebra for analyzing security protocols, such that parallel contexts coincide with probabilistic polynomial-time adversaries. In this framework, further extended by Mitchell, Ramanathan, Scedrov, and Teague [17], they develop an equational theory and bisimulation-based proof techniques.

2 Low-Level Target Model

Before presenting our language design and implementation, we specify the target systems. We rely on standard notions of security for cryptographic primitives (CCA2 for encryption [18], CMA for signing [13]) recalled in the technical report.

We consider systems that consist of a finite number of communicating principals $a, b, c, e, u, v, \ldots \in \mathsf{Prin}$. Each principal runs its own program, written in our high-level language and executed by the PPT machine outlined in Section 5. Each machine M_a has two wires, $\mathbf{in}_a$ and $\mathbf{out}_a$, representing a basic network interface. When activated, the machine reads a bitstring from $\mathbf{in}_a$, performs some local computation, then writes a bitstring on $\mathbf{out}_a$ and yields. The machine embeds probabilistic algorithms for encryption, signing, and random-number generation—thus the machine outputs are random variables. The machine is also parameterized by a security parameter $\eta \in \mathbb{N}$—intuitively, the length for all keys—thus these outputs are ensembles of probabilities.

Some of these machines may be corrupted, under the control of the attacker; their implementation is then unspecified and treated as part of the attacker. We let $a, b \in \mathcal{H}$ with $\mathcal{H} \subset \mathsf{Prin}$ range over principals that comply with our implementation, and let $\mathsf{M} = (\mathsf{M}_a)_{a \in \mathcal{H}}$ describe our whole system. Of course, when a interacts with $u \in \mathsf{Prin}$, its implementation M_a does not know whether $u \in \mathcal{H}$ or not.

The adversary, A, is a PPT algorithm that controls the network, the global scheduler, and some compromised principals. At each moment, only one machine is active: whenever an adversary delivers a message to a principal, this principal is activated, runs until completion, and yields an output to the adversary.

Definition 1 (Run). *A run of* A *and* M *with security parameter* $\eta \in \mathbb{N}$ *goes as follows:*

1. *key materials are generated for every principal* $a \in \mathsf{Prin}$*;*
2. *every* M_a *is activated with* 1^η*, the keys for* a*, and the public keys for all* $u \in \mathsf{Prin}$*;*
3. A *is activated with* 1^η*, the keys for* $e \in \mathsf{Prin} \setminus \mathcal{H}$*, and the public keys for* $a \in \mathcal{H}$*;*
4. A *performs a series of low-level exchanges:*
   - A *writes a bitstring on wire* $\mathbf{in}_a$ *and activates* M_a *for some* $a \in \mathcal{H}$*;*
   - *upon completion of* M_a*,* A *reads a bitstring on* $\mathbf{out}_a$*;*
5. A *returns a bitstring* s*, written* $s \longleftarrow \mathsf{A}[\mathsf{M}]$.

To study their security properties, we compare systems that consist of machines running on behalf of the same principals $\mathcal{H} \subseteq \mathsf{Prin}$, but with different internal programs and states. Intuitively, two systems are equivalent when no adversary, starting with the information normally given to the principals $e \in \mathsf{Prin} \setminus \mathcal{H}$, can distinguish between their two behaviors, except with negligible probability (written $\operatorname{neg}(\eta)$). This is the notion of *computational indistinguishability* introduced by Goldwasser and Micali [12]. Our goal is to develop a simpler, higher-level semantics that entails indistinguishability.

Definition 2. *Two systems* M^0 *and* M^1 *are indistinguishable, written* $\mathsf{M}^0 \approx \mathsf{M}^1$*, when for every PPT adversary* A*, we have* $|\Pr[1 \longleftarrow \mathsf{A}[\mathsf{M}^0]] - \Pr[1 \longleftarrow \mathsf{A}[\mathsf{M}^1]]| \leq \operatorname{neg}(\eta)$.

3 A Distributed Calculus with Principals and Authentication

We now present our high-level language. We successively define terms, patterns, processes, configurations, and systems. We then give their operational semantics. Although some aspects of the design are unusual, the resulting calculus is still reasonably abstract and convenient for distributed programming.

Syntax and Informal Semantics. Let Name be a countable set of *names* disjoint from Prin. Let $\mathtt{f}$ range over a finite number of function symbols, each with a fixed arity $k \geq 0$. Terms and patterns are defined by the following grammar:

$V, W ::=$	Terms
x, y	variable
$m, n \in \mathsf{Name}$	name
$a, b, e, u, v \in \mathsf{Prin}$	principal identity
$\mathtt{f}(V_1, \ldots, V_k)$	constructed term (when $\mathtt{f}$ has arity k)
$T, U ::=$	Patterns
$?x$	variable (binds x)
$T\ as\ ?x$	alias (binds x to the term that matches T)
V	constant pattern
$\mathtt{f}(T_1, \ldots, T_k)$	constructed pattern (when $\mathtt{f}$ has arity k)

Names and principals identities are atoms, or "pure names", which may be compared with one another but otherwise do not have any structure. Constructed terms represent structured data, much like algebraic data types in ML or discriminated unions in C. They can represent constants and tags (when $k = 0$), tuples, and formatted messages. As usual, we write $\mathtt{tag}$ and (V_1, V_2) instead of $\mathtt{tag}()$ and $\mathtt{pair}(V_1, V_2)$. Patterns are used for analyzing terms and binding selected subterms to variables. For instance, the pattern $(\mathtt{tag}, ?x)$ matches any pair whose first component is $\mathtt{tag}$ and binds x to its second component. We write _ for a variable pattern that binds a fresh variable.

Local processes represent the active state of principals, with the following grammar:

$P, Q, R ::=$	Local processes
V	asynchronous output
$(T).Q$	input (binds $bv(T)$ in Q)
$*(T).Q$	replicated input (binds $bv(T)$ in Q)
$\mathtt{match}\ V\ \mathtt{with}\ T\ \mathtt{in}\ Q\ \mathtt{else}\ Q'$	matching (binds $bv(T)$ in Q)
$\nu n.P$	name restriction ("new", binds n in P)
$P \mid P'$	parallel composition
$\mathbf{0}$	inert process

The asynchronous output V is just a pending message; its data structure is explained below. The input $(T).Q$ waits for an output that matches T then runs Q with the bound variables of T substituted by the matching subterms of the output message. The replicated input $*(T).Q$ behaves similarly but it can consume any number of outputs that match T and fork a copy of Q for each of them. The match process runs Q if V matches T, and runs Q' otherwise. The name restriction creates a fresh name n then runs P. Parallel composition represents processes that run in parallel, with the inert process $\mathbf{0}$ as unit. Free and bound names and variables for terms, patterns, and processes are defined as usual: x is bound in T if $?x$ occurs in T; n is bound in $\nu n.P$; x is free in T if it occurs in T and is not bound in T. An expression is closed when it has no free variables; it may have free names.

Our language features two forms of authentication, represented as constructors plus well-formed conditions on their usage in processes. Due to space constraints, this extended abstract only describes message authentication—the technical report also describes high level certificates that provide transferable data authentication.

Authenticated messages between principals are represented as terms of the form $\mathtt{auth}(V_1, V_2, V_3)$, written $V_1{:}V_2\langle V_3\rangle$, where V_1 is the sender, V_2 the receiver, and V_3 the content. We let M and N range over messages. The message M is *from* a (respectively *to* a) if a is the sender (respectively the receiver) of M. Authenticated messages are delivered at most once, to their designated receiver. As an example, $a{:}b\langle\texttt{Hello}\rangle$ is an (authentic) message from a to b with content `Hello`, a constructor with arity 0.

Finally, configurations represent assemblies of communicating principals, with the following grammar:

$C ::=$	configurations
$a[P]$	principal a with local state P
M/i	intercepted message M with index i
$C \mid C'$	distributed parallel composition
$\nu n.C$	name restriction ("new", binds n in C)

A configuration is an assembly of running principals, each with its own local state, plus an abstract record of the messages intercepted by the environment and not forwarded yet to their intended recipients. A system S is a top-level configuration (plus an abstract record of the certificates available to the adversary, omitted here).

We rely on well-formed conditions. In local processes, P is *well-formed for* $a \in \mathsf{Prin}$ when no pattern used for input in P matches any message from a. This condition prevents that messages sent by P be read back by some local input. In configurations, intercepted messages have distinct indices i and closed content M; principals have distinct identities a and well-formed local processes P_a. In systems, let $\mathcal{H}$ be the set of identities for all defined principals, called *compliant principals*; intercepted messages are from a to b for some $a, b \in \mathcal{H}$ with $a \neq b$.

Operational Semantics—Local Reductions. We define our high-level semantics in two stages: local reductions between processes, then global labeled transitions between systems and their (adverse) environment. Processes, configurations, and systems are considered up to renaming of bound names and variables.

Structural equivalence, written $P \equiv P'$, represents structural rearrangements for local processes. As in the pi calculus, it is defined as the smallest congruence such that $P \equiv P \mid 0$, $P \mid Q \equiv Q \mid P$, $P \mid (Q \mid R) \equiv (P \mid Q) \mid R$, $(\nu n.P) \mid Q \equiv \nu n.(P \mid Q)$ when $n \notin fn(Q)$, $\nu m.\nu n.P \equiv \nu n.\nu m.P$, and $\nu n.\mathbf{0} \equiv \mathbf{0}$. Intuitively, structural rearrangements are not observable (although this is quite hard to implement).

Local reduction step, written $P \rightarrow P'$, represents internal computation between local processes. It is defined as the smallest relation such that

(LCOMM) $\quad (T).Q \mid T\sigma \rightarrow Q\sigma$

(LREPL) $\quad *(T).Q \mid T\sigma \rightarrow Q\sigma \mid *(T).Q$

(LMATCH) $\quad \texttt{match}\ T\sigma\ \texttt{with}\ T\ \texttt{in}\ P\ \texttt{else}\ Q \rightarrow P\sigma$

(LNOMATCH) $\quad \texttt{match}\ V\ \texttt{with}\ T\ \texttt{in}\ P\ \texttt{else}\ Q \rightarrow Q$ when $V \neq T\sigma$ for any σ

(LPARCTX) $\dfrac{P \rightarrow Q}{P \mid R \rightarrow Q \mid R}$ (LNEWCTX) $\dfrac{P \rightarrow Q}{\nu n.P \rightarrow \nu n.Q}$ (LSTRUCT) $\dfrac{P \equiv P' \quad P' \rightarrow Q' \quad Q' \equiv Q}{P \rightarrow Q}$

where σ ranges over substitutions of closed terms for the variables bound in T. The local process P is *stable* when it has no local reduction step, written $P \not\rightarrow$. We write $P \twoheadrightarrow Q$ when $P \rightarrow^* \equiv Q$ and $Q \not\rightarrow$.

Operational Semantics—System Transitions. We define a labeled transition semantics for systems. Each labeled transition, written $S \xrightarrow{\gamma} S'$, represents a single interaction with the adversary. We let α and β range over input and output labels (respectively from and to the adversary), let γ range over labels, and let φ range over series of labels. We write $S \xrightarrow{\varphi} S'$ for a series of transitions with labels φ. Labeled transitions are defined by the following rules on configurations:

$$\text{(CFGOUT)}\frac{u \neq a}{a[a{:}u\langle V\rangle \mid Q] \xrightarrow{a:u\langle V\rangle} a[Q]} \qquad \text{(CFGIN)}\frac{u{:}a\langle V\rangle \mid P \twoheadrightarrow Q \quad u \neq a}{a[P] \xrightarrow{(u:a\langle V\rangle)} a[Q]}$$

$$\text{(CFGBLOCK)}\frac{C \xrightarrow{b:a\langle V\rangle} C' \quad i \text{ not in } C}{C \mid a[P] \xrightarrow{\nu i.b:a} C' \mid b{:}a\langle V\rangle/i \mid a[P]} \qquad \text{(CFGFWD)}\frac{C \xrightarrow{(M)} C'}{C \mid M/i \xrightarrow{(i)} C'}$$

$$\text{(CFGPRINCTX)}\frac{C \xrightarrow{\gamma} C' \quad \gamma \text{ not from/to } a}{C \mid a[P] \xrightarrow{\gamma} C' \mid a[P]} \qquad \text{(CFGMSGCTX)}\frac{C \xrightarrow{\gamma} C' \quad i \text{ not in } \gamma}{C \mid M/i \xrightarrow{\gamma} C' \mid M/i}$$

$$\text{(CFGOPEN)}\frac{C \xrightarrow{\beta} C' \quad n \text{ free in } \beta}{\nu n.C \xrightarrow{\nu n.\beta} C'} \qquad \text{(CFGNEWCTX)}\frac{C \xrightarrow{\gamma} C' \quad n \text{ not in } \gamma}{\nu n.C \xrightarrow{\gamma} \nu n.C'}$$

$$\text{(CFGSTR)}\frac{C \equiv D \quad D \xrightarrow{\gamma} D' \quad D' \equiv C'}{C \xrightarrow{\gamma} C'}$$

where structural equivalence on configurations, written $C \equiv C'$, is defined by the same rules as for processes plus Rule $\nu n.a[P] \equiv a[\nu n.P]$.

Rules (CFGOUT) and (CFGIN) represent "intended" interactions with the environment, as usual. They enable local processes to send messages to other principals, and to receive their messages. The transition label conveys the complete message content.

Rules (CFGBLOCK) and (CFGFWD) reflect the actions of an active attacker that intercepts, then selectively forwards, messages exchanged between compliant principals; unlike the (COMM) rule of the pi calculus, they ensure that the environment mediates all communications between principals. The label produced by (CFGBLOCK) signals the message interception; the label conveys partial information on the message content that can be observed from its wire format: the environment learns that an opaque message is sent by b, with intended recipient a. In addition, the intercepted message content is recorded within the configuration, using a fresh index i. Later on, when the environment performs an input with label (i), Rule (CFGFWD) restores the original message content and consumes M/i; this ensures that intercepted messages are delivered at most once.

The local-reduction hypothesis in Rules (CFGIN) makes local computations atomic, as they must complete immediately upon receiving a message and lead to some updated stable process Q. Intuitively, this enforces a transactional semantics for local steps, and prevents any observation of their transient internal state. (Otherwise, the environment may for instance observe the order of appearance of outgoing messages.) On the other hand, any outgoing messages are kept within Q; the environment can obtain all of them via rules (CFGOUT) and (CFGBLOCK) at any time, since those outputs commute with any subsequent transitions.

The rest of the rules for configurations are standard closure rules with regards to contexts and structural rearrangements: Rule (CFGOPEN) is the scope extrusion rule of the pi calculus that opens the scope of a restricted name included in a message sent to the environment. In contrast with intercepted messages, messages sent to a principal not defined in the configuration are transmitted unchanged to the environment, after applying the context rules. In Rule (CFGPRINCTX), condition γ not from a excludes inputs from the environment that forge a message from a, whereas condition γ not to a excludes outputs that may be transformed by Rule (CFGBLOCK).

We define auxiliary notions of transitions, used to describe our implementation. We say that S is *stable* when all local processes are stable and S has no output transition. (Informally, S is waiting for any input from the environment.) We say that a series of transitions $S \xrightarrow{\varphi} S'$ is *normal* when every input is followed by a maximal series of outputs leading to a stable system, that is, $\varphi = \varphi_1\varphi_2 \ldots \varphi_n$, $\varphi_i = \alpha_i\widetilde{\beta_i}$, and $S = S_0 \xrightarrow{\varphi_1} S_1 \xrightarrow{\varphi_2} S_2 \ldots \xrightarrow{\varphi_n} S_n = S'$ for some stable systems $S_0, \ldots, S_n$.

By design, our semantics is compositional, as its rules are inductively defined on the structure of configurations. For instance, we obtain that interactions with a principal that is implicitly controlled by the environment are *at least* as expressive as those with any principal explicited within the system.

4 High-Level Equivalences and Safety

Now that we have labeled transitions that capture our implementation constraints, we can apply standard definitions and proof techniques from concurrency theory to reason about systems. Our computational soundness results are useful (and non-trivial) inasmuch as transitions are simpler and more abstract than low-level adversaries. In addition to trace properties (used, for instance, to express authentication properties as correspondences between transitions) , we consider equivalences between systems.

Intuitively, two systems are equivalent when their environment observes the same transitions. Looking at immediate observations, we say that two systems S_1 and S_2 *have the same labels* when, if $S_1 \xrightarrow{\gamma} S_1'$ for some S_1' (and the name exported by γ are not free in S_2), then $S_2 \xrightarrow{\gamma} S_2'$ for some S_2', and vice versa. More generally, bisimilarity demands that this remains the case after matching transitions:

Definition 3 (Bisimilarity). *The relation $\mathcal{R}$ on systems is a labeled simulation when, for all $S_1 \mathcal{R} S_2$, if $S_1 \xrightarrow{\gamma} S_1'$ (and the names exported by γ are not free in S_2) then $S_2 \xrightarrow{\gamma} S_2'$ and $S_1' \mathcal{R} S_2'$. Labeled bisimilarity, written $\approx$, is the largest symmetric labeled simulation.*

In particular, if $S \approx S'$, then S and S' define the same principals and have the same intercepted-message indices. We also easily verify some congruence properties: our equivalence is preserved by name restrictions, definitions of additional principals, and deletions of intercepted messages.

Lemma 1. *1. If $C_1 \approx C_2$, then $\nu n.C_1 \approx \nu n.C_2$.*
2. If $C_1 \approx C_2$, then $C_1 \,|\, a[P] \approx C_2 \,|\, a[P]$ if these systems are well-formed.
3. If $\nu\tilde{n}_1.(C_1 \,|\, M_1/i) \approx \nu\tilde{n}_2.(C_2 \,|\, M_2/i)$, then $\nu\tilde{n}_1.C_1 \approx \nu\tilde{n}_2.C_2$.

As we quantify over all local processes, we must at least bound their computational power. Indeed, our language is expressive enough to code Turing machines and, for instance, one can easily write a local process that receives a high-level encoding of the security parameter η (e.g. as a series of η messages) then delays a message output by 2^η reduction steps, or even implements an 'oracle' that performs some brute-force attacks using high level implementations of cryptographic algorithms.

Similarly, we must restrict non-deterministic behaviors. Process calculi often feature non-determinism as a convenience when writing specifications, to express uncertainty as regards the environment. Sources of non determinism include local scheduling, hidden in the associative-commutative laws for parallel composition, and internal choices. Accordingly, abstract properties and equivalences typically only consider the existence of transitions—not their probability. Observable non-determinism is problematic in a computational cryptographic setting, as for instance a non-deterministic process may be used as an oracle to guess every bit of a key in linear time.

We arrive at the following definitions. We let $\lceil\cdot\rceil$ compute the (high level) size of systems, labels, and transitions, with for instance $\lceil S \xrightarrow{\gamma} S'\rceil = \lceil S\rceil + \lceil\gamma\rceil + \lceil S'\rceil + 1$, and let $input(\varphi)$ be the input labels of φ.

Definition 4 (Safe Systems). *A system S is* polynomial *when there exists a polynomial p such that, for any φ, if $S \xrightarrow{\varphi} S'$ then $\lceil S \xrightarrow{\varphi} S'\rceil \leq p(\lceil input(\varphi)\rceil)$.*

A system S is safe *when it is polynomial and, for any φ, if $S \xrightarrow{\varphi} S_1$ and $S \xrightarrow{\varphi} S_2$ then S_1 and S_2 have the same labels.*

Hence, starting from a safe process, a series of labels fully determines any further observation. Safety is preserved by all transitions, and also uniformly bounds (for example) the number of local reductions and new names.

These restrictions are serious, but they are also easily established when writing simple programs and protocols. (Still, it would be interesting to relax them, maybe using a probabilistic process calculus.) Accordingly, our language design prevents trivial sources of non-determinism and divergence (e.g. with pattern matching on values, and replicated inputs instead of full-fledged replication); further, most internal choices can be coded as external choices driven by the inputs of our abstract environment.

We can adapt usual bisimulation proof techniques to establish both equivalences and safety: instead of examining all series of labels φ, it suffices to examine single transitions for the systems in the candidate relation.

Lemma 2 (Bisimulation Proof). *Let $\mathcal{R}$ be a reflexive labeled bisimulation such that, for all related systems $S_1 \mathcal{R} S_2$, if $S_1 \xrightarrow{\gamma} S_1'$ and $S_2 \xrightarrow{\gamma} S_2'$, then $S_1' \mathcal{R} S_2'$.*

Polynomial systems related by $\mathcal{R}$ are safe and bisimilar.

We illustrate our definitions using basic examples of secrecy and authentication stated as equivalences between a protocol and its specification (adapted from [2]). Consider a principal a that sends a single message. In isolation, we have the equivalence $a[a{:}b\langle V\rangle] \approx a[a{:}b\langle V'\rangle]$ if and only if $V = V'$, since the environment observes V on the label of the transition $a[a{:}b\langle V\rangle] \xrightarrow{a:b\langle V\rangle} a[\mathbf{0}]$.

Consider now the system $S(V,W) = a[a{:}b\langle V,W\rangle] \,|\, b[(a{:}\langle ?x, \_\rangle).P]$, with an explicit process for principal b that receives a's message and, assuming the message is a pair, runs P with the first element of the pair substituted for x. For any terms W_1 and W_2, we have $S(V,W_1) \approx S(V,W_2)$. This equivalence states the strong secrecy of W, since its value cannot affect the environment. The system has two transitions $S(V,W) \xrightarrow{\nu i.a:b} \xrightarrow{(i)} a[\mathbf{0}] \,|\, b[P\{V/x\}]$.

Further, the equivalence $S(V,W) \approx a[a{:}b\langle\rangle] \,|\, b[(a{:}\langle\_\rangle).P\{V/x\}]$ captures both the authentication of V and the absence of observable information on V and W in the communicated message, since the protocol $S(V, W)$ behaves just like another protocol that sends a dummy message instead of V, W.

5 A Concrete Implementation (Outline)

We systematically map high-level systems S to the machines of Section 2, mapping each principal $a[P_a]$ of S to a PPT machine M_a that executes P_a. Due to space constraints, we only give an outline of our implementation, defined in the technical report. The implementation mechanisms are simple, but they need to be carefully specified and composed. (As a non-trivial example, when a machine outputs several messages, possibly to the same principals, we must sort the messages after encryption so that their ordering on the wire leaks no information on the computation that produced them.)

We use two concrete representations for terms: a wire format for (signed, encrypted) messages between principals, and an internal representation for local terms. Various bitstrings represent constructors, principal identities, names, and certificates. Marshaling and unmarshaling functions convert between internal and wire representations. When marshaling a locally restricted name n for the first time, we draw a bitstring s of length η uniformly at random, associate it with n, and use it to represent n on the wire. When unmarshaling a bitstring s into a name, if s is not associated with any local name, we create a new internal identifier n for the name, and also associate s with n.

Local processes are represented in normal form for structural equivalence, using internal terms and multisets of local inputs, local outputs, and outgoing messages. We implement reductions using an abstract machine that matches inputs and outputs using an arbitrary deterministic, polynomial-time scheduler.

To keep track of the runtime state for our machines, we supplement high-level systems S with *shadow states* D that record sufficient information so that each machine is a function $\mathsf{M}_a(S, \mathsf{D})$. For instance, D records maps from names and intercepted messages to bitstrings, and from principals to their keys and the content of their anti-replay caches. The shadow D also determines the information available to the attacker, coded as a bitstring $public(\mathsf{D})$. The structure of $public(\mathsf{D})$ sets the interface between attackers and low-level systems, called the *shape* of D. For instance, the shape fixes the free names that may occur in S, and $public(\mathsf{D})$ provides their associated bitstrings.

In general, a system S may contain restricted names shared between local processes and intercepted messages, making it non-trivial to describe a concrete initialization mechanism that produces $\mathsf{M}(S, \mathsf{D})$ and $public(\mathsf{D})$. Instead of explicitly coding low-level initialization, we define it as the run of a high-level initialization protocol $S^\circ \xrightarrow{\varphi^\circ} S$ that lets the principals exchange names and yield intercepted messages to the environment. In the initialization protocol, S° is a system with no intercepted messages and no free names in local processes. For any system S, there are such transitions $S^\circ \xrightarrow{\varphi^\circ} S$ and, applying a variant of Theorem 1, there is a PPT algorithm $\mathsf{A}_{\varphi^\circ}$ that simulates φ° and produces $public(\mathsf{D})$ from some $public(\mathsf{D}^\circ)$, where D° is the shadow produced by Definition 1(1–3). Thus, we define a run of $\mathsf{M}(S, \mathsf{D})$ with adversary A, written $\mathsf{A}[\mathsf{M}(S, \mathsf{D})]$, as a run of $(\mathsf{A}_{\varphi^\circ}; \mathsf{A})[\mathsf{M}(S^\circ, \mathsf{D}^\circ)]$ where $\mathsf{A}_{\varphi^\circ}; \mathsf{A}$ first runs $\mathsf{A}_{\varphi^\circ}$ then starts A with input $public(\mathsf{D})$. We then say that D is a valid shadow for S.

6 Soundness and Completeness Results

In this section we show that properties that hold with the high-level semantics can be carried over to the low-level implementation, and the other way around. Due to space constraints, most auxiliary results and all proofs appear in the technical report [6].

Our first theorem expresses the soundness of the high-level operational semantics: every series of transitions can be executed (and checked) by a low-level attacker. Said otherwise, the high-level semantics does not give too much power to the environment.

Theorem 1. *For any shape of* D *and labels* φ*, there is a PPT algorithm* A_φ *such that, for any safe stable system* S *with valid shadow* D *where the new names of* φ *are not free in* D*, one of the following holds with overwhelming probability:*

- $1 \longleftarrow \mathsf{A}_\varphi[\mathsf{M}(S, \mathsf{D})]$ *and there exists* S' *with normal transitions* $S \xrightarrow{\varphi} S'$*; or*
- $0 \longleftarrow \mathsf{A}_\varphi[\mathsf{M}(S, \mathsf{D})]$ *and there are no normal transitions* $S \xrightarrow{\varphi} S'$*.*

Since we can characterize any trace using an adversary, we also obtain completeness for trace equivalence: low-level equivalence implies high-level trace equivalence.

Theorem 2. *Let* S_1 *and* S_2 *be safe stable systems with valid shadow* D *such that* $\mathsf{M}(S_1, \mathsf{D}) \approx \mathsf{M}(S_2, \mathsf{D})$*. If there are normal transitions* $S_1 \xrightarrow{\varphi} S_1'$ *and the new names of* φ *are not free in* D*, then there are normal transitions* $S_2 \xrightarrow{\varphi} S_2'$*.*

Our next theorem expresses the completeness of our high-level transitions: every low-level attack can be described in terms of high-level transitions. More precisely, the probability that an interaction with a PPT adversary yields a machine state unexplained by any high-level transitions is negligible.

Theorem 3. *Let* S *be a safe stable system with valid shadow* D *and* A *a PPT algorithm.*

The probability that $\mathsf{A}[\mathsf{M}(S, \mathsf{D})]$ *completes and leaves the system in state* M' *with* $\mathsf{M}' \neq \mathsf{M}(S', \mathsf{D}')$ *for any normal transitions* $S \xrightarrow{\varphi} S'$ *with valid shadow* D' *is negligible.*

Finally, our main result states the soundness of equivalence: to show that two stable systems are indistinguishable, it suffices to show that they are safe and bisimilar.

Theorem 4. *Let* S_1 *and* S_2 *be safe stable systems with valid shadow* D*. If* $S_1 \approx S_2$*, then* $\mathsf{M}(S_1, \mathsf{D}) \approx \mathsf{M}(S_2, \mathsf{D})$*.*

7 Conclusions and Future Work

We designed a simple, abstract language for secure distributed communications. Our language provides uniform protection for all messages; it is expressive enough to program a large class of protocols; it also enables simple reasoning about security properties in the presence of active attackers, using labeled traces and equivalences. We implemented this calculus as a collection of concrete PPT machines embedding standard cryptographic algorithms, and established that low-level PPT adversaries that control their scheduling and the network have essentially the same power as (much simpler) high-level environments. To the best of our knowledge, these are the first cryptographic soundness and completeness results for a distributed process calculus.

We also identified and discussed difficulties that stem from the discrepancy between the two models, and developed proofs that combine techniques from process calculi and cryptography. It would be interesting (and hard) to extend the expressiveness of our calculus, for instance with secrecy and probabilistic choices.

Acknowledgments. This paper benefited from discussions with Martín Abadi, Tuomas Aura, Karthik Bhargavan, Andy Gordon, and David Pointcheval.

References

1. M. Abadi and C. Fournet. Private authentication. *Theoretical Computer Science*, 322(3):427–476, 2004. Special issue on Foundations of Wide Area Network Computing.
2. M. Abadi, C. Fournet, and G. Gonthier. Authentication primitives and their compilation. In *POPL 2000*, pages 302–315. ACM, 2000.
3. M. Abadi, C. Fournet, and G. Gonthier. Secure implementation of channel abstractions. *Information and Computation*, 174(1):37–83, 2002.
4. M. Abadi and A. D. Gordon. A calculus for cryptographic protocols: The Spi Calculus. *Information and Computation*, 148(1):1–70, 1999.
5. M. Abadi and P. Rogaway. Reconciling two views of cryptography (the computational soundness of formal encryption). *Journal of Cryptology*, 15(2):103–127, 2002.
6. P. Adão and C. Fournet. Cryptographically sound implementations for communicating processes. Technical report MSR-TR-2006-49. Microsoft Research, 2006.
7. M. Backes and B. Pfitzmann. Symmetric encryption in a simulatable Dolev-Yao style cryptographic library. In *CSFW-17*, pages 204–218. IEEE, 2004.
8. M. Backes, B. Pfitzmann, and M. Waidner. A composable cryptographic library with nested operations. In *CCS 2003*, pages 220–230. ACM, 2003.
9. M. Backes, B. Pfitzmann, and M. Waidner. Symmetric authentication within a simulatable cryptographic library. *International Journal of Information Security*, 4(3):135–154, 2005.
10. B. Blanchet, M. Abadi, and C. Fournet. Automated verification of selected equivalences for security protocols. In *LICS 2005*, pages 331–340. IEEE, 2005.
11. R. Canetti and H. Krawczyk. Analysis of key-exchange protocols and their use for building secure channels. In *EUROCRYPT 2001*, LNCS 2045, pages 453–474. Springer, 2001.
12. S. Goldwasser and S. Micali. Probabilistic encryption. *Journal of Computer and Systems Sciences*, 28(2):270–299, 1984.
13. S. Goldwasser, S. Micali, and R. Rivest. A digital signature scheme secure against adaptive chosen-message attack. *SIAM Journal on Computing*, 17(2):281–308, 1988.
14. P. Laud. Secrecy types for a simulatable cryptographic library. In *CCS 2005*, pages 26–35. ACM, 2005.
15. P. Lincoln, J. Mitchell, M. Mitchell, and A. Scedrov. A probabilistic poly-time framework for protocol analysis. In *CCS 1998*, pages 112–121. ACM, 1998.
16. D. Micciancio and B. Warinschi. Soundness of formal encryption in the presence of active adversaries. In *TCC 2004*, LNCS 2951, pages 133–151. Springer, 2004.
17. J. C. Mitchell, A. Ramanathan, A. Scedrov, and V. Teague. A probabilistic polynomial-time calculus for the analysis of cryptographic protocols. *Theoretical Computer Science*, 2006.
18. C. Rackoff and D. R. Simon. Non-interactive zero-knowledge proof of knowledge and chosen ciphertext attack. In *CRYPTO'91*, LNCS 576, pages 433–444. Springer, 1991.

A Dolev-Yao-Based Definition of Abuse-Free Protocols

Detlef Kähler, Ralf Küsters, and Thomas Wilke

Christian-Albrechts-Universität zu Kiel
{kaehler, kuesters, wilke}@ti.informatik.uni-kiel.de

Abstract. We propose a Dolev-Yao-based definition of abuse freeness for optimistic contract-signing protocols which, unlike other definitions, incorporates a rigorous notion of what it means for an outside party to be convinced by a dishonest party that it has the ability to determine the outcome of the protocol with an honest party, i.e., to determine whether it will obtain a valid contract itself or whether it will prevent the honest party from obtaining a valid contract. Our definition involves a new notion of test (inspired by static equivalence) which the outside party can perform. We show that an optimistic contract-signing protocol proposed by Asokan, Shoup, and Waidner is abusive and that a protocol by Garay, Jakobsson, and MacKenzie is abuse-free according to our definition. Our analysis is based on a synchronous concurrent model in which parties can receive several messages at the same time. This results in new vulnerabilities of the protocols depending on how a trusted third party reacts in case it receives abort and resolve requests at the same time.

1 Introduction

Abuse freeness is a security property introduced in [9] for optimistic contract-signing protocols: An optimistic (two-party) contract-signing protocol is a protocol run by A (Alice), B (Bob), and a trusted third party T (TTP) to exchange signatures on a previously agreed upon contractual text with the additional property that the TTP will only be involved in a run in case of problems. Such a protocol is *not* abuse-free for (honest) Alice if at some point during a protocol run (dishonest) Bob can "convince" an outside party, Charlie, that he is in an unbalanced state, where, following the terminology of [4], *unbalanced* means that Bob has both (i) a strategy to prevent Alice from getting a valid contract and (ii) a strategy to obtain a valid contract. In other words, Alice can be misused by Bob to get leverage for another contract (with Charlie). Obviously, abuse-free contract-signing protocols are highly desirable.

The main goal of the present work is to present a formal definition of abuse freeness which is as protocol-independent as possible. The crucial issue with such a formal definition is that it needs to specify what it means for Bob to *convince* Charlie. One of the first proposals for this was presented by Kremer and Raskin [12]. Roughly, their proposal is the following: To convince Charlie a message is presented to Charlie from which he can deduce that "a protocol run has been

M. Bugliesi et al. (Eds.): ICALP 2006, Part II, LNCS 4052, pp. 95–106, 2006.

started between Alice and Bob". What that means is, however, not specified in a general fashion in [12]. Instead, this is decided on a case by case basis. The objective of this paper is to give a generic definition. The only part which needs to be decided on a case by case basis in our definition is what it means for Alice (or Bob) to have received a valid contract—something which can hardly be described in a generic way—and what the assumptions are that Charlie makes.

Before we explain our approach and the contribution of our work we need to explain the following crucial point: Whether or not Charlie is convinced should be based on evidence provided by Bob. Following [9], we model this evidence as a message that Bob presents to Charlie. (In [9], this is called an off-line attack.) This, however, has an important implication. Since Bob can hold back any message he wants to (he can himself decide which messages he shows to Charlie) and since Charlie is assumed to be an outside party not involved in the protocol, if Bob could convince Charlie to be in some state of the protocol at some point, at any later point he would be able to convince Charlie that he was in the same state, simply by providing the same evidence. Therefore, Bob can only convince Charlie that he is or was and still might be in an unbalanced state. We employ this notion of abuse freeness for our work. (Note that it is stronger than the one described above as Charlie is more easily convinced.)

Contribution of this Work. We provide a formal definition of the version of abuse freeness just explained, apply our definition to the optimistic contract-signing protocols by Asokan, Shoup, and Waidner [3] (ASW protocol) and by Garay, Jakobsson, and MacKenzie [9] (GJM protocol), and show that the ASW protocol is abusive while the GJM protocol is abuse-free according to our definition.

The idea behind our definition of abuse freeness is that Bob presents a message to Charlie and Charlie performs a certain test on this message. If the message passes the test, then Charlie is convinced that Bob is or was and still might be in an unbalanced state. The test is such that from the point of view of Charlie, Bob can only generate messages passing the test in states where Bob is or was in an unbalanced state and where at least one of these states is in fact unbalanced. To describe the power Bob has, we adopt a Dolev-Yao style approach [8] (see also [2, 1, 7]). Our definition of test is inspired by the notion of static equivalence [2].

We use a synchronous concurrent communication model in which principals and the (Dolev-Yao-style) intruder may send several messages to different parties at the same time. This rather realistic model requires to specify the behavior of protocol participants in case several messages are received at the same time (or within one time slot). This leads to new effects that have not been observed in previous works. In the ASW and GJM protocols, one needs to specify the behavior of the TTP in case an abort and a resolve request are received at the same time (from different parties). The question arises whether the TTP should answer with an abort or a resolve acknowledgment. We show that if the TTP does the former, then the ASW and the GJM protocol are unbalanced for the responder, and if it does the latter, the two protocols are unbalanced for the initiator.

Related Work. As mentioned above, Kremer et al. [12] analyzed the ASW and GJM protocol based on finite-state alternating transition systems, using an automatic analysis tool. They explicitly needed to specify the behavior of dishonest principals and which states are the ones that are convincing to Charlie (they use a propositional variable prove2C, which they set manually). This is what our definition makes obsolete.

Chadha et al. [4] introduce a stronger notion than abuse freeness, namely *balance*: For a protocol to be *unbalanced* one does not require Bob to convince Charlie that he is in an unbalanced state. The fact that an unbalanced state exists is sufficient for a protocol to be unbalanced. Hence, balance is a formally stronger notion than abuse freeness. Unfortunately, this notion is too strong in some cases. In fact, as shown by Chadha et al. [6] in an interleaving (rather than real concurrent) model, if principals are optimistic, i.e., they are willing to wait for messages of other parties, balance is impossible to achieve; in this paper, Chadha et al. also sketch a definition of abuse freeness based on epistemic logic, but without going into details. In [5], Chadha et al. study multi-party contract signing protocols.

Shmatikov and Mitchell [13] employ the finite-state model checker Murφ to automatically analyze contract-signing protocols. They, too, approximate the notion of abuse freeness by a notion similar to balance.

Structure of the Paper. The technical part of the paper starts with an informal description of the ASW protocol in Sect. 2, which then serves as a running example for the further definitions. In Sect. 3, we describe our communication and protocol model, with the new definition of abuse freeness presented in Sect. 4. We then treat the ASW and the GJM protocol in our framework in Sect. 5 and Sect. 6. We conclude in Sect. 7. A full version of our paper is available, see [11].

2 The ASW Protocol

In this section, we recall the Asokan-Shoup-Waidner (ASW) protocol from [3], which will serve as a running example; the Garay-Jakobsson-MacKenzie (GJM) protocol from [9] will be explained in Section 6.

The ASW protocol assumes the following scenario: Alice and Bob want to sign a contract and a TTP is present. Further, it is agreed upon that the following two types of messages, the *standard contract (SC)* and the *replacement contract (RC)*, will be recognized as valid contracts between Alice and Bob with contractual text text: $SC = \langle me_1, N_A, me_2, N_B \rangle$ and $RC = \mathsf{sig}_T(\langle me_1, me_2 \rangle)$ where $me_1 = \mathsf{sig}_A(\langle A, B, \mathsf{text}, \mathsf{hash}(N_A) \rangle)$ and $me_2 = \mathsf{sig}_B(\langle me_1, \mathsf{hash}(N_B) \rangle)$, and as usual, N_A and N_B stand for nonces. In addition to SC and RC, the variants of SC and RC which one obtains by exchanging the roles of A and B are regarded as valid contracts.

There are three interdependent parts to the protocol: an exchange protocol, an abort protocol, and a resolve protocol. The *exchange protocol* consists of four steps, which, in Alice-Bob notation, are displayed in Fig. 1. The first two messages, me_1 and me_2, serve as respective *promises* of Alice and Bob to sign the contract, and N_A and N_B serve as *contract authenticators*: After they have been revealed, Alice and Bob can compose the standard contract, SC.

$$A \to B : me_1$$
$$B \to A : me_2$$
$$A \to B : N_A$$
$$B \to A : N_B$$

Fig. 1. ASW exchange protocol

The *abort protocol* is run between Alice and the TTP and is used by Alice to abort the contract signing process when she does not receive Bob's promise. Alice will obtain (from the TTP) an abort receipt or, if the protocol instance has already been resolved (see below), a replacement contract. The first step is $A \to T$: ma_1 where $ma_1 = \mathsf{sig}_A(\langle \mathsf{aborted}, me_1 \rangle)$ is Alice's *abort request*; the second step is the TTP's reply, which is either $\mathsf{sig}_T(\langle \mathsf{aborted}, ma_1 \rangle)$, the *abort receipt*, if the protocol has not been resolved, or the replacement contract, RC.

Similarly, the *resolve protocol* can be used by Alice and Bob to resolve the protocol, which either results in a replacement contract or, if the protocol has already been aborted, in an abort receipt. When Bob runs the protocol (because Alice has not sent her contract authenticator yet), the first step is $B \to T$: $\langle me_1, me_2 \rangle$; the second step is the TTP's reply, which is either the abort receipt $\mathsf{sig}_T(\langle \mathsf{aborted}, ma_1 \rangle)$, if the protocol has already been aborted, or the replacement contract, RC. The same protocol (with roles of A and B exchanged) is also used by Alice.

It is assumed that the communication between Alice and the TTP and between Bob and the TTP goes through a channel that is not under the control of the intruder (the dishonest party), i.e., the intruder cannot delay, modify, or insert messages. We refer to such a channel as secure. Whether or not the intruder can read messages sent on this channel does not effect the results shown in this paper.

3 The Concurrent Protocol and Intruder Model

In this section, we introduce our protocol and intruder model, which, unlike most other Dolev-Yao-based models, captures real concurrent computation. Given sets S, T, U with $U \subseteq T$, we denote by S^T the set of functions from T to S and for $f \in S^T$ we denote by $f|_U$ the restriction of f to U.

3.1 Concurrent System Model

A concurrent system in our framework is made up of several components, which are automata provided with input and output ports for inter-component communication. Each such port can either carry a message from a given set $\mathcal{M}$ of messages or the special symbol '$\circ$' (no message). We use $\mathcal{M}_\circ$ to denote $\mathcal{M} \cup \{\circ\}$. A run of such a system proceeds in rounds: In every round, every component reads the input on all of its input ports, and then, depending on its current state, writes output on its output ports (possibly $\circ$), and goes into a new state.

A message written on an output port is read in the next round by the component with the corresponding input port. Note that all components perform their "receive-send action" at the same time and that a component may receive and send several messages at the same time.

Formally, a *component* of a concurrent system over a set $\mathcal{M}$ of messages is a tuple $\mathcal{A} = (S, \mathbf{In}, \mathbf{Out}, I, \Delta)$ where S is a (possibly infinite) set of *local states*, $\mathbf{In}$ is the set of *input ports*, $\mathbf{Out}$ is the set of *output ports*, disjoint from $\mathbf{In}$, $I \subseteq S$ is the set of *initial states*, and $\Delta \subseteq \mathcal{M}_\circ^{\mathbf{In}} \times S \times S \times \mathcal{M}_\circ^{\mathbf{Out}}$ is the *transition relation*, which, w.l.o.g., is required to be complete: for each $(m, s) \in \mathcal{M}_\circ^{\mathbf{In}} \times S$ there exist s' and m' with $(m, s, s', m') \in \Delta$. A transition (m, s, s', m') is meant to model that if $\mathcal{A}$ is in state s and reads the messages m on its input ports, then it writes m' on its output ports and goes into state s'.

A *concurrent system* over a set $\mathcal{M}$ of messages is a finite family $\{\mathcal{A}_i\}_{i\in P}$ of components over $\mathcal{M}$ of the form $(S_i, \mathbf{In}_i, \mathbf{Out}_i, I_i, \Delta_i)$ such that $\mathbf{In}_i \cap \mathbf{In}_j = \mathbf{Out}_i \cap \mathbf{Out}_j = \emptyset$ for every i and $j \neq i$. Note that an output port of one component may coincide with the input port of another component, which allows the former component to send messages to the latter component.

Given a concurrent system $\mathcal{G} = \{\mathcal{A}_i\}_{i\in P}$ as above, its set of input and output ports is determined by $\mathbf{In} = \bigcup_{i\in P} \mathbf{In}_i$ and $\mathbf{Out} = \bigcup_{i\in P} \mathbf{Out}_i$, respectively, while its state set and its initial state set are defined by $S = \prod_{i\in P} S_i$ and $I = \{s \in S \mid s(i) \in I_i \text{ for } i \in P\}$ where $s(i)$ denotes the entry with index i in s. We set $\mathbf{P} = \mathbf{In} \cup \mathbf{Out}$.

A *concurrent transition* is a tuple of the form (m, s, s', m') satisfying $(m|_{\mathbf{In}_i}, s(i), s'(i), m'|_{\mathbf{Out}_i}) \in \Delta_i$ for every $i \in P$. Note that if $p \in \mathbf{Out}_l \cap \mathbf{In}_r$ for $l \neq r$, then this means that by applying the transition, component $\mathcal{A}_l$ sends message $m'(p)$ to component $\mathcal{A}_r$. A *global state* of $\mathcal{G}$ is a pair (m, s) with $m \in \mathcal{M}_\circ^{\mathbf{P}}$ and $s \in S$, i.e., it contains all current messages on ports and all local states.

An (m, s)*-computation* of $\mathcal{G}$ is an infinite sequence $\rho = m_0 s_0 m_1 s_1 \ldots$ of global states such that $(m_i, s_i, s_{i+1}, m_{i+1})$ is a concurrent transition for every i and $(m_0, s_0) = (m, s)$. *Finite* (m, s)*-computations* are defined in the same way. An infinite (m, s)-computation is called a *run* of $\mathcal{G}$ if $m(p) = \circ$ for every $p \in \mathbf{P}$ and $s \in I$. A finite prefix of a run is called a *run segment.*

A global state (m, s) is called *reachable* if there is a run segment $\rho = m_0 s_0 m_1 s_1 \ldots m_{k-1} s_{k-1}$ such that $(m_{k-1}, s_{k-1}) = (m, s)$. Let (m, s) and (m', s') be global states. We call (m', s') a *descendant of* (m, s) if there is an (m, s)-computation $\rho = m_0 s_0 m_1 s_1 \ldots$ such that $(m', s') = (m_i, s_i)$ for some $i \geq 0$, in particular, (m, s) is a descendant of (m, s).

3.2 Dolev-Yao Systems

To model protocols and the execution of protocols in presence of an intruder, we consider specific concurrent systems, called Dolev-Yao systems. We first introduce messages and terms, along the lines of [1, 2, 7].

Given a signature Σ and a set of variables $\mathcal{V}$, the set of *terms* $\mathcal{T}(\Sigma, \mathcal{V})$ and the set of *ground terms* $\mathcal{T}(\Sigma)$ are defined as usual. Given a set $\mathcal{S} \subseteq \mathcal{T}(\Sigma, \mathcal{V})$, called a set of *basic operations*, we call a term t an $\mathcal{S}$*-term* if $t \in \mathcal{S}$ or if it is obtained from a term in $\mathcal{S}$ by substituting $\mathcal{S}$-terms for variables and renaming variables. We also consider an equational theory $\mathcal{H}$ over Σ, which we assume is convergent, implying that every term t has a unique normal form, which we denote by $t\downarrow$.

For example, to model the ASW protocol we consider the signature Σ_{ASW} consisting of the following symbols: $\mathsf{sig}(\cdot,\cdot)$, $\mathsf{sigcheck}(\cdot,\cdot,\cdot)$, $\mathsf{pk}(\cdot)$, $\mathsf{sk}(\cdot)$, $\langle\cdot,\cdot\rangle$, $\pi_1(\cdot)$, $\pi_2(\cdot)$, $\mathsf{hash}(\cdot)$, A, B, T, text, ok, $\mathsf{initiator}$, $\mathsf{responder}$, $\mathsf{aborted}$, and an infinite number of constants. Further, we choose operations that model pairing, projections, checking a signature, signing, and hashing, that is, $\mathcal{S}_{ASW}$ consists of the following basic operations: $\langle x_1, x_2\rangle$, $\pi_1(x_1)$, $\pi_2(x_1)$, $\mathsf{sigcheck}(x_1, x_2, x_3)$, $\mathsf{sig}(x_1, x_2)$, and $\mathsf{hash}(x_1)$. The semantics of these operations is determined by the equational theory $\mathcal{H}_{\mathsf{ASW}}$, which consists of the following three identities: $\pi_1(\langle x, y\rangle) = x$, $\pi_2(\langle x, y\rangle) = y$, and $\mathsf{sigcheck}(x, \mathsf{sig}(\mathsf{sk}(y), x), \mathsf{pk}(y)) = \mathsf{ok}$.

Using the set $\mathcal{S}$ of basic operations, an intruder can derive messages from a given set $\mathcal{K}$ of messages by forming $(\mathcal{S} \cup \mathcal{K})$-terms. We define $d_{\mathcal{S}}(\mathcal{K}) = \{m\downarrow \mid m \text{ is an } (\mathcal{S} \cup \mathcal{K})\text{-term without variables}\}$ to be the set of messages (in normal form) that can be derived from $\mathcal{K}$ using $\mathcal{S}$. In the ASW example, with $\mathcal{K} = \{\langle \mathsf{contract}, \mathsf{sig}(\mathsf{sk}(A), \mathsf{contract})\rangle, \mathsf{sk}(B)\}$, the following term is an $(\mathcal{S}_{\mathsf{ASW}} \cup \mathcal{K})$-term: $m = \mathsf{sig}(\mathsf{sk}(B), \pi_1(\langle \mathsf{contract}, \mathsf{sig}(\mathsf{sk}(A), \mathsf{contract})\rangle))$. The normal form $m\downarrow = \mathsf{sig}(\mathsf{sk}(B), \mathsf{contract})$ of m belongs to $d_{\mathcal{S}_{\mathsf{ASW}}}(\mathcal{K})$.

To specify a Dolev-Yao system, we partition a given (finite) set *ALL* of all principals into a set *HON* of *honest principals* and a set $\mathit{DIS} = \mathit{ALL} \setminus \mathit{HON}$ of *dishonest principals*. In the Dolev-Yao system, we have a component $\mathcal{A}_\pi$ for every honest principal (*honest components*) and one component $\mathcal{A}_{\mathcal{I}}$, the *intruder component*, subsuming all dishonest principals. Each honest component $\mathcal{A}_\pi$ has ports (i) $\mathrm{netin}^{\pi}_{\pi'}$ and $\mathrm{sec}^{\pi}_{\pi'}$ for sending messages to π' for every π' through the network and the secure channel, respectively, and (ii) ports $\mathrm{netout}^{\pi'}_{\pi}$ and $\mathrm{sec}^{\pi'}_{\pi}$ for receiving messages coming from the network (supposedly from π') and from the secure channel (definitely from π') for every π'. The input and output port sets of the intruder component $\mathcal{A}_{\mathcal{I}}$ are $\mathbf{In}_{\mathcal{I}} = \{\mathrm{netin}^{\pi}_{\pi'} \mid \pi \in \mathit{HON}, \pi' \in \mathit{ALL}\} \cup \{\mathrm{sec}^{\pi}_{\pi'} \mid \pi \in \mathit{HON}, \pi' \in \mathit{DIS}\}$ and $\mathbf{Out}_{\mathcal{I}} = \{\mathrm{netout}^{\pi}_{\pi'} \mid \pi' \in \mathit{HON}, \pi \in \mathit{ALL}\} \cup \{\mathrm{sec}^{\pi}_{\pi'} \mid \pi \in \mathit{DIS}, \pi' \in \mathit{HON}\}$, respectively. Note that one end of a network port is always connected to the intruder (since he controls the network), while secure channel ports directly connect two honest principals or an honest principal and a dishonest principal (i.e., the intruder). Instead of connecting two honest principals directly through a secure channel, one could plug between two honest principals a secure channel component for more flexible scheduling. However, for simplicity and since this does not change our results (if secure channel components between honest principals are not controlled by the adversary), we choose direct secure channel links.

The intruder component acts as a Dolev-Yao intruder in that it may derive arbitrary messages from its initial knowledge and the messages received so far using $\mathcal{S}$-terms as described above. Note, however, that the intruder component (as all other components) may receive and send several messages at the same time.

Given the set *HON* of honest and the set *DIS* of dishonest principals, a family $\{\mathcal{A}_\pi\}_{\pi \in HON}$ of honest components (with ports as specified above), and a set $\mathcal{K}$ of messages (the initial intruder knowledge), we denote by $\mathrm{DY}[\{\mathcal{A}_\pi\}_{\pi \in HON}, HON, DIS, \mathcal{K}]$ the induced Dolev-Yao system where the set $\mathcal{S}$ of operations the intruder may use to derive new messages is understood from the context. If (m, s) is a global state of a run of such a system, we denote by $\mathcal{K}(m, s)$ the initial knowledge of the intruder plus the messages he has seen so far on his input ports (including the messages currently on his input ports). We say that the intruder can *deduce message* m' *at state* (m, s) if $m' \in d_{\mathcal{S}}(\mathcal{K}(m, s))$.

4 Balanced and Abuse-Free Protocols

In this section, we present our formal definition of abuse freeness, based on the notion of balance, which, in turn, is based on the notion of strategy.

4.1 Balanced Protocols

Throughout this subsection, we assume a concurrent system $\mathcal{G} = \{\mathcal{A}_i\}_{i \in P}$ with set of ports $\mathbf{P}$ and state set S to be given.

Strategies in the context of abuse freeness need to be defined with respect to partial information, since Bob will not necessarily know the global state of the entire protocol at any point of the protocol execution. In addition, strategies can be carried out jointly by several components. This motivates the following definitions.

A function with domain $(\mathcal{M}_\circ^{\mathbf{P}} \times S)^+$ is called a *view function* for $\mathcal{G}$. Given a view function view and a run segment ρ, we say that $\mathsf{view}(\rho)$ is the *view of* ρ *w. r. t.* view. Any subset of P is called a *coalition*. Given a coalition J, we write $\mathbf{Out}_J$ for $\bigcup_{j \in J} \mathbf{Out}_j$.

Given both, a coalition J and a view function $\mathsf{view}\colon (\mathcal{M}_\circ^{\mathbf{P}} \times S)^+ \to W$, a *view-strategy for* J is a function σ which determines how the components of J act depending on their current view $w \in W$, which itself is determined by view. More precisely, σ assigns to each $w \in W$ successor states $s_j \in S_j$ (for $j \in J$) and messages m_p (for $p \in \mathbf{Out}_J$) to be written to the output ports of the components of the coalition. Clearly, these choices are required to be consistent with the individual transition relations Δ_j. Given a strategy σ and a global state (m, s), we denote by $\mathsf{out}((m, s), \sigma)$ the set of all infinite (m, s)-computations in which the components of the coalition J follow the strategy σ.

In our formal definition of balance, path properties are used to define what exactly it means to prevent Alice from getting a valid contract or to obtain one. Formally, a set $\varphi \subseteq (\mathcal{M}_\circ^{\mathbf{P}} \times S)^\omega$ is called a *$\mathcal{G}$-property*.

The notion of balance will be defined w.r.t. what we call a *balance specifier*, i.e., a tuple β of the form $(I, \mathsf{view}, \varphi_1, \varphi_2)$ where I is a coalition, view is a view function, and φ_1 and φ_2 are path properties. For instance, assume we want to describe balance for Alice in a concrete contract signing setting. Then we need to check whether there exist certain strategies for Bob, so we may choose $I = \{B\}$.

More precisely, we want to know whether Bob has a strategy for preventing that Alice gets a valid contract and a strategy for making sure Bob gets a valid contract. So we define φ_1 as the set of all runs of the protocol where Alice does not get a valid contract and φ_2 as the set of all runs where Bob gets a valid contract. Finally, we choose view in such a way that at any given point in a run, view returns everything Bob has observed of the system thus far. Similarly to [4], balance is now defined as follows:

Definition 1 (balance). *Let $\mathcal{G}$ be a concurrent system with index set P, (m, s) a reachable state of $\mathcal{G}$, and $\beta = (I, \mathsf{view}, \varphi_1, \varphi_2)$ a balance specifier.*

The state (m, s) is β-unbalanced *if there are* view*-strategies σ_1 and σ_2 for I such that $\rho \in \varphi_i$ for every $\rho \in \mathsf{out}((m, s), \sigma_i)$ and $i \in \{1, 2\}$. The system $\mathcal{G}$ is* β-unbalanced *if there is a reachable state (m, s) of $\mathcal{G}$ that is β-unbalanced.*

4.2 Abuse-Free Protocols

As already explained earlier, when a protocol is considered abuse-free, then this means that from Charlie's point of view Bob has no way of convincing him that he is in an unbalanced state. That is, the property of being abuse-free is relative to the view that Charlie has of the protocol. Technically, such a view is determined by a Dolev-Yao system and a balance specifier. This motivates the following definition. A pair $(\mathcal{G}^e, \beta^e)$ consisting of a Dolev-Yao system $\mathcal{G}^e$ and a balance specifier β^e is called an *external view (with respect to abuse freeness)*.

We use a specific but natural notion of test that Charlie can make use of to verify that Bob is in fact in the position he claims to be in. As a parameter it uses a set $\mathcal{X} \subseteq \mathcal{M}$ of messages, which should be thought of as Charlie's a-priori knowledge, such as his private key.

A pair (M, M') of $(\mathcal{S} \cup \mathcal{X})$-terms (containing exactly one variable x) is called an *atomic $\mathcal{X}$-test*. A message $m \in \mathcal{M}$ *passes the test* (M, M'), denoted $m \models (M, M')$, if $M[m/x] \equiv_{\mathcal{H}} M'[m/x]$. The message m *fails the test* (M, M') if m does not pass it. This is extended to *boolean* and *ω-tests* in a straightforward fashion, where in ω-tests conjunctions and disjunctions with a denumerable number of arguments are allowed (our results hold for both boolean and ω-tests). For instance, if Charlie wants to check whether a message has the form $\langle c, \mathsf{sig}(\mathsf{sk}(A), c)\rangle$, then he can use the boolean test $(\pi_1(x), c) \wedge (\mathsf{sigcheck}(c, \pi_2(x), \mathsf{pk}(A)), \mathsf{ok})$.

As explained above, in our definition Charlie uses a test to distinguish between messages that give evidence for an unbalanced state and messages which don't. In other words, Charlie considers a state unbalanced when Bob could possibly deduce a message in that state which passes the test. Therefore, we say that for a given $\mathcal{X}$-test θ, a state (m, s) of a Dolev-Yao system $\mathcal{G}^e$ is *θ-possible* if there exists $m' \in d_{\mathcal{S}}(\mathcal{K}(m, s))$ such that $m' \models \theta$.

The next definition puts everything together. A protocol is not abuse-free if there exists a convincing test which indicates unbalanced states as explained in the introduction:

Definition 2 (abuse freeness). *Let $\mathcal{X} \subseteq \mathcal{M}$. An external view $(\mathcal{G}^e, \beta^e)$ is $\mathcal{X}$-*abusive *if there exists an $\mathcal{X}$-test θ such that the following two conditions are satisfied:*
1. *There exists a θ-possible and β-unbalanced state in $\mathcal{G}^e$.*
2. *Each θ-possible state (m, s) of $\mathcal{G}^e$ is a descendant of a θ-possible and β-unbalanced state in $\mathcal{G}^e$.*

Such a test is called $(\mathcal{G}^e, \beta^e)$-convincing. *The external view $(\mathcal{G}^e, \beta^e)$ is called* $\mathcal{X}$-abuse-free *if $(\mathcal{G}^e, \beta^e)$ is not $\mathcal{X}$-abusive.*

5 The ASW Protocol Analyzed

In this section, we present our results concerning the analysis of the ASW contract signing protocol. For our formal analysis of the ASW protocol, we let $\sigma = \sigma_{ASW}$, $\mathcal{S} = \mathcal{S}_{ASW}$, and $\mathcal{H} = \mathcal{H}_{ASW}$, as explained in Sect. 3.

5.1 The ASW Protocol Is Not Balanced

First, we note that the ASW protocol (without an optimistic honest party) can be shown to be balanced in an interleaving (as opposed to a real concurrent) model; the proof is along the same lines as the one presented in [4] for the GJM protocol. By contrast, if we consider a concurrent setting and make the assumptions that Bob (the intruder) is (1) as fast as Alice in sending messages and (2) the TTP handles a resolve request first when an abort request is received at the same time (or in the same time slot), we can argue (informally) that the protocol is unbalanced: Bob has (i) a strategy to prevent Alice from getting a valid contract, namely by simply doing nothing, and (ii) a strategy to resolve the contract signing process after Alice has sent the first message of the exchange protocol, namely by sending a resolve request to the TTP. Even if Alice sends an abort request to the TTP at the same time, because of assumption (1) her request cannot reach the TTP before Bob's resolve request, and with assumption (2), we know that Bob's resolve request takes priority over Alice's abort request.

Assumption (2) from above shows that we need to be careful when implementing the TTP, because of simultaneous requests. If the TTP receives a resolve request and an abort request at the same time, it could first serve the resolve request and then the abort request or vice versa. As a consequence, we distinguish two models of the TTP, denoted T and T', with corresponding components $\mathcal{A}_T$ and $\mathcal{A}_{T'}$, respectively, and we show that for both variants of the TTP, the ASW protocol is unbalanced.

We consider two scenarios. In the first one, we have honest Alice, dishonest Bob, and T, and in the second one, we have dishonest Alice, honest Bob, and T'. More precisely, we consider $\mathcal{G}_{ASW} = \mathrm{DY}[\{\mathcal{A}_i\}_{i\in\{A,T\}}, \{A, T\}, \{B\}, \mathcal{K}]$ and $\mathcal{G}'_{ASW} = \mathrm{DY}[\{\mathcal{A}_i\}_{i\in\{B,T'\}}, \{B, T'\}, \{A\}, \mathcal{K}']$ where $\mathcal{K} = \mathcal{K}_0 \cup \{\mathsf{sk}(B)\}$, $\mathcal{K}' = \mathcal{K}_0 \cup \{\mathsf{sk}(A)\}$, and $\mathcal{K}_0 = \{A, B, T, \mathsf{text}, \mathsf{pk}(A), \mathsf{pk}(B), \mathsf{pk}(T), \mathsf{initiator}, \mathsf{responder}, \mathsf{ok}\}$, which means that among other things the intruder's initial knowledge comprises

Bob's and Alice's private key, respectively. The components $\mathcal{A}_A$ and $\mathcal{A}_B$ are easily obtained from the informal description in Sect. 2.

We define, in a straightforward fashion, path properties $\bar{\varphi}_A$, $\bar{\varphi}_B$, and $\varphi_\mathcal{I}$ to describe that Alice does not get a valid contract, that Bob does not get a valid contract, and that the intruder does get a valid contract, respectively.

We assume that the intruder's view of the system is limited to his own history, that is, we use an appropriate view function $\mathsf{view}_\mathcal{I}$, which removes anything the intruder cannot observe.

Finally, we define the balance specifiers $\beta_{\text{ASW}} = (\{\mathcal{I}\}, \mathsf{view}_\mathcal{I}, \bar{\varphi}_A, \varphi_\mathcal{I})$ and $\beta'_{\text{ASW}} = (\{\mathcal{I}\}, \mathsf{view}_\mathcal{I}, \bar{\varphi}_B, \varphi_\mathcal{I})$, which are designed in such a way that they describe being unbalanced for Bob and for Alice, respectively.

Following the informal reasoning from above, we prove that the ASW protocol is unbalanced for either Alice or Bob, depending on which version of the TTP is used:

Theorem 1 (ASW is unbalanced). *The Dolev-Yao system $\mathcal{G}_{\text{ASW}}$ is β_{ASW}-unbalanced, and, similarly, $\mathcal{G}'_{\text{ASW}}$ is β'_{ASW}-unbalanced.*

5.2 The ASW Protocol Is Not Abuse-Free

For abuse freeness, we imagine that Charlie assumes that there is only one instance of the ASW protocol running, but that he does not know whether Alice is the initiator or responder, which is a realistic assumption. Formally, we replace $\mathcal{A}_A$ by a variant of it, denoted $\mathcal{A}_{A'}$, which in the beginning decides whether it wants to play the role of the initiator or the responder and then sends a corresponding message to Bob. We set $\mathcal{G}^e_{\text{ASW}} = \text{DY}[\{\mathcal{A}_i\}_{i\in\{A',T\}}, \{A',T\}, \{B\}, \mathcal{K}]$ with $\mathcal{K}$ as above, $\beta^e_{\text{ASW}}=(\{\mathcal{I}^e\}, \mathsf{view}_\mathcal{I}, \bar{\varphi}_A, \varphi_\mathcal{I})$, and $\mathcal{X}=\{A, B, C, T, \mathsf{pk}(A), \mathsf{pk}(B), \mathsf{pk}(C), \mathsf{pk}(T), \mathsf{sk}(C), \mathsf{text}, \mathsf{ok}\}$. Here, $\mathcal{I}^e$ denotes the intruder of $\mathcal{G}^e$. We prove:

Theorem 2 (ASW not abuse-free). *The external view $(\mathcal{G}^e_{\text{ASW}}, \beta^e_{\text{ASW}})$ is $\mathcal{X}$-abusive. This remains true when T is replaced by T'.*

In the proof we identify a test for checking whether a message is Alice's promise of signature in an instance initiated by her and show that this test is convincing.

6 The GJM Protocol Analyzed

We show that, in a concurrent setting, the GJM protocol is unbalanced but abuse-free. The structure of the GJM protocol is exactly as for the ASW protocol. However, the actual messages exchanged are different. In particular, in the version of the exchange protocol of the GJM protocol the first two messages are so-called private contract signatures [9] and the last two messages are actual signatures (obtained by converting the private contract signatures into universally verifiable signatures).

For the GJM protocol we consider the signature Σ_{GJM} which contains the following individual symbols: $\mathsf{sig}(\cdot,\cdot,\cdot)$, $\mathsf{sigcheck}(\cdot,\cdot,\cdot)$, $\mathsf{pk}(\cdot)$, $\mathsf{sk}(\cdot)$, $\langle\cdot,\cdot\rangle$, $\pi_1(\cdot)$, $\pi_2(\cdot)$,

$\mathsf{fake}(\cdot,\cdot,\cdot,\cdot,\cdot)$, $\mathsf{pcs}(\cdot,\cdot,\cdot,\cdot,\cdot)$, $\mathsf{pcsver}(\cdot,\cdot,\cdot,\cdot,\cdot)$, $\mathsf{sconvert}(\cdot,\cdot,\cdot)$, $\mathsf{tpconvert}(\cdot,\cdot,\cdot)$, $\mathsf{sver}(\cdot,\cdot,\cdot,\cdot)$, $\mathsf{tpver}(\cdot,\cdot,\cdot,\cdot)$, A, B, T, text, initiator, responder, ok, pcsok, sok, tpok, and aborted. In addition, it includes infinite sets $\mathcal{C}$ and $\mathcal{R}$ of constants that stand for nonces and random coins used by the parties.

The equational theory $\mathcal{H}_{GJM}$ that we consider contains (among some obvious identities) the following identities to model private contract signatures (PCS):

$$\mathsf{pcsver}(w, \mathsf{pk}(x), \mathsf{pk}(y), \mathsf{pk}(z), \mathsf{pcs}(u, \mathsf{sk}(x), w, \mathsf{pk}(y), \mathsf{pk}(z))) = \mathsf{pcsok}, \quad (1)$$

$$\mathsf{pcsver}(w, \mathsf{pk}(x), \mathsf{pk}(y), \mathsf{pk}(z), \mathsf{fake}(u, \mathsf{sk}(y), w, \mathsf{pk}(x), \mathsf{pk}(z))) = \mathsf{pcsok}, \quad (2)$$

$$\mathsf{sver}(w, \mathsf{pk}(x), \mathsf{pk}(z), \mathsf{sconvert}(u, \mathsf{sk}(x), \mathsf{pcs}(v, \mathsf{sk}(x), w, \mathsf{pk}(y), \mathsf{pk}(z)))) = \mathsf{sok}, \quad (3)$$

$$\mathsf{tpver}(w, \mathsf{pk}(x), \mathsf{pk}(z), \mathsf{tpconvert}(u, \mathsf{sk}(z), \mathsf{pcs}(v, \mathsf{sk}(x), w, \mathsf{pk}(y), \mathsf{pk}(z)))) = \mathsf{tpok}. \quad (4)$$

A term of the form $\mathsf{pcs}(u, \mathsf{sk}(x), w, \mathsf{pk}(y), \mathsf{pk}(z))$ stands for a PCS computed by x (with $\mathsf{sk}(x)$) involving the text w, the party y, and the TTP z while u models the random coins used to compute the PCS. Everybody can verify the PCS with the public keys involved (identity (1)), but cannot determine whether the PCS was computed by x or y (identity (2)): instead of x computing the "real" PCS, y could have computed a "fake" PCS which would also pass the verification with pcsver. Using sconvert and tpconvert, see (3) and (4), a "real" PCS can be converted by x and the TTP z, respectively, into a universally verifiable signature (verifiable by everyone who possesses $\mathsf{pk}(x)$ and $\mathsf{pk}(z)$).

With the Dolev-Yao systems $\mathcal{G}_{GJM}$, $\mathcal{G}'_{GJM}$ and the balance specifiers β_{GJM} and β'_{GJM} defined as in the case of the ASW protocol but with the messages adapted to the GJM protocol, we obtain:

Theorem 3 (GJM is unbalanced). *The Dolev-Yao systems $\mathcal{G}_{GJM}$ and $\mathcal{G}'_{GJM}$ are β_{GJM}- and β'_{GJM}-unbalanced, respectively, i.e., GJM is unbalanced for the initiator if in the TTP resolve takes priority over abort. Conversely, GJM is unbalanced for the responder if in the TTP abort takes priority over resolve.*

With the external view $(\mathcal{G}^e_{GJM}, \beta^e_{GJM})$ defined analogously to the ASW protocol (Alice may play the role of the initiator or responder and the TTP gives priority to resolve), again with the messages adapted to the GJM protocol, and $\mathcal{X}$ defined as above, we obtain:

Theorem 4 (GJM is abuse-free). *The external view $(\mathcal{G}^e_{GJM}, \beta^e_{GJM})$ is $\mathcal{X}$-abuse free. The same is true if the TTP gives priority over abort.*

We prove for all tests Charlie can perform: If a message m that satisfies the test can be derived at some unbalanced state, then another message m' could have been derived already in a previous state which is not a descendent of an unbalanced state and which also satisfies the test; m' is essentially obtained by replacing every occurrence of a real PCS by a fake one. Interestingly, the proof requires to model random coins. Without such coins, one could deduce that given two different messages both passing the same test `pcsver`, one of the messages must be the real PCS while the other one is the fake one.

7 Conclusion

We have proposed a new definition of abuse freeness which involves as key features (i) a specifically designed notion of test performed by the outside party and (ii) a formalization of the assumptions of the outside party by the notion of external view. We have applied our definition to the ASW and GJM protocol, where for the latter protocol we have developed an equational theory to describe the semantics of private contract signatures.

In view of the results in [1, 7, 10], we are currently investigating decidability of abuse freeness as defined here. Also, we study whether balance can be achieved in a real concurrent communication model, given that both the ASW and the GJM protocol are unbalanced no matter what priority the TTP gives to abort and resolve requests received at the same time.

References

1. M. Abadi and V. Cortier. Deciding knowledge in security protocols under equational theories. In *ICALP 2004*, volume 3142 of *LNCS*, pages 46–58. Springer, 2004.
2. M. Abadi and C. Fournet. Mobile Values, New Names, and Secure Communication. In *POPL 2001*, pages 104–115. ACM Press, 2001.
3. N. Asokan, V. Shoup, and M. Waidner. Asynchronous protocols for optimistic fair exchange. In *IEEE Symposium on Research in Security and Privacy*, pages 86–99, 1998.
4. R. Chadha, M.I. Kanovich, and A.Scedrov. Inductive methods and contract-signing protocols. In *CCS 2001*, pages 176–185. ACM Press, 2001.
5. R. Chadha, S. Kremer, and A. Scedrov. Formal analysis of multi-party contract signing. In *CSFW 2004*, pages 266–279. IEEE Computer Society Press, 2004.
6. R. Chadha, J.C. Mitchell, A. Scedrov, and V. Shmatikov:. Contract Signing, Optimism, and Advantage. In *CONCUR 2003*, volume 2761 of *LNCS*, pages 361–377. Springer, 2003.
7. Y. Chevalier and M. Rusinowitch. Combining Intruder Theories. In *ICALP 2005*, volume 3580 of *LNCS*, pages 639–651. Springer, 2005.
8. D. Dolev and A.C. Yao. On the Security of Public-Key Protocols. *IEEE Transactions on Information Theory*, 29(2):198–208, 1983.
9. J.A. Garay, M. Jakobsson, and P. MacKenzie. Abuse-free optimistic contract signing. In *CRYPTO'99*, volume 1666 of *LNCS*, pages 449–466. Springer-Verlag, 1999.
10. D. Kähler, R. Küsters, and Th. Wilke. Deciding Properties of Contract-Signing Protocols. In *STACS 2005*, volume 3404 of *LNCS*, pages 158–169. Springer-Verlag, 2005.
11. D. Kähler, R. Küsters, and Th. Wilke. A Dolev-Yao-based Definition of Abus-free Protocols. Technical report, IFI 0607, CAU Kiel, Germany, 2006. Available from http://www.informatik.uni-kiel.de/reports/2006/0607.html
12. S. Kremer and J.-F. Raskin. Game analysis of abuse-free contract signing. In *CSFW 2002*, pages 206–220. IEEE Computer Society, 2002.
13. V. Shmatikov and J.C. Mitchell. Finite-state analysis of two contract signing protocols. *Theoretical Computer Science (TCS), special issue on Theoretical Foundations of Security Analysis and Design*, 283(2):419–450, 2002.

Preserving Secrecy Under Refinement*

Rajeev Alur, Pavol Černý, and Steve Zdancewic

University of Pennsylvania

Abstract. We propose a general framework of *secrecy* and *preservation of secrecy* for labeled transition systems. Our definition of secrecy is parameterized by the distinguishing power of the observer, the properties to be kept secret, and the executions of interest, and captures a multitude of definitions in the literature. We define a notion of *secrecy preserving refinement* between systems by strengthening the classical trace-based refinement so that the implementation leaks a secret only when the specification also leaks it. We show that secrecy is in general not definable in μ-calculus, and thus not expressible in specification logics supported by standard model-checkers. However, we develop a simulation-based proof technique for establishing secrecy preserving refinement. This result shows how existing refinement checkers can be used to show correctness of an implementation with respect to a specification.

1 Introduction

Security and confidentiality are growing concerns in software and system development [14]. The question of how to ascertain that an attacker cannot easily get information about classified data is central in this domain. We investigate the possibilities for using automated verification techniques (such as model checking) to answer this question, and in particular, we focus on the notion of *refinements* that preserve secrecy. Stepwise refinement is considered to be the correct approach to system and software construction, since it enables developers to find design errors in earlier stages of development. Refinements are useful for synthesizing implementations from higher level specifications, for instance via compilation or other code transformations. Such refinement based approach has been advocated by, for example, Hoare [6] and Lamport [8]. Our goal is to develop a formal and general framework for refinement that also takes into account secrecy.

Our contributions are two fold. First, we introduce a general framework for reasoning about secrecy requirements in a system. We use the standard verification framework – labeled transition systems. Our notion of secrecy depends on three parameters: (1) the equivalence relation on runs of the system that models the distinctions the observer can make, (2) the properties that are to be kept secret, and (3) the set of runs that are of interest. Intuitively, a property is secret if, for every run of interest, there is an equivalent run such that only one of these

* This research was partially supported by NSF Cybertrust award CNS 0524059.

M. Bugliesi et al. (Eds.): ICALP 2006, Part II, LNCS 4052, pp. 107–118, 2006.

two runs satisfies the property. We show that by varying these three parameters, it is possible to capture *possibilistic* definitions of secrecy found in the literature such as noninterference and perfect security property [13, 16]. We study whether such a general notion of secrecy can be specified using temporal logics. The answer is negative: we prove that secrecy is not expressible in μ-calculus. It has been claimed (see [3]) that it is possible to specify secrecy in temporal logic on self-composition (self-composition is a composition of a program with itself). However, we demonstrate that this too is not possible for the general definition of secrecy.

It is well-known that standard notions of refinement (e.g. trace inclusion) do not preserve secrets, and the refined program may leak more secrets than the original program [10]. Our second main contribution is that we define secrecy-preserving refinement and present a simulation-based technique for proving that one system is a refinement of another. In our definition, an implementation is a refinement of the specification, if for every run r of the implementation, there exists a run r' of the specification such that the observer cannot distinguish r from r', and for every property that the observer can deduce from r in the implementation can also be deduced by observing r' in the specification. Simulation is a standard technique: in order to show that a program P refines a program Q (in the classical sense), one can show that Q simulates P. This can also be part of the simulation based proof of secrecy-preserving refinement, since we require trace inclusion in the usual way. However, in order to show that P does not leak more secrets than Q, one must also show that P simulates Q. The reason is that using this simulation relation, one can prove that if P leaks a secret, then so does Q. This implies that even though secrecy is not specifiable in μ-calculus, and thus cannot be directly checked by existing model-checkers, showing that implementation preserves secrets of the specification can be done using existing tools (such as Mocha [2], CadenceSMV [11], PVS [12]) by establishing a simulation relation.

Related Work

We know of only two notions of secrecy preserving refinements that were defined previously. Mantel [9] assumes that some fixed, strong information-flow properties of the system are enforced and his definition of refinement preserves those properties. Our approach is more flexible because it permits the specification of arbitrary secrecy requirements. This means that if the specification program does not maintain secrecy of a certain property, the implementation program does not need to either. Jürjens [7] considers a different (and weaker) definitions of secrets. In his approach, a secret is leaked if the program (possibly when interacting with an adversary) outputs the secret. This approach thus ignores information-flow leaks, i.e. cases when the adversary can infer something about the secret without explicitly seeing it.

There is a large body of literature in language-based security (see [13] for an overview). Various definitions of secrecy have been considered, but all possibilistic variations (those that ignore probabilistic information about the distribution of system behaviors) can be captured in our framework. More closely related is the work on checking for secrecy using self-composition techniques—the work by Barthe et al. mentioned above and [15, 4], where the authors consider only deterministic programs. Halpern and O'Neill [5] define a notion of secrecy in the context of multiagent systems that is similar to our definitions, but they do not consider secrecy-preserving refinements. The preservation of secrecy has been studied in the context of programming language translation by Abadi [1] using techniques based on full abstraction.

2 Secrecy Requirements

In this section, we introduce a framework in which we can reason about properties of a system being secret, i.e. not inferable by an observer who sees the behavior of the system. The framework we present is general enough to capture all possibilistic definitions of secrecy defined in both programming language and verification literature, to the best of our knowledge.

A *labeled transition system* (LTS) T is a tuple (Q, L, δ, I), where Q is a set of states, L is a set of labels, $\delta \subseteq Q \times L \times Q$ is a transition relation, and $I \subseteq Q$ a set of initial states.

A sequence $r = q_0 l_0 q_1 \ldots$ of alternating states and labels is a *run* of the labeled transition system T iff $q_0 \in I$ and $\forall i : 0 \leq i < |r| \Rightarrow (q_i, l_i, q_{i+1}) \in \delta$. Let $R(T)$ be the set of all runs of the LTS T.

A *property* α is a subset of the set of runs, i.e. $\alpha \subseteq R(T)$. A *state-property* is a property that depends only on the last state of a run. Formally, α is a state-property iff there is a set of states $Q_s \subseteq Q$ such that $r \in \alpha$ iff $r = q_0 l_0 q_1 l_1 \ldots l_{n-1} q_n$ and q_n is in Q_s.

Given this model of systems, we want to define what an observer can see and what he or she can infer based on those observations. The observer cannot see everything about the current run of the system, that is to say, in general, several runs can correspond to the same observation. We model this using an equivalence relation on runs, $\equiv \subseteq R(T) \times R(T)$. For a property α, the observer is able to conclude that α holds, if α holds for all the runs that correspond to his or her observations. He or she is able to conclude that α does not hold, if it does not hold for all the runs that correspond to the observations. The third possibility is that the observer is not able to conclude whether α holds or not. We will thus need to use a three-valued domain, $\{\top, \bot, m\}$ (true, false, maybe), and a partial order that models the knowledge the observer has. $\sqsubseteq$ is the following partial order on $\{\top, \bot, m\}$: $m \sqsubseteq m, m \sqsubseteq \top, m \sqsubseteq \bot, \bot \sqsubseteq \bot, \top \sqsubseteq \top$.

Function *IP* – *inferable properties*, is a function that, given a run r, a property α and an equivalence relation $\equiv$, represents the knowledge of the observer about the property α after the run r. $IP(r, \alpha, \equiv) = \top$ if $\forall r' : r' \equiv r \Rightarrow r' \in \alpha$, $IP(r, \alpha, \equiv) = \bot$ if $\forall r' : r' \equiv r \Rightarrow r' \notin \alpha$ and $IP(r, \alpha, \equiv) = m$ otherwise. Our notion of secrecy depends on one additional parameter: instead of requiring a property α to be secret in every run of the system, we may want to focus only on a subset β of runs that are of interest, e.g. the set of all terminating runs. This leads to the following formalization of secrecy:

Let T be a labeled transition system and α and β be two properties. The property α is a *secret* in β for T w.r.t. $\equiv$ if for all $r \in \beta$, $IP(r, \alpha, \equiv) = m$.

We present the following examples in order to show that our definition is general enough to capture several standard information-flow properties such as noninterference or Perfect Security Property. We can capture these definitions by varying the parameters $\equiv$ and β.

Linear-Time Secrecy

Consider an observer who can see the actions of the system, i.e. the labels in L. These labels, for example, might be the messages sent or received by the system. Assume that L contains a symbol τ, which models internal actions of the system.

We define the strong (time-sensitive) equivalence relation ($\approx$) as follows. Let Tr be an erasing homomorphism defined on runs: $Tr(q) = \epsilon$, $Tr(l) = l$, i.e. Tr erases all states. Two runs r and r' are strongly equivalent ($r \approx r'$) iff $Tr(r) = Tr(r')$. The equivalence class to which a run r belongs can be represented by $Tr(r)$, which corresponds to what the observer sees when r is the current execution of the system. $Tr(r)$ is a sequence of labels, and such sequences are called *traces*. $Tr(T)$ is the set of all traces of the LTS T.

We define the weak (time-insensitive) equivalence relation ($\approx_w$) as follows. Let Tr_w be an erasing homomorphism defined on runs: $Tr_w(q) = \epsilon$, $Tr_w(l) = l$ for $l \neq \tau$ and $Tr_w(\tau) = \epsilon$, i.e. Tr_w erases all states and all internal actions. Two runs r and r' are weakly equivalent ($r \approx_w r'$) iff $Tr_w(r) = Tr_w(r')$. The equivalence class can be represented by $Tr_w(r)$ and is called a *weak trace*. Let $Tr_w(T)$ be the set of all weak traces of the LTS T.

Consider the following two programs.

```
A: x=?; y=0; z=x; send z;
B: x=?; y=0; z=y; send z;
```

It is easy to see how they can be modeled as transition systems in our framework. The states are valuations of variables. The set L contains three labels s_0, s_1, τ. s_0 denotes the fact that 0 was sent, s_1 that 1 was sent and τ denotes all the internal (silent) actions. The input ($x = ?$) is intended not to be seen by an observer and thus is modeled by a silent action. We want to analyze what an observer might infer about whether or not $x = 0$ during the execution of the program if he or she can observe what the program sends. We model this using the strong observational equivalence $\approx$ and the state property $x = 0$. Suppose that the input is 0. Note that the observer sees the same trace for both programs, namely $t = \tau\tau\tau 0$. For the program A, the observer, after having seen the trace t was sent, can conclude that $x = 0$ holds. For the program B, the observer does not know whether $x = 0$ holds or not after having seen the trace t. We can conclude that for program A, the state property $x = 0$ is not a secret in the set of all runs w.r.t. $\approx$ and it is a secret for program B.

Noninterference

Consider the standard formulation of termination insensitive noninterference [13]. It is defined using low and high variables, where low variables are visible to the observer and high variables are not. Noninterference can be then formulated informally as follows: "if two input states share the same values of low variables then the behaviors of the program executed from these states are indistinguishable by the observer".

We define functional equivalence $\approx_f$ as follows. Let $\approx_f^i$ and $\approx_f^o$ be two equivalence relations on states. For all terminating runs r and r' we have $r \approx_f r'$ iff their initial states are related by $\approx_f^i$ and their final states by $\approx_f^o$. We model noninterference by functional equivalence defined above. Two states q and q' are related by $\approx_f^i$ (and $\approx_f^o$) exactly when the valuation of the low variables is the same in q and q'.

The purpose of using noninterference is to determine whether some property α of high variables is inferable by an observer who sees only low variables. We can capture this in our framework as follows. Let $\mathcal{P}$ be the set of all expressible properties of high variables. For example, if every property of high variables is considered to be expressible, $\mathcal{P}$ corresponds to the powerset of the set of valuations of high variables. Consider a classic requirement such as "a secret key should stay secret." In our framework, this can be expressed as "a secret key stays secret with respect to a set of predicates $\mathcal{P}$", i.e. none of the properties of the secret key that are in $\mathcal{P}$ will be revealed.

Let β_t be the set of all terminating runs. We can conclude that the system satisfies the noninterference property w.r.t. $\mathcal{P}$ iff for all $\alpha \in \mathcal{P}$, α is secret in β_t w.r.t. $\approx_f$.

Perfect Security Property

Let us consider the Perfect Security Property (PSP) [16]. It is an information-flow property defined in a trace-based setting. In order to define it, we divide the labels into low-security and high-security categories. The observer knows the specification of the system - i.e. the set of all possible traces (sequences of labels) and he or she can observe low-security labels. PSP ensures that the observer cannot deduce any information about occurrences of high-security events.

We can model the PSP in our framework by choosing an appropriate equivalence relation on runs and a property on runs. Let $Low \subseteq L$ be a set of low-security labels and let $High \subseteq L$ be a set of high-security labels such that Low and $High$ partition L. We use the following equivalence relation. Let Tr_{psp} be an erasing homomorphism defined on runs as follows: $Tr_{psp}(q) = \epsilon$, $Tr_{psp}(l) = l$ for $l \in Low$ and $Tr_w(l) = \epsilon$ for $l \in High$, i.e. Tr_{psp} erases all states and all high-security actions. Two runs r and r' are *psp*-equivalent ($r \approx_{psp} r'$) iff $Tr_{psp}(r) = Tr_{psp}(r')$. For each label $h \in High$, we define the property α_h: a run r is in α_h if h occurs in r. Now we can conclude that PSP holds iff α_h is secret in β_{all} w.r.t. $\approx_{psp}$ for all $h \in High$.

Specifying Secrecy in Temporal Logics

It is well-known that secrecy cannot be expressed as a predicate on a single trace and hence cannot be specified in linear-time specification languages such as linear temporal logic (see, for example, [10], for a proof). We prove that secrecy is not a branching-time property either.

Let us consider finite trees over alphabet Σ. The vertices are labeled by elements of Σ (edges are not labeled). A tree T can be seen as an LTS T', where states correspond to vertices of the tree, edges are the parent-child edges, and all the edges are labeled by the same symbol. For each label $\alpha \in \Sigma$, let α' be a state-property corresponding to the set of all vertices labeled by α.

Theorem 1. *The set S of trees T over $\{\alpha, \beta\}$ such that α' is secret in β' w.r.t. $\approx$ for T' is not a regular tree-language.*

Proof. For a proof by contradiction, suppose that S is regular. Then the following special case, defined by a regular condition that β' is false only for the root of the tree, would also be regular. The fact that α' is secret in β' w.r.t. $\approx$ corresponds to the fact that at each depth d ($d > 0$) of the tree, there is a node in α' and a node not in α'. It is well-known that this is not a regular property. □

Corollary 1. *The set of trees T over $\{\alpha, \beta\}$ such that α' is secret in β' w.r.t. $\approx$ for T' is not definable in μ-calculus.*

Note that it is possible to devise algorithms based on standard model-checking for special cases of our definition of secrecy. For example, Barthe et al. [3] claim that it is possible to use CTL model-checking to check for noninterference in finite-state systems. However, upon examination, this holds only for a specific definition of noninterference, the one based on functional equivalence relation (as opposed to, e.g., strong equivalence relation). Barthe et al. reduce checking for noninterference to model-checking a CTL formula on self-composition. Self-composition can be viewed as a (sequential or parallel) composition of a program with itself (variables are renamed in the other copy of the program). It can be shown, by a proof similar to the one above, that there is no μ-calculus formula that characterizes the general definition of noninterference on self-composition.

3 Secrecy-Preserving Refinements

Let us suppose that we have two labeled transition systems T_{spec} and T_{imp}. We want to establish that T_{imp} does not leak more secrets than T_{spec}.

First, consider the classical notion of refinement, where T_{imp} refines T_{spec} iff all behaviors of T_{imp} are allowed by T_{spec}. This notion of refinement preserves all properties expressible in linear temporal logic, but does not in general preserve the secrecy of properties. Consider two of the systems in Figure 1, (a) as T_{spec} and (b) as T_{imp}. Using the classical notion, T_{imp} is a refinement of T_{spec}, since the behaviors of T_{imp} are included in behaviors of T_{spec}. This holds for both the functional (input-output) and observational (trace-based) view of behaviors.

However, T_{imp} leaks more secrets than T_{spec} does. If the observer of T_{imp} sees a trace s_0, he or she can conclude that α does not hold. On the other hand, for T_{spec}, the observer cannot determine whether α holds or not.

We proceed to introduce a new notion of refinement, one that preserves secrecy of properties. Intuitively, we want to show that for each run r of T_{imp}, there is an equivalent run r' of T_{spec}, such that the observer can deduce less about the properties of interest when observing T_{imp} executing r than when observing T_{spec} executing r'. Hence, let us extend the equivalence relation $\equiv$ to the runs of the two systems, i.e. $\equiv \subseteq (R(T_{spec}) \cup R(T_{imp})) \times (R(T_{spec}) \cup R(T_{imp}))$. Furthermore, we need to relate properties of interest for the two systems. Analogously, a property α now will be a subset of $R(T_{spec}) \cup R(T_{imp})$.

Now we are ready to state when a refined transition system preserves at least as many secrets as the original one:

Secrecy-preserving refinement
Let T_{spec}, T_{imp} be two labeled transition system, let $\mathcal{P}$ be a set of properties and let $\equiv$ be an equivalence relation on $R(T_{spec}) \cup R(T_{imp})$. T_{imp} *$\mathcal{P}$-refines* T_{spec} w.r.t. $\equiv$ iff for all runs $r \in R(T_{imp})$, there exists a run $r' \in R(T_{spec})$ such that $r \equiv r'$ and for all properties $\alpha \in \mathcal{P}$, $IP(r, \alpha, \equiv) \sqsubseteq IP(r', \alpha, \equiv)$.

We present the following observations and an example to illustrate the definition. First, note that secrecy-preserving refinement extends the classical notion: consider the case when the set of properties $\mathcal{P}$ is empty. For strong (weak) equivalence $\mathcal{P}$-refinement corresponds to (weak) trace inclusion. For functional equivalence, $\mathcal{P}$-refinement corresponds to the requirement that the input-output relation of T_{imp} is included in the input-output relation of T_{spec}.

Consider the programs A and B from Section 2 again. As before, suppose that the observer does not see the input, but this time, we fix the input to be 0 in order to simplify the example. We consider the strong observational equivalence and we are interested in the state-property α that is true iff $x = 0$. There is only one run in each of the programs. Those runs are equivalent, since the trace is simply $\tau\tau\tau s_0$ in both cases. Let r_A denote the run of A and let r_B denote the run of B. As we have seen, $IP(r_A, \alpha, \approx) = \top$ and $IP(r_B, \alpha, \approx) = m$. Thus we can conclude that A does not $\mathcal{P}$-refine B w.r.t. $\approx$, but B $\mathcal{P}$-refines A w.r.t. $\approx$.

The following theorem states that the $\mathcal{P}$-refinement preserves the secrets from $\mathcal{P}$, i.e. that if T_{spec} does not leak a secret $\alpha \in \mathcal{P}$ and T_{imp} is a $\mathcal{P}$-refinement of T_{spec}, then also T_{imp} does not leak the secret α. Before stating the theorem, we need to define one more condition on the set of runs that are of interest, β. A property β is $\equiv$-preserving iff for all r and for all r', if $r \in \beta$ and $r \equiv r'$, then $r' \in \beta$.

Theorem 2. *Let T_{spec} and T_{imp} be two transition systems such that T_{imp} $\mathcal{P}$-refines T_{spec} w.r.t. $\equiv$ and let β be a an $\equiv$-preserving property. If $\alpha \in \mathcal{P}$ is a secret in β for T_{spec} w.r.t. $\equiv$, then α is a secret in β for T_{imp} w.r.t. $\equiv$.*

4 Proving Secrecy Using a Simulation Relation

In this section we restrict our attention to the strong (time-sensitive) and weak (time-insensitive) equivalence relations on runs and we consider only state-properties. With appropriate modifications, simulation-based proof techniques can be developed for other equivalences such as the ones used for noninterference and perfect security.

Let $T_{spec} = (Q_{spec}, L, \delta_{spec}, I_{spec})$ and $T_{imp} = (Q_{imp}, L, T_{imp}, I_{imp})$ be labeled transition systems. As above, let $\mathcal{P}$ denote both a set of properties about T_{spec} and a corresponding set of properties about T_{imp}. Note that the two transition systems have the same set of labels, thus the relation $\approx$ (the strong observational equivalence) can be seen as a relation on $R(T_{spec}) \cup R(T_{imp})$.

A binary relation $\lesssim \subseteq Q_{spec} \times Q_{imp}$ is a simulation relation iff for all $q_1 \lesssim q_1'$ for all state properties $\alpha \in \mathcal{P}$, $q_1 \in \alpha$ iff $q_1' \in \alpha$ and for every $q_2 \in Q_{spec}$ and $l \in L$ such that $q_1 \xrightarrow{l} q_2$ there exists $q_2' \in Q_{imp}$ such that $q_1' \xrightarrow{l} q_2'$ and $q_2 \lesssim q_2'$.

We say that T_{imp} simulates T_{spec} ($T_{spec} \lesssim T_{imp}$), if there exists a simulation relation $\lesssim$ such that for every $q_1 \in I_{spec}$ there exists $q_1' \in I_{imp}$ such that $q_1 \lesssim q_1'$.

A binary relation $\lesssim_w \subseteq Q_{spec} \times Q_{imp}$ is a weak simulation relation iff for all $q_1 \lesssim_w q_1'$, for all state properties $\alpha \in \mathcal{P}$, $q_1 \in \alpha$ iff $q_1' \in \alpha$ and we have:

- $q_1' \xrightarrow{\tau} q_2'$ implies that there exists a q_2 such that $q_1 \xrightarrow{\tau}{}^{*} q_2$ and $q_2 \lesssim_w q_2'$
- $q_1' \xrightarrow{l} q_2'$ implies that there exists a q_2 such that $q_1 \xrightarrow{\tau}{}^{*} \xrightarrow{l} \xrightarrow{\tau}{}^{*} q_2$ and $q_2 \lesssim_w q_2'$.

Weak simulation between transition system is defined similarly to simulation between transition systems.

Let us consider the case of strong (time-sensitive) equivalence relation on runs. Firstly, we note that it follows from the definition of $\mathcal{P}$-refinement that the standard refinement condition ($Tr(T_{imp}) \subseteq Tr(T_{spec})$) is a necessary condition for the $\mathcal{P}$-refinement.

Secondly, note that unlike classical refinement, the condition that T_{spec} simulates T_{imp} is not sufficient for $\mathcal{P}$-refinement. To see this consider again two of the systems in Figure 1, (a) as T_{spec} and (b) as T_{imp}. Note that T_{spec} simulates T_{imp}, but T_{imp} leaks information on α on the trace s_0, whereas T_{spec} does not.

The property we are looking for is in fact the simulation in the other direction, i.e. that T_{imp} simulates T_{spec}. The reason is that using this simulation relation one can prove that if T_{imp} leaks a secret, then so does T_{spec}. Note also the condition that T_{imp} simulates T_{spec} is not a sufficient condition. Consider now the system on Figure 1(a) as T_{spec} and the system on Figure 1(c) as T_{imp}. Now T_{imp} refines T_{spec}, but for the trace s_1, T_{imp} leaks more secrets than T_{spec}.

The combination of the two conditions, $Tr(T_{imp}) \subseteq Tr(T_{spec})$ and T_{imp} simulates T_{spec} is sufficient to guarantee that $\mathcal{P}$-refinement holds.

Theorem 3. *If $Tr(T_{imp}) \subseteq Tr(T_{spec})$ and $T_{spec} \lesssim T_{imp}$, then T_{imp} $\mathcal{P}$-refines T_{spec} w.r.t. $\approx$.*

Proof. Let r be a run in $R(T_{imp})$. We have to prove that there exists a run r' in T_{spec} such that $r \approx r'$ and $IP(r, \alpha, \approx) \sqsubseteq IP(r', \alpha, \approx)$. We have that $Tr(T_{imp}) \subseteq$

$Tr(T_{spec})$, therefore there exists a run r' in T_{spec} such that $r \approx r'$. It remains to prove that $IP(r, \alpha, \approx) \sqsubseteq IP(r', \alpha, \approx)$. Let us suppose that $IP(r, \alpha, \approx) = \top$. We have to show that $IP(r', \alpha, \approx) = \top$. Let us suppose that $IP(r', \alpha, \approx) = \bot$ or $IP(r', \alpha, \approx) = m$. In any of the two cases, we know that there exists $r'' \in R(T_{spec})$ such that $r' \approx r''$ and the last state of r'' is not in α. Using the condition $T_{spec} \lesssim T_{imp}$ we prove (by induction on the length of the $Tr(r'')$) that there exists an $r''' \in R(T_{imp})$ such that $r'' \approx r'''$ and the last state of r''' is not in α. By transitivity of $\approx$, we have that $r''' \approx r$. This is a contradiction with the assumption $IP(r, \alpha, \approx) = \top$. Thus we can conclude that if $IP(r, \alpha, \approx) = \top$, then $IP(r', \alpha, \approx) = \top$. The case of $IP(r, \alpha, \approx) = \bot$ is similar. If $IP(r, \alpha, \approx) = m$, then there is nothing to prove, since $m \sqsubseteq IP(r', \alpha, \approx)$. □

For weak equivalence relation on runs, a similar theorem holds.

Theorem 4. *If $Tr_w(T_{imp}) \subseteq Tr_w(T_{spec})$ and $T_{spec} \lesssim_w T_{imp}$, then T_{imp} $\mathcal{P}$-refines T_{spec} w.r.t $\approx_w$.*

Note that Theorem 3 implies that secrecy is preserved by bisimulation, since for bisimilar systems, both conditions are met. Note also that we have shown in Section 2 that secrecy is not a branching time property.

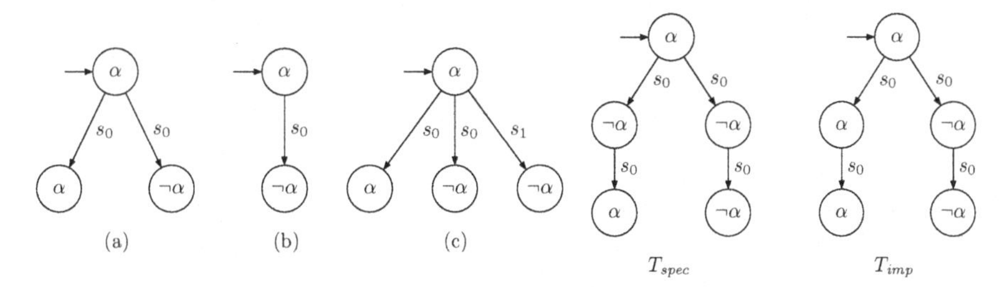

Fig. 1. Refinement by simulation

Fig. 2. Simulation is not a necessary condition

The conjunction of the two conditions of Theorem 3 is not a necessary condition for $\mathcal{P}$ refinement. Consider the two systems in Figure 2 and suppose the set $\mathcal{P}$ of properties is the singleton $\{\alpha\}$. Note that $Tr(T_{imp}) \subseteq Tr(T_{spec})$ and T_{imp} does not simulate T_{spec}. However, T_{imp} does not leak any more secrets T_{spec}.

5 Example

We present an example in order to illustrate the definition of secrecy-preserving refinement and to demonstrate the simulation-based proof method defined above. We will present two implementations, T_{spec} and T_{imp}, of a protocol (more precisely, of one round of a protocol). We will show that while functionally they are equivalent (their input-output relation is the same) and T_{imp} refines T_{spec} in the classical sense (trace inclusion), the implementation T_{imp} leaks some properties that should be secret, whereas T_{spec} does not.

Consider the game Battleship. We will analyze an implementation of one of the players in one round of the protocol, so the following description is sufficient for our purposes[1]. The input (for each round) consists of a grid, where each square is either marked (meaning a ship is there) or unmarked, and of two integers i and j. The output should be *yes* if the square with coordinates i and j is marked.

Let us consider two implementations of this protocol, T_{spec} and T_{imp}. T_{spec} uses the straightforward array representation, T_{imp} uses a list representation. In T_{imp}, the board is represented by a list of rows and each row contains a list of the marked cells. (A possible motivation for the (re-)implementation T_{imp} is that it might be more efficient in case of sparse boards.)

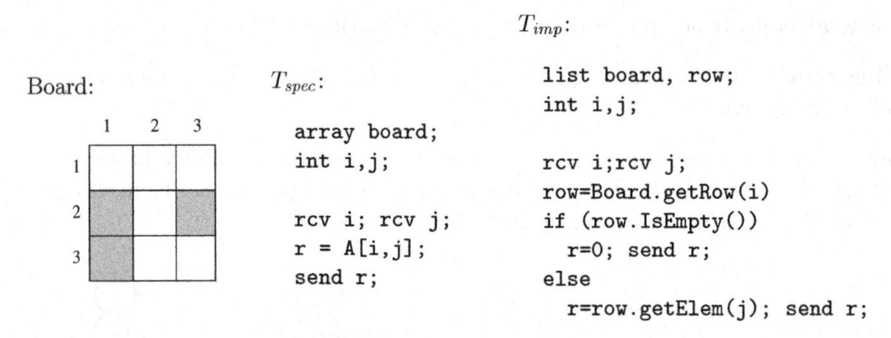

Fig. 3. The Battleship game

We briefly explain how can the programs such as T_{spec} and T_{imp} be modeled in our framework. We use the standard operational semantics approach. The states are valuations of all variables (such as board, row and program counter pc). The label s_0 denotes the fact that 0 was sent, s_1 that 1 was sent and τ denotes all the internal (silent) actions. We model the fact that the board can in general be in any state at the beginning of a round of the battleship protocol by having multiple initial states. For the purposes of this example, we also model *receives* in this way. Thus any valuation of the variables where program counter is equal to 0 is an initial state. An assignment is modeled as an internal action τ. We model the methods (such as getRow()) by one internal action (thus a statement that contains an assignment and a method call is modeled by two internal actions). As an example consider the case shown in Figure 3 and the inputs $i = 1$ (column numbered 1) and $j = 2$ (row numbered 2). The initial state is now determined. The trace produced by T_{spec} is τs_1. The trace produced by T_{imp} is $\tau\tau\tau\tau\tau s_1$. For each cell with coordinates (i, j) of the board, we define a property α_{ij} that is true iff the cell is marked. Let $\mathcal{P}$ be the set of these properties.

We will now show that T_{spec} and T_{imp} are equivalent w.r.t. $\approx_w$, that is, T_{spec} $\mathcal{P}$-refines T_{imp} and T_{imp} $\mathcal{P}$-refines T_{spec} w.r.t. $\approx_w$ and that it can be proven by simulation. We will prove also that T_{imp} does not $\mathcal{P}$-refine T_{spec} w.r.t. $\approx$.

[1] For a full description of the game, google "battleship".

Let us start by proving that for weak refinement, T_{spec} and T_{imp} are equivalent. We show that $Tr(T_{imp}) \subseteq Tr(T_{spec})$ and that $T_{spec} \lesssim_w T_{imp}$. To see that $Tr(T_{imp}) \subseteq Tr(T_{spec})$, note that both T_{spec} and T_{imp} have the same set of weak traces, namely $\{0, 1\}$ (since all the other actions are internal). Now we will show that T_{imp} simulates T_{spec}. We will present a function f from the states of T_{spec} to the states of T_{imp} and show that it defines a simulation relation. Recall that the states of T_{spec} and T_{imp} characterize the current board position, and contain valuations of variables for i, j and the program counter *pc*. In order to be able to define f, we divide the states of T_{spec} into Q^1_{spec} and Q^2_{spec}, where Q^1_{spec} contains those states where the value of *pc* (program counter) indicates that the send instruction has not been executed yet and Q^2_{spec} all the other states. We divide the states of T_{imp} analogously. The function f will relate each state q of T_{spec} to a state q' of T_{imp} that has the same board position, the same valuations of i and j, and $q \in Q^i_{spec}$ iff $q' \in Q^i_{imp}$. It is easy to check that f defines a weak simulation such that T_{spec} simulates T_{imp}. By Theorem 4, we can conclude that T_{imp} is a $\mathcal{P}$-refinement of T_{spec} w.r.t. $\approx$. Note also that $Tr(T_{imp}) = Tr(T_{spec})$ and the simulation we just defined is a bisimulation. Thus we can similarly conclude that T_{spec} is a $\mathcal{P}$-refinement of T_{imp} w.r.t. $\approx$.

We also show that T_{imp} does not $\mathcal{P}$-refine T_{spec} w.r.t. $\approx$ according to our definition, because T_{imp} leaks more secrets in certain situations. Again, consider the case depicted in Figure 3, but this time we fix the inputs to be $i = 1$ (column numbered 1) and $j = 1$ (row numbered 1). We assume the observer knows these inputs (but note that he or she does not see the board.) We analyze what he or she can infer from the execution of T_{spec} (and T_{imp}) on these inputs. As noted above, once the initial state is fixed (by the input values) there is only one possible run of T_{imp} (we denote it by r). The corresponding trace is $t_1 = \tau\tau\tau\tau s_0$. However, after the observer observes the trace t_1, he or she can infer that the j-th row is empty. For example, he or she knows that $IP(r, \alpha_{21}, \approx) = \bot$. This can be inferred because the number of internal actions is 4 (whereas if the j-th row was not empty, 5 internal actions would be observed before the final s_0). The execution of T_{spec} is similar in that there is only one possible run r given the inputs. The corresponding trace is τs_0. Given the program T_{spec}, it is clear that it is not possible to infer information about a property $\alpha_{ij}, i \neq 1, j \neq 1$, i.e. $IP(r', \alpha_{i'j'}, \approx) = m$ for $i \neq 1$ and $j \neq 1$. In particular, $IP(r', \alpha_{21}, \approx) = m$. We can thus conclude that T_{imp} is not a $\mathcal{P}$-refinement of T_{spec} w.r.t. $\approx$.

6 Conclusion

This paper presents a general framework for formal reasoning about secrecy properties. The framework is based on labeled transition systems and is thus suitable for presentation of algorithms for automated verification of secrecy. We presented how different definitions of secrecy can be captured in our framework. We showed also that secrecy is not definable by a μ-calculus formula. The main focus of this work was on defining a notion of refinement that preserves secrecy of properties and providing a method for proving that such a refinement holds. This method is based on simulation and thus can be used for automatic verification using existing tools.

There are several directions for future research. One possibility is to extend this work with static analysis for secrecy-preserving refinements of programs. Second, it would be useful to define program transformations to help designers to transform designs in a way that guarantees the preservation of secrecy. Third, we plan to investigate a logic for secrecy of properties. Fourth, it would be interesting to apply the framework presented here to resource-driven protocol transformation for embedded systems, such as Java cards or smart cards.

References

1. M. Abadi. Protection in programming-language translations. In *Proc. of ICALP'98*, pages 868–883, 1998.
2. R. Alur, T.A. Henzinger, F. Mang, S. Qadeer, S. Rajamani, and S. Tasiran. MOCHA: Modularity in model checking. In *Proc. of CAV'98*, pages 521–525. 1998.
3. G. Barthe, P. D'Argenio, and T. Rezk. Secure Information Flow by Self-Composition. In *Proc. of CSFW'04*, pages 100–114, 2004.
4. Á. Darvas, R. Hähnle, and D. Sands. A Theorem Proving Approach to Analysis of Secure Information Flow. In *Proc. of SPC 2005*, pages 193–208, 2005.
5. J. Halpern and K. O'Neill. Secrecy in multiagent systems. In *Proc. of CSFW'02*, pages 32–46, 2002.
6. C.A.R. Hoare. *Communicating Sequential Processes.* Prentice-Hall, 1985.
7. J. Jürjens. Secrecy-preserving refinement. In *Proc. of FME'01*, pages 135–152, 2001.
8. L. Lamport. The temporal logic of actions. *ACM Transactions on Programming Languages and Systems*, 16(3):872–923, 1994.
9. H. Mantel. Preserving information flow properties under refinement. In *Proc. of SP'01*, pages 78–91, 2001.
10. J. McLean. A general theory of composition for trace sets closed under selective interleaving functions. In *Proc. of SP'94*, pages 79–93, 1994.
11. K.L. McMillan. A compositional rule for hardware design refinement. In *Proc. of CAV'97*, pages 24–35, 1997.
12. S. Owre, J. M. Rushby, and N. Shankar. PVS: A prototype verification system. In *CADE'92*, pages 748–752, 1992.
13. A. Sabelfeld and A. Myers. Language-based information-flow security. *IEEE Journal on Selected Areas in Communications*, 21(1):5–19, 2003.
14. F.B. Schneider, editor. *Trust in Cyberspace.* National Academy Press, 1999.
15. T. Terauchi and A. Aiken. Secure Information Flow as a Safety Problem. In *Proc. of SAS 2005*, pages 352–367, 2005.
16. A. Zakinthinos and E. S. Lee. A general theory of security properties. In *Proc. of SP'97*, pages 94–102, 1997.

Quantifying Information Leakage in Process Calculi*

Michele Boreale

Dipartimento di Sistemi e Informatica,
Università di Firenze
Viale Morgagni 65, I–50134 Firenze, Italy
boreale@dsi.unifi.it

Abstract. We study two quantitative models of information leakage in the pi-calculus. The first model presupposes an attacker with an essentially unlimited computational power. The resulting notion of *absolute leakage*, measured in bits, is in agreement with secrecy as defined by Abadi and Gordon: a process has an absolute leakage of zero precisely when it satisfies secrecy. The second model assumes a restricted observation scenario, inspired by the testing equivalence framework, where the attacker can only conduct repeated success-or-failure experiments on processes. Moreover, each experiment has a cost in terms of communication actions. The resulting notion of leakage *rate*, measured in bits per action, is in agreement with the first model: the maximum information that can be extracted by repeated experiments coincides with the absolute leakage A of the process. Moreover, the overall extraction cost is at least A/R, where R is the rate of the process. Strategies to effectively estimate both absolute leakage and rate are also discussed.

Keywords: process calculi, secrecy, information leakage, information theory.

1 Introduction

Research in language-based security has traditionally focused on qualitative aspects. Recently, a few models have been proposed that allow forms of quantitative reasoning on security properties. For a sequential program, it is natural to quantify leakage by measuring the information flow between secret ("high") and public ("low") variables induced by the computed function. Along these lines, an elegant theory of quantitative non-interference has been recently proposed by Clark et al. [12] (other proposals in the literature are examined in the concluding section.)

In this paper, we study quantitative models of information leakage in process calculi. Processes come with no natural notion of computed function. Rather, one is interested in quantifying the leakage induced by their *observable behaviour*. The difference in intent can be illustrated by the following concrete example. A smart-card implements a function that takes documents as input and releases documents signed with a secret key as output. However, typical attacks targeting the secret key do not focus on the function

* Work partially supported by the EU within the FET-GC2 initiative, project SENSORIA.

M. Bugliesi et al. (Eds.): ICALP 2006, Part II, LNCS 4052, pp. 119–131, 2006.

itself, but rather on the behaviour of the card, in terms e.g. of observed time variance of basic operations [9], or observed power consumption [10].

Our starting point is the notion of *secrecy* as formalized by Abadi and Gordon, originally in the setting of the spi-calculus [1]. In the sequel, we will refer to this notion as AG-secrecy. Informally, AG-secrecy holds for a process P and a parameter x representing a sensible information, if the the observable behaviour of P does not depend on the actual values x takes on. In other words, an attacker cannot infer anything about x by interacting with P. The notion of "observable behaviour" is formalized in terms of behavioural equivalence, such as may testing [4,2].

Although elegant and intuitive, AG-secrecy is in practice too rigid. The behaviour of a typical security application depends nontrivially on the sensible information it protects. Nevertheless, many such applications are considered secure, on the ground that the *amount* of leaked information is, on the average, negligible. Consider a PIN-checking process $P(x)$ that receives a code from a user and checks it against a 5-digits secret code x, in order to authorize or deny a certain operation. Clearly, an attacker may acquire negative information about x by interacting with $P(x)$. However, if $P(x)$ is intended to model, say, an off-line device like a card reader, such small leaks should be of no concern. More generally, one would like to first measure the information leakage of a given system and then decide if it is acceptable or not.

In the present paper, we propose two quantitative models of leakage for processes: one for measuring *absolute* leakage, and one for measuring the *rate* at which information is leaked. As explained below, the two models correspond to different assumptions on the control an attacker may exercise over processes. The connections between these two models will also be clarified.

After quickly reviewing a few notions from information theory that will be used in the paper (Section 2), we introduce our reference language, a pi-calculus with data values (Section 3). In the first model (absolute leakage, Section 4), we presuppose an attacker with full control over the process. Using the language of unconditional security, the model can be phrased as follows. A sensible information is modeled as a random variable, say X. The a priori *uncertainty* of an adversary about X is measured by the Shannon *entropy* $H(X)$, expressed in bits. For full generality, it is assumed that some "side-information" Y, possibly related to X, is publicly available: the conditional entropy $H(X|Y)$ measures the uncertainty about X given that Y is known. The process P, depending in general on both X and Y, induces a random variable $Z = P(X,Y)$: following the discussion above, it is reasonable to stipulate that Z takes as values "observable behaviours", that is, equivalence classes of a fixed behavioral semantics. Now, the conditional entropy $H(X|Y,Z)$ quantifies the uncertainty on X left after observing both Y and Z. Hence the difference $I = H(X|Y) - H(X|Y,Z)$ is the amount of uncertainty about X removed by P, that we take as its absolute leakage. We prove that this notion is in full agreement with the qualitative notion of AG-secrecy. In the special case when there is no side-information, this means that $P(x)$ respects AG-secrecy if and only if $P(X)$ has an absolute leakage of 0 for every random variable X. We also offer two alternative characterizations of zero-leakage, hopefully more amenable to automatic checking.

The second model we consider (rate of leakage, Section 5), refines the previous scenario by introducing a notion of *cost*. Adapting the testing equivalence framework [4], we stipulate that an attacker can only conduct upon P repeated experiments E_1, E_2,... each yielding a binary answer, success or failure. The attacker has "full control" – in the sense of the first model – over the compound systems $P||E$, but not over P itself. The security measure we are interested in is the overall number of *communications* required to extract one bit of information in this scenario. Thus, we define the rate at which P leaks information in terms of the maximal number of bits of information per visible action conveyed by an experiment on P. We then give evidence that this is indeed a reasonable notion. First, we establish a relationship with the first model, showing that absolute leakage A coincides with the maximum information that can be extracted by repeated experiments, and that this costs at least A/R, where R is the rate of P. Second, we establish that, under certain conditions, process iteration ($*P$) leaves the rate of P unchanged, which is what one would expect from a good definition of rate. Finally, in the vein of testing equivalence, we give an experiment-independent characterization of rate in terms of execution traces.

Strategies to effectively estimate rate of leakage (Section 5) and absolute leakage (Section 6) are also discussed. These strategies depend on the use of symbolic semantics in the vein of [7, 3]. Some remarks on further and related work conclude the paper (Section 7). Proofs have been omitted due to lack of space.

2 Preliminary Notions

We quickly recall a few concepts from elementary information theory; see e.g. [15] for full definitions and underlying motivations. We shall consider discrete random variables (r.v.) $X, Y, ...$ defined over a common probabilty space Ω. We say that a r.v. X is of *type* U, and write $X : U$, if $X(\Omega) \subseteq U$. We shall always assume U to be finite. Elements $u \in U$ are called *samples* of X, and $|X|$ is $|\{u \in U | \Pr[X = u] > 0\}|$. The concepts of independent and uniformly distributed (u.d.) random variables, and of expectation of X ($E[X]$, for X real-valued) are defined as usual. As a function, every random variable induces a partition into events of its domain Ω, $\{X^{-1}(u) | u \in X(\Omega)\}$: we say that two random variables X and Y are *equivalent* if they induce on Ω the same partition. A vector of random variables $\tilde{X} = (X_1, ..., X_n)$, with $n \geq 0$ and $X_i : U_i$, is just a random variable of type $U_1 \times \cdots \times U_n$.

Given $X : U$, the *entropy* X of and *conditional entropy of X given $Y : V$* are defined by:

$$H(X) \stackrel{\text{def}}{=} -\Sigma_{u \in U} \Pr[X = u] \cdot \log(\Pr[X = u])$$
$$H(X|Y) \stackrel{\text{def}}{=} \Sigma_{v \in V} H(X|Y = v) \cdot \Pr[Y = v]$$

where $H(X|Y = v) = -\Sigma_{u \in U} \Pr[X = u|Y = v] \cdot \log(\Pr[X = u|Y = v])$, all logarithms are taken to the base of 2 and by convention $0 \cdot \log 0 = 0$. Two equivalent random variables have the same entropy and conditional entropy. The following (in)equalities hold:

$$0 \leq H(X) \leq \log|X| \tag{1}$$
$$H(X,Y) = H(X|Y) + H(Y) \quad \text{(chain rule)} \tag{2}$$
$$H(X_1, ..., X_n) \leq H(X_1) + \cdots + H(X_n) \tag{3}$$

where: in (1), equality on the left holds iff X is a constant, and equality on the right holds iff X is u.d.; in (3), equality holds iff the X_i's are pairwise independent. Note that by (2) and (3), $H(X|Y) = H(X)$ iff X and Y are independent. If $Y = F(X)$ for some function F then $H(Y|X) = 0$. *Information on X conveyed by Y* (aka, *mutual information*) is defined as:

$$I(X;Y) \stackrel{\text{def}}{=} H(X) - H(X|Y).$$

By the chain rule, $I(X;Y) = I(Y;X)$, and $I(X;Y) = 0$ iff X and Y are independent. Mutual information can be generalized by conditioning on another r.v. Z: $I(X;Y|Z) \stackrel{\text{def}}{=} H(X|Z) - H(X|Z,Y)$. Conditioning on Z may in general either increase or decrease mutual information between X and Y. Note that entropy of a r.v. only depends on the underlying probability distribution; thus any probability vector $\tilde{p} = (p_1, ..., p_n)$ ($p_i \geq 0$, $\sum_i p_i = 1$) determines a unique entropy value denoted $H(\tilde{p})$; we shall often abbreviate $H(p, 1-p)$ as $H(p)$.

3 The Model

We assume a countable set of *variables* $\mathcal{V} = \{x, y, ...\}$, a family of non-empty, finite *value-sets* $\mathfrak{U} \stackrel{\text{def}}{=} \{U, V, ...\}$, and a countable set of *names* $\mathcal{N} = \{a, b, ...\}$, partitioned into a family of *sorts* $\mathcal{S}, \mathcal{S}', ...$ We assume a function that maps each x to some $T \in \mathfrak{U} \cup \{\mathcal{S}, \mathcal{S}', ...\}$, written $x : T$, and say that x has *type* T. The inverse image of each T is infinite. These notations are extended to tuples as expected, e.g. for $\tilde{x} = (x_1, ..., x_n)$ and $\tilde{T} = (T_1, ..., T_n)$, $\tilde{x} : \tilde{T}$ means $x_1 : T_1, ..., x_n : T_n$. We let u, v be generic elements of a finite value-set. By slight abuse of notation, we sometimes denote by $\tilde{U}$ the cartesian product $U_1 \times \cdots \times U_n$.

An *evaluation* σ is a map from $\mathcal{V}$ to $\bigcup_{U \in \mathfrak{U}} U \cup \mathcal{N}$ that respects typing, that is, for each $x \in \text{dom}(\sigma)$, $x : T$ implies $\sigma(x) \in T$. We denote by $[\tilde{d}/\tilde{x}]$ the evaluation mapping $\tilde{x}$ to $\tilde{d}$ component-wise. By $t\sigma$, where t is a term over an arbitrary signature with free variables $\text{fv}(t) \subseteq \mathcal{V}$, we denote the result of replacing each free variable $x \in \text{dom}(\sigma) \cap \text{fv}(t)$ with $\sigma(x)$.

We assume a language of logical *formulae* $\phi, \psi, ...$ We leave the language unspecified, but assume it includes a first order calculus with variables $\mathcal{V}$, that function symbols include all values in $\mathfrak{U}$ and names as constants, and that the set of predicates includes equality $[x = y]$. We write $\mathfrak{U}, \mathcal{N} \models \phi$, or simply $\models \phi$, if for all evaluations σ s.t. $\text{dom}(\sigma) \supseteq \text{fv}(\phi)$, $\phi\sigma$ is valid (i.e. a tautology). We will often write $\phi(\tilde{x})$ to indicate that the free variables of ϕ are included in $\tilde{x}$, and in this case, abbreviate $\phi[\tilde{u}/\tilde{x}]$ as $\phi(\tilde{u})$.

The process language is a standard pi-calculus with variables and data values. We assume a countable set of *identifiers* $A, B, ...$ and use $e, e' ...$ to range over an unspecified set of *expressions*, that can be formed starting from variables, values and names. The syntax of processes $P, Q, ...$ is given below.

$$m ::= x \mid a$$

$$P, Q ::= \mathbf{0} \mid \tau.P \mid m(\tilde{x}).P \mid \overline{m}\tilde{e}.P \mid \phi P \mid P + P \mid (\nu b)P \mid P|P \mid A(\tilde{e}).$$

Each identifier A has an associated defining equation of the form $A(\tilde{x}) \stackrel{\text{def}}{=} P$. Input prefix $m(\tilde{x})$. and restriction (νb) are binders for $\tilde{x}$ and b, respectively, thus, notions of free variables (fv) and free names (fn) arise as expected. We identify processes up to alpha-equivalence. We assume a few constraints on the syntax above: $\tilde{x}$ is a tuple of distinct elements in input prefix and in $A(\tilde{x}) \stackrel{\mathrm{def}}{=} P$, and in the latter $\mathrm{fv}(P) \subseteq \tilde{x}$; ϕ is quantifier-free. We assume a fixed sorting system *à la* Milner. In particular, each sort $\mathcal{S}$ has an associated *sort object* $ob(\mathcal{S}) = (T_1,...,T_k)$ $(k \geq 0)$. Here, each T_i is either a sort or a value-set from the universe $\mathfrak{U}$. Informally, a process obeys this sorting system if in every input and output prefix, a name/variable m of sort $\mathcal{S}$ carries a tuple of objects of the sort specified by $ob(\mathcal{S})$; we omit the details that are standard. We let Π^{o} the set of processes (possibly containing free variables) obeying these conditions and Π^{c} the subset of *closed* processes. Notationally, we shall often omit trailing $\mathbf{0}$'s, writing e.g. $a.b.$ instead of $a.b.\mathbf{0}$, we shall write $\sum_{i=1}^{n} P_i$ for nondeterministic choice $P_1 + \cdots + P_n$, and let replication $!P$ denote the process defined by the equation: $!P \stackrel{\mathrm{def}}{=} P|!P$.

We assume over Π^{c} the standard *early* operational semantics of pi-calculus – see e.g. [14]. Let us just remind that in this form of semantics transitions are the form $P \xrightarrow{\mu} P'$, where μ is one of τ (invisible action), $a\tilde{d}$ (input action) or $(\nu\tilde{c})\overline{a}\tilde{d}$ with $\tilde{c} \subseteq \tilde{d}$ (output action) and $d ::= a \mid u$ (name or value). A few standard notations will also be used. In particular, for each *visible* (different from τ) action α, $P \stackrel{\alpha}{\Longrightarrow} P'$ means $P(\xrightarrow{\tau})^* \xrightarrow{\alpha} (\xrightarrow{\tau})^* P'$. This notation is extended to any sequence of visible actions s (i.e. a *trace*), $P \stackrel{s}{\Longrightarrow} P'$, as expected. Finally, $P \stackrel{s}{\Longrightarrow}$ means that there is P' s.t. $P \stackrel{s}{\Longrightarrow} P'$.

We let $\sim$ be a fixed equivalence relation over Π^{c}. We denote by $[Q]$ the equivalence class of a process Q. We assume $\sim$ is included in *trace equivalence* [2], includes *strong bisimulation* [14] and preserves all operators of the calculus, except possibly input prefix and unguarded nondeterministic choice. We introduce now the main concept of this section. An *open process* is a pair $(P,\tilde{x})$, written $P(\tilde{x})$, with $\tilde{x}$ a tuple of distinct variables of type $\tilde{U} \subseteq \mathfrak{U}$ and $P \in \Pi^{\mathrm{o}}$ such that $\mathrm{fv}(P) \subseteq \tilde{x}$; when no confusion arises, we shall abbreviate $P[^{\tilde{u}}/_{\tilde{x}}]$ as $P(\tilde{u})$ and $(P[^{\tilde{y}}/_{\tilde{x}}])(\tilde{y})$ as $P(\tilde{y})$ ($\tilde{y}$ a tuple of distinct variables.)

Definition 1 (open processes as random variables). *Let $P(\tilde{x})$ be an open process and $\tilde{X}$ a vector of random variables, with $\tilde{x} : \tilde{U}$ and $\tilde{X} : \tilde{U}$, for one and the same $\tilde{U}$. We denote by $P(\tilde{X})$ the random variable $F \circ \tilde{X}$, where $F = \lambda\tilde{u} \in \tilde{U}.[P[^{\tilde{u}}/_{\tilde{x}}]]$.*

Note that a sample of $P(\tilde{X})$ is an equivalence class of $\sim$.

Example 1. A PIN*-checking process can be defined as follows. Here, $x,z : 1..k$ for some integer k and x represents the secret code. The situation is modeled where an observer can freely interact with the checking process.*

$$Check(x) \stackrel{\mathrm{def}}{=} a(z).([z=x]\overline{ok}.Check(x) + [z \neq x]\overline{no}.Check(x))\,. \tag{4}$$

The range of the function $F : u \mapsto [Check(u)]$ has k distinct elements, as $u \neq u'$ implies $Check(u) \not\sim Check(u')$. As a consequence, if $X : 1..k$ is a random variable, the distribution of $P(X)$ mirrors exactly that of X. E.g., if X is uniformly distributed, then so is $P(X)$, i.e. the probability of each sample is $1/k$.

Note that, if $P(\tilde{u}) \sim Q(\tilde{u})$ for each $\tilde{u}$, then, for any $\tilde{X}$, $P(\tilde{X})$ and $Q(\tilde{X})$ are the same random variable. Another concept we shall rely upon is that of *most general boolean,* borrowed from [7, 3], that is, the most general condition under which two given open processes are equivalent.

Definition 2 **(mgb).** *Let $P(\tilde{x})$ and $Q(\tilde{y})$ be two open processes, with $\tilde{x} : \tilde{U}$ and $\tilde{y} : \tilde{V}$. We denote by* mgb$(P(\tilde{x}), Q(\tilde{y}))$ *a chosen formula $\phi(\tilde{x}, \tilde{y})$ s.t. for each $\tilde{u} \in \tilde{U}$ and $\tilde{v} \in \tilde{V}$: $P(\tilde{u}) \sim Q(\tilde{v})$ if and only if $\phi(\tilde{u}, \tilde{v})$ is true.*

It is worthwhile to notice that in many cases mgb's for pairs of open pi-processes can be automatically computed relying on *symbolic* transition semantics. Let us recall from [7, 3] that a symbolic transition also carries a logical formula: $P \xrightarrow{\mu,\phi} P'$. In [7], an algorithm is described to compute mgb's for pair of processes both having *finite* symbolic transition systems. Here, we will just assume that the logical language guarantees existence of mgb for any given pair of open processes.

4 Absolute Leakage

Throughout the section and unless otherwise stated, we let $P(\tilde{x}, \tilde{y})$ be an arbitrary open process, with $\tilde{x} : \tilde{U}$ and $\tilde{y} : \tilde{V}$, while $\tilde{X} : \tilde{U}$ and $\tilde{Y} : \tilde{V}$ are two arbitrary vectors of random variables, and $Z \stackrel{\text{def}}{=} P(\tilde{X}, \tilde{Y})$.

Definition 3 **(absolute leakage).** *The (absolute)* information leakage from $\tilde{X}$ to P given $\tilde{Y}$ *is $\mathcal{A}(P; \tilde{X} \mid \tilde{Y}) \stackrel{\text{def}}{=} I(\tilde{X}; Z|\tilde{Y}) = H(\tilde{X}|\tilde{Y}) - H(\tilde{X}|\tilde{Y}, Z)$.*

When $\tilde{Y}$ is empty, we simply write leakage as $\mathcal{A}(P; \tilde{X})$. A first useful fact says that leakage is nothing but the uncertainty about Z after observing $\tilde{Y}$. The proof is a simple application of the chain rule (2).

Lemma 1. *$\mathcal{A}(P; \tilde{X} \mid \tilde{Y}) = H(Z|\tilde{Y})$. In particular, if $\tilde{y}$ is empty, $\mathcal{A}(P; \tilde{X}) = H(Z)$.*

Example 2. The process $Check(x)$ defined in (4) leaks all *information about x. For example, if X is u.d on $1..k$ then $Z = P(X)$ is u.d. over a set of k samples. Hence $\mathcal{A}(Check; X) = H(Z) = \log k = H(X)$.*

Suppose now the adversary cannot interact freely with Check, but rather he observes the result of a user interacting with Check:

$$OneTry(x,y) \stackrel{\text{def}}{=} (\nu a)(Check(x)|\overline{a}y)\,. \tag{5}$$

Clearly, for any $X, Y : 1..k$, the range of the random variable $Z = OneTry(X,Y)$ has only two elements, that is $[\tau.\overline{ok}]$ and $[\tau.\overline{no}]$, that have probabilities $\Pr[X = Y]$ *and* $\Pr[X \neq Y]$, *respectively. In the case where X and Y are uniformly distributed and independent, these probabilities are $1/k$ and $1 - 1/k$, respectively. We are interested in $\mathcal{A}(OneTry; X \mid Y)$. Easy calculations show that Z and Y are in fact independent. For the sake of concreteness, let us assume $k = 10$; then we can compute absolute leakage as*

$$\mathcal{A}(OneTry; X \mid Y) = H(Z|Y) = H(Z) = H(\frac{1}{10}) \approx 0.469\,.$$

In this case, knowledge of Y brings no advantage to the adversary.

The next result is about composing leakage. Let us say that a n-holes context $C[\cdot,...,\cdot]$ preserves $\sim$ if whenever $P_i \sim P_i'$ for $1 \le i \le n$ then $C[P_1,...,P_n] \sim C[P_1',...,P_n']$. The following proposition states that leakage of a compound system cannot be greater than the sum of leakage of individual systems. The (simple) proof is based on inequality (3) plus the so called "data processing" inequality, saying that for any r.v. W and any function F of appropriate type, $H(F(W)) \le H(W)$.

Proposition 1 (compositionality). *Let $C[\cdot,...,\cdot]$ be a n-holes context that preserves $\sim$, and let $Q_i(\tilde{x},\tilde{y})$ be open processes, $1 \le i \le n$. Let $P(\tilde{x},\tilde{y}) = C[Q_1(\tilde{x},\tilde{y}),...,Q_n(\tilde{x},\tilde{y})]$. Then*

$$\mathcal{A}(P;\tilde{X} \mid \tilde{Y}) \le \sum_{i=1}^{n} \mathcal{A}(Q_i;\tilde{X} \mid \tilde{Y}). \quad (6)$$

For example, in the case of parallel composition, inequality (6) specializes to $\mathcal{A}(P|Q;\tilde{X} \mid \tilde{Y}) \le \mathcal{A}(P;\tilde{X} \mid \tilde{Y}) + \mathcal{A}(Q;\tilde{X} \mid \tilde{Y})$. The inequality implies that leakage is never increased by unary operators. In the case of replication !, this leads to the somewhat unexpected conclusion $\mathcal{A}(!P;\tilde{X} \mid \tilde{Y}) \le \mathcal{A}(P;\tilde{X} \mid \tilde{Y})$. Inequalities provided by (6) may hold strict or not, as shown below.

Example 3. *Consider $P(x) = ([x=0]a)|a$, where $x : \{0,1\}$, and X u.d. on the same set. Then $1 = \mathcal{A}(P;X) > \mathcal{A}(!P;X) = 0$. The reason for the latter equality is that for $v \in \{0,1\}$, $!P(v) \sim !a$, that is, the behaviour of $!P(x)$ does not depend on x, so $H(P(X)) = 0$.*

On the other hand, consider $P_1(x) = [x=2]a + [x=4]a$ and $P_2(x) = [x=1]b + [x=2]b$, where this time $x : 1..4$, and X is u.d. on the same set. Then $\mathcal{A}(P_1|P_2;X) = \mathcal{A}(P_1;X) + \mathcal{A}(P_2;X) = H(\frac{1}{2}) + H(\frac{1}{2}) = 2$.

Our next task is to investigate the situation of zero leakage. We start from Abadi and Gordon' definition of Secrecy, originally formulated in the setting of the spi-calculus [1]. According to the latter, a process $P(\tilde{x})$ keeps $\tilde{x}$ secret if the observable behaviour of $P(\tilde{x})$ does not depend on the actual values $\tilde{x}$ may take on. Partly motivated by the non-interference scenario [5, 16], where variables are partitioned into "low" and "high", we find it natural to generalize the definition of [1] to the case where the behaviour of P may also depend on further parameters $\tilde{y}$ known to the adversary.

Definition 4 (generalized secrecy). *We say that $P(\tilde{x},\tilde{y})$* keeps $\tilde{x}$ secret given $\tilde{y}$ *if, for each $\tilde{v} \in \tilde{V}$, and for each $\tilde{u} \in \tilde{U}$ and $\tilde{u}' \in \tilde{U}$, it holds $P(\tilde{u},\tilde{v}) \sim P(\tilde{u}',\tilde{v})$.*

The main result of the section states agreement of diverse notions of secrecy: functional (described above), quantitative (zero leakage) and logical (independence of mgb's from $\tilde{x}$). The latter appears to be more amenable to automatic checking, at least in those cases where the mgb can be computed. We also offer an "optimized" version of the quantitative notion, by which it is sufficient to check zero-leakage relatively to uniformly distributed and independent $\tilde{X}$ and $\tilde{Y}$.

Theorem 1 (secrecy). *Let $P(\tilde{x},\tilde{y})$ be an open process. The following assertions are equivalent:*

1. *$P(\tilde{x},\tilde{y})$ keeps $\tilde{x}$ secret given $\tilde{y}$.*
2. *$\mathcal{A}(P;\tilde{X}^* \mid \tilde{Y}^*) = 0$, for some $\tilde{X}^* : \tilde{U}$ and $\tilde{Y}^* : \tilde{V}$ uniformly distributed and independent.*

3. $\max_{\tilde{X}:\tilde{U},\tilde{Y}:\tilde{V}} \mathcal{A}(P;\tilde{X}\,|\,\tilde{Y}) = 0$.
4. $\phi \Leftrightarrow \exists \tilde{x}\tilde{x}'.\phi$, *where* $\phi = \mathrm{mgb}(P(\tilde{x},\tilde{y}),\ P(\tilde{x}',\tilde{y}'))$, *for* $\tilde{x}'$ *and* $\tilde{y}'$ *tuples of distinct variables disjoint from* $\tilde{x}$ *and* $\tilde{y}$, *but of the same type.*

Example 4. Consider the following process, where $x,y:1..4$*:*

$$Q(x,y) \stackrel{\text{def}}{=} (\nu c)(\overline{c}\,|\,\lfloor y=1\rceil c.\overline{a}) + \lfloor x=2\rceil \tau.\overline{a}\,.$$

It is immediate to see that Q *does not keep* x *secret, given* y*. E.g., if the adversary knows that* $y \neq 1$ *and observes the behaviour* $\lceil \tau.\overline{a}\rceil$ *then he can infer that* $x = 2$*. In fact, the* mgb *given by the theorem above is* $\phi = (\lfloor y=1\rceil \to (\lfloor y'=1\rceil \vee \lfloor x'=2\rceil)) \wedge (\lfloor y'=1\rceil \to (\lfloor y=1\rceil \vee \lfloor x=2\rceil))$*, and clearly,* $\phi \not\Leftrightarrow \exists xx'.\phi$*. As an example, for* X,Y *independent and u.d on* $1..4$*, the leakage from* X *to* Q *given* Y *can be computed as* $H(Z|Y) \approx 0.608$*. The process* $Q'(x,y) = Q(x,y) + \lfloor y \neq 1\rceil \tau.\overline{a}$ *keeps* x *secret given* y*.*

5 Rate of Leakage

We assume now an attacker can only conduct upon P repeated experiments, each yielding a binary[1] answer, say success or failure. We are interested in the number of communications that are *necessary* for the adversary to extract one bit of information about $\tilde{X}$ in this way.

In the rest of the section, we fix $\sim$ to be *weak trace* equivalence (aka *may testing* equivalence [4, 2]) written $\simeq$, and defined as: $P \simeq Q$ iff for each trace s, $P \stackrel{s}{\Longrightarrow}$ iff $Q \stackrel{s}{\Longrightarrow}$. For the sake of simplicity, we shall only consider processes where channels transport tuples of values, i.e. we ban name-passing. For the same reason, we shall assume that no side-information is available to the attacker, i.e. $\tilde{y}$ is empty. We plan to present the treatment of the most general case in a full version of the present paper. Throughout the section and unless otherwise stated, $P(\tilde{x})$, where $\tilde{x}:\tilde{U}$, denotes an arbitrary open pi-process, $\tilde{X}$ an arbitrary vector of random variables of type $\tilde{U}$ and Z is $P(\tilde{X})$. Recall that $\mathcal{A}(P;\tilde{X}) = H(Z)$.

Definitions and basic properties. Consistently with the testing equivalence framework [4, 2], we view an experiment E as a processes that, when composed in parallel with P, may succeed or not. Input on a distinct name ω, carrying no objects, is used to signal *success* to the adversary. Here, it is convenient to adjust the notion of composition ($\|$ below) to ensure that, in case of success, exactly one success action is reported to the adversary.

Definition 5 (experiments). *An* experiment E *is a closed process formed without using recursive definitions and possibly using the distinct success action* ω*.*

We say that a nonempty trace of visible actions s *is* successful *for* E *if* ω *does not occur in* s *and* $E \stackrel{s\cdot\omega}{\Longrightarrow}$.

For each E *and process* Q*, let us define* $Q\|E \stackrel{\text{def}}{=} (\nu\tilde{c},\omega')(P|E[^{\omega'}/_{\omega}]|\overline{\omega'}.\omega)$*, where* $\tilde{c} = \mathrm{fn}(Q,E)\setminus\{\omega\}$ *and* $\omega' \notin \mathrm{fn}(P,Q,\omega)$*.*

---

[1] We expect no significant change in the theory if k-ary answers, with $k > 2$ fixed, were instead considered.

Note that for each Q it must be either $Q \| E \simeq \mathbf{0}$ – meaning that E fails – or $Q \| E \simeq \omega.\mathbf{0}$ – meaning that E succeeds. Hence, for each E, we can define a binary random variable thus[2]

$$E^* \stackrel{\text{def}}{=} P(\tilde{X}) \| E .$$

Information on $\tilde{X}$ conveyed by E^* is $I(\tilde{X}; E^*) = H(E^*) - H(E^*|\tilde{X}) = H(E^*)$. This information is at most one bit. The rate notion of rate we are after should involve a ratio between this quantity of information and the *cost* of E. The following example shows the role played by non-determinism in extracting information, and provides some indications as to what we should intend by cost.

Example 5. Consider again Check(X), where this time X is u.d. over $1..k$, for some fixed even integer $k \geq 2$. An experiment E that extracts one bit out of $E \stackrel{\text{def}}{=} \sum_{d=1}^{k/2} \bar{a}d.ok.\omega$. An attacker can only observe the outcome of the interaction between Check and E, i.e. a sample of the r.v. $E^* = Check(X) \| E$. If action ω is observed, then it must be $X \leq k/2$; if action ω is not observed, then it must be $X > k/2$. Note that $I(X;E^*) = H(E^*) = H(\frac{1}{2}) = 1$.

The above example suggests that different successful traces of an experiment should be counted as different "trials" attempted by the attacker. The cost of each trial can be assumed to be proportional to its length as a trace. These considerations motivate the definition below.

Definition 6 (rate). *For each experiment E, define its* cost *as* $|E| \stackrel{\text{def}}{=} \Sigma\{|s| : s \text{ is succesful for } E\}$. *The* rate of P relative to $\tilde{X}$ *is*

$$\mathcal{R}(P;\tilde{X}) \stackrel{\text{def}}{=} \sup_{|E|>0} \frac{H(E^*)}{|E|} . \tag{7}$$

Our first result is an experiment-independent characterization of rate. In accordance with the may testing approach, this characterization is obtained in terms of observations of single traces. In what follows, given a trace of visible actions s, we consider the r.v. $P(\tilde{X}) \stackrel{s}{\Longrightarrow}$, which may yield *true* or *false*, and denote by p_s the probability[3] $\Pr[(P(\tilde{X}) \stackrel{s}{\Longrightarrow}) = true]$. Recall that for $0 \leq p \leq 1$, we denote by $H(p)$ the entropy of the distribution $(p, 1-p)$.

Proposition 2. *It holds that* $\mathcal{R}(P;\tilde{X}) = \sup_{|s|>0} \frac{H(p_s)}{|s|}$.

Example 6. Consider the process CheckOnce$(x) \stackrel{\text{def}}{=} a(z).([z=x]\overline{ok} + [z \neq x]\overline{no})$, where $x,z : 1..10$, and X u.d. on the same interval. It is immediate to verify that the ratio in the proposition above is maximized by any of $s = ad \cdot \overline{ok}$ or $s = ad \cdot \overline{no}$, for $d \in 1..10$. This yields $\mathcal{R}(CheckOnce;X) = H(\frac{1}{10})/2 \approx 0.234$.

---

[2] We would write $E^*(P)$ should any confusion about P arise.

[3] It is important to note that this definition does *not* induce a probability distribution on the set of traces; rather, it assigns each trace s a binary distribution $(p_s, 1-p_s)$.

The proposition above allows one, at least in principle, to compute the rate of any process having a finite symbolic transition system. In fact, relying on P's symbolic transition system, it is possible to compute, for any given trace s, a logical formula $\phi_s(\tilde{x})$ expressing the exact condition on $\tilde{x}$ under which $P(\tilde{x})$ can perform s (see [7, 3]). From these formulae it is easy to compute, or at least estimate with any degree of precision, the rate of P – we omit the details.

The next result explains the relationship between the notion of rate and absolute leakage. In particular, (a) establishes that $H(Z)$ is the maximal information that can be extracted by *repeated* binary experiments; and (b) provides a lower bound on the cost necessary to extract this information, in terms of the rate of P – thus providing a justification for the name "rate". For $\tilde{E} = (E_1, E_2, ..., E_n)$ a vector of experiments, write $|\tilde{E}| = |E_1| + \cdots + |E_n|$ for its cost, and $\tilde{E}^*$ for the vector of r.v. $(E_1^*, E_2^*, ..., E_n^*)$.

Proposition 3. *It holds that*

$$\begin{array}{ll} (a) & \mathcal{A}(P;\tilde{X}) = H(Z) = \max_{\tilde{E}} I(\tilde{X}\,;\tilde{E}^*) \\ (b) & \text{for each } \tilde{E},\;\; I(\tilde{X}\,;\tilde{E}^*) \leq |\tilde{E}| \cdot \mathcal{R}(P;\tilde{X})\ . \end{array}$$

Note in particular, that the cost of extracting *all* the available information $H(Z)$ cannot be less than $\frac{H(Z)}{\mathcal{R}(P;\tilde{X})}$. It is important to remark that processes with the same absolute leakage may well exhibit different rates. Here is a small example to illustrate this point.

Example 7. Let $P(x)$ and $Q(x)$, where $x : 0..3$, be defined as follows:

$$\begin{array}{l} P(x) = [x=0](a+b) + [x=1](b+c) + [x=2](c+d) + [x=3](d+a) \\ Q(x) = [x=0]a \qquad + [x=1]b \qquad + [x=2]c \qquad + [x=3]d\,. \end{array}$$

Assume X is u.d. over $0..3$. Both $P(X)$ and $Q(X)$ are u.d. on a domain of four elements (the four distinct equivalence classes $[P(i)]$, resp. $[Q(i)]$, for $i \in 0..3$). Hence leakage is $H(P(X)) = H(Q(X)) = H(X) = 2$ bits. On the other hand, each nonempty trace of P occurs with probability $1/2$, while each nonempty trace of Q occurs with probability $1/4$. Thus, by Proposition 2, $\mathcal{R}(P;X) = H(\frac{1}{2}) = 1$ and $\mathcal{R}(Q;X) = H(\frac{1}{4}) \approx 0.811$. Proposition 3(b) implies that gaining all information about X costs the attacker no less than 2 in the case of P, and no less than 3 in the case of Q. Indeed, a sequence of two (resp. three) one-action experiments is sufficient ($\bar{a}.\omega$, $\bar{b}.\omega$ for P and $\bar{a}.\omega$, $\bar{b}.\omega$, $\bar{c}.\omega$ for Q) to determine X.

Compositionality. It is possible to give upper bounds for the rate of a compound process in terms of the component expressions, in the vein of Proposition 1. Some of these upper bounds are rather crude (e.g. in the case of restriction), others are more sophisticated (e.g. $\mathcal{R}(\bar{a}e.P;\tilde{X}) \leq \max\{H([e(\tilde{X}) = v])),\ \mathcal{R}(P;\tilde{X})\})$ – we leave the details for the full version of the paper. Here, we concentrate on the rate of iterated processes. In order to define iteration, we have to first define sequential composition. Output on a distinct name *stop*, not carrying objects, is used to signal termination of a thread. Hence we define sequential composition as $P;Q \stackrel{\text{def}}{=} (\nu\ stop')(P[stop'/stop] \,|\, stop'.Q)$ (with $stop'$ fresh). This is not sequential composition in the usual sense, but it is equivalent in the context we are going to consider – see definition below. For any P, let iteration $*P$ be the process recursively defined by $*P \stackrel{\text{def}}{=} P; *P$. We show that, under suitable conditions, the rate of $*P$ is the same as P's. The condition below requires essentially that termination of a single thread in a process is equivalent to termination of the whole process.

Definition 7 (determinate processes). *Let Q be a closed process. We say that a trace s is* terminating for *Q if $Q \stackrel{s\cdot\overline{stop}}{\Longrightarrow}$. We say that Q is* determinate *if for every terminating trace s, whenever $Q \stackrel{s}{\Longrightarrow} Q'$ then $Q' \simeq \overline{stop}$. Finally, an open process $P(\tilde{x})$ is determinate if $\sum_{\tilde{u}\in\tilde{U}} P(\tilde{u})$ is determinate.*

We need another technical condition: let us say that Q is *stable* if whenever $Q \stackrel{\varepsilon}{\Longrightarrow} Q'$ (ε = empty trace) then $Q' \simeq Q$.

Theorem 2 (iteration rate). *Suppose that $P(\tilde{x})$ is determinate, and that for each $\tilde{u}$, $P(\tilde{u})$ is stable. Then $\mathcal{R}(*P;\tilde{X}) = \mathcal{R}(P;\tilde{X})$.*

Example 8. It is easy to check that $CheckOnceStop(x) \stackrel{\text{def}}{=} a(z).([z = x]\overline{ok}.\overline{stop} + [z \neq x]\overline{no}.\overline{stop})$ is determinate. ($x : 1..10$). Hence, being $Check(d) \simeq *CheckOnceStop(d)$, for every d, by Theorem 2 and Example 6 we have: $\mathcal{R}(Check;X) = \mathcal{R}(CheckOnceStop;X) = H(\frac{1}{10}) \approx 0.234$.

6 Computing Bounds on Absolute Leakage

In this section, we analyze the problem of bounding absolute leakage, from the position of someone – e.g. a developer – who has access to the process' code P, and for whom it is inexpensive to draw independent samples of the data $\tilde{X}$. For simplicity, we shall limit our discussion to the case where the side-information $\tilde{Y}$ is empty, so that absolute leakage reduces to $H(Z)$, where $Z = P(\tilde{X})$. The problem is nontrivial, because even for moderately complex P, the distribution of Z may be extremely difficult to compute or approximate. Methods commonly employed to estimate entropy in absence of an explicit description of distribution involve generation of sample sequences, long enough to let the underlying source's redundancy become appreciable. These methods are not applicable to our case, as operating on samples of Z is extremely expensive. Generation of even a *single* sample of Z – that is, an equivalence class, represented in some form or another – generally takes exponential time and space in the size of P.

We suggest a strategy that may work in practice in a number of cases, but we will not dwell on complexity-theoretical issues. For any discrete random variable W, its *index of coincidence* $IC(W)$ is defined as the probability that two independent experiments yield the same result, that is, denoting by U the type of W:

$$IC(W) \stackrel{\text{def}}{=} \sum_{u\in U} \left(\Pr[X = u]\right)^2.$$

Relationship of IC with Shannon entropy is seen by applying a well-known inequality of convex functions (Jensen's inequality, see e.g. [15]), which yields: $-\log IC(W) \leq H(W)$ (the quantity on the LHS is known as *Renyi's entropy of order 2*.) This inequality has been vastly generalized by Harremoës and Topsøe [8], who provide whole families of lower- and *upper*-bounds of Shannon entropy in terms of IC. These bounds are, in a certain technical sense, the "best possible" and provide fairly good estimates of $H(W)$ in many cases[4]. It remains to be seen how $IC(W)$ can be efficiently estimated in our case ($W = Z$). We show that this can be achieved via mgb's. Let $\phi(\tilde{x},\tilde{x}') \stackrel{\text{def}}{=} \text{mgb}\big(P(\tilde{x}),P(\tilde{x}')\big)$, where $\tilde{x}'$ is a tuple of distinct variables disjoint from $\tilde{x}$. By interpreting the boolean values true and false as 1 and 0, $\phi(\tilde{x},\tilde{x}')$ can be interpreted as a function $\tilde{U} \times \tilde{U} \to \{0,1\}$. We then have the following proposition, based on elementary reasoning on probabilities.

Proposition 4. *Let $\tilde{X}'$ be independent from $\tilde{X}$, but with the same type and distribution as $\tilde{X}$. Then $IC(Z) = E[\phi(\tilde{X},\tilde{X}')]$.*

The expectation $E[\phi(\tilde{X},\tilde{X}')]$ can be estimated with any desired precision via the law of large numbers: in practice, one draws several independent samples of $\phi(\tilde{X},\tilde{X}')$ and then takes the resulting arithmetical mean. The efficiency of this procedure depends on the distribution of $\tilde{X}$ and on the size of ϕ. Therefore, the problem of evaluating $IC(Z)$ can be reduced to the task of computing the formula ϕ, and possibly reducing its size by means of logical simplifications. Dedicated algorithms exist for that (see [7]) which are practical in many cases. Using this methodology, we have conducted some simple but very encouraging experiments on timing-dependent leakage in modular exponentiation algorithms (see e.g. [9]) that will be reported in the extended version of the paper.

7 Conclusions and Related Work

Results and proofs presented here carry over essentially unchanged to other calculi equipped with behavioral equivalences, such as the spi-calculus – except for those that depend on pi's symbolic semantics, like effective computation of leakage. The examples considered in the paper are admittedly a bit artificial. More realistic case-studies, possibly involving cryptography or probabilistic behaviour, are needed for assessing the model's scalability. In the leakage rate scenario, different notions of "cost" are also worthwhile to be investigated.

Early works on quantitative information flow are [13, 17, 6]. Volpano and Smith have later developed a quantified theory of non-interference for imperative programs, also giving a notion of rate [16], albeit not based on information theory. These approaches, like the one by Clark et al. [12], presuppose that computations produce some form or another of "result" , possibly with an associated probability distribution, in the sense already discussed in the introduction. A notable exception is represented by the recent work of Lowe [11]. There, quantitative non-interference for timed CSP is defined as the number of different "low" behaviours that a "high" user can induce on the process. This definition is shown to be in agreement with a qualitative notion of lack of information flow due to Focardi and Gorrieri [5]. A notion of rate is also introduced by taking time explicitly into account. These notions are not easily comparable to ours, due to the different goals and settings (secrecy vs. non-interference, untimed vs. timed.)

[4] As an example, in the case of binary distributions $(p, 1-p)$, an upper bound U can be given s.t. the ratio H/U lies between 1 and 0.9 for all distributions with $p \in [0.03, 0.97]$.

Acknowledgements. The reviewers comments have been very useful to improve on the presentation. Valentina Fedi has conducted several experiments with the metodology described in Section 6 as part of her MSc dissertation.

References

1. M. Abadi and A. Gordon. A calculus for cryptographic protocols: The Spi-calculus. *Information and Computation*, 148(1): 1-70, 1999.
2. M. Boreale and R. De Nicola. Testing equivalence for mobile processes. *Information and Computation*, 120(2): 279-303, 1995.
3. M. Boreale and R. De Nicola. A symbolic semantics for the pi-calculus. *Information and Computation*, 126(1): 34-52, 1996.
4. R. De Nicola and M.C.B. Hennessy. Testing equivalences for processes. *Theoretical Computer Science*, 34:83–133, 1984.
5. R. Focardi and R. Gorrieri. A classification of security properties. *Journal of Computer Security*, 3(1): 5-34, 1995).
6. J.W. Gray, III. Towards a mathematical foundation for information flow security. In *Proc. of 1991 IEEE Symposium on Research in Computer Security and Privacy*, 1991.
7. M.C.B. Hennessy and H. Lin. Symbolic bisimulations. *Theoretical Computer Science*, 138(2): 353-389, 1995.
8. P. Harremoës and F. Topsøe. Inequalities between entropy and index of coincidence derived from information diagrams. *IEEE Transactions on Information Theory*, 47(7): 2944-2960, 2001.
9. P. Kocher. Timing Attacks on Implementations of Diffie-Hellman, RSA, DSS, and Other Systems. *CRYPTO 1996*: 104-113, 1996.
10. P. Kocher, J. Jaffe and B. Jun. Differential Power Analysis. *CRYPTO 1999*: 388-397, 1999.
11. G. Lowe. Defining information flow quantity. *Journal of Computer Security*, 12(3-4): 619-653, 2004).
12. D. Clark, S. Hunt and P. Malacaria. Quantitative Analysis of the Leakage of Confidential Data. *Electr. Notes Theor. Comput. Sci.*, 59(3), 2001.
13. J. Millen. Covert channel capacity. In *Proc. of 1987 IEEE Symposium on Research in Computer Security and Privacy*, 1987.
14. D. Sangiorgi and D. Walker. *The pi-calculus: A Theory of Mobile Processes.* Cambridge University Press, 2001.
15. F. Topsøe. Basic concepts, identities and inequalities – the Toolkit of Information Theory. *Entropy*, 3:162–190, 2001. Also available at `http://www.math.ku.dk/~topsoe/toolkitfinal.pdf`.
16. D. Volpano and G. Smith. Verifying Secrets and Relative Secrecy. In *POPL 2000*, 268-276, 2000.
17. J.T. Wittbold and D. Johnson. Information flow in nondeterministic systems. In *Proc. of 1990 IEEE Symposium on Research in Computer Security and Privacy*, 1990.

Symbolic Protocol Analysis in Presence of a Homomorphism Operator and *Exclusive Or*⋆

Stéphanie Delaune[1,2], Pascal Lafourcade[2,3], Denis Lugiez[3], and Ralf Treinen[2]

[1] France Télécom, Division R&D
[2] LSV, CNRS UMR 8643, ENS de Cachan & INRIA Futurs project SECSI
[3] LIF, Université Aix-Marseille1 & CNRS UMR 6166

Abstract. Security of a cryptographic protocol for a bounded number of sessions is usually expressed as a symbolic trace reachability problem. We show that symbolic trace reachability for *well-defined* protocols is decidable in presence of the *exclusive or* theory in combination with the homomorphism axiom. These theories allow us to model basic properties of important cryptographic operators.

This trace reachability problem can be expressed as a system of symbolic deducibility constraints for a certain inference system describing the capabilities of the attacker. One main step of our proof consists in reducing deducibility constraints to constraints for deducibility in one step of the inference system. This constraint system, in turn, can be expressed as a system of quadratic equations of a particular form over $\mathbb{Z}/2\mathbb{Z}[h]$, the ring of polynomials in one indeterminate over the finite field $\mathbb{Z}/2\mathbb{Z}$. We show that satisfiability of such systems is decidable.

1 Introduction

Cryptographic protocols are small programs designed to ensure secure communication via a network that may be controlled by an attacker. They involve a high level of concurrency and are difficult to analyze by hand. These programs are linear sequences of *receive* and *send* instructions on a public network. A *passive* attacker may only listen to messages, while an *active* attacker may also pretend to be a protocol participant and forge messages according to a certain set of *intruder capabilities*.

The problem of deciding whether a protocol preserves the confidentiality of a message under any active attack is known to be undecidable in general (*e.g.* [11]). Several decidability results have been obtained under the assumption that the number of role instances is bounded, among others NP-completeness due to Rusinowitch and Turuani [17]. The idea of their algorithm is to guess a symbolic trace in which the messages are represented by terms containing variables. This symbolic trace corresponds to a concrete execution trace if the variables can be instantiated in such a way that at every moment a message received by an agent can in fact be deduced by the intruder from the messages seen before. Hence, verifying security of a protocol amounts to a non-deterministic guessing of the symbolic trace plus the resolution of a system of *deducibility constraints*. This result [17], as many others (e.g., [15]), relies on the

⋆ This work has been partly supported by the RNTL project PROUVÉ 03V360 and the ACI-SI Rossignol.

M. Bugliesi et al. (Eds.): ICALP 2006, Part II, LNCS 4052, pp. 132–143, 2006.

so-called *perfect cryptography assumption* which states that the cryptographic primitives (like encryption) are perfect and can be treated as black boxes. This assumption is unrealistic since some attacks exploit in a clever way the interaction between protocol rules and properties of cryptographic primitives. A more realistic approach is to take into account properties of the cryptographic primitives (see [4] for a survey). For the constraint based approach, this has been done for different equational theories [16, 8].

In this paper we study the equational theory ACUNh which is the combination of (h) the homomorphism axiom $h(x+y) = h(x) + h(y)$ with the *exclusive or* (ACUN) theory. These two equational theories model basic properties of important cryptographic primitives. Some protocols relying on these algebraic properties are described in [4]. *Exclusive or* is a basic building block in many symmetric encryption methods like DES or AES, or even used directly as an encryption method (Vernam encryption). Homomorphisms are ubiquitous in cryptography. For instance, the Wired Equivalent Privacy (WEP) protocol uses a checksum function C which has the homomorphism property over $+$, *i.e.* $C(x+y) = C(x) + C(y)$. Moreover, the homomorphism property over some binary operator appears in several encryption schemes (RSA, ElGamal ...) and is crucial in the field of electronic voting protocols [5]. Note that the recent result by Chevalier and Rusinowitch [2] for the combination of intruder theories can not be employed here to simply extend the known decidability result [1, 3] for ACUN since the theories ACUN and h share the symbol $+$. Furthermore, their result relies on a model which is different from ours in that it applies only to a restricted class of protocols.

Some results have already been obtained for the ACUNh theory [13, 6], but only for the case of a passive attacker. This algorithm for passive attacks is an important ingredient to the algorithm for active attacks developed in the present paper. Another important ingredient is ACUNh unification which has been shown decidable in [12]. However, for our procedure, we need to establish that unification in ACUNh is finitary, i.e. that every problem has a finite set of most general solutions. Our work is inspired by Millen and Shmatikov's approach [16] for the equational theory of Abelian groups. However, there are fundamental differences in the technical development.

Outline of the paper. We present our attacker model in Section 2, and the classes of constraint systems that we employ in our algorithm in Section 3. The proof of our main result (Theorem 1) proceeds in two steps: First we reduce satisfiability of deducibility constraints to satisfiability of constraints for one-step deducibility by a particular inference rule (Section 4). Second, we reduce satisfiability of these constraints to the satisfiability of a particular form of quadratic equations over the ring $\mathbb{Z}/2\mathbb{Z}[h]$, which we finally show to be decidable in Section 5 (satisfiability of quadratic equations over $\mathbb{Z}/2\mathbb{Z}[h]$, or for that matter $\mathbb{Z}$, is undecidable in general). Due to lack of space, proofs are omitted and can be found in [9].

2 Attacker Model

2.1 Inference System

The deduction capabilities of the intruder are formalized by the *Dolev-Yao model* [10]. We extend the intruder capabilities by equational reasoning modulo a given set E of equational axioms; we denote this intruder model by $\mathcal{I}_{\mathsf{DY+E}}$. In this paper, we consider the equational theory $\mathsf{E} = \mathsf{ACUNh}$ which consists of the well-known axioms of *exclusive or* in combination with a homomorphism symbol. More formally, ACUNh contains the following equations:

- Associativity, Commutativity (AC): $x + (y + z) = (x + y) + z$, $x + y = y + x$,
- Unit (U): $x + 0 = x$,
- Nilpotence (N): $x + x = 0$,
- homomorphism (h): $h(x + y) = h(x) + h(y)$.

We obtain the inference system described in Figure 1 where equational reasoning is taken into account through the normalization function $\downarrow$ associated to E. In the case of the ACUNh equational theory, the AC-convergent rewrite system is obtained by orienting from left to right the equations (U), (N), (h) and by adding the consequence $h(0) \rightarrow 0$ (see [13] for details). We omit the equality rule for AC and just work with equivalence classes modulo AC.

$$\text{Unpairing (UL)}\ \frac{T \vdash \langle u, v\rangle}{T \vdash u} \qquad \text{Compose (C)}\ \frac{T \vdash u_1\ \ldots\ T \vdash u_n}{T \vdash f(u_1, \ldots, u_n)}\ \text{with } f \in \mathcal{F} \smallsetminus \{+, h, 0\}$$

$$\text{Unpairing (UR)}\ \frac{T \vdash \langle u, v\rangle}{T \vdash v} \qquad \text{Context}(\mathsf{M_E})\ \frac{T \vdash u_1\ \ldots\ T \vdash u_n}{T \vdash C[u_1, \ldots, u_n]\downarrow}\ \text{with } C \text{ an E-context}$$

$$\text{Decryption (D)}\ \frac{T \vdash \{u\}_v \quad T \vdash v}{T \vdash u}$$

Fig. 1. Dolev-Yao Model Extended with an Equational Theory: $\mathcal{I}_{\mathsf{DY+E}}$

The intended meaning of a *sequent* $T \vdash u$ is that the intruder is able to deduce the term $u \in \mathcal{T}(\mathcal{F}, \mathcal{X})$ from the finite set of terms $T \subseteq \mathcal{T}(\mathcal{F}, \mathcal{X})$. As in the standard Dolev-Yao model, the intruder can compose new terms from known terms (C), he can decompose pairs (UL, UR), and he can decrypt ciphertexts provided that he can deduce the decryption key (D). Finally, the intruder may apply ($\mathsf{M_E}$) any E-context, *i.e.* term of the form $C[x_1, \ldots, x_n]$ with $C \in \mathcal{T}(\{0, +, h\}, \{x_1, \ldots, x_n\})$, to terms he already knows. Examples of instances of this rule are

$$\frac{T \vdash a + h(a) \qquad T \vdash b}{T \vdash a + h(h(h(a))) + h(b)}\ (\mathsf{M_E}) \qquad\qquad \frac{}{T \vdash 0}\ (\mathsf{M_E})$$

obtained with $C[x_1, x_2] = x_1 + h(x_1) + h(h(x_1)) + h(x_2)$, resp. $C[] = 0$.

The notation $h^n(t)$ represents the term t if $n = 0$ and $h(h^{n-1}(t))$ otherwise. Along this paper, we consider implicitly that terms are kept in normal form, *i.e.* we write u (resp. $u\sigma$) instead of $u\downarrow$ (resp. $u\sigma\downarrow$).

This deductive system is equivalent in deductive power to a variant of the system in which terms are not automatically normalized, but in which arbitrary equational proofs are allowed at any moment of the deduction (see [6, 13]). The inference system described in Figure 1 deals with symmetric encryption. However, it is not difficult to design a similar deduction system for asymmetric encryption and to extend the results of this paper to this new inference system.

2.2 Factors, Subterms

A term t is *standard* if and only if it is not of the form $f(t_1, \ldots, t_n)$ for some term $t_1, \ldots, t_n$ and some $f \in \{0, h, +\}$. In particular, every variable is a standard term.

Definition 1. *Let t be a term in normal form. We have $t = C[t_1, \ldots, t_n]$ for some standard terms $t_1, \ldots, t_n$ and an* E*-context C. The set $Fact_{\mathsf{E}}(t)$ of* factors *of t is defined by $Fact_{\mathsf{E}}(t) = \{t_1, \ldots, t_n\}$. The set $St_{\mathsf{E}}(t)$ of* subterms *of t is the smallest set such that:*

- $0, t \in St_{\mathsf{E}}(t)$,
- *if $f(t_1, \ldots, t_n) \in St_{\mathsf{E}}(t)$ is standard then $t_1, \ldots t_n \in St_{\mathsf{E}}(t)$,*
- *if $s \in St_{\mathsf{E}}(t)$ is not standard then $Fact_{\mathsf{E}}(s) \subseteq St_{\mathsf{E}}(t)$.*

Note that the set of factors is uniquely defined since equality is taken to be modulo AC. Note also that, by definition, 0 is not a standard term and the factors of any term are necessarily standard. We extend the notations $St_{\mathsf{E}}(\cdot)$ and $Fact_{\mathsf{E}}(\cdot)$ in a natural way to sets of terms.

Example 1. Let $t_1 = h^2(a)+b+x$ and $t_2 = h(\langle a, b\rangle)+x$, we get $Fact_{\mathsf{E}}(t_1) = \{a, b, x\}$, $St_{\mathsf{E}}(t_1) = \{t_1, a, b, x\}$, $Fact_{\mathsf{E}}(t_2) = \{\langle a, b\rangle, x\}$, $St_{\mathsf{E}}(t_2) = \{t_2, \langle a, b\rangle, a, b, x\}$.

2.3 Proofs

Definition 2. *A* proof *P of $T \vdash u$ is a finite tree such that*

- *the root of P is labeled with $T \vdash u$,*
- *every leaf of P labeled with $T \vdash v$ is such that $v \in T$,*
- *for every node of P labeled with $T \vdash v$ having n sons labeled with $T \vdash v_1, \ldots, T \vdash v_n$, there is an instance $\dfrac{T \vdash v_1 \ \ldots \ T \vdash v_n}{T \vdash v}$ (R) of an inference rule. If this node labeled with $T \vdash v$ is the root of P, we say that P* ends *with an instance of* (R).

Note that the terms in the proof are not necessarily ground. A proof P of $T \vdash u$ is *minimal* if there is no proof P' of $T \vdash u$ with less nodes than P.

Definition 3. *A term u is* R-one-step deducible *from a set of terms T in any of the following cases:*

- *$T \vdash u$ is a proof of $T \vdash u$ (i.e, $u \in T$ or $u = 0$),*
- *there exists $u_1, \ldots, u_n$ such that $\dfrac{T \vdash u_1 \ \ldots \ T \vdash u_n}{T \vdash u}$ (R) is a proof of $T \vdash u$.*

The term u is one-step deducible *from T if u is R-one-step deducible from T for some inference rule R.*

The following lemma, due to [6], shows that if there exists a proof of a sequent then there exists a "small" one.

Lemma 1. *A minimal proof P of $T \vdash u$ contains only terms in $St_{\mathsf{E}}(T \cup \{u\})$.*

3 Constraint Systems

3.1 Well-Defined Constraint Systems

It is well-known that the security problem of a protocol for a *fixed* number of parallel sessions reduces to the satisfiability of a constraint system (see, e.g. [1, 15]):

Definition 4. *A* constraint *(resp. one-step constraint,* M_E *constraint) is a sequent of the form* $T \Vdash u$ *(resp.* $T \Vdash_1 u$, $T \Vdash_{\mathsf{M}_\mathsf{E}} u$*) where* T *is a finite subset of* $\mathcal{T}(\mathcal{F}, \mathcal{X})$ *and* $u \in \mathcal{T}(\mathcal{F}, \mathcal{X})$*. We call* T *the* hypothesis set *of the constraint. A* system of constraints *is a sequence of constraints. A solution to a system* C *of constraints is a substitution* σ *such that:*

- *for every* $T \Vdash u \in C$ *there exists a proof of* $T\sigma \vdash u\sigma$*;*
- *for every* $T \Vdash_1 u \in C$ *the term* $u\sigma$ *is one-step deducible from* $T\sigma$*;*
- *for every* $T \Vdash_{\mathsf{M}_\mathsf{E}} u \in C$ *the term* $u\sigma$ *is* M_E*-one-step deducible from* $T\sigma$*.*

A solution σ to C is *non-collapsing* if for all $u, v \in St_\mathsf{E}(C) \setminus \mathcal{X}$ such that $u\sigma =_\mathsf{E} v\sigma$ then $u =_\mathsf{E} v$. If $\mathcal{F}'$ is a sub-signature of $\mathcal{F}$ then a solution σ to a constraint system is called a $\mathcal{F}'$-solution if $x\sigma \in \mathcal{T}(\mathcal{F}', \mathcal{X})$ for every $x \in dom(\sigma)$.

Note that, if σ is solution to a constraint $T \Vdash u$ (resp. one-step constraint, M_E constraint), then $\sigma\theta$ is also a solution to $T \Vdash u$ for every substitution θ.

Definition 5. *A constraint system* $C = \{T_i \Vdash u_i\}_{1 \le i \le k}$ *is* well-defined *if:*

1. (monotonicity) *for all* $i < k$: $T_i \subseteq T_{i+1}$,
2. (origination) *for all substitution* θ: $C\theta$ *satisfies the following requirement:*
   $\forall i \le k, \forall x \in vars(T_i\theta), \exists j < i$ *such that* $x \in vars(u_j\theta)$.

This notion of well-definedness, due to Millen and Shmatikov, is defined in an analogous way on systems of one-step (resp. M_E) constraints. In [16] they show that "reasonable" protocols, in which legitimate protocol participants only execute deterministic steps (up to the generation of random nonces) always lead to a well-defined constraint system. This notion is crucial for several steps of our algorithm.

Theorem 1. *The problem of deciding whether a well-defined constraint system has a solution in* $\mathcal{I}_{\mathsf{DY+E}}$*, where* $\mathsf{E} = \mathsf{ACUNh}$*, is decidable.*

The remainder of the paper is devoted to the proof of this result.

3.2 Conservative Solutions

Intuitively, a *conservative solution* to a constraint system is a solution which does not introduce any new structure. Lemma 2 states that it is sufficient to search for conservative solutions of a constraint system. Moreover, conservative solutions allow us to lift Lemma 1 to deducibility constraints (Lemma 3).

Definition 6. *Let* C *be a constraint system and* σ *a substitution,* σ *is* conservative *w.r.t.* C *if and only if for all* $x \in vars(C)$, $Fact_\mathsf{E}(x\sigma) \subseteq (St_\mathsf{E}(C) \setminus vars(C))\sigma$.

Lemma 2. *Let $\mathcal{C}$ be a well-defined constraint system. If there exists a solution σ to $\mathcal{C}$ then there exists a conservative one.*

Example 2. Consider the following well-defined constraint system $\mathcal{C}$ which is made up of two deducibility constraints: $a, h(b) \Vdash h(x)$ and $a, h(b), x \Vdash \langle a, b\rangle$. One solution is $\sigma = \{x \mapsto \langle a, a\rangle + b\}$. This solution is not conservative w.r.t. $\mathcal{C}$ since $Fact_{\mathsf{E}}(\langle a, a\rangle + b) = \{\langle a, a\rangle, b\}$, and $\langle a, a\rangle$ does not belong to $(St_{\mathsf{E}}(\mathcal{C})\backslash\{x\})\sigma$. However, as it is said in Lemma 2, there is a conservative solution: $\{x \mapsto b\}$.

Lemma 3. *Let σ be a conservative solution to $\mathcal{C} = \{C_1, \ldots, C_k\}$. For each $i \leq k$ there exists a proof of $C_i\sigma$ that involves only terms in $St_{\mathsf{E}}(\mathcal{C})\sigma$.*

4 From Constraints to M_{E} Constraints

We proceed in two non-deterministic steps to reduce the satisfiability of a constraint system to the satisfiability of a M_{E} constraint system:

1. From constraints to one-step constraints (see Lemma 4 and Figure 2).
2. From one-step constraints to M_{E} constraints (see Lemma 5).

```
Input: C = {T1 ⊩ u1, ..., Tk ⊩ uk}
  guess S ⊆ StE(C)
  for all s ∈ S, guess j(s) ∈ {1, ..., k}
  C':= ∅
  for i = 1 to k do
     let Si := {s | j(s) = i}
     choose a total ordering on Si (Si = {si^1, ..., si^ki})
     for j = 1 to ki do
        T := Ti ∪ S1 ... ∪ Si-1 ∪ {si^1,..., si^(j-1)}
        C':= C' ∪ {T ⊩1 si^j}
     end
     C':= C' ∪ {T ⊩1 ui}
  end
return C'
```

Fig. 2. Step 1: from constraints to one-step constraints

The idea of the first step is to guess among the subterms of $\mathcal{C}$ those that are going to be deduced by the intruder, and to insert each of them in some order into the constraint system. The completeness of this reduction step is essentially due to the existence of a conservative solution (Lemma 2) and to Lemma 3. In the resulting constraint system, every constraint can be solved by application of a single inference rule:

Lemma 4. *Let $\mathcal{C}$ be a well-defined system of constraints. Let $\mathscr{C}'$ be the set of constraint systems obtained by applying on $\mathcal{C}$ the algorithm described in Figure 2.*

1. $\mathscr{C}'$ *is a finite set of well-defined systems of one-step constraints.*
2. *If some* $C' \in \mathscr{C}'$ *has a solution then* C *has a solution.*
3. *If* C *has a conservative solution then some* $C' \in \mathscr{C}'$ *has a conservative solution.*

Lemma 5 allows us to reduce the satisfiability of a system of one-step constraints to the satisfiability of a system of $\mathsf{M_E}$ constraints. We first guess a set R of equalities between subterms. Then, we choose an E-unifier of R among the finite number of possibilities given by Theorem 2.

Theorem 2. *Unification in the theory* ACUNh *is finitary, and there exists an algorithm to compute a complete finite set* $mgu_{\mathsf{E}}(R)$ *of unifiers of any unification problem* R.

We write $T \vdash_{\mathsf{DY}} u$ if u is (R)-one step deducible from T where R is one of (D, UL, UR, C). It is trivial to decide whether $T \vdash_{\mathsf{DY}} u$ or not. We can now eliminate all constraints $T \Vdash_1 u$ for which $T \vdash_{\mathsf{DY}} u$ already holds.

Lemma 5. *Let* C *be a well-defined system of one-step constraints. Let*

$$\mathcal{P} = \{\bigwedge_{(s_1,s_2)\in S'} s_1 = s_2 \mid S' \subseteq St_{\mathsf{E}}(C)^2\}.$$

Let $R \in \mathcal{P}$ *and* $\theta \in mgu_{\mathsf{E}}(R)$. *Let* $C_\theta = \{T\theta \Vdash_{\mathsf{M_E}} u\theta \mid T \Vdash_1 u \in C \text{ and } T\theta \not\vdash_{\mathsf{DY}} u\theta\}$. *Let* $\mathscr{C}$ *be the set of constraint systems* C_θ *obtained this way.*

1. $\mathscr{C}$ *is a finite set of well-defined systems of* $\mathsf{M_E}$ *constraints.*
2. *If some* $C_\theta \in \mathscr{C}$ *has a solution then* C *has a solution.*
3. *If* C *has a conservative solution then some* $C_\theta \in \mathscr{C}$ *has a non-collapsing solution.*

Note that we can now restrict our attention to *non-collapsing* solutions, thanks to the fact that we have guessed the subterms that are identified by the solution.

5 Solving $\mathsf{M_E}$ Constraints

Now, we have to solve well-defined $\mathsf{M_E}$ constraint systems, where it is sufficient to look for non-collapsing solutions. In the remainder, we consider a $\mathsf{M_E}$ constraint system $C = \{T_1 \Vdash_{\mathsf{M_E}} u_1, \ldots, T_i \Vdash_{\mathsf{M_E}} u_k\}$ and we assume w.l.o.g. that the set of terms T_i is equal to $\{t_1, \ldots, t_{n+i-1}\}$.

A constraint system is called *factor-preserving* if all its factors appear for the first time in an hypotheses set of a constraint. More formally,

Definition 7. *A* $\mathsf{M_E}$ *constraint system is* factor-preserving *if for all* i, $1 \leq i \leq k$, *we have that* $Fact_{\mathsf{E}}(u_i) \setminus \mathcal{X} \subseteq \bigcup_{j=1}^{j=n+i-1} Fact_{\mathsf{E}}(t_j)$.

Example 3. The systems, $\langle a, b\rangle \Vdash_{\mathsf{M_E}} \langle x_1, x_2\rangle$ and $\langle\langle a, b\rangle, a\rangle \Vdash_{\mathsf{M_E}} \langle a, b\rangle$ are not factor-preserving. Note that the first one has no non-collapsing solution whereas the second one has no solution using the $\mathsf{M_E}$ inference rule only.

This notion is important to ensure that well-definedness is maintained when we abstract a constraint system by replacing factors by new constants (see Lemma 7). Fortunately, requiring factor preservation is not a restriction, since:

Lemma 6. *If a well-defined* $\mathsf{M_E}$*-constraint system* $\mathcal{C}$ *has a non-collapsing solution then it is factor-preserving.*

Factor preservation is of course trivial to check. We can hence suppose that the constraint system under consideration is factor-preserving, since if it is not then we conclude immediately by Lemma 6 that it has no non-collapsing solution.

5.1 Reducing the Signature

We will show in Lemma 7 that we can reduce the satisfiability of $\mathsf{M_E}$ constraint systems to the satisfiability of $\mathsf{M_E}$ constraint systems over a signature consisting only of 0, +, h, and a set of constants.

If $\rho : M \to N$ is a replacement, that is a bijection between two finite sets of terms M and N, then we denote for any term t by t^ρ the term obtained by replacing in t any top-most occurrence of a subterm $s \in M$ by $s\rho$. This extends in a natural way to constraint systems, and to substitutions.

Lemma 7. *Let* $\mathcal{C}$ *be a well-defined factor-preserving* $\mathsf{M_E}$ *constraint system and* $F = Fact_{\mathsf{E}}(\mathcal{C}) \setminus \mathcal{X}$. *Let* $\mathcal{F}_0$ *be a set of new constant symbols of the same cardinality as* F *and* $\rho : F \to \mathcal{F}_0$ *a bijection.*

1. $\mathcal{C}^\rho$ *is well-defined.*
2. $vars(\mathcal{C}^\rho) = vars(\mathcal{C})$.
3. *If* $\mathcal{C}$ *has a non-collapsing solution then* $\mathcal{C}^\rho$ *has a* $\mathcal{F}_0 \cup \{0, h, +\}$*-solution.*
4. *If* $\mathcal{C}^\rho$ *has a* $\mathcal{F}_0 \cup \{0, h, +\}$*-solution then* $\mathcal{C}$ *has a solution.*

As shown by the example below, well-definedness is not necessarily preserved under abstraction when the system is not factor-preserving.

Example 4. Abstraction of the system $a \Vdash_{\mathsf{M_E}} \langle x_1, x_2 \rangle;\ a, x_1, x_2 \Vdash_{\mathsf{M_E}} b$, which is not factor preserving, yields $a \Vdash_{\mathsf{M_E}} c_{new};\ a, x_1, x_2 \Vdash_{\mathsf{M_E}} b$, which is not well-defined.

5.2 Another Characterization of Well-Definedness

Let $\sum_{i=0}^{n} b_i h^i$ where $b_i \in \mathbb{Z}/2\mathbb{Z}$ be a polynomial of $\mathbb{Z}/2\mathbb{Z}[h]$. The product $\odot$ of a polynomial by a term is a term defined as follows:

$$(\sum_{i=0}^{n} b_i h^i) \odot t = \sum_{i=0 \mid b_i \neq 0}^{n} h^i(t)$$

For instance $(h^2+1)\odot(x+a) = h^2(x)+x+h^2(a)+a$. Every $t \in \mathcal{T}(\mathcal{F}, \{x_1, \ldots, x_p\})$ can be written $t^{x_1} \odot x_1 + \ldots t^{x_p} \odot x_p + t^0$ with t^{x_v} in $\mathbb{Z}/2\mathbb{Z}[h]$ and $Fact_{\mathsf{E}}(t^0) \cap \mathcal{X} = \emptyset$. We will denote with $\boldsymbol{t}$ the vector $(t^{x_1}, \ldots, t^{x_p})$.

Definition 8. *Let* $\mathcal{V} = \{v_1, \ldots, v_m\}$ *be a subset of* $\mathbb{Z}/2\mathbb{Z}[h]^n$. $\mathcal{V}$ *is* independent *if whenever there exist* $\alpha_i \in \mathbb{Z}/2\mathbb{Z}[h]$ *such that* $\alpha_1 v_1 + \ldots + \alpha_m v_m = 0$ *then* $\alpha_i = 0$ *for all* $1 \leq i \leq m$. *Otherwise* $\mathcal{V}$ *is* dependent.

Remember that we consider a constraint system $\mathcal{C} = \{t_1, \ldots, t_{n+i-1} \Vdash_{\mathsf{M_E}} u_i\}_{i=1,\ldots,k}$. The set $L = L_k$ of indexes of the so-called *defining constraints* is defined as follow. We set $L_0 = \emptyset$, and we define $L_{i+1} = L_i \cup \{i+1\}$ if $\{\boldsymbol{u_{i+1}}\} \cup \{\boldsymbol{u_j} \mid j \in L_i\}$ is independent, and $L_{i+1} = L_i$ otherwise. We note $\mathcal{B}_i = \{\boldsymbol{u_j} \mid j \in L, j \leq i\}$ and $\mathcal{B} = \mathcal{B}_k$. Lemma 8 gives an algebraic characterization of well-definedness in the special case of the signature $\mathcal{F}_0 \cup \{0, h, +\}$. Now, we have reduced the problem to this restricted signature (Lemma 7), we are going to use the following characterization in Section 5.3 to solve systems of equations over $\mathbb{Z}/2\mathbb{Z}[h]$.

Lemma 8. *A factor-preserving* $\mathsf{M_E}$ *constraint system* $\{t_1, \ldots, t_{n+i-1} \Vdash_{\mathsf{M_E}} u_i\}_{i=1,\ldots,k}$ *over the signature* $\{0, h, +\} \cup \mathcal{F}_0$ *is well-defined if, and only if, for every* $i \leq k$, *the set of vectors* $\{\boldsymbol{t_{n+i-1}}\} \cup \{\boldsymbol{u_j} \mid j \in L_i\}$ *is dependent.*

Intuitively, this is related to the fact that matching modulo ACUNh is essentially linear equation solving.

5.3 Solving $\mathsf{M_E}$ Constraint Systems over $\{0, h, +\} \cup \mathcal{F}_0$

We may by Lemma 6 assume that we have a factor-preserving $\mathsf{M_E}$ constraint system. By Lemma 7 satisfiability of such a system can be reduced to satisfiability of a $\mathsf{M_E}$ constraint system over a signature $\{0, h, +\} \cup \mathcal{F}_0$ where $\mathcal{F}_0$ is a finite set of constants. The characterization of Lemma 8 allows us to use the following well-known fact.

Fact 1. *Let A be a matrix $n \times m$ over $\mathbb{Z}/2\mathbb{Z}[h]$ such that the n row vectors are independent ($n \leq m$) then there exists $Q \in \mathbb{Z}/2\mathbb{Z}[h]$ such that*

$$\forall b \in \mathbb{Z}/2\mathbb{Z}[h]^n, \exists X \in \mathbb{Z}/2\mathbb{Z}[h]^m \;\; A \cdot X = Q \cdot b \tag{1}$$

Moreover, such a coefficient Q is computable as a determinant of a submatrix of A.

We denote Q_{max} the coefficient Q which satisfies the equation (1) for the matrix $\mathcal{B}$.

Example 5. (running example) To illustrate our procedure, we consider the following well-defined $\mathsf{M_E}$ constraint system:

$$\begin{array}{ll} h(a) + a, b + h^2(a) & \Vdash_{\mathsf{M_E}} h(x_1) + h^2(x_2) \\ h(a) + a, b + h^2(a), x_1 + h(x_2) & \Vdash_{\mathsf{M_E}} x_1 + a \\ h(a) + a, b + h^2(a), x_1 + h(x_2), h(x_1) + h(a) & \Vdash_{\mathsf{M_E}} h(x_1) + h^2(x_2) + x_1 + a \end{array}$$

We have $\boldsymbol{u_1} = (h, h^2)$, $\boldsymbol{u_2} = (1, 0)$ and $\boldsymbol{u_3} = (1+h, h^2)$. The algorithm returns $L = \{1, 2\}$ and we obtain $Q_{max} = det(\boldsymbol{u_1}, \boldsymbol{u_2}) = h^2$.

Satisfiability of such an $\mathsf{M_E}$ constraint system $\mathcal{C}$ is equivalent to the satisfiability of the following system $\mathcal{S}$ of equations between terms. The variables $z[i,j]$, called *context variables*, take their value in $\mathbb{Z}/2\mathbb{Z}[h]$. Let $\mathcal{Z} = \{z[i,j] \mid 1 \leq i \leq k, 1 \leq j \leq n+i-1\}$.

$$\begin{array}{l} z[1,1] \odot t_1 + \ldots + z[1,n] \odot t_n = u_1 \\ z[2,1] \odot t_1 + \ldots + z[2,n] \odot t_n + z[2,n+1] \odot t_{n+1} = u_2 \\ \vdots \\ z[p,1] \odot t_1 + \ldots + z[p,n] \odot t_n + \ldots + z[p,n+p-1] \odot t_{n+k-1} = u_k \end{array}$$

Example 6. (running example) Let $t_1 = h(a) + a$ and $t_2 = b + h^2(a)$.

$$\begin{aligned}
&z[1,1] \odot t_1 + z[1,2] \odot t_2 = h(x_1) + h^2(x_2) \\
&z[2,1] \odot t_1 + z[2,2] \odot t_2 + z[2,3] \odot (x_1 + h(x_2)) = x_1 + a \\
&z[3,1] \odot t_1 + z[3,2] \odot t_2 + z[3,3] \odot (x_1 + h(x_2)) + z[3,4] \odot (h(x_1) + h(a)) \\
&\qquad\qquad\qquad\qquad\qquad\qquad\qquad\qquad = h(x_1) + h^2(x_2) + x_1 + a
\end{aligned}$$

Definition 9. *Let $\mathcal{C}$ be a well-defined $\mathsf{M_E}$ constraint system over the signature $\{0, h, +\} \cup \mathcal{F}_0$ and $\mathcal{S}(\mathcal{C})$ be the system of equations obtained from $\mathcal{C}$. A solution to $\mathcal{S}(\mathcal{C})$ is a couple $(\rho : \mathcal{Z} \mapsto \mathbb{Z}/2\mathbb{Z}[h], \theta : vars(\mathcal{C}) \mapsto \mathcal{T}(\{0, h, +\} \cup \mathcal{F}_0))$ such that all the equations of $\mathcal{S}(\mathcal{C})\rho\theta$ are satisfied.*

We split the *context variables* $\mathcal{Z}$ into two parts, those which stem from L and the others. More formally, $\mathcal{Z}_L = \{z[i,j] \mid i \in L \text{ and } 1 \leq j < n + i\}$.

A polynomial $P = \sum_{i=0}^{i=n} p_i h^i$ ($p_n \neq 0$) is *smaller* than $Q = \sum_{i=0}^{i=m} q_i h^i$ ($q_m \neq 0$), written $P < Q$, if either $n < m$, or $P \neq Q$, $n = m$ and $p_i < q_i$ for the greatest i with $p_i \neq q_i$.

Fact 2. *Given any polynomial $P \in \mathbb{Z}/2\mathbb{Z}[h]$, there is only a finite number of polynomials which are smaller (w.r.t. $<$) than P.*

The following Lemma is the crucial point in the proof of Lemma 10.

Lemma 9. *Let $\mathcal{S}(\mathcal{C})$ be a system of equations obtained from a well-defined $\mathsf{M_E}$ constraint system $\mathcal{C}$ over the signature $\{0, h, +\} \cup \mathcal{F}_0$. If $\mathcal{S}(\mathcal{C})$ has a solution then there exists σ a solution to $\mathcal{S}(\mathcal{C})$ such that for all $z \in \mathcal{Z}_L$, $0 \leq z\sigma < Q_{max}$.*

The proof of this lemma proceeds by induction on the number of variables in Z_L.

Lemma 10. *Given $\mathcal{C}$ a well-defined $\mathsf{M_E}$constraint system. It is decidable whether $\mathcal{S}(\mathcal{C})$ has a solution.*

Example 7 (running example). Thanks to Lemma 9, we know that $z[1,1]$, $z[1,2]$, $z[2,1]$, $z[2,2]$ and $z[2,3]$ are bounded by h^2, the value of Q_{max}. We choose $\rho_1 = \{z[1,1] \mapsto 0; z[1,2] \mapsto h; z[2,1] \mapsto h + 1; z[2,2] \mapsto 1; z[2,3] \mapsto 0\}$. We do the replacement on the two first equations:

$$\begin{aligned}
&h \odot (b + h^2(a)) = h(x_1) + h^2(x_2) \\
&(h+1) \odot (h(a) + a) + 1 \odot (b + h^2(a)) = x_1 + a
\end{aligned}$$

This completely determines the value of x_1 and x_2: $\theta = \{x_1 \mapsto b, x_2 \mapsto h(a)\}$. Lastly, we can apply the substitution θ on the third equation to obtain:

$$\begin{aligned}
&z[3,1] \odot (h(a) + a) + z[3,2] \odot (b + h^2(a)) + z[3,3] \odot (b + h^2(a)) + \\
&\qquad\qquad\qquad\qquad z[3,4] \odot (h(b) + h(a)) = h(b) + h^3(a) + b + a
\end{aligned}$$

Since this system is linear it is easy to decide whether it has solution.

Let $\rho_2 = \{z[3,1] \mapsto h+1; z[3,2] \mapsto h+1; z[3,3] \mapsto 0; z[3,4] \mapsto 0\}$. The couple $(\rho_1 \cup \rho_2, \theta)$ is a solution to the system of equations described in Example 6.

Now, we are able to prove our main result as stated in Section 3.

Theorem 1. *The problem of deciding whether a well-defined constraint system has a solution in $\mathcal{I}_{\mathsf{DY+E}}$, where $\mathsf{E} = \mathsf{ACUNh}$, is decidable.*

Proof. The procedure described along the paper is sound and complete.

Soundness. Let $\mathcal{C}_1$ be some factor-preserving $\mathsf{M_E}$-constraint system obtained by applying the first part of our procedure on $\mathcal{C}$, a well-defined constraint system. Thanks to Lemma 4 and 5, $\mathcal{C}_1$ is well-defined since $\mathcal{C}$ is well-defined. Let $\mathcal{C}_2$ be the constraint system obtained from $\mathcal{C}_1$ by replacing all factors by different constants. $\mathcal{C}_2$ is well-defined thanks to Lemma 7. Assume that $\mathcal{S}(\mathcal{C}_2)$ (the system of equations associated to $\mathcal{C}_2$) has a solution. We easily deduce that $\mathcal{C}_2$ has a solution, hence by Lemma 7 that $\mathcal{C}_1$ has a solution, and by Lemma 4 and 5 that $\mathcal{C}$ has a solution.

Completeness. Assume that σ is a solution to $\mathcal{C}$. Thanks to Lemma 2, we can assume that σ is conservative w.r.t. $\mathcal{C}$. Let $\mathscr{C}'$ be the finite set of well-defined one-step constraint systems obtained by applying the algorithm described in Section 4 on $\mathcal{C}$. By Lemma 4, we know that there exists $\mathcal{C}' \in \mathscr{C}'$ such that σ is a conservative solution of $\mathcal{C}'$. By Lemma 5, we know that there exists $\mathcal{C}_\theta$ a well-defined $\mathsf{M_E}$-constraint system which has a non-collapsing solution. Hence, $\mathcal{C}_\theta$ is factor-preserving due to Lemma 6. By Lemma 7, $\mathcal{C}_\theta^\rho$ has solution over $\{0, h, +\} \cup \mathcal{F}_0$. Then, Lemma 10 allows us to conclude. □

6 Conclusion

Our solution for solving deducibility constraints is general enough to hold in related equational theories since it relies on general algebraic concepts. In particular, our technique generalizes previous results for the case of the *exclusive or* equational theory ACUN [1, 3] (context variables take values in $\mathbb{Z}/2\mathbb{Z}$) and the theory of Abelian groups AG [16] (contexts are in $\mathbb{Z}$). However, our technique does not apply to the case AGh of the extension of Abelian groups with a homomorphism since then the contexts are in $\mathbb{Z}[h]$, and Fact 2 does not hold. In fact it has recently been shown that this case is undecidable [7].

Despite a superficial similarity between our algorithm and the one of [16], our procedure to reduce $\mathsf{M_E}$-constraints (*cf.* Section 5) to a special class of quadratic equations is different. In particular it makes use of our novel algebraic characterization of well-defined constraint systems. Furthermore, our procedure to solve a particular form of quadratic equations in polynomials over the finite field $\mathbb{Z}/2\mathbb{Z}[h]$ is different from the one proposed in [16].

An open question is the case of an encryption algorithm distributing over *exclusive or*. Although the case of a passive intruder is decidable in this framework [14], the case of an active intruder seems quite intricate since it amounts to having an infinite number of distinct homomorphisms (one for each term used as a key).

References

1. Y. Chevalier, R. Küsters, M. Rusinowitch, and M. Turuani. An NP decision procedure for protocol insecurity with XOR. In *Proc. of 18th Annual IEEE Symposium on Logic in Computer Science (LICS'03)*, pages 261–270. IEEE Comp. Soc. Press, 2003.
2. Y. Chevalier and M. Rusinowitch. Combining intruder theories. In *Proc. 32nd International Colloquium on Automata, Languages and Programming (ICALP'05)*, volume 3580 of *LNCS*, pages 639–651. Springer, 2005.
3. H. Comon-Lundh and V. Shmatikov. Intruder deductions, constraint solving and insecurity decision in presence of exclusive or. In *Proc. of 18th Annual IEEE Symposium on Logic in Computer Science (LICS'03)*, pages 271–280. IEEE Comp. Soc. Press, 2003.
4. V. Cortier, S. Delaune, and P. Lafourcade. A survey of algebraic properties used in cryptographic protocols. *Journal of Computer Security*, 14(1):1–43, 2006.
5. R. Cramer, R. Gennaro, and B. Schoenmakers. A secure and optimally efficient multi-authority election scheme. In *Proc. International Conference on the Theory and Application of Cryptographic Techniques*, volume 1233 of *LNCS*, pages 103–118, 1997.
6. S. Delaune. Easy intruder deduction problems with homomorphisms. *Information Processing Letters*, 97(6):213–218, 2006.
7. S. Delaune. An undecidability result for AGh. Research Report LSV-06-02, LSV, ENS Cachan, France, 2006.
8. S. Delaune and F. Jacquemard. A decision procedure for the verification of security protocols with explicit destructors. In *Proc. of 11th ACM Conference on Computer and Communications Security (CCS'04)*, pages 278–287. ACM Press, 2004.
9. S. Delaune, P. Lafourcade, D. Lugiez, and R. Treinen. Symbolic protocol analysis in presence of a homomorphism operator and *exclusive or*. Research Report LSV-05-20, Cachan, 2005.
10. D. Dolev and A. Yao. On the security of public key protocols. In *Proc. of the 22nd Symp. on Foundations of Computer Science*, pages 350–357. IEEE Comp. Soc. Press, 1981.
11. N. Durgin, P. Lincoln, J. Mitchell, and A. Scedrov. Undecidability of bounded security protocols. In *Proc. Workshop on formal methods in security protocols*, 1999.
12. Q. Guo, P. Narendran, and D. A. Wolfram. Complexity of nilpotent unification and matching problems. *Information and Computation*, 162(1-2):3–23, 2000.
13. P. Lafourcade, D. Lugiez, and R. Treinen. Intruder deduction for AC-like equational theories with homomorphisms. In *Proc. of 16th Int. Conference on Rewriting Techniques and Applications (RTA'05)*, volume 3467 of *LNCS*, pages 308–322. Springer, 2005.
14. P. Lafourcade, D. Lugiez, and R. Treinen. Intruder deduction for the equational theory of exclusive-or with distributive encryption. Research Report LSV-05-19, ENS Cachan, 2005.
15. J. Millen and V. Shmatikov. Constraint solving for bounded-process cryptographic protocol analysis. In *Proc. of 8th ACM Conference on Computer and Communications Security (CCS'01)*. ACM Press, 2001.
16. J. Millen and V. Shmatikov. Symbolic protocol analysis with an Abelian group operator or Diffie-Hellman exponentiation. *Journal of Computer Security*, 13(3):515 – 564, 2005.
17. M. Rusinowitch and M. Turuani. Protocol insecurity with a finite number of sessions, composed keys is NP-complete. *Theoretical Computer Science*, 1-3(299):451–475, 2003.

Generalized Compact Knapsacks Are Collision Resistant*

Vadim Lyubashevsky and Daniele Micciancio

University of California, San Diego
9500 Gilman Drive, La Jolla, CA 92093-0404, USA
{vlyubash, daniele}@cs.ucsd.edu

Abstract. In (Micciancio, FOCS 2002), it was proved that solving the generalized compact knapsack problem *on the average* is as hard as solving certain *worst-case* problems for cyclic lattices. This result immediately yielded very efficient one-way functions whose security was based on worst-case hardness assumptions. In this work, we show that, while the function proposed by Micciancio is not collision resistant, it can be easily modified to achieve collision resistance under essentially the same complexity assumptions on cyclic lattices. Our modified function is obtained as a special case of a more general result, which yields efficient collision-resistant hash functions based on the worst-case hardness of various new problems. These include new problems from algebraic number theory as well as classic lattice problems (e.g., the shortest vector problem) over *ideal lattices*, a class of lattices that includes cyclic lattices as a special case.

1 Introduction

Ever since Ajtai's discovery of a function whose average-case hardness can be proved based on worst-case complexity assumptions about lattices [2], the possibility of building cryptographic functions whose security is based on worst-case problems has been very alluring. Ajtai's initial discovery [2] and subsequent developments [5, 15, 17] are very interesting from a theoretical point of view because they are essentially the only problems for which such a worst-case / average-case connection is known. Unfortunately, the cryptographic functions proposed in these works are not efficient enough to be practical. The source of impracticality is the use of lattices, which are described as $n \times n$ integer matrices. This results in cryptographic functions with key size and computation time at least quadratic in the security parameter n.

A step in the direction of creating efficient cryptographic functions based on worst-case hardness was taken by Micciancio [14]. He showed how to create a family of efficiently computable *one-way functions*, namely, the generalized

* The full version of this extended abstract appears in ECCC TR05-142. Research supported by NSF CAREER 0093029 and NSF ITR 0313241.

M. Bugliesi et al. (Eds.): ICALP 2006, Part II, LNCS 4052, pp. 144–155, 2006.

compact knapsack functions, whose security is based on a certain problem for a particular class of lattices, called cyclic lattices. These lattices admit a much more compact representation than general ones, and the resulting functions can be described and evaluated in time almost linear in n. However, one-wayness is a rather weak security property, interesting mostly from a theoretical point of view, because it is sufficient to prove the existence (via polynomial time, but rather impractical, constructions) of other cryptographic primitives, like commitment schemes, digital signatures, and private-key encryption. By contrast, the (inefficient) functions based on general lattices considered in [2, 5, 15, 17] are collision-resistant hash functions, a much more useful cryptographic primitive.

In this work, we take the next step in creating efficient cryptographic functions based on worst-case assumptions. We show how to create efficient, collision-resistant *hash functions* whose security is based on standard lattice problems for *ideal lattices* (i.e., lattices that can be described as ideals of certain polynomial rings). With current hash functions that are not based on any hardness assumptions, but used in practice, being broken [23, 24, 4], we believe that it may be an appropriate time to consider using efficient hash functions which do have an underlying hardness assumption, especially worst-case ones.

Our contributions and comparison with related work. The generalized knapsack problem is the following: given m random elements $a_1, \ldots, a_m$ in a ring R, and a target $t \in R$, find $z_1, \ldots, z_m \in D$ such that $\sum a_i z_i = t$, where D is some fixed subset of R. In [14], it was shown that for appropriate choices of R and D, the generalized compact knapsack problem is a one-way function with security based on the worst-case hardness of problems for lattices that can be represented as ideals in the ring $\mathbb{Z}[x]/\langle x^n - 1\rangle$ (i.e. cyclic lattices). In this work, we show how to construct collision-resistant hash functions based on the hardness of problems for lattices that can be represented as ideals in the ring $\mathbb{Z}[x]/\langle f\rangle$, where f can be one of infinitely many polynomials, including $x^n - 1$. Thus our result has two desirable features: it weakens the complexity assumption while strengthening the cryptographic primitive. As in [14], our functions are an instance of the generalized compact knapsack problem, but with ring R and subset D instantiated in a different way. The way we change ring R and subset D is simple, but essential, as we can show that the generalized compact knapsack instances considered in [14] are not collision resistant.

Concurrently with, and independently from our work, Peikert and Rosen [18] have shown, using very similar techniques, that the one-way function in [14] is not collision resistant and showed how to construct collision-resistant hash functions based on the hardness of finding the shortest vector for lattices which correspond to ideals in the ring $\mathbb{Z}[x]/\langle x^n - 1\rangle$. While our more general result is interesting from a purely theoretical standpoint, it turns out that choices of certain f other than $x^n - 1$ result in somewhat better hash functions, making our generalization also of practical use. Also, our hardness assumptions are formulated in a way that leads to natural connections with algebraic number theory, and we are able to relate our complexity assumptions to problems from that area. We believe that this will further our understanding of ideal lattices.

There have been many proposed cryptographic primitives whose hardness relied on the knapsack problem (e.g., [13, 7, 6]), but attacks against them (e.g., [21, 11, 22]) rendered the primitives impractical. These attacks, however, were applied to a group-based knapsack problem, and it is unclear how to apply them to our ring-based one. Also, none of those primitives had a reduction to worst-case instances of lattice problems, and, to the best of our knowledge, there are no known efficient algorithms that are able to solve lattice problems in the worst case (such as shortest vector) for lattices of dimension ≈ 100. Of course, the hardness of our primitive is based on worst-case problems for *ideal* lattices, and very little is known about these. Still, currently there appear to be no algorithms able to take advantage of the ring structure that these lattices possess (see[14] for a discussion of known algorithms for cyclic lattices). Determining the worst-case hardness of lattice problems for ideal lattices is a very interesting open problem.

The ring-based cryptosystem NTRU [10] uses lattices that are similar to ours. While that cryptosystem has no known security proofs (not even one based on average-case assumptions), it has resisted attacks. This is perhaps due to the inherent hardness of ring-based cryptographic constructions that are used in [10] as well as in our work. While we only construct a hash function, our work may be viewed as a strong justification for using such ring based constructions. Our hope is that we have taken another step in the direction of constructing provably secure and *efficient* cryptosystems based on worst case hardness of lattice problems.

The hash function. We now give an informal description of the hash function families that we will be proving collision resistant. Given a ring $R = \mathbb{Z}_p[x]/\langle f \rangle$, where $f \in \mathbb{Z}[x]$ is a monic, irreducible polynomial of degree n and p is an integer of order roughly n^2, generate m random elements $a_1, \ldots, a_m \in R$, where m is a constant. The ordered m-tuple $h = (a_1, \ldots, a_m) \in R^m$ is our hash function. It will map elements in D^m, where D is a strategically chosen subset of R, to R. For an element $b = (b_1, \ldots, b_m) \in D^m$, the hash is $h(b) = \sum_{i=1}^{m} a_i \cdot b_i$. Notice that the size of the key (the hash function) is $O(mn \log p) = O(n \log n)$, and the operation $a_i \cdot b_i$ can be done in time $O(n \log n \log \log n)$ by using the fast Fourier transform, for appropriate choice of the polynomial f. Since m is a constant, hashing requires time $O(n \log n \log \log n)$. To prove that our hash function family is collision resistant, we will show that if there is a polynomial-time algorithm that succeeds with non-negligible probability in finding $b \neq b' \in D^m$ such that $h(b) = h(b')$, for a randomly chosen hash function $h \in R^m$, then a certain problem called the "shortest polynomial problem" is solvable in polynomial time for *every* ideal of the ring $\mathbb{Z}[x]/\langle f \rangle$. We then show that the shortest polynomial problem is equivalent to some lattice and algebraic number theory problems.

Paper outline. Our main result and techniques rely on a connection between lattices and ideals of certain rings, which we describe in section 3. In section 4, we define the worst case problems on which we will be basing the security of our hash function. We formally define the hash function families in section 5.1 and show the worst-case to average-case reduction in section 5.2.

2 Preliminaries

2.1 Algebra

Let $\mathbb{Z}[x]$ and $\mathbb{R}[x]$ be the sets of polynomials with integer and real coefficients respectively. We identify polynomials (of degree $< n$) with the corresponding n-dimensional vectors having the coefficients of the polynomial as coordinates. We define the ℓ_p norm $\|g(x)\|_p$ of $g(x) \in \mathbb{Z}[x]$ as the norm of the corresponding vector, and the product of two n-dimensional vectors $\mathbf{x} \cdot \mathbf{y}$ as the $(2n-1)$-dimensional vector associated to the product of the corresponding polynomials.

Let R be a ring. The smallest ideal of R containing a subset $S \subseteq R$ is denoted $\langle S \rangle$. Much of our work deals with the rings $\mathbb{Z}[x]/\langle f \rangle$ where f is monic and irreducible. When f is a monic polynomial of degree n, every equivalence class $(g + \langle f \rangle) \in (\mathbb{Z}[x]/\langle f \rangle)$ has a unique representative $g' \in (g + \langle f \rangle)$ of degree less than n. This representative is denoted $(g \bmod f)$ and can be efficiently computed using the standard division algorithm. We endow the ring $\mathbb{Z}[x]/\langle f \rangle$ with the (infinity) norm $\|(g + \langle f \rangle)\|_f = \|g \bmod f\|_\infty$. Notice that the function $\|\cdot\|_f$ is well defined (i.e., it does not depend on the choice of representative g) and it is indeed a norm (i.e., it satisfies the positivity and triangle inequality properties). As shorthand, we will sometimes write $\|g\|_f$ instead of $\|g + \langle f \rangle\|_f$. Also, whenever there is no confusion from context, instead of writing $g + \langle f \rangle$ for elements of $\mathbb{Z}[x]/\langle f \rangle$, we just write g.

2.2 Lattices

An n-dimensional *integer lattice* is a subgroup of $\mathbb{Z}^n$ generated by linearly independent vectors $\mathbf{b}_1, \ldots, \mathbf{b}_n \in \mathbb{Z}^n$. The set of vectors $\mathbf{b}_1, \ldots, \mathbf{b}_n$ is called a *basis* for the lattice, and can be compactly represented by the matrix $\mathbf{B}$ having the basis vectors as columns. The lattice generated by $\mathbf{B}$ is denoted $\mathcal{L}(\mathbf{B})$. The dual of this lattice, denoted $\mathcal{L}(\mathbf{B})^*$, is the lattice generated by the matrix $\mathbf{B}^{-T}$, and consists of all vectors that have integer scalar product with all lattice vectors. For any basis $\mathbf{B}$, we define the fundamental parallelepiped $\mathcal{P}(\mathbf{B}) = \{\mathbf{Bx} \colon \forall i.0 \le x_i < 1\}$. Sampling random lattice points from the fundamental parallelepiped associated to a given sublattice can be done in polynomial time [16, Proposition 8.2].

The *minimum distance* of a lattice $\mathcal{L}(\mathbf{B})$ is the minimum distance between any two (distinct) lattice points and equals the length of the shortest nonzero lattice vector. The minimum distance can be defined with respect to any norm. For any $p \ge 1$, the ℓ_p norm of a vector $\mathbf{x}$ is defined by $\|\mathbf{x}\|_p = \sqrt[p]{\sum_i |x_i|^p}$ and the corresponding minimum distance is denoted

$$\lambda_1^p(\mathcal{L}(\mathbf{B})) = \min\{\|\mathbf{x} - \mathbf{y}\|_p : \mathbf{x} \neq \mathbf{y} \in \mathcal{L}(\mathbf{B})\} = \min\{\|\mathbf{x}\|_p : \mathbf{x} \in \mathcal{L}(\mathbf{B}) \setminus \{\mathbf{0}\}\}.$$

Each norm gives rise to a corresponding computational problem SVP_γ^p (the γ-approximate *Shortest Vector Problem* in the ℓ_p norm): given a lattice $\mathcal{L}(\mathbf{B})$, find a nonzero vector $\mathbf{v} \in \mathcal{L}(\mathbf{B})$ such that $\|\mathbf{v}\|_p \le \gamma \lambda_1^p(\mathcal{L}(\mathbf{B}))$. We also consider the restriction of SVP to specific classes of lattices. The restriction of SVP to a class of lattices Λ is denoted Λ-SVP. (E.g, [14] considers *Cyclic-SVP*).

The notion of minimum distance can be generalized to define the ith successive minimum (in the ℓ_p norm) $\lambda_i^p(\mathcal{L}(\mathbf{B}))$ as the smallest radius r such that the closed sphere $\bar{\mathcal{B}}_p(r) = \{\mathbf{x}\colon \|\mathbf{x}\|_p \leq r\}$ contains i linearly independent lattice points: $\lambda_i^p(\mathcal{L}(\mathbf{B})) = \min\{r : \dim(\text{span}(\mathcal{L}(\mathbf{B}) \cap \bar{\mathcal{B}}_p(r))) \geq i\}$.

In this work, we focus on the infinity norm $\|\mathbf{x}\|_\infty = \lim_{p\to\infty} \|\mathbf{x}\|_p = \max_i |x_i|$ since it is the most natural and convenient norm when dealing with polynomials, but most of our results are easily translated to other norms as well. The shortest vector problem in the infinity norm SVP_γ^∞ was shown to be NP-hard for factor up to $\gamma(n) = n^{1/\log\log n}$ by Dinur [8]. The asymptotically fastest algorithm for computing the shortest vector exactly takes time $2^{O(n)}$ [3] and the best polynomial time algorithm approximates the shortest vector to within a factor of $2^{O(\frac{n \log\log n}{\log n})}$ [3],[20],[12]. It is conjectured that approximating SVP to within a polynomial factor is a hard problem, although it is shown that (under standard complexity assumptions) for small polynomial factors it is not NP-hard [1], [9].

2.3 Gaussian Distribution

Let X and Y be random variables over a set A with probability density functions δ_X and δ_Y. We denote the statistical distance between X and Y by $\Delta(X,Y)$.

For any vectors $\mathbf{c}, \mathbf{x}$ and any $s > 0$, let $\rho_{s,\mathbf{c}}(\mathbf{x}) = e^{-\pi\|(\mathbf{x}-\mathbf{c})/s\|^2}$ be a Gaussian function centered in $\mathbf{c}$ scaled by a factor of s. The total measure associated to $\rho_{s,\mathbf{c}}$ is $\int_{\mathbf{x}\in\mathbb{R}^n} \rho_{s,\mathbf{c}}(\mathbf{x})d\mathbf{x} = s^n$. So, $\int_{\mathbf{x}\in\mathbb{R}^n} (\rho_{s,\mathbf{c}}(\mathbf{x})/s^n)d\mathbf{x} = 1$ and $\rho_{s,\mathbf{c}}/s^n$ is a probability density function. The distribution $\rho_{s,\mathbf{c}}/s^n$ can be efficiently approximated using standard techniques (see [17]), so in the rest of the paper we make the simplifying assumption that we can sample from $\rho_{s,\mathbf{c}}/s^n$ exactly and work with real numbers.

Functions are extended to sets in the usual way; e.g., $\rho_{s,\mathbf{c}}(A) = \sum_{\mathbf{x}\in A} \rho_{s,\mathbf{c}}(\mathbf{x})$ for any countable set A. For any s, $\mathbf{c}$ and lattice Λ, define the discrete probability distribution (over the lattice Λ) $D_{\Lambda,s,\mathbf{c}}(\mathbf{x}) = \frac{\rho_{s,\mathbf{c}}(\mathbf{x})}{\rho_{s,\mathbf{c}}(\Lambda)}$, where $\mathbf{x} \in \Lambda$. Intuitively, $D_{\Lambda,s,\mathbf{c}}$ is the conditional probability[1] that $(\rho_{s,\mathbf{c}}/s^n) = \mathbf{x}$ given $(\rho_{s,\mathbf{c}}/s^n) \in \Lambda$. For brevity, we sometimes omit s or $\mathbf{c}$ from the notation $\rho_{s,\mathbf{c}}$ and $D_{\Lambda,s,\mathbf{c}}$. When $\mathbf{c}$ or s are not specified, we assume that they are the origin and 1 respectively.

In [17] Gaussian distributions are used to define a new lattice invariant (called the *smoothing parameter*) defined below, and many important properties of this parameter are established. The following properties will be used in this paper.

Definition 1. *For an n-dimensional lattice Λ, and positive real $\epsilon > 0$, the smoothing parameter $\eta_\epsilon(\Lambda)$ is the smallest s such that $\rho_{1/s}(\Lambda^* \setminus \{\mathbf{0}\}) \leq \epsilon$.*

Lemma 1 ([17, Lemma 4.1]). *Let $\rho_s/s^n \bmod \mathbf{B}$ be the distribution obtained by sampling a point according to the probability density function ρ_s/s^n and reducing the result modulo $\mathbf{B}$. For any lattice $\mathcal{L}(\mathbf{B})$, the statistical distance between $\rho_s/s^n \bmod \mathbf{B}$ and the uniform distribution over $\mathcal{P}(\mathbf{B})$ is at most $\frac{1}{2}\rho_{1/s}(\mathcal{L}(\mathbf{B})^* \setminus \{\mathbf{0}\})$. In particular, if $s \geq \eta_\epsilon(\mathcal{L}(\mathbf{B}))$, then the distance $\Delta(\rho_s/s^n \bmod \mathbf{B}, U(\mathcal{P}(\mathbf{B})))$ is at most $\epsilon/2$.*

[1] We are conditioning on an event that has probability 0; this can be made rigorous by standard techniques.

Lemma 2 ([17, Lemma 3.3]). *For any n-dimensional lattice Λ and positive real $\epsilon > 0$,*

$$\eta_\epsilon(\Lambda) \leq \sqrt{\frac{\ln(2n(1+1/\epsilon))}{\pi}} \cdot \lambda_n^2(\Lambda) \leq \sqrt{\frac{n\ln(2n(1+1/\epsilon))}{\pi}} \cdot \lambda_n^\infty(\Lambda).$$

3 Generalized Compact Knapsacks and Ideal Lattices

In [14], Micciancio introduced the following generalization of the compact knapsack problem. Let R be a ring, $D \subset R$ a subset, and $m \geq 1$ a positive integer. The generalized knapsack function family $\mathcal{H}(R, D, m)$ is the collection of all functions $\mathfrak{h}_\mathbf{a} : D^m \to R$ indexed by $\mathbf{a} \in R^m$ mapping $\mathbf{b} \in D^m$ to $\mathfrak{h}_\mathbf{a}(\mathbf{b}) = \sum_{i=1}^m b_i \cdot a_i \in R$.

For any function family $\mathcal{H}$, define the problem $Col_\mathcal{H}$ as follows: given a function $\mathfrak{h} \in \mathcal{H}$, find a collision, i.e., a pair of inputs $\mathbf{b}, \mathbf{c} \in D^m$ such that $\mathbf{b} \neq \mathbf{c}$ and $\mathfrak{h}(\mathbf{b}) = \mathfrak{h}(\mathbf{c})$. If there is no polynomial time algorithm that can solve $Col_\mathcal{H}$ with non-negligible probability when given an $\mathfrak{h}$ which is distributed uniformly at random in $\mathcal{H}$, then we say that $\mathcal{H}$ is a collision resistant family of hash functions.

Let $f \in \mathbb{Z}[x]$ be a monic polynomial of degree n, and consider the quotient ring $\mathbb{Z}[x]/\langle f \rangle$. Using the standard set of representatives $\{(g \bmod f) : g \in \mathbb{Z}[x]\}$, and our identification of polynomials with vectors, the quotient ring $\mathbb{Z}[x]/\langle f \rangle$ is isomorphic (as an additive group) to the integer lattice $\mathbb{Z}^n$, and any ideal $I \subseteq \mathbb{Z}[x]/\langle f \rangle$ defines a corresponding integer sublattice $\mathcal{L}(I) \subseteq \mathbb{Z}^n$. Notice that not every integer lattice $\mathcal{L}(\mathbf{B}) \subseteq \mathbb{Z}^n$ can be represented this way.[2] We define ideal lattices as lattices that admit such a representation.

Definition 2. *An* ideal lattice *is an integer lattice $\mathcal{L}(\mathbf{B}) \subseteq \mathbb{Z}^n$ such that $\mathcal{L}(\mathbf{B}) = \{g \bmod f : g \in I\}$ for some monic polynomial f of degree n and ideal $I \subseteq \mathbb{Z}[x]/\langle f \rangle$.*

It turns out that the relevant properties of f for the resulting function to be collision resistant are:

- f should be irreducible.
- the ring norm $\|g\|_f$ is not much bigger than $\|g\|_\infty$ for any polynomial g, in a quantitative sense to be explained later.

The first property implies that every ideal of the ring $\mathbb{Z}[x]/\langle f \rangle$ defines a full-rank lattice in $\mathbb{Z}^n$ and plays a fundamental role in our proofs.

Lemma 3. *Every ideal I of $\mathbb{Z}[x]/\langle f \rangle$, where f is a monic, irreducible integer polynomial of degree n, is isomorphic to a full-rank lattice in $\mathbb{Z}^n$.*

The second property affects the strength of our security proofs: the smaller the ratio $\|g\|_f / \|g\|_\infty$ is, the harder to break our functions seems to be. We elaborate

[2] Take, for example, the 2-dimensional lattice generated by the vectors $(2, 0)$ and $(0, 1)$ (or in terms of polynomials, by $2x$ and 1). This lattice cannot be represented by an ideal, because any ideal containing 1 must also contain the polynomial $1 \cdot x$, but the vector $(1, 0)$ (corresponding to the polynomial x) does not belong to the lattice.

on the second property by defining a quantitative parameter (the expansion factor) that captures the relation between $\|\cdot\|_\infty$ and $\|\cdot\|_f$.

3.1 The Expansion Factor

Notice that when we reduce a polynomial g modulo f, the maximum coefficient of g can increase by quite a bit, and thus $||g||_f$ could be a lot bigger than $||g||_\infty$. For example if $f = x^n - 2x^{n-1}$, then $x^{2n} \equiv 2^{n+1}x^{n-1}$ modulo f. On the other hand, if $f = x^n - 1$, we can never have such an exponential growth of coefficients. We capture this property of f by defining the *expansion factor* of f as

$$EF(f,k) = \max_{g\in\mathbb{Z}[x],deg(g)\le k(deg(f)-1)} ||g||_f/||g||_\infty$$

The below theorem gives tight bounds for the expansion factor of certain polynomials that have small expansion factors.

Theorem 1. *(1)$EF(x^{n-1} + x^{n-2} + \ldots + 1, k) \le 2k$. (2)$EF(x^n + 1, k) \le k$.*

In the full version of this work, we also provide some general formulas that upper bound the expansion factors of arbitrary polynomials.

4 Worst Case Problems

In this section we define the worst case problems and provide reductions among them. Because of the correspondence between ideals and integer lattices, we can use the successive minima notation used for lattices for ideals as well. So for any ideal I of $\mathbb{Z}[x]/\langle f\rangle$, where f is a monic integer polynomial, we'll define $\lambda_i^p(I)$ to be $\lambda_i^p(\mathcal{L}(I))$.

Definition 3. *In the approximate Shortest Polynomial Problem ($SPP_\gamma(I)$), we are given an ideal $I \subseteq \mathbb{Z}[x]/\langle f\rangle$ where f is a monic polynomial of degree n, and we are asked to find a $g \in I$ such that $g \neq 0$ and $||g||_f \le \gamma\lambda_1^\infty(I)$.*

As for the shortest vector problem, we can consider the restriction of SPP to specific classes of ideals. We will write f-SPP for SPP restricted to ideals of the ring $\mathbb{Z}[x]/\langle f\rangle$. The f-SPP problem for any monic, irreducible f is the main worst-case problem of this work, as it is the problem upon which the security of our hash functions will be based. Since SPP is a new problem whose hardness has not been explored, we show that other better-known problems can be reduced to it. If we denote by $\mathcal{I}(f)$ the set of lattices that are isomorphic (as additive groups) to ideals of $\mathbb{Z}[x]/\langle f\rangle$ where f is monic, then there's a straightforward reduction from $\mathcal{I}(f)$-SVP_γ to f-SPP_γ (and also the other way around).

Lattices in the class $\mathcal{I}(x^n - 1)$ (cyclic lattices) do not fall into the category of lattices that are isomorphic to ideals of $\mathbb{Z}[x]/\langle f\rangle$ for an irreducible f (since $x^n - 1$ is not irreducible). In the full version, we give a reduction from $(x^n - 1)$-$SPP_{2\gamma}$ to $(x^{n-1} + x^{n-2} + \ldots + 1)$-$SPP_\gamma$, thus establishing the security of hash functions based on the hardness of the shortest vector problem for cyclic lattices of prime dimension. Another problem that we reduce to SPP is the problem of finding complex numbers with small conjugates in ideals of integers of certain number fields. This problem and the reduction is described in detail in the full version.

Now we state a lemma which shows that if I is an ideal of $\mathbb{Z}[x]/\langle f\rangle$ where f is monic and irreducible, then $\lambda_n^\infty(I)$ cannot be much bigger than $\lambda_1^\infty(I)$.

Lemma 4. *For all ideals I of $\mathbb{Z}[x]/\langle f\rangle$ where f is a monic, irreducible polynomial of degree n, we have $\lambda_n^\infty(I) \leq EF(f,2)\lambda_1^\infty(I)$*

Proof. Let g be a polynomial in I of degree less than n such that $||g||_\infty = \lambda_1^\infty(I)$. Then consider the polynomials $g, gx, \ldots, gx^{n-1}$. By lemma 3, the polynomials $g, gx, \ldots, gx^{n-1}$ are linearly independent. And since the maximum degree of any of these polynomials is $2n-2$, $||gx^i||_f \leq EF(f,2)||gx^i||_\infty \leq EF(f,2)||g||_\infty = EF(f,2)\lambda_1^\infty(I)$ for all $0 \leq i \leq n-1$.

We now define the incremental version of SPP. In this version, we are not looking for the shortest polynomial, but for a polynomial that is smaller than the one given to us. We will be reducing this problem to the average-case problem.

Definition 4. *In the approximate Incremental Shortest Polynomial Problem ($IncSPP_\gamma(I,g)$), we are given I and a $g \in I$ such that $||g||_f > \gamma\lambda_1^\infty(I)$ and are asked to return an $h \in I$ such that $||h||_f \neq 0$ and $||h||_f \leq ||g||_f/2$.*

We define the restricted version of $IncSPP$ in the same was as the restricted version for SPP.

Lemma 5. *There is a polynomial time reduction from f-SPP_γ to f-$IncSPP_\gamma$.*

5 Collision Resistant Hash Function Families

In this section, we define families of hash functions which are instances of generalized compact knapsacks and prove that finding collisions in these hash functions is at least as hard as solving the approximate shortest polynomial problem.

5.1 The Hash Function Families

The hash function family $\mathcal{H}(R, D, m)$ we will be considering in this paper will be instances of generalized knapsacks instantiated as follows. Let $f \in \mathbb{Z}[x]$ be an irreducible, monic polynomial of degree n with expansion factor $EF(f,3) \leq \mathcal{E}$. Let the ring R be $\mathbb{Z}_p[x]/\langle f\rangle$ for some integer p, and let $D = \{g \in R : ||g||_f \leq d\}$ for some positive integer d. The family of functions $\mathcal{H}$ is mapping elements from D^m to R where $|D^m| = (2d+1)^{nm}$ and $|R| = p^n$. So if $m > \frac{\log p}{\log 2d}$, then $\mathcal{H}$ will be a family of functions that have collisions. We will only be interested in such families. We will now state the main theorem:

Theorem 2. *Let $\mathcal{H}$ be a hash function family as above with $m > \frac{\log p}{\log 2d}$ and $p > 2\mathcal{E}dmn^{1.5}\log n$. Then, for $\gamma = 8\mathcal{E}^2 dmn\log^2 n$, there is a polynomial time reduction from f-$SPP_\gamma(I)$ for any I to $Col_{\mathcal{H}}(\mathfrak{h})$ where $\mathfrak{h}$ is chosen uniformly at random from $\mathcal{H}$.*

The proof of the theorem is given in the next subsection. To achieve the best approximation factor for f-$SPP_\gamma(I)$, we can set $m = \Theta(\log n, \log \mathcal{E})$ and $d = \Theta(\log n)$. This makes $\gamma = \tilde{O}(n)\mathcal{E}^2$. For purposes of being able to compute the function faster, though, it is useful to have m be smaller than $\Theta(\log n)$. It is possible to make m constant at the expense of being able to approximate f-SPP only to a factor of $\gamma = \tilde{O}(n^{1+\delta})\mathcal{E}^2$. To be able to set m to a constant, we can set $d = n^\delta$ for some $\delta > 0$. Then we can set $m = \frac{\log(\mathcal{E})}{\delta \log n} + \frac{2+\delta}{\delta} + o(1)$.

In order to get the "tightest" reduction, we should pick an f such that the bound $\mathcal{E}$ on f's expansion factor is small. In theorem 1, we show that we can set $\mathcal{E}$ to be 3 and 6 for polynomials of the form $x^n + 1$ and $x^{n-1} + x^{n-2} + \ldots + 1$ respectively. The polynomial $x^n + 1$ is irreducible whenever n is a power of 2 and $x^{n-1} + x^{n-2} + \ldots + 1$ is irreducible for prime n, so those are good choices for f. Among other possible f's with constant bounds for $EF(f, 3)$ are polynomials of the form $x^n \pm x \pm 1$ (see [19, Chapter 2.3.2] for sufficient conditions for the irreducibility of polynomials of this form).

Some sample instantiations of the hash function. If we let $f = x^{126} + \ldots + x + 1, n = 126, d = 8, m = 8$, and $p \approx 2^{23}$, then our hash function is mapping $\log(|2d|^{mn}) = 4032$ bits to $\log|R_p| = \log(p^n) \approx 2900$ bits. If we want to base our hardness assumption on lattices of higher dimension, we can instantiate $f = x^{256} + \ldots + x + 1, n = 126, p \approx 2^{25}, d = 8, m = 8$, and our hash function will be mapping 8192 bits to $\log(p^n) \approx 6400$ bits. If we instead let $f = x^{256} + 1$, we can let p be half as small (because the expansion factor for $x^n + 1$ is half of the expansion factor of $x^n + \ldots + x + 1$) and thus we will be mapping 8192 bits to around 6150 bits.

5.2 Finding Collisions Is Hard

In this section, we will provide the proof of theorem 2. Let $\mathcal{H}$ be the family of hash functions described in the last subsection with $p > 2\mathcal{E}dmn^{1.5}\log n$. We will show that if one can solve in polynomial time, with non-negligible probability, the problem $Col_\mathcal{H}(\mathfrak{h})$ where $\mathfrak{h}$ is chosen uniformly at random from $\mathcal{H}$, then one can also solve f-$IncSPP_\gamma(I, g)$ for any ideal I for $\gamma = 8\mathcal{E}^2 dmn \log^2 n$. And since by lemma 5, f-$SPP_\gamma(I) \leq f$-$IncSPP_\gamma(I, g)$, we will have a reduction from f-$SPP_\gamma(I)$ for any I to $Col_\mathcal{H}(\mathfrak{h})$ for a random $\mathfrak{h}$. Let $\mathcal{C}$ be an oracle such that when given a uniformly random $\mathfrak{h} \in \mathcal{H}$, $\mathcal{C}(\mathfrak{h})$ returns a solution to $Col_\mathcal{H}(\mathfrak{h})$ with non-negligible probability in polynomial time. Now we proceed with giving an algorithm for f-$IncSPP_\gamma$ when given access to oracle $\mathcal{C}$.

Given: I, $g \in I$ such that $g \neq 0$ and $||g||_f > 8\mathcal{E}^2 dmn \log^2 n \lambda_1^\infty(I)$
Find: $h \in I$, such that $h \neq 0$ and $||h||_f \leq ||g||_f/2$.

Without loss of generality, assume that g has degree less than n and thus $||g||_\infty = ||g||_f$. So we are looking for an h such that $||h||_f \leq ||g||_\infty/2$. In this section, it will be helpful to think of ideals I and $\langle g \rangle$ as subgroups of $\mathbb{Z}^n$ (or equivalently, as sublattices of $\mathbb{Z}^n$). Define a number s as

$$s = \frac{||g||_\infty}{8\mathcal{E}\sqrt{n}\log ndm} \geq \mathcal{E}\sqrt{n}(\log n)\lambda_1^\infty(I) \geq \sqrt{n}(\log n)\lambda_n^\infty(I) \geq \eta_\epsilon(I)$$

for $\epsilon = (\log n)^{-2\log n}$, where the last inequality follows by lemma 2, and the inequality before that is due to lemma 4. By lemma 1, it follows that if $y \in \mathbb{R}^n$ where $y \sim \rho_s/s^n$, then $\Delta(y+I, U(\mathbb{R}^n/I)) \leq (\log n)^{-2\log n}/2$. (That is, y is in an almost uniformly random coset of $\mathbb{R}^n/I$). By our definition of s, we have that $||g||_\infty = 8\mathcal{E}dms\sqrt{n}\log n$. Now we will try to create an $h \in I$ which is smaller than g using the procedure below. In the procedure, it may not be obvious how each step is performed, and the reader is referred to lemma 6 for a detailed explanation of each step.

(1) for $i = 1$ to m
  (2) generate a uniformly random coset of $I/\langle g\rangle$ and let v_i be a polynomial in that coset
  (3) generate $y_i \in \mathbb{R}^n$ such that y_i has distribution ρ_s/s^n and consider y_i as a polynomial in $\mathbb{R}[x]$
  (4) let w_i be the unique polynomial in $\mathbb{R}[x]$ of degree less than n with coefficients in the range $[0,p)$ such that $p(v_i + y_i) \equiv gw_i$ in $\mathbb{R}^n/\langle pg\rangle$
  (5) $a_i = [w_i] \bmod p$ (where $[w_i]$ means round each coefficient of w_i to the nearest integer)
(6) call oracle $\mathcal{C}(a_1,\ldots,a_m)$, and using its output, find polynomials $z_1,\ldots,z_m$ such that $||z_i||_f \leq 2d$ and $\sum z_ia_i \equiv 0$ in the ring $\mathbb{Z}_p[x]/\langle f\rangle$.
(7) output $h = \left(\sum\left(\frac{g(w_i-[w_i])}{p} - y_i\right)z_i\right) \bmod f$.

To complete the proof, we will have to show five things: first, we have to prove that the above procedure runs in polynomial time, which is done in lemma 6. Then, in lemma 7, we show that in step (6) we are feeding the oracle $\mathcal{C}$ with an $\mathfrak{h} \in \mathcal{H}$ where the distribution of $\mathfrak{h}$ is statistically close to uniform over $\mathcal{H}$. In lemma 8, we show that the resulting polynomial h is in the ideal I. We then show that if $\mathcal{C}$ outputted a collision, then with non-negligible probability, $||h||_f \leq ||g||_\infty/2$ and that $h \neq 0$. This is done in lemmas 9 and 10 respectively. These five things prove that with non-negligible probability, we will obtain a solution to $IncSPP_\gamma$. If we happen to fail, we repeat the procedure again. Since each run of the procedure is independent, we will obtain a solution to $IncSPP_\gamma$ in polynomial time.

Lemma 6. *The above procedure runs in polynomial time.*

Proof. We will show that each step in the algorithm takes polynomial time. In step (2), we need to generate a random element of $I/\langle g\rangle$. By lemma 3, the ideals I and $\langle g\rangle$ can be thought of as $\mathbb{Z}$-modules of dimension n. Since $\langle g\rangle \subseteq I$, the group $I/\langle g\rangle$ is finite, and we can efficiently generate a random element of $I/\langle g\rangle$. Step (4) of the algorithm will be justified in lemma 7. In step (5), we are just rounding each coefficient of w_i to the nearest integer and then reducing modulo p. Now each a_i can be thought of as an element of $\mathbb{Z}_p[x]/\langle f\rangle$, so in step (6) we can feed $(a_1,\ldots,a_m)$ to the algorithm that solves $Col_\mathcal{H}(a_1,\ldots,a_m)$. The algorithm will return $(\alpha_1,\ldots,\alpha_m),(\beta_1,\ldots,\beta_m)$ where $\alpha_i,\beta_i \in \mathbb{Z}[x]/\langle f\rangle$ such that $||\alpha_i||_f, ||\beta_i||_f \leq d$ and $\sum a_i\alpha_i \equiv \sum a_i\beta_i$ in the ring $\mathbb{Z}_p[x]/\langle f\rangle$. Thus if we set $z_i = \alpha_i - \beta_i$, we will have $||z_i||_f \leq 2d$ and $\sum z_i a_i \equiv 0$ in the ring $\mathbb{Z}_p[x]/\langle f\rangle$.

Lemma 7. *Consider the polynomials a_i as elements in $\mathbb{Z}_p^n$. Then,*

$$\Delta((a_1,\ldots,a_m), U(\mathbb{Z}_p^{n\times m})) \leq m\epsilon/2.$$

Proof. We know that v_i is in a uniformly random coset of $I/\langle g\rangle$ and let's assume for now that y_i is in a uniformly random coset of $\mathbb{R}^n/I$. This means that $v_i + y_i$ is in a uniformly random coset of $\mathbb{R}^n/\langle g\rangle$ and thus the distribution of $p(v_i + y_i)$ is in a uniformly random coset of $\mathbb{R}^n/\langle pg\rangle$. A basis for the additive group $\langle pg\rangle$ is $pg, pgx, \ldots, pgx^{n-1}$, thus every element of $\mathbb{R}^n/\langle pg\rangle$ has a unique representative of the form $\alpha_0 pg + \alpha_1 pgx + \ldots + \alpha_{n-1}pgx^{n-1} = g(p\alpha_0 + p\alpha_1 x + \ldots + p\alpha_{n-1}x^{n-1})$ for $\alpha_i \in [0,1)$. So step (4) of the algorithm is justified, and since $p(v_i + y_i)$ is in a uniformly random coset of $\mathbb{R}^n/\langle pg\rangle$, the coefficients of the polynomial $w_i = p\alpha_0 + p\alpha_1 x + \ldots + p\alpha_{n-1}x^{n-1}$ are uniform over the interval $[0,p)$, and thus the coefficients of $[w_i]$ are uniform over the integers modulo p. The caveat is that y_i is not really in a uniformly random coset of $\mathbb{R}^n/I$, but is very close to it. By our choice of s, we have that $\Delta(\rho_s/s^n + I, U(\mathbb{R}^n/I)) \leq \epsilon/2$, and since a_i is a function of y_i, by a property of statistical distance, we have that $\Delta(a_i, U(\mathbb{Z}_p^n)) \leq \epsilon/2$. And since all the a_i's are independent, we get that $\Delta((a_1,\ldots,a_m), U(\mathbb{Z}_p^{n\times m})) \leq m\epsilon/2$.

Due to space constraints, the proofs of the below lemmas are omitted, and we refer the interested reader to the full version of this work.

Lemma 8. $h \in I$.

Lemma 9. *With probability negligibly different from* 1, $||h||_f \leq \frac{||g||_\infty}{2}$.

Lemma 10. $Pr[h = 0|(a_1,\ldots,a_m),(z_1,\ldots,z_m)] = \Omega(1)$.

6 Conclusions and Open Problems

We gave constructions of efficient collision-resistant hash functions that can be proven secure based on the conjectured worst-case hardness of the shortest vector problem for ideal lattices, i.e., lattices that can be represented as ideals of $\mathbb{Z}[x]/\langle f\rangle$ for some monic, irreducible polynomial f. Moreover, our results can be extended to certain polynomials f that are not irreducible, e.g., the polynomial $f = x^n - 1$ corresponding to the class of cyclic lattices.

The central question raised by our work is the hardness of $\mathcal{I}(f)$-SVP, or equivalently, the hardness of f-SPP for different f's. It is known that SVP is hard in the general case, and it was conjectured in [14] that $\mathcal{I}(x^n-1)$-SVP is hard as well. We show worst-case to average-case reductions that work for many other f's, so, in essence, we are giving more "targets" that can be proved hard.

Almost nothing is currently known about the complexity of problems for ideal lattices. We hope that our constructions of efficient collision-resistant hash functions based on the worst-case hardness of these problems provides motivation for their further study.

References

1. D. Aharonov and O. Regev. Lattice problems in NP ∩ coNP. *Journal of the ACM*, 52(5):749–765, 2005.
2. M. Ajtai. Generating hard instances of lattice problems. In *STOC*, pages 99–108, 1996.
3. M. Ajtai, R. Kumar, and D. Sivakumar. A sieve algorithm for the shortest lattice vector problem. In *STOC*, pages 601–610, 2001.
4. E. Biham, R. Chen, A. Joux, P. Carribault, W. Jalby, and C. Lemuet. Collisions of SHA-0 and reduced SHA-1. In *EUROCRYPT*, 2005.
5. J. Cai and A. Nerurkar. An improved worst-case to average-case connection for lattice problems. In *FOCS*, pages 468–477, 1997.
6. B. Chor and R. L. Rivest. A knapsack type public-key cryptosystem based on arithmetic in finite fields. *IEEE Trans. Inform. Theory*, 34(5):901–909, 1988.
7. I. Damgard. A design principle for hash functions. In *CRYPTO '89*, pages 416–427.
8. I. Dinur. Approximating SVP_∞ to within almost-polynomial factors is NP-hard. *Theor. Comput. Sci.*, 285(1):55–71, 2002.
9. O. Goldreich and S. Goldwasser. On the limits of nonapproximability of lattice problems. *J. Comput. Syst. Sci.*, 60(3), 2000.
10. J. Hoffstein, J. Pipher, and J. H. Silverman. Ntru: A ring-based public key cryptosystem. In *ANTS*, pages 267–288, 1998.
11. A. Joux and L. Granboulan. A practical attack against knapsack based hash functions. In *EUROCRYPT'94*, pages 58–66, 1994.
12. A. K. Lenstra, H. W. Lenstra Jr., and L. Lovasz. Factoring polynomials with rational coefficients. *Mathematische Annalen*, (261):513–534, 1982.
13. R.C. Merkle and M.E. Hellman. Hiding information and signatures in trapdoor knapsacks. *IEEE Transactions on Information Theory*, IT-24:525–530, 1978.
14. D. Micciancio. Generalized compact knapsacks, cyclic lattices, and efficient one-way functions from worst-case complexity assumptions. *Computational Complexity.* (To appear. Preliminary version in FOCS 2002).
15. D. Micciancio. Almost perfect lattices, the covering radius problem, and applications to Ajtai's connection factor. *SIAM J. on Computing*, 34(1):118–169, 2004.
16. D. Micciancio and S. Goldwasser. *Complexity Of Lattice Problems: A Cryptographic Perspective.* Kluwer Academic Publishers, 2002.
17. D. Micciancio and O. Regev. Worst-case to average-case reductions based on Gaussian measures. *SIAM J. on Computing.* (To appear. Preliminary version in FOCS 2004).
18. C. Peikert and A. Rosen. Efficient collision-resistant hashing from worst-case assumptions on cyclic lattices. In *TCC*, 2006.
19. V. V. Prasolov. *Polynomials*, volume 11 of *Algorithms and Computation in Mathematics.* Springer-Verlag Berlin Heidelberg, 2004.
20. C. P. Schnorr. A hierarchy of polynomial time basis reduction algorithms. *Theoretical Computer Science*, 53:201–224, 1987.
21. A. Shamir. A polynomial time algorithm for breaking the basic Merkle-Hellman cryptosystem. *IEEE Transactions on Information Theory*, IT-30(5):699–704, 1984.
22. S. Vaudenay. Cryptanalysis of the Chor–Rivest cryptosystem. *Journal of Cryptology*, 14(2):87–100, 2001.
23. X. Wang, X. Lai, D. Feng, H. Chen, and X. Yu. Cryptanalysis for hash functions MD4 and RIPEMD. In *EUROCRYPT*, 2005.
24. X. Wang and H. Yu. How to break MD5 and other hash functions. In *EUROCRYPT*, 2005.

An Efficient Provable Distinguisher for HFE

Vivien Dubois, Louis Granboulan, and Jacques Stern*

École normale supérieure
Département d'Informatique, 45 rue d'Ulm, 75230 Paris cedex 05, France
{dubois, granboulan, stern}@di.ens.fr

Abstract. The HFE cryptosystem was the subject of several cryptanalytic studies, sometimes successful, but always heuristic. To contrast with this trend, this work goes back to the beginnning and achieves *in a provable way* a first step of cryptanalysis which consists in distinguishing HFE public keys from random systems of quadratic equations. We provide two distinguishers: the first one has polynomial complexity and subexponential advantage; the second has subexponential complexity and advantage close to one. These distinguishers are built on the differential methodology introduced at Eurocrypt'05 by Fouque & *al.* Their rigorous study makes extensive use of combinatorics in binary vector spaces. This combinatorial approach is novel in the context of multivariate schemes. We believe that the alliance of both techniques provides a powerful framework for the mathematical analysis of multivariate schemes.

Keywords: Multivariate cryptography, HFE, differential cryptanalysis.

1 Introduction

While quantum computers, if they are ever built, would threaten most popular public-key cryptosystems such as RSA [17], alternative families of systems are currently designed and evaluated. One such family is based on multivariate quadratic polynomials on finite fields, and demonstrated very fruitful. Initiated in the early 80's by Matsumoto-Imai and Fell-Diffie [19] [5], multivariate cryptography received interest after the work of Shamir [3] and Patarin [10, 11]. Since then, about four basic trapdoors along with a large number of non-exclusive additional modifications have been invented [4]. These modifications, called *variations*, are designed to prevent structural attacks against the trapdoor.

HFE, probably the most promising of these cryptosystems, was proposed by Patarin [11] as a repair of the broken Matsumoto-Imai cryptosystem [20]. A little later, Kipnis and Shamir found a structural attack reducing the recovery of the private key to a MinRank problem [1]. Unfortunately, no known method to solve MinRank problems is practical for usual parameter sizes; still, the attack reveals weaknesses in the hiding of the trapdoor. Next, Courtois discovered that the multivariate quadratic equations coming from an HFE public key satisfy many

* This work is supported in part by the French government through X-Crypt, in part by the European Commission through ECRYPT.

M. Bugliesi et al. (Eds.): ICALP 2006, Part II, LNCS 4052, pp. 156–167, 2006.

low degree polynomial implicit equations [15]. Finally, Faugère and Joux demonstrated experimentally that systems of multivariate quadratic equations coming from HFE keys have good elimination properties that allow much easier Gröbner bases computations [6] — they broke the basic HFE for the first suggested parameters. Nevertheless, the attack did not extend to some major variations, requires a huge workload both in time and memory for the suggested parameter sizes and its complexity is unclear. Also all mentioned cryptanalytic approaches are heuristic and none provides a provable distinguisher.

Recently, Fouque-Granboulan-Stern proposed a new technique of analysis for multivariate schemes [16]. The method consists in studying the rank of the differential of the public key in order to extract information about the internal structure. The *differential* methodology already proved useful by providing an enhanced cryptanalysis of the Matsumoto-Imai cryptosystem and by breaking its Internal Perturbation variation [16] proposed by Ding [7].

Our Results. In this paper, we present a further application of the differential approach. It provides a provable distinguisher of HFE public keys, with polynomial complexity and subexponential advantage. This distinguisher can be improved into an algorithm with subexponential complexity and proven advantage close to one. This is the first cryptanalytic insight into the internal structure of HFE which is both entirely proven and practical for standard parameters. Our study requires combinatorics in finite fields of characteristic 2, which we believe to provide a new powerful approach for the analysis of multivariate schemes.

Organization of the Paper. In Section 2 of this paper, we recall the basic mathematical setting of multivariate cryptography and set up some combinatorial results related to the distribution of ranks of linear maps. In Section 3, we recall the definitions of HFE and its differential, and using the previous combinatorial tools, we show how the HFE internal structure can be detected from a public key with a precisely estimated complexity. A few proofs are sketched in this paper; they appear in details in the appendices of the full paper.

2 Mathematical Setting

2.1 Univariate-Multivariate Correspondence

Finite Fields. [13] We note $\mathbb{F}_2^n$ the n-dimensional vector space over $\mathbb{F}_2$. All fields with 2^n elements are isomorphic, and can be considered as instantiations of the same entity, called *the degree n extension field of* $\mathbb{F}_2$, denoted $\mathbb{F}_{2^n}$. $\mathbb{F}_{2^n}$ is an $\mathbb{F}_2$-vector space of dimension n and every choice of a basis of $\mathbb{F}_{2^n}$ defines a linear isomorphism from $\mathbb{F}_{2^n}$ to $\mathbb{F}_2^n$. Besides, the non-zero elements of $\mathbb{F}_{2^n}$ form a multiplicative group of size $2^n - 1$ and every element a of $\mathbb{F}_{2^n}$ satisfies $a^{2^n} = a$. Last, $\mathbb{F}_{2^n}$ has characteristic 2, that is for all x of $\mathbb{F}_{2^n}$, $x + x = 0$.

$\mathbb{F}_2$-Linear and $\mathbb{F}_2$-Quadratic Polynomials over $\mathbb{F}_{2^n}$. Characteristic 2 implies that for any a, b in $\mathbb{F}_{2^n}$ and any integer i, $(a + b)^{2^i} = a^{2^i} + b^{2^i}$. As a consequence, for any integer i, the polynomial X^{2^i} defines an $\mathbb{F}_2$-linear map from $\mathbb{F}_{2^n}$ to $\mathbb{F}_{2^n}$. Besides, since for all a in $\mathbb{F}_{2^n}$, $a^{2^n} = a$, polynomials X^{2^i} and $X^{2^{i+n}}$ define the same function. Thus, we can focus on monomials X^{2^i} for i restricted to $[0, n-1]$. Next, linear combinations over $\mathbb{F}_{2^n}$ of these monomials again define $\mathbb{F}_2$-linear maps from $\mathbb{F}_{2^n}$ to $\mathbb{F}_{2^n}$ and we define the set

$$\mathcal{L} = \left\{ \sum_{i=0}^{n-1} a_i X^{2^i} ; a_i \in \mathbb{F}_{2^n}, \forall i \in [0, n-1] \right\}$$

that we call the $\mathbb{F}_2$-*linear polynomials over* $\mathbb{F}_{2^n}$. The same way, it is easy to check that linear combinations over $\mathbb{F}_{2^n}$ of monomials in two variables of the form $X^{2^i} Y^{2^j}$ for i, j in $[0, n-1]$ define $\mathbb{F}_2$-bilinear maps from $\mathbb{F}_{2^n} \times \mathbb{F}_{2^n}$ to $\mathbb{F}_{2^n}$. Taking $Y = X$ defines a subset of $\mathbb{F}_{2^n}[X]$

$$\mathcal{Q} = \left\{ \sum_{i,j=0:i\leq j}^{n-1} a_{ij} X^{2^i+2^j} ; a_{ij} \in \mathbb{F}_{2^n}, \forall i, j \in [0, n-1], i \leq j \right\}$$

that we call the $\mathbb{F}_2$-*quadratic polynomials over* $\mathbb{F}_{2^n}$.

Univariate-Multivariate Correspondence. Any function from $\mathbb{F}_{2^n}$ to $\mathbb{F}_{2^n}$ is the evaluation of a polynomial over $\mathbb{F}_{2^n}$, and this polynomial is unique in the quotient ring $\mathbb{F}_{2^n}[X]/(X^{2^n} - X)$. This allows to identify any function from $\mathbb{F}_{2^n}$ to $\mathbb{F}_{2^n}$ to a univariate polynomial in $\mathbb{F}_{2^n}[X]/(X^{2^n} - X)$. The same way, a function from $\mathbb{F}_2^n$ to $\mathbb{F}_2^n$ is defined by n coordinate-functions, which are boolean functions in n variables. Each coordinate-function is the evaluation of a polynomial in $\mathbb{F}_2[x_1, \ldots, x_n]$, which is unique in the quotient-ring $\mathbb{F}_2[x_1, \ldots, x_n]/\{x_1^2 - x_1, \ldots, x_n^2 - x_n\}$. This allows to define any function from $\mathbb{F}_2^n$ to $\mathbb{F}_2^n$ by its multivariate representation in $(\mathbb{F}_2[x_1, \ldots, x_n]/\{x_1^2 - x_1, \ldots, x_n^2 - x_n\})^n$. Further, these two sets are isomorphic, by an extension of the isomorphism between $\mathbb{F}_{2^n}$ and $\mathbb{F}_2^n$. In particular the set of linear maps from $\mathbb{F}_2^n$ to $\mathbb{F}_2^n$, denoted $\mathcal{L}_n$, is in bijection with $\mathcal{L}$. Also, the set of quadratic maps from $\mathbb{F}_2^n$ to $\mathbb{F}_2^n$, denoted $\mathcal{Q}_n$, is in bijection with $\mathcal{Q}$.

2.2 Combinatorics in $\mathbb{F}_2^n$

Linearly Independent Sequences and Subspaces of $\mathbb{F}_2^n$. We denote by $S(n, d)$ the number of linearly independent sequences of length d of vectors of $\mathbb{F}_2^n$; it is easily seen that $S(n, d) = \prod_{i=0}^{n-1}(2^n - 2^i)$. Each such sequence generates a subspace of dimension d which is also generated by $S(d, d)$ other linearly independent sequences of length d. Therefore the number $E(n, d)$ of subspaces of dimension d in $\mathbb{F}_2^n$ is $S(n, d)/S(d, d)$. Defining $\lambda(n) = \prod_{i=1}^{n}\left(1 - \frac{1}{2^i}\right)$, we have

$$S(n, d) = \frac{\lambda(n)}{\lambda(n-d)} 2^{nd} \quad \text{and} \quad E(n, d) = \frac{\lambda(n)}{\lambda(n-d)\lambda(d)} 2^{d(n-d)}$$

$S(n, d)$ is similar to the *number of permutations of size d over n elements*, and $E(n, d)$ is similar to the *number of combinations of size d over n elements*. These quantities sparsely appear in the literature [9, 2, 18, 12], however we could not find any enumerative results dealing with algebraic aspects of binary vector spaces.

Number of Linear Maps of a Given Rank. We consider a fixed integer r in $[0, n]$ and we enumerate the number of linear maps of rank r. Let $\mathcal{K}$ be the kernel of a map of rank r, and let $\mathcal{B}$ a basis of a complement of $\mathcal{K}$. Any linear map of kernel $\mathcal{K}$ is uniquely defined by the image of $\mathcal{B}$, which is a linearly independent sequence of length r. Therefore, the number of linear maps with kernel $\mathcal{K}$ is $S(n, r)$. This depends only on the dimension $n - r$ of $\mathcal{K}$, and there are $E(n, n - r)$ such subspaces. Finally, the number of linear maps of rank r is

$$E(n, n-r)S(n, r) = \frac{\lambda(n)^2}{\lambda(n-r)^2\lambda(r)} 2^{r(n-r)} 2^{nr}$$

Dividing by 2^{n^2} provides the proportion of linear maps of rank r. The collection of these proportions for all ranks defines the distribution of ranks of linear maps.

Distribution of Ranks of $\mathbb{F}_2$-Linear Polynomials of Constrained Degree. We close this section by explaining how to compute the distribution of ranks of a random $\mathbb{F}_2$-linear polynomial of a given degree. While only the easy part of our results will be used in the sequel, it gives an other application of the combinatorial approach, which will later show interesting in the context of HFE.

An $\mathbb{F}_2$-linear polynomial P has as many roots as the number of elements in its kernel. Hence, if r is the rank of the $\mathbb{F}_2$-linear polynomial P considered as a linear map, it is easily seen that P has 2^{n-r} roots. Fixing an integer D in $[0, n-1]$, we denote $\mathcal{L}^D$ the subset of $\mathbb{F}_2$-linear polynomials of degree 2^D. A polynomial of degree 2^D has at most 2^D roots, or is the zero polynomial. Then, the rank of a non-zero $\mathbb{F}_2$-linear polynomial P in $\mathcal{L}^D$ is at least $n - D$. The distribution of ranks of $\mathbb{F}_2$-linear polynomials of degree 2^D is given by the following theorem. Although, the theorem does not provide a closed form for these numbers, it allows to compute them for any choice of the parameters.

Theorem 1. *Let D an integer in the interval $[0, n-1]$. A non-zero $\mathbb{F}_2$-linear polynomial of degree 2^D has rank at least $n-D$. The proportions $p_D(0),\ldots, p_D(D)$ of elements of $\mathcal{L}^D$ of ranks respectively $n, \ldots, n-D$ satisfy the following invertible triangular system*

$$d \in [0, D], \quad E(n, d)2^{-nd} = \sum_{m=d}^{D} E(m, d) p_D(n-d)$$

Sketch of proof. The number of $\mathbb{F}_2$-linear polynomials of degree 2^D is $(2^n - 1)2^{nD}$. Given a subspace of dimension d with d in $[0, D]$, the vanishing of an $\mathbb{F}_2$-linear polynomial of degree 2^D results in d linear constraints over its $D + 1$ coefficients. It implies that for each subspace of dimension d, there are exactly $(2^n - 1)2^{n(D-d)}$ $\mathbb{F}_2$-linear polynomials which vanish on it. In the product $E(n, d)(2^n - 1)2^{n(D-d)}$, the $\mathbb{F}_2$-linear polynomials whose kernel has dimension m with $m \geq d$ are counted

$E(m, d)$ times. Therefore, the proportions $p_D(n-d)$ of $\mathbb{F}_2$-linear polynomials of degree 2^D which have rank $n-d$ satisfy the above invertible triangular system.

3 Distinguishers for HFE

The distinguishers that we provide are built on the observation of the previous section: a $\mathbb{F}_2$-linear polynomial of degree at most 2^D has large rank at least $n-D$, while there is a very small albeit non-zero probability that a random linear map of any rank appears. Applying this observation to the differential yields a distinguisher. Even if the idea appears straightforward, the technicalities required to turn it into a precise mathematical proof and to estimate the advantage of the distinguisher are non-trivial and require the previously introduced combinatorial framework. This is especially true of the enhanced distinguisher, where the advantage is made close to one by iteration: the difficulty here is that we have to play with non pairwise independent random variables, whose precise relationship can only be understood through this combinatorial framework.

3.1 Description of HFE

At the basis of multivariate cryptography is the problem of solving a set of multivariate polynomial equations over a finite field. This problem is proven NP-hard [14] and considered very hard in practice for systems of equations at least quadratic with about the same number of equations and unknowns. For such systems, the best algorithms use Gröbner bases theory, have at least exponential complexity, and are impractical for even a few unknowns (or equations).

Informally, the general construction of multivariate cryptosystems consists in hiding an easily solvable multivariate quadratic system into a random-looking system by a secret transformation. More precisely, one considers a quadratic map $\boldsymbol{P}$ from $\mathbb{F}_2^n$ to $\mathbb{F}_2^n$ defined by n polynomials of degree 2 in n unknowns of a specific form, which allows to easily solve the system $\boldsymbol{P}(x_1, \ldots, x_n) = (a_1, \ldots, a_n)$ for any element $(a_1, \ldots, a_n)$ of $\mathbb{F}_2^n$. Then, one chooses two invertible affine maps $\boldsymbol{S}, \boldsymbol{T}$ from $\mathbb{F}_2^n$ to $\mathbb{F}_2^n$, each defined by n multivariate equations of degree 1. Clearly, the composition $\boldsymbol{T} \circ \boldsymbol{P} \circ \boldsymbol{S}$ is again a multivariate quadratic map $\boldsymbol{P'}$ of $\mathbb{F}_2^n$, and any related system $\boldsymbol{P'}(x_1, \ldots, x_n) = (a_1, \ldots, a_n)$ where $(a_1, \ldots, a_n)$ is an element of $\mathbb{F}_2^n$ is impractical to solve by the dedicated algorithms for a prescribed parameter n. To create an asymmetric cryptosystem, the user randomly picks $\boldsymbol{P}$ of the specific form and two invertible affine maps $\boldsymbol{S}, \boldsymbol{T}$, and keeps them secret. Then, he publishes $\boldsymbol{P'} = \boldsymbol{T} \circ \boldsymbol{P} \circ \boldsymbol{S}$. A message $\boldsymbol{a}$ encrypted into $\boldsymbol{b} = \boldsymbol{P'}(\boldsymbol{a})$ can only be decrypted by the legitimate user since the multivariate quadratic system $\boldsymbol{P'}(x_1, \ldots, x_m) = \boldsymbol{b}$ can only be solved by inverting the secret process.

HFE is a way to generate easily solvable multivariate quadratic systems. As seen in Section 2.1, the set of quadratic maps, called $\mathcal{Q}_n$, is isomorphic to a specific subset of the univariate polynomials over $\mathbb{F}_{2^n}$, namely $\mathcal{Q}$. It implies that solving a given multivariate quadratic system is equivalent to finding the roots of the related univariate polynomial. In HFE, the latter is made easy by generating quadratic systems from *low degree* univariate polynomials of $\mathcal{Q}$. Parameters for the first challenge of HFE are $n = 80$ and degree 96.

3.2 Differential Analysis of Multivariate Quadratic Maps

The Differentials of a Multivariate Quadratic Map. Given a quadratic map $\boldsymbol{P}$, its *differential* at a point $\boldsymbol{a}$ of $\mathbb{F}_2^n$ is the linear map defined by

$$\boldsymbol{DP_a}(\boldsymbol{x}) = \boldsymbol{P}(\boldsymbol{a}+\boldsymbol{x}) + \boldsymbol{P}(\boldsymbol{x}) + \boldsymbol{P}(\boldsymbol{a}) + \boldsymbol{P}(\boldsymbol{0})$$

It vanishes at $\boldsymbol{a}$. If $\boldsymbol{P}$ is seen as a polynomial, $\boldsymbol{DP_a}$ is an $\mathbb{F}_2$-linear polynomial.

For any element $\boldsymbol{a}$, the rank of $\boldsymbol{DP_a}$ can be evaluated. We call *distribution of ranks of the differentials of* $\boldsymbol{P}$ the collection for all rank r in $[0, n]$ of the proportions of elements $\boldsymbol{a}$ at which the rank of $\boldsymbol{DP_a}$ is r. The distribution of ranks of the differentials is *a major element of analysis of multivariate schemes* because it is invariant in the hiding process. Indeed, for $\boldsymbol{P}$ a quadratic map, $\boldsymbol{S}, \boldsymbol{T}$ two affine bijections of linear parts respectively $\underline{\boldsymbol{S}}, \underline{\boldsymbol{T}}$ (bijective), and $\boldsymbol{P}'$ the quadratic map $\boldsymbol{T} \circ \boldsymbol{P} \circ \boldsymbol{S}$, then it can be checked that for any point $\boldsymbol{a}$

$$\boldsymbol{DP}'_{\boldsymbol{a}} = \underline{\boldsymbol{T}} \circ \boldsymbol{DP}_{\underline{\boldsymbol{S}}(\boldsymbol{a})} \circ \underline{\boldsymbol{S}}$$

Consequently, the internal function $\boldsymbol{P}$ and the public key $\boldsymbol{P}'$ have the same distribution of ranks of the differentials. Hence, whenever the distribution of ranks of the differentials of $\boldsymbol{P}$ has some property, it can be seen from $\boldsymbol{P}'$.

Distribution of Ranks of the Differentials of a Random Quadratic Map. We consider a random quadratic map $\boldsymbol{P}$ of $\mathbb{F}_2^n$ and we are interested in the rank $r_{\boldsymbol{a}}$ of its differential $\boldsymbol{DP_a}$ at $\boldsymbol{a}$.

Theorem 2. *Given a non-zero element $\boldsymbol{a}$ of $\mathbb{F}_2^n$, and a random quadratic map $\boldsymbol{P}$, the rank of $\boldsymbol{DP_a}$ follows the distribution of ranks of linear maps vanishing at $\boldsymbol{a}$. Therefore, for any t in $[1, n]$ the probability that $\boldsymbol{DP_a}$ has rank $n - t$ is $\alpha_t 2^{-t(t-1)}$ where α_t is a constant in the interval $[0.16, 3.58]$.*

Proof. Let $\boldsymbol{a} = (a_1, \ldots, a_n)$ a non-zero element of $\mathbb{F}_2^n$ and $\boldsymbol{L}$ a linear map that cancels at $\boldsymbol{a}$: $\sum_{i=1}^n \boldsymbol{l}_i a_i = \boldsymbol{0}$ (Note that $\boldsymbol{l}_i \in \mathbb{F}_2^n$ and $a_i \in \mathbb{F}_2$). A quadratic map $\boldsymbol{P}(x_1, \ldots, x_n) = \sum_{i=1}^n \sum_{j=i+1}^n \boldsymbol{p}_{ij} x_i x_j$ has for differential at $\boldsymbol{a}$

$$\boldsymbol{DP_a}(x_1, \ldots, x_n) = \sum_{i=1}^n \left(\sum_{j=1}^{i-1} \boldsymbol{p}_{ji} a_j + \sum_{j=i+1}^n \boldsymbol{p}_{ij} a_j \right) x_i$$

Therefore, $\boldsymbol{DP_a} = \boldsymbol{L}$ is equivalent to

$$\begin{bmatrix} \boldsymbol{l}_1 \\ \\ \vdots \\ \\ \boldsymbol{l}_n \end{bmatrix} = \begin{bmatrix} 0 & \boldsymbol{p}_{12} & \boldsymbol{p}_{13} & \cdots & \boldsymbol{p}_{1n} \\ \boldsymbol{p}_{12} & 0 & \boldsymbol{p}_{23} & \cdots & \boldsymbol{p}_{2n} \\ \boldsymbol{p}_{13} & \boldsymbol{p}_{23} & 0 & & \boldsymbol{p}_{3n} \\ \vdots & \vdots & & \ddots & \vdots \\ \boldsymbol{p}_{1n} & \boldsymbol{p}_{2n} & \boldsymbol{p}_{3n} & \cdots & 0 \end{bmatrix} \begin{bmatrix} a_1 \\ \\ \vdots \\ \\ a_n \end{bmatrix}$$

Up to a reordering of coordinates, one can assume $a_n \neq 0$. Then any choice of coefficients $\boldsymbol{p}_{ij}$ for $i < j < n$ can be completed in a quadratic map such that $\boldsymbol{DP_a} = \boldsymbol{L}$. Indeed, we define for all i in $[1, n-1]$

$$\boldsymbol{p}_{in} = \boldsymbol{l}_i + \sum_{j=1}^{i-1} \boldsymbol{p}_{ji} a_j + \sum_{j=i+1}^{n-1} \boldsymbol{p}_{ij} a_j$$

and we can check that the last row equation $\sum_{i=1}^{n-1} \boldsymbol{p}_{in} a_i = \boldsymbol{l}_n$ is satisfied, using the vanishing at $\boldsymbol{a}$ of both $\boldsymbol{L}$ and $\boldsymbol{DP_a}$. Hence the number of $\boldsymbol{P}$ in $\mathcal{Q}_n$ such that $\boldsymbol{DP_a} = \boldsymbol{L}$ is independent of $\boldsymbol{a}$ and $\boldsymbol{L}$, and the first point of the theorem follows.

Next, for any t in $[1, n]$, a linear map of rank $n - t$ which vanishes at $\boldsymbol{a}$ is a map whose kernel has dimension t and contains $\boldsymbol{a}$. Since the number of such subspaces is $E(n-1, t-1)$, the number of linear maps of rank $n-t$ vanishing at $\boldsymbol{a}$ is $E(n-1, t-1)S(n, n-t)$. Finally the overall number of linear maps vanishing at $\boldsymbol{a}$ is $2^{n(n-1)}$. Among them, those of rank $n - t$ are in proportion

$$\Pr_{\boldsymbol{L} \in \mathcal{L}_n; \boldsymbol{L}(\boldsymbol{a})=\boldsymbol{0}} \left[rk\, \boldsymbol{L} = (n-t) \right] = \alpha_t 2^{-t(t-1)} \quad \text{with} \quad \alpha_t = \frac{\lambda(n)\lambda(n-1)}{\lambda(t)\lambda(t-1)\lambda(n-t)}$$

Since the sequence λ decreases towards a value over 0.28 [18], α_t lies in [0.16, 3.58].

3.3 A Fast Distinguisher for HFE

A Specific Property of HFE. We denote $\boldsymbol{P}$ the hidden internal function in HFE and we let $D = \lceil \log_2 \deg(\boldsymbol{P}) \rceil$ where $\deg(\boldsymbol{P})$ is the degree of $\boldsymbol{P}$ considered as a polynomial over $\mathbb{F}_{2^n}$. For any element $\boldsymbol{a}$ of $\mathbb{F}_2^n$, $\boldsymbol{DP_a}$ is an $\mathbb{F}_2$-linear polynomial of degree at most 2^D. Unless it is the zero function, its rank is at least $n - D$. In contrast, we saw in the previous paragraph that the differential of a random quadratic system has rank $n - D - 1$ with probability of the order of $2^{-D(D+1)}$.

A Fast Distinguisher for HFE. For any parameter D in $[0, n]$, we define the algorithm T_D which takes as input a quadratic map $\boldsymbol{P}$ and a non-zero point $\boldsymbol{a}$, computes the differential of $\boldsymbol{P}$ at $\boldsymbol{a}$ and evaluates its rank, finally answers 1 when this rank is $n - D - 1$ and 0 otherwise. The running time of this algorithm is polynomial, more precisely it is $\mathcal{O}(n^3)$.

Using algorithm T_D, we can devise a distinguisher for any non-zero arbitrary value $\boldsymbol{a}$, defined the following way

`INPUT`: a quadratic function $\boldsymbol{P}$ which is
- either a HFE function of degree $\leq 2^D$ (probability 1/2)
- or a random quadratic function (probability 1/2)

`DO`: compute $T_D(\boldsymbol{P}, \boldsymbol{a})$
if $T_D(\boldsymbol{P}, \boldsymbol{a}) = 1$ output `random`, else output `HFE`

The distinguisher always answers `HFE` on HFE functions, but it may answer `HFE` on a random quadratic map which is not HFE. Following Theorem 2, the distinguisher answers `random` on a random quadratic maps with a probability of the order of $2^{-D(D+1)}$. This probability is the advantage of the distinguisher and does not depend on $\boldsymbol{a}$. Since 2^D is polynomial in the security parameter to allow decryption of the HFE cryptosystem, $2^{D(D+1)}$ is subexponential. Hence, any non-zero element of $\mathbb{F}_2^n$ yields a distinguisher for HFE with proven subexponential advantage, or more accurately with advantage the inverse of a subexponential function. A test answering 1 when the rank is $\leq n-D-1$ is a little more efficient but its study is more complicated without changing the order of complexity.

3.4 Enhanced Distinguisher

For any parameter D in $[0, n]$ and a fixed integer N, we define the algorithm T_D^N which takes as input a quadratic map $\boldsymbol{P}$ and N distinct non-zero points $\boldsymbol{a}_1, \ldots, \boldsymbol{a}_N$ of $\mathbb{F}_2^n$, computes the values of $T_D(\boldsymbol{P}, \boldsymbol{a}_i)$ for all i, finally answers 1 if $T_D(\boldsymbol{P}, \boldsymbol{a}_i) = 1$ was found for at least one $\boldsymbol{a}_i$, and 0 otherwise. The running time of this algorithm is $\mathcal{O}(Nn^3)$.

The intention behind this algorithm is simple ; it aims at increasing the probability to detect a non-HFE quadratic map by testing for multiple points, yielding a distinguisher with improved advantage. Using algorithm T_D^N, we can devise as before such an improved distinguisher from any arbitrary distinct non-zero values $\boldsymbol{a}_1, \ldots, \boldsymbol{a}_N$.

Let fix N such points $\boldsymbol{a}_1, \ldots, \boldsymbol{a}_N$ and define the random variable

$$S_N^D(\boldsymbol{P}) = \sum_{i=1}^{N} T_D(\boldsymbol{P}, \boldsymbol{a}_i)$$

over the set $\mathcal{Q}_n$ of quadratic maps. All $T_D(\boldsymbol{P}, \boldsymbol{a}_i)$ are $\{0, 1\}$ valued random variables over $\mathcal{Q}_n$ and the advantage of the distinguisher is

$$\Pr_{\boldsymbol{P} \in \mathcal{Q}_n}[S_N^D(\boldsymbol{P}) \geq 1]$$

From Theorem 2, we deduce that all $T_D(\boldsymbol{P}, \boldsymbol{a}_i)$ have the same law, of mean value $\mu_D \simeq 2^{-D(D+1)}$. Hence, we could easily determine the advantage of the distinguisher, if the random variables $T_D(\boldsymbol{P}, \boldsymbol{a}_i)$ were independent; unfortunately these random variables are even not pairwise independent. In the sequel, we give more details about this fact and show that this difficulty can be overcome: using our combinatorial framework, the standard deviation of S_N^D can be actually computed. Next, using Chebychev inequality, we prove that for $N = 2^{D(D+2)}$, *the advantage of the distinguisher is close to one.*

Mean Value and Standard Deviation of S_N^D

Theorem 3. *The mean value and the standard deviation of S_N^D satisfy respectively*

$$\begin{cases} A_N^D &= N\mu_D \\ (\sigma_N^D)^2 &= N\mu_D - N\mu_D^2(1+\epsilon_D) + \epsilon_D N^2 \mu_D^2 \end{cases}$$

where ϵ_D is lower than $2^{2D+2}/(2^n - 1)$ and μ_D is of the order of $2^{-D(D+1)}$.

Proof. For the reader's convenience, we omit the D superscripts and write X_i in place of $T_D(\boldsymbol{P}, \boldsymbol{a}_i)$.

The mean value comes from linearity. The standard deviation satisfies

$$(\sigma_N)^2 = \mathrm{E}_{\boldsymbol{P}\in\mathcal{Q}_n}[(S_N)^2] - (A_N)^2$$

where $\mathrm{E}_{\boldsymbol{P}\in\mathcal{Q}_n}$ denotes the expectation. Further, since the X_i are $\{0,1\}$ valued and the expectation is linear,

$$\mathrm{E}_{\boldsymbol{P}\in\mathcal{Q}_n}[(S_N)^2] = A_N + \sum_{i=1}^{N}\sum_{j\neq i} \mathrm{E}_{\boldsymbol{P}\in\mathcal{Q}_n}[X_i X_j]$$

where for each pair $i \neq j$,

$$\mathrm{E}_{\boldsymbol{P}\in\mathcal{Q}_n}[X_i X_j] = \mathrm{Pr}_{\boldsymbol{P}\in\mathcal{Q}_n}[rk\, DP_{\boldsymbol{a}_i} = n - D - 1\,,\, rk\, DP_{\boldsymbol{a}_j} = n - D - 1] \quad (1)$$

As already mentioned, random variables X_i and X_j are not independent, for any pair $i \neq j$. Indeed, the differentials of $\boldsymbol{P}$ at $\boldsymbol{a}_i$ and $\boldsymbol{a}_j$ satisfy $\boldsymbol{DP}_{\boldsymbol{a}_i}(\boldsymbol{a}_j) = \boldsymbol{DP}_{\boldsymbol{a}_j}(\boldsymbol{a}_i)$. Therefore, the vanishing (or not) of $\boldsymbol{DP}_{\boldsymbol{a}_i}$ at $\boldsymbol{a}_j$ is correlated to the vanishing (or not) of $\boldsymbol{DP}_{\boldsymbol{a}_j}$ at $\boldsymbol{a}_i$. It follows that the ranks of $\boldsymbol{DP}_{\boldsymbol{a}_i}$ and $\boldsymbol{DP}_{\boldsymbol{a}_j}$ are not independent. Fortunately, the distribution of ranks of pairs $(\boldsymbol{DP}_{\boldsymbol{a}_i}, \boldsymbol{DP}_{\boldsymbol{a}_j})$ can be fully understood: defining the set $D(\boldsymbol{a},\boldsymbol{b})$ of pairs of linear maps $(\boldsymbol{L},\boldsymbol{L}')$ such that $\boldsymbol{L}(\boldsymbol{a}) = \boldsymbol{0}, \boldsymbol{L}'(\boldsymbol{b}) = \boldsymbol{0}, \boldsymbol{L}(\boldsymbol{b}) = \boldsymbol{L}'(\boldsymbol{a})$, we can prove the following lemma whose proof is very similar to that of Theorem 2.

Lemma 1. *Given two distinct non-zero elements* $\boldsymbol{a}$ *and* $\boldsymbol{b}$ *of* $\mathbb{F}_2^n$*, and a random quadratic map* $\boldsymbol{P}$*, the rank of the pair* $(\boldsymbol{DP}_{\boldsymbol{a}}, \boldsymbol{DP}_{\boldsymbol{b}})$ *follows the distribution of ranks of pairs of linear maps in* $D(\boldsymbol{a},\boldsymbol{b})$*.*

Lemma 1 implies that

$$\mathrm{Pr}_{\boldsymbol{P}\in\mathcal{Q}_n}\begin{bmatrix} rk\,\boldsymbol{DP}_{\boldsymbol{a}_i} = n - D - 1 \\ rk\,\boldsymbol{DP}_{\boldsymbol{a}_j} = n - D - 1 \end{bmatrix} = \mathrm{Pr}_{(\boldsymbol{L},\boldsymbol{L}')\in D(\boldsymbol{a}_i,\boldsymbol{a}_j)}\begin{bmatrix} rk\,\boldsymbol{L} = n - D - 1 \\ rk\,\boldsymbol{L}' = n - D - 1 \end{bmatrix} \quad (2)$$

It remains to compute the probability on the right hand-side of the above. This probability is part of the distribution of ranks of pairs of linear maps in $D(\boldsymbol{a},\boldsymbol{b})$, which can be computed by the same combinatorial methods.

As a preliminary, let $N_k(r)$ denote the number of linear maps of rank r vanishing on a prescribed subspace of dimension k. The values $N_1(r)$ for all r were computed in the proof of the Theorem 2. In the following, we will need in addition the values $N_2(r)$ for all r, which can be computed the same way. This computation is systematic and can be done at no cost for a general k : for r in $[0, n-k]$, the number of subspaces of dimension $n - r$ containing the prescribed subspace is $E(n-k, n-k-r)$, and the number of linear maps of rank r having one of these subspaces as kernel is $S(n,r)$. Therefore $N_k(r) = E(n-k, n-k-r)S(n,r)$ for r in $[0, n-k]$, and 0 otherwise.

The distribution of ranks of pairs of linear maps in $D(\boldsymbol{a},\boldsymbol{b})$ is given by the following lemma.

Lemma 2. *Given two non-zero distinct points $\boldsymbol{a}, \boldsymbol{b}$ in $\mathbb{F}_2^n$, and for any integers r and s in $[0, n-1]$, the proportion of pairs $(\boldsymbol{L}, \boldsymbol{L}')$ of linear maps in $D(\boldsymbol{a}, \boldsymbol{b})$ which have rank (r, s) is*

$$\frac{1}{2^{n(2n-3)}} \times \left(N_2(r)N_2(s) + \frac{1}{2^n - 1}(N_1(r) - N_2(r))(N_1(s) - N_2(s)) \right)$$

Proof. A pair $(\boldsymbol{L}, \boldsymbol{L}')$ in $D(\boldsymbol{a}, \boldsymbol{b})$ must satisfy $\boldsymbol{L}(\boldsymbol{a}) = 0, \boldsymbol{L}'(\boldsymbol{b}) = 0, \boldsymbol{L}(\boldsymbol{b}) = \boldsymbol{L}'(\boldsymbol{a})$, which are three independent linear constraints over the $2n$ coefficients in $\mathbb{F}_2^n$ defining $\boldsymbol{L}$ and $\boldsymbol{L}'$. Consequently $D(\boldsymbol{a}, \boldsymbol{b})$ has $2^{n(2n-3)}$ elements.

We define $V_{\boldsymbol{a}}$ as the set of linear maps which vanish at $\boldsymbol{a}$ and $V_{[\boldsymbol{a},\boldsymbol{b}]}$ as the set of linear maps which vanish on the subspace generated by $\boldsymbol{a}$ and $\boldsymbol{b}$. Some fraction of functions $\boldsymbol{L} \in V_{\boldsymbol{a}}$ also vanish at $\boldsymbol{b}$, and when it happens, the functions $\boldsymbol{L}'$ such that $(\boldsymbol{L}, \boldsymbol{L}') \in D(\boldsymbol{a}, \boldsymbol{b})$ are those in $V_{[\boldsymbol{a},\boldsymbol{b}]}$. Conversely, for each function $\boldsymbol{L} \in V_{\boldsymbol{a}} \setminus V_{[\boldsymbol{a},\boldsymbol{b}]}$, functions $\boldsymbol{L}'$ such that $(\boldsymbol{L}, \boldsymbol{L}') \in D(\boldsymbol{a}, \boldsymbol{b})$ are those in $V_{\boldsymbol{b}} \setminus V_{[\boldsymbol{a},\boldsymbol{b}]}$ with $\boldsymbol{L}'(\boldsymbol{a}) = \boldsymbol{L}(\boldsymbol{b})$; these functions represent a fraction $1/(2^n - 1)$ of all functions in $V_{\boldsymbol{b}} \setminus V_{[\boldsymbol{a},\boldsymbol{b}]}$ since $\boldsymbol{L}(\boldsymbol{b})$ is one of the $2^n - 1$ equally possible non-zero values for $\boldsymbol{L}'(\boldsymbol{a})$. □

Applying Lemma 2 with $r = s = (n - D - 1)$ provides the probability of equation (2). Using the relation

$$N_1(n - D - 1) = \frac{2^{n-1} - 1}{2^D - 1} N_2(n - D - 1)$$

this probability is

$$\frac{N_1(n - D - 1)^2}{2^{n(2n-3)}} \times \left(\left(\frac{2^D - 1}{2^{n-1} - 1} \right)^2 + \frac{1}{2^n - 1} \left(1 - \frac{2^D - 1}{2^{n-1} - 1} \right)^2 \right) \quad (3)$$

Besides, the proportion of linear maps of rank $n - D - 1$ vanishing at $\boldsymbol{a}$, denoted μ_D, is $N_1(n - D - 1)/2^{n(n-1)}$. Therefore, the factor in (3) equals $\mu_D^2 2^n$ and after a few steps, we get for the above probability

$$\mu_D^2 (1 + \epsilon_D) \qquad \text{with} \quad \epsilon_D = \frac{1}{2^n - 1} \left(\frac{2^n(2^D - 1)}{2^{n-1} - 1} - 1 \right)^2$$

As a remark, since the proportion of pairs of linear maps in $V_{\boldsymbol{a}} \times V_{\boldsymbol{b}}$ of rank $(n - D - 1, n - D - 1)$ is μ_D^2, ϵ_D is a correcting term which measures the distance between the distribution of ranks in $D(\boldsymbol{a}, \boldsymbol{b})$ and in $V_{\boldsymbol{a}} \times V_{\boldsymbol{b}}$ at the pair of ranks $(n - D - 1, n - D - 1)$. From

$$\epsilon_D = \frac{1}{2^n - 1} \left(2^{D+1} - 1 - 2 \left(1 - \frac{2^D - 1}{2^{n-1} - 1} \right) \right)^2$$

we see that the correcting term ϵ_D is less than $2^{2(D+1)}/(2^n - 1)$.

We can now come back to equation (1)

$$\mathrm{E}_{\boldsymbol{P} \in \mathcal{Q}_n}[X_i X_j] = \mu_D^2 (1 + \epsilon_D)$$

to finally obtain

$$(\sigma_N)^2 = N\mu_D - N\mu_D^2(1 + \epsilon_D) + \epsilon_D N^2 \mu_D^2$$

Lower Bound on the Advantage. Using Chebychev inequality, we can upper-bound $\Pr_{\boldsymbol{P}\in\mathcal{Q}}[S_N^D(\boldsymbol{P})=0]$. Indeed, for all t in the interval $(0, A_N^D/\sigma_N^D]$

$$\Pr_{\boldsymbol{P}\in\mathcal{Q}}[S_N^D(\boldsymbol{P})=0] \leq \Pr_{\boldsymbol{P}\in\mathcal{Q}}[|S_N^D(\boldsymbol{P}) - A_N^D| \geq t\,\sigma_N^D] \leq \frac{1}{t^2}$$

We take $t = A_N^D/\sigma_N^D$; then

$$\frac{1}{t^2} = \frac{(\sigma_N^D)^2}{(A_N^D)^2} = \frac{1}{N\mu_D} - \frac{1}{N}(1+\epsilon_D) + \epsilon_D < \frac{1}{N\mu_D} + \epsilon_D$$

Now let fix $N\mu_D = 2^a$, for some integer a. Then

$$\frac{1}{t^2} < \frac{1}{2^a} + \epsilon_D$$

and the advantage is

$$\Pr_{\boldsymbol{P}\in\mathcal{Q}}[S_N^D(\boldsymbol{P}) \geq 1] = 1 - \Pr_{\boldsymbol{P}\in\mathcal{Q}}[S_N^D(\boldsymbol{P})=0] > 1 - \frac{1}{2^a} - \epsilon_D$$

For instance, for $N = 2^D/\mu_D$, our distinguisher has running time $\mathcal{O}(2^{D(D+2)}n^3)$ and advantage at least of the order of

$$1 - \frac{1}{2^D} - \frac{4}{2^{n-2D}}$$

For $N = 2^{D^2}/\mu_D$, the complexity becomes $\mathcal{O}(2^{D(2D+1)}n^3)$ and the advantage is made at least $1 - 2^{-D^2} - 4.2^{-(n-2D)}$.

4 Conclusion

In this paper, we provide two distinguishers of HFE public keys: the first one has polynomial complexity and subexponential advantage; the second has subexponential complexity and advantage close to one. Though the cryptanalytic impact is smaller than the work of Faugere and Joux [6], our work is the first which shows without heuristics how the internal structure of HFE yields some particularities. It aims in particular at initiating a process of mathematical analysis of multivariate primitives, enlightened by the precedent heuristic approachs. The methodology used in this paper is new and widely applicable in the context of multivariate schemes. It should provide a solid framework of analysis for the numerous variations, which mostly escape all previous heuristic approachs. In particular, it is well suited to analyze the Internal Perturbation of HFE [21] suggested by Ding [8].

This study used differential properties of quadratic maps over an $\mathbb{F}_2$-extension $\mathbb{F}_{2^n}$, and combinatorics in $\mathbb{F}_2$-linear spaces. We showed that HFE public keys have very specific differential properties. This raises an interesting open problem: is the set of public keys such that all differentials have rank at least $n-D$ larger than the set of public keys affinely equivalent to an $\mathbb{F}_2$-linear polynomial of degree at most 2^D ? Another open problem is the existence of a polynomial time distinguisher for HFE public keys.

References

1. A.Kipnis and A.Shamir. Cryptanalysis of the HFE Public Key Cryptosystem. In *Crypto'99*, LNCS 1666, pages 19–30. Springer-Verlag, 1999.
2. A.E.Solow A.Nijenhuis and H.S.Wilf. Bijective methods in the theory of finite vector spaces. *J. Combin. Theory (A)*, 37:80–84, 1984.
3. A.Shamir. Efficient signature schemes based on Birational Permutations. In *Crypto'93*, LNCS 773, pages 1–12. Springer-Verlag, 1994.
4. C.Wolf and B.Preneel. Taxonomy of Public Key Schemes based on the problem of Multivariate Quadratic equations. Cryptology ePrint Archive, Report 2005/077, 2005. `http://eprint.iacr.org/`.
5. H.Fell and W.Diffie. Analysis of a Public Key Approach based on Polynomial Substitution. In *Crypto'85*, LNCS 218, pages 340–349. Springer-Verlag, 1985.
6. J-C.Faugère and A.Joux. Algebraic cryptanalysis of Hidden Field Equation (HFE) cryptosystems using Gröbner Bases. In *Crypto'03*, LNCS 2729, pages 44–60. Springer-Verlag, 2003.
7. J.Ding. A new variant of the Matsumoto-Imai Cryptosystem through Perturbation. In *PKC'04*, LNCS 2947, pages 305–318. Springer-Verlag, 2004.
8. J.Ding and D.Schmidt. Cryptanalysis of HFEv and Internal Perturbation of HFE. In *PKC'05*, LNCS 3386, pages 288–301. Springer-Verlag, 2005.
9. J.Goldman and G-C.Rota. The number of subspaces of a vector space. In W.T.Tutte, editor, *Recent progress in Combinatorics*, pages 75–83. Academic Press, 1969.
10. J.Patarin. Cryptanalysis of the Matsumoto and Imai Public Key Scheme of Eurocrypt'88. In *Crypto'95*, LNCS 963, pages 248–261. Springer-Verlag, 1995.
11. J.Patarin. Hidden Field Equations (HFE) and Isomorphisms of Polynomials (IP): two families of asymetric algorithms. In *Eurocrypt'96*, LNCS 1070, pages 33–46. Springer-Verlag, 1996.
12. K.E.Morrison. An introduction to q-species. *The Electronic Jounral of Combinatorics*, 12(R62), 2005.
13. K.Ireland and M.Rosen. *A Classical Introduction to Modern Number Theory*, chapter 7. Springer-Verlag, second edition, 1998.
14. M.Garey and D.Johnson. *Computer and Intractability: A guide to the theory of NP-completeness*. Freeman, 1979.
15. N.Courtois. The security of Hidden Field Equations (HFE). In *CT-RSA'01*, LNCS 2020, pages 266–281. Springer-Verlag, 2001.
16. P-A.Fouque, L.Granboulan, and J.Stern. Differential cryptanalysis for Multivariate Schemes. In *Eurocrypt'05*, LNCS 3386, pages 341–353. Springer-Verlag, 2005.
17. P.Shor. Polynomial-time algorithms for prime factorzation and discrete logarithms on a quantum computer. *SIAM J. Comput.*, 26(5):1484–1509, 1997.
18. S.Finch. *Mathematical Constants*, pages 354–361. Cambridge, 2003.
19. T.Matsumoto and H.Imai. A class of asymetric cryptosystems based on Polynomials over Finite Rings. In *ISIT'83*, pages 131–132, 1983.
20. T.Matsumoto and H.Imai. Public Quadratic Polynomial-tuples for efficient signature-verification and message encryption. In *Eurocrypt'88*, LNCS 330, pages 419–453. Springer-Verlag, 1988.
21. V.Dubois, L.Granboulan, and J.Stern. Cryptanalysis of HFE with Internal Perturbation. work in progress, 2006.

A Tight Bound for EMAC

Krzysztof Pietrzak*

Département d'Informatique, École Normale Supérieure, Paris
pietrzak@di.ens.fr

Abstract. We prove a new upper bound on the advantage of any adversary for distinguishing the encrypted CBC-MAC (EMAC) based on random permutations from a random function. Our proof uses techniques recently introduced in [BPR05], which again were inspired by [DGH$^+$04].

The bound we prove is tight — in the sense that it matches the advantage of known attacks up to a constant factor — for a wide range of the parameters: let n denote the block-size, q the number of queries the adversary is allowed to make and ℓ an upper bound on the length (i.e. number of blocks) of the messages, then for $\ell \leq 2^{n/8}$ and $q \geq \ell^2$ the advantage is in the order of $q^2/2^n$ (and in particular independent of ℓ). This improves on the previous bound of $q^2\ell^{\Theta(1/\ln\ln\ell)}/2^n$ from [BPR05] and matches the trivial attack (which thus is basically optimal) where one simply asks random queries until a collision is found.

1 Introduction

Cipher Block Chaining (CBC) is a popular mode of operation for block ciphers which is used (in some variations) for encryption and message authentication, i.e. as a Message Authentication Code (MAC).

SOME DEFINITIONS. The CBC function with key $\pi : \{0,1\}^n \to \{0,1\}^n$, denoted CBC_π, takes as input a message (whose length must be a multiple of n) $M = M_1 \cdots M_m \in (\{0,1\}^n)^m$ and outputs C_m which is inductively computed as

$$\mathsf{CBC}_\pi(M) = C_m \text{ where } C_0 = 0^n \text{ and } C_i = \pi(C_{i-1} \oplus M_i) \text{ for } i = 1, \ldots, m$$

The ECBC function (E for encrypted) is derived from the CBC function by additionally encrypting the output with an independent permutation[1]

$$\mathsf{ECBC}_{\pi_1,\pi_2}(M) \stackrel{\text{def}}{=} \pi_2(\mathsf{CBC}_{\pi_1}(M))$$

CBC BASED MACs. The CBC and ECBC function, with the π's instantiated by a block-cipher, are popular MACs called CBC-MAC and EMAC respectively.

As for the CBC-MAC, two parties sharing a secret key $K \in \mathcal{K}$ for a block-cipher $E : \mathcal{K} \times \{0,1\}^n \to \{0,1\}^n$ can authenticate their communication by

* Part of this work is supported by the Commission of the European Communities through the IST program under contract IST-2002-507932 ECRYPT.

[1] The ECBC function must not be confused with the ECBC-MAC from [BR00].

M. Bugliesi et al. (Eds.): ICALP 2006, Part II, LNCS 4052, pp. 168–179, 2006.

sending, together with their message M, the authentication tag $\mathsf{CBC}_{E(K,.)}(M)$.
ON THE SECURITY OF CBC BASED MACs. The CBC-MAC as just described is well known to be completely insecure in general,[2] but has been proven secure (under the assumption that the underlying block-cipher is a secure pseudorandom permutation) under the restriction that all messages have the same length in [BKR00], which then has been relaxed to the condition that no message is the prefix of another [PR00]. This means that the CBC-MAC can be safely used for messages of different length, if some prefix free encoding is applied.

The EMAC is a popular variant of the CBC-MAC which was developed by the RACE project [BP95], unlike the "plain" CBC-MAC it is secure without any restriction on the message space [PR00]. The EMAC, along with the UMAC, TTMAC and HMAC, is one of the message authentication codes recommended by NESSIE [NES].

THE MODEL. As nowadays usual, we analyse the security of the construction we are interested in (which is $\mathsf{ECBC}_{\pi_1,\pi_2}$) in a setting where the underlying primitive (here π_1, π_2) are realized by their ideal functionality (here uniformly random permutations), thus separating the analysis of the security of the construction from the security of the underlying primitive.[3] More precisely, we prove an upper bound on $\mathbf{Adv}_{\mathsf{ECBC}}(q,n,\ell)$, by which we denote probability of any adversary making q queries of length at most ℓ blocks, in (existentially) forging $\mathsf{ECBC}_{\pi_1,\pi_2}$.

Following [BR00, BPR05], we view the EMAC as a Carter-Wegman MAC [CW79]. This reduces the task of bounding $\mathbf{Adv}_{\mathsf{ECBC}}(q,n,\ell)$ to the task of bounding the probability that there is a collision amongst the CBC-MACs of q messages of length at most ℓ blocks, we denote this probability by $\mathbf{CP}_{q,n,\ell}$ (see (5)). In practice one would instantiate the π_i's by a block-cipher (and not with uniform random permutations). If this block-cipher is secure in the sense of being a good pseudorandom permutation, then the security of the EMAC is basically $\mathbf{CP}_{q,n,\ell}$, thus proving a good bound on this probability translates into improved security guarantees for the EMAC.

KNOWN LOWER BOUNDS. There is a trivial lower bound $\mathbf{CP}_{q,n,\ell} \in \Omega(q^2/2^n)$ for any q, n and $\ell > 1$ as by the birthday bound we can find a collision with probability $\Omega(q^2/2^n)$ for any input shrinking function by asking random queries.[4]

For $q = 2$ [BPR05] show a lower bound of $\mathbf{CP}_{2,n,\ell} \in \Omega(d(\ell)/2^n)$ where $d(\ell) \stackrel{\text{def}}{=} max_{t\leq\ell} |\{x; 1 \leq x \leq 2^n, x|t\}|$ denotes the maximum number of divisors between 1 and 2^n of any number $\leq \ell$. It is known (Theorem 317 in [HW80]) that $D(\ell) \stackrel{\text{def}}{=} max_{t\leq\ell} |\{x; x|t\}| \in \ell^{\Theta(1/\ln\ln\ell)}$, so the same bound applies for $d(\ell)$ if $\ell \leq 2^n$ as then $d(\ell) = D(\ell)$.

---

[2] In particular, it is not existentially unforgeable as shown by the following simple attack: for any $X \in \{0,1\}^n$, request the MAC $C = CBC_\pi(X) = \pi(X)$, and output a message $X \| X \oplus C$ with tag C. This is a successful forgery as $CBC_\pi(X \| X \oplus C) = \pi(\pi(X) \oplus X \oplus C) = \pi(X) = C$.

[3] See e.g. [Mau02] for more detailed discussion of this concept.

[4] For $\ell = 1$ we have $\mathbf{CP}_{q,n,\ell} = 0$ as a permutation does not have collisions.

Where	Upper Bound, $O(.)$ of	Range restriction	Other restrictions
[PR00]	$\ell^2q^2/2^n$	-	-
[BPR05]	$d(\ell)q^2/2^n$	$\ell \in O(2^{n/4})$	-
[DGH+04]	$q^2/2^n$	$\ell \in O(2^{n/3})$	Equal length messages
Here	$q^2/2^n$	$\ell \in O(2^{n/8}), q \in \Omega(\ell^2)$	-

Where	Lower Bound, $\Omega(.)$ of
Folklore (birthday bound)	$q^2/2^n$
[BPR05]	$d(\ell)q/2^n$

Fig. 1. Upper and lower bounds for $\mathbf{CP}_{q,n,\ell}$ (which then imply basically the same bounds for $\mathbf{Adv}_{\mathsf{ECBC}}(q,n,\ell)$)

KNOWN UPPER BOUNDS. Until now the best known upper bound was a $\mathbf{CP}_{n,q,\ell} \in O(d(\ell)q^2/2^n)$ (for $\ell \leq 2^{n/4}$) due to Bellare et al. [BPR05], this bound improved on the $O(\ell^2q^2/2^n)$ bound of Petrank and Rackoff [PR00]).

TIGHT BOUND FOR EQUAL LENGTH. Dodis et al. [DGH+04] investigated a restricted case where the messages have same length (which is uninteresting for the EMAC construction, but this was not their goal), they state a tight collision probability of $\mathbf{CP}_{2,n,\ell} \in O(q^2/2^n)$ (for $\ell \leq 2^{n/3}$) for the CBC-MAC of two messages, which immediately gives an optimal $\mathbf{CP}_{q,n,\ell} \in \Theta(q^2/2^n)$ bound for the collision probability of q *equal length* messages.

OUR CONTRIBUTION. In this paper we prove the optimal bound $\mathbf{CP}_{q,n,\ell} \in \Theta(q^2/2^n)$ for $q \geq \ell^2$ and $\ell \leq 2^{n/8}$. So for this range the security of ECBC (and thus the EMAC) matches the security of an ideal MAC (i.e. the birthday bound) up to constant factors.

THE TECHNIQUE FROM [BPR05]. Both, the "classical" $O(q^2\ell^2/2^n)$ [PR00] and the $O(d(\ell)q^2/2^n)$ upper bound [BPR05] are achieved by first proving an upper bound on $\mathbf{CP}_{2,n,\ell}$, the collision probability of two messages, and then applying the union bound

$$\mathbf{CP}_{q,n,\ell} \leq \frac{q(q-1)}{2} \cdot \mathbf{CP}_{2,n,\ell} \tag{1}$$

to get a bound for $\mathbf{CP}_{q,n,\ell}$. In particular [BPR05] prove that

$$\mathbf{CP}_{2,n,\ell} \leq 2d(\ell)/2^n + 64\ell^4/2^{2n}. \tag{2}$$

This bound is tight up to the higher order term and a factor 2:

$$\mathbf{CP}_{2,n,\ell} \geq d(\ell)/2^n \tag{3}$$

The proof of (2) uses ideas from [DGH+04, Dod05] and goes roughly as follows: For any two messages M_1, M_2 and a permutation π one maps the computation of $\mathsf{CBC}_\pi(M_1)$ and $\mathsf{CBC}_\pi(M_2)$ to a graph (called structure graph) consisting of two paths associated with the message M_1 and M_2 respectively. In this graph the vertices correspond to the outputs of π during this computation.

Each such graph contains zero or more *accidents*, by which one denotes the "unexpected" collisions in the graph. The main technical lemma (Lemma 2 in this paper) now states that the probability (over the choice of π) that some particular structure graph G will appear is exponentially small in the number of accidents of G. From this lemma one gets that the probability that a random structure graph has at least one accident is in $O(\ell^2/2^n)$. We can now use that $\mathsf{CBC}_\pi(M_1) = \mathsf{CBC}_\pi(M_2)$ implies that there must be at least one accident to get a $O(\ell^2/2^n)$ upper bound on $\mathbf{CP}_{2,n,\ell}$, and further with (1) the "classical" $\mathbf{CP}_{q,n,\ell} \in O(q^2\ell^2/2^n)$ bound. But this bound is not tight as having an accident is only necessary, but not sufficient to have $\mathsf{CBC}_\pi(M_1) = \mathsf{CBC}_\pi(M_2)$. By more carefully upper bounding the number of graphs for which $\mathsf{CBC}_\pi(M_1) = \mathsf{CBC}_\pi(M_2)$ by $O(d(\ell)/2^n + \ell^4/2^{2n})$ one gets the (2) bound. Here the $O(d(\ell)/2^n)$ term bounds the graphs which have *exactly* one accident and $\mathsf{CBC}_\pi(M_1) = \mathsf{CBC}_\pi(M_2)$, whereas all graphs with two or more accidents are "generously" bounded by the "higher order" term $O(\ell^4/2^{2n})$, whech will be dominated by the leading $d(\ell)/2^n$ while ℓ is not too large, $\ell \in O(2^{n/4})$ is small enough.

Unfortunately the bound (3) implies that bounding the collision probability for two messages and then using (1) one cannot prove $\mathbf{CP}_{q,n,\ell} \in o(d(\ell)q^2/2^n)$.

PROOF IDEA. The obvious idea to overcome this barrier is to upper bound the number of structure graphs built by many (and not just two) messages. We prove a lemma (Lemma 4) which states that the number of structure graphs built from any k messages of length at most ℓ blocks, having exactly one accident and a collision on the output for some pair of messages, is at most $k(k+\ell^2)$, this then gives the claimed $\mathbf{CP}_{q,n,\ell} \in O(q^2/\ell^2)$ bound. Unfortunately now the graph is so big (i.e. $q\ell$ vertices) that the higher order term which bounds the cases where we have two or more accidents is in the order $q^4\ell^4/2^{2n}$ (so unless we assume some bound $o(2^{n/2})$ on q, we only achieve a tight $O(q^2/2^n)$ for constant ℓ, but this is already achieved by the classical $q^2\ell^2/2^n$ of [PR00]).

Fortunately one can get out of this apparent cul-de-sac using an approach "between" the one just described and the one given by (1). The q messages are divided into q/ℓ^2 sets of size $r = \ell^2$. Now, if there's a collision, then this collision occurs in the union of two (or maybe just one) such sets. For such a union of two sets (of size $2r$) we can now upper bound the probability that there's a collision amongst any two of the $2r$ messages by $O(r^2/2^n)$ as the sets are sufficiently large (such that applying the before-mentioned Lemma 4 gives a $2r(2r+\ell^2) = \Theta(r^2)$ upper bound on the number of structure graphs), but still small enough for the higher order term to be ignored for a reasonable range of ℓ. Finally we get our $\mathbf{CP}_{q,n,\ell} \in O(q^2/2^n)$ bound (for $\ell \leq 2^{n/8}$) from the union bound applied over all pairs of sets.

ABOUT THE RANGE. The tight upper bound $\mathbf{CP}_{q,n,\ell} \in O(q^2/2^n)$ we prove holds for $q \in \Omega(\ell^2)$ and $\ell \in O(2^{n/8})$. In the next two paragraphs we'll shortly discuss those two bounds.

LOWER BOUND ON q. The $\mathbf{CP}_{2,n,\ell} = \Theta(d(\ell)/2^n)$ bound implies (under a reasonable assumption[5]) $\mathbf{CP}_{q,n,\ell} = \Omega(d(\ell)q/2^n)$. Thus $\mathbf{CP}_{q,n,\ell} \in O(q^2/2^n)$ can only hold if we have a lower bound for q of at least $\Omega(d(\ell))$, the bound we actually require is $q \in \Omega(\ell^2)$.[6] But this lower bound on q is not really relevant as long as there's a upper bound $\ll 2^{n/2}$ on ℓ, as it only means that we don't match the birthday bound $O(q^2/2^n)$ for a range of parameters, where the collision probability given by the classical $q^2\ell^2/2^n$ bound is extremely small anyway.
UPPER BOUND ON ℓ. Wlog. we can assume an upper bound $\ell \leq 2^n!$ as considering longer messages makes no sense: note that every $x, 1 \leq x \leq 2^n$ divides $2^n!$ and thus $\mathbf{CP}_{2,n,2^n!} \geq d(2^n!)/2^n = 1$, i.e. we can find a collision with probability one with only two queries.[7] This (doubly exponential) bound is far from the $\ell \leq 2^{n/8}$ we require, and can probably be relaxed already with the techniques used in this paper. One possibility would be via a better counting argument, which means improving on Lemma 4 from this paper (in particular, Claim 2 from the proof of this lemma seems quite loose). Lowering the $O(q(q+\ell^2))$ bound on the number of graphs given by the lemma to $q(q+o(\ell))$ would already allow a range of $\ell \leq 2^{n/(4+o(1))}$. Further, counting graphs with more than just one (but still constantly many) accidents could have the potential to get the bound to $\ell \leq 2^{n/(2+\epsilon)}$ for any $\epsilon > 0$. Such a bound might still be far from the necessary one, but would be sufficient for any practical application as a length of $2^{n/2}$ is quite big already for small block lengths (say $n = 128$ which is the smallest block-length provided by AES).

2 Definitions and the Main Technical Lemma

NOTATION. If x is a string then $|x|$ denotes its length. We let $B_n \stackrel{\text{def}}{=} \{0,1\}^n$. If $X \subseteq \{0,1\}^*$ then $X^{\leq m}$ denotes the set of all non-empty strings formed by concatenating m or fewer strings from X. If S is a set equipped with some probability distribution then $s \stackrel{\$}{\leftarrow} S$ denotes the operation of picking s from S according to this distribution. If no distribution is explicitly specified, it is understood to be uniform. We denote by $\mathrm{Perm}(n)$ the set of all permutations over $\{0,1\}^n$ and with $\mathrm{Func}(n)$ the set of all functions $\{0,1\}^* \to \{0,1\}^n$.

SECURITY. An *adversary* is a computationally unbounded, randomised oracle-algorithm which finally outputs a bit. $\mathcal{A}_{q,n,\ell}$ denotes the class of adversaries that make at most q oracle queries, each of length at most ℓ n-bit blocks. For a family of functions $F\colon B_n^* \to \{0,1\}^n$, the distinguishing advantage of $\mathcal{A}_{q,n,\ell}$ for F is

$$\mathbf{Adv}_F(q,n,\ell) = \max_{A \in \mathcal{A}_{q,n,\ell}} \{\, \mathbf{Adv}_F(A) \,\} \quad \text{where}$$

[5] We must assume that one can generate $q/2$ pairs of messages where each pair achieves the "worst case" collision probability $\Omega(d(\ell)/2^n)$, and moreover the events that any pair of messages collides are sufficiently independent.

[6] As both, the lower $d(\ell)$ and the upper ℓ^2 bound follow by rather loose arguments, the truth is probably strictly in-between, i.e. in $\omega(d(\ell))$ and $o(\ell^2)$.

[7] In fact, with $\ell = 2^n!$ we can forge a message in a no-query attack as for any $X \in B_n$ and $\pi \in \mathrm{Perm}(n)$ one has $\mathsf{CBC}_\pi(X^{2^n!}) = X$.

$$\mathbf{Adv}_F(A) = \Pr[f \xleftarrow{\$} F: A^f \Rightarrow 1] - \Pr[f \xleftarrow{\$} \mathrm{Func}(n): A^f \Rightarrow 1]$$

CBC AND ECBC. Fix $n \geq 1$. Recall that for $M = M^1 \cdots M^m \in B_n^m$ and $\pi\colon B_n \to B_n$ we defined in the introduction

$$\mathsf{CBC}_\pi(M) = C_m \text{ where } C_0 = 0^n \text{ and } C_i = \pi(C_{i-1} \oplus M_i) \text{ for } i = 1, \ldots, m$$

Let $\mathsf{CBC} = \{\mathsf{CBC}_\pi\colon \pi \in \mathrm{Perm}(n)\}$, this set of functions has the distribution induced by picking π uniformly from $\mathrm{Perm}(n)$. The encrypted CBC MAC is

$$\mathsf{ECBC}_{\pi_1,\pi_2}(M) \stackrel{\text{def}}{=} \pi_2(\mathsf{CBC}_{\pi_1}(M))$$

Let $\mathsf{ECBC} = \{\mathsf{ECBC}_{\pi_1,\pi_2}\colon \pi_1, \pi_2 \in \mathrm{Perm}(n)\}$, with the distribution induced by picking π_1, π_2 independently and uniformly at random from $\mathrm{Perm}(n)$.
COLLISIONS. For q distinct messages $M_1, \ldots, M_q \in B_n^*$ we denote by

$$\mathbf{CP}_n(M_1, \ldots, M_q) \quad \Pr_{\pi \xleftarrow{\$} \mathrm{Perm}(n)} [\exists i, j, M_i \neq M_j : \mathsf{CBC}_\pi(M_i) = \mathsf{CBC}_\pi(M_j)]$$

the probability that the CBC-MACs (based on a uniform random permutation) of any two messages collide. The maximum collision probability for any q messages of length at most ℓ n-bit blocks is denoted by

$$\mathbf{CP}_{q,n,\ell} = \max_{M_1,\ldots,M_q \in B_n^{\leq \ell}} \mathbf{CP}_n(M_1, \ldots, M_q) \tag{4}$$

Following [BR00], we view ECBC as an instance of the Carter-Wegman paradigm [CW79]. This enables us to reduce the problem of bounding $\mathbf{Adv}_{\mathsf{ECBC}}(q, n, \ell)$ to bounding the collision probability $\mathbf{CP}_{q,n,\ell}$ as

$$\mathbf{Adv}_{\mathsf{ECBC}}(q, n, \ell) \leq \mathbf{CP}_{q,n,\ell} + q(q-1)/2^{n+1} \tag{5}$$

We prove the following bound on $\mathbf{CP}_{q,n,\ell}$.

Lemma 1. *For any* $q \geq \ell^2$: $\quad \mathbf{CP}_{q,n,\ell} \leq 16 \cdot q^2/2^n + 128 \cdot q^2\ell^8/2^{2n}$

From this lemma and (5) we get that $\mathbf{Adv}_{\mathsf{ECBC}}(q, n, \ell) \in O(q^2/2^n)$ whenever $q \in \Omega(\ell^2)$ and $\ell \in O(2^{n/8})$, for example

Corollary 1. *For any* $q \geq \ell^2$ *and* $\ell \leq 2^{n/8-1}$: $\mathbf{Adv}_{\mathsf{ECBC}}(q, n, \ell) \leq 18 \cdot q^2/2^n$.

3 A Graph-Based Representation of CBC

In this section we review the graph-based approach to bound collision probabilities from (the full version of) [BPR05]. In this approach the collision probability is related to the number of graphs satisfying some property.

We fix for the rest of this section a blocklength $n \geq 1$, the number of messages $t \geq 1$ and t distinct messages $\mathcal{M} \stackrel{\text{def}}{=} \{M_1, \ldots, M_t\}$, where for $1 \leq i \leq t$ we denote with $m_i \geq 1$ the length (in blocks) of the i'th message $M_i \stackrel{\text{def}}{=} M_i^1 \cdots M_i^{m_i} \in B_n^{m_i}$.

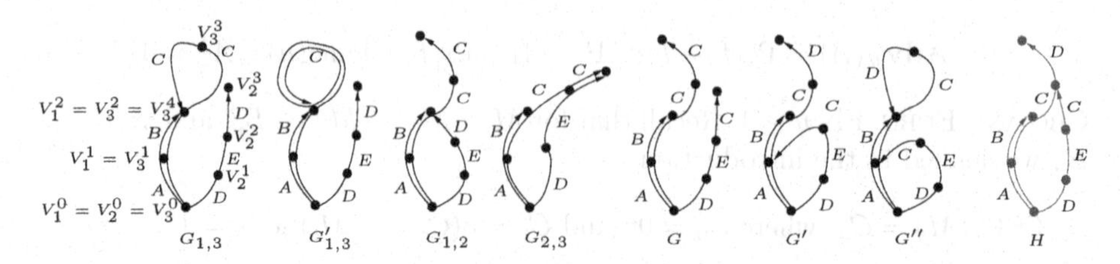

Fig. 2. $\mathcal{G}_{col}(\mathcal{M}) = \{G_{1,3}, G'_{1,3}, G_{1,2}, G_{2,3}\}$ are all structure graphs for $\mathcal{M} = \{AB, DEC, ABCC\}$ which have exactly one accident and a collision on the outputs. Further $\{G, G', G''\} \in \mathcal{G}(\mathcal{M}) \setminus \mathcal{G}_{col}(\mathcal{M})$ are valid structure graphs but not in $\mathcal{G}_{col}(\mathcal{M})$ as: G has 0 and G'' has 2 accidents. G' has exactly one accident but no collision on the outputs. H is not a structure graph as there's a vertex which has two ingoing edges, both labelled C but not being parallel.

For $1 \leq j \leq t$ let $m^j = \sum_{i=1}^{j} m_i$ be the length of the first j messages. It is convenient to set $m^0 \stackrel{\text{def}}{=} 0$ and $m \stackrel{\text{def}}{=} m^t$ to be the total length. Let $M \stackrel{\text{def}}{=} M_1 \| M_2 \| \cdots \| M_t$ denote the concatenation of all messages and M^i the i'th block of M, i.e. $M \stackrel{\text{def}}{=} M^1 \cdots M^m$.

Structure graphs. To $\mathcal{M}$ and any $\pi \in \text{Perm}(n)$ we associate the *structure graph* $G_\pi^{\mathcal{M}}$, which is a directed graph (V, E) where $V \subseteq [0, \ldots, m]$.

The structure graph $G_\pi^{\mathcal{M}} = G = (V, E)$ is defined as follows: We set $C_0 = 0^n$ and for $i = 1, \ldots, m$ we define

$$C_i = \begin{cases} \pi(C_{i-1} \oplus M^i) & \text{if } i \notin [m_0 + 1, \ldots, m_{t-1} + 1] \\ \pi(M^i) & \text{otherwise} \end{cases}$$

From this C_i's we define the mapping $[.]_G : [0, \ldots, m] \to [0, \ldots, m]$ as $[i]_G = \min\{j : C_j = C_i\}$. It is convenient to define a mapping $[.]'_G$ as $[i]'_G = [i]_G$ if $i \notin [m^0, \ldots, m^{t-1}]$ and $[i]'_G = 0$ otherwise. Now the structure graph $G_\pi^{\mathcal{M}} = G = (V, E)$ is given by

$$V = \{[i]_G : 0 \leq i \leq m\} \qquad E = \{([i-1]'_G, [i]_G) : 1 \leq i \leq m\}$$

From this definition it is clear that the mapping $[.]_G$ defines G uniquely and vice versa. Throughout the "i'th edge of G" refers to the edge $([i-1]'_G, [i]_G)$ (note that this not injective) and the "label" of the i'th edge is M^i.

If the C_i's are all distinct, then G is simply a star like tree with t paths leaving the root 0, the i'th path being $0 \to m^{i-1}+1 \to \ldots \to m^{i-1}+m_i = m^i$. In general G is the graph one gets by starting with the tree just described and doing the following while possible: if there are two vertices i, j where $i \neq j$ and $C_i = C_j$ then collapse i and j into one vertex and label it $\min\{i, j\}$.

For a structure graph G we will denote the vertices on the path built by the i'th message by $V_i^0(G), V_i^1(G), \ldots, V_i^{m_i}(G)$, we call this path the i-path, we write V_i^j for $V_i^j(G)$ if G is understood (cf. Figure 2).

Let $\mathcal{G}(\mathcal{M}) = \{G_\pi : \pi \in \text{Perm}(n)\}$ denote the set of all structure graphs associated to messages $\mathcal{M}$. This set has the probability distribution induced by picking π at random from $\text{Perm}(n)$.

COLLISIONS. Suppose a structure graph $G = G_\pi^{\mathcal{M}} \in \mathcal{G}(\mathcal{M})$ is exposed edge by edge (i.e. in step i the value $[i]_G$ is shown to us). We say that G has a *collision* in step i if the edge exposed in step i points to a vertex which is already in the graph. With $\mathsf{Col}(G)$ we denote all collisions, i.e. all pairs (i, j) where in step i there was a collision which hit the vertex computed in step $j < i$:

$$\mathsf{Col}(G) = \{(i, [i]_G) : [i]_G \neq i\}$$

We distinguish between *induced collisions* IndCol and *accidents* Acc where

$$\mathsf{Col}(G) = \mathsf{Acc}(G) \cup \mathsf{IndCol}(G) \qquad \mathsf{Acc}(G) \cap \mathsf{IndCol}(G) = \emptyset$$

Informally, an induced collision in step i is a collision which is implied by the collisions in the first $i-1$ steps, whereas an accident is a "surprising" collision.

The following lemma is the heart of the whole approach, it states that the probability that a randomly sampled structure graph will be some particular graph H is exponentially small in $\mathsf{Acc}(H)$.

Lemma 2. *Let $n \geq 1, t \geq 1, \mathcal{M} = \{M_1, \ldots, M_t\}$ where $M_i \in B_n^{m_i}$ and $m = m_1 + \ldots + m_t$. Then for any structure graph $H \in \mathcal{G}(\mathcal{M})$:*

$$\Pr[G \xleftarrow{\$} \mathcal{G}(\mathcal{M}) :\ G = H] \leq (2^n - m)^{-|\mathsf{Acc}(H)|}$$

Form this lemma we get the following bound on the probability that a random structure graph has two or more accidents:

Lemma 3. *With $\mathcal{M}, m$ as in the previous lemma*

$$\Pr[G \xleftarrow{\$} \mathcal{G}(\mathcal{M}) :\ |\mathsf{Acc}(G)| \geq 2] \leq 4m^4/2^{2n}$$

The proofs of Lemma 2 and 3 can be found in the full version of [BPR05].

SOME USEFUL FACTS. In [BPR05] accidents are formally defined to be exactly those collisions which do *not close a (even length) cycle with alternating edge directions*. It is shown that this are exactly those collisions which are "surprising" in the sense that they are not induced by the already exposed edges. We will not need to work with this formal definition of accidents here, it will be sufficient to consider the more intuitive concept of *true collisions*, which are all collisions except those where no edge is added, or equivalently, we have a true collision in some step i if in this step we add a new edge, but no new vertex (from this definition we see that in a structure graph $G = (V, E)$ the number of true collisions is $|E| - |V| + 1$). Also, it's not hard to see that if G has k accidents, then it has at least k true collisions.[8] Although the converse is not true in general, there are implications in the other direction which will be sufficient for us. In

[8] This follows from the definitions, recall that accidents are those collisions which do *not close a cycle with alternating edge directions*, and true collisions are those which do *not close a cycle with alternating edge directions of length 2* (as such a cycle is given by two parallel edges). So true collisions are just a subset of the accidents.

particular, it's not hard to see that the first true collision that occurs must always be an accident. And if we only consider structure graphs built by at most two paths, then also the second true collision is necessarily an accident (see Lemma 10 in the full version of [BPR05], fact (i) below follows from this).

For $G \in \mathcal{G}(\mathcal{M})$ and i, j, $1 \leq i < j \leq q$ let $G_{[i,j]}$ denote the subgraph of G built by the i-path and the j-path. We will need the following facts:

(i) If G has at most one accident, then for any i, j the $G_{[i,j]}$ has at most one true collision.
(ii) If G has exactly one accident, then G is uniquely determined by $\mathcal{M}$ and any subgraph of G which contains a true collision.

Informally, fact (ii) holds as given the single accident, we know the only "surprising" collision, and thus can deterministically extend the subgraph to G.

4 Bounding $\mathbf{CP}_{q,n,\ell}$

For $i = 1, \ldots, q$, let $M_i \in B_n^{\leq \ell}$ be such that the collision probability is maximised, i.e. with $\mathcal{M} = \{M_1, \ldots, M_q\}$ we have $\mathbf{CP}_{q,n,\ell} = \mathbf{CP}(\mathcal{M})$. To bound $\mathbf{CP}(\mathcal{M})$ we now consider the random experiment where a permutation π is chosen at random and $\mathsf{CBC}_\pi(M_i)$ is computed for $i = 1, \ldots, q$. We can decide whether there was a collision $\mathsf{CBC}_\pi(M_i) = \mathsf{CBC}_\pi(M_j)$ given the structure graph $G_\pi^{\mathcal{M}}$ of this computation. Thus we see $\mathbf{CP}_{q,n,\ell}$ as the probability that $G_\pi^{\mathcal{M}}$ (for a random π) contains such a collision on the outputs of two messages. Let $\mathcal{G}_{col}(\mathcal{M}) \subset \mathcal{G}(\mathcal{M})$ denote the subset of structure graphs where there's a collision on the outputs:

$$\mathcal{G}_{col}(\mathcal{M}) \stackrel{\text{def}}{=} \{G \in \mathcal{G}(\mathcal{M}) \ ; \ \exists i, j, 1 \leq i < j \leq q : V_i^{m_i}(G) = V_j^{m_j}(G)\}$$

As just said, with this definition $\mathbf{CP}_{q,n,\ell} = \Pr_{G \stackrel{\$}{\leftarrow} \mathcal{G}(\mathcal{M})}[G \in \mathcal{G}_{col}(\mathcal{M})]$.

We split this probability into the "single accident" and the "two or more accidents" case. For this let $\mathcal{G}_{col}^i \stackrel{\text{def}}{=} \{G \in \mathcal{G}_{col}(\mathcal{M}) \ ; \ |\mathsf{Acc}(G)| = i\}$, now

$$\mathbf{CP}_{q,n,\ell} = \Pr_{G \stackrel{\$}{\leftarrow} \mathcal{G}(\mathcal{M})}[G \in \mathcal{G}_{col}^1(\mathcal{M})] + \Pr_{G \stackrel{\$}{\leftarrow} \mathcal{G}(\mathcal{M})}[G \in \mathcal{G}_{col}^i(\mathcal{M}) \text{ for some } i \geq 2]. \quad (6)$$

To bound the second term on the rhs. of (6) we can use Lemma 3 and "generously" upper bound the probability that there are two or more accidents.

$$\Pr_{G \stackrel{\$}{\leftarrow} \mathcal{G}(\mathcal{M})}[G \in \mathcal{G}_{col}^i(\mathcal{M}) \text{ for some } i \geq 2] \leq \Pr_{G \stackrel{\$}{\leftarrow} \mathcal{G}(\mathcal{M})}[|\mathsf{Acc}(G)| \geq 2] \leq \frac{4q^4\ell^4}{2^{2n}}. \quad (7)$$

To bound the first term on the rhs. of (6) we can't be so generous any more and simply upper bound the probability of $|\mathsf{Acc}(G)| = 1$ as this would only give a $O(q^2\ell^2/2^n)$ bound. We will more carefully upper bound $|\mathcal{G}_{col}^1(\mathcal{M})|$ (by Lemma 4 below), and then apply Lemma 2 which in our case states that $G \in \mathcal{G}_{col}^1(\mathcal{M})$ appears with

$$\Pr_{G \stackrel{\$}{\leftarrow} \mathcal{G}(\mathcal{M})}[G \in \mathcal{G}_{col}^1(\mathcal{M})] \leq \frac{|\mathcal{G}_{col}^1(\mathcal{M})|}{2^n - \ell q}. \quad (8)$$

Lemma 4. *Let $n, q \geq 1$ and $1 \leq m_1, \ldots, m_q \leq \ell$ and $\mathcal{M} = \{M_1, \ldots, M_q\}$ with $M_i \in B_n^{m_i}$ be distinct messages, then*

$$|\mathcal{G}^1_{col}(\mathcal{M})| \leq q(q + \ell + \ell^2)/2$$

Now combining (6)-(8) and the above Lemma we get:

Lemma 5. $\mathbf{CP}_{q,n,\ell} \leq \frac{q(q+\ell+\ell^2)}{2(2^n - \ell q)} + \frac{4q^4\ell^4}{2^{2n}}$

This already gives $\mathbf{CP}_{q,n,\ell} \in O(q^2/2^n)$ for $q^2\ell^4 \in O(2^n)$ and $q \in \Omega(\ell^2)$. But we can do better. The reason why this bound is not so great is that the term which bounds the "two or more" accident case is of rather large order $q^4\ell^4/2^n$ as we consider a graph (i.e. a total message length) of size $q\ell$. We achieve the bound claimed by Lemma 1 by splitting the messages in chunks of size ℓ^2 (with foresight) and then applying the following lemma which is a generalisation of (1).

Lemma 6. *If r divides q then* $\mathbf{CP}_{q,n,\ell} \leq \mathbf{CP}_{2r,n,\ell} \cdot \frac{q(q-r)}{2 \cdot r^2}$

Proof. Consider q messages $M_1, \ldots, M_q$ where $\mathbf{CP}_{q,n,\ell} = \mathbf{CP}_n(M_1, \ldots, M_q)$. We split the q messages into q/r sets $S_1, \ldots, S_{q/r}$, each containing r messages. If two messages collide, then there are two sets containing this two messages, so using the union bound $\mathbf{CP}_n(M_1, \ldots, M_q) \leq \sum_{i,j,1 \leq i < j \leq q/r} \mathbf{CP}_n(S_i, S_j)$.
The lemma follows as by definition $\mathbf{CP}_n(S_i, S_j) \leq \mathbf{CP}_{2r,n,\ell}$ and the sum has $q(q-r)/2r^2$ terms. □

We now have all ingredients to prove our main result

Proof (of Lemma 1). Let $\widetilde{q}$ be minimal satisfying $\widetilde{q} \geq q$ and $\ell^2 | \widetilde{q}$. Now using Lemma 6 (with $r = \ell^2$) in the second, and Lemma 5 in the third step

$$\mathbf{CP}_{q,n,\ell} \leq \mathbf{CP}_{\widetilde{q},n,\ell} \leq \mathbf{CP}_{2\ell^2,n,\ell} \cdot \frac{\widetilde{q}^2}{2\ell^4} \leq \left(\frac{\ell^2(3\ell^2 + \ell)}{2^n - 2\ell^3} + \frac{4(2\ell^2)^4\ell^4}{2^{2n}} \right) \frac{\widetilde{q}^2}{2\ell^4} \quad (9)$$

We can assume that $2\ell^3 \leq 2^{n-1}$ and $n > 1$ as otherwise the above is ≥ 1 which is a trivial upper bound for $\mathbf{CP}_{q,n,\ell}$. We also have $\widetilde{q} < 2q$ by the $q \geq \ell^2$ precondition an can further simplify (9) to $\mathbf{CP}_{q,n,\ell} \leq \frac{16 \cdot q^2}{2^n} + \frac{128 \cdot q^2 \cdot \ell^8}{2^{2n}}$. □

Proof (of Lemma 4). Wlog. we assume that $m_j \leq m_{j+1}$ for $1 \leq j \leq q-1$. Let

$$\mathcal{G}_{i,j} = \{G \in \mathcal{G}_{col}(\mathcal{M}) \ ; \ V_i^{m_i}(G) = V_j^{m_j}(G) \wedge |\mathsf{Acc}(G) = 1|\}$$

denote the structure graphs with exactly one accident, and where there's a collision on the outputs of the i'th and j'th message. Let $\mathcal{P}_j \subseteq [j-1]$ denote the indices of the messages which are prefixes of M_j after the common suffix has been removed, more formally

$$\mathcal{P}_j = \{i \in [1, \ldots, j-1] \ ; \exists S, X \in B_n^* \ : \ M_j = M'_j \| S, M_i = M'_i \| S, M'_j = M'_i \| X\}$$

Let $\overline{\mathcal{P}} = [1, \ldots, j-1] \setminus \mathcal{P}$. For example if $\mathcal{M} = \{M_1 = A, M_2 = AB, M_3 = ABC, M_4 = ACDB\}$ then $\mathcal{P}_4 = \{1, 2\}$ and $\overline{\mathcal{P}}_4 = \{3\}$.

We will prove two claims, which then will imply the statement of the lemma. The first claim — which is basically Lemma 19 from [BPR05] — states that if

$i \in \overline{\mathcal{P}}_j$, then there's at most one structure graph with exactly one accident where M_i and M_j collide.

The second claim bounds the number of structure graphs having one accident and a collision between M_j and any other message M_i where $i \in \mathcal{P}_j$ by $\ell(\ell+1)/2$ (note that this bound only depends on the length, but not on the number of messages considered). To prove this claim we use the simple observation that if there's a collision between M_j and any M_i where $i \in \mathcal{P}_j$, then it must be the case that the j-path makes a loop. So we can upper bound the number of structure graphs having such a collision by the number of structure graphs having where the j-path loops.

Fig. 3. Figure for proof of Claim 1

Claim 1. *For each* $i \in \overline{\mathcal{P}}_j$, $|\mathcal{G}_{i,j}| \leq 1$.

Proof (of Claim). Let P denote the common prefix and S the common suffix of M_i and M_j. So $M_i = P\|M_i'\|S$ and $M_j = P\|M_j'\|S$ where the M_i' and M_j' are nonempty as $i \in \overline{\mathcal{P}}_j$. Let $p = |P|/n, s = |S|/n$.

By definition $G \in \mathcal{G}_{i,j}$ means $V_i^{m_i} = V_j^{m_j}$, this implies that also $V_i^{m_i-s} = V_j^{m_j-s}$ (as for the last s steps the i and j path must go in parallel). Now as $M_i^{m_i-s-1} \neq M_j^{m_j-s-1}$ (otherwise we could extend the suffix) we have $V_i^{m_i-s-1} \neq V_j^{m_j-s-1}$ (because in a structure graph two edges with distinct labels cannot be parallel). So there's a true collision in $G_{[i,j]}$ which hits the vertex $V_i^{m_i-s}$.

As by fact (i)[9] there can be only one true collision in $G_{[i,j]}$ this means that the "suffix path" $V_i^{m_i-s} = V_j^{m_j-s} \to \ldots \to V_i^{m_i} = V_j^{m_j}$ has no loops. For the same reason the "prefix path" $V_i^1 = V_j^1 \to \ldots \to V_i^p = V_j^p$ makes no loop and also the prefix and suffix paths must be disjoint. So the subgraph of $G_{[i,j]}$ built by the first $p+1$ and the last $s+1$ edges of the i and j path looks like shown on the left in Figure 3. There's only way to extend this subgraph to the full $G_{[i,j]}$ without introducing more true collisions, this is the second graph in Figure 3.

So there's only one possible $G_{[i,j]}$, and by fact (ii) it uniquely determines the whole structure graph, thus there's just one $G \in \mathcal{G}_{i,j}$. △

Claim 2. $\left|\bigcup_{i \in \mathcal{P}} \mathcal{G}_{i,j}\right| \leq \ell(\ell+1)/2.$

Proof (of Claim). Consider any $i \in \mathcal{P}$, and let S denote the common suffix of M_i and M_j. Now, as $i \in \mathcal{P}$, for some P we can write $M_j = P\|M_j'\|S$ and $M_i = P\|S$. Let $p = |P|/n$ and $s = |S|/n$.

[9] We refer to the facts stated at the end of Section 3.

Consider any $G \in \mathcal{G}_{i,j}$, by definition $V_i^{m_i} = V_j^{m_j}$, which implies $V_i^{m_i-s} = V_j^{m_j-s}$ as the last s blocks are equal. And as the first p blocks are equal we have $V_i^p = V_j^p$. Now $V_i^p = V_i^{m_i-s}$ and thus also $V_j^p = V_j^{m_j-s}$, as $p < m_j - s$ there's a true collision on the j-path (i.e. it contains a loop). As there are at most $m_j(m_j+1)/2$ possibilities for the j-path to make a loop[10] and as by fact (ii) the shape of the j path determines G completely, there can be at most $m_j(m_j+1)/2 \leq \ell(\ell+1)/2$ different G's in $\bigcup_{i \in \mathcal{P}} \mathcal{G}_{i,j}$. △

The lemma follows by the two claims as

$$|\mathcal{G}^1_{col}(\mathcal{M})| \leq \sum_{1 \leq i < j \leq q} |\mathcal{G}_{i,j}| \leq \sum_{j=1}^{q} \left(\left| \bigcup_{i \in \mathcal{P}} \mathcal{G}_{i,j} \right| + \sum_{i \in \overline{\mathcal{P}}} |\mathcal{G}_{i,j}| \right)$$

$$\leq \sum_{j=1}^{q} \left(\frac{\ell(\ell+1)}{2} + j - 1 \right) \leq \frac{q(q-1+\ell(\ell+1))}{2}. \qquad \square$$

References

[BKR00] Mihir Bellare, Joe Kilian, and Phillip Rogaway. The security of the cipher block chaining message authentication code. *Journal of Computer and System Sciences*, 61(3):362–399, 2000. Earlier version in *Crypto '94.*

[BP95] Antoon Bosselaers and Bart Preneel, editors. *Integrity Primitives for Secure Information Systems, Final Report of RACE Integrity Primitives Evaluation RIPE-RACE 1040*, volume 1007 of *LNCS* Springer, 1995.

[BPR05] Mihir Bellare, Krzysztof Pietrzak, and Phillip Rogaway. Improved security analyses for CBC MACs. In *Proc. Crypto '05.* Full Version on `www.crypto.ethz.ch/~pietrzak/publications.html`.

[BR00] John Black and Phillip Rogaway. CBC MACs for arbitrary-length messages: The three-key constructions. In*Proc. Crypto '00.*

[CW79] Larry Carter and Mark N. Wegman. Universal classes of hash functions. *Journal of Computer and System Sciences (JCSS)*, 18:143–154, 1979.

[DGH+04] Yevgeniy Dodis, Rosario Gennaro, Johan Håstad, Hugo Krawczyk, and Tal Rabin. Randomness Extraction and Key Derivation Using the CBC, Cascade and HMAC Modes. In *Proc. Crypto '04.*

[Dod05] Yevgeniy Dodis, 2005. Personal Communication.

[HW80] G. Hardy and E. Wright. *An Introduction to the Theory of Numbers.* Oxford University Press, 1980.

[Mau02] Ueli Maurer. Indistinguishability of random systems. In *Proc. Eurocrypt '02.*

[NES] NESSIE. European project ist-1999-12324 on new european schemes for signature, integrity and encryption. `http://www.cryptonessie.org`.

[PR00] Erez Petrank and Charles Rackoff. Cbc mac for real-time data sources. *Journal of Computer and System Sciences*, pages 315–338, 2000.

[10] This observation is trivial, it also follows from the (more general) equation (6) in [BPR05].

Constructing Single- and Multi-output Boolean Functions with Maximal Algebraic Immunity

Frederik Armknecht and Matthias Krause

[1] Network Laboratories
NEC Europe Ltd.
69115 Heidelberg, Germany
[2] Lehrstuhl Theoretische Informatik
Universität Mannheim
68131 Mannheim, Germany

Abstract. The aim of this paper is to construct boolean functions $f : \{0,1\}^n \longrightarrow \{0,1\}^m$, for which the graph $gr(f) = \{(x, f(x)), x \in \{0,1\}^n\} \subseteq \{0,1\}^{n+m}$ has maximal algebraic immunity. This research is motivated by the need for appropriate boolean functions serving as building blocks of symmetric ciphers. Such functions should have large algebraic immunity for preventing vulnerability of the cipher against algebraic attacks. We completely solve the problem of constructing explicitely defined single-output functions for which the graph has maximal algebraic immunity. Concerning multi-output functions, we present an efficient algorithm, based on matroid union, which computes for given m, n, d the table of a function $h : \{0,1\}^n \longrightarrow \{0,1\}^m$ for which the algebraic immunity of the graph is greater than d. To the best of our knowledge, this is the first systematic method for constructing multi-output functions of high algebraic immunity.

Keywords: Cryptographic primitives, boolean functions, algebraic attacks, matroid union algorithm.

1 Introduction

The degree, $\deg(p)$, of a single-output boolean function $p : \{0,1\}^n \longrightarrow \{0,1\}$ is defined as the length of a longest monomial occurring in the ring-sum-expansion of $p = \bigoplus_{\alpha \in \{0,1\}^n} p_\alpha m_\alpha$ of p, i.e. $\deg(p) = \max\{|\alpha|, \alpha \in \{0,1\}^n, p_\alpha \neq 0\}$. (As usual, $|\alpha|$ denotes the number of ones in α, and $m_\alpha = \Pi_{i,\alpha_i=1} x_i$.)

We say that a function $p : \{0,1\}^n \longrightarrow \{0,1\}$ annihilates a subset $S \subseteq \{0,1\}^n$ (or, equivalently, is an annihilator of S) if $p(x) = 0$ for all $x \in S$. The algebraic immunity, $AI(S)$, of S is defined to be the minimal d for which there is degree-d annihilator $p \not\equiv 0$ of S.

Following Meier, Pasalic and Carlet (2004), the algebraic immunity $AI(f)$ of a single output function $f : \{0,1\}^n \longrightarrow \{0,1\}$ is defined to be the minimum of $AI(f^{-1}(0))$ and $AI(f^{-1}(1))$. This definition can be easily generalized to multi-output functions $f : \{0,1\}^n \longrightarrow \{0,1\}^m$, $AI(f)$ is is defined to be the minimum of $AI(f^{-1}(z))$ over all $z \in \{0,1\}^m$.

M. Bugliesi et al. (Eds.): ICALP 2006, Part II, LNCS 4052, pp. 180–191, 2006.

For boolean functions used as building blocks in cryptographic systems (like, e.g., S-Boxes) it is important to know whether there exist nontrivial low degree annihilating relations between input- and output bits. Corresponding to this, the algebraic immunity $AI(gr(f))$ of the graph $gr(f) = \{(x, f(x)), x \in \{0,1\}^n\} \subseteq \{0,1\}^{n+m}$ is a further important design parameter of cryptographic boolean functions $f : \{0,1\}^n \longrightarrow \{0,1\}^m$. Note that for all boolean functions $f : \{0,1\}^n \longrightarrow \{0,1\}^m$ it holds that $AI(f) \leq AI(gr(f)) \leq AI(f) + m$ (see Lemma 1 below).

The aim of this paper is to construct boolean single- and multi-output functions f for which $AI(gr(f))$ is maximal. This is motivated by the necessity of making secret-key cryptosystems immun against algebraic attacks, which are based on defining and solving systems of multivariate equations in the variables corresponding to the bits of a secret key.

Algebraic attacks on secret-key cryptosystems consist in detecting nontrivial low-degree annihilators of relations between secret input- and output bits for building a system of low-degree equations in the keybits, and trying to solve it efficiently.

Algebraic attacks on simple (memoryless) combiners, a special class of keystream generators, have been firstly described by Courtois and Meier (2003), using relations on known outputs and corresponding unknown internal bits. Keystream generators are finite state machines which produce on the basis of a secret key a secret bitstreams of arbitrary lengths. Armknecht and Krause (2003) extended these attacks to the more general class of combiners with memory, including the E_0-generator used in the Bluetooth standard.

In general, solving a system of T degree-d equations over $\mathbf{F}_2$ is NP-hard even for $d = 2$. However, if T is greater than the number of unknowns, there is a certain chance (which is hard to evaluate theoretically) that nontrivial approaches like Gröbner bases succeed (e.g., see Faugère and Ars (2003)). If T even exceeds the number of occurring monomials, efficient strategies exist (Shamir et al. (2000)). In both cases, the effort is heavily influenced by the degree of the relations. In this context, the notion of the "algebraic immunity of a single-output function f" has been introduced by Meier, Pasalic and Carlet (2004), and further developed by Armknecht (2005).

Since the (hypothetical) attack on the Advanced Encryption Standard (AES) presented by Courtois and Pieprzyk (2002), the question of the existence of efficient algebraic attacks on round-based block ciphers attracted a lot of public interest. Contrary to the case of keystream generators, the system of equations obtained here is generally not overdefined but may have a very low degree which is defined by the algebraic immunity of the S-Boxes. For example, quadratic equations exist in the case of AES, although the input/output format of the S-Boxes AES (8/8) would allow an algebraic immunity of 3 for the graph of these S-Boxes. Even though the feasibility of these attacks is still unknown, a huge number of corresponding approaches and results (e.g., see Murphy, Robshaw (2003), Ars et al. (2004) and Cid, Leurent (2005)) shows the interest on this topic.

We think that these developments are motivation enough to study the concepts of the algebraic immunity of boolean functions and graphs of boolean functions in more detail.

In section 2 we completely solve the problem of constructing single-output functions $f : \{0,1\}^n \longrightarrow \{0,1\}$, for which $AI(f)$ and $AI(gr(f))$ are both maximal. We will see that it is quite straightforward to solve this problem for odd n , while it is more nontrivial to construct such functions for even n.

Up to now, for $m > 1$, one is not able to give the explicit definition of a sequence of functions $h_n : \{0,1\}^n \longrightarrow \{0,1\}^m$ of maximal immunity. However, in section 3 we derive a polynomial time algorithm based on matroid union which outputs for given n, m, d the table of a function $h : \{0,1\}^n \longrightarrow \{0,1\}^m$, for which $AI(gr(h))$ is at least d. This implies the first efficient method so far to construct S-boxes of arbitrary input/output format having maximal algebraic immunity. In section 4 we present first experimental results which imply interesting theoretical problems for further research.

Note that so far, our constructions refer only to one out of several important security parameters of boolean functions, the algebraic immunity. Very recently, Carlet (2006) and Carlet, Dalai, Gupta, Maitra (2006) obtained results which relate algebraic immunity of single-output functions to other relevant security parameters like balancedness, nonlinearity and correlation immunity.

For all natural $d \leq n$ let $W_n^{\leq d}$ (resp. $W_n^{=d}$, $W_n^{<d}$, $W_n^{\geq d}$, $W_n^{>d}$) denote the set of all $\alpha \in \{0,1\}^n$ with $|\alpha| \leq d$ (resp. $|\alpha| = d$, $|\alpha| < d$, $|\alpha| \geq d$, $|\alpha| > d$). Further let $\Phi_n(d) = |W_n^{\leq d}| = \sum_{i=0}^{d} \binom{n}{i}$, and $\Phi_n^{-1}(D) = \min\{d, \Phi_n(d) > D\}$.

For all positive integers n we denote by M^n the $2^n \times 2^n$-matrix for which rows and columns are labelled by all $\alpha \in \{0,1\}^n$ and $x \in \{0,1\}^n$, respectively, and for which $M^n_{\alpha,x} = m_\alpha(x)$ (which is 1 iff $\{i, \alpha_i = 1\} \subseteq \{i, x_i = 1\}$). For all $d \leq n$ and $S \subseteq \{0,1\}^n$ we denote by $M^n_{d,S}$ the $\Phi_n(d) \times |S|$-submatrix of M^n corresponding to the rows labelled by elements $\alpha \in W_n^{\leq d}$ and columns labelled by all $x \in S$.

We identify each degree d boolean function $p = \sum_{\alpha \in W_n^{\leq d}} p_\alpha m_\alpha$, with its coefficient vector $\overrightarrow{p} = (p_\alpha)_{\alpha \in W_n^{\leq d}}$. Note that p annihilating S is equivalent to ${}^T\overrightarrow{p} \circ M^n_{d,S} = \overrightarrow{0}$ where ${}^T\overrightarrow{p}$ is the transponent of the vector $\overrightarrow{p}$.

Consequently, for subsets $S \subset \{0,1\}^n$ the set of all degree-d polynomials annihilating S can be computed by solving a system of $|S|$ linear equations in $\Phi_n(d)$ unknowns, which implies that the immunity of S can be at most $\Phi_n^{-1}(|S|)$.

It is quite straightforward to construct sets $S \subseteq \{0,1\}^n$ of maximal possible immunity $\Phi_n^{-1}(|S|)$. Consider the linear ordering $\omega : \{0,1\}^n \longrightarrow \{0, \cdots, 2^n - 1\}$ on $\{0,1\}^n$ defined by $\omega(\overrightarrow{0}) = 0$, and $\omega(x) < \omega(x')$ if $|x| < |x'|$ or, if $|x| = |x'|$, x is lexicographically less than x'.

For all natural $D < 2^n$ we define the set $A_n^D \subseteq \{0,1\}^n$ to consist of the first D elements of $\{0,1\}^n$ ordered with respect to ω. As, if rows and columnes are ordered with respect to w, the matrix M^n_{d,A_n^D} is a triangular matrix with 1's on the diagonal, the set A_n^D is $\Phi_n^{-1}(D)$-immun.

In the following we will deal with the more nontrivial problem to construct boolean functions (i.e. special sets $A \subseteq \{0,1\}^{n+m}$ which correspond to the graph of a boolean function $f : \{0,1\}^n \longrightarrow \{0,1\}^m$) with maximal immunity. We conclude this Introduction with the following lemma stating a basic relation between the immunity parameters $AI(f)$ and $AI(gr(f))$.

Lemma 1. *For all $f : \{0,1\}^n \longrightarrow \{0,1\}^m$ it holds that $AI(f) \leq AI(gr(f)) \leq AI(f) + m$.*

Proof. The upper bound follows from the fact that if p is a nonzero annihilator of $f^{-1}(z)$ for some $z \in \{0,1\}^m$ then $p \cdot \Pi_{i,z_i=1} z_i \cdot \Pi_{i,z_i=0}(z_i \oplus 1)$ is a nonzero annihilator of $gr(f)$. On the other hand, let q be an annihilator of $gr(f)$ of minimal degree, and fix some $z \in \{0,1\}^m$ such that $q(\cdot, z) \not\equiv 0$. Then $q(\cdot, z)$ is a nontrivial annihilator of $f^{-1}(z)$ of degree at most $\deg(q)$.

2 Single Output Boolean Functions of Maximal Immunity

In this section we construct single-output functions $f : \{0,1\}^n \longrightarrow \{0,1\}$ for which $AI(f)$ and $AI(gr(f))$ are both maximal. Note that $\Phi_n^{-1}(2^{n-1})$ is an upper bound for $AI(f)$ and that $\Phi_{n+1}^{-1}(2^n)$ is an upper bound for $AI(gr(f))$.

For n odd it holds $\Phi_n^{-1}(2^{n-1}) = \Phi_{n+1}^{-1}(2^n) = \lceil n/2 \rceil$, and it is easy to construct functions f with $AI(f) = AI(gr(f)) = \lceil n/2 \rceil$. As the complement of $A_n^{2^{n-1}}$ is the affine translation of $A_n^{2^{n-1}}$ by $\vec{1}$, and as the algebraic immunity is invariant under affine translations, the characteristic function of $A_n^{2^{n-1}}$ (which equals negated majority) has this property.

For even n the situation is more complicated as it holds $\Phi_n^{-1}(2^{n-1}) = n/2$, but $\Phi_{n+1}^{-1}(2^n) = n/2 + 1$. The question is how to construct functions f with $AI(gr(f)) = n/2 + 1$, which by Lemma 1 implies that $AI(f) = n/2$.

An intuitive candidate is again the characteristic function of $A_n^{2^{n-1}}$. However, it is quite straightforward to show that if $f^{-1}(0)$ is an affine translation of $f^{-1}(1)$ then $AI(f) = AI(gr(f))$. Indeed, let $f^{-1}(0) = f^{-1}(1) \oplus \vec{v}$ and p be an annihilator of $f^{-1}(1)$ of degree d. Then $p(x \oplus \vec{v})$ and $zp \oplus (z \oplus 1)p(x \oplus \vec{v})$ are degree-d annihilators of $f^{-1}(0)$ and $gr(f)$, respectively. This implies that the graph of the characteristic function of $A_n^{2^{n-1}}$ has algebraic immunity of only $n/2$.

The following approach is more successfull. For n even and subsets $A \subseteq W_n^{=n/2}$ let us consider functions of type f_A defined as $f^A(x) := 0$ iff $x \in W_n^{<n/2} \cup A$. We will see that for certain subsets $A \subseteq W_n^{=n/2}$ it holds that $AI(gr(f^A)) = n/2 + 1$.

Theorem 1. *For all even $n \geq 2$ and all nonempty $A \subseteq W_n^{=n/2}$ it holds that $AI(gr(f^A)) = n/2 + 1$ if and only if $A = A \oplus \vec{1} = \{x \oplus \vec{1}; x \in A\}$.*

Proof. Let $A \subseteq W_n^{=n/2}$ be arbitrarily fixed and denote $B = W_n^{=n/2} \setminus A$.

For a set $T \subseteq \{0,1\}^n$ let us call a boolean function $p = p(x_1, \cdots, x_n)$ to be a T-polynomial, if p can be written as $p = \bigoplus_{\alpha \in T} c_\alpha m_\alpha$.

Let $P = P(x_1, \cdots, x_{n+1})$ be an nonzero annihilator of $gr(f^A)$ of minimal degree. We write $P = p \oplus z \cdot q$, where p and q depend only on $x_1, \cdots, x_n$, p

annihilates $A \cup W_n^{<n/2}$ and $p \oplus q$ annihilates $B \cup W_n^{>n/2}$. Note that if $\deg(P) = n/2$ then $\deg(p) \leq n/2$ and $\deg(q) < n/2$.

Lemma 2. *If $\deg(p) \leq n/2$ then p is a B-polynomial.*

Proof. Let $p = \sum_{\alpha \in W_n^{\leq n/2}} p_\alpha m_\alpha$. We show that $p_\alpha = 0$ for all $\alpha \in W_n^{<n/2} \cup A$ by induction on $|\alpha|$.

As $\vec{0} \in W_n^{<n/2} \cup A$ it follows that $p_{\vec{0}} = 0$.

Now fix $\alpha \in W_n^{<n/2} \cup A$ with $|\alpha| > 0$. As, by induction, $p_\beta = 0$ for all $\beta \subset \alpha$ it holds that $0 = p(\alpha) = p_\alpha m_\alpha(\alpha) = p_\alpha$.

It follows from Lemma 2 that $p \oplus q$ is a $(B \cup W_n^{<n/2})$-polynomial annihilating $B \cup W_n^{>n/2}$ and that $r = p \oplus q(x \oplus \vec{1})$ is a $B \cup W_n^{<n/2}$-polynomial annihilating $(B \oplus \vec{1}) \cup W_n^{<n/2}$. The theorem follows from

Lemma 3. *There exists a nontrivial $(B \cup W_n^{<n/2})$-polynomial annihilating $(B \oplus \vec{1}) \cup W_n^{<n/2}$ if and only if $B \neq B \oplus \vec{1}$.*

Proof. Let $B = B \oplus \vec{1}$. As the submatrix of M^n formed by all rows corresponding to monomials m_α, $\alpha \in B \cup W_n^{<n/2}$ and inputs $\alpha \in B \cup W_n^{<n/2}$ is an upper triangle matrix, nontrivial $(B \cup W_n^{<n/2})$-polynomials annihilating $B \cup W_n^{<n/2}$ do not exist.

Now let $B \neq B \oplus \vec{1}$ and denote $C = B \setminus (B \oplus \vec{1})$. As $|C| + |W_n^{<n/2}| > |W_n^{<n/2}|$ there is a nontrivial $(C \cup W_n^{<n/2})$-polynomial s annihilating $W_n^{<n/2}$. As s annihilates $(B \oplus \vec{1})$, too, it annihilates $(B \oplus \vec{1}) \cup W_n^{<n/2}$.

We have shown that $A = A \oplus \vec{1}$ implies that $AI(gr(f^A)) = n/2 + 1$.

For showing that for $A \neq A \oplus \vec{1}$ it holds $AI(gr(f^A)) < n/2 + 1$ let B, C, s be defined as above. Then the polynomial t, defined by $t(x) = s(x \oplus \vec{1})$, is a nontrivial $(B \cup W_n^{<n/2})$-polynomial annihilating $B \cup W_n^{>n/2}$. Now write $t = p \oplus q$, where p is a B-polynomial and $\deg(q) < n/2$, and define $P = p \oplus z \cdot q$. It is not hard to check that P is a nontrivial degree-$\frac{n}{2}$-polynomial annihilating $A \cup W_n^{<n/2}$.

Note that $A = A \oplus \vec{1}$ implies that $|A|$ has to be even. As $(10) = (01) \oplus (11)$ it is not possible to construct a function of type f^A for $n = 2$ fulfilling that $AI(gr(f^A)) = 2$. However, there are functions f in two variables such that $AI(f)) = 2$, take e.g. $f = x_1 \wedge x_2$. For $n = 4$ there are functions of type f^A fulfilling $AI(gr(f^A)) = 3$, namely if A is one of the sets $\{0011, 1100\}$, $\{1001, 0110\}$, $\{1010, 0101\}$ or the union of two such sets. As $\binom{4}{2} = 6$, all balanced functions of type f^A do not fulfil $AI(gr(f^A)) = 3$. By exhaustive search over all 12,870 balanced function, we could exclude the existence of balanced functions f in four variables with $AI(gr(f)) = 3$ at all. For even $n \geq 6$ $\binom{n}{2}$ is divisible by 4, i.e. there are balanced functions of type f^A with $AI(gr(f^A)) = n/2 + 1$ for all even $n \geq 6$.

Note that very recently Carlet (2006) used functions of type f^A for the construction of balanced and easy computable functions f for which $AI(f)$ is maximal.

3 Constructing Multi-output Boolean Functions of Maximal Algebraic Immunity

In this section we present an algorithm which computes for given n, m, d the table of a function $h : \{0,1\}^n \longrightarrow \{0,1\}^m$ fulfilling $AI(gr(h)) > d$, and runs in polynomial time in the relevant output-size parameter 2^n. The algorithm is based on characterizing the existence of function $h : \{0,1\}^n \longrightarrow \{0,1\}^m$ fulfilling $AI(gr(h)) > d$ by the existence of so-called (n, m, d)-kernels, and the observation that (n, m, d)-kernels correspond to bases of the union of certain matroids.

In the following, we use at several places the relation $\Phi_{n+m}(d) = \sum_{z \in W_m^{\leq d}} \Phi_n(d - |z|)$, which results from partitioning the set of all $(\alpha, \beta) \in W_{n+m}^{\leq d}$ with respect to the β-component.

Definition 1. *We call a collection $\mathcal{U} = (U_z)_{z \in W_m^{\leq d}}$ of pairwise disjoint subsets of $\{0,1\}^n$ to be an (n, m, d)-kernel if for all $z \in W_m^{\leq d}$ it holds that $|U_z| = \Phi_n(d - |z|)$ and U_z has maximal algebraic immunity, i.e. $AI(U_z) = \Phi_n^{-1}(\Phi_n(d - |z|)) = d - |z| + 1$.*

Theorem 2. *For all positive integers n, m, d it holds that there is a function $h : \{0,1\}^n \longrightarrow \{0,1\}^m$ fulfilling $AI(gr(h)) > d$ if and only if there is an (n, m, d)-kernel.*

Proof. We show at first the only-if direction. Let us fix an (n, m, d)-kernel $\mathcal{U} = (U_z)_{z \in W_m^{\leq d}}$ and denote by $U \subseteq \{0,1\}^{n+m}$ the set

$$U = \bigcup_{z \in W_m^{\leq d}} \{(u, z); u \in U_z\}.$$

It is sufficient to show that $AI(U) > d$ as then all functions $h : \{0,1\}^n \longrightarrow \{0,1\}^m$, which fulfill for all $z \in W_m^{\leq d}$ and $u \in U_z$ that $h(u) = z$, fulfil $AI(gr(h)) > d$.

For showing that the rank of $M_{d,U}^{n+m}$ equals $\Phi_{n+m}(d)$ we order rows and columns of $M_{d,U}^{n+m}$ in an appropriate way. The rows of $M_{d,U}^{n+m}$ are labelled with monomials $m_\alpha m_\beta = \Pi_{i,\alpha_i=1} x_i \Pi_{j,\beta_j=1} z_j$ for all $(\alpha, \beta) \in (\{0,1\}^n \times \{0,1\}^m) \cap W_{n+m}^{\leq d}$. We divide the set of rows into β-groups, $\beta \in \{0,1\}^m$, containing all rows labelled $m_\alpha m_\beta$ for some $\alpha \in \{0,1\}^n$, and the set of columns into z-groups, $z \in W_m^{\leq d}$, containing all columns labelled (u, z), $u \in U_z$. Note that each β-group consists of $\Phi_n(d - |\beta|)$ rows, and that each z-group consists of $\Phi_n(d - |z|)$ columns. Place the β-groups and the z-groups from top to bottom and from left to right, respectively, according to the ordering ω on β and z, respectively. Denote by $M_{\beta,z}^{n+m}$ the submatrix of $M_{d,U}^{n+m}$ formed by all rows from the β-group and all columns from the z-group. Note that for all β, z with $\omega(\beta) < \omega(z)$ it holds that $M^{n+m}_{\beta,z} \equiv 0$, i.e. $M^{n+m}_{d,U}$ becomes a stepwise triangle matrix. Further, by definition 1, for all $z \in W^{\leq d}_{m}$ it holds that $M^{n+m}_{z,z}$ is quadratic and regular. This implies that

$$\Phi_{n+m}(d) \geq rank(M^{n+m}_{d,U}) \geq \sum_{z \in W^{\leq d}_{m}} rank(M^{n+m}_{z,z}) =$$

$$= \sum_{z \in W^{\leq d}_{m}} \Phi_n(d - |z|) = \Phi_{n+m}(d).$$

For showing the if-direction fix a function $h : \{0,1\}^n \longrightarrow \{0,1\}^m$ with $AI(gr(h)) > d$. Then, the matrix $M^{n+m}_{d,gr(h)}$ has rank $\Phi_{n+m}(d)$. Given a subset $S \subseteq \{0,1\}^n$ we define $\tilde{S} = \{(s, h(s)), s \in S\} \subseteq \{0,1\}^{n+m}$. We fix a subset $S \subseteq \{0,1\}^n$ of cardinality $\Phi_{n+m}(d)$ such that $M^{n+m}_{d,\tilde{S}}$ is regular. We use the following result from linear algebra.

Lemma 4. *Let r, N be integers fulfilling $1 < r \leq N$ and let A be a regular $N \times N$-matrix over an arbitrary field K of characteristic 2. Further let $I = \{I_s\}_{s=1}^{r}$ be an arbitrary partition of the rows of A into r pairwise disjoint nonempty sets. Then there is a partition $J = \{J_s\}_{s=1}^{r}$ of the columns of A into r pairwise disjoint nonempty sets such that for all $s = 1, \cdots, r$ it holds $|I_s| = |J_s|$ and A_{I_s,J_s} is regular. (Given subsets I of rows and J of columns of A then $A_{I,J}$ denotes the $|I| \times |J|$-submatrix of A formed by the I-rows and the J-columns.)*

Before giving the proof we apply this lemma to the quadratic regular matrix $M^{n+m}_{d,\tilde{S}}$. Observe that this matrix is defined over the finite field $GF(2)$ which has characteristic 2. Consider the partition $\{T_\beta\}_{\beta \in W^{\leq d}_{n+m}}$ of the rows of $M^{n+m}_{d,\tilde{S}}$ into the above described β-blocks. By Lemma 4 there is a partition $\{\tilde{S}_\beta\}_{\beta \in W^{\leq d}_{n+m}}$ of the columns of $M^{n+m}_{d,\tilde{S}}$ such that for all $\beta \in W^{\leq d}_{n+m}$ the submatrix of $M^{n+m}_{d,\tilde{S}}$ corresponding to the T_β-rows and $\tilde{S}_\beta$-columns is regular. It follows directly that the set system $\mathcal{U} = \{U_{\beta \in W^{\leq d}_{n+m}}\}$, defined by $U_\beta = \{u \in \{0,1\}^n; (u, h(u)) \in \tilde{S}_\beta\}$ is an (n, m, d)-kernel.

Proof (Lemma 4). It is known that A is regular iff $\det(A) \neq 0$. The determinant of A is defined as

$$\det(A) = \sum_{\pi \in \mathcal{S}_N} A(\pi)$$

where $A(\pi) = sign(\pi) \Pi_{i=1}^{N} A_{i,\pi(i)}$, and $\mathcal{S}_N$ denotes the set of permutations of $\{1, \cdots, N\}$. As K has characteristic 2, it holds that -1 is equal to 1. Thus, the expression for $A(\pi)$ can be simplified to $A(\pi) = \Pi_{i=1}^{N} A_{i,\pi(i)}$

For all equal size subsets I and J of $\{1, \cdots, N\}$ we denote by $\mathcal{S}(I, J)$ the set of all bijective mappings from I to J. Furthermore, let $\mathcal{J}$ be the set of all partitions $J = \{J_s\}_{s=1}^{r}$ of $\{1, \cdots, N\}$ into r pairwise disjoint sets, which fulfil

$|I_s| = |J_s|$ for all $s = 1, \cdots, r$. Further denote for all $J \in \mathcal{J}$ by $\mathcal{S}_N^J$ the set of all permutations $\pi \in \mathcal{S}_N$ for which $\pi(I_s) = J_s$ for all $s = 1, \cdots, r$. Thus, any $\pi \in \mathcal{S}_N^J$ is equivalently described by $(\pi_1, \ldots, \pi_r)$ with $\pi_s \in \mathcal{S}(I_s, J_s)$. For this case, we introduce the identifier $A(\pi_S) = \prod_{i \in I_s} A_{i,\pi_s(i)}$. Observe that for $s \in \{1, \ldots, r\}$, it holds that $\sum_{\pi_s \in \mathcal{S}(I_s,J_s)} A(\pi_s) = \det(A_{I_s,J_s})$. It follows that

$$\det(A) = \sum_{J \in \mathcal{J}} \sum_{\pi \in \mathcal{S}_N^J} A(\pi) = \sum_{J \in \mathcal{J}} \Pi_{s=1}^r \sum_{\pi_s \in \mathcal{S}(I_s,J_s)} A(\pi_s)$$

$$= \sum_{J \in \mathcal{J}} \Pi_{s=1}^r \det(A_{I_s,J_s}).$$

Hence, if $\det(A) \neq 0$ then there is some $J = \{J_s\}_{s=1}^r \in \mathcal{J}$ fulfilling $\det(A_{I_s,J_s}) \neq 0$ for all $s = 1, \cdots, r$.

We have reduced the problem of computing functions $h : \{0,1\}^n \longrightarrow \{0,1\}^m$ fulfilling $AI(gr(h)) > d$ to the existence of (n, m, d)-kernels. We will see in the following that the theory of matroids provides an efficient algorithm of computing (n, m, d)-kernels.

We list at first those basics on matroids which are relevant for our considerations. For deeper insight into the theory of matroids see e.g. Schrijver (2003). A pair $\mathcal{M} = (S, \mathcal{I})$, where S is some finite ground set and $\mathcal{I}$ denotes a system of subsets of S, is called a matroid if the following three conditions are fulfilled.

(1) $\emptyset \in \mathcal{I}$,
(2) If $I \in \mathcal{I}$ and $I' \subseteq I$ then $I' \in \mathcal{I}$,
(3) If $I_1, I_2 \in \mathcal{I}$ and $|I_1| < |I_2|$ then there is some $e \in I_2 \setminus I_1$ such that $I_1 + e \in \mathcal{I}$.

Elements $I \in \mathcal{I}$ of a matroid $\mathcal{M} = (S, \mathcal{I})$ are called $\mathcal{M}$-independent sets. An $\mathcal{M}$-independent set I, which is maximal in the sense that $I + s \notin \mathcal{I}$ for all $s \in S \setminus I$, is called an $\mathcal{M}$-basis. Conditions (1),(2), and (3) imply that all $\mathcal{M}$-bases in a matroid have the same cardinality, which is defined as the rank $r(\mathcal{M})$ of the matroid $\mathcal{M}$. The standard example for matroids is that S is a finite set of vectors of a K-vector space V, where K denotes some field, and $\mathcal{I}$ denotes the set of all linearly independent subsets of S.

Now let $\mathcal{M}_h = (S, \mathcal{I}_h)$, $h = 1, \cdots, H$, be a family of matroids over the same finite ground set S. Suppose that for all $h = 1, \cdots, H$ the decision if a given subset J of S is $\mathcal{M}_h$-independent can be efficiently computed in $|J|$ and the representation size of elements in S.

Edmonds (1968) showed that $\cup\mathcal{I} = \{I_1 \cup \cdots \cup I_H; I_h \in \mathcal{I}_h\}$ forms the system of independent sets of another matroid

$$\mathcal{M} = \mathcal{M}_1 \vee \cdots \vee \mathcal{M}_H = (S, \cup\mathcal{I}),$$

called the union of the matroids $\mathcal{M}_1, \cdots, \mathcal{M}_H$. Clearly, it holds that the rank of $\mathcal{M}$ is at most $r(\mathcal{M}_1) + \cdots + r(\mathcal{M}_H)$, where equality holds if and only if S contains for all $h = 1, \cdots, H$ a $\mathcal{M}_h$-basis $B_h \subseteq S$ such that $B_1, \cdots, B_H$ are pairwise disjoint.

Definition 2. *The matroid* $\mathcal{M}_1 \vee \cdots \vee \mathcal{M}_H$ *is called to have full rank, if* $r(\mathcal{M}_1 \vee \cdots \vee \mathcal{M}_H) = \sum_{h=1}^{H} r(\mathcal{M}_h)$.

The relation of (n, m, d)-kernels to matroids is as follows. For $1 \leq d \leq n$ and $x \in \{0,1\}^n$ let $V(x,d) \in \mathbf{F}_2^{\Phi_n(d)}$ denote the vector $(m_\alpha(x))_{\alpha \in W_n^{\leq d}}$. For a set $S \subseteq \{0,1\}^n$, we extend this definition to $V(S,d) := \{V(x,d), x \in S\}$. Note that for $I \subseteq \{0,1\}^n$ the matrix $M_{d,I}^n$ consists of the columns $\{V(x,d), x \in I\}$. Thus, a set S of cardinality $\Phi_n(d)$ is d-immune if and only if $V(S,d)$ is a linearly independent subset of the $\mathbf{F}_2$-vector space $\mathbf{F}_2^{\Phi_n(d)}$. Let $\mathcal{I}_n^d$ consist all linearly independent sets $V(I,d)$. Then, $\mathcal{U} = (U_z)_{z \in W_m^{\leq d}}$ is a (n,m,d)-kernel if and only if $U_z \in \mathcal{I}_n^{d-|z|}$ for all $z \in W_m^{\leq d}$.

On the other hand, the definition of $\mathcal{I}_n^d$ gives rise to the following matroid $\mathcal{M}_n^d = (\{0,1\}^n, \mathcal{I}_n^d)$ over the ground set $\{0,1\}^n$. In particular, any (n,m,d)-kernel gives a matroid $\bigvee_{z \in W_m^{\leq d}} \mathcal{M}_n^{d-|z|}$ of full rank and vice versa. Thus, Theorem 2 directly implies

Theorem 3. *For all positive integers* n, m, d *it holds that there is a function* $h : \{0,1\}^n \longrightarrow \{0,1\}^m$ *fulfilling* $AI(gr(h)) > d$ *if and only if the matroid* $\bigvee_{z \in W_m^{\leq d}} \mathcal{M}_n^{d-|z|}$ *has full rank.* □

Now let as above $\mathcal{M}_h = (S, \mathcal{I}_h)$, $h = 1, \cdots, H$, be a family of matroids over the same finite ground set S, where $\mathcal{I}_h \subseteq 2^S$ denotes the set of all $\mathcal{M}_h$-independent sets. What we need is an efficient algorithm (in the representation size of elements in S and H) for constructing a basis in the matroid $\mathcal{M}_1 \vee \cdots \vee \mathcal{M}_H$. Such an algorithm, known as Matroid Union Algorithm, is due to Edmonds (1968). In the following, we describe this algorithm.

We call collections $I = (I_1, \cdots, I_H)$ of pairwise disjoint $\mathcal{M}_h$-independent sets I_h, $1 \leq h \leq H$, a disjoint collections of independent sets (for short, DCIS), and denote $\cup I = \bigcup_{h=1}^{H} I_h$. Further, for subsets $T \subseteq S$ and $s, t \in S$ we write $T + s$ instead of $T \cup \{s\}$ and $T - t$ instead of $T \setminus \{t\}$.

We have to compute a DCIS I^* such that $|\cup I^*|$ is maximal. For this purpose it is sufficient to have an efficient algorithm which decides for given DCIS I if there is some $s \in S \setminus (\cup I)$ such that $(\cup I) \cup \{s\}$ is again $(\mathcal{M}_1 \vee \cdots \vee \mathcal{M}_H)$-independent, and, if yes, computes a new DCIS I' such that $\cup I' = (\cup I) \cup \{s\}$. The matroid properties guarantee that the greedy strategy based on this step yields a basis in $\mathcal{M}_1 \vee \cdots \vee \mathcal{M}_H$ (see, e.g., Schrijver 2003).

Let us fix a DCIS $I = (I_1, \cdots, I_H)$ and some $s \in S \setminus (\cup I)$. For all h, $1 \leq h \leq H$, we denote by $F(I_h)$ the set of all $x \in S \setminus I_h$ such that $I_h + x \in \mathcal{I}_h$, and by $A(I_h)$ the set of all directed arcs (x, y) with $x \in I_h$, $y \in S \setminus I_h$, $I_h - x + y \in \mathcal{I}_h$. Let $F = F(I_1) \cup \cdots \cup F(I_H)$, and let G denote the directed graph $G = (S, A(I_1) \cup \cdots \cup A(I_H))$. Note that F and G can be efficiently computed from $I_1, \cdots, I_H$ and S.

How to find an $s \in S \setminus \cup I$ such that $\cup I + s$ is again $\mathcal{M}_1 \vee \cdots \vee \mathcal{M}_H$-independent? There are two easy cases and one nontrivial case.

Case 1: I_h is $\mathcal{M}_h$-basis (i.e. $F(I_h) = \emptyset$) for all $h = 1, \cdots, H$. In this case, I is obviously a maximal independent set in $\mathcal{M}_1 \vee \cdots \vee \mathcal{M}_H$.

Case 2: It holds $F \not\subseteq \cup I$. Suppose w.l.o.g. that there is some $s \in F(I_1) \setminus \cup I$. Then $s \notin I_h$ for all $1 \leq h \leq H$ and $I' = (I_1 + s, I_2, \cdots, I_h)$ is an DCIS with $\cup I' = \cup I + s$.

Case 3: $F \neq \emptyset$ but $F \subseteq \cup I$, i.e. for all $s \in S \setminus \cup I$ and all $1 \leq h \leq H$ it holds that $I_h + s$ is $\mathcal{M}_h$-dependent.

If a $\bigvee_{h=1}^{H} M_h$-independent set I fulfils the condition of case 3 then let us I call to be locally maximal. We use of the following Theorem 42.4 from Schrijver (2003) that for all $s \in S \setminus \cup I$ it holds that $\cup I + s$ is $\mathcal{M}_1 \vee \cdots \vee \mathcal{M}_H$- independent iff there is a directed path from x to s in G for some $x \in F$.

If our greedy algorithm reaches a situation in which a locally maximal $\bigvee_{h=1}^{H} M_h$-independent set I is constructed then we have to do a more complex step which in the following well be called a *Case-3 step.* A Case-3 step involves first to test whether I is globally maximal by checking the condition of Theorem 42.4 in Edmonds (1968) for all $s \in S \setminus I$. If not then we fix some $s \in S \setminus I$, $x \in F$ and a shortest path $p = (x = s_0, s_1, s_2, \cdots, s_k = s)$ from x to s. Due to the condition of Case 3 it holds $x \in \cup I$, let us suppose w.l.o.g. that $x \in F(I_1)$. The proof of Theorem 42.4 from Schrijver (2003) yields the following efficient algorithm for computing a new partition DCIS I' with $\cup I + s = \cup I'$.

Note that due to the definition of $A(I)$ and the condition of Case 3, it holds for all $1 \leq i \leq k-1$ that $s_i \in \cup I$. Fix $h_1, \cdots, h_{k-1}$ in such a way that $s_i \in I_{h_i}$ for all $1 \leq i \leq k-1$. Note that, by the definition of $A(I)$, $h_i \neq h_{i+1}$ for all $1 \leq i \leq k-2$.

Moreover, p is acyclic, i.e., all s_i, $0 \leq i \leq k$, are pairwise distinct. This implies that for all $1 \leq h \leq H$ the p-edges starting from nodes in I_h form a matching in G. For $1 \leq h \leq H$, let $J_h = p \cap I_h$ and $K_h = \{s_i; 1 \leq i \leq k, s_{i-1} \in J_h\}$.

The DCIS I' is obtained by shifting x into I_1 and then, for $i = 2, \cdots, k$, shifting s_i into $I_{h_{i-1}}$. More formally, for $1 \leq h \leq H$, let $J_h = p \cap I_h$ and $K_h = \{s_i; 1 \leq i \leq k, s_{i-1} \in J_h\}$. Then $I' = (I'_1, \cdots, I'_H)$, where $I'_1 = ((I_1 + x) \setminus J_1) \cup K_1$ and $I'_h = (I_h \setminus J_h) \cup K_h$ for $2 \leq h \leq H$. For all $1 \leq h \leq H$ the p-edges starting from nodes in I_h form a matching in G. This implies (see Theorem 39.13 from Schrijver (2003)) that I_h is $\mathcal{M}_h$-independent for all $1 \leq h \leq H$, i.e. I' is an DCIS. Obviously, $\cup I' = \cup I + s$.

4 Experimental Results and Open Problems

We have seen that the decision of the existence and the computation of an (n, m, d)-kernel $\mathcal{U}$ can be performed by the computation of a maximal independent set $\cup_{z \in W_m^{\leq d}} U_z$ in the matroid union $\bigvee_{z \in W_m^{\leq d}} \mathcal{M}_{d-|z|}^n$, which can be performed by the Matroid Union Algorithm. A trivial necessary condition is that an (n, m, d)-kernel fits into $\{0,1\}^n$, i.e. $|\mathcal{U}| = \Phi_{n+m}(d) \leq 2^n$. We denote by $d_{n,m} = \Phi_{n+m}^{-1}(2^n)$ the corresponding theoretical upper bound for the immunity of the graph of a function $h : \{0,1\}^n \longrightarrow \{0,1\}^m$.

With help of our algorithm, we checked the existence of $(n, m, d_{n,m} - 1)$-kernels for the cases $14 \geq n \geq m \geq 2$ and $(n, m) = (15, 2)$. It turned out that $(n, m, d_{n,m} - 1)$-kernels could be found for all our test cases. This raises the interesting question whether $(n, m, d_{n,m} - 1)$-kernels do exist in general. Note that our results in Section 2 answer this question positively for all n and $m = 1$.

Even if the experiments indicate that functions for which the graph has immunity $d_{n,m}$ may exist for all choices of n and m, one should not forget that a cryptographic reasonable S-box should fulfill additional criteria as high non-linearity or balancedness. Essentially, one hardly finds cryptographic S-boxes in practice which attain maximal algebraic immunity. For example, the Data Encryption Standard (DES) uses eight different S-boxes with $n = 6$ and $m = 4$. The maximal value $d_{6,4} = 3$ is only achieved by five of them (see Shimoyama and Kaneko (1998)). Another example is the S-box ($n = m = 8$) used in the Advanced Encryption Standard (AES). It has been pointed out by Courtois and Pieprzyk (2002) that unless $(8, 8, 2)$-kernels do exist, the immunity of the graph is only 2. This leaves the open task to develop our method further to achieve a tradeoff between algebraic immunity and other conditions.

Acknowledgement

We would like to thank Petr Savicky, Pavel Pudlák, Claude Carlet, Markus Bläser, Rüdiger Reischuk and Hellen Altendorf for helpful comments. This work was supported by DFG-grant KR 1521/7.

References

1. Armknecht, F., M. Krause: *Algebraic attacks on Combiners with Memory*, Proceedings of Crypto 2003, LNCS 2729, pp. 162-176, Springer, 2003.
2. Armknecht, F.: *Algebraic Attacks and Annihilators*, Proceedings of WeWORC 2005, LNI P-74, pp. 13-21, 2005.
3. Ars, G., J.-C. Faugère, H. Imai, M. Kawazoe, M. Sugita: *Comparison Between XL and Gröbner Basis Algorithms.*, Proceedings of Asiacrypt 2004, LNCS 3329, pp. 338-353, Springer, 2004.
4. Carlet, C., D. K. Dalai, K. C. Gupta, S. Maitra: *Algebraic Immunity for Cryptographically Significant Boolean Functions: Analysis and Construction* to appear in IEEE Transactions of Information Theory, 2006.
5. Carlet, C.: *A method of Construction of Balanced Functions with Optimum Algebraic Immunity*, cryptology ePrint archive http://eprint.iacr.org, report 149, 2006.
6. Cid, C., G. Leurent: *An Analysis of the XSL Algorithm*, Proceedings of Asiacrypt 2005, LNCS 3788, pp. 333-352, Springer, 2005.
7. Courtois, N., J. Pieprzyk: *Cryptanalysis of block ciphers with overdefined systems of equations*, Proceedings of Asiacrypt 2002, LNCS 2501, pp. 267-287, Springer, 2002.
8. Courtois, N., W. Meier: *Algebraic attacks on Stream Ciphers with Linear Feedback*, Proceedings of Eurocrypt 2003, LNCS 2656, pp. 345-359, Springer, 2003. An extended version is available at http://www.cryptosystem.net/stream/
9. Courtois, N.: *Fast Algebraic Attacks on Stream Ciphers with Linear Feedback*, Proceedings of CRYPTO 2003, LNCS 2729, pp. 176-194, Springer, 2003.
10. Edmonds, J.: Matroid Partition. Journal of the AMS Vol.11, 1968, 335-345.

11. Faugère, J.-C., G. Ars: *An algebraic cryptanalysis of nonlinear filter generators using Gröbner bases*, 2003. Available at http://www.inria.fr/rrrt/rr-4739.html.
12. Meier, W., E. Pasalic, C. Carlet: *Algebraic attacks and decomposition of Boolean functions*, Proceeding of Eurocrypt 2004, LNCS 3027, pp. 474-491, Springer, 2004.
13. Murphy, S., M. Robshaw: *Comments on the Security of the AES and the XSL Technique*, Electronic Letters, 39:26-38, 2003.
14. Schrijver, A.: *Combinatorial Optimization. Polyhedra and Efficiency*, Springer 2003.
15. Shamir, A., J. Patarin, N. Courtois, A. Klimov: *Efficient Algorithms for Solving Overdefined Systems of Multivariate Polynomial Equations*, Proceedings of Eurocrypt '00, Springer LNCS 1807, pp. 392-407.
16. Shimoyama, T., T. Kaneko: *Quadratic Relation of S-boxes and Its Application to the Linear Attack of Full Round DES*, Proceedings of Crypto 1998, LNCS 1462, pp. 200-211, Springer, 1998.

On Everlasting Security in the *Hybrid* Bounded Storage Model*

Danny Harnik[1,**] and Moni Naor[2,***]

[1] Dept. of Computer Science, Technion, Haifa, Israel
harnik@cs.technion.ac.il
[2] Dept. of Computer Science and Applied Math., Weizmann Institute of Science, Rehovot, Israel
moni.naor@weizmann.ac.il

Abstract. The *bounded storage model* (BSM) bounds the storage space of an adversary rather than its running time. It utilizes the public transmission of a long random string $\mathcal{R}$ of length r, and relies on the assumption that an eavesdropper cannot possibly store all of this string. Encryption schemes in this model achieve the appealing property of *everlasting security*. In short, this means that an encrypted message remains secure even if the adversary eventually gains more storage or gains knowledge of (original) secret keys that may have been used. However, if the honest parties do not share any private information in advance, then achieving everlasting security requires high storage capacity from the honest parties (storage of $\Omega(\sqrt{r})$, as shown in [9]).

We consider the idea of a *hybrid bounded storage model* were computational limitations on the eavesdropper are assumed up until the time that the transmission of $\mathcal{R}$ has ended. For example, can the honest parties run a computationally secure key agreement protocol in order to agree on a shared private key for the BSM, and thus achieve everlasting security with low memory requirements? We study the possibility and impossibility of everlasting security in the hybrid bounded storage model. We start by formally defining the model and everlasting security for this model. We show the equivalence of two flavors of definitions: *indistinguishability of encryptions* and *semantic security*.

On the negative side, we show that everlasting security with low storage requirements cannot be achieved by *black-box* reductions in the hybrid BSM. This serves as a further indication to the hardness of achieving low storage everlasting security, adding to previous results of this nature [9, 15]. On the other hand, we show two augmentations of the model that allow for low storage everlasting security. The first is by adding a random oracle to the model, while the second bounds the accessibility of the adversary to the broadcast string $\mathcal{R}$. Finally, we show that in these two modified models, there also exist bounded storage oblivious transfer protocols with low storage requirements.

* Research supported in part by a grant from the Israel Science Foundation.
** This work was conducted while at the Weizmann Institute.
*** Incumbent of the Judith Kleeman Professorial Chair.

M. Bugliesi et al. (Eds.): ICALP 2006, Part II, LNCS 4052, pp. 192–203, 2006.

1 Introduction

1.1 The Bounded Storage Model

The *bounded storage model*, introduced by Maurer [17] postulates a bound on the *space* (memory size) of dishonest players rather than their running time. The model makes use of a long random string $\mathcal{R}$ of length r that is publicly transmitted and accessible to all parties. One can imagine that $\mathcal{R}$ is broadcast at a very high rate by a trusted party or by some natural source or phenomena. Security in this model relies on the assumption that an adversary cannot possibly store all of the string $\mathcal{R}$ in his memory. For instance, consider the case of honest parties Alice and Bob that want to exchange secret messages in presence of an eavesdropper Charlie. Let the honest parties Alice and Bob use storage of respective size s_A and s_B while an eavesdropper Charlie has a bound of s_C on his storage capacity. Typically we ask that security of the encryption holds in a setting where $s_A, s_B << s_C < r$. That is, the adversary is allowed to have storage space that is much larger than that of the honest players, but still smaller than r. In addition there are no computational restrictions on Charlie.

This model has enjoyed much success for the task of private key encryption. It has been shown that Alice and Bob who share a short private key can exchange messages secretly using only very small storage (a key of length $O(\log r + \log \frac{1}{\varepsilon})$ can be used with storage of size $s_A = s_B = O(\ell + \log r + \log \frac{1}{\varepsilon})$ for an ℓ bit message and ε probability of error). On the other hand an eavesdropper who can store up to a constant fraction of $\mathcal{R}$ (e.g. $\frac{1}{2}r$ bits) cannot learn anything about the messages (this was shown initially in [2] and improved in [1, 8, 10, 16] and ultimately in [22]). These encryption schemes have the important property called *everlasting security* (put forward in [1, 8]), where once the broadcast is over and $\mathcal{R}$ is no longer accessible then the message remains secure even if the private key is exposed and Charlie gains stronger storage capabilities.

In contrast, the situation is far from satisfiable when Alice and Bob do not share any secret information in advance. Cachin and Maurer [4] suggest a method for a key agreement protocol in the bounded storage model. However, this solution requires Alice and Bob to use storage of size at least $\Omega(\sqrt{r})$ which is quite high and renders this approach far less appealing (if not impractical). Dziembowski and Maurer [9] subsequently proved that this is the best one can do.

1.2 Everlasting Security and the Hybrid Bounded Storage Model

The inability to achieve everlasting secure encryption in the bounded storage model with memory requirements smaller than $\sqrt{r}$, has lead to the following appealing suggestion that we call the *hybrid BSM* and is the focus of this paper. Let Alice and Bob agree on their secret key using a computationally secure key agreement protocol (e.g. Diffie-Hellman). The rationale being that while an unbounded eavesdropper will eventually break the key, this is likely to happen only after the broadcast had already occurred. In such a case, the knowledge of the shared key would be useless by that time (this should be expected from the everlasting security property where getting the shared key after the broadcast has ended is useless).

Somewhat surprisingly, Dziembowski and Maurer [9] showed that this rationale may fail. They introduce a specific computationally secure key agreement protocol (containing a non-natural modification based on private information retrieval (PIR) protocols). If this key agreement protocol is used in the hybrid BSM setting with a specific private key scheme, then the eavesdropper can completely decrypt the encrypted message. However, their result does not rule out the possibility that the hybrid idea will work with some other key agreement protocol. For instance, using the plain Diffie Hellman key agreement may still work.

This hybrid model is very natural as we try to achieve everlasting security by adding limitations on the adversary that have a *strict time limit* (expiration date). Assumptions of this sort are generally very reasonable. For instance, in the key agreement example, all that we require is that the computational protocol is not broken in the short time period between its execution and the transmission of $\mathcal{R}$. An assumption such as the Diffie Hellman key agreement cannot be broken within half an hour, can be made with far greater degree of trust than actually assuming the long term security of a computational key agreement protocol.

1.3 This Paper's Contributions

This paper studies the possibility and impossibility of everlasting security in the hybrid bounded storage model. Our contributions are as follows:

- We formally define the model and everlasting security for this model.
- On the negative side we show that everlasting security with low storage requirements cannot be achieved by *black-box* reductions in the hybrid BSM.
- On the other hand, we show two augmentations of the model that allow for low storage everlasting security. The first is by adding a random oracle to the model, while the second bounds the accessibility of the adversary to the broadcast string $\mathcal{R}$.
- Finally, we show that in these two modified models, there also exist oblivious transfer protocols with low storage requirements.

We elaborate on each of these points:

Defining Everlasting Security in the Hybrid BSM: We give rigorous definitions of the type of security we are pursuing. We first define what a hybrid BSM encryption scheme is and then define everlasting security for such a scheme. Following the common practice (stemming from [13]), we give security definitions by indistinguishability of encryptions and by semantic security. We then prove that these two definitions are equivalent.

Regarding the Impossibility of Hybrid schemes: We have more than one indication that proving everlasting security in the hybrid bounded storage model is hard. In fact, it seems quite plausible that everlasting security is not achievable in this model at all. We survey two results (of [9] and [15]) that contribute to this point of view and provide additional evidence in the form of a black-box impossibility result for the general hybrid schemes (this shows the impossibility of semi-black-box constructions in the sense of [21]).

We show a setting (or "world") in which (general) hybrid schemes with everlasting security do not exist. However, if computational key agreement protocols exist in the plain model, then they also exist in this world (this property holds for any computationally secure protocol and not only for key agreement). Thus, we deduce that there can be no black-box proof of everlasting security, basing hybrid BSM schemes on the security of any computational primitive. In particular, this claim rules out the use of computational key agreement protocols via a black-box proof.

Positive Results: We show two modifications of the model that allow for hybrid everlasting security, with small memory requirements (much smaller than $\sqrt{r}$).

Everlasting Security in the presence of Random Oracles: Suppose that the parties are given access to a random oracle. A random oracle is a different type of public random string than the broadcast string $\mathcal{R}$. It is exponentially long and due to its length an efficient algorithm can only access a small fraction of it. This is unlike the broadcast string $\mathcal{R}$ that can be fully accessed by an efficient algorithm. On the other hand the random oracle alone cannot assure everlasting security as it does not eventually disappear and may always be queried at a later stage. We show that combining these two different types of public random strings (the random oracle and $\mathcal{R}$) is sufficient for achieving low storage everlasting security. Everlasting security in this setting means, in particular, that encrypted messages remain secure even if the adversary queries all of the random oracle entries after the broadcast is over.

There are several interpretations to the above result. If one can assemble a random oracle then this presents a methodology for low storage everlasting security. For example, such an oracle may be assembled using natural phenomena (if the broadcast string $\mathcal{R}$ can be implemented this way then why not a random oracle) or using a distributed protocol with partially (and temporarily) trusted parties, e.g. [19]. The emphasis being that such a random oracle need only be secure up until the time of the broadcast. One can also view this result, somewhat optimistically, as a suggested heuristic that achieves everlasting security "for all practical purposes" (when plugging some hash function instead of the random oracle). On the other hand, in light of the negative results for the general hybrid BSM model, one can view the above statement as a testament against relying (blindly) on random oracles to determine whether a task is feasible at all. This is since it shows a task (everlasting security with low storage requirements) that is achievable with a random oracle but might be impossible altogether (or at least very hard to achieve) without it. This is a different statement than previous results regarding random oracles (such as [5, 12, 18]) that show a specific *protocol* (rather than a task) that becomes insecure if the random oracle is replaced by a function with a small representation.

Bounded Accessibility: The *Bounded Accessability Model* assumes that the adversary cannot actually read the whole broadcast string $\mathcal{R}$. Rather, the adversary may choose (adaptively) γr locations of $\mathcal{R}$ that he would like to read and access only these bits during the broadcast. This model was considered by Maurer [17]. The bounded accessibility assumption can be justified, for example, if the broadcast is at a very high rate, and all the adversary can do during a broadcast is to read a bit and write it in his memory (that is of size γr). Another example is if the source consists of a large number of simultaneous transmissions and one can only record a limited number of sources.

We show that the basic hybrid scheme has everlasting security in this model with memory requirements that are substantially smaller than $\sqrt{r}$. An important observation is that this protocol would not have been successful in a plain bounded access model (without introducing the computational limitations on the adversary). This follows since the lower bound of $\sqrt{r}$ [9] applies for this model as well.

On Oblivious Transfer in the Hybrid BSM: We demonstrate that the hybrid BSM in the two augmented models described above can also achieve oblivious transfer (OT) protocols with everlasting security using only low storage. This is in contrast to OT in the standard BSM (see for example [3, 6, 7]) that requires storage of at least $\Omega(\sqrt{r})$ as the lower bound for key agreement [9] applies also for OT. The hybrid scheme is based on the assumption that there exist computationally secure OT protocols.

Paper Organization: We give an overview of the definitions of the hybrid BSM and everlasting security in Section 2 (the rigorous treatment is omitted due to space limitations). The positive results are given in Section 3, while the negative results appear in Section 5. Section 4 discusses the results for OT. Due to the limited space we omit several of the proofs and instead refer the reader to the full version of this paper [14].

2 Hybrid BSM – Setting and Security Definitions

We consider the task of exchanging secret messages between two parties in the presence of an eavesdropper. following is a general description of what a hybrid BSM encryption scheme consists of. We consider the honest parties Alice and Bob (A and B) and an eavesdropper Charlie (C) Charlie is limited to run in polynomial time up to the end of the broadcast, at which point he must store at most s_C bits of information. After this point, Charlie is not limited in any way. In the general hybrid scheme we do not restrict Alice and Bob to a specific behavior, but rather allow them to communicate before and after the broadcast. The first communication is not necessarily a key agreement, (as was suggested in the introduction) but rather any protocol. The point is that this protocol should take advantage of the assumption that at this time Charlie is restricted to being a polynomial time algorithm. After the broadcast the parties may communicate again, though Charlie is no longer bounded.[1]

We then give two equivalent flavors of definitions for everlasting security of a hybrid scheme (following the format in [13]). Roughly, the definitions are:

- **Indistinguishability of encryptions:** A hybrid scheme has everlasting security if for every two messages m_0 and m_1, Charlie cannot tell apart the encryptions of these two messages. That is, the *view* of Charlie has essentially the same distribution (up to a negligible statistical difference) when message m_0 is used or when message m_1 is used. Charlie's view consists of the encryption, the transcripts of all the interaction between Alice and Bob and the s_C bits that Charlie has *efficiently* stored from the broadcast string $\mathcal{R}$ (as well as any prior information such as the messages m_0 and m_1).
- **Semantic security:** A hybrid scheme has everlasting security if an unbounded Charlie cannot compute from his *view* a boolean function of the encrypted message m with a non-negligible advantage.

Due to space limitations, we give the formal definitions as well as the proof of equivalence in the full version of this paper [14].

3 Positive Results

Due to the negative results of Section 5 it seems that proving everlasting security with low storage requirements is out of our reach at this time (or not possible altogether). We next try to modify the model itself in ways that will allow for positive results. We provide two modifications of the standard hybrid model, and prove that under this modifications we can achieve low storage everlasting security.

3.1 Everlasting Security with a Random Oracle

We show a simple scheme that achieves hybrid everlasting security when given access to a random oracle. We view a random oracle RO as an exponentially long list of entries that is publicly available and allows random access. Each entry in the list contains a (relatively short) string of bits. We start by showing a feasibility result that produces just a one bit key and then describe how this can be generalized to give an efficient scheme that outputs a longer key. *Protocol*

Encrypt$_{RO}$

1. Alice and Bob run any computational key agreement (KA) protocol and agree on a key CK of length k.

[1] We note that this is a description of the most general hybrid scheme we consider, and is used for the impossibility results in order to rule out as many potential solutions as possible. On the other hand, we would like actual protocols to be much simpler and in particular, all of our positive results have schemes that have interaction only before the broadcast.

2. Alice and Bob access the random oracle at the location CK, the output $SK = RO(CK)$ consists of $k \cdot \log r$ bits. We view these bits as a list of k indices $i_1, \ldots, i_k$ where each $i_j \in [r]$.[2]
3. For each $j \in [k]$ Alice and bob store the bit $\mathcal{R}(i_j)$.
4. The output bit $K = \bigoplus_{j \in [k]} \mathcal{R}(i_j)$.
5. The key K is the final key and the encryption of the message m is $m \oplus K$.

Theorem 1. *The protocol Encrypt$_{RO}$ has $(\gamma, neg(\cdot))$-everlasting security.*

The above simple scheme demonstrates that everlasting security can be achieved by when a random oracle is available. However, it is not efficient in the sense that the number of bits read from the random oracle is far larger than the actual key. This may be greatly improved by using one of the known schemes for private key encryption in the regular (non-hybrid) bounded storage model, such as the scheme in [16, 22]. In this hybrid scheme, the key $SK = RO(CK)$ serves as the private key of the regular BSM scheme. Such a regular BSM scheme was shown in [22] with private key SK of length $O(\log r + \log \frac{1}{\varepsilon})$ that reads only $O(k + \log \frac{1}{\varepsilon})$ bits of broadcast string $\mathcal{R}$. Thus we get a hybrid scheme with comparable parameters (the storage requirement of Alice and Bob is only $O(k + \log \frac{1}{\varepsilon})$).

3.2 Bounded Accessability and Hybrid Everlasting Security

The crux in the use of the random oracle is that the adversary does not *read* some information. Thus making assumptions on the ability of the adversary to read all of the bits (rather than store them) seems helpful and indeed we show that it is. The *Bounded Accessability Model* (considered already in the original paper of Maurer [17]) assumes that the adversary cannot actually read the whole of the broadcast string $\mathcal{R}$ but rather just a chosen γ fraction of it. That is, the adversary chooses $\gamma \cdot r$ bits that he wishes to read from the string $\mathcal{R}$ and may then store all of these bits.[3] Such an assumption can be justified, for example, if the broadcast is at a very high rate, allowing the parties to read and store just a small fraction that they prepare for in advance. The definitions of everlasting security are the same as in Section 2 only that in the underlying model, the efficient part of Charlie can only decide on $\gamma \cdot r$ locations in $\mathcal{R}$ and store these actual bits from $\mathcal{R}$.

We show that hybrid everlasting security is achievable in this model with memory requirements that are much smaller than $\sqrt{r}$. An important note is that the lower bound of $\Omega(\sqrt{r})$ [9] applies for this model as well. Therefore the hybrid with computational assumptions is essential.

We present a basic hybrid encryption scheme and prove its security. We note that the following example is aimed at showing the feasibility of such a scheme and does not try to optimize the parameters. The output of the scheme is a

[2] We consider the random oracle $RO(\cdot)$ as an exponentially long array broken into cells where each cell contains $k \cdot \log r$ bits.

[3] Note that we allow the choice of locations to store to be an adaptive choice (rely on the answers of the first bits that where read to determine what to read next).

one bit key, that should be pseudorandom. We discuss how to achieve better parameters towards the end of this section.

Protocol $Encrypt_{BAM}$

1. Alice and Bob run a computational key agreement protocol KA with the security parameter k. They agree on a key $\bar{i} = (i_1, \ldots i_k)$ where for each $j \in [k]$ the jth entry is an index $i_j \in \{0,1\}^{\log r}$ (the transcript of KA is denoted by T_{KA}).
2. For each $j \in [k]$ Alice and Bob store the bit $\mathcal{R}(i_j)$.
3. The final key is $K = \bigoplus_{j \in [k]} \mathcal{R}(i_j)$ and a message m is encrypted by $m \oplus K$.

Theorem 2. *The protocol* $Encrypt_{BAM}$ *has* $(\gamma, neg(\cdot))$*-everlasting security.*

As in the case of the random oracle protocol, an efficient protocol can be achieved by using the computational key agreement to agree on a key CK of $O(\log r + \log \frac{1}{\varepsilon})$ bits, that will serve as the private key for a regular BSM scheme of [22].

4 On Hybrid Everlasting Security for Oblivious Transfer

Oblivious transfer (OT) (originally defined by Rabin [20] and presented here using the definition of [11]) is a protocol between Alice holding two secrets s_0 and s_1, and Bob holding a choice bit c. At the end of the protocol Bob should learn the secret of his choice (i.e., s_c) but learn nothing about the other secret. Alice, on the other hand, should learn nothing about Bob's choice c. Oblivious transfer is an important building block for construction of secure computation.Cachin, Crepeau and Marcil [3] showed an implementation of OT in the Bounded Storage model with everlasting security in the sense that once the broadcast is over, no party can learn additional information about the secrets, even if the party has gained more power (or storage space) since. This protocol was subsequently improved in [6] and ultimately in [7]. We refer the reader to [7] for rigorous definitions of OT in the bounded storage model.

The main drawback of all bounded storage OT schemes is that the honest parties are required to use storage of $\sqrt{r}$ bits. This requirement is tight since the lower bound of [9] holds also for OT (as OT implies key agreement). We next show that the idea of the hybrid BSM may be useful for implementing OT with low storage requirements. More precisely, we show such schemes under the two modified models discussed for the case of encryption. That is, if random oracles are allowed and in the bounded accessability model.

The idea, in both cases, is to run a computational string OT between Alice and Bob. At the end of this protocol Bob will have learnt one of the two strings that Alice holds. Alice then uses the two strings as secret keys in a classical BSM private key encryption scheme (e.g., as seeds to a locally computable strong extractor in the scheme of [16, 22]). Alice encrypts her two secrets using the two respective keys. Bob can decrypt one (as he received one of the keys) but not the other.

It is interesting that this protocol does not follow the path of the oblivious transfer in the bounded storage model. Oblivious transfer is inherently an asymmetric protocol, and in the above protocol the asymmetry follows only from the computational protocol. Thus not utilizing the mechanism designed for this task in the bounded storage model.

5 Negative Results

We have more than one indication that proving everlasting security in the hybrid bounded storage model is hard. In fact, it seems quite plausible that everlasting security is not achievable in this model at all. In this section we survey two results contributing to this point of view and provide additional evidence in the form of a black-box impossibility result for the general hybrid schemes.

The [DM04] Example: Dziembowski and Maurer [9] show an example that proves that basic hybrid schemes cannot be blindly trusted. The example takes any combination of a computational key agreement protocol and a private key BSM encryption scheme, and replaces the key agreement with a non-natural, yet secure key agreement protocol, that renders the overall scheme insecure. The new key agreement consists of the old one, and an additional "hint" as to what an adversary should store from the string $\mathcal{R}$. This hint is based on private information retrieval (PIR) protocols, and does not give the computationally bounded adversary any information on the underlying key. It does allow him to store a function of $\mathcal{R}$ that contains the necessary information about what the honest players stored. This information can be extracted at a later stage once the adversary is no longer bounded.

This result can be viewed as saying such a basic hybrid scheme cannot work in a black-box manner, taking any key agreement with any private key BSM scheme. It falls short though of saying that hybrid schemes using specific protocols are insecure. Neither does it rule out a black-box general hybrid scheme (we show such an impossibility result in Section 5.1).

On Compression of $\mathcal{NP}$ instances and Hybrid Schemes: In [15] we define the following problem regarding the compression of $\mathcal{NP}$ instances: Consider $\mathcal{NP}$ problems that have long instances but relatively short witnesses. The question is, can one efficiently compress an instance and store a shorter representation that maintains the information of whether the original input is in the language or not. For example, the compression of the language SAT is formulated as follows: A compression for SAT is an efficient algorithm and a polynomial $p(\cdot,\cdot)$ such that the algorithm takes as input a CNF formula Φ with m clauses over n variables (where $m >> n$). The output should be a formula Ψ of size $p(n, \log m)$ such that Ψ is satisfiable if and only if Φ is satisfiable.

In [15], a family of $\mathcal{NP}$ languages is defined that includes, for example, the languages SAT and Clique. It is shown that if there exists a compression algorithm for any language in this family, then the hybrid BSM is no more powerful (with respect to everlasting security) than the standard BSM. Conversely, in order to prove everlasting security in the hybrid model with low storage requirements, one must prove (or assume) that certain languages are incompressible.

5.1 Black-Box Impossibility

We next show that there exists no black-box *proof* of everlasting security of a hybrid encryption scheme, based on the security of a computational primitive. This forms a wider black-box impossibility than that of the [9] result, as it captures any general hybrid scheme (rather than just a combination of a key agreement and BSM encryption scheme).

We prove the result by introducing a setting (or "world") containing a specialized oracle where any hybrid scheme with low storage requirements (that does not use the oracle) may be broken, but otherwise computational primitives are left unaffected. As a corollary we get that there can be no black-box proof of everlasting security of a hybrid scheme based on a computational protocol (such as key agreement, but also others, e.g., oblivious transfer). The reason is that a security proof is a reduction showing that if the hybrid BSM can be broken then so can the computational scheme. Such a black-box proof would also apply in a world as described above and would thus prove the insecurity of the computational scheme. Therefore, if there existed such a black-box proof based on key agreement (for example) then it would serve as a proof that there exists no computational key agreement protocols altogether. In particular this means that there is no fully black-box reduction (in the terminology of [21]).

To specify the world mentioned above we define an oracle Z. Loosely speaking, the oracle Z takes as input an $\mathcal{NP}$ relation L and a (presumed) instance x of L, and generates several random witnesses to the fact that $x \in L$ (assuming such witnesses exist). However, the oracle does not actually output these witnesses. Instead it returns some form of encryption of them. The point is that a polynomial time adversary essentially gains nothing from the presence of the oracle Z. Thus any primitive that is secure against a polynomial time adversary remains secure even when the adversary is given access to the oracle Z. But a hybrid BSM encryption cannot have everlasting security since a polynomial time adversary with access to Z can save encrypted information for future use.

The Oracle Z**:** For every $\ell \in \mathbb{N}$ let $\pi_\ell : \{0,1\}^\ell \to \{0,1\}^\ell$ be a random permutations. For every $\ell, k \in \mathbb{N}$, let $R_{\ell,k}$ be a $2^\ell \times 2^k$ matrix of entries in $\{0,1\}^\ell$. The entries of the matrix $R_{\ell,k}$ are random strings subject to the sole restriction that $\bigoplus_{i\in[2^k]} R_{\ell,k}[y,i] = \pi_\ell^{-1}(y)$ (where $y \in \{0,1\}^\ell$ is also viewed as an index $y \in [2^\ell]$ and the XOR is performed bit by bit).

1. The oracle takes an input (L, x, n, k) where L is the description of an $\mathcal{NP}$ relation $L(\cdot,\cdot)$, x is a string (presumably in L), n is a parameter and k is a limit on witness length. If there is witness w of size at most k such that $L(x,w) = 1$, then the oracle computes n witnesses $w_1, ..., w_n$ that are randomly chosen under the restriction that $L(x, w_i) = 1$ for all $1 \le i \le n$ (suppose w.l.o.g. that each of the witnesses is of length exactly k). The

output of the oracle is $y = \pi_{nk}(w_1, \ldots, w_n)$. In case there is no witness of size k for x, then the output is simply a random string y.
2. On input (y, i) where $y \in \{0,1\}^{nk}$ and $i \in [2^k]$, the output is $R_{nk,k}[y, i]$.

Thus the oracle Z is defined by the ensemble of permutations π_ℓ and matrices $R_{\ell,k}$ as well as the randomness used to sample the witnesses.

Theorem 3. *In a world containing the Oracle Z (and no other oracle) we have:*

- *Any cryptographic primitive that is computationally secure in the plain model is secure in this world as well. More precisely, for any polynomial time adversary A with access to Z, there exists a polynomial time adversary A' in the plain model (*without *access to Z) so that any polynomial time environment interacting with the adversary without access to Z, cannot distinguish between A and A'.*
- *Any hybrid BSM encryption scheme in which (i) Alice and Bob require storage of size at most $o(\sqrt{r})$, and (ii) Alice and Bob make no calls to Z, cannot have everlasting security.*

The proof idea: For a computationally bounded adversary, the oracle Z is indistinguishable from a random oracle. This is because unless a whole row of R is read, then all of the outputs of Z are truly random strings. Thus, the oracle is of no use to a computationally bounded adversary as such an adversary can simulate the random oracle on his own, simply by tossing random coins whenever it queries the oracle.

On the other hand, the oracle is very handy in breaking any hybrid scheme. This is shown by designing a specific $\mathcal{NP}$ relation L and invoking a Lemma of Dziembowski and Maurer [9], that essentially states that a large enough number of random witnesses to L are sufficient to break a bounded storage scheme with low memory requirements. The adversary Charlie in the hybrid model can query for these witnesses from Z, but gets only an encrypted version via π_{nk} which is useless at the point. However, after the broadcast is over, Charlie may use his unbounded powers to extract the required information from the oracle Z, by decrypting π_{nk} using $R_{nk,k}$, and break the scheme. The actual proof appears in the full version of this paper ([14]).

6 Open Problems

The obvious open problem is to settle the possibility or impossibility of low storage everlasting security in the general hybrid model. This would likely entail resolving the existence or in-existence of relevant compression algorithms (following the formulation of [15]). Another problem is to come up with additional (reasonable) models where the notion of everlasting security may be achieved.

Finally, our solution for oblivious transfer (OT) requires the use of computationally secure oblivious transfer protocols. An interesting question is can one achieve low storage everlasting security OT protocols based on weaker assumptions than computational OT such as key agreement, for instance. It is tempting to think such a protocol exists since bounded storage OT may be achieved with no computational assumption at all (albeit, with high storage requirements).

References

1. Y. Aumann, Y.Z. Ding, and M. O. Rabin. Everlasting security in the bounded storage model. *IEEE Transactions on Information Theory*, 48(6):1668–1680, 2002.
2. Y. Aumann and M. O. Rabin. Information theoretically secure communication in the limited storage space model. In *CRYPTO '99*, volume 1666, pages 65–79, 1999.
3. C. Cachin, C. Crépeau, and J. Marcil. Oblivious transfer with a memory-bound receiver. In *39th IEEE FOCS*, pages 493–502, 1998.
4. C. Cachin and U. Maurer. Unconditional security against memory-bound adversaries. In *CRYPTO '97, LNCS*, volume 1294, pages 292–306. Springer, 1997.
5. R. Canetti, O. Goldreich, and S. Halevi. The random oracle methodology, revisited. *Journal of the ACM*, 51(4):557–594, 2004.
6. Y.Z. Ding. Oblivious transfer in the bounded storage model. In *CRYPTO '01, LNCS*, volume 2139, pages 155–170. Springer, 2001.
7. Y.Z. Ding, D. Harnik, A. Rosen, and R. Shaltiel. Constant round oblivious transfer in the bounded storage model. In *TCC '04*, volume 2951, pages 446–472, 2004.
8. Y.Z. Ding and M.O. Rabin. Hyper-encryption and everlasting security. In *STACS*, pages 1–26, 2002.
9. S. Dziembowski and U. Maurer. On generating the initial key in the bounded-storage model. In *EUROCRYPT ' 2004*, volume 3027, pages 126–137, 2004.
10. S. Dziembowski and U. Maurer. Optimal randomizer efficiency in the bounded-storage model. *Journal of Cryptology*, 17(1):5–26, 2004.
11. S. Even, O. Goldreich, and A. Lempel. A randomized protocol for signing contracts. *Communications of the ACM*, 28(6):637–647, 1985.
12. S. Goldwasser and Y. Tauman Kalai. On the (in)security of the fiat-shamir paradigm. In *44th IEEE FOCS*, pages 102–111, 2003.
13. S. Goldwasser and S. Micali. Probabilistic encryption. *Journal of the ACM*, 28(4):270–299, 1984.
14. D. Harnik and M. Naor. On everlasting security in the *hybrid* bounded storage model. `www.wisdom.weizmann.ac.il/~naor/PAPERS/hybrid_abs.html`, 2006.
15. D. Harnik and M. Naor. On the compressibility of NP instances and cryptographic applications. In *ECCC, TR06-022*, 2006.
16. C. Lu. Encryption against space-bounded adversaries from on-line strong extractors. *Journal of Cryptology*, 17(1):27–42, 2004.
17. U. Maurer. Conditionally-perfect secrecy and a provably-secure randomized cipher. *Journal of Cryptology*, 5(1):53–66, 1992.
18. U. Maurer, R. Renner, and C. Holenstein. Indifferentiability, impossibility results on reductions, and applications to the random oracle methodology. In *TCC '04*, volume 2951 of *LNCS*, pages 21–39, 2004.
19. M. Naor, B. Pinkas, and O. Reingold:. Distributed pseudo-random functions and kdcs. In *EUROCRYPT '99, LNCS*, volume 1592, pages 327–346. Springer, 1999.
20. M. O. Rabin. How to exchange secrets by oblivious transfer. TR-81, Harvard, 1981.
21. O. Reingold, L. Trevisan, and S. Vadhan. Notions of reducibility between cryptographic primitives. In *TCC '04*, volume 2951 of *LNCS*, pages 1–20, 2004.
22. S.P. Vadhan. Constructing locally computable extractors and cryptosystems in the bounded storage model. *Journal of Cryptology*, 17(1):43–77, 2004.

On the Impossibility of Extracting Classical Randomness Using a Quantum Computer

Yevgeniy Dodis[1] and Renato Renner[2]

[1] New York University, USA
dodis@cs.nyu.edu
[2] University of Cambridge, UK
r.renner@damtp.cam.ac.uk

Abstract. In this work we initiate the question of whether quantum computers can provide us with an almost perfect source of classical randomness, and more generally, suffice for classical cryptographic tasks, such as encryption. Indeed, it was observed [SV86, MP91, DOPS04] that classical computers are insufficient for either one of these tasks when all they have access to is a realistic *imperfect* source of randomness, such as the Santha-Vazirani source.

We answer this question in the *negative*, even in the following very restrictive model. We generously assume that quantum computation is error-free, and *all the errors come in the measurements*. We further assume that all the measurement errors are not only small but also *detectable*: namely, all that can happen is that with a small probability $p_\perp \leq \delta$ the (perfectly performed) measurement will result in some distinguished symbol $\perp$ (indicating an "erasure"). Specifically, we assume that if an element x was supposed to be observed with probability p_x, in reality it might be observed with probability $p'_x \in [(1-\delta)p_x, p_x]$, for some small $\delta > 0$ (so that $p_\perp = 1 - \sum_x p'_x \leq \delta$).

1 Introduction

Randomness is important in many areas of computer science, such as algorithms, cryptography and distributed computing. A common abstraction typically used in these applications is that there exists some source of unbiased and independent random bits. However, in practice this assumption seems to be problematic: although there seem to be many ways to obtain somewhat random data, this data is almost never uniformly random, its exact distribution is unknown, and, correspondingly, various algorithms and protocols have to be based on *imperfect sources of randomness*.

Not surprisingly, a large body of work (see below) has attempted to bridge the gap between this convenient theoretical abstraction and the actual reality. So far, however, most of this work concentrated on studying if *classical* computers can effectively use classical imperfect sources of randomness. In this work, we initiate the corresponding study regarding *quantum computation*. To motivate our question, we start by surveying the state of the art in using classical computers, which will demonstrate that such computers are provably incapable of tolerating even "mildly" imperfect random sources.

CLASSICAL APPROACH TO IMPERFECT RANDOMNESS. The most straightforward approach to dealing with an imperfect random source is to *deterministically* (and efficiently) extract nearly-perfect randomness from it. Indeed, many such results were ob-

M. Bugliesi et al. (Eds.): ICALP 2006, Part II, LNCS 4052, pp. 204–215, 2006.

tained for several classes of imperfect random sources. They include various "streaming" sources [Eli72, Blu86, LLS89], "bit-fixing" sources [CGH$^+$85, AL93, DSS01, KZ03], multiple independent imperfect sources [Vaz87b, CG88, DO03, DEOR04, BIW04] and efficiently samplable sources [TV00]. While these results are interesting and non-trivial, the above "deterministically extractable" sources assume a lot of structure or independence in the way they generate randomness. A less restrictive, and arguably more realistic, assumption on the random source would be to assume only that the source contains *some* entropy. We call such sources *entropy sources*. Entropy sources were first introduced by Santha and Vazirani [SV86], and later generalized by Chor and Goldreich [CG88], and Zuckerman [Zuc96].

The entropy sources of Santha and Vazirani [SV86] are the least imperfect (which means it is the hardest to show impossibility results for such sources) among the entropy sources considered so far (e.g., as compared to [CG88, Zuc96]). SV sources, as they are called, require *every bit output by the source* to have almost one bit of entropy, even when conditioned on all the previous bits. Unfortunately, already the original work of [SV86] (see also a simpler proof in [RVW04]) showed that deterministic randomness extraction of even a single bit is *not* possible from all SV sources. This can also be considered as impossibility of pseudo-random generators with access to only an SV source. Moreover, this result was later extended by McInnes and Pinkas [MP91], who showed that in the classical setting of computationally unlimited adversaries, one cannot have secure symmetric encryption if the shared key comes from an SV source. Finally and most generally, Dodis et al. [DOPS04] showed that SV sources in fact cannot be used essentially for any interesting classical cryptographic task involving privacy (such as encryption, commitment, zero-knowledge, multiparty computation), even when restricting to computationally bounded adversaries. Thus, even for the currently most restrictive entropy sources, classical computation does not seem to suffice for applications inherently requiring randomness (such as extraction and cryptography).[1]

We also mention that the impossibility results no longer hold when the extracting party has a small amount of true randomness (this is the study of so called probabilistic *randomness extractors* [NZ96]), or if several *independent* entropy sources are available [Vaz87b, CG88, DO03, DEOR04, BIW04].

QUANTUM COMPUTERS? Given the apparent inadequacy of classical computers to deal with entropy sources — at least for certain important tasks such as cryptography —, it is natural to ask if quantum computers can be of help. More specifically, given that quantum computation is *inherently probabilistic*, can we use quantum computers to generate nearly perfect randomness? (Or maybe just "good enough" randomness for cryptographic tasks like encryption, which, as we know [DS02], do not require *perfect* randomness?) For example, to generate a perfectly random bit from a fixed qubit $|0\rangle$, one can simply apply the Hadamard transform, and then measure the result in the standard basis. Unfortunately, what prevents this simple solution from working in

[1] In contrast, a series of celebrated positive results [VV85, SV86, CG88, Zuc96] show that even very weak entropy sources are enough for simulating probabilistic polynomial-time algorithms — namely, the task which does not *inherently* need randomness. This result was extended to interactive protocols by [DOPS04]. [DOPS04] also show that under certain strong, but reasonable computational assumptions, secure signatures seem to be possible with entropy sources.

practice is the fact that it is virtually impossible to perform the above transformation (in particular, the measurement) precisely, so the resulting bit is likely to be slightly biased. In other words, we must deal with the noise.

But, first, let us explain why there are good reasons to hope for quantum computers to be useful despite the noise. When dealing with classical imperfect sources, we usually assume that the source comes from some family of distributions "outside of our control" (e.g., "nature"), so we would like to make as few assumptions about these distributions as we can. For example, this is why the study of imperfect randomness quickly converged to entropy sources as being the most plausible sources one could get from nature. In contrast, by using a quantum computer to generate our random source for us, we are *proactively designing* a source of randomness which is convenient for use, rather than *passively hoping* that nature will give us such a source. Indeed, if not for the noise, it would be trivial to generate ideal randomness in our setting. Moreover, even with noise we have a lot of freedom in *adapting* our quantum computer to generate and measure quantum states *of our choice*, depending on the computation so far.

OUR MODEL. We first define a natural model for using a (realistically noisy) quantum computer for the task of randomness extraction (or, more generally, any probabilistic computation, such as the one needed in classical cryptography). As we will see shortly, we will prove a *negative* result in our model, despite the optimism we expressed in the previous paragraph. Because of this, we will make the noise as small and as restrictive as we can, even if these restrictions are completely "generous" and unrealistic. Indeed, *we will assume that the actual quantum computation is error-free, and all the errors come in the measurements* (which are necessary to extract some classical result out of the system). Of course, in reality the quantum computation will also be quite noisy, but our assumption will not only allow us to get a stronger result, but also reduce our "quantum" question to a natural "purely classical" question of independent interest.

Moreover, we will further assume that all the measurement errors are not only very small, but also *detectable*: namely, all that can happen is that with a small probability $p_\perp \leq \delta$ the (perfectly performed) measurement will result in some distinguished symbol $\perp$ (indicating an "erasure"). Specifically, we assume that if an element x was supposed to be observed with probability p_x, in reality it might be observed with probability $p'_x \in [(1-\delta)p_x, p_x]$, for some small $\delta > 0$ (so that $p_\perp = 1 - \sum_x p'_x \leq \delta$). Thus, it is guaranteed that no events of small probability can be completely "removed", and the probability of no event can be increased. Moreover, as compared to the classical SV model, in our model the state to be measured can be prepared arbitrarily, irrespective of the computational complexity of preparing this state. Further, such quantum states can even be generated adaptively and based on the measurements so far. For comparison, in the SV model the "ideal" measurement would always correspond to an unbiased bit; additionally, the SV model allows for "errors" while we only allow "erasures".

OUR RESULT. Unfortunately, our main result will show that even in this *extremely restrictive* noise model, one cannot extract even a *single* nearly uniform bit. In other words, if the measurement errors could be *correlated*, quantum computers do not help to extract classical randomness. More generally, we extend the technique of [DOPS04] to our model and show that one cannot generate two (classical) computationally indistinguishable distributions which are not nearly identical to begin with. This can be used to show the impossibility of classical encryption, commitment, zero-knowledge and other tasks exactly as in [DOPS04]. We notice, however, that our result does *not* exclude the possibility of generating perfect entanglement, which might be used to encrypt a message into a *quantum state*. Nevertheless, our result implies that, even with the help of such perfect entanglement, the user will not be able to generate a (shared) classical key that can be used for cryptographic tasks. To summarize, we only rule out the possibility of *classical* cryptography with *quantumly generated randomness*, leaving open the question of (even modeling!) *quantum* cryptography with noise.

Of independent interest, we reduce our "quantum" problem to the study of a new classical source, which is considerably more restrictive than the SV source (and this restriction can really be enforced in our model). We then show a classical impossibility result for our new source, which gives a non-trivial generalization of the corresponding impossibility result for the SV sources [DOPS04, SV86]. From another angle, it also generalizes the impossibility of extraction from the so called "bias-control limited" (BCL) sources of [Dod01]. As with our source, the most general BCL source considered in [Dod01] can adaptively generate samples from arbitrary distributions (and not just random bits). However, the attacker is given significantly more freedom in biasing the "real" distributions. First, all expected "real" distributions can be changed to arbitrary statistically close ones (which gives more power than performing "detectable erasures"), and, second, a small number of "real" distributions can be changed arbitrarily (which we do not allow at all).

To summarize, our main results can be viewed in three areas:

1. A model of using noisy quantum computers for classical probabilistic computation.
2. A reduction from a "quantum" question to the classical question concerning a much more restrictive variant of the SV (or general BCL) source(s).
3. A non-trivial impossibility result for the classical source we define.

RELATION TO QUANTUM ERROR-CORRECTION. What differentiates us from the usual model of quantum computation with noise is the fact that our errors are not assumed independent. In particular, conventional results on fault-tolerant quantum computation (such as the *threshold theorem*; see [NC00] for more details) do not apply in our model (as is apparent from our negative results). From another perspective, our impossibility result is not just a trivial application of the principle that one can always and without loss of generality postpone all the measurements until the end (a useful observation true in the "perfect measurement" case). For example, if all the measurements are postponed to the end, then we might observe a single "useless" $\perp$ symbol with non-trivial probability δ, while with many measurements we are bound to observe a lot of "useful" non-$\perp$ symbols with probability exponentially close to one.

Nevertheless, in our model one can trivially simulate probabilistic algorithms computing deterministic outputs, just as was the case for the classical computation. For example, here we actually *can* postpone all the measurements until the end, and then either obtain an error (with probability at most δ in which case the computation can be repeated), or the desired result (with probability arbitrarily close to $1 - \delta$). Of course, this "positive" result only holds because our noise model was made unrealistically restrictive (since we proved an *impossibility result*). Thus, it would be interesting to define a less restrictive (and more realistic!) error model — for example where the actual quantum computation is not error-free — and see if this feasibility result would still hold.

2 Definition of the Source

A source with n outputs $X_1, X_2, \ldots, X_n$ is specified by a joint probability distribution $P_{X_1 \cdots X_n}$. However, for most realistic sources, the actual distribution $P_{X_1 \cdots X_n}$ can usually not be fully determined. Instead, only a few characteristics of the source are known, e.g., that the conditional probability distributions[2] $P_{X_i|X^{i-1}}$ have certain properties. A well-known example for such a characterization are the Santha-Vazirani sources.

Definition 1 ([SV86]). *A probability distribution $P_{X_1 \cdots X_n}$ on $\{0,1\}^n$ is an α-SV source iff[3] for all $i \in \{1, \ldots, n\}$ and $x^{i-1} \in \{0,1\}^{i-1}$ we have*

$$P_{X_i|X^{i-1}=x^{i-1}}(0) \in [\alpha, 1-\alpha]$$

We will define a more general class of sources which, in some sense, includes the SV sources. The main motivation for our definition is to capture any kind of randomness that can be generated using imperfect (quantum) physical devices. Indeed, we will show in Section 3 that the randomness generated by any imperfect physical device cannot be more useful than the randomness obtained from a source as defined below.

Intuitively, a source can be seen as a device which sequentially outputs symbols $X_1, \ldots, X_n$ from some alphabet $\mathcal{X}$. Each output X_i is chosen according to some fixed probability distribution which might depend on all previous outputs $X_1, \ldots, X_{i-1}$. The "imperfectness" of the source is then modeled as follows. Each output X_i is "erased" with some probability $p_\perp$, i.e., it is replaced by some distinguished symbol $\perp$. This erasure probability might depend on the actual output X_i as well as on all previous outputs $X_1, \ldots, X_{i-1}$, but is upper bounded by some fixed parameter δ.

Before stating the formal definition, let us introduce some notation to be used in the sequel. For any set $\mathcal{X}$, we denote by $\bar{\mathcal{X}}$ the set $\bar{\mathcal{X}} := \mathcal{X} \cup \{\perp\}$ which contains an extra symbol $\perp$. For a probability distribution P_X on $\mathcal{X}$ and $\delta \geq 0$, let $\mathcal{P}^\delta(P_X)$ be the set of probability distributions $\bar{P}_X$ on $\bar{\mathcal{X}}$ such that

$$(1-\delta)P_X(x) \leq \bar{P}_X(x) \leq P_X(x) ,$$

for all $x \in \mathcal{X}$. In particular, the probability of the symbol $\perp$ is bounded by δ, that is, $\bar{P}_X(\perp) \leq \delta$.

Definition 2. *Let $\delta \geq 0$ and let, for any $i \in \{1, \ldots, n\}$, $Q_{X_i|X^{i-1}}$ be a channel[4] from $\bar{\mathcal{X}}^{i-1}$ to $\mathcal{X}$. A probability distribution $P_{X_1 \cdots X_n}$ on $\bar{\mathcal{X}}^n$ is a $(\delta, \{Q_{X_i|X^{i-1}}\})$-source if for all $i \in \{1, \ldots, n\}$ and $x^{i-1} = (x_1, \ldots, x_{i-1}) \in \bar{\mathcal{X}}^{i-1}$ we have*

$$P_{X_i|X^{i-1}=x^{i-1}} \in \mathcal{P}^\delta(Q_{X_i|X^{i-1}=x^{i-1}})$$

[2] We write X^k to denote the k-tuple $(X_1, \ldots, X_k)$.

[3] $P_{X_i|X^{i-1}=x^{i-1}}$ denotes the probability distribution of X_i conditioned on the event that the $(i-1)$-tuple $X^{i-1} = (X_1, \ldots, X_{i-1})$ takes the value $x^{i-1} = (x_1, \ldots, x_{i-1})$.

[4] A *channel* $Q_{Y|X}$ from $\mathcal{X}$ to $\mathcal{Y}$ is a function on $\mathcal{Y} \times \mathcal{X}$ such that, for any $x \in \mathcal{X}$, $Q_{Y|X=x} := Q_{Y|X}(\cdot, x)$ is a probability distribution on $\mathcal{Y}$.

In the full version we show that $(\delta, \{Q_{X_i|X^{i-1}}\})$-sources can be used to simulate α-SV sources, for some appropriately chosen α. This means that $(\delta, \{Q_{X_i|X^{i-1}}\})$-sources are at least as useful as SV sources. The other direction is, however, not true. That is, $(\delta, \{Q_{X_i|X^{i-1}}\})$-sources have a strictly less"malicious" behavior than SV sources (which makes our impossibility proofs stronger).

3 The Quantum Model

In this section, we propose a model that describes the extraction of classical information from imperfect quantum physical devices. Clearly, our considerations also include purely classical systems as a special case.

First, in Section 3.1, we review the situation where the quantum device is perfect. In this case, the process of extracting randomness can most generally be seen as a sequence of perfect quantum operations and perfect measurements. Then, in Section 3.2, we consider the imperfect case where the quantum device is subject to (malicious) noise. As we shall see, in order to get strong impossibility results, it is sufficient to extend the standard notion of perfect measurements by the possibility of detectable failures in the measurement process.

3.1 The Perfect Case

Let us briefly review some basic facts about quantum mechanics. The *state* of a quantum system is specified by a projector $P_{|\psi\rangle}$ onto a vector $|\psi\rangle$ in a Hilbert space $\mathcal{H}$. More generally, if a system is prepared by choosing a state from some family $\{|\psi_z\rangle\}_{z\in\mathcal{Z}}$ according to a probability distribution P_Z on $\mathcal{Z}$, then the behavior of the system is fully described by the *density operator* $\rho := \sum_{z\in\mathcal{Z}} P_Z(z)P_{|\psi_z\rangle}$. The most general *operation* that can be applied on a quantum system is specified by a family $\mathcal{E} = \{E_x\}_{x\in\mathcal{X}}$ of operators on $\mathcal{H}$ such that $\sum_{x\in\mathcal{X}} E_x^\dagger E_x = \mathrm{id}_{\mathcal{H}}$ (see, e.g., [NC00]). When $\mathcal{E}$ is applied to a system which is in state ρ, then, with probability $P_X(x) := \mathrm{tr}(E_x\rho E_x^\dagger)$, the classical output $x \in \mathcal{X}$ is produced and the final state ρ_x of the system is $\rho_x := \frac{1}{P_X(x)}E_x\rho E_x^\dagger$. Hence, when ignoring the classical output x, the state $\mathcal{E}(\rho)$ of the system after applying the operation $\mathcal{E}$ is the average of the states ρ_x, that is, $\mathcal{E}(\rho) := \sum_x P_X(x)\rho_x = \sum_x E_x\rho E_x^\dagger$.

It is important to note that also the action of preparing a quantum system to be in a certain state ρ_0 can be described by a quantum operation $\mathcal{E}$. To see this, let ρ_0 be given by $\rho_0 = \sum_{z\in\mathcal{Z}} P_Z(z)P_{|\psi_z\rangle}$, for some family of vectors $\{|\psi_z\rangle\}_{z\in\mathcal{Z}}$ and a probability distribution P_Z on $\mathcal{Z}$. Additionally, let $\{|i\rangle\}_{i\in\{1,\ldots,d\}}$ be an orthonormal basis of $\mathcal{H}$. It is easy to verify that the quantum operation $\mathcal{E} = \{E_{z,i}\}_{z\in\mathcal{Z},i\in\{1,\ldots,d\}}$ defined by the operators

$$E_{z,i} := \sqrt{P_X(z)}|\psi_z\rangle\langle i|$$

maps any arbitrary state ρ to ρ_0, that is, $\mathcal{E}(\rho) = \rho_0$.

We are now ready to describe the process of randomness extraction from a quantum system. Consider a classical user with access to a quantum physical device. The most general thing he can do is to subsequently apply quantum operations, where each of these operations provides him with classical information which he might use to select the next operation. To describe this on a formal level, let $\mathcal{H}$ be a Hilbert space and let $\mathcal{X}$ be a set. The strategy of the user in each step i is then defined by the quantum operation $\mathcal{E}^{x^{i-1}} = \{E_x^{x^{i-1}}\}_{x\in\mathcal{X}}$ he applies depending on the classical outputs $x^{i-1} \in \mathcal{X}^{i-1}$ obtained in the previous steps. Note that, according to the above discussion, this description also includes the action of preparing (parts of) the quantum system in a certain state. We can thus assume without loss of generality that the initial state of the system is given by some fixed projector $P_{|\psi_0\rangle}$. The probability distribution $P_{X_i|X^{i-1}=x^{i-1}}$ of the classical outcomes in the ith step conditioned on the previous outputs x^{i-1} as well as the quantum state ρ_{x^i} after the ith step given the outputs x^i is then recursively defined by $\rho_{x^0} := P_{|\psi_0\rangle}$ and

$$P_{X_i|X^{i-1}=x^{i-1}}(x) := \mathrm{tr}(E_x^{x^{i-1}} \rho_{x^{i-1}} E_x^{x^{i-1}\dagger}) \tag{1}$$

$$\rho_{x^i} = \rho_{(x^{i-1},x)} := \frac{1}{P_{X_i|X^{i-1}=x^{i-1}}(x)} E_x^{x^{i-1}} \rho_{x^{i-1}} E_x^{x^{i-1}\dagger}. \tag{2}$$

3.2 Quantum Measurements with Malicious Noise

We will now extend the model of the previous section to include situations where the quantum operations are subject to noise. As we are interested in proving the *impossibility* of certain tasks in the presence of noise, our results are stronger if we assume that only parts of the quantum operation are noisy. In particular, we will restrict to systems where only the classical measurements are subject to perturbations.[5]

Formally, we define an imperfect quantum device by its behavior when applying any operation $\mathcal{E}$. Let $\delta \geq 0$ and let $\mathcal{E} = \{E_{x,u}\}_{x\in\mathcal{X},u\in\mathcal{U}}$ be a quantum operation which produces two classical outcomes x and u, where x is the part of the output that is observed by the user. The operation $\mathcal{E}$ acts on the imperfect device as it would in the perfect case, except that each output x is, with some probability $\lambda_x \leq \delta$, replaced by a symbol $\perp$, indicating that something went wrong. Additionally, we assume that, whenever such an error occurs, the state of the system remains unchanged.[6] The resulting probability distribution P_X of the outputs when applying $\mathcal{E}$ to an imperfect device in state ρ is thus given by

$$P_X(x) := \sum_u (1-\lambda_x)\,\mathrm{tr}(E_{x,u}\rho E_{x,u}^\dagger).$$

In particular, the probability of the symbol $\perp$ is $P_X(\perp) = 1 - \sum_{x\in\mathcal{X}} P_X(x) \leq \delta$.

[5] To see that our model leads to strong impossibility results, consider for example an adversary who is allowed to transform the quantum state ρ of the device into a state ρ' which has at most trace distance δ to the original state ρ. Let $\mathcal{M}$ be a fixed measurement and let P be the distribution resulting from applying $\mathcal{M}$ to ρ. It is easy to see that, for any given probability distribution P' which is δ-close to P, the adversary can set the device into a state ρ' such that a measurement $\mathcal{M}$ of ρ' gives raise to the distribution P'. Consequently, such an adversary is at least as powerful as an adversary who can only modify the distribution of the measurement outcomes, as proposed in our model. In particular, our impossibility results also apply to this case.

[6] This means that, even if a measurement error occurs, the state of the quantum system is not destroyed. (Recall that our impossibility results are stronger the closer our model is to a model describing perfect systems.)

Let us now consider the interaction of a user with such an imperfect quantum device. In each step i, he either observes the correct outcome or he gets the output $\perp$, indicating that something went wrong. The user might want to use this information to choose the subsequent operations. His strategy is thus defined by a family $\{\mathcal{E}^{x^{i-1}}\}_{x^{i-1}\in\bar{\mathcal{X}}^{i-1}}$ of quantum operations $\mathcal{E}^{x^{i-1}} = \{E_{x,u}^{x^{i-1}}\}_{x\in\mathcal{X},u\in\mathcal{U}}$. The conditional probability distributions $P_{X_i|X^{i-1}=x^{i-1}}$ of the observed outputs in the ith step, for $x^{i-1}\in\bar{\mathcal{X}}^{i-1}$, and the states ρ_{x^i} after the ith step are recursively defined, analogously to (1) and (2), by

$$P_{X_i|X^{i-1}=x^{i-1}}(x) := (1-\lambda_{x^{i-1},x})Q_{X_i|X^{i-1}=x^{i-1}}(x) \qquad \text{for } x\in\mathcal{X}$$

$$\rho_{x^i} = \rho_{(x^{i-1},x)} := \begin{cases} \frac{1}{Q_{X_i|X^{i-1}=x^{i-1}}(x)}\sum_{u\in\mathcal{U}} E_{x,u}^{x^{i-1}}\rho_{x^{i-1}}E_{x,u}^{x^{i-1}\dagger} & \text{if } x\in\mathcal{X} \\ \rho_{x^{i-1}} & \text{if } x=\perp. \end{cases}$$

for some $\lambda_{x^{i-1},x}\in[0,\delta]$, where $Q_{X_i|X^{i-1}}$ is the channel from $\bar{\mathcal{X}}^{i-1}$ to $\mathcal{X}$ given by $Q_{X_i|X^{i-1}=x^{i-1}}(x) := \sum_{u\in\mathcal{U}}\mathrm{tr}(E_{x,u}^{x^{i-1}}\rho_{x^{i-1}}E_{x,u}^{x^{i-1}\dagger})$.

Let $P_{X^n} = P_{X_1\cdots X_n}$ be the probability distribution of the observed outcomes after n steps. It follows directly from the above formulas that P_{X^n} is a $(\delta,\{Q_{X_i|X^{i-1}}\})$-source. On the other hand, if P_{X^n} is a $(\delta,\{Q_{X_i|X^{i-1}})$-source, then there exist weights $\lambda_{x^{i-1},x}\in[0,\delta]$ such that the conditional probabilities are given by the above formulas. This reduces our "quantum" problem to a totally classical problem for an imperfect source considerably more restrictive than an SV source. The corresponding impossibility result is given in the next section.

4 Main Technical Lemma

Our main technical result can be seen as an extension of a result proved for SV sources (cf. Lemma 3.5 of [DOPS04]). Roughly speaking, Lemma 1 below states that a task g which requires perfect random bits can generally not be replaced by another task f which only uses imperfect bits. Note that this impossibility is particularly interesting for cryptography where many tasks do in fact use randomness.

More precisely, let g be an arbitrary strategy which uses imperfect randomness X^n and, in addition, some perfect randomness Y (whose probability distribution might even depend on the values of X^n). Let f be another strategy which only uses imperfect randomness X^n. Furthermore, assume that, for any $(\delta,\{Q_{X_i|X^{i-1}}\})$-source $P_{X_1\cdots X_n}$, the output distributions of the strategies g and f are (almost) identical. Then the strategy g is (roughly) the same as f, that is, it (virtually) does not use the randomness Y.

Lemma 1. *Let f be a function from $\bar{\mathcal{X}}^n$ to $\mathcal{Z}$, g be a function from $\bar{\mathcal{X}}^n\times\mathcal{Y}$ to $\mathcal{Z}$ and $m=\lceil\log_2(|\mathcal{Z}|)\rceil$. For any $i\in\{1,\ldots,n\}$, let $Q_{X_i|X^{i-1}}$ be a channel from $\bar{\mathcal{X}}^{i-1}$ to $\mathcal{X}$, let $Q_{Y|X^n}$ be a channel from $\bar{\mathcal{X}}^n$ to $\mathcal{Y}$, and let $\delta\geq 0$. Let Γ be the set of all probability distributions P_{X^nY} on $\bar{\mathcal{X}}^n\times\mathcal{Y}$ such that P_{X^n} is a $(\delta,\{Q_{X_i|X^{i-1}}\})$-source*[7]

[7] Similarly to the argument in [DOPS04], the proof can easily be extended to a statement which holds for an even stronger type of sources, where the conditional probability distributions of each X_i given all other source outputs, and not only the previous ones X^{i-1}, is contained in a certain set $\mathcal{P}^\delta$.

and $P_{Y|X^n} = Q_{Y|X^n}$. *If, for all* $P_{X^nY} \in \Gamma$,

$$|P_{f(X^n)} - P_{g(X^n,Y)}|_1 < \varepsilon,$$

then there exists $P_{\bar{X}^n\bar{Y}} \in \Gamma$ *such that*

$$\Pr_{(x^n,y)\leftarrow P_{\bar{X}^n\bar{Y}}}[f(x^n) \neq g(x^n,y)] < 5\varepsilon m\delta^{-1},$$

Proof. Assume first that the functions f and g are binary, i.e., $\mathcal{Z} = \{0,1\}$. The idea is to define two probability distributions $P_{V^nY}, P_{W^nY} \in \Gamma$ such that the output distributions of the function f, $f(V^n)$ and $f(W^n)$, are "maximally different". Then, by assumption, the output distributions of $g(V^n,Y)$ and $g(W^n,Y)$ must be different as well. This will then be used to conclude that the outputs of f and g are actually equal for most inputs.

In order to define the distributions P_{V^nY} and P_{W^nY}, we first consider some "intermediate distribution" $P_{\bar{X}^n\bar{Y}}$. It is defined as the unique probability distribution on $\bar{\mathcal{X}}^n \times \mathcal{Y}$ such that $P_{\bar{Y}|\bar{X}^n} = Q_{Y|X^n}$ and, for any $i \in \{1,\ldots,n\}$ and $x^{i-1} \in \bar{\mathcal{X}}^{i-1}$,

$$P_{\bar{X}_i|\bar{X}^{i-1}=x^{i-1}}(x) := \begin{cases} (1-\frac{\delta}{2})Q_{X_i|X^{i-1}=x^{i-1}}(x) & \text{if } x \in \mathcal{X} \\ \frac{\delta}{2} & \text{if } x = \perp. \end{cases}$$

Note that $P_{\bar{X}_i|\bar{X}^{i-1}=x^{i-1}} \in \mathcal{P}^\delta(Q_{X_i|X^{i-1}=x^{i-1}})$, i.e., $P_{\bar{X}^n}$ is a $(\delta,\{Q_{X_i|X^{i-1}}\})$-source, and thus $P_{\bar{X}^n\bar{Y}} \in \Gamma$.

The distribution P_{V^n} is now defined from $P_{\bar{X}^n}$ by raising the probabilities of all values[8] $x^n \in f^{-1}(0)$ that f maps to 0 and lowering the probabilities of all $x^n \in f^{-1}(1)$. Similarly, P_{W^n} is defined by changing the probabilities of $P_{\bar{X}^n}$ in the other direction. For the formal definition, we assume without loss of generality that $P_{f(\bar{X}^n)}(0) \leq \frac{1}{2}$ and set $\alpha := P_{f(\bar{X}^n)}(0)/P_{f(\bar{X}^n)}(1)$, i.e., $\alpha \leq 1$. P_{V^n} and P_{W^n} are then given by

$$P_{V^n}(x^n) := \begin{cases} P_{\bar{X}^n}(x^n)(1+\tau) & \text{if } x^n \in f^{-1}(0) \\ P_{\bar{X}^n}(x^n)(1-\alpha\tau) & \text{if } x^n \in f^{-1}(1) \end{cases}$$

$$P_{W^n}(x^n) := \begin{cases} P_{\bar{X}^n}(x^n)(1-\tau) & \text{if } x^n \in f^{-1}(0) \\ P_{\bar{X}^n}(x^n)(1+\alpha\tau) & \text{if } x^n \in f^{-1}(1), \end{cases}$$

where $\tau := \frac{\delta}{4}$. Because

$$\begin{aligned}\sum_{x^n\in\bar{\mathcal{X}}^n} P_{V^n}(x^n) &= \sum_{x^n\in f^{-1}(0)} P_{X^n}(x^n)(1+\tau) + \sum_{x^n\in f^{-1}(1)} P_{X^n}(x^n)(1-\alpha\tau) \\ &= P_{f(X^n)}(0)(1+\tau) + P_{f(X^n)}(1)(1-\alpha\tau) = 1,\end{aligned}$$

P_{V^n} and, similarly, P_{W^n}, is indeed a probability distribution.

[8] For $z \in \{0,1\}$, $f^{-1}(z) := \{x \in \bar{\mathcal{X}}^n : f(x) = z\}$ denotes the preimage of z under the mapping f.

We claim that P_{V^n} and P_{W^n} are $(\delta, \{Q_{X_i|X^{i-1}}\})$-sources. To see this, note first that, for any $i \in \{1, \dots n\}$ and $x^i \in \bar{\mathcal{X}}^i$, $(1-\alpha\tau)P_{\tilde{X}^i}(x^i) \leq P_{V^i}(x^i)$ and $P_{V^i}(x^i) \leq (1+\tau)P_{\tilde{X}^i}(x^i)$. Hence, for any $x \in \mathcal{X}$ and $x^{i-1} \in \bar{\mathcal{X}}^{i-1}$,

$$\begin{aligned} P_{V_i|V^{i-1}=x^{i-1}}(x) &= \frac{P_{V_i V^{i-1}}(x, x^{i-1})}{P_{V^{i-1}}(x^{i-1})} \geq \frac{(1-\alpha\tau)P_{\tilde{X}_i \tilde{X}^{i-1}}(x, x^{i-1})}{(1+\tau)P_{\tilde{X}^{i-1}}(x^{i-1})} \\ &= \frac{1-\alpha\tau}{1+\tau} P_{\tilde{X}_i|\tilde{X}^{i-1}=x^{i-1}}(x) = \frac{1-\alpha\tau}{1+\tau}(1-\tfrac{\delta}{2})Q_{X_i|X^{i-1}=x^{i-1}}(x). \end{aligned}$$

Because $\alpha \leq 1$, we have $P_{V_i|V^{i-1}=x^{i-1}}(x) \geq (1-\delta)Q_{X_i|X^{i-1}=x^{i-1}}(x)$. Similarly,

$$P_{V_i|V^{i-1}=x^{i-1}}(x) \leq \frac{1+\tau}{1-\alpha\tau} P_{\tilde{X}_i|\tilde{X}^{i-1}=x^{i-1}}(x) = \frac{1+\tau}{1-\alpha\tau}(1-\tfrac{\delta}{2})Q_{X_i|X^{i-1}=x^{i-1}}(x)$$

which implies $P_{V_i|V^{i-1}=x^{i-1}}(x) \leq Q_{X_i|X^{i-1}=x^{i-1}}(x)$. Combining these inequalities, we conclude $P_{V_i|V^{i-1}=x^{i-1}} \in \mathcal{P}^{\delta}(Q_{X_i|X^{i-1}=x^{i-1}})$, i.e., P_{V^n} is a $(\delta, \{Q_{X_i|X^{i-1}}\})$-source. A similar computation shows that also the distribution P_{W^n} is a $(\delta, \{Q_{X_i|X^{i-1}}\})$-source. Consequently, the distributions $P_{V^n Y}$ and $P_{W^n Y}$ defined by $P_{Y|V^n} = Q_{Y|X^n}$ and $P_{Y|W^n} = Q_{Y|X^n}$, respectively, are contained in the set Γ.

Next, we will analyze the behavior of the function g for inputs chosen according to $P_{V^n Y}$ and $P_{W^n Y}$, respectively, and compare it to f. For this, let q_{x^n} be the probability that, given some fixed $x^n \in \bar{\mathcal{X}}^n$, the output of g is zero, i.e., $q_{x^n} := \Pr_{y \leftarrow Q_{Y|X^n=x^n}}[g(x^n, y) = 0]$. Because $P_{\tilde{Y}|\tilde{X}^n} = P_{Y|V^n} = P_{Y|W^n} = Q_{Y|X^n}$, we get

$$q_{x^n} = P_{g(\tilde{X}^n, \tilde{Y})|\tilde{X}^n=x^n}(0) = P_{g(V^n,Y)|V^n=x^n}(0) = P_{g(W^n,Y)|W^n=x^n}(0) .$$

The probability that the output of f is zero for the distributions P_{V^n} and P_{W^n} can then, obviously, be written as

$$\begin{aligned} P_{f(V^n)}(0) &= \sum_{x^n \in f^{-1}(0)} P_{\tilde{X}^n}(x^n)(1+\tau) \\ P_{f(W^n)}(0) &= \sum_{x^n \in f^{-1}(0)} P_{\tilde{X}^n}(x^n)(1-\tau). \end{aligned}$$

Similarly, for g, we have

$$\begin{aligned} P_{g(V^n,Y)}(0) &= \sum_{x^n \in f^{-1}(0)} P_{\tilde{X}^n}(x^n)(1+\tau)q_{x^n} + \sum_{x^n \in f^{-1}(1)} P_{\tilde{X}^n}(x^n)(1-\alpha\tau)q_{x^n} \\ P_{g(W^n,Y)}(0) &= \sum_{x^n \in f^{-1}(0)} P_{\tilde{X}^n}(x^n)(1-\tau)q_{x^n} + \sum_{x^n \in f^{-1}(1)} P_{\tilde{X}^n}(x^n)(1+\alpha\tau)q_{x^n}. \end{aligned}$$

By assumption of the lemma, because, $P_{V^n Y}$ and $P_{W^n Y}$ are contained in the set Γ, the output distributions of f and g must be close, that is, $|P_{f(V^n)}(0) - P_{g(V^n,Y)}(0)| < \frac{\varepsilon}{2}$ and $|P_{f(W^n)}(0) - P_{g(W^n,Y)}(0)| < \frac{\varepsilon}{2}$, and hence $(P_{f(V^n)}(0) - P_{g(V^n,Y)}(0)) - (P_{f(W^n)}(0) - P_{g(W^n,Y)}(0)) < \varepsilon$. Replacing these probabilities by the above expressions leads to

$$\sum_{x^n \in f^{-1}(0)} P_{\tilde{X}^n}(x^n)2\tau(1-q_{x^n}) + \sum_{x^n \in f^{-1}(1)} P_{\tilde{X}^n}(x^n)2\alpha\tau q_{x^n} < \varepsilon . \qquad (3)$$

Note that this imposes some restrictions on the possible values of q_{x^n}. Roughly speaking, if f maps a certain input x^n to 0, then the probability $1 - q_{x^n}$ that g maps x^n to 1 must be small. In fact, as we shall see, (3) implies a bound on the probability that the outputs of f and g are different.

With the definition $p_{z,w} := P_{f(\tilde{X}^n)g(\tilde{X}^n,Y)}(z,w)$, for $(z,w) \in \{0,1\}^2$ and using again the assumption of the lemma,

$$|p_{0,1} - p_{1,0}| = |(p_{0,0} + p_{0,1}) - (p_{0,0} + p_{1,0})| = |P_{f(\tilde{X}^n)}(0) - P_{g(\tilde{X}^n,\tilde{Y})}(0)| < \frac{\varepsilon}{2},$$

hence,

$$\Pr_{(x^n,y)\leftarrow P_{\tilde{X}^n Y}}[f(x^n) \neq g(x^n,y)] \leq p_{0,1} + p_{0,1} + |p_{1,0} - p_{0,1}| < 2p_{0,1} + \frac{\varepsilon}{2}. \quad (4)$$

Using (3) and the fact that the second sum is nonnegative, we get an upper bound for $p_{0,1}$, that is,

$$\begin{aligned} p_{0,1} &= \sum_{x^n \in \bar{\mathcal{X}}^n} P_{\tilde{X}^n}(x^n) P_{f(\tilde{X}^n)|\tilde{X}^n=x^n}(0) P_{g(\tilde{X}^n,\tilde{Y})|\tilde{X}^n=x^n}(1) \\ &= \sum_{x^n \in f^{-1}(0)} P_{\tilde{X}^n}(x^n)(1 - q_{x^n}) < \frac{\varepsilon}{2\tau} = \frac{2\varepsilon}{\delta}. \end{aligned}$$

Combining this with (4), we conclude $\Pr_{(x^n,y)\leftarrow P_{\tilde{X}^n Y}}[f(x^n) \neq g(x^n,y)] < \frac{4\varepsilon}{\delta} + \frac{\varepsilon}{2} \leq \frac{5\varepsilon}{\delta}$, which proves the lemma for the binary case where $\mathcal{Z} = \{0,1\}$.

To deduce the statement for arbitrary sets $\mathcal{Z}$, consider an (injective) encoding function c which maps each element $z \in \mathcal{Z}$ to an m-tuple $(c_1(z), \ldots, c_m(z))$. Since the L_1-norm $|\cdot|_1$ can only decrease when applying a function, the assumption of the lemma implies that, for all probability distributions $P_{X^n Y} \in \Gamma$, $|P_{f_k(X^n)} - P_{g_k(X^n,Y)}|_1 < \varepsilon$, where $f_k := c_k \circ f$ and $g_k := c_k \circ g$, for any $k \in \{1, \ldots, m\}$. The assertion then follows from the binary version of the lemma and the union bound.

As was shown in [DOPS04], Lemma 1 implies not only impossibility of extracting nearly perfect randomness, but also impossibility of doing almost any classical task involving privacy (such as encryption, commitment, etc.). For illustrative purposes, we give such an argument for extraction, referring to [DOPS04] regarding the other tasks.

Corollary 1. *Let f be a function from $\bar{\mathcal{X}}^n$ to $\{0,1\}$ and P_U be the uniform distribution on $\{0,1\}$. For any $i \in \{1, \ldots, n\}$, let $Q_{X_i|X^{i-1}}$ be a channel from $\bar{\mathcal{X}}^{i-1}$ to $\mathcal{X}$, and let $\delta \geq 0$. Then there exists a $(\delta, \{Q_{X_i|X^{i-1}}\})$-source P_{X^n} such that*

$$|P_{f(X^n)} - P_U|_1 \geq \frac{\delta}{10},$$

Proof. Assume by contradiction that, for any $(\delta, \{Q_{X_i|X^{i-1}}\})$-source P_{X^n}, $|P_{f(X^n)} - P_U|_1 < \frac{\delta}{10}$. Let g be the function on $\mathcal{X}^n \times \{0,1\}$ defined by $g(x^n, u) := u$. Then, for any probability distribution $P_{X^n U} = P_{X^n} \times P_U$, where P_{X^n} is a $(\delta, \{Q_{X_i|X^{i-1}}\})$-source, we have $|P_{f(X^n)} - P_{g(X^n,U)}|_1 < \frac{\delta}{10}$. Lemma 1 thus implies that there exists a $(\delta, \{Q_{X_i|X^{i-1}}\})$-source $P_{\tilde{X}^n}$ with $\Pr_{(x^n,u)\leftarrow P_{\tilde{X}^n} \times P_U}[f(x^n) \neq g(x^n,u)] < \frac{1}{2}$, that is, $\Pr_{(x^n,u)\leftarrow P_{\tilde{X}^n} \times P_U}[f(x^n) \neq u] < \frac{1}{2}$. This is a contradiction because P_U is the uniform distribution on $\{0,1\}$.

References

[AL93] Miklós Ajtai and Nathal Linial. The influence of large coalitions. *Combinatorica*, 13(2):129–145, 1993.

[BIW04] Boaz Barak, Russell Impagliazzo, and Avi Wigderson. Extracting randomness from few independent sources. In *Proc. 45th FOCS*, 2004.

[Blu86] M. Blum. Independent unbiased coin flips from a correlated biased source—a finite state Markov chain. *Combinatorica*, 6(2):97–108, 1986.

[CG88] Benny Chor and Oded Goldreich. Unbiased bits from sources of weak randomness and probabilistic communication complexity. *SIAM J. Comput.*, 17(2):230–261, 1988.

[CGH+85] Benny Chor, Oded Goldreich, Johan Håstad, Joel Friedman, Steven Rudich, and Roman Smolensky. The bit extraction problem of t-resilient functions. In *Proc. 26th FOCS*, pages 396–407. IEEE, 1985.

[Dod01] Yevgeniy Dodis. New Imperfect Random Source with Applications to Coin-Flipping. In *Proc. ICALP '01*, pages 297–309, 2001.

[DEOR04] Yevgeniy Dodis, Ariel Elbaz, Roberto Oliveira, and Ran Raz. Improved randomness extraction from two independent sources. In *Proc. RANDOM '04*, 2004.

[DO03] Yevgeniy Dodis and Roberto Oliveira. On extracting private randomness over a public channel. In *Proc. RANDOM '03*, pages 252–263, 2003.

[DOPS04] Yevgeniy Dodis, Shien Jin Ong, Manoj Prabhakaran, and Amit Sahai. On the (im)possibility of cryptography with imperfect randomness. In *Proc. FOCS '04*, pages 196–205, 2004.

[DSS01] Yevgeniy Dodis, Amit Sahai, and Adam Smith. On perfect and adaptive security in exposure-resilient cryptography. In *Proc. EUROCRYPT '01*, pages 301–324, 2001.

[DS02] Yevgeniy Dodis and Joel Spencer. On the (non)Universality of the One-Time Pad. In *Proc. FOCS '02*, pages 376–385, 2002.

[Eli72] Peter Elias. The efficient construction of an unbiased random sequence. *Ann. Math. Stat.*, 43(2):865–870, 1972.

[KZ03] Jess Kamp and David Zuckerman. Deterministic extractors for bit-fixing sources and exposure-resilient cryptography. In *Proc. 35th FOCS*, pages 92–101, 2003.

[LLS89] D. Lichtenstein, N. Linial, and M. Saks. Some extremal problems arising from discrete control processes. *Combinatorica*, 9(3):269–287, 1989.

[MP91] J. L. McInnes and B. Pinkas. On the impossibility of private key cryptography with weakly random keys. In *Proc. CRYPTO '90*, pages 421–436, 1991.

[NC00] M. A. Nielsen and I. L. Chuang. *Quantum computation and quantum information*. Cambridge University Press, 2000.

[NZ96] Noam Nisan and David Zuckerman. Randomness is linear in space. *J. Comput. Syst. Sci.*, 52(1):43–52, 1996.

[RVW04] Omer Reingold, Salil Vadhan, and Avi Wigderson. A note on extracting randomness from Santha-Vazirani sources. Unpublished manuscript, 2004.

[SV86] M. Santha and U. V. Vazirani. Generating quasi-random sequences from semi-random sources. *J. Comput. Syst. Sci.*, 33(1):75–87, 1986.

[TV00] Luca Trevisan and Salil Vadhan. Extracting randomness from samplable distributions. In *Proc. 41st FOCS*, pages 32–42, 2000.

[Vaz87b] Umesh V. Vazirani. Efficiency considerations in using semi-random sources. In *Proc. 19th STOC*, pages 160–168, 1987.

[VV85] Umesh V. Vazirani and Vijay V. Vazirani. Random polynomial time is equal to slightly-random polynomial time. In *Proc. 26th FOCS*, pages 417–428, 1985.

[Zuc96] David Zuckerman. Simulating BPP using a general weak random source. *Algorithmica*, 16(4/5):367–391, 1996.

Quantum Hardcore Functions by Complexity-Theoretical Quantum List Decoding

Akinori Kawachi[1] and Tomoyuki Yamakami[2]

[1] Graduate School of Information Science and Engineering
Tokyo Institute of Technology
2-12-1 Ookayama, Meguro-ku, Tokyo 152-8552, Japan
[2] ERATO-SORST Quantum Computation and Information Project
Japan Science and Technology Agency
5-28-3 Hongo, Bunkyo-ku, Tokyo 113-0033, Japan

Abstract. We present three new quantum hardcore functions for any quantum one-way function. We also give a "quantum" solution to Damgård's question (CRYPTO'88) on his pseudorandom generator by proving the quantum hardcore property of his generator, which has been unknown to have the classical hardcore property. Our technical tool is quantum list-decoding of "classical" error-correcting codes (rather than "quantum" error-correcting codes), which is defined on the platform of computational complexity theory and cryptography (rather than information theory). In particular, we give a simple but powerful criterion that makes a polynomial-time computable code (seen as a function) a quantum hardcore for any quantum one-way function. On their own interest, we also give quantum list-decoding algorithms for codes whose associated quantum states (called codeword states) are "nearly" orthogonal using the technique of pretty good measurement.

1 Introduction: From Hardcore to List-Decoding

Background: Modern cryptography heavily relies on computational hardness and pseudorandomness. One of its key notions is a *hardcore* bit for a one-way function—a bit that can be completely determined from all the information available to the adversary but still looks random to any "feasible" adversary. A hardcore function transforms the onewayness into pseudorandomness by generating such hardcore bits of a given one-way function. Such a hardcore function is a crucial element of constructing a pseudorandom generator as well as a bit commitment protocol from a one-way permutation. A typical example is the inner product mod 2 function $\mathrm{GL}_x(r)$ of Goldreich and Levin [12], computing the bitwise inner product modulo two $\langle x, r\rangle$, which constitutes a hardcore bit for any (strong) one-way function.[1] Since $\mathrm{GL}_x(r)$ equals the rth bit of the codeword

[1] Literally speaking, this statement is slightly misleading. To be more accurate, such a hard-core function concerns only the one-way function of the form $f'(x,r) = (f(x), r)$ with $|r| = poly(|x|)$ induced from an arbitrary strong one-way function f. See, e.g., [11] for a detailed discussion.

M. Bugliesi et al. (Eds.): ICALP 2006, Part II, LNCS 4052, pp. 216–227, 2006.

$\mathrm{HAD}_x^{(2)} = (\langle x, 0^n\rangle, \langle x, 0^{n-1}1\rangle, \cdots, \langle x, 1^n\rangle)$ of message x of a binary Hadamard code, Goldreich and Levin essentially gave a polynomial-time list-decoding algorithm for this Hadamard code. In the recent literature, list-decoding has kept playing a key role in a general construction of hardcores [2, 17].

Thirteen years later, the "quantum" hardcore property (i.e., a hardcore property against a feasible quantum adversary) of $\mathrm{GL}_x(\cdot)$ was shown by Adcock and Cleve [1], who implicitly gave a simple and efficient quantum algorithm that recovers x from the binary Hadamard code by exploiting the robust nature of a quantum algorithm of Bernstein and Vazirani [6]. The simplicity of the proof of Adcock and Cleve can be best compared to the original proof of Goldreich and Levin, who employed a rather complicated algorithm with powerful techniques: self-correction property of the aforementioned Hadamard code and pairwise independent sampling. This highlights a significant role of robust quantum computation in list-decoding (and thus hardcores); however, it has been vastly unexplored until our work except for a quantum decoder of Barg and Zhou [5] for simplex codes. No other quantum hardcore has been proven so far. The efficiency of robust quantum algorithms with access to biased oracles has been also discussed in a different context [3, 7, 18].

Our Major Contributions: As our main result, we present three new quantum hardcore functions, $\mathrm{HAD}^{(q)}$, SLS^p, and PEQ (see Section 5 for their definition), for any (strongly) quantum one-way function, the latter two of which are not yet known to be hardcores in a classical setting (see, e.g., [13]). In particular, we prove the quantum hardcore property of Damgård's pseudorandom generator [8]. This gives a "quantum" solution to his question of whether his generator has the classical hardcore property (this is also listed as an open problem in [13]). Our proof technique exploits the quantum list-decodability of classical error-correcting codes (rather than quantum error-correcting codes). For our purpose, we formulate the notion of *complexity-theoretical* quantum list-decoding to conduct message-recovery from a quantum-computational error rather than an information-theoretical error which is usually associated with a transmission error. This notion naturally expands the classical framework of list-decoding. Our goal is to present fast quantum list-decoding algorithms for the aforementioned codes.

Proving the quantum hardcore property of a given code C (seen as a function) corresponds to solving the *quantum list-decoding problem* (QLDP) for C via direct access to a *quantum-computationally (or quantumly) corrupted word*, which is given as a black-box oracle. The task of a quantum list-decoder is simply to list all message candidates whose codewords match the quantumly-corrupted word within a certain error rate bound.

The key notion of this paper is a specific quantum state, called a *(k-shuffled) codeword state*, which embodies the full information on a given codeword. Note that similar states have appeared in several quantum algorithms in the recent literature [6, 9, 14, 20]. In our key lemmas, we show (i) how to generate such a codeword state from *any* (even adversarial) quantumly corrupted word and (ii) how to convert a *codeword-state decoder* (i.e., a quantum algorithm that recovers a message x from a codeword state which is given as an input) to a quantum list-decoding algorithm working with a quantumly corrupted word. The robust construction made in the course of our proofs also provides a useful means, known as "hardness" reduction, which is often crucial in the security proof of a quantum cryptosystem. Moreover, using pretty good measurement [10, 16], we present a quantum list-decoding algorithm for any code whose codeword states are "nearly" orthogonal.

Further Implications: Classical list-decodable codes have provided numerous applications in classical computational complexity theory, including proving hardcores for any one-way function, hardness amplification, and derandomization (see, e.g., [19]). Because our formulation of quantum list-decoding naturally extends classical one, many classical list-decoding algorithms work in our quantum setting as well. This will make our quantum list-decoding a powerful tool in quantum complexity theory and quantum computational cryptography.

2 Quantum Hardcore Functions

We begin with the notion of a quantum one-way function, which straightforwardly expands the classical notion of a one-way function. This notion has been studied in the recent literature.

Definition 1 (quantum one-way function). *A function f from $\{0,1\}^*$ to $\{0,1\}^*$ is called* (strongly) quantum one-way *if (i) there exists a polynomial-time* deterministic *algorithm G computing f and (ii) for any polynomial-time quantum algorithm $\mathcal{A}$, for any positive polynomial p, and for any sufficiently large n, $\Pr_{x\in\{0,1\}^n,\mathcal{A}}[f(\mathcal{A}(f(x),1^n)) = f(x)] < 1/p(n)$, where x is uniformly distributed over $\{0,1\}^n$ and the subscript $\mathcal{A}$ is a random variable determined by measuring the final state of $\mathcal{A}$ on the computational basis. We consider only* length-regular *(i.e., $|f(x)| = l(|x|)$ for length function $l(n)$) one-way functions.*

For any quantum one-way function f, the notation f' denotes the function induced from f by the scheme: $f'(x,r) = (f(x),r)$ for all $x,r \in \{0,1\}^*$ with $|r| = poly(|x|)$. Note that f' is also a quantum one-way function. Throughout this paper, we deal only with quantum one-way functions of this form, which is in direct connection to quantum hardcores.

The standard definition of a hardcore function h from $\{0,1\}^n$ to $\{0,1\}^{l(n)}$ is given in terms of the *indistinguishability* between $h(x)$ and a truly random variable over $\{0,1\}^{l(n)}$ whereas a hardcore predicate (i.e., a hardcore function of output length $l(n) = 1$) is usually defined using the notion of *nonapproximability* instead of indistinguishability. It is, nevertheless, well-known that both notions coincide for hardcore functions of output length $O(\log n)$ (see Excise 31 in [11]). In this paper, we conveniently define our quantum hardcores in terms of nonapproximability.

Definition 2 (quantum hardcore function). *Let f be any length-regular function. A polynomial-time computable function h with length function $l(n)$ is* *called a* quantum hardcore *of f if, for any polynomial-time quantum algorithm $\mathcal{A}$, for any polynomial p, and for any sufficiently large number n,*

$$\left| \Pr_{x\in\{0,1\}^n,\mathcal{A}}[\mathcal{A}(f(x),1^n) = h(x)] - 1/2^{l(n)} \right| < 1/p(n),$$

where x is uniformly distributed over $\{0,1\}^n$ and the subscript $\mathcal{A}$ is a random variable determined by measuring the final state of $\mathcal{A}$ on the computational basis.

3 How to Prove Quantum Hardcores

We outline our argument of proving quantum hardcore functions for any quantum one-way function. To prove new quantum hardcores, we exploit the notion of quantum list-decoding as a technical tool. Our approach toward list-decoding is, however, *complexity-theoretical* in nature rather than information-theoretical. Our main objects of quantum list-decoding are "classical" codes and codewords, which are manipulated in a quantum fashion. Generally speaking, a *code* is a set of strings of the same length over a finite alphabet Σ. Each string is indexed by a message and is called a *codeword.* A *code family* is specified by a series $(\Gamma_n, I_n, \Sigma_n)$ of message space Γ_n, index set I_n, and code alphabet Σ_n for each length parameter n. For simplicity, let $\Gamma^* = \bigcup_{n\in\mathbb{N}} \Gamma_n$.

Usually, a code (family) C consists of codewords C_x for each message $x \in \Gamma^*$. As standard now in computational complexity theory, we view the code C as a function that, for each *message length* n (which serves as a *basis parameter* in this paper), maps $\Gamma_n \times I_n$ to Σ_n. Let $N(n) = |\Gamma_n|$ and $q(n) = |\Sigma_n|$. It is convenient to assume that $\Gamma_n \subseteq (\Sigma_n)^n$ so that n actually represents the *length* of a message. By abbreviating $C(x,y)$ as $C_x(y)$, we also treat $C_x(\cdot)$ as a function mapping I_n to Σ_n. Denote by $M(n)$ the *block length* $|I_n|$ of codeword C_x. We simply set $I_n = \{0,1,\ldots,M(n)-1\}$, each element of which can be expressed in $\lceil \log_2 M(n) \rceil$ bits. We freely identify C_x with the vector $(C_x(0), C_x(1), \cdots, C_x(M(n)-1))$ in the *ambient space* $(\Sigma_n)^{M(n)}$ of dimension $M(n)$. We often work on a finite field and it is convenient to regard Σ_n as the finite field $\mathbb{F}_{q(n)}$ of numbers $0,1,\ldots,q(n)-1$. The *(Hamming) distance* $d(C_x, C_y)$ between two codewords C_x and C_y is the number of non-zero components in the vector $C_x - C_y$. The *minimal distance* $d(C)$ of a code C is the smallest distance between any pair of distinct codewords in C. The above-described code is simply called a $(M(n), n)_{q(n)}$-code[2] (or $(M(n), n, d(n))$-code if $d(n)$ is emphasized). We often drop a length parameter n from subscript and argument place whenever we discuss a set of codewords with a "fixed" n (for instance, write Γ and M respectively for Γ_n and $M(n)$).

Now, we wish to prove that a code $C(x,r)$ (seen as a function) is indeed a quantum hardcore for any quantum one-way function of the form $f'(x,r) = (f(x), r)$ with $|r| = poly(|x|)$. First, we assume to the contrary that there exists a feasible quantum algorithm $\mathcal{A}$ that approximates $C_x(r)$ from input $(f(x), r)$

[2] In some literature, the notation $(M(n), N(n))_{q(n)}$ is used instead.

with probability $\geq 1/q(n) + \varepsilon(n)$ (where $\varepsilon(n)$ is a certain *noticeable function*). To be more precise, the outcome of $\mathcal{A}$ on input (y, r), where $r \in I_n$ and $y = f(x)$ for a certain $x \in \Gamma_n$, is of the form:

$$\mathcal{A}(y, r) = \alpha_{y,r,C_x(r)}|r\rangle|C_x(r)\rangle|\phi_{y,r,C_x(r)}\rangle + \sum_{s\in\Sigma_n-\{C_x(r)\}} \alpha_{y,r,s}|r\rangle|s\rangle|\phi_{y,r,s}\rangle$$

for certain amplitudes $\alpha_{y,r,s}$ and ancilla quantum states $|\phi_{x,r,s}\rangle$, where the second register corresponds to the output of the algorithm. For each fixed y, the algorithm $\mathcal{A}_y(\cdot) =_{def} \mathcal{A}(y, \cdot)$ gives rise to the oracle $\tilde{O}_{\mathcal{A}_y}$ (seen as a unitary operator) defined by the map:

$$\tilde{O}_{\mathcal{A}_y}|r\rangle|u\rangle|t\rangle = \sum_{s\in\Sigma} \alpha_{y,r,s}|r\rangle|u \oplus s\rangle|t \oplus \phi_{y,r,s}\rangle$$

for every triplet (r, u, t) of strings, where $\oplus$ is the bitwise XOR and the notation $|t \oplus \phi_{y,r,s}\rangle$ denotes the quantum state $\sum_{v:|v|=|t|}\langle v|\phi_{y,r,s}\rangle|t \oplus v\rangle$. This oracle $\tilde{O}_{\mathcal{A}_y}$ describes *computational error* (not transmission error) occurring during the computation of C_x. This type of erroneous quantum computation is similar to the computational errors (e.g., [1, 3, 4, 18]) dealt with in quantum computational cryptography and quantum algorithm designing.

Similar to the classical notion of a *received word* in coding theory, we introduce our terminology concerning an oracle which represents a "quantum-computationally" corrupted word.

Definition 3 (quantum-computationally corrupted word). *Fix $n \in \mathbb{N}$. We say that an oracle $\tilde{O}$* represents a quantum-computationally (or quantumly) corrupted word *if $\tilde{O}$ satisfies $\tilde{O}|r\rangle|u\rangle|t\rangle = \sum_{s\in\Sigma}\alpha_{r,s}|r\rangle|u \oplus s\rangle|t \oplus \phi_{r,s}\rangle$ for certain unit vectors $|\phi_{r,s}\rangle$ depending only on (r, s). For convenience, we identify a quantumly corrupted word with its representing oracle.*

Remember that $\tilde{O}$ may choose amplitudes $\{\alpha_{r,s}\}_{r,s}$, adversely, not favorably.

To lead to the desired contradiction, we wish to invert f by extracting x from the quantumly corrupted word $\tilde{O}$. Note that the entity $(1/M(n))\sum_{r\in I_n}|\alpha_{r,C_x(r)}|^2$ yields the probability of $\mathcal{A}$'s computing $C_x(\cdot)$ correctly on average. This entity also indicates "closeness" between a codeword C_x and its quantumly corrupted word $\tilde{O}$. In classical list-decoding, for any given oracle $\tilde{O}$ that represents a *received word* and for any error bound e, we need to output a list that include all messages x such that the relative (Hamming) distance between codeword C_x and its received word $\tilde{O}$ is at most $1 - e$ (i.e., $\Pr_{r\in I_n}[\tilde{O}(r) = C_x(r)] \geq 1 - e$). By setting $p_{r,s} = 1$ if $\tilde{O}(r) = s$ and 0 otherwise, the behavior of $\tilde{O}$ can be viewed in a unitary style as $\tilde{O}|r\rangle|0\rangle = \sum_{r\in I_n} p_{r,s}|r\rangle|s\rangle$. The aforementioned entity $(1/M(n))\sum_{r\in I_n}|\alpha_{r,C_x(r)}|^2$ equals the relative distance, $\Pr_{r\in I_n}[\tilde{O}(r) = C_x(r)]$, in a classical setting. For our convenience, we name this entity the *presence* of C_x in $\tilde{O}$ and denote it by $\mathrm{Pre}_{\tilde{O}}(C_x)$. The requirement for the error rate of classical list-decoding is rephrased as $\mathrm{Pre}_{\tilde{O}}(C_x) \geq 1 - e$.

Here, we formulate a quantum version of a classical list-decoding problem using our notions of quantumly corrupted words and presence. Let $C = \{C_x\}_{x\in\Gamma^*}$ be any $(M(n), n, d(n))_{q(n)}$-code. Let $\varepsilon(n)$ be any error bias and let $\delta(n)$ be any confidence parameter (i.e., $0 \leq \varepsilon(n), \delta(n) \leq 1$ for all $n \in \mathbb{N}$).

(ε,δ)-QUANTUM LIST DECODING PROBLEM ((ε,δ)-QLDP) FOR CODE C

> INPUT: a message length n and the values $1/\varepsilon(n)$ and $1/\delta(n)$.
> IMPLICIT INPUT: an oracle $\tilde{O}$ representing a quantumly corrupted word.
> OUTPUT: with success probability at least $1-\delta(n)$, a list of messages that include all messages $x \in \Gamma_n$ such that $\mathrm{Pre}_{\tilde{O}}(C_x) \geq 1/q(n) + \varepsilon(n)$; in other words, codewords C_x have "slightly" higher presence in $\tilde{O}$ than the average.

For any given quantumly corrupted word $\tilde{O}$, how many messages x satisfy the required inequality $\mathrm{Pre}_{\tilde{O}}(C_x) \geq 1/q(n) + \varepsilon(n)$? An upper bound on the number of such messages directly follows from a nice argument of Guruswami and Sudan [15], who gave a q-ary extension of the Johnson bound using a geometric method.

Lemma 1. *Let n be any message length. Let $\varepsilon(n)$, $q(n)$, $d(n)$, and $M(n)$ satisfy that $\varepsilon(n) > \ell(n) =_{def} (1-1/q(n))\sqrt{1-d(n)/M(n)\,(1+1/(q(n)-1))}$. For any $(M(n), n, d(n))_{q(n)}$-code C and for any quantumly corrupted word $\tilde{O}$, there are at most $J_{\varepsilon,q,d,M}(n) =_{def}$*

$$\min\left\{M(n)(q(n)-1), \frac{d(n)\,(1-1/q(n))}{d(n)\,(1-1/q(n)) + M(n)\varepsilon^2(n) - M(n)\,(1-1/q(n))^2}\right\}$$

messages $x \in \Gamma_n$ such that $\mathrm{Pre}_{\tilde{O}}(C_x) \geq 1/q(n) + \varepsilon(n)$. If $\varepsilon(n) = \ell(n)$, then the above bound is replaced by $2M(n)(q(n)-1)-1$.

The proof of Lemma 1 is obtained by an adequate modification of the proof in [15]. As a simple example, consider the $(q^n, n, q^n - q^{n-1})_q$ Hadamard code $\mathrm{HAD}^{(q)} = \{\mathrm{HAD}_x^{(q)}\}_{x\in\Gamma_n}$. Lemma 1 guarantees that, for any quantumly corrupted word $\tilde{O}$, there are only at most $(1-1/q)^2/\varepsilon^2(n)$ messages x that satisfy the inequality $\mathrm{Pre}_{\tilde{O}}(\mathrm{HAD}_x^{(q)}) \geq 1/q + \varepsilon(n)$.

Definition 4 (quantum list-decoding algorithm). *Let C be any code, let $\varepsilon(n)$ be any error bias, and let $\delta(n)$ be any confidence parameter. Any quantum algorithm (i.e., a unitary operator) $\mathcal{A}$ that solves the (ε,δ)-QLDP for C is called a* quantum list-decoding algorithm *for C w.r.t. (ε,δ). If $\mathcal{A}$ further runs in time polynomial in $(n, 1/\varepsilon(n), 1/\delta(n))$, it is called a* polynomial-time quantum list-decoding algorithm *for C w.r.t. (ε,δ).*

To complete our argument (which we started at the beginning of this section), assume that there exists a polynomial-time quantum list-decoding algorithm that solves the $(\varepsilon, 1/poly(n))$-QLDP for $C_x(\cdot)$. Such a list-decoder can output with high probability all possible candidates x' of required presence. Since we can check that $x' \in f^{-1}(x)$ in polynomial time, this list-decoder gives rise to a polynomial-time quantum algorithm that inverts f with high probability.

Clearly, this contradicts the quantum one-wayness of f. Therefore, we obtain the following key theorem that bridges between quantum hardcores and quantum list-decoding.

Theorem 1. *Let $C = \{C_x\}_{x\in\Gamma^*}$ be any $(M(n), n, d(n))_{q(n)}$-code, which is also polynomial-time computable, where $\log_2 M(n) \in n^{O(1)}$ and $\log_2 q(n) \in n^{O(1)}$. If, for any two noticeable functions $\varepsilon(n)$ and $\delta(n)$, there exists a polynomial-time quantum list-decoding algorithm for C w.r.t. (ε, δ), then $C(x, r)$ is a quantum hardcore function for any quantum one-way function of the form $f'(x, r) = (f(x), r)$ with $|x| = \lceil \log_2 |\Gamma_n| \rceil$ and $|r| = \lceil \log_2 M(n) \rceil$.*

4 How to Construct Quantum List-Decoding Algorithms

Theorem 1 makes it suffice to solve the QLDP for any given candidate of a quantum hardcore function. Our goal is now to find a "systematic" way to construct a polynomial-time quantum list-decoder for a wide range of codes. Classically, however, it seems hard to design such list-decoding algorithms in general. Nevertheless, the robust nature of quantum computation enables us to prove that, if we have a decoding algorithm $\mathcal{A}$ from a unique quantum state (called a *codeword state*), then we can construct a list-decoding algorithm by calling $\mathcal{A}$ as a black-box oracle. The notion of such codeword states plays our central role as a technical tool in proving new quantum hardcores in Section 5.

Hereafter, we assume the arithmetic (multiplication, addition, subtraction, etc.) on the finite field $\mathbb{F}_q$ (of numbers $0, 1, \ldots, q-1$), where q is a prime. Denote by ω_q the complex number $e^{2\pi i/q}$.

Definition 5 (k-shuffled codeword state). *Let $C = \{C_x\}_{x\in\Gamma_n}$ be any $(M(n), n)_{q(n)}$-code and let k be any number in $\mathbb{F}_{q(n)}$. A* k-shuffled codeword state *for the codeword C_x that encodes a message $x \in \Gamma_n$ is the quantum state*

$$|C_x^{(k)}\rangle = \frac{1}{\sqrt{M(n)}} \sum_{r\in I_n} \omega_{q(n)}^{k\cdot C_x(r)} |r\rangle.$$

In particular when $k = 1$, we write $|C_x\rangle$ instead of $|C_x^{(1)}\rangle$.

Remark: Codeword states for binary codes have appeared implicitly in several important quantum algorithms. For instance, Grover's search algorithm [14] produces such a codeword state after the first oracle call. In the quantum algorithms of Bernstein and Vazirani [6], of Deutch and Jozsa [9], and of van Dam, Hallgren, and Ip [20], such codeword states were generated to obtain their desired results.

We consider how to generate the k-shuffled codeword state $|C_x^{(k)}\rangle$ for each q-ary codeword C_x with oracle accesses to a quantumly corrupted word $\tilde{O}$. Note that it is easy to generate $|C_x\rangle$ from the oracle O_{C_x} that represents C_x *without* any corruption (behaving as the "standard" oracle). Here, we claim that there exists a generic quantum algorithm that generates codeword states for any q-ary code C. For convenience, write $\mathbb{F}_q^+ = \mathbb{F}_q - \{0\}$ in the rest of this paper.

Lemma 2. *There exists a quantum algorithm $\mathcal{A}$ that, for any quantumly corrupted word $\tilde{O}$, for any message $x \in \Gamma_n$, and for any $k \in \mathbb{F}_q^+$, generates the quantum state*

$$|\psi_k\rangle = \kappa_x^{(k)}|k\rangle|C_x^{(k)}\rangle|\tau\rangle + |\Lambda_x^{(k)}\rangle$$

from the initial state $|\psi_k^{(0)}\rangle = |k\rangle|0^{\lceil \log_2 M(n)\rceil}\rangle|0\rangle|0^{l(n)}\rangle$ with only two queries to $\tilde{O}$ and $\tilde{O}^{-1}$, where $|\tau\rangle$ is a fixed basis vector in $\mathcal{H}_{q^{l(n)+1}}$, and $\kappa_x^{(k)}$ is a complex number, and $|\Lambda_x^{(k)}\rangle$ is a vector satisfying $(\langle k|\langle C_x^{(k)}|\langle\tau|)|\Lambda_x^{(k)}\rangle = 0$ with the following condition: for every $x \in \Gamma_n$, there exists a number $k \in \mathbb{F}_q^+$ with the inequality $|\kappa_x^{(k)}| \geq (q/(q-1))\,|\mathrm{Pre}_{\tilde{O}}(C_x) - 1/q|$.

Isolating simultaneously all individual messages x in Lemma 2 requires a certain type of "orthogonality," which we call *phase-orthogonality.*

Definition 6 (phase-orthogonal code). *A code $C = \{C_x\}_{x\in\Gamma_n}$ is called k-*shuffled phase-orthogonal *if, for any two distinct messages $x, y \in \Gamma_n$, $\langle C_x^{(k)}|C_y^{(k)}\rangle = 0$. If $\langle C_x^{(k)}|C_y^{(k)}\rangle = 0$ holds for every number $k \in \mathbb{F}_q^+$, the code C is simply called* phase-orthogonal.

Note that phase-orthogonality for a binary code, in particular, is naturally induced from the standard *inner product* of two codewords when we translate their binary symbols $\{0,1\}$ into $\{+1,-1\}$. It is not difficult to prove that, for any pair (C_x, C_y) in a given $(M(n), n, d(n))_{q(n)}$-code C, we have $|\langle C_x|C_y\rangle| \geq 1 - 2\cdot d(C_x,C_y)/M(n)$, where the equality holds for any binary code C. Such orthogonality helps us generate $|\psi'\rangle = (1/\sqrt{q-1})\sum_{k\in\mathbb{F}_q^+}\sum_{x\in\Gamma_n}\kappa_x^{(k)}|k\rangle|C_x^{(k)}\rangle|\tau\rangle + |\Lambda'\rangle$.

Now, we give the proof of our key lemma, Lemma 2. Notice that Lemma 2 is true for any $q(n)$-ary code. The binary case ($q = 2$) was implicit in [1]; however, our argument for the general $q(n)$-ary case is more involved because of the introduction of "k-shuffledness."

Proof Sketch of Lemma 2. First, we describe our codeword-state generation algorithm $\mathcal{A}$ in detail. Fix $x \in \Gamma_n$ and $k \in \mathbb{F}_q^+$ and let $m = \lceil \log_2 M(n)\rceil$.

(1) Start with the initial state: $|\psi_k^{(0)}\rangle = |k\rangle|0^m\rangle|0\rangle|0^l\rangle$.
(2) Apply the Fourier transformation $(F_q)^{\otimes m}$ over $\mathbb{F}_q$ to the second register. We then obtain the superposition $|\psi_k^{(1)}\rangle = (1/\sqrt{M})\sum_{r\in I_n}|k\rangle|r\rangle|0\rangle|0^l\rangle$.
(3) Invoke $\tilde{O}$ using the last three registers. The resulting state is $|\psi_k^{(2)}\rangle = (1/\sqrt{M})\sum_{r\in I_n}\sum_{z\in\mathbb{F}_q}\alpha_{r,z}|k\rangle|r\rangle|z\rangle|\phi_{r,z}\rangle$.
(4) Encode the information on the first and the third resisters into the "phase" so that we obtain the state $|\psi_k^{(3)}\rangle = (1/\sqrt{M})\sum_{r\in I_n}\sum_{z\in\mathbb{F}_q}\omega_q^{k\cdot z}\alpha_{r,z}|k\rangle|r\rangle|z\rangle|\phi_{r,z}\rangle$.
(5) Apply $\tilde{O}^{-1}$ to the last three registers. Let $|\psi_k^{(4)}\rangle$ be the resulting state $(I\otimes\tilde{O}^{-1})|\psi_k^{(3)}\rangle$. See, e.g., [1] for how to implement $\tilde{O}^{-1}$ from $\tilde{O}$.
(6) The state $|\psi_k^{(4)}\rangle$ can be expressed in the form $\kappa_x^{(k)}|k\rangle|C_x^{(k)}\rangle|\tau\rangle + |\Lambda_x^{(k)}\rangle$, where $|\tau\rangle = |0\rangle|0^l\rangle$ and $(\langle k|\langle C_x^{(k)}|\langle\tau|)|\Lambda_x^{(k)}\rangle = 0$. The amplitude $\kappa_x^{(k)}$ equals $\mathrm{Pre}_{\tilde{O}}(C_x) + (1/M)\sum_{r\in I_n}\sum_{z:z\neq C_x(r)}\omega_q^{k(z-C_x(r))}|\alpha_{r,z}|^2$.

The non-trivial part of the lemma is to prove the lower-bound of $|\kappa_x^{(k)}|$. For each $j \in \mathbb{F}_q$, let $\beta_j = (1/M)\sum_{r\in I_n} |\alpha_{r,C_x(r)+j}|^2$. By letting $\chi_x^{(k)} = \sum_{j\in\mathbb{F}_q^+} \omega_q^{k\cdot j}\beta_j$, κ_x can be expressed as $\kappa_x^{(k)} = \mathrm{Pre}_{\tilde{O}}(C_x)+\mathrm{Re}(\chi_x^{(k)})+\mathrm{Im}(\chi_x^{(k)})$. To estimate $|\kappa_x^{(k)}|$, it thus suffices to prove that, for each $x \in \Gamma_n$, there exists a number $k \in \mathbb{F}_q^+$ such that $\mathrm{Re}(\chi_x^{(k)}) \geq -(1/(q-1))\,(1 - \mathrm{Pre}_{\tilde{O}}(C_x))$. Since $|\kappa_x^{(k)}|^2 = (\mathrm{Pre}_{\tilde{O}}(C_x) + \mathrm{Re}(\chi_x^{(k)}))^2 + (\mathrm{Im}(\chi_x^{(k)}))^2$, the lemma immediately follows.

To complete the proof, we employ an "adversary" argument. Now, assume that our adversary has cleverly chosen $\tilde{O}$ to make $|\kappa_x^{(k)}|^2$ the smallest for every $k\in\mathbb{F}_q^+$. We argue that the adversary's best choice is to set $\beta_j = \hat{\beta}/(q-1)$ for all $j\in\mathbb{F}_q^+$, where $\hat{\beta}=\sum_{j\in\mathbb{F}_q^+}\beta_j$. This follows directly from the claim below. Let $\hat{\chi}_x = \sum_{k\in\mathbb{F}_q^+}\chi_x^{(k)}$.

Claim 1

1. $\hat{\chi}_x = -\hat{\beta}$.
2. *For his best strategy, the adversary can be assumed to have chosen* $\{\beta_j\}_{j\in\mathbb{F}_q^+}$ *so that* $\beta_j = \beta_{q-j}$ *for any* $j \in \mathbb{F}_q^+$ *and* $\mathrm{Im}(\chi_x^{(k)}) = 0$.

Since $\beta_j = \hat{\beta}/(q-1)$ for every $j \in \mathbb{F}_q^+$ and $\hat{\beta} = 1 - \beta_0$, it easily follows that $\mathrm{Re}(\chi_x^{(k)}) \geq -(1/(q-1))\,(1 - \mathrm{Pre}_{\tilde{O}}(C_x))$, as required. □

The following theorem shows how to convert a *codeword-state decoder* (i.e., a quantum algorithm that recovers x from $|C_x^{(k)}\rangle$ with high probability for any k and x) into a quantum list-decoder. This complements Lemma 2.

Theorem 2. *Let $C = \{C_x\}_{x\in\Gamma_n}$ be any $(M(n), n, d(n))_{q(n)}$-code. Let $\mathcal{A}$ denote the quantum algorithm in Lemma 2 and let $\mathcal{U}$ be any quantum algorithm that, for each $x \in \Gamma_n$ and $k \in \mathbb{F}_{q(n)}^+$, outputs x with probability $\geq 1 - \nu(n)$ from a k-shuffled codeword state $|C_x^{(k)}\rangle \in \mathcal{H}_{M(n)}$ given as an input. For any two real functions $\varepsilon(n)$ and $\delta(n)$ with $(1 - 1/q(n))\sqrt{2\nu(n)} < \varepsilon(n) \leq 1$ and $0 \leq \delta(n) \leq 1$ for all $n \in \mathbb{N}$, there exists a quantum list-decoding algorithm $\mathcal{B}$ for C with oracle accesses to $\tilde{O}$ such that $\mathcal{B}$ produces a list of size at most*

$$\lceil q(n)(\eta_\varepsilon^2(n)/2 - \nu(n))^{-1}(\log_2 J_{\varepsilon,q,d,M}(n) + \log_2(1/\delta(n)))\rceil,$$

where $\eta_\varepsilon(n) = (q(n)/(q(n)-1))\varepsilon(n)$ and $J_{\varepsilon,q,d,M}(n)$ is from Lemma 1. Moreover, if $\mathcal{U}$ is polynomial-time computable and both $q(n)$ and $(\eta_\varepsilon^2(n)/2 - \nu(n))^{-1}$ are polynomially-bounded functions in $(n, 1/\varepsilon(n))$, then $\mathcal{B}$ runs in polynomial time.

Proof Sketch. For any $n \in \mathbb{N}$ and any $\tilde{O}$ as an implicit input, the following algorithm $\mathcal{B}$ solves the (ε,δ)-QLDP for C. Initially, set $k = 1$.

(1) Run the algorithm $\mathcal{A}$ to obtain the quantum state $|\psi_k\rangle$ from $|\psi_k^{(0)}\rangle$.
(2) Apply the algorithm $\mathcal{U}$ to the second register of $|\psi_k\rangle$ using an appropriate number of ancilla qubits, say m. We then obtain the state $\mathcal{U}|\psi_k\rangle|0^m\rangle$.
(3) Measure the obtained state and add this measured result to the list of message candidates.

(4) Repeat Steps (1)–(3) $\lceil(\log_2 J_{\varepsilon,q,d,M}(n) + \log_2(1/\delta(n)))/e\rceil$ times, where $e = 1 - \nu(n) - \sqrt{1 - \eta_\varepsilon^2(n)} \geq \eta_\varepsilon^2(n)/2 - \nu(n)$.
(5) By incrementing k by one until $k = q(n)$, repeat Steps (1)–(4). Finally, output the list.

We claim the following. Let $B_\varepsilon^{(k)} = \{x \in \Gamma_n \mid \mathrm{Pre}_{\tilde{O}}(C_x^{(k)}) \geq 1/q(n) + \varepsilon(n)\}$.

Claim 2

1. *With probability* $\geq e$, *we can observe* x *for a certain* $k \in \mathbb{F}_{q(n)}^+$ *when measuring the quantum state at Step (3) on the computational basis.*
2. *If we perform Steps (1)–(3)* $\lceil e^{-1}(\log_2 |B_\varepsilon^{(k)}| + \log_2(1/\delta(n)))\rceil$ *times for each* $k \in \mathbb{F}_{q(n)}^+$, *then we obtain a list that includes all messages in* $B_\varepsilon^{(k)}$ *with probability at least* $1 - \delta(n)$.

Since $|B_\varepsilon^{(k)}| \leq J_{\varepsilon,q,d,M}(n)$, we obtain the desired list of message candidates at Step (5) with probability at least $1 - \delta(n)$ by the above claim. □

What types of codes satisfy the premise of Theorem 2 and therefore have quantum list-decoders? We show that "nearly" phase-orthogonal codes are indeed quantumly list-decodable (if we ignore the running time of their list-decoders). Our argument uses the notion of pretty-good measurement (known also as square-root measurement or least-squared measurement) [10, 16].

Theorem 3. *Let* $k \in \mathbb{F}_q$ *and let* C *be any* $(M(n), n, d(n))_q$ *code in which there exists a constant* $\xi \in [0, 1/2]$ *satisfying* $|\langle C_x^{(k)}|C_y^{(k)}\rangle| \leq \xi$ *for any distinct pair* $x, y \in \Gamma_n$. *Let* S *be the matrix of the form* $(|C_0^{(k)}\rangle, |C_1^{(k)}\rangle, \ldots, |C_{N-1}^{(k)}\rangle)$. *If* $\xi < 2\varepsilon^2$ *and* $\mathrm{rank}(S) = N$, *then there exists a quantum list-decoding algorithm for* C.

Proof Sketch. From Theorem 2, it suffices to construct a unitary operator U whose success probability $|\langle z|U|C_z\rangle|^2$ of obtaining z from $|C_z\rangle$ is at least $1 - \xi$ whenever $|\langle C_x|C_y\rangle| \leq \xi$ for any distinct $x, y \in \Gamma$ and $\mathrm{rank}(S) = N$.

We want to design U by following an argument of pretty good measurement [10, 16]. Note that, since $\mathrm{rank}(S) = N$, the matrices $S^\dagger S$ and $SS^\dagger$ share the same eigenvalues, say $\lambda_0, \ldots, \lambda_{N-1}$. Perform the *singular-value decomposition* and we obtain $S = PTQ$ for M- and N-dimensional unitary operators P and Q, respectively, and a diagonal matrix $T = \mathrm{diag}(\sqrt{\lambda_0}, \sqrt{\lambda_1}, \ldots, \sqrt{\lambda_{N-1}}, 0, \ldots, 0)$. We therefore have $\langle z|_M US|z\rangle_N = \langle z|_M UPTQ|z\rangle_N$, where $|z\rangle_M$ and $|z\rangle_N$ are, respectively, arbitrary vectors of dimension M and of dimension N.

The desired matrix U is defined as $U = RP^\dagger$, where $R = \left(\begin{smallmatrix} Q^\dagger & 0 \\ 0 & I \end{smallmatrix}\right)$. It immediately follows that $\langle z|_M US|z\rangle_N = \langle z|_M RTQ|z\rangle_N = \langle z|_N Q^\dagger T' Q|z\rangle_N$ with the diagonal matrix $T' = \mathrm{diag}(\sqrt{\lambda_0}, \sqrt{\lambda_1}, \ldots, \sqrt{\lambda_{N-1}})$. The probability of decoding $|C_z^{(k)}\rangle$ to z is therefore lower-bounded by $|\langle z|Q^\dagger T' Q|z\rangle|^2 \geq |\lambda_{\min}|$, where $|\lambda_{\min}|$ denotes $\min\{|\lambda_1|, |\lambda_2|, \ldots, |\lambda_{N-1}|\}$.

Now, the theorem follows from the claim below.

Claim 3. $|\lambda_{\min}| \geq 1 - \xi$.

This completes the proof. □

5 New Quantum Hardcore Functions

Finally, as our main result, we present three new quantum hardcore functions, two of which are unknown to be classically hardcores. We explain them as codes and give polynomial-time list-decoding algorithms for them. From Theorem 2, we need to build only their codeword-state decoders.

Proposition 1. *There exist polynomial-time quantum list-decoding algorithms for the following codes: letting $p(n), q(n)$ be any functions from $\mathbb{N}$ to the primes,*

1. *The* $q(n)$-ary Hadamard code $\mathrm{HAD}^{(q)}$ *with $q(n) \in n^{O(1)}$, whose codeword is defined as* $\mathrm{HAD}_x^{(q)}(r) = \sum_{i=0}^{2^n-1} x_i \cdot r_i \bmod q(n)$.
2. *The* shifted Legendre symbol code SLS^p, *which is a $(p(n), n)_2$-code with $n = \lceil \log p(n) \rceil$, whose codeword is defined by the Legendre symbol*[3] *as* $\mathrm{SLS}_x^p(r) = 1$ *if* $(\frac{x+r}{p(n)}) = -1$, *and* $\mathrm{SLS}_x^p(r) = 0$ *otherwise.*
3. *The* pairwise equality code PEQ *for even $n \in \mathbb{N}$, which is a $(2^n, n)_2$-code, whose codeword is* $\mathrm{PEQ}_x(r) = \oplus_{i=0}^{n/2} \mathrm{EQ}(x_{2i}x_{2i+1}, r_{2i}r_{2i+1})$, *where* EQ *denotes the equality predicate.*

Combining Proposition 1 and Theorem 1, we obtain the quantum hardcore property of all the aforementioned codes.

Theorem 4. *The functions* $\mathrm{HAD}^{(q)}$, SLS^p, *and* PEQ *are all quantum hardcore functions for any quantum one-way function of the form $f'(x,r) = (f(x), r)$ with $|r| = poly(|x|)$, where f is an arbitrary quantum one-way function.*

Remark: Damgård [8] introduced the so-called *Legendre generator*, which produces a bit sequence whose rth bit equals $\mathrm{SLS}^p(r)$. He asked whether his generator possesses the classical hardcore property. (This is also listed as an open problem in [13].) Our result proves the "quantum" hardcore property of Damgård's generator for any quantum one-way function.

Proof Sketch of Proposition 1. It suffices to provide a codeword-state decoder for each of the given codewords because such a decoder satisfies the premise of Theorem 2.

(1) To obtain x from the codeword state $|\mathrm{HAD}^{(q)}\rangle$, we simply apply the Fourier transformation $F_{q(n)}$ over $\mathbb{F}_{q(n)}$ and then extract x deterministically.

(2) Our codeword-state decoder is obtained by an appropriate modification of a quantum algorithm of van Dam, Hallgren, and Ip [20].

(3) Consider the *circulant Hadamard transformation* H_C:

$$H_C =_{def} \begin{pmatrix} -1 & 1 & 1 & 1 \\ 1 & -1 & 1 & 1 \\ 1 & 1 & -1 & 1 \\ 1 & 1 & 1 & -1 \end{pmatrix} = F_4^{-1} \begin{pmatrix} 1 & 0 & 0 & 0 \\ 0 & -1 & 0 & 0 \\ 0 & 0 & -1 & 0 \\ 0 & 0 & 0 & -1 \end{pmatrix} F_4,$$

where F_4 is the quantum Fourier transformation over $\mathbb{F}_4$. We can obtain x from the codeword state $|\mathrm{PEQ}_x\rangle$ by applying $U = H_C^{\otimes n/2}$. □

[3] For any odd prime p, let $(\frac{x}{p}) = 0$ if $p|x$, $(\frac{x}{p}) = 1$ if $p \nmid x$ and x is a quadratic residue modulo p, and $(\frac{x}{p}) = -1$ otherwise.

References

1. M. Adcock and R. Cleve. A quantum Goldreich-Levin theorem with cryptographic applications. In *Proc. STACS 2002*, LNCS Vol.2285, Springer, pp.323–334, 2002.
2. A. Akavia, S. Goldwasser, and S. Safra. Proving hard-core predicates using list decoding. In *Proc. FOCS 2003*, pp.146–157, 2003.
3. A. Ambainis, K. Iwama, A. Kawachi, R. H. Putra, and S. Yamashita. Robust quantum algorithms for oracle identification. Available at http://arxiv.org/abs/quant-ph/0411204, 2004.
4. A. Atici and R. Servedio. Improved bounds on quantum learning algorithms. To appear in *Quantum Information Processing.* Available also at http://arxiv.org/abs/quant-ph/0411140.
5. A. Barg and S. Zhou. A quantum decoding algorithm for the simplex code. In *Proc. Allerton Conference on Communication, Control and Computing*, 1998. Available at http://citeseer.ist.psu.edu/barg98quantum.html.
6. E. Bernstein and U. Vazirani. Quantum complexity theory. *SIAM J. Comput.*, 26(5):1411–1473, 1997.
7. H. Buhrman, I. Newman, H. Röhrig, and R. de Wolf. Robust quantum algorithms and polynomials. In *Proc. STACS 2003*, LNCS Vol.3404, Springer, pp.593–604, 2003.
8. I. B. Damgård. On the randomness of Legendre and Jacobi sequences. In *Proc. CRYPTO '88*, LNCS Vol.403, Springer, pp.163–172, 1988.
9. D. Deutsch and R. Jozsa. Rapid solution of problems by quantum computation. In *Proc. Roy. Soc. London*, A, Vol.439, pp.553–558, 1992.
10. Y. C. Eldar and G. D. Forney, Jr. On quantum detection and the square-root measurement. *IEEE Trans. Inform. Theory*, 47(3):858–872, 2001.
11. O. Goldreich. *Foundations of Cryptography: Basic Tools*, Cambridge University Press, 2001.
12. O. Goldreich and L. A. Levin. A hard-core predicate for all one-way functions. In *Proc. STOC '89*, pp.25–32, 1989.
13. M. I. González Vasco and M. Näslund. A survey of hard core functions. In *Proc. Workshop on Cryptography and Computational Number Theory*, Birkhauser, pp.227–256, 2001.
14. L. K. Grover. Quantum Mechanics helps in searching for a needle in a haystack. Phys. Rev. Lett. 79(2):325–328, 1997.
15. V. Guruswami and M. Sudan. Extensions to the Johnson bound. Manuscript, 2000. Available at http://theory.csail.mit.edu/ madhu/.
16. P. Hausladen and W. K. Wootters. A 'pretty good' measurement for distinguishing quantum states. *J. Mod. Opt.*, 41:2385–2390, 1994.
17. T. Holenstein, U. M. Maurer, and J. Sjödin. Complete classification of bilinear hard-core functions. In *Proc. CRYPTO 2004*, LNCS Vol.3152, Springer, pp.73–91, 2004.
18. P. Høyer, M. Mosca, and R. de Wolf. Quantum search on bounded-error inputs. In *Proc. ICALP 2003*, LNCS Vol.2719, Springer, pp.291–299, 2003.
19. M. Sudan. List decoding: Algorithms and applications. *SIGACT News*, 31(1): 16–27, 2000.
20. W. van Dam, S. Hallgren, and L. Ip. Quantum algorithms for some hidden shift problems. In *Proc. SODA 2003*, pp.489–498, 2003.

Efficient Pseudorandom Generators from Exponentially Hard One-Way Functions

Iftach Haitner[1,*], Danny Harnik[2,**], and Omer Reingold[3,*,***]

[1] Dept. of Computer Science and Applied Math., Weizmann Institute of Science, Rehovot, Israel
iftach.haitner@weizmann.ac.il
[2] Dept. of Computer Science, Technion, Haifa, Israel
harnik@cs.technion.ac.il
[3] Dept. of Computer Science and Applied Math., Weizmann Institute of Science, Rehovot, Israel
omer.reingold@weizmann.ac.il

Abstract. In their seminal paper [HILL99], Håstad, Impagliazzo, Levin and Luby show that a pseudorandom generator can be constructed from any one-way function. This plausibility result is one of the most fundamental theorems in cryptography and helps shape our understanding of hardness and randomness in the field. Unfortunately, the reduction of [HILL99] is not nearly as efficient nor as security preserving as one may desire. The main reason for the security deterioration is the blowup to the size of the input. In particular, given one-way functions on n bits one obtains by [HILL99] pseudorandom generators with seed length $\mathcal{O}(n^8)$. Alternative constructions that are far more efficient exist when assuming the one-way function is of a certain restricted structure (e.g. a permutations or a regular function). Recently, Holenstein [Hol06] addressed a different type of restriction. It is demonstrated in [Hol06] that the blowup in the construction may be reduced when considering one-way functions that have *exponential* hardness. This result generalizes the original construction of [HILL99] and obtains a generator from any exponentially hard one-way function with a blowup of $\mathcal{O}(n^5)$, and even $\mathcal{O}(n^4 \log^2 n)$ if the security of the resulting pseudorandom generator is allowed to have weaker (yet super-polynomial) security.

In this work we show a construction of a pseudorandom generator from any exponentially hard one-way function with a blowup of only $\mathcal{O}(n^2)$ and respectively, only $\mathcal{O}(n \log^2 n)$ if the security of the resulting pseudorandom generator is allowed to have only super-polynomial security. Our technique does not take the path of the original [HILL99] methodology, but rather follows by using the tools recently presented in [HHR05] (for the setting of regular one-way functions) and further developing them.

* Research supported in part by grant no. 1300/05 from the Israel Science Foundation.

** Part of this research was conducted at the Weizmann Institute. Research was supported in part at the Technion by a fellowship from the Lady Davis Foundation.

*** Incumbent of the Walter and Elise Haas Career Development Chair.

M. Bugliesi et al. (Eds.): ICALP 2006, Part II, LNCS 4052, pp. 228–239, 2006.

1 Introduction

Pseudorandom Generators, a notion first introduced by Blum and Micali [BM82] and stated in its current, equivalent form by Yao [Yao82], are one of the cornerstones of cryptography. Informally, a pseudorandom generator is a polynomial-time computable function G that stretches a short random string x into a long string $G(x)$ that "looks" random to any efficient (i.e., polynomial-time) algorithm. Hence, there is no efficient algorithm that can distinguish between $G(x)$ and a truly random string of length $|G(x)|$ with more than a negligible probability. Originally introduced in order to convert a small amount of randomness into a much larger number of effectively random bits, pseudorandom generators have since proved to be valuable components for various cryptographic applications, such as bit commitments [Nao91], pseudorandom functions [GGM86] and pseudorandom permutations [LR88], to name a few.

The seminal paper of Håstad et al. [HILL99] introduced a construction of a pseudorandom generator using any one-way function (called here the HILL generator). This result is one of the most fundamental and influential theorems in cryptography. While the HILL generator fully answers the question of the plausibility of a generator based on any one-way function, the construction is quite involved and very inefficient. This inefficiency also plays a crucial role in the deterioration of the security within the construction.

The seed length and security of the construction: There are various factors involved in determining the security and efficiency of a reduction. In this discussion, however, we focus only on one central parameter, which is the length m of the generator's seed compared to the length n of the input to the underlying one-way function. The HILL construction produces a generator with seed length on the order of $m = \mathcal{O}(n^8)$ (a formal proof of this seed length does not actually appear in [HILL99] and was given in [Hol06]). An alternative construction was recently suggested in [HHR05] which improves the seed length to $\mathcal{O}(n^7)$.

The length of the seed is of great importance to the security of the resulting generator. While it is not the only parameter, it serves as a lower bound to how good the security may be. For instance, the HILL generator on m bits has security that is at best comparable to the security of the underlying one-way function, but on only $\mathcal{O}(\sqrt[8]{m})$ bits. To illustrate the implications of this deterioration in security, consider the following example: Suppose that we only trust a one-way function when applied to inputs of at least 100 bits, then the HILL generator can only be trusted on seed lengths of 10^{16} (ignoring constants) and up (or 10^{14} using the construction of [HHR05]). Thus, trying to improve the seed length towards a linear one is of great importance in making these constructions practical.

Pseudorandom generators from restricted one-way functions: On the other hand, there are known constructions of pseudorandom generators from one-way functions that are by far more efficient when restrictions are made on the type of one-way functions at hand. Most notable is the so called BMY generator (of [BM82, Yao82]) based on any one-way *permutation.* This construction gives a generator with seed length of $\mathcal{O}(n)$ bits. A generator based on any *regular* one-way function of seed length $\mathcal{O}(n \log n)$ was presented in [HHR05] (improving the original such construction of seed length $\mathcal{O}(n^3)$ from [GKL93]). Basing generators on one-way functions with *known preimage-size* [ILL89] also yield constructions that are significantly more efficient than the general case.

The common theme in all of the above mentioned restrictions is that they deal with the *structure* of the one-way function. A different approach was taken by Holenstein [Hol06], that builds a pseudorandom generator from any one-way function with *exponential hardness*. This approach is different as it discusses *raw hardness* as opposed to structure. The result in [Hol06] is essentially a generalization of the HILL generator that also takes into account the parameter stating the hardness of the one-way function. In its extreme case where the hardness is exponential (i.e. 2^{-Cn} for some constant C), then the pseudorandom generator takes a seed length of $\mathcal{O}(n^5)$. Alternatively, the seed length can be reduced to as low as $\mathcal{O}(n^4 \log^2 n)$ when the resulting generator is only required to have super-polynomial security (i.e. security of $n^{\log n}$).

This Work: We give a construction of a pseudorandom generator from any exponentially hard one-way function with seed length $\mathcal{O}(n^2)$. If the resulting generator is allowed to have only super-polynomial security then the construction gives seed length of only $\mathcal{O}(n \log^2 n)$.

Unlike Holenstein's result, our constructions is specialized for one-way functions with exponential hardness. If the security parameter is $2^{-\phi n}$ then the result holds only when $\phi > \Omega(\frac{1}{\log n})$, and does not generalize for use of weaker one-way functions. The core technique of our construction is the *randomized iterate* that was introduced by Goldreich, Krawczyk and Luby [GKL93], and is the focal point in [HHR05].

Paper Organization: Due to space limitations, we provide formal proofs only for the core technique, namely the randomized iterate (in Section 3). In Section 2 we provide an overview of the construction and techniques. Section 4 presents the multiple randomized iterate and its properties, while Section 5 presents the actual construction of the generator. The proofs of the theorems in these sections appear in the full version of the paper.

2 Overview of the Construction

As a motivating example we start by briefly describing the BMY generator. This generator works by iteratively applying the one-way permutation on its own output. More precisely, for a given function f and input x define the kth iterate recursively as $f^k(x) = f(f^{k-1}(x))$ where $f^0(x) = f(x)$.[1] To complete the construction, one needs to take a hardcore-bit at each iteration. If we denote by $b(z)$ the hardcore-bit of z (take for instance the Goldreich-Levin [GL89] predicate), then the BMY generator on seed x outputs the hardcore-bits $b(f^0(x)), \ldots, b(f^\ell(x))$. The rationale behind this technique is that for all k, the kth iteration

[1] We take $f^0(x) = f(x)$ rather than $f^0(x) = x$ for consistency with [HHR05] (see also remark in Section 3).

of f is hard to invert (it is hard to compute $f^{k-1}(x)$ given $f^k(x)$). Indeed, Levin [Lev87] showed that the same generator works with any function that is "one-way on its iterates". However, a general one-way function does not have this guarantee, and in fact, may lose all of its hardness after just one iteration (since there may be too little randomness in the output of f).

The randomized Iterate and regular one-way functions: With the above problem in mind, Goldreich et al. [GKL93] suggested to add a randomizing step between every two iterations. This idea is central in our work and we define it next (following [HHR05]):

Definition (Informal): (The Randomized Iterate). *For function f, input x and random pairwise-independent hash functions $\overline{h} = (h_1, \ldots, h_\ell)$, recursively define the ith randomized iterate (for $i \leq \ell$) by:*

$$f^i(x, \overline{h}) = f(h_i(f^{i-1}(x, \overline{h})))$$

where $f^0(x) = f(x)$.

The rational is that $h_i(f^i(x, \overline{h}))$ is now uniformly distributed, and the challenge is to show that f, when applied to $h_i(f^i(x, \overline{h})$, is hard to invert even when the randomizing hash functions $\overline{h}$ are made public. Indeed, in [HHR05] it was shown that the last randomized iteration is hard to invert even when $\overline{h}$ is known, when the underlying one-way function is *regular*[2] (a regular function is a function such that every element in its image has the same preimage size). Once this is shown, a generator from regular one-way function is similar in nature to the BMY generator, replacing iterations with randomized iterations (the generator outputs $b(f^0(x, \overline{h})), \ldots, b(f^\ell(x, \overline{h})), \overline{h}$).

The randomized iterate and general one-way functions: Unfortunately, the last randomized iteration of a general one-way function is not necessarily hard to invert. It may in fact be easy on a large fraction of the inputs. However, following the proof method presented in [HHR05], we manage to prove the following statement regarding the kth randomized iteration (Lemma 1): There exists a set S^k of inputs to f^k such that the kth randomized iteration is hard to invert over inputs taken from this set. Moreover, the density of S^k is at least $\frac{1}{k}$ of the inputs.

So taking a hard core bit of the kth randomized iteration is beneficial, in the sense that this bit will look random (to a computationally bounded observer) just that this will happen only $\frac{1}{k}$ of the time.

The multiple randomized iterate: Our goal is to get a string of pseudorandom bits, and the idea is to run m independent copies of the randomized iterate (on m independent inputs). We call this the *multiple randomized iterate.* From each of the m copies we output a hardcore bit of the kth iteration. This forms a string of m bits, of which $\frac{m}{k}$ are expected to be random looking. The next

[2] Such a statement was originally proved in [GKL93] for n-wise independent hash functions rather than pairwise independent hash.

step is to run a *randomness extractor* on such a string (where the output of the extractor is of length, say, $\frac{m}{2k}$). This ensures that with very high probability, the output of the extractor is a pseudorandom string of bits.

The use of randomness extractors in a computational setting, was initiated in [HILL99]. We give a general "uniform extraction lemma" for this purpose that is proved using a uniform hardcore Lemma of Holenstein from [Hol05]. Note that similar proofs were given previously [Hol06, HHR05]. In the full paper we give a new version since we require a more careful analysis of the security parameters.

The pseudorandom generator – a first attempt: A first attempt for the pseudorandom generator runs the multiple randomized iterate (on m independent inputs) for ℓ iterations. For each $k \in [\ell]$ we extract $\frac{m}{2k}$ bits at the kth iteration. These bits are guaranteed to be pseudorandom (even when given all of the values at the $(k+1)^{st}$ iterate and all of the randomizing hash functions). Thus outputting the concatenation of the pseudorandom strings for the different values of k forms a long pseudorandom output (by a standard hybrid argument).

However, this concatenation is still not long enough. It is required that the output of the generator is longer than its input, which is not the case here. The input contains m strings $x_1, \ldots, x_m$ and $m \cdot \ell$ hash functions. The hash functions are included in the output, so the rest of the output needs to make up for the mn bits of $x_1, \ldots, x_m$. At each iteration we output $\frac{m}{2k}$ bits which adds up to $\sum_{k=1}^{\ell} \frac{m}{2k}$ bits. This is a harmonic progression that is bounded by $m\frac{\log \ell}{2}$ and in order to exceed the mn lost bits of the input, we need $\ell > 2^n$ which is far from being efficient.

The pseudorandom generator and exponential hardness: The failed generator from above can be remedied when the exponential hardness comes into play. It is known that if a function has hardness 2^{-Cn} (for some constant C), then it has a hardcore function of $C'n$ bits (for another constant C'). Such a general hardcore function appears in the original Goldreich-Levin paper [GL89]. Thus, if the original hardness was exponential, then in the kth iteration we can actually extract $C'n$ random looking strings, each of length $\frac{m}{2k}$. Altogether we get that the output length is $C'n\sum_{k=1}^{\ell} \frac{m}{2k} \geq C'mn \log \ell$. Thus for a choice of ℓ such that $\log \ell > C'$ we get that the overall output is a pseudorandom string of length greater than the input.

The input length of the construction is $\mathcal{O}(nm)$, where m can be taken to be approximately $\mathcal{O}(\log \frac{\ell}{\varepsilon(n)})$ where $\varepsilon(n)$ is the security of the resulting generator. In particular, in order to get an exponentially strong generator, one needs to take a seed of length $\mathcal{O}(n^2)$.

To sum up, we describe the full construction in a slightly different manner: One first creates a matrix of size $m \times \ell$, where each *row* in the matrix is generated by computing the first ℓ randomized iterates of f (each row takes independent inputs). Now from each entry in the matrix $\mathcal{O}(\phi n)$ hardcore bits are computed (thus generating a matrix of hardcore bits). The final stage runs a randomness extractor on each of the *columns* of the hardcore bits matrix.[3] Moreover, the number of pseudorandom bits extracted from a column deteriorates from one iteration to another ($\frac{m}{k}$ pseudorandom bits are taken at the columns associated with the kth randomized iterate).

Some Notes

- Our method works for one-way functions with hardness $2^{-\phi n}$ as long as $\phi > \Omega(\frac{1}{\log n})$. Loosely speaking, this is because for large values of ℓ, the value $\frac{1}{\ell}$ becomes too small to overcome with limited repetition (and thus requires m to grow substantially).
- The paper focuses on length-preserving one-way functions, however, the results may be generalized to use non-length preserving functions (see [HHR05]).

Notations: We denote by $Im(f)$ the image of a function f. Let $y \in Im(f)$, we denote the preimages of y under f by $f^{-1}(y)$. The **degeneracy** of f on y is defined by $D_f(y) \stackrel{\text{def}}{=} \lceil \log |f^{-1}(y)| \rceil$. Due to space limitations we omit standard definitions (provided in the full version).

3 The Randomized Iterate of a One-Way Function

As mentioned in Section 2, the use of randomized iterations lies at the core of our generator. We formally define this notion:

Definition 1 (The kth Randomized Iterate of f). *Let $f : \{0,1\}^n \to \{0,1\}^n$ and let $\mathcal{H}$ be an efficient family of pairwise-independent hash functions*[4] *from $\{0,1\}^n$ to $\{0,1\}^n$. For input $x \in \{0,1\}^n$ and $h^1, \ldots, h^{k-1} \in \mathcal{H}$ define the kth* Randomized Iterate *$f^k : \{0,1\}^n \times \mathcal{H}^k \to Im(f)$ recursively as:*

$$f^k(x, h^1, \ldots, h^k) = f(h^k(f^{k-1}(x, h^1, \ldots, h^{k-1})))$$

where $f^0(x) = f(x)$. For convenience we denote $\overline{h} = (h^1, \ldots, h^k)$.

Another handy notation is the kth explicit randomized iterate $\widehat{f^k} : \{0,1\}^n \times \mathcal{H}^k \to Im(f) \times \mathcal{H}^k$ defined as:

$$\widehat{f^k}(x, \overline{h}) = (f^k(x, \overline{h}), \overline{h})$$

Remark: In the definition randomized iterate we define $f^0(x) = f(x)$. This was chosen for ease of notation and consistency with the results for general OWFs in [HHR05]. For the construction presented in this paper one can also define $f^0(x) = x$, thus saving a single application of the function f.

[3] Note that each execution of the extractor runs on a column in which each entry consists of a single bit (rather than $\mathcal{O}(\phi n)$ bits). This is a requirement of the proof technique.

[4] Pairwise independent hash functions where defined in, e.g. [CW77].

3.1 The Last Randomized Iterate Is (sometimes) Hard to Invert

We now formally state and prove the key observation, that there exists a set of inputs of significant weight for which it is hard to invert the kth randomized iteration even if given access to all of the hash functions leading up to this point.

Lemma 1. *Let $f : \{0,1\}^n \to \{0,1\}^n$ be a one-way function with security $2^{-\phi n}$, and let f^k and $\mathcal{H}$ be as defined in Definition 1.*

Let

$$S^k \stackrel{def}{=} \left\{(x,\overline{h}) \in (\{0,1\}^n \times \mathcal{H}^k) \mid D_f(f^k(x,\overline{h}) = \max_{j\in[k]} D_f(f^j(x,\overline{h}))\right\}$$

Then,

1. *The set S^k has density at least $\frac{1}{k}$.*
2. *For every* PPT *A,*

$$\Pr_{(x,\overline{h}) \leftarrow S^k}[A(f^k(x,\overline{h}),\overline{h}) = f^{k-1}(x,\overline{h})] \leq 2^{-\mathcal{O}(\phi n)}$$

where the probability is also taken over the random coins of A.

More precisely, given a PPT *A that runs in time T_A and inverts the last iteration over S^k with probability $\varepsilon(n)$ one can construct an algorithm that runs in time $T_A + poly(n)$ and inverts the OWF f with probability $\frac{\varepsilon(n)^3}{32k(k+1)n}$.*

Proof

Proving (1). By the pairwise independence of the randomizing hash functions $\overline{h} = (h^1,\ldots,h^k)$ we have that for each $0 \leq i \leq k$, the value $f^i(x,\overline{h})$ is independently and randomly chosen from the distribution $f(U_n)$. Thus, simply by a symmetry argument, the kth (last) iteration is has the heaviest preimage size with probability at least $\frac{1}{k}$. Thus $\Pr_{(x,\overline{h}) \leftarrow (U_n,\mathcal{H}^k)}[(x,\overline{h}) \in S^k] \geq \frac{1}{k}$.

Proving (2). Suppose for sake of contradiction that there exists an efficient algorithm A that given $(f^k(x,\overline{h}),\overline{h})$ computes $f^{k-1}(x,\overline{h})$ with probability $\varepsilon(n)$ over S^k (for simplicity we simply write ε). In particular A inverts the last-iteration of $\widehat{f^k}$ with probability at least ε, that is

$$\Pr_{(x,\overline{h}) \leftarrow S^k}[f(h(A(\widehat{f^k}(x,\overline{h})))) = f^k(x,\overline{h})] \geq \varepsilon$$

Our goal is to use this procedure A in order to break the one-way function f. This goal s achieved by the following procedure:

M^A on input $z \in Im(f)$:

1. Choose a random $\overline{h} = (h^1,\ldots,h^k) \in \mathcal{H}^k$.
2. Apply $A(z,\overline{h})$ to get an output y.
3. If $f(h^k(y)) = z$ output $h^k(y)$, otherwise abort.

We prove that M^A succeeds in inverting f with sufficiently high probability. We focus on the following set of outputs, on which A manages to invert their last-iteration with reasonably high probability.

$$T_A = \left\{(y, \overline{h}) \in Im(\widehat{f^k}) \mid \Pr[f(h^k(A(y, \overline{h})) = y] > \varepsilon/2\right\}$$

A simple Markov argument shows that the set T_A has reasonably large density (the proof is omitted).

Claim 1

$$\Pr_{(x,\overline{h}) \leftarrow (U_n, \mathcal{H}^k)} [\widehat{f^k}(x, \overline{h}) \in T_A] \geq \frac{\varepsilon}{2}.$$

Moreover, since the density of S^k is at least $\frac{1}{k}$, it follows that $\Pr_{(x,\overline{h}) \leftarrow (U_n, \mathcal{H}^k)}$ $[\widehat{f^1}(x, \overline{h}) \in T_A \bigwedge (x, \overline{h}) \in S^k] \geq \varepsilon/2k$. We now make use of the following Lemma, that relates the density of a set with respect to pairs $(f^k(x, \overline{h}), \overline{h})$ where the value of $f^k(x, \overline{h})$ is actually generated using the given randomizing hash functions $\overline{h}$ (i.e. the pair is an output of $\widehat{f^k}$) as opposed to the density of the same set with respect to pairs consisting of a random output of f concatenated with an *independently* chosen hash functions.

Lemma 2. *For every set* $T \subseteq Im(\widehat{f^k})$, *if*

$$\Pr_{(x,\overline{h}) \leftarrow (U_n, \mathcal{H}^k)} [\widehat{f^k}(x, \overline{h}) \in T \bigwedge (x, \overline{h}) \in S^k] \geq \delta$$

then

$$\Pr_{(z,\overline{h}) \leftarrow (f(U_n), \mathcal{H}^k)} [(z, \overline{h}) \in T] \geq \delta^2/2(k+1)n$$

To conclude the proof of Lemma 1, take $T = T_A$ and $\delta = \varepsilon/2k$, and Lemma 2 yields that $\Pr_{(z,\overline{h}) \leftarrow (f(U_n), \mathcal{H}^k)}[(z, \overline{h}) \in T_A] \geq \frac{\varepsilon^2}{16k(k+1)n}$. On each of these inputs A succeeds with probability $\varepsilon/2$, thus altogether M^A manages to invert f with probability $\frac{\varepsilon^3}{32k(k+1)n}$. ∎

Proof. (of Lemma 2) Divide the outputs of the function f into n slices according to their preimage size. The set T is divided accordingly into n subsets. For every $i \in [n]$ define the ith slice $T_i = \left\{(z, \overline{h}) \in T \mid D_f(z) = i\right\}$. We divide S^k into corresponding slices as well, define the ith slice as $S_i^k = \left\{(x, \overline{h}) \in S^k \mid D_f(f^k(x, \overline{h})) = i\right\}$ (note that since $S_i^k \subseteq S^k$, for each $(x, \overline{h}) \in S_i^k$ and thus for each $0 \leq j < k$ it holds that $D_f(f^j(x, \overline{h})) \leq D_f(f^k(x, \overline{h})) = i$). The proof of Lemma 2 follows the methods from [HHR05], used to obtain a similar argument in the case of regular functions. The method follows by studying the collision-probability of $\widehat{f^k}$ when restricted to S_i^k (we work separately on each slice). Denote this as:

$$CP(\widehat{f^k}(U_n, \mathcal{H}^k) \bigwedge S_i^k) = \Pr_{(x_0,\overline{h}_0),(x_1,\overline{h}_1)} [\widehat{f^k}(x_0, \overline{h}_0) = \widehat{f^k}(x_1, \overline{h}_1) \bigwedge (x_0, \overline{h}_0),$$
$$(x_1, \overline{h}_1) \in S_i^k]$$

We first give an upper-bound on this collision-probability (we note that the following upper-bound also holds when only one of the input pairs, e.g. $(x_0, \overline{h}_0)$, is required to be in S_i^k). Recall that $\widehat{f^k}(x, \overline{h})$ includes the hash functions $\overline{h}$ in its output, thus, for every two inputs $(x_0, \overline{h}_0)$ and $(x_1, \overline{h}_1)$, in order to have a collision we must first have that $\overline{h}_0 = \overline{h}_1$ which happens with probability $(1/|\mathcal{H}|)^k$. Now, given that $\overline{h}_0 = \overline{h}_1 = \overline{h}$ (with $\overline{h} \in \mathcal{H}^k$ being uniform), we require also that $f^k(x_0, \overline{h})$ equals $f^k(x_1, \overline{h})$.

If $f(x_0) = f(x_1)$ then a collision is assured. Since it is required that $(x_0, \overline{h}_0) \in S_i^k$ it holds that $D_f(f(x_0)) \leq D_f(f^k(x_1, \overline{h})) = i$ and therefore $\left|f^{-1}(f(x_0))\right| \leq 2^i$. Thus, the probability for that $x_1 \in f^{-1}(f(x_0))$ (and thus of $f(x_0) = f(x_1)$) is at most 2^{i-n}. Otherwise, there must be an $i \in [k]$ for which $f^{i-1}(x_0, \overline{h}) \neq f^{i-1}(x_1, \overline{h})$ but $f^i(x_0, \overline{h}) = f^i(x_1, \overline{h})$. Since $f^{i-1}(x_0, \overline{h}) \neq f^{i-1}(x_1, \overline{h})$, then due to the pairwise-independence of h_i, the values $h_i(f^{i-1}(x_0, \overline{h}))$ and $h_i(f^{i-1}(x_1, \overline{h}))$ are uniformly random values in $\{0,1\}^n$, and thus $f(h_i(f^{i-1}(x_0, \overline{h}))) = f(h_i(f^{i-1}(x_1, \overline{h})))$ also happens with probability at most 2^{i-n}. Altogether:

$$CP(\widehat{f^k}(U_n, \mathcal{H}^k) \bigwedge S_i^k) \leq \frac{1}{|\mathcal{H}|^k} \sum_{i=0}^{k} 2^{i-n} \leq \frac{k+1}{|\mathcal{H}|^k \, 2^{n-i}} \tag{1}$$

On the other hand, we give a lower-bound for the above collision-probability. We seek the probability of getting a collision inside S_i^k and further restrict our calculation to collisions whose output lies in the set T_i (this further restriction may only reduce the collision probability and thus the lower bound holds also without the restriction). For each slice, denote $\delta_i = \Pr[\widehat{f^k}(x, \overline{h}) \in T_i \bigwedge (x, \overline{h}) \in S_i^k]$. In order to have this kind of collision, we first request that both inputs are in S_i^k and generate outputs in T_i, which happens with probability δ_i^2. Then once inside T_i we require that both outputs collide, which happens with probability at least $\frac{1}{|T_i|}$. Altogether:

$$CP(\widehat{f^k}(U_n, \mathcal{H}^k) \bigwedge S_i^k) \geq \delta_i^2 \frac{1}{|T_i|} \tag{2}$$

Combining Equations (1) and (2) we get:

$$\frac{|T_i| \, 2^{i-n-1}}{|\mathcal{H}|^k} \geq \frac{\delta_i^2}{2(k+1)} \tag{3}$$

However, note that when taking a random output z and independent hash functions $\overline{h}$, the probability of hitting an element in T_i is at least $2^{i-n-1}/|\mathcal{H}|^k$ (since each output in T_i has preimage at least 2^{i-1}). But this means that $\Pr[(z,h) \in T_i] \geq |T_i| \, 2^{i-n-1}/|\mathcal{H}|$ and by Equation (3) we deduce that $\Pr[(z,h) \in T_i] \geq \delta_i^2/2(k+1)$. Finally, the probability of hitting T is $\Pr[(z,h) \in T] = \sum_i \Pr[(z,h) \in T_i] \geq \sum_i \delta_i^2/2(k+1)$. Since $\sum_i \delta_i^2 \geq (\sum_i \delta_i)^2/n$ and (by definition) $\sum_i \delta_i = \delta$, it holds that $\Pr[(z,h) \in T] \geq \delta^2/2(k+1)n$ as claimed. ∎

A Hardcore Function for the Randomized Iterate. A hardcore function of the kth randomized iteration is simply taken as the GL hardcore function

([GL89]). The number of bits taken in this construction depends on the hardness of the function at hand (that is the last iteration of the randomized iterate). Thus combining Lemma 1 regarding the hardness of inverting the last iteration, and the Goldreich-Levin Theorem on hardcore functions we get the following lemma:

Lemma 3. *Let* $f : \{0,1\}^n \to \{0,1\}^n$ *be a one-way function with security* $2^{-\phi n}$, *and let* f^k *and* $\mathcal{H}$ *be as defined in Definition 1. Let* $s = \lceil \frac{\phi}{20} n \rceil$ *and take* $hc = gl_s$ *to be the Goldreich-Levin hardcore function which outputs s hardcore bits.*

Then, for every polynomial k, there exist a set $S_k \subseteq \{0,1\}^n \times \mathcal{H}^k$, *of density at least* $\frac{1}{k}$ *such that for any* PPT *A,*

$$\Pr[A(\widehat{f^k}(x,\overline{h}),r) = hc(f^{k-1}(x,\overline{h},r) \mid (x,\overline{h}) \in S_k] < 2^{-\mathcal{O}(\phi n)}.$$

In other words, hc is a hardcore function for the kth *randomized iterate over the set* S^k *with* $\mu_{hc} \leq 2^{-\phi n/20}$ *security.*[5]

4 The Multiple Randomized Iterate

In this section we consider the function $\overline{f^k}$ which consists of m independent copies of the randomized iterate f^k.

Construction 1 (The kth Multiple Randomized Iterate of f). *Let* $m, k \in \mathbb{N}$, *and let* f^k *and* $\mathcal{H}$ *be as in Construction 1. We define the* k^{th} Multiple Randomized Iterate $\overline{f^k} : \{0,1\}^{mn} \times \mathcal{H}^{mk} \to Im(f)^m$ *as:*

$$\overline{f^k}(\overline{x}, H) = f^k(\overline{x}_1, H_1), \ldots, f^k(\overline{x}_m, H_m),$$

where $\overline{x} \in \{0,1\}^{mn}$ *and* $H \in \mathcal{H}^{m \times k}$. *We define the* k^{th} *explicit multi randomized iterate* $\widehat{\overline{f^k}}$ *as:*

$$\widehat{\overline{f^k}}(\overline{x}, H) = \overline{f^k}(\overline{x}, H), H$$

For each of the m outputs of $\overline{f^k}$ we look at its hardcore function hc. By Lemma 3 it holds that m/k of these m hardcore strings are expected to fall inside the "hard-set" of $\widehat{f^k}$ (and thus are indeed pseudorandom given $\widehat{\overline{f^k}}(\overline{x}, H)$). The next step is to invoke a randomness extractor on a concatenation of one bit from each of the different independent hardcore strings. The output of the extractor is taken to be of length $\frac{m}{4k}$. The intuition being that with high probability, the concatenation of single bits from the different outputs of hc contains at least $m/2k$ "pseudoentropy". Thus, the output of the extractor should form a pseudorandom string. Therefore, the output of the extractor serves as a hardcore function of the multiple randomized iterate $\widehat{\overline{f^k}}$.

[5] The constant 1/20 in the security is an arbitrary choice. It was chosen simply as a constant of the form $1/a \cdot b$ where $a > 3$ and $b > 6$ (which are the constants from Lemma 1 and the GL Theorem.

Construction 2 (Hardcore Function for the Multiple Randomized Iterate). *Let f be a one-way function with security μ_f and let $m, k \in \mathbb{N}$. Let s, $\widehat{f^k}$ and hc be as Construction 1 and Lemma 3 yield w.r.t. f, m and k. Let $\varepsilon_{Ext^k} : \mathbb{N} \to [0,1]$ and let $Ext^k : \{0,1\}^n \times \{0,1\}^m \to \{0,1\}^{\lceil \frac{m}{4k} \rceil}$ be a $(\lfloor \frac{m}{k} \rfloor, \varepsilon_{Ext^k})$-strong extractor. We define $\overline{hc^k} : Dom(\widehat{f^k}) \times \{0,1\}^{2n} \to \{0,1\}^{\lceil \frac{m}{4k} \rceil}$ as*

$$\overline{hc^k}(\overline{x}, H, r, y) = w_1^k(\overline{x}, H, r, y), \ldots, w_s^k(\overline{x}, H, r, y),$$

where $\overline{x} \in \{0,1\}^{mn}$, $H \in \mathcal{H}^{m \times k}$, $r \in \{0,1\}^{2n}$ and $y \in \{0,1\}^n$, and for any $i \in [s]$

$$w_i^k(\overline{x}, H, r, y) = Ext^k((hc(f^k(\overline{x}_1, H_1), r)_i, \ldots, (hc(f^k(\overline{x}_m, H_m), r)_i, y).$$

The following lemma implies that, for the proper choice of m and k, it holds that $\overline{hc^{k-1}}$ is a hardcore function of $\widehat{f^k}$. The proof of Lemma 4 appears in the full version. At the heart of this proof lies the uniform extraction Lemma (see discussion in Section 2).

Lemma 4. *Let hc be a hardcore function of the randomized iterate f^k over the set S^k (as in Lemma 3), and denote its security by μ_{hc}. Let $\overline{hc^{k-1}}, \rho^k$ and ε_{Ext^k} be as in Construction 2 and suppose that ρ^k and ε_{Ext^k} are such that $2s(\rho^k + \varepsilon_{Ext^k}) \leq \mu_{hc}$. Then $\overline{hc^{k-1}}$ is a hardcore function of the multiple randomized iterate $\widehat{f^k}$ with security $\mu_{\overline{hc^{k-1}}} < poly(m,n)\mu_{hc}^{\alpha}$ for some constant $\alpha > 0$.*

5 A Pseudorandom Generator from Exponentially Hard One-Way Functions

We are now ready to present our pseudorandom generator. After deriving a hardcore function for the multiple randomized iterate, the generator is similar to the construction from regular one-way function. That is, run randomized iterations and output hardcore bits. The major difference in our construction is that, for starters, it uses hardcore functions rather than hardcore bits. More importantly, the amount of hardcore bits extracted at each iteration is not constant and deteriorates with every additional iteration.

Construction 3 (The Pseudorandom Generator). *Let $m, \ell \in \mathbb{N}$ and let f be a one-way function with security μ_f. Let s and $\overline{hc^k}$ be as Construction 2 yields w.r.t. f and m. We define G as*

$$G(\overline{x}, H, r, y) = \overline{hc^1}(\overline{x}, H, r, y) \ldots, \overline{hc^\ell}(\overline{x}, H, r, y), H, r, y$$

where $\overline{x} \in \{0,1\}^{mn}$, $H \in \mathcal{H}^{m \times s}$, $r \in \{0,1\}^{2n}$ and $y \in \{0,1\}^n$.

Theorem 1. *Let $\phi : \mathbb{N} \to [0,1]$ and let f be a one-way function with security $\mu_f = 2^{-\phi(n)n}$. Let $\delta > 2^{-\frac{\phi(n)n}{20}}$. Let $\ell \in poly(n)$ be such that $\sum_{j=1}^{\ell} \frac{1}{j} > \frac{\phi(n)}{80}$ and let $m = 8\ell \log(\frac{\phi(n)n}{\delta})$. Then G presented in Construction 3 is a pseudorandom generator with input length $\mathcal{O}(n\ell m)$ and security $poly(n)\delta^{\alpha}$, where $\alpha > 0$ is a constant.*

With the appropriate choice of parameters we get the statements mentioned in the introduction, as summarized in the following Corollary:

Corollary 1. *Let $C > 0$ be a constant, Theorem 1 yields the following pseudo-random generators,*

- *For $\delta = 2^{-\frac{C}{20}n}$ and $\mu_f = 2^{-Cn}$ - G is pseudorandom generator with security $2^{-C'n}$ (where $C' > 0$ is a constant) and input length $\mathcal{O}(n^2)$.*
- *For $\delta = 2^{-log^2(n)}$ and $\mu_f = 2^{-Cn}$ - G is pseudorandom generator with security $2^{-\Omega(log^2(n))}$ and input length $\mathcal{O}(n\log(n)^2)$.*
- *For $\delta = 2^{-log^2(n)}$ and $\mu_f = 2^{-Cn/log(n)}$, G is pseudorandom generator with security $2^{-\Omega(log^2(n))}$ and input length $\mathcal{O}(n^{1+\frac{160}{C}}\log(n)^2)$.*[6]

References

[BM82] M. Blum and S. Micali. How to generate cryptographically strong sequences of pseudo random bits. In *23th Annual FOCS*, pages 112–117, 1982.

[CW77] I. Carter and M. Wegman. Universal classes of hash functions. In *9th ACM STOC*, pages 106–112, 1977.

[GGM86] O. Goldreich, S. Goldwasser, and S. Micali. How to construct random functions. *Journal of the ACM*, 33(2):792–807, 1986.

[GKL93] O. Goldreich, H. Krawczyk, and M. Luby. On the existence of pseudorandom generators. *SIAM Journal of Computing*, 22(6):1163–1175, 1993.

[GL89] O. Goldreich and L.A. Levin. A hard-core predicate for all one-way functions. In *21st ACM STOC*, pages 25–32, 1989.

[HHR05] I. Haitner, D. Harnik, and O. Reingold. On the power of the randomized iterate. ECCC, TR05-135, 2005.

[HILL99] J. Håstad, R. Impagliazzo, L. A. Levin, and M. Luby. A pseudorandom generator from any one-way function. *SIAM Journal of Computing*, 29(4): 1364–1396, 1999.

[Hol05] T. Holenstein. Key agreement from weak bit agreement. In *Proceedings of the 37th ACM STOC*, pages 664–673, 2005.

[Hol06] Thomas Holenstein. Pseudorandom generators from one-way functions: A simple construction for any hardness. In *3rd Theory of Cryptography Conference – (TCC '06)*, LNCS. Springer-Verlag, 2006.

[ILL89] R. Impagliazzo, L. A. Levin, and M. Luby. Pseudo-random generation from one-way functions. In *21st ACM STOC*, pages 12–24, 1989.

[Lev87] L. A. Levin. One-way functions and pseudorandom generators. *Combinatorica*, 7:357–363, 1987.

[LR88] M. Luby and C. Rackoff. How to construct pseudorandom permutations from pseudorandom functions. *SIAM Journal of Computing*, 17(2): 373–386, 1988.

[Nao91] M. Naor. Bit commitment using pseudorandomness. *Journal of Cryptology*, 4(2):151–158, 1991.

[Yao82] A. C. Yao. Theory and application of trapdoor functions. In *23rd IEEE FOCS*, pages 80–91, 1982.

[6] Thus this choice of parameters is only useful when $C > \frac{160}{6}$.

Hardness of Distinguishing the MSB or LSB of Secret Keys in Diffie-Hellman Schemes

Pierre-Alain Fouque, David Pointcheval,
Jacques Stern, and Sébastien Zimmer

CNRS-École normale supérieure – Paris, France
{Pierre-Alain.Fouque, David.Pointcheval, Jacques.Stern,
Sebastien.Zimmer}@ens.fr

Abstract. In this paper we introduce very simple deterministic randomness extractors for Diffie-Hellman distributions. More specifically we show that the k most significant bits or the k least significant bits of a random element in a subgroup of $\mathbb{Z}_p^\star$ are indistinguishable from a random bit-string of the same length. This allows us to show that under the Decisional Diffie-Hellman assumption we can deterministically derive a uniformly random bit-string from a Diffie-Hellman exchange in the standard model. Then, we show that it can be used in key exchange or encryption scheme to avoid the leftover hash lemma and universal hash functions.

Keywords: Diffie-Hellman transform, randomness extraction, least significant bits, exponential sums.

1 Introduction

Motivation. The Diffie-Hellman key exchange [15] is a classical tool allowing two entities to agree on a common random element in a group G. It maps a pair of group elements (g^x, g^y) to g^{xy}. Since x and y are randomly chosen, the latter value is uniformly distributed in G. However it is not secret from an information theoretic point of view since x and y are uniquely determined modulo $|G|$ and so is g^{xy}. That is why an additional computational assumption is needed to guarantee that no computationally bounded attacker can find this element with a significant probability. The Computational Diffie-Hellman assumption (CDH) basically expresses this security notion. However, it does not rule out the ability to guess some bits of g^{xy}.

To obtain a cryptographic key from g^{xy} we need that no information leaks and further assumptions are required. Among those, the DDH is perhaps the most popular assumption and allows cryptographers to construct secure protocols [4]. It states the intractability of distinguishing DH-triples (g^x, g^y, g^{xy}) from random triples (g^x, g^y, g^z). Under the decisional Diffie-Hellman assumption (DDH) one can securely agree on a random and private element. However, a problem remains: this element is a random element in G but not a random bit-string as is generally required in further symmetric use. The common secret will indeed thereafter be used as a symmetric key to establish an authentic

M. Bugliesi et al. (Eds.): ICALP 2006, Part II, LNCS 4052, pp. 240–251, 2006.

and private channel. Hence, one has to transform this random element into a random-looking bit-string, i.e. extract the *computational entropy* injected by the DDH assumption in the Diffie-Hellman element. To solve this problem, different methods have been proposed.

Thanks to the Leftover Hash Lemma [23, 25], one can extract entropy hidden within g^z by means of a family of universal hash functions. This solution has the advantage of being proven in the standard model and does not require any cryptographic assumption. One can indeed easily construct such families [10], and they are furthermore quite efficient to compute. However it requires extra randomness which needs to be of good quality (unbiased) and independent of the random secret g^z. Consequently, in a key exchange protocol, this extra randomness either needs to be authenticated or hard-coded in the protocol. This solution is mostly theoretical and is not widely used in standard protocols for the simple reason that families of universal hash functions are not present in cryptographic softwares, while they would be quite efficient [16, 33].

In practice, designers prefer to apply hash functions, such as MD5 or SHA-1, to the Diffie-Hellman element. This solution can be proven secure under the CDH assumption in the random oracle model [2], under the assumption that the compression function acts as a random oracle [13], but not in the standard model (unless one makes additional non-standard assumptions [1, 16, 18]).

In this paper, we analyze a quite simple and efficient randomness extractor for Diffie-Hellman distributions. The security relies on the DDH assumption in the *standard model.*

Related Works. To extract randomness from a Diffie-Hellman secret, one approach is to focus on the distribution induced by the DDH assumption. In [9], Canetti *et al.* show that given the k most significant bits of g^x and g^y, one cannot distinguish, in the statistical sense, the k most significant bits of g^{xy} from a random k bit-string. As Boneh observes [4], this is quite interesting but cannot be applied to practical protocols because an adversary always learns all of g^x and g^y. Chevassut *et al.* [11, 12] review a quite simple and optimal randomness extractor but which can be applied to $\mathbb{Z}_p^\star$, with a safe prime p only. This randomness extractor is very efficient but requires high computational effort to compute g^x, g^y and g^{xy} because of the requirement of a large group. They also presented a new technique (TAU [12]) but which applies to specific elliptic curves only. Independently, Gürel [22] proved that, under the DDH assumption over an elliptic curve, the most significant bits of the Diffie-Hellman transform are statistically close to a random bit-string, when the elliptic curve is defined over a quadratic extension of a finite field. However, $\mathbb{Z}_p^\star$ is one of the most interesting group and in order to speed up the Diffie-Hellman key-exchange, the computations must be performed in a small subgroup. To this end, Gennaro *et al.* [18] prove that a family of universal hash functions can be used even in non-DDH groups, provided that the group contains a large subgroup where the DDH assumption holds. However, this result still requires the use of a family of universal hash functions.

A second line of research is to study usual cryptographic primitives in protocols and prove that they are good randomness extractors. Dodis *et al.* [16]

therefore tried to analyze the security of IPsec. They showed that NMAC, the cascade construction and CBC-MAC are probabilistic randomness extractors. This is the first formal study of the randomness extraction phase of Diffie-Hellman standards in the standard model. These extractors can be applied with several distributions, not only the Diffie-Hellman distributions. However, these results require the assumption that the compression functions of the hash-based constructions under review (the hash functions MD5 or SHA-1) are a family of almost universal hash functions, which is not realistic.

In [5, 6], Boneh and Venkatesan show that the k most significant bits or least significant bits of g^{xy} are hard to compute. Namely, they prove that given an oracle which takes as input (g^x, g^y) and returns the k most significant bits of g^{xy}, one can construct an algorithm to compute g^{ab} given (g^a, g^b). They can take into account faulty oracle which can fail with probability at most $1/\log p$. In order to use these results to show that these bits are hardcore bits, the oracle must correctly answer with probability better than $1/2^k + \epsilon$. Indeed, in this case, the oracle finds the k bits more frequently than by guessing them. However, the techniques used cannot take into account such faulty oracles. Moreover their proof is known to contain a gap which was fixed by Gonzales-Vasco and Shparlinski in [21]. The result of [5, 6] is improved in [21, 20] and in [3]. In the latter, it is shown that under the DDH assumption the two most significant bits of the Diffie-Hellman result are hard to compute. Our main result here tells that under the DDH assumption, a good distinguisher for the two distributions (g^a, g^b, U_k) and $(g^a, g^b, \mathsf{lsb}_k(g^{ab}))$ cannot exist.

Our Result. In this paper, we use the exponential sum techniques to analyze cryptographic schemes. These techniques date back to the beginning of the last century, but we borrowed them from [9, 8] where they are used for cryptographic purposes. They allow us to study a very simple deterministic randomness extractor. Deterministic extractors have been recently introduced in complexity theory by Trevisan and Vadhan [28]. We describe here a deterministic randomness extractor which is provably secure in the standard model, under classical assumptions. We focus on the distribution induced by the DDH in a prime subgroup G of $\mathbb{Z}_p^\star$, where p is prime and $|G| \gg \sqrt{p}$. We prove that the k least significant bits of a random element of G are statistically close to a perfectly random bit-string. In other words, we have a very simple *deterministic* randomness extractor which consists in keeping the k least significant bits of the random element and discarding the others. This extractor can be applied to Diffie-Hellman Key Exchange and El Gamal-based encryption schemes, under the DDH assumption. It does not need any family of universal hash functions neither any extra randomness. We also show that if p is sufficiently close below of a power of 2 by a small enough amount, the k most significant bits are also uniformly distributed.

Organization. In section 2, we present some definitions and results about entropy and randomness extraction. In section 3, we present and analyze our new randomness extractor. In section 4, we compare our extractor with other randomness extractors. In section 5, we present some natural and immediate applications of our extractor. In section 6, we relax the DDH assumption into the weaker CDH assumption and analyze the bit-string we can generate in that case.

2 Entropy and Randomness Extractors

First of all we introduce the notions used in randomness extraction. In the following, a randomness source is viewed as a probability distribution.

2.1 Measures of Randomness

Definition 1 (Min Entropy). *Let X be a random variable with values in a set $\mathcal{X}$ of size N. The* guessing probability *of X, denoted by $\gamma(X)$, is the probability* $\max_{x\in\mathcal{X}}(\Pr[X = x])$. *The min entropy of X is: $H_\infty(X) = -\log_2(\gamma(X))$.*

For example, when X is drawn from the uniform distribution on a set of size N, the min-entropy is $\log_2(N)$. To compare two random variables we use the classical statistical distance:

Definition 2 (Statistical Distance). *Let X and Y be two random variables with values in a set $\mathcal{X}$ of size N. The* statistical distance *between X and Y is the value of the following expression:*

$$\mathbf{SD}(X, Y) = \frac{1}{2}\sum_{x\in\mathcal{X}} |\Pr[X = x] - \Pr[Y = x]| \,.$$

We denote by U_k a random variable uniformly distributed over $\{0,1\}^k$. We say that a random variable X with values in $\{0,1\}^k$ is *δ-uniform* if the statistical distance between X and U_k is upper bounded by δ.

2.2 From Min Entropy to δ-Uniformity

The most common method to obtain a δ-uniform source is to extract randomness from high-entropy bit-string sources. Presumably, the most famous randomness extractor is provided by the Leftover Hash Lemma [23, 25], which requires to introduce the notion of universal hash function families.

Definition 3 (Universal Hash Function Families). *Let $\mathcal{H} = \{h_i\}_i$ be a family of efficiently computable hash functions $h_i : \{0,1\}^n \to \{0,1\}^k$, for $i \in \{0,1\}^d$. We say that $\mathcal{H}$ is a universal hash function family if for every $x \neq y$ in $\{0,1\}^n$,*

$$\Pr_{i\in\{0,1\}^d}[h_i(x) = h_i(y)] \leq 1/2^k.$$

Theorem 4 (Leftover Hash Lemma). *Let $\mathcal{H}$ be a universal hash function family from $\{0,1\}^n$ into $\{0,1\}^k$, keyed by $i \in \{0,1\}^d$. Let i denote a random variable with uniform distribution over $\{0,1\}^d$, let U_k denote a random variable* *uniformly distributed in* $\{0,1\}^k$*, and let* A *denote a random variable taking values in* $\{0,1\}^n$*, with* i *and* A *mutually independent. Let* $\gamma = \gamma(A)$*, then:*

$$\mathbf{SD}(\langle i, h_i(A)\rangle, \langle i, U_k\rangle) \leq \frac{\sqrt{2^k \gamma}}{2}.$$

Proof. See [32]. □

The Leftover Hash Lemma extracts nearly all of the entropy available whatever the randomness sources are, but it needs to invest few additional truly random bits. To overcome this problem, it was proposed to use deterministic functions. They do not need extra random bits, but only exist for some specific randomness sources.

Definition 5 (Deterministic Extractor). *Let* f *be a function from* $\{0,1\}^n$ *into* $\{0,1\}^k$*. Let* $\mathcal{X}$ *be a set of random variables of min entropy* m *taking values in* $\{0,1\}^n$ *and let* U_k *denote a random variable uniformly distributed in* $\{0,1\}^k$*, where* U_k *and* X *are independent for all* $X \in \mathcal{X}$*. We say that* f *is an* (m,ε)*-deterministic extractor for* $\mathcal{X}$ *if for all* $X \in \mathcal{X}$*:*

$$\mathbf{SD}\left(f(X), U_k\right) < \varepsilon.$$

3 Randomness Extractor in a Subgroup of $\mathbb{Z}_p^\star$

In this section, we propose and prove the security of a simple randomness extractor for the Diffie-Hellman exchange in sufficiently large subgroups of $\mathbb{Z}_p^\star$. The main result of this section is theorem 7 which shows that least significant bits of a random element in G are statistically close to truly random bits. To prove this result, we apply the exponential sum techniques in order to find an upper bound on the statistical distance. It is very similar to the results of [27] who studies the distribution of fractional parts of ag^x/p in given intervals of $[0,1]$.

Our result does not require the DDH assumption. However, as it is precised in section 5, to apply it in a cryptographic protocol, the DDH assumption is needed to obtain a random element in the subgroup of $\mathbb{Z}_p^\star$.

3.1 Description of the Deterministic Extractor

Let p be an n-bit prime, that is $2^{n-1} < p < 2^n$, G a subgroup of $\mathbb{Z}_p^\star$ of order q with $q \gg \sqrt{p}$, ℓ the integer such that $2^{\ell-1} \leq q < 2^\ell$ and X a random variable uniformly distributed in G. In the following, we denote by k an integer, by s a k-long bit-string and the associated integer in $[\![0,\, 2^k-1]\!]$, and by U_k a random variable uniformly distributed in $\{0,1\}^k$. If x is an integer, we denote by $\mathsf{lsb}_k(x)$ the k least significant bits of x and by $\mathsf{msb}_k(x)$ the k most significant bits of x.

In this section we show that the k least significant bits of a random element g of G are statistically close to a truly random k-long bit-string provided that G is large enough. A direct consequence of this result is that the function from

$\mathbb{Z}_p^\star$ to $\{0,1\}^k$ which keeps only the k least significant bits of its input is a good deterministic extractor for a G-group source (that is for variables uniformly distributed in the group $G \subset \mathbb{Z}_p^\star$).

Definition 6. *The function* $\mathsf{Ext}_k : \{0,1\}^n \to \{0,1\}^k : c \mapsto \mathsf{lsb}_k(c)$ *is called an* (n,p,q,k)-***extractor*** *for a* G*-group source.*

Theorem 7. *With the above notations of an* (n,p,q,k)*-extractor for a group source, we have:*

$$\mathbf{SD}(\mathsf{lsb}_k(X), U_k) < \frac{2^k}{p} + \frac{2^k\sqrt{p}\log_2(p)}{q} < 2^{k+n/2+\log_2(n)+1-\ell}.$$

This inequality is non trivial only if $k < \ell - n/2 - \log_2(n) - 1$.

Proof. Let us define $K = 2^k$, $H_s = \lfloor \frac{p-1-s}{K} \rfloor$ for $s \in [\![0,\, K-1]\!]$. Let denote by e_p the following character of $\mathbb{Z}_p$: for all $y \in \mathbb{Z}_p$, $e_p(y) = e^{\frac{2i\pi y}{p}} \in \mathbb{C}^*$. The character e_p is an homomorphism from $(\mathbb{Z}_p, +)$ in $(\mathbb{C}^*, \cdot)$. Since

$$\frac{1}{p} \times \sum_{a=0}^{p-1} e_p(a(g^x - s - Ku)) = \mathbb{1}(x,s,u),$$

where $\mathbb{1}(x,s,u)$ is the characteristic function which is equal to 1 if $g^x = s + Ku \bmod p$ and 0 otherwise, we have:

$$\begin{aligned}\Pr_{X\in G}[\mathsf{lsb}_k(X) = s] &= \frac{1}{q} \times \left|\{(x,u) \in [\![0,\, q-1]\!] \times [\![0,\, H_s]\!] \mid g^x = s + Ku \bmod p\}\right| \\ &= \frac{1}{qp} \times \sum_{x=0}^{q-1}\sum_{u=0}^{H_s}\sum_{a=0}^{p-1} e_p(a(g^x - s - Ku)).\end{aligned}$$

Let us change the order of the sums, and split sum on the a's in two terms:

1. the first one comes from the case $a = 0$, and is equal to $(H_s+1)/p$, that is approximately $1/2^k$,
2. the second one comes from the rest, and will be the principal term in the statistical distance in which we can separate sums over x and u.

Twice the statistical distance, that is 2Δ, is equal to:

$$\begin{aligned}&\sum_{s\in\{0,1\}^k} \left| \Pr_{X\in G}[\mathsf{lsb}_k(X) = s] - 1/2^k \right| \\ &\leq \sum_{s\in\{0,1\}^k} \left| \frac{H_s+1}{p} - \frac{1}{2^k} \right| + \sum_{s\in\{0,1\}^k} \frac{1}{qp} \sum_{a=1}^{p-1} \left| \left(\sum_{x=0}^{q-1} e_p(ag^x)\right) \left(\sum_{u=0}^{H_s} e_p(-aKu)\right) \right|.\end{aligned}$$

For the first term, we notice that $\left|(H_s+1)/p - 1/2^k\right| \leq 1/p$, since $K = 2^k$, $H_s = \lfloor \frac{p-1-s}{K} \rfloor$ and:

$$-\frac{1}{p} \leq -\frac{1+s}{Kp} \leq \left(1 + \left\lfloor \frac{p-1-s}{K} \right\rfloor\right)\frac{1}{p} - \frac{1}{K} \leq \frac{K-(1+s)}{Kp} \leq \frac{1}{p}.$$

For the second term, we introduce $M = \max_a \left(\left| \sum_{x=0}^{q-1} e_p(ag^x) \right| \right)$, and show that:

$$\begin{aligned}\sum_{a=1}^{p-1} \left| \sum_{u=0}^{H_s} e_p(-aKu) \right| &= \sum_{a=1}^{p-1} \left| \sum_{u=0}^{H_s} e_p(-au) \right| = \sum_{a=1}^{p-1} \left| \frac{1 - e_p(-a(H_s+1))}{1 - e_p(-a)} \right| \\ &= \sum_{a=1}^{p-1} \left| \frac{\sin(\frac{\pi a(H_s+1)}{p})}{\sin(\frac{\pi a}{p})} \right| = 2 \sum_{a=1}^{\frac{p-1}{2}} \left| \frac{\sin(\frac{\pi a(H_s+1)}{p})}{\sin(\frac{\pi a}{p})} \right| \\ &\leq 2 \sum_{a=1}^{\frac{p-1}{2}} \left| \frac{1}{\sin(\frac{\pi a}{p})} \right| \leq \sum_{a=1}^{\frac{p-1}{2}} \left| \frac{p}{a} \right| \leq p \log_2(p).\end{aligned}$$

The first equality results from a change of variables. The second equality comes from the fact that $[\![0, H_s]\!]$ is an interval, therefore the sum is a geometric sum. We use the inequality $\sin(y) \geq 2y/\pi$ if $0 \leq y \leq \pi/2$ for the second inequality. In summary we have:

$$2\Delta \leq \frac{2^k}{p} + \frac{2^k M \log_2(p)}{q}. \tag{1}$$

Using the bound $M \leq \sqrt{p}$ that can be found in [26], $2^{n-1} < p < 2^n$ and $2^{\ell-1} \leq q < 2^\ell$, we obtain the expected result.

Consequently, since the min entropy of X, as an element of $\mathbb{Z}_p^\star$ but randomly distributed in G, equals $\log_2(|G|) = \log_2(q)$, the previous proposition leads to:

Corollary 8. *Let e be a positive integer and let suppose that we have $\log_2(q) > m = n/2 + k + e + \log_2(n) + 1$. Then the application Ext_k is an $(m, 2^{-e})$-deterministic extractor for the G-group distribution.*

3.2 Improvements

One drawback of the previous result is that we need a subgroup of order at least $\sqrt{p}$. In order to have more efficient Diffie-Hellman key exchange, one prefers to use smaller subgroups. Therefore to improve the results obtained on this random extractor, one idea would be to find a better bound than $\sqrt{p}$ on $M = \max_a \left(\left| \sum_{x=0}^{q-1} e_p(ag^x) \right| \right)$. There are several results which decrease this bound, as these from [7, 24]. Many of them are asymptotic, and do not explicit the constants involved. However, by looking carefully at the proof in [24] or [26] we can find them:

Theorem 9 ([26]). *With the notations of the previous subsection, if $q \geq 256$ then, for all $x \in \mathbb{Z}_p^\star$, we have:*

$$M \leq \begin{cases} p^{1/2} & (\textit{interesting if } p^{2/3} \leq q) \\ 4p^{1/4}q^{3/8} & (\textit{interesting if } p^{1/2} \leq q \leq p^{2/3}) \\ 4p^{1/8}q^{5/8} & (\textit{interesting if } 256 \leq q \leq p^{1/2}) \end{cases}$$

The bound $\sqrt{p}$ is always valid whatever p and q are. Yet, if $\sqrt{p} < q < p^{2/3}$, the second bound is better and similarly to the third bound. For example, with $n = 2048$, $\ell = 1176$ and $e = 80$, theorem 7 says that we can extract 60 bits. Using the second bound given in the theorem above with the equation 1 we obtain that $k \leq 5\ell/8 - (e + n/4 + \log_2(n) + 3)$. It means that we can actually extract 129 bits and obtain a bit-string of reasonable size. However, in most practical cases, the classical bound $\sqrt{p}$ is the most appropriate.

Moreover when G is the group of quadratic residues, Gauss has proven that $\left|\sum_{x=0}^{p-1} e_p(ax)\right| = \sqrt{p}$, for all $a \in \mathbb{Z}_p^\star$. Therefore, $\left|\sum_{x=0}^{q-1} e_p(ag^x)\right| \geq (\sqrt{p} - 1)/2$. This means that in the case of safe primes and with this proof technique, our result is nearly optimal.

3.3 Other Result

The theorem presented in the previous section considers least significant bits. A similar result for most significant bits can be proved with the same techniques. We have the following theorem, whose proof is omitted by lack of space:

Theorem 10. *Let δ be $(2^n - p)/2^n$. If p, m, k and e are integers such that $3\delta < 2^{-e-1}$ and $\log_2(|G|) > m = n/2 + k + e + \log_2(n) + 1$, then the function $\mathsf{msb}_k(\cdot)$ is a $(m, 2^{-e})$-deterministic extractor for the G-group distribution.*

The first assumption on p to be close by below to a power of 2 is easily justified by the fact that the most significant bit is highly biased whenever p is just above a power of 2. Indeed in this case, with high probability, the most significant bit is equal to 0.

4 Comparisons

In the literature other randomness extractors proven secure in the standard model are also available.

4.1 The Leftover Hash Lemma

A famous one is the leftover hash lemma which is presented in subsection 2.2. If one uses a universal hash function family, we can extract up to $\log_2(|G|) - 2e + 2$ bits from a random element in G. With our extractor, the number of random bits extracted is approximately $\log_2(|G|) - (n/2 + \log_2(n) - e + 1)$. However, the leftover hash lemma needs the use of a universal hash function family and extra truly random bits.

In practice we can derandomize it by fixing the key of the hash function. Shoup [32] proved that in this case, there is a linear loss of security in the number of calls of the hash function.

4.2 An Optimal Randomness Extractor for Safe Prime Groups

To extract randomness from a random element of a subgroup of $\mathbb{Z}_p^\star$, where $p = 2q + 1$ is a safe prime (q is also a prime), there is another deterministic extractor reviewed in [11, 12]. Let $G = \langle g \rangle$ denote the subgroup of quadratic residues of $\mathbb{Z}_p^\star$, and let g^x be a random element in G. To extract the randomness of g^x, the extractor needs this function f:

$$f(g^x) = \begin{cases} g^x & \text{if } g^x \leq (p-1)/2 \\ p - 1 - g^x & \text{otherwise} \end{cases}$$

This function is a bijection from G to $\mathbb{Z}_q$. To obtain a random bit-string, one has to truncate the result of f. The composition of f and the truncation is a good deterministic extractor. As f is a bijection, in some sense it is optimal : all the randomness is extracted. However this simple extractor is very restrictive because it can be applied only with a safe prime when our extractor can be used with a significantly larger set of primes. Moreover our extractor is more efficient than this simple one.

5 Applications

The DDH assumption allows to find to our extractor some natural applications in cryptographic protocols. It can indeed be applied in every protocol which generates a random element in a subgroup of $\mathbb{Z}_p^\star$ and where a randomness extractor is needed.

5.1 Key Exchange Protocol

Our extractor is designed to extract entropy from a random element in a group G. It is exactly what is obtained after a Diffie-Hellman key exchange performed in a DDH group G, where G is a subgroup of $\mathbb{Z}_p^\star$.

This means that we have an efficient solution to the problem of agreeing on a random bit-string which is based on the following simple scheme, provably secure in the standard model under the DDH assumption: Alice sends g^x, Bob sends g^y and they compute $\mathsf{lsb}_k(g^{xy})$.

The multiplicative group $\mathbb{Z}_p^\star$ is not a DDH group but if $p = \alpha q + 1$ with q a large prime and α small then the subgroup of $\mathbb{Z}_p^\star$ with q elements may be assumed a DDH group (in such a group, the DDH assumption is reasonable.) Therefore in this case we can extract up to $k = n/2 - (e + \log_2(n) + 2 + \log_2(\alpha))$ bits from an $n - \log_2(\alpha)$ min entropy source.

In practice, the security parameters are often $n = 1024$, $e = 80$. Hence we can extract approximately $420 - \log_2(\alpha)$ bits at the cost of two exponentiations modulo an integer of 1024 bits. It means that if we need a 128-long bit-string, the subgroup should have approximately 2^{731} elements.

5.2 Encryption Schemes

El Gamal Encryption Scheme [17]. In the El Gamal encryption scheme, the message must be an element of a cyclic group G of order q. Alice generates a random element x in $\mathbb{Z}_q$ and publishes $y = g^x$ where g is a generator of G. To encrypt the message m, she generates a random element r of $\mathbb{Z}_q$ and computes (g^r, my^r). This scheme is proven IND-CPA secure if $m \in G$. However in practice messages are often bit-strings and not elements from G. One solution to avoid this problem is to extract the randomness from y^r and xor the generated bit-string with the message. This way, the encryption scheme is still IND-CPA secure. Our extractor can be used in this context to extract randomness.

Cramer-Shoup Encryption Scheme [14, 31]. The Cramer-Shoup encryption scheme is an improvement of the El Gamal encryption scheme which is IND-CCA secure. The principle is the same as in El Gamal, it hides m multiplying it with a random element h^r of G. The security proof requires that m is in G. In order to use bit-string messages, we can use the same solution: extract randomness from h^r with our extractor and xor the result with m.

6 Other Assumptions

In this section, we apply our result under various assumptions, related to the DDH one. First, we make a stronger assumption, the so-called Short Exponent Discrete Logarithm, which allows quite efficient DH-like protocols. Then, we relax the DDH assumption to the CDH one.

6.1 The s-DLSE Assumption

To speed up our randomness extractor, we can use a group in which the additional Short Exponent Discrete Logarithm (DLSE) assumption holds. First introduced in [34], it is formalized in [29] and [18] as follows:

Assumption 1 (s-DLSE [29]). *Let s be an integer, $\mathcal{G} = \{G_n\}_n$ be a family of cyclic groups where each G_n has a generator g_n and $ord(G_n) = q_n > 2^n$. We say that the s-DLSE Assumption holds in $\mathcal{G}$ if for every probabilistic polynomial time Turing machine I, for every polynomial $P(\cdot)$ and for all sufficiently large n we have that* $\Pr_{x \in_R [\![1,\, 2^s]\!]} [I(g_n, q_n, s, g_n^x) = x] \leq 1/P(n)$.

As explained in [18], current knowledge tends to admit that in a group of prime order, for a 2^{-e} security level, we can choose $s \geq 2e$. The usual security parameter of $e = 80$ leads to $s \geq 160$, which is quite reasonable, from a computational cost.

Gennaro *et al.* prove in [18] that under the s-DLSE and the DDH assumption, the two following distributions are computationally indistinguishable:

$$\left\{(g^x, g^y, Z) \middle|\ x, y \in_R [\![1,\, 2^s]\!],\ Z \in_R G\right\} \text{ and } \left\{(g^x, g^y, g^{xy}) \middle|\ x, y \in_R [\![1,\, 2^s]\!]\right\}.$$

This result allows us to use our extractor with the latter distribution and in that way be computationally more efficient.

6.2 The CDH Assumption

In practice, to apply our extractor, we need to work in a group where the DDH assumption is true. It is more difficult to extract entropy in a group where only the CDH assumption is supposed to hold. As precised in the introduction, in the random oracle model, it is possible to extract entropy using hash functions such as MD5 or SHA-1. Yet, in the standard model under the CDH assumption, we currently know how to extract only $O(\log \log p)$ bits and not a fixed fraction of $\log_2(p)$ as we prove in this paper under the DDH assumption. This bound of $O(\log \log p)$ bits is an indirect application of the Goldreich-Levin hard-core predicate [19], using the Shoup's trick [30].

Acknowledgement. The authors would like to thank I. Shparlinski for his helpful remarks and the anonymous reviewers for their comments.

References

1. M. Abdalla, M. Bellare, and P. Rogaway. The Oracle Diffie-Hellman Assumptions and an Analysis of DHIES. In *CT – RSA '01*, LNCS 2020, pages 143–158. Springer-Verlag, Berlin, 2001.
2. M. Bellare and P. Rogaway. Random Oracles Are Practical: a Paradigm for Designing Efficient Protocols. In *Proc. of the 1st CCS*, pages 62–73. ACM Press, 1993.
3. I. F. Blake, T. Garefalakis, and I. E. Shparlinski. On the bit security of the Diffie-Hellman key. In *Appl. Algebra in Engin., Commun. and Computing*, volume 16, pages 397–404, 2006.
4. D. Boneh. The Decision Diffie-Hellman Problem. In J. P. Buhler, editor, *Algorithmic Number Theory Symposium (ANTS III)*, LNCS 1423, pages 48–63. Springer-Verlag, Berlin, 1998.
5. D. Boneh and R. Venkatesan. Hardness of Computing the Most Significant Bits of Secret Keys in Diffie-Hellman and Related Schemes. In *Crypto '96*, LNCS 1109, pages 129–142. Springer-Verlag, Berlin, 1996.
6. D. Boneh and R. Venkatesan. Rounding in Lattices and its Cryptographic applications. In *Proc. of ACM-SIAM SODA'97*, pages 675–681, 1997.
7. J. Bourgain and S. V. Konyagin. Estimates for the Number of Sums and Products and for Exponential Sums Over Subgroups in Fields of Prime Order. *Comptes Rendus Mathmatiques*, 337:75–80, 2003.
8. R. Canetti, J. Friedlander, S. Konyagin, M. Larsen, D. Lieman, and I. Shparlinski. On the Statistical Properties of Diffie-Hellman Distributions. *Israel Journal of Mathematics*, 120:23–46, 2000.
9. R. Canetti, J. Friedlander, and I. Shparlinski. On Certain Exponential Sums and the Distribution of Diffie-Hellman Triples. *Journal of the London Mathematical Society*, 59(2):799–812, 1999.
10. L. Carter and M. Wegman. Universal Hash Functions. *Journal of Computer and System Sciences*, 18:143–154, 1979.
11. O. Chevassut, P. A. Fouque, P. Gaudry, and D. Pointcheval. Key derivation and randomness extraction. Cryptology ePrint Archive, Report 2005/061, 2005. `http://eprint.iacr.org/`.
12. O. Chevassut, P. A. Fouque, P. Gaudry, and D. Pointcheval. The twist-augmented technique for key exchange. In *PKC '06*, LNCS 3958, pages 410–426. Springer-Verlag, Berlin, 2006.
13. J.-S. Coron, Y. Dodis, C. Malinaud, and P. Puniya. Merkle-Damgard Revisited : How to Construct a Hash Function. In *Crypto '05*, LNCS 3621, pages 430–448. Springer-Verlag, Berlin, 2005.

14. R. Cramer and V. Shoup. A Practical Public Key Cryptosystem Provably Secure against Adaptive Chosen Ciphertext Attack. In *Crypto '98*, LNCS 1462, pages 13–25. Springer-Verlag, Berlin, 1998.
15. W. Diffie and M. E. Hellman. New Directions in Cryptography. *IEEE Transactions on Information Theory*, IT–22(6):644–654, November 1976.
16. Y. Dodis, R. Gennaro, J. Håstad, H. Krawczyk, and T. Rabin. Randomness Extraction and Key Derivation Using the CBC, Cascade and HMAC Modes. In *Crypto '04*, LNCS, pages 494–510. Springer-Verlag, Berlin, 2004.
17. T. El Gamal. A Public Key Cryptosystem and a Signature Scheme Based on Discrete Logarithms. *IEEE Transactions on Information Theory*, IT–31(4):469–472, July 1985.
18. R. Gennaro, H. Krawczyk, and T. Rabin. Secure Hashed Diffie-Hellman over Non-DDH Groups. In *Eurocrypt '04*, LNCS 3027, pages 361–381. Springer-Verlag, Berlin, 2004.
19. O. Goldreich and L.A. Levin. A Hard-Core Predicate for all One-Way Functions. In *Proc. of the 21st STOC*, pages 25–32. ACM Press, New York, 1989.
20. M. I. Gonzalez Vasco, M. Näslund, and I. E. Shparlinski. New results on the hardness of Diffie-Hellman bits. In *PKC '04*, LNCS 2947, pages 159–172, 2004.
21. M. I. Gonzalez Vasco and I. E. Shparlinski. On the security of Diffie-Hellman bits. In *Proc. Workshop on Cryptography and Computational Number Theory Singapore, 1999*, pages 331–342. Birkhauser, 2001.
22. N. Gürel. Extracting bits from coordinates of a point of an elliptic curve. Cryptology ePrint Archive, Report 2005/324, 2005. `http://eprint.iacr.org/`.
23. J. Håstad, R. Impagliazzo, L. Levin, and M. Luby. A Pseudorandom Generator from any One-Way Function. *SIAM Journal of Computing*, 28(4):1364–1396, 1999.
24. D. R. Heath-Brown and S. Konyagin. New bounds for Gauss sums derived from k^{th} powers, and for Heilbronn's exponential sum. *Q. J. Math.*, 51(2):221–235, 2000.
25. R. Impagliazzo and D. Zuckerman. How to recycle random bits. In *Proc. of the 30th FOCS*, pages 248–253. IEEE, New York, 1989.
26. S. V. Konyagin and I. Shparlinski. *Character Sums With Exponential Functions and Their Applications.* Cambridge University Press, Cambridge, 1999.
27. N. M. Korobov. The distribution of digits in periodic fractions. *Mat. Sb. (N.S.)*, 89(131):654–670, 672, 1972.
28. L. Trevisan and S. Vadhan. Extracting Randomness from Samplable Distributions. In *Proc. of the 41st FOCS*, pages 32–42. IEEE, New York, 2000.
29. S. Patel and G. Sundaram. An Efficient Discrete Log Pseudo Random Generator. In *Crypto '98*, LNCS 1462. Springer-Verlag, Berlin, 1998.
30. V. Shoup. Lower Bounds for Discrete Logarithms and Related Problems. In *Eurocrypt '97*, LNCS 1233, pages 256–266. Springer-Verlag, Berlin, 1997.
31. V. Shoup. Using Hash Functions as a Hedge against Chosen Ciphertext Attack. In *Eurocrypt '00*, LNCS 1807, pages 275–288. Springer-Verlag, Berlin, 2000.
32. V. Shoup. *A Computational Introduction to Number Theory and Algebra.* Cambridge University Press, Cambridge, 2005.
33. V. Shoup and T. Schweinberger. ACE: The Advanced Cryptographic Engine. Manuscript, March 2000. Revised, August 14, 2000.
34. P. C. van Oorschot and M. J. Wiener. On Diffie-Hellman Key Agreement with Short Exponents. In *Eurocrypt '96*, LNCS 1070, pages 332–343. Springer-Verlag, Berlin, 1996.

A Probabilistic Hoare-style Logic for Game-Based Cryptographic Proofs

Ricardo Corin and Jerry den Hartog

Department of Computer Science, University of Twente, The Netherlands
{ricardo.corin, jerry.denhartog}@cs.utwente.nl

Abstract. We extend a Probabilistic Hoare-style logic to formalize game-based cryptographic proofs. Our approach provides a systematic and rigorous framework, thus preventing errors from being introduced. We illustrate our technique by proving semantic security of ElGamal.

1 Introduction

A typical proof to show that a cryptographic construction is secure uses a reduction from the desired security notion towards some underlying hardness assumption. The security notion is usually represented as a game, in which one proves that the attacker's chance of winning the game is (arbitrarily) small. From a programming language perspective, these games can be thought of as programs whose behaviour is partially known, since the program typically contains invocations to an unknown function representing an arbitrary attacker. In this context, the cryptographic reduction is a sequence of valid program transformations.

Even though cryptographic proofs based on game reductions are powerful, the price one has to pay is high: these proofs are complex, and can easily become involved and intricated. This makes the verification difficult, with subtle errors difficult to spot. Some errors may remain uncovered long after publication, as illustrated for example by Boneh and Franklin's IBE encryption scheme [4], whose cryptographic proof has been recently patched by Galindo [8].

Recently, several papers from the cryptographic community (e.g., the work of Bellare and Rogaway [2], Halevi [9], and Shoup [16]) have recognized the need to tame the complexity of cryptographic proofs. There, the need for (development of) rigorous tools to organize cryptographic proofs in a systematic way is advocated. These tools would prevent subtle easily overlooked mistakes from being introduced in the proof. As another advantage, this precise proof development framework would standardize the proof writing language so that proofs can be checked easily, even perhaps using computer aided verification.

The proposed frameworks [2, 9, 16] provide ad hoc formalisms to reason about the sequences of games, providing useful program transformation rules and illustrating the techniques with several cryptographic proofs from the literature. As we mentioned earlier, the games may be thought of as computer programs, and the reductions thought of as valid program transformations, i.e. transformations that do not change (significantly at least) the "behaviour" of the program. If

M. Bugliesi et al. (Eds.): ICALP 2006, Part II, LNCS 4052, pp. 252–263, 2006.

we represent program behaviour by predicates that establish which states are satisfied by the program before and after its execution, we arrive to a well known setting studied by computer scientists for the past thirty years: program correctness established by a Hoare logic [12]. In Hoare logic, a programming language statement (e.g., value assignment to a variable) is prefixed and postfixed with assertions which state which conditions hold before and after the execution of the statement, respectively. There exists a wealth of papers building on the basic Hoare logic setting, making it one of the most studied subjects for establishing (imperative) program correctness.

This paper's contributions are twofold. First, we adapt and extend our earlier work on Probabilistic Hoare-logic [10, 11] to cope with game-based cryptographic proofs. In particular, we introduce the notion of arbitrary functions, that can be used to model the invocation of an unknown computation (e.g., an arbitrary attacker function). We also include procedures, which are subroutines that can be used to "wrap" function invocations. We provide the associated deduction rules within the logic.We also present a useful program transformation operation, called *orthogonality*, which we use to relate Hoare triples. Orthogonality is our basic "game stepping" operation. Second, to illustrate our approach, we elaborate in full detail a proof of security of ElGamal [6], by reducing the semantic security of the cryptosystem to the hardness of solving the (well-known) Decisional Diffie-Hellman problem.

To the best of our knowledge, ours is the first application of a well known program correctness logic (i.e. Hoare logic) to analyze cryptographic proofs based on transformation of probabilistic imperative programs.

A longer version of this paper [5] contains additional details and proofs.

Related Work. Differently from us, almost all formalisms we know of are directed towards analysing security protocols, thus including concurrency as a main modelling operation. One prime example is in the work of Ramanathan et al. [15], where a probabilistic poly-time process algebraic language is presented. Much effort is paid to measure the computational power of (possibly parallel) processes, so that an environmental context can be precisely regulated to run in probabilistic polynomial time. On the other hand, our logic is fitted for proofs on a simple probabilistic imperative language, without considering parallel systems, nor communication or composition. This simplifies the reasoning and is closer to the original cryptographic proofs which always consider imperative programs (the "games").

Tarento [17] develops machine checkable proofs of signature schemes, focusing on formalizing the semantics of the generic and random oracle models. This differs from the present work, which uses a Hoare-style logic to "derive" the (syntactic) cryptographic algorithms, and then uses the soundness of the logic to obtain the security proofs.

Recently, Blanchet [3] has developed an automated procedure to generate security proofs of protocols; the approach is similar to ours in that also sequences of games are used, although our technique, based on Hoare-logic derivations, can be used to develop proofs manually (however proof checking could be automated); still, it would be interesting to relate the approaches in the future.

2 The Probabilistic Hoare-style Logic *pL*

We shortly recall the probabilistic Hoare style logic *pL* (see [10, 11]). We introduce probabilistic states, and programs which transform such states. Then we introduce probabilistic predicates and a reasoning system to establish Hoare triples which link a precondition and a postcondition to a program.

Probabilistic Programs. We define programs (or statements) s, integer expressions e and Boolean expressions (or conditions) c by:

$$s ::= \texttt{skip} \mid \mathrm{x} := e \mid s\,;\,s \mid \texttt{if}\ c\ \texttt{then}\ s\ \texttt{else}\ s\ \texttt{fi} \mid \texttt{while}\ c\ \texttt{do}\ s\ \texttt{od} \mid s \oplus_\rho s$$
$$e ::= n \mid \mathrm{x} \mid e + e \mid e - e \mid e \cdot e \mid e\,\texttt{div}\,e \mid e\,\texttt{mod}\,e$$
$$c ::= \texttt{true} \mid \texttt{false} \mid b \mid e = e \mid e < e \mid c \wedge c \mid c \vee c \mid \neg c \mid c \rightarrow c$$

where x is a variable of *type* (or 'has *range*') integer, b is a variable of type Boolean and n a number. We assume it is clear how this can be extended with additional operators and to other types and mostly leave the type of variables implicit, assuming that all variables and values are of the correct type.

The basic statements do nothing (`skip`) and assignment ($\mathrm{x} := e$) can be combined with sequential composition (`;`), conditional choice (`if`), iteration (`while`) and probabilistic choice $\oplus_\rho$. In the statement $s \oplus_\rho s'$ a probabilistic decision is made which results in executing s with probability ρ and statement s' with probability $1 - \rho$.

A *deterministic state*, $\sigma \in \mathcal{S}$, is a function that maps each program variable to a value. A *probabilistic state*, $\theta \in \Theta$ gives the probability of being in a given deterministic state. Thus a probabilistic state θ can be seen as a (countable) weighed set of deterministic states which we write as $\rho_1 \cdot \sigma_1 + \rho_2 \cdot \sigma_2 + \ldots$. Here, the probability of being in the (deterministic) state σ_i is ρ_i, $i \geq 0$. For simplicity and without loss of generality we assume that each state σ occurs at most once in θ; multiple occurrences of a single state can be merged into one single occurrence by adding the probabilities, e.g. $1 \cdot \sigma$ rather than $\frac{3}{4} \cdot \sigma + \frac{1}{4} \cdot \sigma$.

The sum of all probabilities is at most 1 but may be less. A probability less than 1 indicates that this execution point may not be always reached (e.g., because of non-termination or because it is part of an '`if`' conditional branch).

To manipulate and combine states we have *scaling* ($\rho \cdot \theta$) which scales the probability of each state in θ, *addition* ($\theta + \theta'$) which unites the two sets and adds probabilities if the same state occurs in both θ and θ', *weighed sum* ($\theta \oplus_\rho \theta' = \rho \cdot \theta + (1 - \rho) \cdot \theta'$) and *conditional selection* ($c?\theta$) which selects the states satisfying c (and removes the rest). For example,

$$\tfrac{1}{2} \cdot (\tfrac{1}{2} \cdot [\mathrm{x} = 1] + \tfrac{1}{2} \cdot [\mathrm{x} = 2]) = \tfrac{1}{4} \cdot [\mathrm{x} = 1] + \tfrac{1}{4} \cdot [\mathrm{x} = 2]$$
$$(\tfrac{1}{4} \cdot [\mathrm{x} = 1] + \tfrac{1}{4} \cdot [\mathrm{x} = 2]) + \tfrac{1}{4} \cdot [\mathrm{x} = 2] = \tfrac{1}{4} \cdot [\mathrm{x} = 1] + \tfrac{1}{2} \cdot [\mathrm{x} = 2]$$
$$(x \leq 2)?(\tfrac{1}{4} \cdot [\mathrm{x} = 1] + \tfrac{1}{2} \cdot [\mathrm{x} = 2] + \tfrac{1}{4} \cdot [\mathrm{x} = 3]) = \tfrac{1}{4} \cdot [\mathrm{x} = 1] + \tfrac{1}{2} \cdot [\mathrm{x} = 2]$$

A program s is interpreted as a transformer of probabilistic states, i.e. its *semantics* $\mathcal{D}(s)$ is a function that maps input states of s to output states. The program transforms the probabilistic state element-wise, with the usual interpretation of the deterministic operations. (See [10] for the fixed point construction used for the semantics of `while`.) For probabilistic choice we use the weighed sum: $\mathcal{D}(s \oplus_\rho s')(\theta) = \mathcal{D}(s)(\theta) \oplus_\rho \mathcal{D}(s')(\theta)$.

Reasoning About Probabilistic Programs. To reason about deterministic states we use *deterministic predicates*, $dp \in DPred$. These are first order logical formulas, i.e. Boolean expressions with the addition of logical variables i, j and the quantification $\forall i :, \exists i :$ over such variables. Similarly, to reason about probabilistic states and programs we introduce *probabilistic predicates*, $p \in Pred$:

$$\begin{aligned} p &::= \mathtt{true} \mid \mathtt{false} \mid b \mid e = e \mid e < e \mid e_r = e_r \mid e_r < e_r \mid p \rightarrow p \mid \neg p \\ &\quad\mid p \wedge p \mid p \vee p \mid \exists j : p \mid \forall j : p \mid \rho \cdot p \mid p + p \mid p \oplus_\rho p \mid c?p \\ e_r &::= \rho \mid \mathtt{r} \mid \mathbb{P}(dp) \mid e_r + e_r \mid e_r - e_r \mid e_r * e_r \mid e_r / e_r \mid \ldots \end{aligned}$$

where e is an expression using logical variables rather than program variables, ρ is a real number and $\mathtt{r}$ a variable with range $[0, 1]$. A *probabilistic expression* e_r is meant to express a probability in $[0, 1]$.

Example 1. We have that $(i < j) \rightarrow (\mathbb{P}(\mathrm{x} = 5 \wedge \mathrm{y} < \mathrm{x} + i) > \mathbb{P}(\mathrm{x} = j) + \frac{1}{4})$ is a probabilistic predicate but $(\mathrm{x} > i)$ is not as the use of program variable x outside of the $\mathbb{P}(\cdot)$ construction is not allowed.

The value of $\mathbb{P}(dp)$, in a given probabilistic state, is the sum of the probabilities for deterministic states that satisfy dp, e.g. in $\frac{1}{4} \cdot [\mathrm{x} = 1] + \frac{1}{4} \cdot [\mathrm{x} = 2] + \frac{1}{4} \cdot [\mathrm{x} = 3] + \frac{1}{4} \cdot [\mathrm{x} = 4]$ we have that $\mathbb{P}(\mathrm{x} \geq 2) = \frac{3}{4}$. Establishing the value of a probabilistic expression e_r and a (basic) predicate p from a probabilistic state θ is standard; the latter is denoted (as usual) by the satisfaction relation $\theta \models p$. The 'arithmetical' operators $+$, $\oplus_\rho$,$\rho\cdot$, ? specific to our probabilistic logic are the logical counterparts of the same operations on states. For example,

$$\theta \models p + p' \text{ when there exists } \theta_1, \theta_2\text{: } \theta = \theta_1 + \theta_2, \theta_1 \models p \text{ and } \theta_2 \models p' \quad (1)$$

$$\theta \models c?p \text{ when there exists } \theta'\text{: } \theta = c?\theta', \theta \models p \quad (2)$$

The satisfaction relation also includes an interpretation function giving values to the logical variables, which we omit from the notation when no confusion is possible. We write $\models p$ if p holds in any probabilistic state.

Hoare triples, also known as *program correctness triples*, give a precondition and a postcondition for a program. A triple is called valid, denoted $\models \{p\}\, s\, \{q\}$, if the precondition guarantees the postcondition after execution of the program, i.e. for all θ with $\theta \models p$ we have $\mathcal{D}(s)(\theta) \models q$.

Our derivation system for Hoare triples adapts and extends the existing Hoare logic calculus. The standard rules for skip, assignment, sequential composition, precondition strengthening and postcondition weakening remain the same. The rule for conditional choice is adjusted and a new rule for probabilistic choice is added, along with some structural rules. We only present the main rules here (see e.g. [10] for a complete overview), noting that the other rules come directly from Hoare logic or from natural deduction.

$$\{p[x/e]\}\ \mathtt{x := e}\ \{p\} \quad \text{(Assign)} \qquad \frac{\{c?p\}\ s\ \{q\} \quad \{\neg c?p\}\ s'\ \{q'\}}{\{p\}\ \mathbf{if}\ c\ \mathbf{then}\ s\ \mathbf{else}\ s'\ \mathbf{fi}\ \{q+q'\}} \text{(If)}$$

$$\frac{\{p\}\ s\ \{p'\} \quad \{p'\}\ s'\ \{q\}}{\{p\}\ s\ ;\ s'\ \{q\}} \text{(Seq)} \qquad \frac{\{p\}\ s\ \{q\} \quad \{p\}\ s'\ \{q'\}}{\{p\}\ s \oplus_\rho s'\ \{q \oplus_\rho q'\}} \text{(Prob)}$$

$$\frac{\{p\}\ s\ \{q\} \quad \{p\}\ s\ \{q'\}}{\{p\}\ s\ \{q \wedge q'\}} \text{(And)} \qquad \frac{\models p' \rightarrow p \quad \{p\}\ s\ \{q\} \quad \models q \rightarrow q'}{\{p'\}\ s\ \{q'\}} \text{(Cons)}$$

These rules are used in the proof of ElGamal in Section 4, but first we extend the language and logic to cover the necessary elements for cryptographic proofs.

3 Extending *pL*

We consider two language extensions and one extension of the reasoning method:

- *Functions* are computations that are a priori unknown. These are useful to reason about arbitrary *attacker* functions, for which we do not know what behavior they will produce.
- *Procedures* allow the specification of subroutines. These are useful to specify cryptographic assumptions that hold 'for every procedure' satisfying some appropriate conditions. Procedures are programs for which its behavior (i.e. the procedure's *body*) is assumed to be partially known (since it may contain an invocation to an arbitrary function).
  We assume that both functions and procedures are deterministic. However this poses no loss of generality as enough "randomness" can be sampled before and then passed to the function or procedure as an extra parameter. We explicitly distinguish functions and procedures for readability and convenience, rather than because there is a fundamental difference between the two; it clarifies the different roles (i.e. procedures are specified routines and functions are unknown attacker functions) directly in the syntax.
- *Orthogonality* allows to reason about independent statements. This is a program transformation operation that is going to be useful when reasoning on cryptographic proofs as sequences of games.

Functions. Functions, as opposed to procedures, are undefined (i.e. we do not provide a body). We use these functions to represent arbitrary attackers, for which we do not know a priori their behaviour.

To include functions in the language we add function symbols to expressions (as defined in the previous section): $e ::= \ldots \mid f(e, \ldots, e)$. We assume that the functions are used correctly, that is functions are always invoked with the right number of arguments and correct types. Also, note that by considering functions to be expressions we allow functions to be used in the (deterministic and probabilistic) predicates. The fact that a function is deterministic is represented in the logic by the following remark.

Remark 2. For any function $f(\cdot)$ (of arity n) and expressions $e_1, \ldots, e_n, e'_1, \ldots, e'_n$ we have $\models (e_1 = e'_1 \wedge \ldots \wedge e_n = e'_n) \rightarrow f(e_1, \ldots, e_n) = f(e'_1, \ldots, e'_n)$.

To deal with functions in the semantics, we assume that any function symbol f has some fixed (albeit unknown) deterministic, type correct interpretation $\hat{f}$. Thus, e.g. the semantics for an assignment using f becomes $\mathcal{D}(\mathtt{x} := f(\mathtt{y}))(\rho_1 \cdot \sigma_1 + \rho_2 \cdot \sigma_2 + \ldots) = \rho_1 \cdot \sigma_1[^{\hat{f}(\sigma_1(\mathtt{y}))} / _{\mathtt{x}}] + \rho_2 \cdot \sigma_2[^{\hat{f}(\sigma_2(\mathtt{y}))} / _{\mathtt{x}}] + \ldots$. The rules given above are also valid for the extended language; extending the correctness proof [10] for the Assign rule is direct, while the proof for the other rules remains the same as it only uses structural properties of the denotational semantics.

Procedures. We now extend the language with procedures, which are used to model (partially) known subprograms. Each procedure has a list of variables, the *formal parameters* (divided in turn into value parameters and variable parameters) and a set of local variables. We assume that none of these variables occur in the main program or in other procedures. The procedure also has a body, $\mathtt{B}_{proc}$, which is a program statement which uses only the formal parameters and local variables, only assigns to variable parameters and local variables, and assigns to a local variable before using its value. We also enforce the procedure to be deterministic by excluding any probabilistic choice statement from $\mathtt{B}_{proc}$. Finally, we require that the procedure is non-recursive (i.e. we can order procedures such that any procedure only calls procedures of a lower order). We use the notation procedure $proc(\textbf{value}\ \mathtt{v}_1, \ldots, \mathtt{v}_n; \textbf{var}\ \mathtt{w}_1, \ldots, \mathtt{w}_m) : \mathtt{B}_{proc}$ to list the value and variable parameters and the body of a procedure (any variables in $\mathtt{B}_{proc}$ that are not formal parameters are local variables).

We add procedures to the language by including *procedure* calls to the statements, $s ::= \ldots \mid proc(\mathrm{e}, \ldots, \mathrm{e}; \mathrm{x}, \ldots, \mathrm{x})$. Here we assume that there is *no aliasing of variables*; i.e. a different variable is used for each variable parameter.

The procedure call $proc(\mathrm{e}_1, \ldots, \mathrm{e}_n, \mathrm{x}_1, \ldots, \mathrm{x}_m)$ (in state σ) corresponds to first assigning the value of the appropriate expression (e_i or x_j) to the formal parameters, running the body of the program and finally assigning the resulting value of the variable arguments $\mathtt{w}_1, \ldots, \mathtt{w}_m$ to $\mathrm{x}_1, \ldots, \mathrm{x}_m$. Thus the semantics is:

$$\begin{aligned}\mathcal{D}(proc(\mathrm{e}_1, \ldots, \mathrm{e}_n; \mathrm{x}_1, \ldots, \mathrm{x}_m))(\theta) = \mathcal{D}(&\mathtt{v}_1 := \mathrm{e}_1; \ldots; \mathtt{v}_n := \mathrm{e}_n;\\ &\mathtt{w}_1 := \mathrm{x}_1; \ldots; \mathtt{w}_m := \mathrm{x}_m;\\ &\mathtt{B}_{proc};\ \mathrm{x}_1 := \mathtt{w}_1; \ldots; \mathrm{x}_n := \mathtt{w}_n)(\theta)\end{aligned}$$

To enable reasoning about a procedure $proc(\mathbf{value}\ \mathtt{v}_1, \ldots, \mathtt{v}_n; \mathbf{var}\ \mathtt{w}_1, \ldots, \mathtt{w}_m)$: $\mathtt{B}_{proc}$, we add the following derivation rule:

$$\frac{\{p\}\ \mathtt{B}_{proc}\ \{q\}}{\{p[^{\mathtt{e}_1,\ldots,\mathtt{e}_n,\mathtt{x}_1,\ldots,\mathtt{x}_n}/_{\mathtt{v}_1,\ldots,\mathtt{v}_n,\mathtt{w}_1,\ldots,\mathtt{w}_m}]\}\ proc(\mathtt{e}_1,\ldots,\mathtt{e}_n;\mathtt{x}_1,\ldots,\mathtt{x}_n)\ \{q[^{\mathtt{x}_1,\ldots,\mathtt{x}_n}/_{\mathtt{w}_1,\ldots,\mathtt{w}_m}]\}} \quad (3)$$

The extended logic including this rule is correct, i.e. any Hoare triple derived from the proof system is valid. Extending the correctness proof for the added rule is again a simple exercise using the definition of the semantics given above and properties of the assignment statement.

Distributions and Independence. We now illustrate how to express the (joint) distribution of variables (and more generally of expressions) in the logic. Then we discuss the issue of independence of variables and expressions.

A commonly used component in (security) games is a variable chosen completely at *random*, which in other words is a variable with a *uniform distribution* over its (finite) range. Suppose that variable x and i have the same range S. Then the following predicate expresses that x is uniformly distributed over S:

$$R_S(\mathtt{x}) \;=\; \forall i : \mathbb{P}(\mathtt{x} = i) = 1/|S|$$

where $|S|$ denotes the size of the set S. The variable x can be given a uniform distribution over $S = \{v_1, \ldots, v_n\}$ by running the program

$$\mathtt{x} := \mathtt{v}_1 \oplus_{1/n} (\mathtt{x} := \mathtt{v}_2 \oplus_{1/(n-1)} (\cdots \oplus_{1/2} \mathtt{x} := \mathtt{v}_n))$$

As this is a commonly used construction we introduce a shorthand notation for this statement: $\mathtt{x} \leftarrow S$. Using our logic, it is straightforward to derive (using repeatedly rule (Prob)) that after running this program x has a uniform distribution over S: $\models \{\, \mathbb{P}(\mathtt{true}) = 1 \,\}\ \mathtt{x} \leftarrow S\ \{\, R_S(\mathtt{x}) \,\}$.

More interestingly, after running the program $\mathtt{x} \leftarrow S; \mathtt{y} \leftarrow S'$ we not only know that x has a uniform distribution over S and y has a uniform distribution over S', but we also know that y has a uniform distribution over S' *independently* from the value of x. In other words, the *joint distribution* of x and y is $R_{S,S'}(\mathtt{x}, y) ::= \forall i, j : \mathbb{P}(\mathtt{x} = i \wedge \mathtt{y} = j) = 1/|S| \cdot 1/|S'|$ (with $i \in S$, $j \in S'$). This is a stronger property than only the information that x and y are uniformly distributed. (The difference is exactly the independence of the variables.) Below we introduce a predicate expressing independence and generalize these results.

Definition 3 (Independent $I(\cdot)$ and Random $R(\cdot)$ expressions). *The predicate $I(e_1, \ldots, e_n)$ states independence of expressions $e_1, \ldots, e_n$, and is defined by (where i_j is of the same type as e_j, $1 \le j \le n$.):*

$$I(e_1, \ldots, e_n) \;=\; \forall i_1, \ldots, i_n : \mathbb{P}(e_1 = i_1 \wedge \ldots \wedge e_n = i_n) = \mathbb{P}(e_1 = i_1) \cdot \ldots \cdot \mathbb{P}(e_n = i_n)$$

The predicate $R_{S_1,\ldots,S_n}(e_1, \ldots, e_n)$ states that $e_1, \ldots, e_n$ are randomly and independently distributed over $S_1, \ldots, S_n$ respectively, is defined as follows:

$$R_{S_1,\ldots,S_n}(e_1, \ldots, e_n) \;=\; \forall i_1, \ldots, i_n : \mathbb{P}(e_1 = i_1 \wedge \ldots \wedge e_n = i_n) = 1/|S_1| \cdot \ldots \cdot 1/|S_n|$$

Lemma 4 (Relations between $R(\cdot)$ and $I(\cdot)$).

1. *An expression list has a joint uniform distribution when they are independent and each has a uniform distribution, i.e.* $\models R_{S_1,\ldots,S_n}(e_1,\ldots,e_n) \leftrightarrow R_{S_1}(e_1) \wedge \ldots \wedge R_{S_n}(e_n) \wedge I(e_1,\ldots,e_n)$.
2. *Separate randomly assigned variables have a joint random distribution:* $\models \{\mathbb{P}(\mathtt{true}) = 1\}\ \mathtt{x}_1 \leftarrow S_1; \ldots; \mathtt{x}_n \leftarrow S_n\ \{R_{S_1,\ldots,S_n}(x_1,\ldots,x_n)\}$.
3. *Independence is maintained by functions; if an expression e is independent from the inputs $e_1,\ldots,e_n$ of a function f, then e is also independent of $f(e_1,\ldots,e_n)$, i.e.,* $\models I(e,e_1,\ldots,e_n) \to I(e, f(e_1,\ldots,e_n))$.

Both (1) and (3) express basic properties, shown easily to hold semantically for any (probabilistic) state. The triple in (2) is shown valid by using the logic.

Example 5. The lemma above can be used in a derivation as follows:

$$\begin{array}{l}
\{\mathbb{P}(\mathtt{true}) = 1\} \\
\quad b \leftarrow Bool\ ; \\
\{R_{Bool}(b)\} \\
\quad \mathrm{x} \leftarrow S; \\
\{R_{Bool,S}(b,\mathrm{x})\} \to \{I(b,\mathrm{x})\} \to \{I(b, f(\mathrm{x}))\} \\
\quad b' := f(\mathrm{x})\ ; \\
\{I(b,b')\}
\end{array}$$

The derivation above is represented as a so called *proof outline*, which is a commonly used way to represent proofs in Hoare logic. Briefly, rather than giving a complete proof tree only the most relevant steps of the proof are given in an intuitively clear format. The predicates in between the program statements give properties that are valid at that point in the execution.

Orthogonality. A (terminating) program that does not change the value of variables in a predicate (i.e. is 'orthogonal to the predicate') will not change its truth value. In this section we make this intuitive property more precise. As we show in the proof of ElGamal cryptosystem in Section 4, orthogonality is a powerful method to reason about programs and Hoare triples yet is easy to use as it only requires a simple syntactical check.

Let $Var(p)$ denote the set of program variables occurring in the probabilistic predicate p, $Var(s)$ the variables occurring in the statement s and let $Var_a(s)$ denote the set of program variables which are *assigned to* (i.e. subject to assignment) in s (x is assigned to in s if $x := e$ occurs in s for some e or when x is used as a variable parameter in a procedure call). We write $s \perp p$ if $Var_a(s) \cap Var(p) = \emptyset$ and $s \perp s'$ if $Var_a(s) \cap Var(s') = \emptyset$. Thus we call a program orthogonal to a predicate (or to another program) if the program does not change the variables used in the predicate (or in the other program).

The following theorem states that we can add and remove orthogonal statements without changing the validity of a Hoare triple. As we shall see in Section 4, this is precisely what is needed to establish the security of ElGamal.

Theorem 6. *If $s' \perp q$ and $s' \perp s''$ then $\{p\}\ s\ ;\ s'\ ;\ s''\ \{q\}$ is valid if and only if $\{p\}\ s\ ;\ s''\ \{q\}$ is valid.*

The notion of orthogonality $\perp$ is a practical and purely syntactically defined relation, and thus easy to check. On the other hand, $\perp$ does not have commonly used properties of relations such as reflexiveness, transitivity and congruence properties. Therefore, care must be taken in reasoning with this relation outside of its intended purpose, that is to add or remove non-relevant program sections in a derivation, so one can transform a program into the exact required form.

4 Application: Security Analysis of ElGamal

We now apply our technique to derive semantic security for ElGamal [6].

ElGamal. Let G be a group of prime order q, and let $\gamma \in G$ be a generator. (The descriptions of G and γ, including q, represent arbitrary "system parameters"). Let $Z_q^* = \{1, \ldots, q-1\}$ denote the usual multiplicative group. A key is created by choosing a number uniformly from Z_q^*, say $x \in Z_q^*$. Then x is the private key and γ^x the public key. To encrypt a message $m \in Z_q^*$, a number $y \in Z_q^*$ is chosen uniformly from Z_q^*. Then (c, k) is the ciphertext, for $c = m \cdot \gamma^{xy}$, and $k = \gamma^y$. To decrypt using the private key x, compute c/k^x, since $\frac{c}{k^x} = \frac{m \cdot \gamma^{xy}}{\gamma^{yx}} = m$.

Security Analysis. The security of ElGamal cryptosystem is shown w.r.t. the Decisional Diffie-Hellman (DDH) assumption. Suppose we sample uniformly the values x, y and z. Fix ε_{ddh} small and RND large w.r.t. the system parameters. Then the DDH assumption (for G) states that no effective procedure $D(\cdot)$ (with randomness given by a uniform sample from $\{1, \ldots, RND\}$, encoding a finite tape of uniformly distributed bits) can distinguish triples of the form $\langle \gamma^x, \gamma^y, \gamma^{xy} \rangle$ from triples of the form $\langle \gamma^x, \gamma^y, \gamma^z \rangle$ with a chance better than ε_{ddh}.

In our formalism we do not precisely define the meanings of "small", "large", "better" and "effective", as they are not required in the actual proof transformations. However, one should keep in mind that these notions need to be defined properly, where e.g. "effective" means time bounded by a polynomial in the security parameter. Moreover, our fixed values (e.g., ε_{ddh}) implicitly depend on the arbitrary system parameters, so asymptotic bounds can be expressed properly (so in fact ε_{ddh} is negligible when the security parameter tends to infinity).

Semantic Security. The semantic security game for ElGamal cryptosystem consists of the following four steps: (1). Setup: x is sampled from Z_q^* and r is sampled from RND. (2). Attacker chooses m_0, m_1 using inputs γ^x, r. (3). y is sampled from Z_q^*, bit b is sampled uniformly, and let $c = \gamma^{xy} \cdot m_b$. (4). Attacker chooses b' using inputs γ^x, γ^y, r, c.

Now, the attacker wins this game if it outputs b' equating b, that is the attacker can guess b with a non-negligible probability (in our case, better than $1/2 + \varepsilon_{ddh}$). A standard proof (e.g., the one given in [16]) reduces the security of this notion (i.e. that the attacker cannot win the game) to the DDH assumption described above. We now describe a similar proof within our formalism.

ElGamal Security Analysis in pL. In our formalism, the DDH assumption ensures that for any effective procedure $D(\mathtt{v1}, \mathtt{v2}, \mathtt{v3}, \mathtt{v4}, \mathtt{v5}; \mathtt{x1})$ with inputs $\mathtt{v1}, \mathtt{v2}, \mathtt{v3}, \mathtt{v4}, \mathtt{v5}$ and output boolean $\mathtt{x1}$, the following is a valid Hoare triple.

$$
\begin{array}{l}
\{\mathbb{P}(\mathtt{true}) = 1\} \\
\quad \mathtt{x} \leftarrow Z_q^*;\ \mathtt{y} \leftarrow Z_q^*;\ \mathtt{r1} \leftarrow RND;\ \mathtt{b1} \leftarrow Bool;\ D(\gamma^{\mathtt{x}}, \gamma^{\mathtt{y}}, \gamma^{\mathtt{xy}}, \mathtt{r1}, \mathtt{b1}; \mathtt{out1}); \\
\quad \mathtt{z} \leftarrow Z_q^*;\ \mathtt{r2} \leftarrow RND;\ \mathtt{b2} \leftarrow Bool;\ D(\gamma^{\mathtt{x}}, \gamma^{\mathtt{y}}, \gamma^{\mathtt{z}}, \mathtt{r2}, \mathtt{b2}; \mathtt{out2}) \\
\{|\mathbb{P}(\mathtt{out1}) - \mathbb{P}(\mathtt{out2})| \leq \varepsilon_{ddh}\}
\end{array}
$$

Here, the extra provided randomness $\mathtt{b1}$ and $\mathtt{b2}$ to procedure $D(\cdot)$ are given solely to ease the exposition (as $\mathtt{r1}$ and $\mathtt{r2}$ already provide enough randomness).

ElGamal Semantic security. We assume three attacker functions $A0(\mathtt{v1}, \mathtt{v4})$, $A1(\mathtt{v1}, \mathtt{v4})$ and $A2(\mathtt{v1}, \mathtt{v2}, \mathtt{v3}, \mathtt{v4})$. Functions $A0(\mathtt{v1}, \mathtt{v4})$ and $A1(\mathtt{v1}, \mathtt{v4})$ return two numbers $\mathtt{m0}$ and $\mathtt{m1}$ from Z_q^*. Similarly, function $A2(\mathtt{v1}, \mathtt{v2}, \mathtt{v3}, \mathtt{v4})$ returns a boolean. From these attacker functions we define another procedure $S(\mathtt{v1}, \mathtt{v2}, \mathtt{v3}, \mathtt{v4}, \mathtt{v5}; \mathtt{x1}) : \mathtt{B}_S$, where the body B_S is defined as follows:

$$
\begin{array}{rl}
\mathtt{B}_S \triangleq & \mathtt{m0} := A0(\mathtt{v1}, \mathtt{v4}); \mathtt{m1} := A1(\mathtt{v1}, \mathtt{v4}); \\
& \text{if } \mathtt{v5} = \mathtt{false} \text{ then } \mathtt{tmp} := \mathtt{v3} \cdot \mathtt{m0} \text{ else } \mathtt{tmp} := \mathtt{v3} \cdot \mathtt{m1} \text{ fi}; \\
& \mathtt{b} := A2(\mathtt{v1}, \mathtt{v2}, \mathtt{tmp}, \mathtt{v4}); \\
& \text{if } \mathtt{v5} = \mathtt{b} \ \mathtt{then}\ \mathtt{x1} := true \ \mathtt{else}\ \mathtt{x1} := false\ \mathtt{fi};
\end{array}
$$

Proving the semantic security of ElGamal amounts to establish:

Theorem 7. *The following is a valid probabilistic Hoare Triple:*

$$
\begin{array}{l}
\{\mathbb{P}(\mathtt{true}) = 1\} \\
\quad \mathtt{x} \leftarrow Z_q^*;\ \mathtt{y} \leftarrow Z_q^*;\ \mathtt{r1} \leftarrow RND;\ \mathtt{b1} \leftarrow Bool;\ S(\gamma^{\mathtt{x}}, \gamma^{\mathtt{y}}, \gamma^{\mathtt{xy}}, \mathtt{r1}, \mathtt{b1}; \mathtt{out1}) \\
\{|\mathbb{P}(\mathtt{out1}) - 1/2| \leq \varepsilon_{ddh}\}
\end{array}
$$

To establish this result, we first show the following lemma.

Lemma 8. *The following is a valid Probabilistic Hoare Triple:*

$$
\{R_{Z_q^{*3}, RND, Bool}(\gamma^{\mathtt{x}}, \gamma^{\mathtt{y}}, \gamma^{\mathtt{z}}, \mathtt{r2}, \mathtt{b2})\}\ S(\gamma^{\mathtt{x}}, \gamma^{\mathtt{y}}, \gamma^{\mathtt{z}}, \mathtt{r2}, \mathtt{b2}; \mathtt{out2})\ \{\mathbb{P}(\mathtt{out2}) = 1/2\}
$$

Proof. (Sketch) We use rule (3) from Section 3 on the definition of procedure $S(\cdot)$, to establish the validity of the following triple (We derive this triple formally in [5]):

$$
\{R_{Z_q^{*3}, RND, Bool}(\mathtt{v1}, \mathtt{v2}, \mathtt{v3}, \mathtt{v4}, \mathtt{v5})\}\ \mathtt{B}_S\ \{\mathbb{P}(\mathtt{x1}) = 1/2\}
$$

Now, to establish Theorem 7, we start by showing the validity of the following Hoare triple:

$$
\begin{array}{l}
\{\mathbb{P}(\mathtt{true}) = 1\} \\
\quad \mathtt{x} \leftarrow Z_q^*;\ \mathtt{y} \leftarrow Z_q^*;\ \mathtt{z} \leftarrow Z_q^*;\ \mathtt{r2} \leftarrow RND;\ \mathtt{b2} \leftarrow Bool; \\
\{R_{Z_q^{*3}, RND, Bool}(\mathtt{x}, \mathtt{y}, \mathtt{z}, \mathtt{r2}, \mathtt{b2})\} \rightarrow \{R_{Z_q^{*3}, RND, Bool}(\gamma^{\mathtt{x}}, \gamma^{\mathtt{y}}, \gamma^{\mathtt{z}}, \mathtt{r2}, \mathtt{b2})\} \\
\quad S(\gamma^{\mathtt{x}}, \gamma^{\mathtt{y}}, \gamma^{\mathtt{z}}, \mathtt{r2}, \mathtt{b2}; \mathtt{out2}) \\
\{\mathbb{P}(\mathtt{out2}) = 1/2\}
\end{array}
$$

The lower part of the triple is given by Lemma 8. For the upper part, we use Lemma 4(1) to obtain to obtain $\{R_{Z_q^{*3},RND,Bool}(\mathtt{x},\mathtt{y},\mathtt{z},\mathtt{r2},\mathtt{b2})\}$ from the random samples. The implication follows from standard properties of the group Z_q^* and the generator γ, which is a permutation of Z_q^* (In the long version [5] we derive formally a similar property). Finally, we combine the two triples using rule (Seq).

The next step consists in adding the orthogonal statements (shown boxed below) between the assignments of y and z of the above triple. Since the added statements are orthogonal (they assign to r1,b1,out1 only, which do not occur in the above triple), by Theorem 6 we get that the following triple is valid:

$$\begin{array}{l}\{\mathbb{P}(\mathtt{true}) = 1\} \\ \quad \mathtt{x} \leftarrow Z_q^*;\ \mathtt{y} \leftarrow Z_q^*;\ \boxed{\mathtt{r1} \leftarrow RND;\ \mathtt{b1} \leftarrow Bool;\ S(\gamma^{\mathtt{x}},\gamma^{\mathtt{y}},\gamma^{\mathtt{xy}},\mathtt{r1},\mathtt{b1};\mathtt{out1});} \\ \quad \mathtt{z} \leftarrow Z_q^*;\ \mathtt{r2} \leftarrow RND;\ \mathtt{b2} \leftarrow Bool;\ S(\gamma^{\mathtt{x}},\gamma^{\mathtt{y}},\gamma^{\mathtt{z}},\mathtt{r2},\mathtt{b2};\mathtt{out2}) \\ \{\mathbb{P}(\mathtt{out2}) = 1/2\}\end{array}$$

This is the DDH assumption when $D(\cdot)$ is instantiated by $S(\cdot)$. We use rule (And) and join the postconditions $\{\mathbb{P}(\mathtt{out2}) = 1/2\}$ and $\{|\mathbb{P}(\mathtt{out1}) - \mathbb{P}(\mathtt{out2})\}| \leq \varepsilon_{ddh}$:

$$\begin{array}{l}\{\mathbb{P}(\mathtt{true} = 1)\} \\ \quad \mathtt{x} \leftarrow Z_q^*;\ \mathtt{y} \leftarrow Z_q^*;\ \mathtt{r1} \leftarrow RND;\ \mathtt{b1} \leftarrow Bool;\ S(\gamma^{\mathtt{x}},\gamma^{\mathtt{y}},\gamma^{\mathtt{xy}},\mathtt{r1},\mathtt{b1};\mathtt{out1}); \\ \quad \mathtt{z} \leftarrow Z_q^*;\ \mathtt{r2} \leftarrow RND;\ \mathtt{b2} \leftarrow Bool;\ S(\gamma^{\mathtt{x}},\gamma^{\mathtt{y}},\gamma^{\mathtt{z}},\mathtt{r2},\mathtt{b2},\mathtt{out2}) \\ \{\mathbb{P}(\mathtt{out2}) = 1/2 \wedge |\mathbb{P}(\mathtt{out1}) - \mathbb{P}(\mathtt{out2})| \leq \varepsilon_{ddh}\} \rightarrow \{|\mathbb{P}(\mathtt{out1}) - 1/2| \leq \varepsilon_{ddh}\}\end{array}$$

The last application of rule (Cons) follows from replacing $\mathbb{P}(\mathtt{out2})$ with $1/2$. Finally, we remove the last line of statements thanks to orthogonality (as the assigned to variables do not occur elsewhere), and obtain the desired theorem:

$$\begin{array}{l}\{\mathbb{P}(\mathtt{true}) = 1\} \\ \quad \mathtt{x} \leftarrow Z_q^*;\ \mathtt{y} \leftarrow Z_q^*;\ \mathtt{r1} \leftarrow RND;\ \mathtt{b1} \leftarrow Bool;\ S(\gamma^{\mathtt{x}},\gamma^{\mathtt{y}},\gamma^{\mathtt{xy}},\mathtt{r1},\mathtt{b1},\mathtt{out1}); \\ \{|\mathbb{P}(\mathtt{out1}) - 1/2| \leq \varepsilon_{ddh}\}\end{array}$$

5 Conclusions and Future Work

Cryptographic proofs are complex constructions that use both cryptography and programming languages concepts. In our opinion, both communities can benefit from our approach: First, Hoare logic is well known in the programming languages community, and has been used to prove algorithm correctness for more than three decades. There are readily available computer aided verification systems that can handle Hoare logic reasoning systems (e.g., HOL [14], PVS [13], Coq [7]). Second, developing cryptographic proofs as games is well known in the cryptographic community [2, 9, 16]. Our logic allows to derive correctness proofs directly from these imperative programs, without code modifications.

Future Work. There are several possible directions for future work. A short term goal is to cover more complex examples [2, 9, 16]. This would probably require to refine the notion of equivalence between Hoare triples to equivalence *up-to* ε, to model transitions based on "bad events unlikely to happen" instead of the standard equivalence that models transitions based on pure indistinguishability.

The price to pay for rigorousity is in proof length, as the detailed proofs can quickly become lengthy. An axiomatization of the logic along with a library of ready-to-use proofs for standard constructions would help into reducing the complexity and proof length (this is a matter of ongoing work). Along the same lines, a longer term goal is to develop an implementation on a theorem prover to provide machine-checkable cryptographic proofs, following e.g. earlier work on (standard) Hoare logic formalization [13, 7, 1]. Here axioms and pre-computed proofs would also greatly increase efficiency and usability.

Acknowledgements. We thank Pieter Hartel, Sandro Etalle, Jeroen Doumen, and the anonymous reviewers for helpful comments.

References

1. P. Audebaud and C. Paulin. Proofs of randomised algorithms in coq. In *MPC'06*.
2. M. Bellare and P. Rogaway. The game-playing technique, December 2004. At `http://www.cs.ucdavis.edu/~rogaway/papers/games.html`.
3. B. Blanchet. A computationally sound mechanized prover for security protocols. In *IEEE Symposium on Security and Privacy*, Oakland, California, 2006.
4. D. Boneh and M. K. Franklin. Identity-based encryption from the weil pairing. In *CRYPTO'01*, pages 213–229. Springer-Verlag, 2001.
5. R. Corin and J. den Hartog. A probabilistic hoare-style logic for game-based cryptographic proofs (long version, `http://eprint.iacr.org/2005/467`). 2006.
6. T. ElGamal. A public-key cryptosystem and a signature scheme based on discrete logarithms. *IEEE Transactions on Information Theory*, IT-31:469–472, 1985.
7. J.-C. Filliâtre. Why: a multi-language multi-prover verification tool. Technical report, LRI, Université Paris Sud, 2003.
8. D. Galindo. Boneh-franklin identity based encryption revisited. In *ICALP*, pages 791–802, 2005.
9. S. Halevi. A plausible approach to computer-aided cryptographic proofs, 2005. At `http://eprint.iacr.org/2005/181/`.
10. J.I. den Hartog. *Probabilistic Extensions of Semantical Models*. PhD thesis, Vrije Universiteit Amsterdam, 2002.
11. J.I. den Hartog and E.P. de Vink. Verifying probabilistic programs using a Hoare like logic. *Int. Journal of Foundations of Computer Science*, 13(3):315–340, 2002.
12. C.A.R. Hoare. An axiomatic basis for computer programming. *Communications of the ACM*, 12:576–580, 1969.
13. J. Hooman. Program design in PVS. In *Workshop on Tool Support for System Development and Verification*, Germany, 1997.
14. M.J.C. Gordon. Mechanizing programming logics in higher-order logic. In *Proc. of the Workshop on Hardware Verification*, pages 387–439. Springer-Verlag, 1988.
15. A. Ramanathan, J. C. Mitchell, A. Scedrov, and V. Teague. Probabilistic bisimulation and equivalence for security analysis of network protocols. In *FoSSaCS*, pages 468–483, 2004.
16. V. Shoup. Sequences of games: a tool for taming complexity in security proofs, May 2005. At `http://www.shoup.net/papers/games.pdf`.
17. S. Tarento. Machine-checked security proofs of cryptographic signature schemes. In *ESORICS*, pages 140–158, 2005.

Generic Construction of Hybrid Public Key Traitor Tracing with Full-Public-Traceability

Duong Hieu Phan[1], Reihaneh Safavi-Naini[2], and Dongvu Tonien[2]

[1] University College London
Adastral Park Postgraduate Research Campus,
Ross Building, Adastral Park, IP5 3RE, Ipswich, United Kingdom
h.phan@adastral.ucl.ac.uk
[2] University of Wollongong,Australia
School of Information Technology and Computer Science,
Wollongong, NSW 2522, Australia
{rei, dong}@uow.edu.au

Abstract. In Eurocrypt 2005, Chabanne, Phan and Pointcheval introduced an interesting property for traitor tracing schemes called *public traceability*, which makes tracing a black-box public operation. However, their proposed scheme only worked for two users and an open question proposed by authors was to provide this property for multi-user systems.

In this paper, we give a comprehensive solution to this problem by giving a *generic construction for a hybrid traitor tracing scheme that provides full-public-traceability*. We follow the Tag KEM/DEM paradigm of hybrid encryption systems and extend it to multi-receiver scenario. We define Tag-Broadcast KEM/DEM and construct a secure Tag-BroadcastKEM from a CCA secure PKE and target-collision resistant hash function. We will then use this Tag-Broadcast KEM together with a semantically secure DEM to give a generic construction for Hybrid Public Key Broadcast Encryption. The scheme has a black box tracing algorithm that *always* correctly identifies a traitor. The hybrid structure makes the system very efficient, both in terms of computation and communication cost. Finally we show a method of reducing the communication cost by using codes with identifiable parent property.

1 Introduction

Broadcast encryption and traitor tracing systems are the main cryptographic primitives for secure distribution of copyrighted digital content. In broadcast encryption systems the user group is dynamic and changes over time and access control is by distributing a new session key to authorised users in each session. The session key is used to securely encrypt (e.g. using AES) the content. The separation of content encryption and session key encryption provides flexibility in choosing encryption algorithms that are suitable for specific content (e.g. MPEG2 streams).

Traitor tracing systems however, aim at providing traceability against colluding users who have constructed a *pirate* decoder that can illegally decrypt the content. Public key traitor tracing schemes allow anyone to send content to members of an authorised group. In the model of tracing proposed in [CFN94], tracing is performed by a trusted authority who has access to the secret key of all users. The tracer is able to identify one of the

M. Bugliesi et al. (Eds.): ICALP 2006, Part II, LNCS 4052, pp. 264–275, 2006.

c traitors who have colluded to construct a pirate decoder. The tracing algorithm may always correctly identify a traitor, or it may have ϵ error in which the tracing algorithm either fails to identify a colluder, or it outputs an innocent user as a traitor.

Full Public Traceability. In Eurocrypt 2005, Chabanne, Phan and Pointcheval (CPP05) introduced the notion of *public traceability* where tracing is a black-box and publicly computable procedure. This is an interesting property that strengthens the overall security of the system as it separates the two tasks of key generation and tracing and allows all users to perform tracing on a pirate decoder. However their construction of fully public traceability system only worked for two users (it also required a new strong computational assumptions). The restriction to two user systems was due to a synthesis method that was inspired by a construction in [KY02] that used c-secure codes. However in using this approach to multi-user case, public traceability will be lost because tracing in c-secure codes is a private operation performed by a trusted centre. Authors raised the construction of a multi-receiver traitor tracing scheme with full public traceability as an interesting open problem.

Efficiency. Efficiency of broadcast encryption systems is often measured in terms of (i) *ciphertext rate* that captures the extra bandwidth that is required for transmission of the ciphertext, measured as the ratio of ciphertext length to plaintext length and, (ii) *computational efficiency* of performing encryption and decryption operations. The two most efficient systems from ciphertext rate view points are KY02 and CPP05 scheme, both with constant ciphertext rate. However both schemes require exponentiation of group elements followed by hashing for encryption and decryption of messages and so compared to symmetric key encryption systems, are inefficient.

In two party systems, Hybrid Tag-KEM/DEM provides a secure, efficient, and flexible method of encrypting messages using public key encryption systems for delivering the key information and using symmetric key systems for the encryption of the actual data. In this approach, Key Encapsulation Mechanism (KEM) encrypts a short random key in a header, and a Data Encapsulation Mechanism (DEM) uses this key to encrypt the message into a ciphertext using a symmetric encryption scheme. It is shown [AGKS05] that strong security for the ciphertext can be guaranteed with a semantically secure DEM and CCA security of KEM.

Our contributions. In this paper, we answer the open question of CCP05 by constructing a generic Hybrid Public Key Traitor Tracing that provides full public traceability and very efficient.

Our approach can be summarised as follows. We first extend the Tag KEM/DEM paradigm of hybrid encryption systems to multi-receiver scenario and define *Hybrid Tag-BroadcastKEM/DEM (Hybrid-PKBE)*. In Hybrid-PKBE the random key that is used for the encryption of a message is encrypted by *TBKEM*. This key is only extractable by authorized receivers. We define security of Hybrid-PKBE and prove a result similar to Hybrid Tag-KEM/DEM. We show that a Replayable CCA secure Hybrid-PKBE can be obtained from a Replayable CCA secure TBKEM and a semantically secure DEM. Replayable CCA (RCCA) security was introduced in [CKN03] to capture a security notion that is strictly weaker than CCA but sufficient for many practical applications.

Next we give a construction of a RCCA secure TBKEM (in above sense) from a CCA secure public key encryption (PKE) system and a *target collision free hash function*. Combining this TBKEM and a semantically secure DEM gives a secure Hybrid-PKBE. Moreover, we will show that this construction of Hybrid-PKBE support a tracing algorithm and hence it is a *Hybrid Public Key Traitor Tracing (Hybrid-PKTT)*. The tracing algorithm is black-box and only uses the public key of the system and so the system has *full public traceability*. This provides an elegant solution to the open problem of CPP05.

The hybrid construction makes the system very efficient. The ciphertext rate for long messages approaches one, and computational efficiency is obtained because of the decoupling of key encapsulation mechanism and data encryption module. The RCCA security of the system makes it *the first* construction of traitor tracing systems with this level of security (compared to previous constructions with constant ciphertext rate).

In the final section of the paper, we focus on increasing efficiency of the system. The communication overhead in the above system is a linear function of the size of the receiver group . Although for long messages and fixed size groups this gives a ciphertext rate of 1 (asymptotically), but it is desirable to reduce the size of the ciphertext overhead to make it more applicable for large groups. We use an approach similar to KY02 and CPP05, replacing collusion-secure codes with IPP codes. This reduces the length of the header and makes it *logarithmic in the number of users*. Interestingly, the composition preserves full public traceability as, unlike collusion-secure codes, IPP codes have public tracing algorithm.

2 Preliminaries

2.1 Public-key Broadcast Encryption (PKBE)

A public-key broadcast encryption (without revocation, without traceability) consists of the following algorithms:

- *Key generation algorithm* $\mathsf{PKBE.Gen}(1^\lambda, n) \to (pk, sk_1, \ldots, sk_n)$:
  An algorithm that generates a public-key pk and n private-keys $sk_1, sk_2, \ldots, sk_n$.
- *Encryption algorithm* $\mathsf{PKBE.Enc}_{pk}(m) \to c$:
  An algorithm that encrypts a message m into c by a public key pk.
- *Decryption algorithm* $\mathsf{PKBE.Dec}_{sk_i}(c) \to m$:
  An algorithm that decrypts a ciphertext c to m by a secret key sk_i.

The security against replayable adaptive chosen ciphertext attack (RCCA) is defined as follows. Let A_{pkbe} be a polytime oracle machine that plays the following game.

[GAME.PKBE]
Step 1. $(pk, sk_1, \ldots, sk_n) \leftarrow \mathsf{PKBE.Gen}(1^\lambda, n)$
Step 2. $(m_0, m_1) \leftarrow A_{\text{pkbe}}^{\mathcal{O}_{\text{pkbe}}}(pk)$
Step 3. $b \leftarrow \{0,1\}$, $c^* \leftarrow \mathsf{PKBE.Enc}_{pk}(m_b)$
Step 4. $\hat{b} \leftarrow A_{\text{pkbe}}^{\mathcal{O}_{\text{pkbe}}}(c^*)$

Here $\mathcal{O}_{\text{pkbe}}$ denotes the decryption oracle. A decryption query is of the form (i, c) where i is an integer $\in [1, n]$ and c is a ciphertext with a constraint that in Step 4 c must be different from c^*. To answer this query, the oracle calculates $\mathsf{PKBE.Dec}_{sk_i}(c) = m$.

In Step 2 the oracle outputs m. Moreover, in Step 4, it checks: if $m = m_0$ or $m = m_1$ then it outputs $\perp$, otherwise it outputs m. We define $\epsilon_{\text{pkbe-rcca},A_{\text{pkbe}}} = |Pr[\hat{b} = b] - 1/2|$, and $\epsilon_{\text{pkbe-rcca}} = \max(\epsilon_{\text{pkbe-rcca},A_{\text{pkbe}}})$, where the maximum is taken over all polytime machines A_{pkbe}. We say that a PKBE is RCCA secure if $\epsilon_{\text{pkbe-rcca}}$ is negligible in λ.

Public Key Traitor Tracing Scheme with Public Traceability. A public key traitor tracing (PKTT) scheme is a public key broadcast encryption with an extra algorithm, the *tracing algorithm*. This tracing algorithm takes as input the tracing information *trace-infor* and a pirate decoder $\mathcal{D}$. It outputs at least one of the users (called *traitors*) who have collaborated in producing the pirate decoder $\mathcal{D}$. In a c-traitor tracing scheme, we assume that there are at most c traitors who created the pirate decoder $\mathcal{D}$. The other obvious assumption on $\mathcal{D}$ is that $\mathcal{D}$ can effectively reverse the encryption, *i.e.* $\mathcal{D}(\mathsf{PKBE.Enc}_{pk}(m)) = m$, with high probability.

In general, tracing information *trace-infor* consists of some public information and some system secret parameters (for example, users' secret keys). Chabanne *et. al* introduced [CPP05] an interesting property of *public traceability* for traitor tracing schemes in which *trace-infor* consists of only public key of the system. Thus, it makes it possible for anyone to execute the tracing algorithm.

2.2 Data Encapsulation Mechanism (DEM)

A data encapsulation mechanism consists of the following algorithms:

- *Setup algorithm* $\mathsf{DEM.Setup}(1^\lambda) \to \mathcal{K}_D$:
  An algorithm that specifies the symmetric key space $\mathcal{K}_D$.
- *Encryption algorithm* $\mathsf{DEM.Enc}_{dk}(m) \to c$:
  An algorithm that encrypts m into c using a symmetric-key $dk \in \mathcal{K}_D$.
- *Decryption algorithm* $\mathsf{DEM.Dec}_{dk}(c) \to m$:
  An algorithm that decrypts c to m using a symmetric-key $dk \in \mathcal{K}_D$.

The IND (indistinguishable against passive attack) security of DEM is defined as follows. Let A_{dem} be a poly-time oracle machine that plays the following game.

[GAME.DEM]
Step 1. $\mathcal{K}_D \leftarrow \mathsf{DEM.Setup}(1^\lambda)$
Step 2. $(m_0, m_1) \leftarrow A_{\text{dem}}$
Step 3. $b \leftarrow \{0,1\}$, $dk \leftarrow \mathcal{K}_D$, $c \leftarrow \mathsf{DEM.Enc}_{dk}(m_b)$
Step 4. $\hat{b} \leftarrow A_{\text{dem}}(c)$

We define $\epsilon_{\text{dem},A_{\text{dem}}} = |Pr[\hat{b} = b] - 1/2|$, and $\epsilon_{\text{dem}} = \max(\epsilon_{\text{dem},A_{\text{dem}}})$, where the maximum is taken over all polytime machines A_{dem}. We say that a DEM is IND secure if ϵ_{dem} is negligible in λ.

2.3 Target Collision Resistant Hash Functions

A family $\mathcal{H} = \{H_k : \mathcal{A} \to \mathcal{B}\}_{k \in K}$ of keyed hash functions is *target collision resistant* if given a random $\tau \in \mathcal{A}$ and a random $H_k \in \mathcal{H}$, it is computationally infeasible to find $\tau' \in \mathcal{A}$ such that $\tau' \neq \tau$ and $H_k(\tau') = H_k(\tau)$. A random function H_k of $\mathcal{H}$ is called a *target collision-free* hash function. Associated with H_k, we define the quantity ϵ_{tch} as $\epsilon_{\text{tch}} = \max Pr[\tau' \in \mathcal{A}, \tau' \neq \tau, H_k(\tau') = H_k(\tau) : \tau \leftarrow \mathcal{A}, \tau' \leftarrow A_{\text{tch}}(\tau)]$ where the maximum is taken over all poly-time machines A_{tch}.

3 Generic Construction of Hybrid PKBE

In this section, we generalize Abe *et. al*'s [AGKS05] Tag-KEM/DEM construction of Hybrid-PKE. We show how to construct a hybrid public key broadcast encryption scheme (without revocation and traitor tracing) using two components: a Tag-BroadcastKEM and a DEM, and with this construction, we prove the following composition theorem,

(relaxed) CCA Tag-BKEM + semantic secure DEM → (relaxed) CCA Hybrid-PKBE.

3.1 Tag-Broadcast Key Encapsulation Mechanism (TBKEM)

A tag-broadcast key encapsulation mechanism consists of the following algorithms:

- *Key generation algorithm* $\mathsf{TBKEM.Gen}(1^\lambda, n) \to (pk, sk_1, \ldots, sk_n)$:
  An algorithm that generates a public-key pk and n private-keys $sk_1, sk_2, \ldots, sk_n$. It also specifies tag space $\mathcal{T}$ and encapsulated key space $\mathcal{K}_D$.
- *Key derivation algorithm* $\mathsf{TBKEM.Key(pk)} \to (\omega, dk)$:
  An algorithm that generates a one-time key dk and internal state information ω.
- *Encryption algorithm* $\mathsf{TBKEM.Enc}(\omega, \tau) \to \psi$:
  An algorithm that encrypts dk (embedded in ω) into ψ using a tag τ.
- *Decryption algorithm* $\mathsf{TBKEM.Dec}_{sk_i}(\psi, \tau) \to dk$:
  An algorithm that recovers dk from ψ and τ using one of the private-key sk_i. It may output a special symbol $\perp \notin \mathcal{K}_D$.

The RCCA security of TBKEM is defined as follows. Let A_{tbkem} be a poly-time oracle machine that plays the following game.

[GAME.TBKEM]
Step 1. $(pk, sk_1, \ldots, sk_n) \leftarrow \mathsf{TBKEM.Gen}(1^\lambda, n)$
Step 2. $(w, dk_1) \leftarrow \mathsf{TBKEM.Key}(pk), dk_0 \leftarrow \mathcal{K}_D, \delta \leftarrow \{0, 1\}$
Step 3. $\tau^* \leftarrow A_{\text{tbkem}}^{\mathcal{O}_{\text{tbkem}}}(pk, dk_\delta)$
Step 4. $\psi^* \leftarrow \mathsf{TBKEM.Enc}(w, \tau^*)$
Step 5. $\hat{\delta} \leftarrow A_{\text{tbkem}}^{\mathcal{O}_{\text{tbkem}}}(\psi^*)$

Here $\mathcal{O}_{\text{tbkem}}$ denotes the decryption oracle. A decryption query is of the form (i, ψ, τ) where i is an integer $\in [1, n]$, τ is a tag and ψ is a ciphertext with a constraint that in Step 5 (ψ, τ) must be different from (ψ^*, τ^*). To answer this query, the oracle calculates $\mathsf{TBKEM.Dec}_{sk_i}(\psi, \tau) = dk$. In Step 3 the oracle outputs dk. However, in Step 5, for a decryption query (i, ψ, τ), the oracle first checks if $\tau = \tau*$ and if $dk = dk_\delta$, then outputs the symbol $\perp \notin \mathcal{K}_D$, otherwise outputs dk. In other words, the sole difference between CCA security and RCCA security in the TBKEM model is that the adversary in the RCCA security is forbidden to ask (ψ, τ) where $\tau = \tau*$ and $dk = dk_\delta$. We remark that in TBKEM game, the adversary only knows dk_δ but not both dk_0 and dk_1.

We define $\epsilon_{\text{tbkem-rcca}, A_{\text{tbkem}}} = |Pr[\hat{\delta} = \delta] - 1/2|$, and $\epsilon_{\text{tbkem-rcca}} = \max(\epsilon_{\text{tbkem-rcca}, A_{\text{tbkem}}})$, where the maximum is taken over all polytime machines A_{tbkem}. We say that a TBKEM is RCCA secure if $\epsilon_{\text{tbkem-rcca}}$ is negligible in λ. We will see later that TBKEM with RCCA security and can be constructed from CCA-secure PKE and target collision-free hash function.

3.2 Hybrid Public Key Broadcast Encryption Scheme (Hybrid-PKBE)

The description of the hybrid public key broadcast encryption scheme is as follows.

- *Algorithm* $\mathsf{Hybrid\text{-}PKBE.Gen}(1^\lambda, n) \rightarrow (pk, sk_1, \ldots, sk_n)$:
  Call $\mathsf{TBKEM.Gen}(1^\lambda, n) \rightarrow (pk, sk_1, \ldots, sk_n)$.
- *Algorithm* $\mathsf{Hybrid\text{-}PKBE.Enc}_{pk}(m) \rightarrow c$: call $\mathsf{TBKEM.Key}(pk) \rightarrow (\omega, dk)$, $\mathsf{DEM.Enc}_{dk}(m) \rightarrow \tau$, $\mathsf{TBKEM.Enc}(\omega, \tau) \rightarrow \psi$ and set $c = (\psi, \tau)$.
- *Algorithm* $\mathsf{Hybrid\text{-}PKBE.Dec}_{sk_i}(c) \rightarrow m$: suppose $c = (\psi, \tau)$.
  Call $\mathsf{TBKEM.Dec}_{sk_i}(\psi, \tau) \rightarrow dk$ and $\mathsf{DEM.Dec}_{dk}(\tau) \rightarrow m$.

Theorem 1. *Hybrid-PKBE is RCCA secure under the assumptions that TBKEM is RCCA secure and DEM is IND secure:* $\epsilon_{pkbe\text{-}rcca} \leq 2\,\epsilon_{tbkem\text{-}rcca} + \epsilon_{dem}$.

Proof. We prove the Hybrid-PKBE is RCCA secure using a sequence of games.

Game 0. Let A_{pkbe} be an adversary that plays the following attack game in the definition of RCCA security (section 2.1).

[$\mathsf{GAME.PKBE}_0$]
Step 1. $(pk, sk_1, \ldots, sk_n) \leftarrow \mathsf{TBKEM.Gen}(1^\lambda, n)$
Step 2. $(m_0, m_1) \leftarrow A_{\text{pkbe}}^{\mathcal{O}_{\text{pkbe}}}(pk)$
Step 3. $b \leftarrow \{0, 1\}$, $(\omega, dk^*) \leftarrow \mathsf{TBKEM.Key}(pk)$,
$\tau^* \leftarrow \mathsf{DEM.Enc}_{dk^*}(m_b)$, $\psi^* \leftarrow \mathsf{TBKEM.Enc}(\omega, \tau^*)$
Step 4. $\hat{b} \leftarrow A_{\text{pkbe}}^{\mathcal{O}_{\text{pkbe}}}(\psi^*, \tau^*)$

Let X_0 be the event that $b = \hat{b}$ in the above game then $\epsilon_{\text{pkbe-rcca}, A_{\text{pkbe}}} = |Pr[X_0] - 1/2|$.
In Step 4, a decryption query is of the form (i, ψ, τ) where $(\psi, \tau) \neq (\psi^*, \tau^*)$. To answer this query, the oracle executes

$\mathsf{Hybrid\text{-}PKBE.Dec}_{sk_i}(\psi, \tau)$:
1. $\mathsf{TBKEM.Dec}_{sk_i}(\psi, \tau) = dk$;
2. $\mathsf{DEM.Dec}_{dk}(\tau) = m$.

If $m = m_0$ or $m = m_1$ then $\mathcal{O}_{\text{pkbe}}$ outputs $\perp$, otherwise, it outputs m.

Game 1. We modify Game 0 in Step 3, instead of encrypting m_b using a key produced by $\mathsf{TBKEM.Key}$, we encrypt m_b using a random key.

[$\mathsf{GAME.PKBE}_1$]
Step 1. $(pk, sk_1, \ldots, sk_n) \leftarrow \mathsf{TBKEM.Gen}(1^\lambda, n)$
Step 2. $(m_0, m_1) \leftarrow A_{\text{pkbe}}^{\mathcal{O}_{\text{pkbe}}}(pk)$
Step 3. $b \leftarrow \{0, 1\}$, $(\omega, dk_1) \leftarrow \mathsf{TBKEM.Key}(pk)$,
$dk_0 \leftarrow \mathcal{K}_D$, $\tau^* \leftarrow \mathsf{DEM.Enc}_{dk_0}(m_b)$, $\psi^* \leftarrow \mathsf{TBKEM.Enc}(\omega, \tau^*)$
Step 4. $\hat{b} \leftarrow A_{\text{pkbe}}^{\mathcal{O}_{\text{pkbe}}}(\psi^*, \tau^*)$

Let X_1 be the event that $b = \hat{b}$ in Game 1.
Claim 1. $|Pr[X_0] - Pr[X_1]| \leq 2\epsilon_{\text{tbkem-rcca}}$.
The proof of this claim can be found in the full version [PST06].

Claim 2. $|Pr[X_1] - 1/2| \leq \epsilon_{\text{dem}}$.
The proof of this claim can be found in the full version [PST06].
Finally, we have

$$\begin{aligned}\epsilon_{\text{pkbe-rcca},A_{\text{pkbe}}} &= |Pr[X_0] - 1/2| \\ &\leq |Pr[X_0] - Pr[X_1]| + |Pr[X_1] - 1/2| \\ &\leq 2\,\epsilon_{\text{tbkem-rcca}} + \epsilon_{\text{dem}},\end{aligned}$$

thus, $\epsilon_{\text{pkbe-rcca}} \leq 2\,\epsilon_{\text{tbkem-rcca}} + \epsilon_{\text{dem}}$, where $\epsilon_{\text{tbkem-rcca}}$ and ϵ_{dem} are assumed to be negligible. ∎

Remark. We can also prove that

CCA Tag-BKEM + semantic secure DEM $\rightarrow$ CCA Hybrid-PKBE.

This is a more natural generalization of the result of Abe *et. al* [AGKS05]. However, we don't know how to construct, in a simple manner, a CCA TBKEM from PKE only. This is the reason we introduce the notion of RCCA TBKEM which will be constructed from CCA PKE.

4 Construction of a Basic Hybrid-PKTT

4.1 How to Construct TBKEM from PKE

In this section, we show a generic construction of a TBKEM that is RCCA secure from a CCA secure PKE and a target collision-free hash function H. The construction is as follows.

- *Algorithm* $\mathsf{TBKEM.Gen}(1^\lambda, n) \rightarrow (pk, sk_1, \ldots, sk_n)$:
  For each $i = 1, \ldots, n$, call $\mathsf{PKE.Gen}(1^\lambda) \rightarrow (pk_i, sk_i)$. Set $pk = (pk_1, \ldots, pk_n)$.
- *Algorithm* $\mathsf{TBKEM.Key}(pk) \rightarrow (\omega, dk)$:
  Choose a random dk and set $\omega = pk||dk$.
- *Algorithm* $\mathsf{TBKEM.Enc}(\omega, \tau) \rightarrow \psi$ (where $\omega = pk||dk$):
  Compute $h = H(\tau)$. Call $\mathsf{PKE.Enc}_{pk_i}(dk||h) \rightarrow \sigma_i$ for $i = 1, \ldots, n$. Output $\psi = (\sigma_1, \ldots, \sigma_n)$.
- *Algorithm* $\mathsf{TBKEM.Dec}_{sk_i}(\psi, \tau) \rightarrow dk$ or $\perp$ (where $\psi = (\sigma_1, \ldots, \sigma_n)$):
  Call $\mathsf{PKE.Dec}_{sk_i}(\sigma_i) \rightarrow dk||h$. If $h = H(\tau)$, return dk. Otherwise, return $\perp$.

Theorem 2. TBKEM *is RCCA secure under the assumptions that* PKE *is CCA secure and H is target collision-free:* $\epsilon_{tbkem\text{-}rcca} \leq \epsilon_{tch} + n\,\epsilon_{pke\text{-}cca}$.

We use the same technique as used for proof of Theorem 1, namely Game approach, to prove this theorem. The proof can be found in the full version [PST06].

4.2 Basic Hybrid-PKTT

In section 3, we show how to construct a Hybrid-PKBE using a TBKEM and a DEM. In section 4, we show how to construct a TBKEM using a PKE and a target collision-free hash function. In this section, we combine the above two constructions. Thus, from a DEM, a PKE and a target collision-free hash function, we can construct a Hybrid-PKBE. We show that this Hybrid-PKBE is very special: it is a Hybrid-PKTT with full-public-traceability. Therefore, we have the following composition:

CCA PKE + semantic secure DEM + target collision-free hash function
$\rightarrow$ RCCA Hybrid-PKTT with full-public-traceability.

The description of the Hybrid-PKTT is as follows.

Algorithm $\mathsf{Hybrid\text{-}PKTT.Gen}(1^\lambda, n) \rightarrow (pk, sk_1, \ldots, sk_n)$:
For each $i = 1, \ldots, n$, call $\mathsf{PKE.Gen}(1^\lambda) \rightarrow (pk_i, sk_i)$. Set $pk = (pk_1, \ldots, pk_n)$.

Algorithm $\mathsf{Hybrid\text{-}PKTT.Enc}_{pk}(m) \rightarrow c$:
Choose a random dk and call $\mathsf{DEM.Enc}_{dk}(m) \rightarrow \tau$. Compute $h = H(\tau)$ and for each $i = 1, \ldots, n$, call $\mathsf{PKE.Enc}_{pk_i}(dk||h) \rightarrow \sigma_i$. Output $c = (\sigma_1, \ldots, \sigma_n, \tau)$.

Algorithm $\mathsf{Hybrid\text{-}PKTT.Dec}_{sk_i}(c) \rightarrow m$ or $\perp$, where $c = (\sigma_1, \ldots, \sigma_n, \tau)$:
Call $\mathsf{PKE.Dec}_{sk_i}(\sigma_i) \rightarrow dk||h$. If $h \neq H(\tau)$, return $\perp$. Otherwise, call $\mathsf{DEM.Dec}_{dk}(\tau) \rightarrow m$ and output m.

Algorithm $\mathsf{Hybrid\text{-}PKTT.Public\text{-}Trace}(pk, \mathcal{D})$: A black-box traitor tracing algorithm that can be executed by anyone using the public-key to find a traitor who had created pirate decoder.

- Choose random dk, m, and call $\mathsf{Hybrid\text{-}PKTT.Enc}_{pk}(m) \rightarrow (\sigma_1, \ldots, \sigma_n, \tau)$.
- For each $i = 1, \ldots, n$, choose random $d'_i \neq dk||h$ so that d'_i has the same length as $dk||h$ and call $\mathsf{PKE.Enc}_{pk_i}(d'_i) \rightarrow \sigma'_i$; modify the ciphertext and give them to $\mathcal{D}$ and check if $\mathcal{D}(\sigma_1, \sigma_2, \ldots, \sigma_n, \tau) \stackrel{?}{=} m$, $\mathcal{D}(\sigma'_1, \sigma_2, \ldots, \sigma_n, \tau) \stackrel{?}{=} m$, $\mathcal{D}(\sigma'_1, \sigma'_2, \ldots, \sigma_n, \tau) \stackrel{?}{=} m, \ldots, \mathcal{D}(\sigma'_1, \sigma'_2, \ldots, \sigma'_n, \tau) \stackrel{?}{=} m$.
- Calculate the following probabilities
$p_0 = Pr[\mathcal{D}(\sigma_1, \sigma_2, \ldots, \sigma_n, \tau) = m], p_1 = Pr[\mathcal{D}(\sigma'_1, \sigma_2, \ldots, \sigma_n, \tau) = m]$,
$p_2 = Pr[\mathcal{D}(\sigma'_1, \sigma'_2, \ldots, \sigma_n, \tau) = m], \ldots, p_n = Pr[\mathcal{D}(\sigma'_1, \sigma'_2, \ldots, \sigma'_n, \tau) = m]$.
We assume that $\mathcal{D}$ is a usable decoder so p_0 is not negligible (indeed $p_0 \approx 1$), and obviously, $p_n \approx 0$. So there must exist i such that $|p_i - p_{i-1}|$ is not negligible, in this case, output i as a traitor.

The above scheme, without tracing algorithm, is a Hybrid-PKBE. The security of encryption of above scheme, denoted by $\epsilon_{\text{pktt-rcca}}$, is evidently independent of the tracing algorithm. Therefore, following theorem is a corollary of Theorem 1 and Theorem 2.

Theorem 3. *Hybrid-PKTT is RCCA secure under the assumptions that PKE is CCA secure, H is target collision-free, and DEM is IND secure:* $\epsilon_{pktt\text{-}rcca} \leq 2(\epsilon_{tch} + n\,\epsilon_{pke\text{-}cca}) + \epsilon_{dem}$.

We obtain thus a generic construction of RCCA Hybrid-PKTT with public traceability. One could doubt about the RCCA model. This is, theoretically, a weaker model than the standard CCA. However, it seem to be sufficiently secure for most practical purposes, as showed in [CKN03]. Moreover, in [CKN03], the authors distinguish two types of RCCA: secretly detectable RCCA (sd-RCCA) and publicly detectable RCCA (pd-RCCA). The former means that the detection of a ciphertext, whose underlying plaintext is identical to the underlying plaintext of the challenge, requires secret information and the latter, which is much less restricted, means that such a detection can be done from the public information only. The RCCA used in our proof is publicly detectable RCCA. In fact, if $c = (\sigma_1^*, \ldots, \sigma_n^*, \tau^*)$ is a valid ciphertext outputed by $\mathsf{Hybrid\text{-}PKTT.Enc}_{pk}(m)$, then anyone can choose random $\sigma_1, \ldots, \sigma_{i-1}, \sigma_{i+1}, \ldots, \sigma_n$ to construct a new ciphertext $c' = (\sigma_1, \ldots, \sigma_{i-1}, \sigma_i^*, \sigma_{i+1}, \ldots, \sigma_n, \tau^*)$ so that the decryption of c' under the key sk_i gives back the original message m, *i.e.* $\mathsf{Hybrid\text{-}PKTT.Enc}_{sk_i}(c) = m$.

We call the above scheme the *basic* Hybrid-PKTT scheme. In this scheme, a ciphertext c consists of a ciphertext body τ and a ciphertext header $(\sigma_1, \ldots, \sigma_n)$. The ciphertext body τ has approximately the same size as the message m. We can say somewhat that the transmission rate of the above scheme is asymptotically 1 because the length of the message to be encrypted could be arbitrary.

The two practically inconveniences of the above scheme is that the size of the ciphertext header is linear in the number of users and that the cost of reduction in the security proof is linear on the number of user.

We can overcome both these problems by using the method in [KY02, CPP05] of using a convenient code, namely the IPP codes. Remark that if we use the collusion secure code [BS98], the scheme does not, unfortunately, support the public traceability anymore. The reason is that, in collusion secure code, for tracing back a traitor from a codeword, one has use the secret permutation in the construction of the code and therefore, only the center can do it. Why we don't use the IPP code with the schemes in [CPP05, KY02]? The obstacle is that the basic scheme in [CPP05, KY02] supports only 2 users and there does not exist binary IPP codes. Fortunately, we can combine our basic scheme above with a q-ary IPP code for any $q \geq 3$. As the tracing procedure in IPP code does not require any secret information, the combined scheme supports full public traceability.

We will present the hybrid scheme $\mathsf{Hybrid_{IPP}\text{-}PKTT}$ based on IPP codes in the next section. We remark that in order to use q-ary code, the parameter n in the above basic scheme must be set to $n = q$. Since we can choose q as small as $q = 3$, the number of users in each basic scheme is small ($n = q = 3$), the ciphertext header in the new scheme become small, the cost of reduction in the security proof become constant, and this make the new scheme $\mathsf{Hybrid_{IPP}\text{-}PKTT}$ become a very efficient scheme.

5 Hybrid-PKTT Based on IPP Codes

This is the most interesting section of our paper. We will show how to combine the basic scheme in the previous section with a q-ary c-IPP code to construct an efficient hybrid traitor tracing scheme with full-public-traceability.

5.1 IPP Codes

Let $\mathcal{Q}$ be an alphabet set containing q symbols. If $C = \{w_1, w_2, \ldots, w_N\} \subset \mathcal{Q}^\ell$, then C is called a q-ary code of size N and length ℓ. Each $w_i \in C$ is called a codeword and we write $w_i = (w_{i,1}, w_{i,2}, \ldots, w_{i,\ell})$ where $w_{i,j} \in \mathcal{Q}$ is called the jth component of the codeword w_i.

We define *descendants* of a subset of codewords as follows. Let $X \subset C$ and $u = (u_1, u_2, \ldots, u_\ell) \in \mathcal{Q}^\ell$. The word u is called a descendant of X if for any $1 \leq j \leq \ell$, the jth component u_j of u is equal to a jth component of a codeword in X. In this case, codewords in X are called *parent codewords* of u. For example, $(3, 2, 1, 3)$ is a descendant of three codewords $(3,1,1,2)$, $(1,2,1,3)$ and $(2,2,2,2)$. We denote by $\mathsf{Desc}(X)$ the set of all descendants of X. For a positive integer c, denote by $\mathsf{Desc}_c(C)$ the set of all descendants of subsets of up to c codewords. Codes with identifiable parent property (IPP codes) are defined as follows.

Definition 1. *A code C is called c-IPP if, for any $u \in \mathsf{Desc}_c(C)$, there exists $w \in C$ such that for any $X \subset C$, if $|X| \leq c$ and $u \in \mathsf{Desc}(X)$ then $w \in X$.*

In a c-IPP code, given a descendant $u \in \mathsf{Desc}_c(C)$, we can always identify at least one of its parent codewords. Binary c-IPP codes (with more than two codewords) do not exists, thus in any c-IPP code, the alphabet size $q \geq 3$. There are many constructions [SW98, SSW01, TM05, TS06] of c-IPP codes. We remark that even both c-IPP codes and c-collusion secure codes can be constructed with large number of codewords of similar length, there are three major differences between them:

- In collusion secure codes, there is an error parameter that specifies the probability that the tracing algorithm fails to output the correct parent codeword, however, in IPP code, there is not such error, thus the tracing is *error-free* and a correct traitor is always identified.
- Collusion secure codes use secret permutation and thus the tracing algorithm cannot be made public, whereas, in IPP codes, everything is public.
- Known collusion secure codes are binary codes, whereas, nontrivial IPP codes have alphabet size at least 3. (Binary IPP code has at most two codewords).

5.2 Description of $\mathsf{Hybrid}_{\mathrm{IPP}}$-PKTT

If a q-ary c-IPP code of size N and length ℓ is used, then constructing ℓ instances of Hybrid-PKTT, in each instance of Hybrid-PKTT set the parameter $n = q$, we have a new public key traitor tracing called $\mathsf{Hybrid}_{\mathrm{IPP}}$-PKTT. In this new scheme, there are $q\ell$ public keys and N users, each user holds ℓ secret keys. The formal construction follows:

Let $C = \{\omega_1, \ldots, \omega_N\}$ be a q-ary c-IPP code that allows collusion of up to c users. The N-user $\mathsf{Hybrid}_{\mathrm{IPP}}$-PKTT scheme is a combination of ℓ basic Hybrid-PKTT schemes $S_1, S_2, \ldots, S_\ell$, each scheme S_i is for q users:

Setup: Given the security parameters λ and c:
For each $j = 1, \ldots, \ell$, call the algorithm Hybrid-PKBE.Gen$(1^\lambda, q)$ to generate an encryption key pk_j and q decryption keys $sk_{j,1}, \ldots, sk_{j,q}$ for the q-user system S_j.

Public key: The tuple $(pk_i)_{i=1,\ldots,\ell}$ and the code C.

Private key of each user: User i (for $i = 1, \ldots, N$) is associated to a codeword w_i in C and the corresponding ℓ-tuple key $sk_{1,w_{i,1}}, sk_{2,w_{i,2}}, \ldots, sk_{\ell,w_{i,\ell}}$ where $w_{i,j} \in \mathcal{Q} = \{1, 2, \ldots, q\}$ is the symbol at the jth position of the codeword w_i.

Encryption algorithm: The plaintext space of the ℓ-key system is $\mathcal{M}^\ell$. On input $(m_1, m_2, \ldots, m_\ell)$, the encryption algorithm outputs the ciphertext $(c_1, c_2, \ldots, c_\ell)$, where $c_j = \text{Hybrid-PKBE.Enc}_{pk_j}(m_j) = (\sigma_{j,1}, \ldots, \sigma_{j,q}, \tau_j)$.

Decryption Algorithm: On the ciphertext $(c_1, c_2, \ldots, c_\ell)$, user i uses his secret key to compute $m_j = \text{Hybrid-PKBE.Dec}_{sk_{j,w_{i,j}}}(c_j)$.

Public Tracing Algorithm: For each $j = 1, \ldots, \ell$, fix $\ell - 1$ valid ciphertexts $c_1, \ldots, c_{j-1}, c_{j+1}, \ldots, c_\ell$ and use the tracing algorithm of the instance S_j to trace a traitor $u_j \in \mathcal{Q} = \{1, 2, \ldots, q\}$ in this instance. From the descendant word $(u_1, \ldots, u_\ell) \in \mathcal{Q}^\ell$, identify one of its parent codewords. The user associated with this codeword is a traitor.

Efficiency: The ciphertext contains two parts: ciphertext body $(\tau_1, \ldots, \tau_\ell)$ and ciphertext header $(\sigma_{j,1}, \ldots, \sigma_{j,q})_{j=1,\ldots,\ell}$. The ciphertext body has approximately the same size as the message and the ciphertext header is proportional to $q\ell$. We can choose IPP code with small alphabet size such as $q = 3$ and the code length ℓ is a logarithmic function of the code size N. Thus, the ciphertext header has fixed size and very small compared to the message size.

For the security analysis, one could use the following assumption, from [KY02]: the *threshold assumption* says that a pirate-decoder that just returns correctly a fraction p of a plaintext of length λ where $1-p$ is a non-negligible function in λ, is useless. However, as already mentioned in [KY02], by employing an all-or-nothing transform [Riv97, CDHKS00], this assumption is not necessary.

Proposition 1. *The leak of the secret keys in the $(\ell - 1)$ q-user systems of ℓ q-user systems does not affect the security of the remained q-user system.*

The proof of this proposition is quite similar to the corresponding ones in [KY02, CPP05] and can be found in the full version [PST06].

This proposition combines with the fact that C is a c-IPP code, leads to following corollary:

Corollary 1. *The above scheme is a N-user, c-traitor tracing scheme with full-public traceability.*

We give an example of a concrete scheme in the full version [PST06]. This scheme is inspired from the Cramer-Shoup scheme [CS03].

6 Conclusion

Motivated by an open problem proposed in CPP05, we extended Tag-KEM/DEM paradigm of hybrid encryption to multi-receiver scenario and constructed a hybrid traitor tracing scheme that has the following properties

- full public traceability, and thus is a comprehensive solution to the open problem of CPP05;
- blackbox traitor tracing algorithm that is *error-free* and always can identify correctly at least one traitor. This is an important advantage of our scheme over [KY02, CPP05], because these schemes use collusion secure code and the tracing algorithm of collusion secure code has error;
- it is a generic construction and provides significant improvement in terms of security and efficiency and this is without resorting to new computational assumptions. In fact security is based on the assumptions underlying security of the public key and symmetric key encryption systems used in KEM and DEM, respectively. In

comparison the scheme in [CPP05] (which supports local public traceability against passive attack only) is based on new assumptions which are all stronger than the standard Bilinear DDH assumption;
- the generic construction provides the following powerful composition

$$\text{IPP code} + \text{CCA PKE} + \text{semantic secure DEM} + \text{target collision-free hash function} \\ \rightarrow \text{(relaxed) CCA Hybrid}_{IPP}\text{-PKTT with full-public-traceability.}$$

Our security proofs are in replayable CCA model. Although all previous schemes with constant transmission rate achieve only semantic security against passive attack and so our scheme has much stronger security level, it is a quite interesting question if similar results can be derived if CCA model is assumed. Finally, combining revocation and public traceability to the Hybrid Traitor tracing scheme is an important open problem.

References

[AGKS05] M. Abe, R. Gennaro, K. Kurosawa and V. Shoup, Tag-KEM/DEM: A New Framework for Hybrid Encryption and A New Analysis of Kurosawa-Desmedt KEM, EUROCRYPT'05, LNCS 3494, 128–146, 2005.

[BS98] D. Boneh and J. Shaw, Collusion secure fingerprinting for digital data, *IEEE Transactions on Information Theory* **44** (1998), 1897–1905.

[CDHKS00] R. Canetti, Y. Dodis, S. Halevi, E. Kushilevitz and A. Sahai, Exposure-resilient functions and all-or-nothing transforms, EUROCRYPT'00, LNCS 1807, 453–469, 2000.

[CKN03] R. Canetti, H. Krawczyk and J.B. Nielsen, Relaxing chosen ciphertext security, CRYPTO'03, LNCS 2729, 565–582, 2003.

[CPP05] H. Chabanne, D.H. Phan and D. Pointcheval, Public traceability in traitor tracing schemes, EUROCRYPT'05, LNCS 3494, 542–558, 2005.

[CFN94] B. Chor, A. Fiat and M. Naor, Tracing traitor, CRYPTO'94, LNCS 839, 257–270, 1994.

[CS03] R. Cramer and V. Shoup, Design and analysis of practical public-key encryption schemes secure against adaptive chosen ciphertext attack, *SIAM J. of Computing* **33** (2003), 167–226.

[KY02] A. Kiayias and M. Yung, Traitor tracing with constant transmission rate, EUROCRYPT'02, LNCS 2332, 450–465, 2002.

[PST06] D.H. Phan, R. Safavi-Naini and D. Tonien, Generic construction of hybrid public key traitor tracing with full-public-traceability. Full version available from `http://www.di.ens.fr/users/phan/`.

[Riv97] R. Rivest, All-or-nothing encryption and the package transform, FSE'97, LNCS 1267, 210–218, 1997.

[SSW01] J.N. Staddon, D.R. Stinson and R. Wei, Combinatorial properties of frameproof and traceability codes, *IEEE Transactions on Information Theory* **47** (2001), 1042–1049.

[SW98] D.R. Stinson and R. Wei, Combinatorial properties and constructions of traceability schemes and frameproof codes, *SIAM Journal on Discrete Mathematics* **11** (1998), 41–53.

[TM05] Tran van Trung and S. Martinosyan, New constructions for IPP codes, *Designs, Codes and Cryptography* **35** (2005), 227–239.

[TS06] D. Tonien and R. Safavi-Naini, Recursive constructions of secure codes and hash families using difference function families, *J. of Combinatorial Theory* A **113** (4)(2006), 664–674.

An Adaptively Secure Mix-Net Without Erasures

Douglas Wikström[1,*] and Jens Groth[2,**]

[1] ETH Zürich, Department of Computer Science
douglas@inf.ethz.ch
[2] UCLA, Computer Science Department
jg@cs.ucla.edu

Abstract. We construct the first mix-net that is secure against *adaptive* adversaries corrupting any minority of the mix-servers and any set of senders. The mix-net is based on the Paillier cryptosystem and analyzed in the universal composability model *without erasures* under the decisional composite residuosity assumption, the strong RSA-assumption, and the discrete logarithm assumption. We assume the existence of ideal functionalities for a bulletin board, key generation, and coin-flipping.

1 Introduction

Suppose a set of senders $S_1, \ldots, S_N$ each have an input m_i, and want to compute the sorted list $(m_{\pi(1)}, \ldots, m_{\pi(N)})$ of messages, but keep the identity of the sender of any particular message m_i secret. A trusted party can provide the service required by the senders. First it collects all messages. Then it sorts the inputs and outputs the result. A protocol, i.e., a list of machines $M_1, \ldots, M_k$, that emulates the service of the trusted party as described above is called a *mix-net*, and the parties $M_1, \ldots, M_k$ are referred to as *mix-servers*. The notion of a mix-net was introduced by Chaum [3].

Many mix-net constructions are proposed in the literature without security proofs, and several of these constructions have in fact been broken. The first rigorous definition of security of a mix-net was given by Abe and Imai [1], but they did not provide any construction that satisfies their definition. Wikström [16] gives the first definition of a universally composable (UC) mix-net, and also the first construction with a complete security proof. He recently presented a more efficient UC-secure scheme [17].

An important tool in the construction of a mix-net is a so called "proof of a shuffle". This allows a mix-server to prove that it behaved as expected without leaking knowledge. The first efficient methods to achieve this were given independently by Neff [12] and Furukawa and Sako [8]. Subsequently, other authors improved and complemented these methods, e.g. [9, 7, 17]. Our results seem largely independent of the method used, but for concreteness we use the method presented in [17].

---

* Part of the work done while at Royal Institute of Technology (KTH), Stockholm, Sweden.

** Supported by NSF Cybertrust ITR grant No. 0456717. Part of the work done while at Cryptomathic, Denmark and BRICS, Dept. of Computer Science, University of Aarhus, Denmark.

M. Bugliesi et al. (Eds.): ICALP 2006, Part II, LNCS 4052, pp. 276–287, 2006.

1.1 Our Contribution

All previous works consider a static adversary that decides which parties to corrupt before the protocol is executed. We provide the first efficient mix-net that is secure against an *adaptive* adversary. The problem of constructing such a scheme has been an open problem since the notion of mix-nets was proposed by Chaum [3] two decades ago. The model we consider is the *non-erasure* model, i.e., every state transition of a party is stored on a special history tape that is handed to the adversary upon corruption. It is well known that it is hard to prove the security of protocols in this model, even more so for efficient protocols. Our analysis is novel in that we show that a mix-net can be proved UC-secure even if the zero-knowledge proofs of knowledge of correct re-encryption-permutations computed by the mix-servers are not zero-knowledge against adaptive adversaries and not even straight-line extractable as is often believed to be necessary in the UC-setting. We prove our claims in the full version of this paper.

1.2 Notation

We use $S_1, \ldots, S_N$ and $M_1, \ldots, M_k$ to denote the senders and the mix-servers. All participants are modeled as interactive Turing machines with a history tape where all state transitions are recorded. Upon corruption the entire execution history is given to the adversary. We abuse notation and use S_i and M_j to denote both the machines themselves and their identity. We denote by $k' = \lceil (k+1)/2 \rceil$ the number of mix-servers needed for majority. We denote the set of permutations of N elements by Σ_N. The main security parameter is κ. The zero-knowledge proofs invoked as subprotocols use two additional security parameters, κ_c and κ_r that determine the number of bits in challenges and the statistical distance between a simulated proof and a real proof. We denote by Sort the algorithm that given a list of strings as input outputs the same set of strings in lexicographical order.

1.3 Cryptographic Model

We use the UC-framework [2], but our notation differs from [2] in that we introduce an explicit "communication model" $\mathcal{C}_\mathcal{I}$ that acts as a router of messages between the parties. We define $\mathcal{M}_l^*$ to be the set of adaptive adversaries that corrupt less than l out of k parties of the mix-server type, and arbitrarily many parties of the sender type. We assume an ideal authenticated bulletin board functionality $\mathcal{F}_{\mathrm{BB}}$. All parties can write to it, but no party can erase any message from it. The adversary can prevent any party from reading or writing. We also need an ideal coin-flipping functionality $\mathcal{F}_{\mathrm{CF}}$ at some points in the protocol. It simply outputs random coins when asked to do so. We take the liberty of interpreting random strings as elements in groups, e.g., in the subgroup $\mathrm{QR}_{\mathbf{N}}$.

We use the discrete logarithm (DL) assumption for safe primes $p = 2q + 1$, which says that it is infeasible to compute a discrete logarithm of a random element $y \in G_q$, where G_q is the group of squares modulo p. We use the decision composite residuosity assumption (DCR), which says that given a product n

of two random safe primes of the same size, it is infeasible to distinguish the uniform distribution on elements in $\mathbb{Z}_{n^2}^*$ from the uniform distribution on nth residues in $\mathbb{Z}_{n^2}^*$. We use the strong RSA-assumption (SRSA) which says that given a product $\mathbf{N}$ of two random safe primes, and $\mathbf{g} \in \mathbb{Z}_{\mathbf{N}}^*$, it is infeasible to compute $(\mathbf{b}, \eta)$ such that $\mathbf{b}^\eta = \mathbf{g} \bmod \mathbf{N}$ and $\eta \neq \pm 1$.

1.4 Distributed Paillier

We use a combination of two threshold versions of the Paillier [13] cryptosystem introduced by Lysyanskaya and Peikert [10] and Damgård et al. [5], and also modify the scheme slightly. On input 1^κ the key generator $\mathsf{KG}^{\mathsf{pai}}$ chooses two $\kappa/2$-bit safe primes p and q randomly and defines the public key $\mathsf{n} = \mathsf{pq}$. We define $\mathsf{g} = \mathsf{n} + 1$ and $\mathsf{f} = (\mathsf{p} - 1)(\mathsf{q} - 1)/4$. Then it chooses a private key d under the restriction $\mathsf{d} = 0 \bmod \mathsf{f}$ and $\mathsf{d} = 1 \bmod \mathsf{n}$ and outputs (n, d). Note that $\mathsf{g}^m = 1 + m\mathsf{n} \bmod \mathsf{n}^2$. We define $L(u) = (u - 1)/\mathsf{n}$ and have $L(\mathsf{g}^m) = m$. To encrypt a message $m \in \mathbb{Z}_\mathsf{n}$ a random $r \in \mathbb{Z}_\mathsf{n}^*$ is chosen and the ciphertext is defined by $u = E_\mathsf{n}(m, r) = \mathsf{g}^m r^{2\mathsf{n}} \bmod \mathsf{n}^2$. The decryption algorithm is defined $D_\mathsf{d}(u) = L(u^\mathsf{d} \bmod \mathsf{n}^2)$. Let $\mathsf{g_f} \in \mathbb{Z}_{\mathsf{n}^2}^*$ be an element of order f. Then there exists a $2\mathsf{n}$th root r_f of $\mathsf{g_f}$ modulo n, and an alternative encryption algorithm is $E_{\mathsf{g_f},\mathsf{n}}(m, s) = \mathsf{g}^m \mathsf{g_f}^s = \mathsf{g}^m (r_\mathsf{f}^s)^{2\mathsf{n}} \bmod \mathsf{n}^2$, where s is chosen randomly in $[0, 2^{\kappa+\kappa_r} - 1]$. Here κ_r is an additional security parameter that is large enough to make $2^{-\kappa_r}$ negligible. It should always be clear from the context what is meant.

The cryptosystem is homomorphic, i.e., $E_\mathsf{n}(m_1)E_\mathsf{n}(m_2) = E_\mathsf{n}(m_1 + m_2)$. As a consequence it is possible to re-encrypt a ciphertext u using randomness $s \in \mathbb{Z}_\mathsf{n}^*$ by computing $uE_\mathsf{n}(0, s) = E_\mathsf{n}(m, rs)$, or alternatively using randomness $s \in [0, 2^{\kappa+\kappa_r} - 1]$ as $uE_{\mathsf{g_f},\mathsf{n}}(0, s) = E_\mathsf{n}(m, rr_\mathsf{f}^s)$. Furthermore, given a ciphertext $K_1 = E_\mathsf{n}(1, R_1) = \mathsf{g}R_1^{2\mathsf{n}} \bmod \mathsf{n}^2$ of 1 an alternative way to encrypt a message m is to compute $E_{K_1,\mathsf{n}}(m, r) = K_1^m r^{2\mathsf{n}} \bmod \mathsf{n}^2$.

The scheme is turned into a distributed cryptosystem with k parties of which a majority k' are needed for decryption as follows. Let g and h be two random generators of a subgroup G_q of prime order q of $\mathbb{Z}_{2q+1}^*$ for a random prime $2q+1$ such that $\log_2 q > 2\kappa + \kappa_r$. Let v be a generator of the group of squares $\mathrm{QR}_{\mathsf{n}^2}$. Each party M_j is assigned a random element $\mathsf{d}_j \in [0, 2^{2\kappa+\kappa_r} - 1]$ under the restriction that $\mathsf{d} = \sum_{j=1}^k \mathsf{d}_j \bmod \mathsf{nf}$, and define $\mathsf{v}_j = \mathsf{v}^{\mathsf{d}_j} \bmod \mathsf{n}^2$. We also compute a Shamir-secret sharing [15] of each d_j to allow reconstruction of this value. More precisely we choose for each j a random $(k' - 1)$-degree polynomial f_j over $\mathbb{Z}_q$ under the restriction that $f_j(0) = \mathsf{d}_j$, and define $\mathsf{d}_{j,l} = f_j(l) \bmod q$. A Pedersen [14] commitment $F_{j,l} = g^{\mathsf{d}_{j,l}} h^{\mathsf{t}_{j,l}}$ of each $\mathsf{d}_{j,l}$ is also computed, where $\mathsf{t}_{j,l} \in \mathbb{Z}_q$ is randomly chosen. The joint public key consists of $(\mathsf{n}, \mathsf{v}, (\mathsf{v}_j)_{j=1}^k, (F_{j,l})_{j,l \in \{1,\ldots,k\}})$. The private key of M_j consists of $(\mathsf{d}_j, (\mathsf{d}_{l,j}, \mathsf{t}_{l,j})_{l=1}^k)$.

To jointly decrypt a ciphertext u, the jth share-holder computes $u_j = u^{\mathsf{d}_j} \bmod \mathsf{n}^2$ and proves in zero-knowledge that $\log_u u_j = \log_\mathsf{v} \mathsf{v}_j$. If the proof fails, each M_l publishes $(\mathsf{d}_{j,l}, \mathsf{t}_{j,l})$. Then each honest party finds a set of $(\mathsf{d}_{j,l}, \mathsf{t}_{j,l})$ such that $F_{j,l} = g^{\mathsf{d}_{j,l}} h^{\mathsf{t}_{j,l}}$, recovers d_j using Lagrange interpolation, and computes $u_j = u^{\mathsf{d}_j} \bmod \mathsf{n}^2$. Finally, the plaintext is given by $L(\prod_{j=1}^k u_j) = m$.

2 The Ideal Relaxed Mix-Net

We use a slightly relaxed definition of the ideal mix-net in that corrupt senders may input messages with κ bits whereas honest senders may only input messages with $\kappa_m = \kappa - \kappa_r - 2$ bits. In the final output all messages are truncated to κ_m bits. The additional security parameter κ_r must be chosen such that $2^{-\kappa_r}$ is negligible. It decides the statistical hiding properties of some subprotocols. It is hard to imagine a situation where the relaxation is a real disadvantage, but if it is, it may be possible to eliminate this beauty flaw by an erasure-free proof of membership in the correct interval in the submission phase of the protocol.

Functionality 1 (Relaxed Mix-Net). The relaxed ideal mix-net, $\mathcal{F}_{\mathrm{RMN}}$, running with mix-servers $M_1, \ldots, M_k$, senders $S_1, \ldots, S_N$, and ideal adversary $\mathcal{S}$ proceeds as follows

1. Initialize a list $L = \emptyset$, a database D, $c = 0$, and $J_S = \emptyset$ and $J_M = \emptyset$.
2. Repeatedly wait for new inputs and do
   - Upon receipt of $(S_i, \texttt{Send}, m_i)$ from $\mathcal{C}_\mathcal{I}$ do the following. If $i \notin J_S$ and S_i is not corrupted and $m_i \in [-2^{\kappa_m} + 1, 2^{\kappa_m} - 1]$ or if S_i is corrupted and $m_i \in [-2^\kappa + 1, 2^\kappa - 1]$ then set $c \leftarrow c + 1$, store this tuple in D under the index c, and hand $(\mathcal{S}, S_i, \texttt{Input}, c)$ to $\mathcal{C}_\mathcal{I}$. Ignore other inputs.
   - Upon receipt of $(M_j, \texttt{Run})$ from $\mathcal{C}_\mathcal{I}$, set $c \leftarrow c + 1$, store $(M_j, \texttt{Run})$ in D under the index c, and hand $(\mathcal{S}, M_j, \texttt{Input}, c)$ to $\mathcal{C}_\mathcal{I}$.
   - Upon receipt of $(\mathcal{S}, \texttt{AcceptInput}, c)$ such that something is stored under the index c in D do
     (a) If $(S_i, \texttt{Send}, m_i)$ is stored under c and $i \notin J_S$, then append m_i to the list L, set $J_S \leftarrow J_S \cup \{i\}$, and hand $(\mathcal{S}, S_i, \texttt{Send})$ to $\mathcal{C}_\mathcal{I}$.
     (b) If $(M_j, \texttt{Run})$ is stored under c, then set $J_M \leftarrow J_M \cup \{j\}$. If $|J_M| > k/2$, then truncate all strings in L to κ_m bits and sort the result lexicographically to form a list L'. Sort the list L to form a list L''. Then hand $((\mathcal{S}, M_j, \texttt{Output}, L''), \{(M_l, \texttt{Output}, L')\}_{l=1}^k)$ to $\mathcal{C}_\mathcal{I}$. Otherwise, hand $\mathcal{C}_\mathcal{I}$ the list $(\mathcal{S}, M_j, \texttt{Run})$.

3 The Adaptively Secure Mix-Net

In this section we first describe the basic structure of our mix-net. Then we explain how we modify this to accommodate adaptive adversaries. We also discuss how and why our construction differs from previous constructions in the literature. This is followed by subsections introducing the subprotocols invoked in an execution of the mix-net. Finally, we give a detailed description of the mix-net.

3.1 Key Generation

The mix-servers use a joint κ-bit Paillier public key n and a corresponding secret shared secret key as described above. The public key n is the main public key in the mix-net, but we do need additional keys. We denote by $(\mathsf{n}', \mathsf{g}', \mathsf{d}')$ Paillier parameters generated as above but such that $\mathsf{n}' > \mathsf{n}$. We also need an RSA modulus $\mathbf{N}$ that is chosen exactly as the Paillier moduli n and n'. Finally, we need two Paillier ciphertexts $K_0 = E_{\mathsf{n}}(0, R_0)$ and $K_1 = E_{\mathsf{n}}(1, R_1)$ of 0 and 1 respectively. Below we summarize key generation as an ideal functionality.

Functionality 2 (Key Generation). The ideal key generation functionality, $\mathcal{F}_{\mathrm{PKG}}$, running with mix-servers $M_1, \ldots, M_k$, senders $S_1, \ldots, S_N$, and ideal adversary $\mathcal{S}$ proceeds as follows. It generates keys as described above and hands $((\mathcal{S}, \texttt{PublicKeys}, (\mathbf{N}, g, h, \mathsf{n}', \mathsf{n}, K_0, K_1, \mathsf{v}, (\mathsf{v}_l)_{l=1}^k, (F_{l,l'})_{l,l' \in \{1,\ldots,k\}})),$
$\{(M_j, \texttt{Keys}, (\mathbf{N}, g, h, \mathsf{n}', \mathsf{n}, K_0, K_1, \mathsf{v}, (\mathsf{v}_l)_{l=1}^k, (F_{l,l'})_{l,l' \in \{1,\ldots,k\}}),$
$(\mathsf{d}_j, (\mathsf{d}_{l,j}, \mathsf{t}_{l,j})_{l=1}^k))\}_{j=1}^k)$ to $\mathcal{C}_{\mathcal{I}}$.

3.2 The Overall Structure

Our mix-net is based on the re-encryption-permutation paradigm. Let $L_0 = \{u_{0,i}\}_{i=1}^N$ be the list of ciphertexts submitted by senders. For $l = 1, \ldots, k$ the lth mix-server M_l re-encrypts each element in $L_{l-1} = \{u_{l-1,i}\}_{i=1}^N$ as explained in Section 1.4, sorts the resulting list and publishes the result as L_l. Then it proves, in zero-knowledge, knowledge of a witness that L_{l-1} and L_l are related in this way. The mix-servers then jointly and verifiably re-encrypt the ciphertexts in L_k. Note that no permutation takes place in this step. The result is denoted by L_{k+1}. Finally, the mix-servers jointly and verifiably decrypt each ciphertext in L_{k+1} and sort the resulting list of plaintexts to form the output. Except for the joint re-encryption step this is similar to several previous constructions.

3.3 Accommodating Adaptive Adversaries

To extract the inputs of corrupt senders, each sender forms two ciphertexts $u_{0,i}$ and $u'_{0,i}$ and proves that the same plaintext is hidden in both. Naor and Yung's [11] double-ciphertext trick then allows extraction. Submissions of honest senders must be simulated without knowing which message they actually hand to $\mathcal{F}_{\mathrm{RMN}}$. A new problem in the adaptive setting is that the adversary may corrupt a simulated honest sender S_i that has already computed fake ciphertexts $u_{0,i}$ and $u'_{0,i}$. The ideal adversary can of course corrupt the corresponding dummy party $\tilde{S}_i$ and retrieve the true value m_i it handed to $\mathcal{F}_{\mathrm{RMN}}$. The problem is that it must provide S_i with a plausible history tape that convinces the adversary that S_i sent m_i already from the beginning. To solve this problem we adapt an idea of Damgård and Nielsen [6]. We have two public keys $K_0 = E_{\mathsf{n}}(0, R_0) = R_0^{2\mathsf{n}} \bmod \mathsf{n}^2$ and $K_1 = E_{\mathsf{n}}(1, R_1) = \mathsf{g}R_1^{2\mathsf{n}} \bmod \mathsf{n}^2$ and each sender is given a unique key $K'_i = E_{\mathsf{n}'}(a_i)$ for a randomly chosen $a_i \in \mathbb{Z}_{\mathsf{n}'}$. The sender of a message m_i chooses $b_i \in \mathbb{Z}_{\mathsf{n}}$, $r_i \in \mathbb{Z}_{\mathsf{n}}^*$ and $r'_i \in \mathbb{Z}_{\mathsf{n}'}^*$ randomly, and computes its two ciphertexts as follows $u_i = E_{K_1,\mathsf{n}}(m_i, r_i)$ and $u'_i = E_{(\mathsf{g}')^{b_i} K'_i,\mathsf{n}'}(m_i, r'_i)$. Then it submits (b_i, u_i, u'_i) and proves in zero-knowledge that the same message m_i is encrypted in both ciphertexts. Note that $D_{\mathsf{d}}(u_i) = m_i$ and $D_{\mathsf{d}'}(u'_i) = (a_i + b_i)m_i$ due to the homomorphic property of the cryptosystem.

During simulation we instead define $K_0 = E_{\mathsf{n}}(1, R_0) = \mathsf{g}R_0^{2\mathsf{n}} \bmod \mathsf{n}^2$ and $K_1 = E_{\mathsf{n}}(0, R_1) = R_1^{2\mathsf{n}} \bmod \mathsf{n}^2$. This means that u_i becomes an encryption of 0 for all senders. Furthermore, simulated senders choose $b_i = -a_i \bmod \mathsf{n}'$ which implies that also u_i' is an encryption of 0. The important property of the simulation is that given m_i and R_1 we can define $\bar{r}_i = r_i/R_1^{m_i}$ such that $u_i = E_{K_1,\mathsf{n}}(m_i, \bar{r}_i)$, i.e., we can open a simulated ciphertext as an encryption of an arbitrary message m_i. The ciphertext u_i' can also be opened as an encryption of m_i in a similar way when $b_i + a_i = 0 \bmod \mathsf{n}'$. Finally, the proof of equality we use can also be "opened" in a convincing way. This allows the simulator to simulate honest senders and produce plausible history tapes as required. Corrupt senders on the other hand have negligible probability of guessing a_i, so the simulator can extract the message submitted by corrupt senders using only the private key d' by computing $m_i = D_{\mathsf{d}'}(u_i')/(a_i + b_i) \bmod \mathsf{n}'$. Before the mix-net simulated by the ideal adversary starts to process the input ciphertexts the ideal mix-net $\mathcal{F}_{\mathrm{RMN}}$ has handed the ideal adversary the list of plaintexts that should be output by the simulation. All plaintexts equal zero in the ciphertexts of the input in the simulation and the correct messages are introduced in the joint re-encryption phase. All mix-servers are simulated honestly during the re-encryption-permutation phase and the decryption phase.

The joint re-encryption is defined as follows. Before the mixing each mix-server is given a random ciphertext $\bar{K}_j'$ using the public key n'. Each mix-server M_j chooses random elements $m_{j,i} \in \mathbb{Z}_{\mathsf{n}}$ and commits to these by choosing $\bar{b}_j \in \mathbb{Z}_{\mathsf{n}'}$ and $s_{j,i}' \in \mathbb{Z}_{\mathsf{n}'}^*$ randomly and computing $w_{j,i}' = E_{(\mathsf{g}')^{\bar{b}_j}\bar{K}_j',\mathsf{n}'}(m_{j,i}, s_{j,i}')$. When all mix-servers have published their commitments, it chooses $s_{j,i} \in \mathbb{Z}_{\mathsf{n}}^*$ randomly and computes $w_{j,i} = E_{K_0,\mathsf{n}}(m_{j,i}, s_{j,i})$. It also proves in zero-knowledge that the same random element $m_{j,i}$ is encrypted in both ciphertexts. The jointly re-encrypted elements $u_{k+1,i}$ are then formed as $u_{k+1,i} = u_{k,i} \prod_{l \in I} w_{l,i}^{\prod_{l' \neq l} \frac{l'}{l'-l}}$ where I is the first set of k' indices j such that the proof of M_j is valid. In the real execution this is an elaborate way to re-encrypt $u_{k,i}$, since K_0 is an encryption of 0. In the simulation on the other hand the ideal adversary chooses $\bar{b}_j = -\bar{a}_j \bmod \mathsf{n}'$ and sets $m_{j,i} = 0$ for simulated mix-servers and extracts the $m_{j,i}$ values of corrupt mix-servers from their commitments. It then redefines the $m_{j,i}$ values of simulated honest mix-servers such that $f_i(j) = m_{j,i}$ for a $(k'-1)$-degree polynomial f_i over $\mathbb{Z}_{\mathsf{n}}$ such that $f_i(0)$ equals $m_{\pi(i)}$ for some random permutation $\pi \in \Sigma_N$. Since $\bar{b}_j + \bar{a}_j = 0 \bmod \mathsf{n}'$ it can compute $\bar{s}_{j,i}'$ such that $w_{j,i}' = E_{(\mathsf{g}')^{\bar{b}_j}\bar{K}_j',\mathsf{n}'}(m_{j,i}, \bar{s}_{j,i}')$. In the simulation K_0 is an encryption of 1 and each $u_{k,i}$ is an encryption of zero, which implies that $u_{k+1,i}$ becomes an encryption of $m_{\pi(i)}$ as required. The adversary can not tell the difference since it can only get its hands on a minority of the $m_{j,i}$ values directly, and the semantic security of the cryptosystem prevents it from knowing these values otherwise.

3.4 Some Intuition Behind Our Analysis

Intuitively, the soundness of the subprotocols ensure that each sender knows the message it submits and that the output of the mix-net is correct. The zero-knowledge properties and the knowledge extraction properties of the subprotocols are not used by the ideal adversary sketched above, but they are essential to prove that the ideal adversary produces an indistinguishable simulation.

The private key d corresponding to the Paillier modulus n is needed both in the ideal model and in the real protocol. Thus, even if the environment can distinguish the ideal model from the real model, we can not use it directly to reach a contradiction to the semantic security of the Paillier cryptosystem. To solve this problem we use the single-honest-player proof strategy to sample each of the two distributions, but without using the secret key d. The knowledge extractor of the proof of a shuffle is needed to be able to simulate the joint decryption, since although the set of plaintexts is known their order in the list of ciphertexts that are jointly decrypted is not. Due to the statistical zero-knowledge property of the proof of a shuffle and the fact that in the ideal model all plaintexts are zero from the start we can use the same type of simulation also when sampling the ideal model without changing its distribution more than negligibly. A hybrid argument allows us to assume that the simulated honest senders use the correct plaintexts already from the start. If there is a gap between the resulting distributions this can be used to distinguish a ciphertext of a zero from a ciphertext of a one, i.e., we can break the semantic security of the Paillier cryptosystem.

3.5 Differences with Previous Constructions

Most previous schemes are based on the ElGamal cryptosystem. We need the Paillier cryptosystem to allow adaptive corruption of the senders in the way explained above. The joint re-encryption step which no previous construction has is needed to insert the correct messages in the simulation and still be able to construct plausible history tapes of any adaptively corrupted mix-server.

In [16, 17] the mix-net is given in a hybrid model with access to ideal zero-knowledge proof of knowledge functionalities. These functionalities are then securely realized, and the composition theorem of the UC-framework invoked. The modular approach simplifies the analysis, but the strong demands on subprotocols make them hard to securely realize efficiently. We avoid this problem by showing that a zero-knowledge proof of knowledge of correct re-encryption-permutation in the classical sense is sufficient, i.e., the protocol can *not be simulated to an adaptive adversary* and extraction is *not straight-line*.

3.6 Subprotocols Invoked by the Main Protocol

Some of our subprotocols satisfy a weaker notion of proof of knowledge called "computationally convincing proof (of knowledge)" introduced by Damgård and Fujisaki [4]. Informally, this means that extraction is possible with overwhelming probability over the randomness of a special input that is given to both parties.

Proof of Knowledge of Re-encryption-Permutation. Denote by $\pi_{\mathrm{prp}} = (P_{\mathrm{prp}}, V_{\mathrm{prp}})$ the 5-move protocol for proving knowledge of a witness of a re-encryption and permutation of a list of Paillier ciphertexts given by Wikström

[17]. The parties accept as special parameters an RSA-modulus $\mathbf{N}$ and random $\mathbf{g}, \mathbf{h} \in \mathrm{QR}_{\mathbf{N}}$, and random $g_1, \ldots, g_N \in G_q$. The re-encryption-permutation relation $R_{\mathrm{RP}}^{\mathsf{n},\mathsf{g}_\mathsf{f}}$ and the security properties of π_{prp} are stated below.

Definition 1 (Knowledge of Correct Re-encryption-Permutation). *Define for each N, n and g_f a relation $R_{\mathrm{RP}}^{\mathsf{n},\mathsf{g}_\mathsf{f}} \subset (\mathrm{QR}_{\mathsf{n}^2}^N \times \mathrm{QR}_{\mathsf{n}^2}^N) \times [-2^{\kappa+\kappa_r} + 1, 2^{\kappa+\kappa_r} - 1]^N$, by $((\{u_i\}_{i=1}^N, \{u_i'\}_{i=1}^N), (a, (x_i)_{i=1}^N)) \in R_{\mathrm{RP}}^{\mathsf{n},\mathsf{g}_\mathsf{f}}$ precisely when $a < \sqrt{\mathsf{n}}/4$ equals one or is prime and $(u_i')^a = \mathsf{g}_\mathsf{f}^{x_{\pi(i)}} u_{\pi(i)}^a \mod \mathsf{n}^2$ for $i = 1, \ldots, N$ and some permutation $\pi \in \Sigma_N$ such that the list $\{u_i'\}_{i=1}^N$ is sorted lexicographically.*

Proposition 1 ([17]). *The protocol π_{prp} is an honest verifier statistical zero-knowledge computationally convincing proof of knowledge for the relation $R_{\mathrm{RP}}^{\mathsf{n},\mathsf{g}_\mathsf{f}}$ with respect to the distribution of $(\mathbf{N}, \mathbf{g}, \mathbf{h})$ and $(g_1, \ldots, g_N)$, and it has overwhelming completeness.*

Proof of Equality of Plaintexts. When a sender submits its ciphertexts u_i and u_i' it must prove that they are encryptions of the same $(\kappa - 2)$-bit integer under two distinct public keys. The protocol $\pi_{eq} = (P_{\mathrm{eq}}, V_{\mathrm{eq}})$ used to do this is given below. The security parameters κ_c and κ_r decide the soundness and statistical zero-knowledge property of the protocol.

Protocol 1 (Proof of Equal Plaintexts Using Distinct Moduli)
COMMON INPUT: $\mathsf{n} \in \mathbb{Z}$, $K, u \in \mathbb{Z}_{\mathsf{n}^2}^*$, $\mathsf{n}' \in \mathbb{Z}$, $K', u' \in \mathbb{Z}_{(\mathsf{n}')^2}^*$, $\mathbf{N} \in \mathbb{N}$, *generators* $\mathbf{g}$ *and* $\mathbf{h}$ *of* $\mathrm{QR}_{\mathbf{N}}$.
PRIVATE INPUT: $m \in [-2^{\kappa_m} + 1, 2^{\kappa_m} - 1]$, $r \in \mathbb{Z}_{\mathsf{n}}^*$, *and* $r' \in \mathbb{Z}_{\mathsf{n}'}^*$ *such that* $u = E_{K,\mathsf{n}}(m, r)$ *and* $u' = E_{K',\mathsf{n}'}(m, r')$.

1. *The prover chooses* $r'' \in [0, 2^{\kappa+\kappa_r} - 1]$, $s_0 \in \mathbb{Z}_{\mathsf{n}^2}^*$ *and* $s_1 \in \mathbb{Z}_{(\mathsf{n}')^2}^*$, *and* $t \in [0, 2^{\kappa_m+\kappa_c+\kappa_r} - 1]$ *and* $s_2, \in [0, 2^{2\kappa+\kappa_c+2\kappa_r} - 1]$ *randomly. Then it computes* $C = \mathbf{g}^m \mathbf{h}^{r''} \mod \mathbf{N}$ *and* $(\alpha_0, \alpha_1, \alpha_2) = (K^t s_0^{2\mathsf{n}} \mod \mathsf{n}^2, (K')^t s_1^{2\mathsf{n}'} \mod (\mathsf{n}')^2, \mathbf{g}^t \mathbf{h}^{s_2} \mod \mathbf{N})$, *and hands* $(C, \alpha_0, \alpha_1, \alpha_2)$ *to the verifier.*
2. *The verifier chooses* $c \in [2^{\kappa_c - 1}, 2^{\kappa_c} - 1]$ *and hands* c *to the prover.*
3. *The prover computes* $(e_0, e_1) = (r^c s_0 \mod \mathsf{n}, (r')^c s_1 \mod \mathsf{n}')$, $(e_2, e_3) = (cr'' + s_2 \mod 2^{\kappa+\kappa_c+2\kappa_r}, cm + t \mod 2^{\kappa_m+\kappa_c+\kappa_r})$ *and hands* (e_0, e_1, e_2, e_3) *to the verifier.*
4. *The verifier checks* $(u^c \alpha_0, (u')^c \alpha_1) = (K^{e_3} e_0^{2\mathsf{n}} \mod \mathsf{n}, (K')^{e_3} e_1^{2\mathsf{n}'} \mod \mathsf{n}')$ *and* $C^c \alpha_2 = \mathbf{g}^{e_3} \mathbf{h}^{e_2} \mod \mathbf{N}$.

The protocol is statistical zero-knowledge, but this is not enough since we must construct plausible history tapes for simulated senders.

Proposition 2 ("Zero-Knowledge"). *Let* $K = R^{2\mathsf{n}} \mod \mathsf{n}^2$ *and* $K' = R'^{2\mathsf{n}'} \mod (\mathsf{n}')^2$ *for some* $R \in \mathbb{Z}_{\mathsf{n}}^*$ *and* $R' \in \mathbb{Z}_{\mathsf{n}'}^*$. *Let* $\mathbf{h}$ *be a generator of* $\mathrm{QR}_{\mathbf{N}}$ *and* $\mathbf{g} = \mathbf{h}^{\mathbf{x}}$. *Let* r, r', *and* (r'', s_0, s_1, t, s_2) *be randomly distributed in the domains described in the protocol, and denote by* $I(m) = (\mathsf{n}, K, u, \mathsf{n}', K', u', \mathbf{N}, \mathbf{g}, \mathbf{h})$ *the common input corresponding to the private input* (m, r, r'). *Denote by* c *the random challenge from the verifier and let* $T(m) = (\alpha, c, e)$ *be the proof transcript induced by* (m, r, r'), c, *and* (r'', s_0, s_1, t, s_2).

There is a deterministic polynomial-time algorithm His *such that for every* $m \in \{0,1\}^{\kappa_m}$ *with* $(\bar{r}, \bar{r}', \bar{r}'', \bar{s}_0, \bar{s}_1, \bar{t}, \bar{s}_2) = \mathsf{His}(R, R', \mathbf{x}, m, r, r', r'', s_0, s_1, t, s_2, c)$ *the distributions of* $[I(m), T(m), (m, r, r'), (r'', s_0, s_1, t, s_2)]$ *and* $[I(0), T(0), (m, \bar{r}, \bar{r}'), (\bar{r}'', \bar{s}_0, \bar{s}_1, \bar{t}, \bar{s}_2))]$ *are statistically close.*

The proposition in itself does not imply statistical zero-knowledge, since it only applies to inputs where K and K' are both encryptions of zero.

Proposition 3. *The protocol is a computationally convincing proof with respect to the distribution of* $(\mathbf{N}, \mathbf{g}, \mathbf{h})$*, and has overwhelming completeness.*

Multiple instances of the protocol can be run in parallel using the same RSA-parameters and same challenge. Thus, we use the protocol also for common inputs on the form $(\mathsf{n}, K, \{u_i\}_{i=1}^N, \mathsf{n}', K', \{u_i'\}_{i=1}^N, \mathbf{N}, \mathbf{g}, \mathbf{h})$ and with corresponding private input $(\{m_i\}_{i=1}^N, \{r_i\}_{i=1}^N, \{r_i'\}_{i=1}^N)$. We extend the notation in the next subsection similarly.

Proof of Equality of Exponents. During joint decryption of a ciphertext u each mix-server computes $u^{\mathsf{d}_j} \bmod \mathsf{n}^2$ using its part d_j of the private key, and proves correctness relative $\mathsf{v}_j = \mathsf{v}^{\mathsf{d}_j} \bmod \mathsf{n}^2$, i.e., that it uses the same exponent d_j for both elements. We denote by $\pi_{\mathrm{exp}} = (P_{\mathrm{exp}}, V_{\mathrm{exp}})$ the 3-move protocol proposed in [5]. It has the following properties.

Proposition 4. *The protocol* π_{exp} *is an honest verifier statistical zero-knowledge proof with overwhelming completeness.*

3.7 The Mix-Net

We are now ready to give a detailed description of the mix-net. Recall that $k' = \lceil (k+1)/2 \rceil$ denotes the number of mix-servers needed for majority. Each entry on the bulletin board is given a sequence number denoted by T below (with different subscripts). To ensure that the ciphertexts in the common inputs to the proofs of a shuffle belong to $\mathrm{QR}_{\mathsf{n}^2}$ the mix-servers square the ciphertexts between each mix-server. The effect of the squaring is eliminated at the end.

Protocol 2 (Mix-Net). The mix-net $\pi_{\mathrm{RMN}} = (S_1, \ldots, S_N, M_1, \ldots, M_k)$ consists of senders S_i, and mix-servers M_j.

SENDER S_i. Each sender S_i proceeds as follows.

1. Wait until $(M_l, \mathsf{n}, K_1, \mathsf{n}', \{K_i'\}_{i=1}^N, \mathbf{N}, \mathbf{g}, \mathbf{h})$ appears on $\mathcal{F}_{\mathrm{BB}}$ for k' distinct indices l.
2. Wait for an input $(\mathtt{Send}, m_i)$, such that $m_i \in [-2^{\kappa_m}+1, 2^{\kappa_m}-1]$. Choose $r_i \in \mathbb{Z}_{\mathsf{n}}^*$, $b_i \in \mathbb{Z}_{\mathsf{n}'}$ and $r_i' \in \mathbb{Z}_{\mathsf{n}'}^*$ randomly and compute $u_i = E_{K_1,\mathsf{n}}(m_i, r_i)$, $u_i' = E_{(\mathsf{g}')^{b_i} K_i', \mathsf{n}'}(m_i, r_i')$, and $(\alpha_i, \mathrm{state}_i) = P_{\mathrm{eq}}((\mathsf{n}, K_1, u_i, \mathsf{n}', (\mathsf{g}')^{b_i} K_i', u_i', \mathbf{N}, \mathbf{g}, \mathbf{h}), (m_i, r_i, r_i'))$. Then hand $(\mathtt{Write}, \mathtt{Submit}, (b_i, u_i, u_i'), \mathtt{Commit}, \alpha_i)$ to $\mathcal{F}_{\mathrm{BB}}$.
3. Wait until $(M_j, \mathtt{Challenge}, S_i, c_i)$ appears on $\mathcal{F}_{\mathrm{BB}}$ for k' distinct j with identical c_i. Then compute $e_i = P_{\mathrm{eq}}(\mathrm{state}_i, c_i)$ and hand $(\mathtt{Write}, \mathtt{Reply}, e_i)$ to $\mathcal{F}_{\mathrm{BB}}$.

MIX-SERVER M_j. Each mix-server M_j proceeds as follows.

Preliminaries

1. Wait for a message on the form $(\texttt{Keys}, (\mathbf{N}, g, h, \mathsf{n}', \mathsf{n}, K_0, K_1, \mathsf{v}, (\mathsf{v}_l)_{l=1}^k, (F_{l,l'})_{l,l' \in \{1,\ldots,k\}}), (\mathsf{d}_j, (\mathsf{d}_{l,j}, \mathsf{t}_{l,j})_{l=1}^k))$ from $\mathcal{F}_{\mathrm{PKG}}$.
2. Hand $(\texttt{GenerateCoins}, (N+k)(\kappa+\kappa_r)+(\kappa+\kappa_r)+2(\kappa+\kappa_r)+N(\kappa+\kappa_r))$ to $\mathcal{F}_{\mathrm{CF}}$ and wait until it returns $(\texttt{Coins}, \{K_i'\}_{i=1}^N, \{\bar{K}_j'\}_{j=1}^k, \mathsf{g}_\mathsf{f}, \mathbf{g}, \mathbf{h}, g_1, \ldots, g_N)$. Then hand $(\texttt{Write}, \mathsf{n}, K_1, \mathsf{n}', \{K_i'\}_{i=1}^N, \mathbf{N}, \mathbf{g}, \mathbf{h})$ to $\mathcal{F}_{\mathrm{BB}}$.

Reception of Inputs

3. Initialize $L_0 = \emptyset$, $J_S = \emptyset$ and $J_M = \emptyset$.
4. Repeat
   (a) When given input $(\texttt{Run})$ hand $(\texttt{Write}, \texttt{Run})$ to $\mathcal{F}_{\mathrm{BB}}$.
   (b) When a new entry $(T, M_l, \texttt{Run})$ appears on $\mathcal{F}_{\mathrm{BB}}$ set $J_M \leftarrow J_M \cup \{l\}$ and if $|J_M| \geq k'$ set $T_{\mathrm{run}} = T$ and go to Step 5.
   (c) When a new entry $(S_i, \texttt{Submit}, (b_i, u_i, u_i'), \texttt{Commit}, \alpha_i)$ appears on $\mathcal{F}_{\mathrm{BB}}$ such that $i \notin J_S$, set $J_S \leftarrow J_S \cup \{i\}$ and hand $(\texttt{GenerateCoins}, \kappa_c)$ to $\mathcal{F}_{\mathrm{CF}}$ and wait until it returns $(\texttt{Coins}, c_i)$. Hand $(\texttt{Write}, \texttt{Challenge}, S_i, c_i)$ to $\mathcal{F}_{\mathrm{BB}}$.
5. Request the contents on $\mathcal{F}_{\mathrm{BB}}$ with index less than T_{run}. Find for each i the first occurrences of entries on the forms $(T_i, \texttt{Submit}, (b_i, u_i, u_i'), \texttt{Commit}, \alpha_i)$, $(T_{j,i}', M_j, \texttt{Challenge}, S_i, c_i)$, and $(T_i'', S_i, \texttt{Reply}, e_i)$. Then form a list L_0 of all ciphertexts $u_i^2 \bmod \mathsf{n}^2$ such that $T_i < T_{j,i}' < T_i'' < T_{\mathrm{run}}$ for at least k' distinct indices j and $V_{\mathrm{eq}}(\mathsf{n}, K_1, u_i, \mathsf{n}', (\mathsf{g}')^{b_i} K_i', u_i', \mathbf{N}, \mathbf{g}, \mathbf{h}, \alpha_i, c_i, e_i) = 1$.

Re-encryption and Permutation

6. Write $L_0 = \{u_{0,i}\}_{i=1}^{N'}$ for some N'. Then for $l = 1, \ldots, k$ do
   (a) If $l = j$, then do
      i. Choose $r_{j,i} \in [0, 2^{\kappa+\kappa_r} - 1]$ randomly, compute

$$L_j = \{u_{j,i}\}_{i=1}^{N'} = \mathsf{Sort}(\{\mathsf{g}_\mathsf{f}^{r_{j,i}} u_{j-1,i}^2 \bmod \mathsf{n}^2\}_{i=1}^{N'}) \ , \text{ and}$$

$$(\alpha_j, \mathrm{state}_j) = P_{\mathrm{prp}}(\mathsf{n}, \mathsf{g}_\mathsf{f}, L_{l-1}^4, L_l^2, \mathbf{N}, \mathbf{g}, \mathbf{h}, g, g_1, \ldots, g_{N'}, \{2r_{j,i}\}_{i=1}^{N'}) \ ,$$

      and hand $(\texttt{Write}, \texttt{List}, L_j, \texttt{Commit1}, \alpha_j)$ to $\mathcal{F}_{\mathrm{BB}}$. The exponentiations L_{l-1}^4 and L_l^2 should be interpreted term-wise.
      ii. Hand $(\texttt{GenerateCoins}, \kappa)$ to $\mathcal{F}_{\mathrm{CF}}$ and wait until it returns $(\texttt{Coins}, c_j)$. Then compute $(\alpha_j', \mathrm{state}_j') = P_{\mathrm{prp}}(\mathrm{state}_j, c_j)$ and hand $(\texttt{Write}, \texttt{Commit2}, \alpha_j')$ to $\mathcal{F}_{\mathrm{BB}}$.
      iii. Hand $(\texttt{GenerateCoins}, \kappa_c)$ to $\mathcal{F}_{\mathrm{CF}}$ and wait until it returns $(\texttt{Coins}, c_j')$. Then compute $e_j = P_{\mathrm{prp}}(\mathrm{state}_j', c_j')$ and hand $(\texttt{Write}, \texttt{Reply}, e_j)$ to $\mathcal{F}_{\mathrm{BB}}$.
   (b) If $l \neq j$, then do
      i. Wait until an entry $(M_l, \texttt{List}, L_l, \texttt{Commit1}, \alpha_l)$ appears on $\mathcal{F}_{\mathrm{BB}}$.
      ii. Hand $(\texttt{GenerateCoins}, \kappa)$ to $\mathcal{F}_{\mathrm{CF}}$ and wait until it returns $(\texttt{Coins}, c_l)$.

iii. Wait for a new entry $(M_l, \mathtt{Commit2}, \alpha'_l)$ on $\mathcal{F}_{\mathrm{BB}}$. Hand $(\mathtt{GenerateCoins}, \kappa_c)$ to $\mathcal{F}_{\mathrm{CF}}$ and wait until it returns $(\mathtt{Coins}, c'_l)$.
iv. Wait for a new entry $(M_l, \mathtt{Reply}, e_l)$ on $\mathcal{F}_{\mathrm{BB}}$ and compute $b_l = V_{\mathrm{prp}}(\mathsf{n}, \mathsf{g}_{\mathsf{f}}, L^4_{l-1}, L^2_l, \mathbf{N}, \mathbf{g}, \mathbf{h}, g, g_1, \ldots, g_{N'}, \alpha_l, c_l, \alpha'_l, c'_l, e_l)$.
v. If $b_l = 0$, then set $L_l = L^2_{l-1}$.

Joint Re-encryption

7. Choose $\bar{b}_j \in \mathbb{Z}_{\mathsf{n}'}$, $m_{j,i} \in \mathbb{Z}_{\mathsf{n}'}$ and $s'_{j,i} \in \mathbb{Z}^*_{\mathsf{n}'}$ randomly and compute $W'_j = \{w'_{j,i}\}_{i=1}^{N'} = \{E_{\mathsf{g}'^{\bar{b}_j} \bar{K}'_j, \mathsf{n}'}(m_{j,i}, s'_{j,i})\}_{i=1}^{N'}$. Hand $(\mathtt{Write}, \mathtt{RandExp}, \bar{b}_j, W'_j)$ to $\mathcal{F}_{\mathrm{BB}}$.
8. Wait until $(\mathtt{RandExp}, \bar{b}_l, W'_l)$ appears on $\mathcal{F}_{\mathrm{BB}}$ for $l = 1, \ldots, k$. Then choose $s_{j,i} \in \mathbb{Z}^*_{\mathsf{n}}$ randomly, compute $W_j = \{w_{j,i}\}_{i=1}^{N'} = \{E_{K_0,\mathsf{n}}(m_{j,i}, s_{j,i})\}_{i=1}^{N'}$, and
$$(\alpha_j, \mathrm{state}_j) = P_{\mathrm{eq}}((\mathsf{n}, K_0, W_j, \mathsf{n}', K', W'_j, \mathbf{N}, \mathbf{g}, \mathbf{h}), (\{m_{j,i}\}_{i=1}^{N'}, \{s_{j,i}\}_{i=1}^{N'}, \{s'_{j,i}\}_{i=1}^{N'})) ,$$
and hand $(\mathtt{Write}, \mathtt{RandExp}, W_j, \mathtt{Commit}, \alpha_j)$ to $\mathcal{F}_{\mathrm{BB}}$.
9. Wait until $(\mathtt{RandExp}, W_l, \mathtt{Commit}, \alpha_l)$ appears on $\mathcal{F}_{\mathrm{BB}}$ for $l = 1, \ldots, k$. Hand $(\mathtt{GenerateCoins}, \kappa_c)$ to $\mathcal{F}_{\mathrm{CF}}$ and wait until it returns $(\mathtt{Coins}, c)$. Compute $e_j = P_{\mathrm{eq}}(\mathrm{state}_j, c)$ and hand $(\mathtt{Write}, \mathtt{Reply}, e_j)$ to $\mathcal{F}_{\mathrm{BB}}$.
10. Wait until $(\mathtt{Reply}, e_l)$ appears on $\mathcal{F}_{\mathrm{BB}}$ for $l = 1, \ldots, k$. Let I be the first set of k' indices with $V_{\mathrm{eq}}(\mathsf{n}, K_0, W_l, \mathsf{n}', K', W'_l, \mathbf{N}, \mathbf{g}, \mathbf{h}, \alpha_l, c, e_l) = 1$.
11. Compute $L_{k+1} = \{u_{k+1,i}\}_{i=1}^{N'} = \left\{ u_{k,i} \prod_{l \in I} w_{l,i}^{\prod_{l' \neq l} \frac{l'}{l'-l}} \right\}_{i=1}^{N'}$.

Joint Decryption

12. Compute $\Gamma_j = \{v_{j,i}\}_{i=1}^{N'} = \{u_{k+1,i}^{2\mathsf{d}_j}\}_{i=1}^{N}$ using d_j and a proof $(\alpha_j, \mathrm{state}_j) = P_{\mathrm{exp}}((\mathsf{n}, \mathsf{v}, \mathsf{v}_j, L^2_{k+1}, \Gamma_j), \mathsf{d}_j)$. Then hand $(\mathtt{Write}, \mathtt{Decrypt}, \Gamma_j, \mathtt{Commit}, \alpha_j)$ to $\mathcal{F}_{\mathrm{BB}}$, where exponentiation is interpreted element-wise.
13. Wait until $(M_l, \mathtt{Decrypt}, \Gamma_l, \mathtt{Commit}, \alpha_l)$ appears on $\mathcal{F}_{\mathrm{BB}}$ for $l = 1, \ldots, k$. Then hand $(\mathtt{GenerateCoins}, \kappa_c)$ to $\mathcal{F}_{\mathrm{CF}}$ and wait until it returns $(\mathtt{Coins}, c)$.
14. Compute $e_j = P_{\mathrm{exp}}(\mathrm{state}_j, c)$ and hand $(\mathtt{Write}, \mathtt{Reply}, e_j)$ to $\mathcal{F}_{\mathrm{BB}}$.
15. Wait until $(\mathtt{Reply}, e_l)$ appears on $\mathcal{F}_{\mathrm{BB}}$ for $l = 1, \ldots, k$. For $l = 1, \ldots, k$ do the following. If $V_{\mathrm{exp}}(\mathsf{n}, \mathsf{v}, v_l, L^2_{k+1}, \Gamma_l, \alpha_l, c, e_l) = 0$ do
    (a) Hand $(\mathtt{Write}, \mathtt{Recover}, M_l, \mathsf{d}_{l,j}, \mathsf{t}_{l,j})$ to $\mathcal{F}_{\mathrm{BB}}$.
    (b) Wait until $(M_{l'}, \mathtt{Recover}, M_l, \mathsf{d}_{l,l'}, \mathsf{t}_{l,l'})$ appears on $\mathcal{F}_{\mathrm{BB}}$ for $l' = 1, \ldots, k$. Then find a subset I of k' indices l' such that $F_{l,l'} = g^{\mathsf{d}_{l,l'}} h^{\mathsf{t}_{l,l'}}$ and Lagrange interpolate $\mathsf{d}_l = \sum_{l' \in I} \mathsf{d}_{l,l'} \prod_{l'' \neq l'} \frac{l''}{l''-l'} \bmod q$.
    (c) Compute $\Gamma_l = \{v_{l,i}\}_{i=1}^{N'} = \{u_{k,i}^{2\mathsf{d}_l}\}_{i=1}^{N}$.
16. Interpret each element in $\{L(\prod_{l=1}^{k} v_{l,i})/2^{k+2}\}_{i=1}^{N'}$ as an integer in $[-2^{\kappa_m+\kappa_r} + 1, 2^{\kappa_m+\kappa_r} - 1]$ (this can be done uniquely, since $\kappa_m + \kappa_r < \kappa - 1$), truncate to κ_m bits, and let L_{out} be the result. Output $(\mathtt{Output}, \mathsf{Sort}(L_{\mathrm{out}}))$.

Theorem 1. *The protocol π_{RMN} above securely realizes $\mathcal{F}_{\mathrm{RMN}}$ in the $(\mathcal{F}_{\mathrm{BB}}, \mathcal{F}_{\mathrm{PKG}}, \mathcal{F}_{\mathrm{CF}})$-hybrid model for $\mathcal{M}^*_{k/2}$-adversaries under the DCR- assumption, the strong RSA-assumption, and the DL-assumption.*

References

1. M. Abe and H. Imai. Flaws in some robust optimistic mix-nets. In *Australasian Conference on Information Security and Privacy – ACISP 2003*, volume 2727 of *LNCS*, pages 39–50. Springer Verlag, 2003.
2. R. Canetti. Universally composable security: A new paradigm for cryptographic protocols. In *42nd IEEE Symposium on Foundations of Computer Science (FOCS)*, pages 136–145. IEEE Computer Society Press, 2001. (Full version at Cryptology ePrint Archive, Report 2000/067, `http://eprint.iacr.org`, October, 2001.).
3. D. Chaum. Untraceable electronic mail, return addresses and digital pseudo-nyms. *Communications of the ACM*, 24(2):84–88, 1981.
4. I. Damgård and E. Fujisaki. A statistically-hiding integer commitment scheme based on groups with hidden order. In *Advances in Cryptology – Asiacrypt 2002*, volume 2501 of *LNCS*, pages 125–142. Springer Verlag, 2002.
5. I. Damgård and M. Jurik. A generalisation, a simplification and some applications of paillier's probabilistic public-key system. In *Public Key Cryptography – PKC 2001*, volume 1992 of *LNCS*, pages 119–136. Springer Verlag, 2001.
6. I. Damgård and J. B. Nielsen. Universally composable efficient multiparty computation from threshold homomorphic encryption. In *Advances in Cryptology – Crypto 2003*, volume 2729 of *LNCS*, pages 247–267. Springer Verlag, 2003.
7. J. Furukawa. Efficient and verifiable shuffling and shuffle-decryption. *IEICE Transactions*, 88-A(1):172–188, 2005.
8. J. Furukawa and K. Sako. An efficient scheme for proving a shuffle. In *Advances in Cryptology – Crypto 2001*, volume 2139 of *LNCS*, pages 368–387. Springer Verlag, 2001.
9. J. Groth. A verifiable secret shuffle of homomorphic encryptions. In *Public Key Cryptography – PKC 2003*, volume 2567 of *LNCS*, pages 145–160. Springer Verlag, 2003. Full version at Cryptology ePrint Archive, Report 2005/246, 2005, `http://eprint.iacr.org/`.
10. A. Lysyanskaya and C. Peikert. Adaptive security in the threshold setting: From cryptosystems to signature schemes. In *Advances in Cryptology – Asiacrypt 2001*, volume 2248 of *LNCS*, pages 331–350. Springer Verlag, 2001.
11. M. Naor and M. Yung. Public-key cryptosystems provably secure against chosen ciphertext attack. In *22th ACM Symposium on the Theory of Computing (STOC)*, pages 427–437. ACM Press, 1990.
12. A. Neff. A verifiable secret shuffle and its application to e-voting. In *8th ACM Conference on Computer and Communications Security (CCS)*, pages 116–125. ACM Press, 2001.
13. P. Paillier. Public-key cryptosystems based on composite degree residuosity classes. In *Advances in Cryptology – Eurocrypt '99*, volume 1592 of *LNCS*, pages 223–238. Springer Verlag, 1999.
14. T. P. Pedersen. Non-interactive and information-theoretic secure verifiable secret sharing. In *Advances in Cryptology – Crypto '91*, volume 576 of *LNCS*, pages 129–140. Springer Verlag, 1992.
15. A. Shamir. How to share a secret. *Communications of the ACM*, 22(11):612–613, 1979.
16. D. Wikström. A universally composable mix-net. In *1st Theory of Cryptography Conference (TCC)*, volume 2951 of *LNCS*, pages 315–335. Springer Verlag, 2004.
17. D. Wikström. A sender verifiable mix-net and a new proof of a shuffle. In *Advances in Cryptology – Asiacrypt 2005*, volume 3788 of *LNCS*, pages 273–292. Springer Verlag, 2005.

Multipartite Secret Sharing by Bivariate Interpolation

Tamir Tassa[1] and Nira Dyn[2]

[1] Division of Computer Science, The Open University, Ra'anana, Israel
[2] Department of Applied Mathematics, Tel Aviv University, Tel Aviv, Israel

Abstract. Given a set of participants that is partitioned into distinct compartments, a multipartite access structure is an access structure that does not distinguish between participants that belong to the same compartment. We examine here three types of such access structures - compartmented access structures with lower bounds, compartmented access structures with upper bounds, and hierarchical threshold access structures. We realize those access structures by ideal perfect secret sharing schemes that are based on bivariate Lagrange interpolation. The main novelty of this paper is the introduction of bivariate interpolation and its potential power in designing schemes for multipartite settings, as different compartments may be associated with different lines in the plane. In particular, we show that the introduction of a second dimension may create the same hierarchical effect as polynomial derivatives and Birkhoff interpolation were shown to do in [13].

Keywords: Secret sharing, multipartite access structures, compartmented access structures, hierarchical threshold access structures, bivariate interpolation, monotone span programs.

1 Introduction

Let $\mathcal{U}$ be a set of participants and assume that it is partitioned into m disjoint subsets,

$$\mathcal{U} = \bigcup_{i=1}^{m} \mathcal{C}_i , \tag{1}$$

to which we refer hereinafter as *compartments*. An m-partite access structure on $\mathcal{U}$ is any access structure that does not distinguish between members of the same compartment. More specifically, an access structure $\Gamma \in 2^{\mathcal{U}}$ is m-partite with respect to partition (1) if for all permutations $\pi : \mathcal{U} \to \mathcal{U}$ such that $\pi(\mathcal{C}_i) = \mathcal{C}_i$, $1 \le i \le m$, $\mathcal{A} \in \Gamma$ if and only if $\pi(\mathcal{A}) \in \Gamma$. Weighted threshold access structures [11, 1], multilevel access structures [12, 3], hierarchical threshold access structures [13], compartmented access structures [3, 8], bipartite access structure [10], and tripartite access structures [1, 5, 8] are typical examples of such multipartite access structures. (Of-course, every access structure may be viewed as a multipartite access structure with singleton compartments; however,

M. Bugliesi et al. (Eds.): ICALP 2006, Part II, LNCS 4052, pp. 288–299, 2006.

that term is reserved to non-degenerate cases where the number of compartments is smaller than the number of participants.)

In this paper we show how to utilize bivariate interpolation in order to realize some multipartite access structures. What makes bivariate interpolation suitable for multipartite settings is the ability to associate each compartment with a different line in the plane. Namely, participants from a given compartment are associated with points that lie on the same line, where each compartment is associated with a different line.

In Section 2 we deal with compartmented access structures. We distinguish between two types of such structures: one that agrees with the type that was presented and studied by Brickell in [3], and another that we present here for the first time. We design for those access structures ideal secret sharing schemes that are based on bivariate Lagrange interpolation with data on parallel lines. In Section 3 we deal with hierarchical threshold access structures and we realize them by bivariate Lagrange interpolation with data on lines in general position. In [13], those access structures were realized by introducing polynomial derivatives and Birkhoff interpolation in order to create the desired hierarchy between the different compartments (that are called *levels* in that context). Here, we show that we may achieve the same hierarchical effect by introducing a second dimension, in lieu of polynomial derivatives. All necessary background from bivariate interpolation theory is provided herein. Finally, in Section 4, we contemplate on the possible advantages of using more involved interpolation settings.

Hereinafter, $\mathbb{F}$ is a finite field of size $q = |\mathbb{F}|$. The field size is large enough so that the domain of all possible secrets may be embedded in $\mathbb{F}$. The secret $S \in \mathbb{F}$ will be encoded by the coefficients of an unknown polynomial $P(x, y) \in \mathbb{F}[x, y]$. We also adopt the following notation convention: vectors are denoted by boldface letters while their components are denoted with the corresponding italic-type indexed letter. In addition, $\mathbb{N}$ stands for the nonnegative integers.

Throughout this study we use the following basic lemma that provides an upper bound for the number of zeros of a multivariate polynomial over a finite field.

Lemma 1. *Let $G(z_1, \ldots, z_k)$ be a nonzero polynomial of k variables over a finite field $\mathbb{F}$ of size q. Assume that the highest degree of each of the variables z_j in G is no larger than d. Then the number of zeros of G in $\mathbb{F}^k$ is bounded from above by kdq^{k-1}.*

All proofs are given in the full version of this paper.

2 Compartmented Access Structures

The original compartmented access structure that was presented in [3] is defined as follows. Let $t_i \in \mathbb{N}$, $1 \le i \le m$, and $t \in \mathbb{N}$ be thresholds such that $t \ge \sum_{i=1}^{m} t_i$. Then

$$\Gamma = \{\mathcal{V} \subseteq \mathcal{U} : \exists \mathcal{W} \subseteq \mathcal{V} \;\text{ such that }\; |\mathcal{W} \cap \mathcal{C}_i| \ge t_i, \;\; 1 \le i \le m, \;\text{and}\; |\mathcal{W}| = t\} \,. \tag{2}$$

Such access structures are suitable for situations in which the size of an authorized subset must be at least some threshold t, but, in addition to that, we wish to guarantee that every compartment is represented by at least some number of participants in the authorized subset. In other situations, however, an opposite demand may occur: while the size of an authorized subset must be at least t, we would like to limit the number of participants that represent each of the compartments; namely,

$$\Delta = \{\mathcal{V} \subseteq \mathcal{U} : \exists \mathcal{W} \subseteq \mathcal{V} \text{ such that } |\mathcal{W} \cap \mathcal{C}_i| \leq s_i, \quad 1 \leq i \leq m, \text{ and } |\mathcal{W}| = s\} , \tag{3}$$

where $s_i, s \in \mathbb{N}$ and $s \leq \sum_{i=1}^{m} s_i$. We refer to Γ as a *compartmented access structure with lower bounds*, while Δ is referred to hereinafter as a *compartmented access structure with upper bounds*.

When $m = 1$ both types of compartmented access structures coincide with the standard threshold access structures of Shamir [11]. When $m = 2$ the two types of access structures agree: a compartmented access structure with lower bounds t_1, t_2 and t is a compartmented access structure with upper bounds $s_1 = t - t_2$, $s_2 = t - t_1$ and $s = t$; conversely, an access structure of type (3) with bounds s_1, s_2 and s may be viewed as an access structure of type (2) with bounds $t_1 = s - s_2$, $t_2 = s - s_1$ and $t = s$. However, when $m \geq 3$, these two types of compartmented access structures differ. As an example, consider an access structure of type (2) where $m = 3$, $t_1 = 1$, $t_2 = 1$, $t_3 = 2$, and $t = 5$. Then the minimal subsets $\mathcal{V}$ are of types $(1, 1, 3)$ (namely, $|\mathcal{V} \cap \mathcal{C}_1| = 1$, $|\mathcal{V} \cap \mathcal{C}_2| = 1$, and $|\mathcal{V} \cap \mathcal{C}_3| = 3$), $(1, 2, 2)$, or $(2, 1, 2)$. This collection of minimal subsets does not fall within the framework (3) for any choice of s_i and s. Indeed, if that collection of subsets was to fall under framework (3) then we should have $s_1 = s_2 = 2, s_3 = 3$, and $s = 5$; but then that collection should have included also subsets of type (0,2,3), which it doesn't. Hence, there is no way of fitting that compartmented access structure with lower bounds within the framework with upper bounds.

Compartmented access structures with lower bounds, (2), are already known to be ideal. We design here ideal linear schemes for these access structures, as well as for the corresponding access structures with upper bounds, (3), that are based on bivariate interpolation.

2.1 Ideal Secret Sharing for Compartmented Access Structures with Upper Bounds

In this section we describe a linear secret sharing scheme for compartmented access structures with upper bounds, (3). Let x_i, $1 \leq i \leq m$, be m distinct points in $\mathbb{F}$ and let $P_i(y)$ be a polynomial of degree $s_i - 1$ over $\mathbb{F}$. Define

$$P(x, y) = \sum_{i=1}^{m} P_i(y) L_i(x) = \sum_{i=1}^{m} \sum_{j=0}^{s_i - 1} a_{i,j} \cdot y^j L_i(x) , \tag{4}$$

where $L_i(x)$ is the Lagrange polynomial of degree $m-1$ over $\{x_i : 1 \le i \le m\}$, namely,

$$L_i(x) = \prod_{\substack{1\le j\le m\\ j\ne i}} \frac{x - x_j}{x_i - x_j} \ . \tag{5}$$

These polynomials are orthogonal in the sense that $L_i(x_j) = \delta_{i,j}$ for all $1 \le i, j \le m$. Then the secret sharing scheme is as follows:

Secret Sharing Scheme 1

1. *The secret is* $S = \sum_{i=1}^{m} \sum_{j=0}^{s_i-1} a_{i,j}$.
2. *Each participant* $u_{i,j}$ *from compartment* $\mathcal{C}_i$ *will be identified by a unique public point* $(x_i, y_{i,j})$, *where* $y_{i,j} \ne 1$, *and his private share will be the value of* P *at that point.*
3. *In addition, we publish the value of* P *at* $k := \sum_{i=1}^{m} s_i - s$ *points* (x_i', z_i), *where* $x_i' \notin \{x_1, \ldots, x_m\}$, $1 \le i \le k$.

Figure 1 illustrates that scheme for the case of $m = 3$ compartments and $k = s_1 + s_2 + s_3 - s = 3$. The $k = 3$ public point values are denoted by full bullets. The point values that correspond to the participants are marked by empty circles along the three random parallel lines, $x = x_i$, $1 \le i \le 3$.

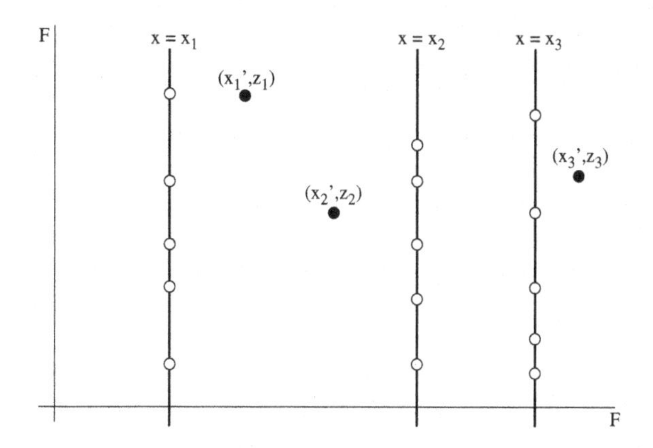

Fig. 1. Secret Sharing Scheme 1

Clearly, this is an ideal scheme since the private shares of all users are taken from the domain of secrets $\mathbb{F}$. The number of unknowns in the polynomial P is $\sum_{i=1}^{m} s_i$ (the coefficients of each of the univariate polynomials $P_i(y)$, $1 \le i \le m$). Since we are given for free $k := \sum_{i=1}^{m} s_i - s$ point values, we need additional s points for full recovery. Moreover, we cannot use more than s_i points from the line $x = x_i$, $1 \le i \le m$, because any s_i points from along that line already fully recover $P_i(y)$, but they do not contribute anything towards the recovery of $P_j(y)$ for $j \ne i$. In view of the above, this scheme agrees with the constraints in (3). We proceed to show that, with high probability, the resulting scheme is perfect.

Theorem 1. *With probability* $1 - O(q^{-1})$*, the ideal Secret Sharing Scheme 1 is a perfect scheme that realizes the compartmented access structure with upper bounds (3).*

We would like to stress that the probability here is with respect to the choices of the points in the plane. Once such a choice was made, the dealer may check that all authorized subsets may recover the secret while all non-authorized subsets may not learn a thing about the secret. If all subsets pass that test then the resulting scheme is perfectly secure. In the rare event (of probability $O(q^{-1})$) that one of the subsets did not pass the test, the dealer has only to try another selection.

2.2 Ideal Secret Sharing for Compartmented Access Structures with Lower Bounds

In this section we describe a linear secret sharing scheme for compartmented access structures with lower bounds, (2). To that end, we construct a scheme for the dual access structure Γ^*. Let us begin with a brief overview of what are dual access structures and recall the main result concerning duality that we need for the design of our scheme.

Karchmer and Wigderson [9] introduced monotone span programs as a linear algebraic model of computation for computing monotone functions. A monotone span program (MSP hereinafter) is a quintuple $\mathcal{M} = (\mathbb{F}, M, \mathcal{U}, \phi, \mathbf{e})$ where $\mathbb{F}$ is a field, M is a matrix of dimensions $a \times b$ over $\mathbb{F}$, $\mathcal{U} = \{u_1, \ldots, u_n\}$ is a finite set, ϕ is a surjective function from $\{1, \ldots, a\}$ to $\mathcal{U}$, and $\mathbf{e}$ is some target row vector from $\mathbb{F}^b$. The MSP $\mathcal{M}$ realizes the monotone access structure $\Gamma \subset 2^{\mathcal{U}}$ when $\mathcal{V} \in \Gamma$ if and only if $\mathbf{e}$ is spanned by the rows of the matrix M whose labels belong to $\mathcal{V}$. The size of $\mathcal{M}$ is a, the number of rows in M. Namely, in the terminology of secret sharing, the size of the MSP is the total number of shares that were distributed to all participants in $\mathcal{U}$. An MSP is ideal if $a = n$.

If Γ is a monotone access structure over $\mathcal{U}$, its dual is defined by $\Gamma^* = \{\mathcal{V} : \mathcal{V}^c \notin \Gamma\}$. It is easy to see that Γ^* is also monotone. In [7] it was shown that if $\mathcal{M} = (\mathbb{F}, M, \mathcal{U}, \phi, \mathbf{e})$ is an MSP that realizes a monotone access structure Γ, then there exists an MSP $\mathcal{M}^* = (\mathbb{F}, M^*, \mathcal{U}, \phi, \mathbf{e}^*)$ of the same size like $\mathcal{M}$ that realizes the dual access structure Γ^*. Hence, an access structure is ideal if and only if its dual is. An efficient construction of the MSP for the dual access structure was proposed in [6].

Realizing the Dual Access Structure. The dual access structure of (2) is given by $\Gamma^* = \{\mathcal{V} : \mathcal{V}^c \notin \Gamma\}$. Hence, $\mathcal{V} \in \Gamma^*$ if and only if $|\mathcal{V}^c| < t$ or $|\mathcal{V}^c \cap C_i| < t_i$ for some $1 \leq i \leq m$. Introducing the notations $n = |\mathcal{U}|$ and $n_i = |C_i|$, $1 \leq i \leq m$, we infer that $\mathcal{V} \in \Gamma^*$ if and only if $|\mathcal{V}| \geq n - t + 1$ or $|\mathcal{V} \cap C_i| \geq n_i - t_i + 1$ for some $1 \leq i \leq m$. Namely,

$$\Gamma^* = \{\mathcal{V} \subseteq \mathcal{U} : |\mathcal{V}| \geq r \text{ or } |\mathcal{V} \cap C_i| \geq r_i \text{ for some } 1 \leq i \leq m\} , \quad (6)$$

where

$$r = n - t + 1 \text{ and } r_i = n_i - t_i + 1 , \quad 1 \leq i \leq m . \quad (7)$$

Since $t \geq \sum_{i=1}^{m} t_i$ and $n = \sum_{i=1}^{m} n_i$, we see that

$$\sum_{i=1}^{m} r_i = \sum_{i=1}^{m} n_i - \sum_{i=1}^{m} t_i + m \geq n - t + m = r + m - 1 \ .$$

Therefore, the thresholds in the dual access structure (6) satisfy

$$\sum_{i=1}^{m} r_i \geq r + m - 1 \ . \tag{8}$$

We proceed to describe a linear ideal secret sharing scheme for realizing such access structures and then prove that, with high probability, it is perfect.

Let x_i, $1 \leq i \leq m$, be m distinct points in $\mathbb{F}$ and let $P_i(y)$ be a polynomial of degree $r_i - 1$ over $\mathbb{F}$, such that

$$P_1(0) = \cdots = P_m(0) \ . \tag{9}$$

Define

$$P(x,y) = \sum_{i=1}^{m} P_i(y) L_i(x) = \sum_{i=1}^{m} \sum_{j=0}^{r_i - 1} a_{i,j} \cdot y^j L_i(x) \ , \tag{10}$$

where $L_i(x)$, $1 \leq i \leq m$, are, as before, the Lagrange polynomials of degree $m-1$ over $\{x_i : 1 \leq i \leq m\}$, (5). Note that condition (9) implies that $a_{1,0} = \cdots = a_{m,0}$ and, consequently, that the number of unknown coefficients in the representation of $P(x,y)$ with respect to the basis $L_i(x)y^j$, $1 \leq i \leq m$, $0 \leq j \leq r_i - 1$, is $g = \sum_{i=1}^{m} r_i - (m-1)$. Note that by (8), $g \geq r$.

Our secret sharing scheme for the realization of the dual access structure Γ^*, (6), is as follows:

Secret Sharing Scheme 2

1. *The secret is* $S = a_{1,0} = \cdots = a_{m,0}$.
2. *Each participant* $u_{i,j}$ *from compartment* C_i *will be identified by a unique public point* $(x_i, y_{i,j})$, *where* $y_{i,j} \neq 0$, *and his private share will be the value of* P *at that point.*
3. *In addition, we publish the value of* P *at* $k = g - r$ *points* (x'_i, z_i), *where* $x'_i \notin \{x_1, \ldots, x_m\}$, $1 \leq i \leq k$.

Theorem 2. *With probability* $1 - O(q^{-1})$, *the ideal Secret Sharing Scheme 2 is a perfect scheme that realizes the access structure (6).*

A Scheme for Compartmented Access Structures with Lower Bounds. Using the results of Section 2.2 we may now easily construct an ideal secret sharing scheme for compartmented access structures with lower bounds, (2). Given such an access structure, Γ, we construct the ideal linear secret sharing scheme for its dual, (6)-(7). Then we translate that ideal scheme (equivalently, MSP) into an ideal scheme (MSP) for $\Gamma = (\Gamma^*)^*$, using the explicit construction that is described in [6]. We omit further details.

3 Hierarchical Threshold Access Structures

3.1 Lagrange Interpolation with Data on Lines in General Positions

Let

$$\{L_i\}_{1\le i\le n}\ ,\quad L_i = \{(x,y)\in\mathbb{F}^2 : L_i(x,y) := a_i x + b_i y + c_i = 0\}\ ,$$

be a collection of n lines in $\mathbb{F}^2$ in general position. Namely, for every pair $1 \le i < j \le n$, L_i and L_j intersect in a point $A_{i,j} = (x_{i,j}, y_{i,j})$ and $A_{i,j} \neq A_{k,\ell}$ whenever $\{i,j\} \neq \{k,\ell\}$ (Figure 2 illustrates the case $n = 4$). Let $f(x,y)$ be a function on $\mathbb{F}^2$. Then there exists a unique polynomial of degree $n-2$,

$$P(x,y) = \sum_{0\le i+j\le n-2} a_{i,j}x^i y^j \in \mathbb{F}[x,y]\ , \tag{11}$$

that satisfies

$$P(x_{i,j}, y_{i,j}) = f(x_{i,j}, y_{i,j}) \quad 1 \le i < j \le n\ . \tag{12}$$

That polynomial is given by

$$P(x,y) = \sum_{1\le i<j\le n} f(x_{i,j}, y_{i,j}) L_{i,j}(x,y) \tag{13}$$

where

$$L_{i,j}(x,y) = \prod_{\substack{1\le k\le n\\ k\neq i,j}} \frac{L_k(x,y)}{L_k(x_{i,j}, y_{i,j})}\ . \tag{14}$$

The bivariate Lagrange polynomials $L_{i,j}(x,y)$ are of degree $n-2$, and they form an orthogonal set in the sense that $L_{i,j}(x_{i,j}, y_{i,j}) = 1$ while $L_{i,j}(x_{k,\ell}, y_{k,\ell}) = 0$ for all $\{k,\ell\} \neq \{i,j\}$ (because the point $(x_{k,\ell}, y_{k,\ell})$ lies on a line other than L_i or L_j, whence the numerator in (14) becomes zero). Note that the number of independent terms (monoms) in (11) agrees with the number of constraints in (12), i.e., $\binom{n}{2}$.

This type of bivariate interpolation was studied first in [4]. We shall be using this bivariate interpolation in a slightly different manner hereinafter. As described above, in order to recover a polynomial $P(x,y)$ of degree k, we need its values at the intersection points of $k+2$ lines in general position. Assume, however, that we have only $k+1$ lines in general position, but we were able to fully recover the restriction of $P(x,y)$ to each of these lines (the restriction of a bivariate polynomial of degree k to a line is the univariate polynomial of degree k that is obtained by replacing x and y in $P(x,y)$ with their linear parameterization along that line). Then that information is also sufficient for the full recovery of $P(x,y)$ since we may add a $(k+2)$th line that intersects all of the original $k+1$ lines and then, as we know the value of P along each of those $k+1$ lines, we know its value in all of the $\binom{k+2}{2}$ intersection points of the $k+2$ lines; this enables the full recovery of $P(x,y)$ through (13)-(14). For example, in order to recover a quadratic polynomial $P(x,y)$ ($k = 2$), we need its values in the 6 intersection points of $k+2=4$ lines in general position (L_1, L_2, L_3 and L_4 in Figure 2); alternatively, we may compute its restriction to only $k+1=3$ of those lines, say L_1, L_2, and L_3, and that is sufficient for finding the value of P in all 6 intersection points of L_1, L_2, L_3 and L_4. Hence, while in this section our setting included n lines and a polynomial $P(x,y)$ of degree $n-2$, in the following sections our settings will include n lines and a polynomial $P(x,y)$ of degree $n-1$.

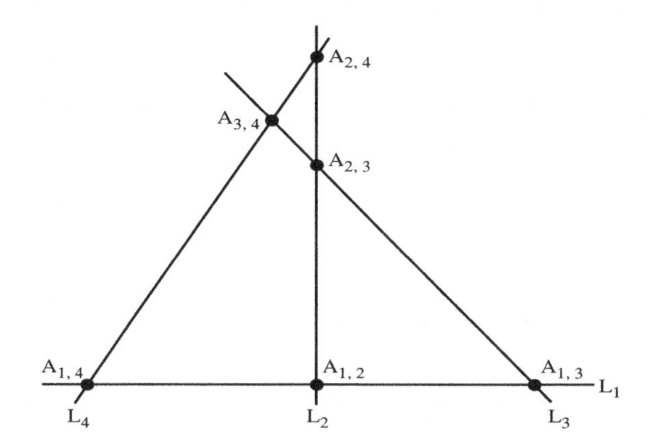

Fig. 2. Four lines in general position and the corresponding interpolation points

3.2 Constructibility and Non-constructibility Results

Let:

- $\mathbb{F}$ be a finite field;
- $\{L_i\}_{1\le i\le n}$, $L_i=\{(x,y)\in\mathbb{F}^2 : L_i(x,y):=a_ix+b_iy+c_i=0\}$, be a collection of n lines in $\mathbb{F}^2$ in general position;
- $P(x,y)=\sum_{0\le i+j\le n-1} a_{i,j}x^iy^j$ be a polynomial of degree (at most) $n-1$ in $\mathbb{F}[x,y]$; and
- $\mathcal{V}\subset\bigcup_{i=1}^n L_i$ be a set of points on the given lines, none of which is an intersection point of two of those lines.

The question that we address here is the amount of information that $D := P|_{\mathcal{V}}$ reveals on $S:=P(0,0)$. Since the underlying model is that of a monotone span program, it is clear that either the given data, D, uniquely determines the unknown, S, or the former does not reveal any information about the latter.

In order to answer that question, we define the *type* of a set $\mathcal{V}$ (Definition 1) and an order on such types (Definition 2).

Definition 1. *Let $\{L_i\}_{1\le i\le n}$ be n lines in general position in $\mathbb{F}^2$. A subset*

$$\mathcal{V}\subset\left(\bigcup_{i=1}^n L_i\right)\setminus\left(\bigcup_{1\le i<j\le n} L_i\cap L_j\right) \tag{15}$$

is said to be of type $\mathbf{v} \in \mathbb{N}^n$*, where* $\mathbf{v}$ *is a monotone vector in the sense that* $0 \le v_1 \le v_2 \le \cdots \le v_n$*, if there exists a permutation* $\pi \in S_n$ *such that* $|\mathcal{V} \cap L_{\pi(i)}| = v_i$ *for all* $1 \le i \le n$*.*

For example, the two subsets depicted in Figure 3 are of type $\mathbf{v} = (2, 2, 3)$.

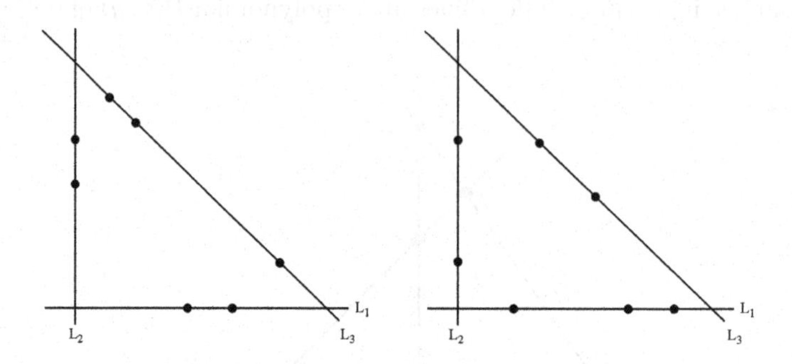

Fig. 3. Two point sets of the same type

Definition 2. *A vector* $\mathbf{u} \in \mathbb{N}^n$ *dominates the vector* $\mathbf{v} \in \mathbb{N}^n$*, denoted* $\mathbf{u} \succeq \mathbf{v}$*, if for all* $1 \le i \le n$*,* $\sum_{j=1}^{i} u_j \ge \sum_{j=1}^{i} v_j$*.*

For example, $(1, 3, 3, 3) \succeq (1, 2, 3, 4)$ while $(1, 1, 4, 5) \not\succeq (1, 2, 3, 4)$.

Definition 3. *Let* $\{L_i\}_{1 \le i \le n}$ *be* n *lines in general position in* $\mathbb{F}^2$*, and let* $\mathcal{V}$ *be a set of points on those lines, (15). The vector set that corresponds to* $\mathcal{V}$ *is the set of vectors*

$$R_{\mathcal{V}} = \{\mathbf{r}_{(x,y),n} : (x, y) \in \mathcal{V}\} \tag{16}$$

where

$$\mathbf{r}_{(x,y),n} := (1, x, y, x^2, xy, y^2, \ldots, x^{n-1}, x^{n-2}y, \ldots, xy^{n-2}, y^{n-1}) \in \mathbb{F}^{n(n+1)/2} \tag{17}$$

Theorem 3. *Let* $\{L_i\}_{1 \le i \le n}$ *be* n *lines in general position in* $\mathbb{F}^2$*, none of which goes through the origin* $(0, 0)$*. Let* $\mathcal{V}$ *be a randomly selected set of points on those lines, (15), and let* $\mathbf{v}$ *be the type of that set. Then the following claim holds in probability* $1 - O(q^{-1})$*:*

1. *If* $\mathbf{v} \succeq (1, 2, \ldots, n)$ *then* $\mathbf{e}_1 \in Span\{R_{\mathcal{V}}\}$*.*
2. *If* $\mathbf{v} \not\succeq (1, 2, \ldots, n)$ *then* $\mathbf{e}_1 \notin Span\{R_{\mathcal{V}}\}$*.*

This theorem actually characterizes the sets of data points that allow the construction of a polynomial $P(x, y) \in \mathbb{F}_{n-1}[x, y]$. Given the values of the polynomial in the points of $\mathcal{V}$, $P|_{\mathcal{V}}$, it is possible (with almost certainty) to reconstruct the entire polynomial (and not just the free coefficient $P(0, 0)$ that corresponds to the vector $\mathbf{e}_1$) if the type of the data set, $\mathbf{v}$, dominates the vector $(1, 2, \ldots, n)$. If, on the other hand, $\mathbf{v} \not\succeq (1, 2, \ldots, n)$, then there is not enough data to reconstruct the polynomial, and this implies (with almost certainty) that we cannot learn any information about $P(0, 0)$.

3.3 Hierarchical Threshold Access Structures

Let $\mathcal{U}$ be a set of participants that is partitioned into m disjoint *levels*, (1), and let $k_1 < k_2 < \cdots < k_m$ be a sequence of thresholds. The corresponding hierarchical threshold access structure is defined by

$$\Gamma = \left\{ \mathcal{V} \subset \mathcal{U} : \left| \mathcal{V} \cap \left(\cup_{j=1}^{i} \mathcal{C}_j \right) \right| \geq k_i \text{ for all } 1 \leq i \leq m \right\} . \tag{18}$$

Those access structures were presented and studied in [13]. They are realized there by an ideal secret sharing scheme that is based on Birkhoff interpolation, namely, interpolation in which the given values of the unknown polynomial, $P(x)$, include also derivative values. Specifically, participants from level $\mathcal{C}_i$, $1 \leq i \leq m$, receive the value of the (k_{i-1})th derivative of P at the point x that identifies them (where hereinafter $k_0 := 0$). As participants from higher levels (namely, $\mathcal{C}_i$ for lower values of i) have shares that equal derivatives of P of lower orders, those shares carry more information on the coefficients of P than shares of participants from lower levels.

Here we show how to realize such hierarchical access structures using bivariate Lagrange interpolation on lines in general position. The scheme that we present here does not use derivatives, as the Birkhoff interpolation-based scheme of [13] did, but instead it adds one more dimension in order to achieve the same hierarchical effect.

Let $\{L_j\}_{1 \leq j \leq n}$ be $n := k_m$ lines in general position in $\mathbb{F}^2$, none of which goes through the origin $(0,0)$. Let

$$P(x,y) = \sum_{0 \leq i+j \leq n-1} a_{i,j} x^i y^j$$

be a random polynomial in $\mathbb{F}_{n-1}[x,y]$.

Secret Sharing Scheme 3

1. *The secret is* $S = P(0,0)$.
2. *Each participant from level* $\mathcal{C}_i$ *will be identified by a unique public point on* $L_{k_i} \setminus \left(\bigcup_{\substack{1 \leq j \leq n \\ j \neq k_i}} L_j \right)$ *and his private share will be the value of* P *at that point.*
3. *In addition, we publish the value of* P *at:*
   - k_{i-1} *additional points on* L_{k_i}, $2 \leq i \leq m$; *and*
   - j *points on* L_j *for all* $j \in \{1, 2, \ldots, n\} \setminus \{k_i : 1 \leq i \leq m\}$.

Example. Assume that there are $m = 3$ levels with thresholds $k_1 = 2$, $k_2 = 4$ and $k_3 = 5$ (namely, $\mathcal{V} \in \Gamma$ if and only if it has at least 2 participants from the highest level $\mathcal{C}_1$, at least 4 participants from the two highest levels $\mathcal{C}_1 \cup \mathcal{C}_2$, and at least 5 participants altogether). Then we select 5 random lines in general position: L_i, $1 \leq i \leq 5$. The allocation of private shares will be as follows:

1. Participants from C_1 will be given polynomial shares on L_2 (since $k_1 = 2$).
2. Participants from C_2 will be given polynomial shares on L_4 (since $k_2 = 4$).
3. Participants from C_3 will be given polynomial shares on L_5 (since $k_3 = 5$).

The corresponding points are marked in Figure 4 by empty circles. The public values will be:

1. 2 point values on L_4, and 4 point values on L_5 (those points are marked by full bullets in Figure 4).
2. 1 point value on L_1 and 3 point values on L_3 (those points are marked by full squares in Figure 4).

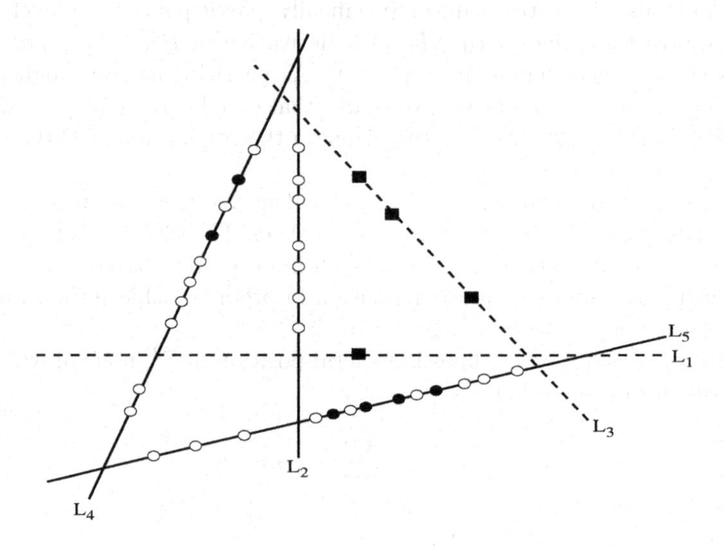

Fig. 4. Secret Sharing Scheme 3

Theorem 4. *With probability* $1 - O(q^{-1})$, *the ideal Secret Sharing Scheme 3 is a perfect scheme that realizes the hierarchical threshold access structure (18).*

4 Epilogue

The advantage of bivariate interpolation over the standard univariate one in designing linear secret sharing schemes for multipartite settings is in the ability to associate different compartments with different lines in the plane. Bivariate interpolation on lines was extended to multivariate interpolation on flats in several dimensions in [2]. By going to higher dimensions and by adequately choosing the flats that represent the compartments, it might be possible to design secret sharing schemes for a wide array of interesting access structures. (In several dimensions we have more flexibility in choosing the dimensions of the flats and their interrelation.) It would be also interesting to explore the possible advantages of using non-linear manifolds instead of flats.

References

1. A. Beimel, T. Tassa and E. Weinreb, Characterizing ideal weighted threshold secret sharing, *The Proceedings of the Second Theory of Cryptography Conference, TCC 2005*, February 2005, MIT, pp. 600-619.
2. C. de Boor, N. Dyn and A. Ron, Polynomial interpolation to data on flats in $\mathcal{R}^d$, *Journal of Approximation Theory*, 105 (2000), pp. 313-343.
3. E.F. Brickell, Some ideal secret sharing schemes, *Journal of Combinatorial Mathematics and Combinatorial Computing*, 9 (1989), pp. 105-113.
4. K.C. Chung and T.H. Yao, On lattices admitting unique Lagrange interpolation, *SIAM Journal on Numerical Analysis*, 14 (1977), pp. 735-743.
5. M.J. Collins, A note on ideal tripartite access structures, available at http://eprint.iacr.org/2002/193/ (2002).
6. S. Fehr Efficient construction of the dual span program, *Manuscript*, May 1999.
7. A. Gál, *Combinatorial methods in Boolean function complexity*, Ph.D. thesis, University of Chicago, 1995.
8. J. Herranz and G. Sáez, New results on multipartite access structures, available at http://eprint.iacr.org/2006/048 (2006).
9. M. Karchmer and A. Wigderson, On span programs, in *The Proceedings of the 8th Structures in Complexity conference*, (1993), pp. 102-111.
10. C. Padró and G. Sáez, Secret sharing schemes with bipartite access structure, *IEEE Transactions on Information Theory*, 46 (2000), pp. 2596-2604.
11. A. Shamir, How to share a secret, *Communications of the ACM*, 22 (1979), pp. 612-613.
12. G.J. Simmons, How to (really) share a secret, *Advances in Cryptology - CRYPTO 88*, LNCS 403 (1990), pp. 390-448.
13. T. Tassa, Hierarchical threshold secret sharing, in *The Proceedings of the First Theory of Cryptography Conference, TCC 2004*, February 2004, MIT, Cambridge, pp. 473-490. (To appear in Journal of Cryptology).

Identity-Based Encryption Gone Wild

Michel Abdalla[1], Dario Catalano[1], Alexander W. Dent[2],
John Malone-Lee[3], Gregory Neven[1,4], and Nigel P. Smart[3]

[1] Département d'Informatique, Ecole Normale Supérieure,
45 rue d'Ulm, 75230 Paris Cedex 05, France
{Michel.Abdalla, Dario.Catalano, Gregory.Neven}@ens.fr
[2] Information Security Group,
Royal Holloway, University of London,
Egham, Surrey, TW20 0EX, United Kingdom
a.dent@rhul.ac.uk
[3] Department of Computer Science, University of Bristol,
Woodland Road,
Bristol, BS8 1UB, United Kingdom
{malone, nigel}@cs.bris.ac.uk
[4] Department of Electrical Engineering,
Katholieke Universiteit Leuven,
Kasteelpark Arenberg 10, B-3001 Heverlee, Belgium
Gregory.Neven@esat.kuleuven.be

Abstract. In this paper we introduce the notion of identity based encryption with wildcards, or WIBE for short. This allows the encryption of messages to multiple parties with common fields in their identity strings, for example email groups in a corporate hierarchy. We propose a full security notion and give efficient implementations meeting this notion in the standard model and in the random oracle model.

1 Introduction

The concept of identity based cryptography was introduced by Shamir as early as in 1984 [12]. However, it took nearly twenty years for an efficient identity based encryption (IBE) scheme to be proposed. In 2000 and 2001 respectively Sakai, Ohgishi and Kasahara [11] and Boneh and Franklin [5] proposed IBE schemes based on elliptic curve pairings. Also, in 2001 Cocks proposed a system based on the quadratic residuosity problem [7].

One of the main application areas proposed for IBE is that of email encryption. In this scenario, given an email address, one can encrypt a message to the owner of the email address without needing to obtain an authentic copy of the owner's public key first. In order to decrypt the email the recipient must authenticate itself to a trusted authority who generates a private key corresponding to the email address used to encrypt the message.

Our work is motivated by the fact that many email addresses correspond to groups of users rather than single individuals. Consider the scenario where

M. Bugliesi et al. (Eds.): ICALP 2006, Part II, LNCS 4052, pp. 300–311, 2006.

there is some kind of organisational hierarchy. Take as an example an organisation called ECRYPT which is divided into virtual labs, AZTEC and STVL for example. In addition, these virtual labs are further subdivided into working groups WG1, WG2 and WG3, and an administrative group ADMIN. Finally, each working group may consist of many individual members. There are several extensions of the IBE primitive to such a hierarchical setting (HIBE) [9, 8]. The idea is that each level can issue keys to users on the level below. For example the owner of the ECRYPT key can issue decryption keys for ECRYPT.AZTEC and ECRYPT.STVL.

Suppose that we wish to send an email to all the members of the AZTEC.WG1 working group, which includes personal addresses ECRYPT.AZTEC.WG1.Nigel, ECRYPT.AZTEC.WG1.Dario and ECRYPT.AZTEC.WG1.John. Given a standard HIBE one would have to encrypt the message to each user individually. To address this limitation we introduce the concept of *identity based encryption with wildcards* (WIBE). The way in which decryption keys are issued is exactly as in a standard HIBE scheme; what differs is encryption. Our primitive allows the encrypter to replace any component of the recipient identity with a *wildcard* so that any identity matching the *pattern* can decrypt. Denoting wildcards by $*$, in the example above the encrypter would use the identity ECRYPT.AZTEC.WG1.$*$ to encrypt to all members of the AZTEC.WG1 group. To send a message to the administrative members of all virtual labs, one can simply encrypt to identity ECRYPT.$*$.ADMIN.$*$.

It is often suggested that identity strings should be appended with the date so as to add timeliness to the message, and so try to mitigate the problems associated with key revocation. Using our technique we can now encrypt to a group of users, with a particular date, by encrypting to an identity of the form ECRYPT.AZTEC.WG1.$*$.22Oct2006 for example. Thus any individual in ECRYPT.AZTEC.WG1 in possession of a decryption key for 22nd October 2006 will be able to decrypt.

Our paper proceeds as follows. In the next section we give an overview of existing material that we will build upon. We formally introduce our new primitive and describe an appropriate security model in Section 3. In Section 4 we describe a generic construction that realises a WIBE from any HIBE. The construction is very simple, yet unsatisfactory as it requires secret keys whose size is exponential in the number of levels of the underlying HIBE.

In Section 5 we turn to the problem of constructing a WIBE scheme with polynomial-size (with respect to all relevant parameters) ciphertexts and keys. We present an efficient WIBE scheme based on Waters' HIBE scheme [13], and prove its security by reducing to the security of Waters' HIBE scheme. The proof, just like that of Waters [13], is in the standard model. In the full version of this paper [1] we give two more efficient constructions, based on the Boneh-Boyen [3] and the Boneh-Boyen-Goh [4] HIBE schemes, and provide security proofs in the random oracle model [2]. We compare the efficiency and security of all our schemes in Section 6, and we also sketch how chosen-ciphertext security can be achieved by adapting the technique of Canetti *et al.* [6].

2 Basic Definitions

In this section we introduce some notation, computational problems and basic primitives that we will use throughout the rest of the paper. Let $\mathbb{N} = \{0, 1, \ldots\}$ be the set of natural numbers. Let ε be the empty string. If $n \in \mathbb{N}$, then $\{0,1\}^n$ denotes the set of n-bit strings, and $\{0,1\}^*$ is the set of all bit strings. More generally, if S is a set, then S^n is the set of n-tuples of elements of S. If S is finite, then $x \stackrel{\$}{\leftarrow} S$ denotes the assignment to x of an element chosen uniformly at random from S. If A is an algorithm, then $y \leftarrow \mathsf{A}(x)$ denotes the assignment to y of the output of A on input x, and if A is randomised, then $y \stackrel{\$}{\leftarrow} \mathsf{A}(x)$ denotes that the output of an execution of $\mathsf{A}(x)$ with fresh coins is assigned to y.

THE DECISIONAL BILINEAR DIFFIE-HELLMAN ASSUMPTION. Let $\mathbb{G}, \mathbb{G}_{\mathrm{T}}$ be multiplicative groups of prime order p with an admissible map $\hat{e} : \mathbb{G} \times \mathbb{G} \to \mathbb{G}_{\mathrm{T}}$. By admissible we mean that the map is bilinear, non-degenerate and efficiently computable. Bilinearity means that for all $a, b \in \mathbb{Z}_p$ and all $g \in \mathbb{G}$ we have $\hat{e}(g^a, g^b) = \hat{e}(g, g)^{ab}$. By non-degenerate we mean that $\hat{e}(g, g) = 1$ if and only if $g = 1$.

In such a setting we can define a number of computational problems. We shall be interested in the following problem, called the bilinear decisional Diffie-Hellman (BDDH) problem: For a generator $g \in \mathbb{G}$, given

$$g \;,\; A = g^a \;,\; B = g^b \;,\; C = g^c \text{ and } Z = \hat{e}(g, g)^z,$$

the problem is to determine whether $Z = \hat{e}(g, g)^{abc}$ for hidden values of a, b, c and z. Formally, we define this via a game between an adversary $\mathcal{A}$ and a challenger $\mathcal{C}$. The challenger first generates random values $a, b, c, z \stackrel{\$}{\leftarrow} \mathbb{Z}_p$ and then it flips a bit β. If $\beta = 1$ it passes $\mathcal{A}$ the tuple $(g, A, B, C, \hat{e}(g, g)^{abc})$, if $\beta = 0$ it passes the tuple $(g, A, B, C, \hat{e}(g, g)^z)$. The adversary $\mathcal{A}$ then must output its guess β' for β. The adversary has advantage ϵ in solving the BDDH problem if

$$\left|\Pr[\mathcal{A}(g, A, B, C, \hat{e}(g, g)^{abc}) = 1] - \Pr[\mathcal{A}(g, A, B, C, \hat{e}(g, g)^z) = 1]\right| \geq 2\epsilon,$$

where the probabilities are over the choice of a, b, c, z and over the random coins consumed by $\mathcal{A}$.

Definition 1. *The (t, ϵ) BDDH assumption holds if no t-time adversary has at least ϵ advantage in the above game.*

We note that throughout this paper we will assume that the time t of an adversary includes its code size, in order to exclude trivial "lookup" adversaries.

IDENTITY-BASED ENCRYPTION SCHEMES. An identity-based encryption (IBE) scheme is a tuple of algorithms $\mathcal{IBE} = (\mathsf{Setup}, \mathsf{KeyDer}, \mathsf{Enc}, \mathsf{Dec})$ providing the following functionality. The trusted authority runs Setup to generate a master key pair (mpk, msk). It publishes the master public key mpk and keeps the master secret key msk private. When a user with identity ID wishes to become part of the system, the trusted authority generates a user decryption key $d_{ID} \stackrel{\$}{\leftarrow}$

$\mathsf{KeyDer}(msk, ID)$, and sends this key over a secure and authenticated channel to the user. To send an encrypted message $\mathfrak{m}$ to the user with identity ID, the sender computes the ciphertext $C \stackrel{\$}{\leftarrow} \mathsf{Enc}(mpk, ID, \mathfrak{m})$, which can be decrypted by the user as $\mathfrak{m} \leftarrow \mathsf{Dec}(d_{ID}, C)$. We refer to [5] for details on the security definitions for IBE schemes.

HIERARCHICAL IBE SCHEMES. In a hierarchical IBE (HIBE) scheme, users are organised in a tree of depth L, with the root being the master trusted authority. The identity of a user at level $0 \le \ell \le L$ in the tree is given by a vector $ID = (ID_1, \ldots, ID_\ell) \in (\{0,1\}^*)^\ell$. A HIBE scheme is a tuple of algorithms $\mathcal{HIBE} = (\mathsf{Setup}, \mathsf{KeyDer}, \mathsf{Enc}, \mathsf{Dec})$ providing the same functionality as in an IBE scheme, except that a user $ID = (ID_1, \ldots, ID_\ell)$ at level ℓ can use its own secret key d_{ID} to generate a secret key for any of its children $ID' = (ID_1, \ldots, ID_\ell, ID_{\ell+1})$ via $d_{ID'} \stackrel{\$}{\leftarrow} \mathsf{KeyDer}(d_{ID}, ID_{\ell+1})$. Note that by iteratively applying the KeyDer algorithm, user ID can derive secret keys for any of its descendants $ID' = (ID_1, \ldots, ID_{\ell+\delta})$, $\delta \ge 0$. We will occasionally use the overloaded notation $d_{ID'} \stackrel{\$}{\leftarrow} \mathsf{KeyDer}(d_{ID}, (ID_{\ell+1}, \ldots, ID_{\ell+\delta}))$ to denote this process. The secret key of the root identity at level 0 is $d_\varepsilon = msk$. Encryption and decryption are the same as for IBE, but with vectors as identities instead of ordinary bit strings. For $1 \le i \le \ell$ and $I \subseteq \{1, \ldots, \ell\}$, we will occasionally use the notations $ID|_{\le i}$ to denote the vector $(ID_1, \ldots, ID_i)$, $ID|_{> i}$ to denote $(ID_{i+1}, \ldots, \ell)$, and $ID|_I$ to denote $(ID_{i_1}, \ldots, ID_{i_{|I|}})$ where $i_1, \ldots, i_{|I|}$ are the elements of I in increasing order. Also, if $S \subset \mathbb{N}$, then we define $S|_{\le i} = \{j \in S : j \le i\}$ and $S|_{> i} = \{j \in S : j > i\}$.

The security of a HIBE scheme is defined through the following game. In a first phase, the adversary is given as input the master public key mpk of a freshly generated key pair $(mpk, msk) \stackrel{\$}{\leftarrow} \mathsf{Setup}$ as input. In a chosen-plaintext attack (IND-ID-CPA), the adversary is given access to a key derivation oracle that on input of an identity $ID = (ID_1, \ldots, ID_\ell)$, returns the secret key $d_{ID} \stackrel{\$}{\leftarrow} \mathsf{KeyDer}(msk, ID)$ corresponding to identity ID. In a chosen-ciphertext attack (IND-ID-CCA), the adversary is additionally given access to a decryption oracle that for a given identity $ID = (ID_1, \ldots, ID_\ell)$ and a given ciphertext C returns the decryption $\mathfrak{m} \leftarrow \mathsf{Dec}(\mathsf{KeyDer}(msk, ID), C)$.

At the end of the first phase, the adversary outputs two equal-length challenge messages $\mathfrak{m}_0^*, \mathfrak{m}_1^* \in \{0,1\}^*$ and a challenge identity $ID^* = (ID_1^*, \ldots, ID_{\ell^*}^*)$, where $0 \le \ell^* \le L$. The game chooses a random bit $b \stackrel{\$}{\leftarrow} \{0,1\}^*$, generates a challenge ciphertext $C^* \stackrel{\$}{\leftarrow} \mathsf{Enc}(mpk, ID^*, \mathfrak{m}_b^*)$ and gives C^* as input to the adversary for the second phase, during which it gets access to the same oracles as during the first phase. The adversary wins the game if it outputs a bit $b' = b$ without ever having queried the key derivation oracle on any ancestor identity $ID = (ID_1^*, \ldots, ID_\ell^*)$ of ID^*, $\ell \le \ell^*$, and, additionally, in the IND-ID-CCA case, without ever having queried (ID^*, C^*) to the decryption oracle.

Definition 2. *A HIBE scheme is (t, q_K, ϵ) IND-ID-CPA-secure if all t-time adversaries making at most q_K queries to the key derivation oracle have at most advantage ϵ in winning the IND-ID-CPA game described above.*

Definition 3. *A HIBE scheme is (t, q_K, q_D, ϵ) IND-ID-CCA-secure if all t-time adversaries making at most q_K queries to the key derivation oracle and at most q_D queries to the decryption oracle have at most advantage ϵ in winning the IND-ID-CCA game described above.*

3 Identity-Based Encryption with Wildcards

SYNTAX. Identity-based encryption with wildcards (WIBE) schemes are essentially a generalisation of HIBE schemes where at the time of encryption, the sender can decide to make the ciphertext decryptable not just by a single target identity ID, but by a whole group of users whose identities match a certain pattern. Such a pattern is described by a vector $P = (P_1, \ldots, P_\ell) \in (\{0,1\}^* \cup \{*\})^\ell$, where $*$ is a special wildcard symbol. We say that identity $ID = (ID_1, \ldots, ID_{\ell'})$ *matches* P, denoted $ID \in_* P$, if and only if $\ell' \leq \ell$ and $\forall\, i = 1 \ldots \ell'$: $ID_i = P_i$ or $P_i = *$. Note that under this definition, any ancestor of a matching identity is also a matching identity. This is reasonable for our purposes because any ancestor can derive the secret key of a matching descendant identity anyway.

More formally, a WIBE scheme is a tuple of algorithms $\mathcal{WIBE}$ = (Setup, KeyDer, Enc, Dec) providing the following functionality. The root authority first generates a master key pair $(mpk, msk) \stackrel{\$}{\leftarrow}$ Setup. A user with identity $ID = (ID_1, \ldots, ID_\ell)$ can use its own decryption key d_{ID} to derive a decryption key for any user $ID' = (ID_1, \ldots, ID_\ell, ID_{\ell+1})$ on the level below by calling $d_{ID'} \stackrel{\$}{\leftarrow}$ KeyDer$(d_{ID}, ID_{\ell+1})$. We will again use the overloaded notation KeyDer$(d_{ID}, (ID_{\ell+1}, \ldots, ID_{\ell+\delta}))$ to denote iterative key derivation for descendants. The secret key of the root identity is $d_\varepsilon = msk$.

To create a ciphertext of message $\mathfrak{m} \in \{0,1\}^*$ intended for all identities matching pattern $P = (P_1, \ldots, P_\ell)$, the sender computes $C \stackrel{\$}{\leftarrow}$ Enc$(mpk, P, \mathfrak{m})$. Any of the intended recipients $ID \in_* P$ can decrypt the ciphertext using its own decryption key as $\mathfrak{m} \leftarrow$ Dec(d_{ID}, C). Correctness requires that for all key pairs (mpk, msk) output by Setup, all messages $\mathfrak{m} \in \{0,1\}^*$, all $0 \leq \ell \leq L$, all patterns $P \in (\{0,1\}^* \cup \{*\})^\ell$, and all identities $ID \in (\{0,1\}^*)^{\ell'}$ such that $ID \in_* P$, Dec(KeyDer(msk, ID) , Enc$(mpk, P, \mathfrak{m})$) $= \mathfrak{m}$ with probability one.

SECURITY. We define the security of WIBE schemes in a way that is very similar to the case of HIBE schemes, but where the adversary chooses a challenge pattern instead of an identity to which the challenge ciphertext will be encrypted. Of course, the adversary is not able to query the key derivation oracle for any identity that matches the challenge pattern, nor is it able to query the decryption oracle with the challenge ciphertext and any identity that matches the challenge pattern.

More specifically, security is defined through the following game with an adversary. In the first phase, the adversary is run on input of the master public key of a freshly generated key pair $(mpk, msk) \stackrel{\$}{\leftarrow}$ Setup. In a chosen-plaintext attack (IND-WID-CPA), the adversary is given access to a key derivation oracle that on input $ID = (ID_1, \ldots, ID_\ell)$ returns $d_{ID} \stackrel{\$}{\leftarrow}$ KeyDer(msk, ID). In a chosen-ciphertext attack (IND-WID-CCA), the adversary additionally has access to a decryption oracle that on input a ciphertext C and an identity $ID = (ID_1, \ldots, ID_\ell)$ returns $\mathfrak{m} \leftarrow \mathsf{Dec}(\mathsf{KeyDer}(msk, ID), C)$.

At the end of the first phase, the adversary outputs two equal-length challenge messages $\mathfrak{m}_0^*, \mathfrak{m}_1^*$ and a challenge pattern $P^* = (P_1^*, \ldots, P_{\ell^*}^*)$ where $0 \leq \ell^* \leq L$. The adversary is given a challenge ciphertext $C^* \stackrel{\$}{\leftarrow} \mathsf{Enc}(mpk, P^*, \mathfrak{m}_b^*)$ for a randomly chosen bit b, and is given access to the same oracles as during the first phase of the attack. The second phase ends when the adversary outputs a bit b'. The adversary is said to win the IND-WID-CPA game if $b' = b$ and if it never queried the key derivation oracle for the keys of any identity that matches the target pattern (i.e., any ID such that $ID \in_* P^*$). Also, in a chosen-ciphertext attack (IND-WID-CCA), the adversary cannot query the decryption oracle on C^* with any matching identity $ID \in_* P^*$.

Definition 4. *A WIBE scheme is (t, q_K, ϵ) IND-WID-CPA-secure if all t-time adversaries making at most q_K queries to the key derivation oracle have at most advantage ϵ in winning the IND-WID-CPA game described above.*

Definition 5. *A WIBE scheme is (t, q_K, q_D, ϵ) IND-WID-CCA-secure if all t-time adversaries making at most q_K queries to the key derivation oracle and at most q_D queries to the decryption oracle have at most advantage ϵ in winning the IND-WID-CCA game described above.*

4 A Generic Construction

We first point out that a WIBE scheme can be constructed from any HIBE scheme, albeit with a secret key size that is exponential in the depth of the hierarchy tree. Let "$*$" be a dedicated bitstring that is not allowed to occur as a user identity. Then the secret key of a user with identity $(ID_1, \ldots, ID_\ell)$ in the WIBE scheme contains the HIBE secret keys of all patterns matching this identity, i.e. the secret keys of all 2^ℓ identities $(ID'_1, \ldots, ID'_\ell)$ such that $ID'_i = ID_i$ or $ID'_i =$ "$*$" for all $i = 1, \ldots, \ell$. To encrypt to a pattern $(P_1, \ldots, P_\ell)$, one uses the HIBE scheme to encrypt to the identity obtained by replacing each wildcard in the pattern with the "$*$" string, i.e. the identity $(ID_1, \ldots, ID_\ell)$ where $ID_i =$ "$*$" if $P_i = *$ and $ID_i = P_i$ otherwise. Decryption is done by selecting the appropriate secret key from the list and using the decryption algorithm of the HIBE scheme.

The efficiency of the WIBE scheme thus obtained is roughly the same as that of the underlying HIBE scheme, except that the size of the secret key is 2^ℓ times that of a secret key in the underlying HIBE scheme. This may be acceptable for some applications, but may not be for others. Moreover, from a theoretical point of view, it is interesting to investigate whether WIBE schemes exist with overhead polynomial in all parameters. We answer this question in the affirmative here by presenting direct schemes with secret key size (and, unfortunately, also ciphertext size) linear in ℓ.

5 A Construction from Waters' HIBE Scheme

5.1 Waters' HIBE Scheme

In [13], Waters argued that his IBE scheme can easily be modified into a L-level HIBE scheme as per [3]. Here we explicitly present this construction as it will be useful in the understanding of our construction of a WIBE scheme.

Setup. The trusted authority chooses random generators g_1 and g_2 from $\mathbb{G}$ and a random value $\alpha \xleftarrow{\$} \mathbb{Z}_p$. For $i = 1, \ldots, L$ and $j = 0, \ldots, n$, it chooses group elements $u_{i,j} \xleftarrow{\$} \mathbb{G}$ where L is the maximum hierarchy depth and n is the length of an identity string. Next, it computes $h_1 \leftarrow g_1^\alpha$ and $h_2 \leftarrow g_2^\alpha$. The master public key is $mpk = (g_1, g_2, h_1, u_{1,0}, \ldots, u_{L,n})$, the corresponding master secret key is $msk = h_2$.

Key Derivation. A user's identity is given by a vector $ID = (ID_1, \ldots, ID_\ell)$ where each ID_i is a n-bit string, applying a collision-resistant hash function if necessary. Let "$j \in ID_i$" denote a variable j iterating over all bit positions $1 \leq j \leq n$ such that the j-th bit of ID_i is one. Using this notation, for $i = 1, \ldots, L$, we define the function

$$F_i(ID_i) = u_{i,0} \prod_{j \in ID} u_{i,j}$$

where the $u_{i,j}$ are the elements in the master public key. To compute the decryption key for identity ID from the master secret key, first random values $r_1, \ldots, r_\ell \xleftarrow{\$} \mathbb{Z}_p$ are chosen, then the private key d_{ID} is constructed as

$$(a_0, a_1, \ldots, a_\ell) = \left(h_2 \prod_{i=1}^{\ell} F_i(ID_i)^{r_i} \ , \ g_1^{r_1}, \ldots, g_1^{r_\ell} \right) .$$

A secret key for identity $ID = (ID_1, \ldots, ID_\ell)$ can be computed by its parent with identity $ID|_{\leq \ell-1}$ as follows. Let $d_{ID|_{\leq \ell-1}} = (a_0, a_1, \ldots, a_{\ell-1})$. The parent chooses $r_\ell \xleftarrow{\$} \mathbb{Z}_p$ and outputs

$$d_{ID} = (a_0 \cdot F_i(ID_i)^{r_\ell} \ , \ a_1, \ldots, a_{\ell-1} \ , \ g_1^{r_\ell}) .$$

Encryption. To encrypt a message $\mathfrak{m} \in \mathbb{G}_T$ for identity $ID = (ID_1, \ldots, ID_\ell)$, the sender chooses $t \xleftarrow{\$} \mathbb{Z}_p$; the ciphertext $C = (C_1, C_2, C_3)$ is computed as

$$C_1 \leftarrow g_1^t \ , \ C_2 \leftarrow \left(C_{2,i} = F_i(ID_i)^t \right)_{i=1,\ldots,\ell} \ , \ C_3 \leftarrow \mathfrak{m} \cdot \hat{e}(h_1, g_2)^t \ .$$

Decryption. If the receiver is the root authority (i.e., the empty identity $ID = \varepsilon$) holding the master key $msk = h_2$, then he can recover the message by computing $C_3 / \hat{e}(C_1, h_2)$. Any other receiver with identity $ID = (ID_1, \ldots, ID_\ell)$ and decryption key $d_{ID} = (a_0, a_1, \ldots, a_\ell)$ decrypts a ciphertext $C = (C_1, C_2, C_3)$ as $C_3 \cdot \prod_{i=1}^{\ell} \hat{e}(a_i, C_{2,i}) / \hat{e}(C_1, a_0)$.

Waters [13] informally states that the above HIBE scheme is IND-ID-CPA secure in the sense that if there is an adversary with advantage ϵ against the HIBE making q_K private key extraction queries, then there is an algorithm solving the BDDH problem with advantage $\epsilon' = O((nq_K)^L \epsilon)$.

5.2 A Waters-Based WIBE Scheme

We first introduce some additional notation. If $P = (P_1, \ldots, P_\ell)$ is a pattern, then let $|P| = \ell$ be the length of P, let $\mathrm{W}(P)$ be the set containing all wildcard indices in P, i.e. the indices $1 \le i \le \ell$ such that $P_i = *$, and let $\overline{\mathrm{W}}(P)$ be the complementary set containing all non-wildcard indices. Clearly $\mathrm{W}(P) \cap \overline{\mathrm{W}}(P) = \emptyset$ and $\mathrm{W}(P) \cup \overline{\mathrm{W}}(P) = \{1, \ldots, \ell\}$. We also extend the notations $P|_{\le i}$, $P|_{> i}$ and $P|_I$ that we introduced for identity vectors to patterns in the natural way.

Let $\mathcal{Wa}\text{-}\mathcal{HIBE}$ = (Setup, KeyDer, Enc, Dec) be the HIBE scheme described in Section 5.1. From $\mathcal{Wa}\text{-}\mathcal{HIBE}$, we can build a WIBE scheme $\mathcal{Wa}\text{-}\mathcal{WIBE}$ = (Setup′, KeyDer′, Enc′, Dec′), where Setup′ and KeyDer′ are equal to those of the $\mathcal{Wa}\text{-}\mathcal{HIBE}$ scheme (i.e., Setup′ = Setup and KeyDer′ = KeyDer), and Enc′ and Dec′ are as follows.

Encryption. To create a ciphertext of message $\mathfrak{m} \in \mathbb{G}_\mathrm{T}$ intended for all identities matching pattern $P = (P_1, \ldots, P_\ell)$, the sender chooses $t \stackrel{\$}{\leftarrow} \mathbb{Z}_p$ and outputs the ciphertext $C = (P, C_1, C_2, C_3, C_4)$, where

$$C_1 \leftarrow g_1^t \qquad C_2 \leftarrow \left(C_{2,i} = F_i(P_i)^t\right)_{i \in \overline{\mathrm{W}}(P)}$$
$$C_3 \leftarrow \mathfrak{m} \cdot \hat{e}(h_1, g_2)^t \qquad C_4 \leftarrow \left(C_{4,i,j} = u_{i,j}^t\right)_{i \in \mathrm{W}(P),\ j=0,\ldots,n}$$

Decryption. If the receiver is the root authority (i.e., the empty identity $ID = \varepsilon$) holding the master key $msk = h_2$, then it can recover the message by computing $C_3/\hat{e}(C_1, h_2)$. Any other receiver with identity $ID = (ID_1, \ldots, ID_\ell)$ matching the pattern P to which the ciphertext was created (i.e., $ID \in_* P$) can decrypt the ciphertext $C = (P, C_1, C_2, C_3, C_4)$ by computing $C_2' = \left(C_{2,i}'\right)_{i=1,\ldots,\ell}$ as

$$C_{2,i}' = F_i(ID_i)^t \leftarrow \begin{cases} C_{2,i} & \text{if } i \in \overline{\mathrm{W}}(P) \\ C_{4,i,0} \cdot \prod_{j \in ID_i} C_{4,i,j} & \text{if } i \in \mathrm{W}(P) \end{cases}$$

and by using his secret key to decrypt the ciphertext $C' = (C_1, C_2', C_3)$ via the Dec algorithm of the $\mathcal{Wa}\text{-}\mathcal{HIBE}$ scheme.

Theorem 6. *Let $\mathcal{Wa}\text{-}\mathcal{HIBE}$ be the HIBE scheme in Section 5.1 and let L be the maximum hierarchy depth. Let $\mathcal{Wa}\text{-}\mathcal{WIBE}$ be the WIBE scheme derived from $\mathcal{Wa}\text{-}\mathcal{HIBE}$ as described in Section 5.2. If $\mathcal{Wa}\text{-}\mathcal{HIBE}$ is $(t, q_\mathrm{K}, \epsilon)$ IND-ID-CPA-secure then $\mathcal{Wa}\text{-}\mathcal{WIBE}$ is $(t', q_\mathrm{K}', \epsilon')$ IND-WID-CPA-secure where*

$$t' = t + t_{\exp} L n (1 + q_\mathrm{K}) \ , \quad q_\mathrm{K}' = q_\mathrm{K} \ , \quad \epsilon' \ge \epsilon / 2^L$$

and $t_{\exp}$ is the time it takes to perform an exponentiation in $\mathbb{G}$.

Proof. The proof of Theorem 6 is by contradiction. That is, we first assume that there exists an adversary $\mathcal{A}$ that breaks the IND-WID-CPA-security of the $\mathcal{Wa}\text{-}\mathcal{WIBE}$ scheme and then we show how to efficiently build another adversary $\mathcal{B}$ which uses $\mathcal{A}$ to break the security of the $\mathcal{Wa}\text{-}\mathcal{HIBE}$ scheme.

Let $mpk_{\mathrm{H}} = (g_1, g_2, h_1, u_{1,0}, \ldots, u_{L,n})$ be the master public key of the *Wa-HIBE* scheme that adversary $\mathcal{B}$ receives as input for its first phase. The idea of the proof is that $\mathcal{B}$ will guess upfront where in the challenge pattern P^* the wildcards are going to be, and "project" the non-wildcard levels of the identity tree of the WIBE scheme onto the first levels of the HIBE scheme. In particular, $\mathcal{B}$ will reuse values $u_{i,j}$ from mpk_{H} for the non-wildcard levels, and will embed new values $u'_{i,j}$ values of which $\mathcal{B}$ knows the discrete logarithms for wildcard levels.

First, $\mathcal{B}$ guesses a random vector $\hat{P} = (\hat{P}_1, \ldots, \hat{P}_L) \stackrel{\$}{\leftarrow} \{\varepsilon, *\}^L$. Define the projection function $\pi : \{1, \ldots, L\} \to \{0, \ldots, L\}$ such that

$$\pi(i) = 0 \text{ if } i \in \mathrm{W}(\hat{P}) \quad \text{and} \quad \pi(i) = i - \left|\mathrm{W}(\hat{P})|_{\leq i}\right| \text{ otherwise.}$$

Intuitively, $\mathcal{B}$ will "project" identities at level i of the WIBE scheme onto level $\pi(i)$ of the HIBE scheme whenever $\pi(i) \neq 0$. Next, the adversary $\mathcal{B}$ runs adversary $\mathcal{A}$ providing it as input for its first phase a public-key $mpk_{\mathrm{W}} = (g_1, g_2, h_1, u'_{1,0}, \ldots, u'_{L,n})$, where for all $1 \leq i \leq L$ and $0 \leq j \leq n$, the elements $u'_{i,j}$ are generated as $u'_{i,j} \leftarrow g_1^{\alpha_{i,j}}$ where $\alpha_{i,j} \stackrel{\$}{\leftarrow} \mathbb{Z}_p$ if $i \in \mathrm{W}(\hat{P})$, and $u'_{i,j} \leftarrow u_{\pi(i),j}$ otherwise. Define functions $F'_i(ID'_i) = u'_{i,0} \prod_{j \in ID'_i} u'_{i,j}$. Notice that $mpk_{\mathcal{A}}$ is distributed exactly as it would be if produced by the setup algorithm described in Section 5.2.

During the first phase, $\mathcal{B}$ has to answer all the key derivation queries $ID' = (ID'_1, \ldots, ID'_\ell)$ that $\mathcal{A}$ is allowed to ask. For that, $\mathcal{B}$ first computes the corresponding identity on the HIBE tree $ID = ID'|_{\overline{\mathrm{W}(\hat{P})}}$, which is the identity obtained by removing from ID' all components at levels where $\hat{P}$ contains a wildcard. That is, the identity ID is obtained from ID' by projecting the component at level i of the WIBE onto level $\pi(i)$ of the HIBE if $\pi(i) \neq 0$. $\mathcal{B}$ then queries its own key derivation oracle for the *Wa-HIBE* scheme on input ID to get the key $d = (a_0, \ldots, a_{\pi(\ell)})$. From this, it computes the key $d' = (a'_0, \ldots, a'_\ell)$ as

$$a'_0 \leftarrow a_0 \cdot \prod_{i \in \mathrm{W}(\hat{P})} F'_i(ID'_i)^{r_i} \ , \ a'_i \leftarrow \begin{cases} g_1^{r_i} & \text{if } i \in \mathrm{W}(\hat{P}) \\ a_{\pi(i)} & \text{if } i \in \overline{\mathrm{W}}(\hat{P}) \end{cases}$$

where $r_i \stackrel{\$}{\leftarrow} \mathbb{Z}_p$ for all $i \in \mathrm{W}(\hat{P})$. At the end of its first phase, $\mathcal{A}$ outputs the challenge pattern $P^* = (P^*_1, \ldots, P^*_{\ell^*})$ and challenge messages $\mathfrak{m}^*_0, \mathfrak{m}^*_1$. If $\mathrm{W}(P^*) \neq \mathrm{W}(\hat{P})$ then $\mathcal{B}$ aborts. Otherwise, $\mathcal{B}$ outputs the corresponding HIBE identity $ID^* = P^*|_{\overline{\mathrm{W}(P^*)}}$ together with challenge messages $\mathfrak{m}^*_0, \mathfrak{m}^*_1$. Let $C^* = (C^*_1, C^*_2, C^*_3)$ be the challenge ciphertext that $\mathcal{B}$ receives in return from its challenger, meaning that C^* is an encryption of $\mathfrak{m}^*_b$ with respect to the identity ID^*, where b is the secret bit chosen at random by the challenger. $\mathcal{B}$ sets $C'^*_1 \leftarrow C^*_1$, $C'^*_2 \leftarrow C^*_2$, $C'^*_3 \leftarrow C^*_3$ and $C'^*_4 \leftarrow ({C^*_1}^{\alpha_{i,j}})_{i \in \mathrm{W}(P^*),\, j=0,\ldots,n}$ and sends to $\mathcal{A}$ the ciphertext $C'^* = (P^*, C'^*_1, C'^*_2, C'^*_3, C'^*_4)$ as the input for its second phase. During the second phase, $\mathcal{A}$ is then allowed to issue more key derivation queries, which are answered by $\mathcal{B}$ exactly as in the first phase. When $\mathcal{A}$ outputs a bit b', $\mathcal{B}$ outputs b' and stops.

In order to analyse the success probability of $\mathcal{B}$, we first need to show that the simulation it provides to $\mathcal{A}$ is correct. The secret key $d' = (a'_0, \ldots, a'_\ell)$ returned for identity $(ID'_1, \ldots, ID'_\ell)$ can be seen to be correctly distributed since if $a'_i = g_1^{r_i}$ for $1 \le i \le \ell$ then

$$\begin{aligned} a'_0 &= h_2 \cdot \prod_{i \in \overline{\mathrm{W}}(\hat{P})} F_{\pi(i)}(ID'_i)^{r_i} \cdot \prod_{i \in \mathrm{W}(\hat{P})} F'_i(ID'_i)^{r_i} \\ &= h_2 \cdot \prod_{i \in \overline{\mathrm{W}}(\hat{P})} \left(u_{\pi(i),0} \prod_{j \in ID'_i} u_{\pi(i),j}\right)^{r_i} \cdot \prod_{i \in \mathrm{W}(\hat{P})} F'_i(ID'_i)^{r_i} \\ &= h_2 \cdot \prod_{i \in \overline{\mathrm{W}}(\hat{P})} \left(u'_{i,0} \prod_{j \in ID'_i} u'_{i,j}\right)^{r_i} \cdot \prod_{i \in \mathrm{W}(\hat{P})} F'_i(ID'_i)^{r_i} \\ &= h_2 \cdot \prod_{i=1}^{\ell} F'_i(ID'_i)^{r_i} \end{aligned}$$

Moreover, the challenge ciphertext $C'^* = (P^*, C'^*_1, C'^*_2, C'^*_3, C'^*_4)$ sent to $\mathcal{A}$ can be seen to be correctly formed when $\mathrm{W}(P^*) = \mathrm{W}(\hat{P})$ as follows. Consider the ciphertext $C^* = (C^*_1, C^*_2, C^*_3)$ that $\mathcal{B}$ receives back from the challenger after outputting $(ID^*, \mathtt{m}^*_0, \mathtt{m}^*_1)$ where $ID^* = P^*|_{\overline{\mathrm{W}}(P^*)}$. We know that, for unknown values $t \in \mathbb{Z}_p$ and $b \in \{0,1\}$, $C^*_1 = g^t$, $C^*_3 = \mathtt{m}^*_b \cdot \hat{e}(h_1, g_2)^t$ and

$$C^*_2 = \left(C^*_{2,i} = F_i(ID^*_i)^t\right)_{i=1,\ldots,\pi(\ell^*)} = \left(C'^*_{2,i} = F'_i(P^*_i)^t\right)_{i \in \overline{\mathrm{W}}(P^*)} .$$

Since $\mathcal{B}$ sets $C'^*_1 = C^*_1$, $C'^*_2 = C^*_2$ and $C'^*_3 = C^*_3$, it follows that C'^*_1, C'^*_2 and C'^*_3 are of the correct form. To show that C^*_4 is correctly formed, notice that $u'_{i,j} = g_1^{\alpha_{i,j}}$ for indices $i \in \mathrm{W}(P^*)$ and $j = 0, \ldots, n$. Thus, $C'^*_{4,i,j} = (C^*_1)^{\alpha_{i,j}} = g_1{}^{t\,\alpha_{i,j}} = (g_1^{\alpha_{i,j}})^t = u'_{i,j}{}^t$ as required.

We also need to argue that $\mathcal{B}$ does not query its key derivation oracle on any identities that are considered illegal in the IND-ID-CPA game when its guess for $\mathrm{W}(P^*)$ is correct. Illegal identities are the challenge identity $ID^* = P^*|_{\overline{\mathrm{W}}(P^*)}$ or any ancestors of it, i.e. any $ID^*|_{\le \ell}$ for $\ell \le \ell^*$. Adversary $\mathcal{B}$ only makes such queries when $\mathcal{A}$ queries its key derivation oracle on an identity $ID' = (ID'_1, \ldots, ID'_{\ell'})$ such that $\ell' \le \ell^*$ and $ID'_i = P^*_i$ for all $i \in \overline{\mathrm{W}}(P^*)|_{\le \ell'}$. By our matching definition, this would mean that $ID' \in_* P^*$, which is illegal in the IND-WID-CPA game as well. Note that, whenever $\ell' > \ell^*$, we always have that $|ID| > |ID^*|$ since $\mathrm{W}(\hat{P})|_{> \ell^*} = \emptyset$.

To conclude the proof, we notice that the success probability of $\mathcal{B}$ is at least that of $\mathcal{A}$ when its guess for $\mathrm{W}(P^*)$ is correct. Let ϵ be the probability that $\mathcal{A}$ wins the IND-WID-CPA game. Thus, it follows that the overall success probability of $\mathcal{B}$ winning the IND-ID-CPA game is at least $\epsilon' \ge \epsilon/2^L$.

Remark 7. The factor of 2^L in the security reduction is not a major drawback given the state of the art in HIBE constructions, which also lose this factor. In addition, we only lose a factor of L^2 when encrypting to patterns with a single sequence of consecutive wildcards, for example $(ID_1, *, *, *, ID_5)$ or $(ID_1, *, *)$.

6 Alternative Constructions and Extensions

In the full version of this paper [1], we present two alternative WIBE implementations, namely the $\mathcal{BB}$-$\mathcal{WIBE}$ scheme based on the Boneh-Boyen HIBE scheme [3] and the $\mathcal{BBG}$-$\mathcal{WIBE}$ scheme based on the Boneh-Boyen-Goh HIBE scheme [4], respectively. We omit them here due to space restrictions. Both of these schemes have security proofs in the standard model under a weaker security notion that can be seen as a variant of selective-ID security with wildcards. Security under the full notion presented in Section 3 can be achieved in the random oracle model [2] at the cost of losing a factor q_{H}^L in the reduction, where q_{H} is the number of an adversary's random oracle queries and L is the maximum depth of the hierarchy. Both schemes have efficiency polynomial in all parameters, unlike the generic construction of Section 4, and offer advantages over the $\mathcal{W}a$-$\mathcal{WIBE}$ scheme in master public key length, ciphertext size and encryption/decryption time. A comparison between all our schemes is provided in Fig. 1.

Scheme	$\|mpk\|$	$\|d\|$	$\|C\|$	Dec	Assumption	RO
Generic	$\|mpk_{\mathcal{HIBE}}\|$	$2^L \cdot \|d_{\mathcal{HIBE}}\|$	$\|C_{\mathcal{HIBE}}\|$	$\mathsf{Dec}_{\mathcal{HIBE}}$	$\mathcal{HIBE}$ is IND-ID-CPA	No
$\mathcal{W}a$-$\mathcal{WIBE}$	$(n+1)L+3$	$L+1$	$(n+1)L+2$	$L+1$	BDDH	No
$\mathcal{BB}$-$\mathcal{WIBE}$	$2L+3$	$L+1$	$2L+2$	$L+1$	BDDH	Yes
$\mathcal{BBG}$-$\mathcal{WIBE}$	$L+4$	$L+2$	$L+3$	2	L-BDHI	Yes

Fig. 1. Efficiency and security comparison between the generic scheme of Section 4, the $\mathcal{W}a$-$\mathcal{WIBE}$ scheme of Section 5.2, and the $\mathcal{BB}$-$\mathcal{WIBE}$ and $\mathcal{BBG}$-$\mathcal{WIBE}$ schemes presented in the full version [1]. The schemes are compared in terms of master public key size ($|mpk|$), user secret key size ($|d|$), ciphertext size ($|C|$), decryption time (Dec), the security assumption under which the scheme is proved secure, and whether this proof is in the random oracle model or not. (The generic construction does not introduce any random oracles, but if the security proof of the HIBE scheme is in the random oracle model, then the WIBE obviously inherits this property.) Values refer to the underlying HIBE scheme for the generic scheme, and to the number of group elements ($|mpk|$, $|d|$, $|C|$) or pairing computations (Dec) for the other schemes. L is the maximal hierarchy depth and n is the bit length of an identity string. Figures are worst-case values, usually occurring for identities at level L with all-wildcard ciphertexts. L-BDHI refers to the decisional bilinear Diffie-Hellman inversion assumption [10, 3].

While the efficiency of our direct schemes is polynomial in all parameters, we stress that their security degrades exponentially with the hierarchy depth L. So just as is the case for the current state of the art in HIBE schemes, we have to leave the construction of a WIBE scheme with polynomial efficiency *and* security in all parameters as an open problem.

Also in the full version of the paper [1], we achieve chosen ciphertext security by adapting the technique of Canetti, Halevi and Katz [6]. In particular, we show that we may use a $(2L+2)$-level CPA-secure WIBE and a strongly unforgeable signature scheme $(\mathsf{SigGen}, \mathsf{Sign}, \mathsf{Verify})$ to construct an L-level CCA-secure WIBE.

Acknowledgments

We would like to thank James Birkett, Jacob Schuldt, Brent Waters and the anonymous referees of ICALP 2006 for their valuable input. This work was supported in part by the European Commission through the IST Programme under Contract IST-2002-507932 ECRYPT. The first two authors were supported in part by France Telecom R&D as part of the contract CIDRE, between France Telecom R&D and École normale supérieure. The fifth author is a Postdoctoral Fellow of the Research Foundation – Flanders (FWO-Vlaanderen), and was supported in part by the Concerted Research Action (GOA) Ambiorics 2005/11 of the Flemish Government.

References

1. M. Abdalla, D. Catalano, A. W. Dent, J. Malone-Lee, G. Neven, and N. P. Smart. Identity-based encryption gone wild. Cryptology ePrint Archive, 2006.
2. M. Bellare and P. Rogaway. Random oracles are practical: A paradigm for designing efficient protocols. In *ACM CCS 1993*, pages 62–73, 1993.
3. D. Boneh and X. Boyen. Efficient selective-ID secure identity based encryption without random oracles. In *EUROCRYPT 2004*, volume 3027 of *LNCS*, pages 223–238. Springer-Verlag, 2004.
4. D. Boneh, X. Boyen, and E.-J. Goh. Hierarchical identity based encryption with constant size ciphertext. In *EUROCRYPT 2005*, volume 3494 of *LNCS*, pages 440–456. Springer-Verlag, 2005.
5. D. Boneh and M. K. Franklin. Identity based encryption from the Weil pairing. *SIAM Journal on Computing*, 32(3):586–615, 2003.
6. R. Canetti, S. Halevi, and J. Katz. Chosen-ciphertext security from identity-based encryption. In *EUROCRYPT 2004*, volume 3027 of *LNCS*, pages 207–222. Springer-Verlag, 2004.
7. C. Cocks. An identity based encryption scheme based on quadratic residues. In *Cryptography and Coding, 8th IMA International Conference*, volume 2260 of *LNCS*, pages 360–363. Springer-Verlag, 2001.
8. C. Gentry and A. Silverberg. Hierarchical ID-based cryptography. In *ASIACRYPT 2002*, volume 2501 of *LNCS*, pages 548–566. Springer-Verlag, 2002.
9. J. Horwitz and B. Lynn. Toward hierarchical identity-based encryption. In *EUROCRYPT 2002*, volume 2332 of *LNCS*, pages 466–481. Springer-Verlag, 2002.
10. S. Mitsunari, R. Saka, and M. Kasahara. A new traitor tracing. *IEICE Transactions*, E85-A(2):481–484, 2002.
11. R. Sakai, K. Ohgishi, and M. Kasahara. Cryptosystems based on pairing. In *SCIS 2000*, Okinawa, Japan, 2000.
12. A. Shamir. Identity-based cryptosystems and signature schemes. In *CRYPTO'84*, volume 196 of *LNCS*, pages 47–53. Springer-Verlag, 1985.
13. B. R. Waters. Efficient identity-based encryption without random oracles. In *EUROCRYPT 2005*, volume 3494 of *LNCS*, pages 114–127. Springer-Verlag, 2005.

Deterministic Priority Mean-Payoff Games as Limits of Discounted Games*

Hugo Gimbert[1] and Wiesław Zielonka[2]

[1] Instytut Informatyki, Warsaw University, Poland
[2] Université Paris 7 and CNRS, LIAFA, case 7014
2, place Jussieu, 75251 Paris Cedex 05, France

Abstract. Inspired by the paper of de Alfaro, Henzinger and Majumdar [1] about discounted μ-calculus we show new surprising links between parity games and different classes of discounted games.

1 Introduction

One of the major results in the theory of stochastic games states that the value of mean-payoff games is the limit of the values of discounted games [2]. Recently de Alfaro, Henzinger and Majumdar [1] presented results that seem to indicate that it is possible to obtain parity games as an appropriate limit of multi-discounted games. In fact, the authors of [1] use the language of the μ-calculus rather than games, but as the links between μ-calculus and parity games are well-known since the advent [3] it is natural to wonder how discounted μ-calculus from [1] can be reflected in games.

Suppose that $\mathcal{A}$ is our arena with each vertex belonging to one of the two players 0 and 1. If the current state s belongs to player P then he chooses an outgoing edge (s, s') and the system moves to the target state s'. Suppose that the states are labeled by priorities from the finite set $\mathbf{D} = \{1, \ldots, k\}$. Inspecting thoroughly the formulas of the discounted μ-calculus from [1] it is not too difficult to discover that it corresponds to the following games. Let us associate with each priority $d \in \mathbf{D}$ a discount factor λ_d from the interval $(0; 1)$. Let $d_0 d_1 d_2 \ldots$ be an infinite sequence of priorities visited during the play. Then we calculate the payoff obtained by player 0 from player 1 using the formula

$$\sum_{i=0}^{\infty} \lambda_{d_0} \cdots \lambda_{d_{i-1}} (1 - \lambda_{d_i}) r_i \tag{1}$$

where

$$r_i = \begin{cases} 0 & \text{if the priority } d_i \text{ is odd,} \\ 1 & \text{if the priority } d_i \text{ is even.} \end{cases}$$

* This research was supported by European Research Training Network: Games and Automata for Synthesis and Validation and ACI Sécurité informatique 2003-22 VERSYDIS.

M. Bugliesi et al. (Eds.): ICALP 2006, Part II, LNCS 4052, pp. 312–323, 2006.

The fact that such games have values and optimal strategies, and in the case of perfect information stochastic games even positional optimal strategies, is known since the seminal paper of Shapley [4]. The results of [1] indicate that

$$\lim_{\lambda_0 \uparrow 1} \dots \lim_{\lambda_{k-1} \uparrow 1} \mathrm{val}_\lambda(s) = \mathrm{val}(s) \tag{2}$$

where $\mathrm{val}_\lambda(s)$ is the value of the multi-discounted game with the payoff (1) for the initial state s and $\mathrm{val}(s)$ is the value of the parity game for the initial state s (more precisely we should take in this case the following version of the parity games: player 0 wins 1 if the smallest priority visited infinitely often is even, otherwise he wins 0).

The first point to note is that if we are in the realm of games rather than μ-calculus then it is completely artificial to limit the numbers r_i appearing in (1) to 0 and 1, it would be much more natural to consider the games with any real valued r_i (and this is of course the point of view adopted by Shapley [4]). Thus now we assume that states are labeled rather by pairs $(d, r) \in \mathbf{D} \times \mathbb{R}$ composed of a priority d and a real number r. If during an infinite play we visit the sequence $(d_0, r_0), (d_1, r_1), \dots$ of labels then we can still calculate the payment obtained by player 0 from player 1 using the formula (1). What about the equation (2) in this case? Does there exists a game that replaces the parity game and such that its value can be put on the right hand side of the equality (2)? As one could expect, such games, *priority mean-payoff games*, exist. In fact priority mean-payoff games were previously introduced in [5], where it was proved that they admit optimal positional strategies. In this paper we show that their values are related to the values of multi-discounted games, generalizing[1] the result of [1].

The formula (2) has a rather limited interest, we would prefer to find a link not only between the game values but also between their optimal strategies. To this end in Section 4 we introduce a new family of discounted games: priority discounted games. They have a considerable advantage over multi-discounted games: their values depend on only one parameter, i.e. to find the limits of their values we do need to use iterated limits. And, what is more important, it is possible to carry out to this framework the concept of Blackwell optimality [6]: for all values of the discount factor sufficiently close to 0, the optimal strategies in priority-discounted games are also optimal for priority mean-payoff games. Note that since the parity games are just a very special subclass of priority mean-payoff games this result establishes a rather unexpected property of parity games. It is an open problem if this be used in practice to calculate optimal strategies for parity games.

[1] This is not really exact, since [1] examines the μ-calculus corresponding to (concurrent) stochastic games while in our paper we limit ourselves to deterministic games. The possibility of generalization of presented results to stochastic games is discussed in Section 5.

2 Games

An *arena* is a tuple $\mathcal{A} = (S_0, S_1, A, \Re)$, where S_0 and S_1 are the sets of *states* controlled by player 0 and player 1 respectively, A is the set of *actions* and $\Re$ is the set of *rewards*.

By $S = S_0 \cup S_1$ we denote the set of all states. Then $A \subseteq S \times \Re \times S$, i.e. each action $a = (s', r, s'') \in A$ is a triple composed of the *source state* $\text{source}(a) = s'$, the *target state* $\text{target}(a) = s''$ and a reward $r = \text{reward}(a) \in \Re$.

An action a is *available* at state s if $a \in A_s$, where A_s denotes the set of actions with source s.

We consider only arenas where the sets of states and actions are finite and such that for each state s the set A_s of available actions is non-empty.

A *path* in arena $\mathcal{A}$ is a finite or infinite sequence $p = a_0a_1a_2\dots$ of actions such that $\forall i, \text{target}(a_i) = \text{source}(a_{i+1})$. The source of the first action a_0 is the source, $\text{source}(p)$, of the path p. If p is finite then the target of the last action is the target, $\text{target}(p)$, of p.

It is convenient to assume that for each state s there is an empty path $\mathbf{1}_s$ with the source and the target s.

Two players 0 and 1 play on $\mathcal{A}$ in the following way. If the current state s is controlled by player $P \in \{0, 1\}$, i.e. $s \in S_P$, then player P chooses an action $a \in A_s$ available at s, this action is executed and the system goes to the state $\text{target}(a)$.

Starting from an initial state s, the infinite sequence of consecutive moves of both players yields an infinite sequence $p = a_0a_1\dots$ of executed actions such that $\text{source}(p) = s$. Such sequences are called *plays*, thus plays in this game are just infinite paths in the underlying arena $\mathcal{A}$.

We shall also use the term "a finite play" as a synonym of "a finite path" but "play" without any qualifier will always denote an infinite play.

An infinite sequence $r_0r_1r_2\dots$ of rewards is *finitely generated* if there exists a finite subset $\Re'$ of $\Re$ such that all elements of this sequence belong to $\Re'$. The set of all infinite finitely generated sequences of $\Re$ is denoted $\Re^\omega$.

By $\Re^*$ we denote the set of all finite sequences of $\Re$ and we set $\Re^\infty = \Re^* \cup \Re^\omega$.

Each path $p = a_0a_1\dots$ yields a sequence of rewards

$$\text{reward}(p) = \text{reward}(a_0)\,\text{reward}(a_1)\dots\ . \tag{3}$$

Note that since our arenas are finite, if p is an infinite path then $\text{reward}(p)$ is finitely generated.

A *utility mapping*

$$u : \Re^\omega \to \mathbb{R} \tag{4}$$

maps each finitely generated infinite reward sequence $x \in \Re^\omega$ to a real number $u(x) \in \mathbb{R}$. The interpretation is that at the end of a play p player 0 receives from player 1 the *payoff* $u(\text{reward}(p))$ (if $u(\text{reward}(p)) < 0$ then it is rather player 1 that receives from player 0 the amount $|u(\text{reward}(p))|$).

A *game* $(\mathcal{A}, u)$ is couple composed of an arena and a utility mapping.

A strategy of a player P is his plan of action that tells him which action to take when the game is at a state $s \in S_P$. The choice of the action can depend on the whole past sequence of moves. Thus a *strategy* for player 0 is a mapping

$$\sigma : \{p \mid p \text{ a finite play with target}(p) \in S_0\} \longrightarrow A \tag{5}$$

such that for each finite play p with $s = \text{target}(p) \in S_0$, $\sigma(p) \in A_s$.

Strategy σ of player 0 is said to be *positional* if for every state $s \in S_0$ and every finite play p such that $\text{target}(p) = s$, $\sigma(p) = \sigma(\mathbf{1}_s)$. Thus the action chosen by a positional strategy depends only on the current state, previously visited states and executed actions are irrelevant. To simplify the notation it is convenient to view a positional strategy as a mapping

$$\sigma : S_0 \to A \tag{6}$$

such that $\sigma(s) \in A_s$.

A finite or infinite play $p = a_0 a_1 \ldots$ is said to be *consistent* with a strategy $\sigma \in \Sigma$ if for each $i \in \mathbb{N}$ such that $\text{target}(a_{i-1}) = \text{source}(a_i) \in S_0$, we have $a_i = \sigma(a_0 \ldots a_{i-1})$. Moreover, if $s = \text{source}(a_0) \in S_0$ then we require that $a_0 = \sigma(\mathbf{1}_s)$.

Strategies, positional strategies and consistent plays are defined in the analogous way for player 1 with S_1 replacing S_0.

In the sequel Σ and $\mathcal{T}$ will stand for the set of strategies for player 0 and player 1, Σ_p and $\mathcal{T}_p$ are the corresponding subsets of positional strategies and finally σ and τ, possibly with subscripts or superscripts, will denote the elements of Σ and $\mathcal{T}$.

Given a pair of strategies $\sigma \in \Sigma$ and $\tau \in \mathcal{T}$, there exists a unique infinite play in arena $\mathcal{A}$, denoted $p(s, \sigma, \tau)$, consistent with σ and τ and such that $s = \text{source}(p(s, \sigma, \tau))$. The corresponding sequence of rewards $\text{reward}(p(s, \sigma, \tau))$ will be denoted $r(s, \sigma, \tau)$.

Definition 1. *Strategies $\sigma^\sharp \in \Sigma$ and $\tau^\sharp \in \mathcal{T}$ are* optimal *in the game $(\mathcal{A}, u)$ if*

$$\forall s \in S, \forall \sigma \in \Sigma, \forall \tau \in \mathcal{T},$$
$$u(r(s, \sigma, \tau^\sharp)) \leq u(r(s, \sigma^\sharp, \tau^\sharp)) \leq u(r(s, \sigma^\sharp, \tau)) \ . \tag{7}$$

We say that a utility mapping u admits optimal positional strategies *if for all games $(\mathcal{A}, u)$ over* finite arenas *there exist positional optimal strategies for both players.*

Thus if both strategies are optimal the players do not have any incentive to change them unilaterally: player 0 cannot increase his gain by switching to another strategy σ while player 1 cannot decrease his loses by switching to τ.

Note that zero-sum games, where the gain of one player is equal to the loss of his adversary, satisfy the exchangeability property for optimal strategies: for any two pairs of optimal strategies $(\sigma^\sharp, \tau^\sharp)$ and $(\sigma^\star, \tau^\star)$, the pairs $(\sigma^\star, \tau^\sharp)$ and $(\sigma^\sharp, \tau^\star)$ are also optimal and, moreover, $u(r(s, \sigma^\sharp, \tau^\sharp)) = u(r(s, \sigma^\star, \tau^\star))$, i.e. the value of the expression $u(r(s, \sigma^\sharp, \tau^\sharp))$ is independent of the choice of the optimal strategies — this is *the value of the game $(\mathcal{A}, u)$ at state s.*

Lemma 2. *Let u be a utility mapping admitting optimal positional strategies for both players.*

(A) Suppose that $\sigma \in \Sigma$ is any strategy while $\tau^\sharp \in \mathcal{T}_p$ is positional. Then there exists a positional strategy $\sigma^\sharp \in \Sigma_p$ such that

$$\forall s \in S, \quad u(r(s,\sigma,\tau^\sharp)) \leq u(r(s,\sigma^\sharp,\tau^\sharp)) \ . \tag{8}$$

(B) Similarly, if $\tau \in \mathcal{T}$ is any strategy and $\sigma^\sharp \in \Sigma_p$ a positional strategy then there exists a positional strategy $\tau^\sharp \in \mathcal{T}_p$ such that

$$\forall s \in S, \quad u(r(s,\sigma^\sharp,\tau^\sharp)) \leq u(r(s,\sigma^\sharp,\tau)) \ .$$

Proof. We prove (A), the proof of (B) is similar. Take any strategies $\sigma \in \Sigma$ and $\tau^\sharp \in \mathcal{T}_p$. Let $\mathcal{A}'$ be a subarena of $\mathcal{A}$ obtained by restricting the actions of player 1 to the actions given by the strategy $\tau^\sharp$, i.e. in $\mathcal{A}'$ the only possible strategy for player 1 is the strategy $\tau^\sharp$. The actions of player 0 are not restricted, i.e. in $\mathcal{A}'$ player 0 has the same available actions as in $\mathcal{A}$. Since $\tau^\sharp$ is positional $\mathcal{A}'$ is a well-defined finite arena and by the assumption concerning u there exists an optimal positional strategy $\sigma^\sharp$ for player 0 in $\mathcal{A}'$; obviously $\tau^\sharp$ is the optimal positional strategy for player 1 in $\mathcal{A}'$. This implies that (8) holds in $\mathcal{A}'$ and therefore also in $\mathcal{A}$. □

Lemma 3. *Suppose that the utility mapping u admits optimal positional strategies. Suppose $\sigma^\sharp \in \Sigma_p$ and $\tau^\sharp \in \mathcal{T}_p$ are positional strategies such that*

$$\forall s \in S, \forall \sigma \in \Sigma_p, \forall \tau \in \mathcal{T}_p,$$
$$u(r(s,\sigma,\tau^\sharp)) \leq u(r(s,\sigma^\sharp,\tau^\sharp) \leq u(r(s,\sigma^\sharp,\tau)) \ , \tag{9}$$

i.e. $\sigma^\sharp$ and $\tau^\sharp$ are optimal in the class of positional strategies. Then $\sigma^\sharp$ and $\tau^\sharp$ are optimal.

Proof. Suppose that

$$\exists \tau \in \mathcal{T}, \quad u(r(s,\sigma^\sharp,\tau)) < u(r(s,\sigma^\sharp,\tau^\sharp)) \ . \tag{10}$$

By Lemma 2 there exists a positional strategy $\tau^\star \in \mathcal{T}_p$ such that $u(r(s,\sigma^\sharp,\tau^\star)) \leq u(r(s,\sigma^\sharp,\tau)) < u(r(s,\sigma^\sharp,\tau^\sharp))$, contradicting (9). Thus $\forall \tau \in \mathcal{T}, u(r(s,\sigma^\sharp,\tau^\sharp)) \leq u(r(s,\sigma^\sharp,\tau))$. The left hand side of (7) can be proved in the similar way. □

3 Priority Mean-Payoff Games as the Limit of Multi-discounted Games

In the sequel of this paper we fix the set of rewards to be

$$\Re = \mathbf{D} \times \mathbb{R} \ , \tag{11}$$

where $\mathbf{D} = \{d \in \mathbb{N} \mid 1 \leq d \leq k\}$ is fixed finite set of *priorities*. We shall note by $|\mathbf{D}| = k$ the cardinality of $\mathbf{D}$.

3.1 Multi-discounted Games

A *discount mapping*

$$\lambda : \mathbf{D} \longrightarrow [0, 1)$$

associates with each priority a real number from the interval $[0, 1)$. The value of λ for a priority $d \in \mathbf{D}$, noted λ_d, is called the discount factor of d.

Given a discount mapping λ we define *multi-discounted utility mapping* u_λ. It is convenient to define u_λ uniformly for infinite as well as for finite reward sequences $t = (d_0, r_0), (d_1, r_1), \ldots \in \Re^\infty$:

$$\begin{aligned} u_\lambda(t) &= (1 - \lambda_{d_0})r_0 + \lambda_{d_0}(1 - \lambda_{d_1})r_1 + \lambda_{d_0}\lambda_{d_1}(1 - \lambda_{d_2})r_2 + \ldots \\ &= \sum_{0 \le i < |t|} \lambda_{d_0} \ldots \lambda_{d_{i-1}}(1 - \lambda_{d_i})r_i \ , \end{aligned} \quad (12)$$

where $|t|$ is the length of t if t is finite and ∞ otherwise.

By an obvious adaptation of the proof of Shapley [4] one can obtain the following theorem which in fact holds even for a more general class of perfect information stochastic games:

Theorem 4 (Shapley). *For each discount mapping $\lambda : \mathbf{D} \to [0; 1)$, the multi-discounted utility mapping u_λ admits optimal positional strategies for both players. In particular each game $(\mathcal{A}, u_\lambda)$ has a value* $\mathrm{val}_\lambda(s)$ *for every initial state s.*

3.2 Priority Mean-Payoff Games

Definition 5. *The* priority *of an infinite reward sequence $t = (d_0, r_0), (d_1, r_1), \ldots \in \Re^\omega$ is the minimal priority occurring infinitely often in t:*

$$\mathrm{priority}(t) = \liminf_{i \to \infty} d_i \ . \quad (13)$$

For any reward sequence $t = (d_0, r_0), (d_1, r_1), \ldots \in \Re^\infty$ and $d \in \mathbb{N}$ let

$$\Pi_d(t) = \{i \in \mathbb{N} \mid 0 \le i < |t| \quad \text{and} \quad d_i = d\}, \quad (14)$$

be the sequence consisting of the indices for which the priority is equal d in t.

Definition 6. *Let $t = (d_0, r_0), (d_1, r_1), \ldots \in \Re^\omega$ be an infinite reward sequence and let $\Pi_d(t) = i_0, i_1, \ldots$ the sequence consisting of the indices i for which $d_i =$* $\mathrm{priority}(t)$. *Then*

$$\mu(t) = \liminf_{n \to \infty} \frac{1}{n} \sum_{j=0}^{n-1} r_{i_j}$$

defines priority mean-payoff utility *mapping* $\mu : \Re^\omega \to \mathbb{R}$.

Thus, intuitively, to calculate $\mu(t)$ we first use the priorities to choose an appropriate subsequence $t' = (d_{i_0}, r_{i_0}), (d_{i_1}, r_{i_1}), (d_{i_1}, r_{i_1}), \ldots$ of t consisting of rewards such that $\mathrm{priority}(t) = d_{i_0} = d_{i_1} = d_{i_1} = \ldots$ and next we apply the usual mean-payoff to the corresponding subsequence $r_{i_0} r_{i_1} r_{i_2} \ldots$ of rewards.

The following result is proved in [5]:

Theorem 7. *The priority mean-payoff utility μ admits optimal positional strategies for both players.*

The *value of the priority mean-payoff game* for an initial state s will be noted $\mathrm{val}(s)$.

The following theorem connects multi-discount and priority mean-payoff games:

Theorem 8. *Let $\mathbf{D} = \{1, \ldots, k\}$ be the set of priorities. Then for each initial state s*

$$\lim_{\lambda_1 \uparrow 1} \lim_{\lambda_2 \uparrow 1} \ldots \lim_{\lambda_k \uparrow 1} \mathrm{val}_\lambda(s) = \mathrm{val}(s) \ , \tag{15}$$

i.e. the value of the priority mean-payoff game is the (iterated) limit of the value of the multi-discounted game ($\lambda_i \uparrow 1$ means that λ_i tends to 1 from below).

The order in which the limits are taken in (15) does matter and is related to the fact that in (13) we have chosen the minimal priority appearing infinitely often as the priority of an infinite sequence of rewards. Let us note that in the particular case when there is only one priority, $|\mathbf{D}| = 1$, Theorem 8 holds in the much larger setting of stochastic games, this is a seminal result of Mertens and Neyman [2]. We skip the proof of Theorem 8 since it is too long to be given here. In fact Theorem 8 will not be used in the sequel and, in our opinion, the subsequent Section 4 contains much more interesting results which are provided with complete proofs.

Before going to the next section let us note however that in fact there exists a whole spectrum of games spanning from multi-discounted to priority mean-payoff games. For an infinite reward sequence $t = (d_0, r_0), (d_1, r_1), \ldots \in \Re^\infty$ we can define an ith partially discounted utility mapping

$$u^i_\lambda(t) = \limsup_{\lambda_{i+1} \uparrow 1} \ldots \limsup_{\lambda_k \uparrow 1} u_\lambda(t) \ ,$$

where $u_\lambda(t)$ is defined by (12). Obviously for $i = k$ this is just the multi-discounted utility but it turns out that for each fixed i, $i = 0, \ldots, k$, the games with the utility u^i_λ have optimal positional strategies and for $i = 0$ the value as well as the optimal strategies of such games are the same as for priority mean-payoff games.

4 Priority-Discounted Games

In Section 3.2 we have established that the value of the priority mean-payoff game is an iterated limit of the multi-discounted game. However, iterated limits are cumbersome so a natural question is if we cannot replace them by a single limit.

Another weakness of multi-discounted games is that they are related to priority mean-payoff games only by their values but not by their optimal strategies.

In this section we introduce the class of priority-discounted games which behave much better in this respect.

Let us take $\beta \in (0;1]$ and, for $d \in \mathbf{D}$, set

$$\lambda_d(\beta) = 1 - \beta^d \ . \tag{16}$$

A *priority-discounted game* is a multi-discounted game in which the discount factor associated with the priority $d \in \mathbf{D}$ is $\lambda_d(\beta)$.

Let $t = (d_0, r_0), (d_1, r_1), \ldots \in \Re^{\infty}$. Then, putting (16) into (12), we get the definition of the *priority-discounted* utility mapping:

$$\begin{aligned} u^{\beta}(t) &= \beta^{d_0} r_0 + (1-\beta^{d_0})\beta^{d_1} r_1 + (1-\beta^{d_0})(1-\beta^{d_1})\beta^{d_2} r_2 + \ldots \\ &= \sum_{0 \le i < |t|} (1-\beta^{d_0})(1-\beta^{d_1}) \ldots (1-\beta^{d_{i-1}})\beta^{d_i} r_i \ . \end{aligned} \tag{17}$$

Let us note that $\lambda_d(\beta) \uparrow 1$ iff $\beta \downarrow 0$. The following theorem is analogous to Theorem 8.

Theorem 9. *Let $\mathcal{A}$ be a finite arena. Then*

(i) For every finite arena $\mathcal{A}$ and for all $\beta \in (0;1]$ both players have optimal positional strategies in the priority discounted game $(\mathcal{A}, u^{\beta})$.

(ii) Let $\mathrm{val}^{\beta}(s)$ be the value of the priority discounted game $(\mathcal{A}, u^{\beta})$ for an initial state s. Then

$$\lim_{\beta \downarrow 0} \mathrm{val}^{\beta}(s) = \mathrm{val}(s) \ , \tag{18}$$

where $\mathrm{val}(s)$ is the value of the priority mean-payoff game.

Proof. (i) obviously is just a special case of Theorem 4. The proof of (ii) will be given at the end of Section 4.1. □

4.1 Blackwell Optimality

The concept known as Blackwell optimality was introduced in [7]. A readable modern presentation can be found in [6]. Roughly speaking, a policy of a Markov decision process with the discounted reward criterion is Blackwell optimal if it is optimal for all discount factors sufficiently close to 1. It turns out that such policies are also automatically optimal for mean-payoff games, hence Blackwell optimality is stronger than the classical concept of optimality in mean-payoff games.

We adapt here the concept of Blackwell optimality to two-person priority-discounted games. We show that corresponding Blackwell optimal strategies exist and that they are optimal for priority mean-payoff games.

Let us fix a finite arena $\mathcal{A}$. Strategies $(\sigma^{\sharp}, \tau^{\sharp}) \in \Sigma \times \mathcal{T}$ are *β-optimal* if they are optimal in the priority-discounted game $(\mathcal{A}, u^{\beta})$ with the discount factor β.

Definition 10. *Strategies $(\sigma^{\sharp}, \tau^{\sharp}) \in \Sigma \times \mathcal{T}$ are Blackwell optimal if they are β-optimal for all values β in an interval $0 < \beta < \beta_0$ for some constant $\beta_0 > 0$.*

The following two lemmas will be useful for establishing the existence of Blackwell optimal strategies in priority-discounted games, stated in Theorem 13. In those Lemmas, we consider the different discounted-priority games obtained when β tends to 0. In Lemma 11 we fix some finite play and describe the asymptotic behavior of the values of this play when β tends to 0. In Lemma 12 we consider the case of ultimately periodic plays.

Lemma 11. *Let $y = (d_0, r_0) \dots (d_n, r_n) \in \Re^*$ be a finite sequence of rewards, $a = \min\{d_0, \dots, d_n\}$ and $I = \{i \mid 0 \le i \le n \quad \text{and} \quad d_i = a\}$. Then*

$$\lim_{\beta \downarrow 0} \frac{u^\beta(y)}{1 - (1-\beta^{d_0}) \cdots (1-\beta^{d_n})} = \frac{1}{|I|} \sum_{i \in I} r_i \ ,$$

where $|I|$ denotes the cardinality of I.

Proof. This is just an elementary exercise: $u^\beta(y) = \beta^{d_0} r_0 + (1-\beta^{d_0})\beta^{d_1} r_1 + (1-\beta^{d_0})(1-\beta^{d_1})\beta^{d_2} r_2 + \dots + (1-\beta^{d_0})(1-\beta^{d_1}) \cdots (1-\beta^{d_{n-1}})\beta^{d_n} r_n = (\sum_{i \in I} r_i)\beta^a + p(\beta)$, where $p(\beta)$ is a polynomial with all monomials having degree $> a$. Similarly, $g(\beta) = 1 - (1-\beta^{d_0}) \cdots (1-\beta^{d_n}) = |I|\beta^a + q(\beta)$, where $q(\beta)$ is a sum of monomials of degree $> a$. Thus

$$u^\beta(\beta)/g(\beta) = ((\sum_{i \in I} r_i) + p(\beta)/\beta^a)/(|I| + q(\beta)/\beta^a) \xrightarrow{\beta \to 0} (\sum_{i \in I} r_i)/|I| \ . \qquad \square$$

Lemma 12. *Given an initial state s and positional strategies $(\sigma, \tau) \in \Sigma_p \times T_p$,*

(i) the function $\beta \mapsto u^\beta(r(s, \sigma, \tau)))$, defined for $0 < \beta < 1$, is a rational function[2] of β.
(ii) $\lim_{\beta \to 0} u^\beta(r(s, \sigma, \tau))) = \mu(r(s, \sigma, \tau)))$, where μ is the priority mean-payoff utility, see Definition 6.

Proof. (i) Since σ and τ are positional, the play $p(s, \sigma, \tau)$ and the resulting sequence $r(s, \sigma, \tau)$ of rewards are ultimately periodic. Thus, for some $x, y \in \Re^*$, $r(s, \sigma, \tau) = xyyy\ldots = xy^\omega$. Then (17) yields

$$\begin{aligned} u^\beta(xy^\omega) &= \\ & u^\beta(x) + (1-\beta^{d_0}) \cdots (1-\beta^{d_l}) u^\beta(y) \sum_{i=0}^{\infty} \left[(1-\beta^{d_{l+1}}) \cdots (1-\beta^{d_m})\right]^i \\ &= u^\beta(x) + \frac{(1-\beta^{d_0}) \cdots (1-\beta^{d_l})}{1 - (1-\beta^{d_{l+1}}) \cdots (1-\beta^{d_m})} u^\beta(y), \quad (19) \end{aligned}$$

where $d_0, \dots, d_l$ is the sequence of priorities of x and $y = (r_{l+1}, d_{l+1}), \dots, (r_m, d_m)$.

Since x and y are finite $u^\beta(x)$ and $u^\beta(y)$ are just polynomials of β.

(ii) It suffices to note that in (19), if $\beta \to 0$ then $u^\beta(x)$ tends to 0 while $(1-\beta^{d_0}) \cdots (1-\beta^{d_l})$ tends to 1. Thus this result is an immediate consequence of Lemma 11. $\square$

[2] A quotient of two polynomials.

We can now state our main result about Blackwell optimality in priority-discounted games.

Theorem 13. *For each finite arena $\mathcal{A}$ there exist Blackwell optimal positional strategies for priority-discounted game $(\mathcal{A}, u^\beta)$.*

Proof. The proof follows very closely the proof given in [6] for Markov decision processes.

Since $\mathcal{A}$ is finite, the set $\Sigma_p \times \mathcal{T}_p$ of pairs of positional strategies is finite. Thus there exists a pair $(\sigma^\sharp, \tau^\sharp) \in \Sigma_p \times \mathcal{T}_p$ of positional β-optimal strategies for all $\beta = \beta_n$, where (β_n) is some sequence such that $\beta_n \downarrow 0$. We claim that $(\sigma^\sharp, \tau^\sharp)$ are Blackwell optimal.

Suppose the contrary. Then there exists a state s and a sequence γ_n tending to 0 with $n \to \infty$ such that

(i) either there exists a sequence $\sigma^\star_n$ of strategies such that $u^{\gamma_n}(r(s, \sigma^\sharp, \tau^\sharp)) < u^{\gamma_n}(r(s, \sigma^\star_n, \tau^\sharp))$,
(ii) or there exists a sequence $\tau^\star_n$ of strategies such that $u^{\gamma_n}(r(s, \sigma^\sharp, \tau^\star_n)) < u^{\gamma_n}(r(s, \sigma^\sharp, \tau^\sharp))$.

Due to Lemma 2, the strategies $\sigma^\star_n$ and $\tau^\star_n$ can be chosen positional and since the number of positional strategies is finite, taking a subsequence if necessary, we can fix one strategy $\sigma^\star$ and one strategy $\tau^\star$ for all n.

Thus we have obtained that

(1) either there exist a state s, a positional strategy $\sigma^\star \in \Sigma_p$ and a sequence (γ_n), $\gamma_n \downarrow 0$, such that for all n

$$u^\beta(r(s, \sigma^\sharp, \tau^\sharp)) < u^\beta(r(s, \sigma^\star, \tau^\sharp)) \quad \text{for all } \beta = \gamma_1, \gamma_2, \dots , \tag{20}$$

(2) or there exist a state s, a positional strategy $\tau^\star \in \mathcal{T}_p$ and a sequence (γ_n), $\gamma_n \downarrow 0$, such that for all n

$$u^\beta(r(s, \sigma^\sharp, \tau^\star)) < u^\beta(r(s, \sigma^\sharp, \tau^\sharp)) \quad \text{for all } \beta = \gamma_1, \gamma_2, \dots . \tag{21}$$

Suppose that (20) holds.

The choice of $(\sigma^\sharp, \tau^\sharp)$ guarantees that

$$u^\beta(r(s, \sigma^\star, \tau^\sharp)) \leq u^\beta(r(s, \sigma^\sharp, \tau^\sharp)) \quad \text{for all } \beta = \beta_1, \beta_2, \dots . \tag{22}$$

Consider the function

$$f(\beta) = u^\beta(r(s, \sigma^\star, \tau^\sharp)) - u^\beta(r(s, \sigma^\sharp, \tau^\sharp)) . \tag{23}$$

By Lemma 12, $f(\beta)$ coincides for $0 < \beta < 1$ with a rational function of the variable β. But from (20) and (22) we can deduce that when $\beta \downarrow 0$ then $f(\beta)$ takes infinitely many times the value 0. This is possible for a rational function only if it is identical to 0, contradicting (20). In a similar way we can prove that (21) entails a contradiction. These contradictions show that $\sigma^\sharp$ and $\tau^\sharp$ are Blackwell optimal. □

Let us note that the proof of Theorem 13 yields in fact a stronger result:

Corollary 14. *For each arena $\mathcal{A}$ there exists $\beta_0 > 0$ such that all priority-discounted games $(\mathcal{A}, u^\beta)$ with $0 < \beta < \beta_0$ have the same optimal positional strategies.*

Now that we know that Blackwell optimal positional strategies exist we are ready to show that they are also optimal for priority mean-payoff games:

Theorem 15. *If $(\sigma^\sharp, \tau^\sharp)$ are Blackwell optimal positional strategies then they are also optimal for the priority mean-payoff game.*

Proof. Suppose the contrary, i.e. that $(\sigma^\sharp, \tau^\sharp)$ is not a pair of optimal strategies for the priority mean-payoff game. This means that there exists a state s such that either

$$\mu(r(s, \sigma^\sharp, \tau^\sharp)) < \mu(r(s, \sigma, \tau^\sharp)) \tag{24}$$

for some strategy σ or

$$\mu(r(s, \sigma^\sharp, \tau)) < \mu(r(s, \sigma^\sharp, \tau^\sharp)) \tag{25}$$

for some strategy τ. Since priority mean-payoff games have optimal positional strategies, by Lemma 2, we can assume without loss of generality that σ and τ are positional. Suppose that (24) holds. By Lemma 12 (B)

$$\lim_{\beta \downarrow 0} u^\beta(r(s, \sigma^\sharp, \tau^\sharp)) = \mu(r(s, \sigma^\sharp, \tau^\sharp)) < \mu(r(s, \sigma, \tau^\sharp)) = \lim_{\beta \downarrow 0} u^\beta(r(s, \sigma, \tau^\sharp)) \ . \tag{26}$$

However inequality (26) implies that there exists $0 < \beta_0$ such that

$$\forall \beta < \beta_0, \quad u^\beta(r(s, \sigma^\sharp, \tau^\sharp)) < u^\beta(r(s, \sigma, \tau^\sharp)) \ ,$$

in contradiction with the Blackwell optimality of $(\sigma^\sharp, \tau^\sharp)$. Similar reasoning shows that also (25) contradicts the Blackwell optimality of $(\sigma^\sharp, \tau^\sharp)$. □

The proof of Theorem 9 (ii) is a direct consequence of Theorems 15 and 13 and Lemma 12.

5 Final Remarks

An interesting open problem is if we can, given an arena $\mathcal{A}$, find the constant β_0 from Corollary 14. If this were possible and β_0 were not too small then we could try to find optimal strategies for priority mean-payoff games (and therefore also parity games) by solving priority-discounted games. And to this end we can adapt the policy improvement algorithm for discounted games given in [8]. Note however that the complexity is not discussed at all in [8] so it is difficult to say if in this way we can outperform the algorithm of Vöge and Jurdziński [9]

Since for (concurrent) stochastic games Theorem 8 holds if there is just one priority [2] as well as for parity games [1] it is reasonable to conjecture that it holds in general. Blackwell optimality fails for stochastic games since stochastic mean-payoff and stochastic parity games do not have optimal positional strategies. For perfect information stochastic games we can preserve some of the results of this paper, in particular concerning optimal positional strategies, this is an ongoing work.

Let us note finally that, as the reviewers pointed out to us judiciously, there is another known link between parity and discounted games: Jurdziński [10] has shown how parity games can be reduced to mean-payoff games and it is well-known that the value of mean-payoff games is a limit of the value of discounted games, see [2] or [11] for the particular case of deterministic games. However, the reduction of [10] does not seem to extend to priority mean-payoff games and, more significantly, it fails also for perfect information stochastic games. Note also that [11] concentrates only on value approximation and the issue of Blackwell optimality in not touched at all.

References

1. de Alfaro, L., Henzinger, T.A., Majumdar, R.: Discounting the future in systems theory. In: ICALP 2003. Volume 2719 of LNCS., Springer (2003) 1022–1037
2. Mertens, J., Neyman, A.: Stochastic games. International Journal of Game Theory **10** (1981) 53–56
3. Emerson, E., Jutla, C.: Tree automata, μ-calculus and determinacy. In: FOCS'91, IEEE Computer Society Press (1991) 368–377
4. Shapley, L.S.: Stochastic games. Proceedings Nat. Acad. of Science USA **39** (1953) 1095–1100
5. Gimbert, H., Zielonka, W.: Games where you can play optimally without any memory. In: CONCUR 2005. Volume 3653 of LNCS., Springer (2005) 428–442
6. Hordijk, A., Yushkevich, A.: Blackwell optimality. In Feinberg, E., Schwartz, A., eds.: Handbook of Markov Decision Processes. Kluwer (2002)
7. Blackwell, D.: Discrete dynamic programming. Annals of Mathematical Statistics **33** (1962) 719–726
8. Raghavan, T., Syed, Z.: A policy-improvement type algorithm for solving zero-sum two-person stochastic games of perfect information. Math. Program. **95**(3) (2003) 513–532
9. Vöge, J., Jurdziński, M.: A discrete strategy improvement algorithm for solving parity games. In: CAV 2000. Volume 1855 of LNCS., Springer (2000) 202–215
10. Jurdziński, M.: Deciding the winner in parity games is in UP $\cap$ co-UP. Information Processing Letters **68**(3) (1998) 119–124
11. Zwick, U., Paterson, M.: The complexity of mean payoff games on graphs. Theor. Computer Science **158**(1-2) (1996) 343–359

Recursive Concurrent Stochastic Games

Kousha Etessami[1] and Mihalis Yannakakis[2]

[1] LFCS, School of Informatics, University of Edinburgh
[2] Department of Computer Science, Columbia University

Abstract. We study Recursive Concurrent Stochastic Games (RCSGs), extending our recent analysis of recursive simple stochastic games [14, 15] to a concurrent setting where the two players choose moves simultaneously and independently at each state. For multi-exit games, our earlier work already showed undecidability for basic questions like termination, thus we focus on the important case of single-exit RCSGs (1-RCSGs).

We first characterize the value of a 1-RCSG termination game as the least fixed point solution of a system of nonlinear minimax functional equations, and use it to show PSPACE decidability for the quantitative termination problem. We then give a strategy improvement technique, which we use to show that player 1 (maximizer) has ϵ-optimal randomized Stackless & Memoryless (r-SM) strategies, while player 2 (minimizer) has optimal r-SM strategies. Thus, such games are r-SM-determined. These results mirror and generalize in a strong sense the randomized memoryless determinacy results for finite stochastic games, and extend the classic Hoffman-Karp [19] strategy improvement approach from the finite to an infinite state setting. The proofs in our infinite-state setting are very different however.

We show that our upper bounds, even for qualitative termination, can not be improved without a major breakthrough, by giving two reductions: first a P-time reduction from the long-standing square-root sum problem to the quantitative termination decision problem for *finite* concurrent stochastic games, and then a P-time reduction from the latter problem to the qualitative termination problem for 1-RCSGs.

1 Introduction

In recent work we have studied Recursive Markov Decision Processes (RMDPs) and turn-based Recursive Simple Stochastic Games (RSSGs) ([14, 15]), providing a number of strong upper and lower bounds for their analysis. These define infinite-state (perfect information) stochastic games that extend Recursive Markov Chains (RMCs) ([12, 13]) with nonprobabilistic actions controlled by players. Here we extend our study to Recursive Concurrent Stochastic Games (RCSGs), where the two players choose moves simultaneously and independently at each state, unlike RSSGs where only one player can move at each state. RCSGs define a class of infinite-state zero-sum (imperfect information) stochastic games that can naturally model probabilistic procedural programs and other systems involving both recursive and probabilistic behavior, as well as concurrent interactions between the system and the environment. Informally, all such

M. Bugliesi et al. (Eds.): ICALP 2006, Part II, LNCS 4052, pp. 324–335, 2006.

recursive models consist of a finite collection of finite state component models (of the same type) that can call each other in a potentially recursive manner. For multi-exit RMDPs and RSSGs, our earlier work already showed that basic questions such as qualitative (i.e. almost sure) termination are already undecidable, whereas we gave strong upper bounds for the important special case of *single-exit* RMDPs and RSSGs (called 1-RMDPs and 1-RSSGs).

Our focus is thus on single-exit RCSGs (1-RCSGs). These models correspond to a concurrent game version of multi-type Branching Processes and Stochastic Context-Free Grammars, both of which are important and extensively studied stochastic processes with many applications including in population genetics, nuclear chain reactions, computational biology, and natural language processing (see, e.g., [18, 20] and other references in [12, 14]). It is very natural to consider game extensions to these stochastic models. Branching processes model the growth of a population of entities of distinct types. In each generation each entity of a given type gives rise, according to a probability distribution, to a multi-set of entities of distinct types. A branching process can be mapped to a 1-RMC such that the probability of eventual extinction of a species is equal to the probability of termination in the 1-RMC. Modeling the process in a context where external agents can influence the evolution to bias it towards extinction or towards survival leads naturally to a game. A 1-RCSG models the process where the evolution of some types is affected by the concurrent actions of external favorable and unfavorable agents (forces).

In [14], we showed that for the 1-RSSG termination game, where the goal of player 1 (2) is to maximize (minimize) the probability of termination starting at a given vertex (in the empty calling context), we can decide in PSPACE whether the value of the game is $\geq p$ for a given probability p, and we can approximate this value (which can be irrational) to within given precision with the same complexity. We also showed that both players have optimal *deterministic Stackless and Memoryless* (SM) strategies in the 1-RSSG termination game; these are strategies that depend neither on the history of the game nor on the call stack at the current state. Thus from each vertex belonging to the player, such a strategy deterministically picks one of the outgoing transitions.

Already for finite-state concurrent stochastic games (CSGs), even under the simple termination objective, the situation is rather different. Memoryless strategies do suffice for both players, but randomization of strategies is necessary, meaning we can't hope for deterministic ϵ-optimal strategies for either player. Moreover, player 1 (the maximizer) can only attain ϵ-optimal strategies, for $\epsilon > 0$, whereas player 2 (the minimizer) does have optimal randomized memoryless strategies (see, e.g., [16, 10]). Another important result for finite CSGs is the classic Hoffman-Karp [19] strategy improvement method, which provides, via simple local improvements, a sequence of randomized memoryless strategies which yield payoffs that converge to the value of the game.

Here we generalize all these results to the infinite-state setting of 1-RCSG termination games. We first characterize values of the 1-RCSG termination game as the least fixed point solution of a system of nonlinear minimax functional equations. We use this to show PSPACE decidability for the *quantitative termination problem* (is the value of the game $\geq r$ for given rational r), as well as PSPACE algorithms for approximating the termination probabilities of 1-RCSGs to within a given number of bits of precision, via results for the existential theory of reals.

We then proceed to our technically most involved result, a strategy improvement technique for 1-RCSG termination games. We use this to show that in these games player 1 (maximizer) has ϵ-optimal randomized-Stackless & Memoryless (r-SM for short) strategies, whereas player 2 (minimizer) has optimal r-SM strategies. Thus, such games are r-SM-determined. These results mirror and generalize in a very strong sense the randomized memoryless determinacy results known for finite stochastic games. Our technique extends Hoffman-Karp's strategy improvement method for finite CSGs to an infinite state setting. However, the proofs in our infinite-state setting are very different. We rely on subtle analytic properties of certain power series that arise from studying 1-RCSGs.

Note that our PSPACE upper bounds for the quantitative termination problem for 1-RCSGs can not be improved to NP without a major breakthrough, since already for 1-RMCs we showed in [12] that the quantitative termination problem is at least as hard as the square-root sum problem (see [12]). In fact, here we show that even the qualitative termination problem for 1-RCSGs, where the problem is to decide whether the value of the game is exactly 1, is already as hard as the square-root sum problem, and moreover, so is the quantitative termination decision problem for *finite* CSGs. We do this via two reductions: we give a P-time reduction from the square-root sum problem to the quantitative termination decision problem for *finite* CSGs, and a P-time reduction from the quantitative finite CSG termination problem to the qualitative 1-RCSG termination problem. Note that this is despite the fact that in recent work Chatterjee et. al. ([6]) have shown that the *approximate* quantitative problems for finite CSGs, including for termination and for more general parity winning conditions, are in NP∩coNP. In other words, we show that quantitative decision problems for finite CSGs will require surmounting significant new difficulties that don't arise for approximation of game values.

We note that, as is known already for finite concurrent games ([5]), probabilistic nodes do not add any power to these games, because the stochastic nature of all the games we consider can in fact be simulated by concurrency alone. The same is true for 1-RCSGs. Specifically, given a finite CSG (or 1-RCSG), G, there is a P-time reduction to a finite concurrent game (or 1-RCG, respectively) $F(G)$, without any probabilistic vertices, such that the value of the game G is exactly the same as the value of the game $F(G)$.

Related work. Stochastic games go back to Shapley [24], who considered finite concurrent stochastic games with (discounted) rewards. See, e.g., [16] for a recent book on stochastic games. Turn-based "simple" finite stochastic games were studied by Condon [8]. As mentioned, we studied RMDPs and (turn-based) RSSGs and their quantitative and qualitative termination problems in [14, 15]. In [15] we showed that the qualitative termination problem for finite 1-RMDPs is in P, and for 1-RSSGs is in NP∩coNP. Our earlier work [12, 13] developed theory and algorithms for Recursive Markov Chains (RMCs), and [11, 3] have studied probabilistic Pushdown Systems which are essentially equivalent to RMCs.

Finite-state concurrent stochastic games have been studied extensively in recent CS literature (see, e.g., [6, 10, 9]). In particular, [6] have shown that for finite CSGs the *approximate* reachability problem and *approximate* parity game problem are in NP∩coNP; however, their results do not resolve the decision problem, which asks whether the value of the game is $\geq r$. (Their approximation theorem (Thm 3.3, part 1.) in its current form is slightly misstated in a way that would actually imply that the decision problem is also in NP∩coNP, but this will be corrected in a journal version of their paper ([5]).) Indeed, we show here that the quantitative decision problem for finite CSGs, as well as the qualitative problem for 1-RCSGs, are as hard as the square-root sum problem, for which containment even in NP is a long standing open problem. Thus our upper bound here, even for the qualitative termination problem for 1-RCSGs, can not be improved to NP without a major breakthrough. Unlike for 1-RCSGs, the qualitative termination problem for finite CSGs is known to be decidable in P-time ([9]). We note that in recent work Allender et. al. [1] have shown that the square-root sum problem is in (the 4th level of) the "Counting Hierarchy" CH, which is inside PSPACE, but it remains a major open problem to bring this complexity down to NP.

2 Basics

Let Γ_1 and Γ_2 be finite sets constituting the *move alphabet* of players 1 and 2, respectively. A *Recursive Concurrent Stochastic Game (RCSG)* is a tuple $A = (A_1, \ldots, A_k)$, where each *component* $A_i = (N_i, B_i, Y_i, En_i, Ex_i, \mathtt{pl}_i, \delta_i)$ consists of:

1. A set N_i of *nodes*, with a distinguished subset En_i of *entry* nodes and a (disjoint) subset Ex_i of *exit* nodes.
2. A set B_i of *boxes*, and a mapping $Y_i : B_i \mapsto \{1, \ldots, k\}$ that assigns to every box (the index of) a component. To each box $b \in B_i$, we associate a set of *call ports*, $Call_b = \{(b, en) \mid en \in En_{Y(b)}\}$, and a set of *return ports*, $Return_b = \{(b, ex) \mid ex \in Ex_{Y(b)}\}$. Let $Call^i = \cup_{b \in B_i} Call_b$, $Return^i = \cup_{b \in B_i} Return_b$, and let $Q_i = N_i \cup Call^i \cup Return^i$ be the set of all nodes, call ports and return ports; we refer to these as the *vertices* of component A_i.
3. A mapping $\mathtt{pl}_i : Q_i \mapsto \{0, play\}$ that assigns to every vertex u a type describing how the next transition is chosen: if $\mathtt{pl}_i(u) = 0$ it is chosen probabilistically and if $\mathtt{pl}_i(u) = play$ it is determined by moves of the two players. Vertices $u \in (Ex_i \cup Call^i)$ have no outgoing transitions; for them we let $\mathtt{pl}_i(u) = 0$.
4. A transition relation $\delta_i \subseteq (Q_i \times (\mathbb{R} \cup (\Gamma_1 \times \Gamma_2)) \times Q_i)$, where for each tuple $(u, x, v) \in \delta_i$, the source $u \in (N_i \setminus Ex_i) \cup Return^i$, the destination $v \in (N_i \setminus En_i) \cup Call^i$, where if $\mathtt{pl}(u) = 0$ then x is a real number $p_{u,v} \in [0, 1]$ (the transition probability), and if $\mathtt{pl}(u) = play$ then $x = (\gamma_1, \gamma_2) \in \Gamma_1 \times \Gamma_2$. We assume that each vertex $u \in Q_i$ has associated with it a set $\Gamma_1^u \subseteq \Gamma_1$ and a set $\Gamma_2^u \subseteq \Gamma_2$, which constitute player 1 and 2's *legal moves* at vertex u. Thus, if $(u, x, v) \in \delta_i$ and $x = (\gamma_1, \gamma_2)$ then $(\gamma_1, \gamma_2) \in \Gamma_1^u \times \Gamma_2^u$. Additionally, for each vertex u and each

$x \in \Gamma_1^u \times \Gamma_2^u$, we assume there is exactly 1 transition of the form (u, x, v) in δ_i. For computational purposes we assume that the given probabilities $p_{u,v}$ are rational. Furthermore they must satisfy the consistency property: for every $u \in \mathtt{pl}^{-1}(0)$, $\sum_{\{v' | (u, p_{u,v'}, v') \in \delta_i\}} p_{u,v'} = 1$, unless u is a call port or exit node, neither of which have outgoing transitions, in which case by default $\sum_{v'} p_{u,v'} = 0$.

We use the symbols $(N, B, Q, \delta$, etc.) without a subscript, to denote the union over all components. Thus, eg. $N = \cup_{i=1}^k N_i$ is the set of all nodes of A, $\delta = \cup_{i=1}^k \delta_i$ the set of all transitions, etc.

An RCSG A defines a global denumerable stochastic game $M_A = (V, \Delta, \mathtt{pl})$ as follows. The global *states* $V \subseteq B^* \times Q$ of M_A are pairs of the form $\langle \beta, u \rangle$, where $\beta \in B^*$ is a (possibly empty) sequence of boxes and $u \in Q$ is a *vertex* of A. More precisely, the states $V \subseteq B^* \times Q$ and transitions Δ are defined inductively as follows: 1. $\langle \epsilon, u \rangle \in V$, for $u \in Q$ (ϵ denotes the empty string.); 2. if $\langle \beta, u \rangle \in V$ & $(u, x, v) \in \delta$, then $\langle \beta, v \rangle \in V$ and $(\langle \beta, u \rangle, x, \langle \beta, v \rangle) \in \Delta$; 3. if $\langle \beta, (b, en) \rangle \in V$, with $(b, en) \in Call_b$, then $\langle \beta b, en \rangle \in V$ & $(\langle \beta, (b, en) \rangle, 1, \langle \beta b, en \rangle) \in \Delta$; 4. if $\langle \beta b, ex \rangle \in V$, & $(b, ex) \in Return_b$, then $\langle \beta, (b, ex) \rangle \in V$ & $(\langle \beta b, ex \rangle, 1, \langle \beta, (b, ex) \rangle) \in \Delta$. Item 1. corresponds to the possible initial states, item 2. corresponds to control staying within a component, item 3. is when a new component is entered via a box, item 4. is when control exits a box and returns to the calling component. The mapping $\mathtt{pl} : V \mapsto \{0, play\}$ is given by $\mathtt{pl}(\langle \beta, u \rangle) = \mathtt{pl}(u)$. The set of vertices V is partitioned into V_0, V_{play}, where $V_0 = \mathtt{pl}^{-1}(0)$ and $V_{play} = \mathtt{pl}^{-1}(play)$.

We consider M_A with various *initial states* of the form $\langle \epsilon, u \rangle$, denoting this by M_A^u. Some states of M_A are *terminating states* and have no outgoing transitions. These are states $\langle \epsilon, ex \rangle$, where ex is an exit node. If we wish to view M_A as a non-terminating CSG, we can consider the terminating states as absorbing states of M_A, with a self-loop of probability 1.

An RCSG where $|\Gamma_2| = 1$ (i.e., where player 2 has only one action) is called a maximizing *Recursive Markov Decision Process* (RMDP), likewise, when $|\Gamma_1| = 1$ is a minimizing RMDP. An RSSG where $|\Gamma_1| = |\Gamma_2| = 1$ is essentially a *Recursive Markov Chain* ([12, 13]).

Our goal is to answer termination questions for RCSGs of the form: *"Does player 1 have a strategy to force the game to terminate (i.e., reach node $\langle \epsilon, ex \rangle$), starting at $\langle \epsilon, u \rangle$, with probability $\geq p$, regardless of how player 2 plays?"*.

First, some definitions: a *strategy* σ for player i, $i \in \{1, 2\}$, is a function $\sigma : V^* V_{play} \mapsto \mathcal{D}(\Gamma_i)$, where $\mathcal{D}(\Gamma_i)$ denotes the set of probability distributions on the finite set of moves Γ_i. In other words, given a history $ws \in V^* V_{play}$, and a strategy σ for, say, player 1, $\sigma(ws)(\gamma)$ defines the probability with which player 1 will play move γ. Moreover, we require that the function σ has the property that for any global state $s = \langle \beta, u \rangle$, with $\mathtt{pl}(u) = play$, $\sigma(ws) \in \mathcal{D}(\Gamma_i^u)$. In other words, the distribution has support only over eligible moves at vertex u.

Let Ψ_i denote the set of all strategies for player i. Given a history $ws \in V^* V_{play}$ of play so far, and given a strategy $\sigma \in \Psi_1$ for player 1, and a strategy $\tau \in \Psi_2$ for player 2, the strategies determine a distribution on the next move of play to a new global state, namely, the transition $(s, (\gamma_1, \gamma_2), s') \in \Delta$ has probability $\sigma(ws)(\gamma_1) * \tau(ws)(\gamma_2)$. This way, given a start node u, a strategy

$\sigma \in \Psi_1$, and a strategy $\tau \in \Psi_2$, we define a new Markov chain (with initial state u) $M_A^{u,\sigma,\tau} = (\mathcal{S}, \Delta')$. The states $\mathcal{S} \subseteq \langle \epsilon, u \rangle V^*$ of $M_A^{u,\sigma,\tau}$ are non-empty sequences of states of M_A, which must begin with $\langle \epsilon, u \rangle$. Inductively, if $ws \in \mathcal{S}$, then: (0) if $s \in V_0$ and $(s, p_{s,s'}, s') \in \Delta$ then $wss' \in \mathcal{S}$ and $(ws, p_{s,s'}, wss') \in \Delta'$; (1) if $s \in V_{play}$, where $(s, (\gamma_1, \gamma_2), s') \in \Delta$, then if $\sigma(ws)(\gamma_1) > 0$ and $\tau(ws)(\gamma_2) > 0$ then $wss' \in \mathcal{S}$ and $(ws, p, wss') \in \Delta'$, where $p = \sigma(ws)(\gamma_1) * \tau(ws)(\gamma_2)$.

Given initial vertex u, and final exit ex in the same component, and given strategies $\sigma \in \Psi_1$ and $\tau \in \Psi_2$, for $k \geq 0$, let $q_{(u,ex)}^{k,\sigma,\tau}$ be the probability that, in $M_A^{u,\sigma,\tau}$, starting at initial state $\langle \epsilon, u \rangle$, we will reach a state $w\langle \epsilon, ex \rangle$ in at most k "steps" (i.e., where $|w| \leq k$). Let $q_{(u,ex)}^{*,\sigma,\tau} = \lim_{k \to \infty} q_{(u,ex)}^{k,\sigma,\tau}$ be the probability of ever terminating at ex, i.e., reaching $\langle \epsilon, ex \rangle$. (Note, the limit exists: it is a monotonically non-decreasing sequence bounded by 1). Let $q_{(u,ex)}^k = \sup_{\sigma \in \Psi_1} \inf_{\tau \in \Psi_2} q_{(u,ex)}^{k,\sigma,\tau}$ and let $q_{(u,ex)}^* = \sup_{\sigma \in \Psi_1} \inf_{\tau \in \Psi_2} q_{(u,ex)}^{*,\sigma,\tau}$. For a strategy $\sigma \in \Psi_1$, let $q_{(u,ex)}^{k,\sigma} = \inf_{\tau \in \Psi_2} q_{(u,ex)}^{k,\sigma,\tau}$, and let $q_{(u,ex)}^{*,\sigma} = \inf_{\tau \in \Psi_2} q_{(u,ex)}^{*,\sigma,\tau}$. Lastly, given a strategy $\tau \in \Psi_2$, let $q_{(u,ex)}^{k,\cdot,\tau} = \sup_{\sigma \in \Psi_1} q_{(u,ex)}^{k,\sigma,\tau}$, and let $q_{(u,ex)}^{*,\cdot,\tau} = \sup_{\sigma \in \Psi_1} q_{(u,ex)}^{*,\sigma,\tau}$.

From, general determinacy results (e.g., "Blackwell determinacy" [22] which applies to all Borel two-player zero-sum stochastic games with countable state spaces; see also [21]) it follows that the games M_A are *determined*, meaning: $\sup_{\sigma \in \Psi_1} \inf_{\tau \in \Psi_2} q_{(u,ex)}^{*,\sigma,\tau} = \inf_{\tau \in \Psi_2} \sup_{\sigma \in \Psi_1} q_{(u,ex)}^{*,\sigma,\tau}$.

We call a strategy σ for either player a (randomized) *Stackless and Memoryless* (*r-SM*) strategy if it neither depends on the history of the game, nor on the current call stack. In other words, a r-SM strategy σ for player i is given by a function $\sigma : Q \mapsto \mathcal{D}(\Gamma_i)$, which maps each vertex u of the RCSG to a probability distribution $\sigma(u) \in \mathcal{D}(\Gamma_i^u)$ on the moves available to player i at vertex u.

We are interested in the following computational problems.

(1) The *qualitative* termination problem: Is $q_{(u,ex)}^* = 1$?
(2) The *quantitative* termination (decision) problem: given $r \in [0,1]$, is $q_{(u,ex)}^* \geq r$? The *approximate* version: approximate $q_{(u,ex)}^*$ to within desired precision.

As mentioned, for multi-exit RCSGs these are all undecidable. Thus we focus on *single-exit* RCSGs (*1-RCSGs*), where every component has one exit. Since for 1-RCSGs it is always clear which exit we wish to terminate at starting at vertex u (there is only one exit in u's component), we abbreviate $q_{(u,ex)}^*$, $q_{(u,ex)}^{*,\sigma}$, etc., as q_u^*, $q_u^{*,\sigma}$, etc., and we likewise abbreviate other subscripts.

3 Nonlinear Minimax Equations for 1-RCSGs

In ([14]) we defined a monotone system S_A of nonlinear min-& -max equations for 1-RSSGs, and showed that its *least fixed point* solution yields the desired probabilities q_u^*. Here we generalize these to nonlinear minimax systems for 1-RCSGs. Let us use a variable x_u for each unknown q_u^*, and let x be the vector of all x_u , $u \in Q$. The system S_A has one equation of the form $x_u = P_u(\mathbf{x})$ for each vertex u. Suppose that u is in component A_i with (unique) exit ex. There are 4 cases based on the "*Type*" of u.

1. $u \in Type_1$: $u = ex$. In this case: $x_u = 1$.
2. $u \in Type_{rand}$: $\mathtt{pl}(u) = 0$ & $u \in (N_i \setminus \{ex\}) \cup Return^i$: $x_u = \sum_{\{v|(u,p_{u,v},v)\in\delta\}} p_{u,v}x_v$. (If u has no outgoing transitions, this equation is by definition $x_u = 0$.)
3. $u \in Type_{call}$: $u = (b, en)$ is a call port: $x_{(b,en)} = x_{en} \cdot x_{(b,ex')}$, where $ex' \in Ex_{Y(b)}$ is the unique exit of $A_{Y(b)}$.
4. $u \in Type_{play}$: $x_u = \mathrm{Val}(A_u(x))$.
   We have to define this case. Given a value vector x, and a play vertex u, consider the zero-sum matrix game given by matrix $A_u(x)$, whose rows are indexed by player 1's moves Γ_1^u from node u, and whose columns are indexed by player 2's moves Γ_2^u. The payoff to player 1 under the pair of deterministic moves $\gamma_1 \in \Gamma_1^u$, and $\gamma_2 \in \Gamma_2^u$, is given by $(A_u(x))_{\gamma_1,\gamma_2} := x_v$, where $(u, (\gamma_1, \gamma_2), v) \in \delta$. Let $\mathrm{Val}(A_u(x))$ be the value of this zero-sum matrix game. By von Neumann's minmax theorem, the value and optimal mixed strategies exist, and they can be obtained by solving a set of linear inequality constraints with coefficients given by the x_i's.

In vector notation, we denote the system S_A by $x = P(x)$. Given 1-exit RCSG A, we can easily construct this system. Note that the operator $P : \mathbb{R}^n_{\geq 0} \mapsto \mathbb{R}^n_{\geq 0}$ is *monotone*: for $x, y \in \mathbb{R}^n_{\geq 0}$, if $x \leq y$ then $P(x) \leq P(y)$. This follows because for two game matrices A and B of the same dimensions, if $A \leq B$ (i.e., $A_{i,j} \leq B_{i,j}$ for all i and j), then $\mathrm{Val}(A) \leq \mathrm{Val}(B)$. Note that by definition of $A_u(x)$, for $x \leq y$, $A_u(x) \leq A_u(y)$. We now identify a particular solution to $x = P(x)$, called the *Least Fixed Point* (LFP) solution, which gives precisely the termination game values. Define $P^1(x) = P(x)$, and define $P^k(x) = P(P^{k-1}(x))$, for $k > 1$. Let $q^* \in \mathbb{R}^n$ denote the n-vector $q^*_u, u \in Q$ (using the same indexing as used for x). For $k \geq 0$, let q^k denote, similarly, the n-vector $q^k_u, u \in Q$.

Theorem 1. *Let $x = P(x)$ be the system S_A associated with 1-RCSG A. Then $q^* = P(q^*)$, and for all $q' \in \mathbb{R}^n_{\geq 0}$, if $q' = P(q')$, then $q^* \leq q'$ (i.e., q^* is the* Least Fixed Point, *of $P : \mathbb{R}^n_{\geq 0} \mapsto \mathbb{R}^n_{\geq 0}$). Moreover, $\lim_{k\to\infty} P^k(\mathbf{0}) \uparrow q^*$, i.e., the "value iteration" sequence $P^k(\mathbf{0})$ converges monotonically to the LFP, q^*.*

The proof is omitted due to space constraints. We will need an important fact established in the proof: suppose for some $q' \in \mathbb{R}^n_{\geq 0}$, $q' = P(q')$. Let τ' be the r-SM strategy for player 2 that always picks, at any state $\langle\beta, u\rangle$, for vertex $u \in \mathtt{pl}^{-1}(play)$, the mixed 1-step strategy which is an optimal minimax strategy in the matrix game $A_u(q')$. Then $q^{*,\cdot,\tau'} \leq q'$. In other words, τ' achieves a value $\leq q'_u$ for the game starting from every vertex u (in the empty context).

Theorem 2. *Given a 1-exit RCSG A and a rational probability p, there is a PSPACE algorithm to decide whether $\mathbf{q}^*_u \leq p$. The running time is $O(|A|^{O(n)})$ where n is the number of variables in $\mathbf{x} = P(\mathbf{x})$. We can also approximate $\mathbf{q}^*$ to within a given number of bits i of precision (i given in unary), in PSPACE and in time $O(i|A|^{O(n)})$.*

Proof. Using the system $x = P(x)$, we can express facts such as $q^*_u \leq c$ as

$$\exists x_1, \ldots, x_n \bigwedge_{i=1}^{n} (x_i = P_i(x_1, \ldots, x_n)) \wedge x_u \leq c$$

We only need to show how to express equations of the form $x_v = \mathrm{Val}(A_v(\mathbf{x}))$ in the existential theory of reals. We can then appeal to well known results for deciding that theory ([4, 23]). But this is a standard fact in game theory (see, e.g., [2, 16, 10] where it is used for finite CSGs). Namely, the minimax theorem and its LP encoding allow the predicate "$y = \mathrm{Val}(A_v(\mathbf{x}))$" to be expressed as an existential formula $\varphi(y, x)$ in the theory of reals with free variables y and $x_1, \ldots, x_n$, such that for every $x \in \mathbb{R}^n$, there exists a unique y (the game value) satisfying $\varphi(y, \mathbf{x})$. To approximate the game values within given precision we can do binary search using such queries. □

4 Strategy Improvement and Randomized-SM-Determinacy

The proof of Theorem 1 implies the following (see discussion after Thm 1):

Corollary 1. *In every 1-RCSG termination game, player 2 (the minimizer) has an optimal r-SM strategy.*

Proof. Consider the strategy τ' in the discussion after Theorem 1, chosen not for just any fixed point $\mathbf{q}'$, but for $\mathbf{q}^*$ itself. That strategy is r-SM. □

Player 1 does not have optimal r-SM strategies, not even in finite concurrent stochastic games (see, e.g., [16, 10]). We next establish that it does have finite r-SM *ϵ-optimal strategies*, meaning that it has, for every $\epsilon > 0$, a r-SM strategy that guarantees a value of at least $\mathbf{q}^*_u - \epsilon$, starting from every vertex u in the termination game. We say that a game is *r-SM-determined* if, letting Ψ'_1 and Ψ'_2 denote the set of r-SM strategies for players 1 and 2, respectively, we have $\sup_{\sigma \in \Psi'_1} \inf_{\tau \in \Psi'_2} q^{*,\sigma,\tau}_u = \inf_{\tau \in \Psi'_2} \sup_{\sigma \in \Psi'_1} q^{*,\sigma,\tau}_u$.

Theorem 3

1. *(Strategy Improvement) Starting at any r-SM strategy σ_0 for player 1, via local strategy improvement steps at individual vertices, we can derive a series of r-SM strategies $\sigma_0, \sigma_1, \sigma_2, \ldots$, such that for all $\epsilon > 0$, there exists $i \geq 0$ such that for all $j \geq i$, σ_j is an ϵ-optimal strategy for player 1 starting at any vertex, i.e., $q^{*,\sigma_j}_u \geq q^*_u - \epsilon$ for all vertices u.*

   Each strategy improvement step involves solving the quantitative termination problem for a corresponding 1-RMDP. Thus, for classes where this problem is known to be in P-time (such as linearly-recursive 1-RMDPs, [14]), strategy improvement steps can be carried out in polynomial time.
2. *Player 1 has ϵ-optimal r-SM strategies, for all $\epsilon > 0$, in 1-RCSG termination games.*
3. *1-RCSG termination games are r-SM-determined.*

Proof. Note that (2.) follows immediately from (1.), and (3.) follows because by Corollary 1, player 2 has an optimal r-SM strategy and thus $\sup_{\sigma \in \Psi'_1} \inf_{\tau \in \Psi'_2} q^{*,\sigma,\tau}_u = \inf_{\tau \in \Psi'_2} \sup_{\sigma \in \Psi'_1} q^{*,\sigma,\tau}_u$.

Let σ be any r-SM strategy for player 1. Consider $q^{*,\sigma}$. First, let us note that if $q^{*,\sigma} = P(q^{*,\sigma})$ then $q^{*,\sigma} = q^*$. This is so because, by Theorem 1, $q^* \leq q^{*,\sigma}$, and on the other hand, σ is just one strategy for player 1, and for every vertex u, $q^*_u = \sup_{\sigma' \in \Psi_1} \inf_{\tau \in \Psi_2} q^{*,\sigma',\tau}_u \geq \inf_{\tau \in \Psi_2} q^{*,\sigma,\tau}_u = q^{*,\sigma}_u$.

Next we claim that, for all vertices $u \notin Type_{play}$, $q^{*,\sigma}_u$ satisfies its equation in $x = P(x)$. In other words, $q^{*,\sigma}_u = P_u(q^{*,\sigma})$. To see this, note that for vertices $u \notin Type_{play}$, no choice of either player is involved, thus the equation holds by definition of $q^{*,\sigma}$. Thus, the only equations that may fail are those for $u \in Type_{play}$, of the form $x_u = \mathrm{Val}(A_u(x))$. We need the following (proof omitted).

Lemma 1. *For any r-SM strategy σ for player 1, and for any $u \in Type_{play}$, $q^{*,\sigma}_u \leq Val(A_u(q^{*,\sigma}))$.*

Now, suppose that for some $u \in Type_{play}$, $q^{*,\sigma}_u \neq Val(A_u(q^{*,\sigma}))$. Thus by the lemma $q^{*,\sigma}_u < Val(A_u(q^{*,\sigma}))$. Consider a revised r-SM strategy for player 1, σ', which is identical to σ, except that locally at vertex u the strategy is changed so that $\sigma'(u) = p^{*,u,\sigma}$, where $p^{*,u,\sigma} \in \mathcal{D}(\Gamma^u_1)$ is an optimal mixed minimax strategy for player 1 in the matrix game $A_u(q^{*,\sigma})$. We will show that switching from σ to σ' will improve player 1's payoff at vertex u, and will not reduce its payoff at any other vertex.

Consider a parameterized 1-RCSG, $A(t)$, which is identical to A, except that u is a randomizing vertex, all edges out of vertex u are removed, and replaced by a single edge labeled by probability variable t to the exit of the same component, and an edge with remaining probability $1-t$ to a dead vertex. Fixing the value t determines an 1-RCSG, $A(t)$. Note that if we restrict the r-SM strategies σ or σ' to all vertices other than u, then they both define the same r-SM strategy for the 1-RCSG $A(t)$. For each vertex z and strategy τ of player 2, define $q^{*,\sigma,\tau,t}_z$ to be the probability of eventually terminating starting from $\langle \epsilon, z \rangle$ in the Markov chain $M^{z,\sigma,\tau}_{A(t)}$. Let $f_z(t) = \inf_{\tau \in \Psi_2} q^{*,\sigma,\tau,t}_z$. Recall that $\sigma'(u) = p^{*,u,\sigma} \in \mathcal{D}(\Gamma^u_1)$ defines a probability distribution on the actions available to player 1 at vertex u. Thus $p^{*,u,\sigma}(\gamma_1)$ is the probability of action $\gamma_1 \in \Gamma_1$. Let $\gamma_2 \in \Gamma_2$ be any action of player 2 for the 1-step zero-sum game with game matrix $A_u(q^{*,\sigma})$. Let $w(\gamma_1, \gamma_2)$ denote the vertex such that $(u, (\gamma_1, \gamma), w(\gamma_1, \gamma_2)) \in \delta$. Let $h_{\gamma_2}(t) = \sum_{\gamma_1 \in \Gamma_1} p^{*,u,\sigma}(\gamma_1) f_{w(\gamma_1,\gamma_2)}(t)$.

Lemma 2. *Fix the vertex u. Let $\varphi : \mathbb{R} \mapsto \mathbb{R}$ be any function $\varphi \in \{f_z \mid z \in Q\} \cup \{h_\gamma \mid \gamma \in \Gamma^u_2\}$. The following properties hold:*

1. *If $\varphi(t) > t$ at some point $t \geq 0$, then $\varphi(t') > t'$ for all $0 \leq t' < t$.*
2. *If $\varphi(t) < t$ at some point $t \geq 0$, then $\varphi(t') < t'$ for all $1 > t' > t$.*

Proof. First, we prove this for $\varphi = f_z$, for some vertex z.

Note that, once player 1 picks a r-SM strategy, a 1-RCSG becomes a 1-RMDP. By a result of [14], player 2 has an optimal deterministic SM response strategy. Furthermore, there is such a strategy that is optimal regardless of the starting vertex. Thus, for any value of t, player 2 has an optimal deterministic SM strategy τ_t, such that for any start vertex z, we have $\tau_t = \arg\min_{\tau \in \Psi_2} q^{*,\sigma,\tau,t}_z$. Let $g_{(z,\tau)}(t) = q^{*,\sigma,\tau,t}_z$, and let $d\Psi_2$ be the (finite) set of deterministic SM strategies of player 2. Then $f_z(t) = \min_{\tau \in d\Psi_2} g_{z,\tau}(t)$. Now, note that the function $g_{z,\tau}(t)$ is the probability of reaching an exit in an RMC starting from a particular vertex. Thus, by [12], $g_{z,\tau}(t) = (\lim_{k\to\infty} R^k(\mathbf{0}))_z$ for a polynomial system $\mathbf{x} = R(\mathbf{x})$ with non-negative coefficients, but with the additional feature that the variable t appears as one of the coefficients. Since this limit can be described by a power series in the variable t with non-negative coefficients, $g_{z,\tau}(t)$ has the following properties: it is a continuous, differentiable, and nondecreasing function of $t \in [0,1]$, with continuous and nondecreasing derivative, $g'_{z,\tau}(t)$, and since the limit defines probabilities we also know that for $t \in [0,1]$, $g_{z,\tau}(t) \in [0,1]$. Thus $g_{z,\tau}(0) \geq 0$ and $g_{z,\tau}(1) \leq 1$.

Hence, since $g'_{z,\tau}(t)$ is non-decreasing, if for some $t \in [0,1]$, $g_{z,\tau}(t) > t$, then for all $t' < t$, $g_{z,\tau}(t') > t'$. To see this, note that if $g_{z,\tau}(t) > t$ and $g'_{z,\tau}(t) \geq 1$, then for all $t'' > t$, $g_{z,\tau}(t'') > t''$, which contradicts the fact that $g_{z,\tau}(1) = 1$. Thus $g'_{z,\tau}(t') < 1$ for all $t' \leq t$, and since $g_{z,\tau}(t) > t$, we also have $g_{z,\tau}(t') > t'$ for all $t' < t$. Similarly, if $g_{z,\tau}(t) < t$ for some t, then $g_{z,\tau}(t'') < t''$ for all $t'' \in [t,1)$. To see this, note that if for some $t'' > t$, $t'' < 1$, $g_{z,\tau}(t'') = t''$, then since $g'_{z,\tau}$ is non-decreasing and $g_{z,\tau}(t) < t$, it must be the case that $g'_{z,\tau}(t'') > 1$. But then $g_{z,\tau}(1) > 1$, which is a contradiction.

It follows that $f_z(t)$ has the same properties, namely: if $f_z(t) > t$ at some point $t \in [0,1]$ then $g_{z,\tau}(t) > t$ for all τ, and hence for all $t' < t$ and for all $\tau \in d\Psi_2$, $g_{z,\tau}(t') > t'$, and thus $f_z(t') > t'$ for all $t' \in [0,t]$. On the other hand, if $f_z(t) < t$ at $t \in [0,1]$, then there must be some $\tau' \in d\Psi_2$ such that $g_{z,\tau'}(t) < t$. Hence $g_{z,\tau'}(t'') < t''$, for all $t'' \in [t,1)$, and hence $f_z(t'') < t''$ for all $t'' \in [t,1)$.

Next we prove the lemma for every $\varphi = h_\gamma$, where $\gamma \in \Gamma_2^u$. For every value of t, there is one SM strategy τ_t of player 2 (depending only on t) that minimizes simultaneously $g_{z,\tau}(t)$ for all nodes z. So $h_\gamma(t) = \min_\tau r_{\gamma,\tau}(t)$, where $r_{\gamma,\tau}(t) = \sum_{\gamma_1 \in \Gamma_1} p^{*,u,\sigma}(\gamma_1) g_{w(\gamma_1,\gamma),\tau}(t)$ is a convex combination (i.e., a "weighted average")of some g functions at the same point t. The function $r_{\gamma,\tau}$ (for any subscript) inherits the same properties as the g's: continuous, differentiable, nondecreasing, with continuous nondecreasing derivatives, and $r_{\gamma,\tau}$ takes value between 0 and 1. As we argued for the g functions, in the same way it follows that $r_{\gamma,\tau}$ has properties 1 and 2. Also, as we argued for f's based on the g's, it follows that h's also have the same properties, based on the r's. □

Let $t_1 = q_u^{*,\sigma}$, and let $t_2 = \mathrm{Val}(A_u(q^{*,\sigma}))$. By assumption $t_2 > t_1$. Observe that $f_z(t_1) = q_z^{*,\sigma}$ for every vertex z. Thus, $h_{\gamma_2}(t_1) = \sum_{\gamma_1 \in \Gamma_1} p^{*,u,\sigma}(\gamma_1) f_{w(\gamma_1,\gamma_2)}(t_1) = \sum_{\gamma_1} p^{*,u,\sigma}(\gamma_1) q^{*,\sigma}_{w(\gamma_1,\gamma_2)}$. But since, by definition, $p^{*,u,\sigma}$ is an optimal strategy for player 1 in the matrix game $A_u(q^{*,\sigma})$, it must be the case that for every $\gamma_2 \in \Gamma_2^u$, $h_{\gamma_2}(t_1) \geq t_2$, for otherwise player 2 could play a strategy against $p^{*,u,\sigma}$ which would force a payoff lower than the value of the game. Thus $h_{\gamma_2}(t_1) \geq t_2 > t_1$, for all γ_2. This implies that $h_{\gamma_2}(t) > t$ for all $t < t_1$ by Lemma 2, and for all $t_1 \leq t < t_2$, because h_{γ_2} is nondecreasing. Thus, $h_{\gamma_2}(t) > t$ for all $t < t_2$.

Let $t_3 = q_u^{*,\sigma'}$. Let τ' be an optimal global strategy for player 2 against σ'; by [14], we may assume τ' is a pure SM strategy. Let γ' be player 2's action in τ' at node u. Then the value of any node z under the pair of strategies σ' and τ' is $f_z(t_3)$, and thus since $h_{\gamma'}(t_3)$ is a weighted average of $f_z(t_3)$'s for some set of z's, we have $h_{\gamma'}(t_3) = t_3$. Thus, by the previous paragraph, it must be that

$t_3 \geq t_2$, and we know $t_2 > t_1$. Thus, $t_3 = q_u^{*,\sigma'} \geq \mathrm{Val}(A_u(q^{*,\sigma})) > t_1 = q_u^{*,\sigma}$. We have shown:

Lemma 3. $q_u^{*,\sigma'} \geq Val(A_u(q^{*,\sigma})) > q_u^{*,\sigma}$.

Note that since $t_3 > t_1$, and f_z is non-decreasing, we have $f_z(t_3) \geq f_z(t_1)$ for all vertices z. But then $q_z^{*,\sigma'} = f_z(t_3) \geq f_z(t_1) = q_z^{*,\sigma}$ for all z. Thus, $q^{*,\sigma'} \geq q^{*,\sigma}$, with strict inequality at u, i.e., $q_u^{*,\sigma'} > q_u^{*,\sigma}$. Thus, we have established that such a "strategy improvement" step does yield a strictly better payoff for player 1.

Suppose we conduct this "strategy improvement" step repeatedly, starting at an arbitrary initial r-SM strategy σ_0, as long as we can. This leads to a (possibly infinite) sequence of r-SM strategies $\sigma_0, \sigma_1, \sigma_2, \ldots$. Suppose moreover, that during these improvement steps we always "prioritize" among vertices at which to improve so that, among all those vertices $u \in Type_{play}$ which can be improved, i.e., such that $q_u^{*,\sigma_i} < \mathrm{Val}(A_u(q^{*,\sigma_i}))$, we choose the vertex which has not been improved for the longest number of steps (or one that has never been improved yet). This insures that, infinitely often, at every vertex at which the local strategy can be improved, it eventually is improved.

Under this strategy improvement regime, we show that $\lim_{i\to\infty} q^{*,\sigma_i} = q^*$, and thus, for all $\epsilon > 0$, there exists a sufficiently large $i \geq 0$ such that σ_i is an ϵ-optimal r-SM strategy for player 1. Note that after every strategy improvement step, i, which improves at a vertex u, by Lemma 3 we will have $q_u^{*,\sigma_{i+1}} \geq \mathrm{Val}(A_u(q^{*,\sigma_i}))$. Since our prioritization assures that every vertex that can be improved at any step i will be improved eventually, for all $i \geq 0$ there exists $k \geq 0$ such that $q^{*,\sigma_i} \leq P(q^{*,\sigma_i}) \leq q^{*,\sigma_{i+k}}$. In fact, there is a uniform bound on k, namely $k \leq |Q|$, the number of vertices. This "sandwiching" property allows us to conclude that, in the limit, this sequence reaches a fixed point of $x = P(x)$. Note that since $q^{*,\sigma_i} \leq q^{*,\sigma_{i+1}}$ for all i, and since $q^{*,\sigma_i} \leq q^*$, we know that the limit $\lim_{i\to\infty} q^{*,\sigma_i}$ exists. Letting this limit be q', we have $q' \leq q^*$. Finally, we have $q' = P(q')$, because letting i go to infinity in all three parts of the "sandwiching" inequalities above, we get $q' \leq \lim_{i\to\infty} P(q^{*,\sigma_i}) \leq q'$. But note that $\lim_{i\to\infty} P(q^{*,\sigma_i}) = P(q')$, because the mapping $P(x)$ is continuous on $\mathbb{R}^n_{\geq 0}$. Thus q' is a fixed point of $x = P(x)$, and $q' \leq q^*$. But since q^* is the *least* fixed point of $x = P(x)$, we have $q' = q^*$. □

Finally, we give the following two reductions (proofs omitted due to space). Recall that the *square-root sum problem* (see, e.g., [17, 12]) is the following: given $(d_1, \ldots, d_n) \in \mathbb{N}^n$ and $k \in \mathbb{N}$, decide whether $\sum_{i=1}^n \sqrt{d_i} \geq k$.

Theorem 4. *There is a P-time reduction from the square-root sum problem to the quantitative termination (decision) problem for finite CSGs.*

Theorem 5. *There is a P-time reduction from the quantitative termination (decision) problem for finite CSGs to the qualitative termination problem for 1-RCSGs.*

Acknowledgement. We thank Krishnendu Chatterjee for clarifying for us several results about finite CSGs obtained by himself and others. This work was partially supported by NSF grant CCF-04-30946.

References

1. E. Allender, P. Bürgisser, J. Kjeldgaard-Pedersen, and P. B. Miltersen. On the complexity of numerical analysis. In *21st IEEE Computational Complexity Conference*, 2006.
2. T. Bewley and E. Kohlberg. The asymptotic theory of stochastic games. *Math. Oper. Res.*, 1(3):197–208, 1976.
3. T. Brázdil, A. Kučera, and O. Stražovský. Decidability of temporal properties of probabilistic pushdown automata. In *Proc. of STACS'05*, 2005.
4. J. Canny. Some algebraic and geometric computations in PSPACE. In *Proc. of 20th ACM STOC*, pages 460–467, 1988.
5. K. Chatterjee, Personal communication.
6. K. Chatterjee, L. de Alfaro, and T. Henzinger. The complexity of quantitative concurrent parity games. In *Proc. of SODA'06*, 2006.
7. K. Chatterjee, R. Majumdar, and M. Jurdzinski. On Nash equilibria in stochastic games. In *CSL'04*, volume LNCS 3210, pages 26–40, 2004.
8. A. Condon. The complexity of stochastic games. *Inf. & Comp.*, 96(2):203–224, 1992.
9. L. de Alfaro, T. A. Henzinger, and O. Kupferman. Concurrent reachability games. In *Proc. of FOCS'98*, pages 564–575, 1998.
10. L. de Alfaro and R. Majumdar. Quantitative solution of omega-regular games. *J. Comput. Syst. Sci.*, 68(2):374–397, 2004.
11. J. Esparza, A. Kučera, and R. Mayr. Model checking probabilistic pushdown automata. In *Proc. of 19th IEEE LICS'04*, 2004.
12. K. Etessami and M. Yannakakis. Recursive markov chains, stochastic grammars, and monotone systems of nonlinear equations. In *Proc. of 22nd STACS'05*. Springer, 2005.
13. K. Etessami and M. Yannakakis. Algorithmic verification of recursive probabilistic state machines. In *Proc. 11th TACAS*, vol. 3440 of LNCS, 2005.
14. K. Etessami and M. Yannakakis. Recursive markov decision processes and recursive stochastic games. In *Proc. of 32nd Int. Coll. on Automata, Languages, and Programming (ICALP'05)*, 2005.
15. K. Etessami and M. Yannakakis. Efficient qualitative analysis of classes of recursive markov decision processes and simple stochastic games. In *Proc. of 23rd STACS'06*. Springer, 2006.
16. J. Filar and K. Vrieze. *Competitive Markov Decision Processes*. Springer, 1997.
17. M. R. Garey, R. L. Graham, and D. S. Johnson. Some NP-complete geometric problems. In *8th ACM STOC*, pages 10–22, 1976.
18. T. E. Harris. *The Theory of Branching Processes*. Springer-Verlag, 1963.
19. A. J. Hoffman and R. M. Karp. On nonterminating stochastic games. *Management Sci.*, 12:359–370, 1966.
20. P. Jagers. *Branching Processes with Biological Applications*. Wiley, 1975.
21. A. Maitra and W. Sudderth. Finitely additive stochastic games with Borel measurable payoffs. *Internat. J. Game Theory*, 27(2):257–267, 1998.
22. D. A. Martin. Determinacy of Blackwell games. *J. Symb. Logic*, 63(4):1565–1581, 1998.
23. J. Renegar. On the computational complexity and geometry of the first-order theory of the reals, parts I-III. *J. Symb. Comp.*, 13(3):255–352, 1992.
24. L.S. Shapley. Stochastic games. *Proc. Nat. Acad. Sci.*, 39:1095–1100, 1953.

Half-Positional Determinacy of Infinite Games

Eryk Kopczyński*

Institute of Informatics, Warsaw University
erykk@mimuw.edu.pl

Abstract. We study infinite games where one of the players always has a positional (memory-less) winning strategy, while the other player may use a history-dependent strategy. We investigate winning conditions which guarantee such a property for all arenas, or all finite arenas. We establish some closure properties of such conditions, and discover some common reasons behind several known and new positional determinacy results. We exhibit several new classes of winning conditions having this property: the class of concave conditions (for finite arenas) and the classes of monotonic conditions and geometrical conditions (for all arenas).

1 Introduction

The theory of infinite games is relevant for computer science because of its potential application to verification of interactive systems. In this approach, the system and environment are modeled as players in an infinite game played on a graph (called *arena*) whose vertices represent possible system states. The players (conventionally called Eve and Adam) decide which edge (state transition, or *move*) to choose; each edge has a specific *color*. The desired system's behavior is expressed as a winning condition of the game — the winner depends on the sequence of colors which appear during an infinite play. If a winning strategy exists in this game, the system which implements it will behave as expected. Positional strategies (i.e. depending only on the position, not on the history of play — also called *memoryless*) are of special interest here, because of their good algorithmic properties which can lead to an efficient implementation.

Among the most often used winning conditions are the parity conditions, which admit positional determinacy ([Mos91], [EJ91], [McN93]). However, not always it is possible to express the desired behavior as a parity condition. An interesting question is, what properties are enough for the winning condition to be positionally determined, i.e. admit positional winning strategies independently on the arena on which the game is played. Recently some interesting characterizations of such positionally determined winning conditions have been found. In [CN06] it has been proven that every (prefix independent) condition which admits positional determinacy for all finite and infinite arenas (with colored moves) is a parity condition (up to renaming colors). There are more such conditions if we only consider finite arenas. In [GZ05] it has been proven that a winning condition is positionally determined for all finite arenas whenever it is so for finite

* Supported by KBN grant 4 T11C 042 25 and the EC RTN GAMES.

M. Bugliesi et al. (Eds.): ICALP 2006, Part II, LNCS 4052, pp. 336–347, 2006.

arenas where only one player is active. Another interesting characterization can be found in [GZ04]. For a survey of recent results on positional determinacy see [Gra04].

Our work attempts to obtain similar characterizations and find interesting properties (e.g. closure properties) of half-positionally determined winning conditions, i.e. ones such that all games using such a winning condition are positionally determined for one of the players (us, say), but the other player (environment) can have an arbitrary strategy. We give uniform arguments to prove several known and several new half-positional determinacy results. As we will see, some results on positional determinacy have natural generalizations to half-positional determinacy, but some do not. This makes the theory of half-positional conditions harder than the theory of positional conditions.

We also exhibit some large classes of half-positionally determined winning conditions. One example is the class of *concave winning conditions*; among examples of such conditions are the parity conditions, Rabin conditions, and the *geometrical condition* associated with convex subsets of $[0,1]^n$. Concavity is sufficient for half-positional determinacy only in the case of games on finite arenas. We investigate to what extent the results on geometrical conditions can be extended to infinite arenas. Another example is the class of monotonic winning conditions, which are defined using a deterministic finite automaton with a monotonic transition function, and includes winning conditions such as $C^\omega - C^*(a^nC^*)^\omega$. Monotonic winning conditions are half-positionally determined on all arenas.

Due to space limitations we had to omit most of proofs and algorithms. They will be presented in the full version of this paper. Its draft can be found at [Kop06].

2 Preliminaries

We consider perfect information antagonistic infinite games played by two players, called conventionally Adam and Eve. Let C be a set of **colors** (possibly infinite).

An **arena** over C is a tuple $G = (\mathrm{Pos}_A, \mathrm{Pos}_E, \mathrm{Mov})$, where:

- Elements of $\mathrm{Pos} = \mathrm{Pos}_E \cup \mathrm{Pos}_A$ are called **positions**; Pos_A and Pos_E are disjoint sets of Adam's positions and Eve's positions, respectively.
- Elements of $\mathrm{Mov} \subseteq \mathrm{Pos} \times \mathrm{Pos} \times C$ are called **moves**; (v_1, v_2, c) is a move from v_1 to v_2 colored by c. We denote $\mathrm{source}(v_1, v_2, c) = v_1$, $\mathrm{target}(v_1, v_2, c) = v_2$, $\mathrm{rank}(v_1, v_2, c) = c$.

A game is a pair (G, W), where G is an arena, and W is a winning condition. A **winning condition** W over C is a subset of C^ω which is *prefix independent*, i.e., $u \in W \iff cu \in W$ for each $c \in C, u \in C^\omega$. We name specific winning conditions WA, WB, ... For example, the **parity condition** of rank n is the winning condition over $C = \{0, 1, \ldots, n\}$ defined with

$$WP_n = \{w \in C^\omega : \limsup_{i\to\infty} w_i \text{ is even}\}. \quad (1)$$

The game (G, W) carries on in the following way. The play starts in some position v_1. The owner of v_1 (e.g. Eve if $v_1 \in \mathrm{Pos}_E$) chooses one of the moves leaving v_1, say (v_1, v_2, c_1). If the player cannot choose because there are no moves leaving v_1, he or she loses. The next move is chosen by the owner of v_2; denote it by (v_2, v_3, c_2). And so on: in the n-th move the owner of v_n chooses a move (v_n, v_{n+1}, c_n). If $c_1c_2c_3 \ldots \in W$, Eve wins the infinite play; otherwise Adam wins.

A **play** in the arena G is any sequence of moves π such that $\mathrm{source}(\pi_{n+1}) = \mathrm{target}(\pi_n)$. By $\mathrm{source}(\pi)$ and $\mathrm{target}(\pi)$ we denote the initial and final position of the play, respectively. The play can be finite ($\pi \in \mathrm{Pos} \cup \mathrm{Mov}^+$, where by $\pi \in \mathrm{Pos}$ we represent the play which has just started in the position π) or infinite ($\pi \in \mathrm{Mov}^\omega$; infinite plays have no target).

A **strategy for player** X is a partial function $s : \mathrm{Pos} \cup \mathrm{Mov}^+ \to \mathrm{Mov}$. For a finite play π such that $\mathrm{target}(\pi) \in \mathrm{Pos}_X$, $s(\pi)$ says what X should do in the next move. A strategy s is **winning** (for X) from the position v if $s(\pi)$ is defined for each finite play π starting in v, consistent with s, and ending in Pos_X, and each infinite play starting in v consistent with s is winning for X.

A strategy s is **positional** if it depends only on $\mathrm{target}(\pi)$, i.e., for each finite play π we have $s(\pi) = s(\mathrm{target}(\pi))$.

A game is **determined** if for each starting position one of the players has a winning strategy. This player may depend on the starting position in the given play. Thus if the game is determined, the set Pos can be split into two sets Win^E and Win^A and there exist strategies s_E and s_A such that each play π with $\mathrm{source}(\pi) \in \mathrm{Win}^X$ and consistent with s_X is winning for X. All games with a Borel winning condition are determined [Mar75], but there exist (exotic) games which are not determined. A winning condition W is **determined** if for each arena G the game (G, W) is determined.

We are interested in games and winning conditions for which one or both of the players have positional winning strategies. Thus, we introduce the notion of a **determinacy type**, given by three parameters: admissible strategies for Eve (positional or arbitrary), admissible strategies for Adam (positional or arbitrary), and admissible arenas (finite or infinite). We say that a winning condition W is (α, β, γ)**-determined** if for every γ-arena G the game (G, W) is (α, β)**-determined**, i.e. for every starting position either Eve has a winning α-strategy, or Adam has a winning β-strategy. Clearly, there are 8 determinacy types in total. For short, we call (positional, positional, infinite)-determined winning conditions **positionally determined** or just **positional**, (positional, arbitrary, infinite)-determined winning conditions **half-positional**, (arbitrary, positional, infinite)-determined winning conditions **co-half-positional**. If we restrict ourselves to finite arenas, we add **finitely**, e.g. (positional, arbitrary, finite)-determined conditions are called **finitely half-positional**. For a determinacy type $D = (\alpha, \beta, \gamma)$, D-arenas mean γ-arenas, and D-strategies mean α-strategies (if they are strategies for Eve) or β-strategies (for Adam).

Note that if a game (G, W) is (α, β)-determined, then its dual game obtained by using the complement winning condition and switching the roles of players is

(β,α)-determined. Thus, W is (α,β,γ)-determined iff its complement is (β,α,γ)-determined.

In the games defined above the moves are colored. In the literature one often studies similar games where positions are colored instead — in this case instead of $\mathrm{Mov} \subseteq \mathrm{Pos}\times\mathrm{Pos}\times C$ we have $\mathrm{Mov} \subseteq \mathrm{Pos}\times\mathrm{Pos}$ and a function $\mathrm{rank} : \mathrm{Pos} \to C$. The winner of a play in such games is defined similarly.

A position-colored game can be easily transformed into a move-colored game — we just have to color each move m with the color $\mathrm{rank}(\mathrm{target}(m))$. Transformation in the other way in general would require splitting positions when they are targets of moves of different colors, which may cause a previously non-positional strategy to become positional. Hence, for position-colored games there are more (half-)positionally determined winning conditions than for move-colored games. The facts proven or cited here do not necessarily hold in the case of position-colored games.

3 Closure Properties of Half-Positional Conditions

Now we will give some closure properties of half-positionally determined winning conditions. We will start with a lemma which is used in many proofs of half-positional determinacy of various winning conditions. This lemma can be proven by transfinite induction.

Lemma 1. *Let D be a determinacy type. Let $W \subseteq C^\omega$ be a winning condition. Suppose that, for each non-empty D-arena G over C, there exists a non-empty subset $M \subseteq G$ such that in game (G, W) one of the players has a D-strategy winning from M. Then W is D-determined.*

Definition 1. *For $S \subseteq C$, WB_S is the set of infinite words where elements of S occur infinitely often, i.e. $(C^*S)^\omega$. Winning conditions of this form are called* Büchi conditions. *Complements of Büchi conditons, $WB'_S = C^*(C-S)^\omega$ are called* co-Büchi conditions.

Theorem 1. *Let D be a determinacy type. Let $W \subseteq C^\omega$ be a winning condition, and $S \subseteq C$. If W is D-determined, then so is $W \cup WB_S$.*

Proof. We will show that the assumption of Lemma 1 holds. Let our arena be $G = (\mathrm{Pos}_E, \mathrm{Pos}_A, \mathrm{Mov})$. *$S$-moves* are moves m such that $\mathrm{rank}(m) \in S$.

Let G' be G with a new position $\top$ added. The position $\top$ belongs to Adam and has no outgoing moves, hence Adam loses here. For each S-move m we change $\mathrm{target}(m)$ to $\top$.

Since Adam immediately loses after doing an S-move in G', the winning conditions W and $W \cup WB_S$ are equivalent for G', thus we can use D-determinacy of W to find the winning sets $\mathrm{Win}'_E, \mathrm{Win}'_A$ and winning D-strategies s'_E, s'_A in G'.

Suppose $\mathrm{Win}'_A \neq \emptyset$. We can see that since Adam's strategy wins in G' from a starting position in Win'_A, he also wins in G from there by using the same strategy (the game G' is „harder" for Adam than G). Thus the assumption of 1 holds (we take $M = \mathrm{Win}'_A$).

Now suppose that $\mathrm{Win}'_A = \emptyset$. We will show that Eve has a winning D-strategy s in Pos everywhere, hence the assumption of Lemma 1 also holds (we take $M = \mathrm{Win}'_E$).

The strategy is as follows. For a finite play π we take $s(\pi) = s_E(\pi')$, where π' is the longest final segment without any S-moves. If s_E tells Eve to make an S-move, Eve makes its counterpart (or one of its counterparts) in G instead. The strategy s is positional if s_E is positional. It can be easily shown that s is indeed a winning strategy. □

Note that, by duality, Thm 1 implies that if W is D-determined, then so is $W \cap WB'_S$. This yields an easy proof of positional determinacy of parity conditions. It is enough to start with an empty winning condition (which is positionally determined) and apply Thm 1 and its dual n times.

A union of co-Büchi and co-half-positional conditions does not need to be co-half-positional ($WB'_{\{a\}} \cup WB'_{\{b\}}$ is not). What about a union of co-Büchi and a half-positional condition, does it have to be half-positional? We have no proof nor counterexample for this yet. This conjecture can be generalized to the following:

Conjecture 1. *Let $\mathcal{W}$ be a (finite, countable, ...) family of half-positionally (finitely) determined winning conditions. Then $\bigcup \mathcal{W}$ is a half-positionally (finitely) determined winning condition.*

Note that we assume prefix independence here. It is very easy to find two prefix dependent winning conditions which are positionally determined, but their union is not half-positionally determined.

This conjecture also fails for non-countable families and infinite arenas, even for such simple conditions as Büchi and co-Büchi conditions:

Theorem 2. *There exists a family of 2^ω Büchi (co-Büchi) conditions such that its union is not a half-positionally determined winning condition.*

Proof. Let $I = \omega^\omega$. Our arena A consists of one Eve's position E and infinitely many Adam's positions $(A_n)_{n \in \omega}$. In E Eve can choose $n \in \omega$ and go to A_n by move $(E, A_n, \star)$. In each A_n Adam can choose $r \in \omega$ and return to E by move $(A_n, E, (n, r))$.

For each $y \in I$, let $S_y = \{(n, y_n) : n \in \omega\} \subseteq C$, and $S'_y = C - S_y - \{\star\}$. Let $WA_1 = \bigcup_{y \in I} WB_{S_y}$, $WA_2 = \bigcup_{y \in I} WB'_{S'_y}$.

The games (A, WA_1) and (A, WA_2) are not half-positionally determined. Let (n_k) and (r_k) be n and r chosen by Eve and Adam in the k-th round, respectively. If Eve always plays $n_k = k$, she will win both the conditions WB_{S_y} and $WB'_{S'_y}$, where $y_k = r_k$. However, if Eve plays with a positional strategy $n_k = n$, Adam can win by playing $r_k = k$. □

There is however a subclass of half-positional winning conditions for which we can prove that it is closed under countable union.

Definition 2. *A* **suspendable winning strategy** *for X is a pair (s, Σ), where $s : \mathrm{Pos} \cup \mathrm{Mov}^+ \to \mathrm{Mov}$ is a strategy, and $\Sigma \subseteq \mathrm{Mov}^*$, such that:*

- *s is defined for every finite play π such that* $\mathrm{target}(\pi) \in \mathrm{Pos}_X$.
- *every infinite play π that is consistent with s from some point t*[1] *has a prefix longer than t which is in Σ;*
- *Every infinite play π that has infinitely many prefixes in Σ is winning for X.*

We say that X has a **suspendable winning strategy in** Win_X *when he has a suspendable winning strategy in the arena* $(\mathrm{Pos}_A \cap \mathrm{Win}_X, \mathrm{Pos}_E \cap \mathrm{Win}_X, \mathrm{Mov} \cap \mathrm{Win}_X \times \mathrm{Win}_X \times C)$.

A winning condition W is **positional/suspendable** *if for each arena G in the game (G, W) Eve has a positional winning strategy in* Win_E *and Adam has a suspendable winning strategy in* Win_A.

Intuitively, if at some moment X decides to play consistently with s, the play will eventually reach Σ; Σ is the set of moments when X can temporarily suspend using the strategy s and return to it later without a risk of ruining his or her victory.

A suspendable winning strategy is a winning strategy, because the conditions above imply that each play which is always consistent with s has infinitely many prefixes in Σ, and thus is winning for X. The co-Büchi condition is positional/suspendable; more examples will be given in Theorems 5 and 6. However, the parity condition WP_2 is positional, but not positional/suspendable, because a suspendable strategy cannot be winning for Adam — it is possible that the play enters state 2 infinitely many times while it is suspended.

Theorem 3. *A union of countably many positional/suspendable conditions is also positional/suspendable.*

If Adam has a suspendable winning strategy for each of given winning conditions and each starting position, then he can use them all in a play — he just has to activate and suspend each of them infinitely many times. Otherwise, we use a lemma similar to Lemma 1 to remove all positions from where Eve can win.

4 Concave Winning Conditions

We will now give some examples of half-positionally determined winning conditions. We will start by giving a simple combinatorial property which guarantees finite half-positional determinacy.

Definition 3. *A word $w \in \Sigma^* \cup \Sigma^\omega$ is a (proper)* **combination** *of words w_1 and w_2, iff for some sequence of words (u_n), $u_n \in \Sigma^*$*

[1] That is, for each prefix u of π which is longer than t and such that $\mathrm{target}(u) \in \mathrm{Pos}_X$, the next move is given by $s(u)$.

– $w = \prod_{k \in \mathbb{N}} u_k = u_0 u_1 u_2 u_3 u_4 u_5 u_6 u_7 u_8 \ldots$,
– $w_1 = \prod_{k \in \mathbb{N}} u_{2k+1} = u_1 u_3 u_5 u_7 \ldots$,
– $w_2 = \prod_{k \in \mathbb{N}} u_{2k} = u_0 u_2 u_4 u_6 \ldots$.

Definition 4. *A winning condition W is* **convex** *if as a subset of C^ω it is closed under combinations, and* **concave** *if its complement is convex.*

Example 1. Parity conditions (including Büchi and co-Büchi conditions) are both convex and concave.

Example 2. Let C be an infinite set. The folowing winning conditions are both convex and concave:

- *Exploration condition*: the set of all v in C^ω such that $\{v_n : n \in \omega\}$ is infinite.
- *Unboundedness condition*: the set of all v in C^ω such that no color appears infinitely often.

Decidability and positional determinacy of these conditions on (infinite) pushdown arenas where each position has a distinct color has been studied in [Gim04] (exploration condition) and [BSW03], [CDT02] (unboundedness condition).

Example 3. Concave winning conditions are closed under union. Convex winning conditions are closed under intersection.

Another example (which justifies the name) is given in Section 6 below.

Theorem 4. *Concave winning conditions are half-positionally finitely determined.*

The proof goes by induction over Mov, and is based on the following idea. Let v be Eve's position, with outgoing moves $m_1, m_2, \ldots$. Suppose that Eve cannot win by using only one of these moves. Then, since the winning condition is concave, she also cannot win by using many of these moves — because it can be written as a combination of subplays that appear after each move $m_1, m_2, \ldots$, and Adam wins all of these plays.

This theorem gives yet another proof of finite positional determinacy of parity games, and also half-positional determinacy of unions of families of parity conditions (where each parity condition may use a different rank for a given color). Half-positional determinacy of Rabin conditions (finite unions of families of parity conditions) over infinite arenas has been proven in [Kla92].

Note that concavity does not imply half-positional determinacy over infinite arenas — for examples see Section 6 below, and also Example 2 and Thm 2. Also, half-positional determinacy (even over infinite arenas) does not imply concavity — examples can be found in Sections 6 and 7.

Concavity does not force any bound on the memory required by Adam. Consider the game (G, W), where G is the arena with one Adam's position A and two moves $A \to A$ colored 0 and 1 respectively, and $W = WF'([0,1] - \{x\})$, where $x \in [0,1]$ is irrational. (WF' is defined in Section 6 below.) Adam requires unbounded memory here.

The following proposition gives some algorithmic properties of concavity, assuming that our winning condition is an ω-regular language.

Proposition 1. *Suppose that a winning condition W is given by a deterministic parity automaton on infinite words using s states and d ranks. Then there exists a polynomial algorithm of determining whether W is concave (or convex). If W is concave and G is an arena with n positions, then the winning sets and Eve's positional strategy can be found in time $O(n(ns)^{d/2} \log s)$.*

5 Weakening the Concavity Condition

In [GZ04] a result similar to Thm 4 has been obtained in the case of full positional determinacy. To present it, we need the following definition:

Definition 5. *A winning condition W is* **weakly convex** *iff for each sequence of words (u_n), $u_n \in C^*$, if*

1. $u_1 u_3 u_5 u_7 \ldots \in W$,
2. $u_2 u_4 u_6 u_8 \ldots \in W$,
3. $(\star)$ $\forall i$ $(u_i)^\omega \in W$,

then $u_1 u_2 u_3 u_4 \ldots \in W$.

A winning condition W is **weakly concave** *iff its complement is weakly convex.*

In the case of normal convexity there is no $(\star)$.

[GZ04] defines *fairly mixing* payoff mappings; in the case of prefix independent winning conditions *fairly mixing* resolves to the conjunction of weak concavity and weak convexity. Theorem 1 from [GZ04] says that games on finite arenas with fairly mixing payoff mappings are positionally determined.

Unfortunately, weak concavity is not enough for half-positional finite determinacy.

Proposition 2. *There exists a weakly concave winning condition WQ which is not half-positionally finitely determined.*

Proof. Let $C = \{0, 1\}$. For $w \in C^\omega$ let $P_n(w)$ be the number of 1's among the first n letters of w, divided by n. The winning condition WQ is a set of w such that $P_n(w)$ is convergent and its limit is rational. It can be easily seen that for each $u \in C^+$ we have $u^\omega \in WQ$. Therefore $(\star)$ is never satisfied for the complement of WQ, hence WQ is a weakly concave winning condition. However, WQ is not half-positionally determined. Consider the arena with two positions $E \in \mathrm{Pos}_E$, $A \in \mathrm{Pos}_A$, and moves $(E, A, 0)$, $(E, A, 1)$, $(A, E, 0)$ and $(A, E, 1)$. If Eve always moves in the same way, Adam can choose the moves 0 and 1 in an irrational proportion, ensuring his victory. However, Eve wins by always moving with the color opposite to Adam's last move — the limit of $P_n(w)$ is then $1/2$. □

Note that the given WQ satisfies the even stronger condition obtained by replacing $\forall i$ by $\exists i$ in $(\star)$ in Definition 5.

6 Geometrical Conditions

In this section we will show some half-positional determinacy results for *geometrical conditions*, which are based on the ideas similar to that used by the *mean payoff game* (also called *Ehrenfeucht-Mycielski game*). We will also show the relations between geometrical conditions and concave winning conditions.

Let $C = [0,1]^n$ (where $[0,1]$ is the real interval; we can also use any compact and convex subset of a normed space). For a word $w \in C^+$, let $P(w)$ be the average color of w, i.e., $\frac{1}{|w|}\sum_{k=1}^{|w|} w_k$. For a word $w \in C^\omega$, let $P_n(w) = P(w_{|n})$ ($w_{|n}$ — an n-letter prefix of w).

Let $A \subseteq C$. We want to construct a winning condition W such that $w \in W$ whenever the limit of $P_n(w)$ belongs to A. Since not every sequence has a limit, we have to define the winner for all other sequences.

Let $WF(A)$ be a set of w such that each cluster point of $P_n(w)$ is an element of A. Let $WF'(A)$ be a set of w such that at least one cluster point of $P_n(w)$ is an element of A. Note that $WF'(A) = C^\omega - WF(C - A)$.

As we will see, for half-positional determinacy the important property of A is whether the complement of A is convex — we will call such sets A *co-convex* (as *concave* usually means "non-convex" in geometry).

Geometrical conditions have a connection with the *mean payoff game*, whose finite positional determinacy has been proven in [EM79]. In the mean payoff game, C is a segment in $\mathbb{R}$ and the payoff mapping is $u(w) = \limsup_{n\to\infty} P_n(w)$ or $u(w) = \liminf_{n\to\infty} P_n(w)$. If $A = \{x : x \geq x_0\}$ then $u^{-1}(A)$ ("Eve wants x_0 or more") is exactly the geometrical condition $WF(A)$ or $WF'(A)$. Geometrical conditions are a generalization of such winning conditions to a larger class of sets A and C.

The following table summarizes what we know about concavity and half-positional determinacy of geometrical conditions. In every point except No. 0 we assume that A is non-trivial, i.e. $\emptyset \neq A \subsetneq C$. The first two columns specify assumptions about A and whether we consider $WF(A)$ or $WF'(A)$, and the last two answer whether the considered condition is concave and whether it has finite and/or infinite half-positional determinacy. Negative answer means that the answer is negative for all sets A in the given class; the question mark means that the given problem has not been solved yet (but we suppose that the answer is positive).

No.	A	condition	concavity	finite	infinite
0	trivial	$WF'(A)$ or $WF(A)$	yes	yes	yes
1	not co-convex	$WF'(A)$ or $WF(A)$	no	no	no
2	co-convex	$WF'(A)$	yes	yes	no
3	co-convex, not open	$WF(A)$	weak only	yes?	no
4	co-convex, open	$WF(A)$	weak only	yes?	yes?

Note that, for any set A which is co-convex and non-trivial, $WF'(A)$ is finitely half-positionally determined, but not infinitely half-positionally determied. This shows a big gap between half-positional determinacy on finite and infinite arenas.

The point 4 remains open in general, but we have a positive result for a special A. Its proof is quite complicated; it uses a similar idea as the proof of Thm 6 below (instead of a state, we use here a real number meaning Eve's „reserve" before falling out of A).

Theorem 5. *Let f be an affine function on C, and $A = f^{-1}(\{x \in \mathbb{R} : x < 0\})$. The condition $WF(A) = \{w : \limsup P_n(f(w)) < 0\}$ is positional/suspendable.*

Note that $WF(A_1) \cup WF(A_2)$ usually is not equal to $WF(A_1 \cup A_2)$, so a union of positional/suspendable conditions given above usually is not of form $WF(A)$ itself.

7 Monotonic Automata

In this section we will show yet another class of half-positionally determined winning conditions. It is based on a different idea than that of concave conditions, and guarantees half-positional determinacy even for infinite arenas. We will need to introduce a special kind of deterministic finite automaton.

Definition 6. *A* **monotonic automaton** $A = (n, \sigma)$ *over an alphabet C is a deterministic finite automaton where:*

- *the set of states is $Q = \{0, \ldots, n\}$;*
- *the initial state is 0, and the accepting state is n;*
- *the transition function $\sigma : Q \times C \to Q$ is monotonic in the first component, i.e., $q \le q'$ implies $\sigma(q, c) \le \sigma(q', c)$.*

Actually, we need not require that the set of states is finite. All the results presented here except for Thm 7 and the remark about finite memory of Adam can be proven with a weaker assumption that Q has a minimum (initial state) and its each non-empty subset has a maximum.

The function σ is extended to C^* as usual: $\sigma^*(q, \epsilon) = q$, $\sigma^*(q, wc) = \sigma(\sigma^*(q, w), c)$ $(w \in C^*, c \in C)$. So defined σ^* is still monotonic. By L_A we denote *the language accepted (recognized) by A*, i.e., the set of words $w \in C^*$ such that $\sigma(0, w) = n$.

Example 4. Monotonic automata can recognize the following languages: $C^*a^nC^*$, $C^*a^{n-1}bC^*$, $C^*ba^{n-1}C^*$. Monotonic automata cannot recognize the following languages: $C^*a^2b^2C^*$, C^*babC^*, C^*bacC^*.

Definition 7. *A* **monotonic condition** *is a winning condition of form $WM_A = C^\omega - L_A^\omega$ for some monotonic automaton A.*

Note that if $w \in L_A$ then $uw \in L_A$ for each $u \in C^*$. Hence $L_A = C^*L_A$. Therefore L_A^ω is equal to $L_A(C^*L_A)^\omega = (L_AC^*)^\omega$. Hence without affecting WM_A we can assume that $\sigma(n, c) = n$ for each c.

Theorem 6. *Any monotonic condition is positional/suspendable.*

The proof is based on the folowing idea. We construct a new game (G',W') where $\mathrm{Pos}' = \mathrm{Pos} \times Q$, moves are natural and $W' = C^\omega - L_A C^\omega$. This game is positionally determined (equivalent to WP_1). From monotonicity we know which position (g,q) for a given g is worst for Eve, and she can always play as if she were in the worst possible state. If Eve can win nowhere, then Adam wins everywhere in the original game; otherwise, Eve wins in some subset of G using a positional strategy, which we can remove using Lemma 1. It is worth to remark that although Adam's strategy given in the proof is not positional, it uses only finite memory (Q is the set of memory states).

From this theorem and the examples of languages recognized by monotonic automata above one can see that e.g. WA_n, the complement of the set of words containing a^n infinitely many times, is monotonic and thus half-positionally determined.

For $n = 1$ the set WA_n is just a co-Büchi condition. However, for $n > 1$ it is easily shown that WA_n is not (fully) positionally determined, and also that it is not concave. For example, for $n = 2$ the word $(bababbabab)^\omega$ is a combination of $(bbbaa)^\omega$ and $(aabbb)^\omega$. However, all monotonic conditions are weakly concave (if $\forall_i w_i^\omega \in WL_A^\omega$ for $A = (n, \sigma)$, then $w_1 w_2 w_3 \ldots \in WL_A^\omega$).

Proposition 3. *Monotonic conditions are closed under finite union.*

A countable union of monotonic conditions is not necessarily defined by a single monotonic automaton, but it is still positional/suspendable; however, a union of cardinality 2^ω of monotonic conditions does not have to be half-positionally determined, since co-Büchi conditions are monotonic. Monotonic conditions are not closed under other Boolean operations.

Theorem 7. *Let $W_1 \subseteq C^\omega$ be a concave winning condition, and A be a monotonic automaton. Then the union $W = W_1 \cup WM_A$ is a half-positionally finitely determined winning condition.*

8 Conclusion and Future Work

We would like to know more closure properties of the class of half-positionally determined winning conditions. Specifically we want to know whether it is closed under finite and countable union (Conjecture 1). In this paper we have proven that it is closed under union with Büchi conditions and intersection with co-Büchi conditions (Thm 1). We have also proven (Theorem 3) that positional/suspendable winning conditions are closed under countable union; and many half-positional winning conditions fall into this category. It seems worthwhile to extend Thm 3 to conditions obtained by using Thm 1 on positional/suspendable winning conditions.

Additionally, some of our results give new proofs of known facts about positional determinacy. Many previous proofs can be simplified by using Lemma 1.

In our opinion the proof of positional determinacy of parity conditions obtained by using Thm 1 is simpler than the proofs previously known to us.

Another direction of further research is to find more examples of half-positional conditions. Theorems 1 and 3 can be used to create new half-positional conditions from old ones. They could be also obtained e.g. by generalizing the results on geometrical conditions and monotonic automata. It would be also interesting to see whether monotonic automata have applications in other areas of automata theory.

References

[BSW03] A. Bouquet, O. Serre, I. Walukiewicz, *Pushdown games with unboundedness and regular conditions*, Proc. of FSTTCS'03, LCNS, volume 2914, pages 88-99, 2003.

[CDT02] T. Cachat, J. Duparc, W. Thomas, *Pushdown games with a Σ_3 winning condition.* Proc. of CSL 2002, LCNS, volume 2471, pages 322-336, 2002.

[CN06] T. Colcombet, D. Niwiński, *On the positional determinacy of edge-labeled games.* Theor. Comput. Sci. 352 (2006), pages 190-196

[EJ91] E. A. Emerson and C. S. Jutla, *Tree automata, mu-calculus and determinacy.* Proceedings 32th Annual IEEE Symp. on Foundations of Comput. Sci., pages 368-377. IEEE Computer Society Press, 1991.

[EM79] A. Ehrenfeucht, J. Mycielski, *Positional strategies for mean payoff games.* IJGT, 8:109-113, 1979.

[Gim04] H. Gimbert, *Parity and Exploration Games on Infinite Graphs.* Proc. of CSL '04, volume 3210 de Lect. Notes Comp. Sci., pages 56-70.

[Gra04] E. Grädel, *Positional Determinacy of Infinite Games.* In STACS 2004, LNCS, volume 2996, pages 4-18, 2004.

[GTW02] E. Grädel, W. Thomas, and T. Wilke, eds., *Automata, Logics, and Infinite Games.* No. 2500 in Lecture Notes in Compter Science, Springer-Verlag, 2002.

[GZ04] H. Gimbert, W. Zielonka, *When can you play positionally?* Proc. of MFCS '04, volume 3153 of Lect. Notes Comp. Sci., pages 686-697. Springer, 2004.

[GZ05] H. Gimbert, W. Zielonka, *Games Where You Can Play Optimally Without Any Memory.* Accepted for CONCUR 2005.

[Kla92] N. Klarlund, *Progress measures, immediate determinacy, and a subset construction for tree automata.* Proc. 7th IEEE Symp. on Logic in Computer Science, 1992.

[Kop06] E. Kopczyński, *Half-positional determinacy of infinite games.* Draft. `http://www.mimuw.edu.pl/~erykk/papers/hpwc.ps`

[Mar75] D. A. Martin, *Borel determinacy.* Ann. Math., 102:363–371, 1975.

[McN93] R. McNaughton, *Infinite games played on finite graphs.* Annals of Pure and Applied Logic, 65:149–184, 1993.

[Mos91] A. W. Mostowski, *Games with forbidden positions.* Technical Report 78, Uniwersytet Gdański, Instytut Matematyki, 1991.

A Game-Theoretic Approach to Deciding Higher-Order Matching

Colin Stirling

School of Informatics
University of Edinburgh
cps@inf.ed.ac.uk

Abstract. We sketch a proof using a game-theoretic argument that the higher-order matching problem is decidable.

Keywords: Games, higher-order matching, typed lambda calculus.

1 Introduction

Higher-order unification is given an equation $t = u$ containing free variables, is there a solution substitution θ such that $t\theta$ and $u\theta$ have the same normal form? Terms t and u are from the simply typed λ-calculus and the same normal form is $\beta\eta$-equality. Higher-order matching is the particular instance when the term u is closed, can t be pattern matched to u? Although higher-order unification is undecidable, higher-order matching was conjectured to be decidable by Huet [4] (and, if so then it has non-elementary complexity [11, 13]). Decidability has been proved for the general problem up to order 4 and for various special cases [7, 8, 9, 10, 2]. Loader showed that matching is undecidable for the variant definition when β-equality is the same normal form [5].

We propose a game-theoretic technique that leads to decidability of matching. It starts with Padovani's reduction to the dual interpolation problem [8]. We then define a game on a closed λ-term t where play moves around it relative to a dual interpolation problem. The game captures the dynamics of β-reduction on t without changing it (using substitution). Small pieces of a solution term, that we call "tiles", can be classified according to their subplays and how they, thereby, contribute to solving it. Two transformations that preserve solution terms are introduced. With these, we show that 3rd-order matching is decidable via the small model property: if there is a solution to a problem then there is a small solution to it. For the general case, the key idea is "tile lowering", copying regions of a term down its branches. A systematic method for tile lowering uses unfolding which is similar to unravelling a model in modal logic. Unfolding requires a non-standard interpretation of game playing where regions of a term are to be understood using suffix subplays. At this point, we step outside terms of typed λ-calculus. Refolding returns us to such terms. The detailed proof of decidability uses unfolding followed by refolding and from their combinatorial properties the small model property follows. However, here we can only outline the method with an example. For all the details and proofs, the reader is invited to access "Decidability of higher-order matching" from the author's web page.

M. Bugliesi et al. (Eds.): ICALP 2006, Part II, LNCS 4052, pp. 348–359, 2006.

2 Matching and Dual Interpolation

Assume simply typed λ-calculus with base type $\mathbf{0}$ and the definitions of α-equivalence, β and η-reduction. A type is $\mathbf{0}$, with *order* 1, or $A_1 \to \ldots \to A_n \to \mathbf{0}$, with *order* $k+1$ where k is the maximum of the orders of the A_is. Assume a countable set of typed variables $x, y, \ldots$ and typed constants, $a, f, \ldots$. The *simply typed terms* is the smallest set T such that if x (f) has type A then $x : A \in T$ ($f : A \in T$); if $t : B \in T$ and $x : A \in T$, then $\lambda x.t : A \to B \in T$; if $t : A \to B \in T$ and $u : A \in T$ then $tu : B \in T$. The *order* of a term is the order of its type and it is *closed* if it does not contain free variables.

A *matching problem* is $v = u$ where $v, u : \mathbf{0}$ and u is closed. The *order* is the maximum of the orders of the free variables $x_1, \ldots, x_n$ in v. A *solution* is a sequence of terms $t_1, \ldots, t_n$ such that $v\{t_1/x_1, \ldots, t_n/x_n\} =_{\beta\eta} u$. Given a matching problem the decision question is, does it have a solution?

We slightly change the syntax of types and terms. $A_1 \to \ldots \to A_n \to \mathbf{0}$ is rewritten $(A_1, \ldots, A_n) \to \mathbf{0}$ and all terms in normal form are in *η-long form*: if $t : \mathbf{0}$ then it is $u : \mathbf{0}$ where u is a constant or a variable, or $u(t_1, \ldots, t_k)$ where $u : (B_1, \ldots, B_k) \to \mathbf{0}$ is a constant or a variable and each $t_i : B_i$ is in η-long form; if $t : (A_1, \ldots, A_n) \to \mathbf{0}$ then t is $\lambda y_1 \ldots y_n.t'$ where each $y_i : A_i$ and $t' : \mathbf{0}$ is in η-long form. A term is *well-named* if each occurrence of a variable y within a λ-abstraction is unique.

Definition 1. *Assume $u : \mathbf{0}$ and $v_i : A_i$, $1 \le i \le n$, are closed terms in normal form and $x : (A_1, \ldots, A_n) \to \mathbf{0}$. $x(v_1, \ldots, v_n) = u$ ($\neq u$) is an* interpolation equation (disequation)*. A* dual interpolation problem *P is a finite family of interpolation equations and disequations, $i : 1 \le i \le m$, $x(v_1^i, \ldots, v_n^i) \approx_i u_i$, with the same free variable x and each $\approx_i \in \{=, \neq\}$. The type and order of P are the type and order of x. A* solution *of P of type A is a closed term $t : A$ in normal form, such that for each equation $t(v_1^i, \ldots, v_n^i) =_\beta u_i$ and for each disequation $t(v_1^i, \ldots, v_n^i) \neq_\beta u_i$. We abbreviate t solves P to $t \models P$.*

Padovani shows that a matching problem of order n reduces to a dual interpolation problem of the same order [8]: given P, is there a solution $t \models P$? We assume a fixed dual interpolation problem P of type A whose order is greater than 1 (as an order 1 problem is easily decided) where the normal form terms v_j^i and u_i are well-named and no pair share bound variables.

A right term u of a (dis)equation may contain bound variables. If $X = \{x_1, \ldots, x_k\}$ are its bound variables then let $C = \{c_1, \ldots, c_k\}$ be a fresh set of constants with corresponding types. The *ground closure* of w with bound variables in X, with respect to C, $\mathrm{Cl}(w, X, C)$, is: if $w = a : \mathbf{0}$, then $\mathrm{Cl}(w, X, C) = \{a\}$; if $w = f(w_1, \ldots, w_n)$, then $\mathrm{Cl}(w, X, C) = \{w\} \cup \bigcup \mathrm{Cl}(w_i, X, C)$; if $w = \lambda x_{j_1} \ldots x_{j_n}.u$ then $\mathrm{Cl}(w, X, C) = \mathrm{Cl}(u\{c_{j_1}/x_{j_1}, \ldots, c_{j_n}/x_{j_n}\}, X, C)$. For $u = f(\lambda x_1 x_2 x_3.x_1(x_2), a)$ with respect to $\{c_1, c_2, c_3\}$, it is $\{u, c_1(c_2), c_2, a\}$.

We also identify subterms of left terms v_j of a (dis)equation relative to a set C: however, these need not be of ground type and may also contain free variables. The *subterms* of w relative to C, $\mathrm{Sub}(w, C)$, is defined using an auxiliary set $\mathrm{Sub}'(w, C)$: if w is a variable or a constant, then $\mathrm{Sub}(w, C) = \mathrm{Sub}'(w, C) = \{w\}$; if w is $x(w_1, \ldots, w_n)$ then $\mathrm{Sub}(w, C) = \mathrm{Sub}'(w, C) = \{w\} \cup \bigcup \mathrm{Sub}(w_i, C)$; if w is $f(w_1, \ldots, w_n)$, then $\mathrm{Sub}(w, C) = \mathrm{Sub}'(w, C) = \{w\} \cup \bigcup \mathrm{Sub}'(w_i, C)$; if w is $\lambda y_1 \ldots y_n.v$, then $\mathrm{Sub}(w, C) = \{w\} \cup \mathrm{Sub}(v, C)$ and $\bigcup\{\mathrm{Sub}(v\{c_{i_1}/y_1, \ldots, c_{i_n}/y_n\}, C) : c_{i_j} \in C$ has the same type as $y_j\}$ is the set $\mathrm{Sub}'(w, C)$. If $v = \lambda z.f(\lambda z_1 z_2 z_3.z_1(z_2), z)$ and $z_2, z_3 : \mathbf{0}$ then $\mathrm{Sub}(v, \{c_1, c_2, c_3\})$ is $\{v, f(\lambda z_1 z_2 z_3.z_1(z_2), z), c_1(c_2), c_1(c_3), c_2, c_3, z\}$.

Given the problem P, let X_i be the (possibly empty) set of bound variables in u_i and let C_i be a corresponding set of new constants (that do not occur in P), the *forbidden* constants.

Definition 2. *Assume P is the fixed problem of type A. T is the set of subtypes of A including A and subterms of u_i. For i, the right subterms are $\mathsf{R}_i = \mathrm{Cl}(u_i, X_i, C_i)$ and $\mathsf{R} = \bigcup \mathsf{R}_i$. For i, the left subterms are $\mathsf{L}_i = \bigcup \mathrm{Sub}(v^i_j, C_i) \cup C_i$ and $\mathsf{L} = \bigcup \mathsf{L}_i$. The* arity, *$\alpha$, of P is the largest k where $(A_1, \ldots, A_k) \to B \in \mathsf{T}$. The right size $\delta(u)$ relative to C is: if $u = a : \mathbf{0}$ then $\delta(u) = 0$; if $u = f(w_1, \ldots, w_k)$ then $\delta(u) = 1 + \sum \delta(w_i)$; if $u = \lambda x_{i_1} \ldots x_{i_k}.w$, then $\delta(u) = \delta(w\{c_{i_1}/x_{i_1}, \ldots, c_{i_k}/x_{i_k}\})$. The* right size *for P, δ, is $\sum \delta(u_i)$ of its right terms.*

So, $\delta(h(a)) = 1$. If δ for P is 0, then each (dis)equation contains a right term that is a constant $a_i : \mathbf{0}$: Padovani proved decidability for this special case [7].

3 Tree-Checking Games

We present a game-theoretic characterization of interpolation inspired by model-checking games (such as in [12]) where a model, a transition graph, is traversed relative to a property. Similarly, in the following game the model is a putative solution term t that is traversed relative to the dual interpolation problem.

A potential solution t for P has the right type, is in normal form, is well-named (with variables that are disjoint from those in P) and does not contain forbidden constants. Term t is represented as a tree, $\mathrm{tree}(t)$. If t is $y : \mathbf{0}$ or $a : \mathbf{0}$ then $\mathrm{tree}(t)$ is the single node labelled with t. For $u(v_1, \ldots, v_k)$ when u is a variable or a constant, a dummy λ with the empty sequence of variables is placed before any subterm $v_i : \mathbf{0}$ in its tree representation. If t is $u(v_1, \ldots, v_n)$, then $\mathrm{tree}(t)$ consists of the root node labelled u and n-successor nodes labelled with $\mathrm{tree}(v_i)$: $u \downarrow_i t'$ represents that t' is the ith successor of u. If t is $\lambda \overline{y}.v$, where $\overline{y}$ could be empty, then $\mathrm{tree}(t)$ consists of the root node labelled $\lambda \overline{y}$ and a single successor node $\mathrm{tree}(v)$: $\lambda \overline{y} \downarrow_1 \mathrm{tree}(v)$. Each node labelled with an occurrence of a variable y_j has a backward arrow $\uparrow^j$ to the $\lambda \overline{y}$ that binds it: the index j tells us which element is pointed at in $\overline{y}$. We use t to be the λ-term t, or its λ-tree or the label (a constant, variable or $\lambda \overline{y}$) at its root node. Dummy λs are central to the analysis in later sections. We also assume that each node of a tree t is uniquely identified.

Example 1. A solution term t from [1] for the problem $x(v) = f(a)$ where $v = \lambda y_1 y_2.y_1(y_2)$ is $\lambda z.z(\lambda x.f(z(\lambda u.x, b)), z(\lambda y.z((\lambda s.s, y), a)))$. The tree for t (without backward edges and indexed forward edges) is in Figure 1. □

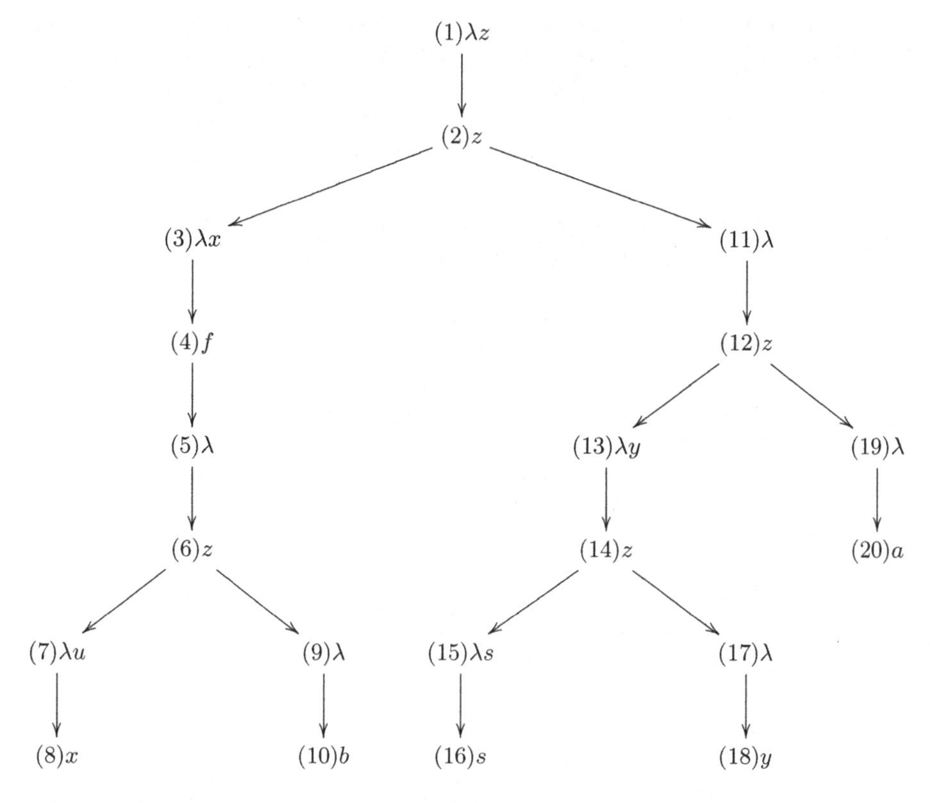

Fig. 1. A term tree

Innocent game semantics following Ong in [6] provides a possible game-theoretic foundation. Given t and a (dis)equation from P, there is the game board $t@(v_1^i, \ldots, v_n^i) \;\; u_i$. Player Opponent chooses a branch of u_i. There is a finite play that starts at the root of t and may repeatedly jump in and out of t and the v_j^i's. At a constant $a : \mathbf{0}$ play ends. At other constants f, player Proponent tries to match Opponent's choice of branch. Proponent wins, when the play finishes, if the sequence of constants encountered matches the chosen branch. Play may reach y in t and then jump to $\lambda\overline{z}$ in v_j^i, as it is this subtree that is applied to $\lambda\overline{y}$, and then when at z in v_j^i play may return to t to a successor of y. Game semantics models β-reduction on the fixed game board without changing it using substitution. This is the rationale for the tree-checking game. However, it starts from the assumption that only t is the common structure for the problem P. So, play will always be in t. Jumping in and out of the v_j^i's is coded using states. The game avoids justification pointers, using iteratively defined look-up tables.

The game $\mathsf{G}(t, P)$ is played by player $\forall$, the *refuter*, who attempts to show that t is not a solution of P. It uses a finite set of states involving elements of L and R from Definition 2. An *argument* state $q[(l_1, \ldots, l_k), r]$ where each $l_j \in \mathsf{L}$

(and k can be 0) and $r \in \mathsf{R}$ occurs at a node labelled $\lambda z_1 \dots z_k$ in t where each l_j has the same type as z_j: $(l_1, \dots, l_k)$ are the subterms applied to $\lambda z_1 \dots z_k$. A *value* state $q[l, r]$ where $l \in \mathsf{L}$ and $r \in \mathsf{R}$ is associated with a node labelled with y in t where y and l share the same type: l is the subterm of some v^i_j that play at y would jump to in game semantics. A *final* state is $q[\forall]$ or $q[\exists]$.

A. $t_m = \lambda y_1 \dots y_j$ and $t_m \downarrow_1 u$ and $q_m = q[(l_1, \dots, l_j), r]$.
So, $t_{m+1} = u$, $\theta_{m+1} = \theta_m\{l_1\eta_m/y_1, \dots, l_j\eta_m/y_j\}$ and q_{m+1}, η_{m+1} are defined by cases on t_{m+1}.
1. $a : \mathbf{0}$. So, $\eta_{m+1} = \eta_m$. If $r = a$ then $q_{m+1} = q[\exists]$ else $q_{m+1} = q[\forall]$.
2. $f : (B_1, \dots, B_k) \to \mathbf{0}$. So, $\eta_{m+1} = \eta_m$. If $r = f(s_1, \dots, s_k)$ then $q_{m+1} = q_m$ else $q_{m+1} = q[\forall]$.
3. $y : B$. If $\theta_{m+1}(y) = l\eta_i$, then $\eta_{m+1} = \eta_i$ and $q_{m+1} = q[l, r]$.

B. $t_m = f : (B_1, \dots, B_k) \to \mathbf{0}$ and $q_m = q[(l_1, \dots, l_j), f(s_1, \dots, s_k)]$.
So, $\theta_{m+1} = \theta_m$, $\eta_{m+1} = \eta_m$ and q_{m+1}, t_{m+1} are decided as follows.
1. $\forall$ chooses a direction $d : 1 \leq d \leq k$ and $t_m \downarrow_d u$. So, $t_{m+1} = u$. If $s_d : \mathbf{0}$, then $q_{m+1} = q[(\ ), s_d]$. If s_d is $\lambda x_{i_1} \dots x_{i_n}.s$ then $q_{m+1} = q[(c_{i_1}, \dots, c_{i_n}), s\{c_{i_1}/x_{i_1}, \dots, c_{i_n}/x_{i_n}\}]$.

C. $t_m = y$ and $q_m = q[l, r]$.
If $l = \lambda z_1 \dots z_j.w$ and $t_m \downarrow_i u_i$, for $i : 1 \leq i \leq j$, then $\eta_{m+1} = \eta_m\{u_1\theta_m/z_1, \dots, u_j\theta_m/z_j\}$ else $\eta_{m+1} = \eta_m$. Elements t_{m+1}, q_{m+1} and θ_{m+1} are by cases on l.
1. $a : \mathbf{0}$ or $\lambda\overline{z}.a$. So, $t_{m+1} = t_m$ and $\theta_{m+1} = \theta_m$. If $r = a$ then $q_{m+1} = q[\exists]$ else $q_{m+1} = q[\forall]$.
2. $c : (B_1, \dots, B_k) \to \mathbf{0}$. So, $\theta_{m+1} = \theta_m$. If $r \neq c(s_1, \dots, s_k)$ then $t_{m+1} = t_m$ and $q_{m+1} = q[\forall]$. If $r = c(s_1, \dots, s_k)$ then $\forall$ chooses a direction $d : 1 \leq d \leq k$ and $t_m \downarrow_d u$. So, $t_{m+1} = u$. If $s_d : \mathbf{0}$, then $q_{m+1} = q[(\ ), s_d]$. If s_d is $\lambda x_{i_1} \dots x_{i_n}.s$ then $q_{m+1} = q[(c_{i_1}, \dots, c_{i_n}), s\{c_{i_1}/x_{i_1}, \dots, c_{i_n}/x_{i_n}\}]$.
3. $f(w_1, \dots, w_k)$ or $\lambda\overline{z}.f(w_1, \dots, w_k)$. So, $t_{m+1} = t_m$ and $\theta_{m+1} = \theta_m$. If $r \neq f(s_1, \dots, s_k)$, then $q_{m+1} = q[\forall]$. If $r = f(s_1, \dots, s_k)$ then $\forall$ chooses a direction $d : 1 \leq d \leq k$. If $s_d : \mathbf{0}$ then $q_{m+1} = q[w_d, s_d]$. If $w_d = \lambda y_1 \dots y_n.w$ and $s_d = \lambda x_{i_1} \dots x_{i_n}.s$, then $q_{m+1} = q[w\{c_{i_1}/y_1, \dots, c_{i_n}/y_n\}, s\{c_{i_1}/x_{i_1}, \dots, c_{i_n}/x_{i_n}\}]$.
4. $x(l_1, \dots, l_k)$ or $\lambda\overline{z}.x(l_1, \dots, l_k)$. If $\eta_{m+1}(x) = t\theta_i$ then $\theta_{m+1} = \theta_i$ and $t_{m+1} = t$ and $q_{m+1} = q[(l_1, \dots, l_k), r]$.

Fig. 2. Game moves

There are two kinds of free variables, in t and in the left terms of states. Free variables in t are associated with left terms and free variables in states are associated with nodes of t. So, the game appeals to a sequence of supplementary look-up tables θ_k and η_k, $k \geq 1$: θ_k is a partial map from variables in t to *pairs* $l\eta_j$ where $l \in \mathsf{L}$ and $j < k$, and η_k is a partial map from variables in elements of L to *pairs* $t'\theta_j$ where t' is a node of the tree t and $j < k$. Initially, θ_1 and η_1 are both empty.

A *play* of $\mathsf{G}(t, P)$ is $t_1q_1\theta_1\eta_1, \dots, t_nq_n\theta_n\eta_n$ where t_i is (the label of) a node of t, $t_1 = \lambda\overline{y}$ is the root node of t, q_i is a state and q_n is a final state. A node t' of t may repeatedly occur in a play. For the initial state, $\forall$ chooses a (dis)equation

$x(v_1^i, \ldots, v_n^i) \approx_i u_i$ from P and $q_1 = q[(v_1^i, \ldots, v_n^i), u_i]$, similar to that in game semantics except v_j^i and u_i are now part of the state (and the choice of branch in u_i happens as play proceeds). If the current position is $t_m q_m \theta_m \eta_m$ and q_m is not final, then $t_{m+1} q_{m+1} \theta_{m+1} \eta_{m+1}$ is determined by a unique move in Figure 2. Moves are divided into groups depending on t_m. Group A covers when it is a $\lambda\overline{y}$, B when it is a constant f (whose type is not $\mathbf{0}$) and C when it is a variable y. In B1, C2 and C3 the constants c_{i_j} belong to the forbidden set C_i: these are also the only rules where $\forall$ can exercise choice (by carving out a branch). The look-up tables are used in A3 and C4 to interpret the two kinds of free variables. If t_m is a λ node, $t_m \downarrow_1 t_{m+1}$ and t_{m+1} is the variable y, then η_{m+1} and q_{m+1} are determined by the entry for y in θ_{m+1}. For C4, if $t_m = y$, $q_m = q[l, r]$ and $l = x(l_1, \ldots, l_k)$ or $\lambda\overline{z}.x(l_1, \ldots, l_k)$, then θ_{m+1} and t_{m+1} are determined by the entry for x in the table η_{m+1}: if the entry is the pair $t'\theta_i$ then $t_{m+1} = t'$ and $\theta_{m+1} = \theta_i$. It is this rule that allows play to jump elsewhere in the term tree (always to a node labelled with a λ). In contrast, for A1-A3, B1 and C2 control passes down the term tree while it remains stationary in the case of C1 and C3.

A play of $\mathsf{G}(t, P)$ finishes with final state $q[\,\forall\,]$ or $q[\,\exists\,]$. Player $\forall$ *wins* it if the final state is $q[\,\forall\,]$ and she *loses* it if it is $q[\,\exists\,]$. $\forall$ *loses the game* $\mathsf{G}(t, P)$ if for each equation she loses every play whose intial state is from it and if for each disequation she wins at least one play whose initial state is from it.

Proposition 1. *$\forall$ loses $\mathsf{G}(t, P)$ if, and only if, $t \models P$.*

Assume $t_0 \models P$, so $\forall$ loses the game $\mathsf{G}(t_0, P)$. The single play for Example 1 is in Figure 3. The number of different plays is at most the sum of the number of branches in the right terms u_i of P. Let $d : \mathbf{0}$ be a constant that is not forbidden and does not occur in any right term of P. We can assume that t_0 only contains d and constants that occur in a right term.

We also allow π to range over *subplays*, consecutive subsequences of positions of any play of $\mathsf{G}(t_0, P)$. The length of π, $|\pi|$, is its number of positions. The ith position of π is $\pi(i)$ and $\pi(i, j)$, $i \leq j$, is the interval $\pi(i), \ldots, \pi(j)$. We write $t \in \pi(i)$, $q \in \pi(i)$, $\theta \in \pi(i)$ and $\eta \in \pi(i)$ if $\pi(i) = tq\theta\eta$ and $t \notin \pi(i)$ if $\pi(i) = t'q\theta\eta$ and $t \neq t'$. If $q = q[(l_1, \ldots, l_k), r]$ or $q[l, r]$ then its *right term* is r.

Definition 3. *A subplay π is* ri, right term invariant, *if $q \in \pi(1)$ and $q' \in \pi(|\pi|)$ share the same right term r. It is* nri *if it is not ri and $q' \in \pi(|\pi|)$ is not final.*

In Figure 3, $\pi(1, 4)$ is ri whereas $\pi(1, 6)$ is nri. Ri subplays are an important ingredient in the decidability proof as they do not immediately contribute to the solution of P.

Proposition 2. *If $t_i q_i \theta_i \eta_i, \ldots, t_n q_n \theta_n \eta_n$ is ri, $t_n = \lambda\overline{y}$ and $q\{r'/r\}$ is state q with right term r' instead of r, then $t_i q_i\{r'/r\}\theta_i \eta_i, \ldots, t_n q_n\{r'/r\}\theta_n \eta_n$ is an ri play.*

Definition 4. *If $\pi \in \mathsf{G}(t_0, P)$ and $\pi(i)$'s look-up table is called when move A3 or C4 produces $\pi(j)$, $j > i$, then position $\pi(j)$ is a* child *of position $\pi(i)$. If $\pi(i+1)$ is the result of move B1 or C2, then $\pi(i+1)$ is a* child *of $\pi(i)$. A look-up table β'* extends *β if for all $x \in \mathrm{dom}(\beta)$, $\beta'(x) = \beta(x)$.*

(1) $q[(v), f(a)]\, \theta_1\, \eta_1$			
(2) $q[v, f(a)]\, \theta_2\eta_2$	$\theta_2 = \theta_1\{(v\eta_1/z\}$	$\eta_2 = \eta_1$	A3
(3) $q[(y_2), f(a)]\, \theta_3\eta_3$	$\theta_3 = \theta_2$	$\eta_3 = \eta_2\{(3)\theta_2/y_1, (11)\theta_2/y_2\}$	C4
(4) $q[(y_2), f(a)]\, \theta_4\eta_4$	$\theta_4 = \theta_3\{y_2\eta_3/x\}$	$\eta_4 = \eta_3$	A2
(5) $q[(\,), a]\, \theta_5\, \eta_5$	$\theta_5 = \theta_4$	$\eta_5 = \eta_4$	B1
(6) $q[v, a]\, \theta_6\, \eta_6$	$\theta_6 = \theta_5$	$\eta_6 = \eta_1$	A3
(7) $q[(y_2), a]\, \theta_7\, \eta_7$	$\theta_7 = \theta_6$	$\eta_7 = \eta_6\{(7)\theta_6/y_1, (9)\theta_6/y_2\}$	C4
(8) $q[y_2, a]\, \theta_8\, \eta_8$	$\theta_8 = \theta_7\{y_2\eta_7/u\}$	$\eta_8 = \eta_3$	A3
(11) $q[(\,), a]\, \theta_9\, \eta_9$	$\theta_9 = \theta_2$	$\eta_9 = \eta_8$	C4
(12) $q[v, a]\, \theta_{10}\eta_{10}$	$\theta_{10} = \theta_9$	$\eta_{10} = \eta_1$	A3
(13) $q[(y_2), a]\, \theta_{11}\eta_{11}$	$\theta_{11} = \theta_{10}$	$\eta_{11} = \eta_{10}\{(13)\theta_{10}/y_1, (19)\theta_{10}/y_2\}$	C4
(14) $q[v, a]\, \theta_{12}\eta_{12}$	$\theta_{12} = \theta_{11}\{y_2\eta_{11}/y\}$	$\eta_{12} = \eta_1$	A3
(15) $q[(y_2), a]\, \theta_{13}\, \eta_{13}$	$\theta_{13} = \theta_{12}$	$\eta_{13} = \eta_{12}\{(15)\theta_{12}/y_1, (17)\theta_{12}/y_2\}$	C4
(16) $q[y_2, a]\, \theta_{14}\, \eta_{14}$	$\theta_{14} = \theta_{13}\{y_2\eta_{13}/s\}$	$\eta_{14} = \eta_{13}$	A3
(17) $q[(\,), a]\, \theta_{15}\, \eta_{15}$	$\theta_{15} = \theta_{12}$	$\eta_{15} = \eta_{14}$	C4
(18) $q[y_2, a]\, \theta_{16}\, \eta_{16}$	$\theta_{16} = \theta_{15}$	$\eta_{16} = \eta_{11}$	A3
(19) $q[(\,), a]\, \theta_{17}\, \eta_{17}$	$\theta_{17} = \theta_{10}$	$\eta_{17} = \eta_{16}$	C4
(20) $q[\,\exists\,]\, \theta_{18}\, \eta_{18}$	$\theta_{18} = \theta_{17}$	$\eta_{18} = \eta_{17}$	A1

Fig. 3. A play

Proposition 3. *If $\pi \in \mathsf{G}(t_0, P)$, $j > 1$, $\pi(j)$ is not a final position and $\lambda\overline{y}$ or y $\in \pi(j)$, then there is a unique $\pi(i)$, $i < j$, such that $\pi(j)$ is a child of $\pi(i)$. If $\pi(j)$ is a child of $\pi(i)$ then $\theta_j \in \pi(j)$ extends $\theta_i \in \pi(i)$ and $\eta_j \in \pi(j)$ extends $\eta_i \in \pi(i)$.*

4 Tiles and Subplays

Assume $t_0 \models P$. The aim is to show there is a *small* $t' \models P$. Although the number of plays in $\mathsf{G}(t_0, P)$ is bounded, there is no bound in terms of P on the length of a play. However, a long play contains ri subplays: across all plays, the right term of a state can change at most δ times, Definition 2. To obtain a small solution term t', ri subplays will be manipulated. First, we need to relate the static structure of t_0 with the dynamics of play.

Definition 5. *Assume $B = (B_1, \ldots, B_k) \to \mathbf{0} \in \mathsf{T}$. λ is an* atomic leaf *of type $\mathbf{0}$. If $x_j : B_j$, $1 \leq j \leq k$, then $\lambda x_1 \ldots x_k$ is an* atomic leaf *of type B. If $u : \mathbf{0}$ is a constant or variable then u is a* simple *tile. If $u : B$ is a constant or a variable and $t_j : B_j$, $1 \leq j \leq k$, are atomic leaves then $u(t_1, \ldots, t_k)$ is a* simple *tile.*

Term t_0 without its very top $\lambda\overline{y}$ consists of simple tile occurrences. Nodes (2),(3) and (11) of Figure 1 form the simple tile $z(\lambda x, \lambda)$ and the leaf (16) is also a simple tile: node (2) by itself and node (2) with (3), are *not* simple tiles. Tiles can be composed to form composite tiles. If $t(\lambda\overline{x})$ is a tile with leaf $\lambda\overline{x}$ and t' is a simple tile, then $t(\lambda\overline{x}.t')$ is a composite tile. A (composite) tile is *basic* if it contains one occurrence of a free variable and no occurrences of constants, or one occurrence of a constant and no occurrences of free variables. The free variable or constant in a basic tile is its head element. Contiguous regions of t_0 are occurrences of basic tiles. In Figure 1 the region $z(\lambda s.s, \lambda)$ is a basic tile rooted at (14). Throughout, we assume our use of tile in t_0 means "tile occurrence" in t_0. We write $t(\lambda\overline{x}_1, \ldots, \lambda\overline{x}_k)$ if t is a basic tile with atomic leaves $\lambda\overline{x}_1, \ldots, \lambda\overline{x}_k$.

Definition 6. *Assume $t = t(\lambda\overline{x}_1, \ldots, \lambda\overline{x}_k)$ is a tile in t_0. t is a* top *tile in t_0 if its free variable y is bound by the initial lambda $\lambda\overline{y}$ of t_0. t is j-*end *in t_0, if every free variable below $\lambda\overline{x}_j$ in t_0 is bound above t. It is an* end *tile in t_0 if it is j-end for all j. t is a constant tile if its head is a constant or its free y is bound by $\lambda\overline{y}$ that is an atomic leaf of a simple constant tile. Two basic tiles t and t' in t_0 are* equivalent, *$t \equiv t'$, if they have the same number and type of atomic leaves and the same free variable y bound to the same $\lambda\overline{y}$ in t_0.*

The tile $z(\lambda x, \lambda)$ in Figure 1 is a top tile which is also 2-end and $z(\lambda u, \lambda)$ is both a top and an end tile: these tiles are equivalent.

We can also classify tiles in terms of their *dynamic* properties.

Definition 7. *π is a* play *on the simple tile $u(\lambda\overline{x}_1, \ldots, \lambda\overline{x}_k)$ in t_0 if $u \in \pi(1)$, $\lambda\overline{x}_i \in \pi(|\pi|)$ for some i and $\pi(|\pi|)$ is a child of $\pi(1)$. It is a* j-play *if $\lambda\overline{x}_j \in \pi(|\pi|)$.*

A play on a simple constant tile $u(\lambda\overline{x}_1, \ldots, \lambda\overline{x}_k)$ is a pair of positions $\pi(i, i+1)$ with $u \in \pi(i)$ and $\lambda\overline{x}_j \in \pi(i+1)$ for some j (by moves B1 or C2 of Figure 2). A play π on a simple non-constant tile $y(\lambda\overline{x}_1, \ldots, \lambda\overline{x}_k)$ in t_0 can be of arbitrary length. It starts at y and finishes at a leaf $\lambda\overline{x}_j$. In between, flow of control can be almost anywhere in t_0. Crucially, the look-up tables of $\pi(|\pi|)$ extend those of $\pi(1)$ by Fact 3: this means that the free variables in the subtree of t_0 rooted at y and the free variables in w when $q[\lambda z_1 \ldots z_k.w, r] \in \pi(1)$ preserve their values.

If $\pi \in \mathsf{G}(t_0, P)$ and $y \in \pi(i)$ then there can be zero or more plays $\pi(i, j)$ on $y(\lambda\overline{x}_1, \ldots, \lambda\overline{x}_k)$ in t_0: simple tiles $u : \mathbf{0}$ have no plays. If $\pi(i, m)$ is a j-play on $y(\lambda\overline{x}_1, \ldots, \lambda\overline{x}_k)$ and $\pi(i, n)$, $n > m$, is also a play on this tile, then there is a position $\pi(m')$, $m < m' < n$, that is a child of $\pi(m)$. In the case of π in Figure 3 on the tree in Figure 1, $\pi(2, 3)$ is a 1-play on $z(\lambda x, \lambda)$ and $\pi(2, 9)$ is also a play on this tile: it is $\pi(8)$ that is the (only) child of $\pi(3)$.

A play π on a basic tile consists of consecutive subplays on the simple tiles that are on the branch between the top of the tile and a leaf $\lambda\overline{x}_i \in \pi(|\pi|)$.

Definition 8. *Assume π is a j-play (play) on tile t in t_0. It is a* shortest *j-play (play) if no proper prefix of π is a j-play (play) on t and it is an* ri *j-play (play) if π is also ri. It is an* internal *j-play (play) when for any i if $t' \in \pi(i)$ then t' is a node of t. Assume $t = t(\lambda\overline{x}_1, \ldots, \lambda\overline{x}_k)$ is in t_0 and π is a subplay. We inductively define when t is j-*directed *in π: if $t \notin \pi(i)$ for all i, then t is j-directed in π; if $\pi(i)$ is the first position with $t \in \pi(i)$ and there is a shortest j-play $\pi(i, m)$ on t and $\pi(i, m)$ is ri and t is j-directed in $\pi(m+1, |\pi|)$, then t is j-directed in π. Tile t is j-*directed *in t_0 if it is j-directed in every $\pi \in \mathsf{G}(t_0, P)$.*

$\pi(2, 3)$ of Figure 3 is a shortest play on tile $z(\lambda x, \lambda)$ of Figure 1: this play is ri, internal and a shortest 1-play. Although $\pi(2, 9)$ is a shortest 2-play, it is neither a shortest play nor an internal play. If t is j-directed in t_0 then each $\pi \in \mathsf{G}(t_0, P)$ contains a (unique) sequence of ri intervals which are shortest j-plays on t.

Assume $t = t(\lambda\overline{x}_1, \ldots, \lambda\overline{x}_k)$ is a top tile in t_0 and $\pi \in \mathsf{G}(t_0, P)$. Consider two positions $t \in \pi(i)$ and $t \in \pi(i')$. The states $q \in \pi(i)$ and $q' \in \pi(i')$ have the form $q[v, r]$ and $q[v, r']$ where v is a closed left term (a v^i_j from a (dis)equation of P). Therefore, a shortest play $\pi(i, i+m)$ on t is internal (as a jump outside t requires there to be a free variable in v via move C4 of Figure 2). If the play $\pi(i, i+m)$ is ri then there is a corresponding ri play $\pi(i', i'+m)$ on t consisting of the same sequence of positions in t and states (except for their right terms r and r'). Tile t is, therefore, j-directed in π when $\lambda\overline{x}_j \in \pi(i+m)$. If the play $\pi(i, i+m)$ is nri then there is a subplay $\pi(i', i'+m')$ where control is never outside t that is either a shortest play on t and nri or $|\pi| \leq i' + m'$. If t is a j-end (end) tile and $t \in \pi(i)$ then there can be at most one j-play (play) $\pi(i, m)$ on t.

Tile t' is *j-below* $t(\lambda\overline{x}_1, \ldots, \lambda\overline{x}_k)$ in t_0 if there is a branch in t_0 from $\lambda\overline{x}_j$ to t'. If two tiles t_1 and t_2 are equivalent, $t_1 \equiv t_2$ and t_2 is j-below t_1 in t_0, then t_2 is an *embedded* tile. Shortest plays on the embedded tile t_2 are constrained by earlier shortest plays on t_1 and in the case of embedded end tiles there is a stronger property that is critical to the decidability proof.

Proposition 4. *If $t_1 \equiv t_2$ are end tiles in t_0 and t_2 is j-below t_1, then either t_2 is j-directed in t_0, or there are $\pi, \pi' \in \mathsf{G}(t_0, P)$, an nri j-play $\pi(m_1, m_1+n_1)$ on t_1 and a subplay $\pi'(m_2, m_2+n_2)$ where $m_2 > m_1+n_1$, $n_2 \leq n_1$ and $\pi'(m_2) = \pi(m_2)$ and either $m_2 = n_1$ and $\pi'(m_2, m_2+n_2)$ is an nri j-play on t_2 or $m_2+n_2 = |\pi'|$.*

5 Outline of the Decision Procedure

A transformation $\mathbf{T}$ converts a tree s into a tree t, written $s\,\mathbf{T}\,t$. Let t' be a subtree of t_0 whose root node is a variable y or a constant $f : B \neq \mathbf{0}$. $\mathsf{G}(t_0, P)$ *avoids* t' if $t' \notin \pi(i)$ for all positions and plays $\pi \in \mathsf{G}(t_0, P)$. Let $t_0[t''/t']$ be the result of replacing t' in t_0 with the tree (of tiles) t''.

T1. If $\mathsf{G}(t_0, P)$ avoids t' and $d : \mathbf{0}$ is a constant then transform t_0 to $t_0[d/t']$

T2. Assume $t(\lambda\overline{x}_1, \ldots, \lambda\overline{x}_k)$ is a j-directed, j-end tile in t_0 and t' is the subtree of t_0 rooted at t. If t_j is the subtree directly beneath $\lambda\overline{x}_j$ then transform t_0 to $t_0[t_j/t']$.

If no play enters a subtree of t_0 then it can be replaced with the constant $d : \mathbf{0}$. If a tile is both j-end and j-directed, Definition 6, then it is redundant and can be removed from t_0. Game-theoretically, the application of **T2** amounts to *omission* of inessential ri subplays that are structurally associated with regions of a term.

Example 2. Consider Example 1 and its single play in Figure 3. The tile $z(\lambda u, \lambda)$ is 1-end and 1-directed because of $\pi(6, 7)$. **T2** allows us to remove it, so node (8) is directly beneath node (5). The *basic* tile $z(\lambda s.s, \lambda)$ is 1-end and 1-directed: the only play $\pi(14, 17)$ is ri. A second application of **T2** places node (18) directly beneath node (13). Consequently, the basic tile $z(\lambda y.y, \lambda)$ is also 1-end and 1-directed because of the play $\pi(12, 19)$. The starting term is therefore reduced to the smaller solution term $\lambda z.z(\lambda x.f(x), a)$. □

Proposition 5. *If* $\mathbf{i} \in \{\mathbf{1}, \mathbf{2}\}$, $s\,\mathbf{Ti}\,t$ *and* $s \models P$ *then* $t \models P$.

If P is 3rd-order and $t_0 \models P$ then t_0 is a tree of simple tiles: each is a constant tile or a top tile that is also an end tile. Assume that $\Pi = \{\pi_1, \ldots, \pi_p\}$ are the plays of $\mathsf{G}(t_0, P)$ and with each such π we associate a unique *colour* $c(\pi)$. We define a partition of each $\pi \in \Pi$ in stages. At stage 1, the initial simple tile t_1 is $u(\lambda\overline{x}_1, \ldots, \lambda\overline{x}_k)$ in t_0, a constant or top tile (where k may be 0). The initial play on t_1, if there is one, is $\pi(i_1, j_1)$ where $i_1 = 2$. If there is no play then $j_1 = |\pi|$ as $q \in \pi(j_1)$ is final, and for all $i > 1$, $u \in \pi(i)$: t_1 is *final* for π and we terminate at this stage. Otherwise, play ends at an atomic leaf of t_1, t_2 is the simple tile directly below it in t_0, $i_2 = j_1 + 1$ and if $\pi(i_1, j_1)$ is nri then t_1 is coloured $c(\pi)$. At stage n and simple tile t_n, $\pi(i_n, j_n)$ is the shortest play on t_n, if there is one. If there is not then $j_n = |\pi|$ and t_n is final for π. If $\pi(i_n, j_n)$ is nri then t_n is coloured $c(\pi)$. If it ends at an atomic leaf of t_n then t_{n+1} is the simple tile directly below it in t_0 and $i_{n+1} = j_n + 1$. The partition of π descends a branch of t_0 until it reaches a final tile.

Consider partitioning with respect to all plays $\pi \in \Pi$. There is a tree of simple tiles, as all plays share the initial tile. Tile t is *coloured* if it has at least one colour and t is *final* if it is final for at least one play. Each play at stage 1 that ends at the same atomic leaf of t_1 shares t_2 at stage 2 and so on. Therefore, branching occurs at a (play) *separator* t_m at stage m if there are plays that end at different atomic leaves of t_m. If a simple tile in t_0 is coloured, final or a separator then it is *special.* A simple tile in t_0 with atomic leaves that is not special is superfluous. Every play avoids it (so, **T1** applies) or every subplay that passes through it is ri and ends at the same atomic leaf (so, **T2** applies). There can be at most δ, the right size for P of Definition 2, coloured tiles, at most p final tiles and at most $p-1$ separators: p is bounded by the number of branches in the right terms of P. Decidability of 3rd-order matching, via the small model property, follows directly from partitioning.

There is just one *level* of simple tile that is not a constant tile in a 3rd-order tree: so, game playing is heavily constrained as control can only descend it. With a 4th or 5th-order tree there are two levels of simple non-constant tiles: top tiles t and end tiles t' where the variable of t' is bound in t. At 8th or 9th-order there are four levels. When there is more than one level, game playing may jump around the tree as Figure 3 illustrates. The mechanism for dealing with these terms hinges on the idea of *tile lowering*, copying tiles down branches.

The mechanism for tile lowering is not a transformation like **T2**. Instead, it uses an intermediate generalized tree, the *unfolding*, analogous to unravelling a model in modal logic, which is then *refolded* into a small tree. Again, a partition of (a subsequence of) π is defined in stages using tiles in t_0. At each stage n, a simple tile t_n in t_0 and a position $\pi(i_n)$ whose control is at the head of t_n are examined. The play $\pi(i_n, j_n)$ is a *suffix* of a play of a constant or *generalized* tile t'_n that contains t_n.

Stage 1 follows the 3rd-order case: t_1 is the first simple tile in t_0, $t'_1 = t_1$, and relative to π, $\pi(i_1, j_1)$ is defined. If t'_1 is not final for π then t_2 is the simple tile directly below it in t_0 and $i_2 = j_1 + 1$. After stage one, the unfolding of t_0

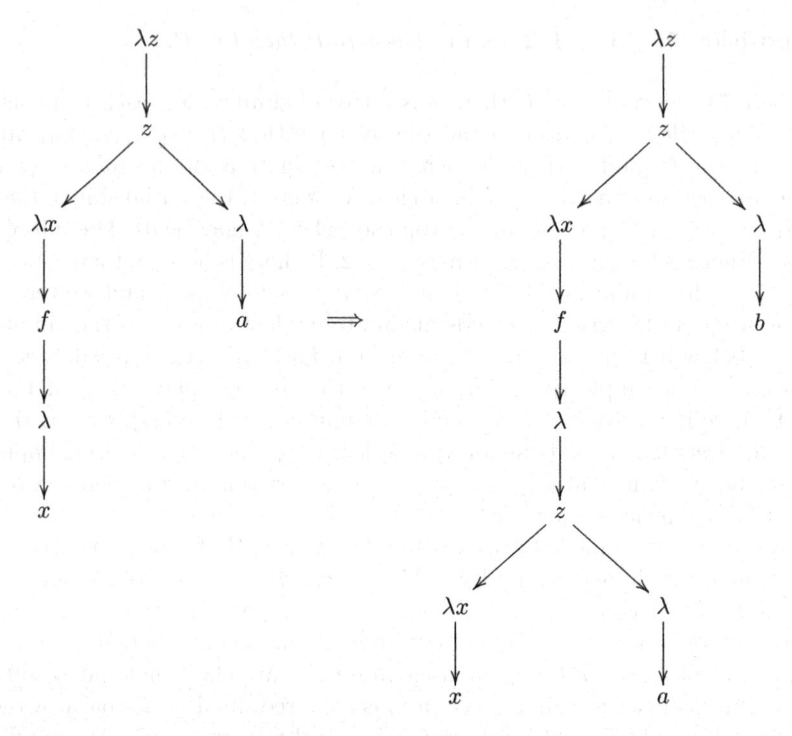

Fig. 4. Illustrating unfolding

can be depicted in linear form, $[t'_1 \ \lambda\overline{x}_j]$ if $\pi(i_1, j_1)$ finishes at $\lambda\overline{x}_j$. Consider the unfolding after stage n for π where t'_n is not final for π. There is the sequence $[t'_1 \ \lambda\overline{x}_1] \ \ldots \ [t'_n \ \lambda\overline{x}_n]$ of (generalized) tiles where each t_{k+1} is directly below $\lambda\overline{x}_k$ in t_0 and there are subplays $\pi(i_k, j_k)$ that start at t_k in t'_k and finish at $\lambda\overline{x}_k$. If t_{n+1} is a top or constant tile, then $t'_{n+1} = t_{n+1}$ and $\pi(i_{n+1}, j_{n+1})$ is either a shortest play on t'_{n+1} or $j_{n+1} = |\pi|$. The other case is that $t_{n+1} = y(\lambda\overline{z}_1, \ldots, \lambda\overline{z}_l)$ is directly below $\lambda\overline{x}_n$ in t_0 and y is bound within an earlier tile t'_k. The position $\pi(i_{n+1})$ is a child of a position in the interpretation of t'_k that is the effect of the suffix play $\pi(i_k, j_k)$. The tile t'_{n+1} is $[\ [t'_k \ \lambda\overline{x}_k] \ [t'_{m_1} \ \lambda\overline{x}_{m_1}] \ldots [t'_{m_l} \ \lambda\overline{x}_{m_l}] \ t_{n+1}\ ]$ where the t'_{m_i} are the minimal number of tiles in $t'_{k+1}, \ldots, t'_n$ that are captured in the sense that they involve extra nri subplays or are final. The interpretation of t'_{n+1} at position $\pi(i_{n+1})$ is $[\ [\pi^1] \ldots [\pi^{l+1}] \, \pi(i_{n+1})]$ where π^1 is the interpretation of the tile t'_k and π^{i+1} is that of t'_{m_i}. The play $\pi(i_{n+1}, j_{n+1})$ is the continuation, assuming (iterated) suffix playing, on t'_{n+1} that starts at t_{n+1} and finishes at an atomic leaf of it or is final. The intention is that unfolding will be true by definition assuming a non-standard interpretation of generalized tiles which includes that their plays are *suffix* plays. As with the 3rd-order case, each play π descends a branch of the unfolded tree. The remainder of the proof, the refolding, is how to extract a small term from the tree of generalized tiles. Game-theoretically, unfolding and refolding is justified by recursive permutations, repetitions and omissions of ri subplays.

Example 3. In Example 2, the term in Figure 1 is reduced to the left tree in Figure 4. We examine its unfolding. Tile $t'_1 = z(\lambda x, \lambda)$, $\pi(i_1, j_1) = \pi(2,3)$, $t'_2 = f(\lambda)$ and $\pi(i_2, j_2) = \pi(4,5)$. Now, $t_3 = x$, so $t'_3 = z(\lambda x.x, \lambda)$ as t'_1 is lowered (and there is no capture) and $\pi(i_3, j_3)$ is the *suffix* play $\pi(8,9)$ that starts at x and finishes at atomic leaf λ. Tile $t'_4 = t_4 = a$ is final for π and $\pi(i_4, j_4) = \pi(20)$. To make the unfolding into a term tree, the initial λ of t_0 is added and the constant $b : \mathbf{0}$ underneath any atomic leaf that does not have a successor, the tree on the right of Figure 4. The issue is t'_3 whose interpretation is a *suffix* play. We can reinterpret it as a complete play on t'_3 because (the prefix play) $\pi(i_1, j_1)$ is ri: the complete play has a different right term in its states, here we use Fact 2. The top $z(\lambda x, \lambda)$ and basic tile $z(\lambda x.x, \lambda)$ are 1-end and 1-directed: so, by **T2**, they are removed. The result is the small term $\lambda z.f(a)$. □

References

1. Comon, H. and Jurski, Y. Higher-order matching and tree automata. *Lecture Notes in Computer Science*, **1414**, 157-176, (1997).
2. Dougherty, D. and Wierzbicki, T. A decidable variant of higher order matching. *Lecture Notes in Computer Science*, **2378**, 340-351, (2002).
3. Dowek, G. Third-order matching is decidable. *Annals of Pure and Applied Logic*, **69**, 135-155, (1994).
4. Huet, G. *Rèsolution d'èquations dans les langages d'ordre* 1, 2, ... ω. Thèse de doctorat d'ètat, Universitè Paris VII, (1976).
5. Loader, R. Higher-order β-matching is undecidable, *Logic Journal of the IGPL*, **11(1)**, 51-68, (2003).
6. Ong, C.-H. L. (2006) On model-checking trees generated by higher-order recursion schemes. Preprint.
7. Padovani, V. Decidability of all minimal models. *Lecture Notes in Computer Science*, **1158**, 201-215, (1996).
8. Padovani, V. Decidability of fourth-order matching. *Mathematical Structures in Computer Science*, **10(3)**, 361-372, (2001).
9. Schubert, A. Linear interpolation for the higher-order matching problem. *Lecture Notes in Computer Science*, **1214**, 441-452, (1997).
10. Schmidt-Schauβ, M. Decidability of arity-bounded higher-order matching. *Lecture Notes in Artificial Intelligence*, **2741**, 488-502, (2003).
11. Statman, R. The typed λ-calculus is not elementary recursive. *Theoretical Computer Science*, **9**, 73-81, (1979).
12. Stirling, C. *Modal and Temporal Properties of Processes*, Texts in Computer Science, Springer, (2001).
13. Wierzbicki, T. Complexity of higher-order matching. *Lecture Notes in Computer Science*, **1632**, 82-96, (1999).

Descriptive and Relative Completeness of Logics for Higher-Order Functions*

Kohei Honda[1], Martin Berger[1], and Nobuko Yoshida[2]

[1] Department of Computer Science, Queen Mary, University of London
[2] Department of Computing, Imperial College London

Abstract. This paper establishes a strong completeness property of compositional program logics for pure and imperative higher-order functions introduced in [18, 16, 17, 19, 3]. This property, called *descriptive completeness*, says that for each program there is an assertion fully describing the program's behaviour up to the standard observational semantics. This formula is inductively calculable from the program text alone. As a consequence we obtain the first relative completeness result for compositional logics of pure and imperative call-by-value higher-order functions in the full type hierarchy.

1 Introduction

Program logics such as Hoare logic are a means to *describe* abstract behaviours of programs as logical assertions; to *verify* that a given program satisfies a specified property; and to *define* axiomatic semantics in the sense that the assertions assign meaning to a program with respect to its observable properties. Because of this strong match with observable and operational semantics of programs in a simple and intuitive manner, many engineering activities ranging from static analyses to program testing increasingly use program logics as their theoretical foundation.

For describing properties of first-order imperative programs, Hoare logic uses a pair of assertions in number theory. For example, in the partial correctness judgement $\{x = i\}x := x+1\{x = i+1\}$, the pair of assertions $x = i$ and $x = i+1$ describes a property of the program $x := x+1$ by saying: *whatever the initial content of x would be, if this program terminates, then the final content of x is the increment of its initial one.* Here a *property* is a subset of programs taken modulo an observational congruence: for example, in `while` programs, we consider programs up to partial functions on store they represent. Since the collection of all properties is uncountable, no standard logical language can represent all properties of any non-trivial programming language. Then what classes of properties should a program logic represent and prove?

In this paper, we focus on *descriptive completeness*, a strong completeness property, which is about representability of behaviour *as a canonical formula*: given a program P, we can always find a unique assertion pair in Hoare logic which represents (pinpoints) P's behaviour. For partial correctness, the best assertion pair for P describes all partial functions equal to or less defined than P. For example, the pair "$x = i$" and "$x = i+1$"

* Work is partially supported by EPSRC GR/R03075/01, GR/T04236/01, GR/S55538/01, GR/T04724/01, GR/T03208/01 and IST-2005-015905 MOBIUS.

M. Bugliesi et al. (Eds.): ICALP 2006, Part II, LNCS 4052, pp. 360–371, 2006.

are also satisfied by a diverging program. Dually for total correctness. A related concept are the *characteristic formulae* of Hennessy-Milner logics, which precisely characterise a CCS process up to bisimilarity [13, 31, 32]. We shift this notion from a process logic to a program logic, establishing descriptive completeness of Hoare logics for pure and imperative higher-order functions introduced in [18, 16, 19, 3].

In first-order Hoare logic, a program defines a partial function from states to states, so that the existence of characteristic formulae is not hard to establish. When we move to higher-order programs, a logic needs to describe how a program *transforms behaviour*. For example $\lambda x^{\mathsf{Nat}\Rightarrow(\alpha\Rightarrow\beta)}.x1$ is a function which receives a function and returns another function. The logics for higher-order functions and their imperative extensions [16, 19, 3, 18] involve direct description of such applicative behaviour. Due to complexity of the underlying semantic universe, it is not immediately obvious if a single pair of formulae can fully describe the behaviour of an arbitrary higher-order program. In the present paper we *construct* a characteristic formula of a program compositionally and algorithmically, following its syntactic structure, and inductively verify that the derived formula has the required properties. The induced algorithm is implemented as a prototype (1,250 LOC in Ocaml) [2]. The size of the resulting formula is asymptotically almost linear to the size of a program under a certain condition.

The generated characteristic assertions clarify the relationship between total and partial correctness for higher-order objects, following early observations [29, 28], but in the context of concrete assertion methods and proof rules. We use the duality between total and partial correctness [28] to derive descriptive completeness for partial correctness from its total variant. A total correctness property denotes an upward closed set of semantic points, representing liveness, while a partial correctness formula stands for a downward closed set, representing safety [28, 26, 23]. This duality subsumes the corresponding notions in the original Hoare logic, and offers a key insight into the nature of assertions for higher-order objects and their derivation. Finally, relative completeness [6] of proof rules is an immediate consequence of descriptive completeness. To our knowledge this work is the first to obtain descriptive and relative completeness in Hoare logics for (imperative) higher-order functions in the full type hierarchy.

In the remainder, Section 2 establishes descriptive and relative completeness for the logic of call-by-value PCF. Section 3 discusses the corresponding results for an imperative extension. Section 4 gives comparisons with related work. Section 5 concludes with further topics. All proofs are omitted, relegated to the long version [1].

2 Descriptive Completeness for PCFv

Call-by-value PCF. The syntax of PCFv is standard [27], and is briefly reviewed below (we can easily treat, but omit, other standard types such as sums and products [18]).

$$\alpha,\beta,\ldots ::= \mathsf{Bool} \mid \mathsf{Nat} \mid \alpha\Rightarrow\beta \qquad V,W,\ldots ::= x^{\alpha} \mid \mathsf{c} \mid \lambda x^{\alpha}.M \mid \mu f^{\alpha\Rightarrow\beta}.\lambda x^{\alpha}.M$$
$$M,N,\ldots ::= V \mid \mathsf{op}(\tilde{M}) \mid MN \mid \mathtt{if}\ M\ \mathtt{then}\ N_1\ \mathtt{else}\ N_2$$

We use numerals (0,1,2,..) and booleans (t and f) as constants (c above) and standard first-order operations ($\mathsf{op}(\tilde{M})$ where $\tilde{M}$ denotes a vector). $V, V', \ldots$ denote values. The typing is standard; henceforth we only consider well-typed programs. A *basis* $(\Gamma, \Delta, \ldots)$

is a finite map from variables to types. If M has type α with its free variables typed following Γ, we write $\Gamma \vdash M : \alpha$. A program is *closed* if it has no free variables. The call-by-value evaluation relation is written $M \Downarrow V$. If M diverges, we write $M \Uparrow$. $M \Downarrow$ means that $M \Downarrow V$ for some V. We use the standard contextual precongruence and congruence [27, 14], written $\lesssim$ and $\cong$, given as: for M and N of the same type, $M \lesssim N$ iff, for each typed closing context $C[\cdot]$, $C[M] \Downarrow$ implies $C[N] \Downarrow$. $\cong$ is the symmetric closure of $\lesssim$.

We list three simple programs. First, the standard recursive factorial program is written $\mathtt{Fact} \stackrel{\mathrm{def}}{=} \mu f^{\mathsf{Nat} \Rightarrow \mathsf{Nat}}.\lambda x^{\mathsf{Nat}}.\mathtt{if}\ x = 0\ \mathtt{then}\ 1\ \mathtt{else}\ x \times f(x-1)$. Second, in each arrow type we find $\Omega^{\alpha \Rightarrow \beta} \stackrel{\mathrm{def}}{=} \mu f^{\alpha \Rightarrow \beta}.\lambda x^{\alpha}.fx$, which diverges whenever invoked. Third, $\omega^{\alpha} \stackrel{\mathrm{def}}{=} \Omega^{\mathsf{Nat} \Rightarrow \alpha} 0$ gives an immediately diverging program (note $\Omega^{\alpha \Rightarrow \beta} \cong \lambda x^{\alpha}.\omega^{\beta}$).

Assertions and their Semantics. We use the following assertion language from [18, 19, 3], common to both total correctness and partial correctness.

$$e ::= \mathsf{c} \mid x^{\alpha} \mid \mathsf{op}(\tilde{e}) \quad A ::= e_1 = e_2 \mid e_1 \bullet e_2 = e_3 \mid A \wedge B \mid A \vee B \mid A \supset B \mid \neg A \mid \forall x^{\alpha}.A \mid \exists x^{\alpha}.A$$

The left definition is for terms, that on the right for formulae. c denotes a constant, either numerals (0,1,2, ...) or booleans ($\mathtt{t}$ and $\mathtt{f}$). Terms are typed as in PCFv. *Henceforth we only consider well-typed terms.* e^{α} indicates e has type α. Constants and first-order operations are from PCFv. We assume the standard bound name convention for formulae. If types of free variables in A follow Γ, we write $\Gamma \vdash A$. We set T as $1 = 1$ and F as its negation. $\equiv$ denotes logical equivalence. The assertion language is first-order, with a ternary predicate $e_1 \bullet e_2 = e_3$, called *evaluation formula*. Intuitively $e_1 \bullet e_2 = e_3$ means:

> *If a function denoted by e_1 is applied to an argument denoted by e_2 then it converges to a value denoted by e_3.*

Note $e_1 \bullet e_2 = e_3$ indicates termination. "=" in $e_1 \bullet e_2 = e_3$ is asymmetric and $\bullet$ is a non-commutative operation like application in an applicative structure. For example, assume f denotes a function which doubles the number n: then the assertion "$f \bullet 5 = 10$" means if we apply that function to 5, then the evaluation terminates and its result is 10.

Meaning of assertions is given by a simple term model. A *model* $(\xi, \xi', \ldots)$ is a finite map from typed variables to closed PCFv-values of the same types. Interpretation of terms is standard, denoted $\xi[\![e]\!]$. The satisfaction relation is written $\xi \models A$, and follows the standard clauses [24, Section 2.2] except the equality is interpreted by the contextual congruence $\cong$ (i.e. $\xi \models e_1 = e_2$ iff $\xi[\![e_1]\!] \cong \xi[\![e_2]\!]$). Further we set:

$$\xi \models e_1 \bullet e_2 = e_3 \quad \text{if} \quad \exists V.(\xi[\![e_1]\!]\xi[\![e_2]\!] \Downarrow V \wedge V \cong \xi[\![e_3]\!]). \tag{2.1}$$

We write $\Gamma \vdash \xi$ if $\mathsf{dom}(\Gamma) = \mathsf{dom}(\xi)$ and the typing of ξ follows Γ. $M\xi$ denotes the term obtained from M by substituting $\xi(x)$ for each free x in M.

Judgements. The judgement for total correctness is written $[A]\,M :_u [B]$, prefixed with $\models$ for validity, and $\vdash$ for provability. It is the standard Hoare triple augmented with an anchor [16, 18, 19, 3]. An anchor is a fresh name denoting the result of evaluation. u may only occur in B. The judgement $[A]\,M :_u [B]$ intuitively says:

> *If a model ξ satisfies A, then $M\xi$ converges and ξ together with the result, named u, satisfy B.*

In $[A]\,M :_u [B]$, we always assume $\Gamma \vdash M : \alpha$, $\Gamma \vdash A$ and $\Gamma \cdot u{:}\alpha \vdash B$ for some Γ and α.

Provability $\vdash [A]M :_u [B]$ is defined by the proof rules [18] listed in Appendix A, which precisely follow the syntax of programs. Validity $\models [A]M :_u [B]$ is defined by the following clause (let Γ be the minimum basis under which M, A and B are typable).

$$\forall \xi.((\Gamma \vdash \xi \wedge \xi \models A) \supset (M\xi \Downarrow V \wedge \xi \cdot u : V \models B)). \tag{2.2}$$

The proof of soundness, $\vdash [A]M :_u [B]$ implies $\models [A]M :_u [B]$, is mechanical. Later we demonstrate the converse. Simple examples of judgements follow.

1. We have $\vdash [\mathsf{T}]\,\mathtt{Fact} :_u [\forall x^{\mathsf{Nat}}.u \bullet x = x!]$, saying Fact computes a factorial whenever invoked. We also have $\vdash [\mathsf{T}]\,\mathtt{Fact} :_u [\forall x^{\mathsf{Nat}}.(Even(x) \supset \exists i.(u \bullet x = i \wedge Even(i))]$ where $Even(n)$ says n is even.
2. We have $\vdash [\mathsf{F}]\,\omega :_u [\mathsf{F}]$, which is the best formulae we can get for ω. Note this judgement holds for arbitrary programs of the same type.
3. From 2 above, we derive $\vdash [\mathsf{T}]\,\lambda x.\omega :_u [\mathsf{T}]$. The judgement contains no information for values, in the sense that all values satisfy it: as it should be, since we had to start from the trivial judgement for ω. Similarly $\vdash [\mathsf{T}]\,\Omega :_u [\mathsf{T}]$ is the best we can get.

Characteristic Formulae. In the last examples of judgements, we have seen the notion of total correctness and compositional verification *demand* that an assertion pair in the present logic cannot directly describe divergence. For this reason the notion of an assertion pair representing a given program pinpoints its behaviour as the *least* element of the described property. We call such a formula a total characteristic assertion pair.

Definition 1. (TCAP) A pair (A,B) is a *total characteristic assertion pair*, or *TCAP, of M at u*, if the following conditions hold (in each clause we assume well-typedness).

1. (soundness) $\models [A]M :_u [B]$.
2. (MTC, minimal terminating condition) $M\xi \Downarrow$ if and only if $\xi \models A$.
3. (closure) Suppose $\models [E]N :_u [B]$ such that $E \supset A$. Then $\xi \models E$ implies $M\xi \lesssim N\xi$.

Proposition 2. *1. If (A,B) is a TCAP of M at u and if $\models [A]N :_u [B]$, then $M \lesssim N$.*
2. *(A,B) is a TCAP of M at u iff (soundness), (MTC) and the following condition hold:* (closure-2): *if $\xi \models A$ and $\xi \cdot u : V \models B$ then $M\xi \lesssim V$.*

By Proposition 2-1, a TCAP of M denotes a collection of behaviours whose minimum element is M, and in that sense characterises that behaviour uniquely.

Descriptive Completeness. We now show all PCFv-terms have TCAPs. The idea is to generate pre/post conditions inductively following the syntax of PCFv-terms. Figure 1 presents the generation rules. They are close to the proof rules in Appendix A, except for having the shape in which once the premise is determined, its conclusion is unique. In $[app]$, the premise says A_1 guarantees M_1's termination, A_2 that of M_2. Hence the conclusion's precondition ought to stipulate $A_{1,2}$ and termination of their application (described by B_1 and B_2). $[rec]$ intuitively says the program now uses itself for the environment f. The size of a formula does not change by applying this rule.

Example 3. 1. We have $\vdash^{\star} [\mathsf{T}]\,\lambda x.x :_u [\forall x.u \bullet x = x]$ (simplified using logical axioms) saying: *whatever value the program receives, it always converges to the same value.*

$$[var]\frac{}{\vdash^\star [\mathsf{T}]\, x :_u [u = x]} \qquad [const]\frac{}{\vdash^\star [\mathsf{T}]\, \mathsf{c} :_u [u = \mathsf{c}]}$$

$$[op]\frac{\vdash^\star [A_i]\, M_i :_{m_i} [B_i]}{\vdash^\star [\bigwedge_i A_i]\, \mathsf{op}(M_1..M_n) :_u [\exists \tilde{m}.(u = \mathsf{op}(m_1..m_n) \wedge \bigwedge_i B_i)]}$$

$$[abs]\frac{\vdash^\star [A]\, M :_m [B]}{\vdash^\star [\mathsf{T}]\, \lambda x.M :_u [\forall x.(A \supset \exists m.(u \bullet x = m \wedge B))]} \qquad [rec]\frac{\vdash^\star [\mathsf{T}]\, \lambda x.M :_u [A]}{\vdash^\star [\mathsf{T}]\, \mu f.\lambda x.M :_u [A[u/f]]}$$

$$[app]\frac{\vdash^\star [A_1]\, M :_m [B_1] \quad \vdash^\star [A_2]\, N :_n [B_2]}{\vdash^\star [A_1 \wedge A_2 \wedge \forall mn.(B_1 \wedge B_2 \supset \exists z.m \bullet n = z)]\, MN :_u [\exists mn.(m \bullet n = u \wedge B_1 \wedge B_2)]}$$

$$[if]\frac{\vdash^\star [A]\, M :_m [B] \quad \vdash^\star [A_i]\, N_i :_u [B_i] \quad \mathsf{b}_1 = \mathsf{t},\ \mathsf{b}_2 = \mathsf{f}}{\vdash^\star [A \wedge \bigwedge_{i=1,2}(B[\mathsf{b}_i/m] \supset A_i)]\, \mathtt{if}\ M\ \mathtt{then}\ N_1\ \mathtt{else}\ N_2 :_u [\bigvee_{i=1,2}(B[\mathsf{b}_i/m] \wedge B_i)]}$$

Fig. 1. Derivation Rules for Total CAPs

2. For $\lambda x.fx$, we get $\vdash^\star [\mathsf{T}]\, \lambda x.fx :_u [\forall xi.(f \bullet x = i \supset u \bullet x = i)]$ (simplification uses axioms for evaluation formulae [19]) which says: *if the application of f to x converges to some value, then the application of u to x converges to the same value.*
3. From 1, we obtain $\vdash^\star [\mathsf{T}]\, \mu f.\lambda x.x :_u [\forall x.u \bullet x = x]$ via vacuous renaming, as expected.
4. From 2, we obtain a TCAP for Ω as $\vdash^\star [\mathsf{T}]\, \Omega :_u [\mathsf{T}]$ by $\forall xi.(u \bullet x = i \supset u \bullet x = i) \equiv \mathsf{T}$. Since Ω is the least defined total behaviour, we cannot say anything better than T for this agent (note T is indeed a TCA of Ω).
5. The factorial program `Fact` is given the following assertion.

$$\vdash^\star [\mathsf{T}]\, \mathtt{Fact} :_u [u \bullet 0 = 1 \wedge \forall xi.(u \bullet x = i \supset u \bullet (x+1) = x \times i)] \qquad (2.3)$$

Note the assertion closely follows the recursive behaviour of the program. Through mathematical induction we obtain $\vdash^\star [\mathsf{T}]\, \mathtt{Fact} :_u [\forall x.(u \bullet x = x!)]$, as expected.

Theorem 4. (descriptive completeness for total correctness) *Assume* $\Gamma \vdash M : \alpha$. *Then* $\vdash^\star [A]\, M :_u [B]$ *implies* (A, B) *is a TCAP of* M *at* u.

Proposition 5. *If* $\vdash^\star [A]\, M :_u [B]$ *then the sum of the size of* A *and* B *is* $O(m \times 2^n)$ *where* m *is the size of* M *and* n *is the number of applications/conditionals in* M.

Definition 6. *Let* x *be fresh in 2 and 3.*

1. *We define* $\sqsubseteq$ *inductively as follows: (1)* $x^\alpha \sqsubseteq y^\alpha$ *iff* $x = y$ *for* $\alpha \in \{\mathsf{Bool}, \mathsf{Nat}\}$; *and (2)* $x^{\alpha \Rightarrow \beta} \sqsubseteq y^{\alpha \Rightarrow \beta}$ *iff* $\forall z^\alpha, v^\beta.(x \bullet z = v \supset \exists w.(y \bullet z = w \wedge v \sqsubseteq w))$.
2. $\mathcal{U}(A, u) \stackrel{def}{=} \forall x.(A[x/u] \supset x \sqsubseteq u)$ *and* $\uparrow(A, u) \stackrel{def}{=} \exists x.(A[x/u] \wedge x \sqsubseteq u)$. *Dually we set* $\mathcal{L}(A, u) \stackrel{def}{=} \forall x.(A[x/u] \supset u \sqsubseteq x)$ *and* $\downarrow(A, u) \stackrel{def}{=} \exists x.(A[x/u] \wedge u \sqsubseteq x)$.
3. *Write* $\models M :_u \{A\}$ *when* $M\xi \Downarrow V$ *implies* $\xi \cdot u : V \models A$ *for each* ξ. *We say* A *is a* PCAP of M at u *when (1)* $\models M :_u \{A\}$ *and (2) whenever* $\models N :_u \{A\}$ *we have* $N \lesssim M$.

Remark. The predicate $\sqsubseteq$ internalises $\lesssim$. $\mathcal{U}(A, u)$ etc. are logical counterparts of the standard order-theoretic operations [7]. A PCAP is the partial counterpart of a TCAP. In partial correctness we do not need a precondition since $\{A\}M :_u \{B\}$ (a partial correctness judgement) is equivalent to $\{\mathsf{T}\}M :_u \{A \supset B\}$, due to statelessness of PCFv.

Corollary 7. *1.* (observational completeness) $M \cong N$ *if and only if, for each A and B, we have* $\models [A]M :_u [B]$ *iff* $\models [A]N :_u [B]$.
2. (relative completeness) *We say B is* upward-closed at u *when* $\uparrow(B,u) \equiv B$. *Then* $\models [A]M :_u [B]$ *such that B is upward-closed at u implies* $\vdash [A]M :_u [B]$.
3. (derivability of PCAP) *If* $\vdash^\star [A]M :_u [B]$ *then* $A \wedge \mathcal{L}(B,u)$ *is a PCAP of M at u.*

For the non-trivial direction of (1), if M and N satisfy the same set of assertion pairs, then each satisfies another's TCAP, hence $M \cong N$ by Proposition 2-1. (2) uses Kleymann's consequence rule (cf. Appendix A) to derive assertion pairs from TCAPs (the restriction to upward-closed formulae is not unduly constraining since upward closure corresponds to total correctness [29, 23]).

3 Descriptive Completeness for Imperative PCFv

The method for deriving TCAPs directly generalises to imperative extensions of the logic [19, 3]. Below we illustrate the key idea, leaving the detailed technical development to the full version [1]. We consider the logic without aliasing [19].

The programming language, imperative PCFv, adds assignment $x := M$ and dereference $!x$, with x of a reference type in both, and $()$ of Unit type (reference types are not carried inside other types to avoid aliasing [19]). Typing is of the form $\Gamma;\Delta \vdash M : \alpha$, where Δ is for free references and Γ for free variables of non-reference types. $\cong$ (resp. $\lesssim$) is a typed congruence (resp. precongruence), relating two programs of a fixed basis. Without loss of generality we restrict "sequentially flattened" forms generated as:

$$
\begin{array}{rcl}
L & ::= & x \mid \mathsf{c} \mid \mathsf{op}(U_1..U_n) \mid \lambda x.L \mid \mathtt{let}\, x = UU' \,\mathtt{in}\, L \mid \mu x.\lambda y.L \\
 & \mid & \mathtt{if}\; U \;\mathtt{then}\, L_1 \;\mathtt{else}\, L_2 \mid x := V; L \mid \mathtt{let}\, x = !y \;\mathtt{in}\, L
\end{array}
$$

where $U, U', \ldots$ range over values in this grammar (i.e. variables, constants, abstraction and recursion). To the assertion language in Section 2 we add dereferencing $!x$ of a reference x. We also replace evaluation formulae with their imperative refinement $[C]\ e_1 \bullet e_2 = x\ [C']$ (x is binding in C') which says:

> *In any state satisfying C, if e_1 is applied to e_2, it converges to a value named x and the resulting state, together satisfying C'.*

A sequent $[C]M :_u [C']$ has a fixed basis, usually left implicit. A judgement for total correctness is written $\models [C]M :_u [C']$ (for validity) and $\vdash [C]M :_u [C']$ (for provability). It is straightforward to define a translation $[\![\cdot]\!]$ which converts all programs of imperative PCFv to their flattened forms so that $[\![M]\!] \cong M$ and $\vdash [C]\, [\![M]\!] :_u [C']$ iff $\vdash [C]M :_u [C']$. Thus it suffices to consider deriving CAPs for flattened programs.

Example 8. 1. The assertion $[!x = i]M :_u [u = i+1]$ says that M reads the content of x and returns the successor of that content. It does not make any guarantee about what is stored in memory after execution of M.
2. Under Δ with domain $\{x,y\}$, the assertion $[!x = i \wedge !y = j]M :_u [u = i+1 \wedge !x = i \wedge !y = j]$ is like (1), but in addition ensures M does not modify any storage cells.

$$[var]\frac{-}{\vdash^{\star\star} [\mathsf{T}]\, y :_u [u = y]} \qquad [const]\frac{-}{\vdash^{\star\star} [\mathsf{T}]\, \mathsf{c} :_u [u = \mathsf{c}]} \qquad [val]\frac{\vdash^{\star\star} [\mathsf{T}]\, U :_u [A] \quad \tilde{i} \text{ fresh}}{\vdash^{\star} [!\tilde{x} = \tilde{i}]\, U :_u [A \wedge !\tilde{x} = \tilde{i}]}$$

$$[op\text{-}val]\frac{\vdash^{\star\star} [\mathsf{T}]\, U_i :_{m_i} [A_i]}{\vdash^{\star} [\mathsf{T}]\, \mathsf{op}(U_1,..,U_n) :_u [\exists \tilde{m}.(u = \mathsf{op}(m_1,..,m_n) \wedge \bigwedge_i A_i)]}$$

$$[abs]\frac{\vdash^{\star} [C]\, L :_m [C'] \quad \tilde{i} = \mathsf{fv}(C, C') \backslash (\mathsf{fv}(L) \cup \{u\tilde{x}\})}{\vdash^{\star\star} [\mathsf{T}]\, \lambda y.L :_u [\forall y\tilde{i}.([C] u \bullet y = m [C'])]} \qquad [rec]\frac{\vdash^{\star\star} [\mathsf{T}]\, \lambda x.L :_u [A]}{\vdash^{\star\star} [\mathsf{T}]\, \mu f.\lambda x.L :_u [A[u/f]]}$$

$$[let\text{-}app]\frac{\vdash^{\star\star} [\mathsf{T}]\, V_1 :_m [A] \quad \vdash^{\star\star} [\mathsf{T}]\, V_2 :_n [B] \quad \vdash^{\star} [C]\, L :_u [C'] \quad \tilde{i} \text{ fresh}}{\vdash^{\star} [!\tilde{x} = \tilde{i} \wedge \forall mn.((A \wedge B) \supset \{!\tilde{x} = \tilde{i}\} m \bullet n = y \{C\})]\, \mathtt{let}\ y = V_1 V_2\ \mathtt{in}\ L :_u [C']}$$

$$[if]\frac{\vdash^{\star\star} [\mathsf{T}]\, U :_m [A] \quad \vdash^{\star} [C_i]\, L_i :_u [C'_i] \quad \mathsf{b}_1 = \mathsf{t},\ \mathsf{b}_2 = \mathsf{f}}{\vdash^{\star} [\bigwedge_{i=1,2} (A[\mathsf{b_i}/m] \supset C_i)]\, \mathtt{if}\ U\ \mathtt{then}\ L_1\ \mathtt{else}\ L_2 :_u [\bigvee_{i=1,2} (A[\mathsf{b_i}/m] \wedge C'_i)]}$$

$$[assign]\frac{\vdash^{\star\star} [\mathsf{T}]\, U :_z [A] \quad \vdash^{\star} [C]\, L :_u [C']}{\vdash^{\star} [\forall z.(A \supset C[z/!y])]\, y := U; L :_u [C']} \qquad [deref]\frac{\vdash^{\star} [C]\, L :_u [C']}{\vdash^{\star} [C[!y/z]]\, \mathtt{let}\ z = !y\ \mathtt{in}\ L :_u [C']}$$

Fig. 2. Derivation Rules for TCAPs for Imperative PCFv

3. Let $A(f) \stackrel{\text{def}}{=} \forall yi.[!y = i]\ f \bullet y = z\ [z = !y = i + 1]$. It characterises a procedure f that increments a reference y and returns the increment.
4. The assertion $[\mathsf{T}]\, \lambda a.(!f)a :_u [\forall ai.[A(f) \wedge !a = i]\ u \bullet a = c\ [!a = i + 1 \wedge c = i + 1]]$ is a procedure that takes a reference, increments it and returns that increment.
5. Finally, just like in the pure functional case, $[\mathsf{F}]\, \omega :_u [\mathsf{F}]$ is the strongest total specification we can derive about ω.

M is *semi-closed* if all its free names are references. A *model* is a pair (ξ, σ) where ξ maps non-reference names to semi-closed values and σ is a store, mapping reference names to semi-closed values. The satisfaction relation $(\xi, \sigma) \vdash C$ is defined as in the logic for PCFv, except that the satisfaction for the evaluation formula is refined to incorporate state change, see [19]. Implicitly assuming a basis, we set $\models [C]\, M :_u [C']$ iff $\forall \xi, \sigma.(\xi, \sigma) \models C \supset ((M\xi, \sigma) \Downarrow (V, \sigma') \wedge (\xi \cdot u : V, \sigma') \models C'$. For $\sigma_{1,2}$ of the same domain and typing, $\sigma_1 \lesssim \sigma_2$ is defined by pointwise ordering.

Definition 9. (TCAP) A pair (C, C') is a *total characteristic assertion pair*, or *TCAP*, *of* $\Gamma; \Delta \vdash M : \alpha$ *at* u, if the following conditions hold.

1. (soundness) $\models [C]\, M :_u [C']$.
2. (MTC, minimal terminating condition) $(\xi \cdot u : M\xi, \sigma) \Downarrow$ iff $(\xi, \sigma) \models C$.
3. (closure) Suppose $\models [E]\, N :_u [C']$, $E \supset C$ and $(\xi, \sigma) \models E$. Then $(M\xi, \sigma) \Downarrow (V, \sigma')$ implies $(N\xi, \sigma) \Downarrow (W, \sigma'')$ such that $V \lesssim W$ and $\sigma' \lesssim \sigma''$.

Figure 2 gives the generation rules for TCAPs. In all rules, we fix, but leave implicit, a reference basis with domain $\tilde{x}$. [*val*] transforms a judgement for values (written with turnstile $\vdash^{\star\star}$) to that for general programs (written with turnstile $\vdash^{\star}$).

Theorem 10. (descriptive completeness for imperative PCFv, total correctness) *If M is typable, then $\vdash^\star [C]M :_u [C']$ implies (C,C') is a TCAP of M at u.*

In 2 and 3 below, we use the notions corresponding to those given in Definition 6 in Section 2, including PCAPs, starting from $\sqsubseteq$ as an intrinsic predicate, see [1].

Corollary 11.
1. (observational completeness) *$M \cong N$ if and only if, for each C and C', we have $\models [C]M :_u [C']$ iff $\models [C]N :_u [C']$.*
2. (relative completeness) *We say C' is* upward closed at u *when $\uparrow(C',u) \equiv C'$. Then $\models [C]M :_u [C']$ such that C' is upward-closed at u implies $\vdash [C]M :_u [C']$.*
3. (derivability of PCAP) *If $\vdash^{\star\star} [\mathsf{T}]V :_u [A]$ then $\mathcal{L}(A,u)$ is a PCAP of V at u.*

In (3) above, any program can be made into a value by vacuous abstraction. We conclude this section with examples.

Example 12.
1. Let us fix a basis for programs and judgements, assuming two imperative variables y and z storing natural numbers. Then we get the following TCAP for $\lambda x.x$ (up to straightforward simplification):

$$\vdash^\star [\mathsf{T}]\ \lambda x.x :_u\ [\forall xnm.([!y = n \wedge !z = m]\ u \bullet x = i\ [i = x \wedge !y = n \wedge !z = m])]$$

Under the assumed basis, the lack of change of the contents of y and z (i.e. n and m) signify that the program has no side effects. For $\lambda x.fx$ we get:

$$\begin{array}{l} \forall xnmn'm'ii'.([!y = n \wedge !z = m]\ f \bullet x = i\ [i = i' \wedge !y = n' \wedge !z = m'] \\ \qquad \supset \quad [!y = n \wedge !z = m]\ u \bullet x = i\ [i = i' \wedge !y = n' \wedge !z = m']) \end{array}$$

Note how causality between the calls to f and $\lambda x.fx$, named u, is described by auxiliary variables n, m, n' and m'. The TCAP for $\mu f.\lambda x.fx$ is again T.
2. As final example, we consider an imperative factorial, that uses a stored procedure to realise recursion.

$$\mathtt{CircFact} \stackrel{\text{def}}{=} w := \lambda x.\mathtt{if}\ x = 0\ \mathtt{then}\ 1\ \mathtt{else}\ x \times !w(x-1)$$

In [19], we have shown that a natural specification for `CircFact` is derivable in the logic for imperative PCFv. For this program, $\vdash^\star$ leads to the following TCAP:

$$[\mathsf{T}]\ \mathtt{CircFact} :_m [m = () \wedge B'(u) \wedge I'(u)]$$

where we set, assuming w constitutes the only store for brevity:

$$\begin{array}{lcl} B'(u) & \stackrel{\text{def}}{=} & \forall f.[!w = f]\ u \bullet 0 = z\ [z = 1 \wedge !w = f] \wedge !w = u \\ I'(u) & \stackrel{\text{def}}{=} & \forall ff'i.\forall x \gneq 0.[!w = f \wedge E'(u)]\ u \bullet x = z\ [z = x \times i \wedge !w = f'] \\ E'(u) & \stackrel{\text{def}}{=} & [!w = f]\ u \bullet (x-1) = z\ [z = i \wedge !w = f'] \end{array}$$

This is the full specification of `CircFact`: it does not directly say the program computes a factorial since the procedure stored in w may change its behaviour depending on what w stores at the time of invocation (note w is not hidden). However through mathematical induction we can justify the following (strict) implication:

$$B'(u) \wedge I'(u) \quad \supset \quad \exists f.(\forall i.[!w = f](!w) \bullet i = z[z = i! \wedge !w = f] \wedge !w = f)$$

arriving at the "natural" specification of `CircFact` given in [19], which says: *after executing* `CircFact`, *w stores a procedure f which would calculate a factorial if w indeed stores that behaviour itself, and that w does store that behaviour.*

4 Related Work

Apart from their usage in verification condition generation [11], weakest preconditions and strongest postconditions [10] help with deriving relative completeness in Hoare logic. Cook's original proof [6] of relative completeness constructs the strongest postcondition for partial correctness. Clarke [5] uses the weakest liberal pre-condition. In both, the pre/post-conditions for loops use Gödel's β-function [24, Section 3.3]. Sokołowski [30] may be the first to give a completeness result for total correctness for the `while` language. De Bakker [8] extends these results to parameterless recursive procedures and concretely constructs what we call MTC (cf. Def. 1). Gorelick [12] seems the first to use most general formulae (MGFs, which correspond to our CAPs) for completeness in Hoare logic. Kleymann [21] introduces a powerful consequence rule and employs MGFs for proving completeness of Hoare logic with parameterless recursive procedures. Halpern [15], Olderog [25] and others establish relative completeness of Hoare logics for sublanguages of Algol (these logics do not include assertions on higher-order behaviours, see [19, Section 8] for a survey). Von Oheimb's recent work [33] gives a mechanised proof of completeness for Hoare logic using MGFs.

Some authors use abstraction on predicates to generate concise verification conditions in the setting of the Floyd-Hoare assertion methods for first-order imperative programs. Blass and Gurevich [4], guided by a detailed study of Cook's completeness result, use an existential fixpoint logic. Leivant [22] uses second-order abstraction (abstraction on first-order predicates), inductively deriving a formula directly representing a partial function defined by a while program with recursive first-order procedure. Once this is done, characteristic assertions for both total and partial correctness for a given program are immediate. We suspect that the use of predicate abstraction in these works may make calculation of validity hard in practice, even for first-order programs.

There are two basic differences between the present work and these preceding studies. First, in the preceding works, generated assertion pairs describe first-order state transformation rather than the behaviour of higher-order programs. Philosophically, our method may be notable in that it extends completeness and related results to assertions which directly talk about (higher-order) behaviour. Second, the presented method for constructing characteristic formulae is different from those employed so far, especially in its treatment of recursion. We need neither the β-predicate, loop annotation, predicate abstraction nor inductively defined formulae for generating TCAPs for recursion. Concretely, this simple treatment for recursion is made possible by evaluation formulae. A deeper reason however may lie in analytical, fine-grained nature of our assertion language, reflecting that of call-by-value higher-order computation. As far as our experience goes, evaluation formulae do not make calculation of validity unduly harder than in first-order Hoare logic: for example, (often implicit) simplifications of assertions in Sections 2 and 3 only use simple syntactic axioms in [16, 19] combined with standard logical axioms and mathematical induction (see Section 5 for further discussions).

The order-theoretic nature of partial and total correctness is observed in early works by Smyth, Plotkin and Stirling [29,28]. The present work differs in that it substantiates these ideas at the level of concrete assertion methods and compositional proof rules (see for example a derivation example of $\lambda x.\omega$ in Section 2). Finally our emphasis on descriptive completeness, and the foundation of the logic itself, comes from Hennessy-Milner logic [16], where Graf, Ingólfsdóttir, Sifakis, Steffen and others [13,31,32] study characteristic formulae for first-order communicating processes.

5 Further Topics

The present work is an inquiry into the descriptive power of program logic for higher-order functions. Through inductive derivation of characteristic formulae, we have shown that the logic allows concise description of full behaviour of programs involving arbitrary higher-order types and recursion. Logics for more complex classes of imperative higher-order functions are studied in [3]. Extensions of the presented results to these logics are important for treating such languages as ML.

Practically speaking, the presented method for TCAP generation, along with its properties, opens a new perspective for program validation based on verification condition generators (VCG) [11,20]. In traditional VCG, we have a target specification $\{C\}P\{C'\}$ and an annotated version of the program. A VCG then automatically generates, usually through backward chaining [20], one or more entailments whose validity entails P's satisfaction of the specification. The presented TCAP generation has the potential to improve this existing scheme. Schematically, our TCAP generation suggests the following framework.

1. Assume given a program V (any program can be made into a value by vacuous abstraction) and a desired specification $[\mathsf{T}]V :_u [A]$.
2. We automatically generate the TCAP (T, A_0).
3. By Theorem 4, if we can validate $A_0 \supset A$, we know V conforms to A.

First, this framework dispenses with the need to annotate programs, which has been one of the obstacles preventing wide-spread adoption of the VCG-based validation methods. Second, at the level of specification, it allows direct treatment of higher-order idioms, opening the use of higher-order languages such as ML and Haskell for program certification (arguably these languages offer a suitable basis for this task through their well-studied semantic foundations). Third, the specification A above can contain assumptions on the environment (say existing libraries, referred to by free variables in V) on which the functionality of V relies. As we discussed in [19, Section 2], this allows specifying complex interplay among the program and library functions beyond the separate treatment of assumptions on procedures in traditional methods. For these reasons, inquiries into the practical potential of TCAP generation for program validation would be worth pursuing. As the first experiment towards this goal, we have developed a prototype implementation of the TCAP generation algorithm [2].

One of the foremost challenges towards practical use of TCAP generation is the development of tractable methods for logical calculation of entailment in Step 3 above, which demands, in addition to first-order logic and mathematical induction, the treatment of logical primitives for (imperative) higher-order functions. It would be especially interesting to extend verification tools like Simplify [9] in this direction, combined with studies on axiom systems for e.g. evaluation formulae (see [16, 19, 3] for a preliminary study). Detailed comparisons with standard VCGs, as well as potential for a combined usage of these two methods, is another interesting further topic.

Finally, we believe that the upper bound in Prop. 5 can be improved upon considerably, at least for large and practically relevant classes of programs.

References

1. A full version of this paper. Available at: http://www.dcs.qmul.ac.uk/~kohei/logics.
2. A prototype implementation of an algorithm deriving characteristic formulae. http://www.dcs.qmul.ac.uk/~martinb/capg, October 2005.
3. M. Berger, K. Honda, and N. Yoshida. A logical analysis of aliasing for higher-order imperative functions. In *ICFP'05*, pages 280–293, 2005.
4. A. Blass and Y. Gurevich. The Underlying Logic of Hoare Logic. In *Current Trends in Theoretical Computer Science*, pages 409–436. 2001.
5. E. M. Clarke. The characterization problem for Hoare logics. In *Proc. Royal Society meeting on Mathematical logic and programming languages*, pages 89–106, 1985.
6. S. A. Cook. Soundness and completeness of an axiom system for program verification. *SIAM J. Comput.*, 7(1):70–90, 1978.
7. B. A. Davey and H. A. Priestley. *Introduction to Lattices and Order*. CUP, 1990.
8. J. W. de Bakker. *Mathematical Theory of Program Correctness*. Prentice-Hall, Inc., Upper Saddle River, NJ, USA, 1980.
9. D. Detlefs, G. Nelson, and J. B. Saxe. Simplify: a theorem prover for program checking. *J. ACM*, 52(3):365–473, 2005.
10. E. W. Dijkstra. *A Discipline of Programming*. Prentice Hall, 1976.
11. R. W. Floyd. Assigning meaning to programs. In *Symp. in Applied Math.*, volume 19, 1967.
12. G. Gorelick. A complete axiomatic system for proving assertions about recursive and nonrecursive programs. Technical Report 75, Univ. of Toronto, 1975.
13. S. Graf and J. Sifakis. A Modal Characterization of Observational Congruence on Finite Terms of CCS. In *ICALP'84*, pages 222–234, London, UK, 1984. Springer-Verlag.
14. C. A. Gunter. *Semantics of Programming Languages*. MIT Press, 1995.
15. J. Y. Halpern. A good Hoare axiom system for an ALGOL-like language. In *11th POPL*, pages 262–271. ACM Press, 1984.
16. K. Honda. From process logic to program logic. In *ICFP'04*, pages 163–174. ACM, 2004.
17. K. Honda. From process logic to program logic (full version of [16]). Available at: www.dcs.qmul.ac.uk/~kohei/logics, November 2004. Typescript, 52 pages.
18. K. Honda and N. Yoshida. A compositional logic for polymorphic higher-order functions. In *PPDP'04*, pages 191–202. ACM, 2004.
19. K. Honda, N. Yoshida, and M. Berger. An observationally complete program logic for imperative higher-order functions. In *LICS'05*, pages 270–279, 2005.
20. J. C. King. A program verifier. In *IFIP Congress (1)*, pages 234–249, 1971.
21. T. Kleymann. Hoare logic and auxiliary variables. Technical report, University of Edinburgh, LFCS ECS-LFCS-98-399, October 1998.
22. D. Leivant. Logical and mathematical reasoning about imperative programs: preliminary report. In *Proc. POPL'85*, pages 132–140, 1985.

23. D. Leivant. Partial correctness assertions provable in dynamic logics. In *FoSSaCS*, volume 2987 of *LNCS*, pages 304–317, 2004.
24. E. Mendelson. *Introduction to Mathematical Logic*. Wadsworth Inc., 1987.
25. E.-R. Olderog. Sound and Complete Hoare-like Calculi Based on Copy Rules. *Acta Inf.*, 16:161–197, 1981.
26. S. Owicki and L. Lamport. Proving liveness properties of concurrent programs. *ACM Trans. Program. Lang. Syst.*, 4(3):455–495, 1982.
27. B. C. Pierce. *Types and Programming Languages*. MIT Press, 2002.
28. G. D. Plotkin and C. Stirling. A framework for intuitionistic modal logics. In *Theor. Aspects of Reasoning about Knowledge*, pages 399–406. Morgan Kaufmann, 1986.
29. M. Smyth. Power domains and predicate transformers: A topological view. In *ICALP'83*, volume 154 of *LNCS*, pages 662–675, 1983.
30. S. Sokołowski. Axioms for total correctness. *Acta Inf.*, 9:61–71, 1977.
31. B. Steffen. Characteristic formulae. In *ICALP'89*, pages 723–732. Springer-Verlag, 1989.
32. B. Steffen and A. Ingólfsdóttir. Characteristic formulae for processes with divergence. *Inf. Comput.*, 110(1):149–163, 1994.
33. D. von Oheimb. Hoare logic for mutual recursion and local variables. In *FSTTCS*, volume 1738 of *LNCS*, pages 168–180, 1999.

A Proof Rules for PCFv and Imperative PCFv (Total Correctness)

The proof rules for PCFv follow. $A^{\text{-}x}$ denotes A in which x does not occur free. See [18, 3, 19] for illustration. The consequence rule comes from Kleymann [21].

$$[\mathit{Var}]\,\frac{-}{[C[x/u]]\,x:_u[C]}\qquad [\mathit{Const}]\,\frac{-}{[C[\mathsf{c}/u]]\,\mathsf{c}:_u[C]}\qquad [\mathit{Add}]\,\frac{[C]\,M:_m[C_0]\quad [C_0]\,N:_n[C'[m+n/u]]}{[C]\,M+N:_u[C']}$$

$$[\mathit{Abs}]\,\frac{[A^{\text{-}x}\wedge C]\,M:_m[C']}{[A]\,\lambda x.M:_u[\forall x.(C\supset \exists m.(u\bullet x=m\ \wedge\ C'))]}$$

$$[\mathit{App}]\,\frac{[C]\,M:_m[C_0]\qquad [C_0]\,N:_n[\exists u.(m\bullet n=u\ \wedge\ C')]}{[C]\,MN:_u[C']}\qquad [\mathit{If}]\,\frac{[C]\,M:_m[C_0]\qquad [C_0[\mathsf{t}/m]]\,N_1:_u[C']\quad [C_0[\mathsf{f}/m]]\,N_2:_u[C']}{[C]\,\mathtt{if}\ M\ \mathtt{then}\ N_1\ \mathtt{else}\ N_2:_u[C']}$$

$$[\mathit{Rec}]\,\frac{[A^{\text{-}xi}\wedge\forall j\precsim i.B(j)[x/u]]\,\lambda y.M:_u[B(i)^{\text{-}x}]}{[A]\,\mu x.\lambda y.M:_u[\forall i.B(i)]}\qquad [\mathit{Conseq}]\,\frac{[A']\,M:_u[B']\quad A\supset(\,A'\wedge(B'\supset B)\,)}{[A]\,M:_u[B]}$$

For Imperative PCFv, the rules for expressions, first-order operators, recursion, and if-then-else are identical with those for pure PCFv.

$$[\mathit{Abs}]\,\frac{[A^{\text{-}x}\wedge C]\,M:_m[C']}{[A]\,\lambda x.M:_u[\forall x.[C]\ u\bullet x=m\ [C']]}\qquad [\mathit{App}]\,\frac{[C]\,M:_m[C_0]\quad [C_0]\,N:_n[C_1\wedge[C_1]\ m\bullet n=u\ [C']]}{[C]\,MN:_u[C']}$$

$$[\mathit{Deref}]\,\frac{-}{[C[!x/u]]\,!x:_u[C]}\qquad [\mathit{Assign}]\,\frac{[C]\,M:_m[C'[m/!x][()/u]]}{[C]\,x:=M:_u[C']}$$

$$[\mathit{Conseq\text{-}Kleymann}]\,\frac{[C_0]M:_u[C_0']\quad C\supset\exists\tilde{j}.(\,C_0[\tilde{j}/\tilde{i}]\wedge(C_0'[\tilde{y}\tilde{j}/\tilde{x}\tilde{i}]\supset C'[\tilde{y}/\tilde{x}])\,)}{[C]\,M:_u[C']}$$

In $[\mathit{Conseq\text{-}Kleymann}]$, we assume a basis Γ (for non-reference) and Δ (for reference) and set $\{\tilde{x}\}=\mathsf{dom}(\Gamma,\Delta)\cup\{u\}$, $\{\tilde{i}\}=\mathsf{fv}(C,C',C_0,C_0')\backslash\{\tilde{x}\}$. In addition, we require the $\tilde{j}$ (resp. $\tilde{y}$) to be fresh and of the same length as $\tilde{i}$ (resp. $\tilde{x}$).

Interpreting Polymorphic FPC into Domain Theoretic Models of Parametric Polymorphism

Rasmus Ejlers Møgelberg*

DISI, Università di Genova
mogelberg@disi.unige.it

Abstract. This paper shows how parametric PILL$_Y$ (Polymorphic Intuitionistic / Linear Lambda calculus with a fixed point combinator Y) can be used as a metalanguage for domain theory, as originally suggested by Plotkin more than a decade ago. Using recent results about solutions to recursive domain equations in parametric models of PILL$_Y$, we show how to interpret FPC in these. Of particular interest is a model based on "admissible" pers over a reflexive domain, the theory of which can be seen as a domain theory for (impredicative) polymorphism. We show how this model gives rise to a parametric and computationally adequate model of PolyFPC, an extension of FPC with impredicative polymorphism. This is the first model of a language with parametric polymorphism, recursive terms and recursive types in a non-linear setting.

1 Introduction

Parametric polymorphism is an important reasoning principle for several reasons. One is that it provides proofs of modularity principles [27] and other results based on "information hiding" such as security principles (see for example [28]). Another is that it can be used to make simple type theories surprisingly expressive by encoding inductive and coinductive types using polymorphism. If further parametric polymorphism is combined with fixed points on the term level, inductive and coinductive types coincide, and Freyd's theory of algebraically compact categories provides solutions to general type equations. However, when introducing fixed points the parametricity principle must be weakened for the theory to be consistent. Plotkin [25, 23] suggested using the calculus PILL$_Y$ (Polymorphic Intuitionistic / Linear Lambda calculus with a fixed point combinator Y), which in combination with parametricity would have inductive, coinductive and recursive types in the linear part of the calculus. This theory was worked out in details, along with a category theoretic treatment by Birkedal, Møgelberg and Petersen [8, 6, 7, 20], see also [10]. In *loc. cit.* a concrete model of PILL$_Y$ is constructed using "admissible" partial equivalence relations (pers) over a reflexive domain. The theory of admissible pers can be seen as a domain theory for (impredicative) polymorphism.

Plotkin suggested using parametric PILL$_Y$ as an axiomatic setup for domain theory. However, as mentioned, the solutions to recursive type equations that PILL$_Y$ provides are in a linear calculus, whereas, as is well known, domain theory also provides models of non-linear lambda calculi with recursive types, such as FPC — a simply typed

* This work is sponsored by Danish Research Agency stipend no. 272-05-0031.

M. Bugliesi et al. (Eds.): ICALP 2006, Part II, LNCS 4052, pp. 372–383, 2006.

lambda calculus with recursive term definitions and general recursive types, equipped with an operational call-by-value semantics. In this paper we test Plotkin's thesis by showing that the solutions to recursive type equations in the linear type theory PILL_Y can be used to model FPC. The interpretation uses a category of coalgebras, and the resulting translation is basically an extension of Girard's interpretation of intuitionistic logic into linear logic presented in [16] and developed on term level in [19]. The new technical contribution in this part of the paper is the treatment of recursive types.

Of particular interest is the example of the model of admissible pers. When writing out the model of FPC in this case, it becomes clear that it also models parametric polymorphism. We use this to show that PolyFPC, an extension of FPC with polymorphism defined below, can be modeled soundly in the per-model. In fact this model is computationally adequate. The model is to the authors knowledge the first model of the combination of parametric polymorphism, recursive terms and recursive types in a non-linear setting. For many readers the construction of this model may be the main result of the paper, but the earlier abstract analysis is needed to show that it models recursive types.

The adequate model of PolyFPC may be used to derive consequences of parametricity, such as modularity proofs, up to ground contextual equivalence along the lines of the proofs of [21], but using denotational methods. The model is also interesting because of the mix of parametricity and partiality, a combination which, as earlier research has shown, requires an alternative formulation of parametricity, such as the one suggested in [17]. This paper sketches the resulting parametricity principle derivable from the model. In future work, the parametric reasoning in the model will be lifted to a logic for parametricity for PolyFPC.

A related paper is [1], in which a model of polymorphism and recursion is constructed using admissible pers (as here) satisfying a uniformity property as well as various other properties ensuring that recursive types may be constructed as in domain theory. The main differences between *loc. cit.* and this paper is that the present model is parametric, and in our model the recursive types are constructed using parametricity.

The paper is organized as follows. Section 2 recalls the language PILL_Y and the theory of models for it, in particular the per-model. The language PolyFPC is defined in Section 3, and Section 4 shows how to model FPC in general models of PILL_Y. In Section 5 the interpretation of PolyFPC in the per-model resulting from the general theory is written out in detail and computational adequacy is formulated. Unfortunately the proof of adequacy is omitted for reasons of space. Finally, Section 6 discusses reasoning about parametric polymorphism for PolyFPC using the per-model.

2 Polymorphic Intuitionistic / Linear Lambda Calculus

The calculus PILL_Y is a Polymorphic dual Intuitionistic / Linear Lambda calculus with a fixed point combinator denoted Y. In other words it is the calculus DILL of [2] extended with polymorphism and fixed points for terms. This section sketches the calculus, but the reader is referred to [9, 20] for full details.

Types of PILL_Y are given by the grammar

$$\sigma ::= \alpha \mid I \mid \sigma \otimes \tau \mid \sigma \multimap \tau \mid !\sigma \mid \textstyle\prod \alpha.\, \sigma,$$

and we use the notation $\alpha_1,\ldots,\alpha_n \vdash \sigma\colon \mathsf{Type}$ to mean that σ is a well-formed type with free type variables among $\alpha_1,\ldots,\alpha_n$. The grammar for terms is

$$t ::= x \mid \star \mid Y \mid \lambda^\circ x\colon \sigma.t \mid t\,t \mid t \otimes t \mid !t \mid \Lambda\alpha\colon \mathsf{Type}.\, t \mid t(\sigma) \mid \\ \text{let } x\colon \sigma \otimes y\colon \tau \text{ be } t \text{ in } t \mid \text{let } !x\colon \sigma \text{ be } t \text{ in } t \mid \text{let } \star \text{ be } t \text{ in } t.$$

Terms have two contexts of ordinary variables — a context of linear variables and a context of intuitionistic variables. We refer the reader to *loc. cit.* for the term formation rules of the calculus and the equality theory for terms. The term constructor $\lambda^\circ x\colon \sigma.t$ constructs terms of type $\sigma \multimap \tau$ by abstracting *linear* term variables. Using the Girard encodings one can define $\sigma \to \tau$ to be $!\sigma \multimap \tau$, and there is a corresponding definable λ-abstraction for *intuitionistic* term variables. Under this convention, the type of the fixed point combinator is $Y\colon \prod\alpha.\,(\alpha \to \alpha) \to \alpha$.

2.1 PILL$_Y$-Models

The most general formulation of models of PILL$_Y$ uses fibred category theory, but here we will just consider a large class of PILL$_Y$-models, which includes all important models known by the author (except some constructed from syntax), since the theory of the next sections is much simpler in this case.

Suppose $\mathbf{C}$ is a linear category, i.e., a symmetric monoidal closed category with a symmetric monoidal comonad ! satisfying a few extra properties as described for example in [18, 20]. We shall write $\sigma \multimap \tau$ for morphisms in $\mathbf{C}$. If further any functor $\mathbf{C}_0^{n+1} \to \mathbf{C}$, where $\mathbf{C}_0$ denotes the objects of $\mathbf{C}$ considered as a discrete category, has a right Kan extension along the projection $\mathbf{C}_0^{n+1} \to \mathbf{C}_0^n$, then one can form a model of PILL, the subset of PILL$_Y$ not including the fixed point combinator (for a full model of PILL$_Y$, a term modeling the fixed point combinator must exist). Types with n free type variables are modeled as functors $\mathbf{C}_0^n \to \mathbf{C}$ (or equivalently maps $\mathbf{C}_0^n \to \mathbf{C}_0$) by modeling $\boldsymbol{\alpha} \vdash \alpha_i$ as the i'th projection, $\otimes, I, \multimap$ using the symmetric monoidal structure, ! using the comonad and polymorphism using the Kan-extensions.

A category theoretic definition of what a *parametric* model of PILL$_Y$ is, is given in [20], but we shall not repeat that now. Instead we mention that the per-model described below is parametric, and we sketch the model theoretic formulations of the consequences of parametricity.

If $\mathbf{C}$ is a parametric model of PILL$_Y$, one can prove that it has, among other type constructions, products and coproducts, and that one can solve a large class of recursive type equations. Syntactically, a recursive type equation is usually given by a type σ with a free variable α and a solution is a type τ such that $\sigma[\tau/\alpha] \cong \tau$. Usually, one is not just interested in any solution, but rather a solution satisfying a universal condition, which in the case of α ocurring only positively in σ (e.g. if $\sigma = \alpha + 1$) means an initial algebra or final coalgebra for the functor induced by σ.

For the more general case of both positive and negative occurences of α in σ (such as $\sigma = (\alpha \to \alpha) + 1$), one can split the occurences of α in σ into positive and negative and obtain a functor of mixed variance. This way any type $\alpha_1, \ldots \alpha_n \vdash \sigma$ in PILL$_Y$ induces a functor $(\mathbf{C}^{\mathrm{op}} \times \mathbf{C})^n \to \mathbf{C}$, which is strong in the sense that there exists a PILL$_Y$ term of type

$$\prod \boldsymbol{\alpha}, \boldsymbol{\alpha}', \boldsymbol{\beta}, \boldsymbol{\beta}'\colon \mathsf{Type}.\, (\boldsymbol{\alpha}' \multimap \boldsymbol{\alpha}) \to (\boldsymbol{\beta} \multimap \boldsymbol{\beta}') \to \sigma(\boldsymbol{\alpha}, \boldsymbol{\beta}) \multimap \sigma(\boldsymbol{\alpha}', \boldsymbol{\beta}')$$

inducing the functor. In general, a functor $F\colon (\mathbf{C}^{\mathrm{op}} \times \mathbf{C})^n \to \mathbf{C}$ is *strong* if there exists a term in the model inducing it. For any strong functor $F\colon (\mathbf{C}^{\mathrm{op}} \times \mathbf{C})^{n+1} \to \mathbf{C}$, there exists a strong functor $\mathrm{Fix}F\colon (\mathbf{C}^{\mathrm{op}} \times \mathbf{C})^n \to \mathbf{C}$ such that

$$F \circ \langle id_{(\mathbf{C}^{\mathrm{op}} \times \mathbf{C})^n}, \breve{\mathrm{Fix}F} \rangle \cong \mathrm{Fix}F$$

where $\breve{\mathrm{Fix}F}\colon (\mathbf{C}^{\mathrm{op}} \times \mathbf{C})^n \to \mathbf{C}^{\mathrm{op}} \times \mathbf{C}$ is the functor that maps $(A_1, B_1, \ldots A_n, B_n)$ to $(\mathrm{Fix}F(B_1, A_1, \ldots B_n, A_n), \mathrm{Fix}F(A_1, B_1, \ldots A_n, B_n))$. The functor $\mathrm{Fix}F$ is encoded in PILL_Y using encodings due to Plotkin. The proof of this proceeds by first showing that any strong functor $\mathbf{C} \to \mathbf{C}$ has an initial algebra whose inverse is a final coalgebra. This phenomena called algebraic compactness has been studied by Freyd [14, 13, 15], who showed how to solve general recursive type equations in this setting. As a consequence of Freyd's theory, the functor $\mathrm{Fix}F$ also satisfies a universal property called the dinaturality condition generalizing at the same time the notion of initial algebra and final coalgebra. See [9, 20] for full details.

2.2 A Per-model

We sketch a model of parametric PILL_Y. For details, see [9, 20]. The model is a variant of the parametric per-model for second order lambda calculus, restricted to a notion of admissible pers to encompass fixed points.

Suppose D is a reflexive domain, i.e., a pointed complete partial order such that $[D \to D]$ is a retract of D. Then D has a combinatory algebra structure with application $x \cdot y$ defined by applying the function corresponding to x by the reflection to y. An *admissible per* is a partial equivalence relation R on D closed under chains and relating $\bot$ to itself. A map of admissible pers from R to S is a map of equivalence classes $f\colon D/R \to D/S$ such that there exists an element $e \in D$ *tracking* f in the sense that $f([x]_R) = [e \cdot x]_S$. This defines a category $\mathbf{AP}$ of admissible pers on D. We also define the subcategory $\mathbf{AP}_\bot$ of $\mathbf{AP}$ of morphisms with strict trackers, i.e., trackers satisfying $e \cdot \bot = \bot$. The category $\mathbf{AP}_\bot$ has products and also a symmetric monoidal closed structure with tensor product defined as a quotient of the product, and $R \multimap S$ as $\{(d, e) \mid d \cdot \bot = e \cdot \bot = \bot \wedge \forall x, y \in D.\, xRy \Rightarrow S(d \cdot x, e \cdot y)\}$. Finally, there is a symmetric monoidal comonad ! definable on $\mathbf{AP}_\bot$, the coKleisli category of which is $\mathbf{AP}$.

By an admissible proposition on a per R we shall mean a subset of the set of equivalence classes for R which itself constitutes an admissible per. An admissible relation between pers R and S is an admissible proposition on $R \times S$. Since $D/(R \times S) \cong D/S \times D/R$, we will often think of such a relation as a subset of the product of equivalence classes. We write $\mathbf{AdmRel}_{\mathbf{AP}_\bot}$ for the category of admissible relations on admissible pers, with as maps pairs of maps from $\mathbf{AP}_\bot$ mapping related elements to related elements. There is a reflexive graph of categories

$$\mathbf{AdmRel}_{\mathbf{AP}_\bot} \;\substack{\longrightarrow \\ \longleftarrow \\ \longrightarrow}\; \mathbf{AP}_\bot \tag{1}$$

where the two maps from left to right map a relation to its domain and codomain respectively, and the last map maps a per to the identity relation on the per. The symmetric monoidal structure of $\mathbf{AP}_\bot$ can be extended to a symmetric monoidal structure on $\mathbf{AdmRel_{AP_\perp}}$ commuting with the functors of (1) and likewise for the symmetric monoidal comonad.

In the parametric variant of the per-model, a type is modeled as a pair (σ^r, σ^p), where $\sigma^p: \mathbf{AP}_0^n \to \mathbf{AP}_0$ is a map as before and σ^r is a map taking an n-vector of admissible relations $(A_i\colon \mathrm{AdmRel}(R_i, S_i))_{i\leq n}$ and producing an admissible relation $\sigma^r(\boldsymbol{A})\colon \mathrm{AdmRel}(\sigma^p(\boldsymbol{R}), \sigma^p(\boldsymbol{S}))$, satisfying $\sigma^r(eq_{R_1}, \ldots, eq_{R_n}) = eq_{\sigma^p(\boldsymbol{R})}$. A term in the model from (σ^r, σ^p) to (τ^r, τ^p) (assumed to be two types with the same number of free variables) is a family of maps $(f_{\boldsymbol{R}}\colon D/\sigma^p(\boldsymbol{R}) \to D/\tau^p(\boldsymbol{R}))_{\boldsymbol{R}}$ with a common tracker, such that for all $(A_i\colon \mathrm{AdmRel}(R_i, S_i))_{i\leq n}$ and all x, y, if $([x], [y]) \in \sigma^r(\boldsymbol{A})$, then $(f_{\boldsymbol{R}}([x]), f_{\boldsymbol{S}}([y])) \in \tau^r(\boldsymbol{A})$. For further details, see [9, 20].

To see how the per-model is an example of the general models described in Section 2.1, notice that (1) describes an internal linear category in a presheaf category over the realizability topos for the combinatory algebra D. Interpreting the general construction in this presheaf category gives the per-model.

3 Polymorphic FPC

In this section we present the language PolyFPC, an extension of the language FPC, first defined by Plotkin [24] (see also [12]), with recursive function definitions and full impredicative polymorphism. This language can be considered a powerful intermediate language to be used in compilers. In later sections we will show how to interpret FPC into any PILL$_Y$-model of the form of Section 2.1 and how to interpret PolyFPC into the per-model sketched in Section 2.2.

Since PolyFPC is a language with polymorphism and general (nested) recursive types, types in the languages may have free type variables (as in PILL$_Y$) and are formed using the grammar

$$\sigma, \tau ::= \alpha \mid 1 \mid \sigma + \tau \mid \sigma \times \tau \mid \sigma \to \tau \mid \mathrm{rec}\ \alpha.\, \sigma \mid \textstyle\prod \alpha.\, \sigma$$

As usual, the constructions $\prod \alpha.\, \sigma$ and rec $\alpha.\, \sigma$ binds the type variable α. The grammar for terms is

$$\begin{aligned} t ::= {} & x \mid \star \mid \mathrm{inl}\, t \mid \mathrm{inr}\, t \mid \mathrm{case}\ t\ \mathrm{of}\ \mathrm{inl}\ x.\, t'\ \mathrm{of}\ \mathrm{inr}\ x.\, t'' \mid \langle t, t'\rangle \mid \pi_1(t) \mid \pi_2(t) \mid \\ & \lambda x\colon \sigma.\, t \mid t(t') \mid \mathrm{intro}\, t \mid \mathrm{elim}\, t \mid \mathrm{let\ rec}\ fx = t\ \mathrm{in}\ f\, t' \mid \Lambda\alpha.\, t \mid t(\tau). \end{aligned}$$

For reasons of space, we shall not repeat the well-known typing rules of FPC, but refer the reader to [12] for them. However, since our version of FPC also includes an explicit recursive term constructions, we mention the typing rule for that:

$$\frac{\boldsymbol{\alpha} \mid \boldsymbol{x}\colon \boldsymbol{\sigma}, f\colon \tau \to \tau', x\colon \tau \vdash t\colon \tau' \qquad \boldsymbol{\alpha} \mid \boldsymbol{x}\colon \boldsymbol{\sigma} \vdash t'\colon \tau}{\boldsymbol{\alpha} \mid \boldsymbol{x}\colon \boldsymbol{\sigma} \vdash \mathrm{let\ rec}\ fx = t\ \mathrm{in}\ f\, t'\colon \tau'}$$

(bold letters denote sequents) and since PolyFPC also includes polymorphism, we mention the two typing rules for that:

$$\frac{\boldsymbol{\alpha}, \alpha' \mid \boldsymbol{x}\colon \boldsymbol{\sigma} \vdash t\colon \tau}{\boldsymbol{\alpha} \mid \boldsymbol{x}\colon \boldsymbol{\sigma} \vdash \Lambda\alpha'.\, t\colon \prod \alpha'.\, \tau}\ \boldsymbol{\alpha} \vdash \boldsymbol{\sigma} \qquad \frac{\boldsymbol{\alpha} \mid \boldsymbol{x}\colon \boldsymbol{\sigma} \vdash t\colon \prod \alpha'.\, \tau \quad \boldsymbol{\alpha} \vdash \tau'}{\boldsymbol{\alpha} \mid \boldsymbol{x}\colon \boldsymbol{\sigma} \vdash t(\tau')\colon \tau[\tau'/\alpha']}$$

The terms intro and elim introduce and eliminate terms of recursive types, e.g. if $t\colon \sigma[\text{rec}\ \alpha.\sigma/\alpha]$ then intro $t\colon \text{rec}\ \alpha.\sigma$.

In the following we shall use the terminology programs, to mean closed typable terms of closed type. The language PolyFPC is equipped with a call-by-value operational semantics. Formally, the operational semantics is a relation $\Downarrow$ relating programs to values, by which we mean programs following the grammar

$$v ::= \star \mid \text{inl}\ v \mid \text{inr}\ v \mid \langle v, v' \rangle \mid \lambda x\colon \sigma.\, t \mid \Lambda\alpha.\, t \mid \text{intro}\ v.$$

Again we refer to [12] for the definition of $\Downarrow$ on FPC (where it is denoted $\rightsquigarrow$), and just mention the two new rules:

$$\frac{e' \Downarrow v' \qquad e[\lambda x\colon \sigma.\, \text{let rec}\ f x' = e\ \text{in}\ f\ x/f, v'/x'] \Downarrow v}{\text{let rec}\ f x' = e\ \text{in}\ f\ e' \Downarrow v}$$

$$\frac{t \Downarrow \Lambda\alpha.\, t' \qquad t'[\tau/\alpha] \Downarrow v}{t(\tau) \Downarrow v}$$

4 Modeling FPC in Categories of Coalgebras

In this section we address the problem of interpreting the intuitionistic calculus FPC into parametric models of the linear calculus PILL_Y. The inspiration for the general case will come from attempting to mimic the usual interpretation of FPC into domain theory (see for example [12, 22]) in the per-model presented in Section 2.2.

In domain theory, types of FPC are interpreted as complete partial orders (cpos) and terms as partial maps between them, i.e., in the Kleisli category for the lifting monad on the category $\mathbf{Cpo}$ of cpos. Neither of the categories $\mathbf{AP}_\perp$ or $\mathbf{AP}$ have the categorical properties needed for playing the role of $\mathbf{Cpo}$ in the adaption of the interpretation of FPC to admissible pers. Instead, the category $\mathbf{CCP}$ of chain closed pers on D and tracked maps between them is a good candidate. As in the category of admissible pers, we will consider as admissible propositions on a chain complete per R, subsets of D/R corresponding to chain complete pers, and an admissible relation is an admissible subset of the product. Admissible relations on chain complete pers form a category $\mathbf{AdmRel}_{\mathbf{CCP}}$ where maps are pairs of maps mapping related elements to related elements. The next proposition shows how to recover $\mathbf{CCP}$ from $\mathbf{AP}_\perp$, and $\mathbf{AdmRel}_{\mathbf{CCP}}$ from $\mathbf{AdmRel}_{\mathbf{AP}_\perp}$.

Proposition 1. *The co-Eilenberg-Moore category* $\mathbf{AP}_\perp^!$ *for the lifting comonad* ! *on* $\mathbf{AP}_\perp$ *is equivalent to* $\mathbf{CCP}$*, and the co-Eilenberg-Moore category for* ! *on* $\mathbf{AdmRel}_{\mathbf{AP}_\perp}$ *is equivalent to* $\mathbf{AdmRel}_{\mathbf{CCP}}$.

Recall that the co-Eilenberg-Moore category for a comonad ! on a category $\mathbf{C}$ is the category whose objects are coalgebras for the monad (maps $\xi\colon \sigma \multimap !\sigma$ satisfying $\epsilon \circ \xi = id$ and $(!\xi) \circ \xi = \delta \circ \xi$, for ϵ, δ are counit and comultiplication) and whose morphisms are maps of coalgebras. Denoting by $\mathbf{C}^!$ the co-Eilenberg-Moore category, one may consider the Kleisli category for the induced monad L on $\mathbf{C}^!$, which we denote

$(\mathbf{C}^!)_L$. This category is isomorphic to the category having the same objects as $\mathbf{C}^!$, but as morphisms from $\xi\colon \sigma \multimap !\sigma$ to $\chi\colon \tau \multimap !\tau$ all morphisms of $\mathbf{C}$ from σ to τ.

For the remainder of this section $\mathbf{C}$ will denote a parametric PILL$_Y$-model. Since $(\mathbf{C}^!)_L$ is a Kleisli category, it is reasonable to think of it as a category of partial maps for $\mathbf{C}^!$. The next lemma shows how these two categories satisfy some of the properties needed for interpreting FPC in categories of partial maps as in Fiore's dissertation [12].

Lemma 1. *The category $\mathbf{C}^!$ is cartesian and has finite coproducts. The category $\mathbf{C}^!$ is partially cartesian closed in the sense that for any object ξ of $\mathbf{C}^!$, the composite of the product functor and the inclusion*

$$\mathbf{C}^! \xrightarrow{\xi\times(-)} \mathbf{C}^! \longrightarrow (\mathbf{C}^!)_L$$

has a right adjoint $\xi \rightharpoonup (-)\colon (\mathbf{C}^!)_L \to \mathbf{C}^!$.

Proof. The first half is well known, the proof can be found in for example [3, Lemma 9]. For $\xi\colon \sigma \multimap !\sigma$, the functor $\xi \rightharpoonup (-)$ maps $\chi\colon \tau \multimap !\tau$ to the free coalgebra $\delta\colon !(\sigma \multimap \tau) \multimap !!(\sigma \multimap \tau)$.

FPC can be interpreted in $(\mathbf{C}^!)_L$ basically as in [12]. A type with n free variables is interpreted as a map $(\mathbf{C}^!_0)^n \to \mathbf{C}^!_0$, with α_i interpreted as the i'th projection and $\times, + \to$ using respectively product, coproduct and partial cartesian structure. Recursive terms is modeled using the fixed point combinator in $\mathbf{C}$. What is different from Fiore's interpretation however, is that here recursive domain equations are solved using parametricity. The next definition defines the class of domain equations which can be solved in $\mathbf{C}^!$.

Definition 1. *A functor $\sigma_{coalg}\colon ((\mathbf{C}^!)^{\mathrm{op}} \times \mathbf{C}^!)^n \to \mathbf{C}^!$ is* induced by a type, *if there exists a strong functor $\sigma\colon ((\mathbf{C})^{\mathrm{op}} \times \mathbf{C})^n \to \mathbf{C}$ making the diagram*

$$\begin{array}{ccc} ((\mathbf{C}^!)^{\mathrm{op}} \times \mathbf{C}^!)^n & \xrightarrow{\sigma_{coalg}} & \mathbf{C}^! \\ \downarrow & & \downarrow \\ (\mathbf{C}^{\mathrm{op}} \times \mathbf{C})^n & \xrightarrow{\sigma} & \mathbf{C} \end{array}$$

commute, where the vertical functors are the obvious forgetful functors. We say that σ induces σ_{coalg}.

If σ induces σ_{coalg} as in Definition 1 above, then σ_{coalg} extends to a functor $(((\mathbf{C}^!)_L)^{\mathrm{op}} \times (C^!)_L)^n \to (C^!)_L$, whose action on morphisms is given by σ.

We show that all recursive domain equations on $(C^!)_L$ corresponding to functors induced by types as in Definition 1 can be solved. The precise formulation of this result is Theorem 1 below. The proof proceeds by first showing that $(C^!)_L$ is algebraically compact as in the next lemma, and then applying Freyd's solution to recursive domain equations in such categories.

Lemma 2. *If the functor $\sigma_{coalg}: \mathbf{C}^! \to \mathbf{C}^!$ is induced by a type σ, then it has an initial algebra. Including the initial algebra into $(\mathbf{C}^!)_L$ gives an initial algebra for the functor $(\mathbf{C}^!)_L \to (\mathbf{C}^!)_L$ induced by (σ, σ_{coalg}), and the inverse of the algebra is a final coalgebra.*

Proof. We just sketch the construction of the initial algebra. As a consequence of parametricity, σ has an initial algebra $in: \sigma(\mu\alpha.\, \sigma) \multimap \mu\alpha.\, \sigma$ whose inverse is a final coalgebra. The object $\mu\alpha.\, \sigma$ of $\mathbf{C}$ has a coalgebra structure for ! defined as the unique map ξ making the diagram

$$\begin{array}{ccccccc}
\sigma(\mu\alpha.\,\sigma) & & \xrightarrow{\quad\quad\cong\quad\quad} & & & & \mu\alpha.\,\sigma \\
\downarrow{\scriptstyle \sigma(\xi)} & & & & & & \downarrow{\scriptstyle \xi} \\
\sigma(!\mu\alpha.\,\sigma) & \xrightarrow{\sigma_{\mathrm{coalg}}(\delta)} & !\sigma(!\mu\alpha.\,\sigma) & \xrightarrow{!\sigma(\epsilon)} & !\sigma(\mu\alpha.\,\sigma) & \xrightarrow{\cong} & !\mu\alpha.\,\sigma
\end{array}$$

commute.

Theorem 1. *For any functor $\sigma_{coalg}: (\mathbf{C}^{!\mathrm{op}} \times \mathbf{C}^!)^{n+1} \to \mathbf{C}^!$ induced by a type, say σ, there exists a functor Fix $\sigma_{coalg}: ((\mathbf{C}^!)^{\mathrm{op}} \times \mathbf{C}^!)^n \to \mathbf{C}^!$ induced by a type Fix σ such that*

$$\sigma_{coalg} \circ \langle id_{((\mathbf{C}^!)^{\mathrm{op}} \times \mathbf{C}^!)^n}, Fix\, \breve{\sigma}_{coalg} \rangle \cong Fix\, \sigma_{coalg} \tag{2}$$

(where the notation $(\breve{-})$ is used as in Section 2.1). The corresponding functors on $(\mathbf{C}^!)_L$ also satisfy (2) and the dinaturality condition. Finally, there exists a general construction of Fix σ_{coalg} such that if $\tau_{coalg}: (\mathbf{C}^{!\mathrm{op}} \times \mathbf{C}^!)^m \to (\mathbf{C}^{!\mathrm{op}} \times \mathbf{C}^!)^n$ is any functor induced by a type, then

$$Fix(\sigma_{coalg} \circ (\tau_{coalg} \times id_{((\mathbf{C}^!)^{\mathrm{op}} \times \mathbf{C}^!)})) = (Fix\, \sigma_{coalg}) \circ \tau_{coalg}$$

Proof (Sketch). The recursive types are constructed as in Freyd's solution to recursive type equations as

$$\begin{aligned}
\omega(\alpha_1, \beta_1, \ldots, \alpha_n, \beta_n, \alpha) &= \mu\beta.\, \sigma(\alpha_1, \beta_1, \ldots, \alpha_n, \beta_n, \alpha, \beta) \\
\tau(\alpha_1, \beta_1, \ldots, \alpha_n, \beta_n) &= \mu\alpha.\, \sigma(\beta_1, \alpha_1, \ldots, \beta_n, \alpha_n, \omega(\alpha_1, \beta_1, \ldots, \alpha_n, \beta_n, \alpha), \alpha) \\
\mathrm{rec}\, \alpha.\, \sigma(\alpha_1, \beta_1, \ldots, \alpha_n, \beta_n, \alpha, \alpha) &= \omega(\alpha_1, \beta_1, \ldots, \alpha_n, \beta_n, \tau(\alpha_1, \beta_1, \ldots, \alpha_n, \beta_n)).
\end{aligned}$$

The dinaturality condition in $(\mathbf{C}^!)_L$ follows from the one in $\mathbf{C}$ since these have the same maps, and the last statement is an easy consequence of the construction.

Of course, to be able to use Theorem 1 for modeling recursive types, one must show that any FPC type $\alpha_1, \ldots, \alpha_n \vdash \sigma$ induces a functor $((\mathbf{C}^!)^{\mathrm{op}} \times \mathbf{C}^!)^n \to \mathbf{C}^!$ induced by a type, by splitting occurrences of free type variables into positive and negative occurrences. This is an easy induction on the structure of σ, and the case of recursive types is simply that Fix σ_{coalg} of Theorem 1 is induced by a type for all σ_{coalg}.

Theorem 2. *FPC can be modeled soundly in $(\mathbf{C}^!)_L$.*

5 Polymorphic FPC in the Per-model

The abstract analysis of Section 4 shows that our main example — the per-model — models recursive types. It also models polymorphism, and Figure 1 shows the interpretation of PolyFPC in the per-model, except the interpretation of recursive types, for which the categorical properties (dinaturality) are more useful than the concrete description as shown for instance in [22].

Types in the per-model of PolyFPC are modeled as pairs $(\llbracket\boldsymbol{\alpha} \vdash \sigma\rrbracket^p, \llbracket\boldsymbol{\alpha} \vdash \sigma\rrbracket^r)$, where $\llbracket\boldsymbol{\alpha} \vdash \sigma\rrbracket^p$ is a map $\mathbf{CCP}^n \to \mathbf{CCP}$ and $\llbracket\boldsymbol{\alpha} \vdash \sigma\rrbracket^r$ is a map taking an n-vector of admissible relations $(A_i\colon \mathbf{AdmRel}(R_i, S_i))$ (admissible in the sense of objects of $\mathbf{AdmRel}_{\mathbf{CCP}}$) on objects of $\mathbf{CCP}$ and produces an admissible relation

$$\llbracket\boldsymbol{\alpha} \vdash \sigma\rrbracket^r(\boldsymbol{A})\colon \mathbf{AdmRel}(\llbracket\boldsymbol{\alpha} \vdash \sigma\rrbracket^p(\boldsymbol{R}), \llbracket\boldsymbol{\alpha} \vdash \sigma\rrbracket^p(\boldsymbol{S}))$$

satisfying $\llbracket\boldsymbol{\alpha} \vdash \sigma\rrbracket^r(eq_{\boldsymbol{R}}) = eq_{\llbracket\boldsymbol{\alpha}\vdash\sigma\rrbracket^p(\boldsymbol{R})}$. In Figure 1 the symbols $1, 2$ denote two incomparable elements of D, and $\langle\cdot,\cdot\rangle$ denotes the pairing function $D \times D \to D$ definable using the combinatory algebra structure on D. The monad induced by the comonad on $\mathbf{AP}_\perp$ and $\mathbf{AdmRel}_{\mathbf{AP}_\perp}$ is denoted L. Explicitly, this monad maps a chain complete per R to $\{(\perp,\perp)\} \cup \{(\langle\iota, x\rangle, \langle\iota, y\rangle) \mid R(x,y)\}$ where ι denotes a code for the identity function on D, and an admissible relation A on chain complete pers R, S is mapped to the relation that on LR, LS that relates $[\perp]$ to $[\perp]$ and $[\langle\iota, x\rangle]$ to $[\langle\iota, y\rangle]$ if $A([x],[y])$.

$$\begin{aligned}
\llbracket\boldsymbol{\alpha} \vdash \alpha_i\rrbracket^p(\boldsymbol{R}) &= R_i\\
\llbracket\boldsymbol{\alpha} \vdash \sigma\times\tau\rrbracket^p(\boldsymbol{R}) &= \{(\langle x,y\rangle, \langle x',y'\rangle) \mid \llbracket\boldsymbol{\alpha} \vdash \sigma\rrbracket^p(\boldsymbol{R})(x,x') \wedge \llbracket\boldsymbol{\alpha} \vdash \tau\rrbracket^p(\boldsymbol{R})(y,y')\}\\
\llbracket\boldsymbol{\alpha} \vdash \sigma+\tau\rrbracket^p(\boldsymbol{R}) &= \{(\langle 1,x\rangle, \langle 1,x'\rangle) \mid \llbracket\boldsymbol{\alpha} \vdash \sigma\rrbracket^p(\boldsymbol{R})(x,x')\}\cup\\
&\quad\ \{(\langle 2,y\rangle, \langle 2,y'\rangle) \mid \llbracket\boldsymbol{\alpha} \vdash \tau\rrbracket^p(\boldsymbol{R})(y,y')\}\\
\llbracket\boldsymbol{\alpha} \vdash \sigma\to\tau\rrbracket^p(\boldsymbol{R}) &= \{(e,f) \mid \forall x,y\in D.\, \llbracket\boldsymbol{\alpha} \vdash \sigma\rrbracket^p(\boldsymbol{R})(x,y) \Rightarrow L\llbracket\boldsymbol{\alpha} \vdash \tau\rrbracket^p(\boldsymbol{R})(e\cdot x, f\cdot y)\}\\
\llbracket\boldsymbol{\alpha} \vdash 1\rrbracket^p(\boldsymbol{R}) &= \{(\perp,\perp)\}\\
\llbracket\boldsymbol{\alpha} \vdash \textstyle\prod\alpha.\,\sigma\rrbracket^p(\boldsymbol{R}) &= \{(x,y) \mid \forall S\colon \mathbf{CCP}_0.\, L\llbracket\boldsymbol{\alpha},\alpha \vdash \sigma\rrbracket^p(\boldsymbol{R},S)(x,y)\wedge\\
&\quad\ \forall S,S'\colon \mathbf{CCP}_0.\, \forall A\colon \mathrm{AdmRel}(S,S').\, L\llbracket\boldsymbol{\alpha},\alpha \vdash \sigma\rrbracket^r(eq_{\boldsymbol{R}}, A)([x],[y])\}
\end{aligned}$$

$$\begin{aligned}
\llbracket\boldsymbol{\alpha} \vdash \alpha_i\rrbracket^r(\boldsymbol{A}) &= A_i\\
\llbracket\boldsymbol{\alpha} \vdash \sigma\times\tau\rrbracket^r(\boldsymbol{A}) &= \{([\langle x,y\rangle], [\langle x',y'\rangle]) \mid \llbracket\boldsymbol{\alpha} \vdash \sigma\rrbracket^r(\boldsymbol{A})([x],[x']) \wedge \llbracket\boldsymbol{\alpha} \vdash \tau\rrbracket^r(\boldsymbol{A})([y],[y'])\}\\
\llbracket\boldsymbol{\alpha} \vdash \sigma+\tau\rrbracket^r(\boldsymbol{A}) &= \{([\langle 1,x\rangle], [\langle 1,x'\rangle]) \mid \llbracket\boldsymbol{\alpha} \vdash \sigma\rrbracket^r(\boldsymbol{A})([x],[x'])\}\cup\\
&\quad\ \{([\langle 2,y\rangle], [\langle 2,y'\rangle]) \mid \llbracket\boldsymbol{\alpha} \vdash \tau\rrbracket^r(\boldsymbol{A})([y],[y'])\}\\
\llbracket\boldsymbol{\alpha} \vdash \sigma\to\tau\rrbracket^r(\boldsymbol{A}) &= \{([e],[f]) \mid \forall([x],[y]) \in \llbracket\boldsymbol{\alpha} \vdash \sigma\rrbracket^r(\boldsymbol{A}).\, ([e\cdot x],[f\cdot y]) \in L\llbracket\boldsymbol{\alpha} \vdash \tau\rrbracket^r(\boldsymbol{A})\}\\
\llbracket\boldsymbol{\alpha} \vdash 1\rrbracket^r(\boldsymbol{A}) &= \{([\perp],[\perp])\}\\
\llbracket\boldsymbol{\alpha} \vdash \textstyle\prod\alpha.\,\sigma\rrbracket^r(\boldsymbol{A}) &= \{([x],[y]) \mid \forall S,S'\colon \mathbf{CCP}_0.\, \forall A\colon \mathrm{AdmRel}(S,S').\, L\llbracket\boldsymbol{\alpha},\alpha \vdash \sigma\rrbracket^r(\boldsymbol{A},A)([x],[y])\}
\end{aligned}$$

Fig. 1. Interpretation of PolyFPC in per-model

Terms of PolyFPC are modeled in the Kleisli category for L. To be more precise, a term $\boldsymbol{\alpha} \mid \boldsymbol{x}\colon \boldsymbol{\sigma} \vdash t\colon \tau$ is modeled as an indexed family of maps

$$(\llbracket t\rrbracket_{\boldsymbol{R}}\colon \textstyle\prod_i \llbracket\boldsymbol{\alpha} \vdash \sigma_i\rrbracket^p(\boldsymbol{R}) \to L\llbracket\boldsymbol{\alpha} \vdash \tau\rrbracket^p(\boldsymbol{R}))_{\boldsymbol{R}}$$

where the product refers to the product in $\mathbf{CCP}$. Such a family must have a common tracker, and must preserve relations, which means that if $\boldsymbol{A}\colon \mathbf{AdmRel}(\boldsymbol{R}, \boldsymbol{S})$, and for each i, $([x_i], [y_i]) \in [\![\boldsymbol{\alpha} \vdash \sigma_i]\!]^r(\boldsymbol{A})$, then

$$([\![t]\!]_{\boldsymbol{R}}([x_1], \ldots, [x_m]), [\![t]\!]_{\boldsymbol{S}}([y_1], \ldots, [y_m])) \in L[\![\boldsymbol{\alpha} \vdash \tau]\!]^r(\boldsymbol{A}).$$

Theorem 3. *The interpretation of PolyFPC types defined in Figure 1 extends to a sound interpretation of PolyFPC.*

5.1 Computational Adequacy

A τ - σ context of PolyFPC for types σ, τ, where τ is closed, is an expression C containing a place holder $-_\tau$ such that whenever an expression t of type τ is substituted for the place holder such that the result $C[t]$ is a closed term, it has type σ. Two terms $t, t'\colon \tau$ of PolyFPC of the same type are called contextually equivalent (written $t \equiv t'$), if for any type σ, and any τ - σ context C,

$$C[t] \Downarrow \text{ iff } C[t'] \Downarrow$$

where $t \Downarrow$ means: There exists a v such that $t \Downarrow v$.

Theorem 4 (Adequacy). *For any program t of PolyFPC, $[\![t]\!] \neq [\bot]$ iff $t \Downarrow$.*

From Theorem 4 the following corollary giving a tight connection between the operational and denotational semantics is easily provable.

Corollary 1. *Suppose t, t' are two PolyFPC terms of the same type. If $[\![t]\!] = [\![t']\!]$ then $t \equiv t'$.*

6 Reasoning Using the Model

The per-model of PolyFPC is parametric by construction, since the interpretations of types have a build-in relational interpretation ($[\![\boldsymbol{\alpha} \vdash \sigma]\!]^r$) satisfying identity extension. This means that the model can be used to verify parametricity arguments about PolyFPC programs. For example, for the usual data abstraction arguments as in [27, 21] proving that two implementations of a data type gives the same final program, one can prove using parametricity of the model, that the two programs denotations are equal, and then use Corollary 1 to prove that the programs are ground contextually equivalent.

In future work, it will be interesting to lift the parametricity of the model to a logic on PolyFPC. Corollary 1 should verify the logic in the sense that two terms that are provably equal in the logic should be ground contextually equivalent. Since the logic reasons about partial functions, it needs to include a termination proposition $(-)\downarrow$. The mix of parametricity and partiality will have the following consequences on the logic.

- Only total functions will have graphs that can be used to instantiate the parametricity principle.
- The relational interpretation of the $\to$ type constructor will relate f to g in $R \to S$ for relations R and S iff $f\downarrow \iff g\downarrow$, and further for all $(x, y) \in R$, $f(x)\downarrow \iff g(y)\downarrow$ and $f(x)\downarrow$ implies $S(f(x), g(y))$.

– The parametricity principle in the logic will say that two terms e, f of, say closed type $\prod\alpha.\,\sigma$, are ground contextually equivalent iff $e\downarrow \iff f\downarrow$ and further, for all pairs of types τ, τ' and any relation R between them $e(\tau)\downarrow \iff f(\tau')\downarrow$ and $e(\tau)\downarrow$ implies $(e(\tau), f(\tau')) \in \sigma[R]$.

Related results can be found in [17], and the interpretation of $\rightarrow$ above is a symmetric version of the one in *loc. cit.*

7 Conclusions

By showing that the solutions to recursive domain equations in the linear part of the calculus PILL_Y can be used to interpret recursive types in languages with no linearity, we have shown that parametric PILL_Y is a useful axiomatic setup for domain theory. The parametric model of PolyFPC constructed by applying the general theory to the case of admissible pers can be used to reason about parametricity for PolyFPC and for example give proofs of modularity properties along the lines of [21], but this time using the denotational semantics.

In recent work, Birkedal, Møgelberg, Petersen and Varming [11] have shown how the programming language lily of Bierman, Pitts and Russo [4] gives rise to a parametric model of PILL_Y. Using this result, the techniques developed here should show how FPC can be translated into lily, but it would be interesting to see if full PolyFPC can be translated into it.

Acknowledgments. The paper contains ideas and creative input from the following people: Lars Birkedal, Eugenio Moggi, Rasmus Lerchedahl Petersen, Pino Rosolini and Alex Simpson.

References

1. M. Abadi and G.D. Plotkin. A per model of polymorphism and recursive types. In *5th Annual IEEE Symposium on Logic in Computer Science*, pages 355–365. IEEE Computer Society Press, 1990.
2. A. Barber. *Linear Type Theories, Semantics and Action Calculi.* PhD thesis, Edinburgh University, 1997.
3. P.N. Benton. A mixed linear and non-linear logic: Proofs, terms and models (preliminary report). Technical report, University of Cambridge, 1995.
4. G. M. Bierman, A. M. Pitts, and C. V. Russo. Operational properties of Lily, a polymorphic linear lambda calculus with recursion. In *Fourth International Workshop on Higher Order Operational Techniques in Semantics, Montréal*, volume 41 of *Electronic Notes in Theoretical Computer Science*. Elsevier, September 2000.
5. L. Birkedal and R. E. Møgelberg. Categorical models of Abadi-Plotkin's logic for parametricity. *Mathematical Structures in Computer Science*, 15(4):709–772, 2005.
6. L. Birkedal, R. E. Møgelberg, and R. L. Petersen. Category theoretic models of linear Abadi & Plotkin logic. Submitted.
7. L. Birkedal, R. E. Møgelberg, and R. L. Petersen. Domain theoretic models of parametric polymorphism. Submitted.
8. L. Birkedal, R. E. Møgelberg, and R. L. Petersen. Linear Abadi & Plotkin logic. Submitted.

9. L. Birkedal, R. E. Møgelberg, and R. L. Petersen. Parametric domain-theoretic models of linear Abadi & Plotkin logic. 2005. Submitted.
10. L. Birkedal, R. E. Møgelberg, and R. L. Petersen. Parametric domain-theoretic models of polymorphic intuitionistic / linear lambda calculus. In *Proceedings of the Twenty-first Conference on the Mathematical Foundations of Programming Semantics*, 2005. To appear.
11. L. Birkedal, R.L. Petersen, R.E. Møgelberg, and C. Varming. Operational semantics and models of linear Abadi-Plotkin logic. Manuscript.
12. M. Fiore. *Axiomatic Domain Theory in Categories of Partial Maps*. Distinguished Dissertations in Computer Science. Cambridge University Press, 1996.
13. P.J. Freyd. Algebraically complete categories. In A. Carboni, M. C. Pedicchio, and G. Rosolini, editors, *Category Theory. Proceedings, Como 1990*, volume 1488 of *Lecture Notes in Mathematics*, pages 95–104. Springer-Verlag, 1990.
14. P.J. Freyd. Recursive types reduced to inductive types. In *Proceedings of the fifth IEEE Conference on Logic in Computer Science*, pages 498–507, 1990.
15. P.J. Freyd. Remarks on algebraically compact categories. In M. P. Fourman, P.T. Johnstone, and A. M. Pitts, editors, *Applications of Categories in Computer Science. Proceedings of the LMS Symposium, Durham 1991*, volume 177 of *London Mathematical Society Lecture Note Series*, pages 95–106. Cambridge University Press, 1991.
16. Jean-Yves Girard. Linear logic. *Theoretical Computer Science*, 50:1–102, 1987.
17. Patricia Johann and Janis Voigtländer. Free theorems in the presence of *seq*. In *Proc. of 31st ACM SIGPLAN-SIGACT Symp. on Principles of Programming Languages, POPL 2004, Venice, Italy, 14–16 Jan. 2004*, pages 99–110. ACM Press, New York, 2004.
18. Paola Maneggia. *Models of Linear Polymorphism*. PhD thesis, University of Birmingham, Feb. 2004.
19. Maraist, Odersky, Turner, and Wadler. Call-by-name, call-by-value, call-by-need and the linear lambda calculus. *TCS: Theoretical Computer Science*, 228:175–210, 1999.
20. R. E. Møgelberg. *Categorical and domain theoretic models of parametric polymorphism*. PhD thesis, IT University of Copenhagen, 2005.
21. A. M. Pitts. Typed operational reasoning. In B. C. Pierce, editor, *Advanced Topics in Types and Programming Languages*, chapter 7, pages 245–289. The MIT Press, 2005.
22. A.M. Pitts. Relational properties of domains. *Information and Computation*, 127:66–90, 1996.
23. G. D. Plotkin. Type theory and recursion (extended abstract). In *Proceedings, Eighth Annual IEEE Symposium on Logic in Computer Science*, page 374, Montreal, Canada, 19–23 June 1993. IEEE Computer Society Press.
24. G.D. Plotkin. Lectures on predomains and partial functions. Notes for a course given at the Center for the Study of Language and Information, Stanford, 1985.
25. G.D. Plotkin. Second order type theory and recursion. Notes for a talk at the Scott Fest, February 1993.
26. Gordon Plotkin and Martín Abadi. A logic for parametric polymorphism. In *Typed lambda calculi and applications (Utrecht, 1993)*, volume 664 of *Lecture Notes in Comput. Sci.*, pages 361–375. Springer, Berlin, 1993.
27. J.C. Reynolds. Types, abstraction, and parametric polymorphism. *Information Processing*, 83:513–523, 1983.
28. Stephen Tse and Steve Zdancewic. Translating dependency into parametricity. *j-SIGPLAN*, 39(9):115–125, September 2004.

Typed GoI for Exponentials

Esfandiar Haghverdi

School of Informatics, Indiana University
Bloomington, IN 47406 USA
ehaghver@indiana.edu
http://xavier.informatics.indiana.edu/~ehaghver

Abstract. In a recent paper we introduced a typed version of Geometry of Interaction, called the Multi-object Geometry of Interaction (MGoI). Using this framework we gave an interpretation for the unit-free multiplicative fragment of linear logic. In this paper, we extend our work to cover the exponentials. We introduce the notion of a *GoI Category* that embodies the necessary ingredients for an MGoI interpretation for unit-free multiplicative and exponential linear logic.

1 Introduction

Geometry of Interaction (GoI) was introduced by Girard in a series of influential papers [6, 7, 8]. This interpretation aims at providing a mathematical model for the cut-elimination process in linear logic. Girard, in [7], gave the first implementation of GoI for system $\mathcal{F}$, through a translation into second order multiplicative and exponential linear logic. For more on the history of the progress made after the inception of the initial ideas see [10] Chapter 5, and [11], and for the most recent advances see the excellent manuscript [9]. The categorical foundations and formulation of GoI started in the unpublished work of M. Hyland, and the work of S. Abramsky and R. Jagadeesan reported in [4]. This approach further led to Abramsky's Program which was completed in [10], see also [3]. It was in this latter work that the notions of a GoI Situation and a reflexive object came to be defined. In a recent paper, [12], we move away from "uni-object GoI" to a typed version that we call multi-object GoI or MGoI for short. The Multi-object GoI (MGoI) interpretation does not require the existence of a reflexive object (i.e., an object with certain retractions, e.g. $U \otimes U \lhd U$, etc), and instead keeps the types as they are suitably interpreted in the underlying category.

This permits a generalization of GoI and axiomatization of its essential features. For example, by removing reflexive objects U, we also unlock the possibilities of generalizing Girard-style GoI to more general tensor categories including cases where the tensor is "product-like" in addition to "sum-like". In particular, in [12], we introduce an axiomatization for partially traced symmetric monoidal categories and give an MGoI interpretation for the multiplicative fragment of linear logic (MLL) without units.

The case of exponentials was not treated in that paper, and that is what we undertake to do in this paper. The main contributions of this paper can be summarized as follows:

M. Bugliesi et al. (Eds.): ICALP 2006, Part II, LNCS 4052, pp. 384–395, 2006.

- We introduce a compatibility notion for the abstract orthogonality relation of [12] in the presence of exponential data, that is the endofunctor T.
- We give an MGoI interpretation for MELL without units, thus extending the work in [12] beyond multiplicatives. This interpretation uses a structure that we define and call a GoI category. A GoI category embodies the necessary structure needed for the categorical multi-object GoI interpretation of MELL. Moreover, we show that interpretations of MELL proofs are partial symmetries in a GoI Category.
- We prove the soundness theorem for our MGoI interpretation.

It seems clear that a treatment of units and additives demands new ideas beyond those that have emerged so far and been crystalized in a GoI category. We shall pursue this direction in future work.

The rest of the paper is organized as follows. In Section 2 we recall the definition of a monoidal $*$-category from [2] with a slight change, we define the notions of Hermitian and partial isometry morphisms in such categories. In Section 3 we recall the notion of partially traced symmetric monoidal categories introduced in [12] and discuss some examples that are new. In Section 4 we recall the definition of abstract orthogonality relation in a partially traced symmetric monoidal category, first introduced in [12]. Furthermore, we introduce the compatibility conditions for such a relation in the presence of exponential data. In Section 5, we recall the MGoI semantics for MLL from [12], and extend it to MELL. We also discuss the execution formula in this section. Section 6 discusses the soundness theorem. Finally, Section 7 contains some thoughts about possible future directions.

2 Monoidal $*$-Categories

In the following we shall recall the definition of monoidal $*$-categories from [2]. Note that we do not require a conjugation functor for the definition.

Definition 1. A monoidal $*$-category $\mathbb{C}$ is a monoidal category with a strict symmetric monoidal functor $(\,)^* : \mathbb{C}^{op} \to \mathbb{C}$ which is strictly involutive and the identity on objects and commutes with the monoidal product, that is, $(f \otimes g)^* = f^* \otimes g^*$.

We say that a morphism $f : A \to A$ is *Hermitian* if $f^* = f$. Also a morphism $f : A \to B$ is called a *partial isometry* if $f^*ff^* = f^*$ and $ff^*f = f$, and a *partial symmetry* if in addition it is Hermitian. That is, if $f^* = f$ and $f^3 = f$. Note that there is no underlying Hilbert space structure on the homsets of $\mathbb{C}$, the terminology here is borrowed from operator algebras to account for the similar properties of such morphisms, which can be expressed in a more general setting of $*$-categories.

An obvious example is the category $\mathbf{Hilb}_{\otimes}$ of Hilbert spaces and bounded linear maps with tensor product of Hilbert spaces as the monoidal product. Given $f : H \to K$, $f^* : K \to H$ is given by the adjoint of f, defined uniquely by $\langle f(x), y\rangle = \langle x, f^*(y)\rangle$. It is easy to see that all the required properties are satisfied. Note that the category $\mathbf{Hilb}_\oplus$ of Hilbert spaces and bounded linear maps but with direct sum as the monoidal product is a $*$-category too, with the same definition for the $()^*$ functor.

Another example is the category $\mathbf{Rel}_\times$ of sets and relations with the cartesian product of sets as the monoidal product. Given $f : X \to Y$, $f^* = \overline{f}$ where $\overline{f}$ is the converse relation. Again, note that the category $\mathbf{Rel}_\oplus$ of sets and relations with monoidal product, the disjoint union (categorical biproduct) is a monoidal $*$-category too, with the same definition for the $()^*$ functor.

Yet another example that shows up frequently in the context of GoI is the category $\mathbf{PInj}_\uplus$ of sets and partial injective maps, with disjoint union as the monoidal product. Given $f : X \to Y$, $f^* = f^{-1}$.

Other examples include $\mathbf{Hilb}_{fd}$ of finite dimensional Hilbert spaces and bounded linear maps, $\mathbf{URep}(G)$, finite representations of a compact group G, etc. For more details, examples and the ways that such categories show up in logic, see [2].

3 Trace Class

The notion of categorical trace was introduced by Joyal, Street and Verity in an influential paper [14]. The motivation for their work arose in algebraic topology and knot theory, although the authors were aware that such traces also have many applications in Computer Science, where they include such notions as feedback, fixedpoints, iteration theories, etc. For references and history, see [1, 3, 11]. In [12], we introduced the notion of partial trace that we shall recall in this section. For details on other approaches to partial trace, examples, and comparison to our definition, see [12].

Recall, following Joyal, Street, and Verity [14], a trace in a symmetric monoidal category $(\mathbb{C}, \otimes, I, s)$ is a family of maps

$$Tr^U_{X,Y} : \mathbb{C}(X \otimes U, Y \otimes U) \to \mathbb{C}(X, Y),$$

satisfying various well-known naturality equations. A *partial* trace requires instead that each $Tr^U_{X,Y}$ be a partial map (with domain denoted $\mathbb{T}^U_{X,Y}$) and satisfy various closure conditions.

Definition 2 (Trace Class). Let $(\mathbb{C}, \otimes, I, s)$ be a symmetric monoidal category. A *trace class* in $\mathbb{C}$ is a choice of a family of subsets, for each object U of $\mathbb{C}$, of the form

$$\mathbb{T}^U_{X,Y} \subseteq \mathbb{C}(X \otimes U, Y \otimes U) \text{ for all objects } X, Y \text{ of } \mathbb{C}$$

together with a family of functions, called a *partial trace*, of the form

$$Tr^U_{X,Y} : \mathbb{T}^U_{X,Y} \to \mathbb{C}(X, Y)$$

subject to the following axioms. A morphism $f \in \mathbb{T}^U_{X,Y}$, by abuse of terminology, is said to be *trace class*.

– **Naturality** in X and Y: For any $f \in \mathbb{T}^U_{X,Y}$ and $g : X' \to X$ and $h : Y \to Y'$,

$$(h \otimes 1_U) f (g \otimes 1_U) \in \mathbb{T}^U_{X',Y'},$$

and $$Tr^U_{X',Y'}((h \otimes 1_U) f (g \otimes 1_U)) = h\, Tr^U_{X,Y}(f)\, g.$$

– **Dinaturality** in U: For any $f : X \otimes U \to Y \otimes U'$, $g : U' \to U$,

$$(1_Y \otimes g) f \in \mathbb{T}^U_{X,Y} \text{ iff } f(1_X \otimes g) \in \mathbb{T}^{U'}_{X,Y},$$

and $$Tr^U_{X,Y}((1_Y \otimes g) f) = Tr^{U'}_{X,Y}(f(1_X \otimes g)).$$

– **Vanishing I:** $\mathbb{T}^I_{X,Y} = \mathbb{C}(X \otimes I, Y \otimes I)$, and for $f \in \mathbb{T}^I_{X,Y}$

$$Tr^I_{X,Y}(f) = \rho_Y f \rho_X^{-1}.$$

Here $\rho_A : A \otimes I \to A$ is the right unit isomorphism of the monoidal category.

– **Vanishing II:** For any $g : X \otimes U \otimes V \to Y \otimes U \otimes V$, if $g \in \mathbb{T}^V_{X \otimes U, Y \otimes U}$, then

$$g \in \mathbb{T}^{U \otimes V}_{X,Y} \text{ iff } Tr^V_{X \otimes U, Y \otimes U}(g) \in \mathbb{T}^U_{X,Y},$$

and $$Tr^{U \otimes V}_{X,Y}(g) = Tr^U_{X,Y}(Tr^V_{X \otimes U, Y \otimes U}(g)).$$

– **Superposing:** For any $f \in \mathbb{T}^U_{X,Y}$ and $g : W \to Z$,

$$g \otimes f \in \mathbb{T}^U_{W \otimes X, Z \otimes Y},$$

and $$Tr^U_{W \otimes X, Z \otimes Y}(g \otimes f) = g \otimes Tr^U_{X,Y}(f).$$

– **Yanking:** $s_{UU} \in \mathbb{T}^U_{U,U}$, and $Tr^U_{U,U}(s_{U,U}) = 1_U$.

A symmetric monoidal category $(\mathbb{C}, \otimes, I, s)$ with such a trace class is called a *partially traced category*, or a *category with a trace class.*

Clearly, all (totally-defined) traces in the usual definition of a traced monoidal category yield a trace class. Here are some more examples of partially traced categories. For more detail on some of these examples see [12].

Example 1. (a) **Finite Dimensional Vector Spaces**
The category $\mathbf{Vec}_{fd}$ of finite dimensional vector spaces and linear transformations is a symmetric monoidal, indeed an additive, category (see [15]), with monoidal product taken to be $\oplus$, the direct sum (biproduct). Hence, given $f : \oplus_I X_i \to \oplus_J Y_j$ with $|I| = n$ and $|J| = m$, we can write f as an $m \times n$ matrix $f = [f_{ij}]$ of its components, where $f_{ij} : X_j \to Y_i$ (notice the switch in the indices i and j). We give a trace class structure on the category $(\mathbf{Vec}_{fd}, \oplus, \mathbf{0})$ as follows. We shall say an $f : X \oplus U \to Y \oplus U$ is *trace class* iff $(Id - f_{22})$ is invertible, where Id is the identity matrix, and Id and f_{22} have size $dim(U)$. In that case, we write

$$Tr^U_{X,Y}(f) = f_{11} + f_{12}(Id - f_{22})^{-1} f_{21} \tag{1}$$

This definition is motivated by a generalization of the fact that for a matrix A, $(Id - A)^{-1} = \sum_i A^i$, whenever the infinite sum converges. If the infinite sum for $(Id - f_{22})^{-1}$ exists, the above formula for $Tr^U_{X,Y}(f)$ becomes the usual "particle-style" trace in [1, 3, 11].

One can also get the analogous result for the category $(\mathbf{Hilb}_{fd}, \oplus)$ of finite-dimensional Hilbert spaces and bounded linear maps. Note also that neither $\mathbf{Vec}_{fd}$ nor $\mathbf{Hilb}_{fd}$ admits nontrivial reflexive objects.

The following example is *new* and does not occur in [12]. It is the first *infinite* dimensional partially traced example.

(b) **Hilbert Spaces**

It can be shown (details appear in the long version of this paper) that the category $(\mathbf{Hilb}, \oplus)$ of (not necessarily finite dimensional) Hilbert spaces and bounded linear maps is also partially traced with the same definition for trace class as above.

(c) **Metric Spaces.** The category **CMet** of complete metric spaces with non-expansive maps has products. We define the trace class structure on **CMet** (where $\otimes = \times$) as follows. We say that a morphism $f : X \times U \to Y \times U$ is in $\mathbb{T}^U_{X,Y}$ iff for every $x \in X$ the induced map $\pi_2 \lambda u.f(x,u) : U \to U$ has a unique fixed point and in that case we define $Tr^U_{X,Y}(f) : X \to Y$ by $Tr^U_{X,Y}(f)(x) = y$, where $f(x,u) = (y,u)$, for the unique u. The category $(\mathbf{Sets}, \times)$ of sets and mappings is partially traced category with the same definition for trace class morphisms as in **CMet**. However, this fails for the category $(\mathbf{Rel}, \times)$ of sets and relations.

4 Orthogonality Relations and GoI Categories

Girard originally introduced orthogonality relations into linear logic to model formulas (or types) as sets equal to their biorthogonal (e.g. in the phase semantics of the original paper [5] and in GoI 1 [6]). Recently M. Hyland and A. Schalk gave an abstract approach to orthogonality relations in symmetric monoidal closed categories [13]. They also point out that an orthogonality on a traced symmetric monoidal category $\mathbb{C}$ can be obtained by first considering their axioms applied to $Int(\mathbb{C})$, the compact closure of $\mathbb{C}$, and then translating them down to $\mathbb{C}$. In [12] we gave this translation (not explicitly calculated in [13]), using the so-called "GoI construction" $\mathcal{G}(\mathbb{C})$ [1, 10] instead of $Int(\mathbb{C})$. The categories $\mathcal{G}(\mathbb{C})$ and $Int(\mathbb{C})$ are both compact closures of $\mathbb{C}$, and are shown to be isomorphic in [10].

As we are dealing with partial traces we need to take extra care in stating the axioms below; namely, an axiom involving a trace should be read with the proviso: "whenever all traces exist". Finally hereafter, without loss of generality and for readability we consider strict monoidal categories. We recall the definition of a strong orthogonality relation from [12] and define the necessary compatibility conditions in the presence of a functor $T : \mathbb{C} \to \mathbb{C}$ with additional monoidal retractions. For more details and discussion on abstract orthogonality relations see [13, 12].

Definition 3. Let $\mathbb{C}$ be a partially traced symmetric monoidal category. A *strong orthogonality relation* on $\mathbb{C}$ is a family of relations $\perp_{UV}$ between maps $u : V \to U$ and $x : U \to V$

$$V \xrightarrow{u} U \perp_{UV} U \xrightarrow{x} V$$

subject to the following axioms:

(i) *Isomorphism* : Let $f : U \otimes V' \to V \otimes U'$ and $\hat{f} : U' \otimes V \to V' \otimes U$ be such that $Tr^{V'}(Tr^{U'}((1 \otimes 1 \otimes s_{U',V'})\alpha^{-1}(f \otimes \hat{f})\alpha)) = s_{U,V}$ and $Tr^{V}(Tr^{U}((1 \otimes 1 \otimes s_{U,V})\alpha^{-1}(\hat{f} \otimes f)\alpha)) = s_{U',V'}$. Here $\alpha = (1 \otimes 1 \otimes s)(1 \otimes s \otimes 1)$ with s at appropriate types. Note that this simply means that $f : (U,V) \to (U',V')$ and $\hat{f} : (U',V') \to (U,V)$ are inverses of each other in $\mathcal{G}(\mathbb{C})$.
Then, for all $u : V \to U$ and $x : U \to V$,

$$u \perp_{UV} x \text{ iff } Tr^{U}_{V',U'}(s_{U,U'}(u \otimes 1_{U'})fs_{V',U}) \perp_{U'V'} Tr^{V}_{U',V'}((1_{V'} \otimes x)\hat{f});$$

that is, orthogonality is invariant under isomorphism.

(ii) *Strong Tensor* : For all $u : V \to U$, $v : V' \to U'$ and $h : U \otimes U' \to V \otimes V'$,

$$v \perp_{U'V'} Tr^{U}_{U',V'}(s_{U,V'}(u \otimes 1_{V'})hs_{U',U}) \quad \text{iff} \quad (u \otimes v) \perp_{U \otimes U', V \otimes V'} h,$$

(iii) *Identity* : For all $u : V \to U$ and $x : U \to V$,

$$u \perp_{UV} x \text{ implies } 1_I \perp_{II} Tr^{V}_{I,I}(xu).$$

(iv) *Symmetry* : For all $u : V \to U$ and $x : U \to V$,

$$u \perp_{UV} x \text{ iff } x \perp_{VU} u.$$

Definition 4. A *GoI category* is a triple $(\mathbb{C}, T, \perp)$ where $\mathbb{C}$ is a partially traced $*$-category as in Section 3, $T = (T, \psi, \psi_I) : \mathbb{C} \to \mathbb{C}$ is a traced symmetric monoidal functor, that is if $f \in \mathbb{T}^{U}_{X,Y}$, then $\psi^{-1}_{Y,U}T(f)\psi_{X,U} \in \mathbb{T}^{TU}_{TX,TY}$ and $Tr^{TU}_{TX,TY}(\psi^{-1}_{Y,U}T(f)\psi_{X,U}) = T(Tr^{U}_{X,Y}(f))$. And $\perp$ is an orthogonality relation on $\mathbb{C}$ as in the above. Furthermore, we require that,

• The following natural retractions exist:

- $\mathcal{K}_I \lhd T\ (w, w^*)$, $\mathcal{K}_I$ denotes the constant I functor.
- $Id \lhd T\ (d, d^*)$
- $T^2 \lhd T\ (e, e^*)$
- $T \otimes T \lhd T\ (c, c^*)$

Above, $T \otimes T \lhd T(c, c^*)$ means that there is a natural transformation $c : T \otimes T \to T$ such that $c^*c = 1$. Similarly for other cases.

• The orthogonality relation be *GoI compatible*, that is, it satisfy the following additional axioms:

(c1) For all $f : V \to U$, $g : U \to V$,

$$f \perp_{U,V} g \text{ implies } d_U f d^*_V \perp_{TU,TV} Tg.$$

(c2) For all $f : U \to U$ and $g : I \to I$,

$$w_U g w_U^* \perp_{TU,TU} Tf.$$

(c3) For all $f : TV \otimes TV \to TU \otimes TU$ and $g : U \to V$,

$$f \perp_{TU \otimes TU, TV \otimes TV} \; Tg \otimes Tg \text{ implies } c_U f c_V^* \perp_{TU,TV} Tg.$$

• The functor T commute with $()^*$, that is $(T(f))^* = T(f^*)$. Moreover, $\psi^* = \psi^{-1}$ and $\psi_I^* = \psi_I^{-1}$.

GoI categories are the main mathematical structures in our semantic interpretation in the following section. Here are a few examples of GoI categories.

Example 2. (a) $(\mathbf{PInj}, \uplus, T, \perp)$ where **PInj** is the category of sets and partial injective functions, monoidal product is the disjoint union with unit the empty set. $T(X) = \mathbb{N} \times X$, where $\mathbb{N}$ is the set of natural numbers, for natural retractions see [10]. $f \perp g$ iff gf is nilpotent.

(b) $(\mathbf{Hilb}, \oplus, T, \perp)$, where **Hilb** is the category of Hilbert spaces and bounded linear maps. The monoidal product is the direct sum of Hilbert spaces. $T(H) = \ell^2 \otimes H$ where ℓ^2 is the space of square summable sequences, for natural retractions see [10]. We define, $f \perp g$ iff $(1 - gf)$ is an invertible linear transformation.

(c) $(\mathbf{Rel}, \oplus, T, \perp)$ using the same definitions for T and $\perp$ as in the case of **PInj**. Note that disjoint union, denoted $\oplus$ is in fact the categorical biproduct in **Rel**.

In [12], we defined an orthogonality relation using the notion of trace class. We recall this definition below and show that it is GoI compatible.

Example 3. Let $(\mathbb{C}, \otimes, I, Tr)$ be a partially traced category where $\otimes$ is the monoidal product with unit I, and Tr is the partial trace operator as in Section 3. Let A and B be objects of $\mathbb{C}$. For $f : A \to B$ and $g : B \to A$, we define, $f \perp_{BA} g$ iff $gf \in \mathbb{T}_{I,I}^A$. It turns out that this is a variation of the notion of *Focussed orthogonality* of Hyland and Schalk [13].

Proposition 1. *Suppose $\mathbb{C}$ is a partially traced *-category that is in addition equipped with an endofunctor T and monoidal retractions as in Definition 4. Then, the orthogonality relation $\perp$ defined as in Example 3 above is GoI compatible.*

5 Multi-object GoI Interpretation

The Multi-object Geometry of Interaction (MGoI) was introduced in [12] and was used to interpret MLL without units. The main idea was to keep the types of the formulas that were defined by a denotational semantic map, during the GoI interpretation. For the multiplicative case this also implied that, in contrast to the usual GoI, there was no need for a reflexive object and this made the interpretation possible in categories like finite dimensional vector spaces. In this section, we generalize MGoI interpretation to cover the exponentials. We will soon observe that infinity forces itself into the framework, it is no longer possible to carry on MGoI interpretation in finite dimension. This transition to infinity occurs, for example when we are forced to admit a retraction $TTA \lhd TA$ for any object A in the relevant category. Note that, although in this way reflexive objects reappear, they are not used to collapse types as in uni-object GoI.

Interpreting the formulas: The interpretation of MLL formulas already appears in [12], however we include them here for the sake of completeness. Given a GoI category $(\mathbb{C}, T, \perp)$, let A be an object of $\mathbb{C}$ and let $f, g \in End(A)$. We say that f is *orthogonal to* g, denoted $f \perp g$, if $(f, g) \in \perp$. Also given $X \subseteq End(A)$ we define

$$X^{\perp} = \{f \in End(A) \,|\, \forall g \in X, f \perp g\}.$$

We now define an operator on the objects of $\mathbb{C}$ as follows: Given an object A, $\mathcal{T}(A) = \{X \subseteq End(A) \,|\, X^{\perp\perp} = X\}$. We shall also need the notion of a denotational interpretation of formulas. We define an interpretation map $[\![-]\!]$ on the formulas of MELL as follows. Given the value of $[\![-]\!]$ on the atomic propositions as objects of $\mathbb{C}$, we extend it to all formulas by:

- $[\![A^{\perp}]\!] = [\![A]\!]$
- $[\![A \parr B]\!] = [\![A \otimes B]\!] = [\![A]\!] \otimes [\![B]\!]$.
- $[\![!A]\!] = [\![?A]\!] = T[\![A]\!]$.

The MGoI-interpretation for formulas is defined as follows.

- $\theta(\alpha) \in \mathcal{T}([\![\alpha]\!])$, where α is an atomic formula.
- $\theta(\alpha^{\perp}) = \theta(\alpha)^{\perp}$, where α is an atomic formula.
- $\theta(A \otimes B) = \{a \otimes b \,|\, a \in \theta(A), b \in \theta(B)\}^{\perp\perp}$
- $\theta(A \parr B) = \{a \otimes b \,|\, a \in \theta(A)^{\perp}, b \in \theta(B)^{\perp}\}^{\perp}$
- $\theta(!A) = \{Ta \,|\, a \in \theta(A)\}^{\perp\perp}$
- $\theta(?A) = \{Ta \,|\, a \in \theta(A^{\perp})\}^{\perp}$

Easy consequences of the definition are: (i) for any formula A, $(\theta A)^{\perp} = \theta A^{\perp}$, (ii) $\theta(A) \subseteq End([\![A]\!])$, and (iii) $\theta(A)^{\perp\perp} = \theta(A)$.

Interpreting the proofs: We define the MGoI interpretation for proofs of MELL without units. The MGoI interpretation for MLL proofs was given in [12], and we refer the reader to that paper for details.

Every MELL sequent will be of the form $\vdash [\Delta], \Gamma$ where Γ is a sequence of formulas and Δ is a sequence of cut formulas that have already been made in the proof of $\vdash \Gamma$ (see [7, 11]). This device is used to keep track of the cuts in a proof of $\vdash \Gamma$. A proof Π of $\vdash [\Delta], \Gamma$ is represented by a morphism $[\![\Pi]\!] \in End(\otimes[\![\Gamma]\!] \otimes [\![\Delta]\!])$. With $\Gamma = A_1, \cdots, A_n$, $\otimes[\![\Gamma]\!]$ stands for $[\![A_1]\!] \otimes \cdots \otimes [\![A_n]\!]$, similarly for Δ. We drop the double brackets wherever there is no danger of confusion. We also define $\sigma = s \otimes \cdots \otimes s$ (m-copies) where s is the symmetry map at different types (omitted for convenience), and $|\Delta| = 2m$. The morphism σ represents the cuts in the proof of $\vdash \Gamma$, i.e. it models Δ. In the case where Δ is empty (that is for a cut-free proof), we define $\sigma : I \to I$ to be 1_I where I is the unit of the monoidal product in $\mathbb{C}$.

Let Π be a proof of $\vdash [\Delta], \Gamma$. We define the MGoI interpretation of Π, denoted by $[\![\Pi]\!]$, by induction on the length of the proof as follows.

1. Π is obtained from Π' by an *of course* rule, that is Π has the form :

$$\frac{\begin{matrix}\Pi' \\ \vdots \\ \vdash [\Delta], ?\Gamma', A\end{matrix}}{\vdash [\Delta], ?\Gamma', !A}\ (of course)$$

Then $[\![\Pi]\!] = (e_{\Gamma'} \otimes 1_{TA} \otimes u_\Delta)\phi^{-1}T([\![\Pi']\!])\phi(e^*_{\Gamma'} \otimes 1_{TA} \otimes v_\Delta)$, where $TT \lhd T(e, e^*)$, with $\Gamma' = A_1, \cdots, A_n$, $e_{\Gamma'} = e_{A_1} \otimes \cdots \otimes e_{A_n}$, similarly for e^*. For any A, $u_A = d^*_A$ and $v_A = d_A$. Finally, with $\Delta = B_1, B_1^\perp, \cdots, B_m, B_m^\perp$, $u_\Delta = u_{B_1} \otimes \cdots \otimes u_{B_m^\perp}$, and ϕ is the canonical isomorphism constructed using $\psi_{X,Y} : TX \otimes TY \to T(X \otimes Y)$.

2. Π is obtained from Π' by the *dereliction* rule, that is, Π is of the form :

$$\frac{\begin{matrix}\Pi' \\ \vdots \\ \vdash [\Delta], \Gamma', A\end{matrix}}{\vdash [\Delta], \Gamma', ?A}\ (dereliction)$$

Then $[\![\Pi]\!] = (1_{\Gamma'} \otimes d_A \otimes 1_\Delta)[\![\Pi']\!](1_{\Gamma'} \otimes d^*_A \otimes 1_\Delta)$ where $Id \lhd T(d, d^*)$.

3. Π is obtained from Π' by the *weakening* rule, that is, Π is of the form:

$$\frac{\begin{matrix}\Pi' \\ \vdots \\ \vdash [\Delta], \Gamma'\end{matrix}}{\vdash [\Delta], \Gamma', ?A}\ (weakening)$$

Then $[\![\Pi]\!] = (1_{\Gamma'} \otimes w_A \otimes 1_\Delta)[\![\Pi']\!](1_{\Gamma'} \otimes w^*_A \otimes 1_\Delta)$, where $\mathcal{K}_I \lhd T(w, w^*)$.

4. Π is obtained from Π' by the *contraction* rule, that is, Π is of the form :

$$\frac{\begin{matrix}\Pi' \\ \vdots \\ \vdash [\Delta], \Gamma', ?A, ?A\end{matrix}}{\vdash [\Delta], \Gamma', ?A}\ (contraction)$$

Then $[\![\Pi]\!] = (1_{\Gamma'} \otimes c_A \otimes 1_\Delta)[\![\Pi']\!](1_{\Gamma'} \otimes c^*_A \otimes 1_\Delta)$, where $T \otimes T \lhd T(c, c^*)$.

Example 4. (a) Let Π be the following proof:

$$\frac{\vdash A, A^\perp \quad \vdash A, A^\perp}{\vdash [A^\perp, A], A, A^\perp}\ cut$$

Then the MGoI semantics of this proof is given by $[\![\Pi]\!] = \tau^{-1}(s \otimes s)\tau = s_{V\otimes V, V\otimes V}$ where $\tau = (1 \otimes 1 \otimes s)(1 \otimes s \otimes 1)$ and $[\![A]\!] = [\![A^\perp]\!] = V$.

(b) Now consider the following proof

$$\dfrac{\dfrac{\dfrac{\vdash A, A^{\perp}}{\vdash A, ?A^{\perp}}}{\vdash !A, ?A^{\perp}} \quad \vdash B, B^{\perp}}{\vdash !A \otimes B, ?A^{\perp} \parr B^{\perp}}$$

Given $[\![A]\!] = V$ and $[\![B]\!] = W$, we have $[\![\Pi]\!] = (1 \otimes s \otimes 1)(1 \otimes e \otimes 1 \otimes 1)(\phi^{-1}T(h)\phi \otimes s)(1 \otimes e^* \otimes 1 \otimes 1)(1 \otimes s \otimes 1)$ where $h = (1 \otimes d_V)s(1 \otimes d_V^*)$.

Proposition 2. *Let Π be an MELL proof of $\vdash [\Delta], \Gamma$. Then $[\![\Pi]\!]$ is a partial symmetry.*

Proof. By induction on the length of proofs. □

5.1 Dynamics

The mathematical model of cut-elimination is given by the so called *execution formula* defined as follows:

$$EX([\![\Pi]\!], \sigma) = Tr^{\otimes \Delta}_{\otimes \Gamma, \otimes \Gamma}((1 \otimes \sigma)[\![\Pi]\!])$$

where Π is a proof of the sequent $\vdash [\Delta], \Gamma$, and $\sigma = s^{\otimes m}$ models Δ, where $|\Delta| = 2m$. Note that $EX([\![\Pi]\!], \sigma)$ is a morphism from $\otimes\Gamma \to \otimes\Gamma$, when it exists. We shall prove below (see Theorem 1) that the execution formula always exists for any MELL proof Π.

Example 5. Consider the proof Π in Example 4(a) above. Recall also that $\sigma = s$ in this case ($m = 1$). Then $EX([\![\Pi]\!], \sigma) = Tr((1 \otimes s_{V,V})s_{V \otimes V, V \otimes V}) = s_{V,V}$.

6 Soundness of the Interpretation

In this section we present one of the main results of this paper: the soundness of the MGoI interpretation. We show that if a proof Π is reduced (via cut-elimination) to another proof Π', then $EX([\![\Pi]\!], \sigma) = EX([\![\Pi']\!], \tau)$; that is, $EX([\![\Pi]\!], \sigma)$ is an invariant of reduction. In particular, if Π' is cut-free (i.e. a normal form) we have $EX([\![\Pi]\!], \sigma) = [\![\Pi']\!]$. The soundness proof for MLL was reported in [12] and we shall not duplicate that here, however we recall a few definitions and lemmas from [12] that are crucial to a clear understanding of our work here. The associativity of cut proven in [12] for MLL holds true for MELL too. Essentially, it's proof relies on properties of categorical trace.

Lemma 1 (Associativity of cut). *Let Π be a proof of $\vdash [\Gamma, \Delta], \Lambda$ and σ and τ be the morphisms representing the cut-formulas in Γ and Δ respectively. Then*

$$EX([\![\Pi]\!], \sigma \otimes \tau) = EX(EX([\![\Pi]\!], \tau), \sigma) = EX(EX((1 \otimes s)[\![\Pi]\!](1 \otimes s), \sigma), \tau),$$

whenever all traces exist.

Definition 5. Let $\Gamma = A_1, \cdots, A_n$ and $V_i = [\![A_i]\!]$.

- A *datum of type* $\theta\Gamma$ is a morphism $M : \otimes_i V_i \to \otimes_i V_i$ such that for any $a_i \in \theta(A_i^\perp)$, $\otimes_i a_i \perp M$ and

$$M{\cdot}a_1 := Tr^{V_1}(s^{-1}_{\otimes_{i\neq 1}V_i, V_1}(a_1 \otimes 1_{V_2} \otimes \cdots \otimes 1_{V_n}) M s_{\otimes_{i\neq 1}V_i, V_1})$$

  exists. (In Girard's notation [7], $M{\cdot}a_1$ corresponds to $ex(CUT(a_1, M))$.)
- An *algorithm of type* $\theta\Gamma$ is a morphism $M : \otimes_i V_i \otimes [\![\Delta]\!] \to \otimes_i V_i \otimes [\![\Delta]\!]$ for some $\Delta = B_1, B_2, \cdots, B_{2m}$ with m a nonnegative integer and $B_{i+1} = B_i^\perp$ for $i = 1, 3, \cdots, 2m-1$, such that if $\sigma : \otimes_{i=1}^{2m} [\![B_i]\!] \to \otimes_{i=1}^{2m} [\![B_i]\!]$ is $\otimes_{i=1, odd}^{2m-1} s_{[\![B_i]\!], [\![B_{i+1}]\!]}$, $EX(M, \sigma)$ exists and is a datum of type $\theta\Gamma$.

Lemma 2. *Let $\widetilde{\Gamma} = A_2, \cdots, A_n$ and $\Gamma = A_1, \widetilde{\Gamma}$. Let $V_i = [\![A_i]\!]$, and $M : \otimes_i V_i \to \otimes_i V_i$, for $i = 1, \cdots, n$. Then, M is a datum of type $\theta(\Gamma)$ iff for every $a_1 \in \theta(A_1^\perp)$, $M{\cdot}a_1$ (defined as above) exists and is in $\theta(\widetilde{\Gamma})$.*

Theorem 1 (Proofs as algorithms). *Let Π be an MELL proof of a sequent $\vdash [\Delta], \Gamma$. Then $[\![\Pi]\!]$ is an algorithm of type $\theta\Gamma$.*

Corollary 1 (Existence of Dynamics). *Let Π be an MELL proof of a sequent $\vdash [\Delta], \Gamma$. Then $EX([\![\Pi]\!], \sigma)$ exists.*

Theorem 2 (EX is an invariant). *Let Π be an MELL proof of a sequent $\vdash [\Delta], \Gamma$ such that $?A$ does not occur in Γ for any formula A. Then,*

- *If Π reduces to Π' by any sequence of cut-eliminations, then $EX([\![\Pi]\!], \sigma) = EX([\![\Pi']\!], \tau)$. So $EX([\![\Pi]\!], \sigma)$ is an invariant of reduction.*
- *In particular, if Π' is any cut-free proof obtained from Π by cut-elimination, then $EX([\![\Pi]\!], \sigma) = [\![\Pi']\!]$.*

7 Conclusions and Future Work

In this paper we extend the scope of Multi-object GoI interpretation introduced in [12] to cover the exponentials. We show that the framework can accommodate the exponentials, given we introduce appropriate compatibility conditions for the orthogonality relation, and properly interpret the formulas involving exponentials. We have shown that the interpretations of MELL proofs are partial symmetries, it is highly desirable to have a converse for this result, that is a completeness theorem for MELL. We have proved the soundness theorem showing that the execution formula is an invariant of cut-elimination reduction for MELL without units. The necessary ingredients for an MGoI interpretation are collected in a compact structure we call a GoI category. There remains the case of units and additives. The work started in [11] and continued through [12] has convinced us that the treatment of units and additives demands new ideas. This is the main current work under progress. Although MGoI interpretation avoided infinity (reflexive object) successfully in the case of multiplicatives, we have seen in this paper that infinity cannot be eliminated when one wishes to treat the exponentials (the retractions $TT \lhd T$ and $T \otimes T \lhd T$). However, at least the notion of partial trace has brought more freedom in the choice of examples. We should take advantage of this situation and look for examples that might bring us closer to the mathematical structures at use in mathematical physics and topology.

References

1. Abramsky, S. (1996), Retracing Some Paths in Process Algebra. In *CONCUR 96, Springer LNCS* **1119**, 1-17.
2. Abramsky, S., Blute, R. and Panangaden, P. (1999), Nuclear and trace ideals in tensored *-categories, *J. Pure and Applied Algebra* vol. 143, 3–47.
3. Abramsky, S., Haghverdi, E. and Scott, P.J. (2002), Geometry of Interaction and Linear Combinatory Algebras. *MSCS*, vol. 12(5), 2002, 625-665, CUP.
4. Abramsky, S. and Jagadeesan, R. (1994), New Foundations for the Geometry of Interaction. *Information and Computation* **111** (1), 53-119.
5. Girard, J.-Y. (1987), Linear Logic. *Theoretical Computer Science* **50** (1), pp. 1-102.
6. Girard, J.-Y. (1988), Geometry of Interaction II: Deadlock-free Algorithms. In *Proc. of COLOG'88, LNCS* **417**, Springer, 76–93.
7. Girard, J.-Y. (1989a) Geometry of Interaction I: Interpretation of System F. In *Proc. Logic Colloquium 88,* North Holland, 221–260.
8. Girard, J.-Y. (1995), Geometry of Interaction III: Accommodating the Additives. In: *Advances in Linear Logic,* LNS **222**,CUP, 329–389,
9. Girard, J.-Y. (2004). Cours de Logique, Rome 2004. Forthcoming. Also available at `http://iml.univ-mrs.fr/~girard/coursang/coursang.html`
10. Haghverdi, E. (2000) *A Categorical Approach to Linear Logic, Geometry of Proofs and Full Completeness*, PhD Thesis, University of Ottawa, Canada.
11. Haghverdi, E. and Scott, P.J. (2006) A Categorical Model for the Geometry of Interaction. *Theoretical Computer Science* vol. 350, pp. 252-274.
12. Haghverdi, E. and Scott, P.J. (2005) Towards a Typed Geometry of Interaction, CSL2005 (Computer Science Logic), Luke Ong, Ed. SLNCS 3634, pp. 216-231.
13. Hyland, M and Schalk, A. (2003), Glueing and Orthogonality for Models of Linear Logic. *Theoretical Computer Science* vol. 294, pp. 183–231.
14. Joyal, A., Street, R. and Verity, D. (1996), Traced Monoidal Categories. *Math. Proc. Camb. Phil. Soc.* **119**, 447-468.
15. Mac Lane, S. (1998), *Categories for the Working Mathematician*, 2nd Ed. Springer.

Commutative Locative Quantifiers for Multiplicative Linear Logic*

Stefano Guerrini[1] and Patrizia Marzuoli[2]

[1] Dipartimento di Informatica
Università degli Studi Roma La Sapienza
Via Salaria, 113
00198 Roma, Italy
guerrini@di.uniroma1.it

[2] Dip. di Scienze Matematiche ed Informatiche "Roberto Magari"
Università degli Studi di Siena
Pian dei Mantellini 44
53100 Siena, Italy
marzuoli@unisi.it

Abstract. The paper presents a solution to the technical problem posed by Girard after the introduction of Ludics of how to define proof nets with quantifiers that commute with multiplicatives. According to the principles of Ludics, the commuting quantifiers have a "locative" nature, in particular, quantified formulas are not defined modulo variable renaming. The solution is given by defining a new correctness criterion for first-order multiplicative proof structures that characterizes the system obtained by adding a congruence implying $\forall x(A \otimes B) = \forall x A \otimes \forall x B$ to first-order multiplicative linear logic with locative quantifiers. In the conclusions we shall briefly discuss the interpretation of locative quantifiers as storage operators.

Keywords: Ludics, linear logic, proof nets.

1 Introduction

Ludics is a logic based on locative principles introduced by Girard in 2001 [Gir01]. In Ludics, the forall quantifier commutes over every connective but the exists. The extension of that general rule of commutativity would force to accept the logical principle

$$\forall x(A \oplus B) \vdash \forall x A \oplus \forall x B$$

or equivalently $\forall x(A \vee B) \vdash \forall x A \vee \forall x B$ that is not even valid in Classical Logic, and the principle

$$\forall x(A \otimes B) \vdash \forall x A \otimes \forall x B \tag{1}$$

that is not valid in Linear Logic, while the corresponding formula $\forall x(A \wedge B) \vdash \forall x A \wedge \forall x B$ is a valid classical principle.

---

* Partially supported by the italian PRIN Project FOLLIA.

M. Bugliesi et al. (Eds.): ICALP 2006, Part II, LNCS 4052, pp. 396–407, 2006.

In this paper we shall analyze what happens if we add the distributivity of the forall quantifier over the tensor in First-Order Multiplicative Linear Logic, MLL^1 for short. We remark that all the other commutativity principles of forall are valid in Linear Logic [Gir87]. In particular, we can derive $\forall xA \otimes \forall xB \vdash \forall x(A \otimes B)$. As a consequence, forcing the distributivity rule 1 in MLL^1 corresponds to add the equivalence $\forall x(A \otimes B) = \forall xA \otimes \forall xB$.

Let $\mathsf{MLL}^1_\sim$ be the logic obtained by adding to MLL^1 a $\sim$-congruence on formulas that contains $\forall x(A \otimes B) \sim \forall xA \otimes \forall xB$. A deductive system for $\mathsf{MLL}^1_\sim$ can be given by means of a subtype relation s.t. $\forall x(A \otimes B) \trianglelefteq \forall xA \otimes \forall xB$. However, the system obtained in this way does not have the subformula property and has some problems with cuts, unless one restricts to locative quantifiers. In fact, since the commutative equivalence may introduce pairs of distinct $\forall$-quantifiers corresponding to the same variable, if the witness of the $\exists$-rule can be any term, we would get that distinct $\forall$-quantifiers binding the same variable should be instantiated by distinct terms, leading to a contradiction.

The problems previously mentioned are solved by moving from sequent calculus to multiplicative proof nets [Gir87, DR89] with first-order quantifiers [Gir91]. In particular, we shall define a subset of MLL^1 proof structures that corresponds to $\mathsf{MLL}^1_\sim$ and for which cut-elimination holds. The proof nets defined in this way are then one of the first syntax for locative principles inspired by Ludics. Anyhow, we remark that the approach followed here is quite different from those presented in [FM05, CF05], in which the authors investigate proof nets with a syntax matching the operators of ludics. In our approach, we stick to the usual syntax of Linear Logic, incorporating into it some principles inspired by locativity.

We also remark that the paper solves the technical problem of a system for locative quantifiers, but does not give any semantic interpretation of them. In the conclusions, section 5, we shall briefly discuss some possible interpretations of locative quantifiers as storage operators that we are investigating.

The reformulation of the results in the paper for second order quantifiers is straightforward.

In section 2 we shall recall the main definitions of the calculus and of the proof nets for MLL^1 (see [Gir91]). In section 3 we shall analyze the system $\mathsf{MLL}^1_\sim$ obtained by forcing that the quantifiers commute over the multiplicatives and we shall give a deductive system based on a notion of subtype. In section 4 we shall extend the proof nets for MLL^1 to $\mathsf{MLL}^1_\sim$ and prove their cut-elimination.

2 First-Order Multiplicative Linear Logic

The formulas of the First-Order Multiplicative Linear Logic without unit MLL^1 [Gir91] are built from atoms p by means of the connectives $\otimes$ and $\parr$, and the quantifiers $\forall$ and $\exists$. For every atom p there is a dual atom $p^\perp$. Duality extends to every formula by $(A \otimes B)^\perp = A^\perp \parr B^\perp$, $(\forall xA)^\perp = \exists xA^\perp$, and $(A^\perp)^\perp = A$. Because of the previous duality rules we can restrict to one-sided sequents $\vdash \Gamma$, where Γ is a finite multiset of formulas.

The rules of the sequent calculus MLL^1 are

$$\frac{}{\vdash A, A^{\perp}}\,\mathsf{ax} \qquad \frac{\vdash \Gamma, A \qquad \vdash \Delta, A^{\perp}}{\vdash \Gamma, \Delta}\,\mathsf{cut}$$

$$\frac{\vdash \Gamma, A \qquad \vdash \Delta, B}{\vdash \Gamma, \Delta, A \otimes B}\,\otimes \qquad \frac{\vdash \Gamma, A, B}{\vdash \Gamma, A \parr B}\,\parr$$

$$\frac{\vdash \Gamma, A}{\vdash \Gamma, \forall x A}\,\forall_x \;\; x \notin \mathsf{FV}(\Gamma) \qquad \frac{\vdash \Gamma, A[t/x]}{\vdash \Gamma, \exists x A}\,\exists_t$$

Let us remark that we do not rename the variable x that we want to bind by a $\forall$-quantifier. The usual property of variable renaming is obtained by assuming that MLL^1 formulas are equated modulo α-congruence.

The variable x in a $\forall$-rule is an *eigenvariable.* W.l.o.g., in the following, we shall assume that the eigenvariables of the universal quantifiers in a proof are distinct, that is, each $\forall$-rule in the proof uses a new eigenvariable. This corresponds to the assumption that the eigenvariables in the proofs combined by a $\otimes$-rule or by a cut are distinct, that is, in some cases, in order to combine two proofs with distinct eigenvariables we may have to replace the eigenvariables that occur in both the proofs with new variables.

2.1 MLL^1 Proof-Nets

A *link* between (occurrences of) formulas is a pair of (possibly empty) sequences of formulas $P \vdash C$: the formulas in P are the *premises* of the link, the formulas in C are its *conclusions.* Every rule of the sequent calculus corresponds to a link whose premises are the main premises of the rule, while the conclusions are the main conclusions of the rule. In particular, the axiom link has two conclusions and no premises, the cut link has no conclusions and two premises, the tensor and par links have two premises and one conclusion, the forall and exists links have one premise and one conclusion.

A *proof structure* is a set of *formulas* and *links* s.t. every formula is conclusion of exactly one link and is premise of at most one link. The formulas that are not premise of any link are the *conclusions of the proof structure.*

Graphically, we can represent axioms/cuts by drawing an edge labeled ax/cut between the conclusions/premises of the link and the other links by drawing an edge between every premise of the link and its conclusion, as explained in the picture below.

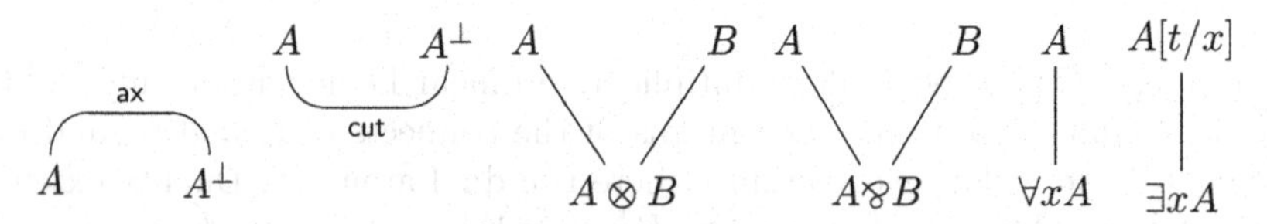

In a forall link, the variable x is the eigenvariable of the link. In an MLL^1 proof structure, the eigenvariables of the $\forall$-links in the proof structure are distinct.

We shall assume that the conclusions of the proof structures are closed. As a consequence, every variable that occurs free in some formula of the proof structure is the eigenvariable of one (and only one) $\forall$-link.

A *switching* of a proof-structure consists in: (i) the choice of a position L or R for each par link; (ii) the choice of a formula B for each $\forall$-link, where B is any formula of the proof structure in which the eigenvariable of the link occurs free or the premise of the link if the eigenvariable does not occur free in any formula.

Given a proof structure S and a switching σ, the *switch* $\sigma(S)$ is the graph defined in the following way: (i) for each axiom or cut link between A and $A^\perp$, there is an edge between A and $A^\perp$; (ii) for each tensor link with conclusion $A \otimes B$ there is an edge between the premise A and $A \otimes B$, and an edge between the premise B and $A \otimes B$; (iii) for each exists link there is an edge between its premise and its conclusion; (iv) for each par link with conclusion $A \parr B$ there is an edge between the premise A and $A \parr B$ if σ chooses L for that link, or an edge between the premise B and $A \parr B$ if σ chooses R; (v) for each forall with conclusion $\forall x A$ there is an edge between the conclusion $\forall x A$ and the formula B selected by σ.

Definition 1 (correctness criterion). *An* MLL^1 *proof structure* N *is* correct *if the switch* $\sigma(N)$ *is connected and acyclic, that is, it is a tree, for every switching* σ*. A correct proof structure* N *is a* proof net.

Theorem 1 (sequentialization). *Given an* MLL^1 *proof* π *of a sequent* Γ*, there is a corresponding* MLL^1 *proof net* $\mathsf{pn}(\pi)$ *with conclusions* Γ *and a link for every rule in* π*. Conversely, for every* MLL^1 *proof net* N*, there is a corresponding* MLL^1 *proof* π *s.t.* $\mathsf{pn}(\pi) = N$*.*

Proof. [Gir91].

2.2 Cut-Elimination

The proof nets of MLL^1 have a terminating cut-elimination defined by the reduction rules

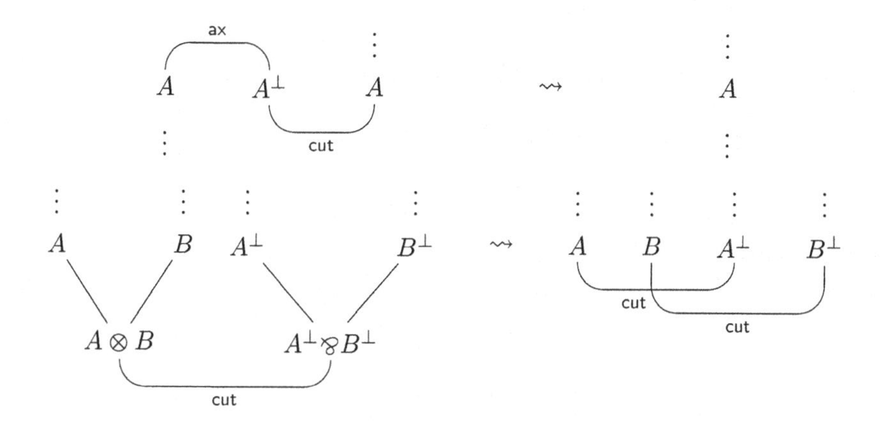

Let us remark that the reduction of a $\forall/\exists$ cut implies the substitution of the witness t of the $\exists$-rule in the cut for the free occurrences of the eigenvariable x of the $\forall$-quantifier introduced by the $\forall$-rule in the cut.

Correctness is preserved by cut-elimination. Therefore, the normal form of any proof net is a proof net too. The reduction rules for cut-elimination are confluent. Hence, the normal form of any proof net is unique.

3 First-Order Commutative Quantifiers

Let $\sim$ be the least congruence over MLL^1 formulas induced by

$$\forall x(A \otimes B) \sim \forall xA \otimes \forall xB \qquad\qquad \forall xA \sim A \quad \text{if } x \notin \mathsf{FV}(A)$$

The congruence $\sim$ naturally extends to sequents: $\Gamma \sim \Gamma'$ if there is a bijection between the formulas in Γ and the formulas in Γ' s.t. every formula $A \in \Gamma$ is mapped to some formula $A' \in \Gamma'$ s.t. $A \sim A'$.

Definition 2 ($\mathsf{MLL}^1_\sim$). *A sequent Γ is derivable in $\mathsf{MLL}^1_\sim$ if there is an MLL^1 proof π of some $\Gamma' \sim \Gamma$ s.t.*

1. *the α-congruence has not been used;*
2. *every $\exists xA$ in π is introduced by an $\exists_t$-rule with $t = x$;*
3. *π is cut-free.*

The restriction on quantifiers and α-congruence are required by cut-elimination. We remark that such a restriction makes the system much weaker, for instance, we cannot prove $\vdash (\forall yA[y/x])^{\perp}, \forall xA$, if y is a variable distinct from x.

The study of cut-elimination will be pursued on proof nets in section 4.6.

3.1 A Deductive System for $\mathsf{MLL}^1_\sim$

$\mathsf{MLL}^1_\sim$ can be also defined as the deductive system obtained from MLL^1 by eliminating the cut rule, by restricting the $\exists$-rule to the case in which the witness t is the variable x bound by the rule

$$\frac{\vdash \Gamma, A}{\vdash \Gamma, \exists xA}\ \exists_x$$

and by equating formulas modulo the $\sim$-equivalence.

Let us recall that the part of the $\sim$ that is missing in MLL^1 is the distributivity of $\forall$ over $\otimes$. We can then define the subtyping relation $\trianglelefteq$ as the least partial order closed by contexts and induced by

$$\forall x(A \otimes B) \trianglelefteq \forall xA \otimes \forall xB \qquad\qquad \forall xA \trianglelefteq A \quad \text{if } x \notin \mathsf{FV}(A)$$

Correspondingly, we have the subtyping rule

$$\frac{\vdash \Gamma, A'}{\vdash \Gamma, A} \triangleleft \quad \text{with } A' \triangleleft A$$

where $\triangleleft$ is the anti-reflexive restriction of $\trianglelefteq$.

It is readily seen that the deductive system $\mathsf{MLL}^1_\triangleleft$ obtained by replacing the condition that formulas are equated modulo $\sim$-equivalence with the previous subtyping rule is equivalent to $\mathsf{MLL}^1_\sim$. In fact, let us observe first that $A \trianglelefteq B$ implies that $\vdash B^\perp, A$ is derivable in MLL^1 (without cuts and using the restricted version of the $\exists$-rule only). Therefore, since for every $A' \sim A$ there is B s.t. $A' \trianglerighteq B \trianglelefteq A$, the formula A is derivable in $\mathsf{MLL}^1_\sim$ iff there is a cut-free MLL^1 proof of some B with $B \trianglelefteq A$ that uses the restricted version of the $\exists$-rule only and does not use the α-congruence.

4 $\mathsf{MLL}^1_\sim$ Proof Nets

In order to define the proof structures for $\mathsf{MLL}^1_\sim$, it is useful to observe that in the deductive system $\mathsf{MLL}^1_\triangleleft$ we can restrict to the case in which the $\triangleleft$-rules are applied to a single variable at a time, restricting the $\triangleleft$-rule to the case

$$\frac{\vdash \Gamma, \forall x\phi(A_1, \dots, A_m, B_1, \dots, B_n)}{\vdash \Gamma, \phi(\forall xA_1, \dots, \forall xA_n, B_1, \dots, B_n)} \triangleleft_\phi \quad x \notin \mathsf{FV}(B_1, \dots, B_n)$$

where ϕ is a $\otimes$-tree, that is, a tree of $\otimes$-connectives, whose leaves are the formulas $A_1, \dots, A_m, B_1, \dots, B_n$. Formally, $\otimes$-trees are defined by: (i) every formula A is a $\otimes$-tree with leaf A; (ii) if ϕ_1 and ϕ_2 are $\otimes$-trees with leaves L_1 and L_2, then $\phi_1 \otimes \phi_2$ is a $\otimes$-tree with leaves L_1, L_2. It is readily seen that, under the proviso of the rule, $\forall x.\phi(A_1, \dots, A_m, B_1, \dots, B_n) \trianglelefteq \phi(\forall x.A_1, \dots, \forall xA_n, B_1, \dots, B_n)$.

If we restrict to atomic axioms, or at least to quantifier free axioms, every $\forall$-quantifier must be introduced by a corresponding $\forall$-rule and we can assume that a $\triangleleft_\phi$-rule that distributes a $\forall x$ is applied immediately after the $\forall_x$-rule that introduces the quantifier. We can then merge the $\forall_x$-rule and the $\triangleleft_\phi$-rule into a unique rule, getting the deductive system $\mathsf{MLL}^1_{\forall\triangleleft}$ defined by

$$\frac{}{\vdash p, p^\perp}\,\mathsf{ax} \qquad \frac{\vdash \Gamma, A \qquad \vdash \Delta, B}{\vdash \Gamma, \Delta, A \otimes B} \otimes \qquad \frac{\vdash \Gamma, A, B}{\vdash \Gamma, A \parr B} \parr \qquad \frac{\vdash \Gamma, A}{\vdash \Gamma, \exists xA} \exists_x$$

$$\frac{\vdash \Gamma, \phi(A_1, \dots, A_m, B_1, \dots, B_n)}{\vdash \Gamma, \phi(\forall xA_1, \dots, \forall xA_m, B_1, \dots, B_n)} \forall_x^\triangleleft \quad x \notin \mathsf{FV}(\Gamma, B_1, \dots, B_n)$$

4.1 Proof Structures

The links of the proof nets for $\mathsf{MLL}^1_\sim$ are the same given for MLL^1, with the restriction that in the $\exists$-link $t = x$. However, in order to define $\mathsf{MLL}^1_\sim$ proof nets, we need to consider proof nets that correspond to $\mathsf{MLL}^1_{\forall\triangleleft}$ proofs too. In particular, we have to replace the $\forall$-link with the link corresponding to the $\forall^\triangleleft$-rule. The $\forall$-link becomes then a particular case of the $\forall^\triangleleft$-link, the case in which the $\otimes$-tree ϕ has only one leaf, that is, $\phi(A) = A$.

Let us recall that in the definition of MLL^1 proof structures we have assumed that the eigenvariables in a proof structure are distinct. Such a condition is too strong in the case of $\mathsf{MLL}^1_\sim$, where the α-congruence does not hold—in the case with cut that we shall analyze in subsection 4.5, such a condition is unsatisfiable as soon as we have a cut formula that contains quantifiers. Therefore, we have to weaken the assumption on the eigenvariables in a proof, reformulating such a condition in a way that allows to define the switching positions of foralll links (i.e., the formulas that in a switch may be connected to the conclusion of the forall) without forcing the renaming of distinct forall links that bind variables with the same name.

4.2 Scope and Eigenvariables

Let us say that two formulas are connected if there is an edge between them (e.g., if they are the two conclusions/premises of an axiom/cut link or one is the premise and the other one is the conclusion of a link). An x-path between two formulas A_0 and A_n is a path from A_0 to A_n in which the variable x occurs free in all the inner formulas of the path; more precisely, an x-path is a sequence of formulas $A_0 A_1 \dots A_{n-1} A_n$ s.t., for every $0 \leq i < n$, the formulas A_i, A_{i+1} are connected and, for every $0 \leq i \leq n$, $x \in \mathsf{FV}(A_i)$ or $A_i = \exists x A$ or A_i is the conclusion of a $\forall^\triangleleft_x$-link.

Let A be the conclusion of a $\forall^\triangleleft_x$-link. A formula B is *x-bound to A*, or x-bound to the $\forall$-link with conclusion A, if there is an x-path between B and A. The *scope* of A, or of the $\forall_x$-link with conclusion A, is the set of the formulas B bound to A s.t. $x \in \mathsf{FV}(B)$.

Let us take a $\forall^\triangleleft_x$-link with conclusion A and a $\forall^\triangleleft_y$-link with conclusion B. (We remark that, since x and y range over the set of the variables, x and y might also denote the same variable.) We shall say that the the two forall links have the same (distinct) *eigenvariable*(s) when they have the same scope (their scopes differ) and we shall write $x \equiv y$ ($x \not\equiv y$). It is readily seen that $x \not\equiv y$ iff $x \neq y$ or, when $x = y$, the scopes of A and B are disjoint.

4.3 Folded and Unfolded Proof Structures

Let us say that an $\mathsf{MLL}^1_{\forall\triangleleft}$ proof structure is *unfolded* when every $\forall^\triangleleft$-link in it is a $\forall$-link. An unfolded $\mathsf{MLL}^1_{\forall\triangleleft}$ proof structure is also an $\mathsf{MLL}^1_\sim$ proof structure.

An $\mathsf{MLL}^1_{\forall\triangleleft}$ proof structure is *folded* when the eigenvariables of all the $\forall^\triangleleft$-links in the proof structure are distinct.

4.4 Cut-Free Proof Nets

Any cut-free folded proof structure S is isomorphic to an MLL^1 proof structure S' obtained by replacing the conclusion $A = \phi(\forall x A_0, \ldots, \forall x A_m, B_1, \ldots, B_n)$ of every $\forall$-link with $A' = \forall x \phi(A_0, \ldots, A_m, B_1, \ldots, B_n)$ and consistently replacing A' for A in every formula that contains A as a subformula. In fact, the restriction on the eigenvariables of S implies that we can find a renaming of the variables in S' s.t. the variables bound by the $\forall$-links in S' are distinct. By construction, if Γ and Γ' are the conclusions of S and S', respectively, then $\Gamma' \trianglelefteq \Gamma$. This immediately implies that, if S' is correct, then Γ is derivable in $\mathsf{MLL}^1_\sim$.

Let us say that the cut-free folded $\mathsf{MLL}^1_{\forall\triangleleft}$ proof structure S is correct when S' is correct.

The definition of correct cut-free folded $\mathsf{MLL}^1_{\forall\triangleleft}$ proof structure can be also given directly, it suffices to adapt the definition of switch, forcing that a $\forall$-quantifier can jump to a formula in its scope only, or to the premise of its $\forall$-link.

Definition 3 **($\mathsf{MLL}^1_{\forall\triangleleft}$ switching).** *An* $\mathsf{MLL}^1_{\forall\triangleleft}$ switching *of a folded* $\mathsf{MLL}^1_{\forall\triangleleft}$ *proof structure is a map that assigns a position* L *or* R *for each par link and a formula B to every $\forall^{\triangleleft}$-link with conclusion A, where B is a formula in the scope of A or the premise of the $\forall^{\triangleleft}$-link.*

Switches and correctness are defined as for MLL^1 proof structures.

Proposition 1. *Given an* $\mathsf{MLL}^1_{\forall\triangleleft}$ *derivation π of a sequent Γ, there is a corresponding* $\mathsf{MLL}^1_{\forall\triangleleft}$ *correct cut-free* $\mathsf{MLL}^1_{\forall\triangleleft}$ *proof structure* $\mathsf{pn}(\pi)$ *with conclusions Γ and a link for every rule in π. Conversely, for every correct cut-free folded* $\mathsf{MLL}^1_{\forall\triangleleft}$ *proof structure S, there is a corresponding* $\mathsf{MLL}^1_{\forall\triangleleft}$ *derivation π s.t.* $\mathsf{pn}(\pi) = S$.

Correctness of unfolded $\mathsf{MLL}^1_{\forall\triangleleft}$ proof structures is defined in terms of correctness for folded ones, in particular, an unfolded structure will be correct only if there is an equivalent correct folded structure.

Let us define the following transformation on $\mathsf{MLL}^1_{\forall\triangleleft}$ proof structures:

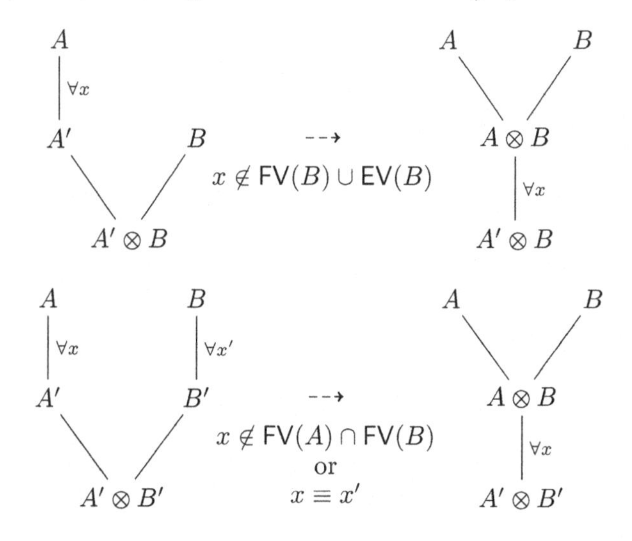

where x and x' denote two occurrences of the same variable, and where $\mathsf{EV}(B)$ denotes the set of the eigenvariables of the quantifier links associated to the quantified subformulas in B (therefore, $x \notin \mathsf{FV}(B) \cup \mathsf{EV}(B)$ means that x cannot occur at all in B, neither free nor bound). Let us remark that the above transformations are well-defined: they transform a proof structure into a proof structure with the same conclusions.

Definition 4. *A cut-free* $\mathsf{MLL}^1_\sim$ *proof structure* N *is correct, it is a* cut-free $\mathsf{MLL}^1_\sim$ proof net, *if there is a correct cut-free folded* $\mathsf{MLL}^1_{\forall\triangleleft}$ *proof structure* S *s.t.* $N \dashrightarrow^* S$.

By reversing the direction of the transformation defined above, we see that, given a folded cut-free $\mathsf{MLL}^1_{\forall\triangleleft}$ proof structure S, there is a (unique) cut-free unfolded $\mathsf{MLL}^1_{\forall\triangleleft}$ proof structure N s.t. $N \dashrightarrow^* S$. Therefore, we can state Proposition 1 for the folded case too.

Proposition 2. *Given an* $\mathsf{MLL}^1_{\forall\triangleleft}$ *derivation* π *of a sequent* Γ, *there is a corresponding cut-free* $\mathsf{MLL}^1_\sim$ *proof net with conclusions* Γ. *Conversely, for every cut-free* $\mathsf{MLL}^1_\sim$ *proof net* N, *there is a corresponding* $\mathsf{MLL}^1_{\forall\triangleleft}$ *derivation* π *s.t.* $\mathsf{pn}(\pi) = N$.

4.5 $\mathsf{MLL}^1_\sim$ Proof Nets with Cut

Unfortunately, the extension to the case with cut is not straightforward. In fact, let $\mathcal{P}$ be the least set of $\mathsf{MLL}^1_\sim$ proof structures (with cut) that contains the cut-free $\mathsf{MLL}^1_\sim$ proof nets and is closed by cut-composition. The reduction rules defined in subsection 2.2 (with $t = x$ in the case of the $\forall/\exists$-cut) defines a terminating and convergent cut-elimination procedure for $\mathcal{P}$. But, $\mathcal{P}$ is not closed by cut-elimination. E.g., by reducing the proof structure obtained by composing the cut-free $\mathsf{MLL}^1_\sim$ proof net N_1 that proves $\vdash (\forall x(A \otimes B))^\perp, \forall x.A \otimes \forall x B$ and the proof net N_2 that proves $\vdash (\forall x A \otimes \forall x B)^\perp, \forall x A \otimes B$ (such a proof net is an MLL^1 proof net indeed), we get the incorrect proof structure

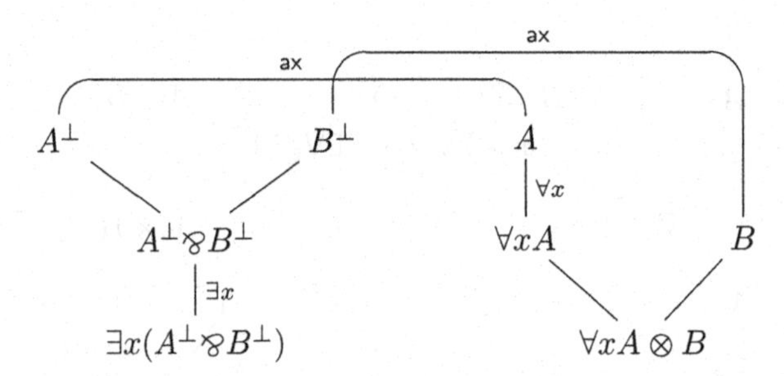

that proves $\vdash (\forall x(A \otimes B))^\perp, \forall x A \otimes B$, which is not derivable in $\mathsf{MLL}^1_\sim$.

The main issue is that in N_1 the $\forall$-links corresponding to the two subformulas $\forall x A$ and $\forall x B$ of the conclusion $\forall x A \otimes \forall x B$ have the same eigenvariable. Therefore, if we want to cut that formula with the conclusion of another proof net N, we have to ask that in N the variables bound by the $\exists$-links corresponding to the subformulas $\exists x A^{\perp}$ and $\exists x B^{\perp}$ of the cut-formula $(\forall x A \otimes \forall x B)^{\perp}$ are bound to the same $\forall$-link, or at least, that the structure of N is consistent with the assumption that the variables bound by $\exists x A^{\perp}$ and $\exists x B^{\perp}$ correspond to the same eigenvariable, in particular, we cannot combine a formula in which the variable x occurs free with a formula in which it occurs bound.

The example shows that, in the presence of cuts, the definition of x-path (sequence of connected formulas in which the variable x occurs free) does not suffice to characterize the dependencies between the occurrences of a variable x in the proof structure. In particular, because of the restriction on quantifiers, the renaming of an eigenvariable may induce the renaming of variables that are not in its scope, creating in this way a connection between formulas in which the variable x occurs and that are not connected by an x-path.

Let us say that two formulas A and B of an $\mathsf{MLL}^1_{\forall\triangleleft}$ proof structure S in which the variable x occurs are *x-correlated* if the renaming of some occurrence of x in A induces a corresponding renaming of some occurrences of x in B.

Let us say that an occurrence α of a variable x in the formula A is *correlated* to an occurrence β of x in the formula B, denoted by $\alpha \doteq \beta$, if the renaming of α induces the renaming of β. It is readily seen that $\doteq$ is an equivalence relation.

Two quantifier links are *correlated* if the occurrence of the variable in the binders in their conclusions are correlated. In particular, when two $\forall$-links are correlated we shall say that their eigenvariables are correlated.

Two correlated $\exists$-links are *siblings* if they are leaves of a tree of $\parr$-links.

Definition 5 (correct folded $\mathsf{MLL}^1_{\forall\triangleleft}$ proof structures). *A folded $\mathsf{MLL}^1_{\forall\triangleleft}$ proof structure S is correct when every switch of S is connected and acyclic and the following sibling condition holds:*

> *if two $\exists$-links of S are siblings, then they are bound to the same $\forall$-link (e.g., to the same eigenvariable).*

Since in a cut-free $\mathsf{MLL}^1_{\forall\triangleleft}$ proof structure two $\exists$-links may be correlated iff they are bound to the same $\forall$-link, the sibling condition trivially holds for cut-free $\mathsf{MLL}^1_{\forall\triangleleft}$ proof structures. Therefore, in the cut-free case, Definition 5 reduces to the definition of correct folded cut-free proof structure given in subsection 4.4.

As in the cut-free case, correctness of unfolded proof structures is defined by means of the transformation rules on page 403 with the addition that the bottom rule can be applied also when x and x' are correlated eigenvariables and not only when they are the same eigenvariable. In any case, let us remark that $\doteq$ contains $\equiv$ and that, in the cut-free case $x \doteq x'$ iff $x \equiv x'$.

Definition 6 ($\mathsf{MLL}^1_{\sim}$ proof nets). *An $\mathsf{MLL}^1_{\sim}$ proof structure N is correct, that is to say it is an $\mathsf{MLL}^1_{\sim}$* proof net, *if there is a correct folded $\mathsf{MLL}^1_{\forall\triangleleft}$ proof structure S s.t. $N \dashrightarrow^* S$.*

By the equivalence between $\doteq$ and $\equiv$ in the cut-free case and by the fact that the sibling condition trivially holds in cut-free $\mathsf{MLL}^1_{\forall\triangleleft}$ proof structures, in the cut-free case, the previous definition reduces to Definition 4.

4.6 Cut-Elimination

Cut-elimination is defined on $\mathsf{MLL}^1_\sim$ proof nets, that is, on unfolded proof structures. Unfortunately, correctness is not preserved by one step of cut-elimination. In fact, let us take the following proof net N.

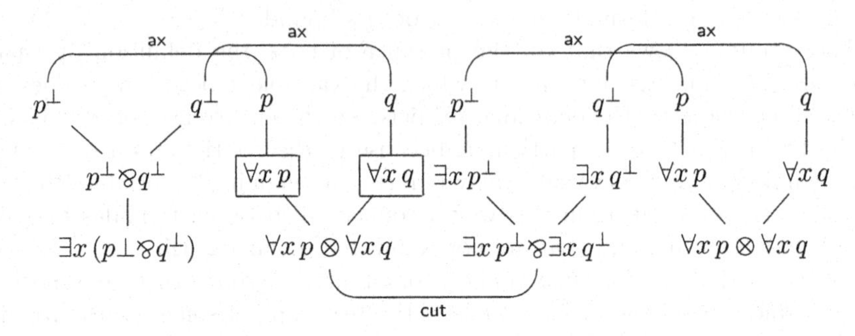

The two framed $\forall$-formulas in N have the same eigenvariable. In one step, N reduces to a proof structure M in which the two framed quantifiers are cut with the matching $\exists$-formulas. M is in normal form for $\dashrightarrow$ but contains two $\forall$-quantifiers with the same eigenvariable. Therefore, M is not a proof net.

Even if correctness is not preserved by a single cut-elimination step, we can define a big-step procedure that reduces an $\mathsf{MLL}^1_\sim$ proof net to another proof net.

Given an $\mathsf{MLL}^1_\sim$ proof net N, let S be the correct folded $\mathsf{MLL}^1_{\forall\triangleleft}$ proof structure s.t. $N \dashrightarrow^* S$. Every conclusion of a $\forall$-link in S is the image of a corresponding formula in N that we shall call a *folding root* of N. Every folding root of N is a $\otimes$-tree $\phi(\forall x A_1, \ldots, \forall x A_m, B_1 \ldots, B_n)$ with $x \notin \bigcup_{1\leq i \leq n} \mathsf{FV}(B_i) \cup \mathsf{EV}(B_i)$. Let us say that ϕ is the *folding tree* of the folding root A.

Let A be a cut formula of an $\mathsf{MLL}^1_\sim$ proof net N. We shall say that N reduces by a *big-step* that eliminates A, written $N \leadsto_b M$, when:

1. A is not a folding root and $N \leadsto M$ by the application of the cut-elimination rule that reduces the cut of A;
2. A is a folding root and $N \leadsto^* M$ by a sequence of cut-elimination rules that eliminate the $\otimes$-links corresponding to the the folding tree of A, moving the cut from the folding root A to the leaves of its folding tree.

Proposition 3 (big-step cut-elimination). *Let N be an $\mathsf{MLL}^1_\sim$ proof net. If $N \leadsto_b M$, then M is an $\mathsf{MLL}^1_\sim$ proof net.*

Big-step cut-elimination is a particular reduction strategy for $\leadsto$. Therefore, by the uniqueness of the normal-form of $\leadsto$, Proposition 3 proves that cut-elimination is sound and that every sequent derivable by an $\mathsf{MLL}^1_\sim$ proof net (with cut) is derivable in $\mathsf{MLL}^1_\sim$.

Theorem 2 (cut-elimination). *For every $\mathsf{MLL}^1_\sim$ proof net N, there is a unique $\mathsf{MLL}^1_\sim$ cut-free $\mathsf{MLL}^1_\sim$ proof net M s.t. $N \leadsto^* M$.*

5 Conclusions and Further Work

The paper solves the problem of how to characterize the proof nets for a logical system with locative quantifiers that commutes over the multiplicative connectives. However, this is just a first step, because in the paper we have not addressed at all any semantics for locative quantifiers. Our ongoing research suggests that locative quantifiers can be interpreted as storage operators. In particular, $\exists x$ states a constraint on or assigns a value to a location x of the store, while $\forall x$ operates on all the values that may be stored in x.

We are also trying to relax the restrictions on the quantifier rules, reintroducing generic terms as witnesses in the $\exists$-rule. A first possibility that we are considering is to associate a store to every proof, assuming that an $\exists_x$-rule with witness t can be applied only if the term t is compatible with the value, if any, in the location x. Another approach is to assume that the language of terms contains an intersection operator s.t. $A(t_1)$ and $A(t_2)$ imply $A(t_1 \cap t_2)$. In this way, we might get rid of the problem of how to instantiate with two distinct terms two $\forall$-quantifiers with the same eigenvariable, as this would correspond to replacing both with the intersection of the two terms.

Acknowledgments

We wish to thank Claudia Faggian and Laurent Regnier who have been the reviewer of the PhD thesis [Mar06] that contains a preliminary version of the work presented in the paper. Their detailed comments and suggestions have been very useful in writing the paper.

References

[CF05] Pierre-Louis Curien and Claudia Faggian. L-nets, strategies and proof-nets. In C.-H. Luke Ong, editor, *Computer Science Logic, 19th International Workshop, CSL 2005, 14th Annual Conference of the EACSL, Oxford, UK, August 22-25, 2005, Proceedings*, pages 167–183. Springer, 2005.

[DR89] V. Danos and L. Regnier. The structure of multiplicatives. *Archive for Mathematical Logic*, 28:181–203, 1989.

[FM05] Claudia Faggian and François Maurel. Ludics nets, a game model of concurrent interaction. In *20th IEEE Symposium on Logic in Computer Science (LICS 2005), 26-29 June 2005, Chicago, IL, USA, Proceedings*, pages 376–385. IEEE Computer Society, 2005.

[Gir87] Jean-Yves Girard. Linear logic. *Theoretical Computer Science*, 50(1):1–102, 1987.

[Gir91] J.-Y. Girard. Quantifiers in linear logic II. Prépublications de l'Équipe de Logique 19, Université Paris VII, Paris, 1991.

[Gir01] Jean-Yves Girard. Locus solum: From the rules of logic to the logic of rules. *Mathematical Structures in Computer Science*, 11(3):301–506, 2001.

[Mar06] Patrizia Marzuoli. *Ludics and Proof-Nets a New Correctness Criterion.* PhD Thesis, Dottorato di Ricerca in Logica Matematica e Informtaica Teorica, Università di Siena, February 2006.

The Wadge Hierarchy of Deterministic Tree Languages

Filip Murlak*

Institute of Informatics, Warsaw University
ul. Banacha 2, 02–097 Warszawa, Poland
fmurlak@mimuw.edu.pl

Abstract. We provide a complete description of the Wadge hierarchy for deterministically recognizable sets of infinite trees. In particular we give an elementary procedure to decide if one deterministic tree language is continuously reducible to another. This extends Wagner's results on the hierarchy of ω-regular languages to the case of trees.

1 Introduction

Two measures of complexity of recognizable languages of infinite words or trees have been considered in literature: the index hierarchy, which reflects the combinatorial complexity of the recognizing automaton and is closely related to μ-calculus, and the Wadge hierarchy, which is the refinement of the Borel/projective hierarchy that gives the deepest insight into the topological complexity of languages. Klaus Wagner was the first to discover remarkable relations between the two hierarchies for finite-state recognizable (ω-regular) sets of infinite words [14]. Subsequently, elementary decision procedures determining an ω-regular language's position in both hierarchies were given [4, 7, 15].

For tree automata the index problem is only solved in the deterministic case [9, 13]. As for topological complexity of recognizable tree languages, it goes much higher than that of ω-regular languages, which are all Δ_3^0. Indeed, co-Büchi automata over trees may recognize Π_1^1-complete languages [8], and Skurczyński [12] proved that there are even weakly recognizable tree languages in every finite level of the Borel hierarchy. This may suggest that in the tree case the topological and combinatorial complexities diverge. On the other hand, the investigations of the Borel/projective hierarchy of deterministic languages [5, 8] reveal some interesting connections with the index hierarchy.

Wagner's results [14, 15], giving rise to what is now called the Wagner hierarchy (see [10]), inspire the search for a complete picture of the two hierarchies and the relations between them for recognizable tree languages. In this paper we solve the Wadge hierarchy problem in the deterministic case. The obtained hierarchy has the height $(\omega^\omega)^3 + 3$, which should be compared with ω^ω for regular ω-languages [15], $(\omega^\omega)^\omega$ for deterministic context-free ω-languages [1], $(\omega_1^{CK})^\omega$ for ω-languages recognized by deterministic Turing machines [11], or an unknown ordinal $\xi > \varepsilon_0$ for nondeterministic context-free ω-languages [2].

The key notion of our argument is an adaptation of the Wadge game to tree languages, redefined entirely in terms of automata. Using this tool we construct a collection of canonical automata representing the Wadge degrees of all deterministic tree

* Supported by KBN Grant 4 T11C 042 25.

M. Bugliesi et al. (Eds.): ICALP 2006, Part II, LNCS 4052, pp. 408–419, 2006.

languages. Finally, we give a procedure calculating the canonical form of a given deterministic automaton, which runs within the time of finding the productive states of the automaton (the exact complexity of this problem is unknown, but not worse than exponential).

In presence of space limitations we omit some of the proofs; they can be found in the full version of the paper (see [6] for a draft).

2 Automata

A *binary tree over* Σ is a partial function $t : \{l,r\}^* \longrightarrow\!\!\!\!\!\circ\;\; \Sigma$ with a prefix closed domain. A tree t is *full* if $\operatorname{dom} t = \{l,r\}^*$. Let T_Σ denote the set of full binary trees over Σ, and let $\tilde{T}_\Sigma$ be the set of all binary trees over Σ.

A *partial deterministic automaton* is a tuple $A = \langle \Sigma, Q, q_I, \delta, \text{rank}\rangle$ where Σ is the input alphabet, Q is the set of states with a specified initial state q_I, the function rank maps states to naturals, and the transition relation δ is a partial function $\delta : Q \times \Sigma \times \{l,r\} \longrightarrow\!\!\!\!\!\circ\;\; Q$. A *run* of a partial automaton over $t \in T_\Sigma$ is a binary tree $\rho_t \in \tilde{T}_Q$ such that $\rho_t(\varepsilon) = q_I$ and for $v \in \operatorname{dom} \rho_t$, $\rho_t(v) = p$, $d = l, r$ it holds that $\delta(p, t(v), d) = q$ if and only if $vd \in \operatorname{dom} \rho_t$ and $\rho_t(vd) = q$. An infinite path $q_1, q_2, \dots$ in a run is *accepting* if $\limsup_i \text{rank}(q_i)$ is even. A run ρ_t is *infinitely accepting* if all its maximal paths are infinite and accepting. A state q is *productive* if there exists an infinitely accepting run starting in that state. A run is *accepting* if all its infinite paths are accepting and all finite maximal paths end in productive states. The *language recognised by* A, in symbols $L(A)$, is the set of trees that admit an accepting run of A. Partial deterministic automata over ω-words are defined analogously.

In classical deterministic automata the transition relation is a complete function. Consequently, all the paths in a run are infinite, and a run is accepting if and only if all the paths are accepting. Our definition is only a slight extension of the classical one. Adding transitions to an all-accepting state $\top$ or an all-rejecting state $\bot$ accordingly, we may transform a partial deterministic automaton A into a classical deterministic automaton $\tilde{A}$ recognizing the same language.

A *branching* transition is a pair $p \xrightarrow{\sigma,l} p_l$, $p \xrightarrow{\sigma,r} p_r$. For branching transitions we will sometimes write $p \xrightarrow{\sigma} p_l, p_r$. Transitions with one branch undefined will be called *non-branching*. We will also say that a partial deterministic automaton is non-branching if it contains only non-branching transitions.

The *head component* of an automaton A is the root of the directed acyclic graph (DAG) of strongly connected components (SCCs) of A. In the definitions of automata we will often specify the *tail components*, always a subset of the leaf SCCs.

3 Reductions and Games

T_Σ and the space of ω-words over Σ are equipped with the standard Cantor-like topology. For trees it is induced by the metric

$$d(s,t) = \begin{cases} 2^{-\min\{|x| \,:\, x \in \{0,1\}^*,\ s(x) \neq t(x)\}} & \text{if } s \neq t\,, \\ 0 & \text{if } s = t\,. \end{cases}$$

L is *Wadge reducible* to M, in symbols $L \leq_W M$, if there exists a continuous function φ such that $L = \varphi^{-1}(M)$. M is *$\mathcal{C}$-hard* if for all $L \in \mathcal{C}$, $L \leq_W M$. If M is $\mathcal{C}$-hard and $M \in \mathcal{C}$, then M is *$\mathcal{C}$-complete.*

Let us introduce a *tree version of Wadge games* (see [10]). For any pair of tree languages L, M the game $G_W(L, M)$ is played by Spoiler and Duplicator. Each player builds a tree, t_S and t_D respectively. In every round, first Spoiler adds at least one level to t_S and then Duplicator can either add some levels to t_D or skip a round (not forever). Duplicator wins the game if $t_S \in L \iff t_D \in M$. Just like for the classical Wadge games, a winning strategy for Duplicator can be easily transformed into a continuous reduction, and vice versa.

Lemma 1. *Duplicator has a winning strategy in $G_W(L, M)$ if and only if $L \leq_W M$.*

For regular languages we find it useful to interpret the game in terms of automata. Let A, B be partial deterministic tree automata. The *automata game* $G(A, B)$ starts with one token put in the initial state of each automaton. In every round players perform a finite number of the following actions:

fire a transition – for a token placed in a state q choose a transition $q \xrightarrow{\sigma} q_1, q_2$, remove the old token from q and put new tokens in q_1 and q_2 (similarly for non-branching transitions),

remove – remove a token placed in a productive state.

Spoiler plays on A and must perform one of these actions at least for the tokens produced in the previous round. Duplicator plays on B and is allowed to postpone performing an action for a token, but not forever. During such a play the paths visited by the tokens of each player define a run of the respective automaton. Removing a token is interpreted as a declaration that this part of the run will be accepting: simply complete the run in construction by any accepting run starting in this state. Duplicator wins the game if both runs are accepting or both are rejecting.

Lemma 2. *Duplicator has a winning strategy in $G(A, B)$ iff $L(A) \leq_W L(B)$.*

Proof. Let $\mathrm{Acc}(C)$ denote the language of accepting runs of an automaton C. For deterministic automata, $\mathrm{Acc}(C) \equiv_W L(C)$. It is enough to observe that $G(A, B)$ is in fact $G_W(\mathrm{Acc}(A), \mathrm{Acc}(B))$. □

We will write $A \leq B$ for $L(A) \leq_W L(B)$, $A \equiv B$ for $L(A) \equiv_W L(B)$, and $A < B$ for $L(A) <_W L(B)$. For $q \in Q^A$ let $A_{q:=B}$ denote the automaton obtained by replacing each A's transition of the form $p \xrightarrow{\sigma,d} q$ with $p \xrightarrow{\sigma,d} q_I^B$. Recall that by A_q we denote the automaton A with the initial state changed to q. Note that $A_{q:=A_q}$ is equivalent to A. The following fact, which will be used implicitly throughout the paper, follows easily from Lemma 2.

Corollary 1. *Let A, B, C be deterministic partial automata and $p \in Q^C$. If $A \leq B$, then $C_{p:=A} \leq C_{p:=B}$.*

4 Gadgets

A partial automaton A' is a *transformation* of A if it was obtained from A by a finite number of the following operations:

relabeling – replacing $p \xrightarrow{\sigma} q_1, q_2$ with $p \xrightarrow{\sigma'} q_1, q_2$ for a fresh letter σ' (analogously for non-branching transitions),

swapping directions – replacing $p \xrightarrow{\sigma} q_1, q_2$ with $p \xrightarrow{\sigma} q_2, q_1$, or replacing a non-branching transition $p \xrightarrow{\sigma,d} q$ with $p \xrightarrow{\sigma,d'} q$.

edge subdivision – replacing $p \xrightarrow{\sigma,d} q$ with $p \xrightarrow{\sigma,d} p' \xrightarrow{\sigma,d} q$, where p' is a fresh state and $\mathrm{rank}(p') = \mathrm{rank}(p)$,

moving entering points in SCC – replacing $p \xrightarrow{\sigma,d} q$ with $p \xrightarrow{\sigma,d} q'$ for $q, q' \in X$, $p \notin X$, where X is a SCC of A (or replacing q_I with a different state from the head component),

moving exit points in SCC – replacing $p \xrightarrow{\sigma} q_1, q_2$ with $p' \xrightarrow{\sigma'} q_1, q_2$ for $p, p' \in X$, $q_1, q_2 \notin X$, and a fresh letter σ' (analogously for non-branching transitions).

Partial automata A and B over Σ are called *isomorphic* if there exists a bijection $\eta : Q^A \to Q^B$, preserving the initial state, such that $p \xrightarrow{\sigma,d} q$ if and only if $\eta(p) \xrightarrow{\sigma,d} \eta(q)$, and for each loop $p_1 \to \ldots \to p_n$, $\max_i \mathrm{rank}^A(p_i)$ and $\max_i \mathrm{rank}^B(\eta(p_i))$ have the same parity. In particular, $L(A) = L(B)$.

We say that A and B are *similar*, in symbols $A \sim B$, if there exist transformations A' and B' that are isomorphic. It can be checked easily that $\sim$ is an equivalence relation. The equivalence class of A will be denoted by $[A]$ and called a *gadget represented by* A.

An easy inductive argument shows that $A \sim B$ implies $L(A) \equiv_W L(B)$ (the converse need not hold). For a gadget Γ, we will denote by $L(\Gamma)$ the Wadge degree of the language recognized by an automaton representing Γ. The meaning of the symbols $<, \leq, \equiv$ introduced for partial automata extends to gadgets in the natural way. By abusing notation we will write $L(\Gamma) \leq_W L(A)$, $\Gamma \leq A$, etc.

Note that the transformations preserve the structure of the DAG of SCCs up to trivial components (singletons with exactly one parent and one child). Therefore, the head component of a gadget is well defined. Analogously, the tail components of a gadget are the tail components of any of the representing automata. Note also that for a single SCC the notion of similarity coincides with the classical notion of graph homeomorphism (up to relabeling and swapping directions).

It follows from the above that non-branching gadgets are well-defined. A non-branching gadget Γ can be represented by a (partial deterministic) ω-automaton over Σ: take any representing tree automaton and ignore the second coordinate of the arrow labels. Obviously, the ω-language recognized by the obtained automaton is Wadge equivalent to $L(\Gamma)$.

B is a *subautomaton of* A if $Q^B \subseteq Q^A$, $\delta^B \subseteq \delta^A$, $\mathrm{rank}^B = (\mathrm{rank}^A)\restriction_{Q^B}$ (the initial states need not be equal). A partial automaton A *contains* a gadget Γ if A contains a (productive) subautomaton B such that $[B] = \Gamma$. A *admits* Γ if it contains a productive subautomaton B which can be obtained from an automaton representing Γ by identifying some states. It follows easily that in both cases $[A] \geq \Gamma$.

5 Operations

The *alternative* $[A_1] \vee [A_2]$ is a gadget represented by an automaton consisting of disjoint copies of A_1 and A_2, and a fresh initial state q_I with the transitions $q_I \stackrel{\sigma_1,d_1}{\longrightarrow} q_I^{A_1}$ and $q_I \stackrel{\sigma_2,d_2}{\longrightarrow} q_I^{A_2}$, with $\sigma_1 \neq \sigma_2$. The tail components are inherited from $[A_1]$ and $[A_2]$. Note that $L([A_1] \vee [A_2])$ is Wadge equivalent to the disjoint sum of $L([A_1])$ and $L([A_2])$. Consequently, $\vee$ is associative and commutative up to Wadge equivalence.

The *parallel composition* $[A_1] \wedge [A_2]$ is defined analogously, only now $\sigma_1 = \sigma_2$ and $d_1 \neq d_2$. Note that, while in $[A_1] \vee [A_2]$ the computation must choose one of the branches, here it continues in both. The language $L([A_1] \wedge [A_2])$ is Wadge equivalent to $\{t : \ t.l \in L([A_1]),\, t.r \in L([A_2])\}$ and $\wedge$ is associative and commutative up to Wadge equivalence. Multiple parallel compositions are performed from left to right: $[A_1] \wedge [A_2] \wedge [A_3] \wedge [A_4] = (([A_1] \wedge [A_2]) \wedge [A_3]) \wedge [A_4]$. We will often write $([A])^n$ to denote $\underbrace{[A] \wedge \ldots \wedge [A]}_{n}$.

A (ι,κ)-*flower* $F_{(\iota,\kappa)}$ is a gadget represented by an automaton $A_{(\iota,\kappa)}$ with states $p, q_\iota, q_{\iota+1}, \ldots, q_\kappa$, $\operatorname{rank}(p) = \iota$, $\operatorname{rank}(q_i) = i$, and transitions $p \stackrel{\sigma_i,d_i}{\longrightarrow} q_i \stackrel{\sigma_i',d_i'}{\longrightarrow} p$ for $i = \iota, \iota+1, \ldots, \kappa$ such that $\sigma_i \neq \sigma_j$ for $i \neq j$. The only SCC of the (ι,κ)-flower is both the head component and the tail component.

The (ι,κ)-*composition* $[A] \stackrel{(\iota,\kappa)}{\longrightarrow} [A_\iota], \ldots, [A_\kappa]$ is a gadget represented by an automaton obtained from the $A_{(\iota,\kappa)}$ above by adding (a single copy of) $A, A_\iota, \ldots, A_\kappa$ and transitions $p \stackrel{\sigma,d}{\longrightarrow} q_I^A$, $p \stackrel{\sigma_i,\bar{d}_i}{\longrightarrow} q_I^{A_i}$ such that $\sigma \neq \sigma_i$ and $\bar{d}_i \neq d_i$ for $i = \iota, \ldots, \kappa$, where d_i, σ_i and p are the ones from the definition of $A_{(\iota,\kappa)}$. The head component is simply the (ι,κ)-flower but for the tail components we choose only the tail components of A. Again, using Corollary 1 it is easy to see that the Wadge degree of the defined automaton depends only on (ι,κ) and the Wadge degrees of $L(A), L(A_\iota), \ldots, L(A_\kappa)$, and so the (ι,κ)-composition also defines an operation on Wadge degrees.

Let $C_1, \ldots, C_k$ be the tail components of $[A_1]$. The *sequential composition* $[A_1] \oplus [A_2]$ is represented by an automaton consisting of (a single copy of) A_1 and A_2 with transitions $p_i \stackrel{\sigma_i,d_i}{\longrightarrow} q_I^{A_2}$, $i = 1, \ldots, k$, such that $p_i \in C_i$ and $\sigma_1, \ldots, \sigma_k$ are fresh letters. $[A_1] \oplus [A_2]$ inherits its head component from A_1 and the tail components from $[A_2]$. The operation $\oplus$ is associative, but it need not be commutative even up to Wadge equivalence. For any gadget Γ and $n < \omega$ let $n\Gamma = \underbrace{\Gamma \oplus \ldots \oplus \Gamma}_{n}$.

6 Canonical Gadgets

Let $C_1 = F_{(0,0)}$, $D_1 = F_{(1,1)}$. Note that in this case we simply get an accepting and a rejecting loop. $L(C_1)$ is Wadge equivalent to the whole space and $L(D_1) \equiv_W \emptyset$. Let $E_1 = C_1 \vee D_1$. Let $C_{\omega^{\omega+k}} = F_{(1,k+2)}$, $D_{\omega^{\omega+k}} = F_{(0,k+1)}$, and $E_{\omega^{\omega+k}} = C_{\omega^{\omega+k}} \vee D_{\omega^{\omega+k}}$. The above gadgets are called *simple non-branching gadgets*.

Let $\Gamma \to \Gamma'$ denote $\Gamma \xrightarrow{(1,1)} \Gamma'$. Let $E_\omega = C_1 \to C_3$ and $E_{\omega^{k+1}} = C_1 \to (C_1 \oplus E_{\omega^k})$ for $k \geq 1$. Let $E_{\omega^{\omega \cdot 2}} = C_1 \to F_{(0,2)}$ and $E_{\omega^{\omega \cdot 2+k+1}} = C_1 \to (C_1 \oplus E_{\omega^{\omega \cdot 2+k}})$ for $k \geq 1$. These in turn are called *simple branching gadgets*.

For every non-zero ordinal $\alpha < \omega^{\omega \cdot 3}$ we have a unique presentation

$$\alpha = \omega^{\omega \cdot 2+k} l_k + \ldots + \omega^{\omega \cdot 2+0} l_0 + \omega^{\omega \cdot +k} m_k + \ldots + \omega^{\omega \cdot +0} m_0 + \omega^k n_k + \ldots + \omega^0 n_0 ,$$

where at least one of l_k, m_k, n_k is non-zero ($\omega^0 = 1$, by convention). Let us define

$$E_\alpha = n_0 E_{\omega^0} \oplus \ldots \oplus n_k E_{\omega^k} \oplus m_0 E_{\omega^{\omega \cdot +0}} \oplus \ldots \oplus m_k E_{\omega^{\omega \cdot +k}} \oplus l_0 E_{\omega^{\omega \cdot 2+0}} \oplus \ldots \oplus l_k E_{\omega^{\omega \cdot 2+k}} .$$

For $\alpha = \omega^{\omega \cdot 2} \alpha_2 + \omega^\omega \alpha_1 + n + 1$, with $\alpha_1, \alpha_2 < \omega^\omega$, $n < \omega$ (at least one non-zero), let

$$C_\alpha = C_1 \oplus E_{\omega^{\omega \cdot 2} \alpha_2 + \omega^\omega \alpha_1 + n} , \quad D_\alpha = D_1 \oplus E_{\omega^{\omega \cdot 2} \alpha_2 + \omega^\omega \alpha_1 + n}$$

and for $\alpha = \omega^{\omega \cdot 2} \alpha_2 + \omega^{\omega + k}(\alpha_1 + 1)$, with $\alpha_1, \alpha_2 < \omega^\omega$ (at least one non-zero) and $k < \omega$, let

$$C_\alpha = C_{\omega^{\omega+k}} \oplus E_{\omega^{\omega \cdot 2} \alpha_2 + \omega^{\omega+k} \alpha_1} , \quad D_\alpha = D_{\omega^{\omega+k}} \oplus E_{\omega^{\omega \cdot 2} \alpha_2 + \omega^{\omega+k} \alpha_1} .$$

Let $\mathcal{G}$ denote the family of all the gadgets defined above.

By $\emptyset \xrightarrow{(\iota,\kappa)} \Gamma_\iota, \ldots, \Gamma_\kappa$ we understand a gadget obtained from $C_1 \xrightarrow{(\iota,\kappa)} \Gamma_\iota, \ldots, \Gamma_\kappa$ by removing the tail loop and the path joining it with the head loop. Let $E_{\omega^{\omega \cdot 3}}$ denote a gadget consisting of an accepting loop λ_0 and $\emptyset \xrightarrow{(1,1)} F_{(0,2)}$, such that λ_0 and the head loop of $\emptyset \xrightarrow{(1,1)} F_{(0,2)}$ form a $(0,1)$-flower. Let $C_{\omega^{\omega \cdot 3}+1} = \emptyset \xrightarrow{(0,0)} F_{(0,1)}$.

The gadgets defined in this section are called *canonical*. Observe that in a play involving any of the canonical gadgets there is always at most one token which can reach a tail component. Let us call such a token *critical*. Whenever a critical token splits in two, exactly one of its children can reach a tail component. We shall identify it with its parent and the other one will be treated as new.

7 Effective Hierarchies

The *index* of an automaton A is a pair $(\min \text{rank}, \max \text{rank})$. Scaling down the ranks by an even integer, we may assume that $\min \text{rank}$ is 0 or 1. A is a (ι, κ)*-automaton* if $L(A)$ can be recognized by a deterministic automaton with index (ι, κ). The (ι, κ)-automata form the *deterministic index hierarchy*. The following theorem shows that the deterministic index hierarchy is effective for tree automata and ω-automata. Let $\overline{(0, m)} = (1, m+1)$, $\overline{(1, m)} = (0, m-1)$.

Theorem 1 (Niwiński & Walukiewicz [7]). *A deterministic automaton over words or trees is a (ι, κ)-automaton iff it does not admit a $\overline{(\iota, \kappa)}$-flower.*

Let Σ^0_n, Π^0_n denote the finite levels of the Borel hierarchy. As usually, $\Delta^0_n = \Sigma^0_n \cap \Pi^0_n$. The class of co-analytical sets (which need not be Borel) is denoted by Π^1_1. We showed in [5] that the Borel hierarchy is effective for deterministic tree languages. The proof relies on two facts, which will also be useful here. By a *split* we will understand a gadget Ω represented by $p \xrightarrow{\sigma,l} p_0 \longrightarrow p$, $p \xrightarrow{\sigma,r} p_1 \longrightarrow p$, with $\text{rank}\, p = \text{rank}\, p_0 = 0$, $\text{rank}\, p_1 = 1$.

Theorem 2 (Niwiński & Walukiewicz [8]). *Let A be a deterministic automaton. If A admits a split, $L(A)$ is Π^1_1-complete (and hence, non-Borel). If A does not admit a split, $L(A) \in \Pi^0_3$.*

Theorem 3 (Murlak [5]). *For a deterministic automaton A, $L(A) \in \Pi^0_2$ iff A does not admit $F_{(0,1)}$, $L(A) \in \Sigma^0_2$ iff A admits neither $F_{(1,2)}$ nor $\emptyset \overset{(0,0)}{\longrightarrow} D_2$, and $L(A) \in \Sigma^0_3$ iff A does not admit $C_{\omega^{\omega\cdot 3}+1}$.*

In fact, $L(F_{(1,2)})$ and $L(\emptyset \overset{(0,0)}{\longrightarrow} D_2)$ are Π^0_2-complete, $L(F_{(1,2)})$ is Σ^0_2-complete, and $L(C_{\omega^{\omega\cdot 3}+1})$ is Π^0_3-complete.

Let $\mathcal{G}'$ denote the set of all non-branching canonical gadgets $\{C_\alpha, D_\alpha, E_\alpha : \alpha = \omega^\omega\gamma + m, \gamma < \omega^\omega, m < \omega\}$. In [15] Wagner showed that the Wadge hierarchy for regular ω-languages is constituted by the Wadge degrees of gadgets from $\mathcal{G}'$ and that the hierarchy is effective.

Theorem 4 (Wagner [15]). *For an ω-automaton A and $C_\alpha, D_\alpha, E_\alpha \in \mathcal{G}'$ it holds that $L(A) \leq_W L(E_\alpha)$ iff A admits neither $C_{\alpha+1}$ nor $D_{\alpha+1}$, and $L(A) \leq_W L(C_\alpha)$ iff A does not admit D_α (and dually).*

Let us state our main result. The remaining of the paper is devoted to the proof of it.

Theorem 5. *The Wadge hierarchy of deterministic tree languages is as follows*

$$\begin{array}{ccccccccccc}
C_1 & & C_2\cdots & & & C_{\omega^\omega} & & C_{\omega^\omega+1}\cdots & & & \\
& \diagdown\ \diagup & & & & & \diagdown\ \diagup & & & & \\
& E_1 & & E_\omega - E_{\omega+1}\cdots & & & E_{\omega^\omega} & & & E_{\omega^\omega+\omega} - E_{\omega^\omega+\omega+1}\cdots & \\
& \diagup\ \diagdown & & & & & \diagup\ \diagdown & & & & \\
D_1 & & D_2\cdots & & & D_{\omega^\omega} & & D_{\omega^\omega+1}\cdots & & &
\end{array}$$

8 Closure Properties

The main result of this section is that the family $\mathcal{G}$ is closed by the basic operations.

Proposition 1. *The family $\mathcal{G}$ is closed by the operations $\vee$, $\wedge$, $\oplus$, $\rightarrow$ up to Wadge equivalence, and the equivalent gadget may be found in polynomial time.*

Instead of giving the whole proof, which is quite technical (see [6]), we present a handful of special cases in which the result of the operation can be given explicitly and which will turn out useful later.

Lemma 3. $E_\gamma \oplus C_{\omega^\omega\alpha} \equiv C_{\omega^\omega\alpha}$ *and* $E_\gamma \oplus D_{\omega^\omega\alpha} \equiv D_{\omega^\omega\alpha}$ *for all* $0 < \gamma, \alpha < \omega^\omega$.

Proof. We shall only consider the case of C_α; the D_α case is dual. A strategy for Duplicator in $G(E_\gamma \oplus C_\alpha, C_\alpha)$ is as follows. While Spoiler keeps inside E_γ, stay in the first flower of C_α. If one of Spoiler's tokens is inside a rejecting loop, loop a rejecting loop in your flower, otherwise loop an accepting one. When he enters C_α, simply copy his actions moving from one flower to another. Only when one of his tokens in E_γ is in a rejecting loop, choose a rejecting loop in your current flower (instead of copying Spoiler's move in C_α). Since a path in E_γ can only be rejecting if one of the tokens stays forever in a rejecting loop, this strategy is winning. □

A pair $(i, i') \in \omega \times \omega$ is called *even* if both i and i' are even. Otherwise (i, i') is *odd*. Let $[\iota, \kappa]$ denote the set $\{\iota, \iota + 1, \ldots, \kappa\} \subseteq \omega$ with the natural order. Consider the set $[\iota, \kappa] \times [\iota', \kappa']$ with the product order: $(x_1, y_1) \leq (x_2, y_2)$ if $x_1 \leq x_2$ and $y_1 \leq y_2$. A (m, n)*-alternating chain* is a sequence $(x_m, y_m) < (x_{m+1}, y_{m+1}) < \ldots < (x_n, y_n)$, such that (x_i, y_i) is even iff i is even. It is enough to consider $(0, n)$ and $(1, n)$ chains. Suppose we have a (m, n)-alternating chain of maximal length in $[\iota, \kappa] \times [\iota', \kappa']$. The parity of n is equal to the parity of (κ, κ'), as defined above, for otherwise we could extend the alternating chain with (κ, κ') and get a $(m, n + 1)$-alternating chain. Consequently, the following operation is well-defined: $(\iota, \kappa) \wedge (\iota', \kappa') = (m, n)$ if the longest alternating chain in $[\iota, \kappa] \times [\iota', \kappa']$ is of the type (m, n).

The following auxiliary gadgets will be called *weak flowers*:

$$WF_{(0,n)} = \underbrace{C_1 \oplus D_1 \oplus C_1 \oplus D_1 \oplus \ldots}_{n+1}, \quad WF_{(1,n+1)} = \underbrace{D_1 \oplus C_1 \oplus D_1 \oplus C_1 \oplus \ldots}_{n+1}.$$

In fact, $WF_{(0,n)} \equiv C_{n+1}$, $WF_{(1,n+1)} \equiv D_{n+1}$, but we find the notation convenient.

Lemma 4. *For all $i, j, m, n < \omega$ it holds that $F_{(i,m)} \wedge F_{(j,n)} \equiv F_{(i,m)\wedge(j,n)}$ and $WF_{(i,m)} \wedge WF_{(j,n)} \equiv WF_{(i,m)\wedge(j,n)}$.*

Proof. For ω-regular languages L, M, let $A_{L\times M}$ be the canonical product automaton recognizing $L \times M = \{(v_1, w_1)(v_2, w_2) \ldots : v_1 v_2 \ldots \in L,\ w_1 w_2 \ldots \in M\}$. Let $L = L(F_{(i,m)})$, $M = L_\omega(F_{(j,n)})$. Since flowers are non-branching gadgets, we may assume that L, M are ω-regular. It holds that $A_{L\times M} \equiv F_{(i,m)} \wedge F_{(j,n)}$. It is easy to see that alternating chains of loops in $A_{L\times M}$ correspond directly to alternating chains in $[i, m] \times [j, n]$. Hence, $A_{L\times M}$ admits a $(i, m) \wedge (j, n)$-flower, and does not admit a $(i, m) \wedge (j, n)$-flower. From Thm. 4 it follows that $A_{L\times M} \equiv F_{(i,m)\wedge(j,n)}$. The proof for weak flowers is analogous. □

Lemma 5. *For all $0 < k, l < \omega$ and all $m < \omega$ it holds that $C_1 \oplus E_{\omega^m k} \wedge C_1 \oplus E_{\omega^m l} = C_1 \oplus E_{\omega^m (k+l)}$ and $C_1 \oplus E_{\omega^{\omega \cdot 2+m} k} \wedge C_1 \oplus E_{\omega^{\omega \cdot 2+m} l} = C_1 \oplus E_{\omega^{\omega \cdot 2+m} (k+l)}$.*

Proof. We will only give a proof of the first equivalence. Let us consider $G(C_1 \oplus E_{\omega^m k} \wedge C_1 \oplus E_{\omega^m l}, C_1 \oplus E_{\omega^m k} \oplus E_{\omega^n l})$. Duplicator has only one critical token which can move along $WF_{(0,2(k+l))}$ formed by the alternating head and tail loops of consecutive copies of E_{ω^m}. Spoiler's starting token splits in the first move into two critical tokens which continue moving along $WF_{(0,2k)}$ and $WF_{(0,2l)}$. The strategy for Duplicator is to loop his critical token inside an accepting loop as long as both Spoiler's critical tokens loop inside accepting loops; if at least one of them moves to a rejecting loop, Spoiler should also move to a rejecting loop, and so on (c. f. Lemma 4). This way, whenever Spoiler produces a new token x using one of the critical tokens, Duplicator can produce its *doppelgänger* y, and let it mimic x. Hence, $C_1 \oplus E_{\omega^m k} \wedge C_1 \oplus E_{\omega^m l} \leq C_1 \oplus E_{\omega^m (k+l)}$. The converse inequality is obvious. □

9 Wadge Ordering

In this section we will investigate the Wadge ordering of the canonical gadgets. First, let us see what the use of the operation $\rightarrow$ is.

Lemma 6. *For all gadgets Γ, Δ and all $0<k<\omega$, $\Gamma \rightarrow \Delta \geq (\Gamma \rightarrow \Delta) \wedge (\Delta)^k$.*

Proof. Consider $G((\Gamma \rightarrow \Delta) \wedge (\Delta)^k, \Gamma \rightarrow \Delta)$. Spoiler's initial moves produce a token x in the head loop of $\Gamma \rightarrow \Delta$, and tokens $x_1, \ldots, x_k$, each in the head component of a different copy of Δ. Duplicator should loop his starting token y around the head loop of $\Gamma \rightarrow \Delta$ exactly k times producing tokens $y_1, \ldots, y_k$ and move them to the head component of Δ. From now on y mimics x, and y_i mimics x_i for $i = 1, \ldots, k$. □

Corollary 2. *For all $k, \iota, \kappa < \omega$ and all $0 < n < \omega$, $E_\omega > WF_{(\iota,\kappa)}$, $E_{\omega^{k+1}} \geq E_{\omega^k n}$, $E_{\omega^{\omega \cdot 2}} > F_{(\iota,\kappa)}$, $E_{\omega^{\omega \cdot 2+k+1}} \geq E_{\omega^{\omega \cdot 2+k} n}$.*

Proof. It is easy to see that $(0, 2m) \wedge (0, 2n) = (0, 2m+2n)$. Consequently, by Lemma 6 and Lemma 4, $E_{\omega^\omega} \geq (WF_{(0,2)})^m \equiv WF_{(0,2m)}$ and by the strictness of the hierarchy for ω-languages $\Gamma_{\omega^\omega} > WF_{(\iota,\kappa)}$. Similarly, using Lemma 6 and Lemma 5 we get $E_{\omega^{k+1}} \geq (C_1 \oplus E_{\omega^k})^n \equiv C_1 \oplus E_{\omega^k n} \geq E_{\omega^k n}$. The remaining inequalities are analogous. □

From the facts above we obtain the following lemma (see [6] for details).

Lemma 7. *If $0 < \alpha \leq \beta \leq \omega^{\omega \cdot 3}$ then $E_\alpha \leq E_\beta$ and whenever C_α and D_α are defined, $C_\alpha \leq E_\beta$, $D_\alpha \leq E_\beta$. If $\alpha < \beta$ then $E_\alpha \leq C_\beta$, $E_\alpha \leq D_\beta$.*

Now we can show that the hierarchy induced on the family of the canonical gadgets by the Wadge ordering actually coincides with the one described in Sect. 7.

Theorem 6. *Let $0 < \alpha \leq \beta \leq \omega^{\omega \cdot 3}$. Whenever the respective gadgets are defined, it holds that $C_\alpha \nleq D_\alpha$, $C_\alpha \ngeq D_\alpha$, $C_\alpha < E_\beta$, $D_\alpha < E_\beta$, and for $\alpha < \beta$, $E_\alpha < E_\beta$, $E_\alpha < C_\beta$, $E_\alpha < D_\beta$.*

Proof. By Lemma 7 it is enough to prove $E_\alpha < E_{\alpha+1}$, $C_\alpha < E_\alpha$, $D_\alpha < E_\alpha$, $C_\alpha \nleq D_\alpha$, $C_\alpha \ngeq D_\alpha$. We will only give a proof of the first inequality; the others can be argued similarly. We will proceed by induction on α. If $\alpha < \omega$, the claim follows by the ω-languages case.

Suppose $\alpha = \omega^k + \alpha'$, $k \geq 1$. Let $\alpha' \geq 1$ (the remaining case is similar). We shall describe a winning strategy for Spoiler in $G = G(E_{\omega^k+\alpha'+1}, E_{\omega^k+\alpha'})$. Spoiler should first follow the winning strategy for $G(E_{\alpha'+1}, E_{\alpha'})$, which exists by the induction hypothesis. When Duplicator enters the head loop of E_{ω^k}, Spoiler removes all his non-critical tokens, moves his critical token to the (accepting) tail loop of $E_{\alpha'+1}$ and loops there until Duplicator leaves the head loop. If Duplicator stays forever in the head loop of E_{ω^k}, he looses. After Duplicator has left the head loop, the play is equivalent to $G' = G(C_1 \oplus E_{\omega^k}, \Gamma)$ for $\Gamma = \Gamma_1 \wedge \ldots \wedge \Gamma_r$, where Γ_j is the part of E_α accessible for the Duplicator's jth token. If $k = 1$, then $\Gamma_j \leq WF_{(0,2)}$ for each j. Hence

$\Gamma \leq WF_{(0,2r)}$ and G' is winning for Spoiler by Corollary 2. Let us suppose $k > 1$. Then $\Gamma_j \leq C_1 \oplus E_{\omega^{k-1}}$ for $j = 1, \ldots, r$. Again, by Corollary 2, $\Gamma \leq E_{\omega^{\omega \cdot 2 + k-1}r+1}$. Since $\omega^{k-1}r + 1 < \omega^{k-1}r + 2 \leq \alpha$, we may use the induction hypothesis to get a winning strategy for Spoiler in G'. In either case Spoiler has a winning strategy in G as well.

Now, assume $\omega^\omega \leq \alpha < \omega^{\omega \cdot 2}$. Let $\alpha = \omega^\omega \alpha_1 + \alpha_0$ with $\alpha_0 < \omega^\omega$, $1 \leq \alpha_1 < \omega^\omega$. Again, we describe a strategy for Spoiler in $G = G(E_{\omega^\omega \alpha_1 + \alpha_0 + 1}, E_{\omega^\omega \alpha_1 + \alpha_0})$ only for $\alpha_0 \geq 1$, leaving the remaining case to the reader. First follow the winning strategy from $G(E_{\alpha_0+1}, E_{\alpha_0})$. If Duplicator does not leave the E_{α_0} component, he will lose. After leaving E_{α_0}, Duplicator has to choose $C_{\omega^\omega \alpha_1}$ or $D_{\omega^\omega \alpha_1}$. Suppose he chooses $C_{\omega^\omega \alpha_1}$. By Lemma 3, $E_{\alpha_0} \oplus C_{\omega^\omega \alpha_1} \equiv C_{\omega^\omega \alpha_1}$, and $E_{\omega^\omega \alpha_1} > C_{\omega^\omega \alpha_1}$. Therefore, Spoiler has a winning strategy in $G_C = G(E_{\omega^\omega \alpha_1}, E_{\alpha_0} \oplus C_{\omega^\omega \alpha_1})$. Imagine a play in G_C in which Duplicator ignores Spoiler's move in the first round and copies the token pattern from the stopped play we were considering above. Of course Spoiler's strategy must work in this case too. The strategy for Spoiler in G is to move the critical token to the first node of the $E_{\omega^\omega \alpha_1}$ component, take all the other tokens away, and then follow the strategy from G_C merging the first two moves into one.

For $\alpha = \omega^{\omega \cdot 2 + k} + \alpha'$ argue like for $\alpha = \omega^k + \alpha'$. □

10 Completeness

In this final section we show that the canonical gadgets represent Wadge degrees of all deterministically recognizable tree languages. To this end we will need the following technical lemma which follows from the closure properties [6].

Lemma 8. *Let A be a deterministic automaton A whose SCCs contain no complete transitions. If A admits neither $E_{\omega^{\omega \cdot 3}}$ nor $C_{\omega^{\omega \cdot 3}+1}$, one can find effectively a gadget $\Gamma \in \mathcal{G}$ such that $\Gamma \equiv A$.*

Theorem 7. *For a deterministic tree automaton A admitting neither $E_{\omega^{\omega \cdot 3}}$ nor $C_{\omega^{\omega \cdot 3}+1}$, one can find effectively an equivalent gadget $\Gamma \in \mathcal{G}$.*

Proof. We may assume that A has a *tree form*, by which we mean that the DAG of SCCs of A is a tree and that the only components containing unproductive states are leaves containing only one all-rejecting state. We will proceed by induction on the structure of this tree. Let X denote the head component of A. Suppose first that X contains a branching transition with one of its branches lying on an accepting loop. Should A admit $F_{(0,1)}$, it would also admit $C_{\omega^{\omega \cdot 3}+1}$, which is excluded by the hypothesis. Consequently, A is a $(1,2)$-automaton. If A admits $F_{(1,2)}$ or $\emptyset \xrightarrow{(0,0)} D_2$, then $A \equiv F_{(1,2)}$. Otherwise, X contains no rejecting loops and canonical forms of the subtrees rooted in the child components of X are at most C_2. If A contains D_1, then $A \equiv C_2$, otherwise $A \equiv C_1$.

Suppose that the above does not happen, but there is a branching transition with one of its branches lying on a rejecting loop and X contains an accepting loop. It follows immediately that X admits $F_{(0,1)}$ and A does not admit $F_{(0,2)}$, which means it is a $(1,3)$ automaton. If A admits neither $F_{(1,2)}$ nor $\emptyset \xrightarrow{(0,0)} D_2$, then $A \equiv F_{(0,1)}$. If A admits one of this two gadgets, it follows easily that $F_{(1,3)} \le A$. Without loss of generality we may assume that only states lying on $(0,1)$-flowers may have rank 3. Let $\text{rank}'(q) = (\text{rank}(q))'$, where $1' = 1$, $2' = 2$, $3' = 1$, and $\text{rank}''(q) = (\text{rank}(q))''$, where $1'' = 0$, $2'' = 0$, $3'' = 1$. Let $A' = \langle \Sigma, Q, q_I, \delta, rank' \rangle$, $A'' = \langle \Sigma, Q, q_I, \delta, rank'' \rangle$. It is obvious that $A \le A' \wedge A''$: a winning strategy for Duplicator in $G(A, A' \wedge A'')$ is to copy Spoilers behaviour both in A' and A''. A' is a $(1,2)$-automaton, so by Theorem 3, $A' \le F_{(1,2)}$. Since A does not admit $\emptyset \xrightarrow{(0,0)} F_{(0,1)}$, A'' will not admit $\emptyset \xrightarrow{(0,0)} D_2$, and hence $A'' \le F_{(0,1)}$. It follows that $A \equiv F_{(1,3)}$.

If X contains a branching transition but contains no accepting loops proceed as follows. Let $q_i \xrightarrow{\sigma_i} q_i', q_i'', i = 1, \ldots, n$ be all the transitions such that $q_i \in X$ and $q_i', q_i'' \notin X$. Let $p_j \xrightarrow{\sigma_i, d} p_j'$ $j = 1, \ldots, m$ be all the remaining transitions such that $p_j \in X$ and $p_j' \notin X$. By the induction hypothesis we may assume that $(A)_{q_i'}$, $(A)_{q_i''}$ and $(A)_{p_j'}$ are in canonical forms. Let $\Delta = ((A)_{q_1'} \wedge (A)_{q_1''}) \vee \ldots \vee ((A)_{q_n'} \wedge (A)_{q_n''}) \vee (A)_{p_1'} \vee \ldots \vee (A)_{p_m'}$. It is not difficult to see that A is equivalent to $C_1 \to \Delta$. By Proposition 1 we get a canonical gadget equivalent to $C_1 \to \Delta$.

Finally, let X contain no branching transitions. By induction hypothesis we may assume that the subtrees rooted in the children of X are in the canonical form. Consequently, no SCC of A contains a branching transition and we may use the lemma. □

Theorem 8. *$L(E_{\omega^\omega \cdot 3})$ is Wadge complete for deterministic Δ_3^0 tree languages.*

Proof. Take a deterministic automaton A recognizing a Δ_3^0-language. By Thm. 3, A does not admit $\emptyset \xrightarrow{(0,0)} F_{(0,1)}$. When a token splits in a branching transition, we will imagine that it goes left, and bubbles a new token to the right. Thus, in every transition only one token is produced. Let us divide the states of A into two categories: a state q is *blue* if there exists a (productive) accepting loop $p \xrightarrow{\sigma,d} p' \to \ldots \to p$ and a (productive) path $p \xrightarrow{\sigma,\bar{d}} p'' \to \ldots \to q$, $d \ne \bar{d}$. The remaining states are *red*. The tokens get the color of their birth state. The essential observation is that during an accepting run the occurrences of red states may be covered by a finite number of infinite paths (see [5], the proof of Thm. 4). Consequently, only finitely many red tokens may be produced during an accepting run.

Let A' be the automaton A with the ranks of red states set to 0, and let A'' be A with the ranks of the blue states set to 0. Like before, $A \le A' \wedge A''$. Since A does not admit $\emptyset \xrightarrow{(0,0)} F_{(0,1)}$, it follows that all $(0,1)$-flowers in A are red. Consequently, A' does not admit $F_{(0,1)}$, and $A' \le F_{(1,2)}$.

Let Λ denote a gadget produced out of $E_{\omega^\omega \cdot 3}$ by replacing $F_{(0,2)}$ with $F_{(\iota,\kappa)}$, where (ι, κ) is the index of A. Consider the game $G(A'', \Lambda)$. In A'' the blue tokens are always in the states with rank 0, so they do not influence the result of the computation. Whenever Spoiler produces a new red token (including the starting token), Duplicator should loop once around the head 1-loop producing a new token in $F_{(\iota,\kappa)}$, and keep looping around the head 0-loop. The new token is to visit states with exactly the same ranks as the token produced by Spoiler. Using the assertion on red tokens, one checks easily that the strategy is winning. Hence $A'' \le \Lambda$. By Lemma 6 and Lemma 4 it follows that $\Lambda \wedge F_{(1,2)} \le E_{\omega^\omega \cdot 3}$. □

The following corollary sums up the results of this section and the whole paper.

Corollary 3. *For a deterministic tree automaton A the exact position of $L(A)$ in the Wadge hierarchy of deterministic tree languages (see Thm. 5) can be calculated within the time of finding the productive states of the automaton.*

Proof. From Thm. 2 it follows that if A admits Ω, $A \equiv \Omega$. If A does not admit Ω, then by Thm. 3 if A admits $C_{\omega^{\omega\cdot 3}+1}$, $A \equiv C_{\omega^{\omega\cdot 3}+1}$. Otherwise $L(A) \in \Delta_3$ and if A admits $E_{\omega^{\omega\cdot 3}}$, then $A \equiv E_{\omega^{\omega\cdot 3}}$ (Thm. 8). The remaining case is settled by Thm. 7.

If the productive states are given, checking if an automaton admits Ω, $C_{\omega^{\omega\cdot 3}+1}$, or $E_{\omega^{\omega\cdot 3}}$ can be done in polynomial time. The algorithm sketched in the proof of Thm. 7 can be implemented polynomially as well, by realizing the described procedure bottom-up on the original DAG of SCCs without constructing the tree form of A explicitly. □

Acknowledgements

The author thanks Damian Niwiński for drawing his attention to the Wadge hierarchy problems and for inspiring discussions, and the anonymous referees for useful comments.

References

1. J. Duparc. A hierarchy of deterministic context-free ω-languages. Theoret. Comput. Sci. **290** (2003) 1253–1300.
2. O. Finkel. Wadge Hierarchy of Omega Context Free Languages. Theoret. Comput. Sci. **269** (2001) 283–315.
3. A. S. Kechris. *Classical Descriptive Set Theory.* Graduate Texts in Mathematics Vol. 156, 1995.
4. O. Kupferman, S. Safra, M. Vardi. Relating Word and Tree Automata. *11th IEEE Symp. on Logic in Comput. Sci.* (1996) 322–332
5. F. Murlak. On deciding topological classes of deterministic tree languages. *Proc. CSL'05*, LNCS 3634 (2005) 428–441.
6. F. Murlak. The Wadge hierarchy of deterministic tree languages. Draft version, `http://www.mimuw.edu.pl/~fmurlak/papers/conred.pdf`.
7. D. Niwiński, I. Walukiewicz. Relating hierarchies of word and tree automata. *Proc. STACS '98*, LNCS 1373 (1998) 320–331.
8. D. Niwiński, I. Walukiewicz. A gap property of deterministic tree languages. Theoret. Comput. Sci. **303** (2003) 215–231.
9. D. Niwiński, I. Walukiewicz. Deciding nondeterministic hierarchy of deterministic tree automata. *Proc. WoLLiC 2004*, Electronic Notes in Theoret. Comp. Sci. 195–208, 2005.
10. D. Perrin, J.-E. Pin. *Infinite Words. Automata, Semigroups, Logic and Games.* Pure and Applied Mathematics Vol. 141, Elsevier, 2004.
11. V. Selivanov. Wadge Degrees of ω-languages of deterministic Turing machines. Theoret. Informatics Appl. **37** (2003) 67-83.
12. J. Skurczyński. The Borel hierarchy is infinite in the class of regular sets of trees. Theoret. Comput. Sci. **112** (1993) 413–418.
13. T. F. Urbański. On deciding if deterministic Rabin language is in Büchi class. *Proc. ICALP 2000*, LNCS 1853 (2000) 663–674.
14. K. Wagner. Eine topologische Charakterisierung einiger Klassen regulärer Folgenmengen. J. Inf. Process. Cybern. EIK **13** (1977) 473–487.
15. K. Wagner. On ω-regular sets. Inform. and Control **43** (1979), 123–177.

Timed Petri Nets and Timed Automata: On the Discriminating Power of Zeno Sequences

Patricia Bouyer[1,*], Serge Haddad[2], and Pierre-Alain Reynier[1,*]

[1] LSV, CNRS & ENS Cachan, France
[2] LAMSADE, CNRS & Université Paris-Dauphine, France
{bouyer, reynier}@lsv.ens-cachan.fr, haddad@lamsade.dauphine.fr

Abstract. Timed Petri nets and timed automata are two standard models for the analysis of real-time systems. In this paper, we prove that they are incomparable for the timed language equivalence. Thus we propose an extension of timed Petri nets with read-arcs (RA-TdPN), whose coverability problem is decidable. We also show that this model unifies timed Petri nets and timed automata. Then, we establish numerous expressiveness results and prove that *Zeno* behaviours discriminate between several sub-classes of RA-TdPNs. This has surprising consequences on timed automata, e.g. on the power of non-deterministic clock resets.

1 Introduction

Timed automata (TA) [3] are a well-accepted model for representing and analyzing real-time systems: they extend finite automata with clock variables which give timing constraints on the behaviour of the system. Another prominent formalism for the design and analysis of discrete-event systems is the model of *Petri nets* (PN) [8]. Thus, in order to model concurrent systems with constraints on time, several timed extensions of PNs have been proposed as a possible alternative to TA.

Time Petri nets (TPN), introduced in the 70's, associate with each transition a time interval [4]. A transition can be fired if its enabling duration lies in its interval and time can elapse only if it does not disable some transition: firing of an enabled transition may depend on other enabled transitions even if they do not share any input or output place, which restricts a lot applicability of partial order methods in this model. Moreover, with this "urgency" requirement, all significant problems become undecidable for unbounded TPNs.

Timed Petri nets (TdPN), also called *timed-arc Petri nets*, associate with each arc an interval (or bag of intervals) [12]. In TdPNs, each token has an age. This age is initially set to a value belonging to the interval of the arc which has produced it or set to zero if it belongs to the initial marking. Afterwards, ages of tokens evolve synchronously with time. A transition may be fired if tokens with age belonging to the intervals of its input arcs may be found in the current configuration. Note that "old" tokens may die (*i.e.* they cannot be used anymore for firing a transition but they remain in the place), and that conditions for firing transitions are thus local and do not depend on the global configuration of the system, like in PNs. This "lazy" behaviour has important consequences. Whereas the reachability problem is undecidable for TdPNs [12], the coverability prob-

* Work partially supported by ACI SI Cortos, a program of the French Ministry of Research.

M. Bugliesi et al. (Eds.): ICALP 2006, Part II, LNCS 4052, pp. 420–431, 2006.

lem [2] and some significant other ones are decidable [1]. Furthermore, TdPNs cannot be transformed into equivalent TA (for the language equivalence), since the untimed languages of the latter model are regular. However the question whether (bounded) TdPNs are more expressive than TA w.r.t. language equivalence was not known.

Our contributions. In this paper, we answer negatively this question, and propose an extension of TdPNs with *read-arcs*[1], yielding the model of *read-arc timed Petri nets* (RA-TdPN). This feature has already been introduced in the untimed framework [10] in order to define a more refined concurrent semantics for nets. However, in the untimed framework, for the interleaving semantics, they do not add any expressive power as they can be replaced by two arcs which check that a token is in the place and replace it immediately. First, we investigate the decidability of the coverability problem for the RA-TdPN model, and we prove that it remains decidable.

We then focus on the expressiveness of read-arcs, and prove quite surprising results. Indeed, we show that read-arcs add expressiveness to the model of TdPNs when considering languages of (possibly *Zeno*) infinite timed words. On the contrary, we also prove that when considering languages of finite or non-*Zeno* infinite timed words, read-arcs can be simulated and thus don't add any expressiveness to TdPNs.

Furthermore we investigate the relative expressiveness of several subclasses of RA-TdPNs, depending on the following restrictions: boundedness of the nets, integrality of constants appearing on the arcs, resets labelling post-arcs. We give a complete picture of their relative expressive power, and distinguish between three timed language equivalences (equivalence over finite words, or infinite words, or non-*Zeno* infinite words) which, as before, lead to different results.

We finally establish that timed automata and bounded RA-TdPNs are language equivalent. From this result and former ones, we deduce several worthwhile expressiveness results, for instance we prove that non-determinism in clock resets adds expressive power to timed automata with integral constants over (possibly *Zeno*) infinite timed words, which contrasts with the finite or non-*Zeno* infinite timed words case [5]. If rational constants are allowed, this is no more the case: it should be emphasized that this latter result implies that the granularity of the automaton has to be refined if we want to remove non-deterministic updates while preserving expressiveness.

Due to lack of space, proofs are omitted, but can be found in [6].

2 Read-Arc Timed Petri Nets

Preliminaries. If A is a set, A^* denotes the set of all finite words over A whereas A^ω denotes the set of infinite words over A. An interval I of $\mathbb{R}_{\geq 0}$ is a $\mathbb{Q}_{\geq 0}$-(resp. $\mathbb{N}$-) *interval* if its left endpoint belongs to $\mathbb{Q}_{\geq 0}$ (resp. $\mathbb{N}$) and its right endpoint belongs to $\mathbb{Q}_{\geq 0} \cup \{\infty\}$ (resp. $\mathbb{N} \cup \{\infty\}$). We denote by $\mathcal{I}$ (resp. $\mathcal{I}_\mathbb{N}$) the set of $\mathbb{Q}_{\geq 0}$-(resp. $\mathbb{N}$-) intervals of $\mathbb{R}_{\geq 0}$.

Bags. Given a set $\mathcal{E}$, $\mathsf{Bag}(\mathcal{E})$ denotes the set of mappings f from $\mathcal{E}$ to $\mathbb{N}$ s.t. the set $\mathsf{dom}(f) = \{x \in \mathcal{E} \mid f(x) \neq 0\}$ is finite. We note $\mathsf{size}(f) = \sum_{x \in \mathcal{E}} f(x)$. Let $x, y \in \mathsf{Bag}(\mathcal{E})$, then $y \leq x$ iff $\forall e \in \mathcal{E}, y(e) \leq x(e)$. If $y \leq x$, then $x - y \in \mathsf{Bag}(\mathcal{E})$ is defined

[1] A similar extension has been proposed independently by Srba in [11].

by: $\forall e \in \mathcal{E}, (x - y)(e) = x(e) - y(e)$. For $d \in \mathbb{R}_{\geq 0}$ and $x \in \mathsf{Bag}(\mathbb{R}_{\geq 0})$ $x + d \in \mathsf{Bag}(\mathbb{R}_{\geq 0})$ is defined by $\forall \tau < d, (x+d)(\tau) = 0$ and $\forall \tau \geq d, (x+d)(\tau) = x(\tau - d)$. Let $x \in \mathsf{Bag}(\mathcal{E}_1 \times \mathcal{E}_2)$. The bags $\pi_i(x) \in \mathsf{Bag}(\mathcal{E}_i)$ for $i = 1, 2$ are defined by: for all $e_1 \in \mathcal{E}_1$, $\pi_1(x)(e_1) = \sum_{e_2 \in \mathcal{E}_2} x(e_1, e_2)$, and similarly for π_2.

Timed words and timed languages. Let Σ be a finite alphabet s.t. $\varepsilon \notin \Sigma$ (ε is the silent action), we note $\Sigma_\varepsilon = \Sigma \cup \{\varepsilon\}$. A *timed word* w over Σ_ε (resp. Σ) is a finite or infinite sequence $w = (a_0, \tau_0)(a_1, \tau_1) \ldots (a_n, \tau_n) \ldots$ s.t. for every $i \geq 0$, $a_i \in \Sigma_\varepsilon$ (resp. $a_i \in \Sigma$), $\tau_i \in \mathbb{R}_{\geq 0}$ and $\tau_{i+1} \geq \tau_i$. The value τ_k gives the date at which action a_k occurs. We write $Duration(w) = \sup_k \tau_k$ for the duration of the timed word w. Since ε is a silent action, it can be removed in timed words over Σ_ε, and it naturally gives timed words over Σ. An infinite timed word w over Σ is said *Zeno* whenever *Duration*(w) is finite. We denote by $\mathcal{TW}^*_\Sigma$ (resp. $\mathcal{TW}^\omega_\Sigma$, $\mathcal{TW}^{\omega_{nz}}_\Sigma$) the set of finite (resp. infinite, non-*Zeno* infinite) timed words over Σ. A *timed language over finite (resp. infinite, non-Zeno infinite) words* is a subset of $\mathcal{TW}^*_\Sigma$ (resp. $\mathcal{TW}^\omega_\Sigma$, $\mathcal{TW}^{\omega_{nz}}_\Sigma$).

The Model of RA-TdPNs. The *qualitative* component of a RA-TdPN is a Petri net extended with read-arcs. A read-arc checks for the presence of tokens in a place without consuming them. The *quantitative* part of a RA-TdPN is described by timing constraints on arcs. Roughly speaking, when firing a transition, tokens are consumed whose ages satisfy the timing constraints specified on the input arcs, and it is checked whether the constraints specified by the read-arcs are satisfied. Tokens are then produced according to the constraints specified on the output arcs.

Definition 1. *A timed Petri net with read-arcs (RA-TdPN for short) $\mathcal{N}$ is a tuple $(P, m_0, T, \mathsf{Pre}, \mathsf{Post}, \mathsf{Read}, \lambda, \mathsf{Acc})$ where:*
- *P is a finite set of places;*
- *$m_0 \in \mathsf{Bag}(P)$ denotes the initial marking of places;*
- *T is a finite set of transitions with $P \cap T = \emptyset$;*
- *Pre, the backward incidence mapping, is a mapping from T to $\mathsf{Bag}(\mathcal{I})^P$;*
- *Post, the forward incidence mapping, is a mapping from T to $\mathsf{Bag}(\mathcal{I})^P$;*
- *Read, the read incidence mapping, is a mapping from T to $\mathsf{Bag}(\mathcal{I})^P$;*
- *$\lambda : P \to \Sigma_\varepsilon$ is a labelling function;*
- *Acc is an accepting condition given as a finite set of formulas generated by the grammar $\mathsf{Acc} ::= \sum_{i=1}^n p_i \bowtie k \mid \mathsf{Acc} \wedge \mathsf{Acc}$, with $p_i \in P$, $k \in \mathbb{N}$ and $\bowtie \in \{\leq, \geq\}$.*

Since $\mathsf{Bag}(\mathcal{I})^P$ is isomorphic to $\mathsf{Bag}(P \times \mathcal{I})$, $\mathsf{Pre}(t)$, $\mathsf{Post}(t)$ and $\mathsf{Read}(t)$ may be also considered as bags. Given a place p and a transition t, if the bag $\mathsf{Pre}(t)(p)$ (resp. $\mathsf{Post}(t)(p)$, $\mathsf{Read}(t)(p)$) is non null then it defines a *pre-arc* (resp. *post-arc*, *read-arc*) of t connected to p.

A *configuration* ν of a RA-TdPN is an item of $\mathsf{Bag}(\mathbb{R}_{\geq 0})^P$ (or equivalently $\mathsf{Bag}(P \times \mathbb{R}_{\geq 0})$). Intuitively, a configuration is a marking extended with age information for the tokens. We will write (p, x) for a token which is in place p and whose age is x. A configuration is then a finite sum of such pairs. Then a token (p, x) belongs to configuration ν whenever $(p, x) \leq \nu$ (in terms of bags). The *initial configuration* $\nu_0 \in \mathsf{Bag}(\mathbb{R}^P_{\geq 0})$ is defined as $\forall p \in P$, $\nu_0(p) = m_0(p) \cdot 0$ (there are $m_0(p)$ tokens of age 0 in place p).

We now describe the semantics of a RA-TdPN in terms of a transition system.

Definition 2 (Semantics of a RA-TdPN). *Let $\mathcal{N} = (P, m_0, T, \mathit{Pre}, \mathit{Post}, \mathit{Read}, \lambda, \mathit{Acc})$ be an RA-TdPN. Its semantics is the transition system $(Q, \Sigma_\varepsilon, \rightarrow)$ where $Q = \mathit{Bag}(\mathbb{R}_{\geq 0})^P$, and $\rightarrow$ is defined by:*

- *For $d \in \mathbb{R}_{\geq 0}$, $\nu \xrightarrow{d} \nu + d$ where the configuration $\nu + d$ is defined by $(\nu + d)(p) = \nu(p) + d$ for every $p \in P$.*
- *A transition t is firable from ν if for all $p \in P$, there exist $x(p), y(p) \in \mathit{Bag}(\mathbb{R}_{\geq 0} \times \mathcal{I})$ such that to Opt$\begin{cases} \pi_1(x(p)) + \pi_1(y(p)) \leq \nu(p), \\ \pi_2(x(p)) = \mathit{Pre}(t)(p) \text{ and } \pi_2(y(p)) = \mathit{Read}(t)(p), \\ \forall(\tau, I) \in \mathit{dom}(x(p)) \cup \mathit{dom}(y(p)),\ \tau \in I. \end{cases}$*

  Let $z(p) \in \mathit{Bag}(\mathbb{R}_{\geq 0} \times \mathcal{I})$ be such that to Opt$\begin{cases} \pi_2(z(p)) = \mathit{Post}(t)(p), \\ \forall(\tau, I) \in \mathit{dom}(z(p)),\ \tau \in I. \end{cases}$

  Define for every $p \in P$, $\nu'(p) = \nu(p) - x(p) + z(p)$. Then $\nu \xrightarrow{\lambda(t)} \nu'$.

A *path* in the RA-TdPN $\mathcal{N}$ is a sequence $\nu_0 \xrightarrow{d_1} \nu_1' \xrightarrow{t_1} \nu_1 \xrightarrow{d_2} \nu_2' \xrightarrow{t_2} \nu_2 \ldots$ in the above transition system. A *timed transition sequence* is a (finite or infinite) timed word over alphabet T, the set of transitions of $\mathcal{N}$. A *firing sequence* is a timed transition sequence $(t_1, \tau_1)(t_2, \tau_2) \ldots$ such that $\nu_0 \xrightarrow{\tau_1} \nu_1' \xrightarrow{t_1} \nu_1 \xrightarrow{\tau_2 - \tau_1} \nu_2' \xrightarrow{t_2} \nu_2 \ldots$ is a path. If $(p, x) \leq \nu$ is a token of a configuration ν, it is a *dead token* whenever for every interval I labelling a pre- or a read-arc of p, x is above I.

Petri nets can be considered as language acceptors. The timed word which is read along a path $\nu_0 \xrightarrow{d_1} \nu_1' \xrightarrow{t_1} \nu_1 \xrightarrow{d_2} \nu_2' \xrightarrow{t_2} \nu_2 \ldots$ is the projection over Σ of the timed word $(\lambda(t_1), d_1)(\lambda(t_2), d_1 + d_2) \ldots$

If ν is a configuration of $\mathcal{N}$, ν satisfies the accepting condition $\sum_{i=1}^{n} p_i \bowtie k$ whenever $\sum_{i=1}^{n} \mathsf{size}(\nu(p_i)) \bowtie k$, and the satisfaction relation for conjunctions of accepting conditions is defined in a natural way. A finite path in $\mathcal{N}$ is accepting if it ends in a configuration satisfying one of the formulas of Acc. An infinite path is accepting if every formula of Acc is satisfied infinitely often along the path (Acc is then viewed as a generalized Büchi condition). We note $\mathcal{L}^*(\mathcal{N})$ (resp. $\mathcal{L}^\omega(\mathcal{N})$, $\mathcal{L}^{\omega_{nz}}(\mathcal{N})$) the set of finite (resp. infinite, non-*Zeno* infinite) timed words accepted by $\mathcal{N}$.

Two RA-TdPNs $\mathcal{N}$ and $\mathcal{N}'$ are $*$-*equivalent* (resp. ω-*equivalent*, ω_{nz}-*equivalent*) whenever $\mathcal{L}^*(\mathcal{N}) = \mathcal{L}^*(\mathcal{N}')$ (resp. $\mathcal{L}^\omega(\mathcal{N}) = \mathcal{L}^\omega(\mathcal{N}')$, $\mathcal{L}^{\omega_{nz}}(\mathcal{N}) = \mathcal{L}^{\omega_{nz}}(\mathcal{N}')$). These equivalences naturally extend to subclasses of RA-TdPNs. In the following, we will use notations like "$\{*, \omega, \omega_{nz}\}$-equivalence" to mean the three equivalences altogether. *Idem* for "$\{*, \omega_{nz}\}$-equivalence" and other combinations.

Notations. Read-arcs are represented by undirected arcs. We use shortcuts to represent bags: for all $I \in \mathcal{I}$, I holds for the bag $1 \cdot I$, $[a]$ is for the interval $[a, a]$. We may write intervals as constraints, *eg* "$\leq a$" is for the interval $[0, a]$. A bag n represents the bag $n \cdot \mathbb{R}_{\geq 0}$, and no bag on an arc means that this arc is labelled by the bag $1 \cdot \mathbb{R}_{\geq 0}$.

Example 1. An example of RA-TdPN is depicted on the next figure. This net models an information provided by a server and asynchronously consulted by clients (transition "read"). Since the information may be obsolete with validity duration "val", the server periodically refreshes the value, but the frequency of this refresh may vary depending on the workload of the server (transition "refresh"). The admission control ensures that at least one time unit elapses between two client arrivals (transition "entry"). Note the interest of the read-arc between "cache" and "read": when transition "read" is fired the age of the token of place "cache" is not reinitialized.

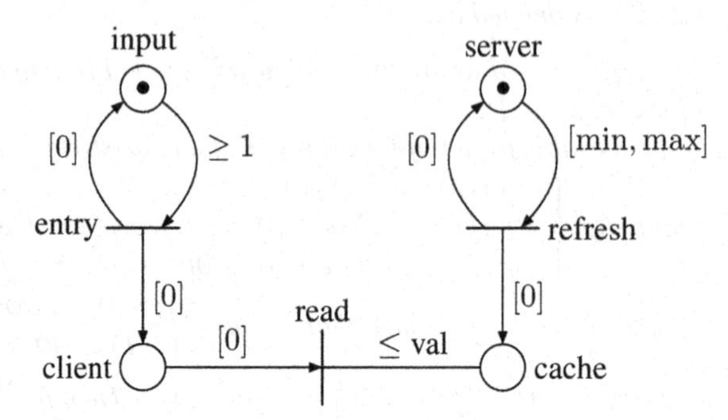

Subclasses of RA-TdPNs. We define several natural subclasses of RA-TdPNs.

Definition 3. *Let $\mathcal{N} = (P, m_0, T, \mathit{Pre}, \mathit{Post}, \mathit{Read}, \lambda, \mathit{Acc})$ be an RA-TdPN. It is*

- *a* timed Petri net *(TdPN for short)*[2] *if for all $t \in T$, $\mathsf{size}(\mathit{Read}(t)) = 0$,*
- integral *if all intervals appearing in bags of $\mathcal{N}$ are in $\mathcal{I}_{\mathbb{N}}$,*
- 0-reset *if for all $t \in T$, for all $p \in P$, $I \neq [0,0] \Rightarrow I \notin \mathit{dom}(\mathit{Post}(t)(p))$,*
- k-bounded *if all configurations ν appearing along a firing sequence of $\mathcal{N}$ are such that for every place $p \in P$, $\mathsf{size}(\nu(p)) \leq k$,*
- bounded *if there exists $k \in \mathbb{N}$ such that $\mathcal{N}$ is k-bounded,*
- safe *if it is 1-bounded.*

The Coverability Problem. Let $\mathcal{N}$ be an RA-TdPN with initial configuration ν_0. Let N be a finite set of configurations of $\mathcal{N}$ where all ages of tokens are rational. We note $N^{\uparrow}$ the upward closure of N, *i.e.* the set $\{\nu \mid \exists \nu' \in N,\ \nu' \leq \nu\}$.

The *coverability problem* for $\mathcal{N}$ and set of configurations N asks whether there exists a path in $\mathcal{N}$ from ν_0 to some $\nu \in N^{\uparrow}$. We obtain the following result.

Theorem 1. *The coverability problem is decidable for RA-TdPNs.*

The proof of this theorem is an extension of the proof done in [9] for TdPNs, based on an extension of classical regions in timed automata [3].

3 Relative Expressiveness of Subclasses of RA-TdPNs

In this section, we thoroughly study the relative expressiveness of subclasses of RA-TdPNs, by distinguishing whether they are bounded, integral, 0-reset, or whether they can be expressed without read-arcs. Surprisingly the results depend on the language equivalence we consider, and whereas finite timed words and non-*Zeno* infinite timed words do not distinguish between (integral, bounded) 0-reset TdPNs and (integral, bounded) RA-TdPNs, *Zeno* infinite timed words lead to a lattice of strict inclusions that will be summarized in Subsection 3.5.

[2] This is the standard model, as defined in [12].

3.1 Two Discriminating Timed Languages

We design two timed languages which distinguish between several subclasses of RA-TdPNs. Notice that these two languages are *Zeno*.

The timed language L_1. The RA-TdPN $\mathcal{N}_1$ of Fig. 1(a) (with a single accepting Büchi condition $p \geq 1$) is a 0-reset, integral and bounded RA-TdPN which recognizes the timed language $L_1 = \{(a, \tau_1)\ldots(a, \tau_n)\ldots \mid 0 \leq \tau_1 \leq \ldots \leq \tau_n \leq \ldots \leq 1\}$. Note that this timed language is also recognized by the TA $\mathcal{A}_1$ of Fig. 1(b).

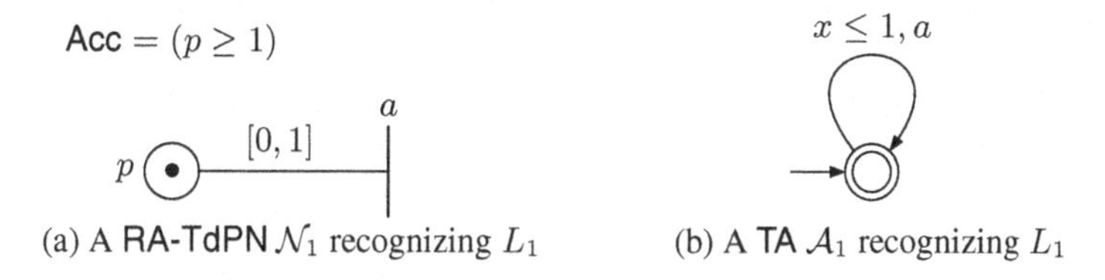

(a) A RA-TdPN $\mathcal{N}_1$ recognizing L_1 (b) A TA $\mathcal{A}_1$ recognizing L_1

Fig. 1. A language L_1 not recognized by any TdPN

Lemma 1. *The timed language L_1 is recognized by no TdPN.*

The timed language L_2. The RA-TdPN $\mathcal{N}_2$ of Fig. 2(a) is an integral bounded RA-TdPN which recognizes the timed language $L_2 = \{(a, 0)(b, \tau_1)\ldots(b, \tau_n)\ldots \mid \exists \tau < 1 \text{ s.t. } 0 \leq \tau_1 \leq \ldots \leq \tau_n \leq \ldots < \tau\}$. Note, and that will be used in Section 4, that the timed language L_2 is also recognized by the TA of Fig. 2(b) (which uses a non-deterministic reset of clock x in the intervals $]0, 1[$).

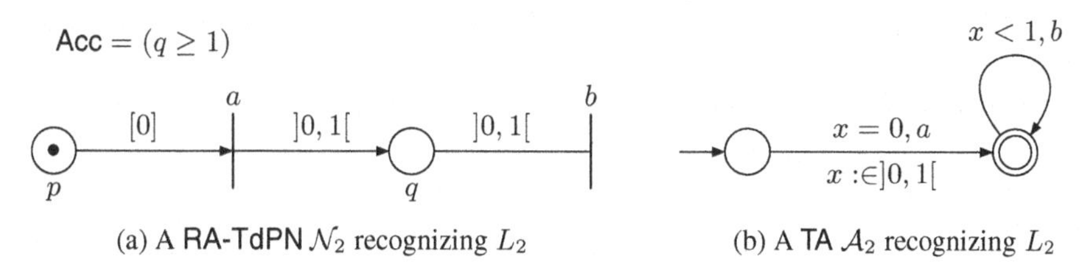

(a) A RA-TdPN $\mathcal{N}_2$ recognizing L_2 (b) A TA $\mathcal{A}_2$ recognizing L_2

Fig. 2. A language L_2 not recognized by any 0-reset integral RA-TdPN

Lemma 2. *The timed language L_2 is recognized by no 0-reset integral RA-TdPN.*

3.2 Normalization of RA-TdPNs

We present a transformation of RA-TdPNs which preserves both languages over finite and (*Zeno* or non-*Zeno*) infinite words, as well as boundedness and integrality of the nets. This construction transforms the net by imposing strong syntactical conditions on places, which will simplify further studies of RA-TdPNs.

Proposition 1. *For any RA-TdPN $\mathcal{N}$, we can effectively construct a RA-TdPN $\mathcal{N}'$ which is $\{*, \omega_{nz}, \omega\}$-equivalent to $\mathcal{N}$, and in which all places are configured as one of the five patterns depicted in Fig. 3, which reads as: "there is an a such that the place is connected to at most one post-arc, at most one pre-arc and possibly several read-arcs, with bags as specified on the figure". Moreover the construction preserves boundedness and integrality properties.*

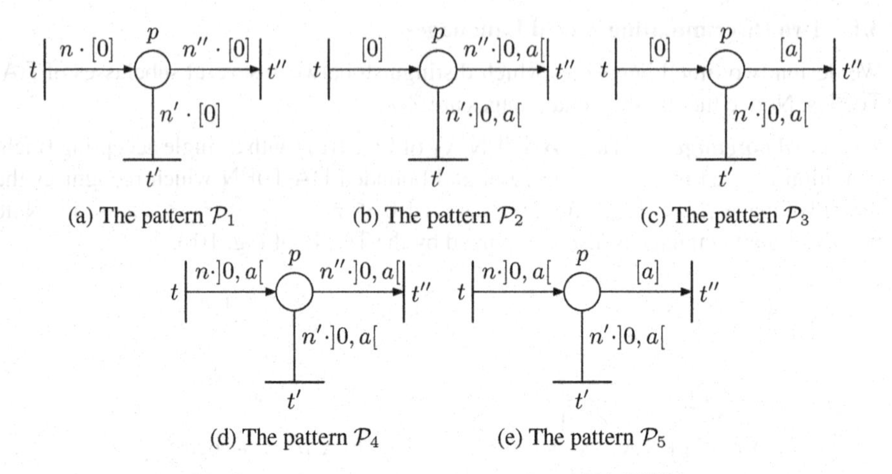

Fig. 3. The five normalized patterns for an RA-TdPN

3.3 Removing the Read-Arcs

In thus subsection, we study the role of read-arcs in RA-TdPNs. Thanks to Lemma 1 (language L_1), we already know that read-arcs add expressive power to TdPNs for the ω-equivalence. We then prove that read-arcs do not add expressiveness to the model of TdPNs when considering finite or infinite non-*Zeno* timed words. We present two different constructions: the first one is correct only for finite timed words, whereas the second one, which extends the first one, is correct for non-*Zeno* infinite timed words. In both correction proofs, we need to assume that places connected to read-arcs do not occur in the acceptance condition. This can be done without loss of generality.

Case of finite words. We state the following result.

Theorem 2. *Let $\mathcal{N}$ be an RA-TdPN, then we can effectively build a TdPN $\mathcal{N}'$, which is $*$-equivalent to $\mathcal{N}$. Note that the construction preserves the boundedness and integrality properties of the nets.*

Proof (Sketch). To prove this result, we first normalize the net. We then distinguish the five possible patterns of Fig. 3 for a place p, and show that we can remove the read-arcs connected to place p. The construction for pattern $\mathcal{P}_4$ is given on the next picture.

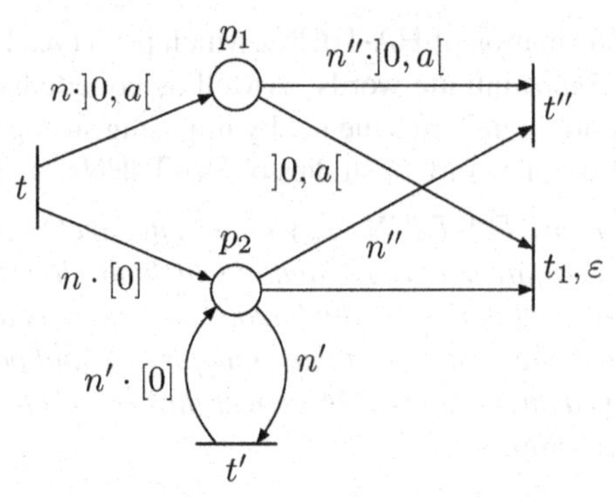

The accepting condition is reinforced by the constraint $p_1 + p_2 \leq 0$, thus imposing to consume (by t'' or t_1) every token produced by t. The idea of this construction is to check pre-arcs with tokens which are in place p_1 and to check read-arcs with tokens in place p_2, but with no timing constraints (there is no sense to check the age of the tokens in p_2 since it is reset each time a read-arc checks the presence of a token in the place). *A posteriori*, before tokens are dead (thus before their age reaches a), they will be consumed by transition t'' or t_1, together with one token in place p_1. □

We illustrate the construction on the RA-TdPN $\mathcal{N}_1$ of Fig. 1(a). It is correct for finite timed words only.

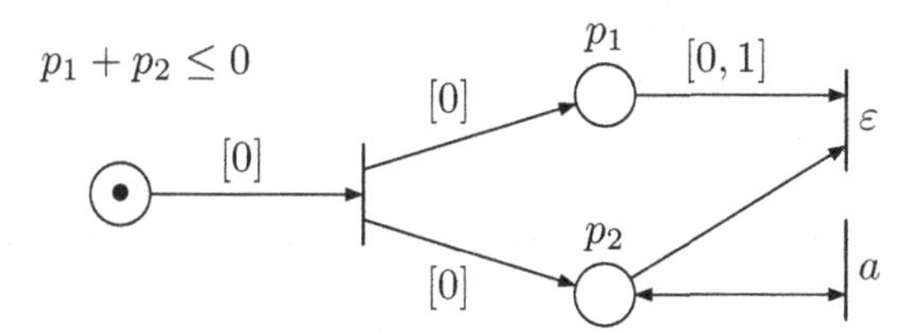

Case of infinite non-Zeno words. The previous construction cannot be applied to languages of infinite words. Indeed, it relies on the following idea. The acceptance condition requires that one empties the places at the end of the sequence in the simulating net in order to check whether the tokens has been appropriately checked.

In the case of infinite timed words, a similar Büchi condition would "eliminate" words accepted by a sequence of the original net in which a place always contains tokens that will be checked in the future. However in the divergent case, we will first apply a transformation of the net that will not change the language, in such a way that in the new net, every infinite non-*Zeno* timed word will be accepted by an appropriate generalized Büchi condition.

Theorem 3. *Let $\mathcal{N}$ be an RA-TdPN, then we can effectively build a TdPN $\mathcal{N}'$, which is ω_{nz}-equivalent to $\mathcal{N}$. Note that the construction preserves the boundedness and the integrality of the nets.*

3.4 Removing General Resets

In this subsection, we study the role of general resets in RA-TdPNs. Thanks to Lemma 2 (language L_2), we know that the class of integral RA-TdPNs is strictly more expressive than the class of 0-reset integral RA-TdPNs for the ω-equivalence. We then prove two results, which show that this is the combination of the presence of read-arcs together with the integrality property which explains the expressiveness gap between 0-reset nets and nets with general resets. Indeed, we design a first construction which holds if there is no read-arc, and which preserves integrality of the net. Then we design a second construction, which holds even for nets with read-arcs, but which does not preserve the integrality of the nets.

Theorem 4. *For every TdPN $\mathcal{N}$, we can effectively build a 0-reset TdPN $\mathcal{N}'$ which is $\{*, \omega, \omega_{nz}\}$-equivalent to $\mathcal{N}$. Moreover, this construction preserves the boundedness and integrality properties of the net.*

This result is not difficult and consists in shifting intervals of pre-arcs connected to a place, depending on the intervals which label post-arcs connected to this place.

The second result is much more involved, and requires to refine the granularity of the net we build. However, it is correct for the whole class of RA-TdPNs.

Theorem 5. *For every RA-TdPN $\mathcal{N}$, we can build a* 0*-reset RA-TdPN $\mathcal{N}'$ which is $\{*, \omega_{nz}, \omega\}$-equivalent to $\mathcal{N}$. The construction preserves the boundedness of the net, but **not** its integrality.*

Proof (Sketch). First, it it worth noticing that in the case of finite words, and non-*Zeno* infinite words, this result is a corollary of previous results (Theorems 2, 3 and 4). This proof, though correct for all finite and infinite timed words, is thus only necessary to deal with *Zeno* infinite timed words.

Let $\mathcal{N}$ be a RA-TdPN which we assume satisfies Proposition 1. The only places of $\mathcal{N}$ which are connected to non 0-reset post-arcs are those which satisfy pattern $\mathcal{P}_4$ or pattern $\mathcal{P}_5$ (Fig. 3(d) and 3(e)). Here, we only present the construction for pattern $\mathcal{P}_4$, it is depicted below.

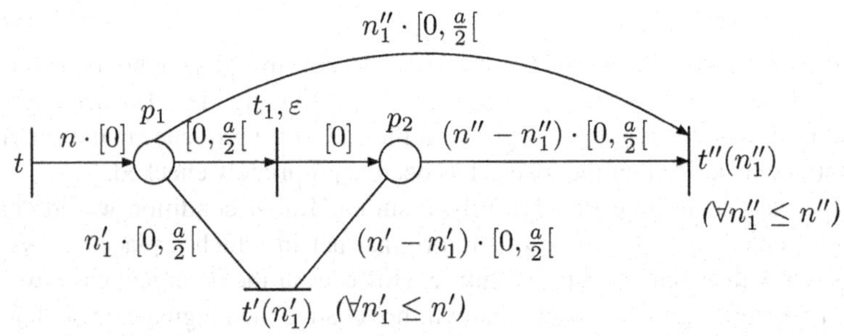

A token which enters place p in the original net (and which will not die) will either be consumed by transition t'' before $\frac{a}{2}$ units of time has elapsed, or after a delay which is greater than $\frac{a}{2}$ but strictly less than 1. In the first case, the token can stay in place p_1 (place in which it can be used by a read-arc) and leave when it is consumed by transition t''. In the second case, the token will stay in place p_1 for some amount of time, and then go to place p_2 where it can also be consumed by transition t''. The read-arc can read tokens in place p_1 or in place p_2 with the constraint that ages of the token are in the interval $[0, \frac{a}{2}[$. □

3.5 Summary of Our Expressiveness Results

Case of finite and infinite non-Zeno words. Applying the results of the two previous subsections, we get equality of all subclasses of RA-TdPNs mentioned on the following picture, for the $\{*, \omega_{nz}\}$-equivalence. Note that this picture is correct for the general classes, for the restriction to integral nets, and also for the restriction to bounded nets.

$$\underbrace{\text{RA-TdPN} = \text{TdPN}}_{\text{Theo. 2,3}} \underbrace{= \text{0-reset TdPN}}_{\text{Theo. 4}}$$

Case of infinite words. The picture in the case of infinite words is much different. Indeed the hierarchy in the previous case collapses, whereas we get here the lattice below. Plain arcs represent strict inclusion, and dashed arcs indicate that the classes are incomparable. Finally note that this picture holds for both bounded and general nets.

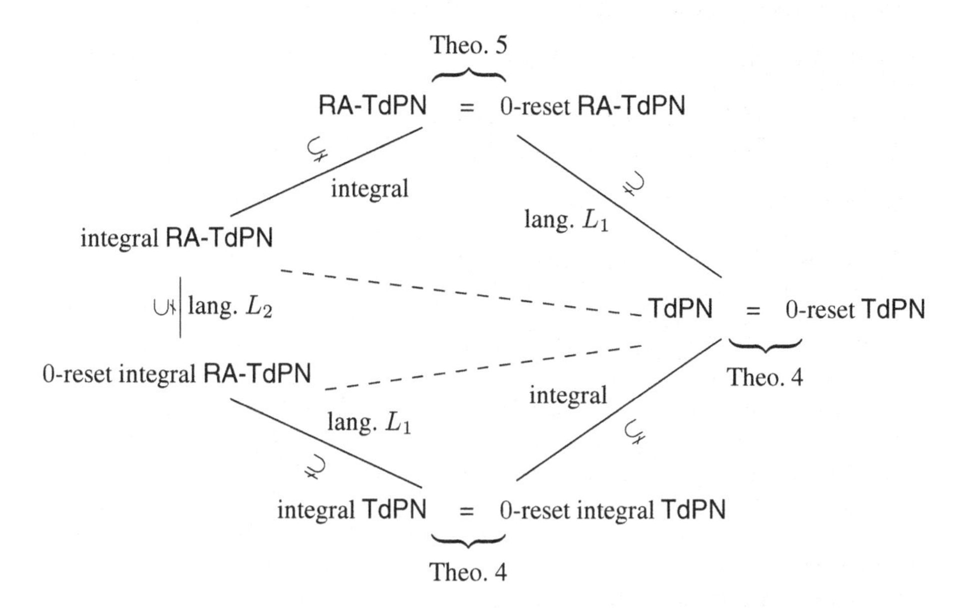

4 Application to Timed Automata

First defined in [3], the model of timed automata (TA) associates with a finite automaton a finite set of non negative real-valued variables called *clocks*. We assume the reader is familiar with TA, and refer to [5] for a formal definition (we allow, in addition to classical resets to 0 of clocks, general resets of the form $x :\in I$ if $I \in \mathcal{I}$ which sets a clock to a value non-deterministically chosen in I). Two examples of TA are given on Fig. 1(b) and 2(b). The following theorem, close to a result by Srba [11], relates TA and bounded RA-TdPNs.

Theorem 6. *Bounded RA-TdPNs and TA are* $\{*, \omega_{nz}, \omega\}$*-equivalent.*

Proof (Sketch). For transforming a bounded RA-TdPNs into an equivalent TA, we first build a safe RA-TdPN, and then a TA, in which a clock is associated with a place and records the age of the token in the place. We illustrate the transformation of a TA into a bounded RA-TdPN on an example.

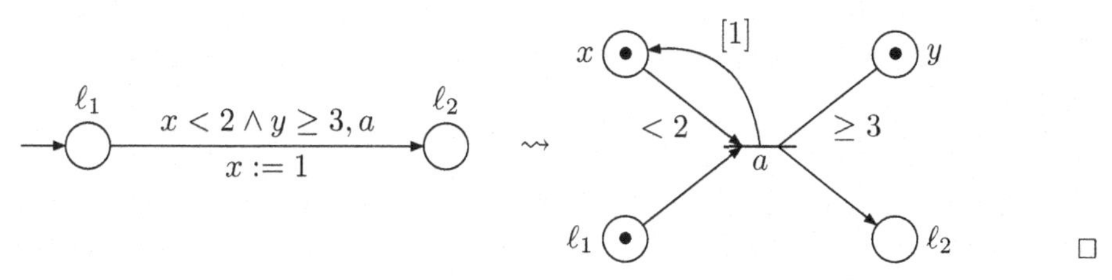

□

***Expressiveness Results for* TA.** Combining this result with the results of the previous section on Petri nets, we get interesting side results on timed automata, and in particular quite surprising results for languages of infinite timed words.

Corollary 1. *For the* $\{*, \omega_{nz}\}$*-equivalence,*

1. *bounded* TdPNs *and* TA *are equally expressive;*
2. *(integral)* TA *and* 0*-reset (integral)* TA *are equally expressive.*

Corollary 2. *For the* ω*-equivalence,*

3. TdPNs *and* TA *are incomparable;*
4. TA *are strictly more expressive than bounded* TdPNs*;*
5. *integral* TA *are strictly more expressive than integral* 0*-reset* TA*;*
6. TA *and* 0*-reset* TA *are equally expressive.*

As a "folk" result, it was thought that TA and bounded TdPNs are equally expressive. We have proved that this is indeed the case for finite and infinite non-*Zeno* timed words (item *1.*), but that it is wrong when considering also *Zeno* behaviours (item *4.*). Indeed, the result is even stronger: even though TdPNs can be somehow seen as timed systems with infinitely many clocks, we have proved that TA and TdPNs are in general incomparable (item *3.*).

The three other results complete the picture of known results about general resets in TA [5]. Item *2.* was already partially proved in the above-mentioned paper, and we provide here a new proof of this result. Items *5.* and *6.* are quite surprising, since they show that refining the granularity of the guards is necessary for removing general resets in TA (and for preserving the languages of infinite timed words). It is one of the first such results in the framework of timed systems (up to our knowledge). Finally, the construction provided in the proof of Theorem 5 applied to TA provides an extension to infinite words of the construction presented in [5] for removing general resets in TA (which is indeed only correct for finite and infinite non-*Zeno* timed words). We illustrate this construction by giving a 0-reset TA ω-equivalent to the timed automaton of Fig. 2(b).

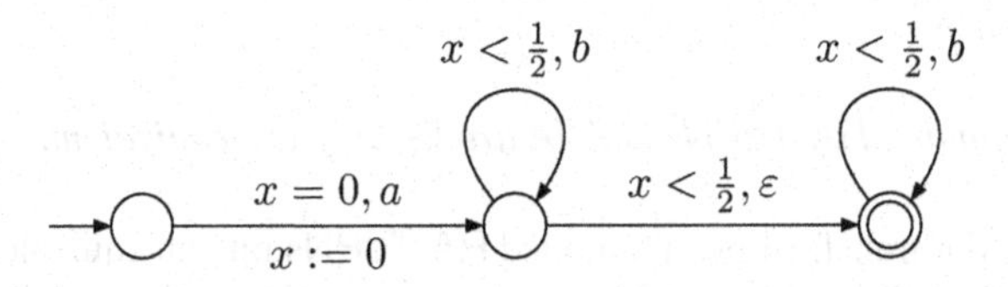

5 Conclusion

In this paper, we have thoroughly studied the relative expressiveness of TdPNs and TA, and we have proved in particular that they are incomparable in general. This has motivated the introduction of read-arcs in TdPNs, yielding the model of RA-TdPNs. This model unifies TA and TdPNs, has a decidable coverability problem, and enjoys pretty surprising expressiveness results.

We have studied the expressive power of read-arcs in RA-TdPNs, and we have proved that, when restricting to finite or infinite non-*Zeno* behaviours, read-arcs do not add expressiveness. On the other hand, we show that *Zeno* behaviours discriminate between several subclasses of RA-TdPNs. For instance, RA-TdPNs are strictly more expressive than TdPNs. Since we also prove that bounded RA-TdPNs and TA are equally expressive, we get the surprising result that TA are strictly more expressive than bounded TdPNs, which is quite counter-intuitive.

Classically, TdPNs use quite general resets, whereas TA use only resets to 0. We have thus studied the expressive power of these general resets, compared with resets to 0. We have shown that they don't add any expressiveness to the above-mentioned models, but that the granularity has to be refined for removing general resets in RA-TdPN when considering *Zeno* behaviours. Up to our knowledge, this is one of the first expressiveness results (at least in the domain of timed systems), which requires to refine the granularity of the model. As side results, we complete the work in [5], and get that it is necessary to refine the granularity of guards in TA for removing general resets, when considering languages of infinite possibly *Zeno* timed words.

Our main further work will be to develop partial-order techniques for RA-TdPNs, taking advantage of the locality of the firing rules (see [7]). Another research direction is to study arcs which do not reset age of tokens.

References

1. P. A. Abdulla, P. Mahata, and R. Mayr. Decidability of Zenoness, syntactic boundedness and token-liveness for dense-timed petri nets. In *Proc. 24th Conf. Foundations of Software Technology and Theoretical Computer Science (FST&TCS'04)*, volume 3328 of *Lecture Notes in Computer Science*, pages 58–70. Springer, 2004.
2. P. A. Abdulla and A. Nylén. Timed Petri nets and bqos. In *Proc. 22nd Int. Conf. Application and Theory of Petri Nets (ICATPN'01)*, volume 2075 of *Lecture Notes in Computer Science*, pages 53–70. Springer, 2001.
3. R. Alur and D. Dill. A theory of timed automata. *Theoretical Computer Science*, 126(2): 183–235, 1994.
4. B. Berthomieu and M. Diaz. Modeling and verification of time dependent systems using time Petri nets. *IEEE Transactions in Software Engineering*, 17(3):259–273, 1991.
5. P. Bouyer, C. Dufourd, E. Fleury, and A. Petit. Updatable timed automata. *Theoretical Computer Science*, 321(2–3):291–345, 2004.
6. P. Bouyer, S. Haddad, and P.-A. Reynier. Timed Petri nets and timed automata: On the discriminating power of Zeno sequences. Research Report LSV-06-06, ENS de Cachan, France, 2006.
7. P. Bouyer, S. Haddad, and P.-A. Reynier. Timed unfoldings of networks of timed automata. Research Report LSV-06-09, ENS de Cachan, France, 2006.
8. C. Girault and R. Valk, editors. *Petri Nets for Systems Engineering*. Springer, 2002.
9. P. Mahata. *Model Checking Parameterized Timed Systems*. PhD thesis, Dept. Information Technology, Uppsala University, Sweden, 2005.
10. U. Montanari and F. Rossi. Contextual nets. *Acta Informatica*, 32(6):545–596, 1995.
11. J. Srba. Timed-arc Petri nets vs. networks of timed automata. In *Proc. 26th International Conference Application and Theory of Petri Nets (ICATPN'05)*, volume 3536 of *Lecture Notes in Computer Science*, pages 385–402. Springer, 2005.
12. V. Valero Ruiz, F. Cuartero Gomez, and D. de Frutos-Escrig. On non-decidability of reachability for timed-arc Petri nets. In *Proc. 8th Int. Work. Petri Nets and Performance Models (PNPM'03)*, pages 188–196. IEEE Computer Society Press, 1999.

On Complexity of Grammars Related to the Safety Problem

Tomasz Jurdziński

Institute of Computer Science, Wrocław University
Przesmyckiego 20, 51151 Wrocław, Poland
tju@ii.uni.wroc.pl

Abstract. Leftist grammars were introduced by Motwani et. al., who established the relationship between the complexity of accessibility problem (or safety problem) for certain general protection system and the membership problem of these grammars. The membership problem for leftist grammars is decidable. This implies the decidability of the accessibility problem. It is shown that the membership problem for leftist grammars is PSPACE-hard. Therefore, the accessibility problem in the appropriate protection systems is PSPACE-hard as well. Furthermore, the PSPACE-hardness result is adopted to very restricted class of leftist grammars, if the grammar is a part of the input.

1 Introduction

Leftist grammars were introduced by Motwani et. al. [9]. They used them as a tool to show decidability of the accessibility problem in certain general protection systems. Those protection systems provide the formal basis for trust management. A protection system is a set of policies that prescribes the ways in which *objects* interact with each other. By objects we mean users, processes or other entities; and interactions can include access rights, information sharing privileges, etc. The accessibility problem (or the safety problem) for a protection system is formulated in the form "Can object p gain (illegal) access to object q by a series of legal moves (as prescribed by the policy)?". A formal treatment of accessibility was first presented by Harrison, Ruzzo, and Ullman [6] who showed that the accessibility problem is undecidable for a general access-matrix model of object-resource interaction. This result prompted extensive research on tradeoffs between expressibility and verifiability in protection systems (see, e.g., [9] for references).

The protection system related to leftist grammars was originally proposed in [4, 10] in the context of Java virtual worlds. The model of this protection system strictly generalizes grammatical protection systems [3] and the take-grant model [8], and it is a special case of the general access-matrix model [6]. In contrast to the general access-matrix model, the accessibility problem for models related to leftist grammars is decidable [9].

We refer the reader to [9, 10] for a description of the protection system related to leftist grammars. For our consideration, it is only important that the accessibility problem of this protection system and the intersection problem of leftist grammars are polynomial time equivalent. Note that the membership problem is a special case of the intersection problem.

M. Bugliesi et al. (Eds.): ICALP 2006, Part II, LNCS 4052, pp. 432–443, 2006.

Leftist grammars can be characterized in terms of rules of the form $a \to ba$ and $cd \to d$ where a, b, c, and d belong to an alphabet Σ. We define the final symbol of a leftist grammar to be some fixed symbol $x \in \Sigma$. We say that a word $w \in \Sigma^*$ belongs to the language defined by a grammar iff there exists a derivation which starts at wx and ends at x.

Although the membership problem for leftist grammars is decidable [9], no efficient algorithm for this problem is known. The only known lower bound states that the class of languages defined by leftist grammars is not included in CFL [7]. However, quite natural restrictions yield context-freenes or even regularity [7, 1]. Motwani et al. designed a sophisticated algorithm for the membership problem of general leftist grammars, which relies on Higman's Lemma. No upper bound for the complexity of this algorithm is known. On the other hand, simplicity of leftist grammars led to the conjecture that there exist efficient algorithms for the membership problem. (Motwani et. al. [9] posed even the question whether all languages defined by leftist grammars were context-free.) We give the first complexity theoretic lower bound for the membership problem by establishing that it is PSPACE-hard. Furthermore, we consider the variable membership problem, i.e., a variant of the membership problem in which not only the tested word but also the grammar is a part of the input. We show that the variable membership problem is PSPACE-hard even in the case of restricted leftist grammars (i.e., with acyclic insert graphs or acyclic delete graphs). Moreover, we obtain EXPSPACE upper bound for this case.

In Section 2 we provide some basic definitions. Section 3 describes the construction which establishes PSPACE-hardness of the membership problem for general leftist grammars. Finally, in Section 4 we analyze complexity of the variable membership problem for restricted leftist grammars. Due to limited space, we omit many details of the proofs.

2 Definitions

For a word x, let $|x|$, $x[i]$ and $x[i, j]$ denote the length of x, the ith symbol of x and the factor $x[i] \dots x[j]$ respectively, where $0 < i \leq j \leq |x|$. Let $[i, j] = \{l \in \mathbb{N} \mid i \leq l \leq j\}$, let $\bar{i} = 1 - i$ for $i \in [0, 1]$. Moreover, let x^R denote the reverse of a word x, that is $x^R = x[n]x[n-1] \dots x[2]x[1]$, where $|x| = n$. Throughout the paper ε denotes the empty word. By $\pi_{i_1,\dots,i_m}(b)$ for $b = \langle b_1, \dots, b_n \rangle \in B_1 \times \dots \times B_n$ we denote the projection of b onto the coordinates $i_1, \dots, i_m$. That is, $\pi_{i_1,\dots,i_m}(b) = \langle b_{i_1}, \dots, b_{i_m} \rangle$.

Definition 1. *A leftist grammar $\mathcal{H} = (\Sigma, P, x)$ consists of the finite alphabet Σ, the final symbol $x \in \Sigma$, and the set of production rules P of the following two types, $ab \to b$ (Delete Rule), $c \to dc$ (Insert Rule) where $a, b, c, d \in \Sigma$. In order to shorten notations, we will describe the above productions as $b \to^D a$ (Delete Rule), and $c \to^I d$ (Insert Rule). We say that $u \Rightarrow_{\mathcal{H}} v$ (or shortly $u \Rightarrow v$) is a derivation step, if $u = u_1yu_2$ and $v = u_1zu_2$ such that $y \to z$ is a production rule in P. A sequence of derivation steps $u_1 \Rightarrow \dots \Rightarrow u_p$ is called a derivation. A word u_i for $i \in [1, p]$ is called a sentential form in this derivation. Finally, the language of $\mathcal{H}$ is defined to be $L(\mathcal{H}) = \{w \in \Sigma^* \mid wx \Rightarrow^* x\}$.*

Throughout the paper, we will implicitely treat symbols of sentential forms as objects which can insert/delete other symbols and can be inserted/deleted. However, in order to simplify notations, we will usually identify the particular occurence of the symbol a in a sentential form with its value a.

We say that the symbol b in the delete rule $ab \to b$ is *active*. Similarly, the symbol c is *active* in the insert rule $c \to dc$. Let $u \Rightarrow v$, where $u = u_1yu_2$ and $v = v_1zv_2$ such that $y \to z$ is a production rule in P. We would like to say that the symbol which is active in the production rule $y \to z$ (that is, the rightmost symbol of the prefix u_1y of u_1yu_2) is also *active* in the derivation step $u \Rightarrow v$. However, it is possible that there are many factorizations $u = u_1yu_2$ such that $v = u_1zu_2$ and $y \to z$ is the production in P, for fixed u, v. Fortunately, one can avoid this ambiguity [7]. So, we will consider only leftist grammars which satisfy the condition that one can determine uniquely which symbol is active in each possible derivation step. The symbol $u_1[i]$ is *active* in u_1 with respect to the fixed derivation $U = (u_1 \Rightarrow^* u_p)$ if it is active in at least one of derivation steps of the derivation U. Otherwise, this symbol is *inactive* in u_1 with respect to U. We say that $u_1[i]$ is *alive* in u_1 with respect to U if $u_1[j]$ is active with respect to U for some $j \leq i$. If $u_1[i]$ is *not alive*, we say that it is *done*. (Note that each active symbol is alive, but the opposite relation is not true. Similarly, each done symbol is inactive.)

We introduce a notion which formally describes the way in which symbols are inserted. Let $U \equiv u_1 \Rightarrow u_2 \Rightarrow \ldots \Rightarrow u_p$ be a derivation. Let b, d be symbols which appear in some sentential forms of this derivation. Then, d is a *descendant* of b in U if (b, d) belongs to the reflexive and transitive closure of the relation $\{(e, f) \mid v\underline{e}w \Rightarrow v\underline{fe}w$ is the derivation step in U for $v, w \in \Sigma^*$ and $e, f \in \Sigma\}$. We say that a word u *eliminates* a word $w \in \Sigma^+$ in the derivation $z_1wuz_2 \Rightarrow^* z'$, if all elements of w (from the sentential form z_1wuz_2) are inactive with respect to this derivation, and all elements of w are deleted (during the derivation $z_1wuz_2 \Rightarrow^* z'$) by the elements of u and their descendants.

A symbol $a \in \Sigma$ is called an *anihilator* (*generator*, resp.) in the grammar $\mathcal{H} = (\Sigma, P, x)$ if at least one production rule $a \to^D b$ ($a \to^I b$, resp.) for some $b \in \Sigma$ belongs to P.

Definition 2 (Interface). *Let $W \equiv (w_1 \Rightarrow w_2 \Rightarrow \ldots \Rightarrow w_m)$ be a derivation of a leftist grammar. Let $a_i = w_i[1]$ if $w_i[1]$ is an anihilator and $a_i = \varepsilon$ otherwise, for $1 \leq i \leq m$. The string $a_1a_2 \ldots a_m$ is called the interface of the derivation W.*

Note that the interface of the derivation $W = (w_1 \Rightarrow^* w_m)$ indirectly describes a set of words which could be eliminated "during" W if we put such a word to the left of w_1.

The derivation $u_1 \Rightarrow u_2 \Rightarrow \ldots \Rightarrow u_p$ is the *leftmost derivation* if the leftmost active symbol with respect to $u_i \Rightarrow^* u_p$ is active in the step $u_i \Rightarrow u_{i+1}$ for $i \in [1, p-1]$. For each $u, v \in \Sigma^*$ such that $u \Rightarrow^*_{\mathcal{H}} v$, there exists a leftmost derivation which starts at u and ends at v [7]. We say that the derivation $U \equiv (w \Rightarrow^* x)$ is *greedy* if:

(a) U is the leftmost derivation;
(b) a symbol a can become done in U only if it is not able to apply any delete rule (that is, it cannot become done in a sentential form $ubav$, where $(a \rightarrow^D b) \in P$);
(c) there are no derivation steps $uav \Rightarrow ubav$ in U such that the inserted symbol b does not eliminate any element of u during U (i.e., if b does not eliminate any element of u, it should not be inserted at all).

Fact 3. *Let $w_1abw_2 \Rightarrow^* x$ be a greedy derivation such that all symbols from the prefix w_1a are done with respect to this derivation. Then, if b is not able to eliminate a, it is done as well.*

Using the following theorem, one can assume in analysis of languages defined by leftist grammars that all derivations of leftist grammars are greedy.

Theorem 1. *Let $\mathcal{H}$ be a leftist grammar. Then, for each w such that $wx \Rightarrow^*_{\mathcal{H}} x$, there exists a greedy derivation $wx \Rightarrow^*_{\mathcal{H}} x$.*

In order to express the influence of the activity of a particular symbol (and its descendants) on the derivation, we define the notion of *trace*.

Definition 4 (Trace). *Let $U \equiv (w_1aw_2 \Rightarrow^* w')$ be a subderivation of the derivation W such that U starts at the first derivation step of W in which the symbol a (following the prefix w_1) is active and finishes at the last derivation step of W in which a or a descendant of a is active. Then, a trace of a in W, $T_W(a)$ or shortly $T(a)$, is equal to va, where v consists of all descendants of a in w', i.e., $w' = w_1'vaw_2'$ for $w_1', w_2' \in \Sigma^*$. (Note that, if W is greedy, all symbols in va are done wrt the remaining part of W.)*

If the symbol a is not active in any step of the derivation W, then $T(a) = a$.

3 PSPACE Hardness for General Grammars

We design a leftist grammar $\mathcal{H}_M$ which corresponds to a linear-bounded automaton (LBA) M, i.e., a one-tape Turing machine which does not leave the part of the tape between two delimiters $\triangleright$ and $\triangleleft$ which appear at the ends of the input, and M does not rewrite the delimiters. LBAs recognize exactly the set of context sensitive languages (CSL), and this class contains some PSPACE-hard languages. Let M be an LBA which recognizes a PSPACE-hard language. Let Γ be the tape alphabet of M ($\triangleright, \triangleleft \in \Gamma$) and let Q be the set of states of M. W.l.o.g., assume that M accepts only in configurations in which all cells between the delimiters are rewritten by a fixed symbol $\diamondsuit$ (where the symbol $\diamondsuit$ is not used at all in non-accepting computations) and the head is located on the right delimiter, $\triangleleft$. Let Accept be the shorthand for such configuration. We encode configurations of M using the alphabet $\Lambda = \Gamma \cup (\Gamma \times Q) \cup \{\flat\}$, where $\flat \notin \Gamma$ is the fixed extra symbol. A configuration with tape content $\triangleright a_1 \ldots a_n \triangleleft$ ($a_i \in \Gamma \setminus \{\triangleright, \triangleleft\}$ for $i \in [1, n]$), the head located at a_i and the state q is encoded as $\flat \triangleright a_1 \ldots a_{i-1} \langle a_i, q \rangle a_{i+1} \ldots a_n \triangleleft \flat$. As usual, $C \vdash_M C'$ denotes that the configuration C' is obtained from the configuration C in one step of M. Transitions

$\delta(q,a) = (q', a', \mathsf{left})$ and $\delta(q,a) = (q', a', \mathsf{right})$ of M can be expressed by the sets of the following rewrite rules applicable on the encodings of the configurations

$$\{c\langle q,a\rangle \to \langle q',c\rangle a' \mid c \in \Gamma\} \text{ for } \delta(q,a) = (q',a',\mathsf{left})$$
$$\{\langle q,a\rangle c \to a'\langle q',c\rangle \mid c \in \Gamma\} \text{ for } \delta(q,a) = (q',a',\mathsf{right})$$

where δ is the transition function of M. Due to limitations of leftist grammars, we will encode configurations using much broader alphabet. Some symbols of our alphabet are tuples, only one coordinate of each tuple corresponds to information stored in configurations of M. Let Φ be a set of all rewrite rules defining transitions of M, let

$$\begin{aligned} \mathcal{G} &= \{\langle G,i,j,a\rangle \mid i,j \in [0,1], a \in \Lambda\} \\ \mathcal{K} &= \{\langle K,i,j,a\rangle \mid i,j \in [0,1], a \in \Lambda\} \\ \mathcal{R} &= \{\langle R,i,a\rangle \mid i \in [0,1], a \in \Lambda\}, \\ \mathcal{G}_\Phi &= \{\langle G_{\alpha,i},j,l,a\rangle \mid i \in [1,2], j,l \in [0,1], a \in \Lambda, \alpha \in \Phi\}, \end{aligned}$$

where G, K, R are some fixed symbols. Finally, the alphabet Σ_M of $\mathcal{H}_M = (\Sigma_M, P_M, x)$ is equal to $\mathcal{G} \cup \mathcal{G}_\Phi \cup \mathcal{R} \cup \mathcal{K} \cup \{H, x\}$, where $\mathcal{H}_M$ is the leftist grammar associated with M (H, x are new symbols). The set P_M consists of the following productions

(10)	$\langle G,i,j,b\rangle \to^I \langle K,i,j,b\rangle$	(100)	$H \to^I \langle G,j,0,\flat\rangle$
(20)	$\langle G,i,j,b\rangle \to^I \langle G,i,\bar{j},b'\rangle$	(110)	$\langle R,0,\flat\rangle \to^D H$
(30)	$\langle K,i,j,b\rangle \to^D \langle K,i,\bar{j},b'\rangle$	(210)	$\langle G,i,j,b\rangle \to^I \langle G_{\alpha,2},i,\bar{j},b_2\rangle$
(40)	$\langle K,i,j,b\rangle \to^D \langle Y,\bar{i},j,b\rangle$	(220)	$\langle G_{\alpha,2},i,j,b_2\rangle \to^I \langle K,i,j,a_2\rangle$
(50)	$\langle R,j,b\rangle \to^D \langle R,\bar{j},c\rangle$	(230)	$\langle G_{\alpha,2},i,j,b_2\rangle \to^I \langle G_{\alpha,1},i,\bar{j},b_1\rangle$
(60)	$\langle R,j,b\rangle \to^D \langle Y,1,j,b\rangle$	(240)	$\langle G_{\alpha,1},i,j,b_1\rangle \to^I \langle K,i,j,a_1\rangle$
(70)	$x \to^D \langle R,i,\flat\rangle$	(250)	$\langle G_{\alpha,1},i,j,b\rangle \to^I \langle G,i,\bar{j},b'\rangle$
(90)	$x \to^D \langle K,i,j,b\rangle$		

where $i,j \in [0,1]$, $b,b',c \in \Lambda$, $\alpha = (a_1a_2 \to b_1b_2)$, $\alpha \in \Phi$, and $Y \in \{G\} \cup \{G_{\alpha,k} \mid \alpha \in \Phi, k \in [1,2]\}$.

We close this section with some additional notations. Let $\widetilde{\mathcal{G}} = \mathcal{G} \cup \mathcal{G}_\Phi$. Let $\mathcal{X}_i = \{a \mid a \in \mathcal{X}, \pi_2(a) = i\}$, and $\mathcal{X}_{i,j} = \{a \mid a \in \mathcal{X}, \pi_2(a) = i, \pi_3(a) = j\}$ for $i,j \in [0,1]$, $\mathcal{X} \in \{\mathcal{G}, \mathcal{K}, \widetilde{\mathcal{G}}, \mathcal{R}\}$. Moreover,

$$A_X(i,j) = \mathcal{X}_{i,j}(\mathcal{X}_{i,\bar{j}}\mathcal{X}_{i,j})^* \cup (\mathcal{X}_{i,\bar{j}}\mathcal{X}_{i,j})^*$$
$$A_X(i) = \mathcal{X}_i(\mathcal{X}_{\bar{i}}\mathcal{X}_i)^* \cup (\mathcal{X}_{\bar{i}}\mathcal{X}_i)^*$$

for $i,j \in [0,1]$, where ($X = G \wedge \mathcal{X} = \widetilde{\mathcal{G}}$), or ($X = K \wedge \mathcal{X} = \mathcal{K}$). Finally, $A_R(i) = (\mathcal{R}_i\mathcal{R}_{\bar{i}})^*\mathcal{R}_i \cup (\mathcal{R}_i\mathcal{R}_{\bar{i}})^*$. We say that a word w is *alternating* if it belongs to the language defined by some of $A_X(i,j)$ or $A_X(i)$ defined above.

3.1 High-Level Description of the Reduction

We say that a word $uwHv$ describes a configuration C of M if:

- $v \in A_R(0)$ (i.e., v is the alternating word over $\mathcal{R}$) such that $\pi_3(v)$ is equal to the reverse of the encoding of the accepting configuration of length $m = |C|$ (i.e., $\pi_3^R(v) = \flat \rhd \Diamond^{m-4}(\lhd, q_A)\flat$, where q_A is the accepting state of M);

- $w \in A_G(i,0)$ for $i \in [0,1]$, $\pi_4(w) = C$ (i.e., w is the alternating word over $\tilde{\mathcal{G}}_i$ which encodes C on its 4th coordinate);
- $u \in A_K(i)$.

In the following, we implicitly give the factorization of $uwHv$ into u, w, v as defined above, by saying that $uwHv$ defines a configuration of M. In order to distinguish the cases that $w \in \widetilde{\mathcal{G}}_0$ or $w \in \widetilde{\mathcal{G}}_1$, we will say that $uwHv$ 0-describes or 1-describes a configuration.

Let C_0 be an (encoding of) initial configuration of M. We reduce the question whether M can accept starting at C_0 to the question whether the language $L(\mathcal{H}_M)$ contains the word u_0w_0Hv which 0-describes C_0, where $u_0 = \varepsilon$, $\pi_1(w_0) = G^{|w_0|}$ (so, u_0w_0Hv is determined uniquely).

Our aim is to show that a greedy derivation of $\mathcal{H}_M$ which starts at w_0Hvx can finish at x only in the following way, which corresponds to the accepting computation of M (and, each accepting computation of M which starts at C_0 determines the appropriate greedy derivation $w_0Hvx \Rightarrow^* x$). Assume that a sentential form u_iw_iHvx appears in this derivation for $i \geq 0$, which j-describes the configuration C_i of M, for $j \in [0,1]$. Then,

(a) if C_i is not the accepting nor rejecting configuration and there exists a computation $C_i \vdash \mathsf{Accept}$, the subderivation $u_iw_iHvx \Rightarrow^* u_{i+1}w_{i+1}Hvx$ appears in the derivation $u_iw_iHvx \Rightarrow^* x$, such that $u_{i+1}w_{i+1}Hv$ $\bar{j}$-describes the configuration C_{i+1} such that $C_i \vdash_M C_{i+1} \vdash^* \mathsf{Accept}$ or $C_i = C_{i+1}$. During this subderivation, w_i (which encodes C_i) is replaced with the alternating word w_{i+1} over $\tilde{\mathcal{G}}_{\bar{j}}$ (which encodes C_{i+1}). We obtain this subderivation in such a way that (first) H inserts the symbol $a \in \mathcal{G}_{\bar{j}}$, then w_i is eliminated by the descendants of a in a subderivation after which a leaves the trace $T(a) = bw_{i+1}$ for $b \in \mathcal{K}_{\bar{j}}$. (See the productions (10)-(40), (100) and (210)-(250).) In the next subsections we concentrate on the proof that this scenario is the only possible in greedy derivations.

(b) if C_i is the accepting configuration and $j = 1$, there exists a subderivation $u_iw_iHvx \Rightarrow^* x$ in which w_iH is eliminated by v and u_i is deleted by x (similarly, if $C_i \vdash^* \mathsf{Accept}$, then there exists the appropriate subderivation). This statement is guaranteed by the fact that the value of the second coordinates in v allow (by the productions (50), (60)) to delete an alternating word over $\widetilde{\mathcal{G}}_1$ iff that word encodes the accepting configuration of M of the length $|v|$ at its 4th coordinate or a subsequence of the appropriate accepting configuration. (See the productions (50)-(90) and (110).)

(c) if there is no accepting computation which starts at C_i, there is no derivation $u_iw_iHvx \Rightarrow^* x$.

Now, we explain the "roles" of the groups of symbols from Σ_M in the grammar. The symbols from $\widetilde{\mathcal{G}}$ are used to store the configurations of M (the elements of $\mathcal{G}_\Phi$ help to introduce changes which reflect the consecutive steps of M; the complication is in ensuring that all derivation steps simulating the changes induced by the step of M are really executed). The anihilators from $\mathcal{K}$ are introduced in order to delete the "previous" configuration and replace it with the new one in the appropriately synchronized way. Moreover, the word over $\mathcal{R}$ is needed in order to verify whether the simulation finishes at the accepting configuration and whether the final configuration has the correct length (note that the elements of $\mathcal{R}$ are not generators). Finally, H is the special symbol which initiates the consecutive stages of the derivation (see (a) above).

Let us point out here why we require that the words w_i and v are alternating (over $\mathcal{G}_{i \bmod 2}$, and $\mathcal{R}$, resp.). This fact, combined with constraints of the production rules, ensures that v should be at least as long as w_i in order to delete w_i (item (b)). Similarly, the word w_{i+1} in the item (a) should (be alternating and) have the length equal or larger than $|w_i|$ in order to eliminate w_i, as w_i is the alternating word (see the productions 10-40). But, in order to finish the derivation at x using the item (b), $|w_{i+1}| = |w_i| = |v|$ for each i. On the other hand, the equality $|w_{i+1}| = |w_i|$ guarantees that $\pi_4(w_{i+1})$ in fact encodes the configuration following $\pi_4(w_i)$ (it ensures that no „artificial" symbols were inserted into the configurations).

3.2 The Formal Proof of the Correctness of the Reduction

In this section, we prove the correctness of the above reduction. All statements formulated below concern the conditions which have to be satisfied by greedy derivations. As we extensively use (implicitly) Fact 3, we collect some properties needed to apply it.

Fact 5. *A symbol $a \in \Sigma_M$ is not able to eliminate a symbol $b \in \Sigma_M$ if*
(a) $a, b \in X$ where $X \in \{\widetilde{\mathcal{G}}_i, \mathcal{K}_{i,j}\}$;
(b) $a \in \mathcal{K}_{i,j}$, $b \in \widetilde{\mathcal{G}}_{\bar{i},\bar{j}}$;
(c) $a \in \Sigma \setminus (\mathcal{K}_i \cup \widetilde{\mathcal{G}}_i \cup \{x\})$, $b \in \mathcal{K}_i$;
(d) $a = x$, $b \in \widetilde{\mathcal{G}}$;
(e) $a \in \mathcal{R}$, $b \in \widetilde{\mathcal{G}}_0 \cup \mathcal{K}$;
for $i, j \in [0, 1]$.

Let us notice that the sets of generators ($\widetilde{\mathcal{G}} \cup \{H\}$) and anihilators ($\mathcal{K} \cup \mathcal{R} \cup \{x\}$) in the grammar $\mathcal{H}_M$ are disjoint. The following lemma specifies conditions under which the alternating word $v \in \mathcal{R}^*$ can eliminate the alternating word $w \in \widetilde{\mathcal{G}}_i^*$ of length $\geq |v|$, where $i \in [0, 1]$.

Lemma 1. *Let $uwHvx$ be a sentential form, where $v \in A_R(0)$, $w \in A_G(i, 0)$, $i \in [0, 1]$ and $|w| \geq |v|$. Then,*
(a) *v is able to eliminate wH iff $|w| = |v|$, $(\pi_4(w))^R = \pi_3(v)$ and $i = 1$.*
(b) *If v does not eliminate the whole w in the derivation $uwHvx \Rightarrow^* x$, it cannot eliminate any symbol of w in this derivation.*

The item (a) of the above lemma follows from the constraints of the production (60) and the fact that v and w are alternating words over $\mathcal{R}$ and $\widetilde{\mathcal{G}}_i$, resp. (so, each symbol in v is able to delete ≤ 1 element of w). The item (b) is based on the observation that, if v does not eliminate whole w, then H or the anihilators from $\mathcal{K}$ will appear between v and w for the whole derivation. Indeed, x cannot eliminate the elements of $\widetilde{\mathcal{G}}$ (see Fact 5(d)), so the remaining part of w (not eliminated by v) should be eliminated by H (before v eliminates any element of w); and it is possible only by inserting the elements of $\mathcal{K}$ which can be deleted only by other elements of $\mathcal{K}$ and by x (not by the elements of v).

Below, we show that a greedy derivation $u_iw_iHvx \Rightarrow^* x$ (where u_iw_iHv j-describes C_i for $j \in [0,1]$) has to contain a subderivation which corresponds to $C_i \vdash C_{i+1}$ such that $C_{i+1} \vdash^* \mathsf{Accept}$ (and conversely, there exists a subderivation which corresponds to each possible step $C_i \vdash C_{i+1}$ of M).

Lemma 2. *Let u_iw_iHvx be a sentential form, which j-describes the non-accepting configuration C_i, where $j \in [0,1]$, let $C_i \vdash_M C_{i+1}$. Then,*
(1) for each $C' \in \{C_i, C_{i+1}\}$, there exists a derivation

$$U \equiv (u_iw_iHvx \Rightarrow^* u_{i+1}w_{i+1}Hvx)$$

such that the symbols from u_iw_i are not active in this derivation, $u_{i+1}w_{i+1}Hv$ is the word which $\bar{j}$-describes C' and $u_{i+1} \in u_i\mathcal{K}_{\bar{j}}$.
(2) if there exists the derivation $U \equiv (u_iw_iHvx \Rightarrow^* x)$ such that the symbols located to the left of H are done in U, it contains the subderivation $u_iw_iHvx \Rightarrow^* u_{i+1}w_{i+1}Hvx$, where $u_{i+1} \in u_iK_{\bar{j}}$, $u_{i+1}w_{i+1}Hv$ $\bar{j}$-describes the configuration C_{i+1} such that $C_i = C_{i+1}$ or $C_i \vdash_M C_{i+1}$. Moreover, all elements of $u_{i+1}w_{i+1}$ are done with respect to each greedy derivation $u_{i+1}w_{i+1}Hvx \Rightarrow^* x$.

The derivation showing correctness of Lemma 2(1) goes as follows. First, H inserts the symbol $w_{i+1}[n]$ (production (100)), where $n = |w_i|$. Then, for $l = n, n-1, \ldots, 1$: $w_{i+1}[l]$ inserts the anihilator b (by (10), (220), or (240)), then b deletes the anihilator inserted by $w_i[l+1]$ (if $l < n$) using (30), b deletes $w_i[l]$ using (40); finally, for $l > 1$, $w_{i+1}[l]$ inserts $w_{i+1}[l-1]$ (productions (20), (210), (230), or (250)). The proof of Lemma 2(2) is presented in Section 3.3.

Theorem 2. *Let u_0w_0Hv be a word which 0-describes C_0, the initial configuration of M on the input word z, let $u_0 = \varepsilon$. Then, there exists a derivation $u_0w_0Hvx \Rightarrow^* x$ if and only if $z \in L(M)$.*

One can prove the direction $\Rightarrow$ by applying Lemma 1 and the induction based on Lemma 2(2). For the opposite direction, the result follows from Lemma 1 and (the inductive application of) Lemma 2(1).

Corollary 1. *The membership problem for leftist grammars is* PSPACE*-hard.*

3.3 Proof of Lemma 2(2)

Our aim is to show that each greedy derivation should satisfy the scenario described in the proof of Lemma 2(1). First, we will see that the behavior of $a \in \widetilde{\mathcal{G}}_i$ (and its descendants) in greedy derivations has to be very regular, close to the scenario from the proof of Lemma 2(1).

Proposition 1. *Let $a \in \widetilde{\mathcal{G}}_{i,j}$ for $i, j \in [0,1]$ be a symbol which appears in the sentential form y. Then, the following conditions are satisfied in each greedy derivation U which starts at y:*

(a) if the symbol a inserts a generator in some derivation step (see (20), (210), (230) or (250)), it becomes inactive directly after this step;
(b) the symbol a inserts at most one anihilator.
(c) $T_U(a) = uw$ for $u \in \mathcal{K}_i$ and $w \in A_G(i,j)$, such that $(w[1] \to^I u) \in P_M$.
(d) Let $T_U(a) = uw$ for u, v as above, let $n = |w|$. Then, the set of interfaces of all possible (sub)derivations $a \Rightarrow^* uw$ is equal to $\{z_n z_{n-1} \ldots z_1 \in \mathcal{K}^+ \mid (w[i] \to^I z_i) \in P_M$ or $z_i = \varepsilon$ for $i \in [1,n]\}$.

The statements (a) and (b) of the above proposition follow from Fact 5(a,b) combined with the restrictions of greedy derivations. The statements (c) and (d) are obtained as a result of the ordering of the applications of the production rules (during the subderivation in which a or its descendants are active) forced by (a), (b) and greadiness. Now, we consider the situation that a symbol $a \in \widetilde{\mathcal{G}}_i$ eliminates an alternating word w over $\widetilde{\mathcal{G}}_{\bar{i}}$.

Proposition 2. *Assume that a symbol $a \in \mathcal{G}_{i,0}$ eliminates $w \in A_G(\bar{i}, 0)$ for $i \in [0,1]$. Then, $wa \Rightarrow^* u'w' = T(a)$ such that $u' \in \mathcal{K}_i$, $w' \in A_G(i,0)$, $(w'[1] \to^I u') \in P_M$, and:*
(a) $|w'| \geq |w|$;
(b) if $\pi_4(w) = C$ for a configuration C of M and $|w| = |w'|$ then:
(i) if w' does not contain any element of $\mathcal{G}_\Phi$ then $\pi_4(w') = \pi_4(w)$.
(ii) otherwise, $\pi_4(w') = C'$ for C' such that $C \vdash_M C'$.

Proof. Proposition 1(c) implies that $T(a) = u'w'$ for u' and w' as above.
(a) Note that the elements of $\widetilde{\mathcal{G}}_{\bar{i},0}$ and $\widetilde{\mathcal{G}}_{\bar{i},1}$ alternate in w. And, the elements of $\mathcal{K}_{i,j}$ cannot eliminate elements of $\mathcal{G}_{\bar{i},\bar{j}}$ (see (40) and Fact 5(b)) for $i, j \in [0,1]$. So, as all possible anihilators which are descendants of a belong to $\mathcal{K}_i$, each such anihilator is able to delete *at most* one element of w. On the other hand, each element of $w' \in \widetilde{\mathcal{G}}_i^*$ inserts at most one anihilator (see Proposition 1(b)). Thus, in order to delete w, the condition $|w'| \geq |w|$ should be satisfied.
(b) Let $|w| = |w'| = n$. We see by the above discussion and the assumption $|w| = |w'|$ that each element of w' has to insert *exactly* one anihilator and this anihilator deletes one element of w. So, by Proposition 1(d), the interface of the subderivation $a \Rightarrow^* u'w'$ is equal to $z = z_n \ldots z_1$ such that z_l is an anihilator inserted by $w'[l]$ for $l \in [1,n]$. And, z_l deletes $w[l]$ for $l \in [1,n]$.
(i) If w' does not contain elements of $\mathcal{G}_\Phi$, then all elements of z are inserted by the production (10), so $z_l = \langle K, i, j, c_l \rangle$ for $w'[l] = \langle G, i, j, c_l \rangle$. Further, the only elements of $\widetilde{\mathcal{G}}_{\bar{i}}$ which can be deleted by z_l are (see (40))

$$\{\langle X, \bar{i}, j, c_l \rangle \mid X \in \{G\} \cup \{G_\alpha \mid \alpha \in \Phi\}\}.$$

So, $\pi_4(w') = (\pi_4(z))^R = \pi_4(w)$, because z_l deletes $w[l]$ for $l \in [1,n]$.
(ii) Intuitively, this statement is guaranteed by the constraints of (210)-(250) which allow to insert symbols of type $G_{\alpha,1}$ and $G_{\alpha,2}$ in pairs and the pair $G_{\alpha,1}G_{\alpha,2}$ can be inserted only if it reflects the change $C \vdash C'$ (otherwise, $|w'| > |w|$). One of the alternatives is that a symbol of type $G_{\alpha,2}$ is inserted as the leftmost symbol of w'. But then it cannot delete the leftmost symbol of w which contains $\flat$ at its 4th coordinate (as $\alpha = (a_1 a_2 \to b_1 b_2)$ for $a_1, a_2 \neq \flat$),

so $|w'| > |w|$. This observation explains why $\flat$ is added as the leftmost and the rightmost symbol of each configuration. Details are presented below.

Let $\alpha \in \Phi$ be a rewrite rule $a_1a_2 \to b_1b_2$ such that w' contains a symbol $\langle G_{\alpha,p}, i, j, b_p\rangle$ for some $j \in [0,1]$ and $p \in [1,2]$. As $\langle G_{\alpha,1}, i, j, b_1\rangle$ can be inserted only by $\langle G_{\alpha,2}, i, j, b_2\rangle$ (and the element of $\widetilde{\mathcal{G}}_i$ cannot eliminate another element of $\widetilde{\mathcal{G}}_i$ by Fact 5(a)), w' has to contain the symbol $\langle G_{\alpha,2}, i, j, b_2\rangle$.

First, we argue that $\langle G_{\alpha,2}, i, j, b_2\rangle$ has to insert $\langle G_{\alpha,1}, i, j, b_1\rangle$. Otherwise, $\langle G_{\alpha,2}, i, j, b_2\rangle$ would be the leftmost element in w'. Indeed, the only element from $\widetilde{\mathcal{G}}$ (and the only generator) which can be inserted by $\langle G_{\alpha,2}, i, j, b_2\rangle$ is $\langle G_{\alpha,1}, i, j, b_1\rangle$. On the other hand, $\langle G_{\alpha,2}, i, j, b_2\rangle$ cannot be the leftmost symbol of w'. In fact, the anihilator inserted by the leftmost element of w' deletes $w[1]$ using (40), so the 4th coordinate of $w'[1]$ and of $w[1]$ should agree. The 4th coordinate of $w[1]$ is equal to $\flat$ (because $\pi_4(w) = C$ for a configuration C) and the 4th coordinate of an anihilator inserted by $\langle G_{\alpha,2}, i, j, b_2\rangle$ is equal to $a_2 \neq \flat$. This follows from the fact that none of symbols a_1, a_2, b_1, b_2 in a rewrite rule $a_1a_2 \to b_1b_2$ from Φ is equal to $\flat$, because the head of M does not move outside of the part of the tape between delimiters $\triangleright$ and $\triangleleft$.

We show that if w' contains exactly one element b such that $\pi_1(b) = G_{\alpha,2}$ for $\alpha \in \Phi$ then $\pi_4(w')$ describes a configuration C' such that $C \vdash_M C'$. Indeed, by the above discussion, the only elements in w' which do not belong to $\mathcal{G}$ form a subword $\langle G_{\alpha,1}, i, j, b_1\rangle\langle G_{\alpha,2}, i, \overline{j}, b_2\rangle$ for some $j \in [0,1]$ and $\alpha \in \Phi$ equal to $a_1a_2 \to b_1b_2$. Assume that this subword appears at positions p and $p+1$ of w'. Then, $z_l = \langle K, i, (n-l) \bmod 2, c_l\rangle$ for $l \notin [p, p+1]$, $z_p = \langle K, i, (n-p) \bmod 2, a_1\rangle$, and $z_{p+1} = \langle K, i, (n-p-1) \bmod 2, a_2\rangle$, where $z = z_nz_{n-1}\dots z_1$ is the interface of $a \Rightarrow^* u'w'$ (see Proposition 1(d)). So, w' is able to delete w if $\pi_4(w')$ is obtained from $\pi_4(w)$ by the application of the rewrite rule $a_1a_2 \to b_1b_2$ (see (40)).

For the sake of contradiction assume that there are (at least) two symbols $\langle G_{\alpha,2}, i, j, b_2\rangle$ and $\langle G_{\alpha',2}, i, j', b_2'\rangle$ in w', where $\alpha = (a_1a_2 \to b_1b_2)$ and $\alpha' = (a_1'a_2' \to b_1'b_2')$ and $\alpha, \alpha' \in \Phi$. According to the above arguments, $\langle G_{\alpha,2}, i, j, b_2\rangle$ is preceded by $\langle G_{\alpha,1}, i, \overline{j}, b_1\rangle$ in w' and $\langle G_{\alpha',2}, i, \overline{j'}, b_2'\rangle$ is preceded by $\langle G_{\alpha',1}, i, j', b_1'\rangle$. And, the anihilators inserted by them can delete the elements with a_1, a_2, a_1', a_2' on the 4th coordinate. However, exactly one of a_1, a_2 and exactly one of a_1', a_2' belongs to $\Gamma \times Q$. But $\pi_4(w)$ contains exactly one element from $\Gamma \times Q$, because it describes a configuration of M. Thus, we obtain contradiction with the fact that each anihilator inserted by a and its descendants deletes one element of w: (at least) one of the inserted anihilators does not delete any element of w. $\square$

Below, we present the key technical argument which helps to see that (and why) a subderivation simulating one step of M, if started, has to be finished. That is, the derivation step $uw\underline{H}vx \Rightarrow uw\underline{aH}vx$ (for $u \in \mathcal{K}^*$, $w \in \widetilde{\mathcal{G}}_i^+$, and $v \in \mathcal{R}^*$) forces the subderivation $uwaHvx \Rightarrow^* u'w'Hvx$, in which a eliminates w.

Proposition 3. *Let z be a sentential form in a derivation $uwHvx \Rightarrow^* x$, where $u \in A_K(j)$, $w \in A_G(j,0)$, $j \in [0,1]$. Then, z cannot contain a subword $z' \in \widetilde{\mathcal{G}}_j^+ \mathcal{K}_{\overline{j}} \widetilde{\mathcal{G}}_{\overline{j}}^+ H$, such that all symbols in z' except H are done with respect to the derivation $z \Rightarrow^* x$.*

Proof. For the sake of contradiction assume that z contains a subword $z' = z_1 d z_2 b H$, where $z_1 \in \widetilde{\mathcal{G}}_j^+$, $d \in \mathcal{K}_{\bar{j}}$, $z_2 \in \widetilde{\mathcal{G}}_{\bar{j}}^*$, $b \in \widetilde{\mathcal{G}}_{\bar{j}}$, and all symbols of $z_1 d z_2 b$ are done. As no production rule can insert H, the symbol H in z' is equal to the only occurence of H in $u_i w_i H v x$. So, $z = \ldots z_1 d z_2 b H v' x$, where v' is a subsequence of v (because vx does not contain any generator). Note that H is active in z with respect to the remaining part of the derivation, because the elements of v cannot eliminate $d \in \mathcal{K}$ (Fact 5(e)) and x is not able to eliminate symbols from $z_1 \in \widetilde{\mathcal{G}}^+$ by Fact 5(d) (recall that $z_1 d z_2 b$ is done). So, if H would not insert anything (note that H is not the anihilator), the factor $z_1 \in \widetilde{\mathcal{G}}_j^+$ has to be eliminated by x what is impossible (Fact 5(e)).

As we consider greedy derivations, H should insert a symbol $c \in \widetilde{\mathcal{G}}_j$ (i.e., the symbol which is able to eliminate b). In order to eliminate b, the element $c' \in \mathcal{K}_j$ should be inserted as a descendant of c: $z_1 d z_2 b \underline{H} \Rightarrow z_1 d z_2 b c \underline{H} \Rightarrow^* z_1 d z_2 b \underline{c'} \ldots c H$. We claim that it makes impossible to delete z_1 during this derivation. Note that $c' \in \mathcal{K}_j$ can be deleted only by x or the elements of $\mathcal{K}_j$ (see the productions). Thus, the leftmost descendant of H during the subderivation which starts at $z_1 d z_2 b \underline{c'} \ldots c H$ will belong to $\mathcal{K}_j$. As the elements of $\mathcal{K}_j$ cannot delete the elements of $\mathcal{K}_{\bar{j}}$ (Fact 5(c)), $d \in \mathcal{K}_{\bar{j}}$ remains undeleted, as long as any symbol to the left of x is active. Finally, if only x is active, it can delete d but there is no possibility to eliminate the elements of z_1 (by Fact 5(d)). □

Now, we apply the above propositions to the derivations that start at the sentential forms which are "similar" to j-descriptions of configurations (for $j \in [0, 1]$).

Proposition 4. *Let $U \equiv (uwHvx \Rightarrow^* x)$ be a derivation for $u \in A_K(i)$, $w \in A_G(i, 0)$, $v \in A_R(0)$, $(\pi_3(v))^R =$ Accept such that the symbols from uw are done wrt U. Moreover,* **(i)** *$|w| > |v|$ or* **(ii)** *$|w| = |v|$ and $\pi_4(w)$ does not describe an accepting configuration.*

Then, H has to insert $a \in \mathcal{G}_{\bar{i},0}$ which eliminates the factor w in the subderivation $wa \Rightarrow^* u'w' = T_U(a)$ for $u' \in K_{\bar{i}}$, and $w' \in A_G(\bar{i}, 0)$ such that $|w'| \geq |w|$.

Let us discuss shortly the proof of Proposition 4. By Lemma 1 and the above condition (i) or (ii), v does not eliminate any symbol of w. So, by Fact 5(d,e), H has to insert $a \in \widetilde{\mathcal{G}}_{i,0}$ which eliminates the whole w. Indeed, if a eliminates only the part of w, the concatenation of the remaining part of w and the trace of a will form the sequence in $\widetilde{\mathcal{G}}_i^+ \mathcal{K}_{\bar{i}} \widetilde{\mathcal{G}}_{\bar{i}}^+$, by Proposition 2. And, this gives contradiction to Proposition 3. (Note that a is able to eliminate w using the strategy from the proof of Lemma 2(1).) The remaining part follows from Proposition 2.

Now, we are ready to prove Lemma 2(2). Proposition 4 guarantees that H has to insert $a \in \mathcal{G}_{\bar{j},0}$ which eliminates w_i in the subderivation $u_i w_i H v y x \Rightarrow^* u_{i+1} w_{i+1} H v y x$, such that $w_{i+1} \in A_G(\bar{j}, 0)$, $u_{i+1} \in u_i K_{\bar{j}}$, $|w_{i+1}| \geq |w_i|$ and $u_{i+1} w_{i+1}$ is done. If $|w_{i+1}| = |w_i|$ then the result holds by Proposition 2(b). Assume by contradiction that $|w_{i+1}| > |w_i|$. Using the induction starting at $u_{i+1} w_{i+1} H v x$, we obtain an infinite sequence of subderivations $u_j w_j H v x \Rightarrow^* u_{j+1} w_{j+1} H v x$ of the derivation $u_{i+1} w_{i+1} H v x \Rightarrow^* x$ for $j \geq i + 1$, by Proposition 4 (as $|w_j| > |v|$ for each j). This implies that there is no derivation $u_i w_i H v x \Rightarrow^* x$, contradiction.

4 Variable Membership Problem

Let $\mathcal{H} = (\Sigma, P, x)$ be a leftist grammar, where $\Sigma = \{a_i\}_{i=1}^{p}$. An *Insert Graph* (*Delete Graph*, resp.) of $\mathcal{H}$ is $G(V, E)$, where $V = \{v_i\}_{i=1}^{p}$, $E = \{(v_i, v_j) \mid (a_i \to a_j a_i) \in P\}$, $(E = \{(v_i, v_j) \mid (a_j a_i \to a_i) \in P\}$, resp.). In this section, we consider the variable membership problem for leftist grammars with acyclic insert/delete graphs. Languages defined by such grammars are context-free [7], so the "standard" membership problem is in P in this case.

Theorem 3. *The variable membership problem for leftist grammars with acyclic insert graphs or acyclic delete graphs is* PSPACE*-hard.*

Proof. (Sketch) As before, we construct a leftist grammars which corresponds to a linear-bounded automaton M. However, for computations on inputs of size n, we design a separate grammar of linear size (with respect to n), $\mathcal{H}_M(n)$. We modify the previous construction in the following way. The values of the third coordinate of each symbol from $\widetilde{\mathcal{G}}, \mathcal{K}$ and the second coordinate of each symbol from $\mathcal{R}$ indicate the position of a symbol in a configuration. It contrasts to the previous construction, where the third coordinate in the elements of $\widetilde{\mathcal{G}}$ and $\mathcal{K}$ and the second coordinate in $\mathcal{R}$ indicated only the parity of the position. All rules in which "odd" symbols insert/delete "even" symbols (i.e., with odd/even values of the third coordinate) are modified such that the jth symbol inserts/deletes the $(j+1)$st or $(j-1)$st symbol. In this way, we avoid cycles in the insert graph (induced by the rule (20)) and in the delete graph (the rules (30), (50)). □

The constructions from [7] give the upper bounds.

Theorem 4. Let $\mathcal{H}$ be a leftist grammar with acyclic insert graph or acyclic delete graph. The variable membership problem for $\mathcal{H}$ is in EXPSPACE.

References

1. S. Bandyopadhyay, M. Mahajan, K.N. Kumar. A non-regular leftist language. Manuscript. 2005.
2. M. Blaze, J. Feigenbaum, J. Ioannidis, A. Keromytis. The role of trust management in distributed security. Secure Internet Programming, LNCS 1603, 185–210.
3. T. Budd. Safety in grammatical protection systems. *International Journal of Computer and Information Sciences*, 12(6):413–430, 1983.
4. O. Cheiner, V. Saraswat. Security Analysis of Matrix. Technical report, AT&T Shannon Laboratory, 1999.
5. M. Harrison. Introduction to Formal Language Theory. Addison-Wesley, 1978.
6. M. Harrison, W. Ruzzo, J. Ullman. Protection in operating systems. *Communications of the ACM*, 19(8):461–470, August 1976.
7. T. Jurdziński, K. Loryś. Leftist Grammars and the Chomsky Hierarchy. Proc. *Fundamentals of Computation Theory (FCT 2005)*, LNCS 3623, 282-293.
8. R. Lipton, L. Snyder. A linear time algorithm for deciding subject security. *Journal of the ACM*, 24(3):455–464, July 1977.
9. R. Motwani, R. Panigrahy, V.A. Saraswat, S. Venkatasubramanian. On the decidability of accessibility problems (extended abstract). *STOC 2000*, 306–315.
10. V. Saraswat. The Matrix Design. Technical report, AT&T Laboratory, April 1997.

Jumbo λ-Calculus

Paul Blain Levy

University of Birmingham

Abstract. We make an argument that, for any study involving computational effects such as divergence or continuations, the traditional syntax of simply typed lambda-calculus cannot be regarded as canonical, because standard arguments for canonicity rely on isomorphisms that may not exist in an effectful setting. To remedy this, we define a "jumbo lambda-calculus" that fuses the traditional connectives together into more general ones, so-called "jumbo connectives". We provide two pieces of evidence for our thesis that the jumbo formulation is advantageous.

Firstly, we show that the jumbo lambda-calculus provides a "complete" range of connectives, in the sense of including every possible connective that, within the beta-eta theory, possesses a reversible rule.

Secondly, in the presence of effects, we show that there is no decomposition of jumbo connectives into non-jumbo ones that is valid in both call-by-value and call-by-name.

1 Canonicity and Connectives

According to many authors [GLT88, LS86, Pit00], the "canonical" simply typed λ-calculus possesses the following types:

$$A ::= \quad 0 \mid A + A \mid 1 \mid A \times A \mid A \to A \tag{1}$$

There are two variants of this calculus. In some texts [GLT88, LS86] the $\times$ connective (type constructor) is a *projection product*, with elimination rules

$$\frac{\Gamma \vdash M : A \times B}{\Gamma \vdash \pi M : A} \qquad \frac{\Gamma \vdash M : A \times B}{\Gamma \vdash \pi' M : B}$$

In other texts [Pit00], $\times$ is a *pattern-match product*, with elimination rule

$$\frac{\Gamma \vdash M : A \times B \quad \Gamma, \mathtt{x} : A, \mathtt{y} : B \vdash N : C}{\Gamma \vdash \mathtt{pm}\ M\ \mathtt{as}\ \langle \mathtt{x}, \mathtt{y} \rangle.\ N : C}$$

This choice of five connectives $0, +, 1, \times, \to$ raises some questions.

1. Why not include a *ternary* sum type $+(A, B, C)$?
2. Why not include a type $(A, B) \to C$ of functions that take *two* arguments?
3. Why not include *both* a pattern-match product $A \times B$ *and* a projection product $A \,\Pi\, B$?

M. Bugliesi et al. (Eds.): ICALP 2006, Part II, LNCS 4052, pp. 444–455, 2006.

In the purely functional setting, these can be answered using Ockham's razor:

1. unnecessary—it would be isomorphic to $(A + B) + C$
2. unnecessary—it would be isomorphic to $(A \times B) \to C$, and to $A \to (B \to C)$
3. unnecessary—they would be isomorphic, so either one suffices.

But these answers are not valid in the presence of effectful constructs, such as recursion or control operators. For example, in a call-by-name language with recursion, $+(A, B, C) \not\cong (A + B) + C$ (a point made in [McC96b]), and $A \times B \not\cong A \sqcap B$. To see this, consider standard semantics that interprets each type by a pointed cpo. Then $+$ denotes lifted disjoint union, $A \sqcap B$ denotes cartesian product, and $A \times B$ denotes lifted product.

This suggests that, to obtain a canonical formulation of simply typed λ-calculus (suitable for subsequent extension with effects), we should—at least *a priori*—replace Ockham's minimalist philosophy with a maximalist one, treating many combinations of the above connectives as primitive. These combinations are called *jumbo connectives.* But how many connectives must we include to obtain a "complete" range?

A first suggestion might be to include *every* possible combination of the original five as primitive, e.g. a ternary connective γ mapping A, B, C to $(A \to B) \to C$. But this seems unwieldy. We need some criterion of reasonableness that excludes γ but includes all the connectives mentioned above.

We obtain this by noting that each of the above connectives possesses, within the $\beta\eta$ equational theory, a *reversible rule.* For example:

$$\frac{\Gamma, A \vdash B}{\overline{\Gamma \vdash A \to B}} \qquad\qquad \frac{\Gamma, A \vdash C \quad \Gamma, B \vdash C}{\overline{\Gamma, A + B \vdash C}}$$

The rule for $A \to B$ means that we can turn each inhabitant of $\Gamma, A \vdash B$ into an inhabitant of $\Gamma \vdash A \to B$, and vice versa, and these two operations are inverse (up to $\beta\eta$-equality). The rule for $A + B$ is understood similarly. Note also that, in these rules, every part of the conclusion other than the type being introduced appears in each premise. Informally, we shall say that a connective is "$\{0, +, 1, \times, \to\}$-like", when, in the presence of $\beta\eta$, it possesses such a reversible rule. In this paper, we introduce a calculus called "jumbo λ-calculus", and show that it contains every $\{0, +, 1, \times, \to\}$-like connective.

As stated above, our main argument for the necessity of jumbo connectives in the effectful setting is that suggested decompositions are not *a priori* valid, but in Sect. 4 we take this further by showing that, *a posteriori*, they do not have a decomposition that is valid in both CBV and CBN.

Related work. Both our arguments for jumbo connectives (invalidity of decompositions, possession of a reversible rule) have arisen in ludics [Gir01].

1.1 Infinitary Variant

Frequently, in semantics, one wishes to study infinitary calculi with countable sum types and countable product types. (The latter are necessarily projection products.) We therefore say that a connective is "$\{0, +, \sum_{i\in\mathbb{N}}, 1, \times, \prod_{i\in\mathbb{N}}, \to\}$-like" when it possesses a reversible rule with countably many premises. By contrast, a $\{0, +, 1, \times, \to\}$-like connective is required to have a reversible rule with finitely many premises.

We shall define an *infinitary* jumbo λ-calculus, as well as the finitary one, and show that the former contains every $\{0, +, \sum_{i\in\mathbb{N}}, 1, \times, \prod_{i\in\mathbb{N}}, \to\}$-like connective.

2 Jumbo λ-Calculus

Jumbo λ-calculus is a calculus of *tuples* and *functions*.

2.1 Tuples

A tuple in jumbo λ-calculus has several components; the first component is a tag and the rest are terms. (We often write tags with a # symbol to avoid confusion with identifiers.) An example of a tuple type is

$$\boxed{\sum}\{ \quad \#\mathtt{a}.\ \mathtt{int}, \mathtt{bool} \quad \#\mathtt{b}.\ \mathtt{bool}, \mathtt{int}, \mathtt{bool} \quad \#\mathtt{c}.\ \mathtt{int} \quad \} \tag{2}$$

This contains tuples such as $\langle \#\mathtt{a}, 17, \mathtt{false}\rangle$ and $\langle \#\mathtt{b}, \mathtt{true}, 5, \mathtt{true}\rangle$. The type (3) can *roughly* be thought of as an indexed sum of finite products:

$$\sum\{ \quad \#\mathtt{a}.\ (\mathtt{int} \times \mathtt{bool}) \quad \#\mathtt{b}.\ (\mathtt{bool} \times \mathtt{int} \times \mathtt{bool}) \quad \#\mathtt{c}.\ \mathtt{int} \quad \} \tag{3}$$

But whether (2) and (3) are actually isomorphic is a matter for investigation below—not something we may assume *a priori.*

If M is a term of the above type, we can pattern-match it:

$$\begin{array}{ll} \mathtt{pm}\ M\ \mathtt{as}\ \{ & \\ \quad \langle \#\mathtt{a}, \mathtt{x}, \mathtt{y}\rangle. & N \\ \quad \langle \#\mathtt{b}, \mathtt{x}, \mathtt{y}, \mathtt{z}\rangle. & P \\ \quad \langle \#\mathtt{c}, \mathtt{w}\rangle. & Q \\ \} & \end{array}$$

where N,P and Q all have the same type.

2.2 Functions

A function in jumbo λ-calculus is applied to several arguments; the first argument is a tag, and the rest are terms. An example of a function type is

$$\boxed{\prod}\{\quad \#\mathsf{a}.\ \mathtt{int},\mathtt{int},\mathtt{int}\vdash\mathtt{bool}\quad \#\mathsf{b}.\ \mathtt{int},\mathtt{bool}\vdash\mathtt{int}\quad \#\mathsf{c}.\ \mathtt{bool},\mathtt{int}\vdash\mathtt{int}\quad\} \tag{4}$$

An example function of this type is

$$\lambda\{\quad (\#\mathsf{a},\mathtt{x},\mathtt{y},\mathtt{z}).\ \mathtt{x}>(\mathtt{y}+\mathtt{z})\quad (\#\mathsf{b},\mathtt{x},\mathtt{y}).\ \mathtt{if}\ \mathtt{y}\ \mathtt{then}\ \mathtt{x}+5\ \mathtt{else}\ \mathtt{x}+7\quad (\#\mathsf{c},\mathtt{x},\mathtt{y}).\ \mathtt{y}+1\quad\} \tag{5}$$

Applying this to arguments $(\#\mathsf{a}, M, N, P)$ gives a boolean, whereas applying it to arguments $(\#\mathsf{b}, N, N')$ gives an integer. (Note the use of () for multiple arguments, and $\langle\rangle$ for tuple formation.) The type (4) can roughly be thought of as an indexed product of function types:

$$\prod\{\quad \#\mathsf{a}.\ (\mathtt{int}\to(\mathtt{int}\to(\mathtt{int}\to\mathtt{bool})))\quad \#\mathsf{b}.\ (\mathtt{int}\to(\mathtt{bool}\to\mathtt{int}))\quad \#\mathsf{c}.\ (\mathtt{bool}\to(\mathtt{int}\to\mathtt{int}))\quad\} \tag{6}$$

But again, we cannot assume *a priori* that (4) and (6) are isomorphic.

2.3 Summary

The types and terms of jumbo λ-calculus are shown in Fig. 1. Here, I ranges over all finite sets (for the finitary variant) or over all countable sets (for the infinitary variant), $\overrightarrow{A}$ indicates a finite sequence of types, $|\overrightarrow{A}|$ is its length, and $\$n$ (for $n \in \mathbb{N}$) is the set $\{0, \dots, n-1\}$. As in, e.g., [Win93], we include a construct `let` to make a binding, although this can be desugared in various ways.

Types $\quad A ::= \boxed{\sum}\{\overrightarrow{A}_i\}_{i\in I} \mid \boxed{\prod}\{\overrightarrow{A}_i \vdash A_i\}_{i\in I}$

Terms

$$\frac{}{\Gamma, \mathtt{x}:A, \Gamma' \vdash \mathtt{x}:A} \qquad \frac{\Gamma\vdash N:A \quad \Gamma,\mathtt{x}:A\vdash M:B}{\Gamma\vdash \mathtt{let}\ N\ \mathtt{be}\ \mathtt{x}.\ M:B}$$

$$\frac{\hat{\imath}\in I \quad \Gamma\vdash N_j : A_{\hat{\imath}j}\ \ (\forall j\in\$|\overrightarrow{A}_{\hat{\imath}}|)}{\Gamma\vdash\langle\hat{\imath},\overrightarrow{N}\rangle : \boxed{\sum}\{\overrightarrow{A}_i\}_{i\in I}} \qquad \frac{\Gamma\vdash N:\boxed{\sum}\{\overrightarrow{A}_i\}_{i\in I} \quad \Gamma,\overrightarrow{\mathtt{x}}:\overrightarrow{A}_i\vdash M_i : B\ \ (\forall i\in I)}{\Gamma\vdash \mathtt{pm}\ N\ \mathtt{as}\ \{\langle i,\overrightarrow{\mathtt{x}}\rangle.M_i\}_{i\in I} : B}$$

$$\frac{\Gamma,\overrightarrow{\mathtt{x}}:\overrightarrow{A}_i\vdash M_i:B_i\ \ (\forall i\in I)}{\Gamma\vdash\lambda\{(i,\overrightarrow{\mathtt{x}}).M_i\}_{i\in I} : \boxed{\prod}\{\overrightarrow{A}_i\vdash B_i\}_{i\in I}} \qquad \frac{\Gamma\vdash M:\boxed{\prod}\{\overrightarrow{A}_i\vdash B_i\}_{i\in I} \quad \hat{\imath}\in I \quad \Gamma\vdash N_j:A_{\hat{\imath}j}\ \ (\forall j\in\$|\overrightarrow{A}_{\hat{\imath}}|)}{\Gamma\vdash M(\hat{\imath},\overrightarrow{N}) : B_{\hat{\imath}}}$$

Fig. 1. Syntax Of Jumbo λ-calculus

2.4 Jumbo-Arities

Many traditional connectives are special cases of the jumbo connectives:

type	comments	expressed as
$A + B$		$\boxed{\sum}\{\#\mathsf{left}.A, \#\mathsf{right}.B\}$
$\sum_{i \in I} A_i$		$\boxed{\sum}\{A_i\}_{i \in I}$
$A \times B$	pattern-match product	$\boxed{\sum}\{\#\mathsf{sole}.A, B\}$
$\times(\overrightarrow{A})$	n-ary pattern-match product	$\boxed{\sum}\{\#\mathsf{sole}.\overrightarrow{A}\}$
$A \sqcap B$	projection product	$\boxed{\prod}\{\#\mathsf{left}. \vdash A, \#\mathsf{right}. \vdash B\}$
$\prod_{i \in I} A_i$	I-ary projection product	$\boxed{\prod}\{\vdash A_i\}_{i \in I}$
$A \to B$	type of functions with one argument	$\boxed{\prod}\{\#\mathsf{sole}.A \vdash B\}$
$(\overrightarrow{A}) \to B$	type of functions with n arguments	$\boxed{\prod}\{\#\mathsf{sole}.\overrightarrow{A} \vdash B\}$
`bool`		$\boxed{\sum}\{\#\mathsf{true}.\epsilon, \#\mathsf{false}.\epsilon\}$
ground_I	ground type with I elements	$\boxed{\sum}\{\epsilon\}_{i \in I}$
TA	studied in call-by-value setting [Mog89]	$\boxed{\prod}\{\#\mathsf{sole}. \vdash A\}$
LA	studied in call-by-name setting [McC96a]	$\boxed{\sum}\{\#\mathsf{sole}.A\}$

To make this more systematic, define a *jumbo-arity* to be a countable family of natural numbers $\{n_i\}_{i \in I}$. Then both $\boxed{\sum}$ and $\boxed{\prod}$ provide a family of connectives, indexed by jumbo-arities, as follows.

- Each jumbo-arity $\{n_i\}_{i \in I}$, determines a connective $\boxed{\sum}_{\{n_i\}_{i \in I}}$ of arity $\sum_{i \in I} n_i$. Given types $\{A_{ij}\}_{i \in I, j \in \$ n_i}$, it constructs the type $\boxed{\sum}\ \{A_{i0}, \ldots, A_{i(n_i-1)}\}_{i \in I}$.
- Each jumbo-arity $\{n_i\}_{i \in I}$, determines a connective $\boxed{\prod}_{\{n_i\}_{i \in I}}$ of arity $\sum_{i \in I}(n_i + 1)$. Given types $\{A_{ij}\}_{i \in I, j \in \$ n_i}$ and types $\{B_i\}_{i \in I}$, it constructs the type $\boxed{\prod}\ \{A_{i0}, \ldots, A_{i(n_i-1)} \vdash B_i\}_{i \in I}$.

Corresponding to the above instances, we have

connective	arity	expressed as
$+$	2	$\boxed{\sum}_{\{\#\mathsf{left}.1, \#\mathsf{right}.1\}}$
$\sum_{i \in I}$	I	$\boxed{\sum}_{\{1\}_{i \in I}}$
$\times$	2	$\boxed{\sum}_{\{\#\mathsf{sole}.2\}}$
$\times$	n	$\boxed{\sum}_{\{\#\mathsf{sole}.n\}}$
$\sqcap$	2	$\boxed{\prod}_{\{\#\mathsf{left}.0, \#\mathsf{right}.0\}}$
$\prod_{i \in I}$	I	$\boxed{\prod}_{\{0\}_{i \in I}}$
$\to$	2	$\boxed{\prod}_{\{\#\mathsf{sole}.1\}}$
$\to$	$n + 1$	$\boxed{\prod}_{\{\#\mathsf{sole}.n\}}$
`bool`	0	$\boxed{\sum}_{\{\#\mathsf{true}.0, \#\mathsf{false}.0\}}$
ground_I	0	$\boxed{\sum}_{\{0\}_{i \in I}}$
T	1	$\boxed{\prod}_{\{\#\mathsf{sole}.0\}}$
L	1	$\boxed{\sum}_{\{\#\mathsf{sole}.1\}}$

3 The $\beta\eta$-Theory of Jumbo λ-Calculus

3.1 Laws and Isomorphisms

In the absence of computational effects, the most natural equational theory for the jumbo λ-calculus is the $\beta\eta$-theory, displayed in Fig. 2.

β-laws

$$\frac{\Gamma \vdash N : A \quad \Gamma, \mathtt{x} : A \vdash M : B}{\Gamma \vdash \mathtt{let}\ N\ \mathtt{be}\ \mathtt{x}.\ M = M[N/\mathtt{x}] : B}$$

$$\frac{\hat{\imath} \in I \quad \Gamma \vdash N_j : A_{\hat{\imath}j} \ (\forall j \in \$|\overrightarrow{A}_{\hat{\imath}}|) \quad \Gamma, \overrightarrow{\mathtt{x}} : \overrightarrow{A}_i \vdash M_i : B \ (\forall i \in I)}{\Gamma \vdash \mathtt{pm}\ \langle \hat{\imath}, \overrightarrow{N} \rangle\ \mathtt{as}\ \{\langle i, \overrightarrow{\mathtt{x}} \rangle.M_i\}_{i \in I} = M_{\hat{\imath}}[\overrightarrow{N/\mathtt{x}}] : B_{\hat{\imath}}}$$

$$\frac{\Gamma, \overrightarrow{\mathtt{x}} : \overrightarrow{A}_i \vdash M : B_i \ (\forall i \in I) \quad \hat{\imath} \in I \quad \Gamma \vdash N_j : A_{\hat{\imath}j} \ (\forall j \in \$|\overrightarrow{A}_{\hat{\imath}}|)}{\Gamma \vdash \lambda\{(i, \overrightarrow{\mathtt{x}}).M_i\}_{i \in I}(\hat{\imath}, \overrightarrow{N}) = M_{\hat{\imath}}[\overrightarrow{N/\mathtt{x}}] : B_{\hat{\imath}}}$$

η-laws

$$\frac{\Gamma \vdash N : \boxed{\sum} \{\overrightarrow{A}_i\}_{i \in I} \quad \Gamma, \mathbf{z} : \boxed{\sum} \{\overrightarrow{A}_i\}_{i \in I} \vdash M : B}{\Gamma \vdash M[N/\mathbf{z}] = \mathtt{pm}\ N\ \mathtt{as}\ \{\langle i, \overrightarrow{\mathtt{x}} \rangle.M[\langle i, \overrightarrow{\mathtt{x}} \rangle/\mathbf{z}]\}_{i \in I} : B} \ \overrightarrow{\mathtt{x}} \text{ fresh for } \Gamma$$

$$\frac{\Gamma \vdash M : \boxed{\prod} \{\overrightarrow{A}_i \vdash B_i\}_{i \in I}}{\Gamma \vdash M = \lambda\{(i, \overrightarrow{\mathtt{x}}).M(i, \overrightarrow{\mathtt{x}})\}_{i \in I} : \boxed{\prod} \{\overrightarrow{A}_i \vdash B_i\}_{i \in I}} \ \overrightarrow{\mathtt{x}} \text{ fresh for } \Gamma$$

Fig. 2. The $\beta\eta$ Equational Theory For Jumbo λ-calculus

A *$\beta\eta$-isomorphism* $A \xrightarrow{\cong} B$ is a pair of terms $\mathtt{y} : A \vdash \alpha : B$ and $\mathbf{z} : B \vdash \alpha^{-1} : A$ such that $\alpha^{-1}[\alpha/\mathbf{z}] = \mathtt{y}$ and $\alpha[\alpha^{-1}/\mathtt{y}] = \mathbf{z}$ is provable up to $\beta\eta$-equality. We identify α and α' when $\alpha = \alpha'$ is provable.

The $\beta\eta$-theory gives non-jumbo decompositions and other isomorphisms, e.g.

$$\boxed{\sum} \{A_{i0}, \ldots, A_{i(n_i-1)}\}_{i \in I} \cong \textstyle\sum_{i \in I}(A_{i0} \times \cdots \times A_{i(n_i-1)})$$
$$\boxed{\prod} \{A_{i0}, \ldots, A_{i(n_i-1)} \vdash B_i\}_{i \in I} \cong \textstyle\prod_{i \in I}(A_{i0} \to \cdots A_{i(n_i-1)} \to B_i)$$
$$\times(\overrightarrow{A}) \cong \sqcap(\overrightarrow{A})$$
$$TA \cong A \cong LA$$

So the $\beta\eta$-theory makes the jumbo λ-calculus equivalent to that of Sect. 1.

3.2 Reversible Rules

Our next task is to make precise the notion of reversible rule from Sect. 1.

Definition 1. 1. For a sequent $s = \Gamma \vdash A$ (i.e. a pair of a context Γ and a type A), we write $\mathsf{inhab}\, s$ for the set of terms (modulo $\beta\eta$-equality) inhabiting s.
2. For a countable family of sequents $S = \{s_i\}_{i \in I}$, we write $\mathsf{inhab}\, S$ for $\prod_{i \in I} s_i$.
3. A *rule* from sequent family S to sequent family S' is a function from $\mathsf{inhab}\, S$ to $\mathsf{inhab}\, S'$. □

The reversible rules for $\to$ and $+$ shown in Sect. 1 are given for all Γ, and, in the case of $+$, for all C. Furthermore, they are "natural", as we now explain.

Definition 2. 1. [Lawvere] A *substitution* from a context $\Gamma = A_0, \ldots, A_{m-1}$ to a context Γ' is a sequence of terms $M_0, \ldots, M_{m-1}$ where $\Gamma' \vdash M_i : A_i$ for each $i \in \$m$. As usual, such a morphism induces a substitution function q^* from terms $\Gamma, \Delta \vdash B$ to terms $\Gamma', \Delta \vdash B$.
2. Any term $\Gamma, \mathtt{y} : C \vdash P : C'$ gives rise to a function $P^\dagger$ from terms inhabiting $\Gamma, \Delta \vdash C$ to terms inhabiting $\Gamma, \Delta \vdash C'$, where $P^\dagger N = P[N/\mathtt{y}]$. □

The $\to$ and $+$ reversible rules are *natural in* Γ in the sense that they commute with q^*, up to $\beta\eta$-equality, for any context morphism $\Gamma' \xrightarrow{q} \Gamma$. (Actually, they commute up to syntactic equality, but that is not significant here.) The $+$ reversible rule is also *natural in* C in the sense that it commutes with $P^\dagger$, up to $\beta\eta$-equality, for any term $\Gamma, \mathtt{y} : C \vdash P : C'$.

Definition 3. A *reversible rule* for a type B, in an equational theory, is a rule r with a single conclusion, such that

- r is a bijection
- the conclusion contains a single occurrence of B (adjacent to $\vdash$, let us say)
- the rest of the conclusion is arbitrary, appears in every premise, and the rule is natural in it.

In detail, either

reversible left rule the conclusion is $\Gamma, B \vdash C$, every premise contains $\Gamma \vdash C$—i.e. is of the form $\Gamma, \Delta \vdash C$—and r is natural in Γ and C, or
reversible right rule the conclusion is $\Gamma \vdash B$, every premise contains $\Gamma \vdash$—i.e. is of the form $\Gamma, \Delta \vdash B'$—and r is natural in Γ. □

Definition 4. We associate to the type $\boxed{\sum}\{\overrightarrow{A}_i\}_{i \in I}$ the reversible left rule

$$\frac{\Gamma, \overrightarrow{\mathtt{x}} : \overrightarrow{A}_i \vdash C \quad (\forall i \in I)}{\Gamma, \mathtt{y} : \boxed{\sum}\{\overrightarrow{A}_i\}_{i \in I} \vdash C} \qquad \begin{array}{ccc} \{M_i\}_{i \in I} & \mapsto & \mathtt{pm}\ \mathtt{y}\ \mathtt{as}\ \{\langle i, \overrightarrow{\mathtt{x}}\rangle.M_i\}_{i \in I} \\ N & \mapsto & \{N[\langle i, \overrightarrow{\mathtt{x}}\rangle/\mathtt{y}]\}_{i \in I} \end{array}$$

We associate to the type $\boxed{\prod}\{\overrightarrow{A}_i \vdash B_i\}_{i \in I}$ the reversible right rule

$$\frac{\Gamma, \overrightarrow{\mathtt{x}} : \overrightarrow{A}_i \vdash B_i \quad (\forall i \in I)}{\Gamma \vdash \boxed{\prod}\{\overrightarrow{A}_i \vdash B_i\}_{i \in I}} \qquad \begin{array}{ccc} \{M_i\}_{i \in I} & \mapsto & \lambda\{(i, \overrightarrow{\mathtt{x}}).M_i\}_{i \in I} \\ N & \mapsto & N(i, \overrightarrow{\mathtt{x}}) \end{array}$$

□

Definition 5. Given a reversible rule r for A, and an $\beta\eta$-isomorphism $A \xrightarrow{\cong} B$ comprised of $\mathtt{y} : A \vdash \alpha : B$ and $\mathtt{z} : B \vdash \alpha^{-1} : A$, we define a reversible rule r_α for B.

- If r is left, with conclusion $\Gamma, \mathtt{y} : A \vdash C$, then r_α has conclusion $\Gamma, \mathtt{z} : B \vdash C$. It maps a to $r(a)[\alpha^{-1}/\mathtt{y}]$, and its inverse maps N to $r^{-1}(N[\alpha/\mathtt{z}])$.
- If r is right, with conclusion $\Gamma \vdash A$, then r_α has conclusion $\Gamma \vdash B$. It maps a to $\alpha[r(a)/\mathtt{y}]$ and its inverse maps N to $r^{-1}(\alpha^{-1}[N/\mathtt{z}])$. □

We can now state the main technical property of jumbo λ-calculus:

Proposition 1. Let s be a reversible rule in the $\beta\eta$-theory of jumbo λ-calculus. Then s is r_α, where r is one of the rules in Def. 4 and α a $\beta\eta$-isomorphism; and r and α are unique. □

Proof. Suppose s is left, with conclusion $\Gamma, \mathtt{z} : B \vdash C$. Call the set indexing its premises I. For each $i \in I$, the ith premise must be of the form $\Gamma, \overrightarrow{\mathtt{x}} : \overrightarrow{A}_i \vdash C$. Set A to be the type $\boxed{\sum}\{\overrightarrow{A}_i\}_{i \in I}$, and r to be the reversible rule that Def. 4 associates to this type. That is clearly is the only possibility for r.

The rest is a syntactic version of the (indexed) Yoneda lemma. Define

- $\mathtt{y} : A \vdash \alpha : B$ to be $rs^{-1}(\mathtt{z} : B \vdash \mathtt{z} : B)$
- $\mathtt{z} : B \vdash \alpha^{-1} : A$ to be $sr^{-1}(\mathtt{y} : A \vdash \mathtt{y} : A)$.

We claim that

$$sr^{-1}(\Gamma, \mathtt{y} : A \vdash M : C) = M[\alpha^{-1}/\mathtt{y}] \tag{7}$$

$$rs^{-1}(\Gamma, \mathtt{z} : B \vdash N : C) = N[\alpha/\mathtt{z}] \tag{8}$$

For (7), we note that $M = M^\dagger k_\Gamma^*(\mathtt{y} : A \vdash \mathtt{y} : A)$. (Here k_Γ means the unique substitution from the empty context to Γ.) Hence the LHS is $sr^{-1}(M^\dagger k_\Gamma^*(\mathtt{y}))$. By naturality of s and r, this is $M^\dagger k_\Gamma^*(sr^{-1}(\mathtt{y}))$, which is $M^\dagger k_\Gamma^*(\alpha^{-1})$, the RHS. (8) is similar. Setting M to be α in (7) gives $\mathtt{z} = \alpha[\alpha^{-1}/\mathtt{y}]$, and similarly $\mathtt{y} = \alpha^{-1}[\alpha/\mathtt{z}]$. Setting M to be $r(a)$ in (7) gives $s = r_\alpha$. For uniqueness, $s = r_\beta$ implies

$$\alpha = rr_\beta^{-1}(\mathtt{z} : B \vdash \mathtt{z} : B) = rr^{-1}(\mathtt{z}[\beta/\mathtt{z}]) = \beta$$

The argument in the case that s is right is similar but easier. □

Thus $\boxed{\sum}$ and $\boxed{\prod}$ are the most general $\{0, +, \sum_{i \in I}, 1, \times, \prod_{i \in I}, \rightarrow\}$-like connectives, and the infinitary jumbo λ-calculus is greatest among calculi consisting of such connectives. Similarly, $\boxed{\sum}$ and $\boxed{\prod}$ with finite tag set are the most general $\{0, +, 1, \times, \rightarrow\}$-like connectives, and the finitary jumbo λ-calculus is greatest among calculi consisting of such connectives.

4 λ-Calculus Plus Computational Effects

4.1 Operational Semantics

In Sect. 4.1–4.2, we adapt standard material from e.g. [Win93] to the setting of jumbo λ-calculus. As a very simple example of a computational effect, let us consider divergence. So we add to the jumbo λ-calculus the typing rule

$$\frac{}{\Gamma \vdash \mathtt{diverge} : B}$$

where B may be any type. The $\beta\eta$-theory is inconsistent in the presence of a closed term of type 0, so we discard it. Our statement that each connective is $\{0, +, \sum_{i\in\mathbb{N}}, 1, \times, \prod_{i\in\mathbb{N}}, \rightarrow\}$-like means that *in the presence of* $\beta\eta$ it has a reversible rule. Since we have now discarded $\beta\eta$, these rules are lost.

We consider two languages with this syntax: call-by-name and call-by-value. As usual, each is defined by an operational semantics that maps closed terms to a special class of closed terms called *terminal terms*. We define this by an interpreter in Fig. 3. The metalanguage for the interpreter (written in italics) is first-order and recursive, containing the following constructs:

rec f lambda	for a recursive definition of a function f
P to D. Q	to mean: first evaluate P, then, if that gives D, evaluate Q
$\overrightarrow{P \textit{ to } D}.\ Q$	to abbreviate P_0 *to* $D_0 . \ldots P_{n-1}$ *to* D_{n-1}. Q.

Terminal Terms
- **CBN** Closed terms of the form $\langle \hat{\imath}, \overrightarrow{M} \rangle$ or $\lambda\{(i, \overrightarrow{\mathtt{x}}).M_i\}_{i\in I}$
- **CBV** Inductively defined by $T ::= \langle \hat{\imath}, \overrightarrow{T} \rangle \mid \lambda\{(i, \overrightarrow{\mathtt{x}}).M_i\}_{i\in I}$

CBN interpreter *rec cbn lambda*{

$\mathtt{let}\ N\ \mathtt{be}\ \mathtt{x}.\ M$	.	*cbn* $M[N/\mathtt{x}]$
$\langle \hat{\imath}, \overrightarrow{N} \rangle$	.	*return* $\langle \hat{\imath}, \overrightarrow{N} \rangle$
$\mathtt{pm}\ N\ \mathtt{as}\ \{\langle i, \overrightarrow{\mathtt{x}} \rangle.M_i\}_{i\in I}$	.	(*cbn* N) *to* $\langle \hat{\imath}, \overrightarrow{N} \rangle$. *cbn* $M_{\hat{\imath}}[\overrightarrow{N/\mathtt{x}}]$
$\lambda\{(i, \overrightarrow{\mathtt{x}}).M_i\}_{i\in I}$	.	*return* $\lambda\{(i, \overrightarrow{\mathtt{x}}).M_i\}_{i\in I}$
$M(\hat{\imath}, \overrightarrow{N})$	.	(*cbn* M) *to* $\lambda\{(i, \overrightarrow{\mathtt{x}}).M_i\}_{i\in I}$. *cbn* $M_{\hat{\imath}}[\overrightarrow{N/\mathtt{x}}]$
$\mathtt{diverge}$	.	*diverge*

}

CBV (left-to-right) interpeter *rec cbv lambda*{

$\mathtt{let}\ N\ \mathtt{be}\ \mathtt{x}.\ M$	.	(*cbv* N) *to* T. *cbv* $M[T/\mathtt{x}]$
$\langle \hat{\imath}, \overrightarrow{N} \rangle$	.	$\overrightarrow{(\textit{cbv } N) \textit{ to } T}$. *return* $\langle \hat{\imath}, \overrightarrow{T} \rangle$
$\mathtt{pm}\ N\ \mathtt{as}\ \{\langle i, \overrightarrow{\mathtt{x}} \rangle.M_i\}_{i\in I}$	.	(*cbv* N) *to* $\langle \hat{\imath}, \overrightarrow{T} \rangle$. *cbv* $M_{\hat{\imath}}[\overrightarrow{T/\mathtt{x}}]$
$\lambda\{(i, \overrightarrow{\mathtt{x}}).M_i\}_{i\in I}$	.	*return* $\lambda\{(i, \overrightarrow{\mathtt{x}}).M_i\}_{i\in I}$
$M(\hat{\imath}, \overrightarrow{N})$	.	(*cbv* M) *to* $\lambda\{(i, \overrightarrow{\mathtt{x}}).M_i\}_{i\in I}$. $\overrightarrow{(\textit{cbv } N) \textit{ to } T}$. *cbv* $M_{\hat{\imath}}[\overrightarrow{T/\mathtt{x}}]$
$\mathtt{diverge}$	.	*diverge*

}

Fig. 3. CBN and (left-to-right) CBV interpreters

Remark 1. Notice the consequences of the call-by-value semantics for the two binary products. A terminal term in $A \times B$ (the pattern-match product) is $\langle T, T' \rangle$, where T and T' are terminal. But, because we do not evaluate under λ, a terminal term in $A \sqcap B$ (the projection product) is $\lambda\{0.M, 1.N\}$, where M and N need not be terminal. This differs from the formulation in [Win93]. □

We write $M \Downarrow_{\mathbf{CBN}} T$ to mean that M evaluates to T in CBN, which can be defined inductively in the usual way. Otherwise M diverges and we write $M \Uparrow_{\mathbf{CBN}}$. Similarly for CBV.

For call-by-value, we inductively define *values*: $V ::= \mathtt{x} \mid \langle i, \overrightarrow{V} \rangle \mid \lambda\{(i, \overrightarrow{\mathtt{x}}).M_i\}_{i\in I}$.

4.2 Denotational Semantics

We extend the cpo semantics for CBN and CBV in [Win93] as follows.

In the call-by-name language, a type denotes a cpo with least element:

$$[\![\boxed{\sum}\{A_{i\,0},\ldots,A_{i\,(n_i-1)}\}_{i\in I}]\!] = (\sum_{i\in I}([\![A_{i\,0}]\!]\times\cdots\times[\![A_{i\,(n_i-1)}]\!]))_\bot$$

$$[\![\boxed{\prod}\{A_{i\,0},\ldots,A_{i\,n_i-1}\vdash B_i\}_{i\in I}]\!] = \prod_{i\in I}([\![A_{i\,0}]\!]\to\cdots\to[\![A_{i\,(n_i-1)}]\!]\to[\![B_i]\!])$$

A context $\Gamma = A_0,\ldots,A_{n-1}$ denotes the cpo $[\![A_0]\!]\times\cdots\times[\![A_{n-1}]\!]$, and a term $\Gamma\vdash M : B$ denotes a continuous function $[\![\Gamma]\!]\xrightarrow{[\![M]\!]}[\![B]\!]$.

In the call-by-value language, a type denotes a cpo:

$$[\![\boxed{\sum}\{A_{i\,0},\ldots,A_{i\,(n_i-1)}\}_{i\in I}]\!] = \sum_{i\in I}([\![A_{i\,0}]\!]\times\cdots\times[\![A_{i\,(n_i-1)}]\!])$$

$$[\![\boxed{\prod}\{A_{i\,0},\ldots,A_{i\,(n_i-1)}\vdash B_i\}_{i\in I}]\!] = \prod_{i\in I}([\![A_{i\,0}]\!]\to\cdots\to[\![A_{i\,(n_i-1)}]\!]\to([\![B_i]\!]_\bot))$$

A context $\Gamma = A_0,\ldots,A_{n-1}$ denotes $[\![A_0]\!]\times\cdots\times[\![A_{n-1}]\!]$, and a term $\Gamma\vdash M : B$ denotes a continuous function $[\![\Gamma]\!]\xrightarrow{[\![M]\!]}[\![B]\!]_\bot$. Each value $\Gamma\vdash V : B$ has another denotation $[\![\Gamma]\!]\xrightarrow{[\![V]\!]^{\mathsf{val}}}[\![B]\!]$ such that $[\![V]\!]\rho = \mathsf{up}\,([\![V]\!]^{\mathsf{val}}\rho)$ for all $\rho\in[\![\Gamma]\!]$.

The detailed semantics of CBN terms and of CBV terms and values are obvious and omitted. For both languages, we prove a substitution lemma, then show that $M \Downarrow T$ implies $[\![M]\!] = [\![T]\!]$, and $M\Uparrow$ implies $[\![M]\!] = \bot$, as in [Win93].

4.3 Invalidity of Decompositions

We say that types A and B are

- *cpo-isomorphic in CBN* when $[\![A]\!]_{\mathbf{CBN}}$ and $[\![B]\!]_{\mathbf{CBN}}$ are isomorphic cpos
- *cpo-isomorphic in CBV* when $[\![A]\!]_{\mathbf{CBV}}$ and $[\![B]\!]_{\mathbf{CBV}}$ are isomorphic cpos.

This is very liberal: e.g., 1_Π and 0 are cpo-isomorphic in CBN, though not isomorphic in other CBN models. But the purpose of this section is to establish *non*-isomorphisms, so that is good enough.

We begin by investigating the most obvious decompositions.

Proposition 2. The following decompositions are cpo-isomorphisms in CBN but not CBV:

$$\Pi(A_0,\ldots,A_{n-1}) \cong A_0\,\Pi\,A_1\cdots\Pi\,A_{n-1}$$

$$\boxed{\sum}\{\overrightarrow{A}_i\}_{i\in I} \cong \textstyle\sum_{i\in I}\Pi\,(\overrightarrow{A}_i)$$

$$(A_0,\ldots,A_{n-1})\to B \cong A_0\to A_1\to\cdots\to A_{n-1}\to B$$

$$(A_0,\ldots,A_{n-1})\to B \cong (A_0\,\Pi\cdots\Pi\,A_{n-1})\to B$$

$$\boxed{\prod}\{\overrightarrow{A}_i\vdash B_i\}_{i\in I} \cong \textstyle\prod_{i\in I}((\overrightarrow{A}_i)\to B_i)$$

The following decompositions are cpo-isomorphisms in CBV but not CBN:

$$+(A_0, \ldots, A_{n-1}) \cong A_0 + A_1 \cdots + A_{n-1}$$
$$\times(A_0, \ldots, A_{n-1}) \cong A_0 \times A_1 \cdots \times A_{n-1}$$
$$\boxed{\sum}\{\overrightarrow{A}_i\}_{i \in I} \cong \sum_{i \in I} \times (\overrightarrow{A}_i)$$
$$(A_0, \ldots, A_{n-1}) \to B \cong (A_0 \times \cdots \times A_{n-1}) \to B$$
$$\boxed{\prod}\{\overrightarrow{A}_i \vdash B_i\}_{i \in \$n} \cong \times_{i \in \$n}((\overrightarrow{A}_i) \to B_i)$$
$$\boxed{\prod}\{\overrightarrow{A}_i \vdash B_i\}_{i \in I} \cong \boxed{\prod}\{\times(\overrightarrow{A}_i) \vdash B_i\}_{i \in I}$$

Some special cases:

			CBV	CBN
$1_\times$	$\cong$	1_Π	yes	no
$\times\overrightarrow{A}$	$\cong$	$\Pi\overrightarrow{A}$	no	no
ground_I	$\cong$	$\sum_{i \in I} 1_\times$	yes	no
ground_I	$\cong$	$\sum_{i \in I} 1_\Pi$	yes	yes
TA	$\cong$	A	no	yes
LA	$\cong$	A	yes	no

□

Proof. For non-isomorphisms: make all the types `bool`, and count elements. □

A stronger statement of non-decomposability is the following. (We omit its proof, which analyzes finite elements.)

Proposition 3. Call the following types of jumbo λ-calculus *non-jumbo.*

$$A ::= \ \mathsf{ground}_I \mid \sum_{i \in I} A_i \mid \times(\overrightarrow{A}) \mid \prod_{i \in I} A_i \mid (\overrightarrow{A}) \to B$$

1. There is no non-jumbo type A such that $\boxed{\sum}\{\#\mathtt{a}.\mathtt{bool}, \mathtt{bool}; \#\mathtt{b}.\mathtt{bool}\}$ is cpo-isomorphic to A in both CBV and CBN.
2. There is no non-jumbo type A such that $\boxed{\prod}\{\#\mathtt{a}.\mathtt{bool} \vdash \mathtt{bool}; \#\mathtt{b}. \vdash \mathtt{bool}\}$ is cpo-isomorphic to A in both CBV and CBN.
3. There is no non-jumbo type A such that $\boxed{\prod}\{T\mathtt{bool} \vdash \mathsf{ground}_{\$n}\}_{n \in \mathbb{N}}$ is cpo-isomorphic to A in CBV. □

Thus, neither $\boxed{\sum}$ nor $\boxed{\prod}$ has a universally valid decomposition. And in the infinitary CBV setting, $\boxed{\prod}$ cannot be decomposed at all.

References

[Gir01] J.-Y. Girard. Locus solum: From the rules of logic to the logic of rules. *Mathematical Structures in Computer Science*, 11(3):301–506, 2001.

[GLT88] J.-Y. Girard, Y. Lafont, and P. Taylor. *Proofs and Types.* Cambridge Tracts in Theoretical Computer Science 7. Cambridge University Press, 1988.

[LS86] J. Lambek and P. Scott. *Introduction to Higher Order Categorical Logic.* Cambridge University Press, Cambridge, 1986.

[McC96a] G. McCusker. Full abstraction by translation. Proc., 3rd Workshop in Theory and Formal Methods, Imperial College, London., 1996.

[McC96b] G. McCusker. *Games and Full Abstraction for a Functional Metalanguage with Recursive Types.* PhD thesis, University of London, 1996.

[Mog89] E. Moggi. Computational lambda-calculus and monads. In *LICS'89, Proc. 4th Ann. Symp. on Logic in Comp. Sci.*, pages 14–23. IEEE, 1989.

[Pit00] A. M. Pitts. Categorical logic. In *Handbook of Logic in Computer Science, Vol. 5.* Oxford University Press, 2000.

[Win93] G. Winskel. *Formal Semantics of Programming Languages.* MIT Press, 1993.

λ-RBAC: Programming with Role-Based Access Control

Radha Jagadeesan[1,*], Alan Jeffrey[2,*], Corin Pitcher[1,**], and James Riely[1,***]

[1] School of CTI, DePaul University
[2] Bell Labs, Lucent Technologies

Abstract. We study mechanisms that permit program components to express role constraints on clients, focusing on programmatic security mechanisms, which permit access controls to be expressed, *in situ*, as part of the code realizing basic functionality. In this setting, two questions immediately arise:

- The user of a component faces the issue of safety: is a particular role sufficient to use the component?
- The component designer faces the dual issue of protection: is a particular role demanded in all execution paths of the component?

We provide a formal calculus and static analysis to answer both questions.

1 Introduction

This paper addresses programmatic security mechanisms as realized in systems such as Java Authentication and Authorization Service (JAAS) and .NET. JAAS and .NET enable two forms of access control mechanisms[1]. First, they permit *declarative* access control to describe security specifications that are orthogonal and separate from descriptions of functionality, e.g. in an interface *I*, a declarative access control mechanism could require the caller to possess a minimum set of rights. Second, JAAS and .NET also permit *programmatic* mechanisms that permit access control code to be intertwined with functionality code, e.g. in the code of a component implementing interface *I*. Why commingle conceptually separate concerns? To enable the programmer to enforce access control that is sensitive to the control and dataflow of the code implementing the functionality.

There is extensive literature on policy languages to specify and implement policies (e.g. [14, 25, 13, 3, 26, 12] to name but a few). This research studies security policies as separate and orthogonal additions to component code, and is thus focused on declarative security in the parlance of JAAS/.NET.

In contrast, we study programmatic security mechanisms. Our motivation is to *extract* the security guarantees provided by access control code which has been written inline with component code. We address this issue from two viewpoints:

* Supported by NSF Cybertrust 0430175.
** Supported by the DePaul University Research Council.
*** Supported by NSF Career 0347542.

[1] In this paper, we discuss only authorization mechanisms, ignoring the authentication mechanisms that are also part of these infrastructures.

M. Bugliesi et al. (Eds.): ICALP 2006, Part II, LNCS 4052, pp. 456–467, 2006.

- The user of a component faces the issue of safety: is a particular set of rights sufficient to use the component? (ie. any greater set of rights will also be allowed to use the component)
- The component designer faces the dual issue of protection: is a particular set of rights demanded in all execution paths of the component? (ie. any lesser set of rights will not be allowed to use the component)

The main contribution of this paper is separate static analyses to calculate approximations to these two questions. An approximate answer to the first question is a set of rights, perhaps bigger than necessary, that is *sufficient* to use the component. On the other hand, an approximate answer to the second question, is a set of rights, perhaps smaller than what is actually enforced, that is *necessary* to use the component.

Related Work. There is extensive literature on Role-Based Access-Control (RBAC) models including NIST standards for RBAC [22, 11]; see [10] for a textbook survey.

The main motivation for RBAC, in software architectures (e.g. [19, 18]) and frameworks such as JAAS/.NET is that it enables the enforcement of security policies at a granularity demanded by the application. In these examples, RBAC allows permissions to be de-coupled from users: Roles are the unit of administration for users and permissions are assigned to roles. Roles are often arranged in a hierarchy for succinct representation of the mapping of permissions. Component programmers design code in terms of a static collection of roles. When the application is deployed, administrators map the roles defined in the application to users in the particular domain.

Unified frameworks encompassing RBAC and trust–management systems have also been studied [23], in part by incorporating history-sensitive ideas into the RBAC model [4]. Our work is close in spirit, if not in technical development, to edit automata [14], which use aspects to avoid the explicit intermingling of security and baseline code.

The papers most closely related to our work are those of Braghin, Gorla and Sassone [6] and Compagnoni, Garalda and Gunter [9]. [6] presents the first concurrent calculus with a notion of RBAC, whereas [9]'s language enables privileges depending upon location. This paper extends the lambda calculus with RBACand internalizes the lattice structure of roles. We present a more detailed comparison at the end of the paper.

An Overview of Our Technical Contributions. We study a lambda calculus enriched with primitives for access control, dubbed λ-RBAC. The underlying lambda calculus serves as an abstraction of the ambient programming framework in a real system. We draw inspiration from the programming idioms in JAASand .NET to determine the expressiveness required for the access control mechanisms.

Roughly, the operation of λ-RBAC is as follows. Program execution takes place in the context of a role, which can be viewed concretely as a set of permissions. Roles are closed under union and intersection operations. The set of roles used in a program is static: we do not allow the dynamic creation of roles. The only run-time operations on roles are as follows. There are combinators to check that the role-context is at least some minimum role: an exception is raised if the check fails. Rights modulation (c.f. "sessions" in RBAC) is achieved by impersonation: this enables an application to operate under the guise of different users at different times.

We assume that roles form a lattice: abstracting the concrete union/intersection operations of the motivating examples. Some of our results assume that the lattice is boolean, i.e. the lattice has a negation operation abstracting the concrete set complement of the motivating examples. Our study is parametric on the underlying role lattice. Our calculus includes a single combinator for role checking and two combinators for impersonation: one for rights weakening and the other for rights amplification. We internalize the right to amplify rights by considering role lattices with an explicit role constructor, *amplify*. This enables the reuse of the access control mechanisms of λ-RBAC to control rights amplification.

We demonstrate the expressiveness of the calculus by building a range of useful combinators and a variety of small illustrative examples. We discuss type systems to perform the two analyses alluded to earlier: (a) an analysis to detect and remove unnecessary role-checks in a piece of code for a caller at a sufficiently high role, and (b) an analysis to determine the (maximal) role that is guaranteed to be required by a piece of code. For both we prove preservation and progress properties.

Rest of the Paper. We begin with a discussion of the dynamic semantics of λ-RBAC in section 2, illustrating the expressiveness of the language with examples in section 3. Section 4 describes the static analyses. The following section 5 provides types for the examples of section 3. We conclude with a summary of related work in section 6.

2 Language and Operational Semantics

2.1 Roles

The language of roles is built up from role constructors. The choice of role constructors is application dependent, but must include at least the six constructors discussed below. We assume that each role constructor κ has an associated arity, $\mathsf{arity}(\kappa)$. Roles P,Q,R,S,T have the form $\kappa(R_1,\ldots,R_n)$.

The semantics of roles is defined by the relation "$\vdash R \geqslant S$" which states that R dominates S. We do not define this relation, but rather assume that it has a suitable, application-specific definition; we impose only the following requirements.

We require that all constructors be monotone with respect to $\geqslant$.

Further we require that roles form a boolean lattice. So, the role lattice is distributive: we require that the set of constructors include the nullary constructors $\bot$ and $\top$ and binary constructors $\sqcup$ and $\sqcap$ (which we write infix). $\bot$ is the least element; $\top$ is the greatest element; $\sqcup$ and $\sqcap$ are idempotent, commutative, associative, and mutually distributive meet and join operations on the lattice of roles. For any R, S, we have $\vdash R \geqslant \bot$ and $\vdash \top \geqslant R$ and $\vdash R \sqcup S \geqslant R$ and $\vdash R \geqslant R \sqcap S$. In addition, there is a complement $R^\star$ for every role R, where R and S are complements if $R \sqcap S = \bot$ and $R \sqcup S = \top$.

Finally, in example 4 we require the unary constructor *amplify*, where $\mathit{amplify}(R)$ represents the right to store R in a piece of code; i.e. if $P = \mathit{amplify}(R)$, then role P stands for the right to provide the role R.

In summary, the syntax is as follows.

$$P,Q,R,S,T ::= \kappa(R_1,\ldots,R_n) \qquad \kappa ::= \cdots \mid \bot \mid \top \mid \sqcup \mid \sqcap \mid {}^\star \mid \mathit{amplify}$$

2.2 Terms

Our goal is to capture the essence of role-based systems, where roles are used to regulate the interaction of components of the system. We have chosen to base our language on the call-by-value lambda calculus[2] because it is simple and well understood, yet rich enough to capture the key concepts. (We expect that our ideas can be adapted to both process and object calculi.) The "components" in a lambda term are abstractions and their calling contexts. Thus it is function calls and returns that we seek to regulate, and, therefore, the language has roles decorating abstractions and applications. Abstraction is written "$\{Q\}\lambda x.M$" where role Q is demanded to execute M. Application is written "$\downarrow P\ U\ V$" where the caller restricts its rights to P during the execution of function U.

We define evaluation using a small-step operational semantics; therefore, we include explicit syntax for the *frame* "$\downarrow P\,[M]$" which represents the execution of term M with rights restricted to P. We use frames additionally for rights escalation, with the form "$\uparrow P\,[M]$." The two forms of frames together allows code to assume any role, which — if entirely uncontrolled — allows code to circumvent an intended policy. We address this issue in example 4, where we describe the use of the *amplify* constructor to control rights escalation.

Let x,y,z,f,g range over variable names. The syntax of values and terms are as follows.

$$U,V ::= \cdots \mid x \mid \{Q\}\lambda x.M$$
$$M,N,L ::= \cdots \mid U \mid \mathsf{let}\,x{=}M;\ N \mid \downarrow P\,U\,V \mid \downarrow P\,[M] \mid \uparrow P\,[M]$$

Evaluation is defined in terms of a *role context*. Formally, we define a judgment $R \vdash M \rightarrow N$, which indicates R is authorized to compute a single step of the initial program M, resulting in the new program N.

EVALUATION ($R \vdash M \rightarrow N$)

$$\frac{\vdash R \geqslant Q}{R \vdash \downarrow P\,(\{Q\}\lambda x.M)\,U \rightarrow \downarrow P\,[M[^U\!/\!_x]]}$$

$$\frac{R \vdash M \rightarrow M'}{R \vdash \mathsf{let}\,x{=}M;\ N \rightarrow \mathsf{let}\,x{=}M';\ N} \qquad \frac{R \sqcap P \vdash M \rightarrow M'}{R \vdash \downarrow P\,[M] \rightarrow \downarrow P\,[M']} \qquad \frac{R \sqcup P \vdash M \rightarrow M'}{R \vdash \uparrow P\,[M] \rightarrow \uparrow P\,[M']}$$

$$\frac{}{R \vdash \mathsf{let}\,x{=}U;\ N \rightarrow N[^U\!/\!_x]} \qquad \frac{}{R \vdash \downarrow P\,[U] \rightarrow U} \qquad \frac{}{R \vdash \uparrow P\,[U] \rightarrow U}$$

The rules are straightforward, but for application; note only that a frame is discarded when the guarded term is fully evaluated. Application involves two participants: the caller (or calling context) and the callee (or abstraction). Each participant may wish to protect itself from the other. When the caller $\downarrow P\ V\ U$ transfers control to V, it may

[2] We have chosen an explicitly sequenced variant (with let). Implicit sequencing can be recovered as follows: $\downarrow R\,M\,N \triangleq \mathsf{let}\,x{=}M;\ \mathsf{let}\,y{=}N;\ \downarrow R\,x\,y$. When x does not appear free in N, we abbreviate $\mathsf{let}\,x{=}M;\ N$ as $M;\ N$. To focus the presentation, we elide base types, indicating them in the syntax using ellipses. In examples, we use base types with the usual operators and semantics, including Int (with values 0, 1, etc), Bool (with values true, false) and Unit (with value ( )). We write $M[^U\!/\!_x]$ for the capture-avoiding substitution of U for x in M.

protect itself by restricting the role context to P while executing V. Symmetrically, the callee $(\{Q\}\lambda x.M)$ may protect itself by demanding that the role context before the call dominates Q. Significantly, the restricting frame created by the caller does not take effect until after the guard is satisfied. In brief, the protocol is "call-test-restrict," with the callee controlling the middle step. This alternation explains why restriction is syntactically fused into application.

A role is *trivial* if it has no effect on evaluation. Thus $\top$ is trivial in restricting frames and applications, whereas $\bot$ is trivial in providing frames and abstractions. We often elide trivial roles and trivial frames; thus, $\lambda x.M$ is read as $\{\bot\}\lambda x.M$ (the check always succeeds), and $U\ V$ is read as $\downarrow\top\ U\ V$ (the role context is unaffected by the resulting frame). In our semantics, these terms evaluate like ordinary lambda terms.

By stringing together a series of small steps, the final value for the program can be determined. Successful termination is written $R \vdash M \Downarrow U$ which indicates that R is authorized to run the program M to completion, with result U. Evaluation can fail because the term diverges or because an inadequate role is provided at some point in the computation; we write the latter as $R \vdash M \Downarrow \mathsf{fail}$[3].

LEMMA 1. *If* $S \vdash M \rightarrow M'$ *and* $\vdash R \geqslant S$ *then* $R \vdash M \rightarrow M'$. □

3 Examples

EXAMPLE 2 (ACCESS CONTROL LISTS). Consider a web server that provides remote access to files protected by Access Control Lists (ACLs) at the filesystem layer. A read-only filesystem can be modeled as:

$$\begin{aligned} \mathit{filesystem} \stackrel{\text{def}}{=}\ & \lambda \mathit{name}.\mathsf{if}\ \mathit{name} = \text{"file1"}\ \mathsf{then\ check}\ \mathit{ADMIN};\ \text{"content1"} \\ & \mathsf{else\ if}\ \mathit{name} = \text{"file2"}\ \mathsf{then\ check}\ \mathit{ALICE} \sqcap \mathit{BOB};\ \text{"content2"} \\ & \mathsf{else}\ \text{"error: file not found"} \end{aligned}$$

Where $\mathsf{check}\ R \triangleq (\{R\}\lambda\_.\,(\,))(\,)$. Assuming incomparable roles *ALICE*, *BOB*, and *CHARLIE* each strictly dominated by *ADMIN*, code running in the *ADMIN* role can access both files:

$$\begin{aligned} &\mathit{ADMIN} \vdash \mathit{filesystem}\ \text{"file1"} \rightarrow^* \mathsf{check}\ \mathit{ADMIN};\ \text{"content1"} \rightarrow^* \text{"content1"} \\ &\mathit{ADMIN} \vdash \mathit{filesystem}\ \text{"file2"} \rightarrow^* \mathsf{check}\ \mathit{ALICE} \sqcap \mathit{BOB};\ \text{"content2"} \rightarrow^* \text{"content2"} \end{aligned}$$

Code running as *ALICE* or *BOB* cannot access the first file but can access the second:

$$\begin{aligned} &\mathit{ALICE} \vdash \mathit{filesystem}\ \text{"file1"} \rightarrow^* \mathsf{check}\ \mathit{ADMIN};\ \text{"content1"} \Downarrow \mathsf{fail} \\ &\mathit{BOB} \vdash \mathit{filesystem}\ \text{"file2"} \rightarrow^* \mathsf{check}\ \mathit{ALICE} \sqcap \mathit{BOB};\ \text{"content2"} \rightarrow^* \text{"content2"} \end{aligned}$$

Finally, assuming that $\mathit{CHARLIE} \not\geqslant \mathit{ALICE} \sqcap \mathit{BOB}$, code running as *CHARLIE* cannot access either file:

$$\begin{aligned} &\mathit{CHARLIE} \vdash \mathit{filesystem}\ \text{"file1"} \rightarrow^* \mathsf{check}\ \mathit{ADMIN};\ \text{"content1"} \Downarrow \mathsf{fail} \\ &\mathit{CHARLIE} \vdash \mathit{filesystem}\ \text{"file2"} \rightarrow^* \mathsf{check}\ \mathit{ALICE} \sqcap \mathit{BOB};\ \text{"content2"} \Downarrow \mathsf{fail} \end{aligned}$$

[3] Write "$R \vdash M_0 \rightarrow^* M_n$" if there exist terms M_i such that $R \vdash M_i \rightarrow M_{i+1}$, for all i $(0 \leq i \leq n-1)$. Write "$R \vdash M \Downarrow U$" if $R \vdash M \rightarrow^* U$. Write "$R \vdash M \Downarrow$" if $R \vdash M \rightarrow^* U$ for some U. Write "$R \vdash M \Downarrow \mathsf{fail}$" if $R \vdash M \rightarrow^* M'$ where $R \vdash M' \not\rightarrow$ and M' is not a value.

Now the web server can use the role assigned to a caller to access the filesystem (unless the web server's caller withholds its role). To prevent an attacker determining the non-existence of files via the web server, the web server fails when an attempt is made to access an unknown file unless the *DEBUG* role is activated.

$$\begin{aligned} webserver \stackrel{\text{def}}{=} \lambda name\,.\,\textsf{if}\ name = \text{"file1"}\ \textsf{then}\ filesystem\ name \\ \textsf{else if}\ name = \text{"file2"}\ \textsf{then}\ filesystem\ name \\ \textsf{else check}\ DEBUG\,;\ \text{"error: file not found"} \end{aligned}$$

For example, code running as Alice can access "file2" via the web server:

$$\begin{aligned} ALICE \vdash webserver\ \text{"file2"} &\rightarrow^* filesystem\ \text{"file2"} \\ &\rightarrow^* \textsf{check}\ ALICE \sqcap BOB\,;\ \text{"content2"} \rightarrow^* \text{"content2"} \end{aligned}$$ □

Example 3 illustrates how the Domain-Type Enforcement (DTE) security mechanism [5, 27], as found in the NSA's Security-Enhanced Linux (SELinux) [15], can be implemented in λ-RBAC. Further discussion of the relationship between RBAC and DTE can be found in [10, 12].

EXAMPLE 3 (DOMAIN-TYPE ENFORCEMENT / SELINUX). DTE grants or denies access requests according to the *domain* of the requesting process and the *type* assigned to the object, e.g., a file or port. The domain of a process only changes when another image is executed. DTE facilitates least privilege by limiting domain transitions based upon the source and target domains, and type assigned to the invoked executable file.

The DTE domain transition from role R to role S (each acting as domains) can be modeled by the function $R \xrightarrow{E} S$ that allows code running at role R to apply a function at role S: $R \xrightarrow{E} S \stackrel{\text{def}}{=} \{R\}\lambda(f,x)\,.\,\downarrow\perp f\ (\{E\}\lambda g\,.\,\uparrow S\,[g\,x]\,)$.

However, the domain transition is only performed when the function is associated with role E, modeling assignment of DTE type E to an executable file. Association of a function g with role E is achieved by accepting a continuation that is called back at role E with the function g. The function $assignType_E$ allows code running at $ADMIN$ to assign DTE types to other code: $assignType_E \stackrel{\text{def}}{=} \{ADMIN\}\lambda g\,.\,\lambda h\,.\,\uparrow E\,[\downarrow\perp h\,g\,]$. For example, for a function value U:

$$ADMIN \vdash \downarrow\perp assignType_E\ U \rightarrow^* \lambda h\,.\,\uparrow E\,[\downarrow\perp h\,U\,]$$

Then given a value V such that $S \vdash U\ V \rightarrow^* W$, we have:

$$R \vdash \downarrow\perp (R \xrightarrow{E} S)\ (\lambda h\,.\,\uparrow E\,[\downarrow\perp h\,U\,], V) \rightarrow^* W$$

With the $R \xrightarrow{E} S$ and $assignType_E$ functions we can adapt the login example from [27] to λ-RBAC. In this example, the DTE mechanism is used to force every invocation of administrative code (running at $ADMIN$) from daemon code (running at $DAEMON$) to occur via trusted login code (running at $LOGIN$). This is achieved by providing domain transitions from $DAEMON$ to $LOGIN$, and $LOGIN$ to $ADMIN$, but no others. Moreover, code permitted to run at $LOGIN$ must be assigned DTE type $LOGINEXE$, and similarly for $ADMIN$ and $ADMINEXE$. Thus a full program running daemon code

M has the following form, where neither M nor the code assigned to g variables contain rights amplification:

$$\text{let } \mathit{daemonToLogin} = \mathit{DAEMON} \xrightarrow{\mathit{LOGINEXE}} \mathit{LOGIN};$$
$$\text{let } \mathit{loginToAdmin} = \mathit{LOGIN} \xrightarrow{\mathit{ADMINEXE}} \mathit{ADMIN};$$
$$\text{let } \mathit{shell} = \text{let } g = \ldots;\ \downarrow\!\bot\ \mathit{assignType}_{\mathit{ADMINEXE}}\ (g);$$
$$\text{let } \mathit{login} = \text{let } g = \lambda(\mathit{password}, \mathit{cmd}).\ \text{if } \mathit{password} = \text{"secret" then} \downarrow\!\bot\ \mathit{loginToAdmin}\ (\mathit{shell}, \mathit{cmd})\ \text{else} \ldots;\ \downarrow\!\bot\ \mathit{assignType}_{\mathit{LOGINEXE}}\ (g);$$
$$\downarrow\!\mathit{DAEMON}\,[M]$$

In the above program, the daemon code M must provide the correct password in order to execute the shell at *ADMIN* because the *login* provides the sole gateway to *ADMIN*. In addition, removal of the domain transition *daemonToLogin* makes it impossible for the daemon code to execute any code at *ADMIN*. □

EXAMPLE 4 (CONTROLLING RIGHTS AMPLIFICATION). The *amplify* constructor provides a flexible dynamic way to control rights amplification. Suppose that M contains no direct rights amplification (subterms of the form $\uparrow P\,[\,\cdot\,]$). Then, in "let $g = U$; $\downarrow R\,[M]$" we may view U as a Trusted Computing Base (TCB) — a privileged function which may escalate rights — and to N as restricted *user* code; g is then an entry point to the TCB and R is the user role. User code is therefore executed at the restricted role R, and rights amplification may only occur through invokation of g.

Non-trivial programs have larger TCBs with more entry points. As the size of the TCB grows, it becomes too difficult to understand the security guarantees offered by a system allowing arbitrary rights amplification in all TCB code. To manage this complexity, one may enforce a coding convention that requires rights increases be justified by earlier checks. As an example, consider the following.

$$\text{let } do_P = \{\mathit{amplify}(P)\}\lambda f.\,\lambda x.\,\uparrow P\,[f\,x];$$

The privileged function do_P that performs rights amplification (for P) is justified by the check for $\mathit{amplify}(P)$ on any caller of do_P. One may also wish to explictly prohibit a term N from direct amplification of some right Q; with such a convention in place, this can be achieved using the frame $\downarrow(\mathit{amplify}(Q)^{\star})\,[N]$.

A formal and systematic general mechanism to enforce such a coding convention — by requiring code to use definable higher order combinators in place of unchecked frames — is omitted from this extended abstract for space reasons. □

4 Statics

We consider two kinds of *static analysis*: (i) a type system to enable removal of unnecessary role-checks in a piece of code for a caller at a sufficiently high role, and (ii) a type system to determine the amount of protection that is enforced by the callee.

To make the issues as clear as possible, we treat a simply-typed calculus with subtyping in the main text. In the rest of this section, we assume that all roles (and therefore all types) are well-formed, in the sense that role constructors have the correct number of arguments.

4.1 A Type System to Calculate Caller Roles

Our first type system attempts to calculate a caller role that guarantees that all execution paths are successful. The judgment $\Gamma \vdash M : \{R\}\ \tau$ asserts that R suffices to evaluate M, i.e. if M is executed with a role that dominates R, then all access checks in M are guaranteed to be successful. Values require no computation to evaluate, thus the value judgment $\Gamma \vdash U : \tau$ includes only the type τ. The syntax of types is as follows.

$$\sigma,\tau ::= \cdots \mid \sigma \rightarrow \{Q \triangleright R\}\ \tau \qquad \Gamma,\Delta ::= x_1 : \sigma_1, \ldots, x_n : \sigma_n$$

The type language includes base types (which we elide in the formal presentation) and function types. The function type $\sigma \rightarrow \{Q \triangleright R\}\ \tau$ is decorated with two latent effects; roughly, these indicate that the role context must dominate Q in order to pass the function's guard, and that the caller must provide a role context of at least R for execution of the function body to succeed. The least role $\bot$ is trivial in function types; thus $\sigma \rightarrow \tau$ abbreviates $\sigma \rightarrow \{\bot \triangleright \bot\}\ \tau$. We also write $\sigma \rightarrow \{R\}\ \tau$ for $\sigma \rightarrow \{\bot \triangleright R\}\ \tau$. If all roles occurring in a term are trivial, our typing rules degenerate to those of the standard simply-typed lambda calculus. The typing judgments are as follows.

VALUE AND TERM TYPING $(\Gamma \vdash U : \tau)$ $(\Gamma \vdash M : \{R\}\ \tau)$

(VAL-VAR)
$$\frac{\Gamma(x) = \tau}{\Gamma \vdash x : \tau}$$

(VAL-ABS)
$$\frac{\Gamma, x{:}\sigma \vdash M : \{R\}\ \tau}{\Gamma \vdash \{Q\}\lambda x.M : \sigma \rightarrow \{Q \triangleright R\}\ \tau}$$

(TERM-VAL)
$$\frac{\Gamma \vdash U : \tau}{\Gamma \vdash U : \{\bot\}\ \tau}$$

(TERM-LET)
$$\frac{\Gamma \vdash M : \{R\}\ \sigma \qquad \Gamma, x{:}\sigma \vdash N : \{S\}\ \tau}{\Gamma \vdash \mathsf{let}\ x{=}M;\ N : \{R \sqcup S\}\ \tau}$$

(TERM-APP)
$$\frac{\Gamma \vdash U : \sigma \rightarrow \{Q \triangleright R \sqcap P\}\ \tau \qquad \Gamma \vdash V : \sigma}{\Gamma \vdash {\downarrow}P\ U\ V : \{Q \sqcup R\}\ \tau}$$

(TERM-RESTRICT)
$$\frac{\Gamma \vdash M : \{R \sqcap P\}\ \tau}{\Gamma \vdash {\downarrow}P[M] : \{R\}\ \tau}$$

(TERM-PROVIDE)
$$\frac{\Gamma \vdash M : \{R \sqcup P\}\ \tau}{\Gamma \vdash {\uparrow}P[M] : \{R\}\ \tau}$$

(TERM-SUBEFFECT)
$$\frac{\Gamma \vdash M : \{S\}\ \tau \qquad \vdash R \geqslant S}{\Gamma \vdash M : \{R\}\ \tau}$$

VAL-ABS simply records the effects that the abstraction will incur when run. By TERM-VAL, a value can be treated as a term that evaluates without error in every role context. TERM-LET indicates that two expressions must succeed sequentially if the current role guarantees success of each individually. TERM-RESTRICT (resp. TERM-PROVIDE) captures the associated rights weakening (resp. amplification). TERM-APP incorporates the role required to evaluate the function to an abstraction and the role required to evaluate the body of the function (while allowing for rights weakening).

Subtyping. A natural notion of subtyping is induced from the role ordering. Formally, subtyping is the least precongruence on types induced by SUBTYPING-BASE.

(SUBTYPING-BASE)
$$\frac{\vdash \sigma' <: \sigma \qquad \vdash Q' \geqslant Q \qquad \vdash S' \geqslant S \qquad \vdash \tau <: \tau'}{\vdash (\sigma \rightarrow \{Q \triangleright S\}\ \tau) <: (\sigma' \rightarrow \{Q' \triangleright S'\}\ \tau')}$$

(VAL-SUBTYPE)
$$\frac{\Gamma \vdash U : \sigma \qquad \vdash \sigma <: \sigma'}{\Gamma \vdash U : \sigma'}$$

(TERM-SUBTYPE)
$$\frac{\Gamma \vdash M : \{R\}\ \tau \qquad \vdash \tau <: \tau'}{\Gamma \vdash M : \{R\}\ \tau'}$$

The following example of Church booleans, illustrates the use of subtyping. The Church booleans, $\lambda t.\lambda f.t$ and $\lambda t.\lambda f.f$ can be given type $(\sigma \to \{R\}\ \tau) \to (\sigma \to \{S\}\ \tau) \to (\sigma \to \{R \sqcup S\}\ \tau)$.

The type system satisfies standard preservation and progress properties.

THEOREM 5. *If* $\Gamma \vdash M : \{R\}\ \tau$ *and* $S \vdash M \to M'$, *then* $\Gamma \vdash M' : \{R\}\ \tau$.
If $\Gamma \vdash M : \{R\}\ \tau$ *then either M is a value, or* $R \vdash M \to M'$, *for some* M'.

4.2 A Type System to Determine Callee Protection

Our second type system ("$\Vdash$") has aims "dual" to the previous type system. Rather than attempting to calculate a caller role that guarantees that all execution paths are successful, we deduce the minimum protection demanded by the callee on all execution paths. View $\Gamma \Vdash M : \{S\}\ \tau$ as asserting that it is not possible for M to evaluate to a value without using a role above S, i.e. the guaranteed protection for M is at least S.

The type system presented below has altered versions of TERM-APP, TERM-RESTRICT, TERM-PROVIDE and inverted versions of TERM-SUBEFFECT and SUBTYPING-BASE.

(TERM-SUBEFFECT)
$$\frac{\Gamma \Vdash M : \{S\}\ \tau \quad \vdash S \geqslant R}{\Gamma \Vdash M : \{R\}\ \tau}$$

(SUBTYPING-BASE)
$$\frac{\Vdash \sigma' <: \sigma \quad \vdash Q \geqslant Q' \quad \vdash S \geqslant S' \quad \Vdash \tau <: \tau'}{\Vdash (\sigma \to \{Q \triangleright S\}\ \tau) <: (\sigma' \to \{Q' \triangleright S'\}\ \tau')}$$

(TERM-APP)
$$\frac{\Gamma \Vdash U : \sigma \to \{Q \triangleright R\}\ \tau \quad \Gamma \Vdash V : \sigma' \quad \vdash \sigma' <: \sigma}{\Gamma \Vdash \downarrow P\ U\ V : \{Q \sqcup R\}\ \tau}$$

(TERM-RESTRICT)
$$\frac{\Gamma \Vdash M : \{R\}\ \tau}{\Gamma \Vdash \downarrow P[M] : \{R\}\ \tau}$$

(TERM-PROVIDE)
$$\frac{\Gamma \Vdash M : \{R\}\ \tau}{\Gamma \Vdash \uparrow P[M] : \{R \sqcap P^{*}\}\ \tau}$$

In TERM-SUBEFFECT above, it is sound to weaken the role since the asserted protection is only reduced. $\uparrow P[M]$ adds P to the role context in the operational semantics. So, in TERM-PROVIDE, the guaranteed protection for $\uparrow P[M]$ removes P from the guaranteed protection for M.

As a consequence of the above rules, if M is a value, we can deduce that it must be typed at role $\bot$; values are already in normal form, so do not enforce any protection. Furthermore, in this system, the Church booleans may be given type: $(\sigma \to \{R\}\ \tau) \to (\sigma \to \{S\}\ \tau) \to (\sigma \to \{R \sqcap S\}\ \tau)$. illustrating the "minimum over all paths" principle via $R \sqcap S$ (to be contrasted with $R \sqcup S$ in the previous typing system.). More generally, the following theorem enables us to understand the invariants established by typing in this system [4].

THEOREM 6. *If* $\Gamma \Vdash M : \{S\}\ \tau$, $\vdash R \not\geqslant S$ *and* $R \vdash M \to M'$, *then* $\Gamma \Vdash M' : \{S\}\ \tau$.

If we start execution in a role context (R in the theorem) that does not suffice to pass the minimum protection guarantee (S in the theorem), then a single step of reduction has only two possibilities: (i) a check, e.g., for S, that is not passed by R occurs and the term gets stuck, or (ii) the check for S does not happen at this step but the invariant that the minimum protection is S continues to get preserved.

[4] The usual form of the type preservation result does not hold for this system. For example $\Vdash (\{\top\}\lambda\_.())\ () : \{\top\}$ Unit and $\top \vdash (\{\top\}\lambda\_.())\ () \to ()$ but $\not\Vdash () : \{\top\}$ Unit.

5 Typing Examples

EXAMPLE 7. Recall the filesystem and web server from example 2. The filesystem code can be assigned the following type, meaning that a caller must possess a role from each of the ACLs in order to guarantee that access checks will not fail:

$$\vdash \mathit{filesystem} : \mathsf{String} \rightarrow \{\bot \rhd \mathit{ADMIN} \sqcup (\mathit{ALICE} \sqcap \mathit{BOB}) \sqcup \bot\}\ \mathsf{String}$$

In the above type, the final role $\bot$ arises from the "unknown file" branch that does not require an access check. The lack of an access check explains the weaker $\Vdash$ type:

$$\Vdash \mathit{filesystem} : \mathsf{String} \rightarrow \{\bot \rhd \mathit{ADMIN} \sqcap (\mathit{ALICE} \sqcap \mathit{BOB}) \sqcap \bot\}\ \mathsf{String}$$

This type indicates that *filesystem* has the potential to expose some information to unprivileged callers with role $\mathit{ADMIN} \sqcap (\mathit{ALICE} \sqcap \mathit{BOB}) \sqcap \bot = \bot$, perhaps causing the code to be flagged for security review.

The access check in the web server does prevent the "unknown file" error message leaking unless the *DEBUG* role is active, but, unfortunately, it is not possible to assign a role strictly greater than $\bot$ to the web server using the $\Vdash$ type system because the *filesystem* type does not record the different roles that must be checked depending upon the filename argument, and hence:

$$\not\Vdash \mathit{webserver} : \mathsf{String} \rightarrow \{\bot \rhd \mathit{ADMIN} \sqcap (\mathit{ALICE} \sqcap \mathit{BOB}) \sqcap \mathit{DEBUG}\}\ \mathsf{String} \qquad \square$$

EXAMPLE 8. Recall the encoding of the DTE/SELinux domain transition mechanism from example 3. Define types for functions running at role S (acting as a domain) and functions that can prove their assigned DTE role is E by calling back with that role:

$$\mathsf{Func}(\sigma,\tau,S) \stackrel{\text{def}}{=} \sigma \rightarrow \{S\}\ \tau$$
$$\mathsf{FuncDTEType}(\sigma,\tau,S,E) \stackrel{\text{def}}{=} (\mathsf{Func}(\sigma,\tau,S) \rightarrow \{E \rhd \bot\}\ \tau) \rightarrow \{\bot\}\ \tau$$

A domain transition will certainly succeed if the caller possesses role R and the function invoked after the domain transition requires at most role S:

$$\vdash R \xrightarrow{E} S : \mathsf{FuncDTEType}(\sigma,\tau,S,E) \times \sigma \rightarrow \{R \rhd \bot\}\ \tau$$

In contrast, the following type guarantees that role R will be demanded from the caller:

$$\Vdash R \xrightarrow{E} S : \mathsf{FuncDTEType}(\sigma,\tau,S,E) \times \sigma \rightarrow \{R \rhd \bot\}\ \tau \qquad \square$$

6 Conclusions

We have presented methods to aid the designer and use of components which include access control code (as permitted in the programmatic RBAC of JAAS/.NET). Our first analysis enables users of code to deduce the role at which code must be run. The other analysis method enables code designers to deduce the protection guarantees of their code by calculating the role that is verified on all execution paths.

In future work, we will explore extensions to role polymorphism and recursive roles following the techniques of [7, 2].

Our paper falls into the broad area of research enlarging the scope of foundational, language-based security methods (see [24, 16, 1] for surveys). The papers that are most directly relevant to the current paper are [6, 9]. Both these papers start off with a mobile process-based computational model. Both calculi have primitives to activate and deactivate roles: these roles are used to prevent undesired mobility and/or communication, and are similar to the primitives for role restriction and amplification in this paper. We expect that our ideas can be adapted to the process calculi framework. In future work, we also hope to integrate the powerful bisimulation principles of these papers.

[6, 9] develop type systems to provide guarantees about the minimal role required for execution to be successful — our first type system occupies the same conceptual space as this static analysis. However, our second type system that calculates minimum access controls does not seem to have an analogue in these papers. More globally, our paper has been influenced by the desire to serve loosely as a metalanguage for programming RBAC mechanisms in examples such as the JAAS/.NET frameworks. Thus, our treatment internalizes rights amplification by program combinators and the amplify role constructor in role lattices. In contrast, the above papers use external — i.e. not part of the process language — mechanisms (namely, user policies in [9], and RBAC-schemes in [6]) to enforce control on rights activation.

Our paper deals with access control, so the extensive work on information flow, e.g., see [20] for a survey, is not directly relevant. However, we note that rights amplification plays the same role in λ-RBAC that declassification and delimited release [8, 21, 17] plays in the context of information flow; namely that of permitting access that would not have been possible otherwise. In addition, by internalizing the ability to amplify code rights into the role lattice, our system permits access control code to actively participate in managing rights amplification.

References

1. M. Abadi, G. Morrisett, and A. Sabelfeld. Language-based security. *J. Funct. Program.*, 15(2):129, 2005.
2. R. M. Amadio and L. Cardelli. Subtyping recursive types. *ACM TOPLAS*, 15(4):575–631, 1993.
3. S. Barker and P. J. Stuckey. Flexible access control policy specification with constraint logic programming. *ACM Trans. Inf. Syst. Secur.*, 6(4):501–546, 2003.
4. E. Bertino, P. A. Bonatti, and E. Ferrari. TRBAC: A temporal role-based access control model. *ACM Trans. Inf. Syst. Secur.*, 4(3):191–233, 2001.
5. W. E. Boebert and R. Y. Kain. A practical alternative to hierarchical integrity policies. In *Proceedings of the Eighth National Computer Security Conference*, 1985.
6. C. Braghin, D. Gorla, and V. Sassone. A distributed calculus for role-based access control. In *CSFW*, pages 48–60, 2004.
7. M. Brandt and F. Henglein. Coinductive axiomatization of recursive type equality and subtyping. *Fundam. Inf.*, 33(4):309–338, 1998.
8. S. Chong and A. C. Myers. Security policies for downgrading. In *ACM Conference on Computer and Communications Security*, pages 198–209, 2004.
9. A. Compagnoni, P. Garralda, and E. Gunter. Role-based access control in a mobile environment. In *Symposium on Trustworthy Global Computing*, 2005.

10. D. F. Ferraiolo, D. R. Kuhn, and R. Chandramouli. *Role-Based Access Control*. Computer Security Series. Artech House, 2003.
11. D. F. Ferraiolo, R. Sandhu, S. Gavrila, D. R. Kuhn, and R. Chandramouli. Proposed NIST standard for role-based access control. *ACM Trans. Inf. Syst. Secur.*, 4(3):224–274, 2001.
12. J. Hoffman. Implementing RBAC on a type enforced system. In *13th Annual Computer Security Applications Conference (ACSAC '97)*, pages 158–163, 1997.
13. S. Jajodia, P. Samarati, M. L. Sapino, and V. S. Subrahmanian. Flexible support for multiple access control policies. *ACM Trans. Database Syst.*, 26(2):214–260, 2001.
14. J. Ligatti, L. Bauer, and D. Walker. Edit automata: enforcement mechanisms for run-time security policies. *Int. J. Inf. Sec.*, 4(1-2):2–16, 2005.
15. P. A. Loscocco and S. D. Smalley. Meeting critical security objectives with Security-Enhanced Linux. In *Proceedings of the 2001 Ottawa Linux Symposium*, 2001.
16. J. C. Mitchell. Programming language methods in computer security. In *POPL*, pages 1–26, 2001.
17. A. C. Myers, A. Sabelfeld, and S. Zdancewic. Enforcing robust declassification. In *CSFW*, pages 172–186, 2004.
18. S. Osborn, R. Sandhu, and Q. Munawer. Configuring role-based access control to enforce mandatory and discretionary access control policies. *ACM Trans. Inf. Syst. Secur.*, 3(2):85–106, 2000.
19. J. S. Park, R. S. Sandhu, and G.-J. Ahn. Role-based access control on the web. *ACM Trans. Inf. Syst. Secur.*, 4(1):37–71, 2001.
20. A. Sabelfeld and A. C. Myers. Language-based information-flow security. *IEEE J. Selected Areas in Communications*, 21(1):5–19, Jan. 2003.
21. A. Sabelfeld and A. C. Myers. A model for delimited information release. In *ISSS*, pages 174–191, 2003.
22. R. Sandhu, E. Coyne, H. Feinstein, and C. Youman. Role-based access control models. *IEEE Computer*, 29(2), 1996.
23. R. S. Sandhu and J. Park. Usage control: A vision for next generation access control. *ACM Trans. Inf. Syst. Secur.*, 2004.
24. F. B. Schneider, G. Morrisett, and R. Harper. A language-based approach to security. In *Informatics—10 Years Back, 10 Years Ahead*, volume 2000 of *LNCS*, pages 86–101, 2000.
25. F. Siewe, A. Cau, and H. Zedan. A compositional framework for access control policies enforcement. In *FMSE*, pages 32–42, 2003.
26. E. G. Sirer and K. Wang. An access control language for web services. In *SACMAT '02: Proceedings of the seventh ACM symposium on Access control models and technologies*, pages 23–30, 2002.
27. K. M. Walker, D. F. Sterne, M. L. Badger, M. J. Petkac, D. L. Shermann, and K. A. Oostendorp. Confining root programs with Domain and Type Enforcement (DTE). In *Proceedings of the Sixth USENIX UNIX Security Symposium*, 1996.

Communication of Two Stacks and Rewriting*

Juhani Karhumäki, Michal Kunc, and Alexander Okhotin

Department of Mathematics, University of Turku, and
Turku Centre for Computer Science, FIN-20014 Turku, Finland
karhumak@utu.fi, kunc@math.muni.cz, alexander.okhotin@utu.fi

Abstract. Rewriting systems working on words with a center marker are considered. The derivation is done by erasing a prefix or a suffix and then adding a prefix or a suffix. This can be naturally viewed as two stacks communicating with each other according to a fixed protocol. The paper systematically considers different cases of these systems and determines their expressiveness. Several cases are identified where very limited communication surprisingly yields universal computation power.

1 Introduction

The earliest evidence of computational power of simple rewriting systems seems to be the following fascinating example of Post [10] dating to 1920's. Given a binary word w, apply to it iteratively the following rule: "omit the three-letter prefix and, if its first symbol was 0 (resp., 1), append a suffix 00 (resp. 1101)". There are three possible outcomes: either the process terminates, or goes into a periodic stage, or proceeds without repetitions. As noticed already by Post, it is not easy to determine what is the case for a given w. In fact, even now no algorithm to decide this is known. Based on the above example Post developed his *canonical systems*, which have universal computing power.

In 1960's Büchi [3] and Kratko [8] independently started to consider simpler rewriting systems of a similar kind. They noticed that if Post's rules are applied locally, that is, a word w is rewritten to $y(x^{-1}w)$ for a rule $x \to y$, then the languages obtained are regular. On the other hand, the power of another simpler variant of Post's rewriting was determined only recently. Given an initial word w and a regular set X, the rule "delete a prefix from X and append any word from X to the end" is iteratively applied. The choice of the word being appended is independent of the word removed, so this was called *uncontrolled one-way rewriting*, and the regularity of the sets generated was established [7].

This process resembles another problem dealing with operations on two ends of a word proposed by Conway [4]. He asked whether for every regular language X the greatest language Z, such that $XZ = ZX$, is regular as well. This problem was recently solved strongly negatively: by a sophisticated construction it was proved that such Z need not be recursively enumerable [9]. This demonstrated that Conway's equation is deeper than it seemed, and motivated the study of its approximate sequential variant, called *uncontrolled two-way rewriting*, defined

* Supported by the Academy of Finland under grants 206039 and 208414.

M. Bugliesi et al. (Eds.): ICALP 2006, Part II, LNCS 4052, pp. 468–479, 2006.

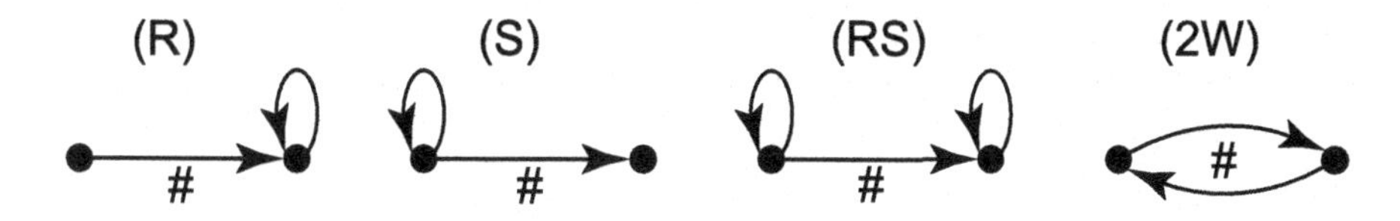

Fig. 1. Modes of rewriting

by the rule "delete an element of X from either of the ends of the word, and at the same time add any element of X to the opposite end". It was found to generate a nonregular language [7], but its exact power was left undetermined.

This paper is dedicated to a systematic study of such rewriting systems and their variants. We assume that the words being rewritten contain a center marker that is never touched: this does not reduce the power of the most general case and leads to interesting special cases. Besides two-way rewriting, we consider three augmented cases of one-way rewriting, which are illustrated by diagrams in Figure 1, where # is the center marker. In these subcases also local Büchi rewriting steps are allowed, i.e., steps where the deletion and adding is done at the same end. If this is allowed on the right, we call the rewriting *one-way R-rewriting*, where R stands for "receiving". If the local rewriting is allowed only at the left end, then it is called *one-way S-rewriting*, where S stands for "sending". Finally, if it is allowed at both ends, then we call it *one-way RS-rewriting*.

In all of the above cases the rewriting may be *controlled* or *uncontrolled*. In the former case a connection between x's and y's is allowed, for example, the pairs (x, y) may be defined by a recognizable or rational relation. In the latter case the words to be deleted and added are chosen independently. Obviously, the computational power of the former is much higher than that of the latter.

We have described our approach in terms of rewriting. However, our systems can be equally interpreted in terms of communication between two (pushdown) stacks. In one-way rewriting we pop from the left stack and simultaneously push onto the right stack, that is, we send a message from left to right. When an uncontrolled relation defines a communication, all messages sent through such a channel are indistinguishable, that is, the fact of sending a message constitutes the entire message. When we pop from and push onto the same stack, this models local processing of data. This interpretation gives a further motivation for our systematic study.

Our presentation is structured as follows. In Section 2 we formally define all variants of our rewriting systems. Then in Section 3 we consider one-way R-rewriting. Here, assuming that the set of initial words is regular, only regular languages can be generated even in the general cases of controlled rewriting. On the other hand, for a context-free initial set the rewriting becomes computationally universal. In the case of one-way S-rewriting studied in Section 4 our essential result is that even quite restricted models generate all context-free languages, but also their essential generalizations do not give anything more. For one-way RS-rewriting, in Section 5 we obtain a greater variety or results: if the one-way rewriting is uncontrolled, then only special cases of context-free languages are generated, while if it is controlled, all recursively enumerable languages can be obtained. Finally, in Section 6 we establish completely unexpected computational universality of uncontrolled two-way rewriting.

2 Formal Definitions and Notation

Let Σ be a finite alphabet and let $\# \notin \Sigma$ be the *center marker*. Let $I \subseteq \Sigma^*\#\Sigma^*$ be a set of words from which the rewriting starts: the *initial set*. Let $\xrightarrow{\ell\ell}, \xrightarrow{rr}, \xrightarrow{\ell r}, \xrightarrow{r\ell} \subseteq \Sigma^* \times \Sigma^*$ be four relations that constitute the *rewriting rules*. Define the corresponding relations of one-step rewriting as follows: for all $w, w' \in \Sigma^*$,

$$xw\#w' \overset{\ell\ell}{\Longrightarrow} yw\#w' \quad ((x,y) \in \xrightarrow{\ell\ell}) \qquad w\#w'x \overset{rr}{\Longrightarrow} w\#w'y \quad ((x,y) \in \xrightarrow{rr})$$
$$xw\#w' \overset{\ell r}{\Longrightarrow} w\#w'y \quad ((x,y) \in \xrightarrow{\ell r}) \qquad w\#w'x \overset{r\ell}{\Longrightarrow} yw\#w' \quad ((x,y) \in \xrightarrow{r\ell})$$

The relation of one-step rewriting, $\Longrightarrow$, is the union of these four relations. Consider the reflexive and transitive closure of $\Longrightarrow$, denoted by $\Longrightarrow^*$. The *language generated* by the rewriting system is defined as $\{w \mid \exists w_0 \in I : \ w_0 \Longrightarrow^* w\}$. The modes of rewriting we consider can be formally defined as follows: if $\xrightarrow{\ell\ell} = \xrightarrow{r\ell} = \varnothing$, this is *one-way R-rewriting*; if $\xrightarrow{rr} = \xrightarrow{r\ell} = \varnothing$, this is *one-way S-rewriting*; if $\xrightarrow{r\ell} = \varnothing$, this is *one-way RS-rewriting*; and if $\xrightarrow{\ell\ell} = \xrightarrow{rr} = \varnothing$, it is *two-way rewriting* (2W).

Let us now turn to the form of relations $\xrightarrow{\ell\ell}, \xrightarrow{rr}, \xrightarrow{\ell r}$ and $\xrightarrow{r\ell}$. A relation $\rightarrow$ is *uncontrolled* using two families of languages $\mathcal{L}$ and $\mathcal{M}$, denoted $Unc(\mathcal{L}, \mathcal{M})$, if $\rightarrow \, = L \times M$ for some $L \in \mathcal{L}$ and $M \in \mathcal{M}$. We consider classes of relations $Unc(Reg, Reg)$ and $Unc(Fin, Fin)$, in which L and M must be regular or finite, respectively. A further restricted form of these relations, where $\rightarrow \, = L \times L$ for a single language $L \in \mathcal{L}$, is denoted by $Unc(\mathcal{L})$.

Finite relations (Fin) are of the form $\rightarrow \, = \{(x_1, y_1), \ldots, (x_n, y_n)\}$; a simple subcase is a *copy relation*, in which $\rightarrow \, = \{(a, a) \mid a \in \Sigma'\}$ for some $\Sigma' \subseteq \Sigma$. *Recognizable* relations (Rec) are those for which the language $\{u\$v \mid (u, v) \in \rightarrow\}$ is regular. *Rational* relations (Rat) are recognized by finite transducers. The most general family we consider are the *regularity preserving relations* ($Reg.Pr$), such that for every regular L the language $\{y \mid \exists x \in L : \ (x, y) \in \rightarrow\}$ is regular.

A class of rewriting systems is defined by a mode of rewriting (R, S, RS or 2W), the families to which the relations belong and the family of languages usable as the initial set. Each class of rewriting systems defines a family of formal languages, and this family characterizes the power of communication modelled by this kind of rewriting. For instance, S-rewriting with recognizable $\xrightarrow{\ell\ell}$, rational $\xrightarrow{\ell r}$ and context-free I, as we shall see, yields exactly the context-free languages.

If either of $\xrightarrow{\ell\ell}, \xrightarrow{rr}$ can be any rational relation, then it is easy to produce an r.e.-complete language from a one-element initial set by using a well-known folklore fact that finite transducers can compute the next configuration of a Turing machine. Therefore the most general relations we are going to consider are regularity-preserving relations $\xrightarrow{\ell r}$ and $\xrightarrow{r\ell}$ and recognizable relations $\xrightarrow{\ell\ell}$ and $\xrightarrow{rr}$.

3 Receiving

In this section we consider *one-way R-rewriting*, which uses relations $\xrightarrow{\ell r}$ and $\xrightarrow{rr}$. We can show that the relation $\Longrightarrow^*$ of such rewriting systems preserves regularity even in the most general case, and that their power is determined by the initial set. Let us organize the study of R-rewriting on the basis of whether I is regular or not. First we consider the case of a regular I.

Theorem 1. *The language L generated by a rewriting system consisting of a regularity-preserving relation $\xrightarrow{\ell r}$, recognizable relation $\xrightarrow{rr}$ and a regular initial set I is always regular. If in addition images of regular languages under $\xrightarrow{\ell r}$ can be algorithmically computed, then L is algorithmically computable too.*

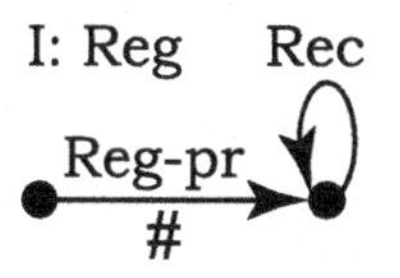

For a given alphabet Σ, take its disjoint copy $\widetilde{\Sigma} = \{\, \widetilde{a} \mid a \in \Sigma \,\}$, where the letter $\widetilde{a}$ will represent deletion of a. For any word $w = a_1 \cdots a_n \in \Sigma^*$, where $n \geqslant 0$ and $a_1, \ldots, a_n \in \Sigma$, denote $\widetilde{w} = \widetilde{a}_n \cdots \widetilde{a}_1$, and extend this notation to languages as $\widetilde{L} = \{\, \widetilde{w} \mid w \in L \,\}$. Define the set of *reduction rules* $\{\, a\widetilde{a} \to \varepsilon \mid a \in \Sigma \,\}$.

The proof consists of three parts. First, we consider *computation histories* of our rewriting, in which, to the right from #, all letters which occurred there during the rewriting are preserved and only symbolically deleted by appending the corresponding "negative" symbols from $\widetilde{\Sigma}$. The language of all such computation histories is defined as follows:

$$L_0 = \big\{ u\#v_0\rho_0 w_1\rho_1 \ldots w_n\rho_n \bigm| n \geqslant 0, \rho_i \in \{\widetilde{x}y \mid (x,y) \in \xrightarrow{rr}\}^*, \\ \exists z_1, \ldots, z_n : z_1 \ldots z_n u\#v_0 \in I, (z_i, w_i) \in \xrightarrow{\ell r} \big\},$$

Lemma 1. *A word $u\#v \in \Sigma^*\#\Sigma^*$ is derivable in the R-rewriting system if and only if there exists $\alpha \in L_0$ reducible to $u\#v$.*

Next, we show regularity of this language.

Lemma 2. *The language L_0 is regular and the corresponding DFA can be effectively constructed (provided that the images under $\xrightarrow{\ell r}$ are computable).*

Finally, we apply a known property of such reductions dating back to Benois [2]; for details, see Sakarovitch [11, Ch. II, Sec. 6], Hofbauer and Waldmann [5], as well as the authors [7]. It states that the image of any regular language under such a reduction is regular. These three results yield a proof of Theorem 1.

If we consider non-regular sets of initial words, then, even for very simple relations, R-rewriting system become computationally universal.

Theorem 2. *For every recursively enumerable language L_0 over an alphabet Σ there exist an alphabet $\Sigma' \supset \Sigma \cup \{\#\}$, a language I over Σ' that is a concatenation of two linear context-free languages, a copy relation $\xrightarrow{\ell r}$ and a finite uncontrolled relation $\xrightarrow{rr}$, such that the language generated by the rewriting system equals $\#L_0$ modulo intersection with $\#\Sigma^*$.*

Unc(Fin,Fin)
Copy
#
I: LinCF·LinCF

Take any two linear context-free languages $K, M \subseteq \Sigma^*$, such that $L_0 = MK^{-1}$; it is easy to construct such languages using the method of Baker and Book [1]. Then define $\Sigma' = \Sigma \cup \widetilde{\Sigma}$, where $\widetilde{\Sigma} = \{\widetilde{a} \mid a \in \Sigma\}$. Let $\overset{\ell r}{\to} = \{(\widetilde{a}, \widetilde{a}) \mid a \in \Sigma\}$ be a copy relation; consider the set $N = \{ a\widetilde{a} \mid a \in \Sigma \} \cup \{\varepsilon\}$ and define $\overset{rr}{\to} = N \times N$. Take the initial set $I = \widetilde{K}\#M$. Then the language L generated by this rewriting system satisfies $L \cap \#\Sigma^* = \#MK^{-1}$.

4 Sending

This section is devoted to *one-way S-rewriting* systems, in which we are allowed to delete and add words on the beginning of the word or delete a word on the beginning and append some word to the end. This has a clear meaning in terms of communication between stacks: the left stack processes information and occasionally sends the results to the right stack as to the output.

Let us first suppose the sending relation $\overset{\ell r}{\to}$ is uncontrolled. Following is an example of a weakest system of this kind, in which the processing relation $\overset{\ell\ell}{\to}$ is also uncontrolled: however, a nonregular language can still be produced.

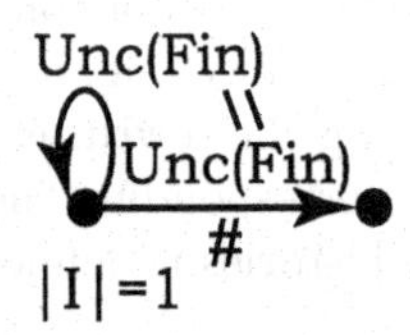

Example 1. Let $\Sigma = \{a, b\}$ and let $\overset{\ell\ell}{\to}, \overset{\ell r}{\to} = \{a, aab\} \times \{a, aab\}$, i.e., $\overset{\ell\ell}{\to}, \overset{\ell r}{\to} \in Unc(Fin)$ and a single two-element set is used for both relations. Let $I = \{ab\#\}$. Then the set L of derivable words satisfies $L \cap b^*\#a^* = \{b^n\#a^n \mid n \geqslant 1\}$.

Notice that every word in L has the same number of occurrences of a and of b, since the initial word has this property and each step of rewriting preserves it. On the other hand, every word $ab^n\#a^{n-1}$, for $n \geqslant 1$, can be inductively derived from $ab\#$, and $b^n\#a^n$ is obtained from it.

Though the language generated in Example 1 is nonregular, it is linear context-free. It turns out that even if the relation $\overset{\ell\ell}{\to}$ is controlled, the generated languages are always linear context-free.

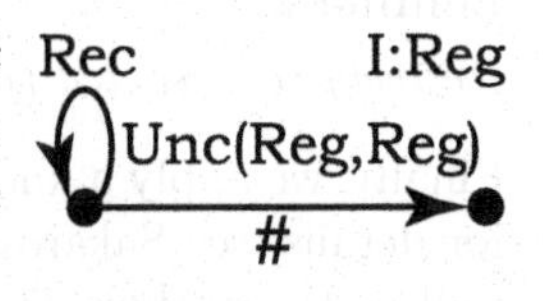

Proposition 1. *Let $\overset{\ell\ell}{\to}$ be a recognizable relation, let $\overset{\ell r}{\to} = X \times Y$ be an uncontrolled relation in $Unc(Reg, Reg)$ and let I be regular. Then the language generated by the system is linear context-free.*

Later we shall prove a more general statement, Theorem 5. For now let us turn to the second case of a controlled sending relation. Here even with an uncontrolled $\overset{\ell\ell}{\to}$ every context-free language can be generated.

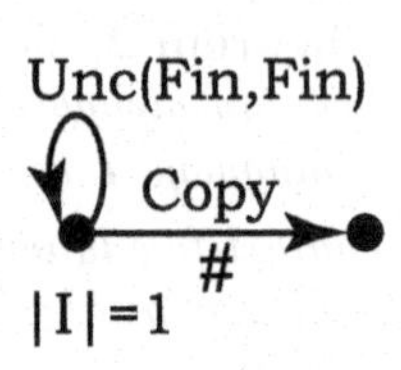

Theorem 3. *For every context-free language $L \subseteq \Sigma^*$ there exist an alphabet $\Gamma \supset \Sigma$, relations $\overset{\ell\ell}{\to} \in Unc(Fin, Fin)$ and $\overset{\ell r}{\to} \in Copy$, and a singleton initial set I, such that the rewriting system generates $\#L$ modulo intersection with $\#\Sigma^*$. Given a context-free grammar for L, this rewriting system can be effectively constructed.*

Assume $G = (\Sigma, N, P, S)$ in Chomsky normal form with possible empty rules. Consider the alphabet $\Gamma = \Sigma \cup N \cup \widehat{N} \cup \widetilde{N}$, where $\widehat{N} = \{\widehat{A} \mid A \in N\}$ and $\widetilde{N} = \{\widetilde{A} \mid A \in N\}$, and construct the following two finite sets:

$$X = N \cup \{\widetilde{A}\widehat{A} \mid A \in N\}$$
$$Y = \{B\widehat{B}C\widehat{C}\widetilde{A} \mid A \to BC \in P\} \cup \{w\widetilde{A} \mid A \to w \in P\} \cup \{\varepsilon\}$$

Define $I = \{S\widehat{S}\#\}$, $\overset{\ell\ell}{\to} = X \times Y$ and let $a \overset{\ell r}{\to} a$ for all $a \in \Sigma$.

The nonterminals always come in pairs, like $A\widehat{A}$. An application of a rule $A \to BC$ is simulated by erasing $A \in X$ and then by writing down $B\widehat{B}C\widehat{C}\widetilde{A} \in Y$, which gives $B\widehat{B}C\widehat{C}\widetilde{A}\widehat{A}$. Later the pair $\widetilde{A}\widehat{A}$ is removed, which verifies that the rule applied here matches the nonterminal erased. It can be proved that if this protocol is violated, then a word $\#w$ can no longer be derived.

Claim. If $uA_1 \ldots A_n$ is a sentential form of G, then $\exists\, \theta_1, \ldots, \theta_n \in \{\widetilde{D}\widehat{D} | D \in N\}^*$, such that $A_1\widehat{A}_1\theta_1 \ldots A_n\widehat{A}_n\theta_n\#u$ is derivable in the rewriting system.

Denote by $\mathrm{d}(\alpha)$ the word obtained from $\alpha \in (\Gamma \cup \{\#\})^*$ by deleting all occurrences of elements of $\widehat{N} \cup \widetilde{N}$.

Claim. Every word α derivable in the rewriting system, from which a word belonging to $\#\Gamma^*$ can be derived, belongs to the language $\Sigma^*\{\, A\widehat{A}, \widetilde{A}\widehat{A} \mid A \in N\,\}^*$ $\#\Sigma^*$. In addition, if $\mathrm{d}(\alpha) = \theta\#u$, then $u\theta$ can be derived in G.

It follows that a word $\#w$ is generated by this rewriting system if and only if $w \in L(G)$, which proves Theorem 3. If the least controlled S-rewriting yields all context-free languages, what could be its expressive power in a fully controlled case? In fact, still nothing but the context-free languages can be generated.

Theorem 4. *Let $\overset{\ell\ell}{\to}$ be a recognizable relation, let $\overset{\ell r}{\to}$ be a rational relation, let I be context-free. Then the language generated by this rewriting system is context-free. Given a finite automaton for $\overset{\ell\ell}{\to}$, a finite transducer for $\overset{\ell r}{\to}$ and a context-free grammar for I, a context-free grammar for the generated language can be effectively constructed.*

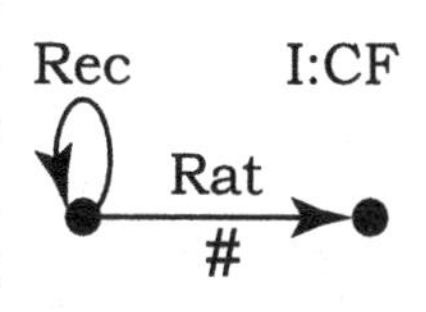

The proof is based upon two results, which are interesting on their own. One of them is the closure of the context-free languages under the cyclic shift operation (see, e.g., Hopcroft and Ullman [6, solution to Ex. 6.4c]).

Lemma 3. *For every context-free language L, the language* $\mathrm{SHIFT}(L) = \{uv \,|\, vu \in L\}$ *is context-free. Given a context-free grammar for L, a context-free grammar for* $\mathrm{SHIFT}(L)$ *can be effectively constructed.*

Another key property is that pushdown automata with initial pushdown contents defined by a context-free language still recognize only context-free languages.

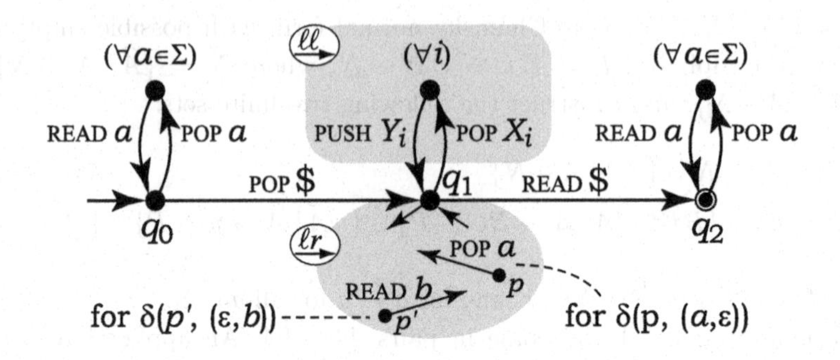

Fig. 2. The PDA in the proof of Theorem 4

Lemma 4. *Let Σ and Γ be two alphabets, and let L_0 be a context-free language over Γ. Let $A = (Q, \Sigma, \Gamma, q_0, \delta, F)$ be a standard PDA without the initial pushdown symbol. Then the language of all words w, for which there exists $x_0 \in L_0$, such that A accepts w given x_0 as the initial pushdown contents, is context-free and a standard PDA recognizing this language can be effectively constructed.*

This can be regarded as a system of two cooperating pushdown automata, in which one PDA supplies the initial contents of the pushdown for the other. We prove such a superposition is no more powerful than a single PDA.

Another interpretation of this result is the following. Let us say that each PDA A defines a relation $R \subseteq \Gamma^* \times \Sigma^*$ between its pushdown and its input. This relation is defined as $R(\alpha, w)$ if and only if $(q_0, \alpha, w) \vdash (q_F, \varepsilon, \varepsilon)$ for some $q_F \in F$. Then our lemma states that any such relation preserves context-freeness.

One more interpretation is that the word given in the pushdown is a "certificate" of the membership of the input word, and we prove that an access to a context-free certificate does not give a PDA any extra computational power.

Lemma 5. *Consider a one-way S-rewriting system over an alphabet Σ as defined in Theorem 4. Let $\overset{\ell\ell}{\to} = \bigcup_{i=1}^{n} X_i \times Y_i$, let a finite transducer for $\overset{\ell r}{\to}$ have a transition function δ, and construct a PDA over a common input and pushdown alphabet $\Sigma \cup \{\$\}$, where $\$ \notin \Sigma$ is a new symbol, as shown in Figure 2. Then the PDA can reach an accepting configuration from a configuration $(q_0, y\$x, v\$u)$, where $y\$x$ is the pushdown and $v\$u$ is the input, if and only if $x\#y \Longrightarrow^* u\#v$.*

Combining Lemmata 3–5, we obtain a succinct proof of our statement on the power of S-rewriting. Let $\$ \notin \Sigma$, let L be the language generated by the rewriting system. Given the initial set I, define $I' = \text{SHIFT}(I\$) \cdot \{\#\}^{-1} = \{v_0\$u_0 \mid u_0\#v_0 \in I\}$. By Lemma 3 and by the well-known closure of the context-free languages under quotient with regular languages, I' is context-free.

Construct a PDA A as in Lemma 5 and let I' be the language of its initial pushdown words. Using Lemma 5, we can prove that $L(A) = \{v\$u \mid u\#v \in L\}$. Next, by Lemma 4, the language $L(A)$ is context-free. It remains to apply the cyclic shift again to obtain $\text{SHIFT}(L(A)\#) \cdot \{\$\}^{-1} = L$, and this language is context-free by the same closure properties as above.

5 Receiving and Sending

We shall now consider *one-way RS-rewriting*, which is the strongest type of one-way rewriting that uses the relations $\overset{\ell\ell}{\rightarrow}$, $\overset{\ell r}{\rightarrow}$ and $\overset{rr}{\rightarrow}$, As in the case of S-rewriting, the expressive power is mainly determined by whether $\overset{\ell r}{\rightarrow}$ is controlled or not. Let us first consider the case of an uncontrolled $\overset{\ell r}{\rightarrow}$.

Theorem 5. *Let $\overset{\ell r}{\rightarrow} \in Unc(Reg, Reg)$ be uncontrollable, let $\overset{\ell\ell}{\rightarrow}$ and $\overset{rr}{\rightarrow}$ be recognizable relations, let I be regular. Then the generated language can be recognized by a one-turn counter automaton, and, given finite automata for $\overset{\ell r}{\rightarrow}$, $\overset{\ell\ell}{\rightarrow}$, $\overset{rr}{\rightarrow}$ and I, this PDA can be constructed.*

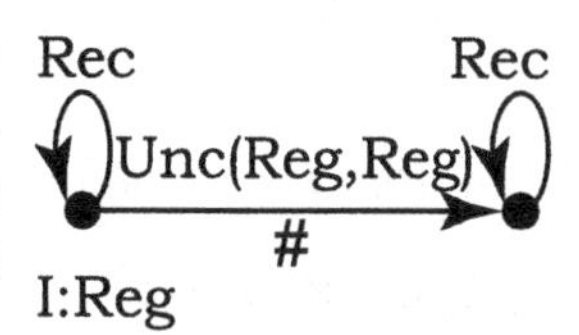

The proof generally follows the scheme used for Theorem 1: first we define the language of computational histories, then use reductions to obtain the language generated by the rewriting system. However, the use of counter one-turn automata instead of finite automata in this context is novel, so we can no longer rely upon any well-known results.

Let Σ be the alphabet and define its two copies: $\overrightarrow{\Sigma} = \{\overrightarrow{a} \mid a \in \Sigma\}$ and $\overleftarrow{\Sigma} = \{\overleftarrow{a} \mid a \in \Sigma\}$. For every word $w = a_1 \dots a_\ell \in \Sigma^*$, with $\ell \geqslant 0$, denote $\overrightarrow{w} = \overrightarrow{a_\ell} \dots \overrightarrow{a_1}$ and $\overleftarrow{w} = \overleftarrow{a_\ell} \dots \overleftarrow{a_1}$. Extend this notation to languages as $\overrightarrow{L} = \{\overrightarrow{w} \mid w \in L\}$ and $\overleftarrow{L} = \{\overleftarrow{w} \mid w \in L\}$ for every $L \subseteq \Sigma^*$. Consider the alphabet $\Sigma_3 = \Sigma \cup \overrightarrow{\Sigma} \cup \overleftarrow{\Sigma}$.

Let $L_\ell = \{y\overrightarrow{x} \mid x \rightarrow y \in \overset{\ell\ell}{\rightarrow}\}$, $L_r = \{\overleftarrow{x}y \mid x \rightarrow y \in \overset{rr}{\rightarrow}\}$ and $\overset{\ell r}{\rightarrow} = Z \times W$, and define the language of computation histories $L_0 \subseteq \Sigma_3$ as follows:

$$L_0 = \{\alpha_n\overrightarrow{z_n} \dots \alpha_1\overrightarrow{z_1}\alpha_0 u_0\#v_0\beta_0 w_1\beta_1 \dots w_n\beta_n \mid \\ u_0\#v_0 \in I, \alpha_i \in L_\ell, \beta_i \in L_r, z_i \in Z, w_i \in W\}$$

Consider the following reduction rules on Σ_3^*: $\overrightarrow{a}a \rightarrow \varepsilon$ and $a\overleftarrow{a} \rightarrow \varepsilon$, for all $a \in \Sigma$. A word $\alpha \in \Sigma_3^*$ is said to be *reducible* to $\beta \in \Sigma_3^*$ if and only if it can be transformed to β by zero or more such reductions.

Lemma 6. *A word $u\#v \in \Sigma^*\#\Sigma^*$ is generated by the system if and only if there exists $\alpha \in L_0$ reducible to $u\#v$.*

In contrast to Theorem 1, here the language of computation histories is, in general, not regular. However, it can be recognized by a pushdown automaton from a simple subclass. Let $I = \bigcup_i I_{\ell,i}\#I_{r,i}$, $\overset{\ell\ell}{\rightarrow} = \bigcup_i X_i \times Y_i$ and $\overset{rr}{\rightarrow} = \bigcup_i U_i \times V_i$.

Lemma 7. *Consider a counter automaton A over the input alphabet $\Sigma_3 \cup \{\#\}$ and with transitions defined according to Figure 3. Then $L(A) = L_0$. Additionally, A makes one reversal of the counter in each computation, and this reversal takes place exactly over the center marker.*

Note that arcs labelled with regular languages specify subautomata that simulate DFAs for these languages without modifying the counter.

It turns out that the reductions we consider can be effectively implemented for PDAs of this restricted form.

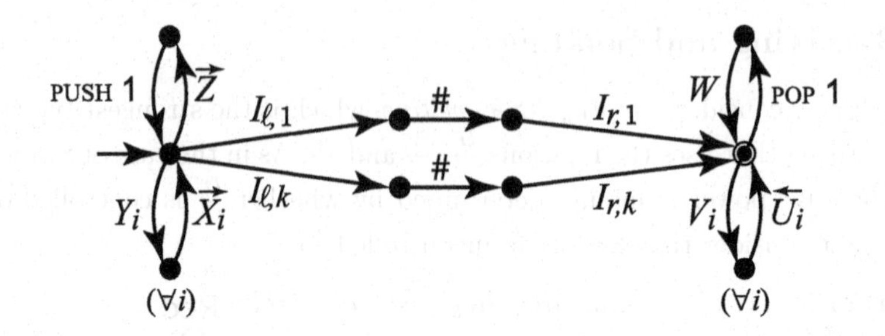

Fig. 3. A one-turn counter automaton recognizing "computation histories"

Lemma 8. *Let A be a one-turn counter automaton over Σ_3, such that $L(A) \subseteq (\Sigma \cup \overrightarrow{\Sigma})^*\#(\Sigma \cup \overleftarrow{\Sigma})^*$, and for every computation of A on an input $u\#v$, the turn of the counter takes place over the center marker. Then the set of words in $\Sigma^*\#\Sigma^*$ that can be obtained by reduction of some words in $L(A)$ is recognized by a one-turn counter automaton B, which, given A, can be effectively constructed.*

These three lemmata yield a proof of Theorem 5.

As we have seen, having an uncontrolled relation $\xrightarrow{\ell r}$ limits the expressive power of RS-rewriting to a subset of the context-free languages. Now let $\xrightarrow{\ell r}$ be a controlled finite relation. We shall see that even if $\xrightarrow{\ell\ell}$ and $\xrightarrow{rr}$ are uncontrolled, the resulting class of systems is still computationally universal.

Theorem 6. *For every recursively enumerable language $L \subseteq \Sigma^*$ there exists a one-way RS-rewriting system formed by relations $\xrightarrow{\ell r} \in Fin$ and $\xrightarrow{\ell\ell}, \xrightarrow{rr} \in Unc(Fin, Fin)$ and a singleton initial set I, such that the language generated by the system, intersected with a regular language R, equals $L\#$. Given a type 0 grammar for L, such a system and a finite automaton for R can be effectively constructed.*

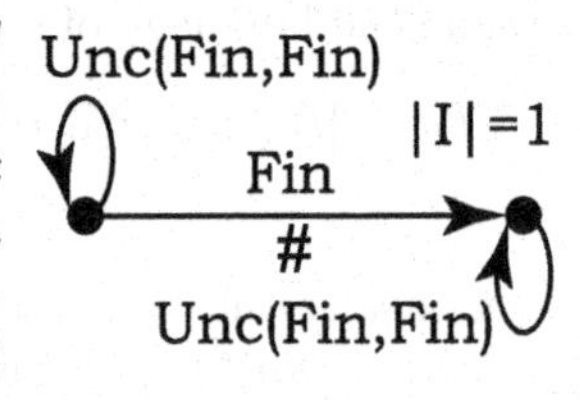

Let $G = (\Sigma, N, P, S)$ be a Chomsky type 0 grammar for L, let $V = \Sigma \cup N$. Consider a new alphabet $V \cup \overline{V}$, where $\overline{V} = \{\overline{s} | s \in V\}$, and define an RS-rewriting system as follows: $I = \{S\#\}$, $\xrightarrow{\ell\ell} = \xrightarrow{rr} = X \times X$, where $X = \{\overline{a}a \mid a \in V\} \cup \{\varepsilon\}$, and $\xrightarrow{\ell r} = \{(a, \overline{a}), (\overline{a}, a) \mid a \in V\} \cup \{(u, \overline{v}) \mid u \to v \in P\}$.

Though this is a one-way rewriting and $\xrightarrow{r\ell} = \varnothing$, a reverse communication channel can be implemented by the following three-step protocol:

$$u\#\overline{v}\overline{a} \overset{\ell\ell}{\Longrightarrow} \overline{a}au\#\overline{v}\overline{a} \overset{\ell r}{\Longrightarrow} au\#\overline{v}\overline{a}a \overset{rr}{\Longrightarrow} au\#\overline{v}$$

First a symbol $a \in \Sigma$ is guessed at the left and a pair $\overline{a}a$ is created. Then one of these symbols is transferred to the right and cancelled there with the existing a. Thus a has effectively been moved from the right to the left.

Claim. If $w \in V^*$ is generated by G from S, then $w\#$ is derivable in the constructed rewriting system.

It can be formally proved that if a wrong symbol is guessed in the above sequence, or the protocol is violated in any other way, then a word of the form $w\#$ can no longer be derived. Denote by $\mathrm{d}(x)$ the word obtained from $x \in (V \cup \overline{V})^*$ by deleting all occurrences of factors of the form $\overline{a}a$ for $a \in V$.

Claim. For every word $\alpha = y\#x$ derivable in the rewriting system, from which any word in $V^*\#$ can be derived, the word $\overline{\mathrm{d}(x)}\,\mathrm{d}(y)$ can be derived in G.

This shows that $w\#$ is derived in the rewriting system if and only if $w \in L(G)$.

6 Two-Way Communication

We shall establish the computational universality of two-way rewriting in its weakest, least controlled form.

Theorem 7. *For every recursively enumerable language $L \subseteq \Sigma^*$ there exists an alphabet $\Gamma \supset \Sigma$, finite uncontrolled relations $\overset{\ell r}{\to}$ and $\overset{r\ell}{\to}$ and a word $w \in \Gamma^*\#\Gamma^*$, such that the language generated by the two-way rewriting system from w equals $\lambda(L)\#$ modulo intersection with $\lambda(\Sigma^*)\#$, for a suitable non-erasing morphism λ.*

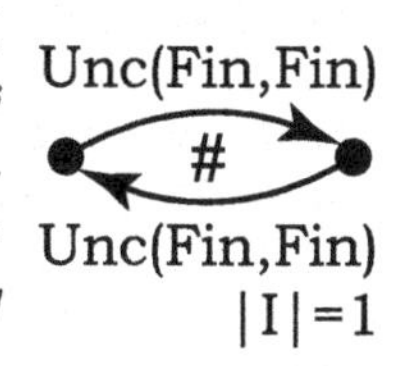

The general idea behind our construction is clear. The two stacks contain a sentential form of Chomsky type 0 grammar, which is redistributed between the stacks before every rewriting step. The symbols and productions are communicated between the stacks in unary notation: in order to send an object number n, $f(n)$ empty messages are sent. The set of uncontrolled rules is constructed in such a way that both parties must faithfully follow a certain rigid protocol, and if they ever divert from it, they will never be able to get back on the right track.

This sounds easy in theory, but if one recalls that the stacks do not even have local states, the task of constructing such a rewriting system will appear impossible. However, there exists quite a sophisticated solution. Let $P \subseteq \Sigma^+ \times \Sigma^+$ be a set of productions generating L from the initial symbol $S \in \Sigma$. We consider the following extended alphabet: $\Gamma = \Sigma \cup \{\pounds, \$, ¢, a, b, c, d, e, g, \bar{g}, \tilde{g}, h, \hbar\}$. Let us denote $n = |\Sigma \cup P| + 3$ and fix an arbitrary bijection $\varphi\colon \Sigma \cup P \to \{3, \ldots, n-1\}$. We define morphisms $\lambda, \rho\colon \Sigma^* \to \Gamma^*$ by setting $\lambda(A) = h\hbar A ¢^n c$ and $\rho(A) = c¢^n A\hbar h$, for every $A \in \Sigma$. Let $\overset{\ell r}{\to} = L_- \times R_+$ and $\overset{r\ell}{\to} = R_- \times L_+$, where

$$\begin{aligned}
L_- &= \{\$, ¢^n c h\hbar, g\bar{g}, \tilde{g}\pounds, \pounds^2 ab\} \cup \{\, \{\varepsilon, h\hbar\} A ¢^{\varphi(A)} \mid A \in \Sigma \,\} \\
&\quad \cup \{\varepsilon, h\hbar\}^{-1} \{\, \lambda(u)(¢^{n-\varphi(u\to v)} c)^{-1} \mid u \to v \in P \,\}, \\
R_+ &= \{\pounds, ¢, g, h, a\pounds^2 \tilde{g}\bar{g}\$, h\hbar b, c¢^n de\hbar\$\} \cup \{\, c¢^n A\hbar\$^{\varphi(A)} \mid A \in \Sigma \,\} \\
&\quad \cup \{\, (\rho(v) h^{-1})\$^{\varphi(u\to v)} \mid u \to v \in P \,\}, \\
R_- &= \{\$, \hbar h c¢^n, \bar{g}g, \pounds\tilde{g}, ba\pounds^2, e\hbar h, ¢^2 d\} \cup \{\, ¢^{\varphi(A)} A \{\varepsilon, \hbar h\} \mid A \in \Sigma \,\}, \\
L_+ &= \{\pounds, ¢, g, h, \$\bar{g}\tilde{g}\pounds^2 a, b\hbar h, \$^2 \hbar S ¢^n c\} \cup \{\, \$^{\varphi(A)} \hbar A ¢^n c \mid A \in \Sigma \,\}.
\end{aligned}$$

The initial set is defined as $I = \{g\bar{g}\tilde{g}\pounds^2 ab\hbar\#\hbar h\hbar\}$. To prove the theorem, it is sufficient to verify the following equivalence:

Main Claim. *A word $w \in \Sigma^*$ can be generated in P if and only if the word $\lambda(w)\#$ is derivable in our rewriting system.*

To prove that every word $\lambda(w)\#$, where $w \in \Sigma^*$ is generated by P, can be derived, we simulate each step $u \overset{P}{\Longrightarrow} v$ of a derivation in P by shifting a factor $\lambda(u)$ from the left to the right, modifying it to $\rho(v)$ during this transfer, and then copying it back to get $\lambda(v)$. During each of these manipulations one pair $h\hbar$ is consumed on the left and one pair $\hbar h$ is consumed on the right, so before starting the actual simulation we have to generate a sufficient amount of these pairs, which we call "fuel".

Claim 1 (Fuel generation). For every $m \in \mathbb{N}$, the word $\lambda(S)(h\hbar)^m\#(\hbar h)^m$ can be derived in the rewriting system.

The proof is by first deriving $g\bar{g}\tilde{g}\pounds^2 ab\hbar(h\hbar)^{m-1}\#(\hbar h)^m\hbar$ for every $m \in \mathbb{N}$, which can be done inductively on m. The required word is obtained from this.

Claim 2 (Copying a letter). Let $x, y \in \Sigma^*$ and $A \in \Sigma$. Then for every $m \in \mathbb{N}$ and $k \in \mathbb{N}_0$, the word $(h\hbar)^{-1}\lambda(x)(h\hbar)^m\#(\hbar h)^k\rho(yA)$ can be derived both from $\lambda(Ax)(h\hbar)^m\#(\hbar h)^k\rho(y)$ and from $(h\hbar)^{-1}\lambda(Ax)(h\hbar)^m\#(\hbar h)^k\rho(y)$. Dually, for every $m \in \mathbb{N}_0$ and $k \in \mathbb{N}$, the word $\lambda(Ax)(h\hbar)^m\#(\hbar h)^k\rho(y)(\hbar h)^{-1}$ can be derived both from $\lambda(x)(h\hbar)^m\#(\hbar h)^k\rho(yA)$ and from $\lambda(x)(h\hbar)^m\#(\hbar h)^k\rho(yA)(\hbar h)^{-1}$.

Claim 3 (Applying a production). For every $x, y \in \Sigma^*$, $m, k \in \mathbb{N}$ and $u \to v \in P$, the word $(h\hbar)^{-1}\lambda(y)(h\hbar)^m\#(\hbar h)^k\rho(xv)$ can be derived both from $\lambda(uy)(h\hbar)^m$ $\#(\hbar h)^k\rho(x)$ and from $(h\hbar)^{-1}\lambda(uy)(h\hbar)^m\#(\hbar h)^k\rho(x)$.

Claim 4 (Simulating generation). For every $m \in \mathbb{N}_0$ and every word w generated by P, the word $\lambda(w)(h\hbar)^m\#(\hbar h)^m$ can be derived in the rewriting system.

The proof is by induction on the length of the derivation of w. The basis (the start symbol) is given by Claim 1. To prove Claim 4 for an n-step derivation and m, we use the inductive hypothesis for the shorter $(n-1)$-step derivation and the number $m+1$. Then we continue simulating the derivation by using Claim 2 an appropriate number of times to move the tape to the desired location, then Claim 3 once to apply the production, and then again redistribute the symbols by Claim 2, consuming one unit of fuel from each side in the process.

Taking $m = 0$ in Claim 4, we obtain that the word $\lambda(w)\#$ can be derived as long as $w \in L$, which proves the forward implication of the Main Claim.

To prove the converse, consider words derived by our rewriting system that belong to one of the following languages:

$$L_1 = \lambda(\Sigma^*)(h\hbar)^*\#(\hbar h)^*\rho(\Sigma^*)(\hbar h)^{-1}$$
$$L_2 = (h\hbar)^{-1}\lambda(\Sigma^*)(h\hbar)^*\#(\hbar h)^*\rho(\Sigma^*)$$

We shall show that if any mistake is made in a derivation, then it can never lead to a word belonging to any of these languages.

Let us first establish that the information about a symbol being moved or a production being applied is always correctly communicated to the other side.

Claim 5 (Soundness of data transfer). Let $m, k \in \mathbb{N}$ and $\alpha \in \Gamma^*\#\Gamma^*$ be arbitrary, and let $j \in \Gamma$ be a letter such that no word from R_- ends with j (no word from L_- starts with j). Then some word from $L_1 \cup L_2$ can be derived from the word $\cent^m c\alpha j\$^k$ ($\$^k j\alpha c\cent^m$) only if $m + k = n$, and every derivation leading to a word from $L_1 \cup L_2$ starts by applying k-times the rule $(\$, \cent) \in \stackrel{r\ell}{\rightarrow}$ ($(\$, \cent) \in \stackrel{\ell r}{\rightarrow}$, respectively). Further, some word from $L_1 \cup L_2$ can be derived from the word $\pounds^m a\alpha j\$^k$ ($\$^k j\alpha a\pounds^m$) only if $m = k = 1$, and every derivation leading to a word from $L_1 \cup L_2$ starts by applying the rule $(\$, \pounds) \in \stackrel{r\ell}{\rightarrow}$ ($(\$, \pounds) \in \stackrel{\ell r}{\rightarrow}$, respectively).

Claim 6 (Soundness of the simulation). Let $x, y \in \Sigma^*$. Then, if any word from $\lambda(y)(h\hbar)^*\#(\hbar h)^*\rho(x)(\hbar h)^{-1} \cup (h\hbar)^{-1}\lambda(y)(h\hbar)^*\#(\hbar h)^*\rho(x)$ can be derived in the rewriting system, this implies that xy can be generated in P.

Now we can verify the converse of the Main Claim: for every $w \in \Sigma^*$, if the word $\lambda(w)\#$ can be derived in our rewriting system, then by Claim 6 the word w can be generated by P. This completes the proof of computational universality of uncontrolled two-way rewriting systems, which in turn concludes our study.

References

1. B. S. Baker, R. V. Book, "Reversal-bounded multipushdown machines", *Journal of Computer and System Sciences*, 8 (1974), 315–332.
2. M. Benois, "Parties rationnelles du groupe libre", *C. R. Acad. Sci. Paris Series A*, 269 (1969) 1188–1190.
3. J. R. Büchi, "Regular canonical systems", *Arch. Math. Logic Grundlagenforsch*, 6 (1964), 91–111.
4. J. H. Conway, *Regular Algebra and Finite Machines*, Chapman and Hall, 1971.
5. D. Hofbauer, J. Waldmann, "Deleting string rewriting systems preserve regularity", *Theoretical Computer Science*, 327:3 (2004), 301–317.
6. J. E. Hopcroft, J. D. Ullman, *Introduction to Automata Theory, Languages and Computation*, Addison-Wesley, 1979.
7. J. Karhumäki, M. Kunc, A. Okhotin, "Computing by commuting", *Theoretical Computer Science*, to appear.
8. M. I. Kratko, "On a certain class of Post calculi", *Soviet Math. Doklady*, 6 (1965), 1544–1545.
9. M. Kunc, "The power of commuting with finite sets of words", *STACS 2005* (Stuttgart, Germany), LNCS 3404, 569–580.
10. E. L. Post, "Formal reductions of the general combinatorial decision problem", *American Journal of Mathematics*, 65:2 (1943), 197–215.
11. J. Sakarovitch, *Elements de théorie des automates*, Vuibert, 2003.

On the Axiomatizability of Priority*

Luca Aceto[1,2], Taolue Chen[3,5], Wan Fokkink[3,4], and Anna Ingolfsdottir[1,2]

[1] Reykjavík University, School of Science and Engineering
Ofanleiti 2, 103 Reykjavík, Iceland
[2] BRICS, Aalborg University, Department of Computer Science
Fr. Bajersvej 7E, 9220 Aalborg Ø, Denmark
[3] CWI, Embedded Systems Group
Kruislaan 413, 1098 SJ Amsterdam, The Netherlands
[4] Vrije Universiteit, Section Theoretical Computer Science
Boelelaan 1081a, 1081 HV Amsterdam, The Netherlands
[5] Nanjing University, State Key Laboratory of Novel Software Technology
Nanjing, Jiangsu, P.R.China, 210093
luca@ru.is, chen@cwi.nl, wanf@cs.vu.nl, annai@ru.is

Abstract. This paper studies the equational theory of bisimulation equivalence over the process algebra BCCSP extended with the priority operator of Baeten, Bergstra and Klop. It is proven that, in the presence of an infinite set of actions, bisimulation equivalence has no finite, sound, ground-complete equational axiomatization over that language. This negative result applies even if the syntax is extended with an arbitrary collection of auxiliary operators, and motivates the study of axiomatizations using conditional equations. In the presence of an infinite set of actions, it is shown that, in general, bisimulation equivalence has no finite, sound, ground-complete axiomatization consisting of conditional equations over BCCSP. Sufficient conditions on the priority structure over actions are identified that lead to a finite, ground-complete axiomatization of bisimulation equivalence using conditional equations.

1 Introduction

Programming and specification languages often include constructs to specify mode switches (see, e.g., [17, 19]). Indeed, some form of mode transfer in computation appears in operating systems in the guise of interrupts, in programming languages as exceptions, and in the behaviour of control programs and embedded systems as discrete "mode switches" triggered by changes in the state of their environment. Such mode changes are often used to encode different levels

* The first and fourth author were partly supported by the project "The Equational Logic of Parallel Processes" (nr. 060013021) of The Icelandic Research Fund. The second and third author were partly supported by the Dutch Bsik project BRICKS (Basic Research in Informatics for Creating the Knowledge Society). The second author was partly supported by 973 Program of China (No. 2002CB312002), NNSFC (No. 60233010, No. 60273034, No. 60403014).

M. Bugliesi et al. (Eds.): ICALP 2006, Part II, LNCS 4052, pp. 480–491, 2006.

of urgency amongst the actions that can be performed by a system as it computes, and implement variations on the notion of pre-emption. Classic process description languages include primitive operators to describe mode changes—for example, LOTOS [9] offers the so-called disruption operator—or have been extended with variations on mode transfer operators. Examples of such operators for the process algebra CCS are discussed by Milner in [18, pp. 192–193].

One of the most widely studied, and natural, notions used to implement different levels of urgency between system actions is priority. (A thorough and clear discussion of the different approaches to the study of priority in process description languages may be found in [12].) In this paper, we consider the well-known priority operator Θ studied by Baeten, Bergstra and Klop [5] in the context of process algebra. (See [10, 11, 12, 13] for later accounts of this operator in the setting of process description languages.) The priority operator Θ gives certain actions priority over others based on an irreflexive partial ordering relation $<$ over the set of actions. Intuitively, $a < b$ is interpreted as "b has priority over a". This means that, in the context of the priority operator Θ, action a is pre-empted by action b. For example, if p is some process that can initially perform both a and b, then $\Theta(p)$ will initially only be able to execute the action b.

In their classic paper [5], Baeten, Bergstra and Klop provided a sound and ground-complete axiomatization for this operator modulo bisimulation equivalence. Their axiomatization uses predicates on actions (to express priorities between actions) and one extra auxiliary operator. Bergstra showed in the earlier paper [6] that, in case of a finite alphabet of actions, there exists a finite equational axiomatization for Θ, without action predicates and help operators. So, if the set of actions is finite, neither conditional equations nor auxiliary operators, as used in [5], are actually necessary to obtain a finite axiomatization of bisimulation equivalence over basic process description languages enriched with the priority operator. But, can Bergstra's positive result be extended to a setting with a countably infinite collection of actions? Or are conditional equations and auxiliary operators necessary to obtain a finite axiomatization of bisimulation equivalence in the presence of an infinite collection of actions? (Note that infinite sets of actions are common in process calculi, and arise, for instance, in the setting of value- or name-passing calculi.) The aim of this paper is to provide a thorough answer to these questions in the setting of the process algebra BCCSP enriched with the priority operator Θ. In case of an infinite alphabet, we permit the occurrence of action variables in axioms.

The process algebra BCCSP contains only basic process algebraic operators from CCS and CSP, but is sufficiently powerful to express all finite synchronization trees. This paper considers the equational theory of BCCSP with the priority operator Θ from [5] modulo bisimulation equivalence. Our first main result is a theorem indicating that the use of conditional equations is indeed inevitable in order to offer a finite axiomatization of bisimulation equivalence over the basic process language we consider in this study. To this end, we prove that, in case of an infinite alphabet and in the presence of at least one priority relation $a < b$ between a pair of actions, there is no finite equational axiomatization for BCCSP enriched with the priority operator (Theorem 2). This result even applies if one is allowed to add an arbitrary collection of help operators to the syntax. Theorem 2 offers a very strong indication that the use of conditional equations, where the conditions consist of action predicates, is essential for axiomatizing Θ, and cannot be circumvented by introducing auxiliary operators. (This is in contrast to the classic positive and negative results on the existence of finite equational axiomatizations for parallel composition offered in [7, 20, 21].)

Having established that conditional equations are necessary in order to obtain a finite, ground-complete equational axiomatization of bisimulation equivalence, we then proceed to investigate whether, in the presence of an infinite set of actions, this equivalence can be finitely axiomatized using conditional equations, but without auxiliary operators like the unless operator used in [5]. We show that, in general, the answer to this question is negative. This we do by exhibiting a priority structure with respect to which bisimulation equivalence affords no finite, sound and ground-complete axiomatization in terms of conditional equations (Theorem 3). This shows that, in general, the use of auxiliary operators is indeed necessary to axiomatize bisimulation equivalence finitely, even using conditional equations and over the simple language considered in this study.

In contrast to the aforementioned negative results, we exhibit a countably infinite, ground-complete axiomatization for bisimulation equivalence over BCCSP with the priority operator in terms of conditional equations (Theorem 4). This axiomatization suggests that infinite collections of pairwise incomparable actions with respect to the priority relation $<$ are the source of our negative result presented in Theorem 3.

Our results add the priority operator to the list of operators whose addition to a process algebra spoils finite axiomatizability modulo bisimulation equivalence; see, e.g., [2, 4, 20, 21, 22] for other examples of non-finite axiomatizability results over process algebras.

Most of the proofs have been omitted from this extended abstract; they can be found in the full version of the paper [1]. Only for the negative result in Section 5.1 do we provide a proof sketch.

2 Preliminaries

We begin by introducing the basic definitions and results on which the technical developments to follow are based.

2.1 The Language BCCSP$_\Theta$

Act denotes a non-empty alphabet of atomic actions, with typical elements a, b, c, d, e. Over *Act* we assume an irreflexive, transitive partial ordering $<$ to express priorities between actions. Intuitively, $a < b$ expresses that the action b has priority over the action a. We say that actions $a_1, \ldots, a_n$ are *incomparable* if they are distinct and $a_i < a_j$ does not hold for all $1 \leq i, j \leq n$.

The language of processes we shall consider in this paper, henceforth referred to as BCCSP$_\Theta$, is obtained by adding the unary priority operator Θ from [5] to the basic process algebra BCCSP [14, 15]. The language is given by the following grammar:

$$t ::= \mathbf{0} \mid a.t \mid t + t \mid \Theta(t) \mid x \mid \alpha.t \ ,$$

where a ranges over Act, x is a process variable and α is an action variable. Process and action variables range over given, disjoint countably infinite sets. We use x, y, z to range over the collection of process variables, and α, β as typical action variables. We use t, u, v to range over the collection of *open* process terms. A process term is *closed* if it does not contain any variables, and p, q, r, range over the set of closed terms T(BCCSP$_\Theta$). The *size* of a term is its length in function symbols.

A substitution maps each process variable to a process term, and each action variable to an action or action variable. A substitution is *closed* if it maps process variables to closed process terms and action variables to actions. For every term t and substitution σ, the term obtained by replacing occurrences of process variables x and action variables α in t with $\sigma(x)$ and $\sigma(\alpha)$ is written $\sigma(t)$.

The semantics of the operators is captured by the transition rules below, which give rise to Act-labelled transitions between closed terms:

$$\frac{}{a.x \xrightarrow{a} x} \qquad \frac{x_1 \xrightarrow{a} y}{x_1 + x_2 \xrightarrow{a} y} \qquad \frac{x_2 \xrightarrow{a} y}{x_1 + x_2 \xrightarrow{a} y} \qquad \frac{x \xrightarrow{a} y \quad x \stackrel{b}{\nrightarrow} \text{ for all } b \text{ such that } a < b}{\Theta(x) \xrightarrow{a} \Theta(y)}$$

where a ranges over Act. Intuitively, closed terms in the language BCCSP$_\Theta$ represent finite process behaviours, where $\mathbf{0}$ does not exhibit any behaviour, $p + q$ is the nondeterministic choice between the behaviours of p and q, and $a.p$ executes action a to transform into p. Furthermore, the process graph of $\Theta(p)$ is obtained by eliminating all transitions $q \xrightarrow{a} q'$ from the process graph of p for which there is a transition $q \xrightarrow{b} q''$ with $a < b$.

We consider the language BCCSP$_\Theta$ modulo bisimulation equivalence.

Definition 1. *A binary symmetric relation* $\mathcal{R}$ *over* T(BCCSP$_\Theta$) *is a* bisimulation *if* $p \mathrel{\mathcal{R}} q$ *together with* $p \xrightarrow{a} p'$ *imply* $q \xrightarrow{a} q'$ *for some* q' *with* $p' \mathrel{\mathcal{R}} q'$*. We write* $p \underline{\leftrightarrow} q$ *if there is a bisimulation relating* p *and* q*. The relation* $\underline{\leftrightarrow}$ *will be referred to as* bisimulation equivalence *or* bisimilarity.

It is well-known that $\underline{\leftrightarrow}$ is an equivalence relation. Moreover, the transition rules are in the GSOS format of [8]. Hence, bisimulation equivalence is a congruence with respect to all the operators in the signature of BCCSP$_\Theta$, meaning that $p \underline{\leftrightarrow} q$ implies $C[p] \underline{\leftrightarrow} C[q]$ for each BCCSP$_\Theta$-context $C[]$.

We can therefore consider the algebra of the closed terms in T(BCCSP$_\Theta$) modulo $\underline{\leftrightarrow}$. In Section 4, we shall offer results that apply to any signature Σ that extends the one of BCCSP$_\Theta$. To this end, we shall tacitly assume that all of the new operators in Σ also preserve bisimulation equivalence, and are semantically interpreted as operations over finite synchronization trees [18].

2.2 Equational Logic

An *axiom system* is a collection of equations $t \approx u$ over the language BCCSP_Θ. An equation $t \approx u$ is derivable from an axiom system E, notation $E \vdash t \approx u$, if it can be proven from the axioms in E using the rules of equational logic (viz. reflexivity, symmetry, transitivity, substitution and closure under BCCSP_Θ contexts). Without loss of generality one may assume that substitutions happen first in equational proofs, i.e., that the rule $\frac{t \approx u}{\sigma(t) \approx \sigma(u)}$ may only be used when $t \approx u \in E$. Moreover, by postulating that for each axiom in E also its symmetric counterpart is present in E, we can disregard applications of symmetry in equational proofs. In the remainder of this paper, we shall tacitly assume that our equational axiom systems are closed with respect to symmetry. Furthermore, it is well-known (cf., e.g., Section 2 in [16]) that if an equation relating two closed terms can be proven from an axiom system E, then there is a closed proof for it. (A proof is *closed* if it only mentions closed terms.) We shall only consider questions related to the provability of closed equations from an axiom system. Therefore, in light of the previous observation, we can restrict ourselves to considering closed proofs.

An equation $t \approx u$ is *sound* with respect to $\underline{\leftrightarrow}$ if $\sigma(t) \underline{\leftrightarrow} \sigma(u)$ holds for each closed substitution σ. An axiom system E is called *sound* over some language modulo $\underline{\leftrightarrow}$ if $E \vdash t \approx u$ implies $t \underline{\leftrightarrow} u$, for all terms t, u in the language. Conversely, E is called *ground-complete* if $p \underline{\leftrightarrow} q$ implies $E \vdash p \approx q$, for all *closed* terms p, q in the language.

Our order of business in the remainder of this paper will be to offer a thorough study of the equational theory of the language BCCSP_Θ modulo bisimulation equivalence. We begin our investigation by considering the case in which the set of actions *Act* is finite. We then move on to investigate the equational properties of bisimulation equivalence over BCCSP_Θ when the set of actions is infinite.

3 $|Act| < \infty$

In this section, we assume that the action set is finite. The axiom system in Table 1 was put forward by Jan Bergstra in [6]. Note that, in the case of a finite action set, this axiom system is finite, since then the axiom schemas PR2–4 give rise to finitely many equations.

Theorem 1 (Bergstra [6]). *The axiom system (A1)–(A4) and (PR1)–(PR4) is sound and ground-complete for* BCCSP_Θ *modulo* $\underline{\leftrightarrow}$.

In the remainder of this paper, process terms are considered modulo associativity and commutativity of $+$. We use $\sum_{i=1}^{n} t_i$ to denote $t_1 + \cdots + t_n$, where the empty sum represents $\mathbf{0}$. Modulo the axioms (A1) and (A2), every term t in the language BCCSP_Θ has the form $\sum_{i=1}^{n} t_i$, where the terms t_i do not have the form $t' + t''$. The terms t_i are called the *summands* of t.

Table 1. Axiomatization in case of $|Act| < \infty$

A1	$x + y$	$\approx$	$y + x$	
A2	$x + (y + z)$	$\approx$	$(x + y) + z$	
A3	$x + x$	$\approx$	x	
A4	$x + \mathbf{0}$	$\approx$	x	
PR1	$\Theta(\mathbf{0})$	$\approx$	$\mathbf{0}$	
PR2	$\Theta(a.x + a.y + z)$	$\approx$	$\Theta(a.x + z) + \Theta(a.y + z)$	
PR3	$\Theta(a.x + b.y + z)$	$\approx$	$\Theta(b.y + z)$	$(a < b)$
PR4	$\Theta(a_1.x_1 + \cdots + a_n.x_n)$	$\approx$	$a_1.\Theta(x_1) + \cdots + a_n.\Theta(x_n)$	$(a_1, \ldots, a_n$ incomparable$)$

4 $|Act| = \infty$

In this section, we deal with the case that the action set is infinite. Our main result is that bisimulation equivalence does *not* afford a finite equational axiomatization over the language BCCSP_Θ, provided that *Act* contains at least two actions a, b with $a < b$. (Otherwise, the equation $\Theta(x) \approx x$ would be sound, and the priority operator could be eliminated from all terms.) This negative result even applies if BCCSP_Θ is extended with an arbitrary collection of operators (over finite synchronization trees) for which bisimulation is a congruence.

The idea behind the proof of our main result of this section is that a finite axiom system E can mention only finitely many action names. So, since *Act* is infinite, we can find a pair c, d of distinct actions that do not occur in E. If c and d are incomparable, then the equation $\Theta(c.\mathbf{0} + d.\mathbf{0}) \approx c.\mathbf{0} + d.\mathbf{0}$ is sound; if $c < d$, then $\Theta(c.\mathbf{0} + d.\mathbf{0}) \approx d.\mathbf{0}$ is sound. In the first case, we show that an equational proof of $\Theta(c.\mathbf{0} + d.\mathbf{0}) \approx c.\mathbf{0} + d.\mathbf{0}$ from E would give rise to a proof of the unsound equation $\Theta(a.\mathbf{0} + b.\mathbf{0}) \approx a.\mathbf{0} + b.\mathbf{0}$ from E. This follows by a simple renaming argument, using that c and d do not occur in E. Likewise, in the second case, a proof of $\Theta(c.\mathbf{0} + d.\mathbf{0}) \approx d.\mathbf{0}$ from E would give rise to a proof of the unsound equation $\Theta(d.\mathbf{0} + c.\mathbf{0}) \approx c.\mathbf{0}$ from E.

Theorem 2. *Let $|Act| = \infty$, and $a < b$ for some $a, b \in Act$. Let Σ be a signature consisting of the operators in BCCSP_Θ, together with auxiliary operators for which bisimulation equivalence is a congruence. Then bisimulation equivalence has no finite, sound and ground-complete axiomatization over $\mathrm{T}(\Sigma)$.*

5 Axiomatizing Priority Conditionally

Theorem 2 offers very strong evidence that, in the presence of an infinite set of actions, equational logic is inherently not sufficiently powerful to achieve a finite axiomatization of bisimilarity over closed terms in the language BCCSP_Θ. Indeed, that result holds true even in the presence of an arbitrary number of auxiliary operators.

In the presence of action variables, it is natural to view our language as consisting of two sorts: one for actions and the other for processes. This is all the more true because the set of actions has the structure of a partial order, and we should like to express axioms over processes that reflect the influence that this poset structure on actions has on the behaviour of processes. In case our set of actions is finite, this can be done by means of a finite number of equations that are instances of (PR3) and (PR4) in Table 1.

In the presence of an infinite action set, however, the axiom schemas (PR3) and (PR4), as well as (PR2), have infinitely many instances. One way to capture their effects finitely is, in the presence of action variables, to phrase the equation schemas (PR3) and (PR4) as conditional equations thus:

$$\text{(CPR3)} \quad (\alpha < \beta) \Rightarrow \Theta(\alpha.x + \beta.y + z) \approx \Theta(\beta.y + z)$$

$$\text{(CPR4)}_n \quad (\bigwedge_{1 \leq i,j \leq n} \neg(\alpha_i < \alpha_j)) \Rightarrow$$

$$\Theta(\alpha_1.x_1 + \cdots + \alpha_n.x_n) \approx \alpha_1.\Theta(x_1) + \cdots + \alpha_n.\Theta(x_n) \quad (n \geq 0) \; .$$

In both of the above conditional equations, we use predicates over actions to restrict the applicability of the equation on the right-hand side of the implication. In general, henceforth in this study we shall consider conditional equations of the form $P \Rightarrow t \approx u$, where P is a predicate over actions, and $t \approx u$ is an equation over the language BCCSP_Θ. In what follows, we shall assume that predicates over actions are expressed using formulae in first-order logic with equality and the binary relation symbol $<$.

The semantics of a predicate P is given by the collection of closed substitutions that satisfy it. If P is a tautology, then we simply write $t \approx u$. For instance, a version of of equation (PR2) with action variables will be written thus:

$$\text{(CPR2)} \quad \Theta(\alpha.x + \alpha.y + z) \approx \Theta(\alpha.x + z) + \Theta(\alpha.y + z) \; .$$

Note that equation (PR1) in Table 1 is just (CPR4)_0. Moreover, since $<$ is irreflexive, the conditional equation (CPR4)_1 reduces to $\Theta(\alpha.x) \approx \alpha.\Theta(x)$. (Note that this equation can be derived from each of the (CPR4)_n with $n \geq 1$ and (A3).)

A conditional equation $P \Rightarrow t \approx u$ is *sound* with respect to bisimilarity, if $\sigma(t) \leftrightarrow \sigma(u)$ holds for each closed substitution σ that satisfies predicate P. It is not hard to see that for each partial order of actions $(Act, <)$, the conditional equations (CPR2), (CPR3) and (CPR4)_n $(n \geq 0)$ are sound modulo bisimilarity over the language BCCSP_Θ.

A natural question to ask at this point, and one that we shall address in the remainder of this study, is whether, unlike standard equational logic, conditional equations suffice to obtain a finite, ground-complete axiomatization of bisimulation equivalence over the language BCCSP_Θ.

In their classic paper [5], Baeten, Bergstra and Klop offered a finite, conditional, ground-complete axiomatization of bisimilarity over the language BPA_δ with the priority operator. Their axiomatization, however, relied upon the

Table 2. Axioms for Θ in the presence of $\triangleleft$

$\Theta(\alpha.x)$	$\approx$	$\alpha.x$
$\Theta(\mathbf{0})$	$\approx$	$\mathbf{0}$
$\Theta(x+y)$	$\approx$	$(\Theta(x) \triangleleft y) + (\Theta(y) \triangleleft x)$
$\neg(\alpha < \beta) \Rightarrow (\alpha.x) \triangleleft (\beta.y)$	$\approx$	$\alpha.x$
$(\alpha < \beta) \Rightarrow (\alpha.x) \triangleleft (\beta.y)$	$\approx$	$\mathbf{0}$
$(\alpha.x) \triangleleft \mathbf{0}$	$\approx$	$\alpha.x$
$\mathbf{0} \triangleleft (\alpha.x)$	$\approx$	$\mathbf{0}$
$(x+y) \triangleleft z$	$\approx$	$(x \triangleleft z) + (y \triangleleft z)$
$x \triangleleft (y+z)$	$\approx$	$(x \triangleleft y) \triangleleft z$

introduction of a binary auxiliary operator, the so-called *unless* operator $\triangleleft$, whose transition rules are:

$$\frac{x \xrightarrow{a} x' \quad y \stackrel{b}{\nrightarrow} \text{ for all } b \text{ such that } a < b}{x \triangleleft y \xrightarrow{a} x'} \;, \text{ where } a \in Act \;.$$

In the setting of BCCSP_Θ, and using action variables in lieu of concrete action names, the relation between the priority operator and the unless operator is expressed by the conditional equations in Table 2. It is not too hard to see that those conditional equations, together with (A1)–(A4) in Table 1, yield a ground-complete, finite, conditional equational axiomatization of bisimulation equivalence. Therefore, even in the presence of an infinite set of actions, bisimulation equivalence affords a finite, ground-complete axiomatization using conditional equations at the price of introducing a single auxiliary operator. But, if the set of actions is infinite, is the use of an auxiliary operator like the unless operator necessary to obtain a finite axiomatizability result for bisimulation equivalence over BCCSP_Θ using conditional equations?

5.1 A Negative Result

Our order of business will now be to prove that, in the presence of an infinite set of actions, in general auxiliary operators are indeed necessary in order to obtain a finite ground-complete axiomatization of bisimulation equivalence over the language BCCSP_Θ. In this section, $Act = \{a_i, b_i \mid i \geq 1\} \cup \{c\}$, where $a_i < b_i < c$ for each $i \geq 1$, and these are the only inequalities. For convenience, we consider terms not only modulo associativity and commutativity of $+$, but also modulo the sound equations $x + \mathbf{0} \approx x$ and $\Theta(\Theta(x) + y) \approx \Theta(x + y)$.

The following lemma is the crux in the proof of Theorem 3. It states a property of closed terms that holds for all of the closed instantiations of axioms in any sound collection of conditional equations. In [3, Section 2.3] this is referred to as a proof-theoretic technique to prove that there is no finite basis for the equational theory. We use Φ_n to abbreviate $\sum_{i=1}^{n} b_i.\mathbf{0}$.

Lemma 1. *Let $P \Rightarrow t \approx u$ be a conditional equation that is sound modulo $\underline{\leftrightarrow}$. Let σ be a closed substitution with $\sigma(P) = true$. Assume that:*

– *n is larger than the size of t, where $n \geq 2$; and*
– *the summands of $\sigma(t)$ are all bisimilar to either Φ_n or $\mathbf{0}$.*

Then the summands of $\sigma(u)$ are all bisimilar to either Φ_n or $\mathbf{0}$.

Proof. The claim is easily seen to hold if $\sigma(t) \leftrightarrow \mathbf{0}$. Assume therefore that some summand of $\sigma(t)$ is bisimilar to Φ_n. Then $\sigma(t) \leftrightarrow \sigma(u) \leftrightarrow \Phi_n$.

Write $t = \sum_{i\in I} t_i$ and $u = \sum_{j\in J} u_j$ for some non-empty, finite index sets I and J, where the terms t_i and u_j are of the form x, $a.v$, $\alpha.v$ or $\Theta(v)$. By the proviso of the lemma, for each $i \in I$, the summands of $\sigma(t_i)$ are all bisimilar to Φ_n or $\mathbf{0}$. Since $n \geq 2$, for each $i \in I$, the term t_i is not of the form $a.v$ or $\alpha.v$. Hence either it is a process variable x, or it is of the form $\Theta(\sum_{\ell\in L_i} d_{i\ell}.t'_{i\ell} + \sum_{m\in M_i} \alpha_m.t''_{im} + \sum_{k\in K_i} z_{ik})$ (modulo $x + \mathbf{0} \approx x$ and $\Theta(\Theta(x) + y) \approx \Theta(x + y)$). Let $I' \subseteq I$ be the set of indices of summands of t that have the above form. Observe that $K_i \neq \emptyset$ for each $i \in I'$ such that $\sigma(t_i)$ is bisimilar to Φ_n (because n is larger than the size of t). Note moreover that summands t_i of t having the above form such that $\sigma(t_i) \leftrightarrow \mathbf{0}$ must have $L_i = M_i = \emptyset$, and for such summands $\sigma(z_{ik}) \leftrightarrow \mathbf{0}$ for each $k \in K_i$.

Let us assume, towards a contradiction, that there is an index $j \in J$ such that $\sigma(u_j)$ has a summand that is bisimilar neither to Φ_n nor to $\mathbf{0}$. We proceed by a case analysis on the form of u_j. The cases where u_j is of the form x, $a.u'_j$ or $\alpha.u'_j$ are easy and are omitted here. We focus on the case where $u_j = \Theta(u')$. Then u_j consists of a single summand, so by assumption, $\sigma(u_j) \not\leftrightarrow \Phi_n$ and $\sigma(u_j) \not\leftrightarrow \mathbf{0}$.

Since $\sigma(u) \leftrightarrow \Phi_n$, u' is of the form $\sum_{\ell\in L} e_\ell.u'_\ell + \sum_{m\in M} \beta_m.u''_m + \sum_{k\in K} y_k$. We distinguish two cases.

1. For each $i \in I'$ with $\sigma(t_i) \not\leftrightarrow \mathbf{0}$ there is a $k_i \in K_i$ such that z_{ik_i} is not a summand of u'.
   Define the substitution σ' as $\sigma'(y) = c.\mathbf{0}$ if either $y = z_{ik_i}$ for some $i \in I'$ with $\sigma(t_i) \not\leftrightarrow \mathbf{0}$ or if y is a summand of t with $\sigma(y) \not\leftrightarrow \mathbf{0}$, and let σ' agree with σ on other process variables and on action variables. It is not hard to see that $\sigma'(t) \overset{b_i}{\nrightarrow}$ for $i = 1, \ldots, n$ (because $c > b_i$ and t has no summand of the form $a.v$ or $\alpha.v$). On the other hand, since $\sigma(u_j) \not\leftrightarrow \mathbf{0}$ and $\sigma(u) \leftrightarrow \Phi_n$, there is an h with $1 \leq h \leq n$ such that $\sigma(u') \overset{b_h}{\rightarrow}$. Furthermore, $\sigma(u') \overset{c}{\nrightarrow}$. By assumption, z_{ik_i} is not a summand of u' for each $i \in I'$ with $\sigma(t_i) \not\leftrightarrow \mathbf{0}$. Moreover, for any variable summand y of t with $\sigma(y) \not\leftrightarrow \mathbf{0}$, y is not a summand of u', because by assumption $\sigma(y) \leftrightarrow \Phi_n$ while $\sigma(u') \not\leftrightarrow \Phi_n$. So $\sigma(u') \overset{b_h}{\rightarrow}$ and $\sigma(u') \overset{c}{\nrightarrow}$ imply $\sigma'(u') \overset{b_h}{\rightarrow}$ and $\sigma'(u') \overset{c}{\nrightarrow}$. It follows that $\sigma'(u_j) \overset{b_h}{\rightarrow}$, and so $\sigma'(u) \overset{b_h}{\rightarrow}$. Hence $\sigma'(t) \not\leftrightarrow \sigma'(u)$. Since $\sigma'(P) = \sigma(P) = \mathit{true}$, this contradicts the fact that $P \Rightarrow t \approx u$ is sound modulo $\leftrightarrow$.
2. $\{z_{i_0k} \mid k \in K_{i_0}\} \subseteq \{y_k \mid k \in K\}$, for some $i_0 \in I'$ with $\sigma(t_{i_0}) \not\leftrightarrow \mathbf{0}$. In this case, K is non-empty since, as previously observed, K_{i_0} is non-empty. By the proviso of the lemma, $\sigma(t_{i_0}) \leftrightarrow \Phi_n$, so (since n is larger than the size of t_{i_0}) there is a $k_0 \in K_{i_0}$ with $\sigma(z_{i_0k_0}) \not\leftrightarrow \mathbf{0}$. Furthermore, by assumption, $\sigma(u_j) \not\leftrightarrow \mathbf{0}$ and $\sigma(u_j) \not\leftrightarrow \Phi_n$. Therefore, there is an h with $1 \leq h \leq n$ such that $\sigma(\Theta(u')) \overset{b_h}{\nrightarrow}$. Define the substitution σ' as $\sigma'(y) = a_h.\mathbf{0}$ if $y = z_{i_0k_0}$, and

let σ' agree with σ on other process variables and on action variables. We argue that $\sigma'(t) \stackrel{a_h}{\nrightarrow}$. To this end, observe, first of all, that, since $\sigma(\Theta(u')) \stackrel{b_h}{\nrightarrow}$, we have $\sigma(\sum_{k \in K} y_k) \stackrel{b_h}{\nrightarrow}$, and so $\sigma(z_{i_0 k_0}) \stackrel{b_h}{\nrightarrow}$. We are now ready to show that no summand of $\sigma'(t)$ affords an a_h-labelled transition. We consider three exhaustive possibilities:

(a) Let $i \in I'$ with $z_{i_0 k_0} \notin \{z_{ik} \mid k \in K_i\}$. Then clearly $\sigma'(t_i) \stackrel{a_h}{\nrightarrow}$.

(b) Let $i \in I'$ with $z_{i_0 k_0} \in \{z_{ik} \mid k \in K_i\}$. Then $\sigma(t_i) \not\leftrightarrow \mathbf{0}$ because $\sigma(z_{i_0 k_0}) \not\leftrightarrow \mathbf{0}$, so by assumption $\sigma(t_i) \leftrightarrow \Phi_n$. This implies $\sigma(t_i) \stackrel{b_h}{\rightarrow}$, so since $\sigma(z_{i_0 k_0}) \stackrel{b_h}{\nrightarrow}$, it follows that $\sigma'(t_i) \stackrel{b_h}{\rightarrow}$. Since the outermost function symbol of t_i is Θ, we can conclude that $\sigma'(t_i) \stackrel{a_h}{\nrightarrow}$.

(c) Finally, since $\sigma(z_{i_0 k_0}) \not\leftrightarrow \mathbf{0}$ and $\sigma(z_{i_0 k_0}) \stackrel{b_h}{\nrightarrow}$, the proviso of the lemma yields that $z_{i_0 k_0}$ cannot be a summand of t.

From the three cases above we can conclude that $\sigma'(t) \stackrel{a_h}{\nrightarrow}$. On the other hand, $\sigma'(\Theta(u')) \stackrel{a_h}{\rightarrow}$, because $\sigma(\Theta(u')) \stackrel{b_h}{\nrightarrow}$ and $z_{i_0 k_0} \in \{y_k \mid k \in K\}$. Hence $\sigma'(u) \stackrel{a_h}{\rightarrow}$, and so $\sigma'(t) \not\leftrightarrow \sigma'(u)$. Since $\sigma'(P) = \sigma(P) = \mathit{true}$, this contradicts the fact that $P \Rightarrow t \approx u$ is sound modulo $\leftrightarrow$.

In summary, the assumption that some $\sigma(u_j)$ has a summand that is bisimilar neither to Φ_n nor to $\mathbf{0}$, leads to a contradiction. This completes the proof. □

The following proposition states that the property of closed instantiations of sound conditional equations mentioned in the above lemma is preserved under equational derivations from a finite collection of sound equations.

Proposition 1. *Let E be a finite collection of conditional equations that is sound modulo $\leftrightarrow$. Let $n \geq 2$ be larger than the size of any term in the equations of E. Assume, furthermore, that*

- *$E \vdash p \approx q$; and*
- *the summands of p are all bisimilar to Φ_n or $\mathbf{0}$.*

Then the summands of q are all bisimilar to Φ_n or $\mathbf{0}$.

Proof. By induction on the depth of the closed proof of the equation $p \approx q$ from E, using Lemma 1. □

Theorem 3. *Let $Act = \{a_i, b_i \mid i \geq 1\} \cup \{c\}$, where $a_i < b_i < c$ for each $i \geq 1$, and these are the only inequalities. Then bisimulation equivalence has no ground-complete axiomatization over* BCCSP$_\Theta$ *consisting of a finite set of sound conditional equations.*

Proof. Let E be a finite collection of conditional equations that is sound modulo $\leftrightarrow$. Let $n \geq 2$ be larger than the size of any term in the equations of E. According to Proposition 1, from E we cannot derive $\Theta(\Phi_n) \approx \Phi_n$. This equation is sound modulo $\leftrightarrow$, and therefore E is not ground-complete. □

5.2 A Positive Result

In the previous section, we offered an example of a priority structure $(Act, <)$ with respect to which it is impossible to give a finite, ground-complete axiomatization of bisimulation equivalence over BCCSP_Θ in terms of conditional equations without auxiliary operators. That result, however, does not imply that auxiliary operators are always necessary to achieve a finite basis of conditional equations for bisimulation equivalence. Our aim in this section is to substantiate this claim by providing some general conditions over the priority structure $(Act, <)$ that are sufficient to guarantee the existence of a finite, ground-complete conditional axiomatization of bisimulation equivalence over BCCSP_Θ.

Definition 2. *An anti-chain in a poset $(Act, <)$ is a subset of Act consisting of pairwise incomparable actions. The* width *of a poset $(Act, <)$ is the least upper bound of the cardinalities of its anti-chains.*

We now offer a countably infinite, ground-complete, conditional axiomatization of bisimulation equivalence over BCCSP_Θ. Such an axiomatization reduces to a finite one if the poset of actions has finite width.

Theorem 4. *Let $(Act, <)$ be an infinite poset of actions.*

1. *The axiom system consisting of (CPR2), (CPR3), $(CPR4)_n$ $(n \geq 0)$ and (A1)–(A4) is ground-complete for bisimilarity over* BCCSP_Θ.
2. *Assume that the width of $(Act, <)$ is k. Then the axiom system consisting of (CPR2), (CPR3), $(CPR4)_k$, (A1)–(A4) and (PR1) is ground-complete for bisimilarity over* BCCSP_Θ.

The sufficient condition over $(Act, <)$ stated in the above theorem applies, for instance, to any poset that has infinitely many finite anti-chains of bounded size. For example, it can be used to show that bisimilarity affords a finite, ground-complete axiomatization consisting of conditional equations over BCCSP_Θ if, for some k, the poset $(Act, <)$ has elements a_{ij} $(i \geq 1, 1 \leq j \leq k)$ ordered thus: $a_{hk} < a_{ij}$ if, and only if, $h < k$. That poset has countably many finite, maximal anti-chains of size k.

A more general sufficient condition over $(Act, <)$ that applies to some posets containing infinite anti-chains, and still guarantees the existence of a finite conditional basis of equations for bisimilarity over BCCSP_Θ may be found in the full version of the paper [1, Section 5.2]. That condition applies, for instance, to the flat priority structure $(\{\bot, a_0, a_1, \ldots\}, <)$, where the only ordering relations are given by $\bot < a_i$ for each $i \geq 0$. Membership of the countably infinite anti-chain $\{a_0, a_1, \ldots\}$ can be characterized syntactically by the predicate $P(\alpha) = \forall\beta.\ \neg(\alpha < \beta)$. We can therefore write the following, sound conditional equation that allows us to reduce the number of summands within the scope of a Θ operator:

$$P(\alpha) \wedge P(\beta) \Rightarrow \Theta(\alpha.x + \beta.y + z) \approx \Theta(\alpha.x + z) + \Theta(\beta.y + z) .$$

The generalization of Theorem 4(2) in the full version of this paper relies on the isolation of conditions on the priority structure that ensure the soundness of the above conditional equation over infinite, maximal anti-chains.

References

1. L. Aceto, T. Chen, W. Fokkink, and A. Ingolfsdottir. On the axiomatizability of priority. BRICS Report Series RS-06-1, BRICS, 2006.
2. L. Aceto, W. Fokkink, A. Ingolfsdottir, and B. Luttik. CCS with Hennessy's merge has no finite equational axiomatization. *Theoretical Computer Science*, 330(3):377–405, 2005.
3. L. Aceto, W. Fokkink, A. Ingolfsdottir, and B. Luttik. Finite equational bases in process algebra: Results and open questions. In *Processes, Terms and Cycles: Steps on the Road to Infinity*, LNCS 3838, pp. 338–367. Springer, 2005.
4. L. Aceto, W. Fokkink, A. Ingolfsdottir, and S. Nain. Bisimilarity is not finitely based over BPA with interrupt. In *Proc. CALCO'05*, LNCS 3629, pp. 52–66. Springer, 2005.
5. J. Baeten, J. Bergstra, and J.W. Klop. Syntax and defining equations for an interrupt mechanism in process algebra. *Fundam. Informat.*, IX(2):127–168, 1986.
6. J. Bergstra. *Put and Get, Primitives for Synchronous Unreliable Message Passing.* Logic Group Preprint Series 3, Utrecht University, Department of Philosophy, 1985.
7. J. Bergstra and J.W. Klop. Process algebra for synchronous communication. *Information & Control*, 60(1/3):109–137, 1984.
8. B. Bloom, S. Istrail, and A.R. Meyer. Bisimulation can't be traced. *Journal of the ACM*, 42(1):232–268, 1995.
9. E. Brinksma. A tutorial on LOTOS. In *Proc. PSTV'85*, pp. 171–194. North-Holland, 1985.
10. J. Camilleri and G. Winskel. CCS with priority choice. *Information and Computation*, 116(1):26–37, 1995.
11. R. Cleaveland and M. Hennessy. Priorities in process algebras. *Information and Computation*, 87(1-2):58–77, 1990.
12. R. Cleaveland, G. Lüttgen, and V. Natarajan. Priorities in process algebra. In *Handbook of Process Algebra*, pp. 711–765. Elsevier, 2001.
13. R. Cleaveland, G. Lüttgen, V. Natarajan, and S. Sims. Priorities for modeling and verifying distributed systems. In *Proc. TACAS'96,* LNCS 1055, pp. 278–297. Springer, 1996.
14. R. van Glabbeek. The linear time-branching time spectrum. In *Proc. CONCUR'90*, LNCS 458, pp. 278–297. Springer, 1990.
15. R. van Glabbeek. The linear time-branching time spectrum I. The semantics of concrete, sequential processes. In *Handbook of Process Algebra*, pp. 3–99. Elsevier, 2001.
16. J.F. Groote. A new strategy for proving ω-completeness with applications in process algebra. In *Proc. CONCUR'90*, LNCS 458, pp. 314–331. Springer, 1990.
17. S. Mauw, PSF – A Process Specification Formalism, PhD thesis, University of Amsterdam, 1991.
18. R. Milner. *Communication and Concurrency.* Prentice-Hall, 1989.
19. R. Milner, M. Tofte, R. Harper, and D. MacQueen. *The Definition of Standard ML (Revised).* MIT Press, 1997.
20. F. Moller. The importance of the left merge operator in process algebras. In *Proc. ICALP'90*, LNCS 443, pp. 752–764. Springer, 1990.
21. F. Moller. The nonexistence of finite axiomatisations for CCS congruences. In *Proc. LICS'90*, pp. 142–153. IEEE, 1990.
22. P. Sewell. Nonaxiomatisability of equivalences over finite state processes. *Annals of Pure and Applied Logic*, 90(1–3), 163–191, 1997.

A Finite Equational Base for CCS with Left Merge and Communication Merge*

Luca Aceto[1,4], Wan Fokkink[2,5], Anna Ingolfsdottir[1,4], and Bas Luttik[3,5]

[1] Department of Computer Science, Reykjavík University, Iceland
luca@ru.is, annai@ru.is
[2] Department of Computer Science, Vrije Universiteit Amsterdam, The Netherlands
wanf@cs.vu.nl
[3] Department of Mathematics and Computer Science, Technische Universiteit Eindhoven, The Netherlands
s.p.luttik@tue.nl
[4] **BRICS**, Department of Computer Science, Aalborg University, Denmark
[5] Department of Software Engineering, CWI, The Netherlands

Abstract. Using the left merge and communication merge from ACP, we present an equational base (i.e., a ground-complete and ω-complete set of valid equations) for the fragment of CCS without restriction and relabelling. Our equational base is finite if the set of actions is finite.

1 Introduction

One of the first detailed studies of the equational theory of a process algebra was performed by Hennessy and Milner [9]. They considered the equational theory of the process algebra that arises from the recursion-free fragment of CCS (see [11]), and presented a set of equational axioms that is complete in the sense that all valid *closed* equations (i.e., equations in which no variables occur) are derivable from it in equational logic [15]. For the elimination of parallel composition from closed terms, Hennessy and Milner proposed the well-known *Expansion Law*, an axiom schema that generates infinitely many axioms. Thus, the question arose whether a finite complete set of axioms exists. With their axiom system ACP, Bergstra and Klop demonstrated in [3] that it does exist if two auxiliary operators are used: the left merge and the communication merge. It was later proved by Moller [13] that without using at least one auxiliary operator a finite complete set of axioms does not exist.

The aforementioned results pertain to the closed fragments of the equational theories discussed, i.e., to the subsets consisting of the closed valid equations only. Many valid equations, such as the equation $(x \parallel y) \parallel z \approx x \parallel (y \parallel z)$ expressing that parallel composition is associative, are not derivable (by means of equational logic) from the axioms in [3] or [9]. In this paper we shall not neglect the variables and contribute to the study of full equational theories of process algebras. We take the fragment of CCS without recursion, restriction and

* The first and third author were partly supported by the project "The Equational Logic of Parallel Processes" (nr. 060013021) of The Icelandic Research Fund.

M. Bugliesi et al. (Eds.): ICALP 2006, Part II, LNCS 4052, pp. 492–503, 2006.

relabelling, and consider the full equational theory of the process algebra that is obtained by taking the syntax modulo bisimilarity [14]. Our goal is then to present an *equational base* (i.e., a set of valid equations from which every other valid equation can be derived) for it, which is finite if the set of actions is finite. Obviously, Moller's result about the unavoidability of the use of auxiliary operations in a finite complete axiomatisation of the closed fragment of the equational theory of CCS a fortiori implies that auxiliary operations are needed to achieve our goal. So we add left merge and communication merge from the start.

Moller [12] considers the equational theory of the same fragment of CCS, except that his parallel operator implements pure interleaving instead of CCS-communication and the communication merge is omitted. He presents a set of valid axiom schemata and proves that it generates an equational base if the set of actions is infinite. Groote [6] does consider the fragment including communication merge, but, instead of the CCS-communication mechanism, he assumes an uninterpreted communication function. His axiom schemata also generate an equational base provided that the set of actions is infinite. We improve on these results by considering the communication mechanism present in CCS, and by proving that our axiom schemata generate an equational base also if the set of actions is finite. Moreover, our axiom schemata generate a finite equational base if the set of actions is finite.

Our equational base consists of axioms that are mostly well-known. For parallel composition ($\|$), left merge ($\lfloor\!\lfloor$) and communication merge ($|$) we adapt the axioms of ACP, adding from Bergstra and Tucker [4] a selection of the axioms for *standard concurrency* and the axiom $(x \mid y) \mid z \approx \mathbf{0}$, which expresses that the communication mechanism is a form of *handshaking communication*.

Our proof follows the classic two-step approach: first we identify a set of normal forms such that every process term has a provably equal normal form, and then we demonstrate that for distinct normal forms there is a distinguishing valuation that proves that they should not be equated. (We refer to the survey [2] for a discussion of proof techniques and an overview of results and open problems in the area. We remark in passing that one of our main results in this paper, viz. Corollary 31, solves the open problem mentioned in [2, p. 362].) Since both associating a normal form with a process term and determining a distinguishing valuation for two distinct normal forms are easily seen to be computable, as a corollary to our proof we get the decidability of the equational theory. Another consequence of our result is that our equational base is complete for the set of valid closed equations as well as ω-complete [7].

The positive result that we obtain in Corollary 31 of this paper stands in contrast with the negative result that we have obtained in [1]. In that article we proved that there does not exist a finite equational base for CCS if the auxiliary operation $/\!\!/$ of Hennessy [8] is added instead of Bergstra and Klop's left merge and communication merge. Furthermore, we conjecture that a finite equational base fails to exist if the unary action prefixes are replaced by binary sequential composition. (We refer to [2] for an infinite family of valid equations that we believe cannot all be derivable from a single finite set of valid equations.)

The paper is organised as follows. In Sect. 2 we introduce a class of algebras of processes arising from a process calculus à la CCS, present a set of equations that is valid in all of them, and establish a few general properties needed in the remainder of the paper. Our class of process algebras is parametrised by a communication function. It is beneficial to proceed in this generality, because it allows us to first consider the simpler case of a process algebra with pure interleaving (i.e., no communication at all) instead of CCS-like parallel composition. In Sect. 3 we prove that an equational base for the process algebra with pure interleaving is obtained by simply adding the axiom $x \mid y \approx \mathbf{0}$ to the set of equations introduced in Sect. 2. The proof in Sect. 3 extends nicely to a proof that, for the more complicated case of CCS-communication, it is enough to replace $x \mid y \approx \mathbf{0}$ by $x \mid (y \mid z) \approx \mathbf{0}$; this is discussed in Sect. 4.

2 Algebras of Processes

We fix a set $\mathcal{A}$ of *actions*, and declare a special action τ that we assume is not in $\mathcal{A}$. We denote by $\mathcal{A}_\tau$ the set $\mathcal{A} \cup \{\tau\}$. Generally, we let a and b range over $\mathcal{A}$ and α over $\mathcal{A}_\tau$. We also fix a countably infinite set $\mathcal{V}$ of *variables*. The set $\mathcal{P}$ of *process terms* is generated by the following grammar:

$$P ::= x \mid \mathbf{0} \mid \alpha.P \mid P + P \mid P \mathbin{\|\!\lfloor} P \mid P | P \mid P \parallel P \ ,$$

with $x \in \mathcal{V}$, and $\alpha \in \mathcal{A}_\tau$. We shall often simply write α instead of $\alpha.\mathbf{0}$. Furthermore, to be able to omit some parentheses when writing terms, we adopt the convention that $\alpha.$ binds stronger, and $+$ binds weaker, than all the other operations.

Table 1. The operational semantics

$$\frac{}{\alpha.P \xrightarrow{\alpha} P} \qquad \frac{P \xrightarrow{\alpha} P'}{P+Q \xrightarrow{\alpha} P'} \qquad \frac{Q \xrightarrow{\alpha} Q'}{P+Q \xrightarrow{\alpha} Q'}$$

$$\frac{P \xrightarrow{\alpha} P'}{P \mathbin{\|\!\lfloor} Q \xrightarrow{\alpha} P' \parallel Q} \qquad \frac{P \xrightarrow{\alpha} P'}{P \parallel Q \xrightarrow{\alpha} P' \parallel Q} \qquad \frac{Q \xrightarrow{\alpha} Q'}{P \parallel Q \xrightarrow{\alpha} P \parallel Q'}$$

$$\frac{P \xrightarrow{a} P',\ Q \xrightarrow{b} Q',\ \gamma(a,b)\downarrow}{P \mid Q \xrightarrow{\gamma(a,b)} P' \parallel Q'} \qquad \frac{P \xrightarrow{a} P',\ Q \xrightarrow{b} Q',\ \gamma(a,b)\downarrow}{P \parallel Q \xrightarrow{\gamma(a,b)} P' \parallel Q'}$$

A process term is *closed* if it does not contain variables; we denote the set of all closed process terms by $\mathcal{P}_0$. We define on $\mathcal{P}_0$ binary relations $\xrightarrow{\alpha}$ $(\alpha \in \mathcal{A}_\tau)$ by means of the transition system specification in Table 1. The last two rules in Table 1 refer to a *communication function* γ, i.e., a commutative and associative partial binary function $\gamma : \mathcal{A} \times \mathcal{A} \rightharpoonup \mathcal{A}_\tau$. We shall abbreviate the statement '$\gamma(a, b)$ is defined' by $\gamma(a, b)\downarrow$ and the statement '$\gamma(a, b)$ is undefined' by $\gamma(a, b)\uparrow$. We shall in particular consider the following communication functions:

1. The *trivial communication function* is the partial function $f : \mathcal{A} \times \mathcal{A} \rightharpoonup \mathcal{A}_\tau$ such that $f(a, b)\uparrow$ for all $a, b \in \mathcal{A}$.
2. The CCS *communication function* $h : \mathcal{A} \times \mathcal{A} \rightharpoonup \mathcal{A}_\tau$ presupposes a bijection $\bar{\cdot}$ on $\mathcal{A}$ such that $\bar{\bar{a}} = a$ and $\bar{a} \neq a$ for all $a \in \mathcal{A}$, and is then defined by $h(a, b) = \tau$ if $\bar{a} = b$ and undefined otherwise.

Definition 1. A *bisimulation* is a symmetric binary relation $\mathcal{R}$ on $\mathcal{P}_0$ such that $P \mathrel{\mathcal{R}} Q$ implies

$$\text{if } P \xrightarrow{\alpha} P', \text{ then there exists } Q' \in \mathcal{P}_0 \text{ such that } Q \xrightarrow{\alpha} Q' \text{ and } P' \mathrel{\mathcal{R}} Q'.$$

Closed process terms $P, Q \in \mathcal{P}_0$ are said to be *bisimilar* (notation: $P \leftrightarroweq_\gamma Q$) if there exists a bisimulation $\mathcal{R}$ such that $P \mathrel{\mathcal{R}} Q$.

The relation $\leftrightarroweq_\gamma$ is an equivalence relation on $\mathcal{P}_0$; we denote the equivalence class containing P by $[P]$, i.e.,

$$[P] = \{Q \in \mathcal{P}_0 : P \leftrightarroweq_\gamma Q\} \ .$$

The rules in Table 1 are all in de Simone's format [5] if P, P', Q and Q' are treated as variables ranging over closed process terms and the last two rules are treated as rule schemata generating a rule for every a, b such that $\gamma(a, b)\downarrow$. Hence, $\leftrightarroweq_\gamma$ has the substitution property for the syntactic constructs of our language of closed process terms, and therefore the constructs induce an algebraic structure on $\mathcal{P}_0/\leftrightarroweq_\gamma$, with a constant $\mathbf{0}$, unary operations $\alpha.$ ($\alpha \in \mathcal{A}_\tau$) and four binary operations $+$, $\mathbin{|\!\lfloor}$, $|$ and $\|$ defined by $\mathbf{0} = [\mathbf{0}]$, $\alpha.[P] = [\alpha.P]$, and $[P]{\star}[Q] = [P{\star}Q]$ for $\star \in \{+, \mathbin{|\!\lfloor}, |, \|\}$.

Henceforth, we denote by $\mathbf{P}_\gamma$ (for γ an arbitrary communication function) the algebra obtained by dividing out $\leftrightarroweq_\gamma$ on $\mathcal{P}_0$ with constant $\mathbf{0}$ and operations $\alpha.$ ($\alpha \in \mathcal{A}_\tau$), $+$, $\mathbin{|\!\lfloor}$, $|$, and $\|$ as defined above. The elements of $\mathbf{P}_\gamma$ are called *processes*, and will be ranged over by p, q and r.

2.1 Equational Reasoning

We can use the full language of process expressions to reason about the elements of $\mathbf{P}_\gamma$. A *valuation* is a mapping $\nu : \mathcal{V} \to \mathbf{P}_\gamma$; it induces an *evaluation mapping*

$$[\![\text{-}]\!]_\nu : \mathcal{P} \to \mathbf{P}_\gamma$$

inductively defined by $[\![x]\!]_\nu = \nu(x)$, $[\![\mathbf{0}]\!]_\nu = \mathbf{0}$, $[\![\alpha.P]\!]_\nu = \alpha.[\![P]\!]_\nu$ and $[\![P \star Q]\!]_\nu = [\![P]\!]_\nu \star [\![Q]\!]_\nu$ for $\star \in \{+, \mathbin{|\!\lfloor}, |, \|\}$. A *process equation* is a formula $P \approx Q$ with P and Q process terms; it is said to be *valid* (in $\mathbf{P}_\gamma$) if $[\![P]\!]_\nu = [\![Q]\!]_\nu$ for all $\nu : \mathcal{V} \to \mathbf{P}_\gamma$. If $P \approx Q$ is valid in $\mathbf{P}_\gamma$, then we shall also write $P \leftrightarroweq_\gamma Q$. The *equational theory* of the algebra $\mathbf{P}_\gamma$ is the set of all valid process equations, i.e.,

$$EqTh(\mathbf{P}_\gamma) = \{P \approx Q : [\![P]\!]_\nu = [\![Q]\!]_\nu \text{ for all } \nu : \mathcal{V} \to \mathbf{P}_\gamma\} \ .$$

The precise contents of the set $EqTh(\mathbf{P}_\gamma)$ depend to some extent on the choice of γ. For instance, the process equation $x \mid y \approx \mathbf{0}$ is only valid in $\mathbf{P}_\gamma$ if γ is the trivial communication function f; if γ is the CCS communication function h, then $\mathbf{P}_\gamma$ satisfies the weaker equation $x \mid (y \mid z) \approx \mathbf{0}$.

Table 2. Process equations valid in every $\mathbf{P}_\gamma$

A1	$x + y \approx y + x$		C1	$\mathbf{0} \mid x \approx \mathbf{0}$	
A2	$(x + y) + z \approx x + (y + z)$		C2	$a.x \mid b.y \approx \gamma(a,b).(x \parallel y)$	if $\gamma(a,b)\downarrow$
A3	$x + x \approx x$		C3	$a.x \mid b.y \approx \mathbf{0}$	if $\gamma(a,b)\uparrow$
A4	$x + \mathbf{0} \approx x$		C4	$(x + y) \mid z \approx x \mid z + y \mid z$	
			C5	$x \mid y \approx y \mid x$	
L1	$\mathbf{0} \mathbin{\|\!\lfloor} x \approx \mathbf{0}$		C6	$(x \mid y) \mid z \approx x \mid (y \mid z)$	
L2	$\alpha.x \mathbin{\|\!\lfloor} y \approx \alpha.(x \parallel y)$		C7	$(x \mathbin{\|\!\lfloor} y) \mid z \approx (x \mid z) \mathbin{\|\!\lfloor} y$	
L3	$(x + y) \mathbin{\|\!\lfloor} z \approx x \mathbin{\|\!\lfloor} z + y \mathbin{\|\!\lfloor} z$				
L4	$(x \mathbin{\|\!\lfloor} y) \mathbin{\|\!\lfloor} z \approx x \mathbin{\|\!\lfloor} (y \parallel z)$		P1	$x \parallel y \approx (x \mathbin{\|\!\lfloor} y + y \mathbin{\|\!\lfloor} x) + x \mid y$	
L5	$x \mathbin{\|\!\lfloor} \mathbf{0} \approx x$				

Table 2 lists process equations that are valid in $\mathbf{P}_\gamma$ independently of the choice of γ. (The equations L2, C2 and C3 are actually axiom schemata; they generate an axiom for all $\alpha \in \mathcal{A}_\tau$ and $a, b \in \mathcal{A}$. Note that if $\mathcal{A}$ is finite, then these axiom schemata generate finitely many axioms.) Henceforth whenever we write an equation $P \approx Q$, we shall mean that it is derivable from the axioms in Table 2 by means of equational logic. It is well-known that the rules of equational logic preserve validity. We therefore obtain the following result.

Proposition 2. For all process terms P and Q, if $P \approx Q$, then $P \leftrightarrows_\gamma Q$.

A set of valid process equations is an *equational base* for $\mathbf{P}_\gamma$ if all other valid process equations are derivable from it by means of equational logic. The purpose of this paper is to prove that if we add to the equations in Table 2 the equation $x|y \approx \mathbf{0}$ we obtain an equational base for $\mathbf{P}_f$, and if, instead, we add $x|(y|z) \approx \mathbf{0}$ we obtain an equational base for $\mathbf{P}_h$. Both these equational bases are finite, if the set of actions $\mathcal{A}$ is finite.

Definition 3. Let P be a process term. We define the *height* of a process term P, denoted $h(P)$, inductively as follows:

$$h(\mathbf{0}) = 0 \ , \qquad h(P + Q) = \max(h(P), h(Q)) \ ,$$
$$h(x) = 1 \ , \qquad h(P \star Q) = h(P) + h(Q) \text{ for } \star \in \{\mathbin{\|\!\lfloor}, |, \parallel\}.$$
$$h(\alpha.P) = h(P) + 1 \ ,$$

Definition 4. We call a process term *simple* if it is not $\mathbf{0}$ and not an alternative composition.

Lemma 5. For every process term P there exists a collection of simple process terms $S_1, \ldots, S_n$ ($n \geq 0$) such that $h(P) \geq h(S_i)$ for all $i = 1, \ldots, n$ and

$$P \approx \sum_{i=1}^{n} S_i \qquad \text{(by A1, A2 and A4)}.$$

We postulate that the summation of an empty collection of terms denotes $\mathbf{0}$. The terms S_i will be called *syntactic summands* of P.

2.2 General Properties of $\mathbf{P}_\gamma$

We collect some general properties of the algebras $\mathbf{P}_\gamma$ that we shall need in the remainder of the paper.

The binary transition relations $\xrightarrow{\alpha}$ ($\alpha \in \mathcal{A}_\tau$) on $\mathcal{P}_0$, which were used to associate an operational semantics with closed process terms, will play an important rôle in the remainder of the paper. They induce binary relations on $\mathbf{P}_\gamma$, also denoted by $\xrightarrow{\alpha}$, and defined as the least relations such that $P \xrightarrow{\alpha} P'$ implies $[P] \xrightarrow{\alpha} [P']$. Note that we then get, directly from the definition of bisimulation, that for all $P, P' \in \mathcal{P}_0$:

$$[P] \xrightarrow{\alpha} [P'] \text{ iff for all } Q \in [P] \text{ there exists } Q' \in [P'] \text{ such that } Q \xrightarrow{\alpha} Q'.$$

Proposition 6. For all $p, q, r \in \mathbf{P}_\gamma$:

(a) $p = \mathbf{0}$ iff there do not exist $p' \in \mathbf{P}_\gamma$ and $\alpha \in \mathcal{A}_\tau$ such that $p \xrightarrow{\alpha} p'$;
(b) $\alpha.p \xrightarrow{\beta} r$ iff $\alpha = \beta$ and $r = p$;
(c) $p + q \xrightarrow{\alpha} r$ iff $p \xrightarrow{\alpha} r$ or $q \xrightarrow{\alpha} r$;
(d) $p \mathbin{\|\!\!\_} q \xrightarrow{\alpha} r$ iff there exists $p' \in \mathbf{P}_\gamma$ such that $p \xrightarrow{\alpha} p'$ and $r = p' \parallel q$; and
(e) $p \mid q \xrightarrow{\alpha} r$ iff there exist actions $a, b \in \mathcal{A}$ and processes $p', q' \in \mathbf{P}_\gamma$ such that $\alpha = \gamma(a, b)$, $p \xrightarrow{a} p'$, $q \xrightarrow{b} q'$, and $r = p' \parallel q'$; and
(f) $p \parallel q \xrightarrow{\alpha} r$ iff $p \mathbin{\|\!\!\_} q \xrightarrow{\alpha} r$ or $q \mathbin{\|\!\!\_} p \xrightarrow{\alpha} r$ or $p \mid q \xrightarrow{\alpha} r$.

Let $p, p' \in \mathbf{P}_\gamma$; we write $p \rightarrow p'$ if $p \xrightarrow{\alpha} p'$ for some $\alpha \in \mathcal{A}_\tau$ and call p' a *residual* of p.

It is easy to see from Table 1 that if $P \xrightarrow{\alpha} P'$, then P' has fewer symbols than P. Consequently, the length of a transition sequence starting with a process $[P]$ is bounded from above by the number of symbols in P.

Definition 7. The *depth* $|p|$ of an element $p \in \mathbf{P}_\gamma$ is defined as

$$|p| = \max\{n \geq 0 : \exists p_n, \ldots, p_0 \in \mathbf{P}_\gamma \text{ s.t. } p = p_n \rightarrow \cdots \rightarrow p_0\}.$$

The *branching degree* $bdeg(p)$ of an element $p \in \mathbf{P}_\gamma$ is defined as

$$bdeg(p) = |\{(\alpha, p') : p \xrightarrow{\alpha} p'\}| \ .$$

We establish some useful properties of parallel composition on $\mathbf{P}_\gamma$.

Lemma 8. For all $p, q \in \mathbf{P}_\gamma$, $|p \parallel q| = |p| + |q|$.

According to the following lemma and Proposition 2, $\mathbf{P}_\gamma$ is a commutative monoid with respect to $\parallel$, with $\mathbf{0}$ as the identity element.

Lemma 9. The following equations are derivable from the axioms in Table 2:

$$\begin{array}{lll} \text{P2} & (x \parallel y) \parallel z & \approx x \parallel (y \parallel z) \\ \text{P3} & x \parallel y & \approx y \parallel x \\ \text{P4} & x \parallel \mathbf{0} & \approx x \ . \end{array}$$

An element $p \in \mathbf{P}_\gamma$ is *parallel prime* if $p \neq \mathbf{0}$, and $p = q \parallel r$ implies $q = \mathbf{0}$ or $r = \mathbf{0}$. Suppose that p is an arbitrary element of $\mathbf{P}_\gamma$; a *parallel decomposition* of p is a finite multiset $[p_1, \ldots, p_n]$ of parallel primes such that $p = p_1 \parallel \cdots \parallel p_n$. (The process $\mathbf{0}$ has as decomposition the empty multiset, and a parallel prime process p has as decomposition the singleton multiset $[p]$.) The following theorem is a straightforward consequence of the main result in [10].

Theorem 10. Every element of $\mathbf{P}_\gamma$ has a unique parallel decomposition.

The following corollary follows easily from the above unique decomposition result.

Corollary 11 (Cancellation). Let $p, q, r \in \mathbf{P}_\gamma$. If $p \parallel q = p \parallel r$, then $q = r$.

Lemma 12. For all $p, q \in \mathbf{P}_\gamma$, $bdeg(p \parallel q) \geq bdeg(p), bdeg(q)$.

We define a sequence of parallel prime processes with special properties that make them very suitable as tools in our proofs in the remainder of the paper:

$$\varphi_i = \tau.\mathbf{0} + \cdots + \tau^i.\mathbf{0} \qquad (i \geq 1). \tag{1}$$

Lemma 13. (i) For all $i \geq 1$, the processes φ_i are parallel prime.
(ii) The processes φ_i are all distinct, i.e., $\varphi_k = \varphi_l$ implies that $k = l$.
(iii) For all $i \geq 1$, the process φ_i has branching degree i.

3 An Equational Base for $\mathbf{P}_f$

In this section, we prove that an equational base for $\mathbf{P}_f$ is obtained if the axiom

$$\text{F} \quad x \mid y \approx \mathbf{0}$$

is added to the set of axioms generated by the axiom schemata in Table 2. The resulting equational base is finite if $\mathcal{A}$ is finite.

Henceforth, whenever we write $P \approx_{\text{F}} Q$, we shall mean that the equation $P \approx Q$ is derivable from the axioms in Table 2 and the axiom F.

Proposition 14. For all process terms P and Q, if $P \approx_{\text{F}} Q$, then $P \leftrightarrows_f Q$.

To prove that adding F to the axioms in Table 2 suffices to obtain an equational base for $\mathbf{P}_f$, we need to establish that $P \leftrightarroweq_f Q$ implies $P \approx_{\mathrm{F}} Q$ for all process terms P and Q. First, we identify a set of normal forms $\mathcal{N}_{\mathrm{F}}$ such that every process term P can be rewritten to a normal form by means of the axioms.

Definition 15. The set $\mathcal{N}_{\mathrm{F}}$ of F-*normal forms* is generated by:

$$N ::= \mathbf{0} \mid N+N \mid \alpha.N \mid x \mathbin{\|\!\underline{\;}} N \qquad (x \in \mathcal{V},\ \alpha \in \mathcal{A}_\tau).$$

Lemma 16. For all $P \in \mathcal{P}$ there is an $N \in \mathcal{N}_{\mathrm{F}}$ s.t. $P \approx_{\mathrm{F}} N$ and $h(P) \geq h(N)$.

It remains to prove that for every two F-normal forms N_1 and N_2 there exists a *distinguishing valuation*, i.e., a valuation $*$ such that if N_1 and N_2 are *not* provably equal, then the $*$-interpretations of N_1 and N_2 are distinct. Stating it contrapositively, for every two F-normal forms N_1 and N_2, it suffices to establish the existence of a valuation $* : \mathcal{V} \to \mathbf{P}_f$ such that

$$\text{if } [\![N_1]\!]_* = [\![N_2]\!]_*, \text{ then } N_1 \approx_{\mathrm{F}} N_2. \tag{2}$$

The idea is to use a valuation $*$ that assigns processes to variables in such a way that much of the original syntactic structure of N_1 and N_2 can be recovered by analysing the behaviour of $[\![N_1]\!]_*$ and $[\![N_2]\!]_*$. To recognize variables, we shall use the special processes φ_i $(i \geq 1)$ defined in Eqn. (1) on p. 498. Recall that the processes φ_i have branching degree i. We are going to assign to every variable a distinct process φ_i. By choosing i larger than the maximal 'branching degrees' occurring in N_1 and N_2, the behaviour contributed by an instantiated variable is distinguished from behaviour already present in the F-normal forms themselves.

Definition 17. We define the *width* $w(N)$ of an F-normal form N as follows:

(i) if $N = \mathbf{0}$, then $w(N) = 0$;
(ii) if $N = N_1 + N_2$, then $w(N) = w(N_1) + w(N_2)$;
(iii) if $N = \alpha.N'$, then $w(N) = \max(w(N'), 1)$;
(iv) if $N = x \mathbin{\|\!\underline{\;}} N'$, then $w(N) = \max(w(N'), 1)$.

The valuation $*$ that we now proceed to define is parametrised with a natural number W; in Theorem 21 we shall prove that it serves as a distinguishing valuation (i.e., satisfies Eqn. (2)) for all F-normal forms N_1 and N_2 such that $w(N_1), w(N_2) \leq W$. Let $\ulcorner \_ \urcorner$ denote an *injective* function

$$\ulcorner \_ \urcorner : \mathcal{V} \to \{n \in \omega : n > W\}$$

that associates with every variable a unique natural number greater than W. We define the valuation $* : \mathcal{V} \to \mathbf{P}_f$ for all $x \in \mathcal{V}$ by

$$*(x) = \tau.\varphi_{\ulcorner x \urcorner} \ .$$

The τ-prefix is to ensure the following property.

Lemma 18. For every F-normal form N, *bdeg*$([\![N]\!]_*) \leq w(N)$.

Lemma 19. Let S be a simple F-normal form, let $\alpha \in \mathcal{A}_\tau$, and let p be a process such that $[\![S]\!]_* \xrightarrow{\alpha} p$. Then the following statements hold:

(i) if $S = \beta.N$, then $\alpha = \beta$ and $p = [\![N]\!]_*$;
(ii) if $S = x \mathbin{\|\!\_} N$, then $\alpha = \tau$ and $p = \varphi_{\ulcorner x \urcorner} \parallel [\![N]\!]_*$.

An important property of $*$ is that it allows us to distinguish the different types of simple F-normal forms by classifying their residuals according to the number of parallel components with a branching degree that exceeds W. Let us say that a process p is of *type* n ($n \geq 0$) if its unique parallel decomposition contains precisely n parallel prime components with a branching degree $> W$.

Corollary 20. Let S be a simple F-normal form such that $w(S) \leq W$.

(i) If $S = \alpha.N$, then the unique residual $[\![N]\!]_*$ of $[\![S]\!]_*$ is of type 0.
(ii) If $S = x \mathbin{\|\!\_} N$, then the unique residual $\varphi_{\ulcorner x \urcorner} \parallel [\![N]\!]_*$ of $[\![S]\!]_*$ is of type 1.

Theorem 21. For every two F-normal forms N_1, N_2 such that $w(N_1), w(N_2) \leq W$ it holds that $[\![N_1]\!]_* = [\![N_2]\!]_*$ only if $N_1 \approx N_2$ modulo A1–A4.

Proof. By Lemma 5 we may assume that N_1 and N_2 are summations of collections of simple F-normal forms. We assume $[\![N_1]\!]_* = [\![N_2]\!]_*$ and prove that then $N_1 \approx N_2$ modulo A1–A4, by induction on the sum of the heights of N_1 and N_2.

We first prove that for every syntactic summand S_1 of N_1 there is a syntactic summand S_2 of N_2 such that $S_1 \approx S_2$ modulo A1–A4. To this end, let S_1 be an arbitrary syntactic summand of N_1; we distinguish cases according to the syntactic form of S_1.

1. Suppose $S_1 = \alpha.N_1'$; then $[\![S_1]\!]_* \xrightarrow{\alpha} [\![N_1']\!]_*$. Hence, since $[\![N_1]\!]_* = [\![N_2]\!]_*$, there exists a syntactic summand S_2 of N_2 such that $[\![S_2]\!]_* \xrightarrow{\alpha} [\![N_1']\!]_*$. By Lemma 18 the branching degree of $[\![N_1']\!]_*$ does not exceed W, so $[\![S_2]\!]_*$ has a residual of type 0, and therefore, by Corollary 20, there exist $\beta \in \mathcal{A}_\tau$ and a normal form N_2' such that $S_2 = \beta.N_2'$. Moreover, since $[\![S_2]\!]_* \xrightarrow{\alpha} [\![N_1']\!]_*$, it follows by Lemma 19(i) that $\alpha = \beta$ and $[\![N_1']\!]_* = [\![N_2']\!]_*$. Hence, by the induction hypothesis, we conclude that $N_1' \approx N_2'$ modulo A1–A4, so $S_1 = \alpha.N_1' \approx \beta.N_2' = S_2$.
2. Suppose $S_1 = x \mathbin{\|\!\_} N_1'$; then $[\![S_1]\!]_* \xrightarrow{\tau} \varphi_{\ulcorner x \urcorner} \parallel [\![N_1']\!]_*$. Hence, since $[\![N_1]\!]_* = [\![N_2]\!]_*$, there exists a summand S_2 of N_2 such that $[\![S_2]\!]_* \xrightarrow{\tau} \varphi_{\ulcorner x \urcorner} \parallel [\![N_1']\!]_*$. Since S_2 has a residual of type 1, by Corollary 20 there exist a variable y and a normal form N_2' such that $S_2 = y \mathbin{\|\!\_} N_2'$. Now, since $[\![S_2]\!]_* \xrightarrow{\tau} \varphi_{\ulcorner x \urcorner} \parallel [\![N_1']\!]_*$, it follows by Lemma 19(ii) that
$$\varphi_{\ulcorner x \urcorner} \parallel [\![N_1']\!]_* = \varphi_{\ulcorner y \urcorner} \parallel [\![N_2']\!]_* \ . \tag{3}$$
Since $[\![N_1']\!]_*$ and $[\![N_2']\!]_*$ are of type 0, we have that the unique decomposition of $[\![N_1']\!]_*$ (see Theorem 10) does not contain $\varphi_{\ulcorner y \urcorner}$ and the unique decomposition of $[\![N_2']\!]_*$ does not contain $\varphi_{\ulcorner x \urcorner}$. Hence, from (3) it follows that $\varphi_{\ulcorner x \urcorner} = \varphi_{\ulcorner y \urcorner}$ and $[\![N_1']\!]_* = [\![N_2']\!]_*$. From the former we conclude, by Lemma 13(ii) and the injectivity of $\ulcorner . \urcorner$, that $x = y$ and from the latter we conclude by the induction hypothesis that $N_1' \approx N_2'$ modulo A1–A4. So $S_1 = x \mathbin{\|\!\_} N_1' \approx y \mathbin{\|\!\_} N_2' = S_2$.

We have established that every syntactic summand of N_1 is provably equal to a syntactic summand of N_2. Similarly, it follows that every syntactic summand of N_2 is provably equal to a syntactic summand of N_2. Hence, modulo A1–A4, $N_1 \approx N_1 + N_2 \approx N_2$, so the proof of the theorem is complete. □

Note that, by instantiating the parameter W with a sufficiently large value, it follows from the preceding theorem that there exists a distinguishing valuation for *every* pair of F-normal forms N_1 and N_2. Thus, we get the following corollary.

Corollary 22. For all process terms P and Q, $P \approx_{\mathrm{F}} Q$ if, and only if, $P \leftrightarroweq_f Q$.

4 An Equational Base for $\mathbf{P}_h$

We now consider the algebra $\mathbf{P}_h$. Note that if $\mathcal{A}$ happens to be the empty set, then $\mathbf{P}_h$ satisfies the axiom F, and it is clear from the proof in the previous section that the axioms generated by the axiom schemata in Table 2 together with F in fact constitute a finite equational base for $\mathbf{P}_h$. We therefore proceed with the assumption that $\mathcal{A}$ is nonempty, and prove that an equational base for $\mathbf{P}_h$ is then obtained if we add the axiom

$$\mathrm{H} \quad x \mid (y \mid z) \approx \mathbf{0}$$

to the set of axioms generated by the axiom schemata in Table 2. Again, the resulting equational base is finite if the set $\mathcal{A}$ is finite.

Henceforth, whenever we write $P \approx_{\mathrm{H}} Q$, we shall mean that the equation $P \approx Q$ is derivable from the axioms in Table 2 and the axiom H.

Proposition 23. For all process terms P and Q, if $P \approx_{\mathrm{H}} Q$, then $P \leftrightarroweq_h Q$.

We proceed to adapt the proof presented in the previous section to establish the converse of Proposition 23. Naturally, with H instead of F not every occurrence of $\mid$ can be eliminated from process terms; we therefore need to adapt the notion of normal form.

Definition 24. The set $\mathcal{N}_{\mathrm{H}}$ of H-*normal forms* is generated by:

$$N \ ::= \ \mathbf{0} \mid N+N \mid \alpha.N \mid x \mathbin{\lfloor\!\lfloor} N \mid (x \mid a) \mathbin{\lfloor\!\lfloor} N \mid (x \mid y) \mathbin{\lfloor\!\lfloor} N \ ,$$

with $x, y \in \mathcal{V}$, $\alpha \in \mathcal{A}_\tau$ and $a \in \mathcal{A}$.

Lemma 25. For every process term P there exists an H-normal form N such that $P \approx_{\mathrm{H}} N$ and $h(P) \geq h(N)$.

We proceed to establish that for every two H-normal forms N_1 and N_2 there exists a valuation $* : \mathcal{V} \to \mathbf{P}_h$ such that

$$\text{if } [\![N_1]\!]_* = [\![N_2]\!]_*, \text{ then } N_1 \approx_{\mathrm{H}} N_2. \tag{4}$$

The distinguishing valuations $*$ will have a slightly more complicated definition than before, because of the more complicated notion of normal form.

As in the previous section, the definition of $*$ is parametrised with a natural number W. Since $|$ may occur in H-normal forms, we now also need to make sure that whatever process $*$ assigns to variables has sufficient communication abilities. To achieve this, we also parametrise $*$ with a finite subset $\mathcal{A}' = \{a_1, \ldots, a_n\}$ of $\mathcal{A}$ that is closed under the bijection $\bar{\cdot}$ on $\mathcal{A}$. (Note that every finite subset of $\mathcal{A}$ has a finite superset with the aforementioned property.) Based on W and $\mathcal{A}'$ we define the valuation $* : \mathcal{V} \to \mathbf{P}_h$ by

$$*(x) = a_1.\varphi_{(1 \cdot \ulcorner x \urcorner)} + \cdots + a_n.\varphi_{(n \cdot \ulcorner x \urcorner)} \ .$$

We shall prove that $*$ satisfies Eqn. (4) provided that the actions occurring in N_1 and N_2 are in $\mathcal{A}' \cup \{\tau\}$ and the width of N_1 and N_2, defined below, does not exceed W. We must also be careful to define the injection $\ulcorner\_\urcorner$ in such a way that the extra factors $1, \ldots, n$ in the definition of $*$ do not interfere with the numbers assigned to variables; we let $\ulcorner\_\urcorner$ denote an injection

$$\ulcorner\_\urcorner : \mathcal{V} \to \{m : m \text{ a prime number such that } m > n \text{ and } m > W\}$$

that associates with every variable a prime number greater than the cardinality of $\mathcal{A}'$ and greater than W.

The definition of width also needs to take into account the cardinality of $\mathcal{A}'$ to maintain that the maximal branching degree in $[\![N]\!]_*$ does not exceed $w(N)$.

Definition 26. We define the *width* $w(N)$ of an H-normal form N as follows:

(i)–(iii) see Definition 17(i–iii).
(iv) if $N = x \lfloor\!\lfloor N'$, then $w(N) = \max(w(N'), n)$;
(v) if $N = (x \mid \alpha) \lfloor\!\lfloor N'$, then $w(N) = \max(w(N'), 1)$; and
(vi) if $N = (x \mid y) \lfloor\!\lfloor N'$, then $w(N) = \max(w(N'), n)$.

Lemma 27. *For every H-normal form N, $bdeg([\![N]\!]_*) \leq w(N)$.*

Lemma 28. *Let S be a simple H-normal form, let $\alpha \in \mathcal{A}_\tau$, and let p be a process such that $[\![S]\!]_* \xrightarrow{\alpha} p$. Then the following statements hold:*

(i) *if $S = \beta.N$, then $\alpha = \beta$ and $p = [\![N]\!]_*$;*
(ii) *if $S = x \lfloor\!\lfloor N$, then $\alpha = a_i$ and $p = \varphi_{i \cdot \ulcorner x \urcorner} \parallel [\![N]\!]_*$ for some $i \in \{1, \ldots, n\}$;*
(iii) *if $S = (x \mid a) \lfloor\!\lfloor N$, then $\alpha = \tau$ and $p = \varphi_{i \cdot \ulcorner x \urcorner} \parallel [\![N]\!]_*$ for the unique $i \in \{1, \ldots, n\}$ such that $\overline{a} = a_i$; and*
(iv) *if $S = (x \mid y) \lfloor\!\lfloor N$, then $\alpha = \tau$ and $p = \varphi_{i \cdot \ulcorner x \urcorner} \parallel \varphi_{j \cdot \ulcorner y \urcorner} \parallel [\![N]\!]_*$ for some $i, j \in \{1, \ldots, n\}$ such that $\overline{a_i} = a_j$.*

As in the previous section, we distinguish H-normal forms by classifying their residuals according to the number of parallel components with a branching degree that exceeds W.

Corollary 29. *Let S be a simple H-normal form such that $w(S) \leq W$ and such that the actions occurring in S are included in $\mathcal{A}' \cup \{\tau\}$.*

(i) If $S = \alpha.N$, then the unique residual of $[\![S]\!]_*$ is of type 0.
(ii) If $S = x \mathbin{|\!\lfloor} N$, then all residuals of $[\![S]\!]_*$ are of type 1.
(iii) If $S = (x \mid a) \mathbin{|\!\lfloor} N$, then the unique residual of $[\![S]\!]_*$ is of type 1.
(iv) If $S = (x \mid y) \mathbin{|\!\lfloor} N$, then all residuals of $[\![S]\!]_*$ are of type 2.

Theorem 30. For every two H-normal forms N_1, N_2 such that $w(N_1), w(N_2) \leq W$ and such that the actions occurring in N_1 and N_2 are included in $\mathcal{A}' \cup \{\tau\}$ it holds that $[\![N_1]\!]_* = [\![N_2]\!]_*$ only if $N_1 \approx N_2$ modulo A1–A4, C5.

Proof. The proof of this theorem is very similar to the proof of Theorem 21, only there are two more cases to consider and the reasoning is slightly more complex due to the more complex definition of $*$. □

Corollary 31. For all process terms P and Q, $P \approx_{\mathrm{H}} Q$ if, and only if, $P \leftrightarroweq_h Q$.

References

1. L. Aceto, W. J. Fokkink, A. Ingolfsdottir, and B. Luttik. CCS with Hennessy's merge has no finite equational axiomatization. *Theor. Comput. Sci.*, 330(3): 377–405, 2005.
2. L. Aceto, W. J. Fokkink, A. Ingolfsdottir, and B. Luttik. Finite equational bases in process algebra: Results and open questions. In A. Middeldorp, V. van Oostrom, F. van Raamsdonk, and R. C. de Vrijer, editors, *Processes, Terms and Cycles: Steps on the Road to Infinity*, LNCS 3838, pages 338–367. Springer, 2005.
3. J. A. Bergstra and J. W. Klop. Process algebra for synchronous communication. *Inform. and Control*, 60(1-3):109–137, 1984.
4. J. A. Bergstra and J. V. Tucker. Top-down design and the algebra of communicating processes. *Sci. Comput. Programming*, 5(2):171–199, 1985.
5. R. de Simone. Higher-level synchronising devices in Meije-SCCS. *Theor. Comput. Sci.*, 37:245–267, 1985.
6. J. F. Groote. A new strategy for proving ω-completeness applied to process algebra. In J. C. M. Baeten and J. W. Klop, editors, *Proceedings of CONCUR'90*, LNCS 458, pages 314–331. Springer, 1990.
7. J. Heering. Partial evaluation and ω-completeness of algebraic specifications. *Theoret. Comput. Sci.*, 43(2-3):149–167, 1986.
8. M. Hennessy. Axiomatising finite concurrent processes. *SIAM J. Comput.*, 17(5):997–1017, 1988.
9. M. Hennessy and R. Milner. Algebraic laws for nondeterminism and concurrency. *J. ACM*, 32(1):137–161, January 1985.
10. B. Luttik and V. van Oostrom. Decomposition orders—another proof of the fundamental theorem of arithmetic. *Theor. Comput. Sci.*, 335(2–3):147–186, 2005.
11. R. Milner. *Communication and Concurrency*. Prentice-Hall International, 1989.
12. F. Moller. *Axioms for Concurrency*. PhD thesis, University of Edinburgh, 1989.
13. F. Moller. The nonexistence of finite axiomatisations for CCS congruences. In *Proceedings of LICS'90*, pages 142–153. IEEE Computer Society Press, 1990.
14. D. M. R. Park. Concurrency and automata on infinite sequences. In P. Deussen, editor, 5th *GI Conference*, LNCS 104, pages 167–183. Springer, 1981.
15. W. Taylor. Equational logic. *Houston J. Math.*, (Survey), 1979.

Theories of HNN-Extensions and Amalgamated Products

Markus Lohrey[1] and Géraud Sénizergues[2]

[1] Universität Stuttgart, FMI, Germany
[2] Université Bordeaux I, LaBRI, France
lohrey@informatik.uni-stuttgart.de, ges@labri.u-bordeaux.fr

Abstract. It is shown that the existential theory of $\mathbb{G}$ with rational constraints, over an HNN-extension $\mathbb{G} = \langle \mathbb{H}, t; t^{-1}at = \varphi(a)(a \in A)\rangle$ is decidable, provided that the same problem is decidable in the base group $\mathbb{H}$ and that A is a finite group. The positive theory of $\mathbb{G}$ is decidable, provided that the existential positive theory of $\mathbb{G}$ is decidable and that A and $\varphi(A)$ are proper subgroups of the base group $\mathbb{H}$ with $A \cap \varphi(A)$ finite. Analogous results are also shown for amalgamated products. As a corollary, the positive theory and the existential theory with rational constraints of any finitely generated virtually-free group is decidable.

1 Introduction

Theories of equations over groups are a classical research topic at the borderline between algebra, mathematical logic, and theoretical computer science. This line of research was initiated by the work of Lyndon, Tarski, and others in the first half of the 20th century. A major driving force for the development of this field was a question that was posed by Tarski around 1945: Is the first-order theory of a free group F of rank two, i.e, the set of all statements of first-order logic with equations as atomic propositions that are true in F, decidable. Decidability results for fragments of this theory were obtained by Makanin (for the existential theory of a free group) [15] and Merzlyakov and Makanin (for the positive theory of a free group) [16, 17]. A complete (positive) solution of Tarski's problem was finally announced in [9]; the complete solution is spread over a series of papers. The complexity of Makanin's algorithm for deciding the existential theory of a free group was shown to be not primitive recursive in [10]. Based on [19], a new PSPACE algorithm for the existential theory of a free group, which also allows to include rational constraints for variables, was presented in [2].

Beside these results for free groups, also extensions to larger classes of groups were obtained in the past: [4, 5, 8, 20]. In [3], a general transfer theorem for existential and positive theories was shown: the decidability of the existential theory is preserved by graph products over groups — a construction that generalizes both free and direct products, see e.g. [7]. Moreover, it is shown in [3] that for a large class of graph products, the positive theory can be reduced to the existential theory. The aim of this paper is to prove similar transfer theorems for HNN-extensions and amalgamated free products. These two operations are of fundamental importance in combinatorial group theory [14]; they are recalled in Section 2 by equations (1) and (3).

M. Bugliesi et al. (Eds.): ICALP 2006, Part II, LNCS 4052, pp. 504–515, 2006.

One of the first important applications of HNN-extensions was a more transparent proof of the celebrated result of Novikov and Boone on the existence of a finitely presented group with an undecidable word problem, see e.g. [14]. Such a group can be constructed by a series of HNN-extensions starting from a free group. This shows that there is no hope to prove a transfer theorem for HNN-extensions, similar to the one for graph products from [3]. Therefore we mainly consider HNN-extensions and amalgamated free products, where the subgroup A in (1) and (3), respectively, is finite. Those groups which can be built up from finite groups using the operations of amalgamated free products and HNN-extensions, both subject to the finiteness restrictions above, are precisely the virtually-free groups [1] (i.e., those groups with a free subgroup of finite index). Virtually-free groups have strong connections to formal language theory and infinite graph theory [18].

In Section 3, we consider existential theories. For an HNN-extension $\mathbb{G}$ of the form (1) where the subgroup A is finite, we prove that the existential theory of $\mathbb{G}$ with rational constraints is decidable if the existential theory of $\mathbb{H}$ with rational constraints is decidable (Thm. 1). In Section 4, we consider positive theories. For an HNN-extension $\mathbb{G}$ where the two isomorphic subgroups A and $\varphi(A)$ have finite intersection, we prove that the positive theory of $\mathbb{G}$ is decidable if the positive existential theory of $\mathbb{G}$ is decidable (Thm. 2). From Thm. 1 and 2 and their analogues for amalgamated free products we deduce that every finitely generated virtually-free group has a decidable existential theory with rational constraints as well as a decidable positive theory (Thm. 4). Our exposition will put emphasis on the case of HNN-extensions and just mention the adaptations to amalgamated free products. Full proofs can be found in the three manuscripts [11, 12, 13].

2 Preliminaries

The powerset of a set A is denoted by $\mathcal{P}(A)$. With $\mathrm{RAT}(\mathbb{M})$ (resp. $\mathcal{B}(\mathrm{RAT}(\mathbb{M}))$) we denote the class of all rational (resp. boolean combinations of rational) subsets of a monoid $\mathbb{M}$. The free product of two monoids $\mathbb{M}_1$ and $\mathbb{M}_2$ is denoted by $\mathbb{M}_1 * \mathbb{M}_2$. For a monoid $\mathbb{M}$, a bijection $h : \mathbb{M} \to \mathbb{M}$ is an *anti-automorphism* if $h(1_\mathbb{M}) = 1_\mathbb{M}$ and $h(a \cdot b) = h(b) \cdot h(a)$ for all $a, b \in \mathbb{M}$. It is called *involutive*, if $h^2(a) = a$ for all $a \in \mathbb{M}$. For two groups A and B, $\mathrm{PGI}(A, B)$ denotes the set of all partial isomorphisms from A to B, i.e., isomorphisms from some subgroup $C \leq A$ to some subgroup $D \leq B$. Let $\mathrm{PGI}\{A, B\} = \mathrm{PGI}(A, B) \cup \mathrm{PGI}(B, A) \cup \mathrm{PGI}(A, A) \cup \mathrm{PGI}(B, B)$.

HNN-extensions and amalgamated free products. See [14] for background in combinatorial group theory. Let Γ be an alphabet and let $\Gamma^{-1} = \{a^{-1} \mid a \in \Gamma\}$ be a disjoint copy of Γ. A pair (Γ, R) with $R \subseteq (\Gamma \cup \Gamma^{-1})^*$ is called a *group presentation*. Elements in R are also called *relations*. The group presented by (Γ, R) is usually denoted by $\langle \Gamma; R \rangle$, and is defined as the quotient monoid $(\Gamma \cup \Gamma^{-1})^*/\rho$, where ρ is the smallest congruence relation on the free monoid $(\Gamma \cup \Gamma^{-1})^*$, which contains all pairs in $\{(aa^{-1}, \varepsilon), (a^{-1}a, \varepsilon) \mid a \in \Gamma\} \cup \{(r, \varepsilon) \mid r \in R\}$; note that this quotient is indeed a group. Instead of $\langle \Gamma; \{r_i \mid i \in I\} \rangle$, we also write $\langle \Gamma; r_i (i \in I) \rangle$. Clearly, every group is isomorphic to a group of the form $\langle \Gamma; R \rangle$ (we do not assume Γ to be finite). For a group $G \simeq \langle \Gamma; R \rangle$, an alphabet Σ with $\Sigma \cap \Gamma = \emptyset$ and a new set of relations $P \subseteq (\Gamma \cup \Sigma \cup \Gamma^{-1} \cup \Sigma^{-1})^*$ we denote with $\langle G, \Sigma; P \rangle$ the group $\langle \Sigma \cup \Gamma;\ P \cup R \rangle$.

Let $\mathbb{H}$ be a group (the base group), together with two proper subgroups $A \leq \mathbb{H}, B \leq \mathbb{H}$ and an isomorphism $\varphi : A \to B$. Let $t \notin \mathbb{H}$ be a new generator. Then, the group

$$\mathbb{G} = \langle \mathbb{H}, t; t^{-1}at = \varphi(a)(a \in A) \rangle \tag{1}$$

is called an *HNN-extension of* $\mathbb{H}$ *by the stable letter* t, where A and B are *associated*. It is well known that $\mathbb{H}$ is a subgroup of $\mathbb{G}$. Clearly, there is a natural projection $\pi_{\mathbb{G}} : \mathbb{H} * \{t, t^{-1}\}^* \to \mathbb{G}$. An element s from the free product $\mathbb{H} * \{t, t^{-1}\}^*$ can be written as

$$s = h_0 t^{\alpha_1} h_1 \cdots t^{\alpha_n} h_n, \tag{2}$$

where $n \in \mathbb{N}, \alpha_i \in \{1, -1\}$, and $h_i \in \mathbb{H}$. It is called a *reduced sequence* iff it has neither a factor of the form $t^{-1}at$ with $a \in A$ nor tbt^{-1} with $b \in B$. We denote by $\mathrm{Red}(\mathbb{H}, t)$ the set of all *reduced t-sequences*; one has $\mathbb{G} = \pi_{\mathbb{G}}(\mathrm{Red}(\mathbb{H}, t))$. Reduced t-sequences turned out to be the right representations for elements from $\mathbb{G}$ for the purpose of deciding $\mathrm{Th}_{\exists}(\mathbb{G}, \mathrm{RAT}(\mathbb{G}))$. Let $\sim$ be the smallest congruence over $\mathbb{H} * \{t, t^{-1}\}^*$ generated by the rules $at \sim t\varphi(a)$ for all $a \in A$ and $bt^{-1} \sim t^{-1}\varphi^{-1}(b)$ for all $b \in B$. The congruence $\approx$ is the kernel of $\pi_{\mathbb{G}} : \mathbb{H} * \{t, t^{-1}\}^* \to \mathbb{G}$. Note that $u \sim v$ implies $u \approx v$. Moreover, if $u, v \in \mathrm{Red}(\mathbb{H}, t)$, then $u \approx v$ iff $u \sim v$. The fact $u \sim v$ for $u, v \in \mathrm{Red}(\mathbb{H}, t)$ can be visualized by a Van Kampen diagram (see [14]) in the group $\mathbb{G}$ of the following form, where $u = h_0 t^{\alpha_1} h_1 t^{\alpha_2} h_2 t^{\alpha_3} h_3 t^{\alpha_4} h_4$, $v = k_0 t^{\alpha_1} k_1 t^{\alpha_2} k_2 t^{\alpha_3} k_3 t^{\alpha_4} k_4$ with $h_0, k_0, \ldots, h_4, k_4 \in \mathbb{H}$ and $c_1, \ldots, c_8 \in A \cup B$. Light-shaded (resp. dark-shaded) areas represent relation in $\mathbb{H}$ (resp. group identities of the form $at = t\varphi(a)$ $(a \in A)$ and $bt^{-1} = t^{-1}\varphi^{-1}(b)$ $(b \in B)$).

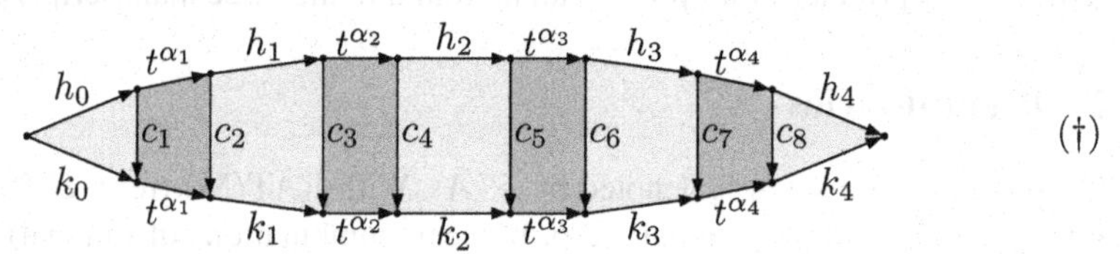

(†)

Now assume that $\mathbb{H}$ and $\mathbb{J}$ are groups with proper subgroups $A < \mathbb{H}$ and $B < \mathbb{J}$ and let $\varphi : A \to B$ be an isomorphism. Then

$$\mathbb{G} = \langle \mathbb{H} * \mathbb{J}, a = \varphi(a)(a \in A) \rangle \tag{3}$$

is called an *amalgamated free product* of $\mathbb{H}$ and $\mathbb{J}$, where A and B are associated.

Logical theories. Let us fix a countable group $\mathbb{G}$, let $\mathcal{C} \subseteq \mathcal{P}(\mathbb{G})$ be a set of *constraints*, and let Ω be an infinite set of variables ranging over $\mathbb{G}$. Formulas of first-order logic over $\mathbb{G}$ with constraints from $\mathcal{C}$ are built up from atomic formulas of the form $x \in L$ $(L \in \mathcal{C}, x \in \Omega)$ and equations $u = v$ $(u, v \in (\Omega \cup \{x^{-1} \mid x \in \Omega\} \cup \mathbb{G})^*)$ using boolean connectives and quantifications over variables. A formula θ is called *positive* if there are no negations in φ, i.e., conjunction and disjunction are the only boolean operators in θ. A formula is called *existential* (resp. existential positive) if it is of the form $\exists x_1 \cdots \exists x_n : \psi(x_1, \ldots, x_n)$, where ψ is a boolean (resp. a positive boolean) combination of atomic formulas. We denote with $\mathrm{Th}_{+}(\mathbb{G}, \mathcal{C})$ (resp. $\mathrm{Th}_{\exists}(\mathbb{G}, \mathcal{C})$, $\mathrm{Th}_{\exists+}(\mathbb{G}, \mathcal{C})$) the set of all positive (resp. existential, existential positive) sentences that are true in $\mathbb{G}$. We briefly write $\mathrm{Th}_X(\mathbb{G})$ for $\mathrm{Th}_X(\mathbb{G}, \emptyset)$ $(X \in \{\exists, +, \exists+\})$.

3 Existential Theories

The following theorem is our main result concerning existential theories:

Theorem 1. $\mathrm{Th}_\exists(\mathbb{G}, \mathrm{RAT}(\mathbb{G}))$ *is decidable in the following two cases:*

(1) $\mathbb{G} = \langle \mathbb{H}, t; t^{-1}at = \varphi(a)(a \in A)\rangle$ *is an HNN-extension, where* A *and* $\varphi(A)$ *are proper subgroups of* $\mathbb{H}$ *with* A *finite, and* $\mathrm{Th}_\exists(\mathbb{H}, \mathrm{RAT}(\mathbb{H}))$ *is decidable.*

(2) $\mathbb{G} = \langle \mathbb{H} * \mathbb{J}, a = \varphi(a)(a \in A)\rangle$ *is an amalgamated free product, where* A *is finite, and* $\mathrm{Th}_\exists(\mathbb{H}, \mathrm{RAT}(\mathbb{H}))$ *and* $\mathrm{Th}_\exists(\mathbb{J}, \mathrm{RAT}(\mathbb{J}))$ *are decidable.*

The statements (1) and (2) in Thm. 1 are orthogonal to the corresponding result for graph products from [3]: none of the three operations (HNN-extensions, amalgamated free products, and graph products) is a special case of another one. At the end of Section 3, we will mention several variants of Thm. 1, which can be obtained by similar techniques. In the following we will sketch a proof of (1) from Thm. 1. Before we go into the details, we will first present some material concerning rational subsets of HNN-extensions, which is of independent interest.

3.1 Rational Subsets of HNN-Extensions

Let us fix throughout this section an HNN-extension $\mathbb{G}$ of a base group $\mathbb{H}$ as described by (1), where A is *finite*. We now define a notion of finite automata which will be well-suited for deciding $\mathrm{Th}_\exists(\mathbb{G}, \mathrm{RAT}(\mathbb{G}))$.

A *finite t-automaton* over $\mathbb{H} * \{t, t^{-1}\}^*$ with labeling set $\mathcal{F} \subseteq \mathcal{P}(\mathbb{H})$ is a 5-tuple

$$\mathcal{A} = \langle \mathcal{L}, \mathsf{Q}, \Delta, \mathsf{I}, \mathsf{T}\rangle, \tag{4}$$

where: (i) $\mathcal{L}$ is a finite subset of $\mathcal{F} \cup \mathcal{P}(A) \cup \mathcal{P}(B) \cup \{\{t\}, \{t^{-1}\}\}$, (ii) Q is a finite set of states, (iii) $\mathsf{I} \subseteq \mathsf{Q}$ is the set of initial states, (iv) $\mathsf{T} \subseteq \mathsf{Q}$ is the set of terminal states, and (v) $\Delta \subseteq \mathsf{Q} \times \mathcal{L} \times \mathsf{Q}$ is the set of transitions. Such an automaton induces a representation map $\mu_\mathcal{A} : \mathbb{H} * \{t, t^{-1}\}^* \to \mathcal{P}(\mathsf{Q} \times \mathsf{Q})$ defined as follows, where $x \in \mathbb{H} \cup \{t, t^{-1}\} \setminus \{1\}$ and $s \in \mathbb{H} * \{t, t^{-1}\}^*$ is of the form (2):

$$\mu_{\mathcal{A},0}(1) = \{(q,q) \mid q \in Q\} \cup \{(q,r) \in Q \times Q \mid \exists (q, L, r) \in \Delta : 1 \in L\}$$
$$\mu_{\mathcal{A},0}(x) = \{(q,r) \in Q \times Q \mid \exists (q, L, r) \in \Delta : x \in L\}$$
$$\mu_\mathcal{A}(s) = \mu_{\mathcal{A},0}(h_0) \circ \mu_{\mathcal{A},0}(t^{\alpha_1}) \circ \mu_{\mathcal{A},0}(h_1) \cdots \mu_{\mathcal{A},0}(t^{\alpha_n}) \circ \mu_{\mathcal{A},0}(h_n).$$

$\mathcal{A}$ recognizes the set $\mathrm{L}(\mathcal{A}) = \{s \in \mathbb{H} * \{t, t^{-1}\}^* \mid (\mathsf{I} \times \mathsf{T}) \cap \mu_\mathcal{A}(s) \neq \emptyset\}$. Let $\mathcal{G}_6 = (\mathcal{T}_6, \mathcal{E}_6)$ and $\mathcal{R}_6 = (\mathcal{T}_6, \mathcal{E}_6')$ be the following two graphs:

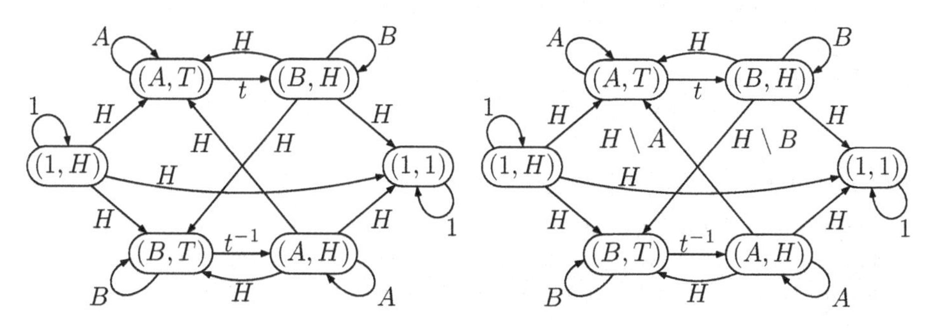

Let $\widehat{\mathcal{E}_6} = \{(p, \ell, q) \mid \exists (p, L, q) \in \mathcal{E}_6, \ell \in L\}$. One can check that $\mathcal{G}_6$ (resp. $\mathcal{R}_6$) endowed with the unique initial state $(1, H)$ and the unique final state $(1, 1)$ is a finite t-automaton recognizing $\mathbb{H} * \{t, t^{-1}\}^*$ (resp. $\mathrm{Red}(\mathbb{H}, t)$). Nodes of $\mathcal{G}_6$, i.e., elements of $\mathcal{T}_6$, are called *vertex-types*. We define a finite partial semigroup $\langle \mathcal{T}, \cdot \rangle$, where $\mathcal{T} = \mathcal{T}_6 \times \mathbb{B} \times \mathcal{T}_6$ and $\mathbb{B} = \langle \{0, 1\}, \vee \rangle$ is the monoid of booleans. The partial product on $\mathcal{T}$ is defined by:

$$\forall (p, b, q), (p', b', q') \in \mathcal{T}_6 \times \mathbb{B} \times \mathcal{T}_6 : (p, b, q) \cdot (p', b', q') = \begin{cases} (p, b \vee b', q') & \text{if } q = p' \\ \text{undefined} & \text{otherwise} \end{cases}$$

The structure $\langle \mathcal{P}(\mathcal{T}), \cdot \rangle$ is thus a (total) monoid. Elements of $\mathcal{T}$ are also called *path-types*. We define an involution $\mathbb{I}_{\mathcal{R}} : \mathcal{T}_6 \to \mathcal{T}_6$ by $(A, T) \leftrightarrow (A, H)$, $(B, T) \leftrightarrow (B, H)$, and $(1, H) \leftrightarrow (1, 1)$. It induces an involution $\mathbb{I}_{\mathcal{T}} : \mathcal{T} \to \mathcal{T}$ defined by: $\mathbb{I}_{\mathcal{T}}(p, b, q) = (\mathbb{I}_{\mathcal{R}}(q), b, \mathbb{I}_{\mathcal{R}}(p))$. This map $\mathbb{I}_{\mathcal{T}}$ is an anti-automorphism of $\mathcal{T}$ and also induces an involutive anti-automorphism of $\langle \mathcal{P}(\mathcal{T}), \cdot \rangle$ that will be denoted by $\mathbb{I}_{\mathcal{T}}$ too. We associate with every element $(p, b, q) \in \mathcal{T}$ its *initial group* $\mathrm{Gi}(p, b, q) = p_1(p) \in \{1, A, B\}$ and its *end group* $\mathrm{Ge}(p, b, q) = p_1(q) \in \{1, A, B\}$. Here p_1 is the projection onto the first component. For $s \in \mathbb{H} * \{t, t^{-1}\}^*$ let $b(s) = 1$ if s contains at least one occurrence of t or t^{-1}, otherwise $b(s) = 0$. Define

$$\gamma_t(s) = \{(p, b(s), q) \in \mathcal{T}_6 \times \mathbb{B} \times \mathcal{T}_6 \mid (p, q) \in \mu_{\mathcal{R}_6}(s)\} \in \mathcal{P}(\mathcal{T}). \tag{5}$$

Let $\mathcal{TA} = \{(\theta, b(x), \theta') \in \mathcal{T} \mid (\theta, x, \theta') \in \widehat{\mathcal{E}_6}\}$ be the set of *atomic path types*. A *normal finite t-automaton* over $\mathcal{F}$ is a 6-tuple $\mathcal{A} = \langle \mathcal{L}, \mathsf{Q}, \tau, \Delta, \mathsf{I}, \mathsf{T} \rangle$, where $\langle \mathcal{L}, \mathsf{Q}, \Delta, \mathsf{I}, \mathsf{T} \rangle$ is as in (4) and $\tau : Q \to \mathcal{T}_6$ maps each state to a vertex-type such that

$$\tau(\mathsf{I}) = \{(1, H)\},\ \tau(\mathsf{T}) = \{(1, 1)\},\ \forall (q, L, r) \in \Delta : \{\tau(q)\} \times L \times \{\tau(r)\} \subseteq \widehat{\mathcal{E}_6},$$

$$[\mathrm{L}(\mathcal{A})]_{\approx} = [\mathrm{L}(\mathcal{A}) \cap \mathrm{Red}(\mathbb{H}, t)]_{\approx}, \tag{6}$$

$$\forall s, s' \in \mathbb{H} * \{t, t^{-1}\}^* : s \sim s' \Rightarrow \mu_{\mathcal{A}}(s) = \mu_{\mathcal{A}}(s'), \tag{7}$$

$$\forall \tilde{\theta} \in \gamma_t(s), \tilde{\theta}' \in \gamma_t(s') : \tilde{\theta} \cdot \tilde{\theta}' \text{ defined in } \mathcal{T} \Rightarrow$$

$$\mu_{\mathcal{A},1}(\tilde{\theta} \cdot \tilde{\theta}', s \cdot s') = \mu_{\mathcal{A},1}(\tilde{\theta}, s) \cdot \mu_{\mathcal{A},1}(\tilde{\theta}', s'),$$

$$\forall \theta \in \mathcal{T}_6 : \mu_{\mathcal{A},1}((\theta, 0, \theta), 1) = \mathrm{id}_{\tau^{-1}(\theta)}.$$

Here, $\mu_{\mathcal{A},1}((\theta, b, \theta'), s) = \mu_{\mathcal{A}}(s) \cap \tau^{-1}(\theta) \times \tau^{-1}(\theta')$; it does not depend on $b \in \{0, 1\}$. $\mathcal{A}$ is said to be *strict* if, instead of (6), it fulfills the condition $\mathrm{L}(\mathcal{A}) \subseteq \mathrm{Red}(\mathbb{H}, t)$.

Lemma 1. *We have:*

- $R \in \mathrm{RAT}(\mathbb{G})$ *iff* $R = \pi_{\mathbb{G}}(\mathrm{L}(\mathcal{A}))$ *for some normal finite t-automaton* $\mathcal{A}$ *with labeling set* $\mathrm{RAT}(\mathbb{H})$.
- *If* $R \in \mathcal{B}(\mathrm{RAT}(\mathbb{G}))$ *then* $R = \pi_{\mathbb{G}}(\mathrm{L}(\mathcal{A}))$ *for some strict normal finite t-automaton* $\mathcal{A}$ *with labeling set* $\mathcal{B}(\mathrm{RAT}(\mathbb{H}))$.

3.2 Deciding $\mathbf{Th}_{\exists}(\mathbb{G}, \mathbf{RAT}(\mathbb{G}))$

AB-Algebras and AB-Homomorphisms. In this section, we introduce an algebraic structure which is devised for handling equations with rational constraints in an HNN-extension. Let A, B be two groups (later, these will be the two subgroups A and

$B = \varphi(A)$ from (1)) and Q be some finite set (it will be the state set of a t-automaton). Let $\mathsf{B}(\mathsf{Q}) = (\mathcal{P}(\mathsf{Q} \times \mathsf{Q}), \cdot)$ be the monoid of binary relations over Q and let $\mathsf{B}^2(\mathsf{Q})$ be the direct product $\mathsf{B}(\mathsf{Q}) \times \mathsf{B}(\mathsf{Q})$. For $m \in \mathsf{B}(\mathsf{Q})$ let $m^{-1} = \{(p,q) \in \mathsf{Q} \times \mathsf{Q} \mid (q,p) \in m\} \in \mathsf{B}(\mathsf{Q})$. Let $\mathbb{I}_{\mathsf{Q}} : \mathsf{B}^2(\mathsf{Q}) \to \mathsf{B}^2(\mathsf{Q})$ be the involutive anti-automorphism defined by $\mathbb{I}_{\mathsf{Q}}(m, m') = (m'^{-1}, m^{-1})$. An *AB-algebra* is a structure $\langle \mathbb{M}, \cdot, 1_{\mathbb{M}}, \mathbb{I}, \iota_A, \iota_B, \gamma, \mu, \delta \rangle$, where $\langle \mathbb{M}, \cdot, 1_{\mathbb{M}} \rangle$ is a monoid, $\iota_A : A \to \mathbb{M}, \iota_B : B \to \mathbb{M}$ are injective monoid homomorphisms, $\mathbb{I} : \mathbb{M} \to \mathbb{M}$ is an involutive anti-automorphism, and $\gamma : \mathbb{M} \to \mathcal{P}(\mathcal{T})$, $\mu : \mathcal{T} \times \mathbb{M} \to \mathsf{B}^2(\mathsf{Q})$, and $\delta : \mathcal{T} \times \mathbb{M} \to \mathrm{PGI}\{A, B\}$ are total mappings fulfilling the axioms (8)–(13) below.

For all $m, m' \in \mathbb{M}$ and all $\tilde{\theta} \in \gamma(m), \tilde{\theta}' \in \gamma(m')$:

$$\gamma(m) \cdot \gamma(m') \subseteq \gamma(m \cdot m') \tag{8}$$

$$\tilde{\theta} \cdot \tilde{\theta}' \text{ defined } \Rightarrow \mu(\tilde{\theta} \cdot \tilde{\theta}', m \cdot m') = \mu(\tilde{\theta}, m) \cdot \mu(\tilde{\theta}', m') \tag{9}$$

$$\mathrm{dom}(\delta(\tilde{\theta}, m)) \subseteq \mathrm{Gi}(\tilde{\theta}), \quad \mathrm{im}(\delta(\tilde{\theta}, m)) \subseteq \mathrm{Ge}(\tilde{\theta}) \tag{10}$$

$$\tilde{\theta} \cdot \tilde{\theta}' \text{ defined } \Rightarrow \delta(\tilde{\theta} \cdot \tilde{\theta}', m \cdot m') = \delta(\tilde{\theta}, m) \circ \delta(\tilde{\theta}', m') \tag{11}$$

For all $a \in A$, $b \in B$, $m \in \mathbb{M}$, and $\tilde{\theta} \in \gamma(m)$:

$$\mathbb{I}(\iota_A(a)) = \iota_A(a^{-1}), \quad \mathbb{I}(\iota_B(b)) = \iota_B(b^{-1}), \tag{12}$$

$$\gamma(\mathbb{I}(m)) = \mathbb{I}_{\mathcal{T}}(\gamma(m)),\ \mu(\mathbb{I}_{\mathcal{T}}(\tilde{\theta}), \mathbb{I}(m)) = \mathbb{I}_{\mathsf{Q}}(\mu(\tilde{\theta}, m)),\ \delta(\mathbb{I}_{\mathcal{T}}(\tilde{\theta}), \mathbb{I}(m)) = \delta(\tilde{\theta}, m)^{-1} \tag{13}$$

Let $\mathcal{M}_i = \langle \mathbb{M}_i, \cdot, 1_{\mathbb{M}_i}, \iota_{A,i}, \iota_{B,i}, \mathbb{I}_i, \gamma_i, \mu_i, \delta_i \rangle$ ($i \in \{1, 2\}$) be two AB-algebras with the same underlying groups A, B and set Q. An *AB-homomorphism* from $\mathcal{M}_1$ to $\mathcal{M}_2$ is a monoid homomorphism $\psi : \mathbb{M}_1 \to \mathbb{M}_2$ fulfilling the five properties (14)–(18) below:

$$\forall a \in A\, \forall b \in B : \psi(\iota_{A,1}(a)) = \iota_{A,2}(a) \ \wedge\ \psi(\iota_{B,1}(b)) = \iota_{B,2}(b) \tag{14}$$

$$\forall m \in \mathbb{M}_1 : \mathbb{I}_2(\psi(m)) = \psi(\mathbb{I}_1(m)) \tag{15}$$

$$\forall m \in \mathbb{M}_1 : \gamma_2(\psi(m)) \supseteq \gamma_1(m) \tag{16}$$

$$\forall m \in \mathbb{M}_1\, \forall \tilde{\theta} \in \gamma_1(m) : \mu_2(\tilde{\theta}, \psi(m)) = \mu_1(\tilde{\theta}, m) \tag{17}$$

$$\forall m \in \mathbb{M}_1\, \forall \tilde{\theta} \in \gamma_1(m) : \delta_2(\tilde{\theta}, \psi(m)) = \delta_1(\tilde{\theta}, m) \tag{18}$$

In the following we will introduce two particular AB-algebras.

The AB-Algebra $\mathbb{H}_t$. From now on, we fix an HNN-extension (1) with A and $B = \varphi(A)$ finite and a strict normal finite t-automaton $\mathcal{A} = \langle \mathcal{L}, \mathsf{Q}, \tau, \Delta, \mathsf{I}, \mathsf{T} \rangle$ with labeling set $\mathcal{B}(\mathrm{RAT}(\mathbb{H}))$. We define an AB-algebra

$$\langle \mathbb{H} * \{t, t^{-1}\}^*, \cdot, 1_{\mathbb{H}}, \iota_A, \iota_B, \mathbb{I}_t, \gamma_t, \mu_t, \delta_t \rangle$$

with underlying monoid $\mathbb{H} * \{t, t^{-1}\}^*$ and set of states Q as follows: ι_A (resp. ι_B) is the natural injection from A (resp. B) into $\mathbb{H} * \{t, t^{-1}\}^*$, and $\mathbb{I}_t$ is the unique involutive anti-automorphism $\mathbb{H} * \{t, t^{-1}\}^* \to \mathbb{H} * \{t, t^{-1}\}^*$ such that $\mathbb{I}_t(h) = h^{-1}$ for $h \in \mathbb{H}$, $\mathbb{I}_t(t) = t^{-1}$, and $\mathbb{I}_t(t^{-1}) = t$. The map γ_t was already defined in (5). The maps μ_t :

$\mathcal{T} \times \mathbb{H} * \{t,t^{-1}\}^* \to \mathsf{B}^2(\mathrm{Q})$ and $\delta_t : \mathcal{T} \times \mathbb{H} * \{t,t^{-1}\}^* \to \mathrm{PGI}\{A,B\}$ are defined as follows, where $s \in \mathbb{H} * \{t,t^{-1}\}^*$ and $\tilde{\theta} \in \mathcal{T}$:

$$\mu_t(\tilde{\theta}, s) = (\mu_{\mathcal{A},1}(\tilde{\theta}, s), (\mu_{\mathcal{A},1}(\mathbb{I}_{\mathcal{T}}(\tilde{\theta}), \mathbb{I}_t(s)))^{-1})$$
$$\delta_t(\tilde{\theta}, s) = \{(c,d) \in \mathrm{Gi}(\tilde{\theta}) \times \mathrm{Ge}(\tilde{\theta}) \mid cs \sim sd\}.$$

Note that $(c,d) \in \delta(\tilde{\theta}, s)$ implies that in the group $\mathbb{G}$ there is a Van Kampen diagram as shown on the right, which (for $s \in \mathrm{Red}(\mathbb{H},t)$) is a diagram of the form (†); note that $c, d \in A \cup B$. E.g., if $\alpha_2 = \alpha_3 = 1$ and $h_2 = k_2$ in (†), then $(c_3, c_6) \in \delta_t(((A,T),1,(B,H)), th_2t)$.

(‡)

One can check that the monoid congruence $\sim$ is compatible with $\mathbb{I}_t$, ι_A, ι_B, γ_t, μ_t, and δ_t (here, (7) is important) so that the quotient $\mathbb{H}_t = \mathbb{H} * \{t,t^{-1}\}^* / \sim$ is naturally endowed with the structure of an AB-algebra (which we denote again with $\mathbb{H}_t$)

$$\mathbb{H}_t = \langle \mathbb{H}_t, \cdot, 1_{\mathbb{H}}, \iota_A, \iota_B, \mathbb{I}_\sim, \gamma_\sim, \mu_\sim, \delta_\sim \rangle. \tag{19}$$

Intuitively, the values $\gamma_\sim(s)$, $\mu_\sim(\tilde{\theta}, s)$, and $\delta_\sim(\tilde{\theta}, s)$ (for $\tilde{\theta} \in \gamma_\sim(s)$) store all information about a sequence s that is relevant when s appears in a solution of a system of equations. Since A, B, and Q are finite, this is only a finite amount of information.

Normal Systems of Equations. A normal system of (dis)equations with constraints from $\mathcal{B}(\mathrm{RAT}(\mathbb{G}))$ is a tuple

$$\mathcal{S}_{\mathbb{G}} = ((u_i = u_i')_{1 \le i \le n}, (u_i \ne u_i')_{n < i \le 2n}, \mu_{\mathcal{A}}, \mu_{\mathcal{U}}), \tag{20}$$

where u_i, u_i' are words over an alphabet of unknowns $\mathcal{U}$, $|u_i| = 1, |u_i'| = 2$ for $1 \le i \le n$, $|u_i| = 1 = |u_i'|$ for $n < i \le 2n$, $\mu_{\mathcal{A}}$ is the representation map associated with the strict normal t-automaton $\mathcal{A}$ from the previous paragraph, and $\mu_{\mathcal{U}} : \mathcal{U} \to \mathsf{B}(\mathrm{Q})$. A solution of the system (20) is any monoid homomorphism $\sigma_{\mathbb{G}} : \mathcal{U}^* \to \mathbb{G}$ such that for all $1 \le i \le n$, $n < j \le 2n$, and $U \in \mathcal{U}$:

$$\sigma_{\mathbb{G}}(u_i) = \sigma_{\mathbb{G}}(u_i'), \quad \sigma_{\mathbb{G}}(u_j) \ne \sigma_{\mathbb{G}}(u_j'), \quad \mu_{\mathcal{A},1}(((1,H),b,(1,1)), \sigma_{\mathbb{G}}(U)) = \mu_{\mathcal{U}}(U),$$

where $b \in \{0,1\}$ ($\mu_{\mathcal{A},1}$ does not depend on the concrete value of b). Since $\mathcal{A}$ is strict normal, $\mu_{\mathcal{A},1}(\tilde{\theta}, g)$ for $g \in \mathbb{G}$ can be defined as $\mu_{\mathcal{A},1}(\tilde{\theta}, s)$ for any $s \in \mathrm{Red}(\mathbb{H},t)$ with $\pi_{\mathbb{G}}(s) = g$. Using Lemma 1, one can reduce $\mathrm{Th}_\exists(\mathbb{G}, \mathrm{RAT}(\mathbb{G}))$ to the question whether a system of the form (20) has a solution. Thus, we may assume to have a system of the form (20) and we aim to decide whether it has a solution.

The AB-Algebra $\mathbb{W}_t$. Whereas our first AB-algebra $\mathbb{H}_t$ from (19) depends on the "concrete" base group $\mathbb{H}$, we now introduce a second "generic" AB-algebra $\mathbb{W}_t$, which depends on our input system (20), but it depends only superficially on $\mathbb{H}$. The idea is to factorize the $\mathbb{G}$-values of a concrete solution of our given system (20) into "generic" symbols, which generate our new AB-algebra $\mathbb{W}_t$. Every generic symbol can be instantiated in $\mathbb{G}$ so that the original solution in $\mathbb{G}$ is recovered.

In order to carry out the above factorization, we introduce for every atomic type $\tilde{\theta} \in \mathcal{TA}$, every $\alpha \in \mathsf{B}^2(\mathrm{Q})$, and every $\beta \in \mathrm{PGI}(\mathrm{Gi}(\tilde{\theta}), \mathrm{Ge}(\tilde{\theta}))$, $54 \cdot n$ (n is from

(20)) many different new "generic" symbols $W_1, \ldots, W_{54n}$ and define: $\gamma(W_i) = \{\tilde{\theta}\}$, $\mu(\tilde{\theta}, W_i) = \alpha$, and $\delta(\tilde{\theta}, W_i) = \beta$. Let $\mathcal{W}$ be the new alphabet obtained in this way. By adding for every $W \in \mathcal{W}$ a new copy to $\mathcal{W}$, we can define on $\mathcal{W}$ an involution $\mathbb{I}$ without fixpoints (i.e., $\mathbb{I}(W) \neq W$) such that (13) holds for every $m = W \in \mathcal{W}$. Let us now consider the free product $\mathcal{W}^* * A * B$. We denote by $\iota_A : A \to \mathcal{W}^* * A * B$ (resp. $\iota_B : B \to \mathcal{W}^* * A * B$) the natural embedding of A (resp. B) into $\mathcal{W}^* * A * B$. We define the AB-algebra

$$\langle \mathcal{W}^* * A * B, \cdot, 1, \iota_A, \iota_B, \mathbb{I}, \mu, \gamma, \delta \rangle$$

with underlying monoid $\mathcal{W}^* * A * B$ and set of states Q as follows: $\mathbb{I}$ is extended as the unique involutive anti-automorphism $\mathcal{W}^* * A * B \to \mathcal{W}^* * A * B$ such that $\mathbb{I}(\iota_A(a)) = \iota_A(a^{-1})$ for $a \in A$ and $\mathbb{I}(\iota_B(b)) = \iota_B(b^{-1})$ for $b \in B$. The mapping $\gamma : \mathcal{W} \to \mathcal{P}(\mathcal{TA})$ is extended to $\iota_A(A) \cup \iota_B(B)$ by

$$\begin{aligned} \forall a \in A \setminus \{1\} : \gamma(\iota_A(a)) &= \{((A,T),0,(A,T)),((A,H),0,(A,H))\}, \\ \forall b \in B \setminus \{1\} : \gamma(\iota_B(b)) &= \{((B,T),0,(B,T)),((B,H),0,(B,H))\}, \\ \gamma(1) &= \{(\theta,0,\theta) \mid \theta \in \mathcal{T}_6\}, \end{aligned}$$

and finally to the full free product $\mathcal{W}^* * A * B$ by

$$\forall g_1, \ldots, g_k \in \mathcal{W} \cup \iota_A(A) \cup \iota_B(B) : \gamma(g_1 \cdots g_k) = \gamma(g_1) \cdots \gamma(g_k).$$

The mappings $\mu : \mathcal{T} \times \mathcal{W} \to \mathsf{B}^2(\mathsf{Q})$ and $\delta : \mathcal{T} \times \mathcal{W} \to \mathrm{PGI}\{A, B\}$ are extended as follows:

$$\begin{aligned} \forall a \in A\, \forall \tilde{\theta} \in \gamma(\iota_A(a)) &: \delta(\tilde{\theta}, \iota_A(a)) = \delta_t(\tilde{\theta}, a), \mu(\tilde{\theta}, \iota_A(a)) = \mu_t(\tilde{\theta}, a) \\ \forall b \in B\, \forall \tilde{\theta} \in \gamma(\iota_B(b)) &: \delta(\tilde{\theta}, \iota_B(b)) = \delta_t(\tilde{\theta}, b),\ \mu(\tilde{\theta}, \iota_B(b)) = \mu_t(\tilde{\theta}, b) \end{aligned}$$

Finally, the maps μ and δ are extended to $\mathcal{W}^* * A * B$ in the only way such that for all $m \in \iota_A(A) \cup \iota_B(B) \cup \mathcal{W}, \tilde{\theta} \in \mathcal{T} \setminus \gamma(m)$: $\mu(\tilde{\theta}, m) = \emptyset$, $\delta(\tilde{\theta}, m) = \{(1,1)\}$ (the trivial partial isomorphism), and axioms (9) and (11) are respected. Let $\equiv$ be the smallest monoid congruence on $\mathcal{W}^* * A * B$ which contains all pairs (cW, Wd) with $W \in \mathcal{W}$ and $(c, d) \in \delta(\tilde{\theta}, W)$ for the unique $\tilde{\theta} \in \gamma(W)$. Let $\mathbb{W} := \mathcal{W}^* * A * B / \equiv$ be the quotient monoid, i.e., we enforce for every $W \in \mathcal{W}$ diagrams of the form (‡) (with $s = W$). One can check that $\equiv$ is compatible with $\mathbb{I}$, ι_A, ι_B, γ, μ, and δ, so that $\mathbb{W}$ inherits from $\mathcal{W}^* * A * B$ the structure of an AB-algebra. Let $\mathcal{W}_t$ be the set of all $W \in \mathcal{W}$ such that for some $s \in \mathbb{H} * \{t, t^{-1}\}^*$: (i) $\gamma(W) \subseteq \gamma_t(s)$ and (ii) the unique $\tilde{\theta} \in \gamma(W)$ fulfills $\mu(\tilde{\theta}, W) = \mu_t(\tilde{\theta}, s)$ and $\delta(\tilde{\theta}, W) = \delta_t(\tilde{\theta}, s)$. Thus, $\mathcal{W}_t$ is the set of all generic symbols that can be realized by a concrete sequence $s \in \mathbb{H} * \{t, t^{-1}\}^*$. With $\mathcal{W}_{\mathbb{H}} \subseteq \mathcal{W}_t$ we denote the set of those $W \in \mathcal{W}_t$ such that moreover $\gamma(W) = \{(\theta, 0, \theta')\}$, where $(\theta, H, \theta') \in \mathcal{E}_6$. Let $\mathbb{W}_t$ (resp. $\mathbb{W}_{\mathbb{H}}$) be the substructure of $\mathbb{W}$ generated by the subset of monoid generators $\iota_A(A) \cup \iota_B(B) \cup \mathcal{W}_t$ (resp. $\iota_A(A) \cup \iota_B(B) \cup \mathcal{W}_{\mathbb{H}}$). It is easy to see that $\psi(\mathbb{W}_{\mathbb{H}}) \subseteq \mathbb{H}$ for every AB-homomorphism $\psi : \mathbb{W}_t \to \mathbb{H}_t$.

The Algorithm. Recall that we have to check, whether the normal system of (dis)equations (20) has a solution.

Step 1. Consider an equation $u_i = u_i'$ from (20), where w.l.o.g. $u_i = U_1$ and $u_i' = U_2U_3$ for $U_1, U_2, U_3 \in \mathcal{U}$; disequations can be treated similarly. Let $\sigma_{\mathbb{G}}$ be a solution for (20). We can choose reduced t-sequences s_1, s_2, and s_3 such that $\sigma_{\mathbb{G}}(U_j) = \pi_{\mathbb{G}}(s_j)$. Then there exists factorizations $s_j = s_{j,1} \cdots s_{j,9}$ and elements $e_{1,2}, e_{2,3}, e_{3,1} \in A \cup B$ such that the Van-Kampen diagram describing the group relation $\pi_{\mathbb{G}}(s_1) = \pi_{\mathbb{G}}(s_2 s_3)$ (i.e., $s_1 \approx s_2 s_3$) decomposes into four pieces, represented by the four relations

$$s_{1,1}\, s_{1,2}\, s_{1,3}\, s_{1,4}\, e_{1,2} \sim s_{2,1}\, s_{2,2}\, s_{2,3}\, s_{2,4} \tag{21}$$

$$s_{2,6}\, s_{2,7}\, s_{2,8}\, s_{2,9} \sim e_{2,3}\, \mathbb{I}_t(s_{3,4})\, \mathbb{I}_t(s_{3,3})\, \mathbb{I}_t(s_{3,2})\, \mathbb{I}_t(s_{3,1}) \tag{22}$$

$$e_{3,1}\, s_{1,6}\, s_{1,7}\, s_{1,8}\, s_{1,9} \sim s_{3,6}\, s_{3,7}\, s_{3,8}\, s_{3,9} \tag{23}$$

$$s_{1,5} = e_{1,2}\, s_{2,5}\, e_{2,3}\, s_{3,5}\, e_{3,1} \text{ in the base group } \mathbb{H}, \tag{24}$$

see the diagram on the right, where the light-shaded area represents a relation in the group $\mathbb{H}$. Dark-shaded areas are diagrams of the form (†) from Section 2. The $s_{j,k}$ ($k \neq 5$) belong to $\mathbb{H} * \{t, t^{-1}\}^*$, while the $s_{j,5} \in \mathbb{H}$. Decomposing e.g. the sequence $s_{1,1}s_{1,2}s_{1,3}s_{1,4}$ into 4 parts allows us to choose all the $s_{1,k}$ ($1 \leq k \leq 4$) either trivial or of some (guessed) atomic type in $\mathcal{TA}$. We now replace every $s_{j,k}$ by a new generic symbol $W_{i,j,k} \in$

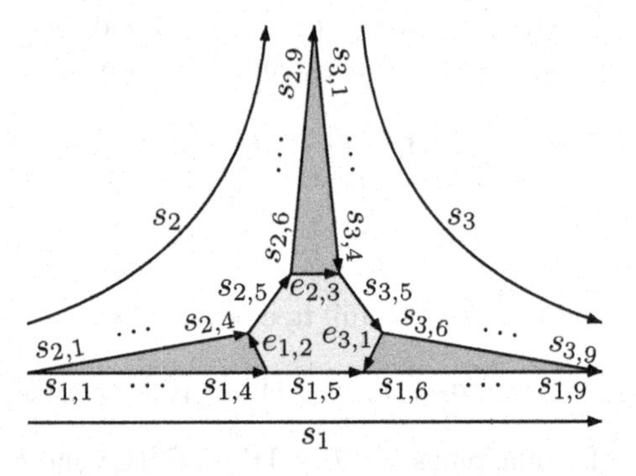

$\mathcal{W}_t$ (or possibly 1); the additional index i refers to the equation $u_i = u_i'$, where $W_{i,j,k}$ comes from. Note that for every i we need 27 symbols $W_{i,j,k}$, this explains the factor $54 = 2 \cdot 27$ in the definition of the alphabet $\mathcal{W}$. The values of the mappings $\mathbb{I}$, γ, μ, and δ on $W_{i,j,k}$ have to be chosen such that the generic symbol $W_{i,j,k}$ captures all the relevant data about the concrete sequence $s_{j,k}$. For instance, $\gamma(W_{i,j,k})$ only contains the guessed type for $s_{j,k}$. In this way, we can translate system (20) into a new system of equations over $\mathbb{W}_t$ (corresponding essentially to (21)–(23)) and another system over $\mathbb{H}$ (corresponding to (24)). Thus, we reduce the problem, whether (20) has a solution, to a finite disjunction of problems of the following form:

INPUT: Finitely many pairs $(v_j, v_j') \in \mathbb{W}_t \times \mathbb{W}_t$ ($j \in J$), with $\gamma(v_j) = \gamma(v_j') \neq \emptyset$, and a (ordinary) system $\mathcal{S}_{\mathbb{H}}$ of equations and disequations in the base group $\mathbb{H}$ and with constraints from $\mathrm{RAT}(\mathbb{H})$; the set of unknowns of $\mathcal{S}_{\mathbb{H}}$ is included in $\mathcal{W}_{\mathbb{H}}$.
QUESTION: Does there exist an AB-homomorphism $\sigma_t : \mathbb{W}_t \to \mathbb{H}_t$ such that

$$\forall j \in J : \sigma_t(v_j) = \sigma_t(v_j') \text{ and simultaneously} \tag{25}$$

$$\sigma_t \text{ solves the system } \mathcal{S}_{\mathbb{H}}? \tag{26}$$

Step 2. We reduce the question, whether (25) and (26) holds for some σ_t to the problem, where the input is the same as above, but the question is:
QUESTION: Do there exist AB-homomorphisms $\sigma_{\mathbb{W}} : \mathbb{W}_t \to \mathbb{W}_t$, $\psi_t : \mathbb{W}_t \to \mathbb{H}_t$ with

$$\forall j \in J : \sigma_{\mathbb{W}}(v_j) = \sigma_{\mathbb{W}}(v_j') \text{ and simultaneously} \tag{27}$$

$$\sigma_{\mathbb{W}} \circ \psi_t \text{ solves the system } \mathcal{S}_{\mathbb{H}}? \tag{28}$$

This reduction is a direct corollary of a *factorization* property for the solutions σ_t of (25): σ_t is a solution iff it can be factorized as $\sigma_{\mathbb{W}} \circ \psi_t$ for AB-homomorphisms $\sigma_{\mathbb{W}} : \mathbb{W}_t \to \mathbb{W}_t$ and $\psi_t : \mathbb{W}_t \to \mathbb{H}_t$. The proof consists in decomposing σ_t into a sort of elementary AB-homomorphisms of the form $W \mapsto cW_1dW_2eW_3f$ ($c, d, e, f \in A \cup B, W_i \in \mathcal{W}_t$), followed by some $\psi_t : \mathbb{W}_t \to \mathbb{H}_t$: we start with the trivial decomposition $\sigma_{\mathbb{W}} = \mathrm{id}_{\mathbb{W}}, \psi_t = \sigma_t$ and then reason by induction over the multiset $\{\!\{d(\sigma_{\mathbb{W}}(v_j), \sigma_{\mathbb{W}}(v'_j)) \mid j \in J\}\!\}$, where d is a kind of distance on $\mathbb{W}$.

Step 3. We introduce the group $\mathbb{U} = \langle \mathbb{W};\ W \cdot \mathbb{I}(W) = 1\ (W \in \mathcal{W})\rangle$. This group turns out to be obtained from a base group $\mathbb{K}$, which is a semi-direct product of the finite group A by a free group of finite rank, by a finite number of HNN-extensions with associated subgroups *strictly smaller* than A. Using the main result of [2], one can show that $\mathrm{Th}_{\exists}(\mathbb{K}, \mathrm{RAT}(\mathbb{K}))$ is decidable. Moreover, by induction on the cardinality of A, also $\mathrm{Th}_{\exists}(\mathbb{U}, \mathrm{RAT}(\mathbb{U}))$ is decidable. One can show that for every AB-homomorphism $\sigma_{\mathbb{W}} : \mathbb{W}_t \to \mathbb{W}_t$ and every generator $W \in \mathcal{W}_{\mathbb{H}}$ one has $\sigma_{\mathbb{W}}(W) \in (A \cup B)\mathcal{W}_{\mathbb{H}}(A \cup B)$. This implies that for the restriction $\sigma_{\mathbb{H}} = \sigma_{\mathbb{W}}\restriction_{\mathcal{W}_{\mathbb{H}}}$ of $\sigma_{\mathbb{W}}$ in (27) there are only finitely many possibilities. By enumerating all these mappings $\sigma_{\mathbb{H}}$ and substituting them into (27) and (28), we reduce the simultaneous satisfiability of (27) and (28) to: (i) on one hand solving finitely many specialized instances of (27), which reduce to the theory $\mathrm{Th}_{\exists}(\mathbb{U}, \mathrm{RAT}(\mathbb{U}))$, and (ii) on the other hand, for every specialized instance from the previous point, solving a corresponding system of the form $\sigma_{\mathbb{H}}(\mathcal{S}_{\mathbb{H}})$, which reduces to $\mathrm{Th}_{\exists}(\mathbb{H}, \mathrm{RAT}(\mathbb{H}))$. Following this strategy we prove Thm. 1.

Using the embedding of an amalgamated free product (3) into the HNN-extension $\langle \mathbb{H} * \mathbb{J}', t; t^{-1}at = \varphi(a)(a \in A)\rangle$ by the map defined by $h \in \mathbb{H} \mapsto t^{-1}ht$, $j \in \mathbb{J} \mapsto j'$ (where $\mathbb{J}' = \{j' \mid j \in \mathbb{J}\}$ is a copy of $\mathbb{J}$, disjoint from $\mathbb{H}$, and φ maps every element of A to its copy in $\mathbb{J}'$, see [14, Thm. 2.6. p. 187]), we obtain statement (2) of Thm. 1. Let us finally discuss some variations of Thm. 1.

Remark 1. Thm. 1(1) remains true when $\mathrm{Th}_{\exists}(X, \mathrm{RAT}(X))$ ($X \in \{\mathbb{H}, \mathbb{G}\}$) is replaced by: $\mathrm{Th}_{\exists+}(X)$ (variant 1), $\mathrm{Th}_{\exists}(X)$ (variant 2), or $\mathrm{Th}_{\exists+}(X, \mathrm{RAT}(X))$ (variant 3). If $\mathrm{Th}_{\exists+}(\mathbb{H}, \{A_1, \ldots, A_n\})$ is decidable, where every A_i is a finitely generated subgroup of $\mathbb{H}$ containing A, then also $\mathrm{Th}_{\exists+}(\mathbb{G}, \{A_i, \langle A_i, t\rangle \mid 1 \le i \le n\})$ is decidable (variant 4), where $\langle A_i, t\rangle$ is the subgroup of $\mathbb{G}$ generated by $A_i \cup \{t\}$. These variants can be also shown if $\mathbb{H}$ is a cancellative monoid instead of a group (only A and B have to be groups). Finally, variant 2 still holds for amalgamated products of cancellative monoids.

4 Positive Theories

The following two theorems are our main results concerning positive theories:

Theorem 2. $\mathrm{Th}_{+}(\mathbb{G})$ *is decidable in the following two cases:*

(1) $\mathbb{G} = \langle \mathbb{H}, t; t^{-1}at = \varphi(a)(a \in A)\rangle$ *is an HNN-extension, where* A *and* $\varphi(A)$ *are proper subgroups of* $\mathbb{H}$ *with* $A \cap \varphi(A)$ *finite, and* $\mathrm{Th}_{\exists+}(\mathbb{G})$ *is decidable.*
(2) $\mathbb{G} = \langle \mathbb{H} * \mathbb{J}, a = \varphi(a)(a \in A)\rangle$ *is an amalgamated free product with* A *finite and* $\mathrm{Th}_{\exists+}(\mathbb{G})$ *is decidable.*

In Thm. 2 we cannot allow a cancellative monoid for $\mathbb{H}$, because the positive theory of $\{a,b\}^* \simeq \mathbb{N} * \mathbb{N}$ is undecidable [6]. For the same reason, we cannot include rational constraints: $\{a,b\}^*$ is a rational subset of the free group of rank 2.

Let us sketch a proof of (1) from Thm. 2. Our strategy for reducing $\mathrm{Th}_+(\mathbb{G})$ to $\mathrm{Th}_{\exists+}(\mathbb{G})$ is similar to [16, 17]: From a positive sentence ψ, which is interpreted over $\mathbb{G}$, we construct an existential positive sentence ψ' with subgroup constraints of a very special form, which is interpreted over a multiple HNN-extension $\mathbb{G}'$ of $\mathbb{G}$, where only finite subgroups of $\mathbb{G}$ are associated. Roughly speaking, ψ' results from ψ by replacing the universally quantified variables by the stable letters of the HNN-extension $\mathbb{G}'$. Let $\mathbb{G}$ be an HNN-extension as in Thm. 2. Let $X \leq A \cap \varphi(A)$ be a (necessarily finite) subgroup of $\mathbb{H}$. With $\mathrm{In}(X)$ we denote the group of all automorphisms f of X such that for some $g \in \mathbb{G}$ we have: $f(c) = g^{-1}cg$ for all $c \in X$. For new constants $k_1, \ldots, k_m \notin \mathbb{G}$ and $f_1, \ldots, f_m \in \mathrm{In}(X)$ we define the multiple HNN-extension

$$\mathbb{G}_{k_1,\ldots,k_m}^{f_1,\ldots,f_m} = \langle \mathbb{G}, k_1, \ldots, k_m; k_i^{-1}ck_i = f_i(c) \ (c \in X, 1 \leq i \leq m) \rangle. \tag{29}$$

The following theorem yields the reduction from $\mathrm{Th}_+(\mathbb{G})$ to $\mathrm{Th}_{\exists+}(\mathbb{G})$.

Theorem 3. *There is a subgroup $X \leq A \cap B \leq \mathbb{H} \leq \mathbb{G}$ such that for every formula $\psi(z_1, \ldots, z_m) \equiv \forall x_1 \exists y_1 \cdots \forall x_n \exists y_n \, \phi(x_1, \ldots, x_n, y_1, \ldots, y_n, z_1, \ldots, z_m)$, where ϕ is a positive boolean combination of equations (with constants) over the group $\mathbb{G}$, and for all $u_1, \ldots, u_m \in \mathbb{G}$ we have: $\psi(u_1, \ldots, u_m)$ in $\mathbb{G}$ iff*

$$\bigwedge_{f_1 \in \mathrm{In}(X)} \exists y_1 \cdots \bigwedge_{f_n \in \mathrm{In}(X)} \exists y_n \left\{ \begin{array}{l} \bigwedge_{1 \leq i \leq n} y_i \in \mathbb{G}_{k_1,\ldots,k_i}^{f_1,\ldots,f_i} \wedge \\ \phi(k_1, \ldots, k_n, y_1, \ldots, y_n, u_1, \ldots, u_m) \text{ in } \mathbb{G}_{k_1,\ldots,k_n}^{f_1,\ldots,f_n} \end{array} \right\} \tag{30}$$

In [3], a result analogous to Theorem 3 for the case that $\mathbb{G}$ is a free product was shown. In this case, the new generators $k_1, \ldots, k_n$ do not interact with the group $\mathbb{G}$, i.e., the HNN-extension $\mathbb{G}_{k_1,\ldots,k_n}^{f_1,\ldots,f_n}$ is replaced by the free product $\mathbb{G} * F_n$, where F_n is the free group generated by $k_1, \ldots, k_n$. For the more general case that $\mathbb{G}$ is an HNN-extension, we cannot avoid some nontrivial interaction between k_i and $\mathbb{G}_i$. This interaction is expressed by the identities $k_i^{-1}ck_i = f_i(c)$ $(c \in X)$ in the HNN-extension $\mathbb{G}_{k_1,\ldots,k_n}^{f_1,\ldots,f_n}$. Note that the sentence in (30) is not interpreted in a single HNN-extension of $\mathbb{G}$. But it is not difficult to construct an HNN-extension $\mathbb{G}'$ of $\mathbb{G}$ such that each of the groups $\mathbb{G}_{k_1,\ldots,k_n}^{f_1,\ldots,f_n}$ can be embedded into $\mathbb{G}'$. Moreover, each single HNN-extension that leads from $\mathbb{G}$ to $\mathbb{G}'$ associates X with itself as in (29). In this way, we can construct from (30) an existential positive sentence $\Psi = (\exists y_\sigma \in \mathbb{G}_\sigma)_{\sigma \in J} \, \chi((k_\sigma)_{\sigma \in J}, (y_\sigma)_{\sigma \in J}, u_1, \ldots, u_m)$ (for some index set J larger than n in (30)) such that (30) iff Ψ is true in $\mathbb{G}'$. Moreover, all constraint-groups $\mathbb{G}_\sigma$ in Ψ are generated by $\mathbb{G}$ and some of the stable letters k_σ. To complete the proof of (1) in Thm. 2, notice that an iterated application of variant 4 from Remark 1 (recall that X is finite) enables us to reduce $\mathrm{Th}_{\exists+}(\mathbb{G}', \{\mathbb{G}_\sigma \mid \sigma \in J\})$ to $\mathrm{Th}_{\exists+}(\mathbb{G})$. A proof of (2) in Thm. 2 follows a similar strategy.

We conclude this paper with an application to virtually-free groups. A finitely generated group $\mathbb{G}$ is *virtually-free*, if it has a free subgroup of finite index. Since these groups have finite decompositions over finite groups by means of the operations (1) and (3) with A finite [1], we obtain from Thm. 1 and 2:

Theorem 4. *If* $\mathbb{G}$ *is virtually-free, then* $\mathrm{Th}_{\exists}(\mathbb{G}, \mathrm{RAT}(\mathbb{G}))$ *and* $\mathrm{Th}_{+}(\mathbb{G})$ *are decidable.*

Thm. 4 immediately leads to the question, whether also the full first-order theory of a virtually-free group is decidable. This is certainly a difficult question. The full proof of Kharlampovich and Myasnikov for the decidability of the theory of a free group (see [9] for an overview) takes several hundred pages. Moreover, there seems to be no obvious reduction from the theory of a virtually-free group to the theory of a free group.

References

1. W. Dicks and M. J. Dunwoody. *Groups Acting on Graphs*. Cambridge Univ. Press, 1989.
2. V. Diekert, C. Gutiérrez, and C. Hagenah. The existential theory of equations with rational constraints in free groups is PSPACE-complete. *Inf. Comput.*, 202(2):105–140, 2005.
3. V. Diekert and M. Lohrey. Word equations over graph products. In *Proc. FSTTCS 2003*, LNCS 2914, pages 156–167. Springer, 2003.
4. V. Diekert and M. Lohrey. Existential and positive theories of equations in graph products. *Theory Comput. Syst.*, 37(1):133–156, 2004.
5. V. Diekert and A. Muscholl. Solvability of equations in free partially commutative groups is decidable. *Int. J. Algebra Comput.*, 2006. to appear.
6. V. G. Durnev. Undecidability of the positive $\forall\exists^3$-theory of a free semi-group. *Sibirsky Matematicheskie Jurnal*, 36(5):1067–1080, 1995. English translation.
7. E. R. Green. *Graph Products of Groups*. PhD thesis, The University of Leeds, 1990.
8. B. Khan, A. G. Myasnikov, and D. E. Serbin. On positive theories of groups with regular free length function. Manuscript, 2005.
9. O. G. Kharlampovich and A. Myasnikov. Tarski's problem about the elementary theory of free groups has a positive solution. *Electron. Res. Announc. AMS*, 4(14):101–108, 1998.
10. A. Kościelski and L. Pacholski. Makanin's algorithm is not primitive recursive. *Theor. Comput. Sci.*, 191(1-2):145–156, 1998.
11. M. Lohrey and G. Sénizergues. Equations in HNN-extensions. Manuscript, 2006.
12. M. Lohrey and G. Sénizergues. Positive theories of HNN-extensions and amalgamated free products. Manuscript, 2006.
13. M. Lohrey and G. Sénizergues. Rational subsets of HNN-extensions. Manuscript, 2006.
14. R. C. Lyndon and P. E. Schupp. *Combinatorial Group Theory*. Springer, 1977.
15. G. S. Makanin. Equations in a free group. *Math. USSR, Izv.*, 21:483–546, 1983. English translation.
16. G. S. Makanin. Decidability of the universal and positive theories of a free group. *Math. USSR, Izv.*, 25:75–88, 1985. English translation.
17. Y. I. Merzlyakov. Positive formulas on free groups. *Algebra i Logika Sem.*, 5(4):25–42, 1966. In Russian.
18. D. E. Muller and P. E. Schupp. Groups, the theory of ends, and context-free languages. *J. Comput. Syst. Sci.*, 26:295–310, 1983.
19. W. Plandowski. Satisfiability of word equations with constants is in PSPACE. *J. ACM*, 51(3): 483–496, 2004.
20. E. Rips and Z. Sela. Canonical representatives and equations in hyperbolic groups. *Invent. Math.*, 120:489–512, 1995.

On Intersection Problems for Polynomially Generated Sets

Wong Karianto[1], Aloys Krieg[2], and Wolfgang Thomas[1]

[1] Lehrstuhl für Informatik 7, RWTH Aachen, Germany
{karianto, thomas}@informatik.rwth-aachen.de
[2] Lehrstuhl A für Mathematik, RWTH Aachen, Germany
krieg@mathA.rwth-aachen.de

Abstract. Some classes of sets of vectors of natural numbers are introduced as generalizations of the semi-linear sets, among them the 'simple semi-polynomial sets.' Motivated by verification problems that involve arithmetical constraints, we show results on the intersection of such generalized sets with semi-linear sets, singling out cases where the non-emptiness of intersection is decidable. Starting from these initial results, we list some problems on solvability of arithmetical constraints beyond the semi-linear ones.

1 Introduction

The study of arithmetical constraints, in particular regarding their effective solvability, is of central interest in several branches of theoretical computer science. One of these fields, which serves as motivation for the present work, is the verification of infinite-state systems where the aspect of infinity arises by including the domain of the natural numbers in the model under consideration.

In the context of infinite-state verification, conditions on vectors of natural numbers usually occur in two roles. First, the considered transition systems $\mathcal{A}$ are assumed to have some mechanism of 'counting' and thus generate, by each run, some vector of $\mathbb{N}^n$; the set of all such vectors is the set $A_{\mathcal{A}} \subseteq \mathbb{N}^n$ generated by $\mathcal{A}$. An example is the computation of the Parikh mapping by an automaton on words over an alphabet with letters $a_1, \ldots, a_n$: the occurrences of the letters a_i are counted by updating a vector from $\mathbb{N}^n$ in each step, incrementing the i-th component by one for an a_i-labeled transition. Taking finite automata or pushdown automata $\mathcal{A}$, the corresponding sets $A_{\mathcal{A}}$ are known to coincide with the semi-linear sets (Parikh's Theorem [12]).

The second role of arithmetical conditions enters when the vectors arising from the runs of the transition systems under consideration are also subject to an 'acceptance condition' φ. In the context of automata, acceptance of an input word w then means that a corresponding run reaches a 'final state' and generates a vector that satisfies φ or, in other words, belongs to the set A_φ defined by φ. As a recent model of this kind, Klaedtke and Rueß [9] proposed 'Parikh automata,' which use more general transitions than those mentioned above: in the update operation an arbitrary vector of $\mathbb{N}^n$ is added (rather than just 1 in a single

M. Bugliesi et al. (Eds.): ICALP 2006, Part II, LNCS 4052, pp. 516–527, 2006.

component), and for the constraints φ formulas of Presburger arithmetic are used (which precisely define the semi-linear sets).

A fundamental property of Parikh automata is the decidability of the nonemptiness problem. This is established easily by observing that the nonemptiness of the language recognized by a Parikh automaton $\mathcal{A}$ with acceptance condition φ is equivalent to the nonemptiness of the intersection $A_{\mathcal{A}} \cap A_{\varphi}$. Since $A_{\mathcal{A}}$ is semi-linear, and since the semi-linear sets are effectively closed under intersection (and their nonemptiness is trivially decidable), one obtains an algorithm for solving the nonemptiness problem.

Many other papers on model-checking infinite-state systems follow similar ideas; see, for example, [1, 3, 7]. Another application area is the study of XML-document specifications. As observed by several authors [2, 10, 13, 14], the automata on unranked trees which capture document type definitions can be extended by counting conditions (on the occurrences of certain data as sons of an XML-tree node). If these arithmetical conditions are restricted to semi-linear sets, then the desired decidability results on type checking can be shown.

The purpose of the present paper is to explore possibilities of extending the framework of semi-linear sets (or, equivalently, Presburger arithmetic or systems of linear equations), while still keeping the fundamental property that nonemptiness of intersection is decidable. As noted above, the two sets of such an intersection may arise differently (e.g., as generated by a system and as specified by an acceptance condition), so it is reasonable to consider intersections $A \cap B$ where A and B are possibly from different classes.

We basically consider two classes extending the semi-linear sets. Firstly, we introduce 'simple semi-polynomial sets' and show initial results on closure properties with respect to intersection (with implications for deciding nonemptiness). Secondly, some variants of 'quadratic' sets are introduced, where a recent result of Grunewald and Segal [5] helps to show the decidability of certain nonemptiness problems. As a conclusion, we suggest some questions motivated by our observations.

In the present paper we do not address applications in detail, for example in concrete verification problems. Instead, we focus on the arithmetical aspects and only remark here that in the scenario above (regarding the sets $A_{\mathcal{A}}$ and A_{φ}) we obtain cases which are substantially more general (or, at least, different) than the existing framework of semi-linear sets and still allow an algorithmic solution.

2 Preliminaries

Recall that a subset A of $\mathbb{N}^n$, $n \geq 1$, is called *linear* if there are vectors $\bar{u}_0, \bar{u}_1, \ldots, \bar{u}_m \in \mathbb{N}^n$, $m \geq 0$, such that

$$A = \{\bar{u}_0 + k_1\bar{u}_1 + \cdots + k_m\bar{u}_m \mid k_1, \ldots, k_m \in \mathbb{N}\} . \quad (1)$$

The vector $\bar{u}_0$ is called the *constant vector*, the vectors $\bar{u}_1, \ldots, \bar{u}_m$ the *periods*, and all of them the *generators* of A. Alternatively, we may replace (1) with

$$A = \{(L_1(k_1, \ldots, k_m), \ldots, L_n(k_1, \ldots, k_m)) \mid k_1, \ldots, k_m \in \mathbb{N}\} , \quad (2)$$

where $L_i(k_1, \dots, k_m) := (\bar{u}_0)_i + k_1(\bar{u}_1)_i + \cdots + k_m(\bar{u}_m)_i$ for $i = 1, \dots, n$.[1] In other words, $L_1, \dots, L_n \in \mathbb{N}[X_1, \dots, X_m]$ are linear forms with nonnegative integer coefficients. A finite union of linear sets is called *semi-linear*.

In [4] Ginsburg and Spanier showed that the solutions of a linear equation $c_0 + \sum_{i=1}^{n} c_i x_i = c'_0 + \sum_{i=1}^{n} c'_i x_i$, where $c_i, c'_i \in \mathbb{N}$, for $i = 0, \dots, n$, form a semi-linear set. Further, if we close the sets defined by linear equations under Boolean operations and projection, the *Presburger-definable sets* (i.e., the first-order-definable sets over $(\mathbb{N}, +)$) are generated. Ginsburg and Spanier [4] also showed that the semi-linear sets coincide with the Presburger-definable ones. Moreover, all the logical closure operations are effective. For instance, given the generators of A and B, generators of $A \cap B$ can be computed. This implies that the nonemptiness of this intersection is decidable.

For a finite, nonempty alphabet $\Sigma = \{a_1, \dots, a_n\}$, the *Parikh mapping* $\Phi \colon \Sigma^* \to \mathbb{N}^n$ is defined by $\Phi(w) := (|w|_{a_1}, \dots, |w|_{a_n})$, for each $w \in \Sigma^*$. Parikh's Theorem [12] asserts that the *Parikh image* $\Phi(L) := \{\Phi(w) \mid w \in L\}$ of a context-free language L over Σ is semi-linear. Conversely, every semi-linear set is the Parikh image of a context-free language (even of a regular language).

3 Simple Semi-polynomial Sets

A natural generalization of semi-linear sets involves general polynomials rather than just linear ones in (2): A subset A of $\mathbb{N}^n$, $n \geq 1$, is a *polynomial set* if there are polynomials $P_1, \dots, P_n \in \mathbb{N}[X_1, \dots, X_m]$ such that

$$A = \{(P_1(k_1, \dots, k_m), \dots, P_n(k_1, \dots, k_m)) \mid k_1, \dots, k_m \in \mathbb{N}\} . \tag{3}$$

A finite union of polynomial sets is called *semi-polynomial*.

Since the polynomials in (3) may have mixed terms, i.e., terms in which more than one variable occur, we get a class which is not manageable. In fact, the nonemptiness of intersection is undecidable even for the case of a two-dimensional polynomial set and a semi-linear one. This is clear by a simple reformulation of Hilbert's Tenth Problem (note that the identity relation $\mathrm{id}_{\mathbb{N}} := \{(k, k) \mid k \in \mathbb{N}\}$ is a linear set):

$$\begin{aligned} & \exists k_1 \dots k_m \; P(k_1, \dots, k_m) = 0 \ , \ \text{ where } P \in \mathbb{Z}[X_1, \dots, X_m] \\ \text{iff} \ \ & \exists k_1 \dots k_m \; Q(k_1, \dots, k_m) = R(k_1, \dots, k_m) \ , \ \text{ where } Q, R \in \mathbb{N}[X_1, \dots, X_m] \\ \text{iff} \ \ & \left\{ \begin{pmatrix} Q(k_1, \dots, k_m) \\ R(k_1, \dots, k_m) \end{pmatrix} \mid k_1, \dots, k_m \in \mathbb{N} \right\} \cap \mathrm{id}_{\mathbb{N}} \neq \emptyset \ . \end{aligned}$$

Therefore, we restrict the polynomials in (3) by disallowing mixed terms and obtain sets of the form

$$\left\{ \begin{pmatrix} c_1 + P_{11}(k_1) + \cdots + P_{1m}(k_m) \\ \vdots \\ c_n + P_{n1}(k_1) + \cdots + P_{nm}(k_m) \end{pmatrix} \mid k_1, \dots, k_m \in \mathbb{N} \right\} , \tag{4}$$

[1] For a vector $\bar{x} \in \mathbb{N}^n$, we write $(\bar{x})_i$ for the i-th component of $\bar{x}$.

where $c_1, \ldots, c_n \in \mathbb{N}$, and $P_{ij} \in \mathbb{N}[X]$ is a (univariate) polynomial without constants, for each $i = 1, \ldots, n$ and $j = 1, \ldots, m$. A set defined in this way is called a *simple polynomial set*, and *simple semi-polynomial sets* are finite unions of simple polynomial sets.

In analogy to (1) for linear sets, a simple polynomial set as in (4) can be represented in terms of its *generators* as follows:

$$\begin{aligned}\{\bar{u}_0 + k_1\bar{u}_{1,1} + k_1^2\bar{u}_{1,2} + \cdots + k_1^{d-1}\bar{u}_{1,d-1} + k_1^d\bar{u}_{1,d} \\ + \cdots + k_m\bar{u}_{m,1} + k_m^2\bar{u}_{m,2} + \cdots + k_m^{d-1}\bar{u}_{m,d-1} + k_m^d\bar{u}_{m,d} \\ \mid k_1, \ldots, k_m \in \mathbb{N}\} \ ,\end{aligned} \tag{5}$$

where $\bar{u}_0$ and $\bar{u}_{i,j}$ ($1 \leq i \leq m$, $1 \leq j \leq d$) are vectors from $\mathbb{N}^n$. In this case, the simple polynomial set is said to be of *degree d*. Note that from a representation (4) one easily obtains (5), and vice versa.

Clearly, each (semi-)linear set is a simple (semi-)polynomial set. An interesting special case is given by the *simple quadratic sets*

$$\{\bar{u}_0 + k_1\bar{u}_{1,1} + k_1^2\bar{u}_{1,2} + \cdots + k_m\bar{u}_{m,1} + k_m^2\bar{u}_{m,2} \mid k_1, \ldots, k_m \in \mathbb{N}\}$$

and finite unions of such sets, the *simple semi-quadratic sets.*

Given the generators of a (simple) polynomial set as in (3) or (5), one can decide whether a given vector $\bar{v} = (v_1, \ldots, v_n)$ belongs to this set; it suffices to check the k_i-values up to $\max\{v_1, \ldots, v_n\}$. Hence, a (simple) semi-polynomial set is decidable.

Example 1. The set $A_1 := \{(u_1, u_2) \in \mathbb{N}^2 \mid u_2 = u_1^2\}$ is a simple quadratic set since $A_1 = \{(0,0) + k(1,0) + k^2(0,1) \mid k \in \mathbb{N}\}$.

Let us verify that A_1 is not semi-linear. Towards a contradiction, suppose that A_1 is a finite union of linear sets, say $A_1 = \bigcup_{i=1}^r B_i$, for some $r \geq 1$. Since A_1 is infinite, there is some linear set B_i that contains at least two elements. Let $\bar{u}_0$ be the constant vector and $\bar{u}_1, \ldots, \bar{u}_m$ be the periods of B_i. If $m = 0$, or if all periods of B_i are $\bar{0}$,[2] then B_i has only one element, namely $\bar{u}_0$, a contradiction. So we can assume that $\bar{u}_1$ is not $\bar{0}$. By definition of linear sets, $\bar{u}_0 + k\bar{u}_1 \in B_i \subseteq A_1$, for all $k \in \mathbb{N}$. Let $\bar{u}_0 = (u_{01}, u_{02})$ and $\bar{u}_1 = (u_{11}, u_{12})$. Then, by definition of A_1, we have for all $k \in \mathbb{N}$

$$(u_{01} + ku_{11})^2 = u_{02} + ku_{12} \ . \tag{6}$$

Since $\bar{u}_1 \neq \bar{0}$, at least one of u_{11} and u_{12} is not zero. Hence, (6) is a polynomial equation of degree one or two in k, which has at most two solutions. Contradiction.

A copy of this argument shows that $\{(u_1, u_2) \in \mathbb{N}^2 \mid u_2 = u_1^{d+1}\}$ is not a simple semi-polynomial set of degree d, for any $d \geq 1$, and that $\{(u_1, u_2) \in \mathbb{N}^2 \mid u_2 = 2^{u_1}\}$ is not a simple semi-polynomial set.

[2] $\bar{0}$ denotes a vector consisting only of zeroes.

Example 2. The product relation $A_{\text{prod}} := \{(u, v, uv) \mid u, v \in \mathbb{N}\} \subseteq \mathbb{N}^3$, which is clearly polynomial, is not a simple semi-polynomial set. To verify this, the simple comparison of growth rates does not suffice, and some structural analysis is needed.

Towards a contradiction, assume that A_{prod} is a finite union of simple polynomial sets, each of them of the form

$$A' = \left\{ \begin{pmatrix} a + P_1(k_1) + \cdots + P_m(k_m) \\ b + Q_1(k_1) + \cdots + Q_m(k_m) \\ c + R_1(k_1) + \cdots + R_m(k_m) \end{pmatrix} \mid k_1, \ldots, k_m \in \mathbb{N} \right\} ,$$

where $P_i, Q_i, R_i \in \mathbb{N}[X]$ are polynomials without constants, and where not all of P_i, Q_i, R_i are zero polynomials, for each $i = 1, \ldots, m$. Since $P_i(0) = Q_i(0) = R_i(0) = 0$ for $i = 1, \ldots, m$, we have, for each $j \geq 2$,

$$\begin{pmatrix} a \\ b \\ c \end{pmatrix} , \begin{pmatrix} a + P_1(1) \\ b + Q_1(1) \\ c + R_1(1) \end{pmatrix} , \begin{pmatrix} a + P_j(1) \\ b + Q_j(1) \\ c + R_j(1) \end{pmatrix} , \begin{pmatrix} a + P_1(1) + P_j(1) \\ b + Q_1(1) + Q_j(1) \\ c + R_1(1) + R_j(1) \end{pmatrix} \in A_{\text{prod}} .$$

Since A_{prod} is the product relation, we have

$$\begin{aligned} c &= ab \\ c + R_1(1) &= (a + P_1(1))(b + Q_1(1)) \\ c + R_j(1) &= (a + P_j(1))(b + Q_j(1)) \\ c + R_1(1) + R_j(1) &= (a + P_1(1) + P_j(1))(b + Q_1(1) + Q_j(1)) \end{aligned}$$

It follows that $P_1(1)Q_j(1) + P_j(1)Q_1(1) = 0$. Since $(P_i(1), Q_i(1)) \neq (0, 0)$, for all $i = 1, \ldots, m$, we have

$$\begin{aligned} P_1(1) = P_j(1) = 0 \quad &\text{and hence} \quad P_1 = P_j \equiv 0 \ , \quad \text{or} \\ Q_1(1) = Q_j(1) = 0 \quad &\text{and hence} \quad Q_1 = Q_j \equiv 0 \ . \end{aligned}$$

Since $j \geq 2$ was arbitrarily chosen, all P_i or all Q_i are zero polynomials, for $i = 1, \ldots, m$, and thus, we have

$$A' \subseteq \{(a, y, ay) \mid y \in \mathbb{N}\} \subseteq A_{\text{prod}} \ , \quad \text{or} \quad A' \subseteq \{(x, b, bx) \mid x \in \mathbb{N}\} \subseteq A_{\text{prod}} \ .$$

Hence, there are s, t and $a_1, \ldots, a_s, b_1, \ldots, b_t$ such that

$$A_{\text{prod}} = \bigcup_{i=1}^{s} \{(a_i, y_i, a_i y_i) \mid y_i \in \mathbb{N}\} \cup \bigcup_{j=1}^{t} \{(x_j, b_j, x_j b_j) \mid x_j \in \mathbb{N}\} \ .$$

Projection to the first two components should yield $\mathbb{N}^2$. However, we obtain the union of the sets $\{(a_i, y_i) \mid y_i \in \mathbb{N}\}$ and $\{(x_j, b_j) \mid x_j \in \mathbb{N}\}$, which is a proper subset of $\mathbb{N}^2$, a contradiction.

As the semi-linear sets can be characterized as the Parikh images of the regular and the context-free languages, one may ask for such a characterization of the simple semi-polynomial sets. In [8] it is shown that all simple semi-polynomial sets (and more sets, e.g., A_{prod} of Example 2) can be obtained as the Parikh images of indexed languages, i.e., those languages which are recognized by level-two pushdown automata (pushdown automata with a stack of stacks).

4 Intersection Problems

We show two results: Simple semi-polynomial sets are not closed under intersection whereas the intersection of a simple semi-polynomial set with a semi-linear set of a 'special kind' is again a simple semi-polynomial set.

Theorem 3. *There exist two simple quadratic sets the intersection of which is not simple semi-polynomial.*

Proof. Consider the simple quadratic sets

$$A = \left\{ \begin{pmatrix} (k_1+1)^2 + (k_2+1)^2 \\ k_3 \end{pmatrix} \mid k_1, k_2, k_3 \in \mathbb{N} \right\} \text{ and } B = \left\{ \begin{pmatrix} k^2 \\ k \end{pmatrix} \mid k \in \mathbb{N} \right\} .$$

The intersection $A \cap B$ consists of the pairs (k^2, k) where $k^2 = (k_1+1)^2 + (k_2+1)^2$, for certain k_1, k_2, is a solution of the Pythagoras equation in positive integers. It is known from elementary number theory (see, e.g., [6]) that these pairs coincide with the pairs $(w^2(u^2+v^2)^2, w(u^2+v^2))$ where u, v, w are positive integers. By the Two-Square Theorem (see, e.g., [6]), the latter pairs coincide with the pairs (n^2, n) of natural numbers where $n \geq 2$ is even or divisible by some prime $p \equiv 1 \pmod 4$.

Suppose that $A \cap B$ is simple semi-polynomial, i.e. a union of sets

$$A_i = \left\{ \begin{pmatrix} \alpha + P_1(k_1) + \cdots + P_m(k_m) \\ \beta + Q_1(k_1) + \cdots + Q_m(k_m) \end{pmatrix} \mid k_1, \ldots k_m \in \mathbb{N} \right\} \quad (i = 1, \ldots, s)$$

where $\alpha, \beta \in \mathbb{N}$ and $P_1, \ldots, P_m, Q_1, \ldots, Q_m \in \mathbb{N}[X]$ are nonzero polynomials without constants. Setting $k_1 = \cdots = k_m = 0$, we obtain $\alpha = \beta^2$. Fixing some $j \in \{1, \ldots, m\}$ and setting $k_r = 0$, for all $1 \leq r \leq s$ with $r \neq j$, we obtain $P_j(k_j) + \beta^2 = (Q_j(k_j) + \beta)^2$ and thus $P_j(k_j) = (Q_j(k_j))^2 + 2\beta Q_j(k_j)$, for each $k_j \in \mathbb{N}$. If $m \geq 2$, we would have for $1 < j \leq m$

$$P_1(k_1) + P_j(k_j) + \beta^2 = (Q_1(k_1) + Q_j(k_j) + \beta)^2 \ ,$$

and hence, $Q_1(k_1)Q_j(k_j) = 0$, for all $k_1, k_j \in \mathbb{N}$, which would imply that one of Q_1 and Q_j is zero, a contradiction. Hence, we have $m = 1$ and can assume

$$A_i = \left\{ \begin{pmatrix} (R_i(k_i))^2 \\ R_i(k_i) \end{pmatrix} \mid k_i \in \mathbb{N} \right\}$$

for some polynomial $R_i \in \mathbb{N}[X]$.

Let $N := 4K$, where K is the least common multiple of the coefficients of $R_1, \ldots, R_s$. Among these polynomials let $R_1, \ldots, R_t$ be the ones of degree 1, say, $R_i = a_i + b_i(X)$, which yields, for $i = 1, \ldots, t$,

$$R_i(\mathbb{N}) = a_i + b_i\mathbb{N} = \bigcup_{j=0}^{m} a_{ij} + N\mathbb{N} \tag{7}$$

as a disjoint union, for some $0 \leq m < N$. By Dirichlet's Prime-Number Theorem, each arithmetic progression $c + N\mathbb{N}$, where $c \equiv 3 \pmod 4$ and where c and N are relatively prime, contains infinitely many primes $q \equiv 3 \pmod 4$. Thus, $c + N\mathbb{N}$ does not occur in the union (7). Now, let p be a prime with $p \equiv 1 \pmod N$. Then, for any $n \in pc + pN\mathbb{N}$, the pair (n^2, n) belongs to $A \cap B$, which means that

$$pc + pN\mathbb{N} \subseteq \bigcup_{i=t+1}^{s} R_i(\mathbb{N}) \ , \tag{8}$$

where, for each $i = t+1, \ldots, s$, R_i is either a constant or of degree ≥ 2. In the latter case we have, $R_i(k_i + 1) - R_i(k_i) \geq 2k_i$, for each $k_i \in \mathbb{N}$. In other words, these differences tend to infinity, which contradicts (8). □

We now exhibit a case of an intersection operation which does not lead out of the simple semi-polynomial sets, respectively the semi-polynomial sets. We consider the intersection with a special form of semi-linear set: A set $A \subseteq \mathbb{N}^n$ is called *componentwise linear* if there are linear sets $A_1, \ldots, A_n \subseteq \mathbb{N}$ such that $A = A_1 \times \cdots \times A_n$. The set A is called *componentwise semi-linear* if it is a finite union of componentwise linear sets.

To simplify notation, in the sequel we do not distinguish between an ordinary natural number and a one-dimensional vector of natural numbers.

Theorem 4. *Let $n \geq 1$. If $A \subseteq \mathbb{N}^n$ is componentwise semi-linear and $B \subseteq \mathbb{N}^n$ is simple semi-polynomial (respectively semi-polynomial) of degree $d \geq 1$, then $A \cap B$ is simple semi-polynomial (respectively semi-polynomial) of degree d. Moreover, if A and B are given by their generators, generators of $A \cap B$ can be computed and hence nonemptiness of $A \cap B$ be checked effectively.*

Proof. We only consider simple semi-polynomial sets; the proof works in the same way for semi-polynomial sets. Furthermore, it suffices to consider the case that A is componentwise linear and B is a simple polynomial set. We construct a simple semi-polynomial representation of $A \cap B$ by a refinement process which successively covers more and more of the n components. For the intersection of the projections of A and B to the first component we obtain a simple semi-polynomial representation, which is then made thinner by taking into account the other components, one by one. To simplify matters, let us first treat the case that B is a simple quadratic set.

Case 1. Let $n = 1$, i.e. $A, B \subseteq \mathbb{N}$. Suppose $A, B \subseteq \mathbb{N}$ are given by

$$A = \{u_0 + k_1 u_1 + \ldots + k_m u_m \mid k_1, \ldots, k_m \in \mathbb{N}\} \ ,$$
$$B = \{v_0 + k_1 v_1 + k_1^2 w_1 + \cdots + k_r v_r + k_r^2 w_r \mid k_1, \ldots, k_r \in \mathbb{N}\} \ ,$$

where the u_i, v_i, w_i are natural numbers and $v_i + w_i \geq 1$ for all $i = 1, \ldots, r$. In order to avoid trivial cases, assume $u_i \geq 1$ for all $i = 1, \ldots, m$.

Let g be the greatest common divisor of $u_1, \dots, u_m$. Clearly, $A \subseteq \{u_0 + kg \mid k \in \mathbb{N}\}$, and for sufficiently large c_0 (we may take $c_0 := u_0 + u_1 \cdots u_m$) the set

$$C := \{c_0 + kg \mid k \in \mathbb{N}\}$$

contains precisely the A-elements from c_0 onwards.

The set $A \setminus C$ is finite; so by decidability of B one computes the set $F := (A \setminus C) \cap B$.

The intersection $A \cap B$ is the union of F with $C \cap B$. Elements in $C \cap B$ have to be solutions of the congruence

$$v_0 + k_1 v_1 + k_1^2 w_1 + \ldots + k_r v_r + k_r^2 w_r \equiv c_0 \pmod{g} \ . \tag{9}$$

It suffices to check the congruence for values $k_i < g$. If no solution exists, we have $C \cap B = \emptyset$ and $A \cap B = F$. If a solution exists, say $\bar{s} = (s_1, \dots, s_r) \in \{0, \dots, g-1\}^r$, it produces the B-elements

$$x = v_0 + m_1 v_1 + m_1^2 w_1 + \ldots + m_r v_r + m_r^2 w_r \ , \tag{10}$$

where $m_i = s_i + n_i g$, $n_i \in \mathbb{N}$.

In order to ensure $x \geq c_0$ (i.e. to obtain $C \cap B$) it suffices to require $\sum_{i=1}^r n_i > \lfloor c_0/g \rfloor$. Only finitely many C-elements are missed by this requirement; we collect them in the finite set $E_{\bar{s}}$. The case $\sum_{i=1}^r n_i > \lfloor c_0/g \rfloor$ is split into finitely many subcases $n_1 \geq l_{1j}, \dots, n_r \geq l_{rj}$ (where j ranges over a finite set J). If we write $l_{ij} + n_i$ ($n_i \geq 0$) instead of $n_i \geq l_{ij}$ and substitute $m_i = s_i + n_i g$ in (10) by $s_i + (l_{ij} + n_i)g$, we obtain the following simple quadratic set in the n_i:

$$C_{\bar{s},j} = \{v_0' + n_1 v_1' + n_1^2 w_1' + \ldots + n_r v_r' + n_r^2 w_r' \mid n_1, \dots, n_r \in \mathbb{N}\} \ . \tag{11}$$

The intersection $C \cap B$ is the union of the finite sets $E_{\bar{s}}$ and the finitely many simple quadratic sets $C_{\bar{s},j}$. Hence, $A \cap B$ is a simple semi-quadratic set. Furthermore, the set is empty iff the finite set F mentioned above is empty and the congruence (9) has no solution.

Case 2. Let $n > 1$. Consider a componentwise linear set $A \subseteq \mathbb{N}^n$ and a simple quadratic set $B \subseteq \mathbb{N}^n$:

$$A = A_1 \times \cdots \times A_n \ ,$$
$$B = \{\bar{v}_0 + k_1 \bar{v}_1 + k_1^2 \bar{w}_1 + \cdots + k_r \bar{v}_r + k_r^2 \bar{w}_r \mid k_1, \dots, k_r \in \mathbb{N}\} \ ,$$

where $\bar{v}_i, \bar{w}_i \in \mathbb{N}^n$.

We analyze the intersection $A \cap B$ for the first component as above. If this intersection is empty, this is also true for $A \cap B$ and we are done. Otherwise we invoke Case 1 for the first components of A and B, which shows that $(A)_1 \cap (B)_1$ is a simple semi-quadratic set.[3] If this intersection is finite (which means

[3] For a set $X \subseteq \mathbb{N}^n$, the set $(X)_i$ denotes $\{(\bar{x})_i \mid \bar{x} \in X\}$. Further, note that $(A)_i = A_i$ since A is componentwise linear.

that (11) above is empty), it suffices to decide for each of the corresponding n-tuples $(k_1, \ldots, k_r)$ whether the second component of the A-element generated by $k_1, \ldots, k_r$ belongs to the simple quadratic set given by the second components of the B-elements.

If $(A)_1 \cap (B)_1$ is infinite, consider a set $C_{\bar{s},j}$ as constructed above. We have to find the $(B)_2$-elements of the form

$$(\bar{v}_0')_2 + n_1(\bar{v}_1')_2 + n_1^2(\bar{w}_1')_2 + \cdots + n_r(\bar{v}_r')_2 + n_r^2(\bar{w}_r')_2 \ .$$

This is a simple quadratic set in the n_i, and the procedure of Case 1 can be invoked to find those vectors $(n_1, \ldots, n_r)$ which describe the second component of an A-element. We obtain a simple semi-quadratic representation of $(A)_{1,2} \cap (B)_{1,2}$ (the intersection of A and B restricted to the first two components), which moreover is testable for nonemptiness. After $n-1$ steps of this kind the procedure terminates with a simple semi-quadratic representation of $A \cap B$, giving also the information whether $A \cap B = \emptyset$.

The same argument is applicable to (simple) polynomial sets B instead of simple quadratic ones. The simple quadratic expressions in (9), (10), (11) change to (simple) polynomial ones, but the form of the solutions $(m_1, \ldots, m_r)$ still is of the form $m_i = s_i + n_i g$ since the component sets $(A)_j$ are (componentwise) linear. So the proof carries over in the obvious way. □

We have shown that for a (simple) semi-polynomial set B the intersection with a componentwise semi-linear set A yields again a (simple) semi-polynomial set whereas this fails in general for a simple semi-polynomial set A. The intermediate case of a semi-linear set A remains open, even for the weaker statement that nonemptiness of $A \cap B$ is decidable. Let us note that this decision problem is as hard as for the intersection of semi-polynomial sets in general. In fact, a decidability proof for semi-linear A and (simple) semi-polynomial B would immediately yield decidability of nonemptiness for intersections of two arbitrary (simple) semi-polynomial sets. The argument resembles the remark at the beginning of Sect. 3. Consider $C = \{(P_1(k_1, \ldots, k_r), \ldots, P_n(k_1, \ldots, k_r)) \mid k_1, \ldots, k_r \in \mathbb{N}\}$ and $D = \{(Q_1(l_1, \ldots, l_s), \ldots, Q_n(l_1, \ldots, l_s)) \mid l_1, \ldots, l_s \in \mathbb{N}\}$. We have $C \cap D \neq \emptyset$ iff there are $i_1, \ldots, i_n$ with $P_1(k_1, \ldots, k_r) = i_1 = Q_1(l_1, \ldots, l_s)$, $\ldots$, $P_n(k_1, \ldots, k_r) = i_n = Q_n(l_1, \ldots, l_s)$. This means that the polynomial set (of dimension $2n$)

$$B = \{(P_1(\bar{k}), Q_1(\bar{l}), \ldots, P_n(\bar{k}), Q_n(\bar{l})) \mid k_1, \ldots, k_r, l_1, \ldots, l_s \in \mathbb{N}\}$$

has a nonempty intersection with the linear set

$$A = \{(i_1(1, 1, 0, \ldots 0) + \cdots + i_n(0, \ldots, 0, 1, 1) \mid i_1, \ldots, i_n \in \mathbb{N}\} \ .$$

5 Quadratic Forms

It is known that the undecidability of Hilbert's Tenth Problem holds for polynomial equations of degree four and for *systems* of polynomial equations of degree two (see [11]).

In this context, Xie, Dang, and Ibarra [16] solved a restricted case regarding pairs of quadratic equations which are generated by products of linear forms.

If only a *single* quadratic form

$$Q(x_1,\ldots,x_n) := \sum_{1\le i,j\le n} a_{ij}x_ix_j + \sum_{1\le i\le n} b_ix_i + c$$

is considered, where $a_{ij}, b_i, c \in \mathbb{Z}$, for $1 \le i,j \le n$, then the solvability of the equation $Q(x_1,\ldots,x_n) = 0$ has also been shown decidable, both in integers and in natural numbers. The solvability in integers follows from Siegel's work [15]. Regarding the solvability in natural numbers, a standard approach is to apply Lagrange's Theorem which characterizes the natural numbers as the sums of four squares (of integers). However, adding this requirement for each variable to a quadratic equation results in a *system* of quadratic equations, where Siegel's analysis does not apply. In a recent paper, Grunewald and Segal [5] show that the solvability of quadratic equations in integers stays decidable even under constraints given by linear inequalities:

Theorem 5 ([5]). *Given a quadratic form $Q \in \mathbb{Z}[X_1,\ldots,X_n]$ and linear forms $L_1,\ldots,L_k \in \mathbb{Z}[X_1,\ldots,X_n]$, it is decidable whether a system*

$$Q(x_1,\ldots,x_n) = 0 \ , \tag{12a}$$

$$L_j(x_1,\ldots,x_n) \ \# \ c_j, \text{ where } c_j \in \mathbb{Z} \text{ and } \# \in \{<,\le\}, \text{ for } j = 1,\ldots,k \ , \tag{12b}$$

$$(x_1,\ldots,x_n) \equiv (h_1,\ldots,h_n) \pmod m, \text{ where } h_1,\ldots,h_n \in \mathbb{Z},\ m \in \mathbb{N} \ , \tag{12c}$$

has a solution in $\mathbb{Z}^n$.

A decision procedure for solvability of quadratic equations in natural numbers can be obtained from Thm. 5 by imposing linear constraints of the form $-x_i \le 0$ for (12b).

The proof of the theorem requires deep number-theoretic constructions and does not come (as yet) with a complexity analysis. Rather than studying the general case it seems more tractable trying to isolate cases where reasonable complexity bounds can be provided.

In the sequel, we demonstrate how the decidability of the solvability of quadratic equations, in particular Thm. 5, can be applied to obtain two kinds of generalizations of semi-linear sets which are yet so modest that decidability results on the intersection problem are retained. The first result is concerned with sets defined via solutions of quadratic equations, and the second one refers to sets which are enumerated by quadratic and linear forms.

As a corollary of Thm. 5, the nonemptiness problem for the intersection of a semi-linear set with the solution set of a quadratic equation is decidable. For this, it suffices to recall that a semi-linear set is the solution set of a linear (in)equation system [4].

Corollary 6. *Nonemptiness of the intersection of a semi-linear set $A \subseteq \mathbb{N}^n$ with the solution set S of a quadratic equation $Q(x_1,\ldots,x_n) = 0$ is decidable.*

This can be applied, for example, to the following scenario indicated in Sect. 1: Given a system that produces a semi-linear set $A \subseteq \mathbb{N}^n$ (for instance, a finite automaton or a pushdown system) and an acceptance constraint given by a quadratic equation $Q(\bar{x}) = 0$ for $Q \in \mathbb{Z}[X_1, \ldots, X_n]$ then it can be decided whether some run of the automaton exists that satisfies the acceptance condition.

Next we introduce sets which refer to the value sets of quadratic forms $Q(\bar{x})$ rather than solutions of the equation $Q(\bar{x}) = 0$. We call a set $A \subseteq \mathbb{N}^n$ *one-quadratic* if A is a polynomial set such that the first component is given by a quadratic form $Q \in \mathbb{N}[X_1, \ldots, X_m]$ and the other components by linear forms $L_2, \ldots, L_n \in \mathbb{N}[X_1, \ldots, X_m]$:

$$A = \{(Q(k_1, \ldots, k_m), L_2(k_1, \ldots, k_m), \ldots, L_n(k_1, \ldots, k_m)) \mid k_1, \ldots, k_m \in \mathbb{N}\} .$$

A *semi-one-quadratic set* is a finite union of one-quadratic sets. The semi-one-quadratic sets encompass the semi-linear ones.

Deciding the nonemptiness of the intersection of one-quadratic sets leads to solving an equation system of the form (12): Given one-quadratic subsets A and B that are defined by $Q, L_2, \ldots, L_n \in \mathbb{N}[X_1, \ldots, X_m]$ and $Q', L'_2, \ldots, L'_n \in \mathbb{N}[X_1, \ldots, X_r]$, respectively, we have that $A \cap B \neq \emptyset$ iff there are $k_1, \ldots, k_m$ and $k'_1, \ldots, k'_r \in \mathbb{N}$ such that $Q(k_1, \ldots, k_m) = Q'(k'_1, \ldots, k'_r)$ and $L_j(k_1, \ldots, k_m) = L'_j(k'_1, \ldots, k'_r)$, for $j = 2, \ldots, n$. Now, we write the equations as $Q - Q' = 0$ and $L_j - L'_j = 0$, and then replace $L_j - L'_j = 0$ by $L_j - L'_j \leq 0$ and $L'_j - L_j \leq 0$. Hence, Thm. 5 can be applied. For the step from one-quadratic sets to semi-one-quadratic sets, we just use the distributivity of union over intersection.

Corollary 7. *Nonemptiness of the intersection of two semi-one-quadratic sets (and hence of a semi-one-quadratic with a semi-linear set) is decidable.*

6 Conclusion

The results of this paper are a small step into a field which is not well explored so far. We have suggested some classes of arithmetical constraints beyond the framework of semi-linear sets where effective solutions are possible. The main purpose of this note is to indicate some perspectives. Let us list some open problems:

1. Study the closure properties of simple semi-polynomial and simple semi-quadratic sets. In particular, does the intersection of a simple semi-polynomial set with a semi-linear set yield again a simple semi-polynomial set? What about the case of a semi-quadratic set?
2. The product relation of Example 2 shows a weakness of the simple semi-polynomial sets. One observes, however, that $2mn = (m+n)^2 - m^2 - n^2$, for any $m, n \in \mathbb{N}$, and thus the product function is (up to a factor) the difference of functions the graphs of which are simple quadratic. This suggests the study of the closure of simple (semi-)quadratic sets under additive operations.
3. Better upper bounds for deciding the membership in a simple semi-polynomial set should be found.

4. A way of extending the simple semi-polynomial sets is to consider the Parikh images of indexed languages. This would cover not only the product relation but also exponential relations like $\{(n, 2^n) \mid n \in \mathbb{N}\}$ (see [8]).
5. Start an algorithmic analysis of [5], and find forms of quadratic equations where reasonable upper bounds for deciding solvability can be established.
6. Study the case of quadratic inequations rather than equations.

References

1. Bruyére, V., Dall'Olio, E., Raskin, J.F.: Durations, parametric model-checking in timed automata with Presburger arithmetic. In Proc. STACS 2003. LNCS 2607. Springer (2003) 687–698
2. Dal Zilio, S., Lugiez, D.: XML schema, tree logic and sheaves automata. In Proc. RTA 2003. LNCS 2706. Springer (2003) 246–263
3. Dang, Z., Ibarra, O.H., Bultan, T., Kemmerer, R.A., Su, J.: Binary reachability analysis of discrete pushdown timed automata. In Proc. CAV 2000. LNCS 1855. Springer (2000) 69–84
4. Ginsburg, S., Spanier, E.H.: Semigroups, Presburger formulas, and languages. Pacific J. Math. **16** (1966) 285–296
5. Grunewald, F., Segal, D.: On the integer solutions of quadratic equations. J. Reine Angew. Math. **569** (2004) 13–45
6. Hardy, G.H., Wright, E.M.: An Introduction to the Theory of Numbers. 5th edn. Oxford University Press (1979)
7. Ibarra, O.H., Bultan, T., Su, J.: Reachability analysis for some models of infinite-state transition systems. In Proc. CONCUR 2000. LNCS 1877. Springer (2000) 183–198
8. Karianto, W.: Parikh automata with pushdown stack. Diploma thesis, RWTH Aachen (2004) Available at http://www-i7.informatik.rwth-aachen.de.
9. Klaedtke, F., Rueß, H.: Monadic second-order logics with cardinalities. In Proc. ICALP 2003. LNCS 2719. Springer (2003) 681–696
10. Lugiez, D.: Counting and equality constraints for multitree automata. In Proc. FOSSACS 2003. LNCS 2620. Springer (2003) 328–342
11. Matiyasevich, Y.V.: Hilbert's Tenth Problem. MIT Press (1993)
12. Parikh, R.J.: On context-free languages. J. ACM **13** (1966) 570–581
13. Seidl, H., Schwentick, T., Muscholl, A.: Numerical document queries. In Proc. PODS 2003. ACM Press (2003) 155–166
14. Seidl, H., Schwentick, T., Muscholl, A., Habermehl, P.: Counting in trees for free. In Proc. ICALP 2004. LNCS 3142. Springer (2004) 1136–1149
15. Siegel, C.L.: Zur Theorie der quadratischen Formen. Nachrichten der Akademie der Wissenschaften in Göttingen, II, Mathematisch-Physikalische Klasse **3** (1972) 21–46
16. Xie, G., Dang, Z., Ibarra, O.H.: A solvable class of quadratic Diophantine equations with applications to verification of infinite-state systems. In Proc. ICALP 2003. LNCS 2719. Springer (2003) 668–680

Invisible Safety of Distributed Protocols*

Ittai Balaban[1], Amir Pnueli[1], and Lenore D. Zuck[2]

[1] New York University, New York
{balaban, amir}@cs.nyu.edu
[2] University of Illinois at Chicago
lenore@cs.uic.edu

Abstract. The method of "Invisible Invariants" has been applied successfully to protocols that assume a "symmetric" underlying topology, be it cliques, stars, or rings. In this paper we show how the method can be applied to proving safety properties of distributed protocols running under arbitrary topologies. Many safety properties of such protocols have reachability predicates, which, at first glance, are beyond the scope of the Invisible Invariants method. To overcome this difficulty, we present a technique, called "coloring," that allows, in many instances, to replace the second order reachability predicates by first order predicates, resulting in properties that are amenable to Invisible Invariants. We demonstrate our techniques on several distributed protocols, including a variant on Luby's Maximal Independent Set protocol, the Leader Election protocol used in the IEEE 1394 (Firewire) distributed bus protocol, and various distributed spanning tree algorithms. All examples have been tested using the symbolic model checker TLV.

1 Introduction

Uniform verification of parameterized systems is one of the most challenging problems in verification today. Given a parameterized system $S(N) : P[1] \| \cdots \| P[N]$ and a property p, uniform verification attempts to verify $S(N) \models p$ for every $N > 1$. One of the most powerful approaches to verification which is not restricted to finite-state systems is *deductive verification*. This approach is based on a set of proof rules in which the user has to establish the validity of a list of premises in order to validate a given property of the system. The two tasks that the user has to perform are:

1. Identify some auxiliary constructs which appear in the premises of the rule;
2. Use the auxiliary constructs to establish the logical validity of the premises.

When performing manual deductive verification, the first task is usually the more difficult, requiring ingenuity, expertise, and a good understanding of the behavior of the program and the techniques for formalizing these insights. The second task is often performed using theorem provers such as PVS[1] or STeP [2], which require user guidance and interaction, placing additional burden on the user. The difficulties in the execution of these two tasks are the main reason why deductive verification is not used more widely.

* This research was supported in part by NSF grant CCR-0205571 and ONR grant N00014-99-1-0131.

M. Bugliesi et al. (Eds.): ICALP 2006, Part II, LNCS 4052, pp. 528–539, 2006.

$$\begin{array}{l} \text{I1. } \Theta \to \varphi \\ \text{I2. } \varphi \wedge \rho \to \varphi' \\ \text{I3. } \varphi \to p \\ \hline \Box p \end{array}$$

Fig. 1. The Proof Rule INV

A representative case is the verification of invariance properties using the *invariance rule* of [3], which is described in Fig. 1. In order to prove that assertion p is an invariant of program P, the rule calls for an *auxiliary assertion* φ that is *inductive* and strengthens (implies) p. Premise I1 requires φ to hold at any initial states, which are characterized by the assertion Θ. Premise I2 requires that every ρ-successor of a φ-state is also φ-state, where ρ is the transition relation, and premise I3 specifies that φ strengthens p. The main challenge in applying INV is identifying a good φ when p itself is not inductive.

In [4, 5] we introduced the method of *Invisible Invariants*, which proposes a method for automatic generation of the auxiliary assertion φ for parameterized systems, as well as an efficient algorithm for checking the validity of the premises of the invariance rule. See [6] for a tool that implements the idea. The generation of invisible auxiliary constructs is based on the observation that frequently an auxiliary assertion φ for a parameterized system has one of the forms $q(i)$, $\forall i.q(i)$ or, more generally, $\forall i \neq j.q(i,j)$. We construct an instance of the parameterized system taking a fixed value N_0 for the parameter N. For the finite-state instantiation $S(N_0)$, we compute, using BDD-techniques, some assertion ψ, which we wish to generalize to an assertion in the required form. Let r_1 be the projection of ψ on process index 1, obtained by discarding references to all variables that are local to all processes other than $P[1]$. We take $q(i)$ to be the generalization of r_1 obtained by replacing each reference to a local variable $P[1].x$ by a reference to $P[i].x$. The obtained $q(i)$ is our candidate for the body of the inductive assertion $\varphi : \forall i.q(i)$. We refer to this part of the process as *proj-gen*. For example, when generating invariants, ψ is the set of reachable states of $S(N_0)$. The process can easily be generalized to generate assertions of the type $\forall i_1, \ldots, i_k.p(\vec{i})$.

Having obtained a candidate for the assertion φ, we still have to check validity of the premises of the proof rule we wish to employ. Under the assumption that our assertional language is restricted to the predicates of equality and inequality between bounded range integer variables (which is adequate for many of the parameterized systems we considered), we proved a *small model* theorem, according to which, for a certain type of assertion, there exists a (small) bound N_0 such that such an assertion is valid for every N iff it is valid for all $N \leq N_0$. This enables using BDD-techniques to check validity of such an assertion. The assertions covered by the theorem are those that can be written in the form $\forall\vec{i}\exists\vec{j}.\psi(\vec{i},\vec{j})$, where $\psi(\vec{i},\vec{j})$ is a quantifier-free assertion that refers only to the global variables and the local variables of $P[i]$ and $P[j]$, where the variables are restricted to be stratified. Thus, for example, if we have a finite domain and an index domain (that ranges over the process id's $[1..N]$), stratification requires that every array maps from the index domain into the finite domain.

Being able to validate the premises on $S[N_0]$ has the additional important advantage that the user never sees the automatically generated auxiliary assertion φ. This assertion is produced as part of the procedure and is immediately consumed in order to validate the premises of the rule. Being generated by symbolic BDD techniques, the representation of the auxiliary assertions is often extremely unreadable and non-intuitive, and will usually not contribute to a better understanding of the program or its proof. Because the user never gets to see it, we refer to this method as the "method of *Invisible Invariants*."

As shown in [4, 5], many concurrent systems are stratified, and the result of embedding a $\forall\vec{i}.q(\vec{i})$ candidate inductive invariant in the main proof rule used for their safety properties results in premises that fall under the small model theorem. In the past we did not study protocols for general topologies, because many of these require reachability analysis, which is not a first order predicate, and therefore was beyond our methods. Thus, all the systems we applied the Invisible Invariants method (or its successors that handle liveness), have a "trivial" topology, be it a star, a clique, or a ring.

In this paper we study applications of the method of invisible invariants to arbitrary fixed topologies. We first present a small-model theorem that applies to such systems. We then study protocols whose specifications include reachability predicates. To handle reachability with an invisible-invariant-like strategy, we augment a given protocol with a coloring scheme that starts at one node (the *initial node*), and propagates colors to adjacent non-colored nodes. At each point in the coloring, only nodes that are reachable from the initial node are colored, and when the coloring terminates, all nodes reachable from the initial node are colored. The coloring allows to replace the second-order reachability predicate with a first order *colored* predicate.

The paper is organized as follows: Section 2 demonstrates how we model the leader election protocol. Section 3 presents the formal model of programs over arbitrary topologies, as well as a small model result. Section 4 formalizes and demonstrates use of the coloring augmentation, Section 5 summarizes runtime and verification results, and Section 6 discusses future work and concludes.

Related Work. We are not aware of any work that deals specifically with automatic verification of distributed algorithms. Most related to the work here is the work on automatic verification of parameterized systems. Our work extends the work surveyed in [7]. The PAX project (e.g., [8]) models parameterized systems in WS1S on which abstractions are computed and checked in MONA. The index predicates (e.g., [9]) combine predicate abstraction with a heuristic, similar to that used here, for constructing quantified invariants.

There have been numerous verification efforts specifically targeted at various aspects of the IEEE 1394 tree identification protocol, among them are [10, 11]. However, none of these works attempt at full automation. The work in [12] deals with the probabilistic aspect of the protocol, which we ignore in the work reported here. (We should, however, state that we have automatically verified the probabilistic aspects of the protocol using methods that are outside the scope of this paper.) For an in depth survey of previous verification efforts of the protocol see [10].

The work in [13] uses a coloring scheme, somewhat different than ours, to obtain over-approximations of reachability predicates for the purpose of shape-analysis. Since we deal with a fixed topology, our coloring scheme is precise with respect to reachability.

2 Leader Election

To demonstrate our techniques, we present the Leader Election protocol [14] (which also serves as the *tree identification* protocol used in the IEEE 1394 bus specification [12]) and its safety properties.

In all of our examples, we assume a network of N processes whose id's are $[1..N]$. The interconnection among the processes is described by the boolean matrix Q, where $Q[i,j]$ denotes a direct link from i to j, and $\neg Q[i,j]$ denotes the absence of such a link. We assume that the communication between neighbors is bi-directional, therefore $Q[i,j] = Q[j,i]$ for all i and j.

The Protocol. IEEE 1394 specifies a network allowing dynamic connection and disconnection of devices. At each point in time, the network is arranged as a tree, with devices as leaves. The *leader election* sub-protocol is invoked during a connection or disconnection event when, based on the new topology, a leader needs to be determined anew. Dynamic aspects of the network need not be modeled here since the leader election sub-protocol itself assumes a static network (i.e., following a connection/disconnection event).

As before, we model communication between nodes by shared variables. We let Q denote the adjacency matrix, and for each process i, we assign a boolean variable $done[i]$ denoting whether i still participates in the protocol or has determined its parent, a boolean $leader[i]$ which is set when i becomes the leader, and a boolean matrix $parent[1..N, 1..N]$ such that $parent[i,j]$ is set when j becomes the parent of i.

In our modeling of the protocol, we assume that each node i, in a single indivisible atomic step, can check all the $parent[1..N, i]$ variables and set $parent[i,j]$ and $leader[i]$ accordingly. This is different from the common synchronous modeling of the protocol that proceeds in send/receive phases, where at a send phase nodes can send "be my parent" requests and at receive phases nodes respond to such requests. There, *contention* may occur when two nodes send one another *be my parent* requests at the same phase. The atomicity assumption here bypasses root contention. As discussed later, the methods proposed here are applicable to less atomic versions that allow for contention.

$$
\begin{array}{l}
Q: \textbf{array } [1..N] \textbf{ of array } [1..N] \textbf{ of bool where } \forall i,j.Q[i,j] \leftrightarrow Q[j,i] \\
parent: \textbf{array } [1..N] \textbf{ of array } [1..N] \textbf{ of bool init } \forall i,j.\neg parent[i,j] \\
leader: \textbf{array } [1..N] \textbf{ of bool init } \forall i.\neg leader[i] \\
done: : \textbf{array } [1..N] \textbf{ of bool init } \forall i.\neg done[i] \\
\displaystyle\Big\|_{i \neq j} P[i,j] :: \left[\begin{array}{l}
\textbf{while } \neg done[i] \textbf{ do} \\
\quad \left[\begin{array}{l}
\textbf{if } \forall k \neq i.Q[i,k] \rightarrow parent[k,i] \textbf{ then } (leader[i], done[i]) := (\mathbf{1},\mathbf{1}) \\
\textbf{elsif } (\neg parent[j,i] \wedge Q[i,j] \wedge \forall k \notin \{i,j\}.Q[i,k] \rightarrow parent[k,i]) \\
\qquad\qquad \textbf{then } (parent[i,j], done[i]) := (\mathbf{1},\mathbf{1})
\end{array} \right]
\end{array} \right]
\end{array}
$$

Fig. 2. Program LEADER-ELECT

The leader election protocol is shown in Fig. 2. For each node, the *parent* matrix identifies which node is the parent of another node. There are $N(N-1)$ processes in the system, each corresponding to a pair $(i,j) \in [1..N]^2$ with $i \neq j$. Each such process, $P[i,j]$, repeatedly performs the following two steps while $done[i] \neq \mathbf{1}$:

1. The first if-statement executes if all nodes directly connected to i have i as their parent. In this case, i becomes the leader and sets $leader[i]$ to **1**.
2. The second if-statement executes if (1) i and j are connected, (2) j has no parent, and (3) all other neighbors of i have i as parent. In this case, j becomes parent of i.

The protocol works as follows: Assume the underlying graph is a tree. Initially, all leaf nodes (and no internal node) can execute the second step. Then, the algorithm climbs up the tree, each node executing the second step, until the root, which executes the first step, is reached.

If the original graph consists of a forest of trees, then a leader will be elected in each tree. If the original graph has non-tree connected components, then no leader will be elected in these components. The safety property of the protocol therefore states that each component contains at most one leader, formally stated by the following property:

$$Unique: \quad \forall i \neq j : reachable(i,j) \rightarrow \neg(leader[i] \wedge leader[j])$$

where for every $i, j \in [1..N]$, $reachable(i,j)$ holds if there is Q-path leading from i to j, i.e., if there are nodes $i_1, \ldots, i_k \in [1..N]$ such that $i_1 = i$, $i_k = j$, and for every $\ell = 1, \ldots, k-1$, $Q[i_\ell, i_{\ell+1}]$.

As discussed in the introduction, none of our old methods can be used to automatically verify this property. The method described in [15] fails since it depends on the *reachable* predicate being based on a relation where each node has at most one successor, and Q, on which our current *reachable* is based, does not satisfy this requirement.

3 Formal Model and Verifying Invariance

In this section we present our computational model, as well as the small model property that forms the basis of the verification method. Both model and property are derived from [5] and only differ in that the version here allows for matrix types (e.g., the Q and *parent* variables in Fig. 2).

Discrete Systems. As our computational model, we take a *discrete system* $S = \langle V, \Theta, \rho \rangle$, where

- V — A set of *system variables*. A *state* of S provides a type-consistent interpretation of the variables V. For a state s and a system variable $v \in V$, we denote by $s[v]$ the value assigned to v by the state s. Let Σ denote the set of all states over V.
- Θ — The *initial condition*: An assertion (state formula) characterizing the initial states.
- $\rho(V, V')$ — The *transition relation*: An assertion, relating the values V of the variables in state $s \in \Sigma$ to the values V' in an S-successor state $s' \in \Sigma$.

For an assertion ψ, we say that $s \in \Sigma$ is a ψ-state if $s \models \psi$.

A *computation* of a system S is an infinite sequence of states $\sigma : s_0, s_1, s_2, \ldots$, satisfying the requirements:

- *Initiality* — s_0 is initial, i.e., $s_0 \models \Theta$.
- *Consecution* — For each $\ell = 0, 1, \ldots$, the state $s_{\ell+1}$ is an S-successor of s_ℓ. That is, $\langle s_\ell, s_{\ell+1} \rangle \models \rho(V, V')$ where, for each $v \in V$, we interpret v as $s_\ell[v]$ and v' as $s_{\ell+1}[v]$.

Finite Network Systems. To allow the automatic decision of validity of assertions, we place further restrictions on the systems we study, leading to what is essentially the model of bounded discrete systems of [5] extended with an additional matrix type. For brevity, we describe here a simplified two-type model; the extension for the general multi-type case is straightforward. We allow the following data types parameterized by the positive integer N, intended to specify the size of the topology:

1. **bool**: boolean and finite-range scalars; With no loss of generality, we assume that all finite domain values are encoded as booleans.
2. **index**: $[1..N]$
3. Arrays of the types **index** $\mapsto$ **bool** (**bool** array) and **index** $\mapsto$ **index** $\mapsto$ **bool** (**bool** matrix)

Constants are introduced as variables with reserved names. Thus, we admit the boolean constants **0** and **1**, and **index** constants such as 1 and N. We often refer to an element of type **index** as a *node*. *Atomic formulas* are defined as follows:

- If x is a boolean variable, B is a **bool** array, and y is an **index** variable, then x and $B[y]$ are atomic formulas.
- If y_1 and y_2 are **index** variables and Q is a **bool** matrix, then $Q[y_1, y_2]$ is an atomic formula.
- If t_1 and t_2 are **index** terms, then $t_1 = t_2$ is an atomic formula.

A *restricted A-assertion* (resp. *restricted E-assertion*) is a formula of the form $\forall \vec{y}.\psi(\vec{x}, \vec{y})$ (resp. $\exists \vec{y}.\psi(\vec{x}, \vec{y})$) where $\vec{x}$ and $\vec{y}$ are lists of **index** variables, and $\psi(\vec{x}, \vec{y})$ is a boolean combination of atomic formulae. A *restricted EA-assertion* is an assertion $\exists \vec{x}.\, \forall \vec{y}.\psi(\vec{x}, \vec{y}, \vec{u})$ where $\vec{u}$ is a list of **index** variables and $\forall \vec{y}.\psi(\vec{x}, \vec{y}, \vec{u})$ is a restricted A-assertion. *Restricted AE-assertions* are similarly defined. As the initial condition Θ and the transition relation ρ we only allow restricted EA-assertions.

Let $\mathcal{V}$ be a *vocabulary* of typed variables, whose types are taken from the restricted type system allowed in a system. A *model* M for $\mathcal{V}$ consists of the following elements:

- A positive integer $N > 0$.
- For each boolean variable $b \in \mathcal{V}$, a boolean value $M[b] \in \{\mathbf{0}, \mathbf{1}\}$. It is required that $M[\mathbf{0}] = \mathbf{0}$ and $M[\mathbf{1}] = \mathbf{1}$.
- For each **index** variable $x \in \mathcal{V}$, a natural value $M[x] \in [1..N]$.
- For each boolean array $B \in \mathcal{V}$, a boolean function $M[B] : [1..N] \mapsto \{\mathbf{0}, \mathbf{1}\}$.
- For each boolean matrix $Q \in V$, a function $M[Q] : [1..N] \mapsto [1..N] \mapsto \{\mathbf{0}, \mathbf{1}\}$

We define the *size* of model M to be N.

The following theorem states that a restricted AE-assertion is valid iff it is valid over all models of a bounded size. It follows from a similar theorem of [5] (which does not deal with the boolean matrix data-type).

Theorem 1 (Small Model Property). *Let* φ: $\forall \vec{y}\ \exists \vec{x}.\psi(\vec{y}, \vec{x})$ *be a closed restricted AE-assertion. Then* φ *is valid iff it is valid over all models of size not exceeding* $|\vec{y}|$.

Checking Invariance. Consider the INV proof rule of Fig. 1. When validating the premises of INV for assertions p and φ, I3 is a boolean combination of A- and E-assertions, while I1 and I2 are AE-assertions. We now compute the cut-off bounds determined by the small model theorem to validate Premises I1 and I2. Assume that the assertions appearing in INV are of the form:

$$p: \ (\forall u_1, \ldots, u_c.p_1(\vec{u})) \otimes \exists \vec{x}.p_2(\vec{x}) \qquad \Theta: \ \exists y_1, \ldots, y_a.\forall \vec{x}.t(\vec{y}, \vec{x})$$
$$\varphi: \ (\forall u_1, \ldots, u_{n_\varphi}.\varphi_1(\vec{u})) \otimes \exists x_1, \ldots, x_{m_\varphi}.\varphi_2(\vec{x}) \quad \rho: \ \exists y_1, \ldots, y_b.\forall \vec{x}.R(\vec{y}, \vec{x})$$

where $\otimes \in \{\vee, \wedge\}$. I.e, p and φ are assertions that are disjunctions or conjunctions of a restricted A-assertion and a restricted E-assertion, and Θ and ρ are restricted EA-assertions. If p has free variables, then let $\hat{c}$ be c plus the number of free variables in p. Define $\hat{n}_\varphi$, $\hat{m}_\varphi$, $\hat{a}$, and $\hat{b}$ similarly. Theorem 1 now implies:

Corollary 1. *The premises of rule* INV *are valid over* $S(N)$ *for all* $N > 1$ *iff they are valid over* $S(N)$ *for all* $N \leq \max\{\hat{a} + \hat{n}_\varphi, \hat{b} + \hat{n}_\varphi + \hat{m}_\varphi, \hat{m}_\varphi + \hat{c}\}$.

In the full version of the paper [16] we describe an automatic verification of the safety properties of Luby's Maximal Independent Set protocol [17].

4 Reachability Avoidance

It is very often the case that safety properties of distributed systems include reachability predicates that are captured neither by Theorem 1 nor by the *proj-gen* heuristic. In this section we define reachability properties we are interested in, and show a methodology that overcomes the challenges they pose to the Invisible Invariants method.

4.1 Safety Properties with Reachability

Let S be a distributed system with an underlying topology described by the adjacency matrix Q. Recall the $\mathit{reachable}(y_1, y_2)$ predicate denoting that y_2 is Q-reachable from y_1. We show how to prove invariant properties of the type $\Box(\alpha \otimes \beta)$, where α is a restricted A-assertion that allows for *reachable* predicates, $\otimes \in \{\vee, \wedge\}$, and β is a restricted E-assertion (without reachability predicates). For simplicity of exposition, we further restrict α to have a single occurrence of a *reachable* predicate, both arguments of which are bound by the universal quantifier. Our results can be easily extended to cases where α has several occurrences of *reachable*, and to cases where some arguments of *reachable* are free. An example of such a property is *Unique* of program LEADER-ELECT in Section 2. There, β is trivial and α has a single *reachable* predicate, both of whose arguments are under the scope of the universal quantification.

For the remaining part of this section we fix a safety property $\Box\, \phi$ we wish to verify over S, where $\phi : \alpha \otimes \beta$ of the form above. Let t be some **index** variable that does not appear free in either ϕ or the transition relation. Without loss of generality, assume that $\alpha\colon \forall i_1, \ldots, i_k.p(i_1, \ldots, i_k)$, where i_k is the first parameter of the (single) reachability predicate in α. Let $\alpha[t]$ be the formula $\forall i_1, \ldots, i_{k-1}.p(i_1, \ldots, i_{k-1}, t)$, and $\phi[t]$ be the formula $\alpha[t] \otimes \beta$. From the choice of t it follows that $S \models \Box\, \phi[t]$ implies that $S \models \Box\, \phi$.

For example, for property *Unique* and $t = 1$, we obtain:

$$\mathit{Unique}[1]: \qquad \forall j.j \neq 1 \wedge \mathit{reachable}(1, j) \rightarrow \neg(\mathit{leader}[1] \wedge \mathit{leader}[j])$$

4.2 Replacing Reachability with a First Order Predicate

The property $\phi[t]$ still contains a reachability predicate and its invariance cannot be handled by the method of Invisible Invariants. We next augment S with a "coloring protocol" and replace ϕ with a new property, ϕ^t, such that ϕ^t is of the form described in Section 3, such that when the augmented system satisfies $\Box\, \phi^t$ we can conclude that $S \models \Box\, \phi[t]$, and therefore $S \models \Box\, \phi$.

The system and coloring protocol alternate once between "protocol" and "coloring" phases. While in the "protocol" phase, the system behaves like S, and the coloring scheme is inactive. Similarly, while in the "coloring" phase, the system is inactive, and the coloring scheme behaves according to its protocol, color_t. An additional component, the "phase changer," determines which phase is first, and switches (once) between them. We shall return to the phase changer and first describe the coloring protocol.

The coloring protocol color_t, described in Fig. 3, propagates a marking starting at the node t. We assume a **boolean** array C_t that does not appear in S, all of whose entries are initially **0**, denoting that all nodes are uncolored. Once activated, the coloring protocol first sets $C_t[t]$, thus marking node t. Thereafter, when an uncolored node i has a colored neighbor j, $C_t[i]$ is set. The correctness of color_t is expressed in the following theorem, whose proof is by induction on the topology of the network:

$$
\begin{array}{l}
\mathsf{color}_t :: \\
\quad \textbf{local } C_t\text{: } \textbf{array } [1..N] \textbf{ of bool init } \forall i. C_t[i] = \mathbf{0} \\
\quad \mathop{\Big\|}_{i \neq j} \Big[\textbf{if } ((i = t) \vee (Q[i,j] \wedge C_t[j] \wedge \neg C_t[i])) \textbf{ then } C_t[i] := 1\Big]
\end{array}
$$

Fig. 3. System color_t

$$
\begin{array}{l}
\text{PHASE}(\Psi) :: \\
\quad \textbf{local } \mathit{phase}, \mathit{init\_phase}\text{: } \{\mathit{protocol}, \mathit{color}\} \textbf{ init } \mathit{phase} = \mathit{init\_phase} \\
\quad \big[\textbf{if } (\Psi \wedge \mathit{phase} = \mathit{init\_phase}) \textbf{ then } \mathit{phase} := \neg \mathit{phase}\big]
\end{array}
$$

Fig. 4. System PHASE(Ψ)

Theorem 2. *Let $S[t] = S \| \mathsf{color}_t$. Then, for every node i, the following all hold:*

1. *$\mathit{reachable}(t,i)$ is S-valid iff it is $S[t]$-valid, i.e., both S and $S[t]$ have the same reachability relations;*
2. *$S[t] \models \Box (C_t[i] \rightarrow \mathit{reachable}(t,i))$, i.e., every colored node is reachable from t;*

Assume phase and $\mathit{init\_phase}$ are variables not in S that can take on the values $\{\mathit{color}, \mathit{protocol}\}$. The *phase changer* is a module which is composed with S and color_t that is allowed to change the phase once, when a condition Ψ, which is an input to PHASE, is met. The module, labeled PHASE, is described in Fig. 4. There, "$\mathit{phase} := \neg \mathit{phase}$" has the obvious meaning. In Subsection 4.3 we discuss how $\mathit{init\_phase}$ and Ψ are initialized.

Let S' be the system S where each instruction is guarded by $(\mathit{phase} = \mathit{protocol})$, i.e., if S is described by $\langle V, \Theta, \rho \rangle$ then S' is defined by $\langle V \cup \{\mathit{phase}\}, \Theta, \rho' \rangle$ where

$\rho' = (phase = protocol \wedge \rho) \vee (phase \neq protocol \wedge \bigwedge_{v \in V} v = v')$. Similarly, let $\mathsf{color}_t{}'$ be the system color_t where each instruction is guarded by $(phase = color)$. Then system S_{aug} is defined by the composition $S' \| \mathsf{color}_t{}' \| \text{PHASE}$. The following claim follows immediately from the definition of $Saug$:

Claim. Let ψ be a safety property over V. Then $S \models \psi$ iff $Saug \models \psi$.

We next construct, from $\phi[t]$, a property ϕ^t such that $S^t \models \square\, \phi^t$ implies that $Saug \models \square\, \phi[t]$ (which, according to the previous claim, implies that $S \models \square\, \phi[t]$). Recall that $\phi[t]$ is of the form $\forall \alpha[t] \otimes \exists \beta$ where the single reachability in $\phi[t]$ appears in α in the form $reachable(t, j)$. We first replace the $reachable(t, j)$ assertion in α by $C_t[j]$. If $reachable(t, j)$ appears in $\alpha[t]$ under positive polarity, we add to the resulting formula the disjunct

$$\exists j \neq k.Q[j, k] \;\wedge\; C_t[j] \;\wedge\; \neg C_t[k]$$

that captures the situation in which the coloring algorithm has not terminated yet. We take ϕ^t to be the resulting formula.

For example, under this transformation, $Unique[1]$ becomes:

$$Unique^1 : \quad \forall j.j \neq 1 \wedge C_1[j] \rightarrow \neg(leader[1] \wedge leader[j]) \tag{1}$$

The following theorem, the proof of which is available in the full version of the paper [16], establishes the soundness of the transformation.

Theorem 3

$$S^t \models \square\, \phi^t \qquad \Longrightarrow \qquad Saug \models \square\, \phi[t]$$

Note that $\square\, \phi^t$ is now of the form covered by Corollary 1. For example, to verify $Unique^1$, we have $a = 0$ (since the initial condition has no existential quantifiers), $b = 3$ since the transition relation of the augmented S^t has i and j under existential quantification, and t appears free in it, and $c = 2$, having j universally quantified and t free. Thus, for an auxiliary invariant φ, we would obtain a cutoff value of $\max\{n_\varphi, 3 + n_\varphi + m_\varphi, m_\varphi + 2\} = 3 + n_\varphi + m_\varphi$. We generated a φ with $n_\varphi = 2$ and $m_\varphi = 0$, and thus verified the premises of INV for every $N_0 \leq 5$.

4.3 Determining the Phase Alternation

There are two main choices to be made, namely, whether *init_phase* is *protocol* or *color*, and whether Ψ is trivially **1** or some non-trivial predicate. In our experiments, we used the trivial $\Psi = \mathbf{1}$ with *init_phase* being both *protocol* or *color*. As to non-trivial Ψ, we had to use it only once, in the verification of LEADER-ELECT, and then *init_phase* was set to *protocol* and Ψ was defined as $leader[t]$. We recommend first trying to use a trivial $\Psi = \mathbf{1}$, and only if it fails under both choices of *init_phase*, to attempt some obvious Ψ's, e.g., predicates that occur in the property.

5 Evaluation

We have evaluated our method on a set of algorithms which, with the exception of Luby's maximal independent set algorithm, are based on versions found in [14]. The

Algorithm	Runtime (seconds)
Leader Election	5
Leader Election (alternate)	54
Spanning Tree	36
MIS	30

Fig. 5. Runtime Results

test cases consist of the leader election protocol used as the running example, a version of leader election that does not assume atomic parent request/acknowledge steps, as well as a distributed spanning tree algorithm. All experiments were evaluated using the TLV symbolic model-checker [18] on a Pentium 3 1GHz PC with 512Mb memory, and can be found at *http://www.cs.nyu.edu/acsys/dist-protocols/index.html*. A summary of runtime results is shown in Fig. 5. The rest of this section summarizes each test case.

The alternate version of leader election allows for contention between nodes. While like the running example it treats the check over all of a node's neighbors as atomic, the assignment of parents is done in 2 phases, a *request* phase and an *acknowledgement* phase. Concretely, the matrix $parent$ is now of type
array $[1..N]$ **of array** $[1..N]$ **of** $\{\text{no}, \text{req}, \text{ack}\}$. Node j is considered the parent of i if $parent[i, j] = \text{ack}$.

For both versions of the leader election protocol, we verified the property $Unique$ defined in Section 2. For the alternate version we proved the additional property of *limited contention*, specifying that if neighboring nodes have requested parenthood from some neighbor, then the request is mutual:

$$\forall i \neq j, k, l : Q[i, j] \ \wedge \ parent[i, k] = \text{req} \ \wedge \ parent[j, l] = \text{req} \rightarrow k = l$$

Since this invariant effectively localizes contention in the protocol to two adjacent nodes, it serves as the basis for a liveness proof showing that any contention eventually converges with probability 1.

The spanning tree algorithm is similar to the coloring protocol color_t in that an arbitrary node is designated as the root, and nodes are added to the tree in a top-down, distributed fashion, starting at the root. For this algorithm we sought to verify the property that any node reachable from the root participates in the tree, unless tree propagation has not yet terminated, expressed as:

$$p\colon (\forall i, t : reachable(t, i) \rightarrow \text{in\_tree}[i]) \vee (\exists j \neq k\colon Q[j][k] \wedge in\_tree[j] \wedge \neg in\_tree[k])$$

where the boolean array in_tree denotes participation of nodes in the tree. However, we failed to generate an inductive auxiliary assertion that also implies this property. Instead, we did successfully verify that $\varphi \ \wedge \ p$ is an inductive invariant, where φ is the generated auxiliary assertion.

6 Conclusion and Discussion

We have described the application of the method of Invisible Invariants to distributed protocols with an arbitrary fixed topology. Contrary to common belief, we found that the

extension of the method to arbitrary, as opposed to trivial, topologies is rather straightforward (as demonstrated by the verification of Luby's MIS protocol). Yet, the correctness of many such protocols is specified by means of reachability predicates, which cannot be captured by the Invisible Invariants method. We present a simple coloring augmentation that allows, in many cases, to replace reachability predicates by first order predicates that can be dealt with by the Invisible Invariants method.
There are several weaknesses to our scheme:

- Many distributed systems are modeled as synchronous, i.e., their transition relation is an AEA-assertion. This is beyond the power of our small model theorem, hence we "de-synchronize" them. We would like to identify the types of synchronous systems our method applies to.
- Our scheme depends on running the "system" and the "coloring," one after the other, switching once from one to the other at some point. Often, this point is nondeterministic and the only choice is which protocol to run first. Yet, it is sometimes the case that the switch can happen only when some condition is attained. Here the method is not fully automatic since the user has to guess the condition, which requires some familiarity with the protocol.
- Our scheme is dependent on the Invisible Invariants method, and is restricted by its power. Being a BDD-based method, the size of the instantiation of the system required may be too large to handle. In addition, *proj-gen* can only generate invariants of certain syntactic type, and it may be the case that the invariants needed are beyond its power (For example, *proj-gen* can generate restricted EA-invariants, but is extremely limited in the AE-invariants it generates).

Yet, in spite of the restrictions, we succeeded to automatically verify, for the first time, some classical examples that have been thoroughly studied in the literature. We are hopeful that our coloring augmentation can be used in verification of other systems too, for example, pointer systems. We are currently working on extending the system to handle mobile networks.

Acknowledgement. We would like to thank Shuvendu Lahiri who brought the Leader Election protocol to our attention, and to Yi Fang who pointed out that our existing small model theorem can be applied to adjacency matrices.

References

1. Shankar, N., Owre, S., Rushby, J.M.: A tutorial on specification and verification using PVS. Technical report (1993)
2. Bjørner, N., Browne, I., Chang, E., Colón, M., Kapur, A., Manna, Z., Sipma, H., Uribe, T.: STeP: The Stanford Temporal Prover, User's Manual. Technical Report STAN-CS-TR-95-1562, Computer Science Department, Stanford University (1995)
3. Manna, Z., Pnueli, A.: Temporal Verification of Reactive Systems: Safety. Springer Verlag, New York (1995)
4. Pnueli, A., Ruah, S., Zuck, L.: Automatic deductive verification with invisible invariants. In: TACAS'01, LNCS 2031 (2001) 82–97

5. Arons, T., Pnueli, A., Ruah, S., Xu, J., Zuck, L.: Parameterized verification with automatically computed inductive assertions. In: CAV'01, LNCS 2102 (2001) 221–234
6. Balaban, I., Fang, Y., Pnueli, A., Zuck, L.: IIV: An invisible invariant verifier. In: Computer Aided Verification (CAV). (2005)
7. Zuck, L., Pnueli, A.: Model checking and abstraction to the aid of parameterized systems. Computer Languages, Systems, and Structures **30**(3–4) (2004) 139–169
8. Baukus, K., Lakhnech, Y., Stahl, K.: Parameterized verification of a cache coherence protocol safety and liveness. In: Proceedings of the 6th International Conference on Verification, Model Checking, and Abstract Interpretation. (2002) 317–330
9. Lahiri, S., Bryant, R.: Constructing quantified invariants via predicate abstraction. In: Proceedings of the 5th International Conference on Verification, Model Checking, and Abstract Interpretation. (2004) 267–281
10. Romijn, J.M.T.: A timed verification of the IEEE 1394 leader election protocol. In Gnesi, S., Latella, D., eds.: Proceedings of the Fourth International ERCIM Workshop on Formal Methods for Industrial Critical Systems (FMICS'99). (1999) pages 3–29
11. Devillers, M., Griffioen, W., Romijn, J., Vaandrager, F.: Verification of a leader election protocol: Formal methods applied to IEEE 1394. Technical Report CSI-R9728, Computing Science Institute, Nijmegen (1997)
12. Daws, C., Kwiatkowska, M., Norman, G.: Automatic verification of the IEEE 1394 root contention protocol with KRONOS and PRISM. In Cleaveland, R., Garavel, H., eds.: Proc. 7th International Workshop on Formal Methods for Industrial Critical Systems (FMICS'02). Volume 66.2 of Electronic Notes in Theoretical Computer Science., Elsevier (2002)
13. Lev-Ami, T., Immerman, N., Reps, T.W., Sagiv, S., Srivastava, S., Yorsh, G.: Simulating reachability using first-order logic with applications to verification of linked data structures. In: CADE. (2005) 99–115
14. Lynch, N.A.: Distributed Algorithms. Morgan Kaufmann Publishers Inc., San Francisco, CA, USA (1996)
15. Balaban, I., Pnueli, A., Zuck, L.: Shape analysis by predicate abstraction. In: Proceedings of the 6th International Conference on Verification, Model Checking, and Abstract Interpretation. (2005) 164–180
16. Balaban, I., Pnueli, A., Zuck, L.: Invisible safety of distributed protocols. Technical report, Computer Science Department, New York University (2006) *http://www.cs.nyu.edu/acsys/pubs/permanent/dist-protocols-icalp06-full.pdf.*
17. Luby, M.: A simple parallel algorithm for the maximal independent set problem. SIAM Journal of Computing **15**(4) (1986) 1036–1053
18. Shahar, E.: The TLV Manual. (2000) *http://www.cs.nyu.edu/acsys/tlv.*

The Complexity of Enriched μ-Calculi

Piero A. Bonatti[1,*], Carsten Lutz[2], Aniello Murano[1,*], and Moshe Y. Vardi[3,**]

[1] Università di Napoli "Federico II", Dipartimento di Scienze Fisiche, 80126 Napoli, Italy
[2] TU Dresden, Institute for Theoretical Computer Science, 01062 Dresden, Germany
[3] Microsoft Research and Rice University, Dept. of Computer Science, TX 77251-1892, USA

Abstract. The *fully enriched μ-calculus* is the extension of the propositional μ-calculus with inverse programs, graded modalities, and nominals. While satisfiability in several expressive fragments of the fully enriched μ-calculus is known to be decidable and EXPTIME-complete, it has recently been proved that the full calculus is undecidable. In this paper, we study the fragments of the fully enriched μ-calculus that are obtained by dropping at least one of the additional constructs. We show that, in all fragments obtained in this way, satisfiability is decidable and EXPTIME-complete. Thus, we identify a family of decidable logics that are maximal (and incomparable) in expressive power. Our results are obtained by introducing two new automata models, showing that their emptiness problems are EXPTIME-complete, and then reducing satisfiability in the relevant logics to this problem. The automata models we introduce are *two-way graded alternating parity automata* over infinite trees (2GAPT) and *fully enriched automata* (FEA) over infinite forests. The former are a common generalization of two incomparable automata models from the literature. The latter extend alternating automata in a similar way as the fully enriched μ-calculus extends the standard μ-calculus.

1 Introduction

The *μ-calculus* is a propositional modal logic augmented with least and greatest fixpoint operators [Koz83]. It is often used as a target formalism for embedding temporal and modal logics with the goal of transferring computational and model theoretic properties such as the EXPTIME upper complexity bound. *Description logics (DLs)* are a family of knowledge representation languages that originated in artificial intelligence [BM+03]. DLs currently receive considerable attention, which is mainly due to their use as an ontology language in prominent applications such as the semantic web [BHS02]. Notably, DLs have recently been standardized as the ontology language OWL by the W3C committee. It has been pointed out by several authors that, by embedding DLs into the μ-calculus, we can identify DLs that are of very high expressive power, but computationally well-behaved [CGL01, SV01, KSV02]. When putting this idea to work, we face the problem that modern DLs such as the ones underlying OWL include several constructs that cannot easily be translated into the μ-calculus. Most importantly, these

* Supported in part by the European Network of Excellence REWERSE, IST-2004-506779.
** Supported in part by NSF grants CCR-0311326 and ANI-0216467, by BSF grant 9800096, and by Texas ATP grant 003604-0058-2003. Work done in part while this author was visiting the Isaac Newton Institute for Mathematical Science, Cambridge, UK, as part of a Special Programme on Logic and Algorithm.

M. Bugliesi et al. (Eds.): ICALP 2006, Part II, LNCS 4052, pp. 540–551, 2006.

	Inverse progr.	Graded mod.	Nominals	Complexity
fully enriched μ-calculus	x	x	x	undecidable
full graded μ-calculus	x	x		EXPTIME (1ary/2ary)
full hybrid μ-calculus	x		x	EXPTIME
hybrid graded μ-calculus		x	x	EXPTIME (1ary/2ary)
graded μ-calculus		x		EXPTIME (1ary/2ary)

Fig. 1. Enriched μ-calculi and previous results

constructs are inverse programs, graded modalities, and nominals. Intuitively, inverse programs allow to travel backwards along accessibility relations [Var98], nominals are propositional variables interpreted as singleton sets [SV01], and graded modalities enable statements about the number of successors and predecessors of a state [KSV02]. All of the mentioned constructs are available in the DLs underlying OWL.

The extension of the μ-calculus with these constructs induces a family of enriched μ-calculi. These calculi may or may not enjoy the attractive computational properties of the original μ-calculus: on the one hand, it has been shown that satisfiability in a number of the enriched calculi is decidable and EXPTIME-complete [CGL01, SV01, KSV02]. On the other hand, it has recently been proved by Bonatti and Peron that satisfiability is undecidable in the *fully enriched μ-calculus*, i.e., the logic obtained by extending the μ-calculus with all of the above constructs simultaneously [BP04]. In computer science logic, it has always been a major research goal to identify decidable logics that are as expressive as possible. Thus, the above results raise the question of maximal decidable fragments of the fully enriched μ-calculus. In this paper, we study this question in a systematic way by considering all fragments of the fully enriched μ-calculus that are obtained by dropping at least one of inverse programs, graded modalities, and nominals. We show that, in all these fragments, satisfiability is decidable and EXPTIME-complete. Thus, we identify a whole family of decidable logics that have maximum (incomparable) expressivity.

The relevant fragments of the fully enriched μ-calculus are shown in Fig. 1 together with the complexity of their satisfiability problem. The results shown in gray are already known from the literature: EXPTIME-completeness of satisfiability in the full hybrid μ-calculus has been shown in [SV01]; under the assumption that the numbers inside graded modalities are coded in unary, the same result was proved for the full graded μ-calculus in [CGL01]; finally, the same was also shown for the (non-full) graded μ-calculus in [KSV02] under the assumption of binary coding. In this paper, we prove EXPTIME-completeness of the full graded μ-calculus and the hybrid graded μ-calculus. In both cases, we allow numbers to be coded in binary (techniques such as those of [CGL01] involve an exponential blow-up when numbers are coded in binary).

Our results are based on the automata-theoretic approach. We introduce *fully enriched automata (FEAs)*, which run on infinite forests and use a parity acceptance condition. Intuitively, these automata generalize alternating automata on infinite trees in a similar way as the fully enriched μ-calculus extends the standard μ-calculus: FEAs can move up to a node's predecessor (by analogy with inverse programs), move down to at least n or all but n successors (by analogy with graded modalities), and jump directly to the roots of the input forest (which are the analogues of nominals). We prove that the emptiness problem is decidable for fully enriched automata and then show how to reduce to this problem satisfiability in the hybrid graded and the full graded μ-calculi, exploiting the forest model property enjoyed by these logics. Observe that decidability of the emptiness problem for FEAs does not contradict the undecidability of the fully enriched μ-calculus: the latter does not enjoy a forest model property [BP04], and hence satisfiability cannot be decided using forest-based FEAs.

To show that the emptiness problem for FEAs is in EXPTIME, we introduce an additional automata model: *two-way graded parity tree automata (2GAPTs)*. These automata are interesting in their own right because they generalize in a natural way two existing, but incomparable automata models: two-way alternating tree automata (2APT) [Var98] and graded parity tree automata (GAPT) [KSV02]. We give a polynomial reduction of the emptiness problem for FEAs to that for 2GAPTs, and then show containment in EXPTIME for the 2GAPT emptiness problem by a reduction to the emptiness of graded nondeterministic parity tree automata (GNPT) as introduced in [KSV02].

Due to space limitations, most of the proofs are omitted. The interested reader can find them in the accompanying technical report [BL+06].

2 Preliminaries

Let AP, Var, $Prog$, and Nom be finite and pairwise disjoint sets of *atomic propositions, propositional variables, atomic programs*, and *nominals*. A *program* is an atomic program a or its converse a^-. The set of *formulas of the fully enriched μ-calculus* is the smallest set such that *(i)* **true** and **false** are formulas; *(ii)* p and $\neg p$, for $p \in AP \cup Nom$, are formulas; *(iii)* $x \in Var$ is a formula; *(iv)* if φ_1 and φ_2 are formulas, α is a program, n is a non-negative integer, and y is a propositional variable, then the following are also formulas: $\varphi_1 \vee \varphi_2$, $\varphi_1 \wedge \varphi_2$, $\langle n, \alpha \rangle \varphi_1$, $[n, \alpha]\varphi_1$, $\mu y.\varphi_1(y)$, and $\nu y.\varphi_1(y)$. Observe that we use positive normal form, i.e., negation is applied only to atomic propositions.

We call μ and ν *fixpoint operators* and use λ to denote a fixpoint operator μ or ν. A propositional variable y occurs *free* in a formula if it is not in the scope of a fixpoint operator, and *bounded* otherwise. A *sentence* is a formula that contains no free variables. For a formula $\lambda y.\varphi(y)$, we write $\varphi(\lambda y.\varphi(y))$ to denote the formula that is obtained by one-step unfolding, i.e. replacing each free occurrence of y in φ with $\lambda y.\varphi(y)$. We refer often to the *graded modalities* $\langle n, \alpha \rangle \varphi_1$ and $[n, \alpha]\varphi_1$ as *atleast formulas* and *allbut formulas* and assume that the integers in these operators are given in binary coding: the contribution of n to the length of the formulas $\langle n, \alpha \rangle \varphi$ and $[n, \alpha]\varphi$ is $\lceil \log n \rceil$ rather than n. We refer to fragments of the fully enriched μ-calculus using the names from Fig. 1.

The semantics of the fully enriched μ-calculus is defined with respect to a *Kripke structure*, i.e., a tuple $K = \langle W, R, L \rangle$ where W is a non-empty set of *states*, $R : Prog \rightarrow 2^{W \times W}$ assigns to each atomic program a transition relation over W, and $L : AP \cup Nom \rightarrow 2^W$ assigns to each atomic proposition and nominal a set of states such that the sets assigned to nominals are singletons. To deal with inverse programs, we extend R as follows: for each $a \in Prog$, set $R(a^-) = \{(v, u) : (u, v) \in R(a)\}$. If $(w, w') \in R(\alpha)$, we say that w' is an α *successor* of w. Informally, an *atleast* formula $\langle n, \alpha \rangle \varphi$ holds at a state w of a Kripke structure K if φ holds at least in $n+1$ α successors of w. Dually, the *allbut* formula $[n, \alpha]\varphi$ holds in a state w of a Kripke structure K

if φ holds in all but at most n α successors of w. Note that $\neg\langle n,\alpha\rangle\varphi$ is equivalent to $[n,\alpha]\neg\varphi$, and that the modalities $\langle\alpha\rangle\varphi$ and $[\alpha]\varphi$ of the standard μ-calculus can be expressed as $\langle 0,\alpha\rangle\varphi$ and $[0,\alpha]\varphi$, respectively.

To formalize semantics, we introduce valuations. Given a Kripke structure $K = \langle W,R,L\rangle$ and a set $\{y_1,\ldots,y_n\}$ of variables in Var, a *valuation* $\mathcal{V} : \{y_1,\ldots,y_n\} \to 2^W$ is an assignment of subsets of W to the variables $y_1,\ldots,y_n$. For a valuation $\mathcal{V}$, a variable y, and a set $W' \subseteq W$, we denote by $\mathcal{V}[y \leftarrow W']$ the valuation obtained from $\mathcal{V}$ by assigning W' to y. A formula φ with free variables among $y_1,\ldots,y_n$ is interpreted over the structure K as a mapping φ^K from valuations to 2^W, i.e., $\varphi^K(\mathcal{V})$ denotes the set of points that satisfy φ under valuation $\mathcal{V}$. The mapping φ^K is defined inductively as follows:

- $\mathbf{true}^K(\mathcal{V}) = W$ and $\mathbf{false}^K(\mathcal{V}) = \emptyset$;
- for $p \in AP \cup Nom$, we have $p^K(\mathcal{V}) = L(p)$ and $(\neg p)^K(\mathcal{V}) = W \setminus L(p)$;
- for $y \in Var$, we have $y^K(\mathcal{V}) = \mathcal{V}(y)$;
- $(\varphi_1 \wedge \varphi_2)^K(\mathcal{V}) = \varphi_1^K(\mathcal{V}) \cap \varphi_2^K(\mathcal{V})$ and $(\varphi_1 \vee \varphi_2)^K(\mathcal{V}) = \varphi_1^K(\mathcal{V}) \cup \varphi_2^K(\mathcal{V})$;
- $(\langle n,\alpha\rangle\varphi)^K(\mathcal{V}) = \{w : |\{w' \in W : (w,w') \in R(\alpha) \text{ and } w' \in \varphi^K(\mathcal{V})\}| \geq n+1\}$;
- $([n,\alpha]\varphi)^K(\mathcal{V}) = \{w : |\{w' \in W : (w,w') \in R(\alpha) \text{ and } w' \notin \varphi^K(\mathcal{V})\}| \leq n\}$;
- $(\mu y.\varphi(y))^k(\mathcal{V}) = \bigcap\{W' \subseteq W : \varphi^K([y \leftarrow W']) \subseteq W'\}$;
- $(\nu y.\varphi(y))^k(\mathcal{V}) = \bigcup\{W' \subseteq W : W' \subseteq \varphi^K([y \leftarrow W'])\}$.

Let $K = \langle W,R,L\rangle$ be a Kripke structure and φ a sentence. For a state $w \in W$, we say that φ *holds* at w in K, denoted $K,w \models \varphi$, if $w \in \varphi^K$. K is a *model* of φ if there is a $w \in W$ such that $K,w \models \varphi$. Finally, φ is *satisfiable* if it has a model.

In the remainder of this section, we show that the full graded μ-calculus has a tree model property, and that the hybrid graded μ-calculus has a forest model property. A *forest* is a set $F \subseteq \mathbb{N}^+$ such that if $x{\cdot}c \in F$ where $x \in \mathbb{N}^+$ and $c \in \mathbb{N}$, then also $x \in F$. The elements of F are called *nodes*, and the strings consisting of a single natural number are the *roots* of F. For each root $r \in F$, the set $T = \{r \cdot x \mid x \in \mathbb{N}^* \text{ and } r \cdot x \in F\}$ is a *tree* of F (the tree *rooted in* r). For every $x \in F$, the nodes $x \cdot c \in F$ where $c \in \mathbb{N}$ are the *successors* of x, and x is their *predecessor*. The number of successors of x is called the *degree* of x, and is denoted by $deg(x)$. The degree of a forest is the maximum of the degrees of a node in the forest and the number of roots.

We call a Kripke structure $K = \langle W,R,L\rangle$ a *forest structure* if (*i*) W is a forest, (*ii*) $(w,v) \in \bigcup_{a\in Prog} R(a)$ iff $(w,v) \in W^2$ and w is either a predecessor or a successor of v, and (*iii*) $R(\alpha) \cap R(\beta) = \emptyset$ for all $\alpha,\beta \in Prog \cup \{a^- \mid a \in Prog\}$ with $\alpha \neq \beta$. K is *directed* if $(w,v) \in \bigcup_{a\in Prog} R(a)$ implies that v is a successor of w. If W consists of a single tree then we call K a *tree structure*.

We call $K = \langle W,R,L\rangle$ a *quasi forest structure* if $\langle W,R',L\rangle$ is a forest structure, where $R'(a) = R(a) \setminus (W \times \mathbb{N})$ for all $a \in Prog$ (i.e., K becomes a forest structure after deleting all the edges entering a root of W). K is *directed* if $\langle W,R',L\rangle$ is. The *degree* of K is the degree of W. Note that forest and tree structures are quasi forest structures. A *forest model* (resp. *tree model*, *quasi forest model*) of φ is a forest (resp. tree, quasi forest) structure $K = \langle W,R,L\rangle$ such that φ and the nominals in φ hold at some (not necessarily different) roots of W. In what follows, a formula φ *counts* up to b if the maximal integer in atleast and allbut restrictions used in φ is $b-1$.

Theorem 1. *Let φ be a sentence of the full graded μ-calculus such that φ has ℓ* atleast *subsentences and counts up to b. If φ is satisfiable, then φ has a tree model whose degree is at most $\ell(b+1)$.*

In contrast to the full graded μ-calculus, the hybrid graded μ-calculus does not enjoy the tree model property. This is for example witnessed by the formula

$$o \wedge \langle 0, a\rangle(p_1 \wedge \langle 0, a\rangle(p_2 \wedge \cdots \langle 0, a\rangle(p_{n-1} \wedge \langle 0, a\rangle o) \cdots))$$

which generates a cycle of length at most n if the atomic propositions are enforced to be mutually disjoint. However, we can follow [SV01] to show that every satisfiable formula of the hybrid graded μ-calculus has a quasi forest model.

Theorem 2. *Let φ be a sentence of the hybrid graded μ-calculus such that φ has k nominals, ℓ* atleast *subsentences and counts up to b. If φ is satisfiable, then φ has a directed quasi forest model K whose degree is at most* $\max\{k+1, \ell(b+1)\}$.

3 Enriched Automata

Nondeterministic automata on infinite trees are a variation of nondeterministic automata on finite and infinite words, see [Tho90] for an introduction. *Alternating automata*, as first introduced in [MS87], are a generalization of nondeterministic automata. Intuitively, while a nondeterministic automaton that visits a node x of the input tree sends one copy of itself to each of the successors of x, an alternating automaton can send several copies of itself to the same successor. In the two-way paradigm [Var98], an automaton can send a copy of itself to its predecessor, too. In graded automata [KSV02], the automaton can specify a number n of successors to which copies of itself are sent, without specifying which successors these exactly are. The fully enriched automata that we are introducing in the next subsection work on infinite forests, include all of the above features, and additionally have the ability to send a copy of themselves to the roots of the forest.

3.1 Fully Enriched Automata

We start with some preliminaries. Let $F \subseteq \mathbb{N}^+$ be a forest and x a node in F. As a convention, we take $x \cdot \varepsilon = x$, $(x \cdot c) \cdot -1 = x$, and $\varepsilon \cdot -1$ as undefined. We call x a *leaf* if it has no successors. A *path* π in F is a minimal set $\pi \subseteq F$ such that some root r of F is contained in π and for every $x \in \pi$, either x is a leaf or there exists a unique $c \in F$ such that $x \cdot c \in \pi$. Given an alphabet Σ, a Σ-labeled forest is a pair $\langle F, V\rangle$, where F is a forest and $V : F \to \Sigma$ maps each node of F to a letter in Σ.

For a given set Y, let $B^+(Y)$ be the set of positive Boolean formulas over Y (i.e., Boolean formulas built from elements in Y using $\wedge$ and $\vee$), where we also allow the formulas **true** and **false** and $\wedge$ has precedence over $\vee$. For a set $X \subseteq Y$ and a formula $\theta \in B^+(Y)$, we say that X satisfies θ iff assigning **true** to elements in X and assigning false to elements in $Y \setminus X$ makes θ true. For $b > 0$, let $\langle[b]\rangle = \{\langle 0\rangle, \langle 1\rangle, \ldots, \langle b\rangle\}$, $[[b]] = \{[0], [1], \ldots, [b]\}$, and $D_b = \langle[b]\rangle \cup [[b]] \cup \{-1, \varepsilon, \langle root\rangle, [root]\}$.

A fully enriched automaton is an automaton in which the transition function δ maps a state q and a letter σ to a formula in $B^+(D_b \times Q)$. Intuitively, an atom $(\langle n\rangle, q)$ (resp.

$([n], q))$ means that the automaton sends copies in state q to $n+1$ (resp. all but n) different successors of the current node, (ε, q) means that the automaton sends a copy (in state q) to the current node, $(-1, q)$ means that the automaton sends a copy to the predecessor of the current node, and $(\langle root \rangle, q)$ and $([root], q)$ mean that the automaton sends a copy to some, respectively all of the roots of the forest. When, for instance, the automaton is in state q, reads a node x, and

$$\delta(q, V(x)) = (-1, q_1) \wedge ((\langle root \rangle, q_2) \vee ([root], q_3)),$$

it sends a copy in state q_1 to the predecessor and either sends a copy in state q_2 to one of the roots or a copy in state q_3 to all roots.

Formally, a *fully enriched automaton* (FEA, for short) is a tuple $A = \langle \Sigma, b, Q, \delta, q_0, \mathcal{F} \rangle$, where Σ is the input alphabet, $b > 0$ is a counting bound, Q is a finite set of states, $\delta : Q \times \Sigma \rightarrow B^+(D_b \times Q)$ is a transition function, $q_0 \in Q$ is an initial state, and $\mathcal{F}$ is the acceptance condition. A *run* of A on an input Σ-labeled forest $\langle F, V \rangle$ is a tree $\langle T_r, r \rangle$ in which each node is labeled by an element of $F \times Q$. Intuitively, a node in T_r labeled by (x, q) describes a copy of the automaton in state q that reads the node x of F. Runs start in the initial state and satisfy the transition relation. Thus, a run $\langle T_r, r \rangle$ with root z has to satisfy the following: (*i*) $r(z) = (c, q_0)$ for some root c of F and (*ii*) for all $y \in T_r$ with $r(y) = (x, q)$ and $\delta(q, V(x)) = \theta$, there is a (possibly empty) set $S \subseteq D_b \times Q$, such that S satisfies θ, and for all $(d, s) \in S$, the following hold:

- If $d \in \{-1, \varepsilon\}$, then $x \cdot d$ is defined and there is $j \in \mathbb{N}$ such that $y \cdot j \in T_r$ and $r(y \cdot j) = (x \cdot d, s)$;
- If $d = \langle n \rangle$, then there are distinct $i_1, \ldots, i_{n+1} \in \mathbb{N}$ such that for all $1 \leq j \leq n+1$, there is $j' \in \mathbb{N}$ such that $y \cdot j' \in T_r$, $x \cdot i_j \in F$, and $r(y \cdot j') = (x \cdot i_j, s)$;
- If $d = [n]$, then there are distinct $i_1 \ldots, i_{deg(x)-n} \in \mathbb{N}$ such that for all $1 \leq j \leq deg(x) - n$, there is $j' \in \mathbb{N}$ such that $y \cdot j' \in T_r$, $x \cdot i_j \in F$, and $r(y \cdot j') = (x \cdot i_j, s)$;
- If $d = \langle root \rangle$, then for some root $c \in F$ and some $j \in \mathbb{N}$ such that $y \cdot j \in T_r$, it holds that $r(y \cdot j) = (c, s)$;
- If $d = [root]$, then for all roots $c \in F$ there exists $j \in \mathbb{N}$ such that $y \cdot j \in T_r$ and $r(y \cdot j) = (c, s)$.

A run $\langle T_r, r \rangle$ is *accepting* if all its infinite paths satisfy the acceptance condition. We consider here the *parity acceptance condition*, where $\mathcal{F} = \{F_1, F_2, \ldots, F_k\}$ is such that $F_1 \subseteq F_2 \subseteq \ldots \subseteq F_k = Q$. The number k of sets in $\mathcal{F}$ is called the *index* of the automaton. Given a run $\langle T_r, r \rangle$ and an infinite path $\pi \subseteq T_r$, let $Inf(\pi) \subseteq Q$ be such that $q \in Inf(\pi)$ iff there are infinitely many $y \in \pi$ for which $r(y) \in F \times \{q\}$. A path π *satisfies* a parity acceptance condition $\mathcal{F} = \{F_1, F_2, \ldots, F_k\}$ iff there is an even i for which $Inf(\pi) \cap F_i \neq \emptyset$ and $Inf(\pi) \cap F_{i-1} = \emptyset$. An automaton *accepts* a forest iff there exists an accepting run of the automaton on the forest. We denote by $\mathcal{L}(\mathcal{A})$ the set of all Σ-labeled forests that A accepts.

The *emptiness problem* for FEAs is to decide, given a FEA A, whether $\mathcal{L}(\mathcal{A}) = \emptyset$. To decide this problem, we first reduce it to the emptiness problem of a more restricted automata model: a *two-way graded alternating parity tree automaton (2GAPT)* is a FEA that accepts trees (instead of forests) and cannot jump to the root of the input tree, i.e., it does not support directions $\langle root \rangle$ and $[root]$ in the transition relation. For each

FEA A, there exists a 2GAPT A' that accepts a tree encoding of A's language. A Σ-labeled forest $\langle F, V\rangle$ is *encoded* by a $\Sigma \cup \{root\}$-labeled tree $\langle T, V'\rangle$ with root $z \notin F$ iff $root \notin \Sigma$, $T = \{z\} \cup \{z \cdot c \mid c \in F\}$, and V' satisfies:

- $V'(z) = \{root\}$,
- $V'(z \cdot x) = V(x)$ for all $x \in F$.

Then, we can prove the following.

Theorem 3. *Let A be a FEA running on Σ-labeled forests with n states, index k and counting bound b. There exists a 2GAPT A' running on $\Sigma \cup \{root\}$-labeled trees ($root \notin \Sigma$) with $3n+1$ states, index k, and counting bound b such that A' accepts a labeled tree $\langle T, V\rangle$ iff A accepts the forest encoded by $\langle T, V\rangle$.*

3.2 Graded Nondeterministic Parity Tree Automata

To decide the emptiness problem of 2GAPTs, we use a reduction to the emptiness problem of graded nondeterministic parity tree automata as introduced in [KSV02]. In the following, we define these automata and state some results concerning them.

For an integer b, a *b-bound* is a pair in $B_b = \{(>, 0), (\leq, 0), (>, 1), (\leq, 1), \ldots, (>, b), (\leq, b)\}$. For a set Y, we use $B(Y)$ to denote the set of all Boolean formulas over atoms in Y. Each formula $\theta \in B(Y)$ induces a set $sat(\theta) \subseteq 2^Y$ such that $x \in sat(\theta)$ iff x satisfies θ. For an integer $b \geq 0$, a *b-counting constraint* for 2^Y is a relation $C \subseteq B(Y) \times B_b$. A tuple $t = \langle x_1, \ldots, x_m\rangle \in (2^Y)^m$ satisfies the b-counting constraint C if for all $\langle \theta, \xi\rangle \in C$, the tuple t satisfies ξ with respect to $sat(\theta)$, that is, when θ is paired with $(>, n)$, at least $n+1$ elements of t should satisfy θ, and when θ is paired with $(\leq, n)$, at most n elements in the tuple satisfy θ. We use $\mathcal{C}(Y, b)$ to denote the set of all b-counting constraints for 2^Y.

A *graded nondeterministic parity tree automaton* (GNPT, for short) is a tuple $A = \langle \Sigma, b, Q, \delta, q_0, \mathcal{F}\rangle$ where Σ, b, q_0, and $\mathcal{F}$ are as in 2GAPT, $Q \subseteq 2^Y$ is the set of states (i.e., Q is encoded by a finite set of variables), and $\delta : Q \times \Sigma \to \mathcal{C}(Y, b)$ maps a state and a letter to a b-counting constraint for 2^Y. Given a GNPT A, a *run* of A on a Σ-labeled tree $\langle T, V\rangle$ rooted in z is a Q-labeled tree $\langle T, r\rangle$ such that $r(z) = q_0$ and for every $x \in T$, the tuple $\langle r(x \cdot 1), \ldots, r(x \cdot deg(x))\rangle$ satisfies $\delta(r(x), V(x))$. The run $\langle T, r\rangle$ is *accepting* if all its infinite paths satisfy the parity acceptance condition.

We need two special cases of GNPT: FORALL automata and SAFETY automata. In FORALL automata, for each $q \in Q$ and $\sigma \in \Sigma$ there is $s \in Q$ such that $\delta(q, \sigma) = \{\langle(\neg\theta_s), (\leq, 0)\rangle\}$, where $\theta_s \in B(Y)$ is such that $sat(\theta_s) = \{s\}$. Thus, a FORALL automaton is a notational variant of a deterministic tree automaton, where the transition function maps q and σ to $\langle s, \ldots, s\rangle$. In SAFETY automata, there is no acceptance condition, and all runs are accepting. Note that this does not mean that SAFETY automata accept all trees, as it may be that on some trees the automaton does not have a run. We will need the following results concerning GNPTs.

Lemma 1. [KSV02] *Given a FORALL GNPT A_1 with n_1 states and index k, and a SAFETY GNPT A_2 with n_2 states and counting bound b, we can define a GNPT A with $n_1 n_2$ states, index k, and counting bound b, such that $\mathcal{L}(A) = \mathcal{L}(A_1) \cap \mathcal{L}(A_2)$.*

Theorem 4. [KSV02] *Given a GNPT $A = \langle \Sigma, b, Q, \delta, q_0, \mathcal{F} \rangle$ with n states, index k, counting bound b, and $|\Sigma| = \ell$, the nonemptiness problem for A can be solved in time* $n^k \ell (b+2)^{O(n(n+2+k \log nk))}$.

4 The Emptiness Problem for 2GAPT

We show that emptiness of the language accepted by a 2GAPT can be decided in EXPTIME. A corresponding lower bound is inherited from alternating tree automata [KVW00].

Let $A = \langle \Sigma, b, Q, \delta, q_0, \mathcal{F} \rangle$ be a 2GAPT. Recall that $D_b = \langle [b] \rangle \cup [[b]] \cup \{-1, \varepsilon\}$ and $\delta : Q \times \Sigma \to B^+(D_b \times Q)$. A *strategy tree* for A is a $2^{Q \times D_b \times Q}$-labeled tree $\langle T, \mathsf{str} \rangle$. Intuitively, the function str (from now on called *strategy*) maps each node of the tree to a set of transitions. For each label $w = \mathsf{str}(x)$, we define $head(w) = \{q : (q, c, q') \in w\}$ as the set of *sources* of w. A strategy tree $\langle T, \mathsf{str} \rangle$ is on a Σ-labeled tree $\langle T, V \rangle$, if $q_0 \in head(\mathsf{str}(root(T)))$ and for each node $x \in T$ and state q, the set $\{(c, q') : (q, c, q') \in \mathsf{str}(x)\}$ satisfies $\delta(q, V(x))$ (where $root(T)$ denotes the root of T). Intuitively, by choosing the atoms that are going to be satisfied for a node x, $\mathsf{str}(x)$ removes the nondeterminism in δ.

A *promise tree* for the automaton A on a Σ-labeled tree $\langle T, V \rangle$ is a $2^{Q \times Q}$-labeled tree $\langle T, \mathsf{pro} \rangle$. Intuitively, in a run that proceeds according to pro (in the following called *promise*), if a node $x \cdot i$ has $(q, q') \in \mathsf{pro}(x \cdot i)$ and the run visits its parent x in state q and proceeds by choosing an atom $(\langle n \rangle, q')$ or $([n], q')$, then $x \cdot i$ is among the successors of x that inherit q'. For each label $w = \mathsf{pro}(x)$, we also define $head(w) = \{q : (q, q') \in w\}$ as the set of *sources* of w.

Consider a 2GAPT A, a Σ-labeled tree $\langle T, V \rangle$, a strategy tree $\langle T, \mathsf{str} \rangle$ and a promise tree $\langle T, \mathsf{pro} \rangle$ for A on $\langle T, V \rangle$. A $(T \times Q)$-labeled tree $\langle T_r, r \rangle$ is *consistent* with str and pro if $\langle T_r, r \rangle$ suggests a possible run of A on $\langle T, V \rangle$ such that whenever the run $\langle T_r, r \rangle$ is in state q as it reads a node $x \in T$, the strategy $\mathsf{str}(x)$ is defined, the run proceeds according to the elements of $\mathsf{str}(x)$ having q as source, and it delivers requirements to each successor $x \cdot j$ according to the elements in $\mathsf{pro}(x \cdot j)$ also having q as source. Formally, $\langle T_r, r \rangle$ is consistent with str and pro iff the following hold:

- $r(root(T_r)) = (root(T), q_0)$;
- for each node y in T_r with $r(y) = (x, q)$, $\mathsf{str}(x)$ is defined and for all $(q, c, q') \in \mathsf{str}(x)$, the following hold:
  - If $c = -1$ or $c = \varepsilon$, then $x \cdot c$ is defined and there is $j \in \mathbb{N}$ such that $y \cdot j \in T_r$ and $r(y \cdot j) = (x \cdot c, q')$;
  - If $c = \langle n \rangle$ or $c = [n]$, then for each $j \in \mathbb{N}$ with $(q, q') \in \mathsf{pro}(x \cdot j)$, there is $j' \in \mathbb{N}$ such that $y \cdot j' \in T_r$ and $r(y \cdot j') = (x \cdot j, q')$.

Note that since the counting constraints in $\mathsf{str}(x)$ may not be satisfied, $\langle T_r, r \rangle$ may not be a legal run.

Consider a strategy tree $\langle T, \mathsf{str} \rangle$ and a promise tree $\langle T, \mathsf{pro} \rangle$ on a Σ-labeled tree $\langle T, V \rangle$. We say that pro *fulfills* str for V if the states promised to be visited by pro satisfy the obligations induced by str as it runs on V. Formally, pro fulfills str for V if for every node $x \in T$, the following hold:

- For every $(q, \langle n \rangle, q') \in \mathsf{str}(x)$, at least $n+1$ successors $x \cdot j$ of x have $(q, q') \in \mathsf{pro}(x \cdot j)$;
- for every $(q, [n], q') \in \mathsf{str}(x)$, at least $deg(x) - n$ successors $x \cdot j$ of x have $(q, q') \in \mathsf{pro}(x \cdot j)$.

Consider a 2GAPT A, a strategy tree $\langle T, \mathsf{str} \rangle$ and promise tree $\langle T, \mathsf{pro} \rangle$ on a Σ-labeled tree $\langle T, V \rangle$. A sequence $(x_0, q_0), (x_1, q_1) \ldots$ is a *trace* induced by str and pro if x_0 is the root of T (notice that q_0 is the initial state of A) and, for each $i \geq 0$, one of the following holds:

- $q_i \notin head(\mathsf{str}(x_i))$ and (x_i, q_i) is the last pair in the trace;
- there is $(q_i, c, q_{i+1}) \in \mathsf{str}(x_i)$ with $c = -1$ or $c = \varepsilon$, $x_i \cdot c$ defined, and $x_{i+1} = x_i \cdot c$;
- $\mathsf{str}(x_i)$ contains $(q_i, \langle n \rangle, q_{i+1})$ or $(q_i, [n], q_{i+1})$, there exists $j \in \mathbb{N}$ with $x_{i+i} = x_i \cdot j$, $x_{i+i} \in T$, and $(q, q') \in \mathsf{pro}(x_{i+1})$.

It is not difficult to see that a sequence of pairs of nodes of T and states of A starting with $(root(T), q_0)$ is a trace induced by a strategy and a promise for A on a Σ-labeled tree $\langle T, V \rangle$ if a run $\langle T_r, r \rangle$ on $\langle T, V \rangle$, which is consistent with both the strategy and the promise, has a path π labeled with the trace. We say that a strategy tree $\langle T, \mathsf{str} \rangle$ and a promise $\langle T, \mathsf{pro} \rangle$ are *good* for $\langle T, V \rangle$ if all the infinite traces induced by str and pro satisfy the acceptance condition $\mathcal{F}$. In [KSV02] it has been shown that a necessary and sufficient condition for a tree to be accepted by a one-way GAPT is to have a strategy tree and a promise tree good for the input tree, with the promise fulfilling the strategy. We establish the same result with respect to the notions of strategy tree and promise tree as introduced above for 2GAPTs.

Theorem 5. *A 2GAPT A accepts $\langle T, V \rangle$ iff there exist a strategy tree $\langle T, \mathsf{str} \rangle$ and a promise tree $\langle T, \mathsf{pro} \rangle$ good for $\langle T, V \rangle$ such that* pro *fulfills* str *for V.*

Strategy and promise trees allow us to define a notion of a run for alternating automata that has the same tree structure as the underlying input tree, unlike the run $\langle T_r, r \rangle$. Since we want to translate 2GAPT into GNPT, we still have the problem that paths in a run can go both up and down. To restrict our attention to unidirectional paths, we extend to our setting the notion of annotation as defined in [Var98]. Annotations allow decomposing a path of a run into a downward path and several finite paths (*detour*) that come back to their origin (possibly in a loop).

Let $A = \langle \Sigma, b, Q, \delta, q_0, \mathcal{F} \rangle$ be a 2GAPT with $\mathcal{F} = \{F_1, \ldots, F_k\}$. Recall that $D_b = \langle [b] \rangle \cup [[b]] \cup \{-1, \varepsilon\}$. For each state $q \in Q$, let $index(q)$ be the minimal i such that $q \in F_i$. Consider a strategy tree $\langle T, \mathsf{str} \rangle$ and a promise tree $\langle T, \mathsf{pro} \rangle$ for A on a Σ-labeled tree $\langle T, V \rangle$, an *annotation tree* for A on $\langle T, \mathsf{str} \rangle$ and $\langle T, \mathsf{pro} \rangle$ is $\langle T, \mathsf{ann} \rangle$ where the *annotation* ann is a mapping $\mathsf{ann} : T \rightarrow 2^{Q \times \{1, \ldots, k\} \times Q}$ such that for every node $x \in T$ the following conditions hold:

- If $(q, \varepsilon, q') \in \mathsf{str}(x)$ then $(q, index(q'), q') \in \mathsf{ann}(x)$;
- if $(q, j', q') \in \mathsf{ann}(x)$ and $(q', j'', q'') \in \mathsf{ann}(x)$, then $(q, min(j', j''), q'') \in \mathsf{ann}(x)$;
- if $x = y \cdot i$, $(q, -1, q') \in \mathsf{str}(x)$, $(q', j, q'') \in \mathsf{ann}(y)$, $\mathsf{str}(y)$ contains $(q'', \langle n \rangle, q''')$ or $(q'', [n], q''')$, and $(q'', q''') \in \mathsf{pro}(x)$, then $(q, min(index(q'), j, index(q''')), q''') \in \mathsf{ann}(x)$;

– if $y = x \cdot i$, $\mathsf{str}(x)$ contains $(q, \langle n \rangle, q')$ or $(q, [n], q')$, $(q, q') \in \mathsf{pro}(y)$, $(q', j, q'') \in \mathsf{ann}(y)$, and $(q'', -1, q''') \in \mathsf{str}(y)$, then $(q, min(index(q'), j, index(q''')), q''') \in \mathsf{ann}(x)$.

Given an annotation tree $\langle T, \mathsf{ann} \rangle$ for A on $\langle T, \mathsf{str} \rangle$ and $\langle T, \mathsf{pro} \rangle$, a *downward path* π induced by str, pro, and ann is a sequence $(x_0, q_0, t_0), (x_1, q_1, t_1), \ldots$ of triples, where $x_0 = root(T)$, q_0 is the initial state of A, and for each i, x_i is in T, q_i is in Q, and t_i is either an element of $\mathsf{str}(x_i)$ or $\mathsf{ann}(x_i)$, such that: (*i*) either t_i is (q_i, c, q_{i+1}) for some $c \in [[b]] \cup [\langle b \rangle]$, $(q_i, q_{i+1}) \in \mathsf{pro}(x_i \cdot d)$ for some $d \in \mathbb{N}$, and $x_{i+1} = x_i \cdot d$; or (*ii*) t_i is (q_i, d, q_{i+1}), for $d \in \{1, \ldots, k\}$, and $x_{i+1} = x_i$. In the first case, we consider $index(t_i)$ as the minimal j such that $q_{i+1} \in F_j$ and, in the second case, $index(t_i) = d$. Moreover, for a downward path π, we consider $index(\pi)$ as the minimal index $index(t_i)$ for all t_i occurring infinitely often in π. We say that a downward path π violates $\mathcal{F}$ if $index(\pi)$ is odd. Given an annotation tree $\langle T, \mathsf{ann} \rangle$ for A on $\langle T, \mathsf{str} \rangle$ and $\langle T, \mathsf{pro} \rangle$, we say that ann is *accepting* if there is no downward path induced by str, pro, and ann that violates $\mathcal{F}$. Notice that a downward path π can also end in a loop where the last t_i is given by ann and π is accepting if $index(t_i)$ is even.

Theorem 6. *A 2GAPT A accepts $\langle T, V \rangle$ iff there exist a strategy tree $\langle T, \mathsf{str} \rangle$ and a promise tree $\langle T, \mathsf{pro} \rangle$ on $\langle T, V \rangle$, and an annotation tree $\langle T, \mathsf{ann} \rangle$ on $\langle T, \mathsf{str} \rangle$ and $\langle T, \mathsf{pro} \rangle$ such that* pro *fulfills* str *for V and* ann *is accepting.*

In the following, we combine the input tree, the strategy, the promise, and the annotation into one tree $\langle T, (V, \mathsf{str}, \mathsf{pro}, \mathsf{ann}) \rangle$. Given a signature Σ for the input tree, let Σ' denote the extended signature for the combined trees, i.e., $\Sigma' = \Sigma \times 2^{Q \times D_b \times Q} \times 2^{Q \times Q} \times 2^{Q \times \{1, \ldots k\} \times Q}$.

Theorem 7. *Let A be a 2GAPT running on Σ-labeled trees with n states, index k and counting bound b. There exists a GNPT A' running on Σ'-labeled trees with $2^{n(2+k \log nk)}$ states, index nk, and b-counting constraints such that A' accepts a tree iff A accepts its projection on Σ.*

5 ExpTime Upper Bounds for Enriched μ-Calculi

We establish ExpTime upper bounds for satisfiability in the full graded μ-calculus and the hybrid graded μ-calculus. p For the full graded μ-calculus, we give a polynomial translation of formulas φ into a 2GAPT A_φ that, roughly speaking, accepts the tree models of φ. By Theorem 1, we can thus decide satisfiability of φ by checking non-emptiness of $L(\mathcal{L}(\mathcal{A}_\varphi)$. There is a minor technical difficulty to be overcome: Kripke structures have labeled edges, while the trees accepted by 2GAPTs do not. This problem can be dealt with by moving the label from each edge to the target node of the edge. For this purpose, we introduce a new propositional symbol p_α for each program α. Let the *tree encoding* of a tree structure $K = \langle W, R, L \rangle$ be the labeled tree $\langle W, L^* \rangle$ such that $L^*(w) = L(w) \cup \{p_\alpha \mid \exists (v, w) \in R(\alpha)$ with w successor of v in $W\}$.

Theorem 8. *Given a sentence φ of the full graded μ-calculus that has ℓ atleast sub-sentences and counts up to b, we can construct a 2GAPT A_φ such that A_φ*

– *accepts exactly the tree encodings of tree models of φ with degree at most $\ell(b+1)$,*
– *has $|\varphi|$ states, index $|\varphi|$, and counting bound b.*

In the case of the hybrid graded μ-calculus, two additional difficulties have to be addressed. First, FEAs accept forests while the hybrid μ-calculus has only a *quasi* forest model property. This problem can be solved by introducing in node labels new propositional symbols $\uparrow_o^\alpha$ (not occurring in the input formula) that represent an α-labeled edge from the current node to the (unique) root node labeled by nominal o. Second, we have to take care of the interaction between graded modalities and the implicit edges encoded via propositions $\uparrow_o^a$. To this end, we need to know the following information before constructing the FEA: which "relevant" formulas are satisfied by each nominal and which nominals are equivalent. This information is provided by a guess, which we define as follows. The *closure* $cl(\varphi)$ of a sentence φ of the full graded μ-calculus is the smallest set of sentences satisfying the following:

– $\varphi \in cl(\varphi)$;
– if $\psi_1 \wedge \psi_2 \in cl(\varphi)$ or $\psi_1 \vee \psi_2 \in cl(\varphi)$, then $\{\psi_1, \psi_2\} \subseteq cl(\varphi)$;
– if $\langle n, \alpha\rangle\psi \in cl(\varphi)$ or $[n, \alpha]\psi \in cl(\varphi)$, then $\psi, \psi \wedge p_\alpha \in cl(\varphi)$;
– if $\lambda x.\psi(x) \in cl(\varphi)$, then $\psi(\lambda x.\psi(x)) \in cl(\varphi)$.
– if $\psi \in cl(\varphi)$, then $\neg\psi \in cl(\varphi)$, where $\neg\psi$ denotes the formula obtained from ψ by dualizing all operators and replacing every literal (i.e., atomic proposition or negation thereof) with its negation.

For a sentence φ, we use $|\varphi|$ to denote the *length* of φ with numbers inside graded modalities coded in binary. Formally, $|\varphi|$ is defined by induction on the structure of φ in a standard way, with $|\langle n, \alpha\rangle\psi| = \lceil log\ n\rceil + 1 + |\psi|$, and similarly for $|[n, \alpha]\psi|$. As proved in [Koz83], for every sentence φ, the number of elements in $cl(\varphi)$ is linear in the length φ.

A *guess* for φ is a pair $(t, \sim)$ where t assigns a subset $t(o) \subseteq cl(\varphi)$ to each $o \in Nom$, and $\sim$ is an equivalence relation on the set of nominals occurring in φ such that the following conditions are satisfied, for all formulas $\psi \in cl(\varphi)$ and nominals o, o' occurring in φ: *(i)* $\psi \in t(o)$ or $\neg\psi \in t(o)$, *(ii)* $o \in t(o)$, and *(iii)* $o \sim o'$ implies $t(o) = t(o')$. We construct a separate FEA $A_{\varphi,G}$ for each guess G for φ. Since the number of guesses is exponential in the length of φ, we get an EXPTIME decision procedure by constructing all of the FEAs and checking whether some of them accept a nonempty language. *Forest encodings* of forest models are defined similar to tree encodings of tree models with the additional property that $\uparrow_o^\alpha \in L^*(w)$ iff there exists $(w, v) \in R(\alpha)$ such that v is a root of W and $o \in L^*(v)$.

Theorem 9. *Given a sentence φ of the hybrid graded μ-calculus that has ℓ* atleast *subsentences, counts up to b, contains k nominals, and a guess $G = (t, \sim)$ for φ, we can construct a FEA $A_{\varphi,G}$ such that $A_{\varphi,G}$*

– *accepts exactly the forest encodings of the quasi forest models of φ having degree at most* $\max\{k+1, \ell(b+1)\}$*, and*
– *has $\mathcal{O}(|\varphi|^2)$ states, index $|\varphi|$, and counting bound b.*

Given a sentence of the full graded μ-calculus with ℓ at-least subformulas, we get by Theorems 7 and 8 a GNPT A_φ with the number of states n and index k bounded by $|\varphi|$, and $|\Sigma|$ and the counting bound b bounded by $2^{|\varphi|}$. While the latter are exponential in

$|\varphi|$, only n and k appear in the exponents in the expression in Theorem 4. This yields the desired EXPTIME upper bound. The lower bound is due to the fact that the μ-calculus is EXPTIME-hard [FL79]. For the hybrid graded μ-calculus, we can argue similarly using Theorems 3, 7, and 9.

Theorem 10. *The satisfiability problems of the full graded μ-calculus and the hybrid graded μ-calculus are* EXPTIME*-complete even if the numbers in the graded modalities are coded in binary.*

References

[BHS02] F. Baader, I. Horrocks, and U. Sattler. Description logics for the semantic web. *KI – Künstliche Intelligenz*, 3, 2002.

[BM+03] F. Baader, D.L. McGuiness, D. Nardi, and P. Patel-Schneider. *The Description Logic Handbook: Theory, implementation and applications.* Cambridge Univ. Press, 2003.

[BC96] G. Bhat and R. Cleaveland. Efficient local model-checking for fragments of the modal mu-calculus. In *Proc. of TACAS'96*, LNCS 1055, pages 107-126, 1996.

[BL+06] P.A. Bonatti, C. Lutz, A. Murano and M.Y. Vardi. The Complexity of Enriched μ-calculi. Chair for Automata Theory, Institute for Theoretical Computer Science, Dresden University of Technology, 2006, LTCS-Report, LTCS-06-02, Germany, see http://lat.inf.tu-dresden.de/research/reports.html.

[BP04] P.A. Bonatti and A. Peron. On the undecidability of logics with converse, nominals, recursion and counting. *Artificial Intelligence*, 158(1):75-96, 2004.

[CGL01] D. Calvanese, G. De Giacomo, and M. Lenzerini. Reasoning in expressive description logics with fixpoints based on automata on infinite trees. In *Proc. of the 16th Int. Joint Conf. on Artificial Intelligence (IJCAI'99)*, pages 84-89, 1999.

[FL79] M.J. Fischer and R.E. Ladner. Propositional dynamic logic of regular programs. *Journal of Computer and Systems Sciences*, Vol.18, pages 194-211, 1979.

[Jut95] C.S. Jutla. Determinization and memoryless winning strategies. *Information and Computation*, 133(2):117–134, 1997.

[Koz83] D. Kozen. Results on the propositional μ-calculus. *Theoretical Computer Science*, Vol.27, pages 333-354, 1983.

[KSV02] O. Kupferman, U. Sattler, and M.Y. Vardi. The complexity of the Graded μ-calculus. In *Proc. of the 18th CADE* LNAI 2392, pages 423-437, 2002.

[KVW00] O. Kupferman, M.Y. Vardi, and P. Wolper. An automata-theoretic approach to branching-time model checking. *Journal of the ACM*, Vol.47(2), pages 312-360, 2000.

[MS87] D.E. Muller and P.E. Schupp. Alternating automata on infinite trees. *Theoretical Computer Science*, Vol.54, pages 267–276, 1987.

[Saf89] S. Safra. Complexity of automata on infinite objects. *PhD thesis*, Weizmann Institute of Science, Rehovot, Israel, 1989.

[SV01] U. Sattler and M. Y. Vardi. The hybrid mu-calculus. In *Proc. of IJCAR'01*, Vol.2083 of LNAI, pages 76-91. Springer Verlag, 2001.

[Tho90] W. Thomas. Automata on Infinite Objects. In *Handbook of Theoretical Computer Science*, pages 133 – 191, 1990.

[Tho97] W. Thomas. Languages, automata, and logic. In *Handbook of Formal Language Theory*, volume III, pages 389-455, G. Rozenberg and A. Salomaa editors, 1997.

[Var98] M.Y. Vardi. Reasoning about the Past with Two-Way Automata. In *Proc. of ICALP'98*, LNCS 1443, pages. 628–641, 1998.

Interpreting Tree-to-Tree Queries

Michael Benedikt[1] and Christoph Koch[2]

[1] Bell Laboratories
[2] Saarland University

Abstract. We establish correspondences between top-down tree building query languages and predicate logics. We consider the expressive power of the query language XQ, a clean core of the practitioner's language XQuery. We show that all queries in XQ with only atomic equality are equivalent to "first-order interpretations", an analog to first-order logic (FO) in the setting of transformations of tree-structured data. When XQ is considered with deep equality, we find that queries can be translated into FO with counting ($FO(\mathbf{Cnt})$). We establish partial converses to this, characterizing the subset of the FO resp. $FO(\mathbf{Cnt})$ interpretations that correspond to XQ. Finally, we study the expressive power of fragments of XQ and obtain partial characterizations in terms of existential FO and a fragment of FO that is two-variable if the tree node labeling alphabet is assumed fixed.

1 Introduction

The formal foundation for relational query languages is well-established. Much of relational querying can be done in the relational calculus, which is equivalent in expressiveness to first-order logic. What about tree-structured data, as exemplified by XML data trees? For Boolean and nodeset queries on pure node-labeled trees, the querying model is fairly well understood. The predominant practitioner language is XPath, whose core corresponds in expressiveness to the two-variable fragment of first-order logic [8]. However, when we turn to queries that produce trees from trees, utilizing not just a fixed labeled structure but also data values that may lie within an unbounded set, the situation becomes less clear. Logics by themselves do not define mappings from structures to structures, and modal and automata based languages do not deal with the data value structure.

The predominant paradigm for practitioners is *top-down non-recursive tree building*, as exemplified in the standard language XQuery. The key feature of such languages is that the output tree is built top-down in parallel, with threads of control at every leaf of the partially-constructed output. Rather than using a general structural recursion mechanism, these languages use explicit nesting of subqueries to build the output up to some fixed depth. In our paper we formalize this model by the query language XQ, an abstraction of XQuery that captures the main tree-formation constructs. We wish to compare the expressiveness of XQ and its fragments to some external benchmark.

We choose here to use logics as a benchmark. We rely on the well-known notion of *first-order interpretation* [1], which associates a collection of first-order

M. Bugliesi et al. (Eds.): ICALP 2006, Part II, LNCS 4052, pp. 552–564, 2006.

formulas with a mapping from structures to structures. Informally, a tree-to-tree transformation T is given by a first-order interpretation if there are a set of relational calculus expressions that produce, from a relational coding of an input data tree D, the relational coding of $T(D)$. Thus queries given by first-order interpretations can be thought of as those that can be implemented relationally using relational algebra or relational calculus.

Contributions. We start by showing that XQ queries using only atomic equality can be transformed into equivalent FO interpretations: this is an analog for tree query languages of the simulations for complex object languages in relational algebra [9]. We then trace how the correspondence filters down to fragments of XQ. Although the transformation for general XQ requires exponential time, it can be made polynomial by restricting to XQ queries in a natural class (the "composition-free" XQ queries). We examine queries that do not use node equality comparisons; for queries in this fragment that use only downward navigation, as well as those that are composition-free, one can map to a restricted fragment of FO, one that is "almost" within two-variable logic. We show that the positive fragment of XQ maps to the existential fragment of first-order logic. We establish lower bounds on the complexity of these transformations.

We turn to the question of mapping back from relational logics to XQ. We observe that XQ is *not FO*-complete, since its ability to construct deep trees is very limited. However, we show that for every tree-to-tree query given by a first-order interpretation, there is a corresponding XQ query that agrees with Q on data trees of fixed depth. We give two different varieties of mappings, one transforming into a composition-free query, but using node equality; another mapping into XQ without node-equality, but relying on composition. We show similar mappings from the existential fragment of first order logic to positive XQ, and the two-variable fragment and composition-free XQ without node equality. We show that the transformations in this direction for Boolean queries can be made in polynomial time, as well as the mappings for interpretations within a certain normal form.

We also study the variant of XQ where queries can compare trees for structural equality (we refer to this as "deep equality", in analogy with the corresponding construct in complex value languages). Not every such query can be transformed into a corresponding first-order interpretation. We hence compare the expressiveness of deep equality queries with interpretations given using first-order logic with counting $FO(\mathbf{Cnt})$, which corresponds to a restricted form of relational calculus with aggregation. Queries given by these interpretations can still be implemented relationally in a fragment of SQL. We give a transformation producing from each XQ query with deep equality a corresponding $FO(\mathbf{Cnt})$ interpretation, and extend the complexity bounds on this transformation from the first-order case.

We derive a variety of consequences of the above results. It follows from our results that the expressiveness of XQ with atomic equality for *Boolean* queries is exactly the same as that of relational calculus on a relational coding of the input. A similar result holds for relational queries – those that take a tree coding of a relation to another tree coding of a relation. We can also derive a characterization of the Boolean expressiveness of XQ fragments: in the case of XQ without node equality, we find a correspondence with two-variable logic. The mappings in both directions give a clean way to show *normal forms* for XQ fragments: they allow us to show, for example, that every XQ query can be converted to its composition-free fragment. Our correspondences are also used to show that node equality is essential for removing composition once both upward and downward navigation in trees is allowed. Finally, our results are applied to the *complexity* of XQ. Using our translations, we give new upper bounds on the data complexity of fragments of XQ, as well as simpler proofs of prior bounds.

Taken together our results make precise the connection between fragments of XQ and classical logical languages on relations, like FO, $FO2$, and $FO(\mathbf{Cnt})$.

Note: Due to space considerations this extended abstract omits boths proofs and an explanation of the (considerable) related work. The reader is encouraged to consult the full paper, to be made availability shortly, for these.

2 Background and Notations

2.1 Data and Query Model

An *unranked ordered tree* is a tree in which nodes may have a variable number of children, with an order among them. A *data tree* is a two-sorted structure of signature

$$\sigma_{nav} = (Node, Lab, \mathsf{lab}(), R_{\mathsf{child}}, R_{\mathsf{next\text{-}sibling}}, R_{\mathsf{descendant}}, R_{\mathsf{following\text{-}sibling}}, =_{atomic}).$$

This represents an unranked ordered tree whose nodes are the elements of sort *Node* and whose labels are the elements of sort *Lab* (that is, we do not assume the labeling alphabet to be fixed); the function $\mathsf{lab}()$: $Node \to Lab$ takes a node to its label; R_{child} is the binary parent-child relation among nodes, $R_{\mathsf{next\text{-}sibling}}$ is the binary immediate right-sibling relation among nodes, $R_{\mathsf{descendant}}$ denotes the descendant relation, $R_{\mathsf{following\text{-}sibling}}$ is the transitive closure of the $R_{\mathsf{next\text{-}sibling}}$ relation, and $=_{atomic}$ denotes equality of labels.

A *data forest* is a relational structure of the same signature σ_{nav}, but with the underlying node structure being an ordered forest; i.e. we allow multiple root nodes. Given a node n in a data forest F, the *subtree of* n refers to the σ_{nav}-substructure of F whose domain consists of the nodes that are descendants of n and the values associated with those nodes by the function $\mathsf{lab}()$ in F.

Equality relativized to nodes will be denoted by $=_{node}$. Isomorphism between data forests will be denoted by $=_{deep}$. For a data tree or forest we can also consider the extended structure with signature $\sigma_{nav+deep} = \sigma_{nav} \cup \{=_{deep}\}$.

By a *query* we mean any function from data trees to data trees, and by a *Boolean query* any function from data trees to $\{true, false\}$. Two queries Q, Q' are said to be *equivalent* if $\forall t\ Q(t) =_{deep} Q'(t)$.

Our tree query language XQ has abstract syntax
$query ::= () \mid query\ query \mid var \mid var/axis :: \nu \mid \langle \mathsf{lab}(var) \rangle\ query\ \langle \mathsf{lab}(var) \rangle$
$\mid \langle a \rangle query \langle /a \rangle \mid$ **for** var **in** $query$ **return** $query \mid$ **if** $cond$ **then** $query$
where $cond ::= var =_{node} var \mid var =_{atomic} var \mid var =_{deep} var \mid query$

In this, a denotes a label, *axis* the tree-structure relations child, descendant, ..., *var* a set of variables $\$x$, $\$x_1$, $\$x_2$, ..., $\$y$, $\$z$, ..., and ν a *label test* (either a label or "*"). Thus, our XQ fragment extends that of [5] by a node equality primitive (which returns true on a pair of nodes iff they are the same node), the power to construct new nodes taking labels from arbitrary input nodes, and the fragment used in the expressiveness results of [6] by all tree structure relations ("axes") of XQuery. ([6] only considers "child" and "descendant".) Our semantics is the usual one for XQuery, as currently undergoing standardization [11], restricted to the subset we present here. The reader can also consult [5] for the semantics of our fragment.

The general semantics for XQ queries is as a function mapping a data forest to a data forest. Since we want to talk about the data tree-to-data tree queries defined by an XQ expression, we will consider the tree query defined by an expression Q with a single variable \$inp which is free (i.e. never bound in a **for** or **let**) in Q, to be the function which, given a data tree, returns the tree formed by applying Q to the forest consisting of that tree with \$inp bound to the root of the tree. We also consider an XQ query Q with free variable \$inp to define a Boolean query Q_{Bool}: this is the query that returns true iff the output of Q is a non-empty list. It is easy to show that alternative definitions of Boolean queries (e.g. by restricting to conditions) yield the same set of queries. We consider an arbitrary XQ query to define a "parameterized Boolean query" – one that gives a Boolean value for every assignment of variables to nodes.

Sublanguages of XQ. By $AtomXQ$ we denote XQ where we restrict general deep equality comparisons between variables, and instead allow only conditions $var =_{deep} \langle a/ \rangle$ for labels a. By $PosXQ$, we refer to the subset where deep equality is not permitted at all. By $AtomXQ^-$ we denote the sublanguage of $AtomXQ$ where comparisons $var =_{node} var$ are forbidden, and by $PosXQ^-$ the corresponding sublanguage of $PosXQ$.

In our definition of the syntax of XQ, we have been economical with operators introduced. For example, we use only simple axis expressions from XPath (i.e. the tree navigation structure): the results of this paper imply that we could have as easily included all of Core XPath [3] as a sublanguage without affecting the expressiveness. The constructs **not** ϕ, ϕ **or** ψ, ϕ **and** ψ, **some** $\$x$ **in** α **satisfies** ϕ can easily be derived from these, a fact which we will use freely in the remainder of this section.

Recalling that an XQ expression Q defines a Boolean query Q_{Bool} holding iff Q is nonempty, we can see that the Boolean query corresponding to ϕ **and** ψ evaluates to true for a given binding of the free variables iff the Boolean queries corresponding to ϕ and of ψ evaluate to true; similarly, the other derived operators give Boolean queries with the semantics corresponding to their usual meaning in logic. Let us just note that the restricted form of deep equality of $AtomXQ$ is sufficient to define negation, but this power is missing from $PosXQ$ (thus the name, suggesting "positive" XQ).

2.2 Tree Structures and First-Order Interpretations

We use the usual notation and semantics of first-order logic FO (e.g. [1]). Given a first-order formula ψ with k free variables over signature σ and a σ-structure $\mathcal{A}$, we abbreviate the relation $\{(x_1, \ldots, x_k) \mid \mathcal{A} \models \psi[x_1, \ldots, x_k]\}$ as $\psi(\mathcal{A})$.

A *first-order interpretation of signature* $\sigma = (Node, Lab, \mathsf{lab}(), R_1, \ldots, R_m)$ (where $R_1, \ldots, R_m$ are binary, e.g. $\sigma = \sigma_{nav}$) consists of a set $\mathcal{C}$ of constants of either Lab or $Node$ sort, and a sequence of formulas

$$I = \langle \phi_{Node}(x_1, \ldots, x_k), \phi_{Lab}(z), \phi_{\mathsf{lab}()}(x_1, \ldots, x_k, z), \\ \phi_{R_1}(x_1, \ldots, x_k, y_1, \ldots, y_k), \ldots, \phi_{R_m}(x_1, \ldots, x_k, y_1, \ldots, y_k) \rangle,$$

where each formula ϕ is over $\sigma \cup \mathcal{C}$. The variable z above is of sort Lab while the x_i and y_i are of sort $Node$. The integer k is the *arity* of the interpretation. The $x_1, \ldots, x_k, y_1, \ldots, y_k$ of the formulas ϕ_{R_i} are implicitly relativized to ϕ_{Node} and ϕ_{Lab}.

An interpretation I is associated with the query $[I]$ that maps a σ-structure $\mathcal{A}$ to the σ-structure $(\phi_{Node}(\mathcal{A} \cup \mathcal{C}), \phi_{Lab}(\mathcal{A} \cup \mathcal{C}), \phi_{\mathsf{lab}()}(\mathcal{A} \cup \mathcal{C}), \phi_{R_{\mathsf{child}}}(\mathcal{A} \cup \mathcal{C}), \ldots)$.

Let σ' be a signature extending σ with new sorts for value constants and node constants, where there are no relations or functions on this sort. If $\mathcal{C} = \mathcal{C}_n \cup \mathcal{C}_v$ is a set of node and value constants, respectively, by $\mathcal{A} \cup \mathcal{C}$ we mean the extension of $\mathcal{A}$ to a σ'-structure whose σ-structure is that of $\mathcal{A}$. Above, the set of k-tuples $\phi_{Node}(\mathcal{A} \cup \mathcal{C})$ is identified as the domain of the $Node$ sort of the output structure, the set $\phi_{Lab}(\mathcal{A} \cup \mathcal{C})$ constitutes the domain of the Lab sort, $\phi_{\mathsf{lab}()}$ restricted to ϕ_{Node} is interpreted as the graph of the $\mathsf{lab}()$ function, and the remaining collections of $2k$-tuples are converted to pairs of k-tuples; that is, to binary relations on the elements satisfying ϕ_{Node}, in the obvious way.

Note that we have restricted our interpretations so that the labels of the output can only come from the input labels or the fixed set of constants.

For a vocabulary σ, the logic $FO(\mathbf{Cnt})(\sigma)$ (or $FO(\mathbf{Cnt})$, when σ is understood) is defined over variables of two sorts Dom and $\mathcal{N}$. Atomic formulas include all atomic formulas of σ, restricted to variables of Dom sort, while there are no atomic predicates or functions on variables of sort $\mathcal{N}$ (other than equality). Formulas are closed under Boolean operations and quantifiers $\exists x, \forall x, \exists i, \forall i$ for x of domain sort and i of number sort. We also have the counting quantifiers $\exists^{=i}\boldsymbol{x}, \forall^{=i}\boldsymbol{x}$ where i is of $\mathcal{N}$ sort and $\boldsymbol{x}$ is a tuple of variables. A formula of the form $\exists^{=i}\boldsymbol{x}\ \phi(\boldsymbol{x}, \boldsymbol{y}, \boldsymbol{j})$ has i and $\boldsymbol{y}$ free.

The semantics of $FO(\mathbf{Cnt})$ formulas is given with respect to a σ structure S and an interpretation mapping variables of Dom sort to elements of the domain of S and variables of $\mathcal{N}$ sort to non-negative integers. A formula $\exists^{=i}\boldsymbol{x}\phi(\boldsymbol{x}, \boldsymbol{y}, \boldsymbol{j})$ holds in structure S in an interpretation assigning $\boldsymbol{x}$ to $\mathbf{c}$, i to integer i_0 and $\boldsymbol{j}$ to integers $\boldsymbol{j}_0$ iff $|\{\boldsymbol{d} \in S \mid (S, \mathbf{d}, \mathbf{c}, \boldsymbol{j}_0) \models \phi\}|$ is exactly i_0.

An $FO(\mathbf{Cnt})$ *interpretation* over one of our data tree signatures (σ_{nav} etc.) is defined as with FO. The evaluation complexity of FO and $FO(\mathbf{Cnt})$ interpretations can be read off from classical results on the logics. Given the standard encoding of a σ_{nav} structure as a string [4],

Proposition 1 (cf. [4]). *Every FO (resp., $FO(\mathbf{Cnt})$) interpretation over σ_{nav} can be evaluated in* AC^0 *(resp.,* TC^0*) data complexity and* PSPACE *combined complexity.*

3 From XQ with Atomic Equality to FO

Our first result is the following.

Theorem 1. *There is an* EXPTIME *function that maps every AtomXQ query to an equivalent FO σ_{nav} interpretation.*

The rather involved proof is omitted — it relies on a composition theorem for interpretations and an inductive construction translating XQ expressions with free variables to FO interpretations with parameters.

Our XQ language allows very limited means of controlling the ordering of the output. Since our concern in this work is not with the ordering capabilities of tree structured query languages (e.g. the "order clause" within an XQuery FLWOR expression [11]), we remain within these limitations. The interpretations resulting from XQ will thus be of special form, which we describe below. For a data tree $\mathcal{T}$ and linear ordering $<_{\mathcal{C}}$ on the node constants $\mathcal{C}$, $<_{\mathcal{T},\mathcal{C}}$ is the ordering on $Node(\mathcal{T}) \cup \mathcal{C}$ formed by placing the elements of $\mathcal{C}$ above all elements of $Node(\mathcal{T})$, ordering $Node(\mathcal{T})$ using "document order", and ordering $\mathcal{C}$ using $<_{\mathcal{C}}$. An interpretation $I = \langle \phi_{Node}(x_1 \dots x_k) \dots \rangle$ of arity k is *document-ordered* if there is some ordering $<_{\mathcal{C}}$ on the node constants $\mathcal{C}$ such that: for every data tree $\mathcal{T}$, for any two sibling nodes $n_1 \dots n_k\, n'_1 \dots n'_k$ in $\phi_{Node}(\mathcal{T})$, $n_1 \dots n_k$ comes before $n'_1 \dots n'_k$ in in the sibling order iff $n_1 \dots n_k$ comes before $n'_1 \dots n'_k$ in the lexicographic ordering based on $<_{\mathcal{T},\mathcal{C}}$. An analysis of the proof of Theorem 1 shows the following: *for every AtomXQ expression Q, there is a document-ordered FO interpretation that can be found in* EXPTIME. In document-ordered interpretations, the $R_{\text{next-sibling}}$ and $R_{\text{following-sibling}}$ relations can be inferred from the ordering on constants and the other relations. From now on, when dealing with a document-ordered interpretation we will assume that its logical representation omits the description of the sibling axes.

Fragments of *AtomXQ***.** We now consider sublanguages of *AtomXQ*, with the goal of seeing what subset of FO they map to.

We first consider the case of *PosXQ*. Let $\exists FO$ be the fragment of FO built up from the atomic formulas of σ_{nav} using positive Boolean operators and existential quantification. The translation that witnesses Theorem 1 only introduces negation in two places: comparisons with $=_{deep}$ and the construction of the sibling axis of the output, which is computed using relativized document order For an interpretation in *PosXQ* we can thus translate into a positive FO query.

Proposition 2. *For every PosXQ expression Q that does not use the sibling axes, one can find (in* EXPTIME*) an equivalent document-ordered interpretation I for which all formulas are in $\exists FO$.*

Complexity of translation and composition-free queries. Theorem 1 gives an EXPTIME translation, due to the need to "inline" the same expression multiple times when translating the **let** clause. It is thus not surprising that XQ queries can be exponentially more succinct than first-order queries. For the translation from $PosXQ$, we can show a stronger result:

Theorem 2. *There is no polynomial time function translating AtomXQ queries to FO interpretations. For a $PosXQ^-$ query, there is no PTime translation to $\exists FO$, even for Boolean queries.*

We can avoid the blow-up by restricting queries in the source of the translation to be in a special form. The language *composition-free XQ* is formed by making the following subtractions and additions to XQ:

- removing **let** from the grammar of XQ, and restricting **for** $\$v$ **in** Q **return** Q' constructs so that Q must be of the form $\$v/axis :: \nu$.
- expanding deep equality comparisons so that they may be of the form $Q_1 =_{deep} Q_2$, not just $\$v_1 =_{deep} \$v_2$, and adding the construct **not** Q with semantics given via (**let** $\$v = \langle A\rangle Q\langle/A\rangle$) $\$v =_{deep} \langle A/\rangle$.

The idea is that we do not have the ability to assign a variable to a query result (or to iterate over a query result) — the ability to do this is what allows XQ to reuse subquery results several times, leading to the exponential blow-up. The language *composition-free AtomXQ* is the language formed from $AtomXQ$ by removing **let**, restricting **for**, adding **not** as above, and also adding $\$v =_{atomic} \langle a\rangle$ for a a label. By going more carefully through the proof of Theorem 1, one can show:

Theorem 3. *There is a polynomial time function producing for every composition-free AtomXQ query an equivalent first-order interpretation. For Q a composition-free query in PosXQ, one can find in* PTIME *an interpretation that is equivalent to Q, in which all formulas are expressible in $\exists FO$.*

We shall see later on that composition-free $AtomXQ$ queries have the same expressiveness as general $AtomXQ$ queries, and composition-free XQ queries have the same expressiveness as general XQ queries. However composition-free $AtomXQ^-$ and full $AtomXQ^-$ have different expressiveness in the presence of upward and sideways axes (if only downward axes are supported, the two languages again coincide [6]).

Node equality and $FO2$. We turn now to queries that are both composition-free and without $=_{node}$. We can show that composition-free $AtomXQ^-$ maps into a small fragment of FO. The logic NFO^2, *navigationally two-variable FO*, is built up from only $=_{atomic}$ (not ordinary $=$ on the domain), using Boolean operations and the limited quantifiers $\exists x\ axis\text{-}pred(x, y)\ \phi(x, y)$ and $\forall x\ axis\text{-}pred(x, y)\ \phi(x, y)$ where *axis-pred* is one of $R_{\mathsf{child}}, R_{\mathsf{descendant}}, \ldots$ We now justify the name "two-variable" for this. For a finite set of labels Σ, an $FO^2(\Sigma)$ formula $\phi, \phi(x)$ is a formula in at most one free variable of first-order logic over the signature including the axis predicates, $=_{atomic}$ comparisons with constants, and unary predicates P_a on the *Node* sort for each $a \in \Sigma$, which hold of a node n iff $\mathsf{lab}(n) = a$. By an $FO^2(\Sigma)$ formula $\phi(x_1 \dots x_n)$ in many variables, we mean a formula that is a Boolean combination of formulas $\phi(x_i)$, each of which is in FO^2 above. We say that an FO formula ϕ is *almost two-variable* if for every finite alphabet Σ, the restriction of ϕ to data trees with labels in Σ is in $FO^2(\Sigma)$. It is then easy to see that *every navigationally two-variable formula is an almost two-variable formula.* That is, navigationally two-variable formulas have two-variable expressive power over any fixed set of labels.

The following result shows that composition-free $AtomXQ^-$ queries translate to navigationally two-variable interpretations:

Theorem 4. *For every composition-free $AtomXQ^-$ query, there is an equivalent FO interpretation where every formula is navigationally two-variable, which can be found in* PTIME. *For every $AtomXQ^-$ query that uses only downward axes, there is an equivalent navigationally two-variable query, that can be found in* PTIME. *For every $PosXQ^-$ composition-free Boolean query there is an equivalent interpretation in which all formulas are $\exists FO$ and navigationally two-variable.*

Applications to Expressiveness and Complexity. The following results on the expressiveness of Boolean queries are an immediate consequence of the results above (in the case of $AtomXQ^-$ below, we also use the equivalence of Core XPath and FO^2 shown in [7]).

Corollary 1. *Let Q be an $AtomXQ$ expression, and Q_{Bool} be the Boolean query defined by Q. Then, (1) Q_{Bool} is expressible in the relational calculus; (2) If $Q \in PosXQ$ and Q does not use the sibling axes, Q_{Bool} is expressible by a union of conjunctive queries (over all atomic formulas); (3) If $Q \in AtomXQ^-$ and Q is composition-free then for every finite set of labels Σ, there is a Core XPath query equivalent to Q_{Bool} on data trees with labels in Σ. In particular (by [7, 2]), there are FO queries that are not expressible in composition-free $AtomXQ^-$.*

The first two results could be generalized to "relational queries" — queries on tree encodings of relational tables. The results on translation to FO immediately give alternative proofs of the following upper bounds:

Corollary 2.
- *All $AtomXQ$ queries can be evaluated in data complexity* AC^0 *on a relational representation of the data, and in* EXPSPACE *combined complexity.*
- *$PosXQ$ queries can be evaluated in combined complexity* NEXPTIME.
- *Composition-free $AtomXQ$ is in* PSPACE *w.r.t. combined complexity.*

Note that in all the above complexity bounds, the input is a σ_{nav} structure representing the data tree $\mathcal{T}$; this is not the default assumption of [5]. However, these results also follow immediately from [5] and the fact that there are LOGSPACE translations back and forth between XML trees and σ_{nav}-structures.

4 Back from *FO* to *XQ*

What about the converse of the above results? It is not hard to see that not every tree-to-tree query given by an *FO* interpretation is expressible in *AtomXQ*. Consider, for example, the query Q_{Reverse} taking data trees consisting of a single chain (i.e., each node has at most one child) and producing a chain in the opposite order (the root of the input becoming the sole leaf of the output, and so forth). On data trees that are not a single chain Q_{Reverse} returns some fixed data tree D_0. It is easy to create a first-order interpretation I_{Reverse} that captures Q_{Reverse} up to isomorphism. The fact that Q_{Reverse} is not expressible in XQ is fairly clear (and will be proved below). The intuition is that XQ can only build trees up to fixed depth and deeper than that can only copy from the input. Let us formalize this intuition.

Definition 1. An FO interpretation $I = (\phi_{Node}, \phi_{Lab}, \phi_{R_{\mathsf{child}}} \ldots)$ is *k-shallow*, for integer $k > 0$, if for every node N of depth k in the output, if N is represented as $(n_1, \ldots, n_k)$ in the interpretation, then for some $i <= k$ the subtree of N is $\{(n_1, \ldots, n_{i-1}, m, n_{i+1}, \ldots, n_k) \mid m \text{ descendant of } n_i\}$ and the mapping $m \rightarrow (n_1, \ldots, n_{i-1}, m, n_{i+1}, \ldots, n_k)$ is an isomorphism of the subtree of n_i onto the subtree of N.

The translation given in the proof of Theorem 1 shows:

Proposition 3. *For every AtomXQ query Q, there is some integer k and a first-order interpretation equivalent to Q that is k-shallow (and document-ordered).*

No I'_{Reverse} capturing Q_{Reverse} up to isomorphism can be k-shallow for any k, hence there can be no XQ query Q'_{Reverse} equivalent to I_{Reverse}.

A second obstacle is the issue of orderings – clearly, we can only mimic document-ordered interpretations. The first result of this subsection is that ordering and shallowness are the only barriers to capturing first-order interpretations in XQ.

Theorem 5. *For every k-shallow first-order interpretation I, there is an AtomXQ query Q equivalent to I modulo ordering. For every k-shallow document-ordered FO interpretation I over σ_{nav} there is an AtomXQ query Q equivalent to I. Furthermore, Q can be calculated from I in* PTIME.

The idea of the proof is to show that for every *FO* formula there is an *AtomXQ* query that "verifies" membership of a tuple of nodes from the input data tree in the formula. This is shown by induction on the formula. Given this result, one can mimic the behavior of an *FO* interpretation up to a given level k by proceeding down from the root, level-by-level, checking whether a given tuple should be placed at that point in the tree.

The PTIME translation above does make use of the "let" construct of *AtomXQ*. We can put the result in composition-free *AtomXQ* as well. We do not know how to do this in PTIME in general. In inductively mapping an interpretation with free variables $x_1 \dots x_n$ to a composition-free *AtomXQ* query that produces the subtree underneath a node corresponding to $x_1 \dots x_n$, our algorithm requires a separate query for every substitution of a subset of $x_1 \dots x_n$ for node constants: since there may be exponentially possibilities for variable-to-node-constant mappings, this leads to an EXPTIME algorithm. We do know how to obtain a composition-free result in PTIME if we deal with interpretations of a more restricted syntactic form. An *FO* interpretation is *unwound* if with every subformula $\phi(w_1 \dots w_j)$ of a formula of I, we can associate a typing that asserts whether a variable is equal to some node constant, and if so which constant. Equivalently, we can require our formulas to be built up with node quantifications $\exists x = c\ \phi(x, \boldsymbol{y})$ for each node constant c and $\exists x \bigwedge_{c \in C} x \neq c\ \phi(x, \boldsymbol{y})$, for C the set of all node constants. Then we have:

Theorem 6. *For every AtomXQ query, an unwound k-shallow document-ordered FO interpretation can be found in* EXPTIME, *and for every composition-free AtomXQ query, such an interpretation can be found in* PTIME. *In the other direction, for every k-shallow document-ordered FO interpretation I, there is a composition-free AtomXQ query Q_I equivalent to $[I]$, which can be found in* PTIME *if I is unwound.*

From fragments of *FO* to fragments of *AtomXQ*. The following result shows that shallow $\exists FO$ interpretations likewise correspond to *AtomXQ* queries.

Theorem 7. *For every k-shallow document-ordered interpretation I with all formulas in $\exists FO$, there is an equivalent query Q_I in PosXQ that can be found in* PTIME. *Furthermore, Q_I does not use $=_{node}$. Q_I can be found to be composition-free; and if I is unwound, a composition-free Q_I can be found in* PTIME.

We now turn to *AtomXQ*$^-$. We do not know if every *FO* interpretation can be transformed into an *AtomXQ*$^-$ query. However, we can show that *AtomXQ*$^-$ queries are complete for *FO* queries with fixed tags. From previous results, this will suffices to separate *AtomXQ*$^-$ from its composition-free subset.

Theorem 8. *For every fixed label alphabet Σ, for every FO boolean query Q, there is an AtomXQ$^-$ query Q' that is equivalent to Q over Σ-labeled tree structures.*

The proof (left for the full paper) proceeds by using **let** abstraction to group the elements of the tree, rather than using $=_{node}$ as in Theorem 5. From the results of the previous section, we know that the resulting query *cannot* in general be taken composition-free, since all composition-free queries map to almost two-variable interpretations.

We now consider the question of FO-completeness of composition-free *AtomXQ*$^-$. From Theorem 4 we know that this language cannot capture all Boolean queries on the navigational structure, since it is restricted to two-variable expressiveness. The following converse shows that composition-free *AtomXQ*$^-$ can capture all "relational" Boolean queries.

Theorem 9. *For every navigationally two-variable sentence ϕ one can find in* PTIME *a composition-free AtomXQ$^-$ query Q with Q_{Bool} equivalent to ϕ. The corresponding result holds for unwound, shallow, document-ordered interpretations. In particular, the Boolean queries expressible in composition-free AtomXQ$^-$ over a fixed alphabet are exactly those expressible in Core XPath.*

Applications to Expressiveness and Complexity. Combining the corollary above with the results of Section 3 we now have:

Corollary 3. *Every AtomXQ query can be converted to an equivalent composition-free query, and every PosXQ query can be converted to an equivalent composition-free PosXQ query.*

We also note the following "conservativity" result, which follows immediately from the results of this section:

Corollary 4. *Every Boolean FO query can be expressed in AtomXQ, and the same for every FO interpretation which outputs only trees of fixed depth k.*

In particular, this implies that the "relational queries" expressible in *AtomXQ* — that is, *AtomXQ* queries restricted to trees that are encodings of flat relations (using one of the standard generic data tree encodings) are exactly those expressible in the relational calculus. Due to space limitations, we do not give this result formally. Note that the corollary above combined with Theorem 4, implies that *i)* Node equality $=_{node}$ cannot be eliminated from composition-free *AtomXQ* (since otherwise every *FO* Boolean query could be converted, over a fixed alphabet, to an FO^2 query, and this is known [2] to be false), *ii)* *AtomXQ* can express Boolean queries not expressible in Core XPath. Because *FO* is known to be PSPACE-complete and $\exists FO$ is known to be NP-complete, we have the following consequences for complexity:

Corollary 5. *The combined complexity of evaluation for composition-free AtomXQ is complete for* PSPACE, *and the combined complexity of evaluation for composition-free PosXQ is* NP*-complete.*

This is of course true because the translations for sentences in this section are all PTIME reductions.

5 *XQ* with Deep Equality vs. *FO*(Cnt)

When we turn to deep equality, first-order logic no longer suffices, even when we restrict to fixed-depth trees. Consider the query (i.e. condition) Q^1 defined by:

$$\textbf{let } \$v_1 := (\textbf{for } \$x \textbf{ in } \$\mathsf{inp/child} :: A \textbf{ return } \langle b/\rangle)$$
$$\textbf{let } \$v_2 := (\textbf{for } \$x \textbf{ in } \$\mathsf{inp/child} :: C \textbf{ return } \langle b/\rangle) \;\; \$v_2 =_{deep} \$v_1$$

Q^1_{Bool} holds iff the number of A's is equal to the number of C's. It is easy to show that there is no first-order interpretation equivalent to Q .

We will see that the absence of the ability to count is, however, the only obstacle.

Theorem 10. *For every XQ query Q, there is an FO*(**Cnt**) σ_{nav} *interpretation equivalent to Q, which can be found from Q in* EXPTIME. *For every composition-free XQ query, an equivalent FO*(**Cnt**) *query Q can be found in* PTIME.

The proof is left for the full paper. It extends the algorithm in Theorem 1, using counting quantifiers to check deep equality of variables.

As in the case of *AtomXQ* and *FO*, we can go back from *FO*(**Cnt**) to *XQ*:

Theorem 11. *For every k-shallow document-ordered FO*(**Cnt**) *interpretation I over* σ_{nav}*, there is an XQ query Q equivalent to I, which can be constructed in polynomial time. Furthermore, the resulting query can be taken to be in Composition-free XQ (in* PTIME *if I is unwound).*

The proof is an extension of the algorithm in Theorem 5. The counting quantifiers are mimicked in *XQ* by forming trees representing the satisfiers of formulas and using $=_{deep}$ comparisons of these sets to do a cardinality comparison of the corresponding sets.

Applications to Expressiveness and Complexity. As in the case of *AtomXQ*, we have an exact characterization for Boolean queries : *FO*(**Cnt**) captures *XQ* for Boolean queries over σ_{nav}. A similar result could be stated for relational queries: the relational queries expressible in *XQ* over the data tree coding of relations (i.e. as flat trees whose attributes match the attributes of the relations) are exactly those that are expressible as *FO*(**Cnt**) interpretations. By combining Theorem 10 and Theorem 11, we also get the result about removal of composition:

Corollary 6. *Composition-free XQ captures XQ.*

Note that in this case, node equality is easy to eliminate:

Proposition 4. *Every XQ query can be rewritten to one that does not use* $=_{node}$.

The proof idea is as follows: for nodes x and y in the same tree, node equality can be mimicked using counting – one requires that for each node x' above x there is a node y' above y with the same distance from the root and the same sibling number. But from above we see that we can translate general *XQ* queries into composition-free queries, and for composition-free queries all nodes can be taken to be in the input tree.

From Theorem 10, Theorem 11, and known results about the complexity of *FO*(**Cnt**) (see [10]), we also get that the data complexity of *XQ* is in TC^0, the combined complexity of *XQ* is in EXPSPACE, while the combined complexity of Composition-free *XQ* is PSPACE-complete.

Acknowledgements. The authors thank Jan van den Bussche and the anonymous reviewers of ICALP 2006 for helpful comments on earlier versions of this paper. Several of the results in this work, including a variant of Theorem 1, have been discovered concurrently by Giovanni Conforti and Thomas Schwentick.

References

1. H.-D. Ebbinghaus and J. Flum. *Finite Model Theory.* Springer, 1999. 2nd edition.
2. K. Etessami, M. Vardi, and T. Wilke. "First Order Logic with Two Variables and Unary Temporal Logic". *Information and Computation*, 179, 2002.
3. G. Gottlob, C. Koch, and R. Pichler. "Efficient Algorithms for Processing XPath Queries". *ACM Transactions on Database Systems*, **30**(2):444–491, June 2005.
4. N. Immerman. *"Descriptive Complexity"*. Springer, 1999.
5. C. Koch. "On the Complexity of Non-recursive XQuery and Functional Languages on Complex Values". In *Proc. PODS*, 2005.
6. C. Koch. "On the Role of Composition in XQuery". In *Proc. WebDB*, 2005.
7. M. Marx. "XPath with Conditional Axis Relations". In *Proc. EDBT*, 2004.
8. M. Marx. "First order paths in ordered trees". In *Proc. of the 10th International Conference on Database Theory (ICDT)*, 2005.
9. J. Paredaens and D. Van Gucht. "Possibilities and Limitations of Using Flat Operators in Nested Algebra Expressions". In *Proc. PODS*, pages 29–38, 1988.
10. N. Schweikardt. "Arithmetic, First-Order Logic, and Counting Quantifiers". *ACM Transactions on Computational Logic*, **6**(3):634–671, 2005.
11. World Wide Web Consortium. "XQuery 1.0 and XPath 2.0 Formal Semantics. W3C Working Draft (Aug. 16th 2002), 2002. `http://www.w3.org/TR/query-algebra/`.

Constructing Exponential-Size Deterministic Zielonka Automata

Blaise Genest[1,2] and Anca Muscholl[1]

[1] LIAFA, Université Paris 7 et CNRS, 75251 Paris Cedex 05, France
[2] IRISA, Université Rennes I et CNRS, 35042 Rennes Cedex, France

Abstract. The well-known algorithm of Zielonka describes how to transform automatically a sequential automaton into a deterministic asynchronous trace automaton. In this paper, we improve the construction of deterministic asynchronous automata from finite state automaton. Our construction improves the well-known construction in that the size of the asynchronous automaton is simply exponential in both the size of the sequential automaton and the number of processes. In contrast, Zielonka's algorithm gives an asynchronous automaton that is doubly exponential in the number of processes (and simply exponential in the size of the automaton).

1 Introduction

A challenging problem concerning concurrent systems is to design distributed algorithms or, even simpler, distributed finite state devices. The problem is that it is easier to think in a sequential rather than a concurrent way, and easier to model the global behavior of a system. In general it is much harder to synthesize local devices, since they only have a local view of the global behavior. Local control of a single process has to deal with partial information, consisting of local behaviors plus information exchanged with other processes.

In this paper we reconsider the problem of synthesizing deterministic asynchronous (trace) automata. These are basically (deterministic) local automata that exchange information using shared (state) variables. The underlying mathematical theory is the theory of Mazurkiewicz traces [10], which has brought a large number of beautiful results in the theory of automata and logics (see [5] for a survey). The basic idea of trace theory is to model actions in a concurrent system by explicitly providing an independence relation between actions that do not share any resource.

A fundamental and difficult result for Mazurkiewicz traces is Zielonka's theorem [16], which states that (diamond) finite state automata can be effectively transformed into deterministic asynchronous automata. This result is all the more fundamental since it has been used for several other closely related problems, as synthesis of communicating automata, with bounded communication channels [13, 7, 8], or existentially-bounded channels [6] or causal memory [1].

The main drawback of Zielonka's theorem is that it yields an asynchronous automaton of doubly exponential size in the size of the alphabet [16, 4]. The

M. Bugliesi et al. (Eds.): ICALP 2006, Part II, LNCS 4052, pp. 565–576, 2006.

paper [14] gives a more direct proof of Zielonka's theorem, with a complexity which is doubly exponential in the number of processes (instead of the size of the alphabet). Similarly, [7] synthesizes bounded communicating automata, and their construction is of size doubly exponential in the number of processes.

We propose here a simple improvement of Zielonka's algorithm in order to lower the complexity by one exponent. We obtain a deterministic asynchronous automaton of size exponential in the size of the input, that is both in the size of the finite state automaton *and* the number of processes. For applications e.g. in verification it is worth to note that the size of the automaton is exponential only because the memory needed by each process is of polynomial size. The time needed to compute any transition of the automaton is also only polynomial, which can be used in practice whenever we need only to simulate the automaton on-the-fly. Moreover, this construction is tight, since the determinization of sequential automata requires exponential size.

Related work. There were several attempts to simplify Zielonka's construction as described in [16, 12]. In some special cases the construction can be indeed simplified (see e.g . [5] Ch. 8), but the complexity is still exponential (the starting point there is a monoid homomorphism in place of an automaton). Our construction also reduces the complexity of other works [12, 15] using Zielonka's theorem or its variant for communicating processes [1, 13, 7] to produce a distributed automaton with local final states or without deadlock. Very recently, [2] proposed a construction of *non-deterministic* asynchronous automata of size $|\mathcal{A}|^{2^{|\Sigma|}}$ while we produce a *deterministic* one. Their complexity is thus polynomial in $|\mathcal{A}|$ but still doubly exponential in Σ, while our construction is of simply exponential complexity in both $|\mathcal{A}|$ and $|\Sigma|$.

Overview of the paper. We first recall some basics of Mazurkiewicz traces in Section 2. Then we recall the main ingredients of Zielonka's construction in Section 3. In Section 4 we present the new idea of decomposing into zones, and in Section 5 we present the new construction of deterministic asynchronous automata.

2 Preliminaries

We assume that there is a set $\mathcal{P}$ of processes and an alphabet Σ which are fixed. Each letter $a \in \Sigma$ is an action associated with the set of processes $\text{dom}(a) \subseteq \mathcal{P}$ involved in a. A pair (Σ, dom) is called *distributed alphabet.* A (non-deterministic) automaton over the alphabet Σ is a tuple $\mathcal{A} = (V, \Sigma, \rightarrow, v^0, F)$ with a finite set of states V, a set of final states F, an initial state v^0 and a non-deterministic transition function $\rightarrow: V \times \Sigma \rightarrow 2^V$. The size of an automaton is the number of states.

Concurrent systems with shared actions given by a distributed alphabet (Σ, dom), are readily modeled by Mazurkiewicz traces. The idea is that the distribution of the alphabet defines an independence relation among actions $I \subseteq \Sigma \times \Sigma$, by setting $(a, b) \in I$ if and only if $\text{dom}(a) \cap \text{dom}(b) = \emptyset$. We call (Σ, I) an *independence alphabet.* The complementary relation $D = \Sigma \times \Sigma \setminus I$ is called a dependence relation. The independence relation induces a congruence $\sim$ on Σ^* by setting $u \sim v$ if there exist words $u_1, \ldots, u_n \in \Sigma^*$ with $u_1 = u$, $u_n = v$ and such that for every $i < n$ we have $u_i = xaby$, $u_{i+1} = xbay$ for some $x, y \in \Sigma^*$ and $(a, b) \in I$. An $\sim$-equivalence class is simply called a *(Mazurkiewicz) trace* [10]. We denote by $[u]$ the trace associated with the word $u \in \Sigma^*$ (for simplicity we do not refer to I, neither in $\sim$ nor in $[u]$, simply because the independence alphabet is fixed). Trace prefixes and trace factors are defined as usual, with $[p]$ a trace prefix (trace factor, resp.) of $[u]$ if p is a word prefix (word factor, resp.) of some $v \sim u$.

A (non-deterministic) automaton $\mathcal{A}$ is called I-diamond if for all $(a, b) \in I$, and all states r, s, t of $\mathcal{A}$ with $r \xrightarrow{a} s$ and $s \xrightarrow{b} t$, there also exists a state s' with $t \xrightarrow{b} s'$ and $s' \xrightarrow{a} t$. Note that the I-diamond property implies that the language $\mathcal{L}(\mathcal{A})$ of $\mathcal{A}$ is *I-closed*: that is, $u \in \mathcal{L}(\mathcal{A})$ if and only if $v \in \mathcal{L}(\mathcal{A})$ for every $u \sim v$.

We use asynchronous automata as distributed models with finite control. Our definition is slightly different from the usual definitions for asynchronous and asynchronous cellular automata [5], see the remark below.

Definition 1. *A deterministic asynchronous automaton over the distributed alphabet (Σ, dom) is a tuple $\mathcal{B} = ((K_p, \delta_p, k_p^0)_{p \in \mathcal{P}}, Acc)$ such that for any $p \in \mathcal{P}$:*

- *K_p is the finite set of local states of process p.*
- *$\delta_p : (\Sigma \times \prod_{q \in \mathcal{P}} K_q) \to K_p$ is the local transition function of process p, satisfying the following conditions for all actions $a \in \Sigma$ and local states $s_q \in K_p$, $q \in \mathcal{P}$:*
  - *for $p \notin dom(a)$, we have $\delta_p(a, s_1, \ldots, s_n) = s_p$.*
  - *for $p \in dom(a)$, the state $\delta_p(a, s_1, \ldots, s_n)$ depends only on $(s_q)_{q \in dom(a)}$, that is $\delta_p(a, s_1, \ldots, s_n) = \delta_p(a, s_1, \ldots, s'_q, \ldots, s_n)$ for $q \notin dom(a)$.*
- *$k_p^0 \in K_p$ is the local initial state of process p.*
- *$Acc \subseteq \prod_{p \in \mathcal{P}} K_p$ is a set of (global) accepting states.*

An asynchronous automaton accepts a regular language with the following global semantics:

Definition 2. *The language of an asynchronous automaton $\mathcal{B} = ((K_p, \delta_p, k_p^0)_{p \in \mathcal{P}}, Acc)$ is defined as $\mathcal{L}(\mathcal{B}) = \mathcal{L}(\mathcal{A})$, where $\mathcal{A} = (K, \delta, k^0, Acc)$ is the following automaton, called the global automaton of $\mathcal{B}$:*

- *The global state space is $K = \prod_{p \in \mathcal{P}} K_p$.*
- *The initial state is $k^0 = (k_p^0)_{p \in \mathcal{P}}$.*
- *The global transition function $\delta : \Sigma \times K \to K$ is defined for all $a \in \Sigma$, $k \in K$ by $\delta(a, k) = (k'_p)_{p \in \mathcal{P}}$ with $k'_p = \delta_p(a, k)$ for all p.*

Remark 1. Our definition of asynchronous automaton differs from the usual one in that we define transitions on processes instead of letters. Moreover, the transitions corresponding to processes are like transitions in a cellular automaton, that is, only the local state associated with the process executing a transition changes.

The definition thus corresponds to a shared-read, owner-write mode (the transition function reads the states of all other processes involved in the current action and writes its local state). However, the difference wrt asynchronous automata is merely syntactical, since in the global behavior of the automaton we synchronize all processes from $\mathrm{dom}(a)$ when executing an action a.

For several purposes it is convenient to represent traces by (labeled) pomsets. Formally, a trace $T = [a_1 \cdots a_n]$ ($a_i \in \Sigma$ for all i) corresponds to a labeled pomset $(E, \lambda, \leq)$ defined as follows: $E = \{e_1, \ldots, e_n\}$ is a set of events (or nodes), one for each position in T. Event e_i is labeled by $\lambda(e_i) = a_i$, for each i. The relation $\leq$ is the least partial order on E with $e_i \leq e_j$ whenever $(a_i, a_j) \in D$ and $i \leq j$. In terms of graphs it is convenient to identify a trace T with its *dependence graph*, by defining an edge from e_i to e_j iff $(a_i, a_j) \in D$ and $i \leq j$. A total order $e_1 \cdots e_n$ that is compatible with $\leq$ is called a *linearization* of T. Since all these formalisms are equivalent, we will refer to graph nodes as events for convenience. Moreover, we will use for convenience set operations on traces, interpreting them on the associated graphs. For instance, assume that T_1, T_2 are both prefixes of some trace T. In other words, each T_i is a downward closed subgraph of T. Then we write $T_1 \cap T_2$ ($T_1 \cup T_2$, resp.) for the least (greatest, resp.) common prefix of T_1, T_2. Also, we write $e_i \in T$ for denoting that e_i is a vertex of (the graph of) T.

For any trace factor T' of T, we denote by $\mathrm{alph}(T') = \bigcup_{e \in T'} \lambda(e)$ the letters occurring in T', resp. by $\mathrm{dom}(T') = \bigcup_{e \in T'} \mathrm{dom}(\lambda(e))$ the processes occurring in T'. For $a \in \Sigma$ we call any event e with $\lambda(e) = a$ an a-event.

We have in Figure 1 a trace T with $(a,b),(a,c) \in D$ and $(a,d) \in I$. Hence, *cbadcbadb* $\sim$ *cbdacbdab* are two representing words of T. Process p can read the state of q when executing action a, but not when executing action c.

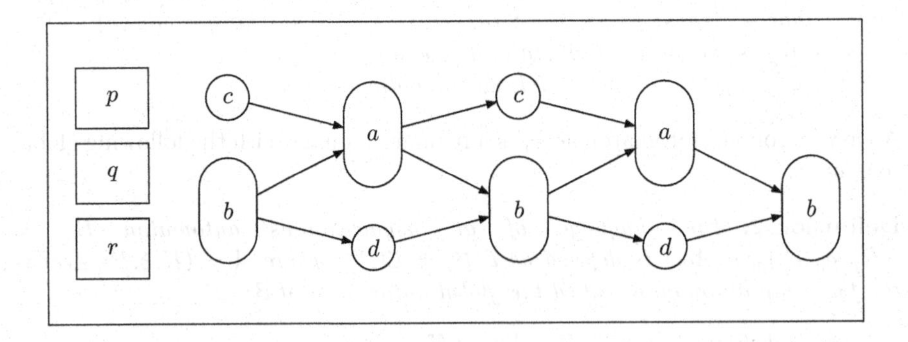

Fig. 1. The pomset associated with the trace $T = [c\,b\,a\,d\,c\,b\,a\,d\,b]$, with $\mathrm{dom}(a) = \{p,q\}$, $\mathrm{dom}(b) = \{q,r\}$, $\mathrm{dom}(c) = \{p\}$, $\mathrm{dom}(d) = \{r\}$

3 Zielonka's Theorem

We recall the main ingredients in Zielonka's construction of deterministic asynchronous automata. Our presentation is based on [5, 14, 4]. Let T be a trace and

$p \in \mathcal{P}$ a process, then we denote by $\text{pref}_p(T)$ the minimal trace prefix of T which contains all events of T on process p. Hence, $\text{pref}_p(T)$ has a unique maximal event which is the last event of T on process p. That is, $\text{pref}_p(T)$ corresponds to the history of process p after executing T, that we also refer to as *p-view*. For instance, in Figure 1 we have $\text{pref}_p(T) = [cbadcba]$. For a set of processes $P \subseteq \mathcal{P}$, let $\text{pref}_P(T) = \cup_{p \in P} \text{pref}_p(T)$ be the P-view of T.

Zielonka's construction starts with a regular, I-closed language, that is presented either through a monoid homomorphism or an automaton satisfying the I-diamond property, [4]. In most applications we are interested in the second case, where we start with a (non-deterministic) I-diamond automaton.

Theorem 1. *[16] Let $\mathcal{A}$ be an I-diamond automaton over the independence alphabet (Σ, I). An equivalent deterministic asynchronous automaton $\mathcal{B}$ with $2^{O(|\mathcal{A}|^2(2^{|\Sigma|}))}$ states can be obtained applying the construction from [4]. An equivalent deterministic asynchronous automaton with $2^{O(|\mathcal{A}|^2(2^{|\mathcal{P}|}))}$ states can be obtained applying [14].*

3.1 General Idea and Timestamping

We first describe informally how Zielonka's construction works. Let $\mathcal{A} = (V, \Sigma, \rightarrow, v^0, F)$ be an I-diamond automaton. When an action $a \in \Sigma$ is executed after T by the processes in $\text{dom}(a)$, each process of $\text{dom}(a)$ reads the states of the other processes in $\text{dom}(a)$ and changes its own state accordingly. At this step, each process $p \in \text{dom}(a)$ computes the events of $\text{dom}(a)$ that were not in its p-view, that is $\text{pref}_p(Ta) \setminus \text{pref}_p(T)$, or equivalently, $\bigcup_{q \in \text{dom}(a)} \text{pref}_q(T) \setminus \text{pref}_p(T)$ plus the last a. First, events are labeled by timestamps in order to recover which events are in $\text{pref}_q(T) \cap \text{pref}_p(T)$ and which events are in $\text{pref}_q(T) \setminus \text{pref}_p(T)$, by simply comparing the timestamps. For instance, in Figure 1, when the last b is executed, then process q reads the state of process r and vice-versa. They find that the second b is the only maximal event in their common past, q discovers that action d was executed in $\text{pref}_r(T) \setminus \text{pref}_q(T)$, and r that ca was executed in $\text{pref}_q(T) \setminus \text{pref}_r(T)$.

The problem is that the number of events that have to be stored is arbitrarily large. Zielonka's construction explains that only a bounded set S_1 of events needs to be timestamped. Finally a finite representation of the behavior of the given sequential automaton $\mathcal{A}$ on $\text{pref}_q(T) \setminus \text{pref}_P(T)$ is needed. To this purpose, transition relations Δ_X are used for $X \in \Sigma^\star$, where $\Delta_X(R) = \{s \in V \mid \exists r \in R, r \xrightarrow{X} s\}$ for any subset R of states of $\mathcal{A}$. That is, $\Delta_X(R)$ is the set of states of $\mathcal{A}$ reached after reading the word X from any state in R. Zielonka's construction explains how process q remembers relations Δ_X only for an exponential number of sequences X. A global state of the asynchronous automaton is then accepting if and only if the Δ relation associated with it satisfies $\Delta_T(v^0) \cap F \neq \emptyset$.

In order to explain the construction in more detail, we define now the sets of events S_1, S_2 used by Zielonka's timestamping and the set of factors X for which we remember the function Δ_X. We will not provide the definition of the timestamping, which is the usual one (the reader is referred to [4, 5, 14] instead). First, the set S_1 simply consists of the last event on process p, for every $p \in \mathcal{P}$:

Definition 3. *Let* $T = (E, \lambda, \leq)$ *be a trace. The primary information of* T *is* $S_1(T) = \{e \in E \mid \exists p \in dom(e), \forall f \in E, p \in dom(f) \implies f \leq e\}$.

The following crucial property of S_1 can be found in [14] (Lemma 1, page 9) and can be quickly obtained from [5] (Proposition 8.3.5, page 259).

Lemma 1. *Let* T *be a trace and* $P, Q \subseteq \mathcal{P}$ *two subsets of processes. Then* $\max(pref_P(T) \cap pref_Q(T)) \subseteq S_1(pref_P(T)) \cap S_1(pref_Q(T))$.

That is, the maximal events of the intersection of the views of the sets $P, Q \subseteq \mathcal{P}$ belong to the primary information of each of these two views.. However, knowing the events in $S_1(\mathrm{pref}_P(T))$ and $S_1(\mathrm{pref}_Q(T))$ is not enough for computing $\max(\mathrm{pref}_P(T) \cap \mathrm{pref}_Q(T))$.

Notice that during the execution of an asynchronous automaton, every event is created as an event of S_1 and can eventually be removed from S_1. Note also that $\{e\} \subseteq S_1(Te) \subseteq S_1(T) \cup \{e\}$ for every event $e \in E$.

After the trace $T' = [cbadcbad]$ is executed, $S_1(\mathrm{pref}_r(T')) = S_1([cbadbd])$ contains the first a (for process p), the second b (for q) and the second d (for r), as shown in Figure 2. The common past of the views of q, r has the second b as unique maximal event. If an event b is then executed (as in Figure 1), we have that $S_1(T'b)$ contains the second a (for p) and the third b (for q and r).

The timestamping function TS in Zielonka's construction [16, 5, 14] labels every event in $S_1(\mathrm{pref}_p(T))$. It allows to compute $S_1(\mathrm{pref}_p(T)) \cap S_1(\mathrm{pref}_q(T))$ for every $p, q \in \mathcal{P}$. For this, the timestamping labels events wrt. the so-called secondary information, that contains the last event on process q before the last event on p for every $p, q \in \mathcal{P}$ (these two events can be equal).

Definition 4. *Let* $T = (E, \lambda, \leq)$. *The secondary information of* T *is the set* $S_2(T) = \{e \in E \mid \exists g \in S_1(T), q \in dom(e), \forall f \leq g, q \in dom(f) \implies f \leq e\}$.

In particular, $S_1(T) \subseteq S_2(T)$. The primary information S_1 is of size $|\mathcal{P}|$, the secondary information has at most $|\mathcal{P}|^2$ events, and the timestamping is of size $2|\mathcal{P}|^3 \log(|\mathcal{P}|)$ [14]. In Figure 2, the secondary information $S_2(\mathrm{pref}_r(T'))$ contains the first two b, the first a and the second d. The first c is not in $S_2(\mathrm{pref}_r(T'))$.

Let $T = (E, \lambda, \leq)$ be a trace. For every subset $W \subseteq S_1(\mathrm{pref}_p(T))$, we define the suffix $X_W(T) = \{e \in E \mid \forall s \in W : e \not\leq s\}$ of T. In particular, $X_\emptyset(T) = T$.

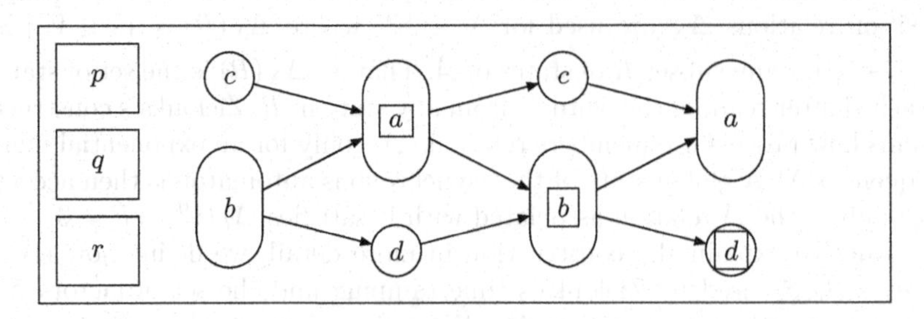

Fig. 2. Trace T', where the distinguished events are those in $S_1(\mathrm{pref}_r(T'))$

For instance, in Figure 2 with $\text{pref}_r(T) = [cbadbd]$, if we fix W to be the first a, then $X_W(\text{pref}_r(T))$ consists of the first and second d and of the second b. For every set $P \subseteq \mathcal{P}$ and every subset $W \subseteq S_1(\text{pref}_P(T))$ of primary events of the P-view, we remember the transition relation Δ_W which associates with every state r of $\mathcal{A}$ the set of states $\Delta_W(r) = \{s \in V \mid r \xrightarrow{X_W} s\}$ that can be reached from r by the trace $X_W(T)$. Process p thus remembers an exponential number (wrt. the number of processes $|\mathcal{P}|$) of relations in $V \times V$. Note that we can compute on-the-fly $\Delta_W(R) = \bigcup_{r \in R} \Delta_W(r)$ for any subset R of states of $\mathcal{A}$.

3.2 The Asynchronous Automaton

We describe now how a deterministic asynchronous automaton updates the state information wrt. the I-diamond automaton, as done e.g. in [16, 5, 14]. A transition of process p in the asynchronous automaton $\mathcal{B}$ consists of reading the states of other processes, and adding a new event e. For instance, if $\lambda(e) = a$ and $\text{dom}(a) = \{p, q, r\}$, then we can decompose the a-transition of process p as first reading the local state of process q and updating its local state, then doing it again with process r, and finally adding the event e.

The state reached on a trace $\text{pref}_P(T)$, $P \subseteq \mathcal{P}$, is a tuple $(S_1, \text{TS}, (\Delta_W)_{W \subseteq S_1})$, containing the primary information S_1 of the P-view $\text{pref}_P(T)$, the timestamping TS, and the transition relations $(\Delta_W)_{W \subseteq S_1}$.

We do not describe here the update of S_1 and TS, since we use the same timestamping algorithm as in [16, 5, 4, 14].

Assume that the current state reached on $\text{pref}_{\text{dom}(a)}(T)$ is $(S_1, \text{TS}, (\Delta_W)\ _{W \subseteq S_1})$ and we add the event e with $\lambda(e) = a$. The new state will be $(S_1', \text{TS}', (\Delta_W')_{W \subseteq S_1'})$. For every $W \subseteq S_1'$ we have two cases, depending on whether or not $e \in W$:

- Either $W \subseteq S_1$, and then $\Delta_W' = \xrightarrow{a} \circ \Delta_W$.
- Else we have $e \in W$, hence $X_W = \emptyset$ and $\Delta_W = \text{Id}$.

We now look at the local state modifications. Assume that $q \in \text{dom}(a)$ and that the local state of process q, say $(S_1^q, \text{TS}^q, (\Delta_W^q)_{W \in S_1^q})$, is added to the state $(S_1, \text{TS}, (\Delta_W)_{W \subseteq S_1})$ reached on $\text{pref}_P(T)$, $P \subseteq \text{dom}(a)$, $q \notin P$. The state reached on $\text{pref}_{P \cup \{q\}}(T)$, say $(S_1', \text{TS}', (\Delta_W')_{W \subseteq S_1'})$ is computed as follows. For every $W' \subseteq S_1'$ we set:

- Let $W = W' \cap S_1$ and $W^q = (W' \cup S_1) \cap S_1^q$.
- Set $\Delta_{W'}' = \Delta_{W^q}^q \circ \Delta_W$. Thus, the $X_{W'}$-suffix of $\text{pref}_{P \cup \{q\}}(T)$ is the union of two suffixes (over disjoint sets of processes), namely the X_W-suffix consisting of events of the P-view that are not below some event in W'; and the X_{W^q}-suffix consisting of events from $\text{pref}_q(T) \setminus \text{pref}_P(T)$ that are not below some event in W'.

A global state$(S_1, \text{TS}, (\Delta_W)_{W \subseteq S_1})$ is accepting if and only if $\Delta_\emptyset(v^0) \cap F \neq \emptyset$.

Let us comment on the complexity of the construction. At each event $e \in E$, updating the primary and secondary information takes polynomial time. Updating the relations Δ may take an exponential time in the number of processes

$|\mathcal{P}|$. Hence, Zielonka's construction gives a transition function which needs exponential time (and space) to compute the next state. Overall there are doubly exponentially many different memory configurations, which is why the automaton is doubly exponential.

The exponential memory comes only from the straightforward use of Δ. In the rest of the paper we explain how we can achieve a better complexity by using new transition relations Δ, still keeping the primary information and the timestamping.

4 Zone Decomposition

The idea of this paper is to define transition relations Δ on *zones* of the p-views $\text{pref}_p(T)$. When the transition relation on some factor of T is needed, we compose the transition relations of the zones included in the factor. Zones are defined as equivalence classes of the following relation:

Definition 5. *Let $T = (E, \leq, \lambda)$ be a trace. For an event $e \in E$ we define the set of events $S(e) = \{f \in S_1(T) \mid e \leq f\}$. We say that two events e, f are equivalent (denoted as $e \equiv f$) if and only if $S(e) = S(f)$. The equivalence classes of $\equiv$ are called zones.*

Let Z be a zone and define $S(Z) = S(e)$ for some event $e \in Z$. Let also Z, Z' be two zones of some trace T. We write $Z < Z'$ if $Z \neq Z'$ and if $e < e'$ for some events $e \in Z$, $e' \in Z'$. By $\text{dom}(Z)$ we denote the set of processes occurring in the zone Z, i.e., $\text{dom}(Z) = \cup_{e \in Z} \text{dom}(e)$. The following lemma is easy to show:

Lemma 2. *1. A zone of T is a factor of T and contains at most one event from the secondary information $S_1(T)$.*
2. The set of zones partitions the set of events of T.
3. The relation $<$ on zones is acyclic. It induces the least partial order such that $S(Z) \supsetneq S(Z')$ and $dom(Z) \cap dom(Z') \neq \emptyset$ implies $Z < Z'$.

Proof of 3). Assume that $Z_1 \leq Z_2 \leq \cdots \leq Z_k = Z_1$, say with $e_i \in Z_i$ and $f_j \in Z_{j+1}$ $(1 \leq i, j < k)$ such that $e_i < f_i$ for all i. Hence, $S(Z_1) \supseteq S(Z_2) \supseteq \cdots \supseteq S(Z_k) = S(Z_1)$, thus $S(Z_i) = S(Z_j)$ and $Z_i = Z_j$ for all i, j.

Figure 3 depicts the same trace $T = [cbadcbad]$ as Figure 2. Recall that $S_1(\text{pref}_r(T))$ consists of the first a, the second b and the second d. There are three zones in $\text{pref}_r(T)$: Z_1 is the first a, b and c, Z_2 is the first d and the second b, and Z_3 is the second d (see also Figure 3). We have $Z_1 < Z_2 < Z_3$. Also, $S(Z_2)$ consists of the second b and the second d.

Zones enjoy some crucial properties, that are stated in the following.

Proposition 1. *Let T be a trace, $P, Q \subseteq \mathcal{P}$ sets of processes, and Z a zone of $\text{pref}_P(T)$. Then either $Z \subseteq \text{pref}_P(T) \cap \text{pref}_Q(T)$, or $Z \cap (\text{pref}_P(T) \cap \text{pref}_Q(T)) = \emptyset$.*

Proof. Assume by contradiction that Z is a zone that violates the statement of the proposition. Consider the factor $Z_1 = Z \cap \text{pref}_P(T) \cap \text{pref}_Q(T)$. By assumption we can write the trace factor Z as $Z \sim Z_1 Z_2$ with Z_1, Z_2 both non-empty. Let $e \in Z_1$ and $f \in Z_2$. Since $e \in \text{pref}_P(T) \cap \text{pref}_Q(T)$, there exists a maximal event g of $\text{pref}_P(T) \cap \text{pref}_Q(T)$ such that $e \leq g$. By Lemma 1, $g \in S_1(\text{pref}_P(T))$. Since $e, f \in Z$, we have $e \equiv f$ and thus $f \leq g$. This implies $f \in \text{pref}_P(T) \cap \text{pref}_Q(T)$, a contradiction. □

Proposition 2. *Let T be a trace. There are at most $|\mathcal{P}|^2 + |\mathcal{P}|$ zones in T.*

Proof. Assume by contradiction that there are more than $|\mathcal{P}|(|S_1(T)|+1)$ zones. Hence we can find a process p involved in at least $k = |S_1(T)| + 2$ zones. Since these zones have intersecting domains, they are strictly ordered, let us say $Z_1 < \cdots < Z_k$. Hence $S(Z_1) \supsetneq \cdots \supsetneq S(Z_k)$. This implies $|S(Z_1)| \geq k-1 = |S_1(T)|+1$. This is a contradiction since there are at most $|S_1(T)|$ events in $S(Z_i)$ for all i. □

The last property is crucial when the zone partition is updated.

Proposition 3. *Let T be a trace, $P \subseteq \mathcal{P}$ a set of processes, $q \notin P$ a process and e, f two events of $\text{pref}_P(T)$. Let us write S for the function of Def. 5 on $\text{pref}_P(T)$ and S' for the function on $\text{pref}_{P \cup q}(T)$. If $S(e) = S(f)$, then $S'(e) = S'(f)$.*

Proof. Assume that $S(e) = S(f)$. Let $g \in S'(e)$. The first case is where $g \in S_1(pref_P(T))$, hence $g \in S(e) = S(f)$ and $f \leq g$ by definition of $S(f)$.

Else, $g \notin S_1(pref_P(T))$, hence $g \in S_1(\text{pref}_q(T))$ and $g \notin \text{pref}_P(T)$. Let h be a maximal event in $\text{pref}_P(T)$ with $e \leq h \leq g$. There exists h' in $\text{pref}_q(T)$ with $h \lessdot h' \leq g$. That is $\text{dom}(h) \cap \text{dom}(h') \neq \emptyset$, and let r be a process in this intersection. By maximality of h, we have $h' \notin \text{pref}_P(T)$. That is, h is the last event on process r of $\text{pref}_P(T)$. By definition of S, we have $h \in S(e) = S(f)$, hence $f \leq h \leq g$. We conclude by symmetry between e and f. □

For each zone Z, the function Δ_Z requires space $\leq |\mathcal{A}|^2$, and there are at most $|\mathcal{P}|^2 + |\mathcal{P}|$ zones. The transition function constructed in the next section gives:

Theorem 2. *Let $\mathcal{A}$ be a (non-deterministic) I-diamond automaton over the independence alphabet (Σ, I). We can construct an equivalent deterministic asynchronous automaton $\mathcal{B}$ with less than $2^{|\mathcal{A}|^2 \times (|\mathcal{P}|^2+|\mathcal{P}|)+2|\mathcal{P}|^4}$ states. Each process has a memory of size $O(|\mathcal{A}|^2 \times |\mathcal{P}|^2 + |\mathcal{P}|^4)$, and computes its next state in time $O(|\mathcal{A}|^2 \times |\mathcal{P}|^2 + |\mathcal{P}|^4)$.*

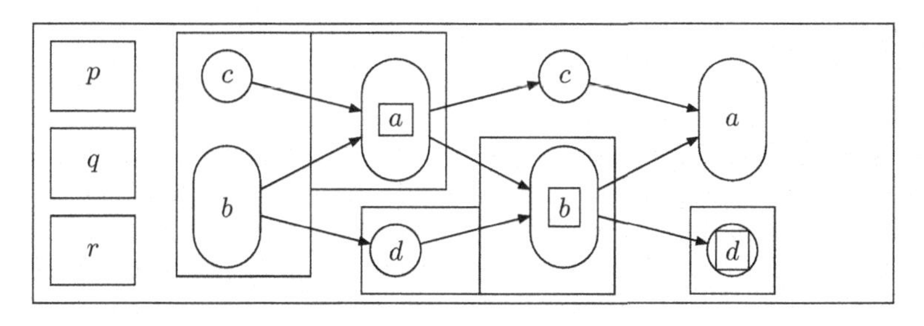

Fig. 3. The three zones of $\text{pref}_r(T)$. The distinguished nodes are those in $S_1(\text{pref}_r(T))$.

5 The New Construction

Our construction follows Zielonka's construction, up to the Δ relations which are remembered only for the zones of $\text{pref}_p(T)$, $p \in \mathcal{P}$. Of course, we have to adapt the transition function of the asynchronous automaton accordingly.

Let $\mathcal{A} = (V, \Sigma, \to, v^0, F)$ be a non-deterministic I-diamond automaton. We explain in the following how to build a deterministic asynchronous automaton $\mathcal{B}$ with the same language. The local p-state of $\mathcal{B}$ reached on a prefix $\text{pref}_p(T)$ is the tuple $(S_1, \text{TS}, \langle \text{dom}(Z_i), S(Z_i), \Delta(Z_i)\rangle_{i=0,\ldots m})$, where:

- $\{Z_1 \ldots, Z_m\}$ is the set of zones of $\text{pref}_p(T)$.
- S_1 is the primary information of $\text{pref}_p(T)$.
- The timestamping TS associates every event of S_1 with its timestamp.
- For a zone Z, $\text{dom}(Z)$ denotes the set of processes occurring in Z.
- For a zone Z, $S(Z) \subseteq \mathcal{P}$ corresponds to Definition 5. That is, $q \in S(Z)$ means that the last event on q in $\text{pref}_p(T)$ is above an event of Z.
- The transition relation Δ_Z gives for each state v of $\mathcal{A}$ the set of states $\Delta_Z(v)$ that can be reached in $\mathcal{A}$ from v by reading some linearization of Z (remember that since $\mathcal{A}$ is I-diamond, this corresponds to the set of states reached by any linearization of Z).

We define now the local transition function δ_p of the asynchronous automaton $\mathcal{B}$. We do not recall how to update the primary information S_1 and the timestamping TS, though the update of S-values includes the update of the secondary information S_2. Recall that the order on zones $Z_i < Z_j$ can be computed from the knowledge of $S(Z_i)$ and of $\text{dom}(Z_i)$, for all zones Z_i (see Lemma 2).

Assume that the action a with $p \in \text{dom}(a)$ is added to the current p-state $(S_1, \text{TS}, \langle \text{dom}(Z_i), S(Z_i), \Delta(Z_i)\rangle_{i=0,\ldots,m})$.

The new p-state is $(S_1', \text{TS}', \langle \text{dom}(Z_i'), S(Z_i'), \Delta(Z_i')\rangle_{i=0,\ldots,m+1})$, with:

1. Let $\text{dom}(Z'_{m+1}) = \text{dom}(a)$, $\Delta_{Z'_{m+1}} = \xrightarrow{a}$, and $S(Z'_{m+1}) = \text{dom}(a)$ (actually $S(Z'_{m+1})$ consists of the unique maximal a event).
2. Let $\langle \text{dom}(Z_j'), S(Z_j'), \Delta_{Z_j'}\rangle_{j=1,\ldots,m} = \langle \text{dom}(Z_j), S(Z_j), \Delta_{Z_j}\rangle_{j=1,\ldots,m}$.
3. For all Z_i', let $S(Z_i') \leftarrow S(Z_i') \cup \{\text{dom}(a)\}$.
4. If $S(Z_i') = S(Z_j')$ with $Z_j \not< Z_i$ then we merge Z_i', Z_j' and let $\Delta_{Z_i'} = \Delta_{Z_j'} \circ \Delta_{Z_i'}$ and $\text{dom}(Z_i') = \text{dom}(Z_j') \cup \text{dom}(Z_i')$ and we delete Z_j'.

That is, a new zone Z'_{m+1} is created representing the new event a (line 1) and the other zones are copied (line 2). We then update the S-values in line 3 and merge zones with equal S-value in line 4.

Assume that the process q is in local state $(S_1^q, \text{TS}^q, \langle \text{dom}(Z_i^q), S(Z_i^q), \Delta_{Z_i^q}\rangle_{i=1,\ldots,n})$ with history $\text{pref}_q(T)$. Moreover, process p in current state $(S_1, \text{TS}, \langle \text{dom}(Z_i), S(Z_i), \Delta_{Z_i}\rangle_{i=1,\ldots,m}\rangle$ and history $\text{pref}_P(T)$ reads the state of q (as usual, $p, q \in \text{dom}(a)$ where a is the new action). The updated state of p is $(S_1', \text{TS}', \langle \text{dom}(Z_i'), S(Z_i'), \Delta_{Z_i'}\rangle_{i=1,\ldots,k})$, with $\text{pref}_{P \cup q}(T)$ as history, where:

1. Let $J = \{i \mid 1 \leq i \leq n, \mathrm{TS}^q(S(Z_i^q)) \cap \mathrm{TS}(S_1) = \emptyset\}$ and $k = m + |J|$,
2. Let $\langle \mathrm{dom}(Z'_j), S(Z'_j), \Delta_{Z'_j} \rangle_{j=1,\dots,m} = \langle \mathrm{dom}(Z_j), S(Z_j), \Delta_{Z_j} \rangle_{j=1,\dots,m}$,
3. $\langle \mathrm{dom}(Z'_{m+j}), S(Z'_{m+j}), \Delta_{Z'_{m+j}} \rangle_{j=1,\dots,|J|} = \langle \mathrm{dom}(Z_i^q), S(Z_i^q), \Delta_{Z_i^q} \rangle_{i \in J}$,
4. The partial order $<'$ on the new zones is given by the transitive closure of the relation $< \cup <^q \cup \{(Z'_i, Z'_j) \mid i \leq m < j, \mathrm{dom}(Z'_i) \cap \mathrm{dom}(Z'_j) \neq \emptyset\}$,
5. For all $Z'_i <' Z'_j$, $S(Z'_i) \leftarrow S'(Z'_i) \cup S'(Z'_j)$,
6. If $S(Z'_i) = S(Z'_j)$ and $Z_j \not< Z_i$, then we merge Z'_i and Z'_j and set $\mathrm{dom}(Z'_i) = \mathrm{dom}(Z'_j) \cup \mathrm{dom}(Z'_i)$ and $\Delta_{Z'_i} = \Delta_{Z'_j} \circ \Delta_{Z'_i}$, and we delete Z'_j.

The update operations consists in copying the zones that form a partition of $\mathrm{pref}_P(T)$ (line 2) and adding in line 3 the zones $(Z_i^q)_{i \in J}$ that partition $\mathrm{pref}_q(T) \setminus \mathrm{pref}_P(T)$. We then update the S-values in line 5. The last line merges zones with equal S-value. We say that a global state with local p-component $S_1^p, \mathrm{TS}^p, \langle \mathrm{dom}(Z_i^p), S(Z_i^p), \Delta_{Z_i^p} \rangle_{i=1,\dots,n_p}$ is accepting if we have $\Delta_{Z_n} \circ \dots \circ \Delta_{Z_1}(v^0) \cap F \neq \emptyset$ for $Z_1 \cdots Z_n$ a linearization of $(Z_i, <)_{i=1,\dots,n}$.

Example: Consider the same trace $T = [cbadcbad]$ as in Figure 3. Then $\mathrm{pref}_q(T) = [cbadcba]$ has two zones (see also Figure 4): zone Z_1^q consisting of the first c, a, d, and the first two b, and zone Z_2^q consisting of the second c and a. Assume that the letter b is now executed, which means that process r can read the state of process q. For instance, we have $S(Z_2^q)) = \{p, q\}$ and with Figure 3, $S(Z_2) = \{q, r\}$, that is Z_2 is not before the last event on p. Also, $S(Z_1) \setminus S(Z_2) = \{p\}$.

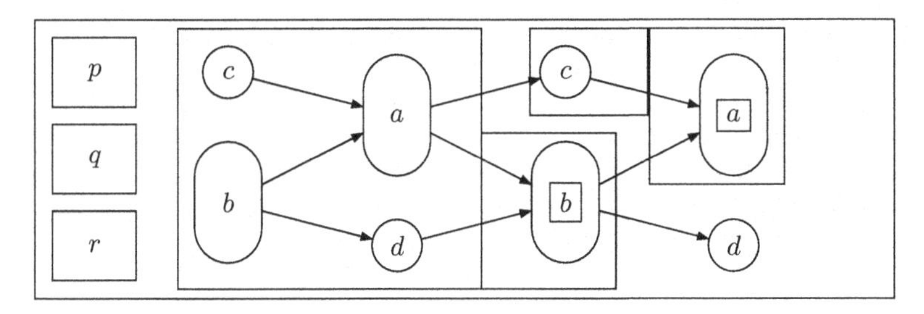

Fig. 4. Zones of $\mathrm{pref}_q(T)$. The distinguished nodes are the primary events of $\mathrm{pref}_q(T)$.

First, with the timestamping, process r computes the maximal event of $\mathrm{pref}_q(T) \cap \mathrm{pref}_r(T)$, which is the second b. Thus $J = \{2\}$, and the new set of zones is $\{Z'_1, Z'_2, Z'_3, Z'_4\}$ with $Z'^4 = Z_2^q$. Process r then computes $Z'_1 < Z'_2 < Z'_4$ and that Z'_3 and Z'_4 are incomparable. Thus $Z'_2 < Z'_4$, then p is added to $S'(Z'_2)$. It means that $S'(Z'_1) = S'(Z'_2)$, and hence it merges Z'_2 with Z'_1. The new set of zones is then $\{Z'_1, Z'_3, Z'_4\}$. Then the letter b is added as a new zone Z''_5, and we get $S''(Z''_1) = \{p, q, r\}, S''(Z''_3) = \{q, r\}, S''(Z''_4) = \{p, q, r\}, S''(Z''_5) = \{q, r\}$, thus we merge zones and we keep two zones $Z''_1 = Z''_1 \cup Z''_4$ and $Z''_3 = Z''_3 \cup Z''_5$.

Acknowledgement. We would like to thank Dietrich Kuske for simplifying the definition of zones.

References

1. B. Adsul, M. Mukund, K. Narayan Kumar and V. Narayanan. Causal closure for MSC languages. Proc. of *FSTTCS'05*, LNCS 3821, pp. 335-347, 2005.
2. N. Baudru and R. Morin. Unfolding Synthesis of Asynchronous Automata. International Computer Science Symposium in Russia, CSR 2006. Available at `http://www.cmi.univ-mrs.fr/~{}morin/papers/CSR.pdf`.
3. D. Brand and P. Zafiropulo. On communicating finite-state machines. In *J. of the ACM*, 30(2):323-342, 1983.
4. R. Cori, Y. Métivier and W. Zielonka. Asynchronous Mappings and Asynchronous Cellular Automata. In *Inf. and Comput.*, 106(2):159-202, 1993.
5. V. Diekert and G. Rozenberg, editors. *The Book of Traces.* Chapter 8 by V. Diekert and A. Muscholl. World Scientific, Singapore, 1995.
6. B. Genest, D. Kuske and A. Muscholl. A Kleene Theorem and Model Checking for a Class of Communicating Automata. Proc. of *DLT'04*, LNCS 3340, pp. 30-48, 2004. Journal version to appear in *Inf. and Comput.*, 2006.
7. J. G. Henriksen, M. Mukund, K. Narayan Kumar, M. Sohoni and P. S. Thiagarajan. A Theory of Regular MSC Languages. In *Inf. and Comput.* 202(1):1-38, 2005.
8. D. Kuske. Regular sets of infinite message sequence charts. In *Inf. and Comput.*, 187(1):80-109, 2003.
9. M. Lohrey and A. Muscholl. Bounded MSC communication. In *Inf. and Comput.*, (189):135–263, 2004.
10. A. Mazurkiewicz. Concurrent program schemes and their interpretation. Technical report, DAIMI Report PB-78, Aarhus University, 1977.
11. P. Madhusudan and B. Meenakshi. Beyond Message Sequence Graphs. Proc. of *FSTTCS'01*, LNCS 2245, pp. 256-267, 2001.
12. M. Mukund. From global specification to local implementations. In *Synthesis and Control of Discrete Event Systems*, Kluwer, pp. 19-34, 2002.
13. M. Mukund, K. Narayan Kumar and M. Sohoni. Synthesizing Distributed Finite-State Systems from MSCs. *TCS* 290(1):221-239 (2003). Extended abstract in *CONCUR'00*, LNCS 1877 , pp. 521-535, 2000.
14. M. Mukund and M. Sohoni. Keeping Track of the Latest Gossip in a Distributed System. In *Distr. Computing* 10(3):137-148, 1997.
15. A. Stefanescu, J. Esparza, and A. Muscholl. Synthesis of distributed algorithms using asynchronous automata. Proc. of *CONCUR'03*, LNCS 2761, pp. 27-41, 2003.
16. W. Zielonka. Note on finite asynchronous automata. In *R.A.I.R.O. - Informatique Théorique et Applications*, 21:99-135, 1987.

Flat Parametric Counter Automata

Marius Bozga[1], Radu Iosif[1], and Yassine Lakhnech[1]

VERIMAG, 2 Avenue de Vignate, 38610 Gières, France
{bozga, iosif, lakhnech}@imag.fr

Abstract. In this paper we study the reachability problem for parametric flat counter automata, in relation with the satisfiability problem of three fragments of integer arithmetic. The equivalence between non-parametric flat counter automata and Presburger arithmetic has been established previously by Comon and Jurski [5]. We simplify their proof by introducing finite state automata defined over alphabets of a special kind of graphs (zigzags). This framework allows one to express also the reachability problem for parametric automata with one control loop as the existence of solutions of a *1-parametric linear Diophantine systems*. The latter problem is shown to be decidable, using a number-theoretic argument. Finally, the general reachability problem for parametric flat counter automata with more than one loops is shown to be undecidable, by reduction from Hilbert's Tenth Problem [9].

1 Introduction

Flat counter automata [5, 6, 3, 4] have been extensively studied, as an important class of infinite-state systems, for which the reachability problem is decidable. The results obtained so far have been used in a number of successful verification tools, like FAST [2], LASH [18] or TREX [1].

Comon and Jurski show in [5] that the reachability problem for a flat counter automaton can be expressed in Presburger arithmetic, given that the automata have transition guards that are conjunctions of relations of the form $x - y \leq c$, where x and y denote either the current or the future (primed) values of the counters, and c is an integer constant. To our knowledge, their result concerns the most general class of flat counter automata, considered so far.

The contributions of the present paper are many fold. First, we give an alternative, easier, proof of the result of [5], using finite state automata defined over alphabets of graphs (zigzags). Second, we consider a more general class of flat counter automata, in which, besides integer constants, parameters are also allowed to occur in transitions. This class is useful in modeling open programs, whose behavior is parameterized by some input values, e.g. procedures in a larger program. The reachability problem in the latter class of automata amounts to checking satisfiability of *Diophantine systems* [12].

Third, we give an effective decision procedure for the following problem: given a linear system with unknowns $x_1, \ldots, x_n$, the coefficients being polynomials of any degree in m, is there a constant $c \in \mathbb{N}$, such that the system resulting from substituting m with c has a positive solution? This result gives an effective algorithm to decide reachability for parametric counter automata with one control loop, whereas in the case of more than one control loop, the reachability problem for such systems is undecidable.

M. Bugliesi et al. (Eds.): ICALP 2006, Part II, LNCS 4052, pp. 577–588, 2006.

1.1 Related Work

Work on the decidability of reachability problems for counter automata starts with the negative result of Minsky [14] regarding two counter machines. The two most studied restrictions of this model are the *reversal bounded* 2-way counter machines [10] and the *flat counter automata* [5, 6, 3]. The class of flat counter automata that is closest to the one considered in this paper is the one studied by Comon and Jurski [5], where the transition relations are conjunctions of inequalities of the form $x - y \leq c$, with $c \in \mathbb{Z}$. Their result is that the set of reachable configurations for such automata is definable in Presburger arithmetic. Our result considers parametric transition relations of the form $x - y \leq f(\mathbf{z})$, and defines the set of reachable configurations as solutions of a linear Diophantine system with one parameter. Decision procedures for this class of systems have been independently found by O. Ibarra and Z. Dang in [11], using a result from the theory of reversal-bounded counter automata, and by Y. Matiyasevich [13]. The latter result uses a similar number theoretic argument, but the proof is based on a more involved case analysis.

2 Preliminaries

Let $\mathbf{x} = \{x_1, \ldots, x_k\}$ be a finite set of variables (counters) ranging over $\mathbb{Z}$, and $\mathbf{x}' = \{y' \mid y \in \mathbf{x}\}$ be the corresponding set of primed variables. For any counter y, we denote by y' its value at the next computational step. In what follows we will abusively use the name of a variable to denote its value also. The (compulsory) occurrence of a set of variables $\mathbf{x}$ in a logical formula φ is denoted as $\varphi(\mathbf{x})$. By $\langle \mathbb{Z}[\mathbf{x}], +, \cdot \rangle$ we denote the ring of polynomials, and by $\langle \mathbf{lin}\mathbb{Z}[\mathbf{x}], + \rangle$ the monoid of linear polynomials, with variables $\mathbf{x}$ and integer coefficients. For a closed formula φ, we write $\models \varphi$ meaning that it is valid, i.e. equivalent to true.

Let $\mathbf{z} = \{z_1, \ldots, z_l\}$ be a set of *parameter* variables, disjoint from $\mathbf{x}$. A relation $\varphi(\mathbf{x}, \mathbf{x}', \mathbf{z})$ that can be written as a finite conjunction of the form:

$$\bigwedge x_i - x_j \leq \alpha_{ij} \wedge \bigwedge x'_m - x_n \leq \beta_{mn} \wedge \bigwedge x_p - x'_q \leq \gamma_{pq} \wedge \bigwedge x'_r - x'_s \leq \delta_{rs}$$

with $1 \leq i, j, m, n, p, q, r, s \leq k$, and $\alpha_{ij}, \beta_{mn}, \gamma_{pq}, \delta_{rs} \in \mathbf{lin}\mathbb{Z}[\mathbf{z}]$, is said to be an *affine relation*. Note the formal difference between variables ($\mathbf{x}$) and parameters ($\mathbf{z}$) in φ: variables are bound to occur both unprimed and primed, whereas parameters can only occur unprimed in formulae.

A *parametric counter automaton* is a tuple $A = \langle \mathbf{x}, \mathbf{z}, Q, \delta, q_0 \rangle$, where $\mathbf{x}$ is the set of working counters, $\mathbf{z}$ is the set of parameters, Q is the set of *control states*, $q_0 \in Q$ is the *initial state*, and δ is the set of *transitions* of the form: $q \xrightarrow{\varphi(\mathbf{x}, \mathbf{x}', \mathbf{z})} q'$, where φ is an affine relation. A *configuration* of A is a tuple $c = \langle q, \mathbf{xz} \rangle$ consisting of a control state, and a set of integer values for the counters and parameters. A *run* of the automaton is a sequence of configurations, $c_0, c_1, c_2, \ldots, c_n$, $c_i = \langle q_i, \mathbf{x_i z} \rangle$, such that $\mathbf{x_0} = \mathbf{0}$, i.e. the counters are initially set to zero, and $q_i \xrightarrow{\varphi(\mathbf{x_i}, \mathbf{x_{i+1}}, \mathbf{z})} q_{i+1}$, for all $0 \leq i < n$. Note that the values of the parameters are not modified throughout the run. A control state q is said to be *reachable* in A if and only if A has a run ending in a configuration $\langle q, \mathbf{xz} \rangle$.

A control state r is said to be the *successor* of a state q if and only if there exists configurations $\langle q, \mathbf{xz} \rangle \rightarrow \langle r, \mathbf{x}'\mathbf{z} \rangle$, for some $\mathbf{x}, \mathbf{x}' \in \mathbb{Z}^k$, $\mathbf{z} \in \mathbb{Z}^l$. A *control path* is a sequence of control states $q_1, q_2, \ldots, q_n$ such that, for all $0 \leq i < n$, q_{i+1} is a successor of q_i. The path is said to be non-trivial if $n > 0$. A *cycle* is a non-trivial control path starting and ending with the same state. A counter automaton is said to be *flat* (FCA) if and only if each control state belongs to at most one cycle. A control state with two or more successors (in the sense mentioned above) is said to be a *branching* state. A branching state with exactly two successors is said to be a *2-branching* state. A FCA is said to be *linear* (LFCA) if and only if the only branching states are 2-branching, and every cycle contains at most one such state. Notice that every FCA can be effectively turned into a finite union of LFCA, the only branching state that is not 2-branching, being the initial state.

It is well-known that the class of affine relations is closed under composition, defined as $(\varphi_1 \circ \varphi_2)(\mathbf{x}, \mathbf{x}', \mathbf{z}) = \exists \mathbf{y}\ \varphi_1(\mathbf{x}, \mathbf{y}, \mathbf{z})\ \wedge\ \varphi_2(\mathbf{y}, \mathbf{x}', \mathbf{z})$. In other words, the existential quantifiers can be eliminated[1], the result being written as another affine relation. As a consequence, we can assume without losing generality, that each control path $q_1 \xrightarrow{\varphi_1} q_2 \ldots q_{n-1} \xrightarrow{\varphi_{n-1}} q_n$, with no incoming edges, is equivalent to a transition $q_1 \xrightarrow{\varphi_1 \circ \ldots \circ \varphi_{n-1}} q_n$. By applying this transformation to the whole counter automaton, we obtain a counter automaton in *normal form*.

Given a counter automaton $A = \langle \mathbf{x}, \mathbf{z}, Q, \delta, q_0 \rangle$ and a control state $q \in Q$, the *reachability problem* asks whether q is reachable in A. As we show in the following, this problem can be defined in various subfragments of the arithmetic of integer numbers. Moreover, we can show equivalence of these logical theories with different subclasses of flat counter automata. The latter are obtained by restricting the number of parameters and loops on a control path. We denote by FCA(p,n) the class of flat counter automata with at most p parameters that occur in the transition relations, and with at most n cycles on each linear component.

3 The Arithmetic of Integers

The undecidability of first-order arithmetic of integers $\langle \mathbb{Z}, +, \cdot, 0, 1 \rangle$ occurs as a consequence of Gödel's Incompleteness Theorem [8]. Moreover, the existential fragment, i.e. *Hilbert's Tenth Problem* [9] was proved undecidable by Y. Matiyasevich [12]. On the positive side, the decidability of the arithmetic of integer numbers with *addition and successor function* $\langle \mathbb{Z}, \geq, +, 0, 1 \rangle$ has been shown by M. Presburger [17].

Let us first introduce the theories of Presburger arithmetic [17] and parametric linear Diophantine systems. Presburger arithmetic $\langle \mathbb{Z}, \geq, +, 0, 1 \rangle$ is the theory of first-order logic of addition and successor function ($S(x) = x + 1$). The interpretation of logical variables is the set of integers $\mathbb{Z}$, and the meaning of the function symbols $0, 1, +$ is the natural one.

A *Diophantine equation* is a formula of the form $P(\mathbf{x}) = 0$, where $P \in \mathbb{Z}[\mathbf{x}]$ is a polynomial of the form $P(\mathbf{x}) = \Sigma_{i=1}^{m} a_i t_i(\mathbf{x}) + a_0$, and t_i are multiplicative terms of the

[1] By e.g. the Fourrier-Motzkin procedure.

form $\Pi_{l=1}^{k} x_l^{i_l}$, with $i_1, \ldots, i_l \in \mathbb{N}$. The equation is said to be *linear with parameter* x_j, $1 \le j \le k$, if for every multiplicative term of the form above, we have $\sum_{l \in \{1 \ldots k\}}^{l \ne j} i_l \le 1$. In other words, the only variable that can occur at a power greater than one is x_j, and moreover, all multiplicative terms contain at most one variable, other than x_j. Note that any Diophantine linear equation with parameter m can be equivalently written as:

$$\sum_{i=1}^{n} p_i(m) x_i + p_0(m) = 0 \tag{1}$$

where $p_i \in \mathbb{Z}[m]$, $0 \le i \le n$ are polynomials of arbitrary degree in m. In the following, we denote by $\mathfrak{D}[m]$ the set of positive boolean combinations of linear Diophantine equations with one parameter, namely m.

In this paper we show that the following problems are inter-reducible:

- the reachability for the class FCA$(0,n)$ (flat counter automata without parameters with any number of loops) and satisfiability of Presburger arithmetic, and
- the reachability for the class FCA$(p,1)$ (flat counter automata with any number of parameters and one loop) and satisfiability of $\mathfrak{D}[m]$.

Notice that the notion of *parameter* changes its meaning, depending on whether we are referring to counter automata, or Diophantine systems.

For the first point, it is already known that, given an arbitrary open Presburger formula $\varphi(\mathbf{x})$, one can build a flat counter automaton that generates exactly the values $\mathbf{x} \in \mathbb{Z}$ satisfying φ. This is a direct consequence of the fact that the set of such values is semilinear [7].

To complete the picture, we show the undecidability of the reachability problem for the class FCA(p,n) with unrestricted number of parameters (p) and loops (n), by reduction from Hilbert's Tenth Problem [9].

4 From FCA to Integer Arithmetic

In this section we develop the framework used to define the reachability problem of a FCA as a formula of either Presburger arithmetic, or $\mathfrak{D}[m]$. Given a FCA $A = \langle \mathbf{x}, \mathbf{z}, Q, \delta, q_0 \rangle$, and a state $q \in Q$, the idea is to build an arithmetic formula $\nu_{A,q}(\mathbf{x}, \mathbf{x}', \mathbf{z})$ such that, for every $\mathbf{x}, \mathbf{x}' \in \mathbb{Z}^k$, $\mathbf{z} \in \mathbb{Z}^l$, there is a run in A from $\langle q_0, \mathbf{x}\mathbf{z} \rangle$ to $\langle q, \mathbf{x}'\mathbf{z} \rangle$ if and only if $\models \nu_{A,q}(\mathbf{x}, \mathbf{x}', \mathbf{z})$. The reachability problem for A and q reduces then to checking the validity of the formula $\exists \mathbf{x} \exists \mathbf{z} \,.\, \nu_{A,q}(\mathbf{0}, \mathbf{x}, \mathbf{z})$.

In order to define $\nu_{A,q}$, we first observe that each $A \in$ FCA(p,n) is a union of disjoint linear flat counter automata, each being composed of a sequence of cycles, connected by non-trivial control paths. Without loss of generality, we will assume that A is in normal form, i.e. each control path with no incoming edges and no branching has been reduced to one transition, by composing the transition relations along the way. It follows that $\nu_{A,q}(\mathbf{x}, \mathbf{x}', \mathbf{z})$ is of the following form:

$$\exists \mathbf{y_{1 \ldots n}} \exists \mathbf{y'_{1 \ldots n}} \bigvee_i \eta_{i1}(\mathbf{x}, \mathbf{y_1}, \mathbf{z}) \wedge \bigwedge_{1 \le j < m_i} [\xi_{ij}(\mathbf{y_j}, \mathbf{y'_j}, \mathbf{z}) \wedge \eta_{ij}(\mathbf{y'_j}, \mathbf{y_{j+1}}, \mathbf{z})] \wedge \mathbf{x}' = \mathbf{y_{m_i}}$$

where $m_i \leq n$, η_{ij} are the affine relations corresponding to the transitions between cycles, and ξ_{ij} represent the transitive closures of the cycle relations, in the following sense: if $q \xrightarrow{\varphi(\mathbf{x},\mathbf{x}',\mathbf{z})} q$ is a cycle, then the transitive closure of φ is the relation between the input and output values of the counters, after *any* number of iterations through the cycle. Since η_{ij} are affine relations, it follows that $\nu_{A,q}$ is a formula in the language of $\langle \mathbb{Z}, \geq, +, 0, 1 \rangle$, if ξ_{ij} belong to the same language. Moreover, for $m_i = 1$, $\nu_{A,q}$ is a formula of $\mathfrak{D}[m]$ if ξ_{ij} are. It is therefore sufficient to analyze the definability of $\nu_{A,q}$ when A has only one transition of the form $q \xrightarrow{\varphi(\mathbf{x},\mathbf{x}',\mathbf{z})} q$. In the following developments, we will silently assume that this is indeed the case.

4.1 Constraint Graph Execution Model

In general, an affine relation $\varphi(\mathbf{x},\mathbf{x}',\mathbf{z})$ can be represented as a directed weighted graph whose set of vertices is the set of variables $\mathbf{x} \cup \mathbf{x}'$, and there is an edge with weight α from x to y if and only if there is an explicit constraint $x - y \leq \alpha$ in φ, where $\alpha \in \mathbf{lin}\mathbb{Z}[\mathbf{z}]$. An n-step execution of $q \xrightarrow{\varphi(\mathbf{x},\mathbf{x}',\mathbf{z})} q$ is represented by a *constraint graph* G^n_φ, defined as the minimal graph whose set of vertices is $\bigcup_{i=0}^n \mathbf{x}^i$, where $\mathbf{x}^i = \{y^i \mid y \in \mathbf{x}\}$ and, for all $0 \leq i < n$, there is an edge labeled α:

- from x^i to y^i, if there is a constraint $x - y \leq \alpha$ in φ.
- from x^{i+1} to y^{i+1}, if there is a constraint $x' - y' \leq \alpha$ in φ.
- from x^i to y^{i+1}, if there is a constraint $x - y' \leq \alpha$ in φ.
- from x^{i+1} to y^i, if there is a constraint $x' - y \leq \alpha$ in φ.

For example, Figure 1 shows the constraint graph for the transition relation $\varphi : x_1 - x_2' \leq z_1 \wedge x_2' - x_3 \leq z_2 \wedge x_3 - x_1' \leq z_3 \wedge x_1 - x_3' \leq z_4$. Intuitively, the nodes $\mathbf{x}^i$ in the execution graph represent the possible values of the counters after i steps of execution. Define $G^\infty_\varphi = \bigcup_{n>0} G^n_\varphi$. We say that a path in G^∞_φ *stretches between n and m*, for some $n \leq m$, if the path contains at least one node from $\mathbf{x}^i$, for each $n \leq i \leq m$.

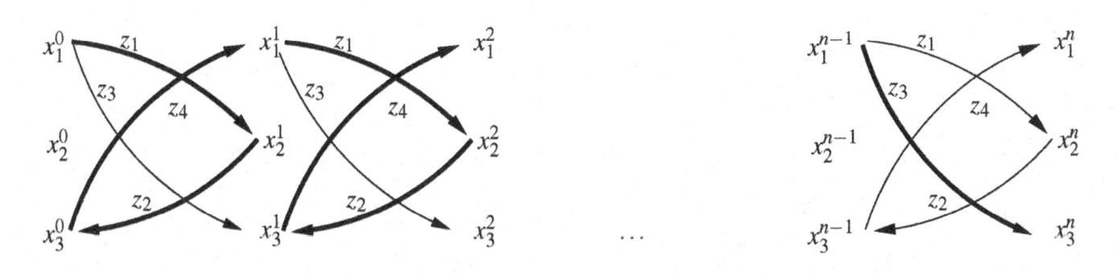

Fig. 1. Constraint Graph for $x_1 - x_2' \leq z_1 \wedge x_2' - x_3 \leq z_2 \wedge x_3 - x_1' \leq z_3 \wedge x_1 - x_3' \leq z_4$

If $\pi : x^i \xrightarrow{\alpha_1} \ldots \xrightarrow{\alpha_m} y^j$, $0 \leq i, j, \leq n$ is a path in G^n_φ, let $\omega(\pi)$ denote the sum of all labels along the path, i.e. $\omega(\pi) = \sum_{k=1}^m \alpha_k$. Notice that $\omega(\pi) \in \mathbf{lin}\mathbb{Z}[\mathbf{z}]$, for any constant $m \in \mathbb{N}$. Clearly, we have $x^i - y^j \leq \omega(\pi)$. We define $\min\{x^i \to y^j\} = \min\{\omega(\pi) \mid \pi : x^i \xrightarrow{\alpha_1} \ldots \xrightarrow{\alpha_m} y^j\}$. By convention, if there are no paths in G^n_φ, between x^i and y^j, we take

$\min\{x^i \to y^j\} = \infty$. On the other hand, if the set of paths between x^i and x^j doesn't have a minimal element, we take $\min\{x^i \to y^j\} = -\infty$. Notice that this can only be the case if G^n_φ has a cycle labeled only with constants, whose sum is less than zero. With the latter notation, we have $x^i - y^j \leq \min\{x^i \to y^j\}$. Moreover, this is the strongest relation involving the values of x and y at the execution times i and j, respectively. Notice that the satisfiability of any constraint between x^i and y^j entails the absence of negative cycles from G^n_φ. The relation between the input and output values of the counters, after n steps is:

$$\bigwedge_{x,y \in \mathbf{x}} x - y \leq \min\{x^0 \to y^0\} \wedge x' - y' \leq \min\{x^n \to y^n\} \wedge$$
$$x - y' \leq \min\{x^0 \to y^n\} \wedge x' - y \leq \min\{x^n \to y^0\} \quad (2)$$

The next step is to define the functions $\min\{x^i \to y^j\}$, $i, j \in \{0, n\}$ using the arithmetic of integers. These functions are definable in $\langle \mathbb{Z}, \geq, +, 0, 1 \rangle$, if φ has no parameters, and in $\mathfrak{D}[m]$, otherwise. The reduction method, based on weighted finite automata, is the same in both cases, and will be presented in the rest of this section.

4.2 The Even and Odd Automata

In the following, we work with a simplified (yet equivalent) form of the transition relation $\varphi(\mathbf{x}, \mathbf{x}', \mathbf{z})$. Namely, all constraints of the form $x - y \leq \alpha$ are replaced by $x - t' \leq \alpha \wedge t' - y \leq 0$, and all constraints of the form $x' - y' \leq \alpha$ are replaced by $x' - t \leq \alpha \wedge t - y' \leq 0$, by introducing fresh variables $t \notin \mathbf{x}$. In other words, we can assume without loss of generality that the constraint graph corresponding to φ is *bipartite*, i.e. it does only contain edges from $\mathbf{x}$ and $\mathbf{x}'$ and viceversa.

As previously mentioned, the presence of any cycle of negative weight within G^n_φ indicates that the constraints represented by G^n_φ are not satisfiable, i.e. the automaton has no run of length n or greater. On the other hand, a path that has a cycle of positive weight is not minimal, as one can obtain a path of smaller weight by eliminating the cycle. So, in principle, we need one tool for recognizing cycles of negative weight, and another one for recognizing acyclic paths within G^∞_φ. Both tools will be finite state automata with weighted transitions, defined on two different alphabets.

Intuitively, a word w of length n represents a path π between, say, x^0 and x^n, with $x, y \in \mathbf{x}$, as follows: the w_i symbol represents *simultaneously* all edges of π that involve only nodes from $\mathbf{x}^i \cup \mathbf{x}^{i+1}$, $0 \leq i < m$. Note that, for a path from x^0 to y^n, coded by a word w, the number of times the w_i symbol is traversed by the path is odd, whereas for a path from x^0 to y^0, or from x^n to y^n, this number is even. Hence the names of *even* and *odd automata*.

Given an affine relation $\varphi(\mathbf{x}, \mathbf{x}', \mathbf{z})$, the *even alphabet* of φ, denoted as Σ^e_φ, is the set of all graphs satisfying the following conditions, for each $G \in \Sigma^e_\varphi$:

1. the set of nodes of G is $\mathbf{x} \cup \mathbf{x}'$,
2. for any $x, y \in \mathbf{x} \cup \mathbf{x}'$, there is an edge with label α from x to y, only if the constraint $x - y \leq \alpha$ occurs in φ.

3. the in-degree and out-degree of each node are at most one.
4. the number of edges from $\mathbf{x}$ to $\mathbf{x}'$ equals the number of edges from $\mathbf{x}'$ to $\mathbf{x}$.

The *odd alphabet* of φ, denoted by Σ_φ^o, is defined in the same way, with the exception of the last condition:

4. the difference between the number of edges from $\mathbf{x}$ to $\mathbf{x}'$ and the number of edges from $\mathbf{x}'$ to $\mathbf{x}$ is either 1 or -1.

Let $\Sigma_\varphi^{e,o} = \Sigma_\varphi^e \cup \Sigma_\varphi^o$. Since, by the previous assumption, no $G \in \Sigma_\varphi^{e,o}$ contains edges of the form $x \xrightarrow{\alpha} y$ or $x' \xrightarrow{\alpha} y'$, the number of edges in all symbols of Σ_φ^e is even, while the number of edges in all symbols of Σ_φ^o is odd. The label of G, is the sum of the weights that occur on its edges. Clearly the weight of a path through G_φ^∞ is the weight of the word it is represented by. We denote by $\omega(w)$ the weight of a word $w \in \Sigma_\varphi^{e,o*}$. Notice that $\omega(w) \in \mathbf{lin}\mathbb{Z}[\mathbf{z}]$, for any given $w \in \Sigma_\varphi^{e,o*}$, where $\mathbf{z}$ is the set of parameters of φ.

Given the set of counters $\mathbf{x} = \{x_1, \ldots, x_k\}$, the even and odd automata share the same transition table, except for the alphabet, which is Σ_φ^e for the former, and Σ_φ^o for the latter. Precisely, we have $A_\varphi^{e,o} = \langle Q, \delta \rangle$, where $Q = \{l, r, lr, rl, \bot\}^k$, and $\mathbf{q} \xrightarrow{G} \mathbf{q}'$ if the following conditions hold, for all $1 \le i \le k$:

- $q_i = l$ iff G has one edge whose destination is x_i, and no other edge involving x_i.
- $q_i' = l$ iff G has one edge whose source is x_i', and no other edge involving x_i.
- $q_i = r$ iff G has one edge whose source is x_i, and no other edge involving x_i.
- $q_i' = r$ iff G has one edge whose destination is x_i', and no other edge involving x_i.
- $q_i = lr$ iff G has exactly two edges involving x_i, one having x_i as source, and another as destination.
- $q_i' = rl$ iff G has exactly two edges involving x_i', one having x_i' as source, and another as destination.
- $q_i' \in \{lr, \bot\}$ iff G has no edge involving x_i'.
- $q_i \in \{rl, \bot\}$ iff G has no edge involving x_i.
- G has at least one edge between $\mathbf{x}$ and $\mathbf{x}'$.

The odd automaton for $\varphi = x_1 - x_2' \le z_1 \wedge x_2' - x_3 \le z_2 \wedge x_3 - x_1' \le z_3 \wedge x_1 - x_3' \le z_4$ is depicted in Figure 2 (a). An example of a run of this automaton is given in Figure 2 (b). Intuitively, $q_{ij} = l$ means that the node x_j^i of G_φ^∞ is traversed from right to left by a path, and no other path comes across this node. Also, $q_{ij} = lr$ means that there is a path coming into x_j^i from $\mathbf{x}^{i+1}$ (left), and leaving also towards $\mathbf{x}^{i+1}$ (right), while no other path comes across this node. The transitions of $A_\varphi^{e,o}$ capture the necessary (yet not sufficient) conditions for a word in $\Sigma_\varphi^{e,o*}$ to represent a path in G_φ^∞. Suppose that $A_\varphi^{e,o}$ has a run $\pi : \mathbf{q}_1 \xrightarrow{G_1} \mathbf{q}_2 \xrightarrow{G_2} \ldots \mathbf{q}_{n-1} \xrightarrow{G_{n-1}} \mathbf{q}_n$. By $G(\pi)$ we shall denote, in the following, the graph associated with the run, i.e. the graph whose nodes are q_{ij}, and there is an edge from q_{ij} to q_{i+1h} if and only if $\mathbf{q}_i \xrightarrow{G_i} \mathbf{q}_{i+1}$ and G_i has an edge from x_j to x_h', for all $1 \le i \le n$, $1 \le j, h \le k$. The edges from q_{i+1h} to q_{ij} are defined symmetrically. Each node in $G(\pi)$ is labeled by a symbol from $\{l, r, lr, rl, \bot\}$, and we write, e.g. $q_{ij} = l$, meaning that q_{ij} is labeled with l. We denote by $\omega(\pi)$ the weight of the run π, defined as $\omega(\pi) = \omega(G(\pi))$.

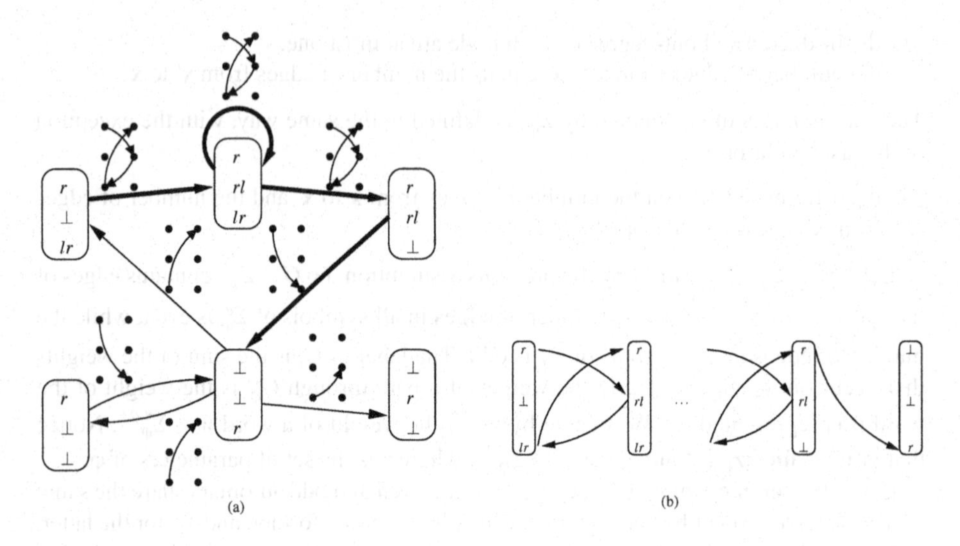

Fig. 2. The Odd Automaton for $x_1 - x_2' \leq z_1 \wedge x_2' - x_3 \leq z_2 \wedge x_3 - x_1' \leq z_3 \wedge x_1 - x_3' \leq z_4$

Lemma 1. *Let* $\pi : \mathbf{q}_1 \xrightarrow{G_1} \mathbf{q}_2 \xrightarrow{G_2} \ldots \mathbf{q}_{n-1} \xrightarrow{G_{n-1}} \mathbf{q}_n$ *be a run of* $A_\varphi^{e,o}$. *Then each node* q_{ij}, $1 \leq i \leq n$, $1 \leq j \leq k$, *from* $G(\pi)$, *has at most one predecessor and at most one successor.*

For some $1 \leq i, j \leq k$, let $A_{ij}^e = \langle A_\varphi^{e,o}, Q_0, F \rangle$ be the (non-deterministic) *even automaton*, defined over Σ_φ^e, where:

$$Q_0 = \begin{cases} \{q \mid q_i = r, q_j = l \text{ and } q_h \in \{lr, \perp\}, 1 \leq h \leq k, h \notin \{i,j\}\} & \text{if } i \neq j \\ \{q \mid q_i = q_j = lr \text{ and } q_h \in \{lr, \perp\}, 1 \leq h \leq k, h \neq i\} & \text{otherwise} \end{cases}$$

is the set of initial states, and $F = \{rl, \perp\}^k$. In the case when $i = j$, we denote A_{ij}^e by A_i^e.

Lemma 2. *For any* $1 \leq i, j \leq k$, $i \neq j$, A_{ij}^e *has an accepting run of length at most m if and only if there exists a path in* G_φ^∞, *from* x_i^0 *to* x_j^0, *that stretches between* 0 *and some* $n \leq m$. *Moreover, if* G_φ^∞ *does not have cycles of negative weight, the minimal weight among all paths from* x_i^0 *to* x_j^0, *stretching from* 0 *to some* $n \leq m$, *equals the minimal weight among all accepting runs of length at most m.*

Lemma 3. *For any* $1 \leq i \leq k$, A_i^e *has an accepting run of negative weight if and only if there exists a cycle of negative weight in* G_φ^∞.

For some $1 \leq i, j \leq k$, let $A_{ij}^o = \langle A_\varphi^{e,o}, Q_0, F \rangle$ be the (non-deterministic) *odd automaton*, defined over Σ_φ^o, where:

$$Q_0 = \{q \mid q_i = r \text{ and } q_h \in \{lr, \perp\}, 1 \leq h \leq k, h \neq i\}$$
$$F = \{q \mid q_j = r \text{ and } q_h \in \{rl, \perp\}, 1 \leq h \leq k, h \neq j\}$$

An example of an odd automaton is given in Figure 2 (a). For $i = 1$ the initial states are $\langle r, \perp, lr \rangle$ and $\langle r, \perp, \perp \rangle$. For $j = 3$ the final state is $\langle \perp, \perp, r \rangle$. An accepting run of A_{13}^o is shown in Figure 2 (b).

Lemma 4. *For any $1 \leq i,j \leq k$, A^o_{ij} has an accepting run of length m if and only if there exists a path in G^{∞}_{φ}, from x^0_i to x^m_j. Moreover, if G^{∞}_{φ} does not have cycles of negative weight, then the minimal weight among all paths from x^0_i to x^m_j equals the minimal weight among all accepting runs of length m.*

4.3 Defining Minimal Accepting Runs

Given a finite automaton with linear weights on transitions, we consider the problem of defining the set of accepting runs of a given length and of minimal weight. This solves the previous problem of defining the functions $\min\{x^i \to y^j\}$, in order to compute the input-output relation for an FCA.

Let $A = \langle Q, q_0, \delta, F\rangle$ be a given finite automaton, and $\omega : Q \times Q \longrightarrow \mathbf{lin}\mathbb{Z}[\mathbf{z}]$ be a weight function associating each transition $q \to r$ a linear expression $\omega(q,r) \in \mathbf{lin}\mathbb{Z}[\mathbf{z}]$. If δ has no transition $q \to r$, we take $\omega(q,r) = 0$. Now associate with any pair of states $q,r \in Q$ a variable x_{qr} and take $\mathbf{x}$ to be the set $\{x_{qr} \mid q,r \in Q\}$. Intuitively, x_{qr} is the number of times the transition $q \to r$ occurs within a run. Hence we take as an implicit condition the fact that all such x_{qr} range over positive integers. The formula characterizing an accepting run of length l and weight w is:

$$\phi_A(l,w) \triangleq \exists \mathbf{x} \bigvee_{q_f \in \mathcal{F}} \varphi_{q_f}(\mathbf{x}) \wedge \sum_{q,r \in Q} x_{qr} = l \wedge \sum_{q,r \in Q} x_{qr}\omega(q,r) = w \tag{3}$$

where $\varphi_{q_f}(\mathbf{x})$ expresses the necessary and sufficient conditions in order for $\mathbf{x}$ to correspond to a valid run of A ending with q_f. The definition of φ_{q_f} in Presburger arithmetic follows a method described in [5], which is based on the fact that the set of states Q of A is finite.

Notice that, if A does not have parameters, ϕ_A is already a formula in the language of $\langle \mathbb{Z}, \geq, +, 0, 1\rangle$, hence we can already define the minimal weight m among all runs of length n by the following formula: $\phi_A(n,m) \wedge \forall z\ [z \leq m \to \neg\phi_A(n,z)]$. However, this is not the case when A has parameters, due to the multiplicative terms of the form $x_{qr}\omega(q,r)$ that occur within ϕ_A. However, it is possible to build from ϕ_A, a formula of $\mathfrak{D}[m]$ defining minimal runs.

Lemma 5. *Given a finite automaton $A = \langle Q, q_0, \delta, F\rangle$, and a weight function $\omega : Q \times Q \longrightarrow \mathbf{lin}\mathbb{Z}[\mathbf{z}]$ associating each transition a linear expression, it is possible build a formula $\psi_A(l,w,\mathbf{z}) \in \mathfrak{D}[m]$ such that, for any values $l \in \mathbb{N}$ and $w, \mathbf{z} \in \mathbb{Z}$, $\models \psi_A$ if and only if w is the weight of the minimal among all accepting runs of length l.*

Intuitively, the parameter m occurring in the formula $\psi_A \in \mathfrak{D}[m]$ above, represents the number of iterations of one control loop in the original parametric FCA. It is thus possible to define the reachability problem for single loop automata in $\mathfrak{D}[m]$. As we show in Section 5, the problem concerning the existence of solutions for such systems is decidable, hence the decidability of the reachability problem for the class of FCA$(p,1)$.

However, for an arbitrary number of loops, one can reduce Hilbert's Tenth Problem to the reachability problem. In the light of [12] The following Lemma entails undecidability of the reachability problem for parametric FCA with unrestricted number of loops.

Lemma 6. *Given a Diophantine system $S(\mathbf{x})$, it is possible to build a parametric FCA $A = \langle \mathbf{y}, \mathbf{z}, Q, \delta, q_0 \rangle$ such that $\mathbf{x} \subseteq \mathbf{z}$, such that, for some control state $q \in Q$, and for all $\mathbf{x} \in \mathbb{Z}$, we have $\models S(\mathbf{x})$ if and only if there exists a run of A $\langle q_0, \mathbf{0z} \rangle \to \ldots \to \langle q, \mathbf{yz} \rangle$.*

5 Solving Parametric Linear Diophantine Systems

In this section we give a proof for the decidability of the class of formulae $\mathfrak{D}[m]$. For a given system, let D denote the maximum degree of all equations, and V is the number of variables in the system. It is known that Diophantine systems become undecidable for $(D \geq 4 \wedge V \geq 2) \vee (D \geq 2 \wedge V \geq 9)$ [15]. For either $D = 1$ or $V = 1$ the systems are decidable. We are unaware of any previously published decidability results for the case $2 \leq D < 4 \wedge 2 \leq V < 9$. The problem considered here has been independently solved by O. Ibarra and Z. Dang in [11], using a property of reversal bounded counter machines. Another proof has been suggested to us by Y. Matiyasevich [13], using a more involved case analysis. Our proof is more concise, due to a result of L. Pottier [16].

Let us fix a linear Diophantine system with parameter m, i.e. a system of the form $\{\sum_{j=1}^{n} p_{ij}(m)x_j + q_i(m) = 0\}_{i=1}^{r}$, with $p_{ij}, q_i \in \mathbb{Z}[m]$. We are interested in the existence of a solution $m, x_1, \ldots, x_n$ in natural numbers, although this is not a restriction.[2] We denote by $A(m)$ the matrix $[p_{ij}(m)]$.

Let us consider first that the system is homogeneous, i.e. $q_i(m)$ is the zero polynomial, for all $1 \leq i \leq n$. The general case will be dealt with in the following, by adding a new variable x_{n+1}, replacing each occurrence of $q_i(m)$ by $q_i(m)x_{n+1}$, and looking only after solutions in which $x_{n+1} = 1$. Let $P(m)$ be the greatest common divisor of all $p_{ij}(m)$ with respect to (symbolic) polynomial division, i.e. obtained by applying Euclid's algorithm in $\mathbb{Z}[m]$. Since $P(m)$ is a polynomial in one variable, its set of roots is finite and effectively computable. If $P(m_0) = 0$ for some $m_0 \in \mathbb{Z}$, then $\langle m_0, x_1, \ldots, x_n \rangle$ is a solution of the system $A(m)\mathbf{x} = 0$, for any choice of $x_1, \ldots, x_n \in \mathbb{Z}$. Thus, we assume in the following that $P(m) \neq 0$, for all $m \in \mathbb{N}$, in other words that, for no value of m, $p_{ij}(m)$ will all become zero at the same time.

Next, we are interested in the minimal solutions of the system. For a given $m \in \mathbb{N}$, a solution $(x_1, \ldots, x_n)$ is said to be *minimal* if it is a least solution with respect to the pointwise ordering on $\mathbb{N}^n$: $(u_1, \ldots, u_n) \preceq (v_1, \ldots, v_n) \iff u_i \leq v_i,\ 1 \leq i \leq n$. The following Theorem has been proved in [16]:

Theorem 1. *For a fixed $m_0 \in \mathbb{N}$, let $x_1, \ldots, x_n$ be any minimal solution of $A(m_0)\mathbf{x} = \mathbf{0}$. Then, for all $1 \leq i \leq n$, we have: $x_i \leq (n - r_0)\left(\frac{\sum_{i,j} a_{ij}(m_0)}{r_0}\right)^{r_0}$, where r_0 is the rank of $A(m_0)$.*

Let $C > 0$ be the maximal absolute value of all coefficients of $a_{ij}(m)$, $1 \leq i \leq r$, $1 \leq j \leq n$, and $K \geq 0$ be the maximum degree of these polynomials. The following is a direct consequence of Theorem 1:

Corollary 1. *For a fixed $m_0 \geq \max(C, n, r)$, let $x_1, \ldots, x_n$ be any minimal solution of $A(m_0)\mathbf{x} = \mathbf{0}$. Then, for all $1 \leq i \leq n$, we have $x_i \leq m_0^{(K+3)r+1}$.*

[2] The satisfiability problem for integers can be reduced to 2^{n+1} instances of the same problem on natural numbers, by performing a case split on the signs of $m, x_1, \ldots, x_n$.

Hence, one can enumerate all $0 \leq m < \max(C,n,r)$, and stop as soon as a solution of the linear Diophantine system $A(m)\mathbf{x} = \mathbf{0}$ has been found. Otherwise, for any $m \geq \max(C,n,r)$ the solution $x_1,\ldots,x_n$ can be represented in base m using at most $M = (K+3)r+1$ digits. Let $(x_i)_m = \sum_{j=0}^{M} \chi_{ij} m^j$, with $0 \leq \chi_{ij} < m$ be the polynomial representing x_i in base m. The entire system $A(m)\mathbf{x} = \mathbf{0}$ can be now represented in base m, as will be explained in the following.

First, we write the system as a set of equations of the form $P(m,x_1,\ldots,x_n) = Q(m,x_1,\ldots,x_n)$, with all coefficients of P and Q being positive. Since m was assumed to be greater that C, the maximal value of all coefficients c of the system, we have $(c)_m = c$. The operations of addition, multiplication by a constant $0 < c < m$, and multiplication by m, respectively, can be defined now using Presburger arithmetic. Let $(d)_m = \sum_{i=0}^{M} \delta_i m^i$, $(e)_m = \sum_{i=0}^{M} \varepsilon_i m^i$ and $(f)_m = \sum_{i=0}^{M} \phi_i m^i$, with $0 \leq \delta_i, \varepsilon_i, \phi_i < m$. We have:

$$(f)_m = (d)_m + (e)_m \iff \bigvee_{\mathbf{r} \in \{0\} \times \{0,1\}^{k-1} \times \{0\}} \bigwedge_{i=0}^{M} \delta_i + \varepsilon_i + r_i = \phi_i + m r_{i+1}$$

$$(e)_m = c(d)_m \iff \bigvee_{\mathbf{r} \in \{0\} \times \{0,\ldots,c-1\}^{k-1} \times \{0\}} \bigwedge_{i=0}^{M} c\delta_i + r_i = \varepsilon_i + m r_{i+1}$$

$$(e)_m = m(d)_m \iff \delta_M = \phi_0 = 0 \wedge \bigwedge_{i=0}^{M-1} \delta_i = \phi_{i+1}$$

The result of applying this transformation to the system $A(m)\mathbf{x} = \mathbf{0}$ is a formula $\Psi_A(m,\chi)$ in Presburger arithmetic, defining all minimal solutions of the original system $(x_i)_m = \sum_{j=0}^{M} \chi_{ij} m^j$, for $m \geq \max(C,n,r)$, with $\chi = \{\chi_{ij} \mid 1 \leq i \leq n,\ 1 \leq j \leq r\}$. The original system has a solution $(m,x_1,\ldots,x_n)$ if and only if, for some $m \in \mathbb{N}$, it has a minimal solution $(x_1^m,\ldots,x_n^m)$. Hence $\Psi_A(m,\chi)$ is satisfiable. Dually, if $\Psi_A(m,\chi)$ is satisfiable, we can construct a solution (not necessarily minimal) of $A(m)\mathbf{x} = \mathbf{0}$.

The non-homogeneous case is handled in the proof of the following:

Theorem 2. *The satisfiability problem for linear parametric Diophantine systems $\mathfrak{D}[m]$ is decidable.*

Theorem 2, together with the results of the previous section entail the main result:

Corollary 2. *The reachability problem for single loop parametric flat counter automata $FCA(p,1)$ is decidable.*

The strength of this result is highlighted by Lemma 6, which entails the undecidability of the reachability problem for $\mathrm{FCA}(p,n)$ with $p > 0$ parameters, and sufficiently many control loops.

6 Conclusions

We have studied a generalization of the flat counter automata considered by Comon and Jurski in [5], obtained by adding parameters to the transition relations. We reduce the reachability problem for these automata to either Presburger arithmetic, in the non-parametric case, and to linear Diophantine systems with one parameter, for single-loop automata with multiple parameters. The existence of solutions for the latter class of systems is shown to be decidable. This entails the decidability of the reachability problem for counter automata with parameters and one control loop, while in general, this problem is undecidable for flat automata with more than one control loop.

Acknowledgements. The authors wish to thank Yuri Matiyasevich and Oscar Ibarra for their enlightening suggestions leading to the proof of Theorem 2.

References

1. A. Annichini, A. Bouajjani, and M.Sighireanu. Trex: A tool for reachability analysis of complex systems. In *Proc.CAV*, volume 2102 of *LNCS*, pages 368 – 372. Springer, 2001.
2. S. Bardin, A. Finkel, J. Leroux, and L. Petrucci. Fast: Fast accelereation of symbolic transition systems. In *Proc. TACAS*, volume 2725 of *LNCS*. Springer, 2004.
3. B. Boigelot. On iterating linear transformations over recognizable sets of integers. *TCS*, 309(2):413–468, 2003.
4. H. Comon and V. Cortier. Flatness is not a weakness. In *Proc. CSL*, volume 1862 of *LNCS*, pages 262 – 276. Springer, 2000.
5. H. Comon and Y. Jurski. Multiple Counters Automata, Safety Analysis and Presburger Arithmetic. In *Proc. CAV*, volume 1427 of *LNCS*, pages 268 – 279. Springer, 1998.
6. A. Finkel and J. Leroux. How to compose presburger-accelerations: Applications to broadcast protocols. In *Proc. FST&TCS*, volume 2556 of *LNCS*, pages 145–156. Springer, 2002.
7. S. Ginsburg and E. H. Spanier. Semigroups, presburger formulas and languages. *Pacific Journal of Mathematics*, 16(2):285–296, 1966.
8. K. Gödel. Über formal unentscheidbare Sätze der Principia Mathematica und verwandter Systeme I. *Monatshefte für Mathematik und Physik*, 38:173 – 198, 1931.
9. D. Hilbert. Mathematische probleme. Vortrag, gehalten auf dem internationalen Mathematiker-Kongress zu Paris. In *Nachrichten von der Königliche Gesellschaft der Wissenschaften zu Göttingen*, pages 253–297, 1900.
10. O. H. Ibarra. Reversal-bounded multicounter machines and their decision problems. *Journal of the Association for Computing Machinery*, 25(1):116 – 133, January 1978.
11. O. H. Ibarra and Z. Dang. On the solvability of a class of diophantine equations and applications. Submitted, 2005.
12. Y. Matiyasevich. Enumerable sets are diophantine. *Journal of Sovietic Mathematics*, 11: 354 – 358, 1970.
13. Y. Matiyasevich. Personal communication, 2005.
14. M. Minsky. *Computation: Finite and Infinite Machines*. Prentice-Hall, 1967.
15. T. Pheidas and K. Zahidi. Undecidability of existential theories of rings and fields: A survey. *Contemporary Mathematics*, 270:49–106, 2000.
16. L. Pottier. Solutions minimales des systemes diophantiens lineaires: bornes et algorithmes. Technical Report 1292, INRIA Sophia Antipolis, 1990.
17. M. Presburger. Über die Vollstandigkeit eines gewissen Systems der Arithmetik. *Comptes rendus du I Congrés des Pays Slaves*, Warsaw 1929.
18. P. Wolper and B. Boigelot. Verifying systems with infinite but regular state spaces. In *Proc. CAV*, volume 1427 of *LNCS*, pages 88–97. Springer, 1998.

Lower Bounds for Complementation of ω-Automata Via the Full Automata Technique

Qiqi Yan*

BASICS Laboratory,
Department of Computer Science and Engineering,
Shanghai Jiao Tong University,
200240, Shanghai, P.R. China
contact@yanqiqi.net
http://www.yanqiqi.net

Abstract. In this paper, we first introduce a new lower bound technique for the state complexity of transformations of automata. Namely we suggest considering the class of full automata in lower bound analysis. Then we apply such technique to the complementation of nondeterministic ω-automata and obtain several lower bound results. Particularly, we prove an $\Omega((0.76n)^n)$ lower bound for Büchi complementation, which also holds for almost every complementation and determinization transformation of nondeterministic ω-automata, and prove an optimal $(\Omega(nk))^n$ lower bound for the complementation of generalized Büchi automata, which holds for Streett automata as well.

1 Introduction

The complementation problem of nondeterministic ω-automata, i.e. nondeterministic automata over infinite words, has various applications in formal verification. For example in automata-theoretic model checking, in order to check whether a system represented by automaton $\mathcal{A}_1$ satisfies a property represented by automaton $\mathcal{A}_2$, one checks that the intersection of $\mathcal{A}_1$ with an automaton that complements $\mathcal{A}_2$ is an automaton accepting the empty language [Kur94, VW94]. In such process, several types of nondeterministic ω-automata are concerned, including Büchi, generalized Büchi, Rabin, Streett etc., and the complexity of complementing these automata has caught great attention.

The complementation of Büchi automata has been investigated for over forty years [Var05]. The first effective construction was given in [Büc62] and the first exponential construction was given in [SVW85] with a $2^{O(n^2)}$ state blow-up (n is the number of states of the input automaton). Even better constructions with $2^{O(n \log n)}$ state blow-up were given in [Saf88, Kla91, KV01], which matches with Michel's $n! = 2^{\Omega(n \log n)}$ lower bound [Mic88], and were thus considered optimal. However, a closer look reveals that the blow-up of the construction in [KV01] is $(6n)^n$ while Michel's lower bound is only roughly $(n/e)^n = (0.36n)^n$, leaving

* Supported by NSFC No. 60273050.

M. Bugliesi et al. (Eds.): ICALP 2006, Part II, LNCS 4052, pp. 589–600, 2006.

a big exponential gap hiding in the asymptotic notation[1]. Motivated by this complexity gap, the construction in [KV01] was further refined in [FKV04] to $(0.97n)^n$. On the other hand, Michel's lower bound was never improved.

For generalized Büchi, Rabin and Streett automata, the best known constructions are in [KV05b, KV05a], which are $2^{O(n \log nk)}$, $2^{O(nk \log n)}$ and $2^{O(nk \log nk)}$ respectively. Here state blow-ups are measured in terms of both n and k, where k is the index of the input automaton. Optimality problems of these constructions have been vastly open, because only $2^{\Omega(n \log n)}$ lower bounds were known by variants of Michel's proof [Löd99], even without k as a factor.

What remains missing are stronger lower bound results. Tighter lower bounds might lead us into better understanding of the intricacy of the complementation of nondeterministic ω-automata, and are the main concern of this paper. Such understanding might suggest ways to further optimize the constructions, or suggest methods to circumvent those difficult cases in practice.

At the core of almost every known lower bound is Michel's result, which was obtained in the traditional way. That is, one first constructs a particular class of automata $\{\mathcal{A}_n\}$, and then proves that complementing $\{\mathcal{A}_n\}$ requires a large state blow-up. Identifying such automata class is usually difficult, and is the main obstacle towards lower bound results. In this paper, we propose a new technique to circumvent such difficulty. Namely, we introduce the notion of full automata, and suggest considering such automata in lower bound analysis.

With the help of full automata, we tighten the state complexity of Büchi complementation from $[(0.36n)^n, (0.97n)^n]$ to $[(0.76n)^n, (0.97n)^n]$. Surprisingly, this $(0.76n)^n$ lower bound also holds for every complementation or determinization transformation concerning Büchi, generalized Büchi, Rabin, Streett, Muller, and parity automata. As to the complementation of generalized Büchi automata, we prove an $(\Omega(nk))^n$ lower bound, matching with the $(O(nk))^n$ bound in [KV05b]. This lower bound also holds for the complementation of Streett automata and the determinization of generalized Büchi automata into Rabin automata. A summary of our lower bounds is given in Section 6.

2 Basic Definitions

A (*nondeterministic*) *automaton* is a tuple $\mathcal{A} = (\Sigma, S, I, \Delta, *)$ with alphabet Σ, finite state set S, initial state set $I \subseteq S$, transition relation $\Delta \subseteq S \times \Sigma \times S$ and $*$ some extra components. Particularly $\mathcal{A}$ is *deterministic* if $|I| = 1$ and for all $p \in S, a \in \Sigma$, $|\{q \in S \mid \langle p, a, q \rangle \in \Delta\}| \leq 1$.

For a *word* $w = a(0)a(1)\ldots a(l-1) \in \Sigma^*$ with $length(w) = l \geq 0$, a *finite run* of $\mathcal{A}$ from state p to q over w is a finite state sequence $\rho = \rho(0)\rho(1)\ldots\rho(l) \in S^*$ such that $\rho(0) = p$, $\rho(l) = q$ and $\langle \rho(i), a(i), \rho(i+1) \rangle \in \Delta$ for all $0 \leq i < l$. We say ρ *visits* a state set T if $\rho(i) \in T$ for some $0 \leq i \leq l$. We write $p \xrightarrow{w} q$ if a finite run from p to q over w exists and $p \xrightarrow[T]{w} q$ if in addition it visits T.

[1] In contrast, for the complementation of nondeterministic finite automata over finite words, the 2^n blow-up of the subset construction [RS59] was justified by a tight lower bound [SS78], which works even if the alphabet concerned is binary [Jir05].

A *(Nondeterministic) Finite Automaton* (NFA for short) is an automaton $\mathcal{A} = (\Sigma, S, I, \Delta, F)$ with final state set $F \subseteq S$. A finite word w is *accepted* by $\mathcal{A}$ if there is a finite run over w from an initial state to a final state. $\mathcal{L}(\mathcal{A})$, the *language* accepted by $\mathcal{A}$, is the set of words accepted by $\mathcal{A}$, and its complement $\Sigma^*\backslash\mathcal{L}(\mathcal{A})$ is denoted by $\mathcal{L}^C(\mathcal{A})$.

For an *ω-word* $\alpha = \alpha(0)\alpha(1)\cdots \in \Sigma^\omega$, i.e., an infinite sequence of letters in Σ, a (infinite) *run* of $\mathcal{A}$ over α is an infinite state sequence $\rho = \rho(0)\rho(1)\cdots \in S^\omega$ such that $\rho(0) \in I$ and $\langle \rho(i), \alpha(i), \rho(i+1)\rangle \in \Delta$ for all $i \geq 0$. Define $Occ(\rho) = \{q \in S \mid \rho(i) = q \text{ for some } i \in \mathbb{N}\}$ and $Inf(\rho) = \{q \in S \mid \rho(i) = q \text{ for infinitely many } i \in \mathbb{N}\}$. We write $\rho[l_1, l_2]$ to denote the infix $\rho(l_1)\rho(l_1+1)\ldots\rho(l_2)$ of ρ.

A *(Nondeterministic) Büchi (Word) automaton* (NBW for short) is an automaton $\mathcal{A} = (\Sigma, S, I, \Delta, F)$ with $F \subseteq S$ the final state set. A run ρ of $\mathcal{A}$ is *successful* if $Inf(\rho) \cap F \neq \emptyset$. An ω-word α is *accepted* by $\mathcal{A}$ if it has a successful run. $\mathcal{L}(\mathcal{A})$, the *ω-language* accepted by $\mathcal{A}$, is the set of ω-words accepted by $\mathcal{A}$, and its complement $\Sigma^\omega\backslash\mathcal{L}(\mathcal{A})$ is denoted by $\mathcal{L}^C(\mathcal{A})$.

The readers are referred to [Tho97] for definitions of other common types of nondeterministic ω-automata, like generalized Büchi, Rabin, Streett, Muller, and parity, which have acronyms NGBW, NRW, NSW etc. We also use acronyms like DRW for the deterministic types.

To visualize the behavior of automata over input words, we introduce Δ-graphs. If $\mathcal{A} = (\Sigma, S, I, \Delta, *)$ is an automaton, then for a finite word $w = a(0)a(1)\ldots a(l-1) \in \Sigma^*$ of length l or for an ω-word $w = a(0)a(1)\cdots \in \Sigma^\omega$ of length $l = \infty$, the *Δ-graph* of w under $\mathcal{A}$ is the directed graph $\mathcal{G}_w^\mathcal{A} = (V_w^\mathcal{A}, E_w^\mathcal{A})$ with vertex set $V_w^\mathcal{A} = \{\langle p, i\rangle \mid p \in S, 0 \leq i \leq l, i \in \mathbb{N}\}$ and edge set $E_w^\mathcal{A}$ defined as: for all $p, q \in S$ and $0 \leq i < l$, $\langle\langle p, i\rangle, \langle q, i+1\rangle\rangle \in E_w^\mathcal{A}$ iff $\langle p, a(i), q\rangle \in \Delta$. For T a subset of S, we say a vertex $\langle p, i\rangle$ is a T-vertex if $p \in T$. By definition $p \xrightarrow{w} q$ iff there is a path (in the directed sense) in $\mathcal{G}_w^\mathcal{A}$ from $\langle p, 0\rangle$ to $\langle q, length(w)\rangle$ and $p \xrightarrow[T]{w} q$ if in addition the path visits some T-vertex.

Finally we define the *state complexity*[2] functions. Assume $\mathcal{T}$ is a type of automata, like NFA or NBW. For a $\mathcal{T}$ automaton $\mathcal{A}$, $C_\mathcal{T}(\mathcal{A})$ is defined as the minimum number of states of a $\mathcal{T}$ automaton that accepts $\mathcal{L}^C(\mathcal{A})$. For $n \geq 1$, $C_\mathcal{T}(n)$ is the maximum of $C_\mathcal{T}(\mathcal{A})$ over all $\mathcal{T}$ automata with n states.

3 The Full Automata Technique

In the recently emerging area of state complexity (see [Yu04] for a survey) or in the theory of ω-automata, we often concern proving theorems of such flavor:

Theorem 1. *[Jir05] For each $n \geq 1$, there exists an* NFA *$\mathcal{A}_n$ with n states over $\{a, b\}$ such that $C_{\text{NFA}}(\mathcal{A}_n) \geq 2^n$.*

[2] In some literature, instead of merely counting the number of states, sizes of transition relations etc. are also taken into account to better measure the sizes of automata. Here we prefer state complexity because it is a measure easier to study, and its lower bound results usually imply lower bounds on the "size" complexity, if the automata witnessing the lower bound are over a not too large alphabet.

That is, we want to prove a lower bound for the state complexity of a transformation (NFA complementation in this case, might be determinization etc.), and further, we hope that the automata witnessing the lower bound ($\mathcal{A}_n$ in this case) are over a fixed small alphabet. Such claims are usually difficult to prove. The apparently easy Theorem 1 was not proved until 2005 by a very technical proof in [Jir05][3], after the efforts in [SS78, Bir93, HK02]. Why is it hard? Let us first review the traditional approach people attempt at such results:

Step I: Identify a class of automata $\{\mathcal{A}_n\}$ with each $\mathcal{A}_n$ having n states.
Step II: Prove that to transform $\{\mathcal{A}_n\}$ needs a large state blow-up.

Almost every known lower bound was obtained this way, including Theorem 1 and the aforementioned Michel's lower bound. In such an approach, Step I is well-known to be difficult. Identifying the suitable $\{\mathcal{A}_n\}$ requires both ingenuity and luck. What is worse, most automata classes that people try are natural ones with simple structures, while the ones witnessing the desired lower bound could be highly unnatural and complex. Finding the right $\{\mathcal{A}_n\}$ seems to be a major obstacle towards lower bound results.

Now let us consider a new kind of automata.

Definition 1. *A* full automaton $\mathcal{A} = (\Sigma, S, I, \Delta, *)$ *is an automaton with* $\Sigma = \mathcal{P}(S \times S)$ *and* Δ *defined as: for all* $p, q \in S, a \in \Sigma$, $\langle p, a, q\rangle \in \Delta$ *iff* $\langle p, q\rangle \in a$.

By definition, a full automaton has a rich alphabet of size $2^{|S|^2}$. With such a rich alphabet, every automaton has some embedding in a full automaton with the same number of states. Let $\mathcal{A}_1 = (\Sigma_1, S_1, I_1, \Delta_1, F_1)$ be an NFA for example. It can be embedded into the full NFA $\mathcal{A}_2 = (\Sigma_2, S_1, I_1, \Delta_2, F_1)$ (so Σ_2 and Δ_2 are determined by S_1) with the same state sets by mapping each letter a_1 in Σ_1 to the letter $\Delta_1(a_1)$ in Σ_2 defined as: for all $p_1, q_1 \in S_1$, $\langle p_1, q_1\rangle \in \Delta_1(a_1)$ iff $\langle p_1, a_1, q_1\rangle \in \Delta_1$. It is then not difficult to see that transforming an automaton can be reduced to transforming a full automaton and so full automata are the automata most difficult to transform. To be specific, if we consider NFA complementation, then:

Theorem 2. *For all* $n \geq 1$, $C_{\mathrm{NFA}}(n) = C_{\mathrm{NFA}}(\mathcal{A})$ *for some full* NFA $\mathcal{A}$ *with* n *states.*

In other words, to prove a lower bound for NFA complementation without restraint on alphabet use, we can simply set $\{\mathcal{A}_n\}$ to be full NFAs in Step I. Similar theorems hold for NBWs etc., and can be verified for determinization and some other kinds of transformations of nondeterministic automata, or even alternating automata. Note that depending on the sizes of the sets $I, F, I \cap F$, there are only $O(n^3)$ essentially different full NFAs of size n (considering symmetry), most of which are likely to be "complex" enough.

[3] The result is actually slightly stronger in that his $\mathcal{A}_n$ has only one initial state. (In some literature NFAs are not allowed to have multiple initial states.)

Remark 1. Sipser's languages $\mathcal{C}_n$ and $\mathcal{B}_n$ [SS78] are similar to full automata in spirit, which are complete with respect to the transformations 2NFA to 2DFA and NFA to 2DFA respectively (2 means two-way). To our knowledge, such languages were never applied to areas other than 2-way automata like we shall do, probably because of the "weird" alphabet. Compared to Sipser's languages, the notion of full automata is slightly more straightforward and extensible.

Now we illustrate how to prove Theorem 1 using full automata.

Proof (of Theorem 1). We first prove a 2^n lower bound for $C_{\mathrm{NFA}}(n)$. For each $n \geq 1$, let $\mathcal{FA}_n = (\Sigma_n, S_n, I_n, \Delta_n, F_n)$ be the full NFA with $S_n = I_n = F_n = \{s_0, \ldots, s_{n-1}\}$. It suffices to prove that $C_{\mathrm{NFA}}(\mathcal{FA}_n) \geq 2^n$. For each subset $T \subseteq$

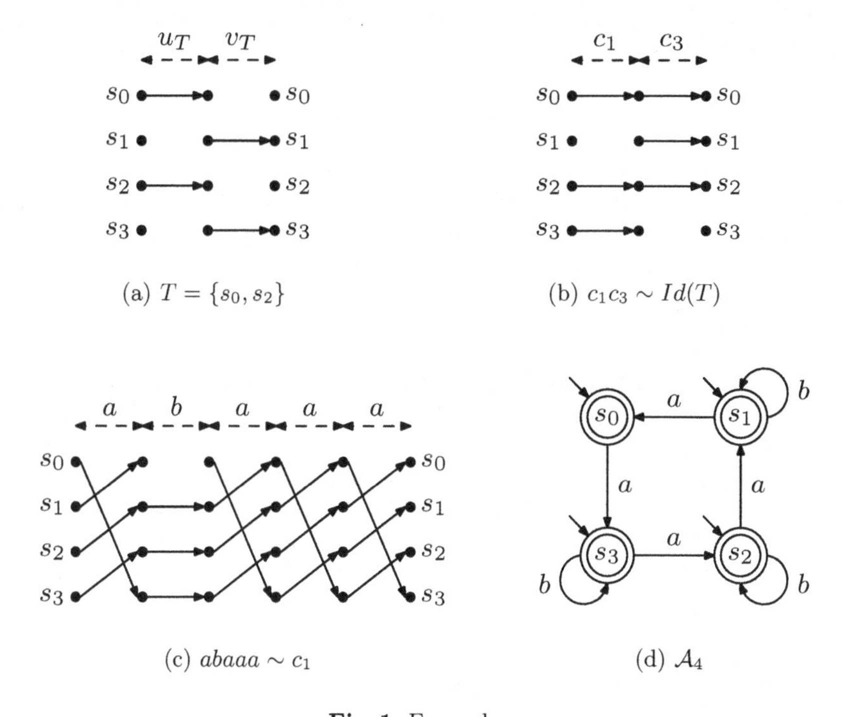

Fig. 1. Examples

S_n, let $Id(T)$ be the letter $\{\langle q, q\rangle \mid q \in T\}$ and let $u_T = Id(T)$, $v_T = Id(S_n \backslash T)$. Figure 1(a) depicts one example of $u_T v_T$ in the language of Δ-graph. Note that, as all states in $\mathcal{FA}_n$ are both initial and final, a word w of length l is accepted by $\mathcal{FA}_n$ iff there is a path from an $\langle s_i, 0\rangle$ vertex to an $\langle s_j, l\rangle$ vertex in the Δ-graph of w under $\mathcal{FA}_n$. So $u_T v_T$ is not accepted by $\mathcal{FA}_n$. Suppose some NFA $\mathcal{CA}$ complements $\mathcal{FA}_n$. So for each $T \subseteq S_n$, there is a state $\hat{q}_T$ of $\mathcal{CA}$ such that $\hat{q}_I \xrightarrow{u_T} \hat{q}_T$ and $\hat{q}_T \xrightarrow{v_T} \hat{q}_F$ for some initial state $\hat{q}_I$ and final state $\hat{q}_F$ of $\mathcal{CA}$. If we prove that $\hat{q}_{T_1} \neq \hat{q}_{T_2}$ whenever $T_1 \neq T_2$, then $\mathcal{CA}$ has at least 2^n states as required. Suppose for contradiction that $\hat{q}_{T_1} = \hat{q}_{T_2}$ for some

$T_1 \neq T_2$. W.l.o.g. there is a state s of $\mathcal{FA}_n$ in $T_1 \backslash T_2$. Then $s \xrightarrow{u_{T_1}} s \xrightarrow{v_{T_2}} s$ and so $u_{T_1}v_{T_2} \in \mathcal{L}(\mathcal{FA}_n)$. On the other hand, for some initial state $\hat{q}_I$ and final state $\hat{q}_F$ of $\mathcal{CA}$, $\hat{q}_I \xrightarrow{u_{T_1}} \hat{q}_{T_1} = \hat{q}_{T_2} \xrightarrow{v_{T_2}} \hat{q}_F$. Thus $u_{T_1}v_{T_2} \in \mathcal{L}(\mathcal{CA})$, contradiction.

The above proof is unpleasant in that the automata witnessing the lower bound are over an exponentially growing alphabet. To fixate the alphabet and prove Theorem 1, we introduce a Step III in which we do "alphabet substitution", as we now illustrate.

We first refine the above proof of $C_{\text{NFA}}(\mathcal{FA}_n) \geq 2^n$ by limiting the number of different letters involved. For two words $u, v \in \Sigma_n^*$, we say that u is *equivalent* to v with respect to $\mathcal{FA}_n$ or simply $u \sim v$ if for all $p, q \in S_n$, $p \xrightarrow{u} q$ iff $p \xrightarrow{v} q$. A little thought shows that if we substitute each $Id(T)$ letter used in the above proof by some equivalent words, the proof still works. First we consider the alphabet $\{c_i\}_{0 \leq i < n}$ with $c_i = Id(S_n \backslash \{s_i\})$. Then for each $T \subseteq S_n$, $Id(T) \sim \Pi_{s_i \notin T} c_i$ (the concatenation of all c_i's with $s_i \notin T$ in arbitrary fixed order), as is clear from Figure 1(b). Further consider the alphabet $\{a, b\}$ with $a = \{\langle s_{i+1}, s_i \rangle \mid 0 \leq i < n-1\} \cup \{\langle s_0, s_{n-1} \rangle\}$ and $b = Id(S_n \backslash \{s_0\})$, then for each $0 \leq i < n$, $c_i \sim a^i b a^{n-i}$, as is clear from Figure 1(c). So if we substitute each letter $Id(T)$ in the above proof by the equivalent word $\Pi_{s_i \notin T} a^i b a^{n-i}$, the proof still works.

After the above refinement of the proof, the part of $\mathcal{FA}_n$ related to letters other than $\{a, b\}$ is in fact irrelevant to the proof. So $\mathcal{A}_n = \mathcal{FA}_n \upharpoonright \{a, b\}$, the restriction of $\mathcal{FA}_n$ to $\{a, b\}$, or formally the NFA $\mathcal{A}_n = (\{a, b\}, S_n, I_n, \Delta_n \cap (S_n \times \{a, b\} \times S_n), F_n)$, also satisfies that $C_{\text{NFA}}(\mathcal{A}_n) \geq 2^n$, as required. ($\mathcal{A}_4$ is depicted in 1(d).) □

We call the above technique of setting $\{\mathcal{A}_n\}$ to be full automata and adding the step for alphabet substitution the "full automata technique". Setting $\{\mathcal{A}_n\}$ to be full automata is crucial here, which in essence delays the trouble of identifying $\{\mathcal{A}_n\}$ to the later analysis of transforming full automata. This makes our life easier because the latter is usually playing with words, which is clearly easier than identifying automata, especially with the rich alphabet of full automata. As to the step of alphabet substitution, our experience is that it could be technical some time, but rarely difficult.

4 Büchi Complementation

4.1 Kupferman and Vardi's Construction

We first briefly introduce the state-of-the-art construction by Kupferman and Vardi in [FKV04], the idea of which is useful for our lower bound. Different from [FKV04], we will continue to work with our Δ-graphs rather than introducing the run graphs. For $x \in \mathbb{N}$, let $[x]$ denote the set $\{0, 1, \ldots, x\}$ and let $[x]^{odd}$ and $[x]^{even}$ denote the sets of odd and even numbers in $[x]$, respectively.

Definition 2. *Given an* NBW $\mathcal{A} = (\Sigma, S, I, \Delta, F)$ *of* n *states, and an* ω*-word* α*, a* co-Büchi ranking *(C-Ranking for short) for* $\mathcal{G}_\alpha^{\mathcal{A}}$ *(i.e. the* Δ*-graph of* α *under* $\mathcal{A}$*) is a partial function* f *from* $V_\alpha^{\mathcal{A}}$ *to the rank set* $[2n-2]$ *such that:*

(i) *For all vertices* $\langle q, l\rangle \in V_\alpha^{\mathcal{A}}$, $f(\langle q, l\rangle)$ *is undefined iff there is no path (in the directed sense) from some* $\langle q_I, 0\rangle$ *with* $q_I \in I$ *to* $\langle q, l\rangle$.
(ii) *For all vertices* $\langle q, l\rangle \in V_\alpha^{\mathcal{A}}$, *if* $f(\langle q, l\rangle)$ *is odd, then* $q \notin F$.
(iii) *For all edges* $\langle\langle q, l\rangle, \langle q', l+1\rangle\rangle \in E_\alpha^{\mathcal{A}}$, *if* $f(\langle q, l\rangle)$ *is defined, then* $f(\langle q, l\rangle) \geq f(\langle q', l+1\rangle)$.

We say that f *is* odd *if for every path in* $\mathcal{G}_\alpha^{\mathcal{A}}$, *there are only finitely many vertices that are assigned even ranks by* f.

Lemma 1. *[KV01]* α *is not accepted by* $\mathcal{A}$ *iff there is an odd C-ranking for* $\mathcal{G}_\alpha^{\mathcal{A}}$.

Proof. We prove the *if* direction here to give a sense of the idea of C-ranking. For every infinite path from a $\langle q_I, 0\rangle$ vertex for some $q_I \in I$, the ranks along the path do not increase by (iii) and so will get trapped in some fixed rank from some point on. Since f is odd, this fixed rank is odd, and thus by (ii), F-vertices are never visited since then. In other words, every run of $\mathcal{A}$ over α visits F finitely often and thus α is not accepted by $\mathcal{A}$. □

A *level ranking*[4] for $\mathcal{A}$ is a partial function $g : S \longrightarrow [2n-2]$ such that if $g(q)$ is odd, then $q \notin F$. Each C-ranking can be "sliced" into such level rankings. It was shown in [KV01] that existence of an odd C-ranking for $\mathcal{G}_\alpha^{\mathcal{A}}$ can be decided by an NBW $\mathcal{CA}$ which guesses an odd C-ranking level by level, and checks the validity in a local manner. By Lemma 1, $\mathcal{CA}$ complements $\mathcal{A}$. In the construction of $\mathcal{CA}$, distinct sets of states are used to handle different level rankings, and the number of such level rankings is the major factor of the $(6n)^n$ blow-up.

We say that a level ranking g for $\mathcal{A}$ is *tight* if (i): the maximum rank in the range of g is some $2m-1$ in $[2n-2]^{odd}$, and (ii): for every $j \in [2m]^{odd}$, there is a state q with $g(q) = j$. In such case, g is also called a TL(m)-ranking. (So $1 \leq m < n$.) It was further showed in [FKV04] that we can restrict attention to tight level rankings and thus use less states in $\mathcal{CA}$. By a careful numerical analysis [FKV04], a $(0.97)^n$ upper bound was proved for the number of states of $\mathcal{CA}$ and thus for Büchi complementation.

4.2 Lower Bound

We turn now to lower bound. It suffices to consider full NBWs. So we define $\mathcal{FB}_n$ for $n > 1$ to be the full NBW $(\Sigma_n, S_n, I_n, \Delta_n, F_n)$ with $I_n = \{s_0, \ldots, s_{n-2}\}$, $F_n = \{s_f\}$ and $S_n = I_n \cup \{s_f\}$. We also let $S'_n = I_n$ denote the "main" states.

We first produce a hard case for Kupferman and Vardi's construction. Since the number of tight level rankings is the major factor of the state blow-up, we try to construct an ω-word $\alpha_n \notin \mathcal{L}(\mathcal{FB}_n)$ such that a great number of tight level rankings would have to be present in every C-ranking for $\mathcal{G}_{\alpha_n}^{\mathcal{FB}_n}$. For such purpose, we introduce Q-rankings for $\mathcal{FB}_n$. We say a tight level ranking g for $\mathcal{FB}_n$ is a *Q-ranking* if $g\,(q)$ is defined for each $q \in S'_n$ and is undefined for $q = s_f$. If g is a $TL(m)$-ranking, then g is also called a $Q(m)$-ranking.

[4] Note that our definitions of level ranking and tight level ranking here are slightly different from [FKV04].

Definition 3. *A word $w \in \Sigma_n^*$ is* compatible *with an ordered pair of $Q(m)$-rankings $\langle f, g\rangle$ for $\mathcal{FB}_n$ if:*

(i) *For all $p, q \in S'_n$, $p \xrightarrow{w} q$ iff $(f(p) > g(q)$ or $f(p) = g(q) \in [2m]^{odd})$.*
(ii) *For all $p, q \in S'_n$, $p \xrightarrow[F_n]{w} q$ iff $f(p) > g(q)$.*
(iii) *For all $p, q \in S_n$, if $p \xrightarrow{w} q$ then $p, q \notin F_n$.*

Lemma 2. *For every two $Q(m)$-rankings f, g for $\mathcal{FB}_n$, there is a word $w_{f,g}$ compatible with $\langle f, g\rangle$.*

Such words are not difficult to construct, see Figure 2 for one construction. Another useful fact is that:

Lemma 3. *Let f, g, h be $Q(m)$-rankings for $\mathcal{FB}_n$. If u and v are compatible with $\langle f, g\rangle$ and $\langle g, h\rangle$ respectively, then uv is compatible with $\langle f, h\rangle$.*

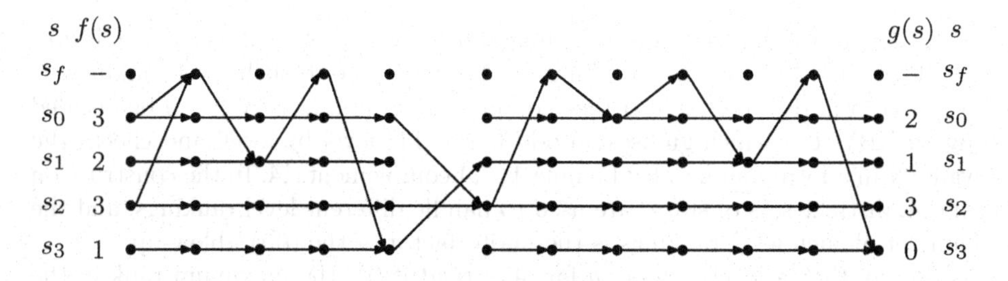

Fig. 2. Δ-graph of $w_{f,g}$

Now we can define our "hard" ω-word α_n. Let $L(n, m)$ be the number of different $Q(m)$-rankings and let $L(n)$ be $\max_{1 \le m < n} L(n, m)$. From now on we fix m such that $L(n) = L(n, m)$ and we may simply write L for $L(n)$. Clearly there exists an infinite looping enumeration $f_0, f_1, \ldots$ of $Q(m)$-rankings such that $f_i \neq f_j$ for all $i \neq j, 0 \le i, j < L$ and $f_i = f_{jL+i}$ for all $i, j \ge 0$. Define α_n to be the ω-word $w_0 w_1 \ldots$ such that $w_i = w_{f_i, f_{i+1}}$ for all $i \ge 0$.

Lemma 4. *The ω-word α_n is not in $\mathcal{L}(\mathcal{FB}_n)$.*

Proof. If there is a successful run ρ of $\mathcal{FB}_n$ over α_n, then there is an infinite state sequence $q_0 q_1 \cdots \in S_n^\omega$ such that $q_i \xrightarrow{w_i} q_{i+1}$ for all $i \ge 0$ and $q_i \xrightarrow[F_n]{w_i} q_{i+1}$ for infinitely many $i \in \mathbb{N}$. As w_i is compatible with $\langle f_i, f_{i+1}\rangle$ for $i \ge 0$, $f_i(q_i) \ge f_{i+1}(q_{i+1})$ for all $i \ge 0$ and $f_i(q_i) > f_{i+1}(q_{i+1})$ for infinitely many $i \in \mathbb{N}$. This is impossible since $f_0(q_0)$ is finite. □

Recall that Kupferman and Vardi's construction uses distinct state sets to handle different TL(m)-rankings. It turns out that if a complement automaton of $\mathcal{FB}_n$ does not have as many states as $Q(m)$-rankings, it would be "confused" by the complex ω-word α_n. We now show this in a slightly broader sense.

Lemma 5. *For each $n > 1$ and each ω-automaton $\mathcal{CA}$ with less than $L(n)$ states, if ρ is a run of $\mathcal{CA}$ over $\alpha_n \notin \mathcal{L}(\mathcal{FB}_n)$, then there is a run ρ' of $\mathcal{CA}$ over some ω-word $\alpha' \in \mathcal{L}(\mathcal{FB}_n)$ with $Occ(\rho') = Occ(\rho)$ and $Inf(\rho') = Inf(\rho)$.*

Proof. Suppose $\mathcal{CA} = (\Sigma_n, \hat{S}, \hat{I}, \hat{\Delta}, Acc)$ is an ω-automaton with less than L states and $\rho = \rho(0)\rho(1)\cdots \in \hat{S}^\omega$ is a run of $\mathcal{CA}$ over α_n, then there is a sequence $k_0, k_1, \ldots$ such that $k_0 = 0$, $k_{i+1} - k_i = length(w_i)$ for all $i \geq 0$, and $\rho(k_i) \xrightarrow{w_i} \rho(k_{i+1})$ for all $i \geq 0$. Define for each $0 \leq i < L$ a nonempty set $\hat{Q}_i = \{\hat{q} \in \hat{S} \mid \rho(k_{jL+i}) = \hat{q}$ for infinitely many $j \in \mathbb{N}\}$. As $\mathcal{CA}$ has less than L states, there exist some $i \neq j, 0 \leq i, j < L$ such that $\hat{Q}_i \cap \hat{Q}_j \neq \emptyset$.

Let $\hat{q} \in \hat{Q}_i \cap \hat{Q}_j$ for some $i \neq j, 0 \leq i, j < L$. So $f_i \neq f_j$. W.l.o.g. there is a $q \in S'_n$ with $f_i(q) > f_j(q)$. By the definitions of $\hat{Q}_i$ and $Occ(\rho)$, there is a $t_1 \in \mathbb{N}$ sufficiently large such that $\rho(k_{t_1 L+i}) = \hat{q}$, every state in $Occ(\rho)$ occurs in $\rho[0, k_{t_1 L+i}]$, and $\rho(t') \in Inf(\rho)$ for all $t' > k_{t_1 L+i}$. By the definitions of $Inf(\rho)$ and $\hat{Q}_j$, there is a sufficiently large $t_2 > t_1$ such that $\rho(k_{t_2 L+j}) = \hat{q}$ and every state in $Inf(\rho)$ occurs in $\rho[k_{t_1 L+i}, k_{t_2 L+j}]$. Let $u = w_0 \ldots w_{t_1 L+i-1}$ and $v = w_{t_1 L+i} \ldots w_{t_2 L+j-1}$. By Lemma 3, u is compatible with $\langle f_0, f_{t_1 L+i}\rangle$, i.e. $\langle f_0, f_i\rangle$, and v is compatible with $\langle f_{t_1 L+i}, f_{t_2 L+j}\rangle$, i.e. $\langle f_i, f_j\rangle$. Let α' be uv^ω.

Let $q_I \in S'_n$ be such that $f_0(q_I) = 2m - 1 \geq f_i(q)$. So $q_I \xrightarrow{u} q$ as u is compatible with $\langle f_0, f_i\rangle$. Also, $q \xrightarrow[F_n]{v} q$ since $f_i(q) > f_j(q)$ and v is compatible with $\langle f_i, f_j\rangle$. Thus $q_I \xrightarrow{u} q \xrightarrow[F_n]{v} q \xrightarrow[F_n]{v} q \ldots$ and α' is accepted by $\mathcal{FB}_n$.

Note that $\rho' = \rho[0, k_{t_1 L+i}] \cdot (\rho[k_{t_1 L+i} + 1, k_{t_2 L+j}])^\omega$ is a run over α', and we have guaranteed that $Occ(\rho') = Occ(\rho)$ and $Inf(\rho') = Inf(\rho)$, as required. □

Theorem 3. *For each $n > 1$, $L(n) \leq C_{\mathrm{NBW}}(\mathcal{FB}_n) \leq C_{\mathrm{NBW}}(n) \leq U(n)$, where $L(n) \approx (c_l n)^n$, $U(n) \approx (c_h n)^n$ with $c_l \approx 0.7645$, $c_h \approx 0.9624$. Here $U(n)$ denotes the state blow-up of Kupferman and Vardi's construction.*

Proof. By Lemma 5, every NBW that complements $\mathcal{FB}_n$ must have at least $L(n)$ states. The estimate for $L(n)$ is from a numerical analysis which is almost the same as the one for the *tight* (n) function in [FKV04].

Thus we have tightened Büchi complementation. By using alphabet substitutions like in the proof of Theorem 1, the NBWs witnessing the lower bound can also be over a fixed alphabet.

Theorem 4. *For each $n > 1$, there exists an* NBW *$\mathcal{B}_n$ of n states over a fixed alphabet such that $L(n) \leq C_{\mathrm{NBW}}(\mathcal{B}_n)$.*

4.3 Other Transformations

Surprisingly, our lower bound on Büchi complementation actually holds for almost every complementation or determinization transformation of nondeterministic ω-automata, via a reduction making use of Lemma 5.

Theorem 5. *For each $n > 1$ and each common type (including nondeterministic Büchi, generalized Büchi, Rabin, Streett, Muller and parity) $\mathcal{T}_1$, there exists a $\mathcal{T}_1$ automaton $\mathcal{A}_n$ with n states over a fixed alphabet such that:*

(i) *For each common type $\mathcal{T}_2$, every $\mathcal{T}_2$ automaton that accepts the complement of $\mathcal{L}(\mathcal{A}_n)$ has at least $L(n)$ states.*
(ii) *For each common type $\mathcal{T}_2$ that is not Büchi or generalized Büchi[5], every deterministic $\mathcal{T}_2$ automaton that accepts $\mathcal{L}(\mathcal{A}_n)$ has at least $L(n)$ states.*

Proof. For each common type $\mathcal{T}_1$, it is easy by definitions that there is a $\mathcal{T}_1$ automaton $\mathcal{A}_n$ equivalent to the NBW $\mathcal{FB}_n$ with also n states [Löd99]. (i) Suppose an automaton $\mathcal{CA}$ of a common type accepts $\mathcal{L}^C(\mathcal{A}_n) = \mathcal{L}^C(\mathcal{FB}_n)$. Since acceptance of ω-automata of a common type only depends on the Occ set and the Inf set of a run, the claim can be obtained by applying Lemma 5. (ii) If some deterministic $\mathcal{T}_2$ automaton with less than $L(n)$ states accepts $\mathcal{L}(\mathcal{A}_n)$ and $\mathcal{T}_2$ is not Büchi or generalized Büchi, then it is simple by definitions that there is a deterministic ω-automaton of a common type (not necessarily $\mathcal{T}_2$) that accepts the complement of $\mathcal{L}(\mathcal{A}_n)$ with also less than $L(n)$ states [Löd99], contrary to (i). Finally, the alphabet of $\mathcal{A}_n$ can be fixated like in the proof of Theorem 4. □

For the transformations involved in this theorem, less than half already had nontrivial lower bounds like $n!$ by Michel's proof or the bunch of proofs by Löding [Löd99], while the others only have trivial or weak $2^{\Omega(n)}$ lower bounds.

5 Complementation of Generalized Büchi Automata

Now we turn to NGBW complementation. For NGBWs, state complexity is preferably measured in terms of both the number of states n and index k, where index measures the size of the acceptance condition. By applying full automata, doing a hard case analysis for the construction in [KV05b] based on GC-ranking, and using a generalization of Michel's technique, we have:

Theorem 6. *For $n > 1$ and $1 < k \leq \binom{n-1}{\lfloor (n-1)/2 \rfloor}$, $C_{\mathrm{NGBW}}(n, k) = (\Omega(nk))^n$.*

This matches neatly[6] with the $(O(nk))^n$ construction in [KV05b], and thus settles the state complexity of NGBW complementation. Like Michel's result, this lower bound can be extended to NSW complementation and the determinization of NGBW into DRW (state complexity denoted by $D_{\mathrm{NGBW}\to\mathrm{DRW}}(n, k)$):

Theorem 7. *For $n > 1$ and $1 < k \leq \binom{n-1}{\lfloor (n-1)/2 \rfloor}$, $C_{\mathrm{NSW}}(n, k) = (\Omega(nk))^n$ and $D_{\mathrm{NGBW}\to\mathrm{DRW}}(n, k) = (\Omega(nk))^n$.*

[5] Deterministic Büchi or generalized Büchi automata are strictly weaker in expressive power than the other common types of ω-automata.

[6] The gap hidden in the notation $(\Theta(nk))^n$ can be at most c^n for some c, while the gap hidden in the more widely used notation $2^{\Theta(n \log nk)}$ can be as large as $(nk)^n$.

Remark 2. For the above lower bounds, the alphabet involved in the proof is of a size polynomial in n. It seems not difficult to fixate the alphabet if we aim at a looser bound in the form $2^{\Omega(n \log nk)}$. But we shall not get into this technical and uninteresting issue.

6 Summary of Lower Bounds

In the following table, we briefly summarize our lower bounds. Here "Any" means any common type of nondeterministic ω-automata (and the two Any's can be different). "co." means complementation and "det." means determinization. "L.B."/"U.B." stands for lower/upper bound. Weak $2^{\Omega(n)}$ lower bounds are considered trivial.

#	Transformation	Previous L.B.	Our L.B.	Known U.B.
1	NBW $\xrightarrow{\text{co.}}$ NBW	$\Omega((0.36n)^n)$ [Mic88]	$\Omega((0.76n)^n)$	$O((0.97n)^n)$ [FKV04]
2	Any $\xrightarrow{\text{co. or det.}}$ Any	trivial or $n!$ [Löd99]	$2^{\Omega(n \log n)}$	-
3	NBW $\xrightarrow{\text{det.}}$ DMW	trivial[7]	$2^{\Omega(n \log n)}$	$2^{O(n \log n)}$ [Saf89]
4	NRW $\xrightarrow{\text{co.}}$ NRW	trivial[8]	$2^{\Omega(n \log n)}$	$2^{O(nk \log n)}$ [KV05a]
5	NGBW $\xrightarrow{\text{co.}}$ NGBW	$\Omega((n/e)^n)$ [Mic88]	$(\Omega(nk))^n$	$(O(nk))^n$ [KV05b]
6	NSW $\xrightarrow{\text{co.}}$ NSW	$\Omega((n/e)^n)$ [Löd99]	$(\Omega(nk))^n$	$2^{O(nk \log(nk))}$ [KV05a]
7	NGBW $\xrightarrow{\text{det.}}$ DRW	$\Omega((n/e)^n)$ [Löd99]	$(\Omega(nk))^n$	$2^{O(nk \log(nk))}$ [Saf89]

Remark 3. Lower bound #2 implies that the $2^{\Omega(n \log n)}$ blow-up is inherent in the complementation and determinization of nondeterministic ω-automata, corresponding to the 2^n blow-up of finite automata. The special case #3 justifies that Safra's construction is optimal in state complexity for the determinization of Büchi automata into Muller automata. We single out this result because this determinization construction is touched in almost every introductory material on ω-automata, and its optimality problem was explicitly left open in [Löd99].

It is hard to believe that the above lower bounds could be obtained in the traditional way. We hope that the full automata technique will stimulate the discovery of new results in automata theory.

Acknowledgment. I thank Orna Kupferman and Moshe Vardi for the insightful discussion and the extremely valuable suggestions. I thank Enshao Shen for his kind support and guidance. I also thank anonymous reviewers for the useful comments.

[7] But if size complexity is concerned, rather than state complexity, then Safra proved that the transformation is inherently doubly exponential [Saf89].

[8] As pointed to us by Moshe Vardi, if size complexity is concerned, then an $2^{\Omega(n \log n)}$ lower bound follows from Michel's lower bound.

References

[Bir93] J.C. Birget. Partial orders on words, minimal elements of regular languages and state complexity (has online erratum). *Theor. Comput. Sci*, 119(2): 267–291, 1993.

[Büc62] J. R. Büchi. On a decision method in restricted second order arithmetic. In *Proceedings of the International Congress on Logic, Method, and Philosophy of Science*, pages 1–12. Stanford University Press, 1962.

[FKV04] E. Friedgut, O. Kupferman, and M.Y. Vardi. Büchi complementation made tighter. In *ATVA*, volume 3299 of *LNCS*, pages 64–78, 2004. refer to http://www.cs.rice.edu/~vardi/papers/index.html.

[HK02] M. Holzer and M. Kutrib. State complexity of basic operations on nondeterministic finite automata. In *7th CIAA*, volume 2608 of *LNCS*, pages 148–157, 2002.

[Jir05] G. Jirásková. State complexity of some operations on binary regular languages. *Theor. Comput. Sci*, 330(2):287–298, 2005.

[Kla91] N. Klarlund. Progress measures for complementation of omega-automata with applications to temporal logic. In *32th FOCS*, pages 358–367, 1991.

[Kur94] R.P. Kurshan. *Computer-Aided Verification of Coordinating Processes: The Automata-Theoretic Approach.* Princeton Univ. Press, 1994.

[KV01] O. Kupferman and M.Y. Vardi. Weak alternating automata are not that weak. *ACM Transactions on Computational Logic*, 2(3):408–429, 2001.

[KV05a] O. Kupferman and M.Y. Vardi. Complementation constructions for nondeterministic automata on infinite words. In *TACAS*, pages 206–221, 2005.

[KV05b] O. Kupferman and M.Y. Vardi. From complementation to certification. *Theor. Comput. Sci*, 345(1):83–100, 2005.

[Löd99] C. Löding. Optimal bounds for transformations of omega-automata. In *19th FSTTCS*, volume 1738 of *LNCS*, pages 97–109, 1999.

[Mic88] M. Michel. Complementation is more difficult with automata on infinite words. CNET, Paris, 1988.

[RS59] M.O. Rabin and D. Scott. Finite automata and their decision problems. *IBM Journal of Research and Development*, 3:114–125, 1959.

[Saf88] S. Safra. On the complexity of ω-automata. In *29th FOCS*, pages 319–327, 1988.

[Saf89] S. Safra. *Complexity of automata on infinite objects.* PhD thesis, Weizmann Institute of Science, 1989.

[SS78] W.J. Sakoda and M. Sipser. Nondeterminism and the size of two way finite automata. In *10th STOC*, pages 275–286, 1978.

[SVW85] A.P. Sistla, M.Y. Vardi, and Pierre Wolper. The complementation problem for Büchi automata with applications to temporal logic (extended abstract). In *12th ICALP*, volume 194 of *LNCS*, pages 465–474, 1985.

[Tho97] W. Thomas. Languages, automata and logic. In *Handbook of Formal Languages*, volume 3, pages 389–455. Springer-Verlag, Berlin, 1997.

[Var05] M.Y. Vardi. Büchi complementation: A 40-year saga. In *The 9th Asian Logic Conference*, 2005.

[VW94] M.Y. Vardi and P. Wolper. Reasoning about infinite computations. *Inf. Comput.*, 115(1):1–37, 1994.

[Yu04] S. Yu. State complexity: Recent results and open problems, 2004. Invited talk at ICALP'04 Workshop in Formal Languages (slides online).

Author Index

Lecture Notes in Computer Science

For information about Vols. 1–3972

please contact your bookseller or Springer

Vol. 4067: D. Thomas (Ed.), ECOOP 2006 – Object-Oriented Programming. XIV, 527 pages. 2006.

Vol. 4063: I. Gorton, G.T. Heineman, I. Crnkovic, H.W. Schmidt, J.A. Stafford, C.A. Szyperski, K. Wallnau (Eds.), Component-Based Software Engineering. XI, 394 pages. 2006.

Vol. 4060: K. Futatsugi, J.-P. Jouannaud, J. Meseguer (Eds.), Algebra, Meaning and Computation. XXXVIII, 643 pages. 2006.

Vol. 4059: L. Arge, R. Freivalds (Eds.), Algorithm Theory – SWAT 2006. XII, 436 pages. 2006.

Vol. 4058: L.M. Batten, R. Safavi-Naini (Eds.), Information Security and Privacy. XII, 446 pages. 2006.

Vol. 4057: J.P. W. Pluim, B. Likar, F.A. Gerritsen (Eds.), Biomedical Image Registration. XII, 324 pages. 2006.

Vol. 4056: P. Flocchini, L. Gąsieniec (Eds.), Structural Information and Communication Complexity. X, 357 pages. 2006.

Vol. 4055: J. Lee, J. Shim, S.-g. Lee, C. Bussler, S. Shim (Eds.), Data Engineering Issues in E-Commerce and Services. IX, 290 pages. 2006.

Vol. 4054: A. Horváth, M. Telek (Eds.), Formal Methods and Stochastic Models for Performance Evaluation. VIII, 239 pages. 2006.

Vol. 4053: M. Ikeda, K.D. Ashley, T.-W. Chan (Eds.), Intelligent Tutoring Systems. XXVI, 821 pages. 2006.

Vol. 4052: M. Bugliesi, B. Preneel, V. Sassone, I. Wegener (Eds.), Automata, Languages and Programming, Part II. XXIV, 603 pages. 2006.

Vol. 4048: L. Goble, J.-J.C.. Meyer (Eds.), Deontic Logic and Artificial Normative Systems. X, 273 pages. 2006. (Sublibrary LNAI).

Vol. 4046: S.M. Astley, M. Brady, C. Rose, R. Zwiggelaar (Eds.), Digital Mammography. XVI, 654 pages. 2006.

Vol. 4045: D. Barker-Plummer, R. Cox, N. Swoboda (Eds.), Diagrammatic Representation and Inference. XII, 301 pages. 2006. (Sublibrary LNAI).

Vol. 4044: P. Abrahamsson, M. Marchesi, G. Succi (Eds.), Extreme Programming and Agile Processes in Software Engineering. XII, 230 pages. 2006.

Vol. 4043: A.S. Atzeni, A. Lioy (Eds.), Public Key Infrastructure. XI, 261 pages. 2006.

Vol. 4041: S.-W. Cheng, C.K. Poon (Eds.), Algorithmic Aspects in Information and Management. XI, 395 pages. 2006.

Vol. 4040: R. Reulke, U. Eckardt, B. Flach, U. Knauer, K. Polthier (Eds.), Combinatorial Image Analysis. XII, 482 pages. 2006.

Vol. 4039: M. Morisio (Ed.), Reuse of Off-the-Shelf Components. XIII, 444 pages. 2006.

Vol. 4038: P. Ciancarini, H. Wiklicky (Eds.), Coordination Models and Languages. VIII, 299 pages. 2006.

Vol. 4037: R. Gorrieri, H. Wehrheim (Eds.), Formal Methods for Open Object-Based Distributed Systems. XVII, 474 pages. 2006.

Vol. 4036: O. H. Ibarra, Z. Dang (Eds.), Developments in Language Theory. XII, 456 pages. 2006.

Vol. 4035: H.-P. Seidel, T. Nishita, Q. Peng (Eds.), Advances in Computer Graphics. XX, 771 pages. 2006.

Vol. 4034: J. Münch, M. Vierimaa (Eds.), Product-Focused Software Process Improvement. XVII, 474 pages. 2006.

Vol. 4033: B. Stiller, P. Reichl, B. Tuffin (Eds.), Performability Has its Price. X, 103 pages. 2006.

Vol. 4032: O. Etzion, T. Kuflik, A. Motro (Eds.), Next Generation Information Technologies and Systems. XIII, 365 pages. 2006.

Vol. 4031: M. Ali, R. Dapoigny (Eds.), Innovations in Applied Artificial Intelligence. XXIII, 1353 pages. 2006. (Sublibrary LNAI).

Vol. 4029: L. Rutkowski, R. Tadeusiewicz, L.A. Zadeh, J. Zurada (Eds.), Artificial Intelligence and Soft Computing – ICAISC 2006. XXI, 1235 pages. 2006. (Sublibrary LNAI).

Vol. 4027: H.L. Larsen, G. Pasi, D. Ortiz-Arroyo, T. Andreasen, H. Christiansen (Eds.), Flexible Query Answering Systems. XVIII, 714 pages. 2006. (Sublibrary LNAI).

Vol. 4026: P.B. Gibbons, T. Abdelzaher, J. Aspnes, R. Rao (Eds.), Distributed Computing in Sensor Systems. XIV, 566 pages. 2006.

Vol. 4025: F. Eliassen, A. Montresor (Eds.), Distributed Applications and Interoperable Systems. XI, 355 pages. 2006.

Vol. 4024: S. Donatelli, P. S. Thiagarajan (Eds.), Petri Nets and Other Models of Concurrency - ICATPN 2006. XI, 441 pages. 2006.

Vol. 4021: E. André, L. Dybkjær, W. Minker, H. Neumann, M. Weber (Eds.), Perception and Interactive Technologies. XI, 217 pages. 2006. (Sublibrary LNAI).

Vol. 4020: A. Bredenfeld, A. Jacoff, I. Noda, Y. Takahashi (Eds.), RoboCup 2005: Robot Soccer World Cup IX. XVII, 727 pages. 2006. (Sublibrary LNAI).

Vol. 4019: M. Johnson, V. Vene (Eds.), Algebraic Methodology and Software Technology. XI, 389 pages. 2006.

Vol. 4018: V. Wade, H. Ashman, B. Smyth (Eds.), Adaptive Hypermedia and Adaptive Web-Based Systems. XVI, 474 pages. 2006.

Vol. 4016: J.X. Yu, M. Kitsuregawa, H.V. Leong (Eds.), Advances in Web-Age Information Management. XVII, 606 pages. 2006.

Vol. 4014: T. Uustalu (Ed.), Mathematics of Program Construction. X, 455 pages. 2006.

Vol. 4013: L. Lamontagne, M. Marchand (Eds.), Advances in Artificial Intelligence. XIII, 564 pages. 2006. (Sublibrary LNAI).

Vol. 4012: T. Washio, A. Sakurai, K. Nakajima, H. Takeda, S. Tojo, M. Yokoo (Eds.), New Frontiers in Artificial Intelligence. XIII, 484 pages. 2006. (Sublibrary LNAI).

Vol. 4011: Y. Sure, J. Domingue (Eds.), The Semantic Web: Research and Applications. XIX, 726 pages. 2006.

Vol. 4010: S. Dunne, B. Stoddart (Eds.), Unifying Theories of Programming. VIII, 257 pages. 2006.

Vol. 4009: M. Lewenstein, G. Valiente (Eds.), Combinatorial Pattern Matching. XII, 414 pages. 2006.

Vol. 4007: C. Àlvarez, M. Serna (Eds.), Experimental Algorithms. XI, 329 pages. 2006.

Vol. 4006: L.M. Pinho, M. González Harbour (Eds.), Reliable Software Technologies – Ada-Europe 2006. XII, 241 pages. 2006.

Vol. 4005: G. Lugosi, H.U. Simon (Eds.), Learning Theory. XI, 656 pages. 2006. (Sublibrary LNAI).

Vol. 4004: S. Vaudenay (Ed.), Advances in Cryptology - EUROCRYPT 2006. XIV, 613 pages. 2006.

Vol. 4003: Y. Koucheryavy, J. Harju, V.B. Iversen (Eds.), Next Generation Teletraffic and Wired/Wireless Advanced Networking. XVI, 582 pages. 2006.

Vol. 4001: E. Dubois, K. Pohl (Eds.), Advanced Information Systems Engineering. XVI, 560 pages. 2006.

Vol. 3999: C. Kop, G. Fliedl, H.C. Mayr, E. Métais (Eds.), Natural Language Processing and Information Systems. XIII, 227 pages. 2006.

Vol. 3998: T. Calamoneri, I. Finocchi, G.F. Italiano (Eds.), Algorithms and Complexity. XII, 394 pages. 2006.

Vol. 3997: W. Grieskamp, C. Weise (Eds.), Formal Approaches to Software Testing. XII, 219 pages. 2006.

Vol. 3996: A. Keller, J.-P. Martin-Flatin (Eds.), Self-Managed Networks, Systems, and Services. X, 185 pages. 2006.

Vol. 3995: G. Müller (Ed.), Emerging Trends in Information and Communication Security. XX, 524 pages. 2006.

Vol. 3994: V.N. Alexandrov, G.D. van Albada, P.M.A. Sloot, J. Dongarra (Eds.), Computational Science – ICCS 2006, Part IV. XXXV, 1096 pages. 2006.

Vol. 3993: V.N. Alexandrov, G.D. van Albada, P.M.A. Sloot, J. Dongarra (Eds.), Computational Science – ICCS 2006, Part III. XXXVI, 1136 pages. 2006.

Vol. 3992: V.N. Alexandrov, G.D. van Albada, P.M.A. Sloot, J. Dongarra (Eds.), Computational Science – ICCS 2006, Part II. XXXV, 1122 pages. 2006.

Vol. 3991: V.N. Alexandrov, G.D. van Albada, P.M.A. Sloot, J. Dongarra (Eds.), Computational Science – ICCS 2006, Part I. LXXXI, 1096 pages. 2006.

Vol. 3990: J. C. Beck, B.M. Smith (Eds.), Integration of AI and OR Techniques in Constraint Programming for Combinatorial Optimization Problems. X, 301 pages. 2006.

Vol. 3989: J. Zhou, M. Yung, F. Bao, Applied Cryptography and Network Security. XIV, 488 pages. 2006.

Vol. 3988: A. Beckmann, U. Berger, B. Löwe, J.V. Tucker (Eds.), Logical Approaches to Computational Barriers. XV, 608 pages. 2006.

Vol. 3987: M. Hazas, J. Krumm, T. Strang (Eds.), Location- and Context-Awareness. X, 289 pages. 2006.

Vol. 3986: K. Stølen, W.H. Winsborough, F. Martinelli, F. Massacci (Eds.), Trust Management. XIV, 474 pages. 2006.

Vol. 3984: M. Gavrilova, O. Gervasi, V. Kumar, C.J. K. Tan, D. Taniar, A. Laganà, Y. Mun, H. Choo (Eds.), Computational Science and Its Applications - ICCSA 2006, Part V. XXV, 1045 pages. 2006.

Vol. 3983: M. Gavrilova, O. Gervasi, V. Kumar, C.J. K. Tan, D. Taniar, A. Laganà, Y. Mun, H. Choo (Eds.), Computational Science and Its Applications - ICCSA 2006, Part IV. XXVI, 1191 pages. 2006.

Vol. 3982: M. Gavrilova, O. Gervasi, V. Kumar, C.J. K. Tan, D. Taniar, A. Laganà, Y. Mun, H. Choo (Eds.), Computational Science and Its Applications - ICCSA 2006, Part III. XXV, 1243 pages. 2006.

Vol. 3981: M. Gavrilova, O. Gervasi, V. Kumar, C.J. K. Tan, D. Taniar, A. Laganà, Y. Mun, H. Choo (Eds.), Computational Science and Its Applications - ICCSA 2006, Part II. XXVI, 1255 pages. 2006.

Vol. 3980: M. Gavrilova, O. Gervasi, V. Kumar, C.J. K. Tan, D. Taniar, A. Laganà, Y. Mun, H. Choo (Eds.), Computational Science and Its Applications - ICCSA 2006, Part I. LXXV, 1199 pages. 2006.

Vol. 3979: T.S. Huang, N. Sebe, M.S. Lew, V. Pavlović, M. Kölsch, A. Galata, B. Kisačanin (Eds.), Computer Vision in Human-Computer Interaction. XII, 121 pages. 2006.

Vol. 3978: B. Hnich, M. Carlsson, F. Fages, F. Rossi (Eds.), Recent Advances in Constraints. VIII, 179 pages. 2006. (Sublibrary LNAI).

Vol. 3977: N. Fuhr, M. Lalmas, S. Malik, G. Kazai (Eds.), Advances in XML Information Retrieval and Evaluation. XII, 556 pages. 2006.

Vol. 3976: F. Boavida, T. Plagemann, B. Stiller, C. Westphal, E. Monteiro (Eds.), NETWORKING 2006. Networking Technologies, Services, and Protocols; Performance of Computer and Communication Networks; Mobile and Wireless Communications Systems. XXVI, 1276 pages. 2006.

Vol. 3975: S. Mehrotra, D.D. Zeng, H. Chen, B. Thuraisingham, F.-Y. Wang (Eds.), Intelligence and Security Informatics. XXII, 772 pages. 2006.

Vol. 3973: J. Wang, Z. Yi, J. Zurada, B.-L. Lu, H. Yin (Eds.), Advances in Neural Networks - ISNN 2006, Part III. XXIX, 1402 pages. 2006.